# ERBIUM-DOPED
# FIBER AMPLIFIERS

# ERBIUM-DOPED FIBER AMPLIFIERS
## Principles and Applications

**EMMANUEL DESURVIRE**
Department of Electrical Engineering
Columbia University

A JOHN WILEY & SONS, INC., PUBLICATION

Copyright © 2002 by John Wiley & Sons, Inc., Hoboken, New Jersey. All rights reserved.

Published simultaneously in Canada.

For ordering and customer service, call 1-800-CALL WILEY.

*Library of Congress Cataloging in Publication Data:*

Desurvire, Emmanuel, 1955–
    Erbium-doped fiber amplifiers : principles and applications /
Emmanuel Desurvire.
        p.   cm.
    Includes bibliographical references and index.
    ISBN 978-0-471-26434-7

    1. Lasers.   2. Optical amplifiers.   3. Optical fibers.   I. Title.
TA1667.D47    1994
621.382′75—dc20                              93-22410
                                                  CIP

# PREFACE

The field of optical fiber communications is now nearly 20 years old, and the maturity of this technology is well reflected by the accelerated rate at which optical fiber links currently are being deployed over the continents and across the oceans. The progress can be measured by considering that existing transoceanic fiber links based on digital communication techniques can transmit 40,000 to 80,000 simultaneous telephone conversations, as compared to 48 in the first telephone cable deployed in 1956 between the United Kingdom and the United States! Formidable technical obstacles had to be overcome in order to reach this performance: essentially, these had to do with manufacturing optical fibers with lowest possible loss and developing reliable laser sources that could survive being turned on and off for several years. A sense of luck may have guided the pioneers of this field because, in spite of such overwhelming difficulties, the technology has never seemed to come to a dead end. In our present information age, where data highways increase in length every day to form gigantic networks on a planetary scale, lightwave technology has a bright future indeed.

Yet, in spite of the tremendous information-carrying capacity offered by lightwave systems, the demand for transoceanic communications is still growing at a constant rate of 25% a year. The demand may become even greater in terrestrial systems, although harder to quantify, as it will follow not only the natural growth of intercity communications, but also the predictable emergence of large-size computer networks and fiber-to-the-home loops. Then, and this will not be the first time, one may wonder whether the limits of lightwave technology are going to be felt soon. What are the limits of its capacity? These are not to be found in the optical fiber itself. Between 1.3 and 1.55-$\mu$m, it offers a comfortable bandwidth of 45 THz still largely underused. Then come the switching electronics used to generate electrical signals and modulate optical sources: for the fastest transistors, the potential bandwidth is up to 100 GHz, only a few times greater than the performance of currently available devices, and greater than the modulation bandwidth of semiconductor optical sources. The real limitation is the need to regenerate optical signals as they undergo attenuation and loss when propagating along a fiber link or network.

The task of signal regeneration traditionally has been that of electronic repeaters in which optical signals are converted into current by a photodiode; the current is then regenerated and converted back into light by a laser diode. Repeaters must be placed at intervals such that the optical signal power never drops below a level where the probability of misreading digital symbols becomes unacceptably high. Current transatlantic fiber links typically have a 70-km repeater spacing—about 100 repeaters over their 7500-km length. The electronic speed of the repeaters is fixed once and for all and cannot be upgraded. This limits the information rate of repeatered systems. Another drawback of electronic regeneration is that wavelength multiplexing is difficult and costly. This is because many parallel repeaters are needed to combine several optical channels into the same fiber. By the end of the 1980s, experts in the field agreed that the maximum capacity of unrepeatered systems, expressed in terms of a bit-rate–length product, had reached its peak. In the future, large systems spanning continental and intercontinental distances would definitely have to rely upon the use of electronic repeaters.

This view has now been radically changed, due to the recent and forceful emergence of optical amplifiers, which have taken the form of laser-diode-pumped erbium-doped fibers. Yet again, luck shines on lightwave technology. In 1985, just when conventional unrepeatered systems had approached their peak performance, a research group at the University of Southampton showed that optical fibers can exhibit laser gain at a wavelength near 1.55 $\mu$m. The fibers were doped with the rare earth erbium and were activated or pumped with low powers of visible light. In 1986, the author and a small group at AT&T Bell Laboratories joined the investigation. In the years to follow, the primary research work of the University of Southampton and Bell Labs would have tremendous consequences in the conception of lightwave systems. Indeed, if practical optical amplifiers were available, they would offer a much wider operating bandwidth (4 THz) and could then replace electronic repeaters used so far in optical communications.

No matter how noble the attempt, any breakthrough in the old field of lightwave systems needs a bit of luck. The lucky nature of erbium ions is that they have some properties of radiative decay with a long excited state lifetime in a glass host, such as the fused silica used to make optical fibers. Their 1.55 $\mu$m lasing wavelength happens to fall in the spectral region where optical fibers have the lowest loss—a narrow window used for long-haul communications. But that's not all. Miniature semiconductor laser diodes (as opposed to cumbersome, water-cooled laser sources) can be used to pump the erbium-doped fibers, which makes them practical as optical amplifier devices. And the gain characteristics of erbium-doped fibers offer all these advantages: polarization insensitivity, temperature stability, quantum-limited noise figure, and immunity to interchannel crosstalk. These advantages could hardly be achieved, all at once, with other optical amplifier approaches.

By early 1989, the first laser-diode-pumped fiber amplifier devices had appeared and system breakthroughs followed at an unprecedented rate. The first experimental systems with lengths of about 1000 km and operations at multigigabit rates without electronic repeaters were demonstrated. They represent a level of performance that no expert could have predicted only a few years ago. At AT&T Bell Laboratories, fiber loop experiments in which data were recirculated many times in order to emulate the characteristics of very long-haul systems based on optical amplifier chains, showed transmission potentials up to 20,000 km. Symbolically, this length is nearly the longest distance possible on earth. Early in 1993, AT&T and the Japanese company KDD reported a straight fiber link incorporating as many as 274 Er-doped fiber amplifiers; the link was 9000 km long and carried optical data at 10 Gbit/s—a record capacity of 90.Tbit/s.km and 100 times the performance level of unrepeatered systems in 1985. Such radical improvements have caused a wide consensus about the importance of optical amplifiers in future optical communications, both in long-haul and network system applications. By 1995, the first transoceanic cables based on optical amplifiers will be deployed over both the Atlantic and Pacific. Each of these high capacity links will have the potential to carry as many as 600,000 voice channels. And we are only witnessing the beginnings of this technology.

The motivation for this book has been to provide the basic materials of a comprehensive introduction to the principles and applications of erbium-doped fiber amplifiers. The literature is abundant on this relatively new subject of many different facets. Each facet reflects issues ranging from the most fundamental to the most applied. For this reason, the student, the communications systems engineer, or the research scientist approaching this field for the first time may feel confused about where to start. This book is intended to answer the most basic and practical questions. How is light amplified in the doped fiber? How much spontaneous emission noise is generated at the output? How does amplification affect the photon statistics and signal-to-noise ratio? What are the ultimate noise limits of fiber amplifiers? Why do detectors with optical preamplifiers perform better than avalanche photodiodes? What are the optimal locations of optical amplifiers in a system? What are the current types and architectures of amplifier-based systems?

The book is organized in three parts that can be read independently. Part A is theoretical. It deals with the fundamentals of light amplification and noise in single-mode fibers, as well as the principles of photodetection of digital signals with optical amplifiers. Part B outlines the characteristics of erbium-doped fibers, from spectroscopic features to fiber amplifier gain and saturation properties. Part C concerns both device and system applications of erbium-doped fibers.

The author is grateful to many coworkers at AT&T Bell laboratories, whose expertise, rigor, and team spirit made the investigation of erbium-doped fiber amplifiers a very challenging and rewarding experience.

EMMANUEL DESURVIRE

*New York*
*April 1993*

# CONTENTS

# APPENDICES

# LIST OF ACRONYMS AND SYMBOLS

## ACRONYMS

| | |
|---|---|
| AFRRG | Amplified fiber ring resonator gyro |
| AGC | Automatic gain control |
| AM | Amplitude modulation |
| AM–VSB | AM vestigial sideband |
| AOD | All-optical device |
| AOM | Acousto-optic modulator |
| AON | All-optical network |
| AOTF | Acousto-optic tunable filter |
| APC | Automatic power control |
| APD | Avalanche photodiode |
| AR | Antireflection coating |
| ARDL | Active recirculating (fiber) delay line |
| ARFG | Amplified reentrant fiber gyro |
| ARQ | Automatic repeat request |
| ASE | Amplified spontaneous emission |
| ASK | Amplitude shift keying |
| ATM | Asynchronous transfer mode |
| BATY | $BaF_2$–$AlF_3$–$ThF_4$–$YF_3$ fluoride glass |
| BE | Bose-Einstein distribution |
| BER | Bit-error rate |
| BFA | Brillouin fiber amplifier |
| BIZYT | $BaF_2$–$InF_3$–$ZnF_2$–$YF_3$–$ThF_4$ fluoride glass |
| CATV | Common antenna television |
| CC | Color center |
| CET | Cooperative energy transfer |
| CFUC | Cooperative frequency upconversion |
| CNR | Carrier-to-noise ratio |

| | |
|---|---|
| CPFSK | Continuous phase frequency shift keying |
| CPM | Colliding pulse mode locking |
| CSO | Composite second order distortion |
| CTB | Composite triple beat |
| CVD | Chemical vapor deposition |
| DBR | Distributed Bragg reflector laser |
| DC | Dispersion compensation |
| DCE | Differential conversion efficiency |
| DCF | Dispersion-compensating fiber |
| DCPBH | Double-channel planar buried heterostructure |
| DFB | Distributed feedback laser |
| DGD | Differential group delay |
| DPL | Diode-pumped laser |
| DPSK | Differential phase shift keying |
| DSF | Dispersion-shifted fiber |
| ECC | Error-correcting code |
| ED | Electric dipole |
| EDD | Electric dipole–dipole |
| EDF | Erbium-doped fiber |
| EDFA | Erbium-doped fiber amplifier |
| EDFL | Erbium-doped fiber laser |
| EDQ | Electric dipole–quadrupole |
| EFS | Error-free seconds |
| ELOS | Enhanced local oscillator system |
| EM | Electromagnetic |
| EQQ | Electric quadrupole–quadrupole |
| ESA | Excited state absorption |
| ESI | Equivalent step-index |
| EYDFA | Erbium–ytterbium-doped fiber amplifier |
| FDM | Frequency division multiplexing |
| FDMA | Frequency division multiple access |
| FEC | Forward error correction |
| FID | Free induction decay |
| FL | Fiber laser |
| FL | Fuchtbauer–Ladenburg relation for cross sections |
| FLN | Fluorescence line narrowing |
| FM | Frequency modulation |
| FP | Fabry–Perot |
| FSK | Frequency shift keying |
| FSR | Free spectral range |
| FTFC | Fused taper fiber coupler |
| FTTC | Fiber to the curb |
| FTTH | Fiber to the home |
| FWHM | Full width at half maximum |
| F8L | Figure-eight laser |
| GRIN–SCH | Linearly graded-index separate confinement heterostructure |
| GS | Gain switching |
| GSA | Ground state absorption |

| | |
|---|---|
| GVD | Group velocity dispersion |
| HDTV | High definition television |
| HEDFA | Hybrid erbium-doped fiber amplifier |
| HMF | Heavy metal fluoride |
| HR | High reflection coating |
| IC | Integrated circuit |
| ICP–AES | Inductively coupled plasma atomic emission spectroscopy |
| IF | Interference filter |
| IFOG | Interferometric fiber optic gyro |
| IM | Intensity modulation |
| IM–DD | Intensity modulation/direct detection |
| JO | Judd–Ofelt theory |
| KdV | Korteveg de Vries |
| KKR | Kramers–Kronig relations |
| LAN | Local area network |
| LASER | Light amplification by stimulated emission of radiation |
| LED | Light-emitting diode |
| LD | Laser diode |
| LO | Local oscillator |
| MAN | Metropolitan area network |
| MC | McCumber relation for cross sections |
| MCVD | Modified chemical vapor deposition |
| MD | Magnetic dipole |
| MDD | Magnetic dipole–dipole |
| MFD | Mode field diameter |
| MKSA | Meter, kilogram, second, Ampere |
| ML | Mode-locking |
| MQW | Multiquantum well |
| NA | Fiber numerical aperture |
| NALM | Nonlinear amplifying loop mirror |
| NDFA | Neodymium-doped fiber amplifier |
| NF | Noise figure |
| NLSE | Nonlinear Schrödinger equation |
| NNB | Noncentral negative binomial distribution |
| NOLM | Nonlinear optical loop mirror |
| NPHB | Nonphotochemical hole burning |
| NRZ | Nonreturn-to-zero data |
| NSIS | Nonlinear Sagnac interferometer switch |
| NTSC | National television system committee |
| OD | Outside fiber diameter |
| OTDR | Optical time domain reflectometry |
| OVD | Outside vapor deposition |
| PA | Optical preamplifier |
| PBS | Polarization beam splitter |
| PCE | Power conversion efficiency |
| PDF | Probability distribution function |
| PDFA | Praseodymium-doped fiber amplifier |
| PDL | Polarization dependent loss |

| | |
|---|---|
| PDM | Polarization division multiplexing |
| PGF | Probability generating function |
| PHB | Photochemical hole burning |
| PM | Phase modulation |
| PM | Polarization-maintaining |
| PMD | Polarization mode dispersion |
| POHB | Polarization hole burning |
| PON | Passive optical network |
| PSK | Phase shift keying |
| PZT | Piezoelectric transducer |
| QCE | Quantum conversion efficiency |
| QW | Quantum well structure |
| RB | Rayleigh backscattering |
| RBS | Rutherford backscattering spectroscopy |
| RE | Rare earth |
| RF | Radio frequency |
| RFA | Raman fiber amplifier |
| RIN | Relative intensity noise |
| RDL | Recirculating delay line |
| RL | Recirculating loop |
| RZ | Return-to-zero data |
| SBS | Stimulated Brillouin scattering |
| SCA | Semiconductor amplifier |
| SCH | Separate confinement heterostructure |
| SCM | Subcarrier multiplexing |
| SES | Severe-errored seconds |
| SFL | Superfluorescent fiber laser |
| SFPM | Stimulated four-photon mixing |
| SHB | Saturation hole burning |
| SHD | Second harmonic distortion |
| SIMS | Secondary ion mass spectrometry |
| SLD | Superluminescent diode |
| SMF | Single-mode fiber |
| SNF | Saturated noise figure |
| SNR | Signal-to-noise ratio |
| SONET | Synchronous optical network |
| SOP | State of polarization |
| SP–EDFA | Single-polarization EDFA |
| SPM | Self-phase modulation |
| SRS | Stimulated Raman scattering |
| SSFS | Soliton self-frequency shift |
| STT | Shimoda, Takahasi, and Townes (linear amplifier theory) |
| TBP | Time–bandwidth product |
| TDM | Time division multiplexing |
| THD | Third harmonic distortion |
| TDFA | Thullium-doped fiber amplifier |
| TDMA | Time division multiple access |
| TLS | Two-level systems |

| | |
|---|---|
| VAD | Vapor axial deposition |
| VIPS | V groove inner stripe, $p$-type substrate |
| WAN | Wide area network |
| WDM | Wavelength division multiplexing |
| WSC | Wavelength selective coupler |
| XM | Cross-modulation |
| XPM | Cross-phase modulation |
| YAG | Yttrium–aluminum garnet |
| YLF | Yttrium–lithium–fluoride |
| ZBLA | Zr–Ba–La–Al fluoride glass |
| ZBLAN | Zr–Ba–La–Al–Na fluoride glass |
| ZBLANP | Zr–Ba–La–Al–Na–Pb fluoride glass |

## ROMAN SYMBOLS

| | |
|---|---|
| $A$ | Effective area |
| $A$ | Einstein's $A$ coefficient |
| $A$ | Polarized cross section ratio |
| $A_{\text{eff}}$ | Mode effective area (EDFA) |
| | Mode effective interaction area (SPM, XPM, SRS, SBS, SFPM/FWM) |
| $A_p, A_s$ | Effective areas at $\lambda_p, \lambda_s$ |
| $A_A^R$ | Radiative decay rate of donor ion $A$ |
| $A_{ij}^{NR}$ | Nonradiative decay rate between energy levels $i$ and $j$ |
| $A_{ij}^R$ | Radiative decay rate between energy levels $i$ and $j$ |
| $A_{NR}^+, A_{NR}^-$ | Nonradiative thermalization rates |
| $A_{S'L'J',SLJ}^{ed}$ | Electric dipole spontaneous emission probability |
| $A_{S'L'J',SLJ}^{md}$ | Magnetic dipole spontaneous emission probability |
| $\mathscr{A}_{pk}$ | Gain dependent pump absorption coefficient |
| $A_{32}$ | Nonradiative decay rate from level 3 to level 2 |
| $A_{21}$ | Spontaneous emission rate |
| $\mathscr{A}_{21}$ | Spontaneous emission rate for Stark split laser systems |
| $a$ | Fiber core radius |
| $a(z)$ | Emission coefficient at fiber coordinate $z$ |
| $a_0$ | Doped core radius |
| $B_e$ | Detector electronic bandwidth |
| $B_o$ | Signal optical bandwidth or optical passband |
| $B_{12}, B_{21}$ | Einstein's $B$ coefficient |
| $b(z)$ | Absorption coefficient at fiber coordinate $z$ |
| $c$ | Speed of light in free space |
| $D$ | Relative laser medium inversion |
| $D$ | Receiver decision level in binary photodetection |
| $D$ | Glass density |
| $D$ | Fiber dispersion parameter |
| $\bar{D}$ | Effective or average fiber dispersion parameter |
| $D(\lambda)$ | Fiber material dispersion |
| $D_H(\lambda)$ | Fiber host dispersion |
| $D_R(\lambda)$ | Fiber resonant dispersion due to RE-doping |

| | |
|---|---|
| $D_{opt}$ | Optimum decision level for which conditional symbol detection errors P(1\|0) and P(0\|1) are equal |
| $E_i$ | Energy of level $i$ |
| $E_{1j}$ | Energy of Stark sublevel $j$ in ground manifold 1 |
| $E_{2k}$ | Energy of Stark sublevel $k$ in excited state manifold 2 |
| $E(r,z,t)$ | Electric field component of electromagnetic wave |
| $e$ | Electric charge |
| $e(z)$ | Amplifier statistical fluctuation at coordinate $z$ |
| $F(N_1,N_2,T)$ | Free energy of atomic system at temperature $T$ |
| $F(x,z)$ | Probability generating function of argument $x$ at coordinate $z$ |
| $F_{ex}$ | Excess noise factor of APD detector |
| $F_i$ | Optical noise figure of individual amplifier $i$ |
| $F_i^{Poisson}$ | Poisson optical noise figure of individual amplifier $i$ |
| $F_o(z)$ | Amplifier optical noise figure at coordinate $z$ |
| $F_o^{min}$ | Minimum optical noise figure |
| $F_o(k)$ | Optical noise figure of amplifier chain with $k$ elements |
| $F_{path}(G)$ | Penalty factor in optical amplifier chains |
| $f(z)$ | Amplifier Fano factor at coordinate $z$ |
| $f$ | Signal frequency in the RF domain |
| $f_i$ | Fano factor of signal input to amplifier $i$ |
| $f_n$ | Saturation function equal to $1/(1+sn)$ |
| $G$ | Fiber amplifier gain |
| $G(z)$ | Fiber amplifier gain at coordinate $z$ |
| $G_k$ | Fiber amplifier gain at wavelength $\lambda_k$ |
| $G_i$ | Fiber amplifier gain of amplifier $i$ in an amplifier chain |
| $G_{max}$ | Small-signal or unsaturated gain |
| $G_{opt}$ | Optimum gain corresponding to optimum EDFA length |
| $g_{em}, g_{am}$ | Emission and absorption coefficients at $\lambda_m$ |
| $g_1, g_2$ | Degeneracies of laser levels 1 and 2 |
| $g_B$ | Brillouin gain coefficient |
| $g_R$ | Raman gain coefficient |
| $g_{inc}$ | Incremental dB gain for additional milliwatt pump power |
| $g_s$ | Small-signal gain coefficient |
| $g(v)$ | Lineshape function at frequency $v$ |
| $h$ | Planck constant $= 6.62 \times 10^{-34}$ J·s |
| $h\varepsilon$ | Free energy required to excite one ion from the ground state |
| $\hbar$ | Planck constant divided by $2\pi$ |
| $I_{a,e}(\lambda)$ | Experimental absorption and fluorescence spectra |
| $I_{a,e}^{peak}$ | Peak intensity of experimental absorption and fluorescence spectra |
| $I_d$ | Total mean photocurrent |
| $I_N$ | Mean ASE noise photocurrent |
| $I_{peak}$ | Pulse peak intensity |
| $I_s$ | Mean unamplified signal photocurrent |
| $I_s(r,\theta)$ | Signal intensity distribution |
| $I'_{sat}(v)$ | Saturation intensity at frequency $v$ for basic laser systems |
| $I_{sat}(v)$ | Saturation intensity at frequency $v$ for Stark split laser systems |
| $i_c$ | Equivalent photodetector circuit noise current density |

| | |
|---|---|
| $i_d$ | Photocurrent corresponding to receiver decision level $D$ |
| $\langle i_0 \rangle$ | Mean photocurrent corresponding to digital symbol 0 |
| $\langle i_1 \rangle$ | Mean photocurrent corresponding to digital symbol 1 |
| $J$ | Total atomic orbital momentum ($J = L + S$) |
| $k$ | Ionization ratio of APD detector |
| $k', k''$ | First and second derivatives of wave vector with respect to $\omega$ |
| $k_B$ | Boltzmann's constant $= 1.38 \times 10^{-23}$ J/K |
| $L$ | Total fiber amplifier length |
| $L$ | Amplifier spacing in amplifier chains |
| $L$ | Total electronic orbital momentum |
| $L_c$ | Soliton collision length |
| $L_{eff}$ | Effective fiber length in fiber Raman and Brillouin amplifiers |
| $L_{opt}$ | Optimal fiber amplifier length |
| $L_{tot}$ | Total length of optical fiber amplifier chain |
| $L_{1/2}$ | Fiber length at which the pump power is half of its input value |
| $\mathscr{L}_{kj}$ | Lorentzian lineshape function with center frequency $\omega_{kj}$, spectral width $\Delta\omega_{kj}$, and unity peak |
| $\mathscr{M}$ | Number of amplifier modes |
| $\hat{M}$ | Optical amplifier transfer matrix |
| $\langle M \rangle$ | Mean APD detector gain |
| $\langle M^2 \rangle = \langle M \rangle^{2+x}$ | Mean-square of APD detector gain |
| $m$ | Analog signal optical modulation index |
| $m_e$ | Electron mass |
| $N$ | Total dopant density (only in Section 1.11) |
| $N(z)$ | Amplified spontaneous emission photon number at coordinate $z$ |
| $N_1, N_2, N_3$ | Atomic population densities of laser system levels 1, 2, and 3 |
| $\bar{N}_1, \bar{N}_2, \bar{N}_3$ | Atomic population densities of Stark split levels 1, 2, and 3 |
| $\mathbf{N}$ | Optical amplifier noise vector |
| $\mathscr{N}$ | Avogadro number $= 6.02 \times 10^{23}$ |
| $n$ | Refractive index |
| $n(T)$ | Bose–Einstein occupation number |
| $n_{sat}$ | Saturation photon number equal to $1/s$ |
| $n_{sp}(z)$ | Spontaneous emission factor at coordinate $z$ |
| $n_{sp}^{min}$ | Minimum spontaneous emission factor |
| $n_{eq}(z)$ | Equivalent input noise factor at coordinate $z$ |
| $n_{eq}^{min}$ | Minimum equivalent input noise factor |
| $n_{eqi}$ | Equivalent input noise factor of amplifier $i$ |
| $n_H$ | Refractive index of host medium |
| $\bar{n}$ | Average number of photons per bit for $10^{-9}$ BER |
| $n_1, n_2$ | Fractional populations of ground and upper level averaged over the fiber length |
| $n_2$ | Nonlinear refractive index coefficient |
| $n_v$ | Boltzmann occupation number |
| $\langle n^k(z) \rangle$ | Photon number of $k$th order moment at coordinate $z$ |
| $\langle n_k \rangle$ | Mean output photon number in an amplifier chain of $k$ elements |
| $\mathscr{P}$ | Degree of polarization |
| $P^{in}$ | Optical power input in fiber |
| $P^{out}$ | Optical power output of fiber |

| | |
|---|---|
| $P_{peak}$ | Pulse peak power |
| $P_p^{in}, P_p^{out}$ | Amplifier input or output pump power |
| $P_s^{in}, P_s^{out}$ | Amplifier input or output signal power |
| $P_p^{trans}$ | Input pump power for EDFA transparency |
| $P_{sat}(v)$ | Saturation power for Stark split laser systems |
| $P_{sat}^{in}$ | Saturation input power at 3dB gain compression |
| $P_{sat}^{out}$ | Saturation output power at 3dB gain compression |
| $\bar{P}$ | Receiver sensitivity (average input signal power per bit for $10^{-9}$ BER) |
| $\bar{P}_k$ | Photon flux at $\lambda_k$ (ratio of optical power to photon energy) |
| $\bar{P}_k^{in,out}$ | Fiber amplifier input and output photon flux at $\lambda_k$ |
| $\bar{P}_{sat}(\lambda_k)$ | Saturation power at $\lambda_k$ normalized to photon energy |
| $\hat{P}_{sat}(\omega)$ | Saturation power at $\omega$ for a single laser transition |
| $P_n^D(k)$ | Output photon statistics probability distribution of amplifier chain of type D = A,C with $k$ elements |
| $P_p^{abs}$ | Absorbed pump power |
| $P_n$ | Photon statistics probability distribution |
| $P_p, P_s$ | Pump and signal powers at fiber coordinate $z$ |
| $P_k$ | Optical power at fiber coordinate $z$ and at wavelength $\lambda_k$ |
| $P_0$ | Equivalent input noise power in bandwidth $\delta v$ |
| $P(1\|0)$ | Conditional error corresponding to the detection of symbol 1 when symbol 0 is transmitted |
| $P(0\|1)$ | Conditional error corresponding to the detection of symbol 0 when symbol 1 is transmitted |
| $\langle P \rangle_{path}$ | Path averaged optical power in optical amplifier chain |
| $\langle P_{ASE} \rangle_{path}$ | Path averaged ASE noise power in optical amplifier chain |
| $\mathscr{P}(r,z,t)$ | Linear polarization of a medium |
| $\mathscr{P}_H(r,z,t)$ | Linear polarization induced by a host medium |
| $\mathscr{P}_A(r,z,t)$ | Linear polarization induced by atomic dopant in a medium |
| $p$ | Signal power normalized to saturation power $P_{sat}$ |
| $p_{0k}$ | Normalized equivalent input noise power in bandwidth $\delta v$ at $\lambda_k$ |
| $p_k^+, p_k^-$ | Normalized forward and backward signal powers at $\lambda_k$ |
| $p_L^+, p_L^-$ | Normalized forward and backward signal powers at $z = L$ |
| $p_0^+, p_0^-$ | Normalized forward and backward signal powers at $z = 0$ |
| $p_{nm}$ | Boltzmann's distribution of main level $n$ and Stark sublevel $m$ |
| $p_{inh}$ | Normalized inhomogeneous density distribution |
| $q$ | Pump power normalized to saturation power $P_{sat}$ |
| $q^+, q^-$ | Normalized forward and backward pump powers |
| $q_L^+, q_L^-$ | Normalized forward and backward pump powers at $z = L$ |
| $q_0^+, q_0^-$ | Normalized forward and backward pump powers at $z = 0$ |
| $R$ | Pumping rate from laser level 1 to level 3 |
| $R_B$ | Bohr radius |
| $R_{bs}$ | Reflection coefficient associated with Rayleigh backscattering |
| $R_{eff}$ | Effective reflection coefficient of Rayleigh backscattering |
| $\mathscr{R}$ | Depolarization ratio |
| $R_{13}, R_{31}$ | Pumping rate and pump emission rate |
| $\mathscr{R}_{13}, \mathscr{R}_{31}$ | Pumping rate and pump emission rate for Stark split laser systems |

| | |
|---|---|
| $\text{RIN}(f)$ | Relative intensity noise at frequency $f$ |
| $S$ | Total electronic spin |
| $S_{a,e}$ | Experimental oscillator strength for absorption or emission |
| $S_{ed}$ | Oscillator strength for electric dipole transition |
| $S_{exp}$ | Experimental oscillator strength for electric dipole transition |
| $S_{md}$ | Oscillator strength for magnetic dipole transition |
| $SNR_e$ | Detector electrical power signal-to-noise ratio |
| $SNR_o(z)$ | Optical signal-to-noise ratio at fiber coordinate $z$ |
| $s$ | Saturation coefficient equal to $1/n_{sat}$ |
| $s$ | Soliton arrival time |
| $T$ | Time constant for unsaturated three-level system equilibrium |
| $T$ | Absolute temperature |
| $T_{a,e}$ | Dimensionless transition parameter for absorption or emission |
| $T_2$ | Atomic dephasing time |
| $T_{2,ij}$ | Effective dephasing time for leaser transition $(ij)$ |
| $\hat{T}$ | Passive fiber transmission matrix |
| $t_{rec}$ | Recovery time of transient gain dynamics |
| $t_{sat}$ | Saturation time of transient gain dynamics |
| $t_{trans}$ | EDFA transit time |
| $t_1, t_2$ | Characteristic time constants of transient gain dynamics |
| $U$ | Eigenvalue parameter for $LP_{01}$ modes $(U^2 + W^2 = V^2)$ |
| $U$ | Parameter defined as $(\eta_s - \eta_p)/(1 + \eta_p)$ |
| $u(z,t), u(\xi,\tau)$ | Soliton envelope |
| $u_k$ | Equals $+1$ or $-1$ for forward or backward signals at $\lambda_k$ |
| $V$ | Fiber V number |
| $V_c$ | V number at cutoff wavelength $\lambda_c$ $(V_c = 2.405)$ |
| $V(\lambda)$ | Voigt profile |
| $v_g$ | Group velocity |
| $X$ | Excess noise factor due to saturation |
| $x(SiO_2), x(Er_2O_3)$ | Molar concentration of $SiO_2$ or $Er_2O_3$ |
| $W$ | Eigenvalue parameter for $LP_{01}$ modes $(U^2 + W^2 = V^2)$ |
| $W$ | Molar weight of RE-doped glass |
| $W_{AB}$ | Rate of cooperative energy transfer between ions A and B |
| $W_{12}, W_{21}$ | Signal absorption rate and stimulated emission rate |
| $Z(SiO_2), Z(Er_2O_3)$ | Molar weights of $SiO_2$ or $Er_2O_3$ |
| $z_c$ | Soliton length |
| $z_0$ | Soliton period |

## GREEK SYMBOLS

| | |
|---|---|
| $\alpha$ | Ratio equal to $\omega_p^2/\omega_s^2$ |
| $\alpha_{GSA}(\lambda_p)$ | Ground state absorption coefficient at $\lambda_p$ |
| $\alpha_{ESA}(\lambda_p)$ | Excited state absorption coefficient at $\lambda_p$ |
| $\alpha_p$ | Er-doping absorption coefficient at $\lambda_p$ |
| $\alpha_s$ | Er-doping absorption coefficient at $\lambda_s$ |
| $\alpha_p'$ | Fiber background loss coefficient at $\lambda_p$ |

| | |
|---|---|
| $\alpha'_s$ | Fiber background loss coefficient at $\lambda_s$ |
| $\beta_{S'L'J',SLJ}$ | Fluorescence branching ratio |
| $\hat{\Gamma}$ | Overlap–inversion factor |
| $\Gamma_p$ | Overlap factor at $\lambda_p$ (equal to $a_0^2/\omega_p^2$ for confined doping) |
| $\Gamma_s$ | Overlap factor at $\lambda_s$ (equal to $a_0^2/\omega_s^2$ for confined doping) |
| $\Gamma_{ek}(z, p_j^{\pm})$ | Power dependent overlap integral factor for emission at $\lambda_k$ corresponding to a two-level system |
| $\Gamma_{ak}(z, p_j^{\pm})$ | Power dependent overlap integral factor for absorption at $\lambda_k$ corresponding to a two-level system |
| $\Gamma_{ap}(z, q^{\pm}, p_j^{\pm}, \alpha)$ | Power dependent overlap integral factor for absorption at $\lambda_p$ corresponding to a three-level system |
| $\Gamma_{ek}(z, q^{\pm}, p_j^{\pm}, \alpha)$ | Power dependent overlap integral factor for emission at $\lambda_k$ corresponding to a three-level system |
| $\Gamma_{ak}(z, q^{\pm}, p_j^{\pm}, \alpha)$ | Power dependent overlap integral factor for absorption at $\lambda_k$ corresponding to a three-level system |
| $\gamma_{ap}$ | Homogeneous pump absorption coefficient with confined Er-doping |
| $\gamma_{ak}$ | Homogeneous absorption coefficient with confined Er-doping |
| $\gamma_{ek}$ | Homogeneous emission coefficient with confined Er-doping |
| $\bar{\gamma}_{am}$ | Unsaturated absorption coefficient at $\lambda_m$ for confined Er-doping |
| $\bar{\gamma}_{em}$ | Unsaturated emission coefficient at $\lambda_m$ for confined Er-doping |
| $\tilde{\gamma}_{ak}$ | Inhomogeneously broadened absorption coefficient at $\lambda_k$ for confined Er-doping |
| $\tilde{\gamma}_{ek}$ | Inhomogeneously broadened emission coefficient at $\lambda_k$ for confined Er-doping |
| $\Delta E$ | Energy separation between two energy levels |
| $\Delta E_{21}$ | Energy separation between lowest energy levels of two Stark manifolds |
| $\Delta g(\omega)$ | Net gain coefficient at frequency $\omega$ ($\Delta g(\omega) = g_e(\omega) - g_a(\omega)$) |
| $\Delta g_k$ | Net gain coefficient at wavelength $\lambda_k$ ($\Delta g_k = g_{ek} - g_{ak}$) |
| $\Delta n$ | Refractive index step |
| $\Delta T$ | Pulse FWHM |
| $\Delta\lambda_{\text{FWHM}}$ | Full wavelength width at half maximum of transition line shape |
| $\Delta\lambda_{a,e}^{\text{eff}}$ | Effective width of transition cross section |
| $\Delta\lambda_{\text{hom}}$ | Homogeneous transition line width |
| $\Delta\lambda_{\text{inh}}$ | Inhomogeneous transition line width |
| $\Delta\nu_A$ | Effective ASE spectral width |
| $\Delta\nu_{\text{FWHM}}$ | Full frequency width at half maximum of transition line shape |
| $\Delta\omega$ | Spectral width of homogeneous laser transition (FWHM) |
| $\Delta\omega_{\text{inh}}$ | Inhomogeneous broadening spectral width |
| $\Delta\omega_{kj}$ | Spectral width of homogeneous laser transition ($jk$) |
| $\Delta\omega_{kj}^{\text{sat}}$ | Saturated spectral width of homogeneous laser transition ($jk$) |
| $\delta$ | Ratio of ground state to excited state absorption cross sections |
| $\delta_{ij}$ | Kronecker delta |
| $\delta E_{1j}$ | Energy difference between Stark sublevel $j$ and lowest energy level in manifold 1 |
| $\delta E_{2k}$ | Energy difference between Stark sublevel $k$ and lowest energy level in manifold 2 |
| $\delta E_1, \delta E_2$ | Average energy difference between Stark levels of manifolds 1 and 2 |

| | |
|---|---|
| $\delta n(\omega)$ | Gain-induced refractive index change at frequency $\omega$ |
| $\delta v_{eq}$ | Equivalent line width broadening |
| $\delta v$ | Elemental signal and noise bandwidth |
| $\varepsilon$ | Electric permittivity in free space |
| $\varepsilon = a_0/\omega_s$ | Er-doping confinement factor |
| $\varepsilon_0$ | Electric permittivity of a medium |
| $\varepsilon_p, \varepsilon_k$ | Ratio of background absorption coefficient to Er-doping absorption coefficient at $\lambda_p$ and $\lambda_k$ |
| $\eta$ | Photodetector quantum efficiency |
| $\eta_A$ | Cooperative energy transfer quantum yield |
| $\eta_c$ | Amplifier to photodetector coupling efficiency |
| $\eta_i$ | Amplifier input coupling efficiency |
| $\eta_{ij}$ | Radiative quantum efficiency of transition ($ij$) |
| $\eta_{in}$ | Amplifier input coupling efficiency |
| $\eta(^4I_{13/2})$ | Total radiative quantum efficiency of energy level $^4I_{13/2}$ |
| $\eta(\lambda)$ | Ratio of emission to absorption cross section at $\lambda$ |
| $\eta_p, \eta_s$ | Ratio of emission to absorption cross section at $\lambda_p$ and $\lambda_s$ |
| $\eta^{peak}$ | Ratio of emission to absorption cross section at peak wavelength $\lambda^{peak}$ |
| $\lambda_c$ | Fiber cutoff wavelength |
| $\lambda_c^{opt}$ | Optimal fiber cutoff wavelength |
| $\lambda_p, \lambda_s$ | Pump and signal wavelengths |
| $\lambda_k$ | Wavelength corresponding to signal at $v_k$ |
| $\lambda_{kj}$ | Wavelength corresponding to laser transition at $\omega_{kj}$ |
| $\lambda_{peak}$ | Peak wavelength of absorption or emission cross section |
| $\lambda^*$ | Wavelength at which absorption and emission cross sections are equal |
| $\mu$ | Atomic dipole operator |
| $\mu_{mn}$ | Atomic dipole matrix element |
| $\mu_0$ | Magnetic permeability in free space |
| $v_p, v_s$ | Pump and signal frequencies |
| $v_B$ | Brillouin frequency shift |
| $v_{peak}$ | Pump frequency of absorption or emission cross section |
| $v_R$ | Raman frequency shift |
| $\bar{v}$ | Effective frequency averaged over the gain spectrum |
| $\rho$ | Density matrix operator (only in Section 1.11) |
| $\hat{\rho}$ | Radiative density matrix operator |
| $\rho_{mn}$ | Density matrix element |
| $\hat{\rho}_{nn}$ | Radiative density matrix diagonal element ($= P_n$) |
| $\bar{\rho}_{11}, \bar{\rho}_{22}$ | Total occupation probabilities of manifolds 1 and 2 |
| $\rho(r)$ | Dopant density distribution |
| $\rho_A$ | Donor density in cooperative energy transfer |
| $\rho_B$ | Acceptor density in cooperative energy transfer |
| $\rho_0$ | Peak dopant density |
| $\rho(Er^{3+})$ | Density of $Er^{3+}$ ions per $cm^3$ or ppm wt |
| $\rho(Er_2O_3)$ | Density of $Er_2O_3$ oxide per ppm mol or wt% |
| $\rho(v, T)$ | Thermal radiation energy density |
| $\sigma(s)$ | Pulse timing jitter |
| $\sigma_{ESA}(\lambda)$ | Excited state absorption cross section at $\lambda$ |
| $\sigma_{GSA}(\lambda)$ | Ground state absorption cross section at $\lambda$ |

| | |
|---|---|
| $\sigma_a(\lambda)$ | Total absorption cross section at $\lambda$ |
| $\sigma_e(\lambda)$ | Total emission cross section at $\lambda$ |
| $\sigma_{am}, \sigma_{em}$ | Absorption and emission cross sections at $\lambda_m$ |
| $\sigma_{kj}(\omega)$ | Cross section of laser transition ($jk$) at frequency $\omega$ |
| $\sigma_{ae}^H$ | Homogeneous cross sections of Stark split two-level laser system |
| $\sigma_{a,e}^I$ | Inhomogeneous cross sections of Stark split two-level laser system |
| $\sigma_{a,e}^{peak}$ | Peak cross section value of absorption or emission |
| $\sigma_{kj}^{peak}$ | Peak cross section value of laser transition |
| $\tilde{\sigma}_{jk}$ | Slowly varying component of density matrix element $\rho_{jk}$ |
| $\sigma_{12}(\lambda_s)$ | Absorption cross section at $\lambda_s$ between levels 1 and 2 |
| $\sigma_{21}(\lambda_s)$ | Emission cross section at $\lambda_s$ between levels 1 and 2 |
| $\sigma^2(z)$ | Photon number variance at coordinate $z$ |
| $\sigma_k^2$ | Photon number output variance in an amplifier chain of $k$ elements |
| $\sigma_{S-ASE}^2$ | Signal–ASE beat noise variance at photodetector |
| $\sigma_{ASE-ASE}^2$ | ASE–ASE beat noise variance at photodetector |
| $\sigma_{LO-ASE}^2$ | Local oscillator–ASE beat noise variance at photodetector |
| $\sigma_{RIN}^2$ | Relative intensity noise variance at photodetector |
| $\hat{\sigma}_{S-ASE}^2$ | Signal–ASE beat noise power spectral density at photodetector |
| $\hat{\sigma}_{ASE-ASE}^2$ | ASE–ASE beat noise power spectral density at photodetector |
| $\tau$ | Spontaneous emission lifetime |
| $\tau_A$ | Radiative lifetime of donor ion A |
| $\tau_k$ | Intensity parameters ($k = 2, 4, 6$) |
| $\tau_{obs}, \tau_i^{obs}$ | Observed radiative lifetime of a given energy level $i$ |
| $\tau_{rad}$ | Calculated radiative lifetime of a given energy level |
| $\tau_{S'L'J',SLJ}$ | Radiative lifetime of transition $S'L'J' \rightarrow SLJ$ |
| $\tau_{S'L'J'}^{rad}$ | Total radiative lifetime of level ($S'L'J'$) |
| $\phi_{ASE}^{out}$ | Total output ASE photon flux |
| $\phi_p^{in}, \phi_p^{out}$ | Input or output pump photon flux |
| $\phi_s^{in}, \phi_s^{out}$ | Input or output signal photon flux |
| $\chi$ | Complex atomic susceptibility of a three-level laser medium |
| $\chi$ | Excess noise factor |
| $\chi'$ | Real part of the complex atomic susceptibility |
| $\chi''$ | Imaginary part of the complex atomic susceptibility |
| $\chi_{ed}$ | Local field correction for electric dipole transitions |
| $\chi_{md}$ | Local field correction for magnetic dipole transitions |
| $\chi_{jk}$ | Complex atomic susceptibility of a Stark-split three-level laser medium corresponding to laser transition ($jk$) |
| $\chi_{jk}^H$ | Complex atomic susceptibility of a Stark split three-level laser medium corresponding to homogeneous laser transition ($jk$) |
| $\chi^I$ | Complex atomic susceptibility of an inhomogeneously broadened, three-level laser medium |
| $\chi_I''$ | Imaginary part of the complex atomic susceptibility of an inhomogeneously broadened, three-level laser medium |
| $\psi_s(r,\theta)$ | Mode envelope distributions at $\lambda_{p,s}$ |
| $\psi_p(r), \psi_s(r)$ | Radial mode envelope distributions at $\lambda_{p,s}$ with unity peak |
| $\psi_k(r)$ | Radial mode envelope distributions at $\lambda_k$ with unity peak |
| $\bar{\psi}_s(r,\theta)$ | Normalized mode envelope distributions at $\lambda_s$ |
| $\xi(r,q)$ | Distribution whose surface integral is the overlap–inversion factor |

| | |
|---|---|
| $\Omega$ | Precession frequency |
| $\Omega$ | Dimensionless relative soliton velocity |
| $\delta\Omega$ | Collision-induced soliton velocity shift |
| $\Omega_k$ | Judd–Ofelt parameters ($k = 2, 4, 6$) |
| $\Omega_{jk}$ | Precession frequency associated with transition ($jk$) |
| $\omega$ | Light frequency ($\omega = 2\pi\nu$) |
| $\omega_p, \omega_s$ | Power mode size or radius at $\lambda_p$ and $\lambda_s$ |

# ERBIUM-DOPED FIBER AMPLIFIERS

# FUNDAMENTALS OF OPTICAL AMPLIFICATION IN ERBIUM-DOPED SINGLE-MODE FIBERS

# MODELING LIGHT AMPLIFICATION IN ERBIUM-DOPED SINGLE-MODE FIBERS

## INTRODUCTION

This first chapter describes the basic formalism used for modeling light amplification in erbium-doped fibers. We prefer to use here the term modeling as opposed to the term theory, as in fact the formal analysis of this type of problem is not based on any novel theory. Instead, modeling Er-doped fibers represents a theoretical understanding that makes use of several fundamental principles borrowed from classical electromagnetism, quantum mechanics, and, most essentially, laser physics. An Er-doped fiber is a system that combines both features of a single-mode waveguide and of a laser glass, for which the basic physics are well understood. However, these combined features produce several new characteristics unobtainable in an undoped fiber or a bulk glass laser. In order to theoretically analyze this particular type of device, a specific model had to be developed.

The first theoretical description of the effect of light amplification is undoubtedly contained in the 1958 proposition by A. L. Schawlow and C. H. Townes on the feasability of optical masers [1]. A review of the literature describing laser theory and travelling-wave amplification of light is beyond the scope of this chapter. But only a few fundamental papers need be studied for the theoretical grounds of this chapter's analysis, most of them published in the early 1960s (see for instance [2]–[4] and references therein).

The first analysis of rare-earth-doped fiber amplifier gain characteristics came with the demonstration of this type of device in 1964 [5]. The effect of modal overlap with the laser medium, initially studied with Gaussian beams in bulk optics, was investigated later on with guided modes in planar and square waveguides, [6] and [7]. The extension of these studies to the case of rare-earth-doped fibers came in the mid 1980s, driven by the interest for laser-diode-pumped Nd single crystal fibers, [8] and [9]. It is only with the emergence, near this period, of the first *Er*-doped fiber lasers and amplifiers, [10] and [11], that new theoretical developments had to be made. The reason for this is that Er:glass lasers behave quite differently from Nd:glass counterparts, owing to their two- and three-level system properties. For its most essential aspects, the theoretical analysis of Er-doped fiber amplifiers was completed by the end of 1990, [12]–[29].

The theoretical analysis developed throughout this chapter represents a unified description based on various models presented in references [12]–[29]. A unified theoretical description was needed because this literature uses a variety of symbols, notations, assumptions, approximations, and experimental data. It was later discovered that Er-doped fibers could be pumped either as a two-level or as a three-level laser system, so some aspects of earlier theoretical descriptions had to be extrapolated in order to be applicable to both schemes.

Also included are theoretical developments and results that were previously unpublished. For instance, using a Gaussian approximation for the guided mode envelopes with $1/e$ width determined by exact $LP_{01}$ mode solutions, accurate analytical expressions for the gain coefficient can be obtained for both two- and three-level systems (Appendix D); a complete density matrix theory for Stark split two- and three-level systems leads to a general expression for the complex atomic susceptibility (Appendix F); an analysis of inhomogeneous broadening, which generalizes results obtained in previous work [27], is made in Section 1.12. The book strives to take account of developments in modeling Er-doped fiber amplifiers during its preparation.

The background needed for understanding this chapter is basic calculus and elementary functional analysis. Only Section 1.11, covering the density matrix model, uses of quantum mechanical operator formalism at its starting point. The reader who is not familiar with such formalism can still follow subsequent developments and acquire a physical interpretation of this model, as it is essentially a semiclassical theory. The whole of this chapter can be skipped in a first reading, though it may be useful for reference. Chapters 2 and 3, which complete this part on the fundamentals of light amplification in Er-doped single-mode fibers, can also be read independently. While various experimental data and parameters are used throughout this chapter for computation and illustration purposes, this chapter is first and foremost theoretical. The justification for experimental data and specific parameter values can be found in Part B, which deals with the phenomenological aspects of Er:glass.

Mathematical notation was kept as simple as possible; various symbols are consistent throughout the book. The symbol list shows precise symbol interpretation. Important definitions, rate equations, and formulae are framed. Some of these theoretical results are associated with specific operating conditions and approximations (e.g., unsaturated regime, amplified spontaneous emission neglected, Gaussian approximation). They should not be applied outside the appropriate context. For easier future reference, maybe the reader can write the corresponding regime and approximations outside each frame. The many appendices detail tedious aspects of some derivations; important results are outlined and interpreted in the main text.

There are several steps to analyze how the signal light is amplified in $Er^{3+}$-doped single-mode fibers, starting as much as possible from the fundamentals. The first step is to derive rate equations for the atomic populations in a basic three-level laser system, as a simplified but conceptually useful representation of the actual $Er^{3+}$ laser glass (Section 1.1). It assumes the ratio of the magnitudes of the radiative and nonradiative rates corresponds to that of $Er^{3+}$:silica glass, the host material most commonly used in the fabrication of Er-doped fibers. Therefore, this basic analysis may not fully apply to the case of other types of glass hosts for erbium or other three-level rare-earth dopants in glass. For simplicity, the two-level pumping scheme, which represents a particular case of this three-level system analysis, will be left until Section 1.5.

We consider next a more accurate description of the actual three-level system, relevant to glass or crystal host surroundings. In such hosts the energy levels are split by the Stark effect induced by the local crystal field (Section 1.2). We show, through a more complex rate equation model, that due to the fast thermalization effect prevailing within the stark manifolds, the multilevel $Er^{3+}$:glass laser system is actually well described by the basis three-level model seen in Section 1.1.

We analyze then the effect of light propagation in a doped fiber waveguide. The results derived previously are applied to the case of a single-mode fiber whose core is doped with $Er^{3+}$ ions. The fundamental laser parameters such as $Er^{3+}$ dopant density, fluorescence lifetime, and cross sections are related to the fiber parameters such as mode envelope, mode radius, and signal gain coefficient (Section 1.3).

The effect of amplification of spontaneous emission (ASE) is introduced phenomenologically in Section 1.4. We can then derive the most general rate equations for the pump, the signal, and the ASE (Section 1.5), using the approximation that the laser transition is homogeneously broadened. A physical interpretation of these equations with respect to mode overlap effects with the doped fiber core is made in the same section. The most general rate equations must be solved by numerical methods, discussed in Section 1.6.

The particular case where the $Er^{3+}$ ion distribution in the fiber core corresponds to a step-like doping profile is analyzed in Section 1.7. The Gaussian envelope approximation makes it possible to integrate the gain coefficient over the transverse plane of the fiber core. This considerably simplifies the rate equations previously derived. In Section 1.8, it is shown that the gain coefficient is maximized for $Er^{3+}$ doping distributions confined in the fiber core; the corresponding rate equations are further simplified. Even in this simplest case, the rate equations must be solved numerically because of the nonlinear coupling between the pump and the signal. Two interesting regimes are the unsaturated gain regime and the low gain regime. Exact analytical solutions for rate equations in these two regimes are derived in Sections 1.9 and 1.10.

Another way to analyse three-level laser systems is through density matrix theory. The basic approach considered here is semiclassical, i.e., the atomic energy levels are quantized, while the electric field inducing the dipole transitions is classical. The interest of such treatment is to derive an expression for the complex atomic susceptibility (corresponding to the complex refractive index). From the susceptibility, both gain coefficient and associated refractive index changes can be calculated (Section 1.11).

We consider penultimately the situation most difficult to analyze. It is the closest approximation to reality. In this situation, the laser transition is inhomogeneously broadened by the random orientation of the crystal field. Modeling the fiber gain coefficient with inhomogeneously broadened Er:glass is analyzed in Section 1.12.

## 1.1 ATOMIC RATE EQUATIONS FOR THREE-LEVEL LASER SYSTEMS

We consider the three-level laser system with energy levels shown in Figure 1.1. By definition, level 1 is the ground level, level 2 is the metastable level characterized by a long lifetime $\tau$, and level 3 is the pump level. The laser transition of interest takes

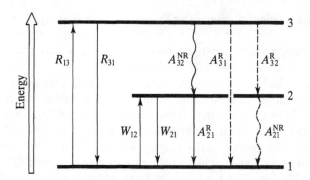

**FIGURE 1.1**   Energy level diagram corresponding to a basic three-level laser system, where the laser transition occurs between levels 1 (ground) and 2 (metastable). The symbols $R$, $W$, and $A$ correspond to pumping rates, stimulated emission rates, and spontaneous decay rates between related levels; superscripts R and NR refer to radiative and nonradiative emission, respectively.

place between levels 1 and 2. Another type of three-level system is possible; the metastable level is level 3 and the terminal level of the laser transition is level 2. To be accurate, we consider here only a three-level system whose ground level is the terminal level of the laser transition; this corresponds to the case of $Er^{3+}$ (Chapter 4).

The pumping rate from levels 1 and 3 is $R_{13}$ and the stimulated emission rate between levels 3 and 1 is $R_{31}$. From the excited state corresponding to level 3, there are two possibilities of decay, i.e., radiative (rate $A_3^R = A_{32}^R + A_{31}^R$) and nonradiative (rate $A_{32}^{NR}$). The spontaneous decay from level 3 is assumed to be predominantly nonradiative, i.e., $A_{32}^{NR} \gg A_3^R$. The stimulated absorption and emission rates between levels 1 and 2 are $W_{12}$ and $W_{21}$, respectively. The spontaneous radiative and nonradiative decay from the excited state corresponding to level 2 is $A_2 = A_{21}^R + A_{21}^{NR}$, with $A_{21}^R = 1/\tau$, where by definition $\tau$ is the fluorescence lifetime. It is assumed that the spontaneous decay is essentially radiative, i.e., $A_{21}^R \gg A_{21}^{NR}$. In the following, the spontaneous decays from levels 2 and 3 will be more simply referred to as $A_{21}$ and $A_{32}$. Let $\rho$ be the laser ion density and $N_1$, $N_2$ and $N_3$ the fractional densities, or populations, of atoms in the energy states 1, 2 and 3, respectively. By definition, $N_1 + N_2 + N_3 = \rho$.

We can now write the atomic rate equations corresponding to these populations:

$$\frac{dN_1}{dt} = -R_{13}N_1 + R_{31}N_3 - W_{12}N_1 + W_{21}N_2 + A_{21}N_2 \tag{1.1}$$

$$\frac{dN_2}{dt} = W_{12}N_1 - W_{21}N_2 - A_{21}N_2 + A_{32}N_3 \tag{1.2}$$

$$\frac{dN_3}{dt} = R_{13}N_1 - R_{31}N_3 - A_{32}N_3 \tag{1.3}$$

We consider now the steady state regime where the populations are time invariant, i.e., $dN_i/dt = 0$ ($i = 1, 2, 3$). Let $a = R_{31} + A_{32}$ and $b = W_{21} + A_{21}$. We obtain from Eqs. (1.2)–(1.3):

$$W_{12}N_1 - bN_2 + A_{32}N_3 = 0 \tag{1.4}$$

$$R_{13}N_1 - aN_3 = 0 \tag{1.5}$$

Relacing $N_3 = \rho - N_1 - N_2$ in the above and solving for $N_1$, $N_2$ yields:

$$N_1 = \rho \frac{ab}{b(a + R_{13}) + aW_{12} + R_{13}A_{32}} \tag{1.6}$$

$$N_2 = \rho \frac{R_{13}A_{32} + aW_{12}}{b(a + R_{13}) + aW_{12} + R_{13}A_{32}} \tag{1.7}$$

We replace then the definitions of $a$ and $b$ and factorize the term $A_{21}A_{32}$ in Eqs. (1.6)–(1.7) to obtain:

$$N_1 = \rho \frac{(1 + W_{21}\tau)\left(1 + \dfrac{R_{13}}{A_{32}}\right)}{(1 + W_{21}\tau)\left(1 + \dfrac{R_{13} + R_{31}}{A_{32}}\right) + W_{12}\tau\left(1 + \dfrac{R_{31}}{A_{32}}\right) + R_{13}\tau} \tag{1.8}$$

$$N_2 = \rho \frac{R_{13}\tau + W_{12}\tau\left(1 + \dfrac{R_{13}}{A_{32}}\right)}{(1 + W_{21}\tau)\left(1 + \dfrac{R_{13} + R_{31}}{A_{32}}\right) + W_{12}\tau\left(1 + \dfrac{R_{31}}{A_{32}}\right) + R_{13}\tau} \tag{1.9}$$

We assume now that the nonradiative decay rate $A_{32}$ dominates over the pumping rates $R_{13,31}$, i.e., $A_{32} \gg R_{13,31}$, and Eqs. (1.8)–(1.9) yield:

$$\boxed{\begin{aligned} N_1 &= \rho \frac{1 + W_{21}\tau}{1 + R\tau + W_{12}\tau + W_{21}\tau} \\[2ex] N_2 &= \rho \frac{R\tau + W_{12}\tau}{1 + R\tau + W_{12}\tau + W_{21}\tau} \end{aligned}} \tag{1.10}$$

with $R = R_{13}$. With the above result, we find that $N_3 = \rho - N_1 - N_2 = 0$, i.e., the pump level population is negligible due to the predominant nonradiative decay $(A_{32})$ toward the metastable level 2.

The steady state populations described by Eqs. (1.10) are central to the calculation of the gain coefficient in Er-doped fibers, as all the assumptions made for this particular three-level laser system entirely apply to the case of $Er^{3+}$ ions in silica glass (see Chapter 4). The effect of pump and signal excited state absorption (ESA), where atoms can be excited to a fourth energy level by absorption of a pump or a signal photon from the metastable level 2, was not described here. The effects of pump and signal ESA, for which the rate equations must be somewhat modified, are analyzed in detail in Chapter 4. The effects of ion/ion energy transfer occurring with ytterbium codoping and cooperative upconversion, to which correspond modified rate equations, are analyzed in Chapter 4.

## 1.2 ATOMIC RATE EQUATIONS IN STARK SPLIT LASER SYSTEMS

The three-level laser system analyzed previously actually corresponds to a simplified representation of the $Er^{3+}$:glass system. As described in more details in Chapter 4, the charge distribution in the glass host generates a permanent electric field, called a crystal or ligand field. A ligand field induces a Stark effect, which results in the splitting of the energy levels. Each of the energy levels we called 1, 2, and 3, is characterized by a total orbital momentum $J$; each splits into a manifold of $g = J + 1/2$ energy sublevels; $g$ is the total level degeneracy. The maximum Stark splitting actually represents a sizeable fraction of the energy separation between the main energy levels 1, 2, and 3.

Considering the multiplicity of levels and the magnitude of the splitting, the previous three-level laser system description seems then to be an oversimplified one. In this section, it is shown that such representation remains accurate, because of the effect of intramanifold thermalization. The effect of thermalization is to maintain a constant population distribution within the manifolds (Boltzmann's distribution), which eventually makes it possible to consider each of them as a single energy level.

The fact that the main energy levels are split with uneven internal subpopulation distributions also makes it possible to pump $Er^{3+}$:glass directly in level 2 (as opposed to level 3) and to achieve overall population inversion between levels 1 and 2. This pumping configuration, which corresponds to a quasi-two-level system, would not be possible if the levels were not split by the Stark effect. Whether the pumping occurs at the second or third levels is not important in the following theoretical analysis.

A schematic diagram of the energy levels with Stark split manifolds is shown in Figure 1.2. Let $\bar{N}_1$, $\bar{N}_2$, and $\bar{N}_3$ be the total population densities of each manifold corresponding to levels 1, 2, and 3, which have degeneracies $g_1$, $g_2$, and $g_3$, respectively. We have $\bar{N}_1 + \bar{N}_2 + \bar{N}_3 = \rho$. energy sublevels of manifolds 1, 2, and 3 are labeled with indexes $j$, $k$, and $l$, respectively, i.e., $j = 1, \ldots, g_1$, $k = 1, \ldots, g_2$, and $l = 1, \ldots, g_3$. The population of each sublevel is then noted as $N_{nm}$ ($n = 1, 2, 3$ and $m = j, k, l$).

The thermalization process occurring within each manifold is characterized by the nonradiative rates $A_{NR}^{+}$ and $A_{NR}^{-}$, which correspond to the excitation or deexcitation of the ions, with absorption or creation of a vibration quantum, or phonon. The condition of overall thermal equilibrium is:

$$A_{NR}^{-} N_{nm} = A_{NR}^{+} N_{n,m-1} \tag{1.11}$$

If $\Delta E_m = E_m - E_{m-1}$ is the energy difference between sublevels $(n, m)$ and $(n, m-1)$, we have the relation $N_{nm}/N_{n,m-1} = A_{NR}^{+}/A_{NR}^{-} = \exp(-\Delta E_m/k_B T)$, where $k_B$ is Boltzmann's constant ($k_B = 1.38 \times 10^{-23}$ $J/K$). A recurrence calculation from Eq. (1.11) gives the well-known relation [30]:

$$N_{nm} = \frac{\exp[-(E_m - E_1)/k_B T]}{\sum\limits_{m=1}^{g_n} \exp[-(E_m - E_1)/k_B T]} \bar{N}_n \equiv p_{nm} \bar{N}_n \tag{1.12}$$

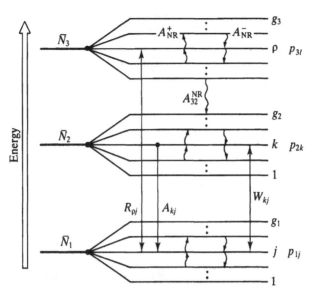

**FIGURE 1.2**   Energy level diagram corresponding to a Stark split three-level laser system. The symbols $A_{NR}^{\pm}$ indicate the thermalization rates between adjacent Stark sublevels.

where $p_{nm}$ is the Boltzmann distribution. The other rates relevant to the quasi-3-level laser system under study and shown in Figure 1.2 are: the pumping rate ($R_{lj}$), the stimulated emission rate ($W_{kj}$), the spontaneous emission rate ($A_{kj}$) and the nonradiative decay rate ($A_{32}$). As a consequence of thermal equilibrium, the absorption rate equals the stimulated emission rate for each sublevel pair, i.e., $R_{ij} = ji$ and $W_{kj} = W_{jk}$ (see derivation of Einstein's coefficients, [30] and [31], Section 4.5).

The rate equations corresponding to the three levels 1, 2, and 3, shown in Appendix A, are far more complex than in the previous section. However, by summing the equations corresponding to each manifold, i.e., Eqs. (A.1)–(A.3), (A.4)–(A.6), and (A.7)–(A.9), and using the relation $\sum_m N_{nm} = \sum_m p_{nm} \bar{N}_n = \bar{N}_n$, we obtain rate equations for the populations $\bar{N}_1$, $\bar{N}_2$, and $\bar{N}_3$ of the main levels:

$$\frac{d\bar{N}_1}{dt} = -\sum_j \sum_l R_{lj}(N_{1j} - N_{3l}) + \sum_j \sum_k \{A_{kj}N_{2k} + W_{kj}(N_{2k} - N_{1j})\} \qquad (1.13)$$

$$\frac{d\bar{N}_2}{dt} = A_{32}^{NR} N_{31} - \sum_j \sum_k \{A_{kj}N_{2k} + W_{kj}(N_{2k} - N_{1j})\} \qquad (1.14)$$

$$\frac{d\bar{N}_3}{dt} = -A_{32}^{NR} N_{31} + \sum_j \sum_l R_{lj}(N_{1j} - N_{3l}) \qquad (1.15)$$

At this point, we can define overall pumping, stimulated and spontaneous emission rates corresponding to the basic three-level system through:

$$\mathscr{R}_{13} = \sum_j \sum_l R_{lj} p_{1j} \qquad (1.16)$$

$$\mathscr{R}_{31} = \sum_j \sum_l R_{lj} p_{3l} \tag{1.17}$$

$$\mathscr{W}_{12} = \sum_j \sum_k W_{kj} p_{1j} \tag{1.18}$$

$$\mathscr{W}_{21} = \sum_j \sum_k W_{kj} p_{2k} \tag{1.19}$$

$$\mathscr{A}_{21} = \sum_j \sum_k A_{kj} p_{2k} \tag{1.20}$$

With the above definitions, and $A_{32}^{\text{NR}} p_{31} \equiv \mathscr{A}_{32}$, we obtain from Eqs. (1.13)–(1.15):

$$\frac{d\bar{N}_1}{dt} = -\mathscr{R}_{13}\bar{N}_1 + \mathscr{R}_{31}\bar{N}_3 - \mathscr{W}_{12}\bar{N}_1 + \mathscr{W}_{21}\bar{N}_2 + \mathscr{A}_{21}\bar{N}_2 \tag{1.21}$$

$$\frac{d\bar{N}_2}{dt} = \mathscr{W}_{12}\bar{N}_1 - \mathscr{W}_{21}\bar{N}_2 - \mathscr{A}_{21}\bar{N}_2 + \mathscr{A}_{32}\bar{N}_3 \tag{1.22}$$

$$\frac{d\bar{N}_3}{dt} = \mathscr{R}_{13}\bar{N}_1 - \mathscr{R}_{31}\bar{N}_3 - \mathscr{A}_{32}\bar{N}_3 \tag{1.23}$$

Such a system of rate equations for $\bar{N}_1$, $\bar{N}_2$, and $\bar{N}_3$ is identical to that of Eqs. (1.1)–(1.3) corresponding to the populations $N_1$, $N_2$, and $N_3$ of the basic three-level laser system seen in Section 1.1. Thus, because of the assumption of thermal equilibrium distributions of populations within each Stark manifold, the Stark split laser system is equivalent to a three-level laser system, but with the pumping, stimulated, and spontaneous emission rates defined in Eqs. (1.16)–(1.20). The solution of system of Eqs. (1.21)–(1.23) is therefore identical to that expressed in Eq. (1.10).

As it will be shown in Chapter 4, it is not necessary to know the exact positions of the Stark levels and associated transition cross sections in order to calculate the overall pumping, stimulated and spontaneous emission rates defined by $\mathscr{R}_{13,31}$, $\mathscr{W}_{12,21}$, and $\mathscr{A}_{21}$. In order to calculate these rates, we shall use in the following the overall phenomenological cross sections that one can determine experimentally, as opposed to the theoretical cross sections between individual Stark sublevels (see section 1.11). Individual Stark sublevels cannot be used in the theory, as a full knowledge of the laser subsystem characteristics, i.e., the Stark level positions, the Boltzmann distribution $p_{nm}$, and the associated radiative decay rates $A_{kj}$, would be necessary and are generally not known. The uncertainty about these parameters is due both to the difficulty of spectrally resolving and characterizing the individual Stark transitions in $Er^{3+}$:glass, and the effect of inhomogeneous broadening, in which the Stark levels randomly vary from one $Er^{3+}$-ion to the next, due to the random orientation of the ligand field in the glass host (see Chapter 4).

## 1.3  GAIN COEFFICIENT AND FIBER AMPLIFIER GAIN

We have derived previously the expressions of steady state atomic populations of a three-level, Stark split laser system which corresponds to $Er^{3+}$:silica glass. We must now take into account the effect of light confinement in the fiber waveguide, and relate the waveguide parameters to the pumping and stimulated emission rates. In this section, we develop the general expression of the signal gain coefficient in a single-mode fiber whose core is doped with $Er^{3+}$-ions of a radial density distribution $\rho(r)$.

When a signal light beam with intensity $I_s$ (power per area) at wavelength $\lambda_s$ traverses a slice of laser medium of infinitesimal thickness $dz$ and atomic population densities $N_1$ (lower level) and $N_2$ (upper level), the intensity change $dI_s$ is given by, [30] and [32]:

$$dI_s = \{\sigma_{21}(\lambda_s)N_2 - \sigma_{12}(\lambda_s)N_1\}I_s dz \qquad (1.24)$$

where $\sigma_{12}(\lambda_s)$ and $\sigma_{21}(\lambda_s)$ are the absorption and emission cross sections of the laser transition at $\lambda_s$, respectively. For simplicity of notation, the dependence in $z$ of $I_s$, $N_1$, and $N_2$ was made implicit. When the transition is degenerate, i.e., there exist $g_1$ and $g_2$ sublevels in the lower and upper states, then we have the relation $g_1\sigma_{12} = g_2\sigma_{21}$ [32], and Eq. (1.24) becomes:

$$\frac{dI_s}{dz} = \sigma_{12}(\lambda_s)\left\{\frac{g_1}{g_2}N_2 - N_1\right\}I_s \qquad (1.25)$$

which is the usual expression found in many textbooks. The relation $g_1\sigma_{12} = g_2\sigma_{21}$, used to obtain this result, assumes that all sublevels are equally populated and all laser transitions between pairs of sublevels are equally probable. Chapter 4 shows this is not the case with $Er^{3+}$:glass, because the magnitude of the Stark splitting within each manifold is such that a significant population difference actually exists between the sublevels. Equation (1.25) is therefore not accurate. It is possible, however, to define absorption and emission cross sections $\sigma_a(\lambda_s)$ and $\sigma_e(\lambda_s)$ for this particular laser system; in this case:

$$\frac{dI_s}{dz} = \{\sigma_e(\lambda_s)N_2 - \sigma_a(\lambda_s)N_1\}I_s = \sigma_a(\lambda_s)\{\eta(\lambda_s)N_2 - N_1\}I_s \qquad (1.26)$$

with, by definition:

$$\eta(\lambda_s) = \frac{\sigma_e(\lambda_s)}{\sigma_a(\lambda_s)} \qquad (1.27)$$

We see that Eqs. (1.25) and (1.26) are similar, but the concept of laser level degeneracy is now included in the parameter $\eta(\lambda_s)$, which can be viewed as a phenomenological cross section ratio. The parameter $\eta$ is of central importance in the modeling of Er-doped fiber amplifiers, as will be shown throughout the theoretical part of this book. From now on, the degeneracy factors $g_{1,2}$ will no longer be used.

In Eq. (1.26), the term $g = \sigma_a(\lambda_s)\{\eta(\lambda_s)N_2 - N_1\}$ represents the signal gain coefficient. The definition, magnitude, and spectral characteristics of the cross sections $\sigma_{a,e}(\lambda_s)$ involved in this coefficient will not be discussed here, but in Chapter 4. However, it is useful at this point to consider the dependence of the gain coefficient with wavelength and relative medium inversion. The relative inversion can be defined as

$D = (N_2 - N_1)/\rho = (2N_2 - \rho)/\rho$, with $D = -1$ when all the laser ions are in the ground state, and $D = 1$ when they are all in the excited state.

Figure 1.3 shows plots of the gain coefficient $g = \rho\{\sigma_e(\lambda_s)(1 + D) - \sigma_a(\lambda_s)(1 - D)\}/2$ around the 1.5 $\mu$m laser transition of a typical $Er^{3+}$:glass fiber (Type B fiber, see Chapter 4) for different values of the relative inversion $D$, by incremental steps of 0.2. The cross sections used in this example correspond to an alumino-germanosilicate $Er^{3+}$:glass type (Chapter 4), with peak values of $\sigma_a(\lambda_s = 1.53 \ \mu m) = 7 \times 10^{-25} \ m^2$ and $\sigma_e(\lambda_s = 1.53 \ \mu m) = 0.92\sigma_a(\lambda_s = 1.53 \ \mu m)$. The $Er^{3+}$ density is assumed to be $\rho = 1 \times 10^{+25} \ m^{-3}$. It is seen from the figure that for $D = -1$ all ions are in the ground state and the medium is absorbing at all signal wavelengths, as the gain coefficient is negative. As the relative inversion increases, however, a spectral region near the long wavelength side of the transition is characterized by a positive gain coefficient. For these wavelengths, the medium is amplifying, while for the rest of the spectrum, the medium is still absorbing. As $D$ increases toward the maximum value $(+1)$, representing complete medium inversion, the region of positive gain coefficient widens to spread eventually over the whole spectral range.

The above example illustrates a characteristic feature of three-level systems having a laser transition ending at the ground level, i.e., the corresponding laser medium is absorbing, or lossy, when not pumped or fully inverted. The medium becomes transparent (or loss-free) only when a certain ratio of excited to ground state ions is achieved, depending on the wavelength considered. Such a feature is not found with four-level laser systems, for which positive gain, in the absence of background loss, occurs as soon as a single atom is in the excited state (see for instance [31]).

It is important to note, however, that the relative inversion and the gain coefficient also vary with the fiber longitudinal coordinate $z$, as the pump energy at the origin

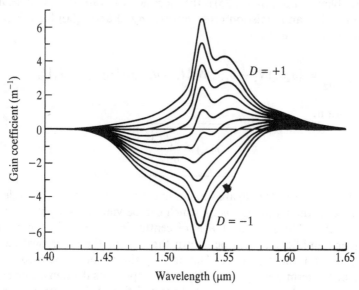

**FIGURE 1.3** Signal gain coefficient as a function of wavelength for different relative medium inversion $D = (N_2 - N_1)/\rho$, for an alumino-germanosilicate $Er^{3+}$:glass. $D = -1$ correspond to the case where all $Er^{3+}$ ions are in the ground state, and $D = +1$ to the case where they are all in the excited state.

of the inversion is absorbed along the fiber length. The actual net signal gain is given by the integral of the gain coefficient over the whole fiber length traversed. Because of the pump power decay along the length, an Er-doped fiber can have a portion of positive gain followed by a portion of negative gain, resulting in signal reabsorption. This feature will be analyzed later on, when discussing fiber length optimization.

We consider now the fact that the signal light is actually guided in the fiber. In a single-mode fiber, the signal power coupled into the mode has a finite spatial distribution over the fiber transverse plane. If the mode envelope is defined as $\psi_s(r, \theta)$, where $(r, \theta)$ represent the cylindrical transverse coordinates ($z$ is the fiber longitudinal coordinate), and if the optical power coupled into the mode is $P_s$, we can define the light intensity distribution $I_s(r, \theta)$ in the fiber transverse plane as:

$$I_s(r, \theta) = P_s \frac{\psi_s(r, \theta)}{\displaystyle\int_S \psi_s(r, \theta) r \, dr \, d\theta} \tag{1.28}$$

It is clear from the above definition that we have $\int_S I_s(r, \theta) r \, dr \, d\theta = P_s$. We can now write an equation for the rate of optical power change, as light propagates along the fiber in the guided mode, through:

$$\frac{dP_s}{dz} = \sigma_a(\lambda_s) \int_S \left\{ \eta(\lambda_s) N_2(r, \theta) - N_1(r, \theta) \right\} I_s(r, \theta) r \, dr \, d\theta$$

$$= \sigma_a(\lambda_s) P_s \int_S \left\{ \eta(\lambda_s) N_2(r, \theta) - N_1(r, \theta) \right\} \bar{\psi}_s(r, \theta) r \, dr \, d\theta \tag{1.29}$$

where the normalized mode envelope $\bar{\psi}_s$ is defined as:

$$\bar{\psi}_s(r, \theta) = \frac{\psi_s(r, \theta)}{\displaystyle\int_S \psi_s(r, \theta) r \, dr \, d\theta} \tag{1.30}$$

With this definition, the light intensity distribution is now given by $I_s(r, \theta) = P_s \bar{\psi}_s(r, \theta)$, which will be useful in the following.

The dependence in $(r, \theta)$ that was introduced in the populations $N_{1,2}$ in Eq. (1.29) is implicit in the definitions of the pumping and stimulated emission rates, as these vary across the fiber transverse plane, since both pump and signal powers are coupled into guided modes. The rates will be defined explicitly later on. Equation (1.29) shows that the effect of transverse distributions of both atomic populations and optical intensity in the transverse plane are actually averaged out, and that one can consider the evolution of the power coupled into the mode as a whole.

We focus now on the case of single-mode fibers. We define a mode *power* radius $\omega_s$ corresponding to the mode *power* distribution through [27]:

$$\omega_s = \left\{ \frac{1}{\pi} \int_S \psi_s(r, \theta) r \, dr \, d\theta \right\}^{1/2} = \left\{ 2 \int_S \psi_s(r) r \, dr \right\}^{1/2} \tag{1.31}$$

where we have used the radial symmetry property of the fundamental mode, and have assumed $\psi_s(0) = 1$. Note that the mode power radius defined here should not be confused with the mode spot size, which conventionally corresponds to the mode field distribution [33]. As the peak value of $\psi_s(r)$ is unity, the intensity distribution is simply given by $I_s(r) = P_s \psi_s(r)/\pi \omega_s^2 = P_s \bar{\psi}_s(r)$.

The above definitions apply to any type of single-mode fiber waveguide, i.e., of the step-index or graded-index types, if one assumes that the Er-doping is light enough so that the index profile and other waveguide parameters are not perturbed. In this case, the optical power evolution along the fiber only depends upon the interaction between the guided mode and the Er-doped region in the fiber core. As the light/atom interaction is an effect of *local* intensity rather than of the whole power guided into the mode, the actual mode envelope $\bar{\psi}_s$ must be precisely known. Without loss of generality, we can consider here the case of weakly guiding step-index fibers, for which the fundamental mode is well approximated by the $LP_{01}$ solution [33], [34].

The $LP_{01}$ mode envelope is defined through Bessel functions, detailed in Appendix B. The simpler Gaussian mode approximation, based on optimum power coupling efficiency into the actual Bessel solution, is also widely used, [33] and [35]. A relevant question is whether such Gaussian approximation is accurate enough when dealing with active waveguides such as Er-doped fibers. This question is addressed in detail in Appendix B. We show that in the general case, the Gaussian mode approximation is actually not accurate, as the corresponding intensity distribution can depart significantly from that of the exact Bessel solution, as discussed in [33]. Appendix B shows that the agreement is improved as the mode wavelength is close to the fiber cutoff wavelength. It is also shown in this appendix that the Bessel solution can be well approximated by a Gaussian *envelope*, provided its $1/e$ mode radius matches the power mode radius of the Bessel solution defined in Eq. (1.31). This approximation should not be confused with the usual Gaussian mode approximation previously mentioned (Appendix B).

In the case of passive (undoped) waveguides, the discrepancy between the intensity profiles between Gaussian and Bessel envelopes is not important; the important parameter is generally considered to be the total optical power launched into the mode. In active or doped fibers, the mode power evolution strongly depends upon the overlap between the mode intensity distribution and the doped core, and such discrepancy introduced by the Gaussian approximation can introduce significant errors in theoretical modeling. However, the advantage of the Gaussian envelope approximation, which we will use at different places in this book, is to provide closed-form expressions that are simpler to handle, but in some cases at the expense of accuracy. Equation (1.31) for the power mode radius applies to both exact and approximated cases.

When we choose for $\psi_s$ the actual $LP_{01}$ mode envelope, the mode radius (1.31) takes the form [36]:

$$\omega_s = a \frac{V K_1(W)}{U K_0(W)} J_0(U) \qquad (1.32)$$

where $a$ is the core radius, $V$, $U$ and $W$ are the conventional normalized frequency and eigenvalue parameters, and $J_0$, $K_{0,1}$ the Bessel solutions describing the LP modes

[33]. In the Gaussian approximation, i.e., $\psi_s(r) = \exp(-r^2/\omega_s^2)$, the mode radius definition (1.31) reduces to the usual $1/e$ mode size parameter.

The above analysis also applies to the pump power, when guided in the fundamental fiber mode. Equations (1.28)–(1.32) are then entirely applicable to the case of the pump at wavelength $\lambda_p$. When the laser system is pumped at level 3, Eq. (1.29) takes the form:

$$\frac{dP_p}{dz} = -\sigma_a(\lambda_p)P_p 2\pi \int_S N_1(r)\bar{\psi}_p(r)r\,dr \tag{1.33}$$

where $P_p$ is the pump power guided in the fundamental mode, and $\bar{\psi}_p(r)$ the corresponding normalized mode profile, defined similarly to Eq. (1.30). We have used the fact that $\eta(\lambda_p) = 0$ or $\sigma_e(\lambda_p) = 0$, as from level 3 the decay is essentially nonradiative and there is negligible stimulated emission at the pump wavelength (see Section 1.1). Equation (1.33) shows that in this case the pump is absorbed along the fiber, as the coefficient in the right-hand side is always negative.

In the above, we have assumed that the pump propagates in the same direction as the signal, i.e., towards positive $z$. This configuration is referred to as forward pumping. Instead, the pump could be launched from the output end of the fiber, and propagate in the direction opposite to the signal. It could also be launched in both directions in the fiber. These cases will be referred to as backward pumping and bidirectional pumping, respectively. For backward pumping, the pump rate equation is identical to Eq. (1.33), with an opposite sign in the right-hand side. For bidirectional pumping, the two types of pump equations must be solved simultaneously, with the population $N_1$ involving pumping rates from both contributions. Such specific pumping configurations will be analyzed in detail later on.

We have studied the case of a three-level laser system pumped at level 3. Another possibility is to pump the system directly at level 2. In this case, the laser medium operates as a two-level system. Contrary to the three-level pumping case, we have $\eta(\lambda_p) \neq 0$ in this case, as the pump wavelength falls into the laser transition, or signal band. For two-level pumping, Eq. (1.33) does not apply. Instead, the evolution of pump power $P_p(z)$ with fiber coordinate $z$ must be described by Eq. (1.29), with all subscripts s changed into p. The rate equations corresponding to the two pumping schemes will be detailed in Section 1.5.

A major assumption was made in the above rate equations for the pump and the signal powers. Indeed, we have assumed that all the ions present in the laser medium at any point $(z, r, \theta)$ are characterized by identical cross sections. This is equivalent to assuming homogeneous broadening, i.e., all ions occupy identical atomic sites in the glass host. This implies that the Stark effect induces identical energy level splitting for each ion. In reality, such is not the case, as the crystal field associated with each site in the glass host is random. Further, the site coordination by neighboring atoms in the host (such as codopants) can randomly vary from one ion to the next. These effects induce inhomogeneous broadening, which makes the analysis of signal gain more complex, as all the variations in cross section must be averaged at the macroscopic scale. The issue of modeling gain coefficient with inhomogeneous broadening is addressed in Section 1.12. Until then, we shall consider only the results associated with the homogeneous broadening approximation.

Having the definitions of the populations $N_1$ and $N_2$ in terms of pumping and stimulated emission rates, and the optical intensity distributions across the fiber core, we are now able to derive the basic rate equations for the pump and the signal. Before moving to this stage, it is useful to introduce the effect of amplified spontaneous emission, or ASE, so that we obtain the most general rate equations.

## 1.4 AMPLIFIED SPONTANEOUS EMISSION

The generation of noise in optical amplifiers is, in essence, an effect of the spontaneous deexcitation of the laser ions. As the ions have a finite excited state lifetime ($\tau = 10$ ms in the case of $Er^{3+}$:glass, see Chapter 4), some of the ions spontaneously return to the ground state, thereby emitting a photon. This photon has no coherence characteristics with respect to the incoming signal light, as opposed to a photon generated by stimulated emission. Therefore the collection of such spontaneously generated photons, being multiplied by the fiber amplifier, forms a background noise adding to the signal light. We refer to this background noise as amplified spontaneous emission, or ASE.

In this section, we shall not be concerned in the first principles analysis of the generation of spontaneous emission in doped fibers. This is the object of the next chapter. We shall simply use basic considerations and results for deriving a complete rate equation for the signal including the effect of ASE.

The number of randomly polarized photons at frequency between $v$ and $v + \delta v$ that are spontaneously generated in the direction of positive $z$, within an infinitesimal volume $dV$ of the laser medium, and coupled into the fiber mode, is given by [37]:

$$dn(v) = A_{21}g(v)\delta v \frac{\Delta\Omega}{4\pi} dV \int_S N_2(r, \theta)\bar{\psi}_s(r, \theta)r \, dr \, d\theta \qquad (1.34)$$

where $g(v)$ is the lineshape function [30], [31], $A_{21} = 1/\tau$ is the spontaneous decay rate, $\Delta\Omega/4\pi$ is the fraction of spontaneous light captured by the fiber, and the integral term accounts for the overlap between the density distribution of excited ions and the guided mode. The lineshape function is defined as $g(v) = 8\pi n^2 \tau \sigma_e(v)/\lambda_s^2$, where $n$ is the medium refractive index [30]. The capture solid angle $\Delta\Omega$ can be defined in a way similar to that corresponding to a blackbody mode in an unguided laser cavity [37], i.e., $\Delta\Omega = \lambda_s^2/n^2\pi\omega_s^2$, and the volume element can be defined as $dV = \pi\omega_s^2 \, dz$. The corresponding spontaneous emission power per unit frequency is given by $dP_{SE} = hv \, dn(v)$. With these definitions, the rate of creation of spontaneous emission power in bandwidth $\delta v$ becomes:

$$\frac{dP_{SE}}{dz} = 2P_0\sigma_e(v) \int_S N_2(r, \theta)\bar{\psi}_s(r, \theta)r \, dr \, d\theta \qquad (1.35)$$

where $P_0 = hv\delta v$ is the power of one spontaneous noise photon in bandwidth $\delta v$. The term $P_0$ is often referred to as equivalent input noise, as in high gain conditions the total spontaneous emission generated along the fiber is equivalent to the amplification of one fictitious input signal photon per mode in bandwidth $\delta v$, as detailed in Chapter 2. The factor of 2 in Eq. (1.35) reflects that spontaneous emission occurs in both polarization modes of the fiber.

Having obtained the rate of creation of spontaneous emission power at fiber coordinate $z$, and propagating in the direction of positive $z$, we can now express the evolution of the total signal power at wavelength $\lambda_s$ and in bandwidth $\delta \nu$ through Eqs. (1.29) and (1.35):

$$\frac{dP_s(\lambda_s)}{dz} = \sigma_a(\lambda_s) 2\pi \int_S \{\eta(\lambda_s) N_2(r)[P_s(\lambda_s) + 2P_0] - N_1(r)P_s(\lambda_s)\} \bar{\psi}_s(r) r \, dr \quad (1.36)$$

The above equation describes both phenomena of *amplification of signal* and *amplification of spontaneous noise*. If the excited state population $N_2$ is different from zero, the constant term $2P_0$ in the right-hand side of Eq. (1.36) causes the generation of optical noise along the fiber, whether optical signals are input to the fiber or not, which represents the total guided ASE in bandwidth $\delta \nu$. ASE is also generated in the direction of negative $z$, propagating in the direction opposite to the signal. We shall refer in the following to *forward* and *backward ASE*, with the forward case corresponding to the direction of positive $z$.

## 1.5 GENERAL RATE EQUATIONS FOR PUMP, SIGNAL, AND ASE

In the previous section, we have obtained rate equations for both pump and signal, the latter including ASE. These equations are expressed as a function of the atomic populations $N_1$ and $N_2$, which must now be written explicitly in terms of pump and signal intensity distributions, and associated saturation powers.

The steady state atomic populations $N_1$ and $N_2$ given in Eq. (1.10) are functions of the pumping rate $R = R_{13}$, which represents the pump absorption rate between level 1 and 3, and of the absorption and stimulated emission rates $W_{12,21}$ between levels 1 and 2. We do not distinguish anymore these rates with the rates $\mathscr{R}_{13}$ and $\mathscr{W}_{12,21}$ defined in Eqs. (1.16)–(1.19), which correspond to the Stark split three-level system, as we have seen that such a system is characterized by identical rate equations, Eqs. (1.21)–(1.23) when the Stark manifold populations are considered as a whole.

The emission rate $W_{21}$ at fiber coordinate $z$ and distance $r$ to the core axis is proportional to the signal intensity $I_s(r, z)$ and is given by [30]:

$$W_{21}(r, z) = \frac{\lambda_s^2}{8\pi n^2 h \nu_s \tau} I_s(r, z) g(\nu_s) \quad (1.37)$$

with $\nu_s = c/\lambda_s$. We introduce a new mode envelope definition $\psi_s(r)$, which has with unity peak value, i.e.,

$$\psi_s(r) = \pi \omega_s^2 \bar{\psi}_s(r) \quad (1.38)$$

which makes possible, after definitions in Eqs. (1.30) and (1.31) to express the signal intensity as $I_s(r, z) = P_s(z)\psi_s(r)/\pi \omega_s^2$, where $P_s(z)$ is the signal power coupled into the mode.

Replacing both Eq. (1.38), and the expression $g(v_s) = 8\pi n^2 \tau \sigma_e(v_s)/\lambda_s^2$ [30] for the lineshape function in Eq. (1.37) we obtain:

$$W_{21}(r, z)\tau = \frac{\sigma_e(v_s)\tau}{hv_s} I_s(r, \dot{z}) = \frac{\sigma_e(v_s)\tau}{hv_s \pi \omega_s^2} P_s(z)\psi_s(r) \tag{1.39}$$

We have seen in Section 1.3 that for the case of Stark split laser systems, the total or effective absorption and stimulated emission rates are given by a summation over all the possible transitions, Eqs. (1.18) and (1.19). Because of the differences in Stark level positions which give different Boltzmann distributions $p_{nm}$, we have therefore:

$$\mathcal{W}_{21} = \sum_j \sum_k W_{kj} p_{2k} \neq \sum_j \sum_k W_{jk} p_{1j} = \mathcal{W}_{12}. \tag{1.40}$$

whereas $W_{kj} = W_{jk}$, according to the Einstein coefficients, [30] and [31]. We must then consistently introduce the following definition for the absorption rate:

$$W_{12}(r, z)\tau = \frac{\sigma_a(v_s)\tau}{hv_s \pi \omega_s^2} P_s(z)\psi_s(r) \tag{1.41}$$

where $\sigma_a(v_s)$ is the phenomenological $E_r^{3+}$:glass absorption cross section at $v_s$. From eqs. (1.39) and (1.41), we have $W_{21} = (\sigma_e/\sigma_a)W_{12} = \eta W_{12}$, consistent with the usual relation $W_{21} = (g_1/g_2)W_{12}$ applying to basic multilevel laser systems [30]. The modified definitions in Eqs. (1.39) and (1.41) for absorption and emission rates in the case of Stark split laser systems can be more rigorously justified, as shown in Section 1.11, which describes a density matrix model for $Er^{3+}$:glass [27], and Chapter 4.

For the pumping rate $R(r, z)$ we introduce a definition similar to that of $W_{12}$:

$$R(r, z)\tau = \frac{\sigma_a(v_p)\tau}{hv\pi\omega_p^2} P_p(z)\psi_p(r) \tag{1.42}$$

where $\sigma_a(v_p)$ is the pump absorption cross section at $v_p$, and $P_p(z)$ the power at $v_p$ coupled into the pump mode of envelope $\psi_p(r)$ and radius $\omega_p$.

The saturation intensity for basic laser systems is by definition [30]:

$$I'_{sat}(v) = \frac{4\pi n^2 hv}{\lambda^2 g(v)} = \frac{hv}{\sigma_e(v)\tau} \tag{1.43}$$

The emission and absorption rates in Eqs. (1.39) and (1.41) can then be written as: $W_{21} = I_s/(\tau I'_{sat})$ and $W_{12} = I_s/(\eta \tau I'_{sat})$. Instead of using the definition (1.43) it is convenient to introduce a different saturation intensity $I_{sat}$ through:

$$I_{sat}(v) = \frac{hv}{[\sigma_a(v) + \sigma_e(v)]\tau} = \frac{hv}{\sigma_a(v)[1 + \eta(v)]\tau} = \frac{\eta(v)}{1 + \eta(v)} I'_{sat}(v) \tag{1.44}$$

The corresponding saturation power at signal frequency $v_s$ is then:

$$P_{sat}(v_s) = \frac{h v_s \pi \omega_s^2}{[\sigma_a(v_s) + \sigma_e(v_s)]\tau} = \frac{h v_s \pi \omega_s^2}{\sigma_a(v_s)[1 + \eta(v_s)]\tau} \qquad (1.45)$$

with similar definition for the pump frequency $v_p$. With these definitions, the pumping, absorption, and emission rates in Eqs. (1.40)–(1.42) take the form:

$$W_{12}\tau = \frac{1}{1 + \eta_s} \frac{P_s(z)}{P_{sat}(v_s)} \psi_s(r) \qquad (1.46)$$

$$W_{21}\tau = \frac{\eta_s}{1 + \eta_s} \frac{P_s(z)}{P_{sat}(v_s)} \psi_s(r) \qquad (1.47)$$

$$R\tau = \frac{P_p(z)}{P_{sat}(v_p)} \psi_p(r) \qquad (1.48)$$

with $\eta_s = \eta(\lambda_s)$. For the pumping rate $R$ in Eq. (1.48), we used the fact that $\eta(\lambda_p) = 0$ or $\sigma_e(\lambda_p) = 0$, as the laser system is pumped at level 3, from which decay is essentially nonradiative (see Section 1.1). If the system is pumped at level 2, i.e., the pump wavelength falls into the signal band, we have $R = 0$ and the rates defined in Eqs. (1.46) and (1.47) must be written explicitly in terms of both pump and signal contributions, as detailed later on.

Using Eqs. (1.46)–(1.48) we can now express the steady state populations $N_{1,2}$ in Eq. (1.10) as explicit functions of the optical powers and transverse mode profiles:

$$N_1(r, z) = \rho(r) \frac{1 + \dfrac{\eta_s}{1 + \eta_s} \dfrac{P_s(z)}{P_{sat}(v_s)} \psi_s(r)}{1 + \dfrac{P_s(z)}{P_{sat}(v_p)} \psi_p(r) + \dfrac{P_s(z)}{P_{sat}(v_s)} \psi_s(r)} \qquad (1.49)$$

$$N_2(r, z) = \rho(r) \frac{\dfrac{P_p(z)}{P_{sat}(v_p)} \psi_p(r) + \dfrac{1}{1 + \eta_s} \dfrac{P_s(z)}{P_{sat}(v_s)} \psi_s(r)}{1 + \dfrac{P_p(z)}{P_{sat}(v_p)} \psi_p(r) + \dfrac{P_s(z)}{P_{sat}(v_s)} \psi_s(r)} \qquad (1.50)$$

In order to simplify the above notations, we can normalize all optical powers with respect to their corresponding saturation powers, through:

$$\begin{aligned} p &= \frac{P_s(z)}{P_{sat}(v_s)} \\ q &= \frac{P_p(z)}{P_{sat}(v_p)} \end{aligned} \qquad (1.51)$$

and

$$p_0 = \frac{P_0}{P_{sat}(v_s)} = \frac{\sigma_a(v_s)(1 + \eta_s)}{\pi \omega_s^2} \delta v \tau$$

(1.52)

From now on, the use of lower case letters (e.g., $p$, $q$, $p_0$) for optical powers will indicate values normalized to their respective saturation powers. Note that $q$ will always be used in the following to refer to the normalized pump power.

For a *three*-level pumping scheme, the pump and signal rate equation (1.33) and (1.36) combined with Eqs. (1.49)–(1.52) take now the form:

$$\frac{dq}{dz} = -q\rho_0\sigma_a(v_p)\frac{2}{\omega_p^2}\int_S \frac{\rho(r)}{\rho_0}\psi_p(r)\left\{\frac{1 + \dfrac{\eta_s}{1 + \eta_s}p\psi_s(r)}{1 + q\psi_p(r) + p\psi_s(r)}\right\} r\,dr$$

(1.53)

$$\frac{dp}{dz} = \rho_0\sigma_a(v_s)\frac{2}{\omega_s^2}\int_S \frac{\rho(r)}{\rho_0}\psi_s(r)$$

$$\times \left\{\frac{\eta_s\left[q\psi_p(r) + \dfrac{1}{1 + \eta_s}p\psi_s(r)\right][p + 2p_0] - \left[1 + \dfrac{\eta_s}{1 + \eta_s}p\psi_s(r)\right]p}{1 + q\psi_p(r) + p\psi_s(r)}\right\} r\,dr$$

(1.54)

For a *two*-level pumping scheme, we obtain instead:

$$\frac{dq}{dz} = -q\rho_0\sigma_a(v_p)\frac{2}{\omega_p^2}\int_S \frac{\rho(r)}{\rho_0}\psi_p(r)\left\{\frac{\dfrac{\eta_s - \eta_p}{1 + \eta_s}p\psi_s(r) + 1}{1 + q\psi_p(r) + p\psi_s(r)}\right\} r\,dr$$

(1.55)

$$\frac{dp}{dz} = \rho_0\sigma_a(v_s)\frac{2}{\omega_s^2}\int_S \frac{\rho(r)}{\rho_0}\psi_s(r)$$

$$\times \left\{\frac{\eta_s\left[\dfrac{1}{1 + \eta_p}q\psi_p(r) + \dfrac{1}{1 + \eta_s}p\psi_s(r)\right][p + 2p_0] - \left[1 - \dfrac{\eta_p}{1 + \eta_p}q\psi_p(r) + \dfrac{\eta_s}{1 + \eta_s}p\psi_s(r)\right]p}{1 + q\psi_p(r) + p\psi_s(r)}\right\} r\,dr$$

(1.56)

Justification for these last two equations can be made through the following. Since in the case of the two-level system the pump falls into the signal band, we set $q = 0$ in eq. (1.54) and replace $p$ by $p'$. The remaining signal power, which we call $p'$, now represents the sum of both pump and signal at $v_p$ and $v_s$, respectively. Thus, to obtain a signal equation at $v_s$ only, the terms proportional to power $p'$ in Eq. (1.54) are rewritten under the form of two separate components: $p$, the power at $v_s$, and $q$, the power at $v_p$. For instance, $p'\psi_s(r)$ becomes $p\psi_s(r) + q\psi_p(r)$, which yields Eq. (1.56).

For the pump at $v_p$, the same equation as (1.56) is used, but the subscripts p, s are interchanged, and the noise source term $p_0$ is taken as zero. The spontaneous emission generated at $v_p$ can indeed be neglected, since it is not amplified. The same equations can be derived more formally by considering separately the absorption and emission rates $W_{12,21}(\nu_p)$ and $W_{12,21}(\nu_s)$, and redoing the analysis leading to Eq. (1.10) for the atomic populations. It is straightforward.

In Eqs. (1.53)–(1.56) we have introduced the additional parameter $\rho_0$, which represents the peak value of the $Er^{3+}$ concentration. With this definition, the term $\rho(r)/\rho_0$ thus represents a normalized $Er^{3+}$ concentration distribution and the terms $\rho_0 \sigma_a(\nu_{p,s})$ represent absorption coefficients. For the two-level pumping scheme, as the condition $\eta_s > \eta_p$ is usually met (Chapter 4), the coefficient in the right-hand side of Eq. (1.55) is negative. This means the pump power is absorbed along the fiber, as in the three-level case.

It is worth observing that when the coefficient $\eta_p$ is set to zero in Eqs. (1.55) and (1.56) for the two-level pumping scheme, these equations reduce exactly to Eqs. (1.53) and (1.54) corresponding to the three-level pumping scheme, which is expected. Thus, rate Eqs. (1.55) and (1.56) can be used for both *two-* and *three-*level pumping schemes, with $\eta_p = 0$ in the last case.

A second observation is that there exists actually no restriction in the validity of Eqs. (1.55) and (1.56) with respect to the choice of pump and signal frequencies (except $\nu_p = \nu_s$, when some terms are wrongly counted twice).

To give a physical interpretation to such a complex signal rate equation, let us first assume that the pump is turned off, i.e., $q = 0$, and that the ASE generated by the signal $p$ is negligible. We obtain from Eq. (1.56):

$$\frac{dp}{dz} = -p\rho_0 \sigma_a(\nu_s) \frac{2}{\omega_s^2} \int_S \frac{\rho(r)}{\rho_0} \frac{\psi_s(r)}{1 + p\psi_s(r)} r\, dr \tag{1.57}$$

This last result shows that when the doped fiber is unpumped, it is absorbing the signal. For input signal powers $P_s(0)$ much higher than the saturation power $P_{sat}(\nu_s)$, i.e., $p(0) = P_s(0)/P_{sat}(\nu_s) \gg 1$, the absorption coefficient in Eq. (1.57) vanishes, which means that the fiber becomes transparent, or bleached. In such a regime, equalization of level populations is achieved according to $\eta_s N_2 = N_1$. This can be obtained from Eq. (1.10) in the limit $\tau W_{12,21} \gg 1$ and $R = 0$, using the relation $\eta_s W_{12} = W_{21}$.

If now we turn the pump on, while keeping the input signal low so that $p \ll q$ at any fiber coordinate $z$ and neglect ASE, we obtain from Eq. (1.56):

$$\frac{dp}{dz} = p\rho_0 \sigma_a(\nu_s) \frac{2}{\omega_s^2} \int_S \frac{\rho(r)}{\rho_0} \psi_s(r) \frac{\dfrac{\eta_s - \eta_p}{1 + \eta_p} q\psi_p(r) - 1}{1 + q\psi_p(r)} r\, dr \equiv p\rho_0 \sigma_a(\nu_s) \hat{\Gamma}(q) \tag{1.58}$$

with

$$\hat{\Gamma}(q) = 2\pi \int_S \xi(r, q) r\, dr \tag{1.59}$$

and

$$\xi(r, q) = \frac{\rho(r)}{\rho_0} \bar{\psi}_s(r) \frac{Uq\psi_p(r) - 1}{1 + q\psi_p(r)} \tag{1.60}$$

and

$$U = \frac{\eta_s - \eta_p}{1 + \eta_p} \qquad (1.61)$$

The function $\xi(r, q)$ defined above represents a normalized gain coefficient density. The surface integral $\hat{\Gamma}(q)$ represents the averaging of population inversion changes across the core, and takes into account the finite overlap existing between the signal mode and the dopant distribution. This parameter $\hat{\Gamma}(q)$ represents then both an overlap factor and a degree of inversion, which can theoretically take values between $+1$ and $-1$. Therefore we call it the *overlap–inversion factor*. The quantity $g(\lambda_s) = \rho_0 \sigma_a(\lambda_s)\hat{\Gamma}(q)$ in eq. (1.58) represents the overall gain coefficient for the signal mode.

In the case of a uniform dopant distribution extending in the fiber cladding (i.e., $\rho(r) = \text{const} = \rho_0$) and of a very high pump power ($q \gg 1$), it can be seen from Eqs. (1.59) and (1.60) that the overlap–inversion factor $\hat{\Gamma}(q)$ takes the maximum value of $\hat{\Gamma}(q \gg 1) = U$, meaning that maximum possible inversion and full overlap between the laser medium and the signal mode are simultaneously achieved. On the other hand, when the pump is off, i.e., $q = 0$, and the doping distribution still extends uniformly over the cladding, we have $\hat{\Gamma}(0) = -1$, which is the case of a uniformly doped, unpumped fiber.

In practice, the overlap–inversion factor $\hat{\Gamma}(q)$ takes maximum values that are generally lower than $U$ at the highest pump powers, as the fiber cladding is usually undoped.

In order to illustrate the effect of finite overlap between the signal mode and the inverted medium, and how such overlap is affected by the total pump power launched into the mode, we show in Figure 1.4 plots of the function $\xi(r, q)$ against radius $r$, for different values of the normalized pump power $q$, as well as of the pump and signal envelopes $\bar{\psi}_{p,s}(r)$, which are exact Bessel mode solutions (see Appendix B). The corresponding overlap–inversion factors $\hat{\Gamma}(q)$ are also shown in the figure. In this example, the pump is chosen at $\lambda_p = 980$ nm ($\eta_p = 0$), and the signal is at $\lambda_s = 1.53$ $\mu$m, corresponding to $U = \eta_s = 0.92$; the fiber cutoff wavelength and numerical aperture are $\lambda_c = 980$ nm and NA $= 0.2$, respectively, corresponding to a fiber diameter of $2a = 3.75$ $\mu$m; the fiber core is assumed to be uniformly doped, and the cladding is undoped.

Figure 1.4 shows that, in this example, the function $\xi(r, q)$ is negative or zero over the whole fiber cross section for pump powers between $q = 0$ and $q = 1$. Thus, the overlap–inversion factor is negative ($\hat{\Gamma}(q) = -0.6$ to $-0.2$), which corresponds to a negative signal gain coefficient, or absorption. At higher pump power ($q = 2$), the function $\xi(r, q)$ is positive over the central region of the doped core, while it is negative or zero outside. The integrated factor $\hat{\Gamma}(q)$ is exactly zero in this case, meaning that the fiber is transparent for the signal. At higher pumps ($q = 5$ to $100$), the function $\xi(r, q)$ is positive over the whole doped region, giving overlap–inversion factors of $\hat{\Gamma}(q) = +0.2$ to $+0.5$. As previously mentioned, the maximum value of $0.5$ found for $\hat{\Gamma}(q)$ at the highest pump power is lower than $U = 0.92$, which illustrates the effect of finite overlap between the doped core and the signal mode.

More generally, Eq. (1.58) shows that a positive net gain coefficient at fiber coordinate $z$ can be achieved only if the integral over the core $\hat{\Gamma}(q)$ of the function

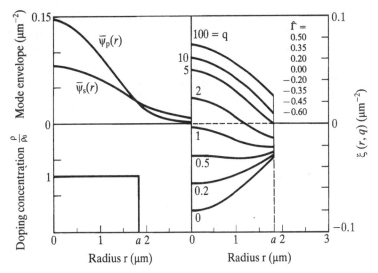

**FIGURE 1.4** Normalized exact Bessel solutions for the pump and signal mode envelopes as a function of radius $r$ for $\lambda_p = 980$ nm pump and $\lambda_p = 1530$ nm signal for a fiber with cutoff wavelength $\lambda_c = 980$ nm and core radius $a = 1.875$ $\mu$m (upper left quadrant). The lower left quadrant shows the relative Er-doping distribution $\rho(r)/\rho_0$. The right-hand side shows plots of the function $\xi(r, q)$ for different values of the normalized pump power $q$. The overlap–inversion factors $\hat{\Gamma}$ corresponding to the surface integral of $\xi(r, q)$ are also shown.

$\xi(r, q)$ is positive. A necessary, but not sufficient condition for the integral to be positive is that the integrand $\xi(r, q)$ be positive within a certain region centered around the fiber core. This condition is equivalent to $Uq\psi_p(r) - 1 > 0$, which is met only if: (1) $U$ is positive, or $\eta_s > \eta_p$, and (2) $q\psi_p(r) > 1/U$. From Eqs. (1.46) and (1.47) the first condition amounts to:

$$\eta_s = \frac{W_{21}(\lambda_s)}{W_{12}(\lambda_s)} > \frac{W_{21}(\lambda_p)}{W_{12}(\lambda_p)} = \eta_p \qquad (1.62)$$

The above relation involves only the ratio of cross sections and is therefore the most fundamental. It shows that if the ratio of stimulated emission to absorption at the signal wavelength is not greater than that at the pump wavelength, a positive signal gain coefficient can never be achieved. For $Er^{3+}$:glass pumped as a three-level system, we have $\eta_p = 0$, and condition (1.62) is always verified. With the two-level pumping scheme, this condition is also verified when the pump is near $\lambda_p = 1480$ nm and the signal is at longer wavelengths (see Chapter 4).

Obviously, the second condition, i.e., $q\psi_p(r) > 1/U$, cannot be achieved, at any distance $r$ from the fiber core axis, as the mode envelope $\psi_p(r)$ decreases with distance. Thus, if the Er-doping density distribution $\rho(r)/\rho_0$ is of finite dimension with respect to the pump mode envelope $\psi_p(r)$, the function $\xi(r, q)$ is zero in the evanescent tails of the pump mode, while it is maximum near the center of the doped core. This fact suggests that in order to achieve maximum signal gain, the Er-doping distribution should ideally be confined near the center of the fiber core. This effect of Er-doping confinement on the gain coefficient is analyzed in detail in Section 1.8 and Chapter 5.

As the pump is progressively absorbed along the fiber, the condition $\bar{\Gamma}(q) > 0$ corresponding to a positive gain coefficient, cannot be realized over any length. Assuming a gain coefficient initially positive at the fiber input end, there exists then a length $L_0$ at which the overlap–inversion factor $\bar{\Gamma}(q)$ is exactly zero. At this point, the signal is no longer amplified, and beyond, it is reabsorbed, as $\bar{\Gamma}(q)$ turns negative. The length $L_0$ corresponds to a maximum achieveable signal gain, and we shall refer to it as the optimum length. Such optimum length critically depends on the input power conditions, the pump and signal wavelengths, and the fiber waveguide parameters; its theoretical determination is most important. This aspect will be discussed in detail in Chapter 5.

Having completed a physical interpretation of the signal rate equation, we can now return to the general case. The ASE power was assumed to be at a single frequency $\nu_s$, identical to that of the signal, and to be confined spectrally in a narrow bandwidth $\delta\nu$. In reality, ASE is generated over a continuum of wavelengths spanning the entire $Er^{3+}$:glass gain spectrum. Likewise, in the most general case, the input signal can also be spread over a set of discrete lines, or optical channels, or form a spectral continuum.

In order to model this continuum, assuming given total spectral width $\Delta\nu$, one must then solve as many separate equations of the form Eq. (1.56), each corresponding to one signal + ASE spectral slot of width $\delta\nu$ and center wavelength $\lambda_k$. If the desired spectral resolution is $\delta\nu$, the number of equations to solve is thus $k = \Delta\nu/\delta\nu$. Thus, the set of these differential equations describes the whole signal and ASE spectrum. It is coupled and nonlinear, as each band is subject to saturation from all other spectral components, due to the homogeneous characteristic of the gain medium.

In the most general case of a wavelength continuum for the signal, one must redefine the absorption and stimulated emission rates in eqs. (1.46) and (1.47) through:

$$W_{12,21}(z, \nu) = \int_{\Delta\nu} \sigma_{a,e}(\nu) n_s(z, \nu) \bar{\psi}_s(r, \nu) \, d\nu \qquad (1.63)$$

where $n_s(z, \nu)$ is the signal photon number density at frequency $\nu$ and coordinate $z$. If we decompose the continuum into a set of $k$ discrete frequency slots of narrow width $\delta\nu \ll \Delta\nu$, with center frequency $\nu_k$ and optical power $P(z, \nu_k) = n_s(z, \nu_k) h\nu_k \delta\nu$, the integral in Eq. (1.63) can be approximated by a discrete sum, and we obtain, using definition (1.45):

$$W_{12}(z, r, \nu) = \frac{1}{\tau} \sum_k \frac{P_s(z, \nu_k)}{(1 + \eta_k) P_{sat}(\nu_k)} \psi_{sk}(r) \qquad (1.64)$$

$$W_{21}(z, r, \nu) = \frac{1}{\tau} \sum_k \frac{\eta_k P_s(z, \nu_k)}{(1 + \eta_k) P_{sat}(\nu_k)} \psi_{sk}(r) \qquad (1.65)$$

with $\psi_{sk}(r) = \psi_s(r, \nu_k)$ and $\eta_k = \eta(\nu_k)$. If the pump also forms a continuum, it is straightforward to extend the same considerations for the pumping rate $R$ defined in Eq. (1.48). For simplicity, we shall assume here that the pump is monochromatic, i.e., it is characterized by a single frequency $\nu_p$. The mode envelope at $\lambda_p$ is defined as $\psi_p(r)$.

Finally, we take into account the two possible propagation directions for the ASE, which brings the total number of signal + ASE equations to $2k$. As mentioned previously, the pump can also be launched in either or both directions in the fiber (forward, backward, or bidirectional pumping), so the most general set of equations should consider two pump equations as well.

Let $q^{\pm}$ be the normalized pump power and $p_k^{\pm}$ the normalized signal + ASE powers at wavelength $\lambda_k$, propagating in the direction of positive $(+)$ and negative $(-)$ fiber coordinate $z$, respectively. Thus $q^{+}$, $p_k^{+}$ represent the powers moving in the forward direction, and $q^{-}$, $q_k^{-}$ the powers moving in the backward direction.

We define $p_{0k}$ as the equivalent input noise power normalized to the saturation power at $\lambda_k$, i.e., $p_{0k} = P_0/P_{\mathrm{sat}}(\lambda_k) = h\nu\delta\nu/P_{\mathrm{sat}}(\lambda_k)$, where $P_{\mathrm{sat}}(\lambda_k)$ is given by Eq. (1.45). The mode radius $\omega_s(\nu_k)$ at frequency $\nu_k$ is noted for short $\omega_k$.

For the *three-level pumping scheme* we can generalize Eqs. (1.53) and (1.54) into:

$$-\frac{dq^{\pm}}{dz} = \pm\rho_0\sigma_a(\lambda_p)\int_S 2\pi r\,dr\,\frac{\rho(r)}{\rho_0}\bar{\psi}_p(r)\left\{\frac{1 + \sum_j \dfrac{\eta_j}{1+\eta_j}(p_j^{+} + p_j^{-})\psi_{sj}(r)}{1 + (q^{+} + q^{-})\psi_p(r) + \sum_j (p_j^{+} + p_j^{-})\psi_{sj}(r)}\right\}q^{\pm}$$

$$\tag{1.66}$$

$$\frac{dp_k^{\pm}}{dz} = \pm\rho_0\sigma_a(\lambda_k)\int_S 2\pi r\,dr\,\frac{\rho(r)}{\rho_0}\bar{\psi}_{sk}(r)$$

$$\times\left\{\frac{\eta_k\left[\dfrac{(q^{+} + q^{-})\psi_p(r)}{1+\eta_p} + \sum_j \dfrac{(p_j^{+} + p_j^{-})}{1+\eta_j}\psi_{sj}(r)\right][p_k + 2p_{0k}] - \left[1 + \dfrac{\eta_p(q^{+} + q^{-})\psi_p(r)}{1+\eta_p} + \sum_j \dfrac{\eta_j(p_j^{+} + p_j^{-})}{1+\eta_j}\psi_{sj}(r)\right]p_k}{1 + (q^{+} + q^{-})\psi_p(r) + \sum_j (p_j^{+} + p_j^{-})\psi_{sj}(r)}\right\}$$

$$\tag{1.67}$$

For the *two-level pumping scheme*, only one equation for each propagation direction is necessary to describe both pump and signal, and Eq. (1.56) can be generalized into:

$$\frac{dp_k^{\pm}}{dz} = \pm\rho_0\sigma_a(\nu_k)\int_S 2\pi r\,dr\,\frac{\rho(r)}{\rho_0}\bar{\psi}_{sk}(r)$$

$$\times\left\{\frac{\eta_k\left[\sum_j \dfrac{(p_j^{+} + p_j^{-})}{1+\eta_j}\psi_{sj}(r)\right][p_k + 2p_{0k}] - \left[1 + \sum_j \dfrac{\eta_j(p_j^{+} + p_j^{-})}{1+\eta_j}\psi_{sj}(r)\right]p_k}{1 + \sum_j (p_j^{+} + p_j^{-})\psi_{sj}(r)}\right\}$$

$$\tag{1.68}$$

The three main assumptions implied in the above general rate equations (1.66)–(1.68) can be restated as follows: (1) pump, signal, and ASE propagate in the fundamental fiber mode; (2) the gain medium is   homogeneously broadened; and (3) ASE is generated in both polarizations.

Another important assumption that was made is that the effect of fiber background loss is negligible. The justification for such assumption is that fiber amplifier lengths are generally short, i.e., they range from one meter to less than one hundred meters, due to the high Er dopant concentrations that can be achieved (Chapter 4). Thus, the background loss coefficients at both pump and signal wavelengths are negligible in comparison to the emission and absorption coefficients due to the Er-doping. However, two cases require the inclusion of such background loss coefficients, i.e., when the background loss is much higher than that of standard single-mode fibers (due to the presence of various impurities), and when the fiber length is made deliberately very long, i.e., from one kilometer to several tens of kilometers, which is the case of distributed fiber amplifiers (Chapter 6). In these particular cases, one must add to the right-hand side of Eq. (1.66) the term $\pm(-\alpha'_p q^\pm)$ and to the right-hand side of Eqs. (1.67)–(1.68) the term $\pm(-\alpha'_s p_k^\pm)$, where $\alpha'_p$ and $\alpha'_s$ represent the background loss coefficients at the pump and the signal wavelengths. In general, we shall omit these additional terms in the rate equations, except when required in the analysis of specific cases. We analyze in the next section the issues associated with the numerical resolution of such a complex differential equation set.

## 1.6  NUMERICAL RESOLUTION

In the most general case, the complexity of the pump, signal and ASE rate equations derived in the previous section obviously requires numerical solving. Only in some particular cases, which will be detailed in Sections 1.9 and 1.10 can such equations be solved exactly, with analytical expressions for the output pump, signal and ASE.

Numerical resolution of Eqs. (1.66)–(1.68) involves three basic degrees of difficulty: evaluation of boundary conditions; integration in the transverse plane to obtain the absorption and gain coefficients for pump and signal; integration over the fiber length.

The first issue is certainly the most delicate to address. For simplicity, we consider here only forward pumping, but the same considerations apply to the case of bidirectional pumping. Regardless of the pumping configuration, the forward and backward ASE distributions along the fiber are not known a priori. In the high gain regime, and in absence of strong signal input, the two ASE power distributions along the fiber mutually saturate the gain. While the forward ASE power spectrum is uniformly zero at the fiber input end, the corresponding backward ASE is maximum at this point. Thus, in order to perform the integration of the equations starting at $z = 0$, one must make some assumptions for the unknown backward ASE solution. Essentially there are two ways to solve such a two boundary value problem: the shooting method and the relaxation method [38].

The shooting method is a trial and error approach in which the unknown boundary value at $z = 0$ is first guessed. The system is then integrated from $z = 0$ to $z = L$, the corresponding boundary value at $z = L$ is compared to the desired value, and the difference is used to correct the guess in subsequent trials [38]. In our problem, we would first guess the normalized power spectrum of the backward ASE at $z = 0$, i.e.,

$p_j^-(z = 0)$. After several integration trials, the backward ASE spectrum at $z = L$ should eventually vanish, i.e., $p_j^-(z = L) = 0$. This approach actually presents considerable difficulties, since having $k$ differential equations for the backward ASE, we would have to first guess $k$ trial values for $p_j^-(z = 0)$. Since we do not know the actual shape of the spectrum, we would have to assume some arbitrary spectral distribution. We could use for this first guess, for instance, a uniform distribution; the shape of the forward ASE spectrum obtained at $z = L$ could then be used for the second guess. But no matter how sophisticated the guessing algorithm, convergence towards the actual solution would be very slow. In some cases, convergence might lead to solutions that arbitrarily depend upon the guessing algorithm. For these reasons, and owing to its complexity, the shooting method is not recommended in the general case.

The relaxation method makes iterative adjustments to the solution. An initial set of boundary values is chosen for the first integration. The system is then integrated again, for instance in the reverse direction, using corrected boundary values. The relaxation method offers many possibilities for solving Eqs. (1.66)–(1.68).

For instance, we can perform the first integration from $z = 0$ to $z = L$ without considering backward ASE, i.e., setting $p_j^-(z = 0) = p_j^+(z = 0) = 0$ as the initial boundary value set for the ASE spectrum. Then only the forward pump and forward ASE are integrated to $z = L$. The whole set of equation (including backward ASE) is then integrated from $z = L$ to $z = 0$, with $q^+(z = L)$, $p_j^+(z = L)$ as previously determined, for the starting conditions. After such a second integration, one has obtained a set of quasisolutions both the pump, forward, and backward ASE distributions along the fiber. These quasisolutions are not accurate, as the saturating effect of the backward ASE was neglected when performing the first integration run. In order to improve the accuracy, the system can then be integrated back and forth (i.e., from $z = 0$ to $z = L$, then from $z = L$ to $z = 0$) using for the unknown boundary values the quasisolutions obtained from the previous trial. With this algorithm, convergence towards the actual solutions can be rapidly achieved, due to a particular property of the rate equations involved. Indeed, overlooking backward ASE in the first integration results in an overestimation of the signal gain and forward ASE power, since the saturation effect by backward ASE is neglected. In the next integration from $z = L$ to $z = 0$, the saturation effect by the signal and forward ASE is then also overestimated, resulting in an underestimated backward ASE solution at $z = 0$. Using this quasisolution as a boundary condition for the next integration from $z = 0$ to $z = L$ is going to yield a more accurate solution for both pump and forward ASE, since saturation by backward ASE, while underestimated, is now taken into account. Each integration round-trip refines the mutual saturation effect and progressively the accuracy of the solutions. Iteration of this integration routine may be stopped when, for instance, the difference between successive solutions is less than 1%.

Obviously, the relaxation method is the most adequate for solving numerically Eqs. (1.66)–(1.68). The approach outlined above represents only one possibility. Improvements in convergence speed, i.e., the number of iterations required for a given accuracy, can be made in a variety of ways, depending on the actual pump and signal inputs. In the case of relatively high input signals at $z = 0$, the saturation effect will be essentially caused by the amplified signals. The terms corresponding to the ASE power spectra will not play a significant role therefore rapid convergence can be obtained with the relaxation algorithm. In the most difficult case, where input signals are weak and saturation by ASE is dominant (amplifier self-saturation), many iterations are required. In order to speed up the process of convergence, one may first assume a trial gain spectrum profile $G(\lambda_k)$, based on experimentally measured data, with a given peak value, e.g., $G(\lambda_{\text{peak}}) = 1000$. The boundary conditions for the backward ASE at $z = 0$ can then be set to the values $p_j^-(z = 0) = n_{\text{sp}}(G(\lambda_j) - 1)p_{0j}$, where $n_{\text{sp}}$ is the spontaneous emission factor (see Chapter 2). The success of this method depends on the distance between the first guess for $G(\lambda_{\text{peak}})$ and the actual solution. An exact analytical solution for $G(\lambda)$ can also be used with an approximate value for $n_{\text{sp}}$, as shown in Section 5.5.

In the most general case described by Eqs. (1.66)–(1.68), the gain coefficient must be integrated over the transverse plane at each fiber step, i.e., between $z$ and $z + dz$; the relaxation routine may consume much computer time. Any possible increase in convergence speed through improved boundary evaluation algorithms is therefore of great importance. But some specific problems in Er-doped fiber amplifiers do not require such a complex description. They can be analyzed through simplified numerical models (Sections 1.7 and 1.8) or exact analytical solutions (Sections 1.9 and 1.10).

An example of a computer program organization and subroutines for numerical integration of general rate Eq. (1.68), based on a Runge–Kutta algorithm, is given in Appendix C.

Various theoretical papers found in the literature claim to present comprehensive models. Such models are actually based on the same equations (1.66)–(1.68) as described above, and differ for the most in minor approximations, notations, and parameter definitions. Comprehensive models may be interpreted as accurate compared with earlier theoretical models that use simplifying approximations to solve practically Eqs. (1.66)–(1.68) [15], [17]. But the theoretical model leading to Eqs. (1.66)–(1.68) is based on the main assumption of *homogeneous gain broadening*. This assumption prevents even the most accurate implementation of the model from comprehensively describing the system, as it does not take into account the small effects of gain inhomogeneity observed in Er-doped fibers (see Section 1.12 and Chapters 4 and 5).

Accurate predictions of experimental data through numerical integration of most general Eqs. (1.66)–(1.68) strongly depend upon knowledge of experimental parameters such as mode envelopes, Er-doping radial distribution, and Er:glass cross sections. The characterization of these essential laser parameters is discussed in Chapter 4, while comparisons between accurate theoretical modeling and experimental data are discussed in Chapter 5.

In the next section, we derive simplified rate equations in which integration over the transverse plane can be performed analytically under certain approximations. This simplification yields closed-form expressions for the absorption and gain coefficients $g_a$ and $g_e$. The use of such closed-form expressions in solving the pump and signal rate equations is fully justified when the Er-doping distribution corresponds to a step profile.

## 1.7 RATE EQUATIONS WITH STEP Er-DOPING

We introduce the $z$-dependent pump emission and absorption coefficients at $\lambda_p$, i.e., $g_{ep}(z), g_{ap}(z)$, and the signal emission and absorption coefficients at $\lambda_k$, i.e., $g_{ek}(z), g_{ak}(z)$, by rewriting Eqs. (1.66)–(1.68) under a more compact form, which applies to both two- and three-level pumping schemes:

$$\pm \frac{dq^{\pm}}{dz} = (g_{ep} - g_{ap})q^{\pm} \tag{1.69}$$

$$\pm \frac{dp_k^{\pm}}{dz} = (g_{ek} - g_{ak})p_k^{\pm} + 2g_{ek}p_{0k} \tag{1.70}$$

with

$$g_{em}(z) = \eta_m \rho_0 \sigma_{am} \int_S 2\pi r \, dr \frac{\rho(r)}{\rho_0} \bar{\psi}_m(r) \frac{\dfrac{(q^+ + q^-)}{1 + \eta_p}\psi_p(r) + \sum_j \dfrac{(p_j^+ + p_j^-)}{1 + \eta_j}\psi_j(r)}{1 + (q^+ + q^-)\psi_p(r) + \sum_j (p_j^+ + p_j^-)\psi_j(r)} \tag{1.71}$$

$$g_{am}(z) = \rho_0 \sigma_{am} \int_S 2\pi r \, dr \frac{\rho(r)}{\rho_0} \bar{\psi}_m(r) \frac{1 + \dfrac{\eta_p(q^+ + q^-)}{1 + \eta_p}\psi_p(r) + \sum_j \dfrac{\eta_j(p_j^+ + p_j^-)}{1 + \eta_j}\psi_j(r)}{1 + (q^+ + q^-)\psi_p(r) + \sum_j (p_j^+ + p_j^-)\psi_j(r)} \tag{1.72}$$

where $m = p$ or $k$ indicates the pump or the signal parameters, and $\sigma_{em} = \sigma_e(\lambda_m)$ and $\sigma_{am} = \sigma_a(\lambda_m)$. For the three-level pumping scheme, and for the pump ($m = p$) the emission coefficient $g_{ep}$ is zero, as $\eta_p = 0$ in Eq. (1.71), and the absorption coefficient $g_{ap}$ is given by Eq. (1.72).

The first simplification that can be done in the computation of the coefficients $g_{em}$, $g_{am}$ in Eqs. (1.71) and (1.72) is to assume that the envelopes $\psi_j(r)$ are equal for all the signal modes, i.e., $\psi_j(r) = \psi_s(r)$. It is useful to keep a separate definition for the mode envelope $\psi_p(r)$ at $\lambda_p$, in order to obtain a definition for the coefficients that would also apply to the case of the three-level pumping scheme, where the approximation $\psi_p(r) \approx \psi_s(r)$ is not valid. Using the ratio

$$f(r) = \frac{\psi_p(r)}{\psi_s(r)} \tag{1.73}$$

We obtain, in the case of three-level systems, the pump and signal coefficients from Eqs. (1.71)–(1.73):

$$g_{ap}(z) = \rho_0 \sigma_{ap} \frac{2}{\omega_p^2} \int_S r \, dr \frac{\rho(r)}{\rho_0} f(r)\psi_s(r) \frac{\psi_s(r)^{-1} + \sum_j \dfrac{\eta_j(p_j^+ + p_j^-)}{1 + \eta_j}}{\psi_s(r)^{-1} + (q^+ + q^-)f(r) + \sum_j (p_j^+ + p_j^-)} \tag{1.74}$$

$$g_{ek}(z) = \eta_k \rho_0 \sigma_{ak} \frac{2}{\omega_s^2} \int_S r \, dr \frac{\rho(r)}{\rho_0} \psi_s(r) \frac{(q^+ + q^-)f(r) + \sum_j \dfrac{(p_j^+ + p_j^-)}{1 + \eta_j}}{\psi_s(r)^{-1} + (q^+ + q^-)f(r) + \sum_j (p_j^+ + p_j^-)} \tag{1.75}$$

$$g_{ak}(z) = \rho_0 \sigma_{ak} \frac{2}{\omega_s^2} \int_S r\, dr \frac{\rho(r)}{\rho_0} \psi_s(r) \frac{\psi_s(r)^{-1} + \sum_j \dfrac{\eta_j(p_j^+ + p_j^-)}{1 + \eta_j}}{\psi_s(r)^{-1} + (q^+ + q^-)f(r) + \sum_j (p_j^+ + p_j^-)} \tag{1.76}$$

with $g_{ep}(z) = 0$.

In the case of two-level systems, we have $f(r) \approx 1$ as $\psi_p(r) \approx \psi_s(r)$, and setting $q^+ = q^- = 0$ we obtain from Eqs. (1.75)–(1.76):

$$g_{ek}(z) = \eta_k \rho_0 \sigma_{ak} \frac{2}{\omega_s^2} \int_S r\, dr \frac{\rho(r)}{\rho_0} \psi_s(r) \frac{\sum_j \dfrac{(p_j^+ + p_j^-)}{1 + \eta_j}}{\psi_s(r)^{-1} + \sum_j (p_j^+ + p_j^-)} \tag{1.77}$$

$$g_{ak}(z) = \rho_0 \sigma_{ak} \frac{2}{\omega_s^2} \int_S r\, dr \frac{\rho(r)}{\rho_0} \psi_s(r) \frac{\psi_s(r)^{-1} + \sum_j \dfrac{\eta_j(p_j^+ + p_j^-)}{1 + \eta_j}}{\psi_s(r)^{-1} + \sum_j (p_j^+ + p_j^-)} \tag{1.78}$$

The integrals involved in Eqs. (1.74)–(1.76) for three-level systems and Eqs. (1.77) and (1.78) for two-level systems cannot be performed to yield closed-form expressions without some approximations. The first approximation that can be done is to use Gaussian envelopes for $f(r)$ and $\psi(r)$. As the normalized Er-doping density $\rho(r)/\rho_0$ can follow any distribution, e.g., Gaussian, in the general case, such an approximation for the mode envelopes is not sufficient. A second type of approximation assumes the Er-doping density distribution to be uniform over a given doping radius. The doping radius may be comparable to the mode sizes $\omega_{p,s}$, or somewhat smaller. As shown below, the integrals involved in Eqs. (1.74)–(1.78) can be reduced in either case to elementary functions; in the three-level pumping scheme, further approximations will be necessary).

The Gaussian *envelope* approximation for $\psi_{p,s}(r)$ should not be confused with the Gaussian *mode* approximation. In this latter, the exact Bessel mode solution is replaced by a Gaussian mode solution whose power launching efficiency into the exact mode is optimized (see Appendix B). In contrast, the Gaussian envelope approximation, considered here, replaces the exact Bessel envelope by a Gaussian envelope with a $1/e$ radius equal to the power radius of the Bessel solution. The advantage of this Gaussian envelope approximation is that the intensity distributions $\bar{\psi}_{p,s}(r) = \psi_{p,s}(r)/\pi\omega_{p,s}^2$ of exact and approximated functions are better matched, in comparison to the case of the Gaussian mode approximation (see Appendix B for discussion).

The Gaussian envelopes are then defined through

$$\psi_{p,s}(r) \approx \exp\left(-\frac{r^2}{\omega_{p,s}^2}\right) \tag{1.80}$$

(with $\omega_{p,s}$ representing the power mode size corresponding to that of the exact Bessel solution, according to Eqs. (1.31) and (1.32).

We assume next that the Er-doping density $\rho(r)$ has a step-like distribution which is uniform across a doping radius $a_0$, i.e.,

$$\rho(r) = \begin{cases} \rho_0 & \text{for} \quad r \leqslant a_0 \\ 0 & \text{for} \quad r > a_0 \end{cases} \tag{1.81}$$

Considering first the simplest case, i.e., the two-level system, we obtain after elementary integration with change of variables $u = \psi_s(r)$ in Eqs. (1.77) and (1.78):

$$g_{ek}(z) = \eta_k \rho_0 \sigma_{ak} \Gamma_{ek}(z, p_j^{\pm}) \tag{1.82}$$

$$g_{ak}(z) = \rho_0 \sigma_{ak} \Gamma_{ak}(z, p_j^{\pm}) \tag{1.83}$$

$$\Gamma_{ek}(z, p_j^{\pm}) = \Gamma_s \left\{ \mathscr{D}_e + \frac{\mathscr{D}_e}{\kappa \Gamma_s} \log \left( 1 - \frac{\kappa}{1 + \kappa} \Gamma_s \right) \right\} \tag{1.84}$$

$$\Gamma_{ak}(z, p_j^{\pm}) = \Gamma_s \left\{ \mathscr{D}_a + \frac{\mathscr{D}_a - 1}{\kappa \Gamma_s} \log \left( 1 - \frac{\kappa}{1 + \kappa} \Gamma_s \right) \right\} \tag{1.85}$$

with

$$\boxed{\Gamma_s = 1 - \exp\left\{ -\left( \frac{a_0}{\omega_s} \right)^2 \right\}} \tag{1.86}$$

$$\kappa = \sum_j (p_j^+ + p_j^-) \tag{1.87}$$

$$\mathscr{D}_e = \frac{1}{\kappa} \sum_j \frac{p_j^+ + p_j^-}{1 + \eta_j} \tag{1.88}$$

$$\mathscr{D}_a = \frac{1}{\kappa} \sum_j \frac{\eta_j}{1 + \eta_j} (p_j^+ + p_j^-) \tag{1.89}$$

In the three-level system case, Eqs. (1.74)–(1.76) with Eqs. (1.73) and (1.80) cannot be integrated without further approximations. Indeed, considering for instance the case of the pump absorption coefficient in Eq. (1.74), we obtain after a change of variables, and using previous definitions (1.86)–(1.89):

$$g_{ap}(z) = \rho_0 \sigma_{ap} \int_{1-\Gamma_p}^1 \frac{1 + \kappa \mathscr{D}_a x^\alpha}{1 + (q^+ + q^-)x + \kappa x^\alpha} \, dx = \rho_0 \sigma_{ap} \Gamma_{ap}(z, q^{\pm}, p_j^{\pm}, \alpha) \tag{1.90}$$

with $\alpha = \omega_p^2 / \omega_s^2$. As $\alpha$ is not generally an integer, it is not possible to directly reduce Eq. (1.90) to a closed-form expression. But one can make the approximation $\alpha \approx 1$ in the integrand of Eq. (1.90) to yield an elementary integral; the validity of this approximation is then studied a posteriori. The same type of approximation is made for the signal coefficients in Eqs. (1.75) and (1.76); we obtain after elementary integration (Appendix D):

$$g_{ap}(z) \approx \rho_0 \sigma_{ap} \Gamma_{ap}(z, q^{\pm}, p_j^{\pm}, 1) \tag{1.91}$$

$$g_{ek}(z) \approx \eta_k \rho_0 \sigma_{ak} \Gamma_{ek}(z, q^{\pm}, p_j^{\pm}, 1) \tag{1.92}$$

$$g_{ak}(z) \approx \rho_0 \sigma_{ak} \Gamma_{ak}(z, q^{\pm}, p_j^{\pm}, 1) \tag{1.93}$$

$$\Gamma_{ap}(z, q^{\pm}, p_j^{\pm}, 1) = \Gamma_p \left\{ \mathscr{D}_a' + \frac{\mathscr{D}_a' - 1}{\kappa' \Gamma_p} \log \left( 1 - \frac{\kappa'}{1 + \kappa'} \Gamma_p \right) \right\} \tag{1.94}$$

$$\Gamma_{ek}(z, q^{\pm}, p_j^{\pm}, 1) = \Gamma_s \left\{ \mathscr{D}_e' + \frac{\mathscr{D}_e'}{\kappa' \Gamma_s} \log \left( 1 - \frac{\kappa'}{1 + \kappa'} \Gamma_s \right) \right\} \tag{1.95}$$

$$\Gamma_{ak}(z, q^{\pm}, p_j^{\pm}, 1) = \Gamma_s \left\{ \mathscr{D}_a' + \frac{\mathscr{D}_a' - 1}{\kappa' \Gamma_s} \log \left( 1 - \frac{\kappa'}{1 + \kappa'} \Gamma_s \right) \right\} \tag{1.96}$$

with

$$\boxed{\Gamma_p = 1 - \exp\left\{ -\left( \frac{a_0}{\omega_p} \right)^2 \right\}} \tag{1.97}$$

$$\kappa' = q^+ + q^- + \sum_j (p_j^+ + p_j^-) \tag{1.98}$$

$$\mathscr{D}_e' = \frac{1}{\kappa'} \left\{ q^+ + q^- + \sum_j \frac{p_j^+ + p_j^-}{1 + \eta_j} \right\} \tag{1.99}$$

$$\mathscr{D}_a' = \frac{1}{\kappa'} \sum_j \frac{\eta_j (p_j^+ + p_j^-)}{1 + \eta_j} \tag{1.100}$$

The validity of the approximation $\alpha \approx 1$, which was necessary to obtain the closed-form expressions for the overlap integrals in Eqs. (1.94)–(1.96) corresponding to the three-level system, is analyzed in detail in Appendix D.

The analysis shows that for all signal powers $p < 50$ the discrepancy between the approximated and exact power-dependent overlap integrals is at maximum 6%. This result assumes that the confinement factor is unity or less, i.e., $a_0/\omega_s < 1$, and that the fiber is locally pumped with power $q \geqslant 2$. In the case where the fiber is underpumped ($q = 2$) with no input signal ($p = 0$), the error is slightly higher, or 15%, but this regime is of limited interest for practical applications.

In the high pump, or unsaturated gain regime, i.e., $q \gg p$, Eqs. (1.94)–(1.96) are 100% accurate, as the dependence of the overlap integrals in the mode size ratio $\alpha$ vanishes. In the high signal, or highly saturated regime, i.e., $p > q$, the approximate equations are very accurate (94%) if the confinement factor is unity or less.

The terms $\Gamma_{ek, ak}(z, p_j^{\pm})$ in Eqs. (1.82)–(1.85) for two-level systems, and $\Gamma_{ap, ak, ek}(z, q^{\pm}, p_j^{\pm}, 1)$ in Eqs. (1.91)–(1.96) three-level systems, represent *power-dependent overlap integral factors*. Such factors include all the effects of pump and signal mode overlap with the doped core, as well as integrated variations of the degree of inversion across it. The notion of power-dependent overlap factors was developed in previous work (C. R. Giles and E. Desurvire [39]) for the two-level pumping scheme, and the above formulae represent a generalization of this analysis to both two- and three-level systems.

Such power-dependent overlap integral factors should not be confused with the constant overlap factors $\Gamma_{s,p}$ defined in Eqs. (1.86) and (1.97), which only reflects the effect of finite overlap between the signal or pump and the doped core.

In the case of an infinite pump power at wavelength $\lambda_p$, i.e., $\kappa \gg 1$, the power-dependent overlap integrals in Eqs. (1.82)–(1.85) and (1.91)–(1.96) reduce to:

$$\Gamma_{ek}(z, p_p \to \infty) = \Gamma_s \frac{1}{1 + \eta_p} \tag{1.101}$$

$$\Gamma_{ak}(z, p_p \to \infty) = \Gamma_s \frac{\eta_p}{1 + \eta_p} \tag{1.102}$$

$$\Gamma_{ap}(z, q \to \infty, p_j^\pm, 1) = 0 \tag{1.103}$$

$$\Gamma_{ek}(z, q \to \infty, p_j^\pm, 1) = \Gamma_s \tag{1.104}$$

$$\Gamma_{ak}(z, p \to \infty, p_j^\pm, 1) = 0 \tag{1.105}$$

These results show that the maximum achievable signal gain coefficients $\Delta g_k$, corresponding to the maximum possible degree of inversion, are, for two- and three-level systems,

$$\Delta g_k(q \to \infty) = (g_{ek} - g_{ak})(q \to \infty) = \rho_0 \sigma_{ak} \Gamma_s \frac{\eta_s - \eta_p}{1 + \eta_p} \text{ (two-level system)} \tag{1.106}$$

$$\Delta g_k(q \to \infty) = (g_{ek} - g_{ak})(q \to \infty) = \rho_0 \sigma_{ak} \Gamma_s \eta_s \text{ (three-level system)} \tag{1.107}$$

Thus, the effect of the waveguide is to affect the gain coefficient by a factor $\Gamma_s < 1$, which reflects the effect of incomplete overlap between the mode and the doped core. Note that $\Gamma_s$ becomes unity when the doped core extends largely into the cladding, i.e., $a_0 \gg \omega_s$. The difference between the other factors involved in Eqs. (1.106) and (1.107) reflects the effect of pumping schemes, discussed in Section 1.5, with the concept of overlap–inversion parameter $\hat{\Gamma}$. The fact that both pump and signal absorption coefficients vanish for three-level systems, as shown in Eqs. (1.103) and (1.105) is consistent with a high degree of inversion, i.e., no ionic population exists at the ground level in this regime.

## 1.8 RATE EQUATIONS WITH CONFINED Er-DOPING

In the previous section, we derived in Eqs. (1.106) and (1.107) the highest limit $\Delta g_k(q \to \infty)$ for the signal gain coefficient in two- and three-level doped fiber amplifiers, corresponding to the case of infinite pump power. We can wonder now for the general case, i.e., for any finite pump power level, whether it is possible to optimize the gain coefficient to make it as close as possible to the highest limit $\Delta g_k(q \to \infty)$.

Taking the example of the two-level system, we express the net gain coefficient $\Delta g_k = g_{ek} - g_{ak}$ from Eqs. (1.82)–(1.85):

$$\Delta g_k = \rho_0 \sigma_{ak}\{\eta_k \Gamma_{ek}(z, p_j^{\pm}) - \Gamma_{ak}(z, p_j^{\pm})\} = \frac{\rho_0 \sigma_{ak} \Gamma_s}{1 + \kappa}\{\kappa(\eta_k \mathscr{D}_e - \mathscr{D}_a) - 1\}(1 - u) \quad (1.108)$$

with

$$u = \frac{1 + \eta_k \mathscr{D}_e - \mathscr{D}_a}{1 - \kappa(\eta_k \mathscr{D}_e - \mathscr{D}_a)}\left\{1 + \frac{1 + \kappa}{\kappa \Gamma_s}\log\left(1 - \frac{\kappa}{1 + \kappa}\Gamma_s\right)\right\} \quad (1.109)$$

We analyze then the different terms involved in the function $u$ in Eq. (1.109). First, with $\Gamma_s$ having values between 0 and 1, it is straightforward to verify that the term between braces in this equation is always negative, i.e., for any positive value of $\kappa$. On the other hand, the term $1 + \eta_k \mathscr{D}_e - \mathscr{D}_a$ in Eq. (1.109) is always positive. This can be shown by writing $1 + \eta_k \mathscr{D}_e - \mathscr{D}_a$ explicitly:

$$1 + \eta_k \mathscr{D}_e - \mathscr{D}_a = 1 + \frac{\displaystyle\sum_j \frac{\eta_k - \eta_j}{1 + \eta_j}(p_j^+ + p_j^-)}{\displaystyle\sum_j (p_j^+ + p_j^-)}$$

$$= 1 + \frac{\displaystyle\sum_{\eta_j < \eta_k} \frac{\eta_k - \eta_j}{1 + \eta_j}(p_j^+ + p_j^-)}{\displaystyle\sum_j (p_j^+ + p_j^-)} - \frac{\displaystyle\sum_{\eta_j > \eta_k} \left|\frac{\eta_k - \eta_j}{1 + \eta_j}\right|(p_j^+ + p_j^-)}{\displaystyle\sum_j (p_j^+ + p_j^-)} \quad (1.110)$$

As we have $|(\eta_k - \eta_j)/(1 + \eta_j)| < 1$ for $\eta_j > \eta_k$, the last term in Eq. (1.110) is less than unity, which gives $1 + \eta_k \mathscr{D}_e - \mathscr{D}_a > 0$, regardless of the value of $\kappa$.

Thus, we have found that the sign of $u$ in Eq. (1.109) is always the same as of $\kappa(\eta_k \mathscr{D}_e - \mathscr{D}_a) - 1$, which is the same factor as involved in Eq. (1.108). The important result is that for $\kappa(\eta_k \mathscr{D}_e - \mathscr{D}_a) - 1 > 0$ we have $u > 0$, which represents a decrease of the positive net gain coefficient $\Delta g_k$ in Eq. (1.108). On the other hand, for $\kappa(\eta_k \mathscr{D}_e - \mathscr{D}_a) - 1 < 0$ we have $u < 0$, which represents an increase in absorption, as the net gain coefficient is negative. It is clear then that the function $u$ should be made the smallest possible. Through a Taylor expansion of the exponential function in $\Gamma_s$ near the origin, it is found from Eq. (1.109) that the condition $u \approx 0$ is achieved for $\Gamma_s \ll 1$, which corresponds to *a confined Er-doping distribution in the core*, or $a_0 \ll \omega_s$. We find then that the optimum net gain coefficient for the two-level system is:

$$\Delta g_k(\text{confined}) = \frac{\rho_0 \sigma_{ak} \Gamma_s}{1 + \kappa}\{\kappa(\eta_k \mathscr{D}_e - \mathscr{D}_a) - 1\}$$

$$= \rho_0 \sigma_{ak} \Gamma_s \frac{\displaystyle\sum_j \frac{\eta_k - \eta_j}{1 + \eta_j}(p_j^+ + p_j^-) - 1}{1 + \displaystyle\sum_j (p_j^+ + p_j^-)} \quad (1.111)$$

The same analysis yields for the three-level system:

$$\Delta g_k(\text{confined}) = \frac{\rho_0 \sigma_{ak} \Gamma_s}{1 + \kappa'} \{\kappa'(\eta_k \mathscr{D}'_e - \mathscr{D}'_a) - 1\}$$

$$= \rho_0 \sigma_{ak} \Gamma_s \frac{\eta_k(q^+ + q^-) + \sum_j \frac{\eta_k - \eta_j}{1 + \eta_j}(p_j^+ + p_j^-) - 1}{1 + q^+ + q^- + \sum_j (p_j^+ + p_j^-)} \qquad (1.112)$$

As the overlap factor $\Gamma_s \ll 1$ remains as a multiplicative term in the definition of the gain coefficients in Eqs. (1.111) and (1.112), confinement of Er-doping results in a sizeable decrease of these coefficients, which is power independent. This decrease in the coefficients can be compensated for by increasing either the dopant density $\rho_0$, or the fiber length $L$. But it is not possible to increase the dopant density above a certain level, due to the detrimental effect of cooperative energy transfer, which results in a drastic reduction in the net gain coefficient (Chapter 4). It is not possible either to increase the fiber length indefinitely, as the effect of background loss from the silica glass material would affect both pump and signal. The effect of Er-doping confinement in the determination of Er-density and fiber length requirements is analyzed in Chapter 5.

A second observation is that the approach of Er-doping confinement is not sufficient to yield the highest possible gain coefficients. There is still a degree of freedom left in the choice of the overlap factors $\Gamma_p$ and $\Gamma_s$ corresponding to the pump and the signal. This issue, which concerns the optimization of the fiber waveguide design, is analyzed in detail Chapter 5.

We now express the rate equations corresponding to the specific case of confined Er-doping, which are the same as Eqs. (1.69) and (1.70) but involving the emission and absorption coefficients $\gamma_{ap}$, $\gamma_{ek,ak}$:

$$\pm \frac{dq^\pm}{dz} = (\gamma_{ep} - \gamma_{ap})q^\pm \qquad (1.113)$$

$$\pm \frac{dp_k^\pm}{dz} = (\gamma_{ek} - \gamma_{ak})p_k^\pm + 2\gamma_{ek}p_{0k} \qquad (1.114)$$

The coefficients $\gamma_{ap}$, $\gamma_{ek,ak}$ can be obtained by taking the limit $\Gamma_{s,p} \ll 1$ in the overlap integral factors defined in Eqs. (1.84) and (1.85) and Eqs. (1.94)–(1.96) and by making explicit all the terms involved in these definitions, which yields, for the two-level pumping scheme:

$$\gamma_{ek} = \alpha_k \frac{\sum_j \frac{\eta_k}{1 + \eta_j}(p_j^+ + p_j^-)}{1 + \sum_j (p_j^+ + p_j^-)} \qquad (1.115)$$

$$\gamma_{ak} = \alpha_k \frac{1 + \sum_j \frac{\eta_j}{1 + \eta_j}(p_j^+ + p_j^-)}{1 + \sum_j (p_j^+ + p_j^-)} \tag{1.116}$$

and for the three-level pumping scheme:

$$\gamma_{ap} = \alpha_p \frac{1 + \sum_j \frac{\eta_j}{1 + \eta_j}(p_j^+ + p_j^-)}{1 + q^+ + q^- + \sum_j (p_j^+ + p_j^-)}; \gamma_{ep} = 0 \tag{1.117}$$

$$\gamma_{ek} = \alpha_k \frac{\eta_k(q^+ + q^-) + \sum_j \frac{\eta_k}{1 + \eta_j}(p_j^+ + p_j^-)}{1 + q^+ + q^- + \sum_j (p_j^+ + p_j^-)} \tag{1.118}$$

$$\gamma_{ak} = \alpha_k \frac{1 + \sum_j \frac{\eta_j}{1 + \eta_j}(p_j^+ + p_j^-)}{1 + q^+ + q^- + \sum_j (p_j^+ + p_j^-)} \tag{1.119}$$

where

$$\boxed{\begin{aligned} \alpha_p &= \rho_0 \sigma_{ap} \Gamma_p \\ \alpha_k &= \rho_0 \sigma_{ak} \Gamma_s \end{aligned}} \tag{1.120), (1.121}$$

are the pump and signal Er-doping absorption coefficients.

Rate equations (1.113) and (1.114) with coefficients (1.115)–(1.121) are considerably simplified in comparison to Eqs. (1.69)–(1.72) which correspond to the general case. Yet the simplified equations remain relatively complex as they form a coupled nonlinear set; a numerical solution requires an approach similar to that described in Appendix C, except for integration over the transverse plane.

But fortunately there are two regimes where these nonlinear equations have exact analytical solutions which can be expressed in closed-form. These are the *unsaturated gain regime* and the *low gain regime*, respectively, which are considered in the next two sections.

## 1.9 ANALYTICAL MODEL FOR UNSATURATED GAIN REGIME

For the time being, we shall define the unsaturated gain regime as a condition where the sum of all normalized signal + ASE powers $\sum (p_j^+ + p_j^-)$ (excluding the pump for two-level systems) at any fiber coordinate $z$ is much smaller than the total normalized pump power $q^+ + q^-$. This definition should not be confused with another notion of unsaturated gain regime discussed later in this book. The latter refers to the net signal gain as integrated along the fiber, and corresponds to a situation where this signal gain is independent of the input signal power. Indeed, if the input signal power is very weak and the gain coefficient high enough, the possibility exists that the sum $\sum (p_j^+ + p_j^-)$ at any point in the fiber is dominated by ASE, not by the amplified signal. This regime is referred to as amplifier self-saturation (Chapter 5). In this case, and as long as ASE dominates, the net integrated gain for the signal is independent of input signal power. Thus, while ASE saturates the signal absorption and emission coefficients, the signal gain remains constant over a certain range of input signal power. This amplification regime is conventionally referred to as the *unsaturated gain regime*. In this chapter, we shall refer to the unsaturated gain regime as the condition where all signal + ASE terms $p_j^+ + p_j^-$ are negligible in front of the pump terms $q^+ + q^-$.

With this approximation, we obtain from Eqs. (1.115)–(1.119) the unsaturated absorption and emission coefficients $\bar{\gamma}_{ap}$ and $\bar{\gamma}_{ek,ak}$ corresponding to confined Er-doping:

$$\bar{\gamma}_{ep} - \bar{\gamma}_{ap} = -\alpha_p \frac{1}{1 + q^+ + q^-} \tag{1.122}$$

for both two- and three-level systems, and

$$\bar{\gamma}_{ek} = \alpha_k \frac{\dfrac{\eta_k}{1 + \eta_p}(q^+ + q^-)}{1 + q^+ + q^-} \tag{1.123}$$

$$\bar{\gamma}_{ak} = \alpha_k \frac{1 + \dfrac{\eta_p}{1 + \eta_p}(q^+ + q^-)}{1 + q^+ + q^-} \tag{1.124}$$

for two-level systems.

The pump and signal equations (1.113) and (1.114) with coefficients defined in (1.122)–(1.124) now take the following form, valid for both two- and three-level systems:

$$\pm \frac{dq^\pm}{dz} = -\alpha_p \frac{1}{1 + q^+ + q^-} q^\pm - \alpha_p' q^\pm \tag{1.125}$$

$$\pm \frac{dp_k^\pm}{dz} = \alpha_k \frac{1}{1 + q^+ + q^-}\left[ \left\{ \frac{\eta_k - \eta_p}{1 + \eta_p}(q^+ + q^-) - 1 \right\} p_k^\pm \right.$$
$$\left. + \frac{\eta_k}{1 + \eta_p}(q^+ + q^-)2p_{0k} \right] - \alpha_s' p_k^\pm \tag{1.126}$$

We have also introduced in each of the above equations an additional term reflecting the effect of fiber background loss, corresponding to absorption coefficients $\alpha_{p,s}'$. This effect is important in the case of *distributed amplifiers*, where the fiber length can be several kilometers (Chapter 6). The above equations are fully appropriate to describe the evolution of pump and signal in such distributed amplifiers, as they are usually operated under unsaturated gain conditions.

While the equation system (1.125) and (1.126) is linear in signal at $\lambda_k$, it is nonlinear in forward and backward pumps, therefore numerical integration is required to obtain the dependence of pump and signal powers with fiber coordinate $z$. But for a given fiber length $L$, it is possible to derive a simple analytical relation between the input $(q_0^+, q_L^-)$ and output $(q_L^+, q_0^-)$ pump powers as a function of the fiber gain $G_k$ at $\lambda_k$. Indeed, Appendix E shows that, in the general case of bidirectional pumping, these powers are related through (E. Desurvire *et al.*, [40] and [41]):

$$\boxed{\begin{aligned} q_L^+ &= q_0^+ \exp(-\mathscr{A}_{\mathrm{p}k} L) \\ q_0^- &= q_L^- \exp(-\mathscr{A}_{\mathrm{p}k} L) \end{aligned}} \qquad (1.127)$$

with

$$\mathscr{A}_{\mathrm{p}k} = \alpha_{\mathrm{p}} \frac{1 + \eta_{\mathrm{p}}}{1 + \eta_k} \left\{ (1 + \varepsilon_k)(1 - bC)\varepsilon_{\mathrm{p}} - \frac{\log G_k}{\alpha_k L} \right\} \qquad (1.128)$$

$$C = \frac{\dfrac{\eta_{\mathrm{p}} - \eta_k}{1 + \eta_{\mathrm{p}}} + \varepsilon_k}{1 + \varepsilon_k} \qquad (1.129)$$

where $\varepsilon_{\mathrm{p},k} = \alpha'_{\mathrm{p},k}/\alpha_{\mathrm{p},k}$ is the ratio of background absorption to ionic absorption coefficients at $\lambda_{\mathrm{p},k}$, and $b = (1 + \varepsilon_{\mathrm{p}})/\varepsilon_{\mathrm{p}}$. By definition, we call the parameter $\mathscr{A}_{\mathrm{p}k}$ the *gain-dependent pump absorption coefficient*, as justified by the form of expression (1.127) and by the dependence in gain $G_k$ of $\mathscr{A}_{\mathrm{p}k}$ in Eq. (1.128).

As simple as the above relations seem, the general case of bidirectional pumping is yet not so easily tractable. Appendix E shows that the required pump powers for a given gain $G_k$ and fiber length $L$ are related through a complicated transcendental equation, (E.21), which must be solved numerically. Unless a specific gain value is required, there is no real advantage in numerically solving this particular equation compared to a more direct numerical solution of Eqs. (1.125) and (1.126). However, the difficulty is removed in the case of unidirectional pumping (forward or backward pump only).

Considering for instance the case of forward pumping ($q_0^+ \neq 0$, $q_L^- = 0$), we find from Appendix E that the required input pump power $q_0^+$, for a fiber with length $L$ and gain $G_k$, is simply given by [40]:

$$\boxed{q_0^+ = b e^{BL} \frac{e^{AL} - 1}{e^{BL} - 1}} \qquad (1.130)$$

with

$$A = \mathscr{A}_{\mathrm{p}k} - B = \varepsilon_{\mathrm{p}} \alpha_{\mathrm{p}} \frac{1 + \eta_{\mathrm{p}}}{1 + \eta_k} \left\{ 1 + \varepsilon_k + \frac{\log G_k}{\alpha_k L} \right\} \qquad (1.131)$$

$$B = w(q_L^- = 0) = b\varepsilon_{\mathrm{p}}(\mathscr{A}_{\mathrm{p}k} - \alpha'_{\mathrm{p}}) = \alpha_{\mathrm{p}}(1 + \varepsilon_{\mathrm{p}}) \frac{1 + \eta_{\mathrm{p}}}{1 + \eta_k} \left\{ \frac{\eta_k - \eta_{\mathrm{p}}}{1 + \eta_{\mathrm{p}}} - \varepsilon_k - \frac{\log G_k}{\alpha_k L} \right\} \qquad (1.132)$$

with $C$ defined in Eq. (1.129).

Thus, in the simplest case of unidirectional pumping and an unsaturated gain regime, the signal gain at $\lambda_k$ as a function of pump power has, through Eq. (1.130), an explicit form convenient to analyze. Plotting the fiber gain characteristics $G_k = f(q_0^+)$ can be done directly from Eq. (1.130), even if this equation is actually in the form $q_0^+ = f^{-1}(G_k)$; in plotting Eq. (1.130), only the region $q_0^+ \geqslant 0$ must then be considered. The above results are used in Chapters 2 and 5 for a detailed analysis of pump power requirements, effects of Er concentration, fiber background loss, and noise in distributed amplifiers. These results can also be used to study fiber amplifiers of higher Er concentrations and shorter lengths, operating only in the unsaturated gain regime. In this case, the fiber background loss can be neglected, and the limit $\varepsilon_{p,k} \to 0$ can be taken in Eqs. (1.130)–(1.32) using a Taylor expansion for $\exp(BL)$ in eq. (1.130). The results for input and output pumps are given in Chapter 2, Eqs. (2.135)–(2.137).

Another interesting feature of the unsaturated gain regime analyzed in this section is, in the case of unidirectional pumping, the existence of exact solutions for the amplified spontaneous emission or ASE.

It was shown previously, that in the unsaturated gain regime, the forward and backward ASE power spectra are only functions of the input and output pump powers (E. Desurvire et al. [42], for which we have derived here explicit solutions. The details of the derivation of the ASE power spectra in the forward ($p_{k,\mathrm{ASE}}^+(q_L^+)$) and the backward ($p_{k,\mathrm{ASE}}^-(q_0^+)$) directions are outlined in Appendix E. These take the form:

$$p_{k,\mathrm{ASE}}^+(q_L^+) = 2p_{0k} \frac{\alpha_k}{\alpha_p} \frac{\eta_k}{1+\eta_p} I^+(q_0^+, q_L^+) \exp\left\{ \frac{\alpha_k}{\alpha_p} \frac{1+\varepsilon_k}{1+\varepsilon_p} [(bC-1)\log(b+q_L^+) + \log(q_L^+)] \right\}$$

(1.133)

$$p_{k,\mathrm{ASE}}^-(q_0^+) = 2p_{0k} \frac{\alpha_k}{\alpha_p} \frac{\eta_k}{1+\eta_p} I^-(q_0^+, q_L^+) \exp\left\{ -\frac{\alpha_k}{\alpha_p} \frac{1+\varepsilon_k}{1+\varepsilon_p} [(bC-1)\log(b+q_0^+) + \log(q_0^+)] \right\}$$

(1.134)

where $I^\pm(q_0^+, q_L^+)$ are elementary integrals which, having no analytical form in the general case, must be computed numerically or approximated through Taylor series expansions. Expressions of the ASE spectra in the limit of negligible background loss ($\varepsilon_{p,k} \to 0$) are given in Appendix E, Eqs. (E.37)–(E.39).

Equations (1.133) and (1.134) defining the ASE spectra, along with Eqs. (1.27) and (1.130) for the input and output pump powers, provide a complete description of the input/output fiber amplifier characteristics as a function of gain $G_k$, in the case of unidirectional pumping and an unsaturated gain regime. As the gain $G_k$ is directly related to the input pump power through Eq. (1.130), these characteristics can also be studied as a function of input pump power (see the analysis of noise in EDFAs, Chapter 2).

Finally, the analytical solutions provided by this unsaturated gain model also make it possible to study the EDFA characteristics as a function of fiber coordinate $z$, i.e., the distributions of the pump, signals, and ASE along the fiber, but this requires solving a transcendental equation. Given the input pump power $q_0^+$, one must solve Eq. (1.130) for each coordinate $z$ (replacing length $L$ by $z$ in Eqs. (1.130)–(1.32)), in order to obtain the corresponding gain $G_k$. The pump power at coordinate $z$ is then given by Eq. (1.127). However, Eq. (1.130) giving $G_k$ is transcendental and therefore must be solved numerically; this fact removes the interest of the approach, as compared to the direct numerical integration of Eqs. (1.125)–(1.126).

## 1.10 ANALYTICAL MODELS FOR LOW GAIN REGIME

The previous section focussed upon the unsaturated gain regime, for which analytical solutions can be derived for pump, signal, and ASE output powers. Also of interest is the regime of low EDFA gains (i.e., $G_k < 20$ dB), where there is no amplifier self-saturation by ASE. As a result, both forward and backward ASE powers can be ignored, the EDFA can be analyzed as a noise-free amplifier, and the effect of gain saturation by high input signal(s) can be taken into account. This is of interest for power amplifier applications (Chapter 5).

We shall outline here the main results of two independent models based on the noise-free assumption (M. Peroni and M. Tamburrini [23], A. A. M. Saleh *et al.* [24]). These models lead to transcendental equations requiring numerical solution. In this section, we present an original graphical method, which may be used to simplify the numerical algorithms.

We consider the first model described in [23]. This model, valid for unidirectional pumping (forward or backward) with only one signal input, was developed for a three-level pumping scheme (e.g., a pump at $\lambda_p = 980$ nm). Next we generalize this model for application to three- and two-level pumping schemes (for two-level schemes, the corresponding pump wavelength is near $\lambda_p = 1480$ nm). The model assumes confined Er-doping, as studied in Section 1.8.

Starting from Eqs. (1.113) and (1.114) and overlooking the ASE source term $p_{0k}$, we obtain the rate equations for two signals at wavelengths $\lambda_{k_1}$ and $\lambda_{k_2}$, propagating in the forward direction, corresponding to the two-level pumping scheme:

$$\frac{dp_{k_1}}{dz} = -\alpha_{k_1} \frac{\dfrac{\eta_{k_2} - \eta_{k_1}}{1 + \eta_{k_2}} p_{k_2} + 1}{1 + p_{k_1} + p_{k_2}} p_{k_1} \tag{1.135}$$

$$\frac{dp_{k_2}}{dz} = \alpha_{k_2} \frac{\dfrac{\eta_{k_2} - \eta_{k_1}}{1 + \eta_{k_1}} p_{k_1} - 1}{1 + p_{k_1} + p_{k_2}} p_{k_2} \tag{1.136}$$

Taking the ratio of the two equations above and integrating the results yields the transcendental equation:

$$\frac{1}{\alpha_{k_2}} \left\{ \frac{\eta_{k_2} - \eta_{k_1}}{1 + \eta_{k_2}} (G_{k_2} - 1) p_{k_2}^{\text{in}} + \log G_{k_2} \right\} = \frac{1}{\alpha_{k_1}} \left\{ \frac{\eta_{k_2} - \eta_{k_1}}{1 + \eta_{k_1}} (1 - G_{k_1}) p_{k_1}^{\text{in}} + \log G_{k_1} \right\}$$

$$\tag{1.137}$$

where superscripts in and out mean input and output, respectively, and the gains are defined by $G_{k_m} = p_{k_m}^{\text{out}} / p_{k_m}^{\text{in}}$ ($m = 1, 2$). In the case of counterpropagating signals, e.g., the signals at $\lambda_{k_1}$ and at $\lambda_{k_2}$ propagate in the backward and forward directions, respectively, the corresponding equation is obtained by changing the absorption coefficient $\alpha_{k_2}$ into $-\alpha_{k_2}$ in Eq. (1.137). For a three-level pumping scheme, and assuming the pump to be at $\lambda_p = \lambda_{k_1}$, and the signal at $\lambda_s = \lambda_{k_2}$, we have $\eta_{k_1} = 0$ (see Section 1.5) and eq. (1.137) gives:

$$\frac{1}{\alpha_s}\left\{\frac{\eta_s}{1+\eta_s}(G_s - 1)p_s^{in} + \log G_s\right\} = \frac{1}{\alpha_p}\{\eta_s(1 - G_p)p_p^{in} + \log G_p\} \tag{1.138}$$

which is the equation obtained in [23] when $\eta_s = 1$ is assumed. The powers $p_p^{in}$ and $p_s^{in}$ are powers normalized to saturation powers, as defined in Eq. (1.51).

Given the two input power conditions $p_{k_1}^{in}$, $p_{k_2}^{in}$, the two unknowns involved in Eqs. (1.137) and (1.138) are the gains $G_{k_1} = G_p$ and $G_{k_2} = G_s$. An assumption that can be made is that the fiber length is chosen optimum for the signal wavelength $\lambda_{k_2}$. At this optimum length, the net gain coefficient is minimal, and Eq. (1.136) gives the condition $p_{k_1}^{out} = (1 + \eta_{k_1})/(\eta_{k_2} - \eta_{k_1})$. This condition implies that $\eta_{k_2} > \eta_{k_1}$, i.e., the signal at $\lambda_{k_1}$ is attenuated rather than amplified, Eq. (1.135), and acts as a pump for the signal at $\lambda_{k_2}$. With this assumption, Eq. (1.137) becomes:

$$\boxed{\begin{aligned}\frac{1}{\alpha_{k_2}}\left\{\frac{\eta_{k_2}-\eta_{k_1}}{1+\eta_{k_2}}(G_{k_2}-1)p_{k_2}^{in} + \log G_{k_2}\right\} = \\ \frac{1}{\alpha_{k_1}}\left\{\frac{\eta_{k_2}-\eta_{k_1}}{1+\eta_{k_1}}p_{k_1}^{in} - 1 - \log\left(\frac{\eta_{k_2}-\eta_{k_1}}{1+\eta_{k_1}}\right) - \log(p_{k_1}^{in})\right\}\end{aligned}}$$

$$\tag{1.139}$$

Similar equations corresponding to the cases of counterpropagating pump and/or three-level pumping scheme are straightforward to derive from the above result.

The signal gain $G_{k_2}$ at $\lambda_{k_2}$ can then be found as a function of the input power conditions $p_{k_1}^{in}$, $p_{k_2}^{in}$ by numerically solving Eq. (1.139). *Graphical* solution of Eq. (1.139) is also possible; this may be implemented as an efficient and straightforward algorithm in the numerical resolution of the transcendental equation. First developed by Volterra in 1931[43], it consists of the following steps. We take the exponential function of both sides of Eq. (1.137) to obtain an equation of the form $f(p_2^{out}) = \mathscr{C}g(p_1^{out})$:

$$f(p_2^{out}) = p_2^{out}\exp(U_2 p_2^{out}) = \mathscr{C}\{p_1^{out}\exp(-U_1 p_1^{out})\}^{\alpha_2/\alpha_1} = \mathscr{C}g(p_1^{out}) \tag{1.140}$$

where $\mathscr{C}$ is a constant defined as:

$$\mathscr{C} = p_2^{in}(p_1^{in})^{-\alpha_2/\alpha_1}\exp(U_2 p_2^{in} + \frac{\alpha_2}{\alpha_1}U_1 p_1^{in}) \tag{1.141}$$

and $U_1 = (\eta_{k_2} - \eta_{k_1})/(1 + \eta_{k_1})$, $U_2 = (\eta_{k_2} - \eta_{k_1})/(1 + \eta_{k_2})$. For clarity, the subscript $k$ was omitted in Eqs. (1.140) and (1.141).

The graphical method, illustrated in Figure 1.5, consists then in plotting the three functions: $f = \mathscr{C}g$ (quadrant I), $f = f(p_2)$ (quadrant II), and $g = g(p_1)$ (quadrant IV),

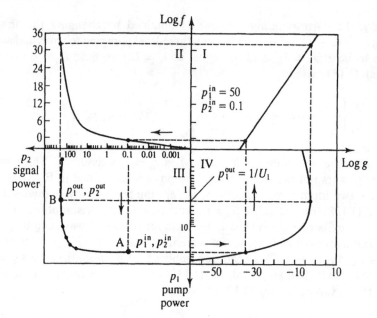

**FIGURE 1.5**   Graphical solution $p_2 = h(p_1)$ of transcendental equation (1.140) corresponding to a forward pumped EDFA with input power conditions of $p_1^{in} = 50$, $p_2^{in} = 0.1$, and the pump and signal wavelengths $\lambda_1 = \lambda_p = 1480$ nm and $\lambda_2 = \lambda_s = 1531$ nm, respectively (see text for description).

as defined in Eqs. (1.140) and (1.141). Only the function $f = \mathscr{C}g$ in quadrant I depends on the initial conditions ($p_1^{in}, p_2^{in}$). The constant $\mathscr{C}$ is calculated in this example assuming forward pumping with the parameter values $p_1^{in} = 50$, $p_2^i = 0.1$, $\alpha_2/\alpha_1 = 3.5$, $U_1 = 0.52$, $U_2 = 0.34$, corresponding to $\eta_{k_1} = 0.26$, $\eta_{k_2} = 0.92$, $\lambda_{k_1} = \lambda_p = 1480$ nm, $\lambda_{k_2} = \lambda_s = 1531$ nm, which apply to a typical aluminosilicate Er-doped fiber (Chapter 4).

The graphical solution $p_2^{out} = h(p_1^{out})$ of the equation $f(p_2^{out}) = \mathscr{C}g(p_1^{out})$ is found in quadrant III by drawing lines parallel to the axes and intercepting the curves in each of the other quadrants, starting from a given value of $p_1$, as shown in Figure 1.5. The points A and B in the figure were determined this way, starting from horizontal lines with ordinates $p_1 = p_1^{in} = 50$ and $p_1 = p_1^{out} = 1/U_1 = 1.92$; this last value corresponds to the output pump power when gain at $\lambda_{k_2}$ is optimum. The coordinates of points A and B on the $p_2$ axis correspond to the input ($p_2^{in}$) and output ($p_2^{out}$) signal powers at $\lambda_{k_2}$, respectively. The complete solution of eqs. (1.137) and (1.140) for any intermediate output pump power value $p_1^{out}$ is also shown in Figure 1.5, as a full line joining a few other points determined through this method.

Figure 1.6 illustrates the graphical solution method for different input pump power conditions ($p_2^{in} = 2$ to 50) and constant input signal power ($p_1^{in} = 1/U_1$ provide the output signal powers $p_2^{out}$ on the $p_2$ axis, which correspond to optimal or maximized gains. As seen from the figure, the signal gain $G_{k_2}$ decreases as the output pump power becomes less than $p_1^{out} = 1/U_1$.

Only the functions $f = \mathscr{C}g$ (straight lines) in quadrant I had to be drawn for each input pump conditions in order to obtain the output signal solutions, since functions in quadrants II and IV do not depend on the input conditions. This property suggests

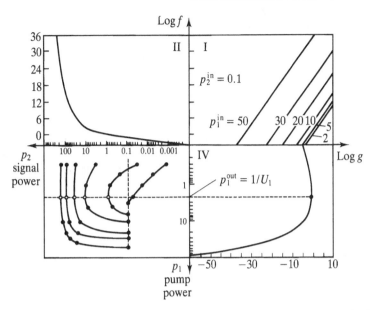

**FIGURE 1.6**  Graphical solution $p_2 = h(p_1)$ following the method of Figure 1.5, for input pump powers $p_1^{in} = 2$ to 50, corresponding to a forward pumped EDFA with pump and signal wavelengths $\lambda_1 = \lambda_p = 1480$ nm and $\lambda_2 = \lambda_s = 1531$ nm, respectively. The maximum output signal powers corresponding to optimal fiber lengths are obtained along the line $p_1^{out} = 1/U_1$ (open circles).

an efficient algorithm for the numerical solution of Eq. (1.137). This algorithm is a simple computer implementation of the graphical method, in which the points corresponding to functions in quadrants I, II, and IV can be entered in three arrays. Only the array corresponding to the function in quadrant I varies as a function of the input conditions, as defined by elementary functions in Eq. (1.141). The computer solution is simply given by comparing these arrays two by two, following the sequence illustrated in Figure 1.5. This method also applies to the case of backward pumping; the input and output conditions at $z = 0$ and $z = L$ become $(p_1^{out} = 1/U_1, p_2^{in})$ and $(p_1^{in}, p_2^{out})$, respectively, and the sign of $\alpha_1$ is reversed in Eqs. (1.140) and (1.141).

In Section 5.3 it is shown that the solution $p_2^{out} = F(p_1^{in}, p_1^{out}, p_2^{in})$ of the transcendental equation (1.140) can actually be expressed explicitly, without numerical or graphical resolution, as given in Eqs. (5.26)–(5.28).

We consider next another low gain, ASE noise-free model described in [24]. This model makes it possible to analyze a fiber amplifier saturated by several signals at different wavelengths; the pumping can be unidirectional or bidirectional, and distributed over several wavelengths or pump bands simultaneously. The Er-doping is also assumed to be confined, as studied in Section 1.8. As such, this model can be viewed as a conceptual extention of the previous one, which corresponded to the simplest case of one input signal with unidirectional pumping.

Let $P_k \equiv P(\nu_k)$ be the optical power at $\lambda_k$ (as opposed to $p_k$ which is the power normalized to the saturation power $P_{sat}(\nu_k)$, which can represent either a pump or a signal in the two-level pumping scheme. First, we start from the definitions (1.64) and (1.65) of the stimulated emission rates $W_{12,21}$. Into these equations we substitute the definition (1.45) for the saturation power $P_{sat}(v_k)$:

$$W_{12,21} = \frac{1}{\pi a_0^2} \sum_k \Gamma_k P_k \frac{\sigma_{ak,ek}}{hv_k} \tag{1.142}$$

In the above equation, we used the definition $\Gamma_k = a_0^2/\omega_k^2$, the limit for small Er-doping radii $a_0$ of the overlap integral defined in Eq. (1.86). We have also assumed consistently $\psi_k = 1$ in Eqs. (1.64) and (1.65), as the doping region is confined.

Now we rewrite the upper level population $N_2$ in Eq. (1.2), assuming steady-state conditions and $N_3 = 0$, from Eq. (1.142) for the stimulated emission rates, as:

$$N_2 = -\frac{\tau}{\pi a_0^2} \sum_k \frac{1}{hv_k} \Gamma_k P_k \sigma_{ak}(\eta_k N_2 - N_1) \tag{1.143}$$

The rate equation for the power at $\lambda_k$ is, from Eqs. (1.114)–(1.116) and (1.121):

$$u_k \frac{dP_k}{dz} = \Gamma_k P_k \sigma_{ak}(\eta_k N_2 - N_1) \tag{1.144}$$

where $u_k = 1$ if the power at $\lambda_k$ propagates in the forward direction and $u_k = -1$ if the power propagates in the backward direction. Summing Eq. (1.144) over $k$ and comparing the result with Eq. (1.143) yields:

$$N_2 = -\frac{\tau}{\pi a_0^2} \sum_k \frac{u_k}{hv_k} \frac{dP_k}{dz} \tag{1.145}$$

Finally, using the relation $N_1 + N_2 = \rho$ and the definition of the absorption coefficient $\alpha_k$ in Eq. (1.121) and of the saturation power $P_{sat}(v_k)$ in eq. (1.45), we obtain from Eqs. (1.144) and (1.145):

$$u_j \frac{dP_j}{P_j} = -\left\{\alpha_j + \frac{hv_j}{P_{sat}(\lambda_j)} \sum_k \frac{u_k}{hv_k} \frac{dP_k}{dz}\right\} dz \tag{1.146}$$

The above equation integrates from $z = 0$ to $z = L$ into:

$$\frac{P_j(L)}{P_j(0)} = \exp\left\{-u_j \alpha_j L + u_j \frac{hv_j}{P_{sat}(\lambda_j)} \sum_k \frac{u_k}{hv_k}[P_k(0) - P_k(L)]\right\} \tag{1.147}$$

With the definitions $P_m(0) = P_m^{in}$, $P_m(L) = P_m^{out}$ for forward signals, and $P_m(0) = P_m^{out}$, $P_m(L) = P_m^{in}$ for backward signals at $\lambda_m$ ($m = j, k$), it is easily verified that Eq. (1.147) takes finally the form [24]:

$$\bar{P}_j^{out} = \bar{P}_j^{in} \exp\left\{-\alpha_j L + \frac{\bar{P}^{in} - \bar{P}^{out}}{\bar{P}_{sat}(\lambda_j)}\right\} \tag{1.148}$$

with

$$\bar{P}^{\text{in, out}} = \sum_k \bar{P}_k^{\text{in, out}} \tag{1.149}$$

In Eqs. (1.148)–(1.149) the power $P_k$ with a bar symbol placed on top ($\bar{P}_k$) represents the photon flux, or the optical power at $\lambda_k$ divided by the corresponding photon energy $h\nu_k$, i.e.,

$$\bar{P}_k = \frac{P_k}{h\nu_k} \tag{1.150}$$

$$\bar{P}_{\text{sat}}(\lambda_k) = \frac{P_{\text{sat}}(\lambda_k)}{h\nu_k} \tag{1.151}$$

Thus, Eq. (1.149) for $\bar{P}^{\text{in, out}}$ defines the fiber amplifier's total input and output fluxes.

By summing Eq. (1.148) over $j$, we obtain a transcendental equation in $\bar{P}^{\text{out}}$:

$$\bar{P}^{\text{out}} = \sum_j a_j \exp(-b_j \bar{P}^{\text{out}}) \tag{1.152}$$

where $a_j$ is only a function of the input powers and defined by:

$$a_j = \bar{P}_j^{\text{in}} \exp(-\alpha_j L + b_j \bar{P}^{\text{in}}) \tag{1.153}$$

and

$$b_j = \frac{1}{\bar{P}_{\text{sat}}(\lambda_j)} \tag{1.154}$$

The solution $\bar{P}^{\text{out}}$ of eq. (1.152) substituted in Eq. (1.148) provides the output power, hence the fiber gain, for each wavelength $\lambda_k$. As the subscript $k$ can represent either a pump or a signal wavelength, Eqs. (1.148) and (1.152) are most general, and can describe the case of a fiber amplifier pumped at different wavelengths simultaneously (distributed over several pump bands), and input with several signals at different wavelengths, with each pump and signal propagating in arbitrary directions.

Now we must show that this model represents a conceptual extension of the previous one, which considered only one signal and one pump in the unidirectional pumping scheme. This can be done by writing Eq. (1.148) for two wavelengths $\lambda_{k_1}$ and $\lambda_{k_2}$:

$$\log\left(\frac{\bar{P}_{k_1}^{\text{out}}}{\bar{P}_{k_1}^{\text{in}}}\right) \equiv \log G_{k_1} = -\alpha_{k_1} L + p_{k_1}^{\text{in}}(1 - G_{k_1}) + p_{k_2}^{\text{in}}(1 - G_{k_2}) \frac{\bar{P}_{\text{sat}}(\lambda_{k_2})}{\bar{P}_{\text{sat}}(\lambda_{k_1})} \tag{1.155}$$

$$\log\left(\frac{\bar{P}_{k_2}^{\text{out}}}{\bar{P}_{k_2}^{\text{in}}}\right) \equiv \log G_{k_2} = -\alpha_{k_2} L + p_{k_2}^{\text{in}}(1 - G_{k_2}) + p_{k_1}^{\text{in}}(1 - G_{k_1}) \frac{\bar{P}_{\text{sat}}(\lambda_{k_1})}{\bar{P}_{\text{sat}}(\lambda_{k_2})} \tag{1.156}$$

Eliminating then the parameter $L$ between these two equations, and using the definition (1.151) for $\bar{P}_{sat}(\lambda_{k_m})$, it is easily found that the resulting equation is identical to Eq. (1.137), which corresponds to the previous model.

An important advantage of this conceptually extended model is, in the case of one signal and one pump in the unidirectional pumping scheme, the possibility to predict the optimal length for which signal gain is maximized. Indeed, after solving transcendental Eq. (1.139), which yields the maximum or optimal gain $G_{k_2}$ at signal wavelength $\lambda_{k_2}$, the corresponding optimal length $L_{opt}$ is determined from Eq. (1.156) with $p_{k_1}^{out} = 1/U_1$ or $G_{k_1} = 1/(U_1 p_{k_1}^{in})$, and $U_1 = (\eta_{k_2} - \eta_{k_1})/(1 + \eta_{k_1})$, for the forward pumping case (application to backward pumping is straightforward). This equation can then be put into the form:

$$L_{opt} = \frac{1}{\alpha_{k_1}} \frac{1 + \eta_{k_2}}{1 + \eta_{k_1}} \left( p_{k_1}^{in} - \frac{1 + \eta_{k_1}}{\eta_{k_2} - \eta_{k_1}} \right) - \frac{1}{\alpha_{k_2}} \{\log G_{k_2} + p_{k_2}^{in}(G_{k_2} - 1)\} \quad (1.157)$$

To summarize, the two low gain, ASE noise-free models corresponding to [23] and [24], described and compared in this section, make it possible to analyze EDFA characteristics under a saturated gain regime. Both models are applicable to two- or three-level pumping schemes, and apply to the case of confined Er-doping. The solutions from both models are in the form of a transcendental equation that requires numerical solution. The first model applies only to the case of unidirectional pumping with one signal wavelength. The second model makes it possible to solve the most general problem of multiple signal and pump wavelengths with any propagation direction. It also permits the prediction of optimal fiber length in the case of a unidirectionally pumped EDFA with a single input signal. An extension of this model to the more general case of arbitrary mode profiles and Er-dopant distributions is treated in [44].

The two models apply to the low gain regime ($G < 20$ dB), where ASE can be ignored. At higher gains ($G > 20$ dB), ASE causes amplifier self-saturation, and these models are not valid to predict the amplifier gain characteristics. But we show in Chapter 5 that the most general model of [24] can also be extended to predict gains and ASE in the case of a self-saturation regime [45]. In this case, ASE is taken into account in the model by assuming two broad spectrum, equivalent input noise signals at both amplifier ends. The power of such equivalent input noise in a frequency bin of width $\delta \nu$ corresponds to $2n_{sp}$ photons, where $n_{sp}$ is the spontaneous emission factor (see Chapter 2). For a high gain amplifier, the value of $n_{sp}$ is independent of wavelength and close to unity. Equations (1.148) and (1.152), which can describe the power evolution of a virtually unlimited number of optical channels, may then be solved to determine the signal gain and output ASE spectral in both propagation directions. Numerical applications of this method of modeling self-saturated amplifiers are presented in Chapter 5.

## 1.11 DENSITY MATRIX DESCRIPTION

So far, the theory outlined in Sections 1.1 through 1.10 has relied upon a rate equation description in which the optical power evolution with fiber coordinate at the pump and signal wavelengths could be determined from the steady state ionic populations along the fiber. The so-called *density matrix description* outlined in this section, as applied to the specific case of doped fibers, represents a different and more powerful approach. We focus here only on a *semiclassical description*, in which the atomic system is quantized and the electrical field interacting with matter is classical. Although we assume various spontaneous relaxation effects between the quantized atomic states (to match a phenomenological three-level system), spontaneous emission and ASE noise are not taken into account. In the rate equation model previously described, spontaneous emission was introduced phenomenologically. For a rigorous description of spontaneous emission, a fully quantized field theory is necessary; this is outlined in detail in Chapter 2. The real advantage of the semiclassical density matrix model is to provide a full description of the *complex atomic susceptibility* and *complex refractive index* associated with the gain medium [27]. The complex refractive index is an important parameter in the study of dispersion and short pulse propagation in doped fibers, as discussed in Chapter 4.

A description of light/matter interaction through the density matrix formalism can be found for instance in [30], [46], and [47]. However, the well-known results from these references cannot be directly applied to the case of Er:glass for several reasons. The first reason is that these descriptions apply to the case of laser systems having two or three discrete energy levels, as opposed to the Stark split manifolds of energy levels found in Er:glass. The second reason is that the results in [30] and [46] apply to the case of a fundamental two-level system, while those of [47] apply to a three-level system where the terminal level of the transition is not the ground level, unlike Er:glass lasers. The results for a three-level system with transition ending at the ground level are somewhat different in saturation terms.

We shall consider first the case of a basic three-level laser system, following strictly the same approach as in [30], [45], and [47]. The case of a Stark split three-level system is analyzed next, using a generalization of the density matrix model. The central assumption made in this case is that all off-diagonal elements of the density matrix, which concern intramanifold transitions, vanish due to the effect of thermal equilibrium. Consequently all diagonal elements are determined by Boltzmann's Law. The basis for such assumption is that thermal equilibrium within the Stark manifolds is not perturbed by the pumping and stimulated emission process, due to the comparatively fast thermalization process between the sublevels.

The result of this analysis is that *the total complex atomic susceptibility of the Stark split system is given by the sum of all susceptibilities associated with three-level subsystems, to each of which corresponds an individual laser transition.* This fundamental result, used only as a postulate in previous work [27], is actually demonstrated here (Appendix F).

Because the derivation through the density matrix theory of the atomic susceptibility in the case of a Stark split laser system is quite tedious, we first analyze the simplest case: the basic, nondegenerate three-level system.

Let the atomic quantum states be defined by $|E_n\rangle$. The matrix elements of the density and the atomic dipole moment operators are then $\rho_{mn} = \langle E_m | \rho | E_n \rangle$ and $\mu_{mn} = \langle E_m | \mu | E_n \rangle$, respectively (the symbol $\rho$ should not be confused with the density of atoms, which in this section we call $N$). let $E(z, t) = E_0[\exp(i\omega t) + cc.]/2$ be the transverse electric field. The system hamiltonian operator is $\mathscr{H} = \mathscr{H}_0 + \mathscr{H}'$, where $\mathscr{H}_0$ is the unperturbed Hamiltonian with $\mathscr{H}_0 | E_n \rangle = E_n | E_n \rangle$ ($E_n$ = atomic state energy).

The Hamiltonian $\mathcal{H}'$ is the electric dipole operator defined by $\mathcal{H}' = -\mu E(z, t)$; $E$ is a classical variable. By definition, $E_m - E_n = \hbar\omega_{mn}$. We use three properties: (1) the diagonal elements of the operator $\mu$ vanish, i.e., $\mu_{11} = \mu_{22} = 0$, [47]; (2) the off-diagonal elements are chosen so that $\mu_{12} = \mu_{21} = \mu$, [30]; (3) the normalization conditions $\text{tr}(\rho) = \rho_{11} + \rho_{22} + \rho_{33} = 1$, and the closure relation $\sum |E_k\rangle\langle E_k| = 1$. The Heisenberg equation of motion [30] for the density matrix is then:

$$\frac{d\rho_{mn}}{dt} = -\frac{i}{\hbar}[\mathcal{H}, \rho]_{mn} = -\frac{i}{\hbar}[\mathcal{H}_0, \rho]_{mn} - \frac{i}{\hbar}[\mathcal{H}', \rho]_{mn} \qquad (1.158)$$

where $[A, B] = AB - BA$ is the commutator of operators $A$ and $B$. Equation (1.158) gives explicitly:

$$\frac{d\rho_{mn}}{dt} = -\frac{i}{\hbar}\langle E_m|(\mathcal{H}_0\rho - \rho\mathcal{H}_0)|E_n\rangle + \frac{i}{\hbar}E(z, t)\langle E_m|\mu\rho - \rho\mu|E_n\rangle$$

$$= -\frac{i}{\hbar}(E_m - E_n)\rho_{mn} + \frac{i}{\hbar}E(z, t)\sum_k \{\langle E_m|\mu|E_k\rangle\langle E_k|\rho|E_n\rangle - \langle E_m|\rho|E_k\rangle\langle E_k|\mu|E_n\rangle\}$$

$$= -i\omega_{mn}\rho_{mn} + \frac{i}{\hbar}E(z, t)\sum_k \{\mu_{mk}\rho_{kn} - \rho_{mk}\mu_{kn}\} \qquad (1.159)$$

From Eq. (1.159), and using the above properties, we can derive three relevant equations:

$$\frac{d\rho_{11}}{dt} = \frac{i}{\hbar}\mu E(z, t)(\rho_{21} - \rho_{12}) - R(\rho_{11} - \rho_{33}) + A_{21}\rho_{22} \qquad (1.160)$$

$$\frac{d\rho_{22}}{dt} = -\frac{i}{\hbar}\mu E(z, t)(\rho_{21} - \rho_{12}) - A_{21}\rho_{22} + A_{32}\rho_{33} \qquad (1.161)$$

$$\frac{d\rho_{12}}{dt} = i\omega_{21}\rho_{12} + \frac{i}{\hbar}\mu E(z, t)(\rho_{22} - \rho_{11}) - A_r\rho_{12} \qquad (1.162)$$

A certain number of additional terms were also included in these equations, which is justified through the following [47]. The term in $A_{21}\rho_{22}$ causes $\rho_{22}$ to decrease and $\rho_{11}$ to increase in the absence of any other forces, and accounts for the effect of spontaneous relaxation from state $|E_2\rangle$ to $|E_1\rangle$. The term in $R(\rho_{11} - \rho_{33})$ effects a transport of atoms from state $|E_1\rangle$ to $|E_3\rangle$, which causes $\rho_{33}$ to grow in absence of the electric field $E(z, t)$ and phenomenologically accounts for the pumping process. The term in $A_{32}\rho_{33}$ causes $\rho_{22}$ to increase, which accounts for the spontaneous (nonradiative) relaxation from state $|E_3\rangle$ to $|E_2\rangle$. Without these terms, no force would move the system away from the thermal equilibrium given by Boltmann's Law. Finally, a decay term in $-A_r\rho_{12}$ is introduced so that the off-diagonal term $\rho_{12}$

vanishes in the absence of all forces, as required by equilibrium conditions [47]. As $\rho_{nn}$ represents the probability of finding the system in state $|E_n\rangle$ [30], the quantity $N\rho_{nn}$ ($N$ = atomic density) represents the atomic population in energy level $n$. Equations (1.160a–c) is thus the density matrix counterpart of rate equations (1.1)–(1.3), which describe the three-level laser system of Section 1.1, illustrated in Figure 1.1.

The rapid time variation in Eqs. (1.160a–c) can be eliminated by the substitution $\rho_{12} = \tilde{\sigma}_{12} \exp(i\omega t)$ ($\tilde{\sigma}_{12}$ represents the slowly varying part of the matrix element $\rho_{12}$, and should not be confused with the cross section symbol $\sigma_{12}$). We obtain from Eq. (1.160c):

$$\frac{d\tilde{\sigma}_{12}}{dt} = i(\omega_{21} - \omega)\tilde{\sigma}_{12} + i\Omega(\rho_{22} - \rho_{11}) - A_r\tilde{\sigma}_{12} \tag{1.161}$$

A term in $\exp(-2i\omega t)$ was ignored in the above derivation, as such nonsynchronous contributions average out to zero in the time scale of interest [30]; the same applies to the following derivations. In Eq. (1.161), $\Omega = \mu E_0/2\hbar$ is called the precession frequency [30]. Using $\rho_{21} = \tilde{\sigma}_{21} \exp(-i\omega t)$ and $\rho_{33} = 1 - (\rho_{11} + \rho_{22})$, we obtain from Eq. (1.160)–(1.161):

$$\frac{d(\rho_{22} - \rho_{11})}{dt} = 2i\Omega(\tilde{\sigma}_{12} - \tilde{\sigma}_{21}) + \rho_{11}(2R - A_{32}) + \rho_{22}(R - 2A_{21} - A_{32}) + A_{32} - R \tag{1.162}$$

We assume now a steady state regime for $\rho_{33}$, i.e., $d(\rho_{33})/dt = 0$ or $-d(\rho_{11} + \rho_{22})/dt = 0$. This condition yields from the sum of Eqs. (1.160) and (1.161) the relation:

$$\rho_{22} = 1 - \rho_{11}\frac{2R + A_{32}}{R + A_{32}} \tag{1.163}$$

With Eq. (1.163), Eq. (1.162) can then be put into form:

$$\frac{d(\rho_{22} - \rho_{11})}{dt} = 2i\Omega(\tilde{\sigma}_{12} - \tilde{\sigma}_{21}) - 2A_{21} + 2\rho_{11}\frac{R(1 - \varepsilon) + A_{21}(1 + 2\varepsilon)}{1 + \varepsilon} \tag{1.164}$$

with $\varepsilon = R/A_{32}$. If we consider that in experimental conditions the nonradiative decay rate $A_{32}$ is considerably greater than the pumping rate, we have $\varepsilon \approx 0$ and from Eq. (1.163) we have $2\rho_{11} = 1 - (\rho_{22} - \rho_{11})$. These conditions, replaced in Eq. (1.164) yield:

$$\frac{d(\rho_{22} - \rho_{11})}{dt} = 2i\Omega(\tilde{\sigma}_{12} - \tilde{\sigma}_{21}) - \frac{1}{\tau}\left\{(\rho_{22} - \rho_{11}) - \frac{R\tau - 1}{R\tau + 1}\right\} \tag{1.165}$$

where $\tau = 1/A_{21}$ and $T = \tau/(1 + R\tau)$ is a time constant. The physical meaning of Eq. (1.165) is that when the external field is turned off $(E_0 = 0)$, the difference $\rho_{22} - \rho_{11}$ relaxes to an equilibrium value $(\rho_{22} - \rho_{11})_{eq} = (R\tau - 1)(R\tau + 1)$ with time constant $T$. This result is consistent with the transient dynamics of a three-level system where the transition terminates to the ground level, with the condition $R/A_{32} \ll 1$ [19].

The steady state solution for the difference can be found by setting Eq. (1.161), its complex conjugate, and Eq. (1.165) to zero. After some algebra, this yields:

$$\tilde{\sigma}_{12,21} = \frac{R\tau - 1}{R\tau + 1} \frac{\dfrac{4\Omega}{\Delta\omega^2}(\omega - \omega_{21}) + 2i\dfrac{\Omega}{\Delta\omega}}{1 + 4\left(\dfrac{\omega - \omega_{21}}{\Delta\omega}\right)^2 + \dfrac{8\Omega^2 T}{\Delta\omega}} \tag{1.166}$$

In Eq. (1.166), the parameter $\Delta\omega = 2A_r$ corresponds to the homogeneous spectral width (FWHM) of the laser transition [47].

The next step is to derive the macroscopic polarization induced by the collection of atoms resonating with the electrical field $E(z, t)$ (i.e., the erbium ions). This macroscopic polarization is given by $P = N\langle\mu\rangle = \mathscr{R}e\,[\varepsilon_0\chi E_0\exp(i\omega t)]$, where $\langle\mu\rangle$ is the average value for the dipole moment, which is given by $\langle\mu\rangle = \mathrm{tr}(\rho\mu) = \mu(\rho_{12} + \rho_{21})$, and $\chi$ is the complex atomic susceptibility[30]. Expressing $\rho_{12,21}$ as functions of $\tilde{\sigma}_{12,21}$ found in Eq. (1.166) we obtain the atomic susceptibility:

$$\chi(\omega) = \rho\frac{R\tau - 1}{R\tau + 1}\frac{2\mu^2}{\varepsilon_0\hbar\Delta\omega}\frac{2\dfrac{\omega - \omega_{21}}{\Delta\omega} + i}{1 + 4\left(\dfrac{\omega - \omega_{21}}{\Delta\omega}\right)^2 + \dfrac{2\mu^2 E_0^2 T}{\hbar^2\Delta\omega}} \tag{1.167}$$

Note that in Eq. (1.167) we have used the symbol $\rho$ instead of $N$ for the atomic density, for consistency with our previous notations; there is no longer any risk of confusing it with the density matrix operator. The dipole moment $\mu$ in Eq. (1.167) can be replaced by $\mu^2 = \pi\varepsilon_0\hbar c^3/\tau n\omega_{21}^3$ and we can convert the term $E_0^2$ into an optical intensity through $cn\varepsilon_0 E_0^2/2 = P_s/A$, where $P_s$ is a signal power and $A$ is an effective area [30]. After these transformations, we obtain from Eq. (1.167):

$$\boxed{\chi(\omega) = n\frac{c}{\omega}\sigma(\omega)\left\{2\frac{\omega - \omega_{21}}{\Delta\omega} + i\right\} \cdot \rho\frac{R\tau - 1}{1 + R\tau + 2\dfrac{P_s}{\hat{P}_{sat}(\omega)}}} \tag{1.168}$$

with

$$\sigma(\omega) = \frac{\sigma_{peak}}{1 + 4\left(\dfrac{\omega - \omega_{21}}{\Delta\omega}\right)^2} \tag{1.169}$$

and

$$\sigma_{\text{peak}} = \frac{\lambda_s^2}{2\pi n^2 \tau \Delta\omega} \qquad (1.170)$$

where $\lambda_s = 2\pi c/\omega_{21}$, and:

$$\hat{P}_{\text{sat}}(\omega) = \frac{\hbar\omega_{21}A}{\sigma(\omega)\tau} \qquad (1.171)$$

By definition, the parameters $\sigma$ and $\hat{P}_{\text{sat}}(\omega)/A$ are the cross section and saturation intensity, respectively, corresponding to a single laser transition [30].

The main result obtained in the above derivation is Eq. (1.68) for the complex atomic susceptibility $\chi$, which applies to the same basic three-level laser system as previously described with the rate equations model (Section 1.1). Referring to Eqs. (1.46)–(1.48) and Eqs. (1.49) and (1.50) but ignoring the radial dependence ($\psi_{p,s}(r) = 1$), and assuming equal absorption and emission cross sections ($\sigma_a = \sigma_e$) in these equations, we may recognize that the multiplicative factor on the right-hand side of Eq. (1.68) is the three-level system population inversion density $N_2 - N_1$. The expression is different for two-level and four-level systems, which are usually found in textbooks [30] and [31].

Another important result of this density matrix analysis that the spectral line assumes the shape of a Lorentzian, as shown in Eq. (1.169) for the transition cross section. This fundamental result is the consequence of the relaxation process introduced in Eq. (1.162) for the off-diagonal density matrix element, through the parameter $A_r$. This relaxation term can be interpreted as a loss of phase coherence from the ensemble of atoms oscillating with the electric field, which causes line broadening [30], [46], and [47].

In the case of actual Er:glass, the fluorescence and absorption characteristics are observed to depart significantly from a Lorentzian shape, and the corresponding cross sections for absorption and emission are different in shape and magnitude, as seen in Chapter 4. This fact is obviously a consequence of the Stark splitting effect of the upper and ground manifolds. Because of this Stark splitting, many different laser transitions can take place between the two manifolds, each one being characterized by its individual line parameters (i.e., fluorescence lifetime, cross section, peak wavelength, and line width). Furthermore, each Stark sublevel is populated according to Boltzmann's distribution, Eq. (1.12), which acts as a weighting function for each of the possible transitions. the laser transitions originating from the Stark sublevels of lowest energy are likely to play a dominant role in shaping the overall cross section spectrum, as these sublevels are the most populated. Thus, the overall cross sections for absorption and emission are made up of a superimposition of overlapping Lorentzian line shapes with different characteristics. This situation is made even more complicated by the phenomenon of *inhomogeneous broadening*, in which the Stark splitting varies from site to site for each lasing ion. The effect of inhomogeneous broadening is overlooked here for simplicity, but will be introduced and analyzed in the next section.

The derivation of the complex atomic susceptibility in the case of the Stark split, three-level laser system strictly follows the same procedure as above, using an approach similar to that of the rate equation model treated in Section 1.2. Because this derivation is tedious, we shall only recall here the main results; the details are given in Appendix F.

The three equations for the relevant density matrix elements are found to take the form:

$$\frac{d\bar{\rho}_{11}}{dt} = \frac{i}{\hbar} E(z, t) \sum_j \sum_k \mu_{kj}(\rho_{kj} - \rho_{jk}) - \mathcal{R}_{13}\bar{\rho}_{11} + \mathcal{R}_{31}\bar{\rho}_{33} + \mathcal{A}_{21}\bar{\rho}_{22} \quad (1.172)$$

$$\frac{d\bar{\rho}_{22}}{dt} = -\frac{i}{\hbar} E(z, t) \sum_j \sum_k \mu_{kj}(\rho_{kj} - \rho_{jk}) - \mathcal{A}_{21}\bar{\rho}_{22} + \mathcal{A}_{32} \quad (1.173)$$

$$\frac{d\rho_{jk}}{dt} = i\omega_{kj}\rho_{jk} + \frac{i}{\hbar} E(z, t)\mu_{kj}\{p_{2k}\bar{\rho}_{22} - p_{1j}\bar{\rho}_{11}\} - A'_{jk}\rho_{jk} \quad (1.174)$$

In Eqs. (1.172)–(1.174) the diagonal elements are summed up through $\bar{\rho}_{11} = \sum \rho_{jj}$ and $\bar{\rho}_{22} = \sum \rho_{kk}$ which represent the total probability of occupation of manifolds 1 and 2, respectively; the matrix elements corresponding to the individual Stark sublevels are given by $\rho_{jj} = p_{1j}\bar{\rho}_{11}, \rho_{kk} = p_{2k}\bar{\rho}_{22}, \rho_{ll} = p_{31}\bar{\rho}_{33}$. These relations express that the occupation probability of sublevels $(1j)$ or $(2k)$ is proportional to the Boltzmann distribution ($p_{1j}$ or $p_{2k}$, see definition in Eq. (1.12)), corresponding to thermal equilibrium, and this equilibrium is not disturbed by the pumping process. This assumption is justified by the fact that the thermalization rates are significantly greater than any other rates, due to the relatively small energy separation ($\Delta E < k_B T$) between the Stark sublevels (Chapter 4). The total pumping rates $\mathcal{R}_{13,31}$ and spontaneous emission rate $\mathcal{A}_{21}$ were already defined in Eqs. (1.16), (1.17), and (1.20) in the rate equation model. The total or effective fluorescence lifetime $\tau$ of the upper level manifold 2 is thus given by:

$$\frac{1}{\tau} \equiv \mathcal{A}_{21} = \sum_j \sum_k A_{kj} p_{2k} = \sum_j \sum_k \frac{1}{\tau_{kj}} p_{2k} \quad (1.175)$$

where $\tau_{kj}$ represents the radiative lifetime associated with the laser transition $(jk)$. The effect of thermal equilibrium is to weight each transition probbability $(A_{kj})$ by the corresponding sublevel occupation probability $p_{2k}$. By thermal equilibrium, we actually mean here *thermal equilibrium in a system excited by an external pump source*; the occupancy distribution in the Stark level manifolds are given by Boltzmann's law, Eq. (1.12), and both ground and upper manifolds are populated. This condition is distinguished from the case of thermal equilibrium with the pump turned off, in which all atoms end up in their ground state, because of the effect of spontaneous decay. Thermal equilibrium is nonetheless appropriate whether the atomic system is driven by an external pump source or whether it is isolated from any external sources. The physical interpretation of Eq. (F.19) in Appendix F clearly illustrates this point.

When solving Eqs. (1.172)–(1.174) in the steady state regime, we find that the manifold occupation probabilities take the form:

$$\bar{\rho}_{11} = \frac{1 + \dfrac{P_s}{A} \sigma_e(\omega_s)\tau}{1 + \mathcal{R}_{13}\tau + \dfrac{P_s}{h\bar{\nu}A}[\sigma_e(\omega_s) + \sigma_a(\omega_s)]\tau} \quad (1.177)$$

$$\bar{\rho}_{22} = \cfrac{\mathscr{R}_{13}\tau + \cfrac{P_s}{A}\sigma_a(\omega_s)\tau}{1 + \mathscr{R}_{13}\tau + \cfrac{P_s}{h\bar{v}A}[\sigma_e(\omega_s) + \sigma_a(\omega_s)]\tau} \tag{1.177}$$

with the following definitions for the total absorption and emission cross sections $\sigma_{a,e}(\omega)$:

$$\boxed{\frac{\sigma_a(\omega)}{h\bar{v}} = \sum_{jk}\frac{\sigma_{kj}(\omega)}{hv_{kj}}p_{1j}} \tag{1.178}$$

$$\boxed{\frac{\sigma_e(\omega)}{h\bar{v}} = \sum_{jk}\frac{\sigma_{kj}(\omega)}{hv_{kj}}p_{2k}} \tag{1.179}$$

In Eqs. (1.178) and (1.179), $\sigma_{kj}(\omega)$ represents the cross section of Lorentzian line shape associated with the individual laser transition $(jk)$, defined similarly to Eqs. (1.169) and (1.170), i.e.,

$$\sigma_{kj}(\omega) = \mathscr{L}_{kj}\sigma_{kj}^{\text{peak}} = \cfrac{1}{1 + 4\left(\cfrac{\omega - \omega_{kj}}{\Delta\omega_{kj}}\right)^2}\frac{\lambda_{kj}^2}{2\pi n^2\tau_{kj}\Delta\omega_{kj}} \tag{1.180}$$

where $\lambda_{kj} = 2\pi c/\omega_{kj}$ is the transition wavelength and $\omega_{kj}$ the associated spectral width (FWHM). The term $\bar{v}$ in definitions (1.176)–(1.179) is an average frequency arbitrarily taken as any value of $c/\lambda_{kj}$ as $\lambda_{kj}$ varies by only 1% over the wavelength range of interest (1.53–1.55 $\mu$m). In Eqs. (1.176) and (1.177) we used the symbol $\omega_s$, representing the fixed saturating signal frequency, in order to express the fact that the quantities $\bar{\rho}_{11}$, $\bar{\rho}_{22}$ are not intrinsically frequency dependent.

Equations (1.178) and (1.179) correspond to a generalized definition of the absorption and emission cross sections applicable to Stark split laser systems. These are made from the superimposition of all individual laser lines bridging each pair of Stark sublevels that can be selected from manifolds 1 and 2, their sum weighted by the Boltzmann distribution. The Stark level energies and the Boltzmann distribution energies, $p_{1j,2k}$ are generally different for each manifold (Chapter 4). Consequently, the absorption and emission cross sections generally have different magnitudes and spectral shapes, as seen previously in Section 1.3 and illustrated in Figure 1.3. As the total cross sections $\sigma_{a,e}(\lambda)$ and fluorescence lifetime $\tau$ can be measured experimentally (Chapter 4), the knowledge of the actual Stark level positions, and individual laser transition characteristics ($\lambda_{kj}$, $\sigma_{kj}$, $\tau_{kj}$, $\Delta\omega_{kj}$) is not actually necessary for modeling the Stark split system. The only difficulty remaining is in the theoretical definition of the peak value for these cross sections. Using this equation, the peak values can be approximated only:

$$\sigma_{a,e}(\omega) \approx \frac{\lambda^2}{2\pi n^2\tau\Delta\omega_{a,e}} \tag{1.181}$$

where $\Delta\omega_{a,e}$ are effective spectral line widths of:

$$\Delta\omega_{a,e} = \int I_{a,e}(\omega)\,d\omega \qquad (1.182)$$

and $I_{a,e}$ represents the experimental (or phenomenological) absorption and emission spectra normalized to a unity peak value. Such an approximate definition represents the Fuchtbauer–Ladenburg (FL) relation. Chapter 4 outlines a method for the theoretical determination of $\sigma_{a,e}(\lambda)$ more accurate than the FL relation.

We recognize for the occupation probabilities $\bar{\rho}_{11}$, $\bar{\rho}_{22}$ in Eqs. (1.176) and (1.177) the same form as in Eq. (1.10) that apply to the rate equation model of Sections 1.1–1.3. The density matrix theory developed here thus validates Eq. (1.26) for the gain coefficient, in which absorption and emission cross sections were introduced phenomenologically. It also shows that *the Stark split three-level system properties are fully identical to that of a nondegenerate three-level system in which the absorption and emission processes would be characterized by different cross sections.*

Finally, the total complex atomic susceptibility $\chi$ is found to take the form (Appendix F):

$$\boxed{\chi = \sum_{jk} \chi_{jk}} \qquad (1.183)$$

where $\chi_{jk}$ is the complex atomic susceptibility associated with the individual laser transition $(jk)$ of expression:

$$\chi_{jk}(\omega) = -nc\rho\,\frac{\sigma_{kj}(\omega)}{\omega_{kj}}\left(2\frac{\omega_{kj}-\omega}{\Delta\omega_{kj}}-i\right)\frac{\mathcal{R}_{13}\tau p_{2k}-p_{1j}+\dfrac{P_s}{h\bar{v}A}[p_{2k}\sigma_a(\omega_s)-p_{1j}\sigma_e(\omega_s)]\tau}{1+\mathcal{R}_{13}\tau+\dfrac{P_s}{h\bar{v}A}[\sigma_e(\omega_s)+\sigma_a(\omega_s)]\tau}$$

$$(1.184)$$

The fundamental result of eq. (1.183) was used only as a postulate in previous work on the analysis of Stark split three-level systems (E. Desurvire [27]), and also utilized to describe laser systems with degenerate energy levels (A. E. Siegman [32]). The linear superposition principle, fully demonstrated in Appendix F, expresses that the total susceptibility of the Stark split three-level system is given by the sum of all susceptibilities associated with three-level subsystems, to each of which correspond an individual laser transition. This linear superposition principle remains true in the saturation regime, due to the fact that the effect of saturation is independent of the index $j$, $k$, as shown by the form taken by the denominator in Eq. (1.184). This property is a consequence of a central assumption, i.e., the thermal equilibrium defining the Boltzmann distributions $p_{1j,2k}$ within the manifolds remains unaffected by the lasing process. In other words, the pumping and stimulated emission rates remain negligible in comparison to the thermalization rates, due to the small energy spacing ($\Delta E < k_B T$) between the Stark sublevels.

Another important result contained in Eqs. (1.183) and (1.184) is that the imaginary part of the total susceptibility of the Stark split system saturates homogeneously, discussed later when considering $\chi''$ in more detail.

If the complex susceptibility $\chi$ has no poles in either upper or lower parts of the complex plane, its real and imaginary parts are related by the Kramers–Kronig relations (KKR), [30] and [31], i.e.,

$$\chi'(\omega) = \frac{1}{\pi} \, \text{P.V.} \int_{-\infty}^{+\infty} \frac{\chi''(\omega')}{\omega' - \omega} \, d\omega' = \frac{1}{\pi} \sum_{jk} \text{P.V.} \int_{-\infty}^{+\infty} \frac{\chi''_{jk}(\omega')}{\omega' - \omega} \, d\omega' \quad (1.185)$$

$$\chi''(\omega) = -\frac{1}{\pi} \, \text{P.V.} \int_{-\infty}^{+\infty} \frac{\chi'(\omega')}{\omega' - \omega} \, d\omega' = -\frac{1}{\pi} \sum_{jk} \text{P.V.} \int_{-\infty}^{+\infty} \frac{\chi'_{jk}(\omega')}{\omega' - \omega} \, d\omega' \quad (1.186)$$

where P.V. means the Cauchy principal value of the integral. Thus, if the KKR apply, it is possible to predict theoretically the real or imaginary part of $\chi$ from its other part. This represents a useful property, as the imaginary part can be easily deduced from an experimental measurement of the gain coefficient (see below), while the real part, representing the refractive index, is more difficult to evaluate experimentally.

Does the total susceptibility $\chi$ of Stark split, three-level systems satisfy the KKR? Rewrite using Eqs. (1.180), (1.183), and (1.184) in the form:

$$\chi = \frac{\pi c^3 \rho}{n(1 + \mathscr{R}_{13}\tau)} \sum_{jk} \frac{\mathscr{U}_{kj}(\omega)}{\tau_{kj}\omega_{kj}^3} \frac{\omega - \left(\omega_{kj} - i\dfrac{\Delta\omega_{kj}}{2}\right)}{\left[\omega - \left(\omega_{kj} - i\dfrac{\Delta\omega_{kj}^{\text{sat}}}{2}\right)\right]\left[\omega - \left(\omega_{kj} + i\dfrac{\Delta\omega_{kj}^{\text{sat}}}{3}\right)\right]} \quad (1.187)$$

with

$$\mathscr{U}_{kj}(\omega) = \mathscr{R}_{13}\tau p_{2k} - p_{1j} + \frac{P_s}{h\bar{v}A}\left[p_{2k}\sigma_a(\omega_s) - p_{1j}\sigma_e(\omega_s)\right]\tau \quad (1.188)$$

and

$$\Delta\omega_{kj}^{\text{sat}} = \Delta\omega_{kj}\sqrt{1 + \frac{P_s\tau}{h\bar{v}A(1 + \mathscr{R}_{13}\tau)} \frac{\sigma_e(\omega_s) + \sigma_a(\omega_s)}{\mathscr{L}_{kj}(\omega)}} \quad (1.189)$$

The parameter $\Delta\omega_{kj}^{\text{sat}}$ defined in Eq. (1.189) represents the saturated homogeneous width of the $(jk)$ laser transition, and is equal to $\Delta\omega_{kj}$ in the absence of the saturating signal ($P_s = 0$). The effect of homogeneous saturation is, according to eq. (1.189), to broaden the line width by a factor that varies as the square root of the saturating power.

From Eq. (1.187), when no saturation occurs, the poles of $\chi$ are all in the upper half of the complex plane, i.e., $\omega = \omega_{kj} + i\Delta\omega_{kj}/2$. In this case, the KKR apply; a physical interpretation of this situation is given in [30]. When saturation occurs, eq. (1.187) shows that the poles are distributed in both sides of the complex plane, i.e., $\omega = \omega_{kj} \pm i\Delta\omega_{kj}/2$, and in this case the KKR do not apply.

We consider next that the summation rules in eqs. (1.178)–(1.179) lead to a relatively simple expression for the imaginary part $\chi''$ of the susceptibility, valid for any regime:

$$\chi''(\omega) = -\frac{nc\rho}{2\pi\bar{v}} \frac{\mathcal{R}_{13}\tau\sigma_e(\omega) - \sigma_a(\omega) + \dfrac{P_s}{h\bar{v}A}[\sigma_e(\omega)\sigma_a(\omega_s) - \sigma_a(\omega)\sigma_e(\omega_s)]\tau}{1 + \mathcal{R}_{13}\tau + \dfrac{P_s}{h\bar{v}A}[\sigma_e(\omega_s) + \sigma_a(\omega_s)]\tau} \qquad (1.190)$$

From the above equation, it is clear that $\chi''(\omega)$ can be theoretically calculated even in the saturation regime, given the experimental data on $\sigma_{a,e}(\omega)$. The form of Eq. (1.190) suggests that $\chi''(\omega)$ saturates homogeneously, *as if* the Stark split laser system is characterized by one single homogeneous laser transition with a non-Lorentzian line shape and with different absorption and emission cross sections. Saturation is homogeneous because thermalization occurs fast within the Stark manifolds, so the overall cross section line shapes remain unchanged, even under high saturation conditions. What is affected by saturation is the overall population of each manifold.

Homogeneous saturation is important for modeling EDFAs. Knowledge of the optical powers and the spectral distributions of $\sigma_{a,e}(\omega)$ allows the fiber gain coefficient (proportional to $\chi''(\omega)$, see below) to be calculated in any regime using Eq. (1.190). Consider now the real part $\chi'(\omega)$ of the complex susceptibility. Direct calculation of $\chi'(\omega)$ comes from Eqs. (1.183) and (1.184), and is given by:

$$\chi'(\omega) = -2nc\rho \sum_{jk} \frac{\sigma_{kj}(\omega)}{\omega_{kj}} \frac{\omega_{kj} - \omega}{\Delta\omega_{kj}} \frac{\mathcal{R}_{13}\tau p_{2k} - p_{1j} + \dfrac{P_s}{h\bar{v}A}[p_{2k}\sigma_a(\omega_s) - p_{1j}\sigma_e(\omega_s)]\tau}{1 + \mathcal{R}_{13}\tau + \dfrac{P_s}{h\bar{v}A}[\sigma_e(\omega_s) + \sigma_a(\omega_s)]\tau}$$

$$(1.191)$$

In contrast to the calculation of $\chi''(\omega)$, this requires the *full knowledge of all individual laser transition characteristics* ($\omega_{kj}$, $\sigma_{kj}$, $\tau_{kj}$, $\Delta\omega_{kj}$). As such knowledge is hardly obtainable experimentally (Chapter 4), the only technique left for the evaluating $\chi'(\omega)$ is the KKR method. *In the saturation regime,* however, the KKR do not apply, so *the real part $\chi'(\omega)$ cannot be theoretically inferred.*

Generalization of the above results to the case of a Stark split, two-level system is straightforward; there is no need to go through the whole density matrix treatment. Instead, express the probabilities $\bar{p}_{11}$, $\bar{p}_{22}$ in Eqs. (1.176) and (1.177) with the pumping rate $\mathcal{R}_{31}$ set to zero, and with an additional term proportional to $P_p$, representing a second saturating signal, or pump, at wavelength $\lambda_p$:

$$\bar{p}_{11} = \frac{1 + \dfrac{P_p}{A_p}\sigma_e(\omega_p)\tau + \dfrac{P_s}{A_s}\sigma_e(\omega_s)\tau}{1 + \dfrac{P_p}{h\bar{v}A_p}[\sigma_e(\omega_p) + \sigma_a(\omega_p)]\tau + \dfrac{P_s}{h\bar{v}A_s}[\sigma_e(\omega_s) + \sigma_a(\omega_s)]\tau} \qquad (1.192)$$

$$\bar{\rho}_{22} = \frac{\dfrac{P_p}{A_p}\sigma_a(\omega_p)\tau + \dfrac{P_s}{A_s}\sigma_a(\omega_s)\tau}{1 + \dfrac{P_p}{h\bar{\nu}A_p}[\sigma_e(\omega_p) + \sigma_a(\omega_p)]\tau + \dfrac{P_s}{h\bar{\nu}A_s}[\sigma_e(\omega_s) + \sigma_a(\omega_s)]\tau} \tag{1.193}$$

Note that different effective areas $A_p$ and $A_s$ were introduced for the pump and the signal, respectively. This derivation of $\bar{\rho}_{11}$, $\bar{\rho}_{22}$ does not stem from a rigorous density matrix analysis, but is consistent with the rate equation model previously described. On the other hand, we can use the rigorous Eq. (F.26) derived for the slowly varying density matrix element $\tilde{\sigma}_{jk}(\omega)$, and replace in this expression the quantities $\bar{\rho}_{11}$, $\bar{\rho}_{22}$ defined in Eqs. (1.192) and (1.193). This substitution gives the same expression as Eq. (F.34) with $\mathscr{R}_{31} = 0$ and additional terms proportional to $P_p$, which repeat the terms proportional to $P_s$. Following the procedure in Appendix F, we obtain for the complex susceptibility of the transition $(jk)$ in the two-level system:

$$
\begin{aligned}
\chi_{jk}(\omega) = &-nc\rho \frac{\sigma_{kj}(\omega)}{\omega_{kj}}\left(2\frac{\omega_{kj} - \omega}{\Delta\omega_{kj}} - i\right) \\
&\times \frac{-p_{1j} + \dfrac{P_p}{h\bar{\nu}A_p}[p_{2k}\sigma_a(\omega_p) - p_{1j}\sigma_e(\omega_p)]\tau + \dfrac{P_s}{h\bar{\nu}A_s}[p_{2k}\sigma_a(\omega_s) - p_{1j}\sigma_e(\omega_s)]\tau}{1 + \dfrac{P_p}{h\bar{\nu}A_p}[\sigma_e(\omega_p) + \sigma_a(\omega_p)]\tau + \dfrac{P_s}{h\bar{\nu}A_s}[\sigma_e(\omega_s) + \sigma_a(\omega_s)]\tau}
\end{aligned}
$$

$$\tag{1.194}$$

Using Eq. (1.183) and the summing rules (1.178) and (1.179), the imaginary part of the total susceptibility of a two-level system takes from Eq. (1.194) the form:

$$
\begin{aligned}
\chi''(\omega) = &-\frac{nc\rho}{2\pi\bar{\nu}} \\
&\times \frac{-\sigma_a(\omega) + \dfrac{P_p}{h\bar{\nu}A_p}[\sigma_e(\omega)\sigma_a(\omega_p) - \sigma_a(\omega)\sigma_e(\omega_p)]\tau + \dfrac{P_s}{h\bar{\nu}A_s}[\sigma_e(\omega)\sigma_a(\omega_s) - \sigma_a(\omega)\sigma_e(\omega_s)]\tau}{1 + \dfrac{P_p}{h\bar{\nu}A_p}[\sigma_e(\omega_p) + \sigma_a(\omega_p)]\tau + \dfrac{P_s}{h\bar{\nu}A_s}[\sigma_e(\omega_s) + \sigma_a(\omega_s)]\tau}
\end{aligned}
$$

$$\tag{1.195}$$

Equation (1.194) for $\chi_{jk}(\omega)$ for the two-level system actually coincides with that of (1.184) found for the three-level system, when the emission cross section at the pump frequency $\sigma_e(\omega_p)$ in Eq. (1.194) is set to zero, since we have $P_p\sigma_a(\omega_p)/h\bar{\nu}A_p = \mathscr{R}_{13}$ from the eq. (1.42). Therefore, the above expressions for $\chi_{jk}(\omega)$ and $\chi''(\omega)$ corresponding to the two-level system are most general and readily applicable to three-level systems.

Considering the imaginary parts $\chi''$ corresponding to both two-level and three-level systems, we note that the term proportional to $P_s$ in the numerator in Eqs. (1.190) and (1.195) vanishes when $\omega = \omega_s$, but is finite for $\omega \neq \omega_s$. In this last case, the effect of saturation in $\chi''$ is partly compensated by the effect of population near-equalization between the two manifolds. This near-equalization effect stems from the fact that, in both two- and three-level systems considered here, the laser transition terminates to the ground level. As a result, the upward and downward stimulated emission rates, which are responsible for saturation, nearly compensate each other in a ratio given by $W_{21}/W_{12} = \eta_s = \sigma_e/\sigma_a = N_2/N_1$, according to Eqs. (1.46) and (1.47), Eqs. (1.49) and (1.50).

The conclusions we derived for the three-level system complex susceptibility also apply to the two-level system, i.e., the imaginary part $\chi''$ can be theoretically predicted for all regimes; the real part $\chi'$ can be theoretically inferred by the KKR only in the unsaturated regime, unless we have full knowledge of the individual laser transition characteristics; the imaginary part of the susceptibility saturates homogeneously.

A useful relation between the real and imaginary parts of the complex atomic susceptibility $\chi(\omega)$ on one hand, and the gain coefficient $\Delta g(\omega)$ and associated refractive index change $\delta n(\omega)$ on the other, can be established through Maxwell's classical electromagnetism.

The density matrix theory developed in this section is semiclassical as opposed to fully quantum mechanical, i.e., the atomic states are quantized but the interacting field is classical. It is therefore appropriate to express the macroscopic linear polarization of the medium $\mathscr{P}(r, z, t)$ through the complex atomic susceptibility found above, and to use this polarization as a driving term in Maxwell's wave equation for the electric field $E(r, z, t)$. The linear polarization has two components, i.e., $\mathscr{P}(r, z, t) = \mathscr{P}_H(r, z, t) + \mathscr{P}_A(r, z, t)$, where $\mathscr{P}_H(r, z, t)$ is the polarization induced by the host medium, and $\mathscr{P}_A(r, z, t)$ the polarization induced by the atomic dopant in this host. We have $\mathscr{P}_H(r, z, t) = \varepsilon_0(1 + \chi_H)E(r, z, t) = \varepsilon_0 n^2 E(r, z, t)$, where $\chi_H$ is the host linear susceptibility, and $\mathscr{P}_A(r, z, t) = \varepsilon_0 \chi E(r, z, t)$ where $\chi$ is the complex susceptibility associated with the atomic dopant. For simplicity, we consider here only a simple unidimensional wave equation and assume a linearly polarized electric field confined in an envelope of radial dependence $\psi(r)$, which corresponds to the fundamental guided mode of fibers. The wave equation then takes the form, [31] and [47]:

$$\frac{\partial^2 E(r, z, t)}{\partial z^2} - \mu_0 \varepsilon \frac{\partial^2 E(r, z, t)}{\partial t^2} = \mu_0 \frac{\partial^2 \mathscr{P}_A(r, z, t)}{\partial t^2} \tag{1.196}$$

where $\mu_0$ is the magnetic permeability in free space, with $\varepsilon = \varepsilon_0 n^2$ and $\varepsilon_0 \mu_0 c^2 = 1$. The electric field can be expressed as $E(r, z, t) = \sqrt{\psi(r)}E_0(z)\exp[i(\omega t - \beta z)]$ where $\psi(r)$ represents the guided power mode envelope and $\beta = n\omega/c$ the wave propagation constant.

Using the slowly varying envelope approximation [47] for Eq. (1.196), multiplying this equation by $\psi(r)$, integrating over the transverse plane, and replacing the expression for $\mathscr{P}_A(r, z, t)$ while assuming a step-like dopant density distribution, we obtain:

$$\frac{\partial E_0(z)}{\partial z} = -i\Gamma_s \frac{\omega}{2nc} \chi E_0(z) = -\Gamma_s \frac{\omega}{2nc}(\chi'' + i\chi')E_0(z) \tag{1.197}$$

where is $\Gamma_s$ the mode/doped core overlap factor, defined in Section 1.7. Integration from $z = 0$ to $z = L$ of eq. (1.197), assuming no gain saturation (which does not affect the generality of the result) yields $E_0(z) = E_0(z)\sqrt{G}\exp(-i\delta n\omega L/c)$, where the power gain $G(\omega)$ and the gain-induced refractive index change $\delta n(\omega)$ are given by:

$$G(\omega) = \exp\left\{-\Gamma_s\frac{\omega}{nc}\int_0^L\chi''(\omega, z)\,dz\right\}$$ (1.198)

$$\delta n(\omega) = \Gamma_s\frac{1}{2nL}\int_0^L\chi'(\omega, z)\,dz$$ (1.199)

Using Eqs. (1.190), (1.176), and (1.177) and the populations of Stark split three-level systems $\bar{N}_i = \rho\bar{\rho}_{ii}$ ($i = 1, 2$), we find that, consistently, the gain in Eq. (1.198) takes the form:

$$G(\omega) = \exp\left\{\Gamma_s\int_0^L[\sigma_e(\omega)\bar{N}_2(z) - \sigma_a(\omega)\bar{N}_1(z)]\,dz\right\}$$ (1.200)

which has been analyzed in the previous sections. Note that the saturation term in the denominator of $\chi''$ in Eq. (1.190) or eq. (1.195) can be put in the form $P_S/P_{sat}(\omega)$ where $P_{sat}(\omega)$ is the saturation power defined in Eq. (1.45). Thus, all the results found in this section are consistent with the previous rate equation analysis for Stark split two- and three-level systems. A study of gain-induced refractive index change $\delta n(\omega)$ under different pumping regimes and its effect on fiber dispersion is given in Chapter 4.

## 1.12  MODELING INHOMOGENEOUS BROADENING

Throughout the previous sections, we assumed the laser transitions were homogeneously broadened, i.e., all atoms in the host medium had the same laser line characteristics: center frequency, line width, peak cross section, and fluorescence lifetime. In the case of a Stark split three-level system, the overall laser line width is broadened by the superposition of all possible homogeneous transitions ($jk$) between the Stark manifolds. This broadening effect can still be considered homogeneous, as the resulting cross sections homogeneously saturate, due to the fast intramanifold thermalization effect, as seen in the previous section.

In an ordered structure material, such as a perfect crystal, each dopant atom (or activator impurity) occupies the same type of site, which has well-determined surroundings. each activator experiences the same electric field generated by the site's surrounding charges. This crystal field causes equal amounts of Stark splitting in each activator. As a result, the spectral properties of the collections of all such activators in the host coincide with that of any single activator in the site [48]. The resulting spectral lines are thus homogeneously broadened.

Defects, dislocations, or lattice impurities in host crystals and the amorphous structure of glass hosts lead to site-to-site variations of the surrounding field and cause a randomization of the Stark effect, changing the energy positions of the Stark sublevels. the surrounding electric field also affects the electric dipole transition characteristics, as outlined in Chapter 4. Finally, some glasses have more than one type of activator site, depending on the various possible coordinations by codopant ions involved in the glass composition. For all the above reasons, site-to-site variations and existence of multiple types of sites in the host cause the activators to exhibit different laser line characteristics. The spectral lines resulting from the superimposition of all activator contributions in such a host are inhomogeneously broadened.

While the effect of inhomogeneous broadening in Er:glass and its importance relative to homogeneous characteristics will be analyzed in detail in Chapter 4, the goal of this section is to describe how it can be taken into account in theoretical modeling. This classical issue is fully addressed in many textbooks for degenerate laser systems (see for instance [30], [32], and [49]), and for Stark split, three-level systems see [25] and [27].

We can generalize the definition of the inhomogeneously broadened atomic susceptibility $\chi^I(\omega)$ of [32] to the case of a Stark split system through [27]:

$$\chi^I(\omega) = \sum_{jk} \int_{-\infty}^{\infty} p_{\mathrm{inh}}(\omega') \chi^H_{jk}(\omega - \omega') \, d\omega' \qquad (1.201)$$

where $\chi^H_{jk}(\omega)$ is the homogeneous atomic susceptibility of laser transition $(jk)$ corresponding to definitions in Eq. (1.184) or Eq. (1.194), and $p_{\mathrm{inh}}(\omega')$ is a normalized density distribution for the resonance frequency $\omega_{kj}$, centered on $\omega' = 0$, and characterized by a width $\Delta\omega_{\mathrm{inh}}$. As $\chi^H_{jk}(\omega)$ is an explicit function of $\omega' - \omega_{kj}$, the function $\chi^H_{jk}(\omega - \omega')$ actually corresponds to a transition of center frequency $\Omega_{kj} = \omega_{kj} \pm |\omega'|$. In this simplified model, only the resonance frequencies $\omega_{kj}$ are made to randomly vary. This is equivalent to assuming that all the line characteristics are conserved from site to site, except for the center frequency $\omega_{jk}$, which is determined by the random positions of the Stark sublevels. In reality, the host can randomly change all line parameters to some extent and the theory becomes enormously complex. In order to account for increased randomness, the total susceptibility could be given by an expression of the type:

$$\chi^I(\omega) = \sum_{jk} \int_{-\infty}^{\infty} p_{\mathrm{inh}}(\omega') p_2(\sigma) p_3(\tau) \chi^H_{jk}(\omega - \omega', \sigma^{\mathrm{peak}}_{kj} - \sigma, \tau_{kj} - \tau) \, d\omega' \, d\sigma \, d\tau \qquad (1.202)$$

where $p_2$ and $p_3$ are density distributions for other line parameters such as peak cross section $\sigma^{\mathrm{peak}}_{kj}$ and fluorescence lifetime $\tau_{kj}$. For simplicity, we shall assume here that the main effect of inhomogeneous broadening is the randomization of the center frequencies.

The density distribution $p_{\mathrm{inh}}(\omega')$ for inhomogeneous broadening of line frequencies in solids is generally Gaussian, [32] and [48], i.e.,

$$p_{\mathrm{inh}}(\omega') = \sqrt{\frac{4\log(2)}{\pi\Delta\omega^2_{\mathrm{inh}}}} \exp\left\{ -4\log(2) \cdot \left(\frac{\omega'}{\Delta\omega_{\mathrm{inh}}}\right)^2 \right\} \qquad (1.203)$$

where $\Delta\omega_{\mathrm{inh}}$ represents an inhomogeneous broadening spectral width. The distribution defined in Eq. (1.203) is normalized (i.e., $\int p_{\mathrm{inh}}(\omega) \, d\omega = 1$). In the limit of negligible inhomogeneous broadening, or $\Delta\omega_{\mathrm{inh}} \to 0$, the distribution takes the form of a delta distribution, i.e., $p_{inh}(\omega') = \delta(\omega')$ and, from Eq. (1.201) we find $\chi^I = \sum \chi^H_{jk}$, the same relation as Eq. (1.183) obtained for homogeneous systems.

In order to evaluate the imaginary part $\chi''_I(\omega)$ of the inhomogeneously broadened susceptibility (relevant to the analysis of the gain coefficient), we consider first the simplest case of the three-level system with an unsaturated regime. using Eq. (1.190) with $P_s = 0$ and Eq. (1.201), we find:

$$\chi''_I(\omega) = -\frac{nc\rho}{2\pi\bar{v}}\frac{\mathscr{R}_{13}\tau\sigma^I_e(\omega) - \sigma^I_a(\omega)}{1 + \mathscr{R}_{13}\tau} \tag{1.204}$$

where $\sigma^I_{a,e}(\omega)$ are the inhomogeneous absorption and emission cross sections of definitions:

$$\sigma^I_a(\omega) = \int_{-\infty}^{\infty} p_{inh}(\omega - \omega')\sigma^H_a(\omega')\,d\omega' \tag{1.205}$$

$$\sigma^I_e(\omega) = \int_{-\infty}^{\infty} p_{inh}(\omega - \omega')\sigma^H_e(\omega')\,d\omega' \tag{1.206}$$

The cross sections $\sigma^H_{a,e}(\omega)$ in Eqs. (1.205) and (1.206) correspond to the homogeneous absorption and emission cross sections, previously defined in Eqs. (1.178) and (1.179). If we substitute the definitions (1.178) and (1.179) for $\sigma^H_{a,e}(\omega)$ into Eqs. (1.205) and (1.206), and carry out the integration with the change of variable $y = \sqrt{4\log(2)}(\omega - \omega')/\Delta\omega_{inh}$ and the definitions $x_{kj} = \sqrt{4\log(2)}(\omega_{kj} - \omega)/\Delta\omega_{inh}$, $b_{kj} = \sqrt{\log(2)}\Delta\omega_{kj}/\Delta\omega_{inh}$ we obtain [49]:

$$\sigma^I_{a,e}(\omega) = 2\pi\bar{v}\sqrt{\pi\log(2)}\sum_{jk} p_{1j,2k}\frac{\sigma^{peak}_{kj}}{\omega_{kj}}\frac{\Delta\omega_{kj}}{\Delta\omega_{inh}}\mathscr{R}e\{\mathscr{W}(x_{kj} + ib_{kj})\} \tag{1.207}$$

where $\mathscr{W}(x + ib)$ is the error function of complex argument as defined by:

$$\mathscr{W}(x + ib) = \frac{i}{\pi}\int_{-\infty}^{\infty} dy\,\frac{\exp(-y^2)}{x + ib + y} \tag{1.208}$$

The real part of $\mathscr{W}(x + ib)$ is called the Voigt profile [49], and its value for the argument $x + ib$ can be found in mathematical functions handbooks [50]. The Voigt integral in Eq. (1.208) can also be calculated exactly in the limits where homogeneous or inhomogeneous broadening effects are dominant [49]. However, as seen in Chapter 4, such is not the case with Er:glass, where both broadening effects are of comparable magnitudes.

As Eq. (1.207) shows, a theoretical calculation of the inhomogeneous cross section $\sigma^I_{a,e}(\omega)$ through this inhomogeneous broadening model would require a full knowledge of all the $(jk)$ line parameters, as well as the Stark sublevel distributions $p_{1j,2k}$. But this calculation is not necessary because $\sigma^I_{a,e}(\omega)$ can be regarded as the actual cross sections that are measurable experimentally. The inhomogeneous width $\Delta\omega_{inh}$ can also be determined experimentally, as shown in Chapter 4. Therefore, the homogeneous cross sections $\sigma^H_{a,e}(\omega)$ involved in the convolution integrals in Eqs. (1.205) and (1.206) can be calculated exactly. Assuming the Gaussian distribution defined in Eq. (1.203) and using the convolution theorem [51], we find:

$$\sigma_{e,a}^{H}(\omega) = \mathscr{F}^{-1}\left[\exp\left(\frac{\Delta\omega_{inh}^2 x^2}{16\log(2)}\right) \cdot \mathscr{F}[\sigma_{e,a}^{I}(\omega); x]; \omega\right] \tag{1.209}$$

where $\mathscr{F}$, $\mathscr{F}^{-1}$ are the Fourier and inverse Fourier transforms, respectively, i.e., $G(x) = \mathscr{F}[g(\omega); x]$, $g(\omega) = \mathscr{F}^{-1}[G(x); \omega]$. Equation (1.209) is an exact solution for the deconvolution of eqs. (1.205) and (1.206) only if the Fourier transform $\mathscr{F}[\sigma_{a,e}^{I}(\omega); x]$ is accurately defined. But the deconvolution operation is exact if the functions $\sigma_{a,e}^{I}(\omega)$ are fitted through a sum of elementary functions (e.g., Gaussian) of known Fourier transforms. An analysis of the homogeneous cross sections $\sigma_{a,e}^{H}(\omega)$ for various Er:glass types and a straightforward method for the computation of the Fourier transforms in Eq. (1.209) based on Gaussian fits of $\sigma_{a,e}^{I}(\omega)$ are presented in Chapter 4.

Knowledge of the homogeneous cross sections $\sigma_{a,e}^{H}(\omega)$, which can be obtained from eq. (1.209), allows calculation of the inhomogeneous gain coefficient in the saturation regime. The terms proportional to the pump and signal power $P_m$ ($m = $ p, s) which cause saturation in Eq. (1.194) must be modified in order to take into account the effect of inhomogeneous broadening. This modification consists of the following substitution:

$$\frac{P_m}{h\bar{\nu}A_m}\sigma_{a,e}(\omega_m)\tau \Rightarrow \frac{1}{h\bar{\nu}A_m}\frac{\tau}{\delta\omega}\int_{-\infty}^{\infty} P_m(\omega'')\sigma_{a,e}^{H}(\omega'' - \omega')\,d\omega'' \tag{1.210}$$

where $\delta\omega$ is the integration increment used in the numerical computation of Eq. (1.210), and $P_m(\omega)$ is the power spectral density of the saturating source with peak frequency $\omega_m$. This transformation is justified because the saturating source is affecting individual atoms differently, depending on the amount of detuning between its peak frequency $\omega_m$ and the atomic homogeneous cross section $\sigma_{a,e}^{H}(\omega)$. Therefore, the resulting effect represents a summation over all possible detunings, which is done by the convolution integral in eq. (1.210). In the case where the saturating source is strongly peaked at the frequency $\omega_m$, i.e., $P_m(\omega) = P_m\delta(\omega - \omega_m)/\delta\omega$ where $\delta(\omega - \omega_m)$ represents the delta distribution and $P_m$ the total power, Eq. (1.210) takes the form $P_m\sigma_{a,e}^{H}(\omega_m - \omega')\tau/h\bar{\nu}A_m$. We can use this result to express now the homogeneous complex susceptibility for the transition ($jk$) as:

$$\chi_{jk}^{H}(\omega - \omega') = -nc\rho\,\frac{\sigma_{kj}(\omega - \omega')}{\omega_{kj}}\left(2\,\frac{\omega_{kj} - \omega + \omega'}{\Delta\omega_{kj}} - i\right)$$

$$\times\;\frac{-p_{1j} + \sum_m \dfrac{P_m}{h\bar{\nu}A_m}[p_{2k}\sigma_a^{H}(\omega_m - \omega') - p_{1j}\sigma_e^{H}(\omega_m - \omega')]\tau}{1 + \sum_m \dfrac{P_m}{h\bar{\nu}A_m}[\sigma_e^{H}(\omega_m - \omega') + \sigma_a^{H}(\omega_m - \omega')]\tau}$$

$$\tag{1.211}$$

and the inhomogeneous complex susceptibility $\chi^I(\omega)$ is given by eq. (1.201). For the imaginary part $\chi_I''(\omega)$, we find then from this equation and using the summation rules (1.178) and (1.179):

$$
\chi_I''(\omega) = -\frac{nc\rho}{2\pi\bar{v}} \int_{-\infty}^{\infty} d\omega'\, p_{\text{inh}}(\omega')
$$
$$
\times \frac{-\sigma_a^H(\omega - \omega') + \sum_m \dfrac{P_m}{h\bar{v}A_m}[\sigma_e^H(\omega - \omega')\sigma_a^H(\omega_m - \omega') - \sigma_a^H(\omega - \omega')\sigma_e^H(\omega_m - \omega')]\tau}{1 + \sum_m \dfrac{P_m}{h\bar{v}A_m}[\sigma_e^H(\omega_m - \omega') + \sigma_a^H(\omega_m - \omega')]\tau}
$$

$$(1.212)$$

Multiplying both sides of the complex wave equation (1.197) for the electric field by $E_0(z)$ to convert it into a rate equation for the optical power $P(\omega_k)$ at frequency $\omega_k$, we find from Eq. (1.212):

$$
\frac{dP(\omega_k)}{dz} = \rho\Gamma_k P(\omega_k) \int_{-\infty}^{\infty} d\omega'\, p_{\text{inh}}(\omega')
$$
$$
\times \frac{-\sigma_a^H(\omega_k - \omega') + \sum_m \dfrac{P_m}{h\bar{v}A_m}[\sigma_e^H(\omega_k - \omega')\sigma_a^H(\omega_m - \omega') - \sigma_a^H(\omega_k - \omega')\sigma_e^H(\omega_m - \omega')]\tau}{1 + \sum_m \dfrac{P_m}{h\bar{v}A_m}[\sigma_e^H(\omega_m - \omega') + \sigma_a^H(\omega_m - \omega')]\tau}
$$

$$(1.213)$$

The above equation can be used to calculate both pump ($k = p$) and signal ($k = s$) powers in the two-level pumping scheme. It can also be used for the same purpose in the three-level pumping scheme, by setting $\sigma_e^H(\omega_m - \omega') = 0$.

In the homogeneous limit, the distribution $p_{\text{inh}}(\omega')$ is the delta distribution $\delta(\omega')$, and it is easily verified that Eq. (1.213) exactly reduces to Eq. (1.114), with $p_{0k} = 0$ (no ASE assumed), and using $\alpha_k = \rho\sigma_a^H(\omega_k)\Gamma_k \equiv \rho\sigma_{ak}\Gamma_s$, $p_j = P_j/P_{\text{sat}}(\omega_j)$, $P_j\sigma_a(\omega_j)\tau/h\bar{v}A_j = p_j/(1 + \eta_j)$, $P_j\sigma_e(\omega_j)\tau/h\bar{v}A_j = p_j\eta_j/(1 + \eta_j)$, and $\eta_j = \sigma_e^H(\omega_j)/\sigma_a^H(\omega_j)$.

When inhomogeneous broadening cannot be neglected, Eq. (1.213) must be used instead. ASE noise can also be introduced by adding to Eq. (1.213) the noise source term proportional to $2p_{0k}$, which gives, considering both propagation directions in the fiber:

$$
\pm\frac{dp_k^\pm}{dz} = (\tilde{\gamma}_{ek} - \tilde{\gamma}_{ak})p_k^\pm + 2\tilde{\gamma}_{ek}p_{0k}
$$

$$(1.214)$$

with the following definitions for the inhomogeneous absorption and emission coefficients $\tilde{\gamma}_{ak}$, $\tilde{\gamma}_{ek}$:

$$\tilde{\gamma}_{ak} = \alpha_k \int_{-\infty}^{\infty} d\omega' p_{\text{inh}}(\omega') \frac{\sigma_a^H(\omega_k - \omega')}{\sigma_a^H(\omega_k)} \frac{1 + \sum_m (p_m^+ + p_m^-) \dfrac{\eta_m}{1 + \eta_m} \dfrac{\sigma_e^H(\omega_m - \omega')}{\sigma_e^H(\omega_m)}}{1 + \sum_m (p_m^+ + p_m^-)} \tag{1.215}$$

$$\tilde{\gamma}_{ek} = \alpha_k \int_{-\infty}^{\infty} d\omega' p_{\text{inh}}(\omega') \frac{\sigma_e^H(\omega_k - \omega')}{\sigma_e^H(\omega_k)} \frac{\sum_m (p_m^+ + p_m^-) \dfrac{\eta_k}{1 + \eta_m} \dfrac{\sigma_a^H(\omega_m - \omega')}{\sigma_a^H(\omega_m)}}{1 + \sum_m (p_m^+ + p_m^-)} \tag{1.216}$$

Because of the convolution integrals involved in Eqs. (1.215) and (.216), computation of the entire signal gain and ASE spectrum, considering iterative integrations for reaching convergence (see Section 1.6) is quite heavy and computer time-consuming. This computation was done in previous work [25] for comparing experimental data to the results of homogeneous and inhomogeneous models. The results of this study are discussed in Section 5.3.

In deriving the expression for the inhomogeneous gain coefficient, we have considered for simplicity (though without loss of generality) the case of confined Er-doping. When the Er-doping assumes any transverse distribution, integration over the transverse plane, as used in Eqs. (1.66)–(1.68), increases the complexity even more. These results would correspond only to a fixed set of input powers $P_k(z = 0)$ and a given fiber length $L$. In modeling EDFAs, however, these parameters and many others must be continuously varied in order to obtain specific EDFA characteristics. Such a complete parametric study, based upon this most comprehensive model, in fact requires the use of a supercomputer.

The above considerations show that *modeling an optical amplifier based on an inhomogeneous laser medium such as Er:glass is actually a highly complex task.* Fortunately, comparison of experimental data with results predicted by the homogeneous approximation remains fairly accurate in the saturation regime, as discussed in Chapters 4 and 5. Investigation of other types of glass hosts (e.g., fluoride glasses, see Chapter 4) for erbium or other rare-earth doping may reveal greater importance for inhomogeneous broadening. Accurately modeling the characteristics of amplifiers based on such materials would then represent a real challenge.

## CHAPTER 2

# FUNDAMENTALS OF NOISE IN OPTICAL FIBER AMPLIFIERS

## INTRODUCTION

In our modern civilization, the notion of noise is quite familiar to anyone; it is experienced nearly everyday when perceiving sound or pictures relayed by radio waves and electrical signals. Intuitive as it may be, the concept of noise escapes any simple definition and, by itself, noise is a notion generally difficult to describe or quantify. Coarsely, noise can be viewed as the random deviation of a physical parameter from an expected value. Considering light, this physical parameter can be for instance: the electrical field amplitude, the optical power, the frequency, the phase, or the polarization. With multimode laser sources, noise is associated with the random excitation in time of the longitudinal modes; with multimode fibers, the same applies to the fiber's transverse modes.

Noise can thus be viewed as the consequence of some randomness associated with various physical processes, but this definition is hardly complete. Indeed, from a quantum-mechanical viewpoint, another and most important contribution to noise is the uncertainty in the measurement of any physical quantity. This notion is well illustrated by the example of photodetection. There is randomness in the physical process where a photon is converted into an electron–hole pair, and there is uncertainty in the photon location, momentum, energy, or arrival time. Thus, there cannot exist any deterministic or noise-free measurement of optical power. As we shall see in this chapter, the notions of randomness and uncertainty inherent to photodetection also apply to optical amplification. Furthermore, an ideal, noiseless amplifier cannot exist physically, as such a device would violate Heisenberg's uncertainty principle.

In communication systems, where electrical, radio, or optical signals are transmitted, noise can be viewed as an impairment resulting in the degradation of the information contained in the signal. As pointed out by D. Marcuse in [1]: If noiseless communication channels and noiseless amplifiers and detectors were possible, we could communicate over arbitrary distances with arbitrary small amounts of power, for there is certainly no limit to the amount of possible signal amplification. In this perspective, the noise is defined by A. Yariv as: what exists in a given communication channel when no signal is present [2].

Since the notion of noise is highly complex and rests upon various concepts borrowed from classical and quantum physics, as well as electrical engineering and information theory, any analysis of noise is necessarily incomplete and of arbitrary depth. A main question is: if the description of noise is possible semiclassically, does one need a further quantum mechanical analysis? This question applies to many other fields in physics. In other words, what does quantum mechanics add to the understanding of noise, from an engineering standpoint?

How far and how deep one should progress in the analysis of optical amplifier noise, from quantum theory's first principles to basic engineering formulae, remains an academic question. When considering optical communications systems based on photodetectors and optical amplifiers, the goal of any noise analysis is at least to quantify the signal degradation from transmitting to receiving ends. It can be shown that this goal can be reached without the help of quantum theory, but this can be done only by a priori postulating the effect of spontaneous emission. No matter how far back to first principles one goes, the end result of the analysis remains unchanged, and relatively simple to implement for the analysis of communications systems.

With this main goal in mind, we have attempted to concentrate into this chapter all the notions that are central to a fundamental understanding of noise, from both engineering and physics perspectives.

The fundamentals of noise described in this chapter are not concerned with nonclassical or squeezed light. In squeezed light, the uncertainties associated with either photon number, phase, or quadrature amplitudes fall below the corresponding limits found for the so-called coherent states [3], [4], and [5], and can reach the minima imposed by Heisenberg's relations. A reason for not considering here the advantagous low noise characteristics of squeezed states is their apparently limited interest in the field of long-haul communication systems. Nonclassical light is indeed highly susceptible to both processes of loss and amplification, which take place in such systems. The randomness associated with loss and amplification actually suppresses the effect of squeezing, and hence the reduced noise characteristics of the optical signals [6].

This chapter will not be concerned either with parametric amplifiers and wave-mixing amplifiers [7]. Likewise, the reason for this restriction is that these types of amplifiers do not have at present a clear potential of application to communications systems, in spite of their remarkable noise properties. Finally, the case of Raman and Brillouin amplifiers (or scattering amplifiers), which might have some application potential, will be overlooked in this chapter, as their relevant noise characteristics can eventually be described in a way similar to that of the Er-doped fiber amplifier. These characteristics are given by comparison in Chapter 5.

The noise analysis description outlined in this chapter concerns exclusively the physical characteristics of light. The process whereby light and its associated noise are converted into photocurrent and noise photocurrent are analyzed in Chapter 3. The results of Chapter 2 can be interpreted as applying to the case of signals measured through an ideal detector. Such an ideal detector also introduces noise in signal measurements, but the uncertainty is minimal (according to the Heisenberg principle and the ideal matching of error characteristics with the amplifier); the quantum efficiency of the ideal detector is unity.

The distinction made between noise associated with light and noise associated with detector photocurrent is useful when nonideal or realistic photodetectors are considered. Nonideal photodetection, whether based on an avalanche gain mechanism or not, introduces additional noise impairments considered in Chapter 3; they are distinct from the fundamental noise characteristics of light.

As complex as the formal descriptions in some sections in this chapter may seem to some readers, the results concerning the amplifier noise output characteristics are quite simple to understand and to handle in practical applications. The reader who is not familiar with the quantum mechanical analysis used in some sections may refer for instance to [1], [2], and [3] for a basic introduction to quantum mechanical principles and formalism. The quantum theory outlined in this chapter actually represents a simplified and straightforward treatment of the light/matter interaction in the optical amplifier; theories of much higher complexity and reach can be found in the literature. The reader who is exclusively interested in the communications system aspects can also skip this chapter, using its results as starting postulates in Chapter 3. The same approach is also possible within this chapter, as some sections can be understood by a reader without a quantum mechanics background.

In Section 2.1, the question of unavoidability of amplifier noise is addressed. It is shown that as ideal noiseless amplifiers violate the Heisenberg uncertainty principle, a minimum amount of noise must be generated by the amplification process, in order to satisfy this principle.

A full quantum description of noise is given in Section 2.2. The quantum description assumes a second-quantized Hamiltonian which represents the electric dipole interaction between the quantized electric field and the atomic system. The interacting atom–light system is described through atom–photon number states, and the amplification regime is assumed linear. The derivation of transition probabilities leads to the photon statistics master equation, which describes the evolution along the amplifier of the photon number probability distribution.

In Section 2.3, the photon statistics and associated probability distribution of the amplifier output light are derived for the linear amplification regime through the probability generating function (PGF) method. The variance of the output statistics is shown to be made of different noise components called excess noise, shot noise, and beat noise. It is shown in particular that the beat noise component lends itself to different physical interpretations, according to the two perspectives of quantum statistics and semiclassical theory.

In Section 2.4, the optical signal-to-noise ratio (SNR) and the amplifier noise figure (NF) are studied from the results of the photon statistics description. It is shown that in the high gain limit, the optical amplifier noise figure reaches a 3 dB minimum value, commonly referred to as the quantum limit. This quantum limit can be reached when the EDFA is pumped as a three-level laser system, but is slightly higher in the case of a two-level pumping scheme.

The evolution of noise throughout a chain of lumped amplifiers and the resulting system noise figure is studied in Section 2.5. The specific case of distributed amplifiers where the gain is distributed over a relatively long strand of lossy fiber, is analyzed in Section 2.6, and the results are compared to that of lumped amplifier chains.

The photon statistics of amplifier chains are analyzed through the PGF method in Section 2.7. The nonlinear photon statistics corresponding to the case of a saturated amplifier are analyzed and discussed in Section 2.8.

## 2.1 MINIMUM AMPLIFIER NOISE AND TEMPERATURE

Why is amplifier noise unavoidable? A complete treatment of this fundamental question could probably fill up several volumes; it is not our purpose to attempt a full answer. However complex the issue may be, it is possible to get at least a satisfactory answer through very basic considerations and physical principles. This question is addressed in a fundamental paper by H. Heffner [8], which shows that *an ideal, noise-free amplifier actually violates Heisenberg's uncertainty principle*. We shall restate here the essential features of this demonstration, after a brief introduction of the uncertainty principle.

According to the uncertainty principle, there exists a fundamental limit in the accuracy of the simultaneous measurement of a particle's momentum $p$ and position $x$ ($p$ and $x$ represent vector projections along a reference axis). The associated uncertainties $\Delta p$ and $\Delta x$ are related through:

$$\Delta p \Delta x \geqslant \frac{\hbar}{2} \tag{2.1}$$

The inequality implies that exact knowledge of either momentum or position variable ($\Delta p = 0$ or $\Delta x = 0$) is accompanied with absolute uncertainty on the conjugate variable ($\Delta x = \infty$ or $\Delta p = \infty$). Considering now photons propagating along the reference axis, the momentum is given by $p = \hbar k = \hbar \omega / c = E/c$, where $k$ is the wave vector of the associated electromagnetic field, $E$ the photon energy, and $c$ the speed of light. Substituting this expression into Eq. (2.1) with $\Delta t = \Delta x / c$ as the photon arrival time at a specified location, we obtain a similar relation linking energy and time:

$$\Delta E \Delta t \geqslant \frac{\hbar}{2} \tag{2.2}$$

If the uncertainty associated with energy is due to an uncertainty in the actual number of photons $n$ associated with the electromagnetic wave, while the light frequency $v$ is known, we have $\Delta E = h v \Delta n$. The wave can be observed at a location of arbitrary precision, so that the phase uncertainty $\Delta \varphi$ is related to $\Delta t$ through $\Delta \varphi = 2 \pi v \Delta t$. Equation (2.2) then yields the *number–phase uncertainty relation*:

$$\Delta n \Delta \varphi \geqslant \frac{1}{2} \tag{2.3}$$

This relation implies that exact knowledge of the phase precludes exact knowledge of the number of photons, and vice versa.

We consider now an ideal linear optical amplifier that is noiseless. Given an input photon stream with number of photons $n_0 \pm \Delta n_0$, the output of such an ideal amplifier has a number of photons $n \pm \Delta n = G n_0 \pm G \Delta n_0$, which means that the input signal is linearly amplified with gain $G$, without any added noise. Given the phase $\varphi_0 \pm \Delta \varphi_0$ of the input signal, the output phase is $\varphi \pm \Delta \varphi = \varphi_0 + \theta \pm \Delta \varphi_0$, where $\theta$ is an additive phase shift accounting for the propagation through the amplifier and any other possible physical effect. The output phase uncertainty is then unchanged, i.e., $\Delta\varphi = \Delta\varphi_0$, while the output photon number uncertainty is increased, i.e., $\Delta n = G\Delta n_0$.

We assume now that we have an ideal detector which is sensitive to both photon number and phase and has unity quantum efficiency. This last condition means that every photon incident onto the detector gives one photocount, while the first condition means that the detector has the smallest possible uncertainty, i.e., $\Delta n\Delta\varphi = 1/2$. This uncertainty in the detected signal after amplification corresponds to an uncertainty in the input signal such that

$$\Delta n_0 \Delta\varphi_0 = \frac{1}{2G} \tag{2.4}$$

As the gain $G$ is by definition greater than unity, Eq. (2.4) represents a violation of the uncertainty principle stated in Eq. (2.3). Therefore, *the ideal amplifier described here cannot exist, and the missing uncertainty must be compensated by some source of noise that is intrinsic to the amplifier.* The physical interpretation of this important result is that optical amplifiers cannot be used as a way to improve the a detector sensitivity beyond the limit imposed by the Heisenberg uncertainty principle.

The next step is to evaluate the minimum noise contribution of the amplifier that would make both input and output signal uncertainties satisfy the Heisenberg principle. This evaluation was also made by H. Heffner [8], and the amplifier minimum noise power output $P_N$ is found to be:

$$\boxed{P_N = h\nu B(G-1)}$$

where $B$ is one half the amplifier full bandwidth.

In the following, we restate the demonstration and the assumption leading to this fundamental result. We consider now a linear amplifier that is not noiseless, followed by an ideal detector of properties identical to the previous case. We label $\Delta n_a^2$, $\Delta\varphi_a^2$ and $\Delta n_d^2$, $\Delta\varphi_d^2$ the uncertainties associated with the amplifier and the detector, respectively. As these are uncorrelated, the photon number and phase uncertainties of the output signal are given by:

$$\Delta n^2 = \Delta n_a^2 + \Delta n_d^2$$
$$\Delta\varphi^2 = \Delta\varphi_a^2 + \Delta\varphi_d^2 \tag{2.6}$$

The uncertainties $\Delta n^2$, $\Delta\varphi^2$ obtained at the detector imply uncertainties $\Delta n_0^2$, $\Delta\varphi_0^2$ in the input signal measurement, as related through:

$$\Delta n^2 = G^2\Delta n_0^2$$
$$\Delta\varphi^2 = \Delta\varphi_0^2 \tag{2.7}$$

Combining Eqs. (2.6) and (2.7), the output number–phase uncertainty product verifies:

$$\Delta n_0^2\Delta\varphi_0^2 = \frac{1}{G^2}\left\{\Delta n_a^2\Delta\varphi_a^2 + \Delta n_d^2\Delta\varphi_d^2 + \Delta\varphi_a^2\Delta\varphi_d^2\left(\frac{\Delta n_a^2}{\Delta\varphi_a^2} + \frac{\Delta n_d^2}{\Delta\varphi_d^2}\right)\right\} \tag{2.8}$$

If we impose that the detector has minimal uncertainty, i.e., $\Delta n_d \Delta \varphi_d = 1/2$, Eq. (2.8) becomes:

$$\Delta n_0^2 \Delta \varphi_0^2 = \frac{1}{G^2} \left\{ \Delta n_a^2 \Delta \varphi_a^2 + \frac{1}{4} + \frac{\Delta \varphi_a^2}{2} \frac{\Delta \varphi_d}{\Delta n_d} \left( \frac{\Delta n_a^2}{\Delta \varphi_a^2} + \frac{\Delta n_d^2}{\Delta \varphi_d^2} \right) \right\} \tag{2.9}$$

In order to optimize the apparatus performance, we can make the additional requirement that the detector uncertainty ratio $\Delta n_d / \Delta \varphi_d$ is chosen to minimize the output number–phase uncertainty. This condition corresponds to a best matching of the detector to the amplifier, as the ratio can be chosen arbitrarily. Minimization of $\Delta n_0^2 \Delta \varphi_0^2$ with respect to $\Delta n_d / \Delta \varphi_d$ leads to the relation:

$$\frac{\Delta n_a}{\Delta \varphi_a} = \frac{\Delta n_d}{\Delta \varphi_d} \tag{2.10}$$

We impose now that the number–phase uncertainty product inferred for the input signal is minimal, i.e., $\Delta n_0 \Delta \varphi_0 = 1/2$. Using this last condition and Eqs. (2.9) and (2.10), we obtain the equation:

$$\Delta n_a^2 \Delta \varphi_a^2 + \Delta n_a \Delta \varphi_a - \frac{G-1}{4} = 0 \tag{2.11}$$

of unique physical solution ($\Delta n_{a,d}$ and $\Delta \varphi_{a,d}$ are assumed positive):

$$\Delta n_a \Delta \varphi_a = \frac{G-1}{2} \tag{2.12}$$

At this point, we can assume that the nature of the uncertainty associated with the amplifier is additive white noise with Gaussian statistics. If the signal-to-noise ratio (SNR) is large, the output statistics of the signal phase are nearly Gaussian [9] and the phase variance is $\Delta \varphi_a^2 = P_N / 2P_S$, where $P_S$ and $P_N$ are the signal and noise mean powers, respectively. In the same conditions, the noise power variance $\Delta P_N^2$ is given by $\Delta P_N^2 = 2P_S P_N$. The noise power variance is related to the photon number variance $\Delta n_a^2$ through $\Delta P_N^2 = (h\nu/T)^2 \Delta n_a^2$, where $T$ is the detector sampling interval. Combining these different relations into Eq. (2.12), we find that the minimum amplifier noise power is $P_N = h\nu(G-1)/2T$, or with a sampling interval being equal to $T = 1/2B$ ($2B = B_0$ = amplifier FWHM bandwidth):

$$\boxed{P_N = \frac{1}{2} h\nu \frac{1}{T} (G-1) = h\nu B(G-1)} \tag{2.13}$$

This fundamental result can be restated in the following way: assuming that the amplifier uncertainty is due to an additive white noise process, and that the SNR and the gain $G$ are large, *the minimum achievable amplifier output noise power corresponds to the amplification of one photon in bandwidth $B$, one half the amplifier bandwidth $B_0$*. The amplifier output noise $P_N = h\nu B(G-1)$ is such that the minimum uncertainty relation $\Delta n \Delta \varphi = 1/2$ is satisfied for both amplifier input signal and detected output signal.

Another equally valid way to interpret Eq. (2.13) is the following: the minimum uncertainty in output signal with a high gain amplifier corresponds to the energy of half a photon in a given observation time T. The same conclusion was also reached independently by H. A. Haus and J. A. Mullen [10], through a different approach, which considered the amplifier noise in the two light field quadratures without restricting assumptions on their statistics.

The quantity $h v/2$ can also be interpreted as being the minimum value for the quantized electromagnetic field energy, which is given by $E_n = (n + \frac{1}{2})h v$ [3]. The energy $h v/2$ thus corresponds to that of a radiation mode in the vacuum state having $n = 0$ photons, and is called zero-point energy. The minimal noise power deduced previously from the uncertainty principle is in fact a consequence of the presence of such zero-point energy, whose origin can be traced to the quantization of the field. The minimum detectable noise power $h v B_0/2$ is often referred to as quantum noise, justified by its fundamental quantum origin (see discussion by D. Marcuse in [1], for instance). Quantum noise can be opposed to thermal noise, which has a power $h v B_0/[\exp(h v/k_B T) - 1]$. At room temperature, thermal and quantum noise reach equal magnitude at a frequency near $v = 6.8\,\text{THz}$, corresponding to a wavelength near $\lambda = 44\,\mu\text{m}$. At shorter wavelengths, quantum noise is greater than thermal noise by several orders of magnitude.

The minimum noise power output of the amplifier, as expressed in Eq. (2.13), is a fundamental result not based on any restrictive assumption regarding the light photon statistics. In the next section, which deals with the quantum description of the amplifier, we shall see that without any restrictive assumptions regarding photon statistics, we obtain the same result as Eq. (2.13) for the mean output noise power $P_N$. In a subsequent section, it is shown that the noise power variance $\Delta P_N^2$ is given by $\Delta P_N^2 = 2 P_S P_N + f(P_S, P_N)$, where additional terms contained in $f(P_S, P_N)$ are negligible at high gains and SNRs, which is consistent with our previous assumption for $\Delta P_N^2$.

We consider next the notion of amplifier effective temperature. The effective temperature $T_a$ corresponds to a figure of merit for the amplifier noise [11], and is defined through:

$$\frac{h v B}{\exp\left(\dfrac{h v}{k_B T_a}\right) - 1} = \frac{P_N}{G} \tag{2.14}$$

The left-hand side in Eq. (2.14) contains the thermal power density distribution, or Bose–Einstein factor; this corresponds to a blackbody source so Eq. (2.14) gives the temperature $T_a$ at which a blackbody source would produce the noise power $P_N/G$, corresponding to an equivalent noise power input to the amplifier. The Nyquist formula $k_B T_a B = P_N/G$ [12] defining the effective temperature for thermal noise in microwave circuits is the same as Eq. (2.14) in the limit $h v \ll k_B T_a$. At infrared frequencies ($\lambda = 1.55\,\mu\text{m}$) there is no reason to assume a priori that $h v \ll k_B T_a$, and only Eq. (2.14) applies, which represents the Nyquist formula generalized to the quantum noise case. Solving this equation for $T_a$ using Eq. (2.13) yields the effective temperature corresponding to the minimum amplifier noise:

$$T_a = \frac{h\nu}{k_B} \frac{1}{\log\left(\dfrac{2 - 1/G}{1 - 1/G}\right)}$$

(2.15)

In the high gain limit, the effective temperature reduces to $T_a^{min} = h\nu/k_B \log(2)$, which corresponds to the amplifier minimum effective temperature. For optical signals with wavelength near $\lambda = 1.55 \ \mu m$, this minimum temperature is $T_a = 13,381$ K or $13,108°C$. By contrast, a fictitious noise-free amplifier would have an effective noise temperature of zero, according to Eq. (2.14).

In the following section, it is shown that both minimum noise power $P_N$ and minimum effective temperature $T_a^{min}$ correspond to a regime where full population inversion is achieved in the amplifying medium.

## 2.2  QUANTUM DESCRIPTION OF NOISE

The quantum theory of amplitude noise in optical amplifiers represents only one limited aspect of the more general quantum theories of coherent light, [13]; coherent light/matter interaction, [14] and [15]; noise and laser oscillation, [16]–[22], whose formalism, developments, and implications are very complex. For a comprehensive description of this field, the reader may consult reference books [1]–[3], [14] and [15], and [23]–[26], for instance. In this section, the quantum treatment of amplifier amplitude noise will be kept to a simplified, though accurate, basic description.

The interaction of an electromagnetic quantized field with two-level atoms results in their polarization and magnetization, which can be described through a multipolar interaction Hamiltonian [3]. Out of the four possible interactions (i.e., electric dipole, electric quadrupole, magnetic dipole and diamagnetic), only the electric dipole contribution dominates, described by the interaction Hamiltonian:

$$\mathcal{H}_{ed} = e\hat{D}.\hat{E}$$

(2.16)

where $\hat{D}$ and $\hat{E}$ represent the atomic dipole and electric field operators, respectively, and $e$ the electric charge. The quantized electric field can be expressed in terms of the photon creation and annihilation operators $a^+$, $a$ through:

$$\hat{E} = \mathcal{C}\hat{\varepsilon}\{ae^{-i(\omega t - kz)} - a^+ e^{i(\omega t - kz)}\}$$

(2.17)

where $\mathcal{C}$ is a constant and $\hat{\varepsilon}$ is a unit vector indicating the field polarization; the field is assumed to be a plane wave.

We consider now the atomic quantum states $|1\rangle$ and $|2\rangle$, corresponding to the ground and excited levels, with associated energy 0 and $\hbar\omega$, respectively. These states are orthonormal, i.e., $\langle i|j\rangle = \delta_{ij}$, where $\delta_{ij}$ is the Kronecker symbol ($\delta_{ij} = 1$ for $i = j$ and $\delta_{ij} = 0$ otherwise). The atomic dipole operator can be expressed then in the form [3]:

$$\hat{D} = d\hat{\varepsilon}'\{|1\rangle \langle 2| + |2\rangle \langle 1|\}$$

(2.18)

where $d$ is a constant and $\hat{\varepsilon}'$ is a unit vector corresponding to the dipole orientation. The symbols $|1\rangle\langle2|$ and $|2\rangle\langle1|$ represent atomic lowering and raising operators, respectively, as we have $|1\rangle\langle2|i\rangle = \delta_{2i}|1\rangle$ and $|2\rangle\langle1|i\rangle = \delta_{1i}|2\rangle$. Using Eqs. (2.16)–(2.18) the electric dipole Hamiltonian takes the form:

$$\mathscr{H}_{ed} = ed\mathscr{C}\hat{\varepsilon}.\hat{\varepsilon}'\{a\,e^{-i(\omega t - kz)}|2\rangle\langle1| - a^+\,e^{i(\omega t - kz)}|1\rangle\langle2|$$
$$+ a\,e^{-i(\omega t - kz)}|1\rangle\langle2| - a^+\,e^{i(\omega - kz)}|2\rangle\langle1|\} \qquad (2.19)$$

The events represented by each of the four terms in Eq. (2.19) correspond to absorption of a photon (terms proportional to $a$) or emission of a photon (terms proportional to $a^+$), with associated changes in atomic states. The events described by the last two terms in Eq. (2.19) are now allowed, since it is not possible for the initial and final states to have the same energy $\hbar\omega$. The neglect of these two terms is equivalent to the so-called rotating wave approximation [3].

The electric field can be represented by a statistical mixture of photon number states. These quantum states correspond to the harmonic oscillator's energy eigenstates [1]–[3], and verify the usual relations:

$$a|n\rangle = \sqrt{n}|n - 1\rangle$$
$$a^+|n\rangle = \sqrt{n + 1}|n + 1\rangle \qquad (2.20)$$

With photon number states, the number of photons is exactly defined ($\Delta n = 0$) but the phase is indefinite. The corresponding uncertainty relations are $\Delta n \Delta \cos\varphi \geqslant |\langle\widehat{\sin\varphi}\rangle|/2$ and $\Delta n \Delta \sin\varphi \geqslant |\langle\widehat{\cos\varphi}\rangle|/2$, where $\widehat{\sin\varphi}$, $\widehat{\cos\varphi}$ are phase measurement operators [3]. With pure photon number states, we have $\langle\widehat{\sin\varphi}\rangle = \langle\widehat{\cos\varphi}\rangle = 0$, and $\Delta\sin\varphi = \Delta\cos\varphi = 1/\sqrt{2}$, consistent with the uncertainty relations, and shows that the phase is equally likely to take any value between $0$ and $2\pi$ [3].

A pure photon number state $|n\rangle$ does not have any counterpart in systems based on conventional laser sources, as electromagnetic waves produced by these sources do not have a definite number of photons. However, photon number states can be generated experimentally by anticorrelation or feedback/feedforward schemes (see discussion in [5]). Such states, called photon number squeezed states, represent nonclassical electromagnetic waves and will not be considered here.

A quantum state representation which corresponds more closely to nonsqueezed light consists of a statistical mixture of photon number states. This mixture or statistical superposition of states can be expressed as $|S\rangle = \sum u_n|n\rangle$, with the probability of finding exactly $n$ photons in the system being given by $P_n = \langle n|S\rangle = |u_n|^2$. The so-called coherent states, whose probability distribution $P_n$ corresponds to Poisson statistics, represent a particular case of this mixture of photon number states. One interesting property of the coherent states is that they approach a close representation of the classical electromagnetic field for large photon numbers [3]. In the following, we will not need to make any a priori assumptions regarding the initial field state $|S\rangle$ and its associated photon statistics $P_n$.

A practical tool to describe the evolution of field photon statistics is the radiative density matrix operator defined through $\hat{\rho} = \sum P_n|n\rangle\langle n|$. The diagonal elements $\hat{\rho}_{nn} = \langle n|\hat{\rho}|n\rangle$ of this operator thus correspond to the photon probability distribution $P_n$. The equation describing the temporal evolution of $\hat{\rho}$ is given by the theory of

M. O. Scully and W. E. Lamb, [21] and [23]. As we are concerned here only with the diagonal elements $\hat{\rho}_{nn}$, we shall consider only a simplified derivation for the corresponding equation, strictly identical to that given by the theory.

This simplified approach considers pure photon number states for the interacting field, as opposed to a statistical superimposition. We define the atom–photon number states $|i, n\rangle$ ($i = 1, 2$) which represent the whole atom/field system. These states verify the relations (2.20) and are orthogonal, i.e., $\langle j, m|i, n\rangle = \delta_{ij}\delta_{mn}P_{in}$, where $P_{in}$ is the probability that the initial state of the system is $|i, n\rangle$. Assuming $|i, n\rangle$ for the initial state (i.e., before interaction), the probability of finding the system after the interaction, in the final state $|j, m\rangle$ is given by:

$$P_{jm;in} = |\langle j, m|\mathscr{H}_{ed}|i, n\rangle|^2 \tag{2.21}$$

With an initial state having $k$ photons, the process of photon absorption is then characterized by the probability $P_{2,k-1;1,k}$ and the process of photon emission by the probability $P_{1,k+1;2,k}$.

Figure 2.1 represents an energy diagram for the field with four different possible transitions associated with the absorption or the emission of one photon in a single radiation mode of the field [3]. Using the definition (2.19), the relations (2.20), and the orthogonality properties of the atom–photon number states, we find the probabilities corresponding to the events represented in Figure 2.1:

$$\begin{aligned}
P_{2,n;1,n+1} &= (n + 1)P_{n+1} \\
P_{2,n-1;1,n} &= nP_n \\
P_{1,n;2,n-1} &= nP_{n-1} \\
P_{1,n+1;2,n} &= (n + 1)P_n
\end{aligned} \tag{2.22}$$

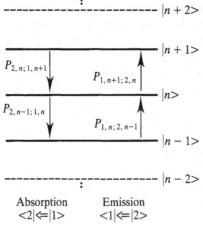

**FIGURE 2.1** Energy level diagram corresponding to the photon number states $|n\rangle$. The arrows indicate four possible transitions, with corresponding probabilities, in which one photon is absorbed (atomic state shifting from $|1\rangle$ to $|2\rangle$) or emitted (atomic state shifting from $|2\rangle$ to $|1\rangle$).

where the constants involved in Eqs. (2.18) and (2.19) were ignored, and the atomic dipole was assumed to be colinear with the interacting field. In Eqs. (2.22), the probabilities that are proportional to the initial photon number in the system $n$ are associated with the processes of absorption and stimulated emission. The last two terms in Eq. (2.22) corresponding to emission also include the effect of spontaneous emission. The action of the photon creation operator $a^+$ on any system quantum state $|n\rangle$ is physically equivalent to stimulated and spontaneous emission processes.

We consider now the collection of atoms in their ground and excited states, with density at coordinate $z$ $N_1(z)$ and $N_2(z)$, respectively, in a slice of medium with infinitesimal width $dz$. The phenomenological cross sections for the absorption and emission processes are $\sigma_a$ and $\sigma_e$, respectively. The change of the photon number probability $dP_n$, between $z$ and $z + dz$ is then given by:

$$dP_n = \{\sigma_a N_1(P_{2,n;1,n+1} - P_{2,n-1;1,n}) + \sigma_e N_2(P_{1,n;2,n-1} - P_{1,n+1;2,n})\}\, dz \quad (2.23)$$

or using Eq. (2.22) and the definitions $a = \sigma_e N_2$, $b = \sigma_a N_1$:

$$\boxed{\frac{dP_n}{dz} = a\{nP_{n-1} - (n+1)P_n\} + b\{(n+1)P_{n+1} - nP_n\}} \quad (2.24)$$

The above equation, called the *photon statistics master equation*, constitutes the starting point of the fundamental 1957 paper by K. Shimoda, H. Takahasi, and C. H. Townes (STT) [27]. The STT solution of Eq. (2.24), obtained for the linear regime, provides an accurate model for describing quantum noise in optical amplifiers.

Equation (2.24) is fundamental; it is also called the *forward Kolmogorov equation* [28]–[30]. By definition, a Kolmogorov equation is of the form:

$$\frac{dP_n}{dz} = \sum_m \lambda_{mn} P_m \quad (2.25)$$

with the two conditions $\lambda_{nn} < 0$ and $\sum_n \lambda_{mn} = 0$. This particular type of equation governs a Markoff process [30], defined most generally as the random time evolution of a set $\{P_n\}$ of $n$ variables. A Markoff process governed by forward or backward Kolmogorov equations is also called a birth–death–immigration (BDI) process; it is used in the analysis of many stochastic phenomena [30].

The first condition ($\lambda_{nn} < 0$) for the Kolmogorov equation guarantees that $0 \leqslant P_n \leqslant 1$ is maintained at all coordinates $z$; the second condition guarantees the probability is conserved, i.e., $d(\sum_n P_n)/dz = 0$, or $\sum_n P_n(z) = 1$. The first condition is satisfied as we have from Eq. (2.24) $\lambda_{mn} = -\{\sigma_a N_1 n + \sigma_e N_2(n+1)\}$. The second condition ($\sum_n \lambda_{mn} = 0$) is verified by performing the following sum with the use of Eq. (2.24):

$$\frac{d}{dz}\sum_{n=1}^{M} P_n = \sum_{n=1}^{M}\sum_m \lambda_{mn} P_m = \sum_m P_m \sum_{n=1}^{M} \lambda_{mn} = -\sigma_a N_1 P_0 - \sigma_e N_2(M+1)P_M = 0 \quad (2.26)$$

The last part of Eq. (2.26) is due to the boundary conditions: $P_0 = 0$ (there is at least one photon in the system) and $MP_M \to 0$ for large $M$ (the probability distribution $P_n$ must exponentially vanish beyond some large photon number).

Before studying the photon statistics, it is useful to evaluate the average noise power predicted by the photon statistics maser equation (2.24), and to relate the result to that of Section 2.1.

The rate of change of the average photon number $\langle n \rangle$ throughout the amplifier is given by the statistical mean of Eq. (2.24):

$$\sum_n n \frac{dP_n}{dz} = \frac{d\langle n \rangle}{dz} = \sum_n [a\{n^2 P_{n-1} - n(n+1)P_n\} + b\{n(n+1)P_{n+1} - n^2 P_n\}] \quad (2.27)$$

where $a = \sigma_e N_2(z)$ and $b = \sigma_a N_1(z)$ are functions of $z$ only. Assuming that the photon number is large enough, we can compute Eq. (2.27) using the substitutions $f(n)P_{n-1} \approx f(n+1)P_n, f(n)P_{n+1} \approx f(n-1)P_n$ [27] and the normalization condition $\Sigma_n P_n = 1$, to find the linear equation

$$\frac{d\langle n \rangle}{dz} = a(\langle n \rangle + 1) - b\langle n \rangle \quad (2.28)$$

whose solution at coordinate $z$ is

$$\boxed{\langle n(z) \rangle = G(z)\langle n(0) \rangle + N(z)} \quad (2.29)$$

where $G(z)$ is the amplifier gain, defined by:

$$\boxed{G(z) = \exp\left\{ \int_0^z [a(z') - b(z')] \, dz' \right\}} \quad (2.30)$$

and

$$\boxed{N(z) = G(z) \int_0^z \frac{a(z')}{G(z')} \, dz'} \quad (2.31)$$

The second term $N(z)$ in eq. (2.29), as defined above, can be interpreted as amplified noise, since it appears in the absence of input signal ($\langle n(0) \rangle = 0$). This noise is due to the factor $+1$ in Eq. (2.28) and in the photon statistics master equation (2.24), previously related to the effect of spontaneous emission. We therefore interpret this noise contribution as the *amplified spontaneous emission* or ASE.

Assuming for simplicity that coefficients $a$ and $b$ are constant with amplifier length, i.e. a uniformly pumped laser medium, we obtain from Eq. (2.31):

$$N(z) = \frac{a}{a - b}(G - 1) = \frac{\eta N_2}{\eta N_2 - N_1}(G - 1) \quad (2.32)$$

where by definition $\eta = \sigma_e/\sigma_a$ and $G = G(z) = \exp[(a - b)z]$. From this result, we obtain the mean output noise power in bandwidth $B$:

$$P_N = n_{sp} h\nu B(G - 1) \tag{2.33}$$

where we have introduced the amplifier *spontaneous emission factor*:

$$n_{sp} = \frac{\eta N_2}{\eta N_2 - N_1} \tag{2.34}$$

In the general case where the coefficients $a$ and $b$ depend on the coordinate $z$, the spontaneous emission factor can be defined through $N(z) = n_{sp}(z)[G(z) - 1]$, with:

$$n_{sp}(z) = \frac{N(z)}{G(z) - 1} = \frac{G(z)}{G(z) - 1} \int_0^z \frac{a(z')}{G(z')} dz' \tag{2.35}$$

With this general definition for $n_{sp}$, Eq. (2.33) holds, and the conclusions that follow are also valid.

The important result of Eq. (2.32) is that at high amplifier gains $(G \gg 1)$ the mean ASE output photon number corresponds to the amplification of $n_{sp}$ photons, which represent an *equivalent input noise*.

In the case of negative medium inversion, i.e., $\eta N_2 - N_1 < 0$, the gain $G$ is less than unity and $n_{sp}$ is also negative but the noise power in Eq. (2.33) is always positive and equal to $P_N = |n_{sp}|(1 - G)$.

At the threshold of medium inversion, i.e., $\eta N_2 - N_1 = 0$, $n_{sp}$ is infinite, but the noise power is finite and equal to $P_N = \eta N_2 z$. At positive medium inversion, i.e., $\eta N_2 - N_1 > 0$, we have $n_{sp} = 1/(1 - N_1/\eta N_2) > 1$.

At full medium inversion, where all the atoms are in the excited state, i.e., $N_1 = 0$ and $N_2 = \rho$, the spontaneous emission factor $n_{sp}$ reaches its minimum value, unity. The output noise power reduces in this case to the amplified quantum noise value $P_N = h\nu B(G - 1)$ found in the previous section in Eq. (2.13). This fundamental result can be restated as follows: *the minimum amplifier noise output is obtained when complete population inversion is achieved in the amplifying medium.* This important conclusion was reached without knowledge of the actual noise statistics of the ASE, as governed by the photon statistics master equation (2.24).

All the results derived in this section concern the case of optical amplifiers in which light is confined into a single propagation mode. As Er-doped fiber amplifiers are made from single-mode fibers, which actually support two degenerate polarization modes, we must account for the generation of amplified spontaneous noise in the polarization mode orthogonal to that of the signal, as well as its effect on the output statistics. For simplicity, this may be ignored by first considering only a single propagation mode.

## 2.3  PHOTON STATISTICS IN LINEAR GAIN REGIME

The photon statistics master equation (2.24) can be solved exactly through the probability generating function (PGF) method [27]–[29]. The PGF is defined by:

$$F(x, z) = \sum_n x^n P_n(z) \tag{2.36}$$

The probability distribution $P_n(z)$ can be calculated from the PGF $F(x, z)$ through the relation:

$$P_n(z) = \frac{1}{n!} \left( \frac{\partial^n F(x, z)}{\partial x^n} \right)_{x=0} \tag{2.37}$$

easily verified from Eq. (2.36).

The differential equation for the amplifier PGF is obtained from Eq. (2.24) through the summation [27]:

$$\frac{\partial F(x, z)}{\partial z} = \sum_n x^n \frac{dP_n(z)}{dz} = \sum_n [a\{nx^n P_{n-1} - (n+1)x^n P_n\} + b\{(n+1)x^n P_{n+1} - nx^n P_n\}]$$

$$= \sum_m [a\{(m+1)x^{m+1} P_m - (m+1)x^m P_m\} + b\{mx^{m-1} P_m - mx^m P_m\}] \tag{2.38}$$

where summations over $m = n + 1$ and $m = n - 1$ were introduced where appropriate, in order to obtain a right-hand side function only of $P_m$. Using the property $\sum mx^{m-1} P_m = x \, \partial F / \partial x$, Eq. (2.38) becomes:

$$\frac{\partial F}{\partial z} = (x - 1) \left\{ (ax - b) \frac{\partial F}{\partial x} + aF \right\} \tag{2.39}$$

The above equation was solved exactly by K. Shimoda, H. Takahasi and C. H. Townes [27] in the general case where the coefficients $a$ and $b$ are functions of $z$, which leads to an exact expression for the output photon number distribution $P_n(z)$. A solution method leading to the general case solution for the PGF, but expressed under a different and more tractable form compared to that of the STT analysis, was also derived by P. Diament and M. C. Teich [29]. The two resolution methods are described in detail in Appendix G.

In this appendix, it is found from the STT analysis that the output photon number distribution $P_m(z)$ is given by the expression:

$$\boxed{P_m(z) = \sum_n P_{n,m}(z) P_n(0)} \tag{2.40}$$

where the function $P_{n,m}(z)$ represents the probability of finding $m$ output photons assuming that $n$ photons are input to the amplifier, and is defined by:

$$P_{n,m}(z) = \frac{(1 + N - G)^n N^m}{(1 + N)^{m+n+1}} \sum_{j=0}^{\min(m,n)} \frac{m! \, n!}{j! j! (m-j)! (n-j)!} \left[ \frac{G}{N(1 + N - G)} \right]^j \tag{2.41}$$

where the terms $G$ and $N$ are functions of $z$ and correspond to the amplifier gain and output ASE noise photon number as defined in Eqs. (2.30) and (2.31).

It is shown in the appendix that the distributions $P_{n,m}(z)$ are related to *Jacobi polynomials*, defined by [31]:

$$P_N^{(\alpha,\beta)}(X) = \frac{(X-1)^N}{2^N} \sum_{j=0}^{N} \binom{N+\alpha}{j}\binom{N+\beta}{N-j}\left(\frac{X+1}{X-1}\right)^j \tag{2.42}$$

In the two cases $m < n$ and $n < m$ the distributions $P_{n,m}(z)$ take the form:

$$P_{n,m}(z) = \frac{(N+1-G)^{n-m}}{(N+1)^{n+1}}(G-N)^m P_m^{(n-m,0)}(X); \quad m < n \tag{2.43}$$

$$P_{n,m}(z) = \frac{N^{m-n}}{(N+1)^{m+1}}(G-N)^n P_n^{(m-n,0)}(X); \quad n < m \tag{2.44}$$

with

$$X = \frac{1}{N-G}\left(G\frac{N-1}{N+1} - N\right) \tag{2.45}$$

The Jacobi polynomials can also be expressed as *hypergeometric functions* $F(a, b, c, x)$ [31].

A case of interest considered in [27] is that of large amplifier gain, i.e., $G \gg 1$, $m \gg 1$ and $m \gg n$. This last condition corresponds to the assumption of a small number of input photons for which no amplifier gain saturation occurs. It is shown in Appendix G that in such a case, the output distribution $P_{n,m}(z)$ can also be put under the form:

$$P_{n,m}(z) \approx \frac{(1+N-G)^n N^m}{(1+N)^{m+n+1}}L_n\left[-m\frac{G}{N(1+N-G)}\right] \tag{2.46}$$

and in the case where the coefficients $a$ and $b$ do not depend on the coordinate $z$:

$$P_{n,m}(z) \approx \frac{a-b}{aG}\left(\frac{b}{a}\right)^n \exp\left(-m\frac{a-b}{aG}\right)L_n\left[-m\frac{(a-b)^2}{abG}\right] \tag{2.47}$$

In Eqs. (2.46) and (2.47) the functions $L_n(X) = L_n^{(0)}(X)$ are *Laguerre polynomials*, with definition [31]:

$$L_n(X) = \sum_{j=0}^{n} \frac{n!}{j!j!(n-j)!}(-X)^j \tag{2.48}$$

The fundamental and accurate result obtained in Eq. (2.40)–(2.45) for the amplifier output photon statistics distribution $P_m(z)$ as a function of the input photon statistics distribution $P_n(0)$ requires, in the general case, a lengthy numerical computation, due to the needs of evaluating the orthogonal polynomials through recursive methods [31] and of carrying on the summation in Eq. (2.40). However, this last summation is not required in the case of interest where the amplifier has exactly $n_0$ input photons, i.e., $P_n(0) = \delta_{n,n_0}$ for which $P_m(z) = P_{n_0,m}(z)$. The approximation $P_m(z) \approx P_{n_0,m}(z)$ can also be made in the case where the input signal has a large number of photons, and for which the statistics are fairly close to the Kronecker function $\delta_{n,n_0}$.

The output photon statistics distributions $P_m(z)$ corresponding to different cases of interest for $P_n(0)$ are actually difficult to study from the above results. It is shown in the following that these cases can be studied more practically by considering only the associated PGFs.

The treatment by P. Diament and M. C. Teich [29] for solving Eq. (2.39) leads to a general expression for the PGF more tractable than in the previous analysis. The details of the derivation, which we have generalized here to the case of an amplifier with variable coefficients $a(z)$ and $b(z)$, are outlined in Appendix G, with notation consistent with the previous analysis. The output PGF is shown to take the form:

$$F(x, z) = F[X(x, z; 0), 0]F_1(x, z) \tag{2.49}$$

with

$$F_1(x, z) = \frac{1}{1 - (x - 1)N(z)} \tag{2.50}$$

and

$$X(x, z; 0) = 1 + \frac{(x - 1)G(z)}{1 - (x - 1)N(z)} = 1 + \frac{(x - 1)G(z)}{1 - (x - 1)n_{\mathrm{sp}}(z)[G(z) - 1]} \tag{2.51}$$

The functions $G(z)$, $N(z)$, and $n_{\mathrm{sp}}(z)$ are defined in Eqs. (2.30), (2.31), and (2.35).

The multiplicative factor $F[Z(x, z; 0), 0]$ in Eq. (2.49) is, according to the definitions of the input PGF and Eq. (2.51):

$$F[X(x, z; 0), 0] = \sum_m \left[ 1 + \frac{(x - 1)G(z)}{1 - (x - 1)N(z)} \right]^m P_m(0) \tag{2.52}$$

The output PGF solution defined in Eqs. (2.49)–(2.51) makes it possible to study the evolution of the output photon statistics $P_m(z)$ and its corresponding moments $\langle n^k \rangle$ ($k = 1, 2, \ldots$). The first and second moments $\langle n(z) \rangle$ and $\langle n^2(z) \rangle$ are given by:

$$\langle n(z) \rangle = \left( \frac{\partial F(x, z)}{\partial x} \right)_{x=1} \tag{2.53}$$

$$\langle n^2(z) \rangle = \left( \frac{\partial^2 F(x, z)}{\partial x^2} \right)_{x=1} + \langle n(z) \rangle \tag{2.54}$$

which stem from the definition for the PGF in Eq. (2.36). The computation of Eqs. (2.53) and (2.54) is detailed in Appendix G. The moments are found to take the form:

$$\langle n(z) \rangle = G(z)\langle n(0) \rangle + N(z) \qquad (2.55)$$

$$\langle n^2(z) \rangle = G^2(z)[\langle n^2(0) \rangle - \langle n(0) \rangle] + 4G(z)N(z)\langle n(0) \rangle + G(z)\langle n(0) \rangle + 2N^2(z) + N(z)$$

$$(2.56)$$

The output photon statistics variance is defined by $\sigma^2(z) = \langle n^2(z) \rangle - [\langle n(z) \rangle]^2$, and is from Eqs. (2.55) and (2.56)

$$\sigma^2(z) = G^2(z)[\sigma^2(0) - \langle n(0) \rangle] + G(z)\langle n(0) \rangle + N(z) + 2G(z)N(z)\langle n(0) \rangle + N^2(z)$$

$$(2.57)$$

where $\sigma^2(0) = \langle n^2(0) \rangle - [\langle n(0) \rangle]^2$ is the input photon statistics variance.

The fundamental results of Eq. (2.55) and (2.57) obtained for the linear amplifier output mean and variance with coefficients $a$ and $b$ depending on fiber coordinate, are most general. Consistently, the output mean photon number $\langle n(z) \rangle$ defined in Eq. (2.55) is identical to that obtained previously by averaging the photon statistics master equation, Eqs. (2.27)–(2.29). The same averaging makes it possible to obtain the higher order moments $\langle n^k \rangle (k > 1)$, as shown later on in this section. This method leads to coupled rate equations for each $\langle n^k \rangle$, which must be integrated to yield the moments. The alternate way to obtain $\langle n^k \rangle$ is to derive $k$ times the PGF in Eq. (2.49), as done in the above for the determination of $\langle n^2(z) \rangle$.

The different terms contributing to the output variance in Eq. (2.57) are usually called *excess noise, shot noises,* and *beat noises,* respectively [32]. These noise contributions combine to decrease the signal-to-noise ratio (SNR), of fundamental significance to optical communications (Chapter 3). *This output noise is inherent to the statistics of amplified light*; it is not associated per se with the effect of photodetection. But an ideal photodetector following the optical amplifier would still exhibit the noise statistics moments of Eqs. (2.55) and (2.57), as shown in Chapter 3.

The above remark is justified through a physical interpretation of the different terms involved in the amplifier output variance.

The first term proportional to $\sigma^2(0) - \langle n(0) \rangle$ is the only contribution to the output noise that actually depends on the input signal photon statistics. An input signal with Poisson statistics, has a photon probability distribution defined by:

$$P_n(0) = \frac{\langle n(0) \rangle^n}{n!} e^{-\langle n(0) \rangle} \qquad (2.58)$$

whose variance is [4]:

$$\sigma^2(0) = \langle n(0) \rangle \qquad (2.59)$$

Thus, for Poisson input statistics, the output noise contribution proportional to $\sigma^2(0) - \langle n(0) \rangle$ vanishes. On the other hand, for input signals made of chaotic or thermal light, the probability distribution is given by the Bose–Einstein statistics:

$$P_n(0) = U_0(1 - U_0)^n = \frac{1}{1 + \langle n(0) \rangle} \frac{1}{(1 + 1/\langle n(0) \rangle)^n} \tag{2.60}$$

with $U_0 = 1/(\langle n(0) \rangle + 1)$, whose variance is [4]:

$$\sigma^2(0) = \langle n(0) \rangle^2 + \langle n(0) \rangle \tag{2.61}$$

In such a case, the contribution to the output noise $\sigma^2(0) - \langle n(0) \rangle$ is nonvanishing, as being proportional to mean input $\langle n(0) \rangle$. The term excess noise for this contribution is justified, as it represents a certain amount of additional noise generated by the amplifier when the input statistics deviate from the Poisson law. The Poisson statistics characterize the photon counting of a superimposition of pure sine waves corresponding to the classical electromagnetic field [1]. They also characterize the quantum coherent states [3], which represent coherent light sources.

The second noise contribution in Eq. (2.57) is equal to the amplifier mean output power $G\langle n(0) \rangle + N$. The corresponding term shot noise is justified by the effect of this noise contribution in photodetection. In photodetectors and vacuum tubes, the process of photon-to-electron conversion is subject to the so-called shot effect. This effect produces a photocurrent noise with mean-square fluctuation $\langle i^2 \rangle = 2eB_e\langle i \rangle (B_e = $ electronic bandwidth), which is called shot noise [1]. Thus, the shot noise is proportional to the photon number of the detected signal; for ideal detectors $\langle i \rangle = P_s/hv = \langle n \rangle B_o (B_o = $ optical signal bandwidth, $P_s = $ light average power). Assuming the photoelectron statistics to equal that of the incident light, the noise contribution proportional to the total light power $G\langle n(0) \rangle + N$ would correspond to the detector photocurrent shot noise. Therefore, the term shot noise for the contribution $G\langle n(0) \rangle + N$ to the light statistics is justified in view of its effect in the photodetection process, but is purely conventional.

The last two terms in Eq. (2.57), i.e., $2G(z)\langle n(0) \rangle N(z) + N^2(z)$, are conventionally called beat noises. The first term, i.e., $2G\langle n(0) \rangle N$, is proportional to the double product $2P_s P_N$ of the output signal and noise powers. This product can be interpreted as being the variance of a light signal with Gaussian noise statistics and large SNR, as seen in Section 2.1. It can also be interpreted as the beat power between the electromagnetic components of the signal and the amplified spontaneous emission noise reaching a photodetector [33]. This interpretation is accurate from a classical standpoint but inaccurate from the quantum standpoint. It can be shown indeed that *a noise term taking the form* $2G\langle n(0) \rangle N$ *is always present in the amplifier output photon statistics, even in the absence of amplified spontaneous emission.*

This important conclusion can be reached by considering an ideal optical amplifier without spontaneous emission. The probabilities associated with absorption and pure stimulated emission processes in Figure 2.1 are:

$$
\begin{aligned}
P_{2,n;1,n+1} &= (n + 1)P_{n+1} \\
P_{2,n-1;1,n} &= nP_n \\
P_{1,n;2,n-1} &= (n - 1)P_{n-1} \\
P_{1,n+1;2,n} &= nP_n
\end{aligned}
\tag{2.62}
$$

and the corresponding photon statistics master equation is:

$$\frac{dP_n}{dz} = b\{(n + 1)P_{n+1} - nP_n\} + a\{(n - 1)P_{n-1} - nP_n\} \qquad (2.63)$$

Such an equation also corresponds to a BDI process where immigration (corresponding to spontaneous emission in optical amplifiers) is absent [29]. The first order and second order moment rate equations are from Eq. (2.63):

$$\frac{d\langle n(z)\rangle}{dz} = \sum_n n \frac{dP_n(z)}{dz} = (a - b)\langle n(z)\rangle \qquad (2.64)$$

$$\frac{d\langle n^2(z)\rangle}{dz} = \sum_n n^2 \frac{dP_n(z)}{dz} = 2(a - b)\langle n^2(z)\rangle + (a + b)\langle n(z)\rangle \qquad (2.65)$$

Integrating Eqs. (2.64) and (2.65) yields the solutions:

$$\langle n(z)\rangle = G(z)\langle n(0)\rangle \qquad (2.66)$$

$$\langle n^2(z)\rangle = G^2(z)\langle n^2(0)\rangle + G(z)N(z)\langle n(0)\rangle + G^2(z)\langle n(0)\rangle \int_0^z \frac{b(z')}{G(z')} dz' \qquad (2.67)$$

where $G(z)$ and $N(z)$ have the usual definitions of Eqs. (2.30) and (2.31). Taking the derivative of the integrals in Eqs. (2.30) and (2.67), and using Eq. (6.48) gives:

$$\int_0^z \frac{b(z')}{G(z')} dz' = \frac{N(z) + 1 - G(z)}{G(z)} \qquad (2.68)$$

Substituting Eq. (2.68) into Eq. (2.67) yields $\langle n^2(z)\rangle$, from which we find the variance:

$$\sigma^2(z) = G^2(z)[\sigma^2(0) - \langle n(0)\rangle] + G(z)\langle n(0)\rangle + 2G(z)N(z)\langle n(0)\rangle \qquad (2.69)$$

The above result can also be obtained directly from the PGF solution of Eq. (2.63) [29].

The output variance in Eq. (2.69) is seen to be identical to that defined in Eq. (2.57), except for missing terms $N(z)$ and $N^2(z)$. According to Eq. (2.69), the three contributions to the output noise variance of the amplifier free from spontaneous emission are the excess noise $G^2[\sigma(0) - \langle n(0)\rangle]$, the amplified signal shot noise $G\langle n(0)\rangle$, and the beat noise term $2G\langle n(0)\rangle N$. Thus, the term $2G\langle n(0)\rangle N$ is predicted by the amplifier photon statistics, even in absence of amplified spontaneous emission. In this case, *the quantity $2G\langle n(0)\rangle N$ cannot be interpreted as a beat component, but rather as reflecting the statistical nature of the stimulated emission process.* The quantity $2N$ in this noise component corresponds to a multiplicative factor, which vanishes if the stimulated emission coefficient $a(z)$ is zero, as seen in Eq. (2.31). The statistical nature of stimulated emission is inherent in Eq. (2.63), a true Kolmogorov equation.

Considering now an amplifier with spontaneous emission but no input signal, i.e., $\langle n(0) \rangle = 0$, the output variance in Eq. (2.57) is given by:

$$\sigma^2(z) = N(z) + N^2(z) \tag{2.70}$$

The noise variance defined above can be viewed as reflecting the *statistical nature of the amplified spontaneous emission process.*

From these considerations, it is clear now that the overall noise contribution $2\langle n(0) \rangle N + N^2$, previously called beat noise, should rather be viewed as the result of *two independent statistical processes* pertaining to either amplified spontaneous emission or stimulated emission. The unsaturated amplification regime causes the two processes to be independent. In such both population inversion density $N_2 - N_1$ and excited atoms density $N_2$, which govern the stimulated and spontaneous emission processes, are independent of the total photon number population. This is not true in the case of saturated amplification, as the atomic densities $N_1$ and $N_2$ depend upon the signal and ASE photon populations, nonlinearly coupled through the atomic system.

Because the two noise contributions relevant to stimulated and spontaneous emission amplification appear independently in the linear regime statistics, the name beat noise is in principle an incorrect term. But remarkably, the semiclassical description predicts a beating effect between signal and noise at the detector of exactly the same amount $2\langle n(0) \rangle N + N^2$, as shown in Chapter 3. Therefore, the physical interpretation of this beat noise component in the amplifier output statistics depends on whether light is analyzed from the perspectives of either quantum statistics or semiclassical theory.

Multiple propagation modes are relevant to Er-doped fibers and were not accounted for in the previous analysis. Let $\mathcal{M}$ be the number of possible modes, and $n$ the total photon population distributed among these modes. Assuming that spontaneous emission events are equally likely for all modes, the transition probabilities in Eq. (2.22) emission must be modified according to:

$$P_{1,n;2,n-1} = (n - 1 + \mathcal{M})P_{n-1}$$
$$P_{1,n+1;2,n} = (n + \mathcal{M})P_n \tag{2.71}$$

The photon statistics master equation (2.23) with Eq. (2.71) now takes the form:

$$\frac{dP_n}{dz} = a\{(n - 1 + \mathcal{M})P_{n-1} - (n + \mathcal{M})P_n\} + b\{(n + 1)P_{n+1} - nP_n\} \tag{2.72}$$

which corresponds to the forward Kolmogorov equation for a BDI process where the immigration rate is $\mathcal{M}$ [29]. Summing as in Eqs. (2.64) and (2.65) then iterating the resulting differential equations, we obtain the output mean, mean-square and variance:

$$\langle n(z) \rangle = G(z)\langle n(0) \rangle + \mathcal{M}N(z) \tag{2.73}$$

$$\langle n^2(z) \rangle = G^2(z)[\langle n^2(0) \rangle - \langle n(0) \rangle] + G(z)\langle n(0) \rangle + \mathcal{M}N(z)$$
$$+ 2(\mathcal{M} + 1)G(z)\langle n(0) \rangle N(z) + \mathcal{M}(\mathcal{M} + 1)N^2(z) \tag{2.74}$$

$$\sigma^2(z) = G^2(z)[\sigma^2(0) - \langle n(0)\rangle] + G(z)\langle n(0)\rangle + \mathcal{M}N(z) + 2G(z)\langle n(0)\rangle N(z) + \mathcal{M}N^2(z)$$

(2.75)

To obtain the result of Eq. (2.74), we must use Eq. (2.68) and the following relations (straightforward to prove using the property $d(N/G)/dz = a/G$):

$$\int_0^z \frac{a(z')N(z')}{G^2(z)} dz' = \frac{1}{2}\left[\frac{N(z)}{G(z)}\right]^2$$

(2.76)

$$\int_0^z \frac{a(z') + b(z')N(z')}{G^2(z)} dz' = \frac{N(z)[N(z) + 2]}{2G^2(z)}$$

(2.77)

The mean output photon number obtained for an amplifier with $\mathcal{M}$ modes in Eq. (2.73) is, consistently, the sum of the amplified signal photons $G\langle n(0)\rangle$ and $\mathcal{M}$ times the ASE noise photon number $N$ in each mode.

Comparing the variance obtained in Eq. (2.75) for the $\mathcal{M}$-mode amplifier with that obtained in Eq. (2.57) for the single-mode amplifier, several observations can be made. First, the shot noise is consistently given by $G\langle n(0)\rangle + \mathcal{M}N$. Second, the beat noise involving the signal is $2G\langle n(0)\rangle N$, independent of the number of modes $\mathcal{M}$. This can be interpreted by the fact that the statistical fluctuations associated with stimulated emission occur only in the mode occupied by the signal. If, on the other hand, the signal were distributed equally into all modes, the signal photon number per mode would be $\langle n(0)\rangle/\mathcal{M}$, and the total statistical fluctuation associated with independent stimulated emission processes in each mode would be $\mathcal{M}$ times $2G(\langle n(0)\rangle/\mathcal{M})N$ or $2G\langle n(0)\rangle N$, the result obtained when the signal is in one mode. Classically, we can interpret this result by considering that there is no possible beating effect between modes, as these are orthogonal. The third observation from Eq. (2.75) is that the beat noise associated with ASE is $\mathcal{M}$ times $N^2$, which is the corresponding beat noise in each mode. This term thus reflects the combined effect of independent statistical fluctuations for the ASE pertaining to each individual mode.

When considering the $\mathcal{M}$-mode amplifier, the previous equations and corresponding results for the PGF and the probability distributions must be modified. As these modifications are straightforward to do from the previous analysis, we shall only summarize the main results.

The output distribution $P_{n,m}(z)$ for the $\mathcal{M}$-mode amplifier is [27]:

$$P_{n,m}(z) = \frac{(1 + N - G)^n N^m}{(1 + N)^{m+n+\mathcal{M}}} \sum_{j=0}^{\min(m,n)} \frac{n!(m + \mathcal{M} - 1)!}{j!(m-j)!(n-j)!(j + \mathcal{M} - 1)!}\left[\frac{G}{N(1 + N - G)}\right]^j$$

(2.78)

which can be written as a function of the Jacobi polynomials, defined in Eq. (2.42):

$$P_{n,m}(z) = \frac{(N + 1 - G)^{n-m}}{(N + 1)^{n+\mathcal{M}}}(G - N)^m P_m^{(n-m,\mathcal{M}-1)}(X); \quad m < n$$

(2.79)

$$P_{n,m}(z) = \frac{N^{m-n}}{(N+1)^{m+\mathcal{M}}} (G-N)^n P_n^{(m-n,\mathcal{M}-1)}(X); \quad n < m \tag{2.80}$$

where $X$ is defined in Eq. (2.45) and where $\mathcal{M}$ is assumed to be an integer.

In the limit of large output photon numbers $m$, we can make the approximation $(m + \mathcal{M} - 1)! \approx m^{j+\mathcal{M}-1}$ and we find:

$$P_{n,m}(z) \approx \frac{m^{\mathcal{M}-1}(1 + N - G)^n N^m}{(1+N)^{m+n+\mathcal{M}}} \sum_j \frac{n!}{j!(n-j)!(j+\mathcal{M}-1)!} \left[ -m \frac{G}{N(1+N-G)} \right]^j \tag{2.81}$$

The PGF solution corresponding to the photon statistics master equation (2.72) for the $\mathcal{M}$-mode amplifier is found to be [29]:

$$F(x,z) = F[X(x,z;0),0]F_1(x,z)^{\mathcal{M}} \tag{2.82}$$

where $X(x,z;0)$ and $F[X(x,z;0),0]$ are defined in Eqs. (2.51) and (2.52). From the definition of $F_1(x,z)$ in Eq. (2.50), Eq. (2.82) can be put into the form:

$$F(x,z) = \frac{1}{[1 - (x-1)N(z)]^{\mathcal{M}}} F[X(x,z;0),0] \tag{2.83}$$

Note that the above result is valid for the general case where the coefficients $a(z)$ and $b(z)$ are not constant, which is demonstrated in Appendix G.

For either the single-mode ($\mathcal{M} = 1$) or the $\mathcal{M}$-mode amplifier cases, the higher order moments $\langle n^k \rangle (k > 2)$ can be deduced by two possible methods. The first calculates the $k$th derivatives of the PGF solution in Eq. (2.83). The second integrates the set of equations $d\langle n^k \rangle/dz$. In the first approach, each of the moments $\langle n^k \rangle$ must be evaluated iteratively, as $\langle n^k \rangle$ depends upon all moments of lower order. The set of rate equations $d\langle n^k \rangle/dz$ can be obtained by summing the photon statistics master equation (2.72) multiplied by $n^k$, as was done in Eqs. (2.64) and (2.65) for $\langle n \rangle$ and $\langle n^2 \rangle$. A recurrence method shows that the rate equation for $\langle n^k \rangle$ is given by, [23] and [34]:

$$\frac{d\langle n^k \rangle}{dz} = \sum_{j=1}^{k} \binom{k}{j-1} \{ a\langle (n+\mathcal{M})n^{j-1} \rangle + (-1)^{k+j-1} b\langle n^j \rangle \} \tag{2.84}$$

A set of $M$ equations (2.84) ($k = 1, ..., M$) can be integrated iteratively to yield closed-form expressions (though difficult to simplify for $a$, $b$ not constant and $k > 2$), or simultaneously by numerical methods.

In the linear amplification regime, the knowledge of moments of order higher than 2 is actually not of practical use, considering that we already have an expression for the output photon statistics distribution $P_n(z)$. But in the nonlinear amplification regime, the photon statistics distribution solution is generally difficult to obtain (see

Section 2.8). In this regime, knowledge of the higher order moments $\langle n^k \rangle$ through integration of Eq. (2.84) ($k = 1, \ldots, M$) is needed to evaluate $\langle n \rangle$ and $\langle n^2 \rangle$, as analyzed by S. Ruíz-Moreno *et al.* in [34], and discussed in Section 2.8.

We consider next different issues involved in the computation of the amplifier photon statistics in the linear regime.

Section 1.9 shows that for unsaturated EDFAs with forward pumping (for instance), the coefficients $a(z)$ and $b(z)$ for the signal wavelength $\lambda_k$ can be expressed as Eqs. (1.123) and (1.124):

$$a(z) = \bar{\gamma}_{ek}(z) = \alpha_k \frac{\dfrac{\eta_k}{1 + \eta_p} q(z)}{1 + q(z)} \tag{2.85}$$

$$b(z) = \bar{\gamma}_{ak}(z) = \alpha_k \frac{1 + \dfrac{\eta_p}{1 + \eta_p} q(z)}{1 + q(z)} \tag{2.86}$$

where $q(z)$ is the pump power normalized to the saturation power at wavelength $\lambda_p$. The pump power distribution $q(z)$ along the fiber is a solution of the rate equation (1.125):

$$\frac{dq(z)}{dz} = -\alpha_p \frac{q(z)}{1 + q(z)} \tag{2.87}$$

For the calculation of the PGF, knowledge of the pump distribution $q(z)$, the emission coefficient $a(z)$, the gain $G[a(z), b(z), z]$, and the ASE noise $N[a(z), b(z), z]$ are required, as seen from Eqs. (2.49)–(2.51). From definitions (2.30) and (2.31), and the above equations for $a(z)$ and $b(z)$, the gain $G(z)$ and mean noise output $N(z)$ are described by the following differential equations:

$$\frac{d}{dz} \{ \log G(z) \} = \frac{\alpha_k}{1 + q(z)} \left\{ \frac{\eta_k - \eta_p}{1 + \eta_p} q(z) - 1 \right\} \tag{2.88}$$

and

$$\frac{d}{dz} \left\{ \frac{N(z)}{G(z)} \right\} = \alpha_k \mathcal{M} \frac{\eta_k}{1 + \eta_p} \frac{q(z)}{G(z)[1 + q(z)]} \tag{2.89}$$

The set of coupled equations (2.87)–(2.89) can be solved numerically from the initial point $z = 0$ to the fiber coordinate $z$, given an input value $q(0)$ and the EDFA parameters $\alpha_p$, $\alpha_k$, $\eta_p$, and $\eta_s$. The PGF solution $F(x, z)$ in Eq. (2.83) is then determined from the solutions $G(z)$ and $N(z)$ and a summation in Eq. (2.52). Finally, the probability distribution $P_n(z)$ is obtained by performing $n$ successive derivations of the PGF function $F(x, z)$ with respect to $x$, according to Eq. (2.37).

When the output photon number is small (e.g., $\langle n(z) \rangle \leqslant 100$), the complete determination of the output probability distribution $P_n(z)$ by the PGF method is then relatively simple to implement. However, this method becomes quite computer time-consuming, if not prohibitive, when the output photon number is large (e.g., $\langle n(z) \rangle \geqslant 1000$).

If the PGF method is excluded in the case of large photon numbers, the remaining possibilities for the determination of $P_m(z)$ are: (1) to integrate numerically the photon statistics master equation (2.24), as done for instance in [2.35]; or (2) use the solution $P_m(z)$ given by the Jacobi molynomials in Eqs. (2.79) and (2.80). This solution requires knowledge of $G(z)$ and $N(z)$, determined by numerical integration of Eqs. (2.87)–(2.89) and two discrete summations, as shown in Eqs. (2.40) and (2.42).

There are several cases of interest where the PGF method remains the most practical. Indeed, for particular input photon statistics conditions, the output photon number distributions $P_m(z)$ can be directly identified through the PGF without performing iterative derivations of the PGF [29].

We observe first that the function $F_1(x, z)^{\mathcal{M}}$ in Eq. (2.83) is the PGF of the *negative binomial (NB) probability distribution* $B_n[\mathcal{M}, N(z)]$ with mean $\mathcal{M}N(z)$ and parameter $\mathcal{M}$. Indeed, the NB distribution is defined most generally by:

$$B_n[\mathcal{M}, N] = \frac{\mathcal{M}^{\mathcal{M}}(n + \mathcal{M} - 1)!}{n!(\mathcal{M} - 1)!} \frac{N^n}{(N + \mathcal{M})^{n + \mathcal{M}}} \tag{2.90}$$

and its PGF is

$$F(x, \mathcal{M}, N) = \frac{1}{\left[ 1 - (x - 1)\dfrac{N}{\mathcal{M}} \right]^{\mathcal{M}}} \tag{2.91}$$

When $\mathcal{M} \to \infty$, $B_n(1, N)$ reduces to the Poisson distribution, when $\mathcal{M} = 1$, the case of a single-mode amplifier, the NB distribution reduces to the Bose–Einstein (BE) distribution. The BE distribution is associated with thermal or chaotic light, and is defined by:

$$B_n[1, N(z)] = \frac{N(z)^n}{[1 + N(z)]^{n + 1}} \tag{2.92}$$

Four amplifier input conditions of interest, for which the photon statistics $P_m(z)$ are straightforward to derive from the output PGF [29], are: (1) no photons, (2) exactly one photon, (3) a deterministic number of photons, and (4) a signal with Poisson statistics.

*Amplifier with no input photons* The input distribution is $P_n(0) = \delta_{n,0}$, for which $F[X(x, z; 0), 0] = 1$ after Eq. (2.52), and the output distribution corresponds to the NB distribution $B_n[\mathcal{M}, \mathcal{M}N(z)]$ with mean $\mathcal{M}N$, defined in Eq. (2.90) by:

$$P_n(z) = \frac{(n + \mathcal{M} - 1)!}{n!(\mathcal{M} - 1)!} \frac{N(z)^n}{[1 + N(z)]^{n + \mathcal{M}}} \tag{2.93}$$

For the single-mode amplifier ($\mathscr{M} - 1$), the above distribution reduces to BE with mean $N(z)$.

*Amplifier with exactly one input photon*   The input distribution is $P_n(0) = \delta_{n,1}$, for which $F[X(x, z; 0), 0] = [1 + (x - 1)(G - N)]/[1 - (x - 1)N]$ after Eq. (2.52), and the output PGF is

$$F(x, z) = \frac{1 + (x - 1)[G(z) - N(z)]}{[1 - (x - 1)N(z)]^{\mathscr{M}+1}} \qquad (2.94)$$

The corresponding distribution is the sum of an NB distribution and a shifted one [29]:

$$P_n(z) = (1 + N - G)B_n[\mathscr{M} + 1, (\mathscr{M} + 1)N] + (G - N)B_{n-1}[\mathscr{M} + 1, (\mathscr{M} + 1)N] \qquad (2.95)$$

where the z-dependence was omitted for clarity. From Eqs. (2.90) and (2.95), we obtain:

$$\boxed{P_n(z) = \frac{(n + \mathscr{M})!}{\mathscr{M}!\, n!} \frac{N^n}{(N + 1)^{n+\mathscr{M}}} \left\{ 1 - \frac{G}{N + 1} - \left(1 - \frac{G}{N}\right)\frac{n}{n + \mathscr{M}} \right\}} \qquad (2.96)$$

*Amplifier with deterministic number of input photons*   The input distribution is $P_n(0) = \delta_{n,n_0}$, where $n_0$ is the exact number of input photons, and the output PGF takes the form:

$$F(x, z) = \frac{\{1 + (x - 1)[G(z) - N(z)]\}^{n_0}}{[1 - (x - 1)N(z)]^{n_0+\mathscr{M}}} \qquad (2.97)$$

This PGF corresponds to the convolution of a positive binomial (PB) distribution and a negative binomial (NB) distribution [29], i.e.,

$$P_n(z) = B_n[n_0 + \mathscr{M}, (n_0 + \mathscr{M})N] \otimes b_n[n_0, n_0(G - N)] \qquad (2.98)$$

where the PB distribution $b_n(\alpha, N)$ is defined by:

$$b_n(\alpha, N) = B_n(-\alpha, N) \qquad (2.99)$$

The mean and PGF for the PB distribution are $N$ and $F = [1 + (x - 1)N/\alpha]^\alpha$, respectively. Another expression for $P_n(z)$ is directly given by the solution $P_n(z) = P_{n_0,n}(z)$ in Eq. (2.78):

$$\boxed{P_n(z) = \frac{(1 + N - G)^{n_0}N^n}{(1 + N)^{n_0+n+\mathscr{M}}} \sum_{j=0}^{\min(n_0,n)} \frac{n_0!(n + \mathscr{M} - 1)!}{j!(n_0 - j)!(n - j)!(j + \mathscr{M} - 1)!} \left[\frac{G}{N(1 + N - G)}\right]^j}$$

$$(2.100)$$

Note that setting $n_0 = 0$ and $n_0 = 1$ in Eq. (2.100) yields the distributions found in Eqs. (2.93) and (2.96), respectively, as can be easily verified.

*Amplifier with input signal having Poisson statistics*    The input signal probability distribution is $P_n(0) = \exp(-\langle n_0 \rangle)\langle n_0 \rangle^n/n!$ From Eqs. (2.52) and (2.83), we find the PGF:

$$F(x, z) = \frac{\exp\left\{\dfrac{\langle n_0 \rangle(x - 1)G}{1 - (x - 1)N(z)}\right\}}{[1 - (x - 1)N(z)]^{\mathscr{M}}} \tag{2.101}$$

This PGF corresponds to the noncentral negative binomial (NNB) distribution, also called the Laguerre distribution, with expression [29]:

$$P_n(z) = \frac{N^n}{(N + 1)^{n+\mathscr{M}}} \exp\left\{-\frac{G\langle n_0 \rangle}{N + 1}\right\} L_n^{(\mathscr{M}-1)}\left\{-\frac{G\langle n_0 \rangle}{N(N + 1)}\right\} \tag{2.102}$$

where $L_n^{(\alpha)}(U)$ is the generalized Laguerre polynomial defined by [31]:

$$L_n^{(\alpha)}(U) = \sum_{j=0}^{n} \frac{(n + \alpha)!}{j!(n - j)!(j + \alpha)!}(-U)^j \tag{2.103}$$

From Eq. (2.101), it is seen that the most general case of the output PGF does not correspond to Poisson statistics; a Poisson distribution with mean $\langle n \rangle$ has PGF $= \exp[\langle n \rangle(x - 1)]$. But if we set $\mathscr{M} = N(z) = 0$ in Eq. (2.101), the PGF becomes that of a Poisson distribution with mean $G(z)\langle n(0) \rangle = \langle n(0) \rangle$, since $N(z) = 0$ implies $G(z) = 1$. The condition $\mathscr{M} = 0$ corresponds to an amplifier free from spontaneous emission, as seen in Eqs. (2.71) and (2.73). Assume now that the amplifier is free from spontaneous emission, but has positive gain. In this case, the the output PGF is $\exp\{G\langle n(0) \rangle(x - 1)/[1 - (x - 1)N]\}$, unrelated to Poisson statistics. Thus, the important result is that *even in the absence of spontaneous emission, light amplification does not conserve Poisson statistics for the signal.* This can be attributed to the term $N(z)$ in the PGF, which reflects the statistical fluctuations in the stimulated emission process, as previously discussed. Note that if the amplifier is free from spontaneous emission and does not have stimulated emission, we have $a(z) = 0$, $N(z) = 0$, and $G(z) = T(z) < 1$. The output PGF is $\exp\{T\langle n(0) \rangle(x - 1)\}$, a Poisson distribution with mean $T\langle n(0) \rangle$. This confirms the well-known property that *Poisson statistics are conserved when light is passing through a passive attenuator,* [3] and [4].

Another way to reach the above conclusion is to consider the amplifier output mean and variance in Eq. (2.75), assuming an input signal with Poisson statistics, for which $\sigma^2(0) - \langle n(0) \rangle = 0$. From these equations, we find for the quantity $\sigma^2(z) - \langle n(z) \rangle$:

$$\sigma^2(z) - \langle n(z) \rangle = 2G(z)\langle n(0) \rangle N(z) + \mathscr{M}N^2(z) \tag{2.104}$$

which vanishes only for $N(z) = 0$, corresponding to the case where no stimulated emission occurs. In the general case, we have $\sigma^2(z) - \langle n(z) \rangle \neq 0$ so the amplifier output statistics are not Poisson.

The probability distribution (2.102) actually describes the statistics of a system where coherent and chaotic (thermal) fields are superimposed, [36] and [37]. A deeper analysis of the statistics associated with such systems can be found in fundamental work on coherent and incoherent states of the radiation field, by R. J. Glauber [13], and in [25], [26], and [38], for instance. This deeper analysis confirms that for the linear regime, stimulated emission by itself causes changes in the light statistics, in addition to the changes introduced by the effect of spontaneous emission. It also shows that the contribution of spontaneous emission is always characterized by *Bose–Einstein statistics*, not by *additive Gaussian noise statistics*, as predicted by earlier quantum theories [39].

An analysis using more elaborate tools in the quantum theory of laser amplifiers shows that changes in the photon statistics due to superposition of coherent and chaotic fields correspond to a diffusion process of the photon probability distribution (associated with the density matrix of the system). Such a diffusion process is characterized by the *Fokker–Planck equation*, [13] and [28].

Without considering the details of this deeper analysis, which requires a more sophisticated formalism (namely the so-called P-representation for the density matrix [13]), it is still possible to show that the photon statistics master equation (2.72) can be transformed into a Fokker–Planck equation [25].

If we consider $n$ as being a continuous variable $x$, make the substitutions $n \to x$, $nP_n \to xp(x)$, and $(n \pm 1)P_{n\pm1} \to (x \pm 1)p(x \pm 1)$ in Eq. (2.72), and use the second order Taylor approximation:

$$(x \pm 1)p(x \pm 1) \approx xp \pm \frac{\partial xp}{\partial x} + \frac{1}{2}\frac{\partial^2 xp}{\partial x^2} \tag{2.105}$$

then we obtain from Eq. (2.72):

$$\boxed{\frac{dp(x, z)}{dz} = -\frac{\partial[M_1(x)p(x, z)]}{\partial x} + \frac{\partial^2[M_2(x)p(x, z)]}{\partial x^2}} \tag{2.106}$$

with $M_1(x) = (a - b)x$ and $M_2(x) = (a + b)x/2$. The Fokker–Planck equation (2.106) describes *a one-dimensional diffusion process* for the probability distribution $p(x)$. More specifically, the first term in Eq. (2.106) including the coefficient $M_1$ corresponds to a drift effect of the peak of the distribution $p(x)$ towards larger values of $x$, in the case of amplification. This can be shown by observing that for $a - b > 0$ (amplification or $G > 1$) we have $-\partial[M_1(x)p]/\partial x < 0$ for the left-hand side tail of $p(x)$, and $-\partial[M_1(x)p]/\partial x > 0$ for the right-hand side tail, causing the left tail to decrease and the right tail to increase, as illustrated in Figure 2.2a. The second term containing $M_2$ characterizes a diffusion effect, as it causes both tails to increase (as $\partial^2[M_2(x)p]/\partial x^2 > 0$ on the left- and right-hand sides of the two inflexion points of $p(x)$) and the peak of $p(x)$ to decrease (as $\partial^2[M_2(x)p]/\partial x^2 < 0$ in the region within the two inflexion points), as shown in Figure 2.2b. The Fokker–Planck equation obtained from the theory of coherent states $|\alpha\rangle$ actually describes a two-dimensional random walk and diffusion process in the $\alpha$-complex plane [3], [13], and [28]. This drift–diffusion effect for the photon probability distribution $P_m(z)$ along the amplifier is well illustrated in the numerical examples shown below.

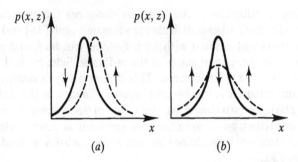

**FIGURE 2.2** Effect of the Fokker–Planck equation on the probability distribution $p(x, z)$: (a) drift caused by coefficient $M_1$, (b) diffusion caused by coefficient $M_2$.

Two functions that are useful in the charactetization of the amplifier output statistics are the Fano factor $f(z)$ and the statistical fluctuation $e(z)$. These are defined through:

$$f(z) = \frac{\sigma^2(z)}{\langle n(z) \rangle} \qquad (2.107)$$

and

$$e(z) = \frac{\sigma(z)}{\langle n(z) \rangle} \qquad (2.108)$$

For Poisson statistics, we have $\sigma^2 = \langle n \rangle$ and the Fano factor is $f = 1$, while for BE statistics with $\sigma^2 = \langle n \rangle^2 + \langle n \rangle$, it is $f = \langle n \rangle + 1$. For Poisson statistics, the fluctuation decreases with the mean photon number as $e = 1/\sqrt{\langle n \rangle}$ with a zero limit for $\langle n \rangle \to \infty$, while for BE statistics, the fluctuation decreases as $e = \sqrt{1 + 1/\langle n \rangle}$ to reach a minimum limit of unity. The Fano factor and fluctuation at the output of the amplifier can be found from Eqs. (2.73) and (2.75) for the mean and variance and the above definitions. For an $\mathcal{M}$-mode amplifier with no input signal ($\langle n(0) \rangle = 0$), the output mean and variance are $\langle n \rangle = \mathcal{M}N$ and $\sigma^2 = \mathcal{M}N(N + 1)$, respectively, and the Fano factor is $f = N + 1 = 1 + \langle n \rangle / \mathcal{M}$. If we assume an input signal with Poisson statistics, we find in the limit of high gains and high input signals ($n_0 \gg n_{sp}$):

$$f(z) \approx 1 + 2n_{sp}(z)[G(z) - 1] \qquad (2.109)$$

$$e(z) \approx \frac{1}{\sqrt{G(z)\langle n(0) \rangle}} \{1 + 2n_{sp}(z)[G(z) - 1]\}^{1/2} \qquad (2.110)$$

where $n_{sp}(z)$ is defined in Eq. (2.35). In this limit, the Fano factor at the amplifier output is the sum of two factors, one corresponding to the Poisson statistics ($f = 1$), the other to BE statistics with mean $2N$. On the other hand, the output fluctuation $e(z)$ is equal to that of Poisson statistics for a signal of mean $G\langle n \rangle$, i.e., $e = 1/\sqrt{G\langle n(0) \rangle}$, times a factor $\sqrt{1 + 2n_{sp}(z)[G(z) - 1]}$, which increases like the square root of the amplifier ASE noise $N = n_{sp}(G - 1)$. This last factor thus characterizes the deviation of the output statistics from Poisson. For large input signals ($n_0 \gg n_{sp}$) and very high gains ($G \gg 1$) the statistical fluctuation reaches an upper limit of $e(G \gg 1) = \sqrt{2n_{sp}/\langle n(0) \rangle}$.

In order to illustrate the above results, we consider next a few examples. The first is an ideal, lossless amplifier with complete and uniform population inversion, i.e., $a = \text{const.}$ and $b = 0$. The value of the emission coefficient is chosen, without loss of generality, to be $a = 0.4065 \, \text{m}^{-1}$, or $a = 2 \, \text{dB/m}$, which gives a gain of $G = +20 \, \text{dB}$ for a 10 m long fiber. In these conditions, we have $G(z) = \exp(az)$ and $N(z) = [G(z) - 1]$.

The probability distributions $P_m(z)$ corresponding to a fiber with $\mathcal{M} = 2$ polarization modes are plotted in a linear scale in Figure 2.3 for two different cases: (a) no input photon, or $n_0 = 0$, Eq. (2.93), and (b) exactly one input photon, or $n_0 = 1$, Eq. (2.96). The mean output $\langle n(z) \rangle$, standard deviation $\sigma(z)$, Fano factor $f(z)$, and statistical fluctuation $e(z)$ corresponding to each gain $G(z)$ are also shown in the figure. The probability distribution $P_m(z)$ broadens with fiber coordinate $z$, while its peak decreases and shifts towards higher values of $m$. This is the drift–diffusion effect predicted by the Fokker–Planck equation (2.106). The NNB distributions for these two cases, i.e., $n_0 = 0$ and $n_0 = 1$ are very similar, with little differences in Fano factor and fluctuation.

Figure 2.4 shows plots of $P_m(z)$ corresponding to a signal input with Poisson statistics and mean $\langle n(0) \rangle = 1$ (Figure 2.4a) and $\langle N(0) \rangle = 10$ (Figure 2.4b), as calculated from Eq. (2.102), with the same conditions as previously, i.e., $a = 2 \, \text{dB/m}$ and $b = 0$, for different amplifier lengths up to $z = 10 \, \text{m}$. For large values of the photon number $n$, the factorial function in Eq. (2.102) needs to be approximated by Stirling's formula:

$$n! \approx n^n \exp(-n)\sqrt{2\pi n} \qquad (2.111)$$

which is 99.9% accurate for $n > 50$.

The values of $\langle n \rangle$, $\sigma(z)$, $f(z)$ and $e(z)$ shown in Figure 2.4a, which correspond to $\langle n(0) \rangle = 1$ with Poisson statistics, are identical as in Figure 2.3b, which correspond to the case of a single photon deterministic input, as the output mean and variance in Eq. (2.75) take the same form in both cases. With the weak input mean photon number $\langle n(0) \rangle = 1$, the output fluctuation $e(z)$ is seen to decrease with length (Figure 2.4a), which shows that the amplifier statistics are dominated by thermal noise with the NNB distribution. With the higher input mean $\langle n(0) \rangle = 10$, the output fluctuation $e(z)$ is seen to increase with length (Figure 2.4b), from its initial value $e(0) = 1/\sqrt{n(0)} = 1/\sqrt{10} = 0.316$. This fact reflects that the photon statistics departs significantly from Poisson, for which we would have instead a decrease of $e(z)$, according to $e(z) = 1/\sqrt{\langle n(z) \rangle}$. The amplifier output photon statistics given in Eq. (2.102) correspond to a mixture of chaotic light (BE statistics) and coherent light (Poisson statistics) and its characteristic fluctuation increases like the square root of the ASE noise, as seen previously. Thus, for a Poisson statistics input, the statistical fluctuation $e = \sigma/\langle n \rangle = \Delta n/\langle n \rangle$ increases; this is consistent with the uncertainty principle, as discussed in Section 2.1. If the amplifier were to conserve Poisson statistics, the uncertainty in the output photon number would be $\sigma = \Delta n = \sqrt{G\langle n(0) \rangle}$, corresponding to an uncertainty in input signal photon number of $\Delta n(0) = \Delta n/\sqrt{G}$, which can be made arbitrary small with high gains, or $G \gg 1$. In this case, the optical amplifier would have reduced the

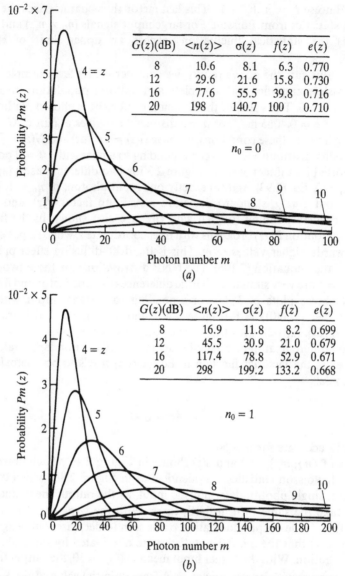

| $G(z)$(dB) | $\langle n(z)\rangle$ | $\sigma(z)$ | $f(z)$ | $e(z)$ |
|---|---|---|---|---|
| 8 | 10.6 | 8.1 | 6.3 | 0.770 |
| 12 | 29.6 | 21.6 | 15.8 | 0.730 |
| 16 | 77.6 | 55.5 | 39.8 | 0.716 |
| 20 | 198 | 140.7 | 100 | 0.710 |

$n_0 = 0$

| $G(z)$(dB) | $\langle n(z)\rangle$ | $\sigma(z)$ | $f(z)$ | $e(z)$ |
|---|---|---|---|---|
| 8 | 16.9 | 11.8 | 8.2 | 0.699 |
| 12 | 45.5 | 30.9 | 21.0 | 0.679 |
| 16 | 117.4 | 78.8 | 52.9 | 0.671 |
| 20 | 298 | 199.2 | 133.2 | 0.668 |

$n_0 = 1$

**FIGURE 2.3**   Output photon probability distribution for different amplifier lengths ($z = 4$ m to $z = 10$ m), in an ideal case of full and uniform inversion with $a = 0.4605\,\text{m}^{-1} = 2\,\text{dB/m}$ and $b = 0$, for different input conditions: (a) no input signal, (b) exactly one photon input signal. The gain $G(z)$ mean output photon number $\langle n(z)\rangle$, standard deviation $\sigma(z)$, Fano factor $f(z)$, and statistical fluctuation $e(z)$ are inset for different amplifier lengths.

photon number uncertainty of the input signals, in violation of the uncertainty principle, as shown in Section 2.1.

A second example to consider is a fiber amplifier with nonuniform and incomplete population inversion, where the coefficients $a$ and $b$ depend upon the coordinate $z$, according to Eqs. (2.85) and (2.86). The evolution of the relative pump power $q(z)$, the gain $G(z)$, and the ASE noise $N(z)$ with fiber length can be calculated by

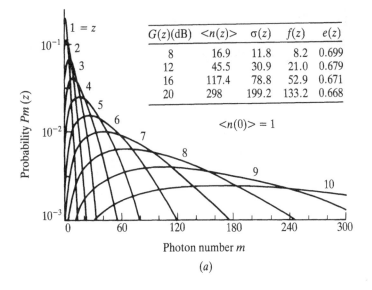

| $G(z)$(dB) | $<n(z)>$ | $\sigma(z)$ | $f(z)$ | $e(z)$ |
|---|---|---|---|---|
| 8 | 16.9 | 11.8 | 8.2 | 0.699 |
| 12 | 45.5 | 30.9 | 21.0 | 0.679 |
| 16 | 117.4 | 78.8 | 52.9 | 0.671 |
| 20 | 298 | 199.2 | 133.2 | 0.668 |

$<n(0)> = 1$

(a)

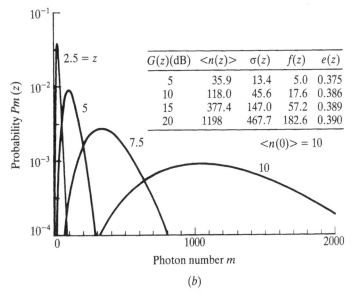

| $G(z)$(dB) | $<n(z)>$ | $\sigma(z)$ | $f(z)$ | $e(z)$ |
|---|---|---|---|---|
| 5 | 35.9 | 13.4 | 5.0 | 0.375 |
| 10 | 118.0 | 45.6 | 17.6 | 0.386 |
| 15 | 377.4 | 147.0 | 57.2 | 0.389 |
| 20 | 1198 | 467.7 | 182.6 | 0.390 |

$<n(0)> = 10$

(b)

**FIGURE 2.4** Output photon probability distribution for different amplifier lengths for $z = 1$ m to $z = 10$ m, in an ideal case of full and uniform inversion ($a = 2$ dB/m and $b = 0$), assuming an input signal with Poisson statistics and a mean of (a) $\langle n(0) \rangle = 1$, and (b) $\langle n(0) \rangle = 10$. The gain $G(z)$, mean output photon number $\langle n(z) \rangle$, standard deviation $\sigma(z)$, Fano factor $f(z)$, and statistical fluctuation $e(z)$ are inset for different amplifier lengths.

simultaneous integration of Eqs. (2.87)–(2.89). These equations correspond to the case of forward pumping. Consider a realistic Er-doped fiber amplifier: $q(z = 0) = q_0 = 6.65$, $\alpha_p = 2$ dB/m, $\alpha_s = 4$ dB/m, $\eta_p = 0.26$ and $\eta_s = 0.92$. The input signal is assumed to have Poisson statistics with mean $\langle n(0) \rangle = 10$. The number of modes $\mathcal{M} = 2$.

Figure 2.5a shows plots of the relative pump power $q(z)$, and mean $\langle n(z) \rangle$ versus fiber coordinate $z$, as integrated from Eqs. (2.87)–(2.89) with the above parameters.

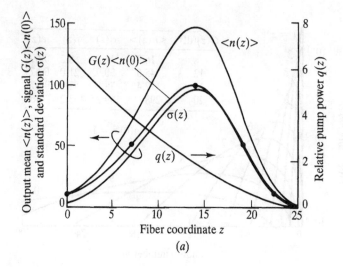

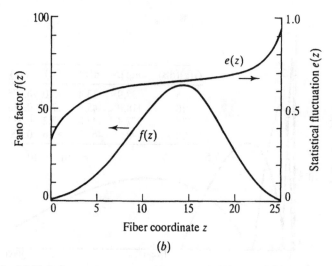

**FIGURE 2.5**   (a) Relative pump power $q(z)$, mean $\langle n(z) \rangle$, amplified signal $G(z)\langle n(0) \rangle$ and standard deviation $\sigma(z)$ versus fiber coordinate $z$, as integrated from Eqs. (2.87)–(2.89), and (b) corresponding plots of the Fano factor $f(z)$ and statistical fluctuation $e(z)$ for $q_0 = 6.65$, $\alpha_p = 2$ dB/m, $\alpha_s = 4$ dB/m, $\langle n(0) \rangle = 10$, $\eta_p = 0.26$, $\eta_s = 0.92$, and $\mathcal{M} = 2$. The points marked in the curve $G(z)\langle n(0) \rangle$ in Figure 2.5a correspond to five fiber lengths ($z = 0$, 7, 14, 19, and 23 m) for which the photon statistics were calculated, using Eq. (2.102), and plotted in Figure 2.6.

The amplified signal $G(z)\langle n(0) \rangle$ and standard deviation $\sigma(z)$ are also shown for reference. The figure shows that a maximum gain of $G = 10$ dB is achieved at $z = 14$ m, corresponding to the optimum amplifier length $L_{opt}$. Beyond this length, the fiber is lossy, and the mean and deviation decrease.

How do the photon statistics evolve with the effect of reabsorption by the fiber, past the optimal length $L_{opt}$? Figure 2.5b shows plots of the Fano factor $f(z)$ and

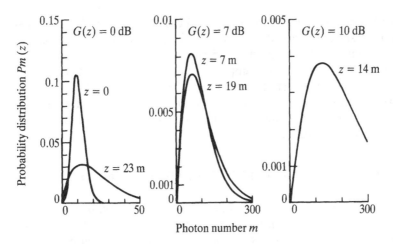

**FIGURE 2.6**  Probability distribution $P_m(z)$ for different amplifier lengths $z = 0$, $z = 7$ m, $z = 14$ m, $z = 19$ m, and $z = 23$ m, corresponding to gains $G = 0$, $G = 7$ dB, $G = 14$ dB, $G = 7$ dB, and $G = 0$ dB, respectively.

statistical fluctuation $e(z)$. The fluctuation $e(z)$ increases from its initial value $e(0) = 1/\sqrt{\langle n(0) \rangle} = 1/\sqrt{10} = 0.316$ but this increase becomes much steeper when the fiber is reabsorbing. This is explained by the factor $1/\sqrt{G(z)}$ in Eq. (2.110), which increases with $z$ when $z > L_{opt}$.

The points marked in the curve $G(z)\langle n(0) \rangle$ in Figure 2.5$a$ correspond to five fiber lengths ($z = 0, 7, 14, 19$, and 23 m) for which the photon statistics were calculated, using Eq. (2.102). The lengths $z = 7$ and $z = 19$ m correspond to the same net gain of $G = 7$ dB, the length $z = 14$ m corresponds to the maximum gain of 10 dB, and the length $z = 23$ m corresponds to unity gain, or fiber transparency. The probability distributions $P_m(z)$ corresponding to each of these cases are shown in Figure 2.6. The curves corresponding to $z = 0$, $z = 7$, and $z = 14$ m show the diffusion effect for $P_m(z)$ previously discussed. The interesting features of the figure concern the photon statistics at $z = 19$ and $z = 23$ m; they correspond to a mean signal photon number $G(z)\langle n(0) \rangle$ identical to that at $z = 7$ and $z = 0$ m, respectively. It is seen that for lengths $z > L_{opt}$ the distributions $P_m(z)$ have experienced broadening. This broadening can be attributed to the statistical fluctuation of stimulated emission (occurring even in the case $a(z) - b(z) < 0$) and the effect of additional spontaneous emission noise, which take place in the lossy portion of the amplifier. In the lossy portion, the drift–diffusion effect continues, but the drift moves the peak of the distribution towards lower values of $m$. Thus, an important conclusion is that *the effect of amplification followed by reabsorption to the same mean signal level does not conserve the photon statistics higher moments.*

The description of photon statistics made in this section assumed that the absorption and emission coefficients $a(z)$ and $b(z)$ are deterministic, not subject to statistical fluctuations. But such fluctuations are likely to occur in a three-level system such as an Er-doped fiber amplifier. There are three main causes: (1) pump intensity noise, which directly affects the density $N_1$ of atoms in the ground state, and indirectly affects the density $N_2$ of atoms in the excited state; (2) statistical fluctuations with the multiphonon decay from the pump to the metastable levels; and (3) statistical fluctuations associated with the thermalization process occuring within the upper and lower manifolds. When the amplifier is operated as a two-level system, the pump intensity noise affects directly both upper and lower populations. A complete statistical treatment including these effects remains to be developed. In the absence of this model, we can assume that possible corrections including the effects of pump noise and thermal population noise only represent small perturbations of the photon statistics previously developed.

## 2.4   OPTICAL SIGNAL-TO-NOISE RATIO AND NOISE FIGURE

Equations (2.73) and (2.75) are general expressions for the amplifier output mean and variance, valid *regardless of the particular input photon statistics* $P_m(0)$. Chapter 3 shows how they are essential to determine characteristics of photodetected signals. Now we need to define some more quantities to use in the statistical analysis of such photodetected signals.

Consider digital communications systems. The signal input to the amplifier with mean $\langle n(0) \rangle$ is actually modulated in time, frequency, or phase for conveying information at a bit rate $B = 1/T$. At the amplifier output, the quantity $\langle n(z) \rangle - \langle n(z) \rangle_{ASE}$. where $\langle n(z) \rangle_{ASE} = \mathcal{M} N(z)$ is the unmodulated background ASE noise, thus represents the mean photon number that contains signal information. At the receiver end, the modulated signal power $P_{sig} \approx \langle n(z) \rangle - \langle n(z) \rangle_{ASE}$ is converted into photocurrent, with electrical power $P_{phot} \approx P_{sig}^2$. On the other hand, the associated noise power is given by $P_{noise} \approx \sigma^2$ [10]. It is therefore appropriate to define an optical signal-to-noise ratio $SNR_o$ in the form:

$$SNR_o(z) = \frac{\langle \langle n(z) \rangle - \langle n(z) \rangle_{ASE} \rangle_T^2}{\sigma^2(z)} = \frac{G^2(z) \langle n(0) \rangle^2}{\sigma^2(z)} \qquad (2.112)$$

In the above definition, the brackets $\langle \cdots \rangle_T$ indicate a time average over the bit period $T$. The quantity $\sigma^2(z)$ is the noise power measured over this period. Equation (2.112) therefore corresponds to the SNR of a single signal bit, which at this point remains an unspecified symbol (assumed to contain energy or $\langle n(0) \rangle \neq 0$). A more elaborate definition for the SNR, which will take into account the difference between bits representing 0 and 1 symbols, is introduced in Chapter 3. The term optical in the above definition for the SNR is justified by the fact that the factors in Eq. (2.112) involve optical powers, as opposed to electrical powers. An ideal detector of unity quantum efficiency, without considering the effect of electronic bandwidth, has an electrical SNR equal to its optical SNR because the electrical quantities are proportional to the squared optical quantities. Optical SNR is useful for characterizing optical amplifier performance, even if in real communications systems only electrical SNR is relevant.

In particular, for input signals with Poisson statistics, we have $\sigma^2(0) = \langle n(0) \rangle$, and the $SNR_o(z)$ at $z = 0$ is equal to the mean, i.e.,

$$SNR_o(0) = \langle n(0) \rangle \qquad (2.113)$$

Using Eqs. (2.73) and (2.75) in Eq. (2.112), we obtain at coordinate $z$:

$$\frac{1}{\mathrm{SNR}_o(z)} = \frac{1}{\langle n(0) \rangle} \left\{ f(0) - 1 + \frac{1 + 2N(z)}{G(z)} + \frac{\mathcal{M}N(z)[N(z) + 1]}{G^2(z)\langle n(0) \rangle} \right\} \quad (2.114)$$

where $f(0)$ is the input Fano factor, defined in Eq. (2.107).

We define the amplifier optical noise figure $F_o(z)$ through the ratio [2.1] and [2.10]:

$$F_o(z) = \frac{\mathrm{SNR}_o(0)}{\mathrm{SNR}_o(z)} \quad (2.115)$$

where $\mathrm{SNR}_o(0)$ is the amplifier input optical SNR. According to Eq. (2.115), the optical noise figure represents a measure of the SNR degradation experienced by the signal after passing through the amplifier. As the amplifier introduces noise, it is expected that $\mathrm{SNR}_o(z) < \mathrm{SNR}_o(0)$, so *the amplifier optical noise figure is always greater than or equal to unity*. This important conclusion can be reached otherwise by considering the contrary case of an ideal, fictitious amplifier that would have no output noise, with gain $G > 1$. The uncertainty in output signal power would then be given by the minimum detector noise, or shot noise, proportional to $\sigma^2 = G\langle n(0) \rangle$, as shown in Section 2.1. The output optical SNR for such an amplifier would then be $\mathrm{SNR}_o = [G\langle n((0)\rangle)]^2/\sigma^2 = G\langle n(0) \rangle$; this shows that the amplifier would conserve Poisson statistics. As a result, the noise figure for this amplifier would be $F_o = 1/G < 1$, which contradicts the previous conclusion. The optical noise figure is always greater than unity; this reflects the property that *optical amplifiers cannot improve the optical signal SNR*. If the contrary were true, the potential accuracy of signal power measurements would violate Heisenberg's uncertainty principle, as discussed in Section 2.1.

The above conclusion also applies to the case where the amplifier is attenuating, as such an amplifier always has a finite ASE noise output. The effect of attenuation itself is a cause of optical SNR degradation. Consider indeed the case of a signal with Poisson statistics passing through an attenuator having neither spontaneous nor stimulated emission, with a transmission $T < 1$. As Poisson statistics are conserved after pure attenuation of light (see previous section), the output optical SNR is then $\mathrm{SNR}_o = T\langle n(0) \rangle$. Given the input optical SNR of $\mathrm{SNR}_o = \langle n(0) \rangle$, the characteristic noise figure of the attenuator is therefore $F_o = 1/T$, which is greater than unity.

We consider now the general case. Using Eq. (2.114) in Eq. (2.115), we find for the optical noise figure:

$$F_o(z) = \frac{\mathrm{SNR}_o(0)}{\langle n(0) \rangle} \left\{ f(0) - 1 + \frac{1 + 2N(z)}{G(z)} + \frac{\mathcal{M}N(z)[N(z) + 1]}{G^2(z)\langle n(0) \rangle} \right\} \quad (2.116)$$

For input signals with Poisson statistics, we have $f(0) = 1$ and $\mathrm{SNR}_o(0) = \langle n(0) \rangle$, Eq. (2.113), and the optical noise figure in Eq. (2.116) becomes:

$$F_o(z) = \frac{1 + 2N(z)}{G(z)} + \frac{\mathcal{M}N^2(z)}{G^2(z)\langle n(0) \rangle} + \frac{\mathcal{M}N(z)}{G^2(z)\langle n(0) \rangle} \quad (2.117)$$

The three contributions involved in the noise figure in Eq. (2.117) can be traced back to the various noise contributions shown in Eq. (2.75): signal shot noise $G\langle n(0)\rangle$ and beat noise $2G\langle n(0)\rangle N$ (term $[1 + 2N]/G$), ASE beat noise $\mathcal{M}N^2$ (term $\mathcal{M}N^2/G^2\langle n(0)\rangle$), and ASE shot noise $\mathcal{M}N$ (term $\mathcal{M}N/G^2\langle n(0)\rangle$). If we assume an input high enough such that $G\langle n(0)\rangle \gg N$, the optical noise figure in Eq. (2.117) reduces to [40]:

$$F_o(z) = \frac{1 + 2N(z)}{G(z)} = \frac{1 + 2n_{sp}(z)[G(z) - 1]}{G(z)} \qquad (2.118)$$

where $n_{sp}(z)$ is the spontaneous emission factor defined in Eq. (2.35).

In the high gain limit $G \gg 1$, the optical noise figure reduces to

$$F_o(z, G \gg 1) = 2n_{sp}(z) \qquad (2.119)$$

The spontaneous emission factor is always greater than or equal to unity; it equals unity for complete medium inversion, Eq. (2.34). The fundamental result therefore is that *the noise figure of a high gain amplifier is always greater than 2, or $F_o > 3$ dB.*

Now consider the case of an actual Er-doped fiber amplifiers (EDFA), for which the spontaneous emission factor depends upon the cross section ratios $\eta = \sigma_e/\sigma_a$ at both pump and signal wavelengths. Using Eq. (2.34) for $n_{sp}$, and Eqs. (1.123) and (1.124) for the unsaturated absorption and emission coefficients at wavelength $\lambda_k$, i.e., $a = \bar{\gamma}_{ek} = \alpha_k \eta_k N_2$ and $b = \bar{\gamma}_{ak} = \alpha_k N_1$, we obtain:

$$n_{sp}(z, \lambda_p, \lambda_k) = \frac{1}{1 - \dfrac{\eta_p}{\eta_k} - \dfrac{1 + \eta_p}{\eta_k}\dfrac{1}{q(z)}} \qquad (2.120)$$

where $q(z)$ is the total (forward + backward) relative pump power at coordinate $z$. In the limit of high pump powers ($q(z) \gg 1$), we obtain the minimum spontaneous emission factor value $n_{sp}^{min}$, [40] and [41]:

$$n_{sp}^{min}(\lambda_p, \lambda_k) = \frac{1}{1 - \dfrac{\eta_p}{\eta_k}} = \frac{1}{1 - \dfrac{\sigma_e(\lambda_p)\sigma_a(\lambda_k)}{\sigma_a(\lambda_p)\sigma_e(\lambda_k)}} \qquad (2.121)$$

which corresponds to the minimum amplifier noise figure $F_o^{min}$:

$$F_o^{min}(\lambda_p, \lambda_k) = \frac{2}{1 - \dfrac{\eta_p}{\eta_k}} = \frac{2}{1 - \dfrac{\sigma_e(\lambda_p)\sigma_a(\lambda_k)}{\sigma_a(\lambda_p)\sigma_e(\lambda_k)}} \qquad (2.122)$$

In the case where the EDFA is pumped as a three-level system (see Chapter 1), we have $\sigma_e(\lambda_p) = 0$ and from Eq. (2.122) the minimum achieveable noise figure is $F_o = 3$ dB.

The 3 dB lower limit obtained for the optical noise figure at high amplifier gains is commonly referred to as the *quantum limit*. The existence of such a limit can also be demonstrated from a rigorous quantum mechanical treatment, where the amplifier input SNR is defined with respect to the uncertainty noise [10]. *The quantum limit condition $F_o \geqslant 2$ only applies to the case of high amplifier gains.* On the other hand, for low gain amplifiers, the optical noise figure can be such that $1 \leqslant F_o \leqslant 2$. An example is given by the case of a fully inverted amplifier with $a = 0.693\,\mathrm{m}^{-1} = 3.01\,\mathrm{dB/m}$ and $b = 0$. The gain at $z = 1\,\mathrm{m}$ is $G = 2.0$, and the spontaneous emission factor is $n_{sp} = 1$. In this case, the optical noise figure is from Eq. (2.118): $F_o(z = 1\,\mathrm{m}) = [1 + 2n_{sp}(G - 1)]/G = 1.5$ or $F_o = 1.76\,\mathrm{dB}$, which is below the quantum limit. Thus, a reduced SNR degradation as compared to the quantum limit is possible, but at the expense of amplifier gain. Since $n_{sp} = 1$ for fully inverted amplifiers, the optical noise figure is actually always strictly lower than 3 dB even at high gains, as in this case $F_o = 2 - 1/G < 2$. Thus, for fully inverted amplifiers, the optical noise figure approaches at high gains the quantum limit by lower values. But in experimental conditions, it is difficult to achieve full amplifier inversion with $n_{sp}$ equal to exactly unity; in practice $n_{sp} > 1$. As an example, $n_{sp} = 1.05$ yields from Eq. (2.118) a high gain limit for the optical noise figure of $F_o = 2.1$.

In the case where the EDFA is pumped as a two-level system, we have $\sigma_e(\lambda_p) \neq 0$ in Eqs. (2.121) and (2.122), consequently the minimum achievable noise figure is always greater than 3 dB. This property can be illustrated by considering the example of an alumino-germanosilicate Er:glass (see Chapter 4), whose absorption and emission cross sections and ratio $\eta$ are plotted against wavelength in Figure 2.7. The spectra of the minimum spontaneous emission factor $n_{sp}^{min}(\lambda)$ and corresponding optical noise figure $F_o^{min}(\lambda)$, defined in Eqs. (2.121) and (2.122) are plotted for different pump wavelengths in Figure 2.8. Figure 2.8a shows a decrease in the minimum spontaneous emission factor from short signal wavelengths ($\lambda_s = 1.52\,\mu\mathrm{m}$) to long signal wavelengths ($\lambda_s = 1.57\,\mu\mathrm{m}$), with an intermediate peak near $\lambda_s = 1.535\,\mu\mathrm{m}$. This feature is independent of the pump wavelength $\lambda_p$. The highest value reached by $n_{sp}$ at short signal wavelengths steadily increases with the pump wavelength. For the type of Er:glass shown in this example, the minimal achievable noise figure $F_o^{min}$, corresponding to the minimal achievable value $n_{sp}^{min}$ for the spontaneous emission factor is, according

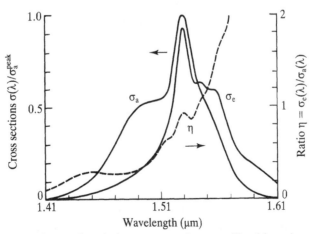

**FIGURE 2.7** Absorption and emission cross sections $\sigma_{a,e}(\lambda)$ with ratio $\eta = \sigma_e/\sigma_a$ versus wavelength, corresponding to a typical alumino-germanosilicate Er-doped fiber.

to Figure 2.8b, $F_0^{\min} = 3.8$ dB, obtained at the pump and signal wavelengths of $\lambda_p = 1.46 \ \mu$m and $\lambda_s = 1.57 \ \mu$m, respectively.

This dependence of $n_{sp}^{\min}$ on pump and signal wavelengths can be interpreted if we consider first the unsaturated absorption and emission coefficients $\bar{\gamma}_{ak}$ and $\bar{\gamma}_{ek}$ in Eqs. (1.123) and (2.124). In the high pump regime ($q(z) = q^+ + q^- \gg 1$), $a(\lambda_k) = \bar{\gamma}_{ek} = \alpha_k \eta_k/(1 + \eta_p)$ and $b(\lambda_k) = \bar{\gamma}_{ak} = \alpha_k \eta_p/(1 + \eta_p)$. We define a dimensionless population inversion experienced by the laser transition at wavelength $\lambda_k$ through $U(\lambda_k) = [a(\lambda_k) - b(\lambda_k)]/\alpha_k$, or $U(\lambda_k) = (\eta_k - \eta_p)/(1 + \eta_p)$. The parameter $U$ is also defined in Eq. (1.61), for the analysis of the degree of inversion in all pump power regimes. The parameter $U$ thus represents the maximum inversion possible at wavelength $\lambda_k$, determined by the cross section ratios $\eta_p$ and $\eta_k$. The highest value of $U$ corresponds to $\eta_p = 0$, representing full inversion. Figure 2.7 shows that this condition never occurs with two-level amplifiers, as we have $\eta_p > 0$ at all possible pump wavelengths ($\lambda_p = 1.45$–$1.53 \ \mu$m). A low value for $\eta_p$ indicates the amplifier operates almost as a three-level system (as there is negligible stimulated emission at the pump wavelength); a finite value for $\eta_p$ indicates the amplifier operates nearly as a two-level system (as there exists sizeable stimulated emission at the pump wavelength). There is obviously no clear-cut transition between the two- and three-level regimes, as the coefficient $\eta_p$ varies continuously (see Figure 2.7). The higher the pump wavelength $\lambda_p$, the greater the coefficient $\eta_p$, and the more the amplifier operates as a two-level system. This regime also corresponds to the lowest value for the inversion $U(\lambda_k)$ at the signal wavelength $\lambda_k$, which corresponds to the highest value for the spontaneous emission factor, as $n_{sp}^{\min} = a(\lambda_k)/U(\lambda_k)$. The above considerations explain the increase with pump wave-length of $n_{sp}^{\min}$ in Figure 2.8a. The decrease of $n_{sp}^{\min}$ with increasing signal wavelengths and fixed pump wavelengths (or fixed $\eta_p$) is due to the increase of $\eta_k$, which results in a higher inversion parameter $U(\lambda_k)$.

The laser system involved in the Er-doped glass medium is neither two-level nor three-level. It is a laser system quasicontinuum. The wavelength independent textbook definition for the spontaneous emission factor, i.e., $n_{sp} = N_2/[N_2 - (g_2/g_2)N_1]$, where $g_1, g_2$ represent the lower and upper energy level degeneracies, must be replaced by the wavelength dependent definition $n_{sp} = N_2/(N_2 - N_1/\eta_k)$. Further discussion is provided in Section 1.3.

Spontaneous emission factor becomes irrelevant when the amplifier gain is less than or equal to unity. For amplifier transparency ($G = 1$), we have $a - b = 0$ and $n_{sp} = a/(a - b) \to \infty$; ASE and optical noise remain finite as $N(z) \to az$ and $F_o(z) \to 1 + 2az$. For $G < 1$, the spontaneous emission factor becomes negative. For this reason, it is convenient to introduce the definition of an *equivalent input noise factor* $n_{eq}(z)$ through [42]:

$$n_{eq}(z) = \frac{N(z)}{G(z)} = \frac{n_{sp}(z)[G(z) - 1]}{G(z)} = \int_0^z \frac{a(z')}{G(z')} \, dz' \qquad (2.123)$$

This equivalent input noise factor can be regarded as an equivalent input noise signal. Using Eqs. (2.73) and (2.123), the total ASE output is $\mathcal{M}N = \mathcal{M}n_{eq}G$. At high gains, the equivalent input noise factor matches the spontaneous emission factor; at transparency, it is finite: $n_{eq}(z) = az/G(z)$, for $a = $ const; below transparency it is

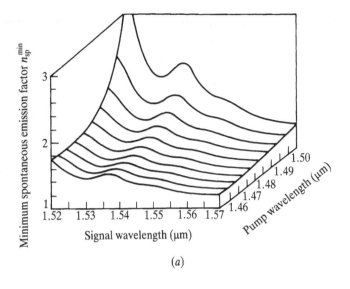

(a)

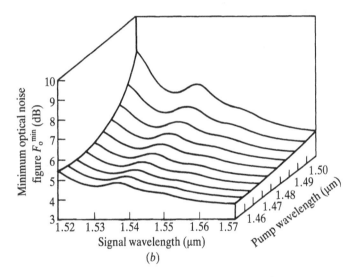

(b)

**FIGURE 2.8** (a) Minimum spontaneous emission factor $n_{sp}^{min}(\lambda)$, and (b) corresponding minimum optical noise figure $F_o^{min}(\lambda)$, plotted versus signal wavelength $\lambda_s = 1.52\ \mu m$ to $1.57\ \mu m$, for different pump wavelengths $\lambda_p = 1.46\ \mu m$ to $1.50\ \mu m$.

positive, unlike $n_{sp}$. It is therefore appropriate to consider $n_{eq}$ instead of $n_{sp}$ for analyzing ASE an equivalent input noise over the whole gain spectrum. This includes two spectral regions where $G \leqslant 1$. From Eqs. (2.118) and (2.123), the optical noise figure can then be defined as:

$$F_o(z) = 2n_{eq}(z) + \frac{1}{G(z)} \qquad (2.124)$$

We have considered so far only the high pump regime, which corresponds to the lowest possible amplifier noise figure $F_\circ^{min}$. In the general case, the pump power decreases along the length, and the condition $q(z) \gg 1$ is not realized over the whole amplifier length. Forward pumping and backward pumping are the two possible pumping configurations for which the pump power either increases or decreases with length. The effect of having a finite pump power distribution along the fiber and the effect of pumping direction with respect to the ASE noise can be analyzed through the exact analytical solutions obtained for the unsaturated gain regime (Section 1.9).

From the results of Appendix E, and assuming the fiber background loss is negligible (i.e., $\alpha'_{p,k} = 0$), the gain and ASE noise photon numbers in the two fiber propagation directions can be put in the form (E. Desurvire [43]):

$$G_k \equiv G_k^+(q_L) = G_k^-(q_0) = \exp\left\{\frac{\alpha_k}{\alpha_p}\left[\frac{\eta_k - \eta_p}{1 + \eta_p}(q_0 - q_L) - \log\left(\frac{q_0}{q_L}\right)\right]\right\} \quad (2.125)$$

$$N^+(\lambda_k, q_L) = \frac{\alpha_k}{\alpha_p}\frac{\eta_k}{1 + \eta_p}I^+(q_0, q_L)\exp\left\{\frac{\alpha_k}{\alpha_p}\left[\log(q_L) - \frac{\eta_k - \eta_p}{1 + \eta_p}q_L\right]\right\} \quad (2.126)$$

$$N^-(\lambda_k, q_0) = \frac{\alpha_k}{\alpha_p}\frac{\eta_k}{1 + \eta_p}I^-(q_0, q_L)\exp\left\{-\frac{\alpha_k}{\alpha_p}\left[\log(q_0) - \frac{\eta_k - \eta_p}{1 + \eta_p}q_0\right]\right\} \quad (2.127)$$

with

$$I^\pm(q_0, q_L) = \int_{q_L}^{q_0} dx \frac{\exp\left(\pm\frac{\alpha_k}{\alpha_p}\frac{\eta_k - \eta_p}{1 + \eta_p}x\right)}{X^{\pm\alpha_k/\alpha_p}} \quad (2.128)$$

In the above equations, $q_0$ and $q_L$ represent the relative input and output pump powers, respectively, assuming $q_0 > q_L$, which corresponds to the forward pumping case. The quantities $G_k^+(q_L)$ and $N^+(\lambda_k, q_L)$ represent the gain and forward ASE photon number at $z = L$, for forward pumping. The quantities $G_k^-(q_0)$ and $N^-(\lambda_k, q_0)$ represent the gain and backward ASE photon number at $z = 0$, for forward pumping. The forward and backward ASEs in the forward pumping case respectively correspond to the backward and forward ASEs in the backward pumping case. The above results are therefore most general. Note also that the amplifier gain is the same in both directions, i.e., $G_k^+(q_L) = G_k^-(q_0) = G_k$, while in the general case the ASE noises are not equal, with the relation $N^-(\lambda_k, q_0) > N^+(\lambda_k, q_L)$, as shown below [43].

From Eqs. (1.127) and (1.128), with the limit $\alpha'_{p,k} \to 0$ (or $\varepsilon_{p,k} \to 0$), the output pump power $q_L$ is related to the input pump $q_0$ through:

$$q_L = q_0 \exp(-\mathscr{A}_{pk}L) \quad (2.129)$$

with

$$\mathscr{A}_{pk} = \alpha_p\frac{1 + \eta_p}{1 + \eta_k}\left\{\frac{\eta_k - \eta_p}{1 + \eta_p} - \frac{\log G_k}{\alpha_k L}\right\} \quad (2.130)$$

Elimination of the parameter $G_k$ by combining Eqs. (2.125), (2.129), and (2.130) yields the pump transcendental equation

$$q_0 - q_L = \alpha_p L - \log\left(\frac{q_0}{q_L}\right) \tag{2.131}$$

which is the solution of Eq. (1.125) for forward pumping and $\alpha'_p = 0$.

Using Eqs. (2.123), (2.124), (2.126), (2.127), and (2.132), we obtain the equivalent input noise factors $n_{eq}^{\pm}(\lambda_k) = N^{\pm}(\lambda_k)/G_k$ and the optical noise figures $F_o^{\pm}(\lambda_k)$ at wavelength $\lambda_k$ at $z = L$ and $z = 0$, corresponding to the two propagation directions, i.e.,

$$n_{eq}^{+}(\lambda_k) = \frac{\alpha_k}{\alpha_p}\frac{\eta_k}{1 + \eta_p} I^{+}(q_0, q_L)\exp\left\{\frac{\alpha_k}{\alpha_p}\left[\log(q_0) - \frac{\eta_k - \eta_p}{1 + \eta_p}q_0\right]\right\} \tag{2.132}$$

$$n_{eq}^{-}(\lambda_k) = \frac{\alpha_k}{\alpha_p}\frac{\eta_k}{1 + \eta_p} I^{-}(q_0, q_L)\exp\left\{-\frac{\alpha_k}{\alpha_p}\left[\log(q_0) - \frac{\eta_k - \eta_p}{1 + \eta_p}q_L\right]\right\} \tag{2.133}$$

and

$$F_o^{\pm}(\lambda_k) = 2n_{eq}^{\pm}(\lambda_k) + \frac{1}{G_k} \tag{2.134}$$

The above relations giving the equivalent input noise factor and the optical noise figure uniquely depend upon the input and output pump powers $q_0$, $q_L$, and the Er-doped fiber parameters ($\alpha_p$, $\alpha_k$, $\eta_p$ and $\eta_k$). However, $q_0$, $q_L$ are related to the fiber length $L$ and gain $G_k$ at $\lambda_k$ through Eqs. (2.129) and (2.130), which is of the form $q_L/q_0 = f[\log(G_k)/L]$. Finally, the input pump $q_0$ is related to $L$ and $G_k$ through (see Eqs. (E.23)–(E.25) with the limit $\alpha'_{p,k}, \varepsilon_{p,k} \to 0$):

$$q_0 = \alpha_p L \frac{Q_k}{1 - \exp\{-\alpha_p L(1 - Q_k)\}} \tag{2.135}$$

with

$$Q_k = \frac{1 + \eta_p}{1 + \eta_k}\left(1 + \frac{\log G_K}{\alpha_k L}\right) \tag{2.136}$$

with Eqs. (2.129) and (2.135) we can now rewrite the output pump $q_L$ through:

$$q_L = q_0 \exp\{-\alpha_p L(1 - Q_k)\} \tag{2.137}$$

For a numerical application, we must first assume values for the pump and signal absorption coefficients $\alpha_p$, $\alpha_k$ and the fiber length $L$. The maximum possible EDFA gain, is determined by $L$ and the fiber parameters, independently of the input pump

$q_0$. Indeed, the condition $(1 - Q_k) > 0$ in Eq. (2.137), which implies that the pump is absorbed in the fiber or $q_L < q_0$, gives:

$$G_K < \exp\left\{\frac{\eta_K - \eta_p}{1 + \eta_p} \alpha_k L\right\} \qquad (2.138)$$

As a numerical example, we consider an EDFA pumped at $\lambda_p = 980$ nm, with cross sections shown in Figure 2.7. We have $\eta_p = 0$, since with that pump wavelength the EDFA operates as a three-level laser system (see Chapter 1), and from the figure we find $\eta_s(\lambda_{peak} = 1.53 \ \mu m) = 0.92$. We choose for the pump and peak signal absorption coefficients the values of $\alpha_p = 0.5$ dB/m and $\alpha_s(\lambda_{peak}) = 1$ dB/m, respectively, typical in Er-doped fibers (Chapter 4).

For an amplification regime free from amplifier self-saturation, we must simulate experimental conditions where the maximum gain is not greater than 25 dB. From Eq. (2.138) and the above values for $\eta_p$, $\eta_s(\lambda_{peak})$, $\alpha_p$ and $\alpha_s(\lambda_{peak})$, we find that a fiber length of $L = 27$ m gives a maximum gain $G_k(peak) = 25$ dB, therefore this set of parameters represents a valid example. The gain $G_k(peak)$ can be chosen as the only variable parameter; the corresponding values for the input/output pump powers $q_0$, $q_L$ are found from Eq. (2.135)–(2.137).

The EDFA gain spectrum $G_k(\lambda_k)$ corresponding to this example is plotted in Figure 2.9, for different values of the peak gain $G_k(peak) = -27$ dB to $+24$ dB, or input pump power $q_0 = 0$ to 64, as calculated analytically in Eq. (2.125). These gain spectra show that given a pump power $q_0$, there exists a spectral range over which the fiber is amplifying $(G > 1)$ and outside which the fiber is lossy $(G < 1)$. To the unpumped fiber $(q_0 = 0)$ corresponds the absorption spectrum $T(\lambda_k) = \exp(-\alpha_k L)$, which repre-

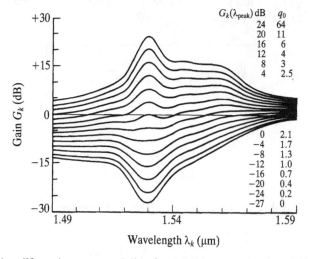

**FIGURE 2.9** Amplifier gain spectrum $G_k(\lambda_k)$ of an EDFA pumped at $\lambda_p = 980$ nm, with length $L = 27$ m, as calculated as calculated for $G_k(\lambda_{peak}) = -27$ dB to $+24$ dB from analytical solutions shown in Eqs. (2.125) and (2.135)–(2.137). Fiber parameters: $\eta_p = 0$, $\eta_s(\lambda_{peak} = 1.53 \ \mu m) = 0.92$, $\alpha_p = 0.5$ dB/m, $\alpha_p(\lambda_{peak}) = 1$ dB/m, absorption and emission cross sections shown in Figure 2.7.

sents the characteristic fiber transmission. As the pump power is increased, fiber transparency is followed by a regime of fiber amplification, starting from long wavelengths. This fact can be explained by the effect of ground level absorption, which at long wavelengths is smaller than at short wavelengths. Thus, at any point in the fiber, when the pump is gradually increased, the condition for a positive net gain coefficient $\gamma(z, \lambda_k) = a(z, \lambda_k) - b(z, \lambda_k) > 0$, is realized first for long wavelengths, as also shown in Figure 1.3. The net fiber gain $G_k$ is determined by integration of $\gamma(z, \lambda_k)$ along the fiber. Depending upon the signal wavelength $\lambda_k$, $G_k$ is positive (amplifying fiber); null (transparent fiber); or negative (lossy fiber). Figure 2.9 shows for instance that for a pump power $q_0 = 2.1$, the fiber is amplifying at all wavelengths $\lambda_k > 1.54\,\mu m$, and lossy at other wavelengths, except at $\lambda_{peak} = 1.53\,\mu m$ for which fiber transparency is achieved. A small increase of pump power to $q_0 = 2.5$ shifts the transparency point to the shorter wavelength $\lambda_k = 1.52\,\mu m$; the fiber amplifies at longer wavelengths. The transparency condition for a given wavelength $\lambda_k$ means that signal amplification along the fiber $(\gamma(z, \lambda_k) > 0)$ occurs up to a certain point $z = L_{opt}$, past which the signal is reabsorbed $(\gamma(z, \lambda_k) > 0)$. This results from the effect of pump decay or absorption along the fiber. At the highest pump power $(q_0 \to \infty)$, complete medium inversion is eventually realized at all wavelengths $\lambda_k$ over the whole fiber length $L$, and the fiber is amplifying $(G > 1)$ over the entire spectrum, as shown in Figure 2.9. Significant for the amplifier noise figure at intermediate pump powers $(0 < q_0 < \infty)$ is the fact that population inversion decays with the pump along the fiber.

The equivalent input noise factor $n_{eq}(\lambda_k)$ and optical noise figure $F_o(\lambda_k)$, as calculated analytically in Eqs. (2.132)–(2.137) and corresponding to the above example, are plotted in Figures 2.10, 2.11, and 2.12 for forward and backward propagation directions.

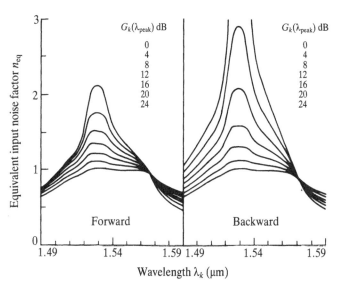

**FIGURE 2.10** Amplifier equivalent input noise factor spectrum $n_{eq}(\lambda_k)$ corresponding to the example in Figure 2.9 ($\lambda_p = 980\,nm$, $L = 27\,m$) shown for the forward and backward propagation directions, calculated for $G_k(\lambda_{peak}) = 0\,dB$ to $+24\,dB$ from analytical solutions shown in Eqs. (2.132), and (2.133), and (2.135)–(2.137).

Figure 2.10, shows plots of $n_{eq}(\lambda_k)$ for the cases where $G_k(\lambda_{peak}) > 1$. For both propagation directions, as the gain increases the equivalent input noise factor becomes nearly independent of signal wavelength, and tends towards unity over a broad (50 nm) spectral range, [40] and [44]. The limit $n_{eq}(\lambda_k) \approx n_{sp}(\lambda_k) \to 1$ corresponds to a regime of complete inversion along the fiber ($b(z, \lambda_k) = 0$). Such a regime can be

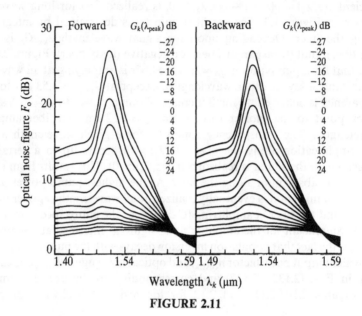

**FIGURE 2.11**

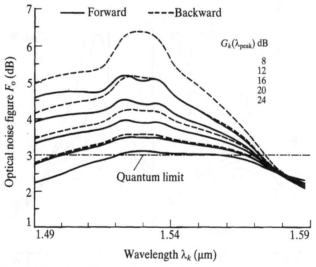

**FIGURES 2.11 and 2.12** Amplifier optical noise figure spectrum $F_o(\lambda_k)$ corresponding to the example in Figures 2.9 and 2.10 ($\lambda_p = 980$ nm, $L = 27$ m) for the forward and backward propagation directions, as calculated for $G_k(\lambda_{peak}) = -27$ dB to $+24$ dB from analytical solutions shown in Eqs. (2.132)–(2.137). The magnified scale shows the optical noise figure spectra for the highest gains $G_k(\lambda_{peak}) = +8$ dB to $+24$ dB, with values near the 3 dB quantum limit: (———) forward (– – –) backward.

achieved with a pump wavelength of 980 nm, for which the EDFA operates as a three-level system, as seen previously. In this regime, the output ASE noise is given by $N(\lambda_k) = n_{eq}(\lambda_k)G_k \approx G_k$ meaning that the output ASE noise corresponds, in average, to the amplification of a single photon. Figure 2.9 shows that the 50 nm spectral range over which $n_{eq}(\lambda_k) \approx 1$ correspond to fiber gains $G_k$ greater or nearly equal to 10 dB. In the case where the gain is lower ($G_k < 10$ dB), the equivalent input noise factor $n_{eq}(\lambda_k)$ vanishes as the signal wavelength is detuned from the gain peak. This may be expected as, at wavelengths away from the laser resonance, the doped fiber core is actually a passive medium.

A second observation from Figure 2.10 is the important difference between equivalent input noise factors in the two propagation directions. Indeed $n_{eq}(\lambda_k,$ forward) $< n_{eq}(\lambda_k,$ backward), i.e., the equivalent input noise is greater in the propagation direction opposite the pump. In the 50 nm wide region of interest, this difference is observed to vanish at high gains, i.e., when complete inversion is reached. This property, common to all types of end-pumped fiber amplifiers [42], is a consequence of the effect of pump decay and nonuniform inversion along the fiber, as was shown specifically for EDFAs in [45].

The plots of the optical noise figure $F_o(\lambda_k) = 2n_{eq}(\lambda_k) + 1/G_k(\lambda_k)$ shown in Figure 2.11 for $G_k(\lambda_{peak}) = -27$ dB to $+24$ dB reveal sizeable changes of $F_o$ with pumping conditions. For the unpumped fiber ($q_0 = 0$), the optical noise figure is the reciprocal of the fiber transmission, i.e., $F_o(\lambda_k) = 1/T(\lambda_k) = \exp(+\alpha_k L)$. As the fiber is pumped with increasing input power, corresponding to increasing gains and overall fiber inversion, the optical noise figure is seen to decrease for both forward and backward propagation directions. Plots of $F_o(\lambda_k)$ corresponding to the highest gains ($G_k(\lambda_{peak}) = +8$ dB to $+24$ dB) are shown in Figure 2.12. The figure indicates that for the highest gains the optical noise figure reaches a lower limit of 3 dB (quantum limit) over a broad spectral range (50 nm), and is very nearly wavelength independent, [44] and [45]. For this high gain regime, the difference between forward and backward optical noise figures is seen to vanish. Figures 2.11 and 2.12 show that in the pumping regime where $F_o \leqslant 3.5$ dB, the difference in noise figure between the two propagation directions remains small ($<0.1$ dB); at lower gains, it reaches several decibels (up to 4 dB). The difference also vanishes when $q_0 \to 0$, as for the unpumped fiber the optical noise figure is $F_o(\lambda_k) = 1/T(\lambda_k)$ for both propagation directions.

This leads to the conclusion that *in EDFAs, the optical noise figure is higher when the signal propagates in the direction opposite to the pump, compared to the case where signal and pump are copropagating.* Before analyzing in more detail this fundamental property of EDFAs, consider a second example where the same Er-doped fiber is pumped at $\lambda_p = 1.48 \mu$m, corresponding to the case where the EDFA is operated as a two-level laser system.

From the cross sections in Figure 2.7, we find $\eta_p(\lambda_p = 1.48 \mu$m$) = 0.28$, $\eta_s(\lambda_{peak}) = 0.92$, and, choosing the same peak absorption coefficient as previously, i.e., $\alpha_s(\lambda_{peak}) = 1$ dB/m, we find $\alpha_p = \alpha_s(\lambda_{peak})\sigma_a(\lambda_p)/\sigma_a(\lambda_{peak}) = 0.44$ dB/m. This last relation takes into account the fact that the mode overlaps with the doped core are nearly the same for pump and signal, according to Eqs. (1.120), and (1.121). As before, we choose a fiber length which yields a maximum gain of $G_k(peak) = 25$ dB; with the above parameters in Eq. (2.138), this length is $L = 50$ m.

The EDFA gain spectrum $G_k(\lambda_k)$ for this second example, as calculated analytically in Eq. (2.125), is plotted in Figure 2.13 for different values of the peak gain

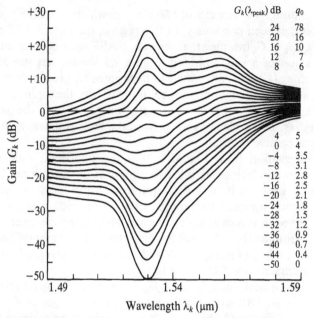

**FIGURE 2.13**   Amplifier gain spectrum $G_k(\lambda_k)$ of an EDFA pumped at $\lambda_p = 1480\,$nm, with length $L = 50\,$m, calculated for $G_k(\lambda_{peak}) = -50\,$dB to $+24\,$dB from analytical solutions shown in Eqs. (2.125) and (2.135)–(2.137). Fiber parameters: $\eta_p = 0.28$, $\eta_s(\lambda_{peak} = 1.53\,\mu m) = 0.92$, $\alpha_p = 0.44\,$dB/m, $\alpha_p(\lambda_{peak}) = 1\,$dB/m, with cross sections shown in Figure 2.7.

$G_k(\lambda_{peak}) = -50\,$dB to $+24\,$dB or input pump power $q_0 = 0$ to 78. The corresponding equivalent input noise factor $n_{eq}(\lambda_k)$ and optical noise figure $F_o(\lambda_k)$, as calculated analytically in Eqs. (2.132)–(2.137) are plotted in Figure 2.14, 2.15, and 2.16, for forward and backward propagation directions. The same observations as in the $\lambda_p = 980\,$nm pump case apply to this $\lambda_p = 1.48\,\mu m$ case, except for a few features. Comparison of Figures 2.9 and 2.13 shows that if the gain spectra are given by the same relation $G_k(\lambda_k) = \exp(-\alpha_k L)$ for both pump wavelengths when $q_0 = 0$, they have dissimilar features for $q_0 \to \infty$, corresponding in both cases to $G_k(\lambda_{peak}) = +24\,$dB. This is due to the fact that the population inversion at any point in the fiber, and at all pump power levels, is reduced by the effect of stimulated emission at the pump wavelength, proportional to $\sigma_e(\lambda_p) \approx \eta_p(\lambda_p) \neq 0$. The existence of stimulated emission at $\lambda_p$ results in a spontaneous emission factor always higher than unity, as imposed by Eq. (2.121). Consequently, the minimum possible optical noise figure obtained for $q_0 \to \infty$ is expected to be higher than the 3 dB quantum limit, as shown in Figures 2.15 and 2.16. The detailed plots in Figure 2.16 indicate that in the range of interest of $\lambda_s = 1.52$–$1.57\,\mu m$ for signal wavelengths where $G_k > 10\,$dB (Figure 2.13), the EDFA noise figure is $F_o = 3.5$ to 5 dB, with a value near 5 dB at the peak gain wavelength $\lambda_{peak} = 1.53\,\mu m$. The important result is that *in a fiber amplifier pumped as a two-level system, the minimum achievable optical noise figure is, in the high gain spectral region of interest, higher than the 3 dB quantum limit.*

Figures 2.11 and 2.15 also show that for signal wavelengths away from the peak gain region or laser resonance, the optical noise figure decreases to a limit of unity

($F_o = 0$ dB), which is the property of a passive optical medium. This fact illustrates that the notion of a 3 dB quantum limit for the optical noise figure only applies to the spectral region where high gains can be achieved, i.e., near the laser resonance, and is irrelevant for signal wavelengths in the region where the laser medium is passive. However, the notion of optical noise figure, as derived from the previous

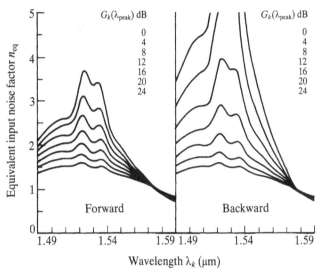

**FIGURE 2.14**  Amplifier equivalent input noise factor spectrum $n_{eq}(\lambda_k)$ corresponding to the example in Figure 2.13 ($\lambda_p = 1480$ nm, $L = 50$ m) for the forward and backward propagation directions, calculated for $G_k(\lambda_{peak}) = 0$ dB to $+24$ dB from analytical solutions shown in Eqs. [1] (2.132), (2.133) and (2.135)–(2.137).

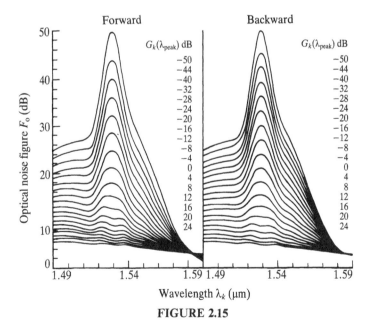

**FIGURE 2.15**

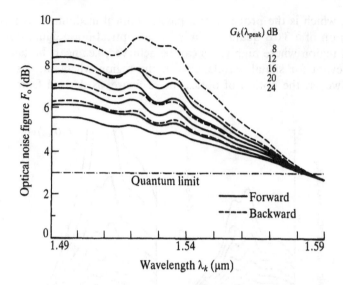

**FIGURE 2.15 and 2.16** Amplifier optical noise figure spectrum $F_o(\lambda_k)$ corresponding to the example in Figures 2.13 and 2.14 ($\lambda_p = 1480$ nm, $L = 50$ m) for the forward and backward propagation directions, as calculated for $G_k(\lambda_{peak}) = -50$ dB to $+24$ dB from analytical solutions shown in Eqs. (2.132)–(2.137). The magnified scale shows the optical noise figure spectra for the highest gains $G_k(\lambda_{peak}) = +8$ dB to $+24$ dB, with values near the 3 dB quantum limit: (———) forward (---) backward.

photon statistics analysis, is relevant to any optical medium, whether amplifying, transparent, or lossy at wavelengths near or away from a laser resonance. Although the optical noise figure can in fact take any value greater than unity, the 3 dB quantum limit is an important notion because most fiber amplifier systems operate in the high gain regime.

There is a physical explanation of the higher amplifier noise figure for propagation in the direction opposite to the pump compared to propagation in the same direction as the pump. Instead of studying the optical noise in a fiber amplifier where the emission and absorption coefficients $a(z)$, $b(z)$ continuously vary along the fiber length, it may be simpler to consider a chain of two amplifiers having different medium inversion.

Let $G_1$, $G_2$ be the gains of amplifiers 1 and 2, respectively, and $n_{eq1}$, $n_{eq2}$ the equivalent input noises when operated separately. We assume that the gain in either amplifier remains unsaturated. If amplifier 1 is the first in the chain, the respective noise outputs $N$ and $N'$ of amplifiers 1 and 2 are, in one polarization mode:

$$N = n_{eq1} G_1 \tag{2.139}$$

$$N' = G_2 N + n_{eq2} G_2 = n_{eq1} G_1 G_2 + n_{eq2} G_2 \tag{2.140}$$

The net gain of the amplifier chain is $G' = G_1 G_2$, and from definitions in Eqs. (2.118) and (2.123), the corresponding optical noise figure is $F' = (1 + 2N')/G'$, which from

Eqs. (2.139) and (2.140) takes the form:

$$F' = F_1 + \frac{F_2 - 1}{G_1} \tag{2.141}$$

where $F_1 = 2n_{eq1} + 1/G_1$ and $F_2 = 2n_{eq2} + 1/G_2$ are the optical noise figures of amplifiers 1 and 2, when operated separately. If the gain in the first amplifier is high enough (such that $G_1 \gg F_2 - 1$), the noise figure $F'$ of the amplifier chain is determined only by the noise figure of the first amplifier, i.e., $F' \approx F_1$. This result represents a well-known engineering theorem for amplifiers operated in tandem [1]. An important consequence of this theorem is that if the two amplifiers have different noise characteristics, i.e., $F_1 \neq F_2$, the resulting noise figure depends on the order in which the amplifiers are placed in the chain. *The lowest noise figure for the chain is obtained when the amplifier of lowest noise figure is placed first in the sequence.*

Equation (2.141) for tandem amplifiers is consistent with definition (2.118) for the amplifier noise figure, which takes the form $F_o = (1 + 2N)/G$. Equations (2.118) and (2.141) are valid only in the limit of large amplified signals, i.e., $G\langle n(0)\rangle \gg N$, for which the signal shot noise and signal–ASE beat noise are the dominant contributions to the total amplifier noise. In the general case, where the other noise contributions must be taken into account, an equation more accurate than (2.141) must be used, as shown in Section 2.5. But for simplicity and without loss of generality, we shall use Eq. (2.141) for a comparison of noise figures in the forward and backward propagation directions; this corresponds to the limit of large amplified signals.

We can now consider the case of a fiber amplifier in which the pump power decays with length. This condition can be viewed as being similar to that of two identical amplifiers operating in tandem, each having a different input pump power, i.e., $q_{01}$ and $q_{02}$, with $q_{01} > q_{02}$. The first amplifier has the lowest noise figure, since it has a higher input pump, and consequently, experiences a greater medium inversion, i.e., $n_{sp1}(q_{01}) < n_{sp2}(q_{02})$ and $F_1(q_{01}) < F_2(q_{02})$.

For a signal propagating along with the pump, the resulting noise figure of the tandem is given by Eq. (2.141), i.e., $F' = F_1(q_{01}) + \{F_2(q_{02}) - 1\}/G_1(q_{01}) = F_1 + (F_2 - 1)/G_1$. On the other hand, for a signal that propagates in the direction opposite to the pump, the noise figure is $F'' = F_2(q_{02}) + \{F_1(q_{01}) - 1\}/G_2(q_{02}) = F_2 + (F_1 - 1)/G_2$. The difference in noise figure $\Delta F = F'' - F'$ between the two propagation schemes can then be put in the form:

$$\begin{aligned}
\Delta F = F'' - F' &= F_2 + \frac{F_1 - 1}{G_2} - F_1 - \frac{F_2 - 1}{G_1} \\
&= F_2 \frac{G_1 - 1}{G_1} - F_1 \frac{G_2 - 1}{G_2} + \frac{1}{G_1} - \frac{1}{G_2} \\
&= \frac{2}{G_1 G_2} \{N_2(G_1 - 1) - N_1(G_2 - 1)\} \\
&= \frac{2N_1(G_2 - 1)}{G_1 G_2} \left\{ \frac{N_2(G_1 - 1)}{N_1(G_2 - 1)} - 1 \right\}
\end{aligned} \tag{2.142}$$

By replacing $N_1 = n_{sp1}(G_1 - 1)$ and $N_2 = n_{sp2}(G_2 - 1)$ in Eq. (2.142) we obtain finally:

$$\Delta F = \frac{2n_{sp1}(G_1 - 1)(G_2 - 1)}{G_1 G_2}\left(\frac{n_{sp2}}{n_{sp1}} - 1\right) \tag{2.143}$$

Since we have $n_{sp2} > n_{sp1}$, Eq. (2.143) shows that $\Delta F$ is always positive or $F'' > F'$, regardless of the values of $G_1$ and $G_2$. This result means that *the noise figure is always higher for the signal propagating in the direction opposite to the pump.*

An actual EDFA, in which the pump is continuously decaying along the fiber, can be viewed as a chain of discrete amplifiers with infinitesimal length $dz$, and for which $q_0(z + dz) < q_0(z)$ or $n_{sp}(z + dz) > n_{sp}(z)$. The difference in noise figure between the two propagation directions $\delta(dF) = dF'' - dF'$ is therefore always positive, and so is the noise figure difference $\Delta F$ as integrated along the fiber path; this can be checked numerically.

## 2.5   LUMPED AMPLIFIER CHAINS

In certain system applications, several optical amplifiers are operating in cascade, providing periodic signal regeneration. This section analyzes the optical noise figure characteristics of such a chain of discrete or lumped amplifiers. The case of distributed amplification, which represents the continuous limit of a lumped amplifier chain, is analyzed in the next section.

An important observation is that all results previously derived for the amplifier characteristics, namely the gain $G(z)$, ASE photon number $N(z)$, and spontaneous emission factor $n_{sp}(z)$, in Eqs. (2.30), (2.31), and (2.35); the output mean $\langle n(z) \rangle$ and variance $\sigma^2(z)$, in Eqs. (2.73) and (2.75), do apply to the most general case, without any restriction in the actual distribution along the fiber path of the emission and absorption coeffficients $a(z)$ and $b(z)$. Therefore, any chain of lumped amplifiers, whether or not it incorporates lossy elements between amplifiers, can be viewed as a single fiber amplifier whose emission and absorption coefficients follow a periodic (and even discontinuous) distribution along the length. Therefore, the optical noise figure of the amplifier chain at coordinate $z = L$ is given by the same equations, Eqs. (2.118) and (2.124), i.e.,

$$F_o(L) = \frac{1 + 2\mathcal{N}(L)}{\mathscr{G}(L)} = 2n_{eq}(L) + \frac{1}{\mathscr{G}(L)} \tag{2.144}$$

where $\mathscr{G}(L)$, $\mathcal{N}(L)$ and $n_{eq}(L)$ have their usual meaning but represent quantities integrated along the amplifier chain. Equation (2.144) is valid only in the high signal limit $\mathscr{G}(L)\langle n(0) \rangle \gg \mathcal{N}(L)$. The general case is derived later.

We consider first the two basic chain elements of length $L$ in Figure 2.17. The first (Type A) consist of 2 fiber amplifier followed by a strand of passive fiber. The second (Type C) consists of a strand of passive fiber followed by a fiber amplifier. We shall keep the term Type B for a chain element where amplification is distributed along the whole element length, as analyzed in the next section.

Let $l_0 \ll L$ be the fiber amplifier length, $a(z)$, $b(z)$ the emission and absorption coefficients at wavelength $\lambda_k$, and $G$, $N$ the gain and ASE photon number at this

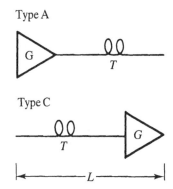

**FIGURE 2.17**  Two basic elements of an optical fiber amplifier chain: Type A, where an amplifier of gain $G$ precedes a strand of passive fiber with length $L$ and transmission $T = \exp(-\alpha_k L)$ at signal wavelength $\lambda_k$; Type C, where the fiber strand precedes the optical amplifier.

wavelength. In the passive or undoped portion of the chain, we have $a(z) = 0$ and $b(z) = \alpha_k'$, where $\alpha_k'$ is the background loss coefficient at $\lambda_k$. For Type A and C elements, the gain $\mathscr{G}(L)$ is found from Eq. (2.30) to be:

$$\mathscr{G}(L) = GT \tag{2.145}$$

where $T = \exp\{-\alpha_k'(L - l_0)\} \approx \exp\{-\alpha_k' L\}$ is the transmission of the passive fiber. For the ASE photon number $\mathscr{N}(L)$ it is found from Eq. (2.31):

$$\mathscr{N}(L) = TN \tag{2.146}$$

for Type A, and

$$\mathscr{N}(L) = N \tag{2.147}$$

for Type C. Equations (2.145)–(2.147) could also have been derived intuitively, considering the effect of fiber loss on the signal and ASE before or after amplification.

When the above results are replaced into the general formulae (2.73), (2.75), and (2.144), we obtain the output mean, output variance, and optical noise figure of both chain types, i.e., for Type A:

$$\langle n(L) \rangle = GT\langle n(0) \rangle + \mathscr{M}TN \tag{2.148}$$

$$\sigma^2(L) = G^2 T^2 [\sigma^2(0) - \langle n(0) \rangle] + GT\langle n(0) \rangle + \mathscr{M}TN + 2GT^2\langle n(0) \rangle N + \mathscr{M}T^2N^2 \tag{2.149}$$

$$\boxed{F_o(\text{Type A}) = \frac{1 + 2TN}{GT}} \tag{2.150}$$

and for Type C:

$$\langle n(L) \rangle = GT\langle n(0) \rangle + \mathcal{M}N \tag{2.151}$$

$$\sigma^2(L) = G^2 T^2 [\sigma^2(0) - \langle n(0) \rangle] + GT\langle n(0) \rangle + \mathcal{M}N + 2GT\langle n(0) \rangle N + \mathcal{M}N^2 \tag{2.152}$$

$$\boxed{F_o(\text{Type C}) = \frac{1 + 2N}{GT}} \tag{2.153}$$

The difference in optical noise figure between the two types can be expressed from Eqs. (2.150) and (2.153) as:

$$F_o(\text{Type C}) - F_o(\text{Type A}) = \frac{2N}{G}\left(\frac{1}{T} - 1\right) = 2n_{eq}(e^{\alpha'_k L} - 1) \tag{2.154}$$

This difference is always positive and increases exponentially with the element length $L$. It indicates the advantage of the Type A configuration, due to the absorbed fraction $(1 - T)$ of the ASE noise in the Type A chain. This yields a better output SNR for the same output signal $GT\langle n(0) \rangle$. A detailed comparison of optical noise figures of Types A, B (distributed amplifier chain), and C is made in the next section.

For the determination of the optical noise figure in the general case, where all noise terms in Eqs. (2.149) and (2.152) are taken into account, and in amplifier chains made of a succession of several elements of Types A and C, it is useful to obtain a formula that can be easily iterated. For this purpose, the output noise and variance in Eqs. (2.148), (2.149), (2.151), and (2.152) can be put into a more practical vector form (E. Desurvire *et al.* [46]):

$$\begin{pmatrix} \langle n(L) \rangle \\ \sigma^2(L) \end{pmatrix} = \hat{X}\begin{pmatrix} \langle n(0) \rangle \\ \sigma^2(0) \end{pmatrix} + \mathbf{Y} \tag{2.155}$$

In the above relation, $\hat{X}$ is a matrix and $\mathbf{Y}$ a vector, with the following definitions:

$$\hat{X}(\text{Type A}) = GT\begin{pmatrix} 1 & 0 \\ 1 - GT + 2NT & GT \end{pmatrix} \tag{2.156}$$

$$\hat{X}(\text{Type C}) = GT\begin{pmatrix} 1 & 0 \\ 1 - GT + 2N & GT \end{pmatrix} \tag{2.157}$$

$$\mathbf{Y}(\text{Type A}) = \mathcal{M}TN\begin{pmatrix} 1 \\ 1 + TN \end{pmatrix} \tag{2.158}$$

$$\mathbf{Y}(\text{Type C}) = \mathcal{M}N\begin{pmatrix} 1 \\ 1 + N \end{pmatrix} \tag{2.159}$$

We introduce the passive fiber transmission matrix $\hat{T}$, the amplifier transfer matrix $\hat{M}$, and the amplifier noise vector **N** through the definitions:

$$\hat{T} = T\begin{pmatrix} 1 & 0 \\ 1 - T & T \end{pmatrix} \tag{2.160}$$

$$\hat{M} = G\begin{pmatrix} 1 & 0 \\ 1 - G + 2N & G \end{pmatrix} \tag{2.161}$$

$$\mathbf{N} = \mathcal{M}N\begin{pmatrix} 1 \\ N + 1 \end{pmatrix} \tag{2.162}$$

With definitions (2.160)–(2.162), we can rewrite Eq. (2.155) in the form:

$$\begin{pmatrix} \langle n(L) \rangle \\ \sigma^2(L) \end{pmatrix} = \hat{T}\hat{M}\begin{pmatrix} \langle n(0) \rangle \\ \sigma^2(0) \end{pmatrix} + \hat{T}\mathbf{N} \tag{2.163}$$

for a Type A chain, and

$$\begin{pmatrix} \langle n(L) \rangle \\ \sigma^2(L) \end{pmatrix} = \hat{M}\hat{T}\begin{pmatrix} \langle n(0) \rangle \\ \sigma^2(0) \end{pmatrix} + \mathbf{N} \tag{2.164}$$

for a Type C chain.

With the above notations, the action of the amplifier is to transform the input vector $\mathbf{x} = (\langle n \rangle, \sigma^2)$ into the output $\mathbf{y} = \hat{M}x + N$, while the action of the passive fiber follows the transformation $\mathbf{y} = \hat{T}\mathbf{x}$. The matrices $\hat{M}$ and $\hat{T}$ do not commute, i.e., $\hat{M}\hat{T} \neq \hat{T}\hat{M}$.

Using definitions (2.160)–(2.162), the vector equations (2.163) and (2.164) can be easily iterated to yield:

$$\begin{pmatrix} \langle n_k \rangle \\ \sigma_k^2 \end{pmatrix}_{\text{Type A}} = [\hat{T}\hat{M}]^k\begin{pmatrix} \langle n_0 \rangle \\ \sigma_0^2 \end{pmatrix} + \sum_{j=1}^{k} [\hat{T}\hat{M}]^{j-1}\hat{T}\mathbf{N} \tag{2.165}$$

$$\begin{pmatrix} \langle n_k \rangle \\ \sigma_k^2 \end{pmatrix}_{\text{Type C}} = [\hat{M}\hat{T}]^k\begin{pmatrix} \langle n_0 \rangle \\ \sigma_0^2 \end{pmatrix} + \sum_{j=1}^{k} [\hat{M}\hat{T}]^{j-1}\mathbf{N} \tag{2.166}$$

where $(\langle n_0 \rangle, \sigma_0^2)$ and $(\langle n_k \rangle, \sigma_k^2)$ are the input vector at $z = 0$ and output vector at $z = kL$, respectively. Details of the computations of the sums in Eqs. (2.165) and (2.166) are shown in Appendix H. From the appendix, we find:

$$\left\{ \begin{array}{c} \langle n_k \rangle = (GT)^k\langle n_0 \rangle + \mathcal{M}TNx_k \\ \sigma_k^2 = (GT)^k[\sigma_0^2 - \langle n_0 \rangle] + (GT)^k[1 + 2TNx_k]\langle n_0 \rangle \\ + \dfrac{\mathcal{M}TNx_k}{1 + GT}\left\{ \dfrac{1 - GT + 2TN}{1 - GT}GT[1 - (GT)^{k-1}] + (1 + TN)[1 + (GT)^k] \right\} \end{array} \right\}_{\text{Type A}} \tag{2.167}$$

and

$$\left.\begin{cases} \langle n_k \rangle = (GT)^k \langle n_0 \rangle + \mathcal{M} N x_k \\ \sigma_k^2 = (GT)^k [\sigma_0^2 - \langle n_0 \rangle] + (GT)^k [1 + 2N x_k] \langle n_0 \rangle \\ + \dfrac{\mathcal{M} N x_k}{1 + GT} \left\{ \dfrac{1 - GT + 2N}{1 - GT} GT [1 - (GT)^{k-1}] + (1 + N)[1 + (GT)^k] \right\} \end{cases}\right\}_{\text{Type C}}$$

(2.168)

with

$$x_k = \frac{1 - (GT)^k}{1 - GT}$$

(2.169)

In the case $k = 1$ (only one element of Type A or C in the amplifier chain), for which $x_k = 1$, Eqs. (2.167) and (2.168) reduce to Eqs. (2.148) and (2.149), and Eqs. (2.151) and (2.152), respectively.

In the case $k > 1$ ($k$ elements of Type A or C in the amplifier chain), there exist several possibilities, depending on the magnitude of the factor $GT$. For $GT < 1$ or $GT > 1$, the amplified signal rapidly decreases or increases along the amplifier chain as $(GT)^k$; this is of limited interest. For $GT = 1$, each amplifier in the chain exactly compensates the fiber loss in the preceding or following section; this is of major interest. Taking the limit $GT \to 1$ or $GT = 1 + \varepsilon$ with $\varepsilon \to 0$ in Eqs. (2.167)–(2.169) yields:

$$\boxed{\begin{aligned} \langle n_k \rangle &= \langle n_0 \rangle + \mathcal{M} k n_{\text{eq}} \\ \sigma_k^2 &= \sigma_0^2 - \langle n_0 \rangle + (1 + 2k n_{\text{eq}}) \langle n_0 \rangle + \mathcal{M} k n_{\text{eq}} (1 + k n_{\text{eq}}) \end{aligned}} \quad \text{Type A} \quad (2.170)$$

and

$$\boxed{\begin{aligned} \langle n_k \rangle &= \langle n_0 \rangle + \mathcal{M} k G n_{\text{eq}} \\ \sigma_k^2 &= \sigma_0^2 - \langle n_0 \rangle + (1 + 2k G n_{\text{eq}}) \langle n_0 \rangle + \mathcal{M} k G n_{\text{eq}} (1 + k G n_{\text{eq}}) \end{aligned}} \quad \text{Type C} \quad (2.171)$$

where $n_{\text{eq}} = N/G$ is the equivalent input noise factor.

The output SNR of the amplifier chain is given by $\text{SNR}_o = \langle n_0 \rangle^2 / \sigma_k^2$, since the output signal is $\langle n_0 \rangle$. Then, according to Eq. (2.115) with the input SNR being $\text{SNR}_o(0) = \langle n_0 \rangle$ (input signal with Poisson statistics) and using Eqs. (2.170) and (2.171), optical noise figures $F_0(k)$ of Type A and Type C amplifier chains with $k$ elements take the form:

$$\boxed{F_0(k) = 1 + 2k n_{\text{eq}} + \mathcal{M} k \frac{n_{\text{eq}}}{\langle n_0 \rangle} (1 + k n_{\text{eq}})}$$

(2.172)

For Type A, and

$$F_o(k) = 1 + 2kGn_{eq} + \mathcal{M}kG \frac{n_{eq}}{\langle n_0 \rangle}(1 + kGn_{eq})$$    (2.173)

for Type C.

The important result in Eqs. (2.172) and (2.173) is that for amplifier chains operating at transparency ($GT = 1$), the optical noise figure increases with the number of elements $k$. The increase is linear as long as $k/\langle n_0 \rangle \ll 1$ for Type A chains, and $kG/\langle n_0 \rangle \ll 1$ for Type C chains, and is due to the signal–ASE beat noise component (equal to $2kn_{eq}$ or $2kGn_{eq}$, respectively). The increase becomes quadratic in the region of the chain past a certain number of elements where these conditions are no longer met, and is due to the ASE–ASE beat noise component (equal to $\mathcal{M}(kn_{eq})^2/\langle n_0 \rangle$ for Type A, and $\mathcal{M}(kGn_{eq})^2/\langle n_0 \rangle$ for Type C).

Another important result is that the optical noise figure of a Type C chain is always greater than that of a Type A chain, because of the multiplicative factor $G$ in Eq. (2.173), which is not in Eq. (2.172). Comparing these two equations, it can be stated that *the noise figure of a Type C chain with k elements is the same as that of a Type A chain with kG identical elements (assuming G integer), or with k elements including amplifiers of equivalent input noise G times greater.* Equation (2.123) shows that $n_{eq} = n_{sp}(1 - 1/G)$, the rate of increase in noise figure depends upon the amplifier gain $G$, and more specifically, of the *amplifier spacing* $L$, as $G = 1/T = \exp(+\alpha_k L)$. The next section gives more details on noise figure increase versus amplifier spacing for all three types of amplifier chains.

So far, we have considered the case of optical amplifier chains operating at transparency, i.e., $GT = 1$, where the characteristics of all elements in the chain are identical. Consider the more general case where the chain element $i$ can have different gain and fiber length characteristics $G_i$ and $T_i$. To each element $i$ can then be associated a fiber transmission matrix $\hat{T}_i$, and amplifier transfer matrix $\hat{M}_i$ and a noise vector $\mathbf{N}_i$, defined in Eqs. (2.160)–(2.162). Computation of the output mean and variance ($\langle n_k \rangle, \sigma_k^2$) can be done by iteration of Eq. (2.163) or Eq. (2.164) for each type of amplifier chain. As the output signal is $G_1 T_1 \ldots G_k T_k \langle n_0 \rangle$, the optical noise figure is then given by $F_o(k) = \sigma_k^2/[(G_1 T_1 \ldots G_k T_k)^2 \langle n_0 \rangle]$. This iteration is difficult to perform and does not lead to a simple formula for ($\langle n_k \rangle, \sigma_k^2$) and the noise figure $F_o(k)$. A general formula for $F_o(k)$ can be obtained by a different method detailed in Appendix I. The corresponding definitions and results are summarized below.

From Eqs. (2.112) and (2.115), the optical noise figure $F_i$ of an individual amplifier $i$ in a chain of amplifiers can be first expressed as:

$$F_i = \frac{SNR_{in}(i)}{SNR_{out}(i)} = \frac{\langle n_{i-1}(\text{signal}) \rangle^2}{\sigma_{i-1}^2} \frac{\sigma_i^2}{\langle n_i(\text{signal}) \rangle^2} = \frac{\sigma_i^2}{\sigma_{i-1}^2 G_i^2}$$    (2.174)

where we used the property $\langle n_i(\text{signal}) \rangle = G_i \langle n_{i-1}(\text{signal}) \rangle$ for the signal amplified throughout the chain. *The noise figure of an individual amplifier in the chain is not known a priori,* since according to Eq. (2.174) it is a function of the variance $\sigma_{i-1}^2$ of its input signal. For a chain of $k$ amplifiers, we obtain from Eq. (2.174):

$$F_o(k) = \frac{\langle n_0(\text{signal})\rangle^2}{\sigma_0^2} \frac{\sigma_k^2}{\langle n_k(\text{signal})\rangle^2} = \frac{\sigma_1^2}{G_1^2 \sigma_0^2} \cdots \frac{\sigma_{k-1}^2}{G_{k-1}^2 \sigma_{k-2}^2} \frac{\sigma_k^2}{G_k^2 \sigma_{k-1}^2} = F_1 \ldots F_{k-1} F_k$$

(2.175)

Appendix I shows that the optical noise figure of each amplifier $i$ can be put into the form

$$F_i = \frac{\langle n_{i-1}\rangle}{\sigma_{i-1}^2} \{ f_{i-1} - 1 + F_i^{\text{Poisson}} \}$$

(2.176)

where $f_{i-1} = \sigma_{i-1}^2/\langle n_{i-1}\rangle$ is the Fano factor, Eq. (2.107), and $F_i^{\text{Poisson}}$ is what we define as the *Poisson optical noise figure*, with expression:

$$F_i^{\text{Poisson}} = \frac{1 + 2n_{\text{eq}i} G_i}{G_i} + \mathcal{M} \frac{n_{\text{eq}i}}{\langle n_{i-1}\rangle} \frac{1 + n_{\text{eq}i} G_i}{G_i}$$

(2.177)

where $n_{\text{eq}i}$ is the equivalent input noise factor of amplifier $i$.

In the case where the signal input to amplifier $i$ has Poisson statistics, we have $f_i = 1$ and $F_i = F_i^{\text{Poisson}}$ and Eqs. (2.176) and (2.77) are identical. But in the case of an amplifier chain for $i > 1$, this condition does not hold, since the statistics of the signal output from each amplifier depart from Poisson, as shown in Section 2.3.

Using the above definitions, the optical noise figure $F_o(2)$ of a chain of two amplifiers in tandem is given by (Appendix I):

$$F_o(2) = \frac{\langle n_0\rangle}{\sigma_0^2} \left\{ f_0 - 1 + F_1^{\text{Poisson}} + \left(1 + \mathcal{M} \frac{n_{\text{eq}1}}{\langle n_0\rangle}\right) \frac{F_2^{\text{Poisson}} - 1}{G_1} \right\}$$

(2.178)

The optical noise figure $F_o(k)$ of a chain of $k$ amplifiers is given by the following general formula:

$$F_o(k) = \frac{\langle n_0\rangle}{\sigma_0^2} \left\{ f_0 - 1 + F_1^{\text{Poisson}} + g_1 \frac{F_2^{\text{Poisson}} - 1}{G_1} + g_1 g_2 \frac{F_3^{\text{Poisson}} - 1}{G_1 G_2} \right.$$
$$\left. + \cdots + g_1 g_2 \cdots g_{k-1} \frac{F_k^{\text{Poisson}} - 1}{G_1 G_2 \cdots G_{k-1}} \right\}$$

(2.179)

with

$$g_i = \left(1 + \mathcal{M} \frac{n_{\text{eq}i}}{\langle n_{i-1}\rangle}\right)$$

(2.180)

and

$$\langle n_i \rangle = G_1 G_2 \cdots G_i \langle n_0 \rangle + \mathcal{M}\{G_1 G_2 \cdots G_i n_{\mathrm{eq}1} + G_2 \cdots G_i n_{\mathrm{eq}2} + \cdots + G_i n_{\mathrm{eq}i}\}$$

(2.181)

In the case where the total signal input to each amplifier is large, i.e., $\langle n_{i-1} \rangle \gg n_{\mathrm{eq}i}$, we have $g_i \approx 1$ and Eq. (2.179) reduces to:

$$F_o(k) = \frac{\langle n_0 \rangle}{\sigma_0^2}\left\{ f_0 - 1 + F_1^{\mathrm{Poisson}} + \frac{F_2^{\mathrm{Poisson}} - 1}{G_1} + \frac{F_3^{\mathrm{Poisson}} - 1}{G_1 G_2} + \cdots + \frac{F_k^{\mathrm{Poisson}} - 1}{G_1 G_2 \cdots G_{k-1}} \right\}$$

(2.182)

In the case where (1) the total signal input to each amplifier is large ($\langle n_{i-1} \rangle \gg n_{\mathrm{eq}i}$) and (2) the signal input to the amplifier chain has Poisson statistics ($f_0 = 1$), we obtain from Eq. (2.182):

$$F_o(k) = F_1^{\mathrm{Poisson}} + \frac{F_2^{\mathrm{Poisson}} - 1}{G_1} + \frac{F_3^{\mathrm{Poisson}} - 1}{G_1 G_2} + \cdots + \frac{F_k^{\mathrm{Poisson}} - 1}{G_1 G_2 \cdots G_{k-1}}$$

(2.183)

which is the usual formula found in the literature [7], used in Section 2.5 for $k = 2$.

The above results apply to any type of amplifier chain, including the case where the amplifiers are separated by strands of lossy fiber (Types A or C). Each fiber strand can then be viewed as a special case of an optical amplifier having a gain less than unity ($G_i < 1$) and no spontaneous emission noise ($n_{\mathrm{eq}i} = 0$). In the case of a Type A chain, all the terms with odd indices ($i$) correspond to an amplifier with gain $G$ and noise factor $n_{\mathrm{eq}}$, while the even index terms correspond to a fiber strand with gain or transmission $T = 1/G$ and noise factor $n_{\mathrm{eq}} = 0$. For Type A and Type C chains, the most general Eq. (2.179) for the noise figure $F_o(k)$ reduces with a Poisson input signal to eqs. (2.172) and (2.173), respectively, obtained through the transfer matrix method.

Whether considering Eq. (2.179), which corresponds to the most general case, or considering the specific cases of a large throughput signal with or without Poisson statistics input, Eq. (2.182) and (2.183), it is clear that the optical noise figure of the chain strongly depends on the noise figure $F_1^{\mathrm{Poisson}}$ of the first element, as the contributions to the noise figure from the other amplifier elements are proportional $1/G_1 \ldots G_i$. For Type A and Type C chains made of identical elements alternating with gain $G$ and transmission $T = 1/G$, we have from Eq. (2.177) $F_1^{\mathrm{Poisson}}$(Type A) $= 2n_{\mathrm{eq}} + 1/G + (n_{\mathrm{eq}} + 1/G)\mathcal{M}n_{\mathrm{eq}}/\langle n_0 \rangle$ and $F_1^{\mathrm{Poisson}}$(Type C) $= 1/T = G$, respectively. It is clear then that if the amplifier spacing is sufficiently large, such that $G \gg 1$, the optical noise figure of a Type C chain is much greater than that of a Type A chain.

## 2.6  DISTRIBUTED AMPLIFIERS

In the previous section, we defined the type A chain as a chain made of identical elements where the amplifier precedes the lossy fiber strand, and the type C chain as that where the amplifier follows the lossy fiber strand. Both types were referred to as lumped amplifier chains. Consider the possibility that the length of the fiber amplifier is relatively long, e.g., a few kilometers. In this case, it can also be used as the transmission fiber itself. Such a fiber, for which the gain is distributed along the whole transmission length, is called a *distributed amplifier*. The gain of a distributed amplifier is ideally unity. Distributed amplifiers may be cascaded to form a chain of longer span. A chain of distributed amplifiers has a Type B configuration that represents the limit of Type A and Type C lumped amplifier chains where the amplifiers are closely spaced so the lumped gain in each amplifier is small. A Type B amplifier chain can also be viewed as a special case of a single distributed fiber amplifier in which the pump varies periodically, so that to maintain a constant average gain ($G = 1$) along the fiber. We show in this section that the noise figure characteristics of distributed amplifiers and Type B chains are actually intermediate to that of Type A and Type C elements and associated chains.

A distributed amplifier can be viewed as being a strand of lossless fiber. But this view is not quite accurate, since there are two ways to achieve fiber transparency (no net loss). One way is to cancel the fiber background loss at any point along the fiber. This implies uniform medium inversion along the fiber length, which requires (1) either a constant pumping power, $P_p(z) = P_p(0)$; or (2) a pump power high enough such that the effect of pump absorption along the fiber on the medium inversion is negligible. The other way to achieve transparency is to allow the gain increase with length to reach a maximum then decrease, or vice versa, so the net gain at $z = L$ is unity. A condition of gain decrease or fiber absorption (in addition to the background loss) can be realized by pumping the fiber in the regime that is below population inversion. As the pump power is absorbed along the fiber length, regimes of increasing or decreasing population inversion (corresponding to increasing or decreasing gain) can be respectively achieved by forward or backward pumping, as illustrated in Figure 2.18.

The configuration of gain increase followed by gain decrease or signal reabsorption (forward pumping) is similar to that of Type A, in which loss follows amplification, while that of gain decrease followed by gain increase (backward pumping) is similar to that of Type C, in which loss precedes amplification. Another possibility is bidirectional pumping, in which the pump is supplied from both fiber ends. In this case, assuming equal pump powers at both ends, the net gain increases, decreases, then increases again. The bidirectionally pumped configuration is therefore similar to a combination of Type A and Type C elements. Independently of the pumping configuration, the noise characteristics of distributed amplifiers fall between those of Type A and Type C elements [47]–[49].

The signal power excursion represents, by definition, the maximum signal power reached at some point in the chain element (whether lumped or distributed) normalized to the input power. If $P(z)$ is the signal power at fiber coordinate $z$, the path averaged signal power $\langle P \rangle_{\text{path}}$ represents the mean value:

$$\langle P \rangle_{\text{path}} = \frac{1}{L} \int_0^L P(z)\, dz \tag{2.184}$$

With lumped chain elements, we have $P(z) = GP(0)T(z)$ for Type A and $P(z) = P(0)T(z)$ for Type C, with $T(z) = \exp(-\alpha_k' z)$ at signal wavelength $\lambda_k$, and the

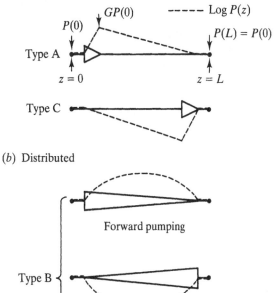

*(a)* Lumped

*(b)* Distributed

**FIGURE 2.18**  Five basic elements for transparent optical fiber amplifier chains: *(a)* lumped amplifier chain elements: Type A, where the amplifier (gain $G$) precedes the strand of passive fiber (transmission $1/G$), and Type C, where the transmission fiber precedes the amplifier; *(b)* distributed amplifier chain elements of Type B, with forward, backward or bidirectional pumping. The dashed line shows the log of the signal power versus length.

transparency condition $GT(L) = 1$. With distributed chain elements (Type B), we have $P(z) = G(z)P(0)$, where $G(z)$ is the gain distribution along the fiber, which depends on the pumping configuration, and verifies the boundary conditions $G(0) = G(L) = 1$. Using Eq. (2.184) we find for Type A and Type C the path averaged signal powers:

$$\langle P \rangle_{\text{path}} = P(0)\,\frac{G-1}{\log G} \quad \text{(Type A)} \tag{2.185}$$

$$\langle P \rangle_{\text{path}} = P(0)\,\frac{G-1}{G \log G} \quad \text{(Type C)} \tag{2.186}$$

The above result shows that for Type C chains, the path averaged signal power is smaller than for Type A chains by a factor $1/G$. If the amplifier spacing is large (e.g., $L > 50$ km, which gives $G = 1/T > 10$ dB for $\alpha'_k(\lambda_k = 1.5 \ \mu\text{m}) = 0.2$ dB/km) the difference in path averaged signal powers between the two types can be important.

Detrimental effects of Fiber nonlinearities increase nonlinearly with signal power (Chapter 7) so it is desirable from a system standpoint to reduce the path averaged power to the lowest level possible. This requirement is particularly stringent in the case of verly long-haul transmission systems as the weak effect of nonlinearity is enhanced by the considerable length (e.g., $kL = 10,000$ km) of transmission fiber. Equations (2.185) and (2.186) show that while for lumped amplifier chains, Type C offers the lowest possible path averaged power; the difference with a Type A chain is not very important when the gain $G$ is small ($G < 10$ dB). This condition corresponds to the case where the amplifier spacing is relatively short ($L < 50$ km).

Equations (2.185) and (2.186) show that with lumped amplifier chains, the path averaged power exclusively depends upon the gain $G$, which is itself both a function of the length $L$ of the chain element and the fiber background loss coefficient $\alpha'_k = \log G/L$. For actual distributed amplifiers, the path averaged power is not exclusively determined by the parameters $\alpha'_k$ and $L$, as the gain distribution along the fiber also depends upon the Er-doping concentration and pumping power [47]. The relation between pump power for fiber transparency and Er-concentration, [49] and [50], is analyzed below. As expected, both Er concentration and pumping power for transparency strongly affect the noise characteristics of actual distributed amplifiers.

For simplicity, we shall first consider an ideal distributed amplifier in which medium inversion along fiber length is uniform. For such an amplifier, the signal power excursion is zero, and the path averaged power is $P(0)$, which falls in between Type A and Type C. Taking into account the signal background loss, the constant net gain coefficient is $a - b - \alpha'_k$; Eqs. (2.30) and (2.31) give the gain and ASE noise at coordinate $z$: $G(z) = \exp[(a - b - \alpha'_k)z]$ and

$$N(z) = \{\exp[(a - b - \alpha'_k)z] - 1\}a/(a - b - \alpha'_k).$$

As $G(z) = 1$ for all $z$, we have $a - b - \alpha'_k = 0$ and $N(L) = aL$. We can also assume that the medium inversion is complete, so that $b = 0$ (no ground level population), and in this case $N(L) = \alpha'_k L$. From Eq. (2.118), the optical noise figure is:

$$F_o = 1 + 2\alpha'_k L \qquad \text{(Type B, ideal)} \tag{2.187}$$

Compare this result with those for Type A and Type C chain elements, [48] and [49], assuming the ideal case of full medium inversion ($n_{sp} = 1$). For lumped amplifiers, the ASE noise output is $N = n_{sp}(G - 1) = G - 1$. For Type A, we have $N(L) = TN = N/G = 1 - 1/G = 1 - T$; for Type C, we have $N(L) = G - 1$. The corresponding optical noise figures are $F_o = [1 + 2N(L)]/GT$, or

$$F_o = 1 + 2[1 - \exp(-\alpha'_k L)] \qquad \text{(Type A, ideal)} \tag{2.188}$$

$$F_o = 1 + 2[\exp(\alpha'_k L) - 1] \qquad \text{(Type C, ideal)} \tag{2.189}$$

For this type of study, instead of the optical noise figure $F_o$, we can consider the equivalent input noise factor $n_{eq}$ defined in Eq. (2.123), also called the excess noise factor [51], as was done in [48].

The optical noise figures corresponding to Types A, B, and C, as defined in Eqs. (2.187)–(2.189), are plotted in Figure 2.19 [49] assuming a fiber background loss of $\alpha'_k = 0.2$ dB/km. The Type A configuration yields the lowest optical noise figure, while for the distributed amplifier (Type B) the noise figure is intermediate between Type A and Type C. For fiber lengths shorter than $L = 10$ km, the noise figure difference between the three types is less than 0.5 dB and vanishes for $L \to 0$, which for Types A and C corresponds to the distributed amplifier limit. At the absorption length $L_{abs} = 1/\alpha'_k = 22$ km, the noise figure difference is about 1 dB between Types B and A, and 1.5 dB between Types C and B [52].

The noise figure of Type A is always smaller than the noise figure of Type C. This is because the lossy fiber following the amplifier attenuates the ASE noise. For Type C, the lossy fiber is placed before the amplifier and the ASE noise is not attenuated. For the distributed amplifier, the ASE noise generated in the beginning portion of the fiber length experiences the most attenuation, while the noise generated in the last portion experiences the least attenuation. For the same total fiber length, the noise figure of the distributed amplifier must therefore fall in between the noise figure of Types A and C. *The notion of a minimum 3 dB quantum limit for the amplifier noise figure,* discussed in Section 2.4, *is not relevant in the case of amplifier chains.* The term excess noise factor for the parameter $n_{eq}$ is actually not justified for distributed amplifiers, as it can take any value greater than zero, as opposed to the case of high gain amplifiers where $n_{eq} \geqslant 1$ (Section 2.4).

Another type of distributed amplifier is the *Raman fiber amplifier* (RFA). The RFA is based on the nonlinear effect of stimulated Raman scattering (SRS) in fibers [53] (for a review of RFAs, see [54], for instance). In RFAs, the undoped glass fiber core represents the gain medium; the stimulated emission coefficient is proportional to the pump power [53]. In the unsaturated gain regime, and for forward pumping, the pump power decays exponentially with length, i.e., $P_p(z) = P_p(0)\exp(-\alpha'_p z)$, and the net gain coefficient at the signal wavelength can be expressed as $g(z) =$

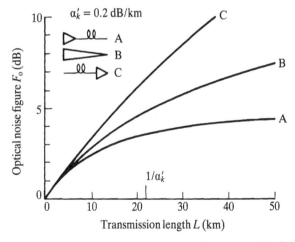

**FIGURE 2.19** Optical noise figure $F_o$ of Type A (lumped), Type B (distributed), and Type C (lumped) amplifier chain elements versus fiber length $L$, assuming the ideal case of complete medium inversion along the fiber, and a background loss coefficient $\alpha'_k = 0.2$ dB/km. From [49] © 1992 IEEE.

$a_0 \exp(-\alpha'_k z) - \alpha'_k$, assuming $\alpha'_p \approx \alpha'_k$. For backward pumping, we have likewise $g(z) = a_0 \exp[-\alpha'_k(L - z)] - \alpha'_k$. After integration, the RFA signal gain is given by $G = \exp(a_0 L_{\text{eff}} - \alpha'_k L)$, where $L_{\text{eff}} = [1 - \exp(-\alpha'_k L)]/\alpha'_k$ is an effective length accounting for pump decay, [53] and [54]. A calculation of the RFA output ASE noise yields the equivalent noise factors $n^{\pm}_{\text{eq}}$ corresponding to the two pumping configurations [42]:

$$n^+_{\text{eq}} = \frac{a_0}{\alpha'_k} \exp\left(-\frac{a_0}{\alpha'_k}\right)\left\{ Ei\left(\frac{a_0}{\alpha'_k}\right) - Ei\left[\frac{a_0}{\alpha'_k} \exp(-\alpha'_k L)\right]\right\} \tag{2.190}$$

$$n^-_{\text{eq}} = 1 - \frac{1}{G} + \frac{\alpha'_k}{a_0}\left\{ \exp(\alpha'_k L) - \frac{1}{G}\right\} \tag{2.191}$$

where $Ei(x)$ is the exponential integral function [31]. With the transparency condition $G = 1$, or $a_0 = \alpha'_k L/L_{\text{eff}}$, we obtain from Eqs. (2.190) and (2.191) the optical noise figures $F^{\pm}_o = 1 + 2n^{\pm}_{\text{eq}}$:

$$F^+_o = 1 + 2\frac{L}{L_{\text{eff}}} \exp\left(-\frac{L}{L_{\text{eff}}}\right)\left\{ Ei\left(\frac{L}{L_{\text{eff}}}\right) - Ei\left[\frac{L}{L_{\text{eff}}} \exp(-\alpha'_k L)\right]\right\} \tag{2.192}$$

$$F^-_o = 1 + 2\frac{L_{\text{eff}}}{L}\left\{ \exp(\alpha'_k L) - 1\right\} \tag{2.193}$$

The optical noise figures of forward pumped and backward pumped Raman fiber amplifiers are plotted versus fiber length in Figure 2.20, along with those of Types A, B, and C chain elements. As expected, the backward pumped RFA has a noise figure between that of Type C and the ideal distributed amplifier (Type B); the forward

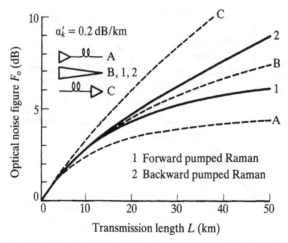

**FIGURE 2.20**   Optical noise figure $F_o$ of forward pumped and backward pumped Raman fiber amplifiers versus fiber length $L$, assuming a background loss coefficient $\alpha'_k = 0.2$ dB/km. The noise figures of Type A, B, and C ideal amplifier chain elements of Figure 2.19 are also shown for comparison.

pumped RFA noise figure is between that of Type A and Type B. These facts are explained by the Raman gain dependence with length; for the forward pumped RFA, the gain coefficient decays with length, and the gain coefficient is the greatest at the beginning of the fiber; the contrary applies to the backward pumped amplifier. Thus, the forward pumped RFA noise characteristics must fall between those of the ideal distributed amplifier and those of a lumped amplifier followed by a lossy fiber; the same type of reasoning applies to the backward pumped RFA.

The preceding has shown that noise figure increases with fiber length for all amplifier types and finally depends upon the net gain coefficient variation along the fiber length. We focus now upon distributed EDFAs and consider the actual case where the pump power, and consequently the gain coefficient, vary along the fiber.

Chapter 1 describes a model to account for fiber background loss in EDFAs. It is adequate to describe distributed EDFAs provided the Er-doping is confined near the center of the fiber core. The relations between the net fiber gain $G_k$ and $\lambda_k$ and the input or output pump powers $q_0^{\pm}, q_L^{\pm}$ are provided in Eqs. (1.130)–(1.32). Rewriting these relations with the condition $G_k = 1$ yields the pump power required for transparency:

$$q_0^+ = q_L^- = \frac{1 + \varepsilon_p}{\varepsilon_p} e^{BL} \frac{e^{AL} - 1}{e^{BL} - 1} \tag{2.194}$$

with

$$A = \varepsilon_p \alpha_p \frac{1 + \eta_p}{1 + \eta_k} (1 + \varepsilon_k) \tag{2.195}$$

$$B = \alpha_p (1 + \varepsilon_p) \frac{1 + \eta_p}{1 + \eta_k} \left( \frac{\eta_k - \eta_p}{1 + \eta_p} - \varepsilon_k \right) \tag{2.196}$$

and $\varepsilon_{p,k} = \alpha'_{p,k} / \alpha_{p,k}$ are the ratio of the background loss coefficient to the Er-doping absorption coefficient at the pump and the signal wavelengths.

The parameter $A$ is always positive. From Eq. (2.194), if $q_0^+, q_L^- > 0$, then $B > 0$. And $B > 0$ imposes a condition on the Er-doping absorption coefficient $\alpha_k$; it must be strictly greater than an asymptotical value $\alpha_k^*$ [50], i.e.,

$$\boxed{\alpha_k > \alpha_k^* = \alpha'_k \frac{1 + \eta_p}{\eta_k - \eta_p} = \frac{\alpha'_k}{U_k}} \tag{2.197}$$

In the above equation, $U_k = (\eta_k - \eta_p)/(1 + \eta_p)$ is a fundamental EDFA parameter, previously defined in Eq. (1.61). The existence of this an asymptotical value for $\alpha_k$ can be explained by integrating Eq. (1.126) for the signal at $\lambda_k$ in the case of unidirectional pumping to obtain another expression for the transparency condition $G_k(L) = 1$:

$$\alpha_k \int_0^L \frac{U_k q(z) - 1}{1 + q(z)} dz - \alpha'_k L = 0 \tag{2.198}$$

or

$$\alpha_k = \frac{\alpha'_k}{\frac{1}{L}\int_0^L \frac{U_k q(z) - 1}{1 + q(z)}\, dz} \tag{2-199}$$

The minimum value for the Er-doping coefficient $\alpha_k$ corresponds then to the maximum possible value for the integral in Eq. (2.199), which is equal to $U_k$ when $q(z) \to \infty$. Therefore, we have $\alpha_k(\text{min}) = \alpha_k^* = \alpha'_k/U_k$, which is the result of Eq. (2.197). The limit $q(z) \to \infty$ corresponds to a condition of uniform inversion along the fiber length, for which the constant gain coefficient is $\alpha_k U_k - \alpha'_k$, which vanishes for $\alpha_k = \alpha'_k/U_k$. In the limit where full inversion can be achieved (three-level system, or $\eta_p = 0$), we have $U_k = \eta_k$, and $\eta_k \alpha_k = \alpha'_k$; in this case, the stimulated emission coefficient $a = \eta_k \alpha_k$ exactly equals the fiber background loss coefficient. The minimum value $\alpha_k^*$ is independent of fiber length.

Plots of the input pump power required for transparency versus Er-doping absorption coefficient are shown in Figure 2.21, for different lengths $L = 10$–$25$ km [50]. The pump and signal wavelengths are $\lambda_p = 1.48\ \mu$m, $\lambda_k = \lambda_s = 1.53\ \mu$m, the fiber background loss coefficient $\alpha'_k = \alpha'_p = 0.5$ dB/km, and typical Er fiber parameters are assumed, i.e., $\eta_p = 0.37, \eta_s = 1$, for which $U_k = 0.46, \alpha_k^* = 1.09$ dB/km, and $\alpha_p = 0.4\alpha_s$.

As expected, the pump increases to very large values ($> 50$) as the Er-doping absorption coeffficient $\alpha_k$ becomes close to $\alpha_k^*$. On the other hand, for large Er-doping coefficients ($\alpha_k > 10$ dB/km), the required pump power also to increases; this becomes more important at longer fiber lengths $L$. The pump is more rapidly absorbed along the length of fibers with a high Er-doping absorption coefficient, therefore more pump power is necessary to achieve a given degree of medium inversion. For a given Er-doping absorption, the required pump power starts to increase when the total fiber length $L$ represents more than ten times the absorption length $1/\alpha_k$. A very long length ($L \gg 1/\alpha_k$) of highly doped fiber would therefore require a very high pump power ($q \gg 10$). Figure 2.21 shows that in the intermediate case, the required input

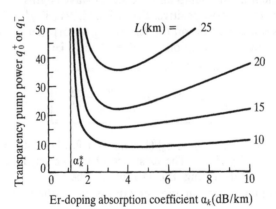

**FIGURE 2.21** Normalized input pump power required for transparency in unidirectionally pumped, distributed EDFAs, as a function of Er-doping absorption coefficient $\alpha_k$ at the signal wavelength $\lambda_s = 1.531\ \mu$m, assuming a pump at $\lambda_p = 1.48\ \mu$m, a background loss coefficient $\alpha'_k = 0.5$ dB/km, $\eta_p = 0.37$, $\eta_s = 1$, and $\alpha_p/\alpha_s = 0.4$. From [50] © 1992 IEEE.

pump power goes through a minimum value $q_{0,L}^{\pm}(\min)$. Therefore, for a given distributed amplifier length $L$, there exists an optimal doping concentration $\alpha_k^{\mathrm{opt}}$ for which the required pump power is minimized [50]. This optimum operating point nearly corresponds to a maximum in the amplifier noise figure, when forward pumped [49].

The optical noise figure of distributed EDFAs can be calculated from the unsaturated gain model developed in Chapter 1. From Eqs. (1.133)–(1.134) and with the transparency condition $G_k = 1$, we find the ASE noise photon number output at $z = L$, corresponding to forward pumping:

$$
n_{k,\mathrm{ASE}}^{\mathrm{fwd\ pump}} = \frac{\alpha_k}{\alpha_p} \frac{\eta_k}{1 + \eta_p} I^+(q_{\mathrm{in}}, q_{\mathrm{out}})
$$

$$
\times \exp\left\{ \frac{\alpha_k}{\alpha_p} \frac{1 + \varepsilon_k}{1 + \varepsilon_p} \left[ \log(q_{\mathrm{out}}) - \frac{1 - \varepsilon_k/\varepsilon_p + bU_k}{1 + \varepsilon_k} \log(b + q_{\mathrm{out}}) \right] \right\} \qquad (2.200)
$$

and the ASE noise photon number output at $z = L$, corresponding to backward pumping:

$$
n_{k,\mathrm{ASE}}^{\mathrm{bkwd\ pump}} = \frac{\alpha_k}{\alpha_p} \frac{\eta_k}{1 + \eta_p} I^-(q_{\mathrm{in}}, q_{\mathrm{out}})
$$

$$
\times \exp\left\{ -\frac{\alpha_k}{\alpha_p} \frac{1 + \varepsilon_k}{1 + \varepsilon_p} \left[ \log(q_{\mathrm{in}}) - \frac{1 - \varepsilon_k/\varepsilon_p + bU_k}{1 + \varepsilon_k} \log(b + q_{\mathrm{in}}) \right] \right\} \qquad (2.201)
$$

where $b = (1 + \varepsilon_p)/\varepsilon_p$ and, for simpler notation, we have introduced $q_{\mathrm{in}} = q_0^+$, $q_{\mathrm{out}} = q_L^+$ for forward pumping, and $q_{\mathrm{in}} = q_L^-$, $q_{\mathrm{out}} = q_0^-$ for backward pumping. The integral factors $I^\pm(q_{\mathrm{in}}, q_{\mathrm{out}})$ in Eqs. (2.200) and (2.201) are defined in Eqs. (E.34)–(E.36). The input pump power for transparency $q_{\mathrm{in}}$ is given by Eq. (2.194), and the relation between input and output pump power $q_{\mathrm{out}}$ is given by Eq. (1.127), which takes the form:

$$
q_{\mathrm{out}} = q_{\mathrm{in}} \exp\left\{ -\alpha_p L \frac{1 + \eta_p}{1 + \eta_k} \varepsilon_p \frac{1 - \varepsilon_k/\varepsilon_p + bU_k}{1 + \varepsilon_k} \right\} \qquad (2.202)
$$

At transparency ($G_k = 1$), the optical noise figures corresponding to either of the pumping schemes are given by $F_o = 1 + 2n_{k,\mathrm{ASE}}$; $n_{k,\mathrm{ASE}}$ is defined in Eq. (2.200) or Eq. (2.201). Finally, the ASE photon numbers $n_{k,\mathrm{ASE}}$ can be expressed as a function of the signal Er-doping absorption $\alpha_k$ only, as the pump Er-doping absorption is related to $\alpha_k$ through $\alpha_p/\alpha_k = \sigma_a(\lambda_p)/\sigma_a(\lambda_k)$, which represents a characteristic parameter of the doped fiber.

Figure 2.22 shows the optical noise figure of distributed EDFAs with forward and backward pumping, plotted as a function of the signal absorption coefficient, and for different fiber lengths. The pump and signal wavelengths are $\lambda_p = 1.48\ \mu m$, $\lambda_k = \lambda_s = 1.5\ \mu m$, and the same parameters as in Figure 2.21 were assumed. The figure shows that, for any Er-doping concentration, the noise figure is always lower with forward pumping than with backward pumping, as expected from previous analysis. With forward pumping, the noise figure reaches a maximum weakly dependent upon fiber

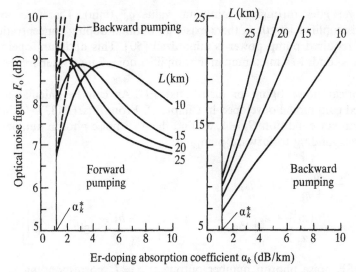

**FIGURE 2.22** Optical noise figure $F_o$ of distributed EDFA versus Er-doping absorption coefficient $\alpha_k$ for forward and backward pumping and different fiber lengths $L$ ($\lambda_p = 1.48\ \mu$m, $\lambda_s = 1.531\ \mu$m, background loss coefficient $\alpha'_k = 0.5$ dB/km, $\eta_p = 0.37$, $\eta_s = 1$, and $\alpha_p/\alpha_s = 0.4$).

length; with backward pumping, the noise figure steadily increases with doping absorption and is strongly dependent upon fiber length. For distributed amplifiers with unidirectional pumping, the lowest noise figure is always achieved with forward pumping. Comparison with figure 2.21 shows that the Er-doping absorption coefficient for which the noise figure is maximum in forward pumping, is close to that which minimizes the pump power required for transparency. Therefore, *with distributed EDFAs a trade-off exists between noise figure and required pump power* (this conclusion also applies to bidirectionally pumped distributed EDFAs).

The dependence of the optical noise figure of distributed EDFAs with length is shown in Figure 2.23, for both pumping configurations and different Er-doping absorption coefficients. The same observations as in the previous figure, regarding the difference between forward and backward pumping apply. The difference in noise figure between the two pumping schemes vanishes at short lengths ($L < 5$ km), as expected.

For comparison with the forward pumped case, the noise figure of an ideal distributed amplifier with complete medium inversion (Type B), as calculated from Eq. (2.187) with the same fiber background loss $\alpha'_k = 0.5$ dB/km, is also shown. Interestingly, for the shorter lengths ($L < 20$ km), the noise figures of the actual distributed EDFAs are up to twice those of the ideal distributed amplifier; for the longer lengths ($L > 20$–30 km), the actual distributed EDFAs have a noise figure lower than ideal. For an actual distributed EDFA pumped at $\lambda_p = 1.48\ \mu$m, complete medium inversion at any point in the fiber is not possible (Section 2.4), therefore the equivalent input noise factor $n_{eq}(z)$ is always greater than unity. This increases the noise figure. The noise factor is also increased by pump decay along the fiber. Pump decay results in lower medium inversion and increases the noise figure with respect to the ideal case. At long lengths, however, the effect of pump decay causes the ending portion of the fiber to be lossy, while the amplifying portion of the fiber is at the

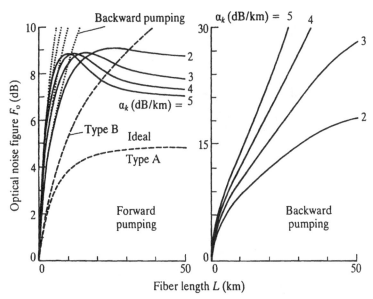

**FIGURE 2.23** Optical noise figure $F_o$ of distributed EDFA versus fiber lengths $L$ for forward and backward pumping and different Er-doping absorption coefficients $\alpha_k$. The noise figures $F_o$ of an ideal distributed EDFA with full medium inversion (Type B), and an ideal lumped EDFA with full inversion followed by a strand of lossy fiber (Type A), are also shown. Parameters: $\lambda_p = 1.48$ $\mu$m, $\lambda_s = 1.531$ $\mu$m, background loss coefficient $\alpha'_k = 0.5$ dB/km, $\eta_p = 0.37$, $\eta_s = 1$, and $\alpha_p/\alpha_s = 0.4$.

beginning. In this case, the actual distributed EDFA is similar to a Type A ideal chain element, where amplification is followed by loss, which has a noise figure lower than that of Type B. Figure 2.23, shows that for long lengths, the actual distributed EDFA noise figure falls between those of ideal Type A and Type B amplifier chain elements.

Bidirectional pumping offers the advantage of a more uniform medium inversion along the fiber, along with the possibility to reduce the required input pump power at both fiber ends. How much pump power reduction can be achieved? Does the bidirectionally pumped configuration have a better noise figure characteristics compared to the unidirectionally pumped configuration? The analytical model previously used cannot be applied to the bidirectional pump as each pump affects the absorption experienced by the other. But numerical calculation of the noise figure only requires two elementary integrations [49].

The total pump power $(q_0^+ + q_L^-)$ required for transparency and optical noise figure of bidirectionally pumped distributed EDFAs as functions of the signal Er-doping absorption coefficient are shown in Figure 2.24, assuming equal input pump powers at each fiber end $(q_0^+ = q_L^-)$, $\lambda_p = 1.48$ $\mu$m, $\lambda_p = 1.53$ $\mu$m and a background loss of $\alpha'_k = 0.2$ dB/km (D. Chen and E. Desurvire [49]). For comparison, the results obtained with unidirectional forward pumping with the same input power are also shown in the figure. For short lengths ($L < 10$ km) the difference in required pump power between the two schemes is negligible, while it increases somewhat for longer lengths (up to 25% difference for $L = 25$ km. For distributed amplifiers of moderate lengths (i.e., $L < 30$ km), the bidirectional pumping scheme represents a marginal performance

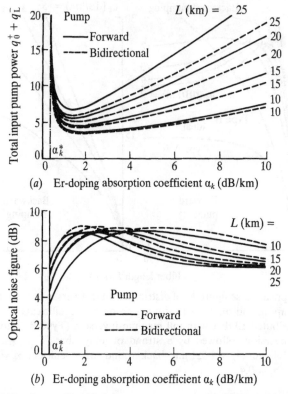

**FIGURE 2.24** (a) Total normalized input pump power required for transparency in distributed EDFAs with unidirectional forward pumping (full lines) or bidirectional pumping (dashed lines), as a function of Er-doping absorption coefficient $\alpha_k$, for different fiber lengths $L$; (b) Corresponding optical noise figures $F_o$. Parameters: $\lambda_p = 1.48\ \mu m$, $\lambda_s = 1.531\ \mu m$, background loss coefficient $\alpha'_k = 0.2\ dB/km$, $\eta_p = 0.26$, $\eta_s = 1$, and $\alpha_p/\alpha_s = 0.4$. From [49] © 1992 IEEE.

improvement compared to the unidirectional, forward pumping scheme. If longer lengths are considered (e.g., $L = 100\ km$), the noise figure difference increases; this favors the unidirectional forward pumping scheme. The lowest noise figure is always achieved by the ideal, lumped amplifier chain element of Type A. The ideal conditions corresponding to a Type A element can be realized by using a pump at the 980 nm wavelength, for which the EDFA operates as a three-level system, with the possibility of full medium inversion. Therefore, the forward pumped distributed EDFA does not represent an optimum configuration, unless other system requirements such as minimization of signal excursion or minimization of path averaged signal power have to be taken into account.

One of the main result of this study is that for 1.48 $\mu m$ pumped distributed EDFAs with lengths 20–50 km and background loss 0.2 dB/km, the optical noise figure at the peak gain wavelength 1.53 $\mu m$ is close to 10 dB, which represents about twice the value obtained with a Type A lumped amplifier chain element. Whether this result actually depends upon the signal wavelength and is valid at longer amplifier lengths ($L = 100\ km$) remains to be addressed.

The spectral dependence over the amplifier gain bandwidth of the transparency pump power and noise figure is described in a study by K. Rottwitt *et al.* [55]. This study considers the case of a 100 km long, bidirectionally pumped EDFA. Figure 2.25 shows the results for four different Er-doping concentrations. The transparency pump power has two minima at $\lambda_s = 1.535\,\mu m$ and $\lambda_s = 1.554\,\mu m$, corresponding to maxima for the noise figure. These wavelengths correspond to the two peaks of the emission and absorption cross section spectra of the germanosilicate Er-doped fiber type (see Chapter 4). When the concentration is fixed, transparency cannot be achieved necessarily at all wavelength (e.g., curves corresponding to concentrations $\rho_1$, $\rho_2$ in Figure 2.25), because of the existence of a minimum absorption coefficient $\alpha_k^*$ (or a minimum Er-doping concentration) for each wavelength $\lambda_k$. Transparency pump power and noise figure are strongly dependent upon signal wavelength and concentration. But the determination of an optimum signal wavelength for minimal noise figure should be nade on the assumption of constant input pump power. The transparency pump power can be equalized at all wavelengths by adjusting the doping concentration. The result of such an optimization is the following: for the unsaturated gain regime, the noise figure is minimal ($F_o = 13$ dB) at the optimal signal wavelength $\lambda_s = 1.554\,\mu m$, while it is 1 dB higher at the peak wavelength $\lambda_s = 1.534\,\mu m$ [55]. The same optimal signal wavelength is also found in the case of a 50 km long distributed amplifier.

Extension of this study to the case where the amplifier gain is saturated shows a sizeable increase of the noise figure [55]. But these results cannot be interpreted in terms of signal-to-noise ratio and signal photon statistics, as the formula used for the noise figure assumes the linear amplification regime (see Section 2.3). An accurate treatment of saturated distributed EDFAs requires a nonlinear photon statistics theory, outlined in Section 2.8. It can be reasonably inferred that the results predicted by the linear theory are fairly accurate at the onset of the saturation regime, but this remains to be established.

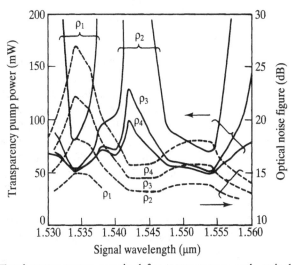

**FIGURE 2.25**  Total pump power required for transparency and optical noise figure in a 100 km bidirectionally pumped, distributed EDFA, for different Er-doping concentrations $\rho_1 < \cdots < \rho_4$. From [55] © 1992 IEEE.

The total pumping power at $\lambda = 1.48\,\mu m$ necessary to achieve transparency in 100 km long, bidirectionally pumped distributed EDFAs is typically between 50 and 100 mW, [55] and [56]. For this power level and fiber length, the potential effect of stimulated Raman scattering (SRS) [53] must be considered. In the SRS process, a fraction of the pump light is spontaneously converted by scattering into Stokes light; this light is then amplified in the forward and backward directions of the fiber. In silica glass fibers, the Stokes power is frequency shifted with respect to the pump by an amount varying from zero to $490\,cm^{-1}$ [53]. When the pump is at $\lambda_s = 1.48\,\mu m$, the Stokes power spectrum spreads from $\lambda = 1.48\,\mu m$ to $1.6\,\mu m$, with a peak at $\lambda = 1.58\,\mu m$, which strongly overlaps with the EDFA gain spectrum (see [56], for comparison). Therefore, both SRS and stimulated emission by the erbium ions contribute to the signal gain. A numerical study [56] of 100 km long bidirectionally pumped EDFAs shows that the effect of SRS is negligible when operated at transparency. When high gains ($G > 25\,dB$) are achieved, the gain is enhanced by SRS by an amount up to 3 dB, which depends upon the Er-doping concentration. This can be explained by the fact that in such a configuration, SRS gain, pump power, and doping concentration are all interdependent.

We have considered so far the optical noise figures of lumped (Types A, C) and distributed (Type B) amplifier chain elements, and their dependence with fiber length. In an optical amplifier chain, where such elements are cascaded, the amplifier spacing, or the length $L$ of each repeated element, is a very important parameter in the evolution of the path averaged ASE power $\langle P_{ASE} \rangle_{path}$ along the chain. The quantity $\langle P_{ASE} \rangle_{path}$ is given by the simple expression [57]:

$$\langle P_{ASE} \rangle_{path} = \alpha'_s L_{tot} n_{sp} F_{path}(G) \tag{2.203}$$

where $F(G)$ is a penalty factor defined as:

$$F_{path}(G) = \frac{1}{G}\left(\frac{G-1}{\log G}\right)^2 \tag{2.204}$$

In Eq. (2.203), $\alpha'_s$ is the loss coefficient of the transmission fiber, $G = \exp(+\alpha'_s L)$ the lumped amplifier gain, and $n_{sp}$ the corresponding spontaneous emission factor. The result of Eqs. (2.203) and (2.204) can be demonstrated through the following. We consider first the case of Type C chains. From Eq. (2.171), the ASE output power in one polarization mode at the element $k$ of the chain is $N = k n_{sp}(G - 1)$. If the chain has $k$ elements of length $L$, the total length is $L_{tot} = kL = k \log(G)/\alpha'_s$ and the ASE noise is $N = \alpha'_s L_{tot} n_{sp}(G - 1)/\log(G)$. From Eq. (2.186), the path averaged power normalized to the chain element output is $(G - 1)/G \log(G)$. Multiplying the output ASE $N$ by this ratio gives the result of Eqs. (2.203) and (2.204). Similarly, for a Type A chain element, the ASE noise is $N = \alpha'_s L_{tot} n_{sp}(G - 1)/G \log(G)$ while the path averaged power normalized to the chain element output is $(G - 1)/\log(G)$, after Eq. (2.185). Thus, the penalty $F_{path}(G)$ for Type A is the same as for Type C.

For a given amplifier chain length $L_{tot}$, the maximum path averaged ASE power $\langle P_{ASE} \rangle_{path}$ along the chain, given by Eq. (2.203), is determined by the penalty factor $F_{path}(G)$, which increases with $G$ or the amplifier spacing (J. P. Gordon and L. F. Mollenauer [57]). For both Types A and C lumped amplifier chains, the penalty factor $F_{path}(G)$ reaches unity in the limit $G \to 1$, which corresponds to the distributed amplifier chain of Type B. In this limit, we find from Eq. (2.203) $\langle P_{ASE} \rangle_{path} = \alpha'_s L_{tot} n_{sp}$, which represents the minimum value of the path averaged ASE power at the chain output. On the other hand, the noise figure at the chain output is given by $F_o = 1 + 2N$, which gives $F_o(G) = 1 + 2\alpha'_s L_{tot} n_{sp}(G-1)/G \log(G)$ for a Type A chain, and $F_o(G) = 1 + 2\alpha'_s L_{tot} n_{sp}(G-1)/\log(G)$ for a Type C chain.

Figure 2.26 shows plots of the penalty factor $F_{path}(G)$ and the quantity $x(G) = [F_o(G) - 1]/2\alpha'_s L_{tot}$ as a function of the gain $G$ or the amplifier spacing $L$ ($\alpha'_s = 0.2$ dB/km), for both Type A and C chains. The figure shows that for a maximum spacing of $L = 150$ km ($G = 30$ dB), the penalty factor is about 13 dB; reducing the amplifier spacing to $L = 100$ km ($G = 20$ dB) and $L = 50$ km brings the penalty factor to about 6.5 dB and 2 dB, respectively. For system applications where the ASE path averaged power must be minimized, the amplifier spacing must be the shortest possible. If the amplifiers are lumped, the limit $L < 1$ km is not realistic as a very large number of amplifiers $k = L_{tot}/L$ and related optical components would be required for realizing a long chain ($L_{tot} > 100$ km). We have also shown that in practice, the continuity limit $G \to 1$ of ideal distributed amplification, which gives the smallest penalty ($F_{path} = 1$) is not possible with realistic EDFAs. Therefore, the solution of reducing amplifier spacing is, in fact, limited by considerations of system complexity and economic sense. Figure 2.26 also shows that the factor $x$ increases for Type C chains and decreases for Type A chains. The noise figure is given by $F_o = 1 + 2\alpha'_s L_{tot} n_{sp} x$, *Type A chains have the lowest noise figures for any amplifier spacings.* We expected this result, as in a chain of lumped amplifiers the total optical noise figure $F_o(k)$ is essentially determined by that of the first element $F_o(1)$ in the chain, Eq. (2.182). For Type A, $F_o(1, \text{Type A}) = (1 + 2N)/G$; for Type C, $F_o(1, \text{Type C}) = 1/T = G \gg F_o(1, \text{Type A})$. Figure 2.27, may tempt us to conclude that Type A chains are the

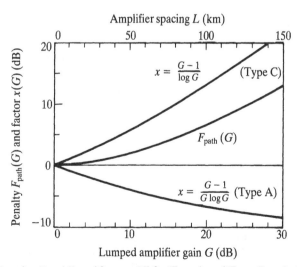

**FIGURE 2.26** Penalty $F_{path}(G)$ and factor $x(G)$ for Type A and Type C optical amplifier chains as a function of lumped amplifier gain $G$ or amplifier spacing $L$ (loss coefficient $\alpha'_s = 0.2$ dB/km). The path averaged ASE power at the chain output is given by $\langle P_{ASE} \rangle_{path}(G) = \alpha'_s L_{tot} n_{sp} F_{path}(G)$ and the optical noise figure by $F_o = 1 + 2\alpha'_s L_{tot} n_{sp} x(G)$.

most advantageous, since their noise figure is lower and their penalty factor is the same as in Type C. However, the signal path averaged power $\langle P_{\text{signal}}\rangle_{\text{path}}$ is greater for Type A by a factor $G$, in comparison to Type C, Eqs. (2.185) and (2.186), which represents an important difference for large amplifier spacings ($G \gg 1$). In some system applications, and particularly for very long-haul systems, a large signal path averaged power is not desirable because of limitations imposed by amplifier gain saturation (Chapter 5) and fiber nonlinearities (Chapter 7). As a matter of fact, Type A and Type C chains differ by the existence of an optical amplifier at the beginning of the chain (generally called a power amplifier) and determining which of the two configurations is advantageous is eventually a matter of careful system analysis.

A fiber nonlinearity that must be taken into account when considering very long amplifier chains ($L > 1000$ km) is the optical Kerr effect [53]. The Kerr effect causes a modulation of the fiber refractive index proportional to the optical power. If the ASE noise power is sufficiently high, random variations of the refractive index are generated in the fiber, causing random fluctuations of the signal group velocity. For short optical pulses ($T < 100$ ps), these fluctuations are experienced as random walk, and at the detector end, as timing jitter. The timing jitter induced by ASE on optical pulses called solitons is known as the Gordon–Haus effect [58], and can be an important cause of bit-error rate (Chapters 3 and 5). This unwanted effect can be reduced by reducing the path averaged power $\langle P_{\text{ASE}}\rangle_{\text{path}}$, defined in Eq. (2.203). As seen previously, the reduction of $\langle P_{\text{ASE}}\rangle_{\text{path}}$ can be done by decreasing the amplifier spacing, whether the chain elements are lumped or distributed. For a fixed amplifier spacing, this reduction can also be also achieved through an appropriate fiber design (cutoff wavelength, numerical aperture). A study by K. Rottwitt *et al.* [59] shows that for a 9000 km long transmission system based on a chain of distributed, bidirectionally pumped 100 km long EDFAs, there exists an optimum fiber design for which the ASE path averaged power and hence the bit-error rate are reduced. The limitations induced by fiber nonlinearities on the performance of very long-haul transmission systems are discussed in Chapter 7.

The effects of amplifier spacing and path averaged signal power on the performance of long-haul communication systems (e.g., SNR, required signal power, bit error rate) are considered in more detail in Chapter 3. The basic experimental characteristics of 1.48 $\mu$m pumped distributed EDFAs (e.g., pump power for transparency, signal excursion, and noise figure) are reviewed in Chapter 6.

## 2.7   PHOTON STATISTICS OF OPTICAL AMPLIFIER CHAINS

Now we have an expression for optical noise figure $F_o(k)$ as a function of the chain element parameters (i.e., $G$, $N$, $\alpha_s$ and $L$) and the number $k$ of such elements. But we have no informations regarding the output probability distributions of the amplified signal and the ASE noise. In communications systems applications, we need to know these probability distributions to determine the bit-error rate (BER). If a Gaussian approximation is made for the received signal and ASE noise statistics, then the knowledge of the mean and variance is sufficient for the BER determination. The validity of this approximation is studied in Chapter 3. In this section, we analyze the exact photon statistics of optical amplifier chains, stemming directly from the linear theory (STT) we have used throughout this chapter.

The photon statistics of lumped amplifier chains can be analyzed through the PGF method (T. Li and M. C. Teich [60], which is generalized here for Types A and C chains. Rewrite Eq. (2.83) to give the amplifier output PGF $F_{out}(x)$ through the simpler form:

$$F_{out}(x) = F_{in}[F_{BD}(x)]F_I(x) \tag{2.205}$$

where $F_{BD}(x) = X(x, z; 0)$ is the PGF of a birth–death (emission–absorption) process, defined in Eq. (2.51), $F_I(x) = F_I^{\mathscr{M}}(x, z)$ is the PGF of the immigration (spontaneous emission) process, defined in Eq. (2.50), and $F_{in}(x) = F(x, 0) = \sum P_m(0)x^m$, is the PGF of the input signal, Eq. (2.36).

We consider first the case of two amplifiers operating in tandem, with characteristic PGFs $F_{BD}^{(1)}(x)$, $F_I^{(1)}(x)$ and $F_{BD}^{(2)}(x)$, $F_I^{(2)}(x)$, respectively. From Eq. (2.205), we find for the PGF at the second amplifier output:

$$F_{out}^{(2)}(x) = F_{in}^{(2)}[F_{BD}^{(2)}(x)]F_I^{(2)}(x) \tag{2.206}$$

with

$$F_{in}^{(2)}(x) = F_{out}^{(1)}(x) = F_{in}^{(1)}[F_{BD}^{(1)}(x)]F_I^{(1)}(x) \tag{2.207}$$

and $F_{in}^{(k)}(x) = \sum P_m^{(1-1)}(0)x^m$. In the general case of arbitrary input photon statistics $P_m(0) = P_m^{(0)}(0)$, calculation of Eq. (2.206) is complex because of the summation involved in the definitions of $F_{in}^{(k)}(x)$. Therefore, we consider for now the simplest case where the amplifier chain input is a single photon, i.e., $P_m(0) = \delta_{m,1}$. In this case, we have $F_{in}^{(1)}(x) = x$, and we find from Eqs. (2.206) and (2.207):

$$F_{in}^{(2)}(x) = F_{BD}^{(1)}(x)F_I^{(1)}(x) \tag{2.208}$$

$$F_{out}^{(2)}(x) = F_{BD}^{(1)}[F_{BD}^{(2)}(x)]F_I^{(1)}[F_{BD}^{(2)}(x)]F_I^{(2)}(x) \tag{2.209}$$

For a Type A amplifier chain, we have from Eqs. (2.50) and (2.51), and the above definitions:

$$F_{BD}^{(1)}(x) = \frac{1 + (x - 1)(G - N)}{1 - (x - 1)N} \tag{2.210}$$

$$F_I^{(1)}(x) = [1 - (x - 1)N]^{-\mathscr{M}} \tag{2.211}$$

$$F_{BD}^{(2)}(x) = 1 + (x - 1)T \tag{2.212}$$

$$F_I^{(2)}(x) = 1 \tag{2.213}$$

while for a Type C amplifier chain, Eqs. (2.210)–(2.213) apply, with superscripts (1) and (2) permuted. Note that $F_{BD}^{(2)}(x) = 1 + (x - 1)T$ is the characteristic PGF of the *Bernoulli random deletion process* [61], which is associated with the effect of light attenuation in the transmission fiber. Substituting these definitions into Eq. (2.209) yields:

$$F_{out}^{(2)}(x, \text{Type A}) = \frac{1 + (x - 1)(TG - TN)}{1 - (x - 1)TN}[1 - (x - 1)TN]^{-\mathscr{M}} \tag{2.214}$$

$$F_{\text{out}}^{(2)}(x, \text{Type C}) = \frac{1 + (x - 1)(TG - N)}{1 - (x - 1)N}[1 - (x - 1)N]^{\mathcal{M}} \qquad (2.215)$$

Using the definitions $G_A = TG$, $N_A = TN$, and $G_C = TG$, $N_C = N$, we can then rewrite the results of Eqs. (2.214) and (2.215) in the form:

$$F_{\text{out}}^A(x) = F_{\text{out}}^{(2)}(x, \text{Type A}) = F_{\text{BD}}^A(x)F_I^A(x) \qquad (2.216)$$

$$F_{\text{out}}^C(x) = F_{\text{out}}^{(2)}(x, \text{Type C}) = F_{\text{BD}}^C(x)F_I^C(x) \qquad (2.217)$$

with

$$F_{\text{BD}}^D(x) = \frac{1 + (x - 1)(G_D - N_D)}{1 - (x - 1)N_D} \qquad (2.218)$$

$$F_I^D(x) = [1 - (x - 1)N_D]^{\mathcal{M}} \qquad (2.219)$$

and D = A or C.

We consider next the case where two chain elements of types A and C are placed in tandem. From Eqs. (2.209), (2.216), and (2.217), the output PGF of a chain of two elements of Type D is:

$$F_D(x, k = 2) = F_{\text{BD}}^D[F_{\text{BD}}^D(x)]F_I^D[F_{\text{BD}}^D(x)]F_I^D(x) \qquad (2.220)$$

Substituting Eqs. (2.218) and (2.219) into Eq. (2.220) yields:

$$F_D(x, k = 2) = \frac{1 + (x - 1)(G_D^2 - N_2)}{1 - (x - 1)N_2}[1 - (x - 1)N_2]^{\mathcal{M}} \equiv F_{\text{BD}}^D(x, k = 2)F_I^D(x, k = 2) \qquad (2.221)$$

with $N_2 = G_D N_D + N_D$. By induction, we find the output PGF $F_D(x, k)$ of a chain of $k$ elements of Type D [60]:

$$F_D(x, k) = \frac{1 + (x - 1)(G_D^k - N_k)}{1 - (x - 1)N_k}[1 - (x - 1)N_k]^{\mathcal{M}} \equiv F_{\text{BD}}^D(x, k)F_I^D(x, k) \qquad (2.222)$$

with

$$N_k = N_D \sum_{n=1}^{k-1} G_D^n = N_D \frac{G_D^k - 1}{G_D - 1} \qquad (2.223)$$

In the transparency limit $G_D \to 1$, we find from Eq. (2.223) $N_k = kN_D$, and the output PGF of the $k$ element amplifier chain in eq. (2.222) becomes:

$$F_D(x, k) = \frac{1 + (x - 1)(1 - kN_D)}{1 - (x - 1)kN_D}[1 - (x - 1)kN_D]^{\mathcal{M}} \qquad (2.224)$$

The important result of Eqs. (2.222) and (2.223) is that *a chain of k identical amplifiers with gain G and ASE noise N has the same output PGF as a single amplifier with gain $G^k$ and ASE noise $N_k$; $N_k$* is defined in Eq. (2.223). A more general output PGF formula for $k$ nonidentical amplifiers can be found in [60].

We have seen in Section 2.3 that if the signal input to a single amplifier of gain $G$ and noise $N$ has Poisson statistics, the output photon statistics correspond to the noncentral negative binomial (NNB) or Laguerre probability distribution defined in Eq. (2.102). If the amplifier has no input, the output photon statistics are also characterized by an NNB distribution, defined in Eq. (2.93). At transparency, the single amplifier equivalent to the chain has unity gain and ASE noise $kN_D$. In this case, the output probability distribution $P_n^D(z, k)$ for a Type D chain (D = A or C) with $k$ elements comes from Eqs. (2.93) and (2.102):

$$P_n^D(k) = \frac{(kN_D)^n}{(kN_D + 1)^{n+\mathcal{M}}} \exp\left\{ -\frac{\langle n_0 \rangle}{kN_D + 1} \right\} L_n^{(\mathcal{M}-1)}\left\{ -\frac{\langle n_0 \rangle}{kN_D(kN_D + 1)} \right\} \qquad (2.225)$$

for an input signal with Poisson statistics, and

$$P_n^D(k) = \frac{(n + \mathcal{M} - 1)!}{n!(\mathcal{M} - 1)!} \frac{(kN_D)^n}{(kN_D + 1)^{n+\mathcal{M}}} \qquad (2.226)$$

in the case where there is no input signal. For Type A and C chains, we have specifically:

$$P_n^A(k) = \frac{(kn_{eq})^n}{(kn_{eq} + 1)^{n+\mathcal{M}}} \exp\left\{ -\frac{\langle n_0 \rangle}{kn_{eq} + 1} \right\} L_n^{(\mathcal{M}-1)}\left\{ -\frac{\langle n_0 \rangle}{kn_{eq}(kn_{eq} + 1)} \right\} \quad (2.227)$$

$$P_n^C(k) = \frac{(kGn_{eq})^n}{(kGn_{eq} + 1)^{n+\mathcal{M}}} \exp\left\{ -\frac{\langle n_0 \rangle}{kGn_{eq} + 1} \right\} L_n^{(\mathcal{M}-1)}\left\{ -\frac{\langle n_0 \rangle}{kGn_{eq}(kGn_{eq} + 1)} \right\} \quad (2.228)$$

for an input signal with Poisson statistics, and

$$P_n^A(k) = \frac{(n + \mathcal{M} - 1)!}{n!(\mathcal{M} - 1)!} \frac{(kn_{eq})^n}{(kn_{eq} + 1)^{n+\mathcal{M}}} \qquad (2.229)$$

$$P_n^C(k) = \frac{(n + \mathcal{M} - 1)!}{n!(\mathcal{M} - 1)!} \frac{(kGn_{eq})^n}{(kGn_{eq} + 1)^{n+\mathcal{M}}} \qquad (2.230)$$

in the case where there is no input signal, with $n_{eq} = N/G$.

Figure 2.27 shows plots of the photon probability distribution $P_n^A(k)$ for a Type A lumped amplifier chain with $L = 100$ km amplifier spacing ($n_{eq} = 1$, $\alpha_s = 0.2$ dB/km, $\mathcal{M} = 2$), and for two cases: (a) there is no input signal, Eq. (2.229), which we call the 0 symbol case; and (b) where the input signal has Poisson statistics with $\langle n_0 \rangle = 100$, Eq. (2.227), which we call the 1 symbol case. In Figure 2.27b, the distribution corresponding to the 0 symbol is plotted on the same scale, for comparison. The figure shows that after a propagation length greater than 500 km ($k > 5$), the overlap

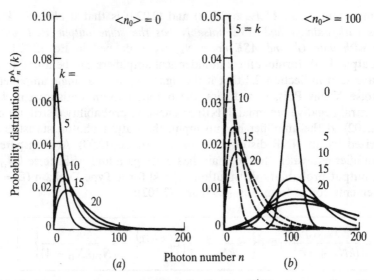

**FIGURE 2.27**   Photon statistics probability distribution $P_n^A(k)$ corresponding to a Type A lumped amplifier chain of $k$ elements with $L = 100$ km amplifier spacing ($n_{eq} = 1$, $\alpha_s = .2$ dB/km), with: (a) no signal input to the chain (0 symbol), and (b) the input signal has Poisson statistics with $\langle n_0 \rangle = 100$ (1 symbol). Distributions of 0 and 1 symbols are plotted at the same scale in (b), for comparison.

between the two becomes significant. For instance, at $z = 1000$ km ($k = 10$) and $n = 60$ the two distributions tails are nearly equal. In this case, the probability of misreading a symbol for the other would be quite high, as discussed in detail in Chapter 3. It is clear then that some knowledge of the light photon statistics at the output of the amplifier chain is necessary for the evaluation of such a symbol reading error (or bit-error rate). It is possible to approximate the distributions of the 0 and the 1 symbols by a Gaussian law, in which case only the knowledge of the corresponding output mean and variance is required. On the other hand, if such an approximation turns out to be inaccurate, knowledge of the exact photon statistics is required.

## 2.8   NONLINEAR PHOTON STATISTICS

An underlying assumption throughout this chapter so far is that the atomic populations $N_1$ and $N_2$ of the upper and lower laser levels are deterministic or statistically constant. This assumption allowed us to derive the photon statistics at the output of the amplifier, using $N_1$, $N_2$ as fixed parameters. If the signal power propagating along the fiber amplifier were to increase to the point where the gain became saturated, our assumption would no longer be valid. The atomic population would exhibit statistical fluctuations in response to fluctuations of the amplified light. The atomic and photon population systems are statistically coupled, each having its own dynamics. The statistics of the amplified light not only depend on the input signal statistics, but also on the input signal power. In such a regime of gain saturation, the determination of the amplified light noise characteristics requires a specific model of *nonlinear photon statistics*, as opposed to the linear amplifier STT model [27] used so far.

A nonlinear photon statistics theory applying to the case of EDFAs is needed for accurate determination of bit-error rate in communications systems where optical amplifiers operate at or near saturation. But a complete theory that would take into account the specific and highly complex features of the Er:glass laser system still remains to be developed. In addition, different types of nonlinear amplification regime can exist with EDFAs, among them: signal self-saturation, amplifier self-saturation by ASE, a combination of the two, cross-saturation between different optical channels, and nonlinear coupling between pump and signal photon statistics. Currently, none of these complex issues have been addressed and their full treatment remains a challenge for the future. At the time of publication, the whole issue of nonlinear photon statistics in EDFAs is actually open to speculation. The purpose of this section is to discuss some difficulties associated with a full statistical analysis of signal self-saturation in EDFAs. We shall also develop a theory based upon a heuristic nonlinear photon statistics master equation for three-level laser systems [62], which could represent the basis of a future and more complete analysis of the problem.

What can be tentatively concluded from this heuristic theory is shown to agree with previous work on nonlinear amplifier statistics in four-level systems, [34] and [63]–[66]. The main conclusion of these referenced investigations is that amplifier gain saturation causes an important reduction of the light statistical fluctuations; the associated probability distribution, however, is always broader than that of coherent light. Therefore, when an optical amplifier operates in the nonlinear regime, no improvement but rather degradation in signal detection can be expected, as was shown theoretically in [67], and experimentally through BER measurements in [68]. The issue of BER with saturated EDFAs is addressed in detail in Section 7.1.

The first photon statistics theory taking into account nonlinearity in light amplification was developed in 1967 by M. O. Scully and W. E. Lamb [20]. This theory made it possible to analyze the transient buildup of laser oscillation from quantum noise [69], as well as the laser output photon statistics and line width, [21] and [22].

The full analysis of saturated laser light statistics turns out to be highly complex, and rests upon a vast array of quantum theoretical treatments. These are based on Langevin, density matrix, or Fokker–Planck equations (see, for instance, the family tree of the quantum theory of the laser, by H. Haken [24]). A derivation of the light field reduced density matrix equations in the photon number representation can be found in the fundamental book by M. Sargent, M. O. Scully, and W. E. Lamb [23]. The theory by R. Loudon [3] considers rate equations for the atomic and photon statistical populations, under the assumption of an adiabatic decoupling in the light/atom system. Both derivations lead to the same *nonlinear photon statistics master equation for* the probability distribution $P_n(z)$, which is our main interest here.

These earliest treatments, however, did not include the effect of saturation by the lower atomic level which, in typical four-level laser systems, is highly damped by nonradiative decay. In the three-level system case of Er:glass, the lower energy level, made of a manifold of Stark sublevels, generally has a finite population, which increases during gain saturation. An extension of the theory of [23] to the case where both upper and lower levels experience saturation (also considered in Chapter 17 of [23]), was extensively studied by Vorobev and Sokolovskii [70], G. Oliver and C.

Benjaballah, [63] and [64], and N. B. Abraham [65]. The nonlinear photon statistics master equation at the base of these studies takes the form:

$$\frac{dP_n}{dz} = a\{nf_nP_{n-1} - (n+1)f_{n+1}P_n\} + b\{(n+1)f_{n+1}P_{n+1} - nf_nP_n\} \quad (2.231)$$

where $a$ and $b$ are the unsaturated emission and absorption coefficients of the four-level laser system, and $f_n$ is the saturation function defined by:

$$f_n = \frac{1}{1 + sn} = \frac{1}{1 + \dfrac{n}{n_{sat}}} \quad (2.232)$$

where $s = 1/n_{sat}$ is the *saturation coefficient*. Resolution of Eq. (2.231) can be done by several techniques and simplifying assumptions, discussed later in this section. Before considering the master equation corresponding to three-level systems, it is useful to analyze some general properties of Eq. (2.231).

In the small signal limit $sn \ll 1$ or $n \ll n_{sat}$ the nonlinear master equation (2.231) reduces to eq. (2.24), which corresponds to the linear amplifier ($\mathscr{M} = 1$) used in the linear STT theory. The rate equations for the statistical moments of order $j$, i.e., $\langle n^j \rangle$, can be obtained as in the linear case through the summation:

$$\langle n^j \rangle = \sum_n n^j \frac{dP_n}{dz} \quad (2.233)$$

which, from eq. (2.231) and using the usual translation property $\sum g(n)P_{n\pm1} = \sum g[n-(\pm1)]P_n$, yields for the first two moments:

$$\frac{d\langle n \rangle}{dz} = a\left\langle \frac{n+1}{1+s(n+1)} \right\rangle - b\left\langle \frac{n}{1+sn} \right\rangle \quad (2.234)$$

$$\frac{d\langle n^2 \rangle}{dz} = a\left\langle \frac{2n^2+3n+1}{1+s(n+1)} \right\rangle - b\left\langle \frac{2n^2-n}{1+sn} \right\rangle \quad (2.235)$$

where the brackets indicate statistical average. A general expression for the rate equations for the higher order moments $\langle n^j \rangle$ can be obtained by the same procedure, which gives (S. Ruiz-Moreno et al. [34]):

$$\frac{d\langle n^j \rangle}{dz} = \sum_{k=1}^{j} \binom{j}{k-1}\left\{ a\left\langle \frac{(n+1)n^{k-1}}{1+s(n+1)} \right\rangle + (-1)^{j+k-1}b\left\langle \frac{n^k}{1+sn} \right\rangle \right\} \quad (2.236)$$

The numerator and denominators involved in the brackets in eqs. (2.234)–(2.236) are statistically correlated, and averaging the corresponding fractions cannot be made without the knowledge of the photon statistics distribution $P_n(z)$. This problem can be alleviated by using various decorrelation methods [64].

A first method is to use the *all order decorrelation*:

$$\left\langle \frac{n^j}{1 + sn} \right\rangle \approx \frac{\langle n^j \rangle}{1 + s\langle n \rangle} \qquad (2.237)$$

Such a decorrelation is valid only for coherent (Glauber) fields, i.e., where the statistics are given by a Poisson law [64]. Using this technique to decorrelate and solve Eqs. (2.234)–(2.236) in the general case could lead to erroneous results if its validity is not carefully checked, by performing for instance the numerical integration of the master equation (2.231). Furthermore, we have seen in the previous section that even when the signal input to the amplifier is coherent, the amplified signal statistics significantly depart from Poisson statistics. Intuitively, we can also infer that, in the general case, Eq. (2.237) represents a coarse approximation. Indeed, considering for instance $j = 1$, the approximation $\langle n/(1 + sn) \rangle \approx \langle n \rangle/(1 + s\langle n \rangle)$ is equivalent to assuming that the signal saturation is, in practice, an effect of the mean photon number $\langle n \rangle$. If on the other hand, we consider instantaneous photon number fluctuations about $\langle n \rangle$, saturation is less important for $n < \langle n \rangle$ than for $n > \langle n \rangle$. Thus, the mean $\langle n/(1 + sn) \rangle$ is actually lower than $\langle n \rangle/(1 + s\langle n \rangle)$ if the probability distribution is symmetrical about $\langle n \rangle$, and the decorrelation (2.237) is not accurate. A case where such a decorrelation would be accurate is a probability distribution having a standard deviation $\sigma$ small compared to the mean $\langle n \rangle$, but this is not applicable to amplified light where $\sigma > G\sqrt{2\langle n_0 \rangle n_{eq}}$.

Another decorrelation method is a Taylor expansion of the saturation function $f_n$ about the mean value $n = \langle n \rangle$, i.e.,

$$f_n = f_{\langle n \rangle} + (n - \langle n \rangle)\left(\frac{df_n}{dn}\right)_{\langle n \rangle} + \cdots + \frac{(n - \langle n \rangle)^k}{k!}\left(\frac{d^k f_n}{dn^k}\right)_{\langle n \rangle}$$

$$= f_{\langle n \rangle} - s\frac{n - \langle n \rangle}{(1 + s\langle n \rangle)^2} + \cdots + (-1)^k s^k \frac{(n - \langle n \rangle)^k}{(1 + s\langle n \rangle)^{k+1}} \qquad (2.238)$$

The advantage of expanding the function $f_n$ about $\langle n \rangle$, instead of about the origin, is that the approximation tracks the increase of $\langle n \rangle$ with coordinate $z$. This ensures improved convergence. With Eq. (2.238), the following decorrelation is obtained:

$$\langle n^j f_n \rangle = \left\langle \frac{n^j}{1 + sn} \right\rangle = \frac{\langle n^j \rangle}{1 + s\langle n \rangle} - s\frac{\langle n^{j+1} \rangle - \langle n^j \rangle\langle n \rangle}{(1 + s\langle n \rangle)^2} + \cdots$$

$$+ (-1)^k s^k \frac{\langle n^j(n - \langle n \rangle)^k \rangle}{(1 + s\langle n \rangle)^{k+1}} \qquad (2.239)$$

Expansion (2.239), found in the right-hand side of the rate equation $d\langle n^j \rangle/dz$ in Eq. (2.236), is seen to involve the higher order moments $\langle n^{j+k} \rangle$. At the first order ($k = 1$), the moment $\langle n^{j+1} \rangle$ is therefore necessary to compute $d\langle n^j \rangle/dz$. Thus, with the decorrelation method of Eq. (2.239), even a first order approximation would require solving all the moment equations $d\langle n^j \rangle/dz$ (up to a certain order), unless the higher order moments in each equation were neglected. This is equivalent to introducing an all order decorrelation at some point, neither accurate nor practical.

The only decorrelation method to alleviate this problem is Taylor expansion of the function $n^j f_n$ [34]. In this case, Eq. (2.239) takes the alternate form:

$$
\begin{aligned}
\langle n^j f_n \rangle &= \left\langle \frac{n^j}{1 + sn} \right\rangle = \frac{\langle n \rangle^j}{1 + s\langle n \rangle} + \sum_k \frac{\langle (n - \langle n \rangle)^k \rangle}{k!} \left[ \frac{d^k}{dn^k} \left( \frac{n^j}{1 + sn} \right) \right]_{\langle n \rangle} \\
&= \frac{\langle n \rangle^j}{1 + s\langle n \rangle} + s^2 \frac{\sigma^2}{2!} \left[ \frac{1}{s^2} \frac{d^2}{dn^2} \left( \frac{n^j}{1 + sn} \right) \right]_{\langle n \rangle} \\
&\quad + s^3 \frac{\langle (n - \langle n \rangle)^3 \rangle}{3!} \left[ \frac{1}{s^3} \frac{d^3}{dn^3} \left( \frac{n^j}{1 + sn} \right) \right]_{\langle n \rangle} + \cdots
\end{aligned}
\tag{2.240}
$$

with $\sigma^2 = \langle n^2 \rangle - \langle n \rangle$. The first order ($k = 1$) term in Eq. (2.240) is always zero and the terms of order $k$ involve only the moments of same rank, i.e., $\langle n^k \rangle$, and of lower order. For $j = 0, 1$ the decorrelation (2.240) yields:

$$
\left\langle \frac{1}{1 + sn} \right\rangle = f_{\langle n \rangle} + s^2 \sigma^2 f_{\langle n \rangle}^3 + \cdots
\tag{2.241}
$$

$$
\left\langle \frac{n}{1 + sn} \right\rangle = \langle n \rangle f_{\langle n \rangle} - s\sigma^2 f_{\langle n \rangle}^3 + \cdots
\tag{2.242}
$$

The same type of development as in Eq. (2.240) also yields:

$$
\left\langle \frac{n + 1}{1 + s(n + 1)} \right\rangle = (\langle n \rangle + 1) f_{\langle n \rangle + 1} - s\sigma^2 f_{\langle n \rangle + 1}^3 + \cdots
\tag{2.243}
$$

Substituting Eqs. (2.242) and (2.243) into Eq. (2.234), keeping only the second order terms, and assuming $\langle n \rangle \gg 1$ in the denominators gives an approximate rate equation for the mean $\langle n(z) \rangle$:

$$
\frac{d\langle n(z) \rangle}{dz} \approx a \frac{\langle n \rangle + 1}{1 + s(\langle n \rangle + 1)} - b \frac{\langle n \rangle}{1 + s\langle n \rangle} - s f_{\langle n \rangle}^3 (a - b) \sigma^2
\tag{2.244}
$$

When reduced to the first two terms on the right-hand side, the above equation is recognized as the usual amplifier rate equation $d\langle n \rangle / dz = g_e(\langle n \rangle + 1) - g_a \langle n \rangle$ seen in Chapter 1, except in the definitions of the coefficients $g_e = a/[1 + s(\langle n \rangle + 1)]$, $g_a = b/(1 + s\langle n \rangle)$ which apply here to four-level systems.

The negative correction in $s\sigma^2 f_{\langle n \rangle}^3$ appearing in Eq. (2.244) reflects the effect of the statistical decorrelation. It is easily verified that such a correction is negligible for coherent fields. Indeed, the condition $s\sigma^2 f_{\langle n \rangle}^3 \ll \langle n \rangle f_{\langle n \rangle}$ with $\sigma^2 = \langle n \rangle$ is equivalent to $\langle n \rangle (2 + \langle n \rangle / n_{sat}) + n_{sat} \gg f = 1$ ($f$ = Fano factor), which is satisfied for any value of $\langle n \rangle$ as $n_{sat} \gg 1$. On the other hand, with (noncoherent) amplified light for which $\sigma^2 \approx 2\langle n \rangle N$ (signal–ASE beat noise limit), the Fano factor is $f = \sigma^2 / \langle n \rangle \approx 2N = 2n_{sp}(G - 1) \approx 2G$, and it is found that the condition $s\sigma^2 f_{\langle n \rangle}^3 \ll \langle n \rangle f_{\langle n \rangle}$ with $\sigma^2 / \langle n \rangle = 2G$ and $\langle n \rangle = G\langle n_0 \rangle$ is equivalent to $\langle n_0 \rangle (1 + G\langle n_0 \rangle / 2n_{sat}) + n_{sat}/2G \gg 1$. Using typical parameters $B_0 = 1$ GHz, $P_{sat} = 1$ mW with $\lambda_s = 1.5 \, \mu m$, we find from the definition of $n_{sat}$, Eq. (2.249), $n_{sat} \ll q \cdot 10^7$; as $q \gg 1$ and $G \leqslant 10^4$, we have $n_{sat}/G \gg 1$ and the condition $s\sigma^2 f^3_{\langle n \rangle} \ll \langle n \rangle f_{\langle n \rangle}$ is actually always verified. These considerations show that in any regime of interest for optical amplifiers, the second order correlation in Eq. (2.244) is small.*

The next step is to derive moment rate equations from a nonlinear photon statistics equation relevant to three-level laser systems, in contrast with Eq. (2.231) which applies to four-level systems. This is done in Appendix J. The resulting equation takes the form:

$$\frac{dP_n}{dz} = -\{A_0(n+1) + B_0 n\}P_n + A_0 n P_{n-1} + B_0(n+1)P_{n+1}$$
$$-\{A_1 f_{n+1}(n+1) + B_1 f_n n\}P_n + A_1 f_n n P_{n-1} + B_1 f_{n+1}(n+1)P_{n+1} \quad (2.245)$$

with

$$A_0 = B_0 = \alpha_s \frac{\eta_s}{1 + \eta_s} \quad (2.246)$$

---

* An important remark can be made regarding stimulated emission in the presence of fluctuations of signal power. The basic amplifier equation (1.24) for the light power $P_s$, i.e.,

$$\frac{dP_s}{dz} \approx (\sigma_e N_2 - \sigma_a N_1)P_s$$

is actually not strictly accurate for describing an average signal power. The correct rate equation is:

$$\frac{d\langle P_s \rangle}{dz} \approx \langle (\sigma_e N_2 - \sigma_a N_1)P_s \rangle$$

which stems from averaging the previous equation with a continuous probability distribution $p(P_s)$ describing the signal light statistics. The laser level populations $N_1$ and $N_2$ are functions of the instantaneous value of $P_s$. The resulting correlation effect can be shown explicitly by replacing $N_1$, $N_2$ from their definitions in Eq. (1.10):

$$\frac{d\langle P_s \rangle}{dz} \approx \frac{\rho(\sigma_e q - \sigma_a)}{1 + q} \left\langle \frac{P_s}{1 + \dfrac{P_s}{P_{sat}^*}} \right\rangle$$

with $P_{sat}^* = P_{sat}/(1 + q)$ and $q = R\tau$. Solving the above equation involves the same decorrelation problem as discussed in the main text. The all order decorrelation shown in Eq. (2.237) applied for $j = 1$ to this equation yields:

$$\frac{d\langle P_s \rangle}{dz} \approx \frac{\rho(\sigma_e q - \sigma_a)}{1 + q} \frac{\langle P_s \rangle}{1 + \dfrac{\langle P_s \rangle}{P_{sat}^*}}$$

which is the rate equation for $\langle P_s \rangle$ widely used in the literature. We have seen in the main text that for the mean value $\langle P_s \rangle$ the correction due to this statistical decorrelation is generally small. This indicates that the above rate equation is a good approximation to the effect of signal amplification by stimulated emission.

$$A_1 = \alpha_s \frac{\eta_s}{1 + \eta_s} \frac{Uq - 1}{1 + q} \tag{2.247}$$

$$B_1 = -\alpha_s \frac{1}{1 + \eta_s} \frac{Uq - 1}{1 + q} \tag{2.248}$$

with $U = (\eta_s - \eta_p)/(1 + \eta_p)$. In Eq. (2.245), $f_n$ is the saturation function defined in Eq. (2.232) and the saturation photon number $n_{sat}$ is defined by:

$$n_{sat} = \frac{1}{s} = (1 + q) \frac{P_{sat}(v_s)}{hv_s B_o} = (1 + q) \frac{\pi \omega_s^2}{\sigma_a(v_s)(1 + \eta_s) B_o \tau} \tag{2.249}$$

where $B_o$ is the signal optical bandwidth. Note that $n_{sat}$ is a function of the normalized pump power $q$. The effect of saturation is reduced with increasing pump. This stems from the fact that in the EDFA laser system, the pump contributes to deplete the population of the transition's terminal level.

Several important remarks can be made about the nonlinear photon statistics master equation (2.245). First, the derivation of this equation is heuristic, not rigorous. However, this derivation can be justified by analogy with the case of four-level systems, for which a complete theory leading to a similar master equation is available.

Second, we have considered, for simplicity, the single-mode case ($\mathcal{M} = 1$), equivalent to assuming the effect of saturation by all the ASE modes is negligible, while saturation is exclusively due to the amplified signal.

Third, in the case $sn \ll 1$ the nonlinear equation (2.245) reduces to the linear photon statistics master equation (2.24) of the STT theory. Thus, nonlinear Eq. (2.245) represents the nonlinear continuation of Eq. (2.24).

Finally, the pump power $q$ was used as a fixed parameter with no statistical fluctuation nor correlation with the signal. This assumption represents an approximation, as in three-level systems where the ground level is the terminal level of the transition, both pump and signal cause fluctuations in the ground level population $N_1$ (for two-level systems, both level populations $N_1$, $N_2$ are affected by the pump and the signal). A complete theory taking into account the effect of pump photon statistics remains to be developed.

In the nonlinear master equation (2.245), we recognize a linear part (coefficients $A_0$, $B_0$) and a nonlinear part (coefficients $A_1$, $B_1$). The linear part corresponds to Eq. (2.24), general form of the linear photon statistics master equation; the nonlinear part corresponds to Eq. (2.231), general form of the nonlinear photon statistics master equation. The main difference between three-level and four-level master equations is the linear part involving $A_0$ and $B_0$. Therefore, we might expect the properties of the solution $P_n(z)$ corresponding to the three-level case to be very similar to those corresponding to the four-level case, [63]–[65].

The rate equations for the first two moments, as derived from the master equation (2.245), are found to be (Appendix J):

$$\boxed{\frac{d\langle n(z) \rangle}{dz} = A_0 + A_1 \left\langle \frac{n + 1}{1 + s(n + 1)} \right\rangle - B_1 \left\langle \frac{n}{1 + sn} \right\rangle} \tag{2.250}$$

$$\frac{d\langle n^2(z)\rangle}{dz} = A_0 + (3A_0 + B_0)\langle n\rangle + A_1\left\langle\frac{2n^2 + 3n + 1}{1 + s(n + 1)}\right\rangle - B_1\left\langle\frac{2n^2 - n}{1 + sn}\right\rangle \qquad (2.251)$$

The moment equations (2.250) and (2.251) are the counterparts for two- and three-level systems of the general moment equations for four-level systems [34]. These equations can be solved without knowledge of the probability distribution $P_n(z)$ by means of a decorrelation method for the bracketed terms. With the decorrelation shown in Eq. (2.240), one obtains up to the second order (Appendix J):

$$\frac{d\langle n(z)\rangle}{dz} \approx A_0 + \frac{(A_1 - B_1)\langle n\rangle + A_1}{1 + s\langle n\rangle} - sf_{\langle n\rangle}^3(A_1 - B_1)\sigma^2 \qquad (2.252)$$

$$\frac{d\langle n^2(z)\rangle}{dz} \approx A_0 + 4A_0\langle n\rangle + \frac{2(A_1 - B_1)\langle n\rangle^2 + (3A_1 + B_1)\langle n\rangle + A_1}{1 + s\langle n\rangle}$$

$$+ (2 + s)f_{\langle n\rangle}^3(A_1 - B_1)\sigma^2 \qquad (2.253)$$

where $\sigma^2 = \langle n^2\rangle - \langle n\rangle^2$ and $f_{\langle n\rangle} = 1/(1 + s\langle n\rangle)$. Using the relation $d\sigma^2/dz = d\langle n^2\rangle/dz - 2\langle n\rangle d\langle n\rangle/dz$, we can also combine Eqs. (2.252) and (2.253), to get a rate equation for $\sigma^2(z)$:

$$\frac{d\sigma^2(z)}{dz} \approx A_0 + 2A_0\langle n\rangle + \frac{(A_1 + B_1)\langle n\rangle + A_1}{1 + s\langle n\rangle} + (2 + s + 2s\langle n\rangle)f_{\langle n\rangle}^3(A_1 - B_1)\sigma^2$$

$$(2.254)$$

Numerical resolution of Eqs. (2.252) and (2.254) in the case of an actual EDFA requires an additional equation for the normalized pump power $q(z)$, involved in the definitions of the coefficients $A_1$, $B_1$, and $s$, Eqs. (2.247)–(2.249). As the pump is assumed to be statistically constant, the corresponding rate equation is, from Eqs. (1.113) and (2.249):

$$\frac{dq(z)}{dz} = -\alpha_p \frac{1 + \dfrac{\eta_p}{1 + \eta_p}q + \dfrac{\eta_s}{1 + \eta_s}(1 + q)s\langle n\rangle}{1 + s\langle n\rangle}\frac{q}{1 + q} \qquad (2.255)$$

An example of numerical resolution of Eqs. (2.252), (2.254), and (2.255) can be considered with the parameters $q_0 = 100$, $\alpha_p = 1\,\mathrm{dB/m}$, $\alpha_p = 2\,\mathrm{dB/m}$, $\eta_p = 0$, $\eta_s = 1$ and $n_{sat} = 4.5 \times 10^6(1 + q)$, for two different relative input signal powers $p_0 < 10^{-4}$ and $p_0 = 10^{-2}$. The choice of $n_{sat}$ (as defined in Eq. (2.249) corresponds to typical EDFA parameters $\omega_s = 4\,\mu\mathrm{m}$, $\tau = 10\,\mathrm{ms}$, $B_o = 1\,\mathrm{GHz}$ and $\sigma_a(v_s) = 6.10^{-25}\,\mathrm{m}^2$. The choice of a relatively high input pump ($q_0 = 100$) generates the condition of full medium inversion for which the optical noise figure is 3 dB at high gains.

In order to evaluate how the signal self-saturation effect progresses with fiber length, the saturation term $s\langle n\rangle$ involved in the saturation function $f_{\langle n\rangle} = 1/(1 + s\langle n\rangle)$ is plotted in Figure 2.28a for the two input signals $p_0 = 10^{-4}$ and $p_0 = 10^{-2}$. It shows that for the high signal $p_0 = 10^{-2}$, the onset of gain saturation is near $z = 15\,\mathrm{m}$, at

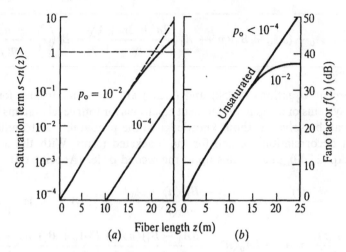

**FIGURE 2.28**   Study of EDFA noise characteristics in the nonlinear amplification regime $q_0 = 100$, $n_{sat} = 4.5 \times 10^6(1 + q)$, based upon a second order decorrelation theory. (a) Saturation term $s\langle n \rangle = \langle n \rangle/n_{sat}$ as a function of fiber length for different relative input signal powers $p_0 = 10^{-4}$ and $p_0 = 10^{-2}$, with relative input pump power $q_0 = 100$. (b) Corresponding Fano factor $f = \sigma^2/\langle n \rangle$ versus fiber length.

which point the term $s\langle n \rangle$ reaches the value of 0.1. For $z > 25$ m and $p_0 < 10^{-4}$ the gain remains unsaturated as $s\langle n \rangle < 0.1$. At $z = 15$ m, for both signal inputs, the gain is near 30 dB. In such a high gain regime, the effect of amplifier self-saturation would be important, but is overlooked here, for simplicity.

The Fano factor $f = \sigma^2/\langle n \rangle$ corresponding to this example is plotted against length in Figure 2.28b. Compared with Figure 2.28a, the unsaturated gain regime ($z < 15$ m for $p_0 = 10^{-2}$) has a Fano factor $f(z)$ nearly equal to the signal gain $G(z)$. This indicates the amplifier noise variance $\sigma^2$ is dominated by the signal–ASE beat noise of expression $\sigma^2_{s-ASE} = 2\langle n \rangle N = 2\langle n \rangle n_{eq} G$, in which case $\sigma^2/\langle n \rangle \approx G$. But at high input signals ($p_0 = 10^{-2}$), the Fano factor decreases relative to the unsaturated value, indicating that the effects of saturation is a reduction of the noise variance with respect to its signal–ASE beat noise limit. This fact can be physically interpreted by the reduction of statistical fluctuations caused by the saturable gain. The effect of gain saturation is thus to reduce the photon number fluctuations above the mean value $\langle n \rangle$, which causes the noise variance to decrease. Likewise, a saturable absorber would reduce the photon number fluctuations below the mean, which would also result in a decrease of noise variance. Signal-to-noise ratio considerations: in the saturable amplifier, when the mean increases, the variance decreases; in the saturable absorber, when the mean decreases, the variance also decreases. From this observation, *the effect of gain saturation is therefore to increase the SNR*. This SNR increase is relative to the case of an unsaturated amplifier with the same gain, not relative to the input SNR.

To further illustrate this point, the corresponding optical noise figure $F_o(z)$ versus fiber length is shown in Figure 2.29. The noise figure is defined here as $F_o(z) = \langle n_0 \rangle \sigma^2(z)/\langle n(z) \rangle^2$, somewhat different from Eq. (2.115), which involves the signal-to-noise ratio $SNR_o(z) = [\langle n(z) \rangle - \langle n(z) \rangle_{ASE}]^2/\sigma^2(z)$, as opposed to the quantity $\langle n(z) \rangle^2/\sigma^2(z)$. But the difference between the two definitions is negligible when the

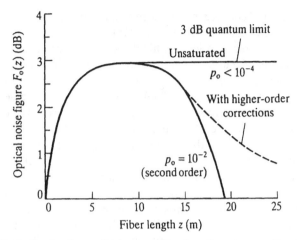

**FIGURE 2.29**    Optical noise figure $F_o(z)$ versus fiber length corresponding to the example of Figure 2.28 ($q_0 = 100, p_0 < 10^{-4}$ and $p_0 = 10^{-2}$) with second order decorrelation approximation. The dashed line approximately sketches the trend followed by $F_o(z)$ with higher order corrections.

input signal photon number is large in comparison to the equivalent input noise $n_{eq}$. After $z = 10$ m ($G > 20$ dB) in the small signal case ($p_0 < 10^{-4}$), the noise figure reaches the 3 dB quantum limit predicted by the linear theory and corresponding to the signal–ASE beat noise regime. For higher input signals ($p_0 = 10^{-2}$), however, the noise figure drops below the 3 dB quantum limit and steadily decreases with length, consistent with the decrease previously observed in the Fano factor.

As the saturation effect increases, Figure 2.29 shows that the noise figure drops to unity (0 dB) somewhere along the fiber ($z \approx 20$ m), with predicted values below unity at longer lengths. This erroneous prediction clearly shows that the second order approximation made in this example for the calculation of the noise variance $\sigma^2$ is no longer valid past a certain point. In order to accurately predict the actual noise figure corresponding to this region in the fiber, a higher order development involving the computations of the moments of higher rank $\langle n^j \rangle$ ($j > 2$) is necessary. The dashed line in Figure 2.29 approximately sketches the trend followed by the noise figure with higher order corrections.

The inadequacy of the second order development for the computation of the noise figure can be attributed to the large number of photons necessary to saturate the EDFA, as opposed to other types of optical amplifiers. Indeed, the value $s\langle n \rangle = 0.1$, which represents the onset of saturation, corresponds with $q = 100$ to $\langle n \rangle = 4.5 \times 10^7$ photons, with the typical EDFA parameters chosen in the previous example. With such a high photon number, the convergence of the Taylor expansion (2.240) is observed to be slow. By contrast, in the example considered in [34], the regime where $s\langle n \rangle = 0.1$ is reached for $\langle n \rangle = 10^3$, and the second order development is far more accurate. From these observations, we emphasize that *the accuracy of results obtained through the decorrelation method (2.240) must be carefully checked by comparison with higher order solutions*. The Taylor development of the brackets in Eq. (2.236) becomes rapidly fastidious and impractical as the order increases. Note that the decorrelation method is not unique, and that there exists only few criteria to determine a priori whichever is best suited, [63] and [64]. If no method in a specific instance makes it possible to achieve rapid convergence at the lowest orders, then a numerical integration of the photon statistics master equation (2.245) remains the only alternative. This approach, and related computational difficulties, are discussed later on.

The above analysis of the decorrelation problem has shown in particular, that it is difficult to predict the amplifier noise characteristics in the saturated gain regime. The all order decorrelation method (2.236) is inaccurate in the case of noncoherent or amplified signal light and could lead to significant errors in the evaluation of the amplifier noise variance. Some of the literature gives the formula $F_o = (1 + 2N)/G$, with $G$ and $N$ determined from the solution of the general rate equations (1.56) and (1.114) in the saturation regime. Such an assumption is equivalent to the all order decorrelation and is inaccurate for describing the saturated EDFA noise characteristics. *The noise figure formula $F_o = (1 + 2N)/G$ stems from the STT linear theory, and its validity as an approximation in the saturation regime has not yet been established.*

The numerical resolution of the nonlinear photon statistics master equation or (2.31), or (2.245) for EDFAs, can be done in several ways. The infinite number of equations need to be truncated from some number $n$ to some other number $n + M$ in order to compute the solution. The truncation implies that some estimation or extrapolation must be made in the two boundary equations in $dP_n/dz$ and $dP_{n+M}/dz$, for the functions $P_{n-1}$ and $P_{n+M+1}$. A possible extrapolation is given by the relations $P_{n-1} \approx (P_n/P_{n+1})P_n$ and $P_{n+M+1} \approx (P_{n+M}/P_{n+M-1})P_{n+M}$ [35]. This approximation means that, during integration, beyond the distribution peak, some functions $P_n$ can acquire negative values or values greater than a predecessor. These cause computational instability. As the solution of the equation system is sensitive to the extrapolation procedure, it is recommended to use for the numerical integration the first order approximation $P_n(z + h) \approx P_n(z) + h\,dP_n/dz$, as opposed to other integration algorithms [35]. The truncation can then be varied with fiber coordinate, so that the domain in which the solution is computed remains centered about the increasing mean $\langle n \rangle$.

Another approach is to treat the equation as a linear system of $M$ coupled equation and to solve it for each fiber coordinate using linear systems algorithms. For instance, in [64] a Laplace transform technique is used to reduce Eq. (2.231) into an eigenvalue problem. The drawback of this method is the number of equations that can be solved simultaneously is limited by the maximum allowable $M \times M$ matrix size. Furthermore, the Laplace transform cannot be made if the coefficients $a$, $b$ involved in the equation depend upon the coordinate $z$, which is the case if the amplifier is not pumped highly above inversion threshold, i.e., $q_0 \gg 1$. An extrapolation of the boundary conditions and a truncation of the photon number domain is also required at each integration step.

Another method is to reduce the set of discrete photon statistics master equations to a single partial differential equation. This method is valid when the mean photon number is large [22]. For large photon numbers $n$, the function $P_n$ slowly varies with $n$ and the following Taylor approximations can be made:

$$P_{n-1} \approx P_n - \frac{\partial P_n}{\partial n} + \frac{1}{2}\frac{\partial^2 P_n}{\partial n^2} \tag{2.256}$$

$$P_{n+1} \approx P_n + \frac{\partial P_n}{\partial n} + \frac{1}{2}\frac{\partial^2 P_n}{\partial n^2} \tag{2.257}$$

Using these approximations in master equation (2.245) yields:

$$\frac{dP_n}{dz} = -\{A_0(2n + 1) + A_1[(n + 1)f_{n+1} + nf_n]\}P_n$$

$$+ \{-n(A_0 + A_1f_n) + (n + 1)(B_0 + f_{n+1}B_1)\}\frac{\partial P_n}{\partial n}$$

$$+ \frac{1}{2}\{n(A_0 + A_1f_n) + (n + 1)(B_0 + f_{n+1}B_1)\}\frac{\partial^2 P_n}{\partial n^2} \qquad (2.258)$$

We can introduce then the function $S(z, n)$, as defined through $P_n(z) = \exp\{S(z, n)\}$, which varies even more slowly with $n$ [22]. Substitution of this definition into Eq. (2.258) yields:

$$\frac{dS}{dz} = -\{A_0(2n + 1) + A_1[(n + 1)f_{n+1} + nf_n]\}S$$

$$+ \{-n(A_0 + A_1f_n) + (n + 1)(B_0 + f_{n+1}B_1)\}\frac{\partial S}{\partial n}$$

$$+ \frac{1}{2}\{n(A_0 + A_1f_n) + (n + 1)(B_0 + f_{n+1}B_1)\}\left[\frac{\partial^2 S}{\partial n^2} + \left(\frac{\partial S}{\partial n}\right)^2\right] \qquad (2.259)$$

Considering now the discrete values $n_k = k\delta n$, where $\delta n$ is an arbitrary photon number increment, the partial derivatives with respect to $n$ involved in Eq. (2.259) can be evaluated through the finite difference approximations (Euler method):

$$\left(\frac{\partial S(z, n)}{\partial n}\right)_{n=n_k} \approx \frac{S(z, n_{k+1}) - S(z, n_{k-1})}{2\delta n} \qquad (2.260)$$

$$\left(\frac{\partial^2 S(z, n)}{\partial n^2}\right)_{n=n_k} \approx \frac{S(z, n_{k+1}) - 2S(z, n_k) + S(z, n_{k-1})}{\delta n^2} \qquad (2.261)$$

Substitution of Eqs. (2.260) and (2.261) into Eq. (2.259) makes possible numerical integration of the function $S(z, n)$ along $z$ over a large photon number range. For instance, the study in [22] shows solutions of Eq. (2.231), corresponding to four-level systems, calculated up to $n = 10^5$. For higher photon number values, a truncation of the photon number interval is necessary. At each integration step, this interval can be shifted by a quantity $j\delta n$ depending upon the mean photon number $\langle n(z)\rangle$. However, the main difficulty is to extrapolate the right tail of the probability distribution as the mean increases. This requires evaluation of $S(z, n + \delta n)$ and $S(z, n + j\delta n)$, where $S(z, n + \delta n)$ represents the last solution known, to be used as input conditions for the next integration step. Several methods are possible for this extrapolation, based for instance on ad hoc extrapolation functions whose coefficients can be determined through least-squares fitting [63] or based upon more general nonlinear curve-fitting algorithms, e.g., Levenberg–Marquardt [71]. The drawback of such extrapolation methods is that slight fitting inaccuracies can result in large errors in the evaluation of the probability distribution tail, therefore the method's accuracy must be carefully checked at each integration step [63]. Another difficulty is the additional computation time required by fitting algorithms, particularly if a large number of reference points are used for the nonlinear extrapolation. The above arguments show that when large photon numbers are considered, as in the case of saturated EDFAs for which the onset of saturation is typically near $n \approx 10^7$, numerical resolution of the nonlinear photon statistics master equation turns out to be a complex task likely to require powerful computers.

The reduction of light statistical fluctuations in saturated amplifiers is a known feature that was analyzed in previous studies of four-level amplifier systems with saturable emission and absorption coefficients, [34], [63], and [65]. This effect is also predicted here for EDFAs and stems from the similarity between the photon statistics master equations used in the theoretical descriptions. As an SNR improvement is predicted with saturated EDFAs, we may conclude that the effect could be advantageously applied to reduce the BER in communications systems. But this conclusion cannot be drawn by exclusively considering the amplifier SNR characteristics. Indeed, Chapter 3 shows that the BER depends upon the photon number probability distribution $P_n$ associated with the digital symbols 0 and 1. These distributions contain more information about the photon statistics than merely the statistical moments $\langle n \rangle$ and $\langle n^2 \rangle$. Intuitively, we might conclude that the reduction of statistical fluctuations in the region $n > \langle n \rangle$ caused by gain saturation must be accompanied by some increase of the probability distribution tails, because of probability conservation. The study [64] shows that *the probability distribution associated with saturated light is characterized by tails that are higher than those of a Poisson distribution with the same mean* $\langle n \rangle$. This is consistent with the increase of the Fano factor with amplifier gain, observed in the previous numerical example, and shows that the probability distribution of amplified light always spreads over a range larger than the Poisson distribution. From [65], the limiting solution corresponding to the case of a highly saturated amplifier (i.e., for which $dP_n/dz \approx 0$) is found to be broader than the Poisson distribution with same mean by the factor:

$$x = \sqrt{\frac{a}{a-b}} = \sqrt{n_{\text{sp}}} \qquad (2.262)$$

This result assumes constant coefficients $a$, $b$ with $a \gg b$. In the case of near-complete medium inversion, $n_{\text{sp}} \approx 1$, the distribution of the output light is nearly equal to Poisson. Thus, the effect of strong saturation is to reduce the fluctuations of amplified light to a limit where the photon statistics approach the Poisson distribution. Assuming a gain $G$ and a mean output $\langle n \rangle$ for such a highly saturated amplifier, the corresponding noise variance in the Poisson limit is $\sigma^2 = \langle n \rangle = G \langle n_0 \rangle$, and the noise figure is $F_o = 1/G$. This result indicates that, consistent with the Heisenberg uncertainty principle (Section 2.1) such a limiting solution must correspond to the case where the amplifier gain is saturated to the point of transparency ($G = 1$), for which the noise figure $F_o$ is unity. The uncertainty principle would be violated if both input and output statistics were Poisson while the amplifier gain were greater than unity. Indeed, in this assumption, the uncertainty $\Delta n$ in the output photon number would be given by $\Delta n = \sigma = \sqrt{G \langle n_0 \rangle} = \sqrt{G} \sigma_0 = \sqrt{G} \Delta n_0$, and assuming a minimum uncertainty $\Delta n_0 \Delta\varphi_0 = 1/2$ for the input signal, we would find $\Delta n\Delta\varphi_0 = 1/(2\sqrt{G}) < 1/2$, in violation of the uncertainty principle. The main conclusion of this demonstration is that *both linear and nonlinear amplification cannot reduce the photon number uncertainty of the input signal.*

EDFA operating conditions close to the limiting regime discussed above are unlikely to prevail. The reason is that amplifier saturation increases pump absorption, which results in reduced inversion along the fiber and consequently in an increase of the spontaneous emission factor $n_{\rm sp}$, as observed experimentally when there is no amplifier self-saturation [68]. In the regime where self-saturation by ASE occurs, the spontaneous emission factor is observed to remain nearly constant during saturation, as discussed in Section 5.4. In any experimental conditions, the output probability distribution of highly saturated EDFAs is always expected to be broader than the Poisson distribution, even in the near-transparency case.

We have seen that photon statistics distributions of the same mean but corresponding to linear and nonlinear amplifiers differ in the extent of their tails. The tails are broader in the nonlinear amplifier so the overlap between distributions associated with symbols 0 and 1 in the nonlinear amplifier must be greater than in the linear case, causing an increase in BER. The theoretical study [67], based on the numerical computation of the corresponding distributions, has proven this statement, i.e., that *the BER is not improved by saturated amplification, when compared to unsaturated amplification.* This conclusion applies to EDFAs, as phenomenologically observed so far [68]. Full theoretical analysis of nonlinear photon statistics in EDFAs, of associated BERs, and of their consequence in communications systems are yet to be accomplished.

# CHAPTER 3

# PHOTODETECTION OF OPTICALLY AMPLIFIED SIGNALS

## INTRODUCTION

Among the most important aspects of physical light that were discussed in the previous chapter, is the fact that the number of photons or energy quanta is not deterministic. On the contrary, it has a definite statistical nature. Light amplification is more specifically a process of photon multiplication and it increases the uncertainty in the number of quanta. This is because there is randomness associated with stimulated emission and because spontaneous emission generates additional noise.

In this chapter, we consider another fundamental process, the conversion of light into electrical current – or photodetection. In photodetection, incident photons are captured to release an electron from a previously bound state. The generation of such a photoelectron obeys statistical laws intimately related to the incident photon stream. This chapter analyzes the relation.

Photodetection can be viewed as one of the means to study some specific characteristics of light (e.g. photon number statistics, power spectrum), providing some information about the physics of light-generating processes (e.g. laser oscillation, atomic transitions), which has opened up the 30 years old field of *photon counting* [1]–[4]. The study of photodetection and signal multiplication statistics in biological systems, such as the eye's retina, provides clues about the complex functioning of visual perception [5]. Photodetection also plays a central role in lightwave communications systems; it is the first step in the retrieval of the information conveyed by light signals. Our main focus will be on this last type of application.

We shall not discuss the deep meanings involved in the notion of information and its vast implications in communications. For insights in communication theory however, the reader may refer to the pioneering work by C. S. Shannon [6], D. Gabor [7], J. P. Gordon [8] and H. Takahasi [9] (see also reviews [10] and [11], for instance, and references therein). We keep here the simplified point of view that if information is contained in a light signal, the goal of photodetection is to convey this information with the smallest loss or degradation possible. While photodetection represents only a step in information retrieval, it is the most critical. In digital systems, for instance, the information is physically reduced to a stream of elementary *mark* and *space symbols*, emulated by pulses of modulated light with different amplitude, frequency, or phase. Considering the simplest case of digital amplitude modulation, where the light is basically turned on and off, the function of the photodetector is to translate these light pulses into current pulses. Interpretation of these current pulses in terms of digital symbols is a matter of subsequent electronic signal processing. But the process of detection is statistical in nature and introduces noise. This can lead to errors in the interpretation of symbols. The main goal of this chapter is to describe the photoelectron statistics, its relation with that of amplified light, and, more importantly, to analyze *the role that can be played by optical amplifiers to improve photodetection performance.*

We have previously discussed the notion of an ideal photodetector, a device whose current output statistics faithfully replicate those of an incoming light signal. In reality, there exists no such ideal detector. This is for two principal reasons: first, detectors phenomenologically exhibit a background noise current, or thermal noise, in the absence of any input optical signal; second, the conversion efficiency from an incident photon to a photoelectron is generally less than 100%. Other limitations also contribute to degrade the light signal characteristics, these relate to the electronic bandwidth and linearity of the detector and its associated circuitry (receiver). It is beyond the scope of this book to provide insights into the characteristics of different types of photodetectors and receiver designs. For a study of this vast subject, the reader may refer to textbooks, [12]–[15]; for an overview of state-of-the-art technology in communication systems, to [16]; and for familiarization with the fundamental physics and technology of semiconductor detectors, to [17]. In this chapter, we shall be concerned only with the basic notions of photodetection necessary for a minimum understanding of the role played by optical amplifiers in communications systems. The concept of noise, which was analyzed in the previous chapter in the case of optical fiber amplifiers, is extended here to include the contribution of the receiver, without which we could not assess the performance of a communications systems using such optical amplifiers.

In Section 3.1, the theory of quantum photodetection statistics for single-mode radiation is analyzed. The photoelectron count is related to the light photon statistics through the *Bernoulli sampling formula.*

The semiclassical photodetection theory outlined in Section 3.2 makes it possible to express the photocurrent variance, which accounts for the power spectral density of the different photocurrent noise components, as well as the effect of the finite electronics bandwidth of the detector.

It is shown in Section 3.3 that optical amplification prior to photodetection, or *signal preamplification,* results in the enhancement of signal-to-noise ratio at the receiver end, which also corresponds to an increase of detector sensitivity.

In Section 3.4, the *detector sensitivity enhancement* obtained with optical pre-amplification is compared to the sensitivity of *avalanche photodetectors.* The combined effect of optical preamplification and avalanche gain is also analyzed.

The signal *bit-error rate* (BER) and *receiver sensitivity* (number of photons/bit to achieve BER $= 10^{-9}$) are studied in Sections 3.5 and 3.6 for the case of digital signals in direct detection systems and coherent detection systems, respectively. It is shown that the effect of optical preamplification is an increase in receiver sensitivity for direct detection, while the effect is of limited consequence for coherent detection.

Section 3.7 analyzes the ultimate bit-error rate performance limits with *direct detection* or *coherent detection systems* based on optical amplifier chains. The amplification and detection of *analog signals* is studied in Section 3.8.

## 3.1  QUANTUM PHOTODETECTION STATISTICS

The statistics of quantum photodetection, or the counting of photoelectrons generated by laser light incident onto a detector, have been quantum mechanically analyzed by M. O. Scully and W. E. Lamb [18]. The purpose of such a theoretical investigation was to determine a relation between the laser output photon statistics $P_n(0)$ and that of the photoelectron count $P'_m$. However complex the analysis, the result turns out to be remarkably simple and intuitive. Namely, if $\eta$ represents the detector quantum efficiency, or the probability of capturing an incident photon ion by the photodetecting material, the photoelectron statistics distribution $P'_m$ is given by the *Bernoulli sampling formula*:

$$P'_m = \sum_{n=m}^{\infty} \binom{n}{m} \eta^m (1 - \eta)^{n-m} P_n(0) \qquad (3.1)$$

where

$$\binom{n}{m} = \frac{n!}{m!\,(n-m)!} \qquad (3.2)$$

represents the combinational factor.

The physical interpretation of the Bernoulli sampling formula is the following. First, the probability of observing $m$ photoelectrons, corresponding to the capture of $m$ photons, is $\eta^m$. Assuming there are $n$ incident photons, this probability is to be multiplied by the probability $\eta^{n-m}$, corresponding to $n-m$ failures to capture a photon. The combinational factor $(n, m)$, accounts for the number of possible ways $m$ photons can be captured among the group of $n$ photons. The overall probability of this event is found from multiplying by the input probability $P_n(0)$, and summing over all possibilities (as the number of photons $n$ is not deterministic), with the requirement $m \leqslant n$ since only $m$ photons out of $n$ are captured.

This important result could also be derived with the photon statistics formalism previously used in Chapter 2 for analyzing optical amplifiers. Indeed, given an initial photon population, the random process of photon capture can be viewed as a *Markoff process*, fully characterized by a *photon statistics master equation*, or forward Kolmogorov equation Eq. (2.72). In the *birth–death–immigration* (BDI) picture, the random capture of photons by the detecting medium represents a pure death (D) process, or a restricted case of BDI interaction. Thus, the master equation for the D case is given by Eq. (2.72) with coefficient $a = 0$ and $b$ the absorption coefficient:

$$\frac{dP_n}{dz} = b\{(n + 1)P_{n+1} - nP_n\} \qquad (3.3)$$

Multiplying the above equation by $n$, summing with the translation rule $\sum f(n)P_{n+1} = \sum f(n-1)P_n$, and integrating over the depth $L$ of the detector yields for the mean number of the photon population having escaped detection or capture

$$\langle n(L) \rangle = \langle n(0) \rangle \exp(-bL) = T \langle n(0) \rangle \tag{3.4}$$

where $\langle n(0) \rangle$ is the mean photon input. The mean number of captured photons or (photoelectrons) is therefore $\langle n_e \rangle = (1 - T)\langle n(0) \rangle$, so that $\langle n(L) \rangle + \langle n_e \rangle = \langle n(0) \rangle$ for conservation of the number of particles. The factor $\eta = (1 - T)$ is the detector quantum efficiency, or the capture probability.

Given $n$ photons incident upon the detector, the probability $P'_m$ of generating $m$ photoelectrons or succeeding in $m$ photon captures is equal to the probability $P_{n-m}$ of observing $n - m$ output photons. Thus, the photon statistics of the light escaping the detector and that of the generated photoelectrons are directly related. In order to determine the statistics of the output light, we can use the analysis by K. Shimoda, H. Takahasi, and C. H. Townes (STT) [19] described in the previous chapter. The detector is a special case of an optical absorber having no spontaneous nor stimulated emission ($a = 0$), which is described by the master equation (3.3). From the STT analysis, the photon statistics distribution $P_k(L)$ of the light output from an optical amplifier with length $L$, gain $G = T$ and output ASE noise $N = 0$, is given by Eqs. (2.40) and (2.41):

$$P_k(L) = \sum_{n=0}^{\infty} P_{n,k}(L)P_n(0)$$

$$= \sum_{n=0}^{\infty} \left\{ (1 - T)^n \sum_{j=0}^{\min(k,n)} \frac{k!\, n!}{j!\, j!\, (k-j)!\, (n-j)!} \left[ \frac{T}{1-T} \right]^j N^{k-j} \right\} P_n(0) \tag{3.5}$$

Two cases can be considered from the above result: for $k > n$, we find $P_k(L) = 0$, since $N^{k-j} = 0$, meaning that the output photon number $k$ cannot be greater than the input $n$. For $k < n$, we have $N^{k-j} = \delta_{k,j}$, and we find

$$P_k(L) = \sum_{n=0}^{\infty} \left\{ (1 - T)^n \sum_{j=0}^{k} \frac{k!\, n!}{j!\, j!\, (k-j)!\, (n-j)!} \left[ \frac{T}{1-T} \right]^j N^{k-j} \right\} P_n(0)$$

$$= \sum_{n=0}^{\infty} \binom{n}{k} (1 - T)^{n-k} T^k P_n(0) \tag{3.6}$$

Using the relation $P'_m = P_{n-m}(L)$ linking the photoelectron and output photon count probabilities, we find from Eq. (3.6) with $k = n - m$ and $T = 1 - \eta$, the photoelectron count:

$$P'_m(L) = \sum_{n=m}^{\infty} \binom{n}{m} \eta^m (1 - \eta)^{n-m} P_n(0) \tag{3.7}$$

which is identical to the Bernoulli sampling formula (3.1).

Considering that the incoming light to be detected is a superimposition of coherent and chaotic (thermal) light, an important property is that the Bernoulli sampling formula conserves the functional form of the statistics [20]. A straightforward example is given by the Bernoulli sampling of purely coherent light (Poisson statistics). Indeed, substituting $P_n(0) = \langle n(0) \rangle \exp\{-\langle n(0) \rangle\}/n!$ into Eq. (3.7) gives:

$$
\begin{aligned}
P'_m(L) &= \sum_{n=m}^{\infty} \frac{n!}{m!(n-m)!} \eta^m (1-\eta)^{n-m} \frac{\langle n(0) \rangle^n \exp\{-\langle n(0) \rangle\}}{n!} \\
&= \frac{(\eta \langle n(0) \rangle)^m \exp\{-\eta \langle n(0) \rangle\}}{m!} \sum_{n=m}^{\infty} \frac{\langle n(0) \rangle^{n-m}(1-\eta)^{n-m}}{(n-m)!} \exp\{-(1-\eta)\langle n(0) \rangle\} \\
&= \frac{(\eta \langle n(0) \rangle)^m \exp\{-\eta \langle n(0) \rangle\}}{m!}
\end{aligned}
\tag{3.8}
$$

which is the Poisson distribution of mean $\eta \langle n(0) \rangle$.

In the more complex case of a superimposition of coherent and chaotic light, such as light output from an optical amplifier, the property of conservation of statistics functional form can be shown easily through the probability generating function (PGF) method. Section 2.3 shows that in the case of an amplifier, 1, followed by a lossy element, 2, (the whole is called a Type A chain element), the output light PGF is given by Eq. (2.206):

$$
F_{\text{out}}^{(2)}(x) = F_{\text{in}}[F_{\text{BD}}^{(2)}(x)]F_{\text{I}}^{(2)}(x)
\tag{3.9}
$$

where $F_{\text{in}}(x)$ is the PGF of the light input to the element 2, and $F_{\text{BD}}^{(2)}(x)$ and $F_{\text{I}}^{(2)}(x)$ are the PGFs corresponding to BD and I processes in this second element, respectively. Assume now that the lossy element 2 is a photodetector, with transmission $T = 1 - \eta$. Then, the PGF characterizing the element 2 is that of a pure D process, i.e., $F_{\text{out}}^{(2)}(x) = 1 + (x - 1)T$, associated with *Bernoulli random deletion* [21]. Since there is no spontaneous emission (immigration) in the detector, we also have $F_{\text{I}}^{(2)}(x) = 1$. Assuming now an input signal with poisson statistics, the light output from the amplifier 1 and input to detector 2 has for PGF Eq. (2.101):

$$
F_{\text{in}}(y) = \frac{\exp\left\{\dfrac{\langle n(0) \rangle(y - 1)G}{1 - (y - 1)N}\right\}}{[1 - (y - 1)N]^{\mathscr{M}}}
\tag{3.10}
$$

Using the above relations, we find from Eqs. (3.9)–(3.10):

$$
F_{\text{out}}^{(2)}(x) = \frac{\exp\left\{\dfrac{\langle n(0) \rangle(x - 1)(1 - \eta)G}{1 - (x - 1)(1 - \eta)N}\right\}}{[1 - (x - 1)(1 - \eta)N]^{\mathscr{M}}}
\tag{3.11}
$$

The PGF in Eq. (3.11) corresponds to the *noncentral negative binomial* (NNB) probability distribution Eq. (2.102):

$$P_k(L) = \frac{[(1 - \eta)N]^k}{[(1 - \eta)N + 1]^{k+\mathcal{M}}} \exp\left\{ -\frac{G(1 - \eta)\langle n_0 \rangle}{(1 - \eta)N + 1} \right\}$$

$$\times L_k^{(\mathcal{M}-1)}\left\{ -\frac{G(1 - \eta)\langle n_0 \rangle}{(1 - \eta)N[(1 - \eta)N + 1]} \right\} \tag{3.12}$$

The distribution (3.12) corresponds to the light output from the detector, i.e., having escaped photon capture. In the case $\eta = 0$ (no photon capture), the distribution is identical to that of the light input to the detector, Eq. (2.102). Comparison of the light statistics before and after passing through the detector shows that in the case of a statistical mixture (coherent + chaotic), the Bernoulli random deletion process conserves the functional form of the statistics.

On the other hand, if we are interested in the statistics of the photoelectron count, we can use the PGF $F_{\text{out}}^{(2)}(x) = 1 + (x - 1)\eta$, which characterizes the random capture of photons with probability $\eta$. From Eqs. (3.9) and (3.10) we obtain the photoelectron count PGF:

$$F_{\text{out}}^{(2)}(x) = \frac{\exp\left\{ \dfrac{\langle n(0) \rangle(x - 1)\eta G}{1 - (x - 1)\eta N} \right\}}{[1 - (x - 1)\eta N]^{\mathcal{M}}} \tag{3.13}$$

which corresponds to the *photoelectron count* probability distribution $P_m'$:

$$\boxed{P_m' = \frac{(\eta N)^m}{(\eta N + 1)^{m+\mathcal{M}}} \exp\left\{ -\frac{G\eta\langle n_0 \rangle}{\eta N + 1} \right\} L_m^{(\mathcal{M}-1)}\left\{ -\frac{G\eta\langle n_0 \rangle}{\eta N(\eta N + 1)} \right\}} \tag{3.14}$$

This last result shows that the photoelectron count statistics also take the same functional form as that of the input light. In the case of an ideal, 100% efficient photodetector, i.e., $\eta = 1$, the distribution (3.14) is identical to that of the input light, Eq. (2.202). *With ideal photodetectors, the photoelectron statistics are identical to those of incident light,* a result only postulated in Chapter 2. In the case where $\eta < 1$, the photoelectron count distribution $P_m'$ in Eq. (3.14) is identical to that of light first amplified with gain $G$ then attenuated with transmission $\eta$, with Poisson statistics input. This last property is very useful; the transfer matrix formulae can be used to determine the photoelectron mean and variance given the input light mean $\langle n \rangle$, its variance $\sigma^2$, and the detector efficiency $\eta$. The photodetector can indeed be characterized by the transfer matrix $\hat{D}$:

$$\hat{D} = \eta\begin{pmatrix} 1 & 0 \\ 1 - \eta & \eta \end{pmatrix} \tag{3.15}$$

The photoelectron count mean $\langle n_e \rangle$ and variance $\sigma_e^2$ are then given by the simple matrix relation:

$$\begin{pmatrix} \langle n_e \rangle \\ \sigma_e^2 \end{pmatrix} = \hat{D} \begin{pmatrix} \langle n \rangle \\ \sigma^2 \end{pmatrix} \tag{3.16}$$

Considering the case of *amplified light detection* (with no particular statistics assumed for the signal input to the amplifier), we can use Eqs. (2.73) and (2.75) for $\langle n \rangle$ and $\sigma^2$, and the above Eq. (3.16) to obtain the photoelectron mean and variance:

$$\boxed{\langle n_e \rangle = \eta G \langle n(0) \rangle + \mathcal{M}\eta N} \tag{3.17}$$

$$\boxed{\sigma_e^2 = \eta^2 G^2[\sigma^2(0) - \langle n(0) \rangle] + \eta G \langle n(0) \rangle + \mathcal{M}\eta N(z) + 2\eta^2 G \langle n(0) \rangle N + \mathcal{M}\eta^2 N^2}$$

$$\tag{3.18}$$

where $\langle n(0) \rangle$ and $\sigma^2(0)$ are the mean and variance of the signal input to the amplifier. In the derivation of Eqs. (3.17) and (3.18), we assumed there were no coupling losses at the optical amplifier input and between the amplifier output and the detector. Losses can be introduced by multiplying $[\langle n(0) \rangle, \sigma^2(0)]$ and $[\langle n \rangle, \sigma^2]$ by coupling matrices $\hat{D}$ of the type (3.15), with $\eta$ representing either the input coupling efficiency or the amplifier/detector coupling efficiency, respectively, and replacing the results in Eqs. (3.17) and (3.18). The use of such coupling or transfer matrices at any point in the amplifier/detector chain is equivalent to assuming modifications of the photon statistics related to the Bernoulli random partitioning or random deletion processes [3].

Equations (3.17) and (3.18) express the mean and variance of the photoelectron count. To put these fundamental results in terms of mean electrical photocurrent and electrical noise power, consider first the mean photocurrent and define signal and noise components $I_s$ and $I_N$, [3] and [22]:

$$I_s = \frac{eP_s(0)}{h\nu_s} = \frac{e\langle n(0) \rangle h\nu_s B_o}{h\nu_s} = e\langle n(0) \rangle B_o \tag{3.19}$$

$$I_N = \frac{eP_N}{h\nu_s} = \frac{eN h\nu_s B_o}{h\nu_s} = eN B_o \tag{3.20}$$

where $e$ is the electric charge, $B_o$ the optical bandwidth, and $P_s(0)$ the signal input to the amplifier. Note that the above definitions for $I_s$, $I_N$ do not involve the quantum efficiency $\eta$ and the amplifier gain $G$, as a matter of convention. From Eqs. (3.17), (3.19), and (3.20) we find the total detector mean photocurrent:

$$\boxed{I_d = \eta(G I_s + \mathcal{M} I_N)} \tag{3.21}$$

Considering next the total electrical noise power $\sigma_d^2$ (normalized to a unit load resistance), we define the *shot noise* and *beat noise* components through:

$$\sigma_{shot}^2 = 2\eta e B_e(GI_s + \mathcal{M}I_N) \tag{3.22}$$

$$\sigma_{beat}^2 = \eta^2\{2GI_sI_N f(B_e, B_o) + \mathcal{M}I_N^2 f'(B_e, B_o)\} \tag{3.23}$$

where $B_e$ is the detector electronic bandwidth. The justification for the factor $2\eta_e B_e$ in the shot noise component (3.22) stems from the physics of the shot effect. The theory [1] shows that the noise variance related to the shot effect, is proportional to the mean photocurrent $\langle I \rangle$ and is approximately given by $\sigma^2 = 2eB_e\langle I \rangle$ (the approximation is valid if the electron transit time from the detector's cathode to anode is small compared to $1/B_{max}$, where $B_{max}$ is the highest frequency in the electronic bandwidth). The beat noise component in Eq. (3.23) was expressed directly as a function of the products of photocurrents defined in Eqs. (3.19) and (3.20). The justification is that the received optical power has an intrinsic beat noise component of standard deviation $\sigma = h v_s B_o \sqrt{(2\eta^2 G\langle n(0)\rangle N + \mathcal{M}\eta^2 N^2)}$. This noise generates a beat photocurrent $I_{beat}$ of zero mean and of variance $\sigma_{beat}^2 = (e\sigma/hv_s)^2 f(B_e, B_o)$, where $f(B_e, B_o)$ is a factor accounting for the fraction of beat noise power that effectively falls into the detector bandwidth. As we do not know a priori whether this fraction is the same for the two beat noise components (i.e., signal–ASE and ASE–ASE), two different functions $f(B_e, B_o)$ and $f'(B_e, B_o)$ were introduced in Eq. (3.23). The form of these functions cannot be predicted at this point, but will be specified in the next section.

In addition to the shot and beat noise, the detector and associated electronic circuitry contribute to a thermal noise component $\sigma_{th}^2$ (also called Johnson or Nyquist noise) with definition [3]:

$$\sigma_{th}^2 = \frac{4k_B T B_e}{R} \tag{3.24}$$

where $k_B$ is Boltzmann's constant ($k_B = 1.38 \times 10^{-23}$ J/K), $T$ an effective (absolute) temperature, and $R$ the load resistor (here $R = 1$).

The total detector electrical noise power $\sigma_d^2$ is then from Eqs. (3.22)–(3.24):

$$\sigma_d^2 = \sigma_{shot}^2 + \sigma_{beat}^2 + \sigma_{th}^2 = 2\eta e B_e(GI_s + \mathcal{M}I_N)$$

$$+ \eta^2\{2GI_sI_N f(B_e, B_o) + \mathcal{M}I_N^2 f'(B_e, B_o)\} + \frac{4k_B T B_e}{R} \tag{3.25}$$

The excess noise component involved in the signal light variance in Eq. (2.74), and proportional to $\sigma^2(0) - \langle n(0)\rangle$, is not accounted for in the Eq. (3.25) for the detector noise power. We therefore assume the input signal light is coherent, i.e., with Poisson statistics. In this case we may ignore the excess noise component.

This derivation of the photodetector mean and variance is based upon a simplified quantum photodetection theory, in which the signal light incident upon the detector is treated as a stochastic population of photon particles. Photodetection is a Bernoulli random deletion (sampling) of the signal photon population [3]. The central assumption made in this theory is that the optical signal to be detected corresponds to a single mode of a radiation field. Chapter 2 shows that the photon statistics master equation, which represents the starting point of the theory developed in this section, describes the photon probability distribution $P_n$ associated with a single radiation mode. The quantum mechanical representation of this mode is the photon number state $|n\rangle$, and the transition probabilities between the different number states $|n\rangle$ are determined from the electric dipole Hamiltonian $\mathcal{H}_{ed}$ defined in Eq. (2.19). This interaction Hamiltonian assumes a single frequency electric field of operator representation Eq. (2.17):

$$\hat{E} = \mathcal{C}\hat{\varepsilon}\{ae^{-i(\omega t - kz)} - a^+ e^{i(\omega t - kz)}\} \tag{3.26}$$

The annihilation and creation operators $a$, $a^+$ can be expressed as a product of amplitude and phase operators, with $\hat{n} = aa^+ - 1$ according to [2]:

$$a = (\hat{n} + 1)^{1/2}\widehat{\exp}(i\varphi), \quad a^+ = \widehat{\exp}(-i\varphi)(\hat{n} + 1)^{1/2} \tag{3.27}$$

An important observation is that in the description of noise and photon statistics made in Chapter 2, the action of quantum mechanical phase operators involved in Eq. (3.27) has been overlooked. This is equivalent to assuming that the quantum fluctuations of the amplified light, which lead to the photon number uncertainty $\sigma$, are determined only by the amplitude fluctuations of the electric field, not by phase fluctuations too. Chapter 5 shows that this approximation is quite accurate at the signal power levels (or photon numbers) used in realistic fiber amplifiers. But a complete theory applicable to the case of small photon numbers should include the effect of phase change associated with light amplification. Then it could predict the statistics of both phase and amplitude fluctuations of the amplified field, which contains more information than the mere photon number statistics.

The introduction of a time-varying phase in the electric field operator (3.26) and the additional assumption that the optical signal is not single-mode but consists of a superposition of radiation modes at discrete frequencies, lead to a far more complex analysis for photocounting statistics [2]. A full quantum analysis of this problem, developed by P. L. Kelley and W. H. Kleiner [23], shows that in the most general case, the photocount distribution $P'_m$ is given by [2]:

$$P'_m = \left\langle \mathcal{N} \frac{\{\zeta\hat{I}(T)T\}^m}{m!} \exp\{-\zeta\hat{I}(T)T\} \right\rangle \tag{3.28}$$

where the brackets denote *quantum mechanical averaging* (i.e., $\langle\hat{X}\rangle = tr\{\hat{\rho}\hat{X}\}$, $\hat{\rho}$ is the density matrix operator), $\mathcal{N}$ the normal ordering operator, $T$ the photon counting interval, $\zeta$ a factor related to the detector quantum efficiency, and $\hat{I}(T)$ the intensity operator time averaged over the photon counting interval, defined by:

$$\hat{I}(T) = \frac{1}{T}\int_0^T 2\varepsilon_0 c\hat{E}(z, t)\hat{E}^+(z, t)\, dt \tag{3.29}$$

It is shown in [2] that for a single-mode initial field, such as defined in Eq. (3.26), the photon counting statistics distribution $P'_m$ in Eq. (3.28) exactly reduces to the

Bernoulli sampling formula of Eq. (3.1). Equation (3.28) shows that in the general case, however, the photocount distribution predicted by the full quantum theory is not easily tractable. For this practical reason, we shall focus in the next section on a *semiclassical theory of photodetection*, which alleviates the difficulty by considering classical multimode fields as opposed to quantized fields. This theory, where the photoelectrons are quantized and the light is classical with continuous statistics, can also predict the photocount probability distribution $P'_m$, through a classical counterpart of Eq. (3.28). This theory, first derived by L. Mandel [24] is also analyzed in [2] and [25]. But if we are only interested in the mean and variance of the photocount, these quantities can be calculated by time averaging as opposed to statistical averaging. Time averaging can be done without knowledge of the probability distribution $P'_m$.

In the multimode signal case, a rigorous computation of the noise power spectral density through quantum photodetection theory would also be very complex. Information about the power spectral density is of great importance, as we expect that the fraction of noise power carried by the photocurrent must be a function of both the incident signal and the detector electronic bandwidths. Equation (3.25) for the photocount variance was derived under the assumption that the detected light (signal + ASE) is confined into a single radiation mode of bandwidth $B_o$, but even in this simplified assumption the theory does not provide any information about the factors $f(B_e, B_o)$ and $f'(B_e, B_o)$ involved in its beat noise components. Determination of these factors requires knowledge of the power spectral densities in the detector bandwith interval $[0, B_e]$ of the corresponding beat noises. While a quantum mechanical determination of this power spectral density would be quite complex, the semiclassical theory of photodetection developed in the next section makes it comparatively straightforward.

## 3.2  SEMICLASSICAL DESCRIPTION OF PHOTODETECTION

In this semiclassical theory, light output from the amplifier is represented as a superposition of classical signal and noise electric fields of deterministic amplitudes; the noise has a random phase. The photodetector is a squaring-law device that yields a photocurrent proportional to the mean optical power. The noise of mean Power $P_N = Nh\nu B_o$ ($N$ = ASE photon number output from the amplifier) that is added to the signal is not predicted by any classical theory, and the detailed expression of $N$ (i.e. $N = n_{sp}(G - 1) = n_{eq}G$) is based upon the quantum theory developed in Chapter 2. As this description borrows from both theories, it is said to be semiclassical.

The process of square-law detection and the determination of photocurrent noise power spectral density is a well-known problem in electrical engineering [26]. The theory outlined in this section restates with notation consistent with this book a generic analysis of square-law detection of amplified light (P. S. Henry and N. A. Olsson [22]).

The signal and noise classical fields amplitudes $E_s$, $E_N$ incident upon the detector are first expressed as follows:

$$E_s(t) = \sqrt{2GP_s^{in}} \cos(2\pi\nu_s t) \tag{3.30}$$

$$E_N(t) = \sqrt{2Nh\nu_s \delta\nu} \sum_{k=-M}^{M} \cos\{2\pi(\nu_s + k\delta\nu)t + \varphi_k\} \tag{3.31}$$

In the above equations, $P_s^{in}$ is the mean signal power input to the amplifier, $Nh\nu_s\delta\nu$ is the mean ASE power in bandwidth $\delta\nu$, $M = B_o/2\delta\nu$ is an integer. The total ASE noise field is assumed to be made of the superimposition of $2M$ independent radiation modes at optical frequency $\nu_s + k\delta\nu$, with a random phase $\varphi_k$. Both fields are assumed to be in the same optical polarization. The mean optical powers associated with signal and ASE noise in bandwidth $\delta\nu$ are given by $|E_s(t)|^2 = GP_s^{in}$ and $|E_N(t, \delta\nu)|^2 = Nh\nu_s\delta\nu$, respectively.

Assuming a detector with unity quantum efficiency, the instantaneous photocurrent $i(t)$ is given by the relation:

$$i(t) = \frac{e}{h\nu_s}[E_s(t) + E_N(t)]^2 \tag{3.32}$$

Substituting Eqs. (3.30) and (3.31) into Eq. (3.32) yields:

$$i(t) = \frac{2e}{h\nu_s} GP_s^{in} \cos^2(2\pi\nu_s t) + \frac{2e}{h\nu_s} \sqrt{GP_s^{in} Nh\nu_s\delta\nu}$$

$$\times \sum_{k=-M}^{M} \cos\{2\pi(\nu_s + k\delta\nu)t + \varphi_k\}\cos(2\pi\nu_s t)$$

$$+ 2eN\delta\nu \left[\sum_{k=-M}^{M} \cos\{2\pi(\nu_s + k\delta\nu)t + \varphi_k\}\right]^2 \tag{3.33}$$

A detailed analysis of the different components involved in Eq. (3.33) is made in [22] and is also given in Appendix K. It shows that the mean signal and ASE photocurrents are given by the same expressions as Eqs. (3.19) and (3.20), the sum of which, accounting for $\mathscr{M}$ ASE modes and the finite detector quantum efficiency $\eta$, is given by Eq. (3.21). On the other hand, the mean-square development $\langle i^2(t)\rangle$ of Eq. (3.33) yields signal–ASE and ASE–ASE power spectral densities with the following expressions:

$$\hat{\sigma}_{s\text{-ASE}}^2(f) = \begin{cases} 4GI_sI_N\dfrac{1}{B_o}; & f \leqslant B_o \\ 0; & f > B_o \end{cases} \tag{3.34}$$

$$\hat{\sigma}_{\text{ASE-ASE}}^2(f) = \begin{cases} \dfrac{2I_N^2}{B_o}\left(1 - \dfrac{f}{B_o}\right); & f \leqslant B_o \\ 0; & f > B_o \end{cases} \tag{3.35}$$

The spectral densities of the two beat noise components (s–ASE and ASE–ASE) in the RF domain, as defined in Eqs. (3.34) and (3.35) and their sum are plotted in Figure 3.1, with as an example $I_N > 2GI_s$. The s–ASE beat noise is uniform over the frequency interval $[0, B_o/2]$, while the ASE–ASE beat noise linearly decays with frequency to vanish at $f = B_o$.

The total beat noise powers falling into the electronic bandwidth $B_e$ (assuming

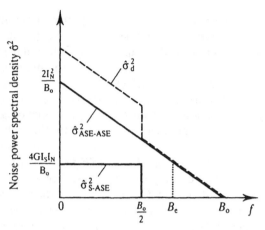

**FIGURE 3.1**   Plots of the noise power spectral densities in the RF domain of photodetector signal–ASE and ASE–ASE beat noise components.

uniform electronic response and no filltering) are given by (Appendix K):

$$\sigma^2_{\text{s-ASE}} = 4GI_sI_N \frac{B_e}{B_o} \tag{3.36}$$

$$\sigma^2_{\text{ASE-ASE}} = I_N^2 \frac{B_e}{B_o^2}(2B_o - B_e) \tag{3.37}$$

From the results of Eqs. (3.36) and (3.37), and accounting for: (1) the phenomenological shot noise, defined in Eq. (3.22); (2) the existence of $\mathcal{M}$ ASE modes ($\mathcal{M} = 2$ with single-mode EDFAs if no polarizer is used between the EDFA and the detector); and (3) the finite detector quantum efficiency $\eta$, we can express the total noise power falling into the detector electronic bandwidth $B_e$ [22]:

$$\sigma_d^2 = \sigma^2_{\text{shot}} + \sigma^2_{\text{s-ASE}} + \sigma^2_{\text{ASE-ASE}} + \sigma^2_{\text{th}}$$

$$= \eta e 2B_e(GI_s + \mathcal{M}I_N) + 2\eta^2 GI_sI_N \frac{2B_e}{B_o} + \mathcal{M}\eta^2 I_N^2 \frac{2B_e}{B_o^2}\left(B_o - \frac{B_e}{2}\right) + \frac{4k_B T B_e}{R}$$

$$\tag{3.38}$$

Equation (3.38) for the detector noise variance assumes a detector and associated electronics having a uniform frequency response in the interval $[0, B_e]$. Likewise, no optical filtering prior to photodetection is assumed. In actual receivers used in communications systems, the overall frequency response of the electronics circuitry is generally not uniform, and also includes the effect of RF filters and equalizers. In communications based on optical frequency modulation (FSK) and direct detection, signal demodulation also requires the use of Fabry–Perot filters prior to photodetection. In all these cases, the noise power spectral density must be evaluated by taking into account both electronics and optical filter responses, [27] and [28], for which leads to a photodetector noise variance different from Eq. (3.38).

The detector noise variance defined in Eq. (3.38), as obtained from the semiclassical theory, completes the result of Eq. (3.25) obtained in the previous section from the quantum photodetection theory, and in which some factors remained undetermined. Let us compare both theories and associated physical interpretations. In the semiclassical theory, the shot noise was introduced phenomenologically, while in the quantum theory this noise component is intrinsic to the light statistics. *In the semiclassical theory, the signal–spontaneous beat noise component (s–ASE) stems from the effect of mixing amplified signal and ASE fields at the detector.* This beat noise would vanish in the case of an ideal, ASE noise-free amplifier. But Section 2.3 shows that for an ASE noise-free amplifier the quantum theory predicts a beat noise component having an identical definition. *In this quantum description, the beat noise only represents the effect of fluctuations in stimulated emission of photons.* No mixing of fields takes place at the detector, as only one signal field (or photon stream) with intrinsic amplitude noise is incident upon the detector. When the amplifier has spontaneous emission, we predict an additional ASE–ASE beat noise component, which takes the same form as its counterpart in the semiclassical theory. Likewise, this component is not due to an effect of field mixing, but is intrinsic to the statistics of amplified light.

We can summarize the difference between the two theories as follows. Semiclassical: an amplified signal field of deterministic amplitude is linearly superimposed on an ASE noise field of random phase; the beating effect between these two is a consequence of square-law detection of their mixing. Quantum theory, i.e., STT + quantum photodetection: the amplified light is a photon stream representing a statistical superimposition of amplified signal and ASE. The s–ASE beat noise component reflects fluctuations in stimulated emission, rather than an effect of spontaneous emission, contrary to the case of the ASE–ASE beat noise, which is present only when spontaneous emission exists. Such beat noises, intrinsic to the amplified light statistics, produce counterparts in the detector photocurrent. Remarkably, while the two physical interpretations are fundamentally different, the theories predict essentially the same result. But note that the quantum theory outlined in this book represents a simplified quantum description of photodetection of amplified light (photon number state, single-mode radiation field), which explains why the semiclassical description apparently predicts more information about the photodetector noise (power spectral density).

This semiclassical theory of direct signal photodetection does not represent the most complete description of noise possible. Many additional effects can be included, for instance, optical and electronic filtering, [27] and [28]; detector impulse response and signal phase noise, [29] and [30]; signal timing jitter [31]; and electronic intersymbol interference [32]. The fundamental studies of receiver design by S. D. Personick [33] and Y. Yamamoto, [34] and [35], developed some 10 to 20 years ago, included not only a detailed analysis of the receiver electronics response but also of avalanche multiplication in the detector. Section 3.4 discusses the effect of avalanche photodetection on receiver sensitivity with optically preamplified signals. Regardless of the accuracy and the sophistication of available theories of receiver noise there is a consensus in the literature to make use of results derived in this section for a basic evaluation of signal-to-noise ratio and overall system performance.

## 3.3 ENHANCEMENT OF SIGNAL-TO-NOISE RATIO BY OPTICAL PREAMPLIFICATION

Optical preamplification as a means to enhance the signal-to-noise ratio (SNR) at the detector and hence the minimum detectable signal emerged in the early 1960s. In this period, several theoretical studies addressed the issue of SNR improvement: H. Steinberg, [36] and [37]; H. Kogelnick and A. Yariv [38]; F. Arams and M. Wang [39]. Contemporary with these studies are the first experimental proofs of the concept achieved with gas lasers, [39]–[41]. The first two experiments demonstrated 16 dB and 32 dB improvements in detector SNR using xenon or helium–neon laser amplifiers, [39] and [40]. The first fiber amplifiers based on neodymium-doping were also investigated during that period by C. J. Koester and E. Snitzer [42].

The concept that a maximum achievable SNR (quantum limit) is imposed by the effect of the beat noise between signal and spontaneous emission was derived semiclassically by J. Arnaud [41]. This conclusion agreed with the previous quantum mechanical analysis of the minimum optical amplifier noise figure made by H. A. Haus and J. A. Mullen [43], discussed in Chapter 2.

Chapter 2 shows that the noise figure $F_o$ of an optical amplifier, which represents a measure of SNR degradation, is always greater than unity, and reaches a 3 dB quantum limit in the high gain regime. Since the optical SNR is degraded by 3 dB after passing through the preamplifier, how can an improvement in electrical SNR be measured?

First we clarify this apparent contradiction. Then a detailed analysis will show the SNR improvement actually has two causes: (1) the enhancement of received signal power by the preamplifier gain; and (2) the domination of the overall receiver noise power by signal–ASE beat noise.

Assume that the receiver is characterized by an electrical noise power level $N$. The minimum detectable signal $S_0$ can then be arbitrarily defined as the electrical signal power $S$ for which the signal-to-noise ratio SNR = $S/N$ reaches a minimum value $S_0/N$ representing a system standard (e.g., $S_0/N = 20$ dB). Using now an optical preamplifier with gain $G$, the received signal power is $G^2S'$, and the SNR is given by:

$$\text{SNR}' = \frac{G^2S'}{N + N'} \tag{3.39}$$

where $N'$ represents the amplified spontaneous emission generated in the optical amplifier and converted by the detector into an additional noise background. The new minimum detectable signal $S_0'$ is then given by the relation:

$$\frac{G^2S_0'}{N + N'} = \frac{S_0}{N} \tag{3.40}$$

If the value $S_0'$ is found to be lower than $S_0$, this corresponds to an *improvement of minimum detectable signal or detector sensitivity* of:

$$\frac{S_0}{S_0'} = G^2 \frac{N}{N + N'} \tag{3.41}$$

From Eq. (3.41), if the amplifier gain $G$ is high enough such that $G^2 > 1 + N'/N$, then such an improvement is possible.

We assume now that the noise $N'$ introduced by the preamplifier is dominated by signal–ASE beat noise, i.e., $N' \approx \sigma_{\text{s-ASE}}^2$, defined in Eq. (3.36). If the detector noise is of thermal origin, i.e., $N = 4k_{\text{B}}TB_{\text{e}}/R$, the condition $G^2 > 1 + N'/N$ for sensitivity improvement is equivalent at high gains to $P_{\text{s}} < h\nu_{\text{s}}k_{\text{B}}T/R\eta^2 e^2 n_{\text{eq}}$, or $P_{\text{s}} < 400\,\mu\text{W}$ ($\eta = n_{\text{eq}} = 1$, $T = 300\,\text{K}$, $R = 50\,\Omega$, and $\lambda_{\text{s}} = 1.5\,\mu\text{m}$). This condition is always satisfied in the usual operating conditions of optical preamplifiers. If we assume now $S_0' = S_0$, i.e., the input signals with and without preamplifier are identical, Eq. (3.40) shows that $G^2 > 1 + N'/N$ is equivalent to SNR$'$ > SNR, which means an improvement in the detector's *electrical* SNR.

If we consider now an ideal detector, no thermal noise and unity quantum efficiency, the detector noise without preamplifier is the shot noise $N = 2eB_{\text{e}}I_{\text{s}}$. In the signal–ASE beat noise limit, i.e., $N' \approx \sigma_{\text{s-ASE}}^2$, it is straightforward to find from Eq. (3.41) that the ratio of minimum detectable signals becomes

$$\frac{S_0}{S_0'} = \frac{1}{2} \tag{3.42}$$

This last result shows that *in the case of an ideal detector, no improvement of detector sensitivity or electrical SNR by optical preamplification is possible, and more specifically, the SNR is degraded by 3 dB*. Since the detector is ideal, there is no difference between electrical and optical SNRs, and this degradation actually represents a measurement of the amplifier optical noise figure $F_{\text{o}}$.

These considerations have shown that *with nonideal detectors (characterized by thermal noise), optical preamplification can improve the detector electrical SNR, in spite of the fact that the optical SNR is necessarily degraded by 3 dB in the process.*

For the sole purpose of analyzing the enhancement in detector electrical SNR and sensitivity, we shall use in this section a simplified SNR definition not applicable to the case of two-level digital signals. Section 3.5 uses the same type of analysis for this more complex case, which requires further discussion about the notion of signal decision level.

The simple definition chosen here for the photodetector SNR is:

$$\text{SNR}_{\text{e}} = \frac{I_{\text{signal}}^2}{\sigma_{\text{d}}^2} \tag{3.43}$$

where $I_{\text{signal}} = \eta G I_{\text{s}}$ represents the electrical signal photocurrent and $\sigma_{\text{d}}^2$ the total detector noise variance, defined in Eq. (3.38). The subscript e in SNR$_{\text{e}}$ indicates an electrical power SNR, by contrast with previous definitions of optical power SNR seen in Chapter 2. Replacing the definitions of $I_{\text{signal}}$ and $\sigma_{\text{d}}^2$ in Eq. (3.43) yields:

$$\text{SNR}_{\text{e}} = \frac{(\eta G I_{\text{s}})^2}{\eta e 2 B_{\text{e}}(G I_{\text{s}} + \mathcal{M} I_{\text{N}}) + 2\eta^2 G I_{\text{s}} I_{\text{N}} \dfrac{2B_{\text{e}}}{B_{\text{o}}} + \mathcal{M}\eta^2 I_{\text{N}}^2 \dfrac{2B_{\text{e}}}{B_{\text{o}}^2}\left(B_{\text{o}} - \dfrac{B_{\text{e}}}{2}\right) + \dfrac{4k_{\text{B}}TB_{\text{e}}}{R}} \tag{3.44}$$

Using definitions $I_{\text{s}} = eP_{\text{s}}/h\nu_{\text{s}}$ and $I_{\text{N}} = eN/h\nu_{\text{s}} = eGP_{\text{o}}/h\nu_{\text{s}}$ ($P_{\text{o}} = n_{\text{eq}}h\nu_{\text{s}}B_{\text{o}}$ is an equivalent ASE noise power), the electrical SNR in Eq. (3.44) can also be expressed explicitly as a function of the preamplifier gain $G$:

$$\text{SNR}_e = \frac{P_s^2}{\dfrac{h\nu_s}{\eta G}2B_e(P_s + \mathcal{M}P_o) + 4P_sP_o\dfrac{B_e}{B_o} + \mathcal{M}P_o^2\dfrac{2B_e}{B_o^2}\left(B_o - \dfrac{B_e}{2}\right) + \dfrac{4k_B TB_e}{R}\left(\dfrac{h\nu_s}{e\eta G}\right)^2}$$

(3.45)

The different noise power components in the denominator in Eq. (3.44) and the corresponding SNR in Eq. (3.45) are plotted as functions of preamplifier gain $G$ in Figure 3.2 (parameters: $\eta = n_{eq} = 1$, $T = 300\,\text{K}$, $\lambda_s = 1.5\,\mu\text{m}$, $P_s = 100\,\text{nW}$, $B_o = 2B_e = 10\,\text{GHz}$, $R = 50\,\Omega$ and $\mathcal{M} = 2$). The figure shows that the total detector noise power is characterized by two regimes: at low gains, the thermal noise component $\sigma_{th}^2$ dominates; at high gains the signal–ASE beat noise component $\sigma_{s\text{-ASE}}^2$ dominates. The figure also shows that the electrical SNR steadily increases with gain until the detector is in the s–ASE beat noise regime. In this regime, the SNR reaches an upper limit $\text{SNR}_e^{max}$. The existence of this limit is explained by the fact that in this regime, both received signal and s–ASE beat noise increase quadratically with the gain. Figure 3.2 shows that in this example, optical preamplification results in about a 36 dB increase in electrical SNR. Without preamplification, the detector could not be used to detect input signals as low as $P_s = 100\,\text{nW}$, since the corresponding electrical SNR would be too low ($-20.5$ dB), while with preamplification the maximum possible SNR of $\text{SNR}_e^{max} = +16$ dB could be high enough in some system applications.

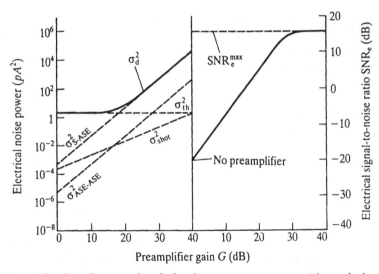

**FIGURE 3.2**   Left: photodetector electrical noise power components (thermal, shot, s–ASE and ASE–ASE) and total electrical noise power $\sigma_d^2$, plotted as a function of optical preamplifier gain $G$. Right: photodetector electrical signal-to-noise ratio as a function of optical preamplifier gain $G$ (parameters: $\eta = 1$, $n_{eq} = 1$, $T = 300\,\text{K}$, $\lambda_s = 1.5\,\mu\text{m}$, $B_0 = 2B_e = 10\,\text{GHz}$, $R = 50\,\Omega$, and $P_s = 100\,\text{nW}$, and $\mathcal{M} = 2$).

We obtain the upper limit $\text{SNR}_e^{\max}$ of the electrical SNR from Eq. (3.45) by assuming $G \gg 1$ and a large input signal $(P_s \gg P_o)$, i.e.,

$$\text{SNR}_e^{\max} = \frac{P_s}{4P_o \dfrac{B_e}{B_0}} = \frac{P_s}{4n_{eq} h v_s B_e} \tag{3.46}$$

This limit is to be compared with the electrical SNR obtained without preamplifier, setting $G = 1$ and $P_o = 0$ in Eq. (3.45):

$$\text{SNR}_e = \frac{P_s}{\dfrac{2h v_s B_e}{\eta} + \dfrac{4k_B T B_e}{R P_s}\left(\dfrac{h v_s}{e\eta}\right)^2} \tag{3.47}$$

Taking the ratio of Eqs. (3.46) and (3.47), we obtain an expression for the SNR enhancement factor:

$$x = \frac{1}{2\eta n_{eq}}\left(1 + \frac{2k_B T h v_s}{R\eta e^2}\frac{1}{P_s}\right) \tag{3.48}$$

At zero temperature $(T = 0)$, for which the detector would be free from thermal noise, the enhancement factor in Eq. (3.47) is given by:

$$x = \frac{1}{2\eta n_{eq}} \tag{3.49}$$

Equation (3.49) shows thar SNR enhancement $(x > 1)$ is possible if the detector quantum efficiency is poor, i.e., $\eta < 1/(2n_{eq}) < 1/2$ (since in the high gain limit $n_{eq} \geqslant 1$), which is of limited interest. In the case of an ideal detector with unity quantum efficiency and a fully inverted amplifier $(n_{eq} = 1)$, there is no enhancement since $x = 1/2$, which corresponds to the $-3$ dB SNR degradation predicted by the quantum theory (Chapter 2).

At finite temperatures where thermal noise is present, Eq. (3.48) shows that the enhancement factor increases as the reciprocal of the signal optical power $P_s$. The enhancement cannot be arbitrarily large, as the formula is valid only in the case where $P_s \gg P_o$, corresponding to a large input photon numbber such that $\langle n \rangle \gg n_{eq} \geqslant 1$. But with a realistic example, the enhancement is generally very large. Consider for instance the case $\eta = n_{eq} = 1$, for which both SNRs in Eqs. (3.46) and (3.47) are maximized, the temperature $T = 300$ K, and the wavelength $\lambda_s = c/v_s = 1.5\ \mu\text{m}$. We find that for an incident optical power of $P_s = 100$ nW, the enhancement is $x = 36$ dB. For an optical bandwidth $B_o = 1$ GHz, such an optical signal power corresponds to a large photon number of $\langle n \rangle \approx 750$, for which Eq. (3.48) is valid.

We have seen that with the optical preamplifier the maximum electrical SNR $(\text{SNR}_e^{\max})$ increases linearly with $P_s$, Eq. (3.46), while the SNR enhancement with respect to detection without amplifier increases as $1/P_s$. The maximum enhancement corresponds to the lowest $\text{SNR}_e^{\max}$ value, so the SNR enhancement factor does not represent a useful criterion.

A better figure of merit is the enhancement of *minimum detectable power*, or *detector sensitivity*. The minimum detectable power $P_s$ is defined from Eq. (3.26):

$$\text{SNR}_e = \frac{P_s}{\dfrac{2hv_sB_e}{\eta} + \dfrac{4k_BTB_e}{RP_s}\left(\dfrac{hv_s}{e\eta}\right)^2} = \text{SNR}_e^{st} \qquad (3.50)$$

where $\text{SNR}_e^{st}$ is a standard SNR value to be used in a system, e.g. $\text{SNR}_e^{st} = 20\,\text{dB}$. Solving Eq. (3.50) for $P_s$ yields the minimum detectable power:

$$P_s = \text{SNR}_e^{st}\frac{4hv_sB_e}{\eta}\left(1 + \sqrt{1 + \frac{4k_BT}{Re^2B_e\text{SNR}_e^{st}}}\right) \approx \text{SNR}_e^{st}\frac{4hv_sB_e}{\eta}\sqrt{\frac{4k_BT}{Re^2B_e\text{SNR}_e^{st}}}$$

$$(3.51)$$

On the other hand, if an optical preamplifier is used, while operating to yield the maximum possible SNR defined in Eq. (3.46), the minimum detectable power $P_s'$ is given by $\text{SNR}_e^{max} = P_s'/4n_{eq}hv_sB_e = \text{SNR}_e^{st}$, which, with Eq. (3.51), gives the sensitivity enhancement:

$$y = \frac{P_s}{P_s'} \approx \frac{1}{\eta n_{eq}}\sqrt{\frac{4k_BT}{Re^2B_e\text{SNR}_e^{st}}} \qquad (3.52)$$

As an example, taking the parameter values $\text{SNR}_e^{st} = 20\,\text{dB}$, $\eta = n_{eq} = 1$, $B_e = 1\,\text{GHZ}$, $R = 50\,\Omega$ and $T = 300\,\text{K}$, we find a sensitivity enhancement of $y = 25.5\,\text{dB}$.

The fact that optical preamplifiers make it possible to increase the detector sensitivity is of considerable impact in communications systems. Indeed, if smaller signals can be detected at the preamplifier/detector end while providing the same standard electrical SNR as a detector without preamplifier, then additional loss or longer fiber spans can be supported by the system. Taking the previous example, and assuming a fiber loss of 0.2 dB/km, a sensitivity improvement of $y = 25.5\,\text{dB}$ corresponds to a possible additional loss of 25.5 dB or a supplementary fiber span of $L = y/0.2 = 127\,\text{km}$.

The detector SNR and sensitivity improvements by optical preamplification discussed in this section assumed a detector with no internal gain. The case where photocarrier multiplication occurs in the detector (such as with avalanche photodiodes, or APD) leads to different conclusions. It would be interesting to compare the sensitivity of APD detectors to that of detectors with no internal gain but optically preamplified signals (OPA + D), and also to determine whether optical preamplification can improve the sensitivity of APD detectors themselves. These issues are addressed in the next section.

The formulae derived in this section correspond to a simplified definition for the receiver electrical SNR. In the case of digital communication systems, we must take into account the two possible levels of symbols mark and space in the SNR definition. A reference level for the electronics decision circuit which interprets these symbols must also be introduced. The analysis of SNR, sensitivity, and bit-error rate in digital direct detection systems with optical preamplifiers is detailed in Section 3.5.

## 3.4 OPTICAL PREAMPLIFICATION VERSUS AVALANCHE PHOTODETECTION

In the phenomenon of avalanche multiplication, the individual photocarriers (electrons and holes) generated in a detector are accelerated under a high bias voltage (kV) to create secondary pairs by impact ionization, [3] and [12]–[17]. Both types of carriers can contribute to the process with different ionization efficiencies, which is a source of electrical signal feedback [3]. As a single input photon can give birth to $M$ photocarriers of a given type, the detector is characterized by an intrinsic signal gain $M$.

Because avalanche multiplication is a random process by itself, and also because feedback adds further randomness, the avalanche gain is probabilistic, i.e., it is characterized by a finite variance $\sigma_M^2 = \langle M^2 \rangle - \langle M \rangle^2$, unlike the case of the optical amplifier where the gain $G$ is deterministic. The excess noise factor of avalanche photodiodes (APD) is defined by the ratio:

$$F_{ex} = \frac{\langle M^2 \rangle}{\langle M \rangle^2} \tag{3.53}$$

The excess noise factor is a function of the mean gain $\langle M \rangle$ and the ionization ratio $k = \alpha_h/\alpha_e$ (assuming electron injection) [3]:

$$F_{ex} = k\langle M \rangle + (1 - k)\left(2 - \frac{1}{\langle M \rangle}\right) \tag{3.54}$$

Equation (3.54) shows that when the ionization ratio is zero, i.e. in the absence of feedback, the excess noise factor is $F_{ex} = 1$ or 0 dB at $\langle M \rangle = 1$ and increases to reach a 3 dB limit for $\langle M \rangle \gg 1$.

The mean-square of the avalanche gain is usually given as $\langle M^2 \rangle = \langle M \rangle^{2+x} = F_{ex}\langle M \rangle^2$, where $x$ is an excess exponent defined as:

$$x = \frac{\log(F_{ex})}{\log(\langle M \rangle)} \tag{3.55}$$

The photodetection statistics of APDs and the fundamental limitations caused by APD noise in communications systems were studied by S. D. Personick [44], Y. Yamamoto, [34] and [35], and J. C. Simon [45]. When an optical preamplifier is used in conjunction with an APD detector (we shall refer to this as the OPA + APD configuration) the mean electrical signal power and noise variance are given by, [34] and [35]:

$$I_d = \langle M \rangle \eta G I_s \tag{3.56}$$

$$\sigma_d^2 = \sigma_{shot}^2 \langle M \rangle^{2+x} + \sigma_{beat}^2 \langle M \rangle^2 + \sigma_{th}^2 \tag{3.57}$$

where $\sigma_{shot}^2$ and $\sigma_{beat}^2 = \eta^2(\sigma_{s\text{-}ASE}^2 + \sigma_{ASE\text{-}ASE}^2)$ are defined in Eqs. (3.22), (3.36), and (3.37). The electrical SNR of an OPA + APD receiver is then, from Eqs. (3.56) and

(3.57):

$$\mathrm{SNR}_e = \frac{(\langle M \rangle \eta G I_s)^2}{2\eta e B_e (G I_s + \mathcal{M} I_N)\langle M \rangle^{2+x} + \left\{4\eta^2 G I_s I_N \dfrac{B_e}{B_o} + \eta^2 I_N^2 \dfrac{B_e}{B_o^2}(2B_o - B_e)\right\}\langle M \rangle^2 + \sigma_{th}^2}$$

$$= \frac{P_s^2}{\dfrac{h\nu_s}{\eta G} 2B_e (P_s + \mathcal{M} P_o) F_{ex} + 4 P_s P_o \dfrac{B_e}{B_o} + \mathcal{M} P_o^2 \dfrac{2B_e}{B_o^2}\left(B_o - \dfrac{B_e}{2}\right) + \dfrac{4 k_B T B_e}{R \langle M \rangle^2}\left(\dfrac{h\nu_s}{e\eta G}\right)^2}$$

(3.58)

In the case where there is no preamplifier ($G = 1$, $P_o = n_{eq} = 0$), the electrical SNR of a single APD detector is, from Eq. (3.58):

$$\mathrm{SNR}_e = \frac{P_s}{\dfrac{2 h\nu_s B_e}{\eta} F_{ex} + \dfrac{4 k_B T B_e}{R P_s}\left(\dfrac{h\nu_s}{e\eta \langle M \rangle}\right)^2}$$

(3.59)

We focus first on the case of a single APD detector, to compare it with the results previously obtained for the optical preamplifier + detector configuration (OPA + D). Taking the same parameters as before, i.e. $\eta = 1$, $T = 300\,\mathrm{K}$, $\lambda_s = 1.5\,\mu\mathrm{m}$, $P_s = 100\,\mathrm{nW}$, $B_o = 2B_e = 10\,\mathrm{GHz}$, $R = 50\,\Omega$, we can plot the electrical SNR as a function of avalanche gain $\langle M \rangle$ for different values of the ionization ratio $k$, as done in Figure 3.3. The figure shows that, as expected, the electrical SNR increases with avalanche gain $\langle M \rangle$, with the increase being maximized for a null ionization ratio ($k = 0$). When $k = 0$ and the gain is high ($\langle M \rangle \gg 1$), the excess noise factor in Eq. (3.54) reaches a value of 2. In this case, the SNR of an APD, Eq. (3.59), takes the same form as the SNR of an OPA + D, Eq. (3.45), having the same gain $G = \langle M \rangle$ and an equivalent

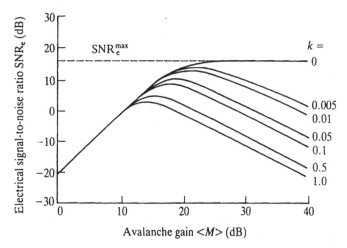

**FIGURE 3.3** Electrical SNR of APD photodiode as a function of mean avalanche gain $\langle M \rangle$, for different values of the ionization ratio $k$ (parameters: $\eta = 1$, $T = 300\,\mathrm{K}$, $\lambda_s = 1.5\,\mu\mathrm{m}$, $B_0 = 2B_e = 10\,\mathrm{GHz}$, $R = 50\,\Omega$, and $P_s = 100\,\mathrm{nW}$).

input noise factor $n_{eq} = 1/\eta$ ($P_0 = n_{eq} h v_s B_0$), and in which only the terms $\sigma^2_{\text{s-ASE}}$ and $\sigma^2_{\text{th}}$ are present. At low gains where the thermal noise $\sigma^2_{\text{th}}$ dominates, the two types of receivers (APD and OPA + D) also have the same electrical SNR. In this example where $\eta = n_{eq} = 1$, therefore, the curve $k = 0$ in Figure 3.3, which gives the maximum possible SNR increase for an APD, is identical to that of Figure 3.2 for an OPA + D. The maximum achievable SNR for an ideal APD is, from Eq. (3.59) with $k = 0$ and $\langle M \rangle \gg 1$:

$$\text{SNR}_e^{\max} = \frac{P_s}{4 h v_s (1/\eta) B_e} \tag{3.60}$$

This result is identical to Eq. (3.46) for an OPA + D with $\eta = 1/n_{eq}$, i.e., *the maximum SNR improvement with an ideal APD is identical to that of an OPA + D having an equivalent input noise $n_{eq} = 1/\eta$.*

The similarity between an APD and an OPA + D in electrical SNR improvement disappears when the ionization ratio $k$ is greater than zero, as the improvement drops if the avalanche multiplication is too high, as shown in Figure 3.3. In this case, the maximum electrical SNR that can be achieved decreases as the ionization ratio increases. As typical values for the ionization ratio are $k = 0.7$–$1.0$ for Ge-APDs and $k = 0.3$–$0.5$ for InGaAs-APDs [16], *realistic APDs do not offer the same potential of electrical SNR improvement as the case of optical preamplification.* Furthermore, for an OPA + D receiver, the maximum possible improvement is independent of the detector quantum efficiency, Eq. (3.46), contrary to the APD case, Eq. (3.60).

Now consider an APD with optical preamplifier (OPA + APD). The electrical SNR of this configuration is given in Eq. (3.58). In the high gain regime ($G \gg 1$ and $\langle M \rangle \gg 1$) and for input signals large enough ($P_s \gg P_o$), we find the OPA + APD maximum achievable electrical SNR:

$$\text{SNR}_e^{\max} = \frac{P_s}{\dfrac{2 h v_s B_e}{\eta}\left(k\,\dfrac{\langle M \rangle}{G} + 2(1-k)\right) + 4 n_{eq} h v_s B_e} \tag{3.61}$$

Equation (3.61) shows that with an ideal APD ($k = 0$), that the maximum SNR of an OPA + APD is identical to that of an OPA + D configuration with equivalent input noise $N'_{eq} = n_{eq} + 1/\eta$. For $\eta = n_{eq} = 1$, the maximum SNR of an OPA + APD is therefore 3 dB lower than the maximum SNR of an OPA + D. In the nonideal case ($k \neq 0$), the maximum SNR of an OPA + APD is lower than that of an OPA + D. For $\eta = n_{eq} = 1$ and $\langle M \rangle = 2G$, the maximum SNR of an OPA + APD is 3 dB lower than that of an OPA + D. This shows that for the purpose of enhancing detector sensitivity, *the combination of optical preamplification and avalanche multiplication in the detector is not advantageous, as it entails an SNR penalty of at least 3 dB.*

## 3.5  BIT-ERROR RATE AND RECEIVER SENSITIVITY IN DIGITAL DIRECT DETECTION

Knowledge of the electrical signal-to-noise ratio at the detector is sufficient to assess whether the quality of the received signal meets a given system standard, as in the case of basic analog transmission of voice or video information (Section 3.8). For digital communications, however, the electrical SNR is not by itself a meaningful figure of merit. This can be shown by considering the basic principle of digital detection.

In digital communication systems, the optical signals are made of mark and space symbols, carried by modulated light. These symbols are generally called mark and space, rather than 1 and 0, as several modulation formats are possible. The most basic of these is the ON/OFF format where a light pulse represents 1 and absence of a pulse represents 0. In direct detection systems, the mark and space symbols can be represented by either amplitude modulated light (ASK, for amplitude shift keying), frequency modulated light (FSK, for frequency shift keying), or phase modulated light (DPSK, for differential phase shift keying). Basic modulation formats used in digital communications are discussed in Section 3.6, which deals with coherent systems. In direct detection systems, FSK and PSK signals are demodulated to be converted into ASK signals, or 1 and 0 light pulses, prior to photodetection. The FSK signals are optically demodulated through a narrowband optical filter (Fabry–Perot), while the DPSK signals are demodulated through a delayed Mach–Zehnder interferometer.

The basic principle of digital photodetection is illustrated in Figure 3.4. After demodulation (for FSK or PSK), the light pulses are photodetected and converted into electrical pulses. After electronic preamplification, a circuit integrates the bit energy over the bit period. The resulting signal is compared to some reference level by a following decision circuit, which assesses whether a 0 or a 1 symbol has been detected. The signal is interpreted as a 1 if it is above the decision level and as a 0 otherwise. In this simplified description, other electronic functions that are not relevant to the present discussion were overlooked.

If randomness is present in the incident optical signals, then the process of digital photodetection outlined above is characterized by a finite probability of error. The error occurs each time the integrated energy of the bit randomly fluctuates about the decision level. In the case where the detector electrical SNR is very high, this is unlikely, i.e., the probability of error is negligible. In the general case where the received optical signals are weak, decision errors are more likely to occur. The photodetector's electrical SNR does not represent a meaningful figure of merit for this type of signal detection.

A valid figure of merit for digital detection is the probability of error associated with a given signal and noise power level. Knowledge of this error probability makes possible to infer how many bits on average can be successfully transmitted through the system over a given period of time.

The error probability, called bit-error rate (BER) can be defined as [45]:

$$\mathrm{BER} = P(1|0)p(0) + P(0|1)p(1) \tag{3.62}$$

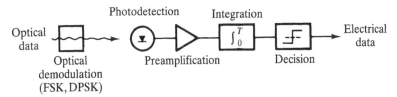

**FIGURE 3.4** Basic principle of digital direct photodetection.

where $P(1|0)$ represents the conditional probability for the electronics decision circuit of falsely detecting a 1 symbol when a 0 symbol was transmitted, and $P(0|1)$ represents the contrary event. The terms $p(0)$ and $p(1)$ represent the probabilities that 0 or a 1 symbol is transmitted, respectively. In the case where the number of bits $N$ in the message sequence is large (e.g., $n \geqslant 2^{15}$), the 0 and 1 bit transmissions are equiprobable, i.e., $p(0) \approx p(1) \approx 1/2$, and we have from Eq. (3.62) BER $= [P(1|0) + P(0|1)]/2$. The integration process is equivalent to counting the photoelectrons generated in the bit duration. Assuming that the probability distributions of the photoelectron count of symbols 0 and 1 are $P_n(0)$ and $P_n(1)$, respectively, and that the decision level is $D$ counts, the conditional probabilities in Eq. (3.62) are given by:

$$P(1|0) = \sum_{n=D}^{\infty} P_n(0) \tag{3.63}$$

$$P(0|1) = \sum_{n=0}^{D-1} P_n(1) \tag{3.64}$$

If the number of counts is large, the probability distributions $P_n(0)$, $P_n(1)$ can be approximated by continuous functions $P_x(0)$, $P_x(1)$ and the sums in Eqs. (3.63) and (3.64) by integrals. Figure 3.5 graphically illustrates the notion of BER. The distributions $P_n(0)$ and $P_n(1)$ are represented in this example by Gaussian functions with mean and standard deviation $x_0$, $\sigma_0$ and $x_1$, $\sigma_1$, respectively. The conditional probabilities $P(1|0)$ and $P(0|1)$ are equal to the shaded areas shown in the figure, and the BER is given by the half-sum of the total shaded area. The probability of successful transmission $1 -$ BER corresponds to half the total nonshaded area below the two curves $P_x(0)$, $P_x(1)$. In this example, the BER is comparatively high, as the variances for 0 and 1 were chosen, for illustration purposes, to be relatively large with respect to the mean separation $x_1 - x_0$.

In actual communications systems, for both short-haul (400 km) and long-haul (6400 km), the bit-error rate standard for error-free operation is BER $\leqslant 10^{-9}$, which

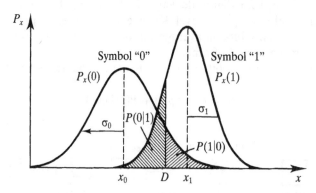

**FIGURE 3.5**    Probability distributions corresponding to the photodetection of digital symbols 0 and 1, with continuous Gaussian approximation. The shaded areas $P(1|0)$ and $P(0|1)$ represent the probabilities of error in symbol interpretation. The total shaded area is equal to twice the bit-error rate (BER).

means an average of less than one error per second at 1 Gbit/s data rate. Other figures of merit are the percentage of *error-free seconds* (EFS) and the *severe-errored seconds* (SES) [16]. A system that operates at 1 Gbit/s with BER $= 10^{-8}$ has zero EFS, as each second contains 10 errors on average. The SES is defined as a second where more than 100 errors occur. A system outage occurs when the BER is higher than $10^{-3}$ for a period of 10 consecutive seconds [16].

This shows that the determination of BER at the receiver end is of major importance in the design of any digital communications system. In particular, when the system includes an optical amplifier chain (Section 3.7) or an optical preamplifier at the receiver, the additional noise generated by these elements directly affects the BER. The computation of BER from Eq. (3.62) requires knowledge of the photocurrent statistics, related to those of the incident light, and of other noise sources present in the detector.

The different theoretical models found in the literature for the BER analysis of optically amplified signals can be put into two general categories. The first category assumes square-law detection (similar to that of microwave systems) with a semi-classical noise field. The corresponding models differ by various assumptions; the most used in the literature is the Gaussian approximation for the photocurrent probability distributions, [22], [27]–[30], [33], [34], [45], and [46]–[48]. Other models provide exact probability distributions when different dominating noise sources in the amplifier and the receiver are assumed, [31], and [49]–[56]. The second category considers amplified light as a quantized radiation field and quantum mechanically analyzes the photoelectron count assuming, for instance, coherent light superimposed to a thermal noise background, [57] and [58], or the exact probability distribution solutions of the photon statistics master equation, [59]–[62].

The BER analysis of optically amplified digital signals lends itself to many different models, approximations, and levels of complexity. The basic results associated with the gaussian approximation and the square-law detection models are best adapted to the analysis of problems in communications systems. For comparison, we apply the quantum photon counting model to ASK signals; it is also applicable to demodulated FSK and PSK signals if additional noise factors caused by demodulation (e.g., signal phase noise, frequency noise, polarization noise, jitter, chirp, and dispersion) are neglected. At low BERs ($10^{-9}$), all three models generally give similar predictions.

We focus fist on a BER analysis based upon the Gaussian approximation. In this analysis, developed by S. D. Personick [33], the conditional error probabilities $P(1|0)$ and $P(0|1)$ are defined by:

$$P(1|0) = \frac{1}{\sigma_0 \sqrt{2\pi}} \int_D^\infty \exp\left\{\frac{(\langle i_0 \rangle - x)^2}{2\sigma_0^2}\right\} dx \tag{3.65}$$

$$P(0|1) = \frac{1}{\sigma_1 \sqrt{2\pi}} \int_{-\infty}^D \exp\left\{\frac{(\langle i_1 \rangle - x)^2}{2\sigma_1^2}\right\} dx \tag{3.66}$$

In Eqs. (3.65) and (3.66) the quantities $\langle i_0 \rangle$, $\langle i_1 \rangle$ are the mean photocurrents associated with symbols 0 and 1; $\sigma_0$, $\sigma_1$ are the average derivations from the mean photocurrents; and $D$ is the decision level. Through a change of variables, we obtain:

$$P(1|0) = \frac{1}{\sqrt{2\pi}} \int_{Q_0}^{\infty} \exp\left(-\frac{x^2}{2}\right) dx \tag{3.67}$$

$$P(0|1) = \frac{1}{\sqrt{2\pi}} \int_{Q_1}^{\infty} \exp\left(-\frac{x^2}{2}\right) dx \tag{3.68}$$

where $Q_0$, $Q_1$ are defined by:

$$Q_0 = \frac{D - \langle i_0 \rangle}{\sigma_0} \tag{3.69}$$

$$Q_1 = \frac{\langle i_1 \rangle - D}{\sigma_1} \tag{3.70}$$

If two error probabilities in Eqs. (3.67) and (3.68) are set equal for minimal error, since symbols 0 and 1 are equiprobable. This implies $Q_0 = Q_1 = Q$, which yields from Eqs. (3.69) and (3.70) the decision level $D$:

$$D = \frac{\sigma_1 \langle i_0 \rangle + \sigma_0 \langle i_1 \rangle}{\sigma_0 + \sigma_1} \tag{3.71}$$

The corresponding parameter $Q$ (which we shall refer to as Personick's $Q$ factor) is then from Eq. (3.70) and (3.71):

$$Q = \frac{\langle i_1 \rangle - \langle i_0 \rangle}{\sqrt{\sigma_0^2} + \sqrt{\sigma_1^2}} \tag{3.72}$$

The BER is given by Eq. (3.62) with $p(0) = p(1) = 1/2$ and $P(1|0) = P(0|1)$, or from eq. (3.68):

$$\text{BER} = \frac{1}{\sqrt{2\pi}} \int_{Q}^{\infty} \exp\left(-\frac{x^2}{2}\right) dx = \frac{1}{2} \text{erfc}\left(\frac{Q}{\sqrt{2}}\right) \tag{3.73}$$

where $\text{erfc}(x) = 1 - \text{erf}(x)$ is the complementary error function, and $\text{erf}(x)$ is the error function [63]. A series definition for $\text{erf}(x)$ is [63]:

$$\text{erf}(x) = \frac{2}{\sqrt{\pi}} \sum_{n=0}^{\infty} \frac{(-1)^n}{n!(2n+1)} x^{2n+1} \tag{3.74}$$

Substituting Eq. (3.74) into Eq. (3.73) yields after elementary manipulation:

$$\text{BER} = \frac{1}{Q\sqrt{2\pi}} \exp\left(-\frac{Q^2}{2}\right) + \frac{Q}{2\sqrt{2\pi}} \left[\frac{\sqrt{2\pi}}{Q} - 2 - \sum_{n=1}^{\infty} \left(-\frac{Q^2}{2}\right)^n \frac{1}{(n+1)!(2n+1)}\right] \tag{3.75}$$

The above expression can be approximated by keeping only the first term, i.e.,

$$\text{BER} \approx \frac{1}{Q\sqrt{2\pi}} \exp\left(-\frac{Q^2}{2}\right) \qquad (3.76)$$

Figure 3.6 shows a plot of BER as a function of $Q$, defined by exact Eq. (3.73) and by approximation Eq. (3.76). Equation (3.76) is very accurate for $Q > 3$. Figure 3.6 also shows that a BER of $10^{-9}$ is obtained for $Q = 6$.

We assume identical noise variances for the 0 and 1 symbols, i.e., $\sigma_0^2 = \sigma_1^2 = \sigma^2$, therefore the electrical signal-to-noise ratio is given by Eq. (3.72) as $\text{SNR} = 4Q^2$, which for $\text{BER} = 10^{-9}$ yields $\text{SNR} = 144 = +21.6\,\text{dB}$. This result, found in many textbooks, applies only to the case where the noise variances corresponding to both symbols are nearly identical, which occurs in particular when thermal noise is dominant. In the case of amplified light, the existence of the signal–ASE beat noise component gives the condition $\sigma_1^2 > \sigma_0^2$, yielding for 1 symbols and for $\text{BER} = 10^{-9}$ a higher electrical SNR. This can be shown by the following: define the SNR corresponding to symbol 1 with $\sigma_1^2 > \sigma_0^2$ by $\text{SNR}(1) = (\langle i_1 \rangle - \langle i_0 \rangle)^2/\sigma_1^2$, and the SNR corresponding to the case $\sigma_1^2 = \sigma_0^2 = \sigma^2$ by $\text{SNR}^* = (\langle i_1 \rangle - \langle i_0 \rangle)^2/\sigma^2$. Letting $x = \sigma_1^2/\sigma_0^2$, we have $\text{SNR}(1) = \text{SNR}^*/x$, and from Eq. (3.72), $Q = \sqrt{\text{SNR}^*}/(1 + x)$. These relations yield the SNR for symbol 1:

$$\text{SNR}(1) = \frac{(1 + x)^2}{x} Q^2 = \left(\frac{\sigma_0}{\sigma_1} + \frac{\sigma_1}{\sigma_0}\right)^2 Q^2 \qquad (3.77)$$

Since $x > 1$ in the case of amplified light ($\sigma_1^2 > \sigma_0^2$), we have $(1 + x)^2/x > 4$ and $\text{SNR}(1) > 4Q^2$. For instance, with $x = 10$ ($\sigma_1^2 = 10\sigma_0^2$), we find $\text{SNR}(1) = 435 = +26.4\,\text{dB}$ for $\text{BER} = 10^{-9}$, which is about 5 dB higher than for $\sigma_1^2 = \sigma_0^2$.

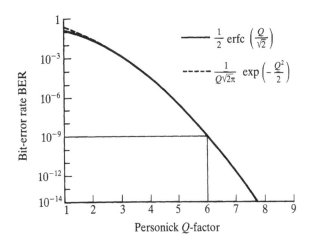

**FIGURE 3.6** Bit-error rate as a function of Personick's $Q$ factor, showing exact expression (full line) and approximate expression (dashed line).

The receiver sensitivity is, by definition, the input signal power $P_s$ for which BER $= 10^{-9}$ is achieved; the value is expressed as an average power per bit, i.e., $\bar{P} = P_s/2$. With an OPA + D receiver, the sensitivity refers to the signal power input to the optical preamplifier. From Eqs. (3.19)–(3.21), the 1 and 0 photocurrents are $\langle i_1 \rangle = \eta(GI_s + I_N)$ and $\langle i_0 \rangle = \eta I_N$, with $I_s$, $I_N$ defined in Eqs. (3.19) and (3.20), and from Eq. (3.38) the corresponding noise variances are:

$$\sigma_1^2 = \eta 2 B_e (GI_s + \mathcal{M} I_N) + 2\eta^2 GI_s I_N \frac{2B_e}{B_o} + \mathcal{M}\eta^2 I_N^2 \frac{2B_e}{B_o^2}\left(B_o - \frac{B_e}{2}\right) + \frac{4k_B T B_e}{R} \quad (3.78)$$

$$\sigma_0^2 = \eta 2 B_e \mathcal{M} I_N + \mathcal{M}\eta^2 I_N^2 \frac{2B_e}{B_o^2}\left(B_o - \frac{B_e}{2}\right) + \frac{4k_B T B_e}{R} \quad (3.79)$$

Substituting the photocurrents and noise variances into Eq. (3.72) and solving for $P_s$ with $Q = 6$ yields the receiver sensitivity $\bar{P}$:

$$\bar{P} = 36 h\nu_s B_e \left\{ F_o + \frac{1}{6}\sqrt{\mathcal{M} n_{eq}^2 \left(2\frac{B_o}{B_e} - 1\right) + \mathcal{M} n_{eq}\frac{2}{\eta G}\frac{B_o}{B_e} + \frac{4k_B T}{R e^2 \eta^2 G^2 B_e}} \right\} \quad (3.80)$$

where $F_o$, the amplifier optical noise figure (see Chapters 2 and 5), defined as:

$$F_o = \frac{1/\eta + 2n_{eq}G}{G} \quad (3.81)$$

Since the electronic bandwidth $B_e$ of photodetectors, i.e., of the $p$–$i$–$n$ type, is fixed by carrier transit time and intrinsic capacitance [16], a way to enhance the sensitivity $\bar{P}$ is to reduce the optical pass band $B_o$. The smallest possible value for $B_o$ is the data spectral width, i.e., $B_o = B$, where $B$ is the bit rate. On the other hand, the minimum electrical bandwidth $B_e$ should be equal to about one half the bit rate, i.e., $B_e = B_o/2$. The combined effect of the preamplifier gain $G$ and the optical pass band $B_o$ on receiver sensitivity $\bar{P}$, given an electrical bandwidth $B_e$, is illustrated in Figure 3.7 ($B_e = 1.25$ GHz, $\eta = n_{eq} = 1$, $\mathcal{M} = 2$, $R = 50\,\Omega$, $T = 300$ K and $\lambda_s = 1.55\,\mu$m). As the figure shows, the receiver sensitivity increases, i.e., $\bar{P}$ decreases, with preamplifier gain, and the ultimate value is reached when $B_e \approx B_o/2$, corresponding to a bit rate $B = B_o = 2.5$ GHz. A penalty in receiver sensitivity is introduced by the effect of nonvanishing extinction ratio in ASK, as shown in [22] and [28].

In the limit of high preamplifier gains ($G \gg 1$), the optical noise figure is $F_o = 2n_{eq}$ and the ultimate receiver sensitivity is given by Eq. (3.80) with $B_e = B_o/2$:

$$\bar{P} = 18 n_{eq} h\nu_s B_o \left(2 + \sqrt{\frac{\mathcal{M}}{12}}\right) \quad (3.82)$$

For a fully inverted amplifier ($n_{eq} = 1$), the result of Eq. (3.82) corresponds to a receiver sensitivity $\bar{P} = 41.2$ photons/bit for $\mathcal{M} = 1$ (ASE filtered by a polarizer), and $\bar{P} = 43.3$ photons/bit for $\mathcal{M} = 2$ (no polarizer). Other BER analyses based upon the Gaussian approximation predict other ideal sensitivities. By including an elaborate

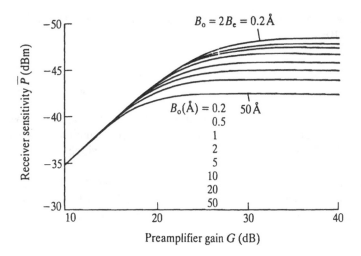

**FIGURE 3.7** Receiver sensitivity as a function of preamplifier gain $G$ for different values of the optical bandwidth $B_o$, assuming an electronic bandwidth $B_e = 1.25$ GHz. Other parameters: $\eta = 1$, $\eta_{eq} = 1$, $\mathcal{M} = 2$, $T = 300$ K, $R = 50\,\Omega$, and $\lambda_s = 1.5\,\mu m$.

analysis of optical filtering (no polarizer), $\bar{P} = 44.5$ photons/bit, [28]; by including the effect of laser phase noise, $\bar{P} = 42.3$ photons/bit (with polarizer) [30]. These sensitivities are within one photon, or 0.1 dB, of the previously calculated values.

We focus now on a BER analysis based on exact probability distributions derived from square-law detection theory. In the detection of binary ASK signals, a basic analysis [64] shows that the probability distribution (PDF) of symbol $q(q = 0$ or $1)$ is given by:

$$P_x(q) = \frac{x}{\sigma_q^2} \exp\left(-\frac{x^2 + \langle i_q \rangle^2}{2\sigma_q^2}\right) I_0\left(\frac{x\langle i_q \rangle}{\sigma_q^2}\right) \tag{3.83}$$

where $I_0(u)$ is the modified Bessel function of the first kind and zero order [63], and $\langle i_q \rangle$, $\sigma_q^2$ are the mean photocurrent and variance.

The PDF defined in Eq. (3.83) is called the Rician distribution, named after S. O. Rice who pioneered the field of random noise analysis [65]. For a 0 symbol having zero mean photocurrent, i.e., $\langle i_0 \rangle = 0$, the Rician PDF reduces to the Rayleigh distribution [64]:

$$P_x(0) = \frac{x}{\sigma_0^2} \exp\left(-\frac{x^2}{2\sigma_0^2}\right) \tag{3.84}$$

In the case of amplified light, and unless a DC block filter is used, the mean 0 photocurrent $\langle i_0 \rangle$ is not zero and the 0 symbol PDF is, as for the 1 symbol, a Rician distribution.

Best receiver sensitivity can be computed numerically through the general expression of BER:

$$\text{BER} = \frac{1}{2}\{P(1|0) + P(0|1)\} = \frac{1}{2}\left\{\int_D^\infty P_x(0)\,dx + \int_0^D P_x(1)\,dx\right\} \tag{3.85}$$

and the PDF definitions in Eqs. (3.83) and (3.84). In this evaluation, we must also determine numerically the decision level $D = D_{\text{opt}}$ for which the two conditional probabilities $P(0|1)$ and $P(1|0)$ in Eq. (3.85) are equal, which yields the minimum achievable BER, as shown below.

For the numerical calculation of Eq. (3.85), we consider the ideal case where thermal and shot noises are negligible in front of s–ASE and ASE–ASE beat noises, with $B_e = B_o/2$ and $n_{\text{eq}} = \eta = 1$. In this case, the noise variances in Eqs. (3.78) and (3.79) can be put in the form:

$$\sigma_0^2 = \left(\frac{eG}{h\nu_s}\right)^2 \frac{3}{4} \mathcal{M} P_o^2 \tag{3.86}$$

$$\sigma_1^2 = \left(\frac{eG}{h\nu_s}\right)^2 \left(4\bar{P}_s P_o + \frac{3}{4} \mathcal{M} P_o^2\right) \tag{3.87}$$

while the mean photocurrents are given by:

$$\langle i_0 \rangle = \frac{eG}{h\nu_s} \mathcal{M} P_0 \quad \text{or} \quad \langle i_0 \rangle = 0 \tag{3.88}$$

$$\langle i_1 \rangle = \frac{eG}{h\nu_s} 2\bar{P}_s \tag{3.89}$$

with $P_o = h\nu_s B_o$. Using Eqs. (3.86)–(3.89) and typical parameters, the argument of the modified Bessel function $I_0(u)$ in Eq. (3.83) is large, and this function is well approximated by [64]:

$$I_0(u) \approx \frac{e^u}{\sqrt{2\pi u}} \tag{3.90}$$

Using Eq. (3.90), the Rician distribution in Eq. (3.83) becomes:

$$P_x(q) \approx \sqrt{\frac{x}{2\pi\langle i_q \rangle \sigma_q^2}} \exp\left(-\frac{x^2 - \langle i_q \rangle^2}{2\sigma_q^2}\right) \tag{3.91}$$

which for values of $x$ near $\langle i_q \rangle$ is a function very similar to the Gaussian or normal distribution previously used.

Figure 3.8 shows plots of the conditional probabilities $P(1|0)$ and $P(0|1)$ and BER as a function of the dimensionless decision leve $D(D = i_d/eB_o G)$, as determined numerically from Eq. (3.85) with Eqs. (3.83), (3.84), and (3.86)–(3.91) for different values of the average photon number per bit $\bar{n}(\bar{P} = \bar{n}h\nu_s B_o)$, and in the case $\langle i_0 \rangle = 0$

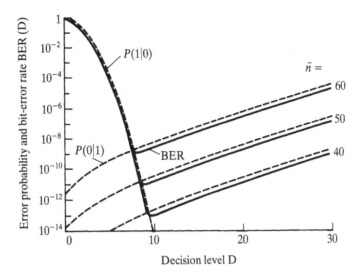

**FIGURE 3.8** Conditional error probabilities $P(1|0)$, $P(0|1)$, and BER versus decision level $D$ ($D = i_s/eB_oG$), assuming Rayleigh and Rician distributions for the 0 and 1 symbols, respectively, no ASE polarizing filter ($\mathcal{M} = 2$), and $\lambda_s = 1.5\,\mu m$, $n_{eq} = \eta = 1$, $G = 30\,dB$, $B_o = 1\,GHz$, The minimum bir-error rate BER ($D_{opt}$) is found at the decision level $D_{opt}$ for which $P(1|0) = P(0|1)$.

($P_x(0)$ = Rayleigh PDF), $\mathcal{M} = 2$ (no ASE polarizer). Figure 3.8 shows that to each value of $\bar{n}$ corresponds an optimum decision level $D_{opt}$ for which $P(1|0) = P(0|1)$ and the BER is minimized.

Figure 3.9 shows plots of the minimal bit-error rate BER($D_{opt}$) versus average number of photons per bit $\bar{n}$. Four cases of interest are considered, which correspond to the following assumptions: $\mathcal{M} = 1$ (ASE filtered by a polarizer) or $\mathcal{M} = 2$ (no polarizer), and either a Rician PDF ($\langle i_0 \rangle \neq 0$) or a Rayleigh PDF ($\langle i_0 \rangle = 0$) for the symbol 0 are assumed. It shows that the best sensitivity is $\bar{n} = 38.0$ photons/bit, corresponding to the case of Rayleigh PDF for the 0 symbol and $\mathcal{M} = 1$ (polarizer). If no polarizer is used ($\mathcal{M} = 2$), the sensitivity is then $\bar{n} = 40.5$ photons/bit. With Rician distributions for the symbol 0, we obtain $\bar{n} = 38.8$ photons/bit and $\bar{n} = 42.1$ photons/bit, corresponding to $\mathcal{M} = 1$ and $\mathcal{M} = 2$, respectively.

The best theoretical sensitivity $\bar{n} = 38$ photons/bit with ASE polarizer (P. S. Henry [66]) is 3 photons/bit or 0.3 dB better than the previous result $n = 41.2$ photons/bit, obtained with the Gaussian PDF approximation. But if the ASE noise photocurrent is assumed finite (as in the case of the Gaussian approximation), the accurate model gives the closer result $\bar{n} = 40.5$ photons/bit.

A more detailed analysis of BER by P. A. Humblet and M. Azizoglu [52] takes into account the number of degrees of freedom of the noise polarization components, i.e.,

$$M = \mathcal{M} \frac{B_o}{B} \qquad (3.92)$$

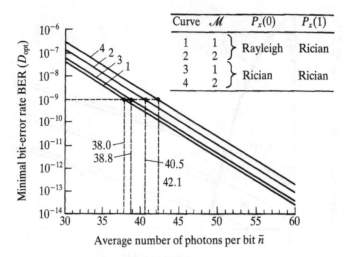

| Curve | $\mathcal{M}$ | $P_x(0)$ | $P_x(1)$ |
|-------|---------------|----------|----------|
| 1 | 1 | Rayleigh | Rician |
| 2 | 2 | | |
| 3 | 1 | Rician | Rician |
| 4 | 2 | | |

**FIGURE 3.9**   Ideal receiver BER versus average number of photons per bit, assuming Rayleigh or Rician probability distributions for the 0 symbol, Rician distribution for the 1 symbol, with ASE polarizing filter ($\mathcal{M} = 1$) or without ASE polarizing filter ($\mathcal{M} = 2$), and other parameters identical to that of Figure 3.8 ($\lambda_s = 1.5\,\mu m$, $n_{eq} = \eta = 1$, $G = 30$ dB, and $B_o = 1$ GHz).

where $B$ is the bit rate [50]. So far, we have considered only the case $M = \mathcal{M}$ or $B_o/B = 1$. In the case where $M > 1$, the PDF can be expressed as a noncentral, chi-square distribution with $2M$ degrees of freedom [52], while $M = 1$ corresponds to the Rician distribution (3.83). The BER corresponding to modulation formats other than ASK (i.e., FSK and DPSK) are also compared with the Gaussian approximation [52]. The study shows that for $M = 1$, the ideal receiver sensitivity with exact PDF is $\bar{n} = 38$ photons/bit for ASK and FSK, while for DPSK it is $\bar{n} \approx 20$ photons/bit; the sensitivities decrease with $M$ for all modulation formats. Another analysis of square-law detection by D. Marcuse, [49]–[51], also provides closed-form expressions for the conditional error probabilities $P(1|0)$ and $P(0|1)$ and the optimum decision level, the accuracy of which is discussed in [52].

So far the analysis of BER is based upon the semiclassical theory of square-law detection. Could a quantum photodetection theory also provide BER predictions in agreement with those of the semiclassical theory? A full quantum mechanical description of photodetection, to account for multiple radiation modes and the effects of optical/electronic filtering, would present computational difficulties. In actual communications systems, photodetection of amplified signals involves a large number of photons per bit (even at low BERs), so we might expect a semiclassical description to be accurate. Perhaps a full quantum theory of photodetection is only of academic interest but we might be interested to compare the results of the semiclassical, square-law detection theory and the results of the simplified quantum photodetection theory presented here.

This quantum photodetection theory requires knowledge of the amplified light photon statistics, analyzed in Chapter 2. The quantum statistics of amplified light and photoelectron count are both described by discrete PDFs. In the case of a coherent signal input to the amplifier (Poisson PDF), the photon statistics at the amplifier output have a noncentral negative binomial distribution (NNB) or Laguerre distribution, defined in Eq. (2.102). For an ideal detector, free from additive thermal noise and with finite quantum efficiency $\eta$, the photoelectron count PDF is given by Eq. (3.14). Likewise, in the case where there is no input signal to the amplifier, the photon statistics at the amplifier output are described by the negative binomial distribution (NB), defined in Eq. (2.93), which reduces to the Bose–Einstein (BE) distribution for a single polarization mode ($\mathscr{M} = 1$).

The photoelectron counts PDFs corresponding to symbols 0 and 1 can be expressed as:

$$P_n(0) = \frac{(n + \mathscr{M} - 1)!}{n!\,(\mathscr{M} - 1)!} \frac{(\eta N)^n}{[1 + \eta N]^{n+\mathscr{M}}} \tag{3.93}$$

$$P_n(1) = \frac{(\eta N)^m}{(\eta N + 1)^{m+\mathscr{M}}} \exp\left\{ -\frac{G\eta\langle n_0\rangle}{\eta N + 1} \right\} L_n^{(\mathscr{M}-1)}\left\{ -\frac{G\eta\langle n_0\rangle}{\eta N(\eta N + 1)} \right\} \tag{3.94}$$

where $L_n^{(\mathscr{M}-1)}$ is the generalized Laguerre polynomial defined in Eq. (2.103). Using Eqs. (3.93) and (3.94), the conditional error probabilities $P(1|0)$ and $P(0|1)$ can be numerically evaluated through the discrete summations in Eqs. (3.63) and (3.64), and the BER through Eq. (3.62) with $p(1) = p(0) = 1/2$.

For $10^{-9}$ BER, and for the 1 symbol, the amplifier input photon number is close to $2\bar{n} = 80$. Thus, with a high gain amplifier, e.g., $G = 30$ dB, the photoelectron count is very large, e.g., $n = 8.10^4$ at $\langle n\rangle = 2G\bar{n}$. The numerical evaluation of Laguerre polynomials in Eq. (3.94) up to order 80,0000 for each value of $n$, which determines the distribution $P_n(1)$, is therefore a time-consuming process. The distribution $P_n(1)$ must also be determined for different input signal mean photon numbers $\langle n\rangle$. Finally, the evaluation of minimal BER requires repeated summations in Eqs. (3.63) and (3.64) for different decision levels $D$, until an optimum $D_{\text{opt}}$ is found.

The summation involved in Eq. (3.64), however, only requires the knowledge of $P_n(1)$ up to the optimum decision level $D_{\text{opt}}$, which is well below $2G\bar{n}/4$ (as shown in Figure 3.8, with $D_{\text{opt}}$ normalized to $G$). Therefore, it is not necessary to compute the whole distribution $P_n(1)$ up to the order $n = 2G\bar{n}$. Additionally, when the mean photon number $2G\bar{n}$ is large, the distributions $P_n(q)$ involved in the sums in Eq. (3.63) and (3.64) are slowly varying and can be assumed constant over bins of $N$ photons (e.g. $N = 10$). This approximation reduces by a factor $N$ the number of points for the distributions $P_n(q)$ required for the BER evaluation.

A BER calculation using the discrete PDFs defined in Eqs. (3.93) and (3.94) and a comparison with a Gaussian approximation is shown in a study by T. Li and M. C. Teich [61]. A $10^{-9}$ BER sensitivity $\bar{n} \approx 38.5$ photons/bit was obtained, assuming an amplifier gain of $G = 22$ db and $\mathscr{M} = 1$ (ASE polarizer), while with a Gaussian PDF approximation, the sensitivity was $\bar{n} \approx 42$ photons/bit. This result indicates that some BER overestimation is introduced by the approximation, in comparison to the exact quantum PDF; also observed with the square-law detection theory and discussed previously.

The $10^{-9}$ BER sensitivities calculated from Eqs. (3.62)–(3.64) and (3.93)–(3.94) for the cases $\mathscr{M} = 1$ (ASE polarizer) and $\mathscr{M} = 2$ (no polarizer), and for a gain value of $G = 15$ dB are found to be $\bar{n} = 38.7$ photons/bit and $\bar{n} = 39.5$ photons/bit, respectively. These values are within one photon, or 0.1 dB, of the previous results found from the square-law detection theory and Rician PDFs.

**TABLE 3.1** **Sensitivity of ideal ASK direct detection receiver for $10^{-9}$ BER in average photons/bit**

| Theory | PDF for 0 | PDF for 1 | Polarizer ($\mathcal{M} = 1$) | No polarizer ($\mathcal{M} = 2$) | Refs. |
|---|---|---|---|---|---|
| | Gaussian | Gaussian | 41.2 | 43.3 | Chapter 3 |
| Square-law | Gaussian | Gaussian | 42.3 | – | [30] |
| detection | Gaussian | Gaussian | – | 44.5 | [28] |
| | Rayleigh | Rician | 38.0 | 40.5 | [66], Chapter 3 |
| | Rician | Rician | 38.8 | 42.1 | Chapter 3 |
| Quantum photodetection | Negative binomial | Noncentral negative binomial | 38.7 | 39.5 | [61], Chapter 3 |

The different values for $10^{-9}$ BER sensitivity obtained in this section from various theories, which correspond to the case of an ideal direct detection ASK receiver with optical preamplifier, are summarized in Table 3.1. Experimental results with these values are compared in Section 7.1.

The results of Table 3.1 suggest that for the prediction of minimal BER of ideal direct detection ASK receivers, the square-law detection theory and the quantum photodetection theory are in good agreement. Furthermore, we have seen that *the Gaussian approximation provides a fair estimate of the BER, with the advantage of closed-form expressions*. We expect good agreement between the semiclassical theory and the quantum theory as the mean photon number incident upon the detector for $10^{-9}$ BER is very large, i.e., $2G\bar{n} = 10^4 - 10^5$. For such large photon numbers, the PDFs corresponding to both theories are nearly identical and approach the Gaussian distribution. The semiclassical theory of square-law detection is not complete, as the amplifier ASE noise is phenomenologically introduced in the form $N = n_{eq}G$, a result borrowed from quantum theory.

Clearly, a pure quantum theory of photodetection is relevant to the statistical description of amplified light having a mean output photon number $\langle n \rangle = 1 - 10^2$, a number considerably lower than in optical communications. For the noise statistics of signals with such low photon numbers and, furthermore, single photon count events, the semiclassical square-law detection theory, which essentially assumes Gaussian random noise, is clearly irrelevant.

## 3.6 BIT-ERROR RATE AND RECEIVER SENSITIVITY IN DIGITAL COHERENT DETECTION

The theory of photodetection described in the previous section considered the case of direct detection receivers. In such receivers, the ASK optical signal is directly converted into a photocurrent, which requires prior optical demodulation if the signal is initially transmitted in FSK or DPSK formats. This principle is generally called incoherent detection, by contrast with coherent detection.

The principle of digital coherent detection is illustrated in Figure 3.10. Prior to photodetection, the optical signals at frequency $\omega_s$, which can be modulated with

ASK, FSK, or PSK, are mixed with the continuous wave (CW) light of a local oscillator (LO) with frequency $\omega_{LO}$ and matched state of polarization. The mixing produces an interference tone at the beat or intermediate frequency (IF) $\omega_{IF} = \omega_{LO} - \omega_s$. The scheme is said to be homodyne if the LO frequency $\omega_{LO}$ matches that of the optical signals to be detected, i.e. $\omega_{IF} = 0$, and heterodyne in the contrary case, i.e., $\omega_{IF} \neq 0$. Photodetection converts the optical beat into a photocurrent in the IF electronic bandwidth. The corresponding receiver sensitivity is greatly enhanced in comparison to the case of incoherent detection. Basic principles and theoretical aspects of coherent detection are in textbooks [14] and [15], and review paper [16] and [67]–[69].

There are two applications for an optical amplifier in coherent receivers: to amplify the signal data (a preamplifier); to amplify the local oscillator (a power booster). The power booster configuration is referred to as ELOS – enhanced LO system [70].

An ideal coherent receiver operates near the shot noise limit; this corresponds to the highest possible sensitivity. Since optical amplifiers generate s–ASE beat noise, the inclusion of a preamplifier in a coherent receiver should, in principle, cause a sensitivity degradation [71]. In actual receivers operating at high data rates (Gbit/s), however, near-shot-noise-limited operation is difficult to achieve, due to electronics limitations, and sensitivity improvement with preamplifiers is possible [70].

Likewise, the best coherent receiver sensitivity is achieved when the LO shot noise overcomes the detector thermal noise. This regime is difficult to achieve because of the limited power available from the LO source and loss introduced by the mixer or other optical components. Therefore, using an optical amplifier to boost the LO power (ELOS) can also improve the receiver sensitivity [70].

Optical amplifiers can also be used in coherent systems for long-haul applications, as in-line repeaters, [72] and [73]. In long-haul systems with in-line amplifiers, the buildup of broadband ASE requires optical filtering at the receiver. In these systems, the advantage of coherent detection over direct detection is that optical homodyning or heterodyning provides the required filtering function, as the unwanted ASE noise is naturally rejected from the IF band. Some form of optical filtering prior to detection may still be necessary for detector saturation considerations, but with a reduced requirement in the filter bandwidth. If amplifier and detector saturation by ASE are negligible, the whole amplifier bandwidth can be exploited for transmission and detection of multiple optical channels. In this scheme, coherent detection instead of direct detection means that the optical bandwidth and channel density may be considerably higher (see [74] and references therein). Hence we analyze the performance of coherent receivers for the detection of optically amplified signals.

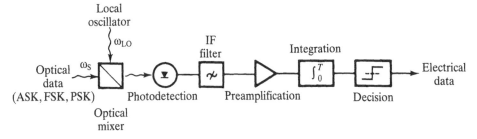

**FIGURE 3.10**  Basic principle of digital coherent photodetection.

The analysis of photodetection with coherent receivers has been the object of considerable theoretical work, based on classical or quantum mechanical descriptions (see [69], [75], and [76] plus references therein). We cannot provide a comparative insight into the various descriptions, not even a limited overview of receiver and system configurations; so many are possible for coherent communications. Instead we outline basic theoretical results that make possible a simplified evaluation of the SNR and BER in digital systems using heterodyne coherent detection.

The photocurrent generated by the detector after mixing signal and LO can be expressed as:

$$I_{\text{phot}} = \frac{e\eta}{h\nu_s} \{E_{\text{LO}}^2 + E_s^2 + 2E_s E_{\text{LO}} \cos(\omega_{\text{IF}} t + \Delta\varphi_{\text{LO}} - \varphi_s)\} \tag{3.95}$$

where $\Delta\varphi = \varphi_{\text{LO}} - \varphi_s$ is the phase difference between signal and LO. For heterodyne detection where $\omega_{\text{IF}} \neq 0$, the time averaged signal power in the IF bandwidth is then:

$$I_{\text{sig}}^2 = \left(\frac{e\eta}{h\nu_s}\right)^2 2P_s P_{\text{LO}} = 2\eta^2 I_s I_{\text{LO}} \tag{3.96}$$

where $P_s$ and $P_{\text{LO}}$ are the signal and the LO optical powers, respectively, $\eta$ is the detector quantum efficiency, and $I_s$, $I_{\text{LO}}$ are the photocurrents, defined as:

$$I_s = \frac{eP_s}{h\nu_s} \tag{3.97}$$

$$I_{\text{LO}} = \frac{eP_{\text{LO}}}{h\nu_s} \tag{3.98}$$

If an optical preamplifier of gain $G$ is included in the receiver, Eq. (3.96) becomes:

$$I_{\text{sig}}^2 = \left(\frac{e\eta}{h\nu_s}\right)^2 2GP_s P_{\text{LO}} = 2\eta^2 G I_s I_{\text{LO}} \tag{3.99}$$

On the other hand, the total noise power $\sigma_d^2$ falling into the receiver IF bandwidth $B_e$ can be expressed as a sum of various components [73]:

$$\sigma_d^2 = \sigma_{\text{shot}}^2 + \sigma_{\text{LO-ASE}}^2 + \sigma_{\text{s-ASE}}^2 + \sigma_{\text{ASE-ASE}}^2 + \sigma_{\text{th}}^2 \tag{3.100}$$

In Eq. (3.100), the shot noise component $\sigma_{\text{shot}}^2$ is defined by:

$$\sigma_{\text{shot}}^2 = \eta e 2 B_e (I_{\text{LO}} + G I_s + \mathcal{M} I_{\text{N}}) \tag{3.101}$$

and the LO–ASE, s–ASE and ASE–ASE beat noise components are defined by:

$$\sigma_{\text{LO-ASE}}^2 = 4\eta^2 I_{\text{LO}} I_{\text{N}} \frac{B_e}{B_o} \tag{3.102}$$

$$\sigma^2_{\text{s-ASE}} = 4\eta^2 G I_s I_N \frac{B_e}{B_o} \tag{3.103}$$

$$\sigma^2_{\text{ASE-ASE}} = \mathcal{M} \eta^2 I_N^2 \frac{2B_e}{B_o} \tag{3.104}$$

where the photocurrent $I_N$ is defined by:

$$I_N = e n_{\text{eq}} G B_o \tag{3.105}$$

Equations (3.101)–(3.104) for the various noise components in coherent heterodyne detection are identical to their counterparts in direct detection, as derived in Section 3.2, except for the ASE–ASE beat noise. Since $I_N/B_o = e n_{\text{eq}} G$ from Eq. (3.105), the LO–ASE and the s–ASE beat noises are functions only of $B_e$, while the ASE–ASE beat noise is a function of $B_o B_e$.

We consider first the case where there is no optical preamplifier, i.e., $G = 1$ and $I_N = 0$. The receiver electrical $\text{SNR}_e$ is then, from Eqs. (3.97)–(3.101):

$$\text{SNR}_e = \frac{I^2_{\text{sig}}}{\sigma^2_d} = \frac{2\eta^2 I_s I_{\text{LO}}}{\eta e 2 B_e I_{\text{LO}} + \sigma^2_{\text{th}}} = \frac{\eta P_s}{h v_s B_e} \frac{1}{1 + \dfrac{h v_s \sigma^2_{\text{th}}}{2 e^2 B_e P_{\text{LO}}}} \tag{3.106}$$

If the LO power $P_{\text{LO}}$ is high enough so that the LO shot noise overcomes the detector thermal noise $\sigma^2_{\text{th}}$, we obtain from Eq. (3.106):

$$\boxed{\text{SNR}_e = \text{SNR}_e^{\text{shot}} = \frac{\eta P_s}{h v_s B_e}} \tag{3.107}$$

The electrical SNR value $\text{SNR}_e^{\text{shot}}$ expressed in Eq. (3.107) corresponds to the so-called *detector shot noise limit* [1], and is maximum for a detector with unity quantum efficiency ($\eta = 1$).

We consider next the case where an optical preamplifier is included in the receiver. The receiver electrical $\text{SNR}_e$ is now:

$$\text{SNR}_e = \frac{I^2_{\text{sig}}}{\sigma^2_d}$$

$$= \frac{2\eta^2 G I_s I_{\text{LO}}}{\eta e 2 B_e (I_{\text{LO}} + G I_s + \mathcal{M} I_N) + 4\eta^2 I_{\text{LO}} I_N \dfrac{B_e}{B_o} + 4\eta^2 G I_s I_N \dfrac{B_e}{B_o} + \mathcal{M} \eta^2 I_N^2 \dfrac{2B_e}{B_o} + \sigma^2_{\text{th}}}$$

$$= \frac{\text{SNR}_e^{\text{shot}}}{\eta F_o \left(1 + \dfrac{G I_s}{I_{\text{LO}}}\right) + \mathcal{M} \left(\dfrac{1}{G} + \eta n_{\text{eq}}\right) \dfrac{I_N}{I_{\text{LO}}} + \dfrac{\sigma^2_{\text{th}}}{\eta e 2 B_e G I_{\text{LO}}}}$$

$$= \frac{\text{SNR}_e^{\text{shot}}}{\eta F_o \left(1 + \dfrac{G P_s}{P_{\text{LO}}}\right) + \mathcal{M} \left(\dfrac{1}{G} + \eta n_{\text{eq}}\right) \dfrac{G P_o}{P_{\text{LO}}} + \dfrac{h v_s \sigma^2_{\text{th}}}{\eta e^2 2 B_e G P_{\text{LO}}}} \tag{3.108}$$

where $\text{SNR}_e^{\text{shot}}$ is the shot-noise-limited SNR defined in Eq. (3.107), $F_o$ is the optical noise figure defined in Eq. (3.81), i.e., $F_o = 2n_{\text{eq}} + 1/\eta G$, and $P_o = n_{\text{eq}} h v_s B_o$ is the amplifier equivalent input noise power.

In the case where the preamplifier gain is high ($G \gg 1$) and the LO power is significantly greater than the amplified signal power ($I_{\text{LO}} \gg GI_s$), with the additional condition $GI_{\text{LO}} \gg \sigma_{\text{th}}^2/eB_e$ the dominating noise component in the receiver is the LO–ASE beat noise, and SNR in Eq. (3.108) becomes:

$$\text{SNR}_e = \frac{\text{SNR}_e^{\text{shot}}}{\eta F_o} = \frac{\text{SNR}_e^{\text{shot}}}{2\eta n_{\text{eq}}} = \frac{P_s}{2n_{\text{eq}} h v_s B_e} \qquad (3.109)$$

Equation (3.109) shows that in the ideal case of a fully inverted preamplifier ($n_{\text{eq}} = 1$) and of a detector with unity quantum efficiency ($\eta = 1$), *the best SNR achievable with a preamplifier/coherent receiver is 3 dB below the shot noise limit $SNR_e^{\text{shot}}$ [70]*. If the detector quantum efficiency is less than unity (as is generally the case), Eq. (3.109) shows that the degradation with respect to the shot-noise-limited SNR that is introduced by the preamplifier is less than 3 dB. In all cases, a minimum degradation of 3 dB always exists with respect to the maximum achievable shot-noise-limited SNR, which is equal to $P_s/h v_s B_e$. The sensitivity improvements introduced by the preamplifier in the case of a nonideal coherent receiver ($\eta < 1$) are studied in [70]. This study also shows that boosting the LO power by optical amplification makes it possible to approach the conditions of shot-noise-limited detection.

The formal derivation of BER for coherent heterodyne detection can be found in many books on communications, [77]–[80], and fundamental papers, [81]–[85]. Receiver sensitivity comparisons between various modulation schemes (i.e., ASK, FSK, PSK), receiver configurations (i.e., synchronous or asynchronous) and homodyning versus heterodyning can be found in [14], [16], and [69]. In heterodyne, asynchronous receivers, for which line width requirements for both signal and LO sources are the least stringent [14], the BER is given by the expression [79]:

$$\text{BER} = \tfrac{1}{2}\exp\{-\xi\text{SNR}_e\} \qquad (3.110)$$

where $\text{SNR}_e$ is defined in Eq. (3.108) and $\xi$ is a parameter reflecting the effect of various modulation/demodulation schemes. The signal power required for achieving $10^{-9}$ BER is therefore given by $\text{SNR}_e = -\log(2.10^{-9})/\xi = 20/\xi$. Combining this result with Eq. (3.108), with the assumptions $B_e = 2B_o$ and $P_{\text{LO}} \gg GP_s$, we obtain the relation:

$$\bar{P} = \frac{20}{\eta\xi} h v_s B_o \left(\frac{1}{G} + 2\eta n_{\text{eq}}\right) \qquad (3.111)$$

where $\bar{P}$ is the average signal power per bit required to achieve $10^{-9}$ BER, which is the receiver sensitivity.

When no amplifier is used ($G = 1$, $n_{eq} = 0$), the sensitivity in photons/bit is given by $\bar{n} = \bar{P}/hv_s B_o = 20/\eta\xi$. For ideal receivers ($\eta = 1$) with heterodyne asynchronous detection, the best sensitivities are $\bar{n} = 40$ photons/bit (ASK, FSK), and $\bar{n} = 20$ photons/bit (PSK), [14]–[16], which correspond to the parameter where $\bar{P}$ is the average signal power per bit required to achieve $10^{-9}$ BER, which is the receiver sensitivity.

When no amplifier is used ($G = 1$, $n_{eq} = 0$), the sensitivity in photons/bit is given by $\bar{n} = \bar{P}/hv_s B_o = 20/\eta\xi$. For ideal receivers ($\eta = 1$) with heterodyne asynchronous detection, the best sensitivities are $\bar{n} = 40$ photons/bit (ASK, FSK), and $\bar{n} = 20$ photons/bit (PSK), [14]–[16], which correspond to the parameter values $\xi = 1/2$ (ASK, FSK) and $\xi = 1$ (PSK), respectively.

If a high gain amplifier is used ($G \gg 1$), the sensitivity in photons/bit is given by $\bar{n} = 40 n_{eq}/\xi$. With an ideal amplifier ($n_{eq} = 1$), the best sensitivities are then $\bar{n} = 80$ photons/bit (ASK, FSK), and $\bar{n} = 40$ photons/bit (PSK).

## 3.7  DIGITAL PHOTODETECTION WITH OPTICAL AMPLIFIER CHAINS

One of the most important application of optical amplifiers in lightwave communications is in long-haul systems, in which distances are typically greater than 1000 km. In these systems, optical amplifiers are advantageously used as in-line optical repeaters, replacing conventional electronic repeaters. Communications systems based on optical amplifier chains are also (paradoxically) called *unrepeatered* or *nonregenerative systems*, as the only function of the amplifiers is to keep the signal at a constant power level, without any other form of data processing or regeneration.

Maybe those are the most important questions we can answer from the analysis of SNR and BER developed throughout this chapter. *What is the ultimate distance over which optical data can be transmitted without regeneration? How is the maximum transmission distance affected by the amplifier spacing, data modulation format, and receiver type?* They have been answered in several theoretical studies and experimental demonstrations using both direct and coherent detection systems, [72]–[74] and [86]–[89]. In this section, we shall only provide a basic understanding of how the maximum transmission distance is affected by the constraints of amplifier spacing, modulation format, and type of receiver. Other system constraints and various technological issues in unrepeatered systems are discussed in Chapter 7.

In Chapter 2, we analyzed the effect of ASE noise buildup in optical amplifier chains and derived closed-form expressions for the output noise variance. We have considered Type A and C chains, which are made of a repeating sequence of lumped elements. These elements consist of an amplifier of gain $G$ followed by a strand of fiber with loss $T = 1/G$ (Type A) or the same elements in reverse order (Type C), as illustrated in Figure 2.17. The output noise variance of an amplifier chain with $k$ elements is identical to that of a single amplifier having unity gain and an equivalent input noise $n'_{eq} = kn_{eq}$ (Type A chain) or $n'_{eq} = kGn_{eq}$ (Type C chain), as shown in Eqs. (2.170) and (2.171).

In this chapter, we analyzed various aspects of photodetection of optically amplified light for both direct detection and heterodyne coherent detection receivers. The equivalence between a $k$ element amplifier chain and a single amplifier makes it possible to generalize the results obtained in this chapter to the case of unrepeatered systems.

For coherent detection systems, Eq. (3.108) is modified to include the effect of ASE noise accumulation along the amplifier chain.

We focus first on direct detection systems. We can take Eq. (3.80) and substitute $n'_{eq} = kn_{eq}$ (Type A) or $n'_{eq} = kG_{eq}$ (Type C). Assuming nonpolarization maintaining systems ($\mathcal{M} = 2$, or no ASE polarization filter at the receiver), assuming $B_e = B/2$ ($B$ = bit rate), and assuming each amplifier has an optical filter of passband $B_o$ (usually much larger than $B/2$), we obtain then for Type A and Type C amplifier chains:

$$\bar{P}(k, \text{Type A}) = 18hv_s B \left\{ \frac{1}{\eta} + 2kn_{eq} + \frac{1}{6}\sqrt{2k^2 n_{eq}^2 \left(\frac{4B_o}{B} - 1\right) + \frac{8kn_{eq}B_o}{\eta B} + \frac{4\sigma_{th}^2}{e^2 \eta^2 B^2}} \right\}$$

(3.112)

$$\bar{P}(k, \text{Type C}) = 18hv_s B \left\{ \frac{1}{\eta} + 2kGn_{eq} + \frac{1}{6}\sqrt{2k^2 G^2 n_{eq}^2 \left(\frac{4B_o}{B} - 1\right) + \frac{8kGn_{eq}B_o}{\eta B} + \frac{4\sigma_{th}^2}{e^2 \eta^2 B^2}} \right\}$$

(3.113)

The above results show that the required power $\bar{P}$ for $10^{-9}$ BER increases linearly with the number of amplifiers $k$, past a certain point in the chain where $k > (\sigma_{th}/e\eta Bn_{eq})$. For the same amplifier gain $G$ in both Type A and Type C configurations, the increase is $G$ times higher for the Type C configuration, which represents an important factor in the case of large amplifier spacings (e.g., 100 km, or $G = 20$–$25$ dB). For this reason, only the Type A configuration will be considered for very long systems.

By definition, the *system penalty* is the ratio $\bar{P}(k)/\bar{P}(0)$, where $\bar{P}(k)$ is the receiver sensitivity after $k$ amplifier chain elements, and $\bar{P}(0)$ is the receiver sensitivity when the chain is bypassed, also called back-to-back sensitivity. For direct detection, the back-to-back sensitivity is given by setting $k = 0$ in Eq. (3.112), which gives:

$$\bar{P}(0, \text{direct det.}) = \frac{18}{\eta} hv_s B \left(1 + \sqrt{\frac{\sigma_{th}^2}{9e^2 B^2}}\right)$$

(3.114)

Figure 3.11 shows plots of the theoretical penalty as a function of total system length, corresponding to an ASK direct detection configuration with Type A optical amplifier chain. The corresponding parameters are: $\eta = 0.8$, $B = 1$ Gbit/s, $B_o = 0.1$ nm to 5 nm, $\lambda_s = 1.55$ $\mu$m, $\sigma_{th}^2 = i_c^2 B$ with $i_c = 10$ pA/$\sqrt{\text{Hz}}$ [73], which give a back-to-back sensitivity of $\bar{P}(0) = -27.2$ dBm. The amplifier gain and equivalent input noise factor are assumed to be $G = 25$ dB and $n_{eq} = n_{sp}(1 - 1/G) = 1.6$, respectively, corresponding to 100 km amplifier spacing with a fiber loss coefficient of $\alpha_s = 0.25$ dB/lm, and an amplifier noise figure of $F_o = 2n_{sp} = 5$ dB. As expected, Figure 3.11 shows the penalty increases with both length and optical filter width $B_o$. For a total system length of 10,000 km, this penalty is near 5 dB with a filter width $B_o = 5$ nm, and near 2 dB with a filter width $B_o = 0.5$ nm. We conclude from these results that *for systems based on ASK direct detection, the penalty caused by ASE noise accumulation is generally small, even at lengths up to 10,000 km* [73]. But in actual systems, the prevailing

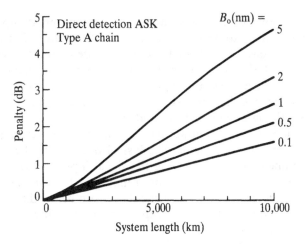

**FIGURE 3.11** System penalty at $B = 1$ Gbit/s versus total transmission length for a Type A amplifier chain with ASK direct detection receiver, and for various optical filter bandwidths $B_o$. The back-to-back receiver sensitivity is $\bar{P}(0) = -27.2$ dBm. $B = 1$ Gbit/s, $\lambda_s = 1.55 \mu m$, $n_{eq} = 1.6$, and BER $= 10^{-9}$.

penalty may be dispersion, fiber nonlinearities, or amplifier saturation, for instance, so the above results only represent ultimate performance limits determined by ASE accumulation. Other system penalties are discussed in detail in Chapter 7.

We consider now the case of a coherent system based on a Type A amplifier chain configuration. We substitute the equivalent amplifier parameters in the SNR expression Eq. (3.108) and solve Eq. (3.110) for $10^{-9}$ BER, assuming $I_s \ll I_{LO}$, $B_e = 2B$ [73]. This gives the sensitivity after $k$ chain elements:

$$\bar{P}(k) = \frac{20}{\eta\xi} h v_s B \left\{ 1 + 2k\eta n_{eq} + 2k n_{eq} \frac{h v_s B_o}{P_{LO}} (1 + k\eta n_{eq}) + \frac{h v_s}{P_{LO}} \frac{\sigma_{th}^2}{4e^2 \eta B} \right\}$$

$$= \bar{P}(0, \text{coherent det.}) + k \frac{40 n_{eq}}{\eta\xi} h v_s B \left\{ \eta + \frac{h v_s B_o}{P_{LO}} (1 + k\eta n_{eq}) \right\} \qquad (3.115)$$

where $\bar{P}(0, \text{coherent det.})$ is the system back-to-back sensitivity, defined by:

$$\bar{P}(0, \text{coherent det.}) = \frac{20}{\eta\xi} h v_s B \left( 1 + \frac{h v_s}{P_{LO}} \frac{\sigma_{th}^2}{4\eta e^2 B} \right) \qquad (3.116)$$

Using the same direct detection parameters as before, except with a thermal noise $\sigma_{th}^2 = i_c^2(2B)$ [73], and an LO power $P_{LO} = 1$ mW, we obtain from Eq. (3.116) back-to-back sensitivities of $\bar{P} = -50.7$ dBm (ASK, FSK) and $\bar{P} = -53.7$ dBm (PSK), corresponding to 66 photons/bit and 33 photons/bit, respectively.

Equation (3.115) shows that the required power $\bar{P}$ for $10^{-9}$ BER increases linearly with the number of amplifiers $k$. The numerical values of the example below show this increase is actually of the order of $k$ times the back-to-back sensitivity $\bar{P}(0, \text{coherent det.})$, much faster than in the case of direct detection. Using Eqs. (3.115) and (3.116) with the above parameters, we find that for system lengths up to 10,000 km, the penalties corresponding to ASK, FSK, and PSK modulation formats are about 20 dB, and the system sensitivities $\bar{P}$ are better than those of direct detection by 3 dB (ASK, FSK) or 6 dB (PSK). Thus, with the Type A amplifier chain configuration there is no significant advantage in using the coherent detection scheme.

It is possible to improve the sensitivity of this system by including an extra loss at its end, between the last chain element and the receiver [73]. The improvement comes from the reduction in ASE power incident upon the detector, which reduces the LO–ASE noise by the same factor. The loss also decreases the received signal but this can be compensated by increasing the signal input to the amplifier chain. However, three constraints for the signal power input are: (1) it should remain below the amplifier saturation input power; (2) it should remain below the threshold where fiber nonlinearities are observed; and (3) at the receiver level, it should be negligible in comparison to the LO power.

With this modified system configuration, the receiver sensitivity is given by the following expression, similar to Eq. (3.115):

$$T'\bar{P}(k) = \bar{P}(0, \text{coherent det.}) + k\,\frac{40T'n_{eq}}{\eta\xi}\,hv_s B\left\{\eta + \frac{hv_s B_o}{P_{LO}}(1 + k\eta T'n_{eq})\right\} \qquad (3.117)$$

In Eq. (3.117), $T'$ is the transmission of the lossy element added to the end of the amplifier chain, $\bar{P}(k)$ is the signal power input to each amplifier in the chain, and $T'\bar{P}(k)$ is the signal power received at the detector. The penalty originating from the LO–ASE noise, proportional to $kT'n_{eq}$, can be considerably reduced by choosing a low value for the transmission $T'$. The maximum sensitivity corresponding to a negligible penalty is given by $T'\bar{P}(k) \approx \bar{P}(0)$, which occurs for $T' \ll 1$ or $\bar{P}(k) \gg \bar{P}(0)$, which must be realized under the three aforementioned restrictions.

There are two ways to achieve the operating conditions described by Eq. (3.117). One is to set the input signal power $\bar{P}(k)$ to a fixed value (e.g., $-20$ dBm), then to adjust the transmission $T'$ given a number $k$ of amplifiers [73]. Another is to fix the value of the transmission $T'$ (e.g., $T' = -20$ dB) then to adjust the input signal power $\bar{P}(k)$. This approach has the advantage that the lossy element at the end of the chain can be a fixed length of fiber, representing additional transmission distance.

Figure 3.12 shows plots of theoretical penalties as a function of the total system length, corresponding to asynchronous heterodyne coherent systems with ASK, FSK, or PSK modulation, and based on a Type A amplifier chain followed by a lossy element of transmission $T'$. As no optical filters are used, we assume the optical bandwidth to be the full EDFA width, i.e., $B_o = 25$ nm. Other parameters are identical to those of Figure 3.11.

Figure 3.12 shows that with coherent systems, the penalty introduced by ASE noise accumulation can be effectively compensated by the effect of additional loss between the end of the amplifier chain and the receiver [73]. With a loss $T' = -30$ dB, which represents 120 km of additional fiber, the penalty at system length 10,000 km is about 0.8 dB for ASK and FSK modulation, and about 0.5 dB for PSK modulation. The signal power $\bar{P}(k)$ required at the input to the chain to achieve $10^{-9}$ BER at the receiver is given by the relation $\bar{P}(k) = \bar{P}(0) + \text{penalty}(k) + |T'|$, where all terms are expressed in dB or dBm. For the longest length ($k = 100$), this relation gives $\bar{P}(100) = -50.7$ dBm $+ 0.8$ dB $+ 30$ dB $= -19.9$ dBm for (ASK, FSK), and $\bar{P}(100) =$

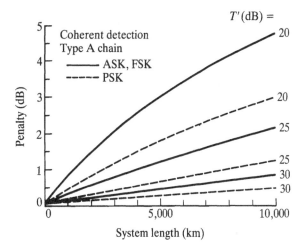

**FIGURE 3.12** System penalty at $B = 1$ Gbit/s versus total transmission length for a Type A amplifier chain with coherent, heterodyne asynchronous detection receiver modulated by (———) ASK, FSK and (– – –) PSK. A lossy element of transmission $T'$ is assumed between the end of the chain and the receiver. The back-to-back receiver sensitivities are $\bar{P}(0) = -50.7$ dBm (ASK, FSK) and $\bar{P}(0) = -53.7$ dBm (PSK). $B_o = 25$ nm, $\lambda_s = 1.55$ μm, $n_{eq} = 1.6$, and BER = $10^{-9}$.

$-53.7$ dBm $+ 0.5$ dB $+ 30$ dB $= -23.2$ dBm for (PSK). The required signal powers in this example are below or near $-20$ dBm, below the saturation input power of typical EDFAs (Chapter 5).

## 3.8 ANALOG SIGNAL PHOTODETECTION

Lightwave communications systems can be divided into analog or digital according to how the light signals are modulated. Digital communications concern the large majority of newly developed lightwave systems, but many applications are currently based on analog communications. Indeed, several fiber optic applications require an analog format to achieve compatibility with existing microwave systems, or compliance with existing standards for voice and video distribution, for their specific advantages over digital modulation, or simply because they may cost less.

Different types of analog systems are reviewed in [12]–[15]. In some analog systems, such as cable TV or CATV (common antenna television) networks, optical fibers find applications as low loss trunk elements between various network nodes; this reduces the number of electronic repeaters. For this type of application, optical fibers also offer the advantages of a virtually unlimited bandwidth, as well as an inherent immunity to electromagnetic interference. Fiber optic systems can also be used as intermediate networks or concentrators for wireless microwave communications and computer networks, [13] and [15].

The recent emergence of optical fiber amplifiers (EDFAs), has opened the perspective of all-optical analog communications systems, characterized by a range and bandwidth capacity far superior to conventional microwave systems, [90] and references therein. Various applications of EDFAs to analog systems, in particular AM and FM video broadcast networks, are described in Chapter 7.

In this section, we provide a basic description of the characteristics of amplified light photodetection in the case where light is used to convey analog signals. We consider the basic analog broadcast transmission system shown in Figure 3.13. In this simple configuration, an optical amplifier (gain $G$) is used as a power booster for the analog signal. This power booster can compensate the loss $L$ of the $1 \times N$ optical splitter (or tree coupler), used to broadcast the signal to $N$ receivers. For analog systems, the *carrier-to-noise ratio* (CNR) is, by definition, the ratio of the electrical carrier power $I_{\text{sig}}^2$ to the sum of all noise powers $\sigma_d^2$ [14], i.e.,

$$\text{CNR} = \frac{I_{\text{sig}}^2}{\sigma_d^2} \tag{3.118}$$

Given an average optical power $P_s$ and a photodetector quantum efficiency $\eta$, the received carrier power is given by $I_{\text{sig}}^2 = (m\eta I_s)^2/2$, where $m$ is the optical modulation index [14]. The carrier power corresponding to the system configuration of Figure 3.13 is then given by:

$$I_{\text{sig}}^2 = \tfrac{1}{2}(m\eta GLI_s)^2 \tag{3.119}$$

$I_s = eP_s/h\nu_s$ is the mean photocurrent defined in Eq. (3.97) and $P_s$ is the average optical power input to the amplifier.

The total receiver noise power corresponding to the configuration of Figure 3.13 can be formally expressed as [90] and [91]:

$$\sigma_d^2 = \sigma_{\text{RIN}}^2 + \sigma_{\text{shot}}^2 + \sigma_{\text{s-ASE}}^2 + \sigma_{\text{ASE-ASE}}^2 + \sigma_{\text{th}}^2 \tag{3.120}$$

In Eq. (3.120), $\sigma_{\text{RIN}}^2$ is the *relative intensity noise* power:

$$\sigma_{\text{RIN}}^2 = \text{RIN} \cdot (\eta GLI_s)^2 B_e \tag{3.121}$$

in which RIN is the relative intensity noise spectral density of the optical transmitter (defined as $\text{RIN} = \langle I_s^2 \rangle / I_s^2$ in dB/Hz) and $B_e$ is the receiver electronic bandwidth.

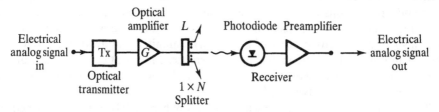

**FIGURE 3.13** Basic system configuration for analog signal broadcast distribution using an optical amplifier at the transmitter.

Consistent with Eqs. (3.101), (3.103), and (3.104), the shot noise ($\sigma_{shot}^2$) and beat noise components ($\sigma_{s\text{-}ASE}^2$, $\sigma_{ASE\text{-}ASE}^2$) are defined by:

$$\sigma_{shot}^2 = 2eB_e\eta GLI_s \tag{3.122}$$

$$\sigma_{s\text{-}ASE}^2 = 4\eta^2 L^2 GI_s I_N \frac{B_e}{B_o} \tag{3.123}$$

$$\sigma_{ASE\text{-}ASE}^2 = \mathcal{M}\eta^2 L^2 I_N^2 \frac{2B_e}{B_o} \tag{3.124}$$

where $I_N$ is the mean ASE noise photocurrent, defined in Eq. (3.105) as $I_N = en_{eq}GB_o$ ($B_o$ = optical pass band) and $\mathcal{M}$ is the number of polarization modes ($\mathcal{M} = 2$ if no ASE polarizer is used at the detector, $\mathcal{M} = 1$ otherwise). Finally, the receiver thermal noise can be expressed as $\sigma_{th}^2 = i_c^2 B_e$, where $i_c$ is the equivalent circuit noise current density (expressed usually in $pA/\sqrt{Hz}$).

The specifications for CNR depend on the system considered. For CATV applications, the required CNR is $+50\,dB$ to $+55\,dB$ for AM TV and $+17\,dB$ for FM TV, [90] and [91]. While such a CNR is required at the main trunk level, it is usually lower by 10 dB at the subscriber level, i.e., CNR = $+40\,dB$ to $+45\,dB$ for AM TV [92]. As the information is retrieved at the subscriber level in the form of a picture, the notion of CNR is mainly subjective. take AM TV, for instance: a CNR of $+45\,dB$ corresponds to a picture subjectively judged excellent, whereas a CNR of $+25\,dB$ corresponds to a picture judged objectional [92].

The optical signal power $P_s$ required at the transmitter end to achieve a desired CNR (which we shall refer to as $CNR_d$) at the receiver end can be derived from the above formulae. Substituting Eqs. (3.119)–(3.124) into Eq. (3.118) with $X = \eta GLI_s$, we obtain the quadratic equation:

$$\left(\frac{m^2}{2CNR_d} - B_e RIN\right)X^2 - 2eB_e(1 + 2\eta n_{eq}LG)X - [\sigma_{th}^2 + (2\eta n_{eq}LG)^2 e^2 B_o B_e] = 0 \tag{3.125}$$

Equation (3.125) shows that a solution exists ($X > 0$) only under the condition:

$$B_e RIN < \frac{m^2}{2CNR_d} \tag{3.126}$$

Considering typical numerical values $B_e = 5\,MHz$, RIN = $-155$ to $-152\,dB/Hz$, $m = 0.05$ and $CNR_d = +50$ to $+55\,dB$ used in state-of-the-art AM TV analog systems [91], we can check that Eq. (3.126) is always satisfied. It is also satisfied for FM TV with the same parameters and $CNR_d = +17\,dB$.

Since Eq. (3.126) applies, the only positive solution of Eq. (3.125) can be put into the form:

$$\boxed{P_s = \frac{\kappa}{\eta} h\nu_s B_e \frac{1 + 2\eta n_{eq}LG}{LG}\left(1 + \sqrt{1 + \frac{\sigma_{th}^2/e^2 B_e B_o + (2\eta n_{eq}LG)^2 B_o}{\kappa(1 + 2\eta n_{eq}LG)^2} \frac{B_o}{B_e}}\right)} \tag{3.127}$$

with

$$\kappa = \left( \frac{m^2}{2\text{CNR}_\text{d}} - B_\text{e}\,\text{RIN} \right)^{-1}$$

The general result obtained in Eq. (3.127) makes it possible to study a variety of cases. Consider first the case where there is no optical amplifier and splitting loss, i.e., $G = L = 1$ and $n_\text{eq} = 0$, which gives the optical power $P_\text{s}$ necessary to achieve a desired CNR. We obtain then from Eq. (3.127):

$$P_\text{s} = \frac{\kappa}{\eta} h\nu_\text{s} B_\text{e} \left( 1 + \sqrt{1 + \frac{\sigma_\text{th}^2}{e^2 B_\text{e}^2}} \right) \approx \frac{\kappa}{\eta} h\nu_\text{s} B_\text{e} \sqrt{\frac{\sigma_\text{th}^2}{e^2 B_\text{e}^2}} \tag{3.128}$$

In Eq. (3.128), we used the fact that for typical parameters $\sigma_\text{th}^2 \gg e^2 B_\text{e}^2$. With the values RIN $= -152\,\text{dB/Hz}$, $B_\text{e} = 5\,\text{MHz}$, $m = 0.05$, $\sigma_\text{th}^2 = 10^{-22}(\text{A}^2/\text{Hz}) \cdot B_\text{e}$ ($i_\text{c} = 10\,\text{pA}/\sqrt{\text{Hz}}$), $\eta = 0.8$ and $\lambda_\text{s} = 1.55\,\mu\text{m}$, we obtain from Eq. (3.128) a required average optical power $P_\text{s} = +33\,\text{dBm}$ (2W) for $\text{CNR}_\text{d} = +50\,\text{dB}$, $P_\text{s} + 28\,\text{dBm}$ (600 mW) for $\text{CNR}_\text{d} = +45\,\text{dB}$ and $P_\text{s} = -0.6\,\text{dBm}$ (0.9 mW) for $\text{CNR}_\text{d} = +17\,\text{dB}$. The optical power required for AM ($\text{CNR}_\text{d} = +45\,\text{dB}$ to $+50\,\text{dB}$) is much higher than that available from conventional long wavelength, single-mode laser diodes, about 100–500 mW. The required power can be decreased to the 100 mW power range by using a modulation index higher than $m = 0.05$ but at the expense of significant clipping distortion [93]. On the other hand, for FM ($\text{CNR}_\text{d} = +17\,\text{dB}$) the required optical power matches that of conventional laser diodes. However, if the power has to be split to 10–100 users, the power requirement increases by the same factor. Optical amplifiers can therefore play an important role in the implementation of AM and FM analog lightwave broadcast systems.

We consider therefore the case where an optical amplifier with gain $G > 1$ is used in the system. If no splitting loss and a high gain amplifier are assumed ($L = 1$, $G \gg 1$), the required power in Eq. (3.127) takes the form:

$$P_\text{s} = 2n_\text{eq} \kappa h\nu_\text{s} B_\text{e} \left( 1 + \sqrt{1 + \frac{B_\text{o}}{\kappa B_\text{e}}} \right) \tag{3.129}$$

When no optical filter is used, $B_\text{o}$ is the optical amplifier pass band (i.e., $B_\text{o} = 25\,\text{nm}$ for an EDFA). In this case, the typical parameters RIN $= -152\,\text{dB/Hz}$, $B_\text{e} = 5\,\text{MHz}$, $m = 0.05$ give $B_\text{o}/\kappa B_\text{e} \ll 1$ and $\kappa = 1.25 \times 10^9$. With an equivalent input noise factor $n_\text{eq} = 1$, we obtain a required power $P_\text{s} = 3.2\,\text{mW}$ for $\text{CNR}_\text{d} = +55\,\text{dB}$. The effect of using a high gain amplifier prior to transmission (post-amplification) is a considerable improvement in receiver CNR. Since the required power for $\text{CNR}_\text{d} = 55\,\text{dB}$ is near 1 mW, conventional infrared laser diodes can be used for the system application considered.

For the actual broadcast system shown in Figure 3.13, the loss caused by the $1 \times N$ optical splitter is required to adjust the transmitter signal power and the amplifier gain for the desired CNR at the receiving stations. In Figure 3.14, the required signal power $P_\text{s}$ is plotted from Eq. (3.127) as a function of amplifier gain $G$ for different values of the loss $L$, two required CNRs of $+45\,\text{dB}$, and $\text{CNR}_\text{d} = +55\,\text{dB}$

corresponding to AM TV standards. The parameters assumed are $B_e = 5\,\text{MHz}$, $B_o = 25\,\text{nm}$, $m = 0.05$, $\text{RIN} = -152\,\text{dB/Hz}$, $\sigma_{th}^2 = 10^{-22}\,(\text{A}^2/\text{Hz}) \cdot B_e$, $\eta = 0.8$ and $\lambda_s = 1.55\,\mu\text{m}$.

Figure 3.14 shows that for $\text{CNR}_d = +45\,\text{dB}$, the required signal powers are between $-10\,\text{dBm}$ and $0\,\text{dBm}$ for a gain $G > 10\,\text{dB}$ and a splitting loss $L \leqslant 20\,\text{dB}$. For $\text{CNR}_d = +55\,\text{dB}$, the required powers are between $+7\,\text{dBm}$ and $+15\,\text{dBm}$ in the same conditions. The figure also shows a plot of the amplifier power limit. This power limit corresponds to the signal power that can be input to the EDFA for a given gain $G$ and, in this example, a maximum amplified signal power $P_{out} = 100\,\text{mW}$. The figure shows that for $\text{CNR}_d = +45\,\text{dB}$ and a splitting loss $L \leqslant 20\,\text{dB}$, the required signal powers are below the amplifier power limit; for $\text{CNR}_d = +55\,\text{dB}$, the system can be realized only for a splitting loss $L \leqslant 10\,\text{dB}$. These results correspond to a number of potential users between 10 and 100. Note that this performance cannot be improved by using a narrow bandpass filter, as can be checked numerically.

In Figure 3.15, the required signal power $P_s$ is plotted as a function of amplifier gain $G$ for different values of the loss $L$, assuming a required $\text{CNR}_d$ of $+17\,\text{dB}$ (FM TV) with modulation index $m = 0.1$ and other parameters identical to those in the previous example. The required powers are considerably lower than in the previous example, with a high gain limit near $-35\,\text{dBm}$. The figure also plots the amplifier power limit assuming a maximum output of $10\,\text{mW}$. In all cases ($L = 0$ to $-30\,\text{dB}$), the required powers are below the amplifier power limit. From these results, we may conclude that an FM TV broadcast system using a conventional laser source ($P_s \leqslant 0\,\text{dB}$) and a single-stage amplifier/splitter could have up to 1000 users.

It should be noted that the excess loss of the $1 \times N$ splitter was overlooked in the AM and FM examples. This loss is not significant, even if the $1 \times N$ splitter has a large number of outputs. Assuming the $1 \times N$ splitter is made of a passive $N \times N$ star topology, the splitter excess loss is given by $L' = \gamma \log_2 N$, where $\gamma$ represents

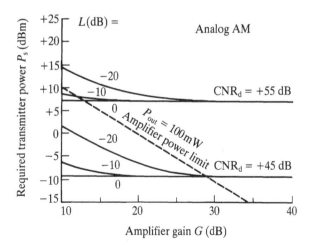

**FIGURE 3.14** Required transmitter power $P_s$ for analog AM TV broadcast systems as a function of amplifier gain $G$ and shown for different $1 \times N$ splitter loss $L$, assuming a receiver CNR of $+45\,\text{dB}$ and $+55\,\text{dB}$. The dashed line indicates the amplifier input power limit, assuming a maximum output $P_{out} = 100\,\text{mW}$. $B_e = 5\,\text{MHz}$, $B_o = 25\,\text{nm}$, $\eta = 0.8$, $\lambda_s = 1.55\,\mu\text{m}$, $\text{RIN} = -152\,\text{dB/Hz}$, $i_c = 10\,\text{pA}/\sqrt{\text{Hz}}$, and $m = 0.05$.

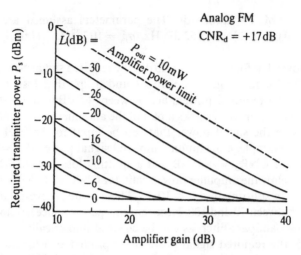

**FIGURE 3.15**   Required transmitter power $P_s$ for analog FM TV broadcast systems as a function of amplifier gain $G$ and shown for different $1 \times N$ splitter loss $L$, assuming a receiver CNR of $+17$ dB. The dashed line indicates the amplifier input power limit, assuming a maximum output $P_{out} = 10$ mW. $B_e = 5$ MHz, $B_o = 25$ nm, $\eta = 0.8$, $\lambda_s = 1.55\ \mu$m, RIN $= -152$ dB/Hz, $i_c = 10$ pA/$\sqrt{\text{Hz}}$, and $m = 0.1$.

the excess loss of a single 3 dB coupler [16]. For $N = 1000$, assuming $\gamma = 0.1$ dB, the splitter excess loss would be only $L' = 1$ dB.

The examples of Figures 3.14 and 3.15 have shown that *for a broadcast system based on one-stage amplification of transmitter power, realistic parameters give a maximum number of users between 100 and 1000.*

The maximum number of users can be increased by using a multiple stage amplification and splitting configuration. Figure 3.16. The configuration in this example includes $k$ amplifiers with identical gain $G$ and $k$ splittings with identical loss $L$, corresponding to a maximum number of users $k/L$. Where $M = k/L$ is large, the amplifier gain should exactly compensate the splitting loss at each stage, i.e., $GL = 1$, otherwise the optical power will rapidly decay. With this transparency condition, the optical path from the transmitter to any receiver in the broadcast system is identical to that of a Type A amplifier chain configuration, in which the fiber loss $T$ is replaced by the splitting loss $L$. Chapter 2 shows that at transparency, a Type A chain made of $k$ amplifiers with equivalent noise factor $n_{eq}$ and gain $G$ is equivalent to a single optical amplifier with unity gain and noise factor $n'_{eq} = k n_{eq}$.

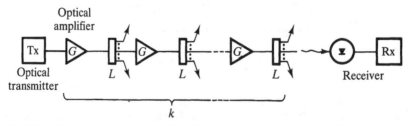

**FIGURE 3.16**   Basic system configuration for analog signal broadcast distribution using $k$ stage optical amplification.

Using this property and Eq. (3.127), we obtain the power $P_s(k)$ required for a desired CNR in a $k$ stage broadcast system:

$$P_s(k) = \frac{\kappa}{\eta} h\nu_s B_e(1 + 2\eta k n_{eq})\left(1 + \sqrt{1 + \frac{\sigma_{th}^2/e^2 B_e B_o + (2\eta k n_{eq})^2}{\kappa(1 + 2\eta k n_{eq})^2}\frac{B_o}{B_e}}\right) \quad (3.130)$$

Equation (3.130) shows that in the $k$ stage broadcast system operated at transparency, the required power $P_s(k)$ is theoretically independent of the amplifier gain $G$ and splitting loss $L$. For a given number of users $M$, the system can be realized with $k$ stages and splitting loss given by $M = 1/L^k$ (or amplifier gain given by $G = M^{1/k}$). The number of stages $k$ can be decreased by increasing both gain and splitting loss. As in reality, the maximum gain achieveable is limited by pump power and input signal power, a trade-off must be made between the splitting loss and the number of stages $k$; a high splitting loss corresponds to a low value of $k$.

We consider in Figure 3.17 examples of AM and FM signals with the same parameters as before. The required transmitter powers $P_s(k)$ are plotted as a function of the maximum number of users $M$, for different values of amplifier gain $G(G = 1/L)$. The bottom graph shows the number of stages $k$ that corresponds to a given number of users $M$ and gain $G$, defined by $k = \log(M)/\log(G)$. It shows, for instance, that a broadcast system of 1000 users can be realized with five amplifier stages with $1 \times 4$ splitters $(G = 1/L = 6 \text{ dB})$, using a transmitter power of $-2$ dBm for AM TV and $-25$ dBm for FM TV. With a 2–3 dB increase in transmitter power for both AM and FM, the number of potential users in this configuration reaches $10^6$. The number of stages can be reduced by using amplifiers with higher gain. With a gain $G = 20$ dB, the number of $10^6$ users can be achieved using only three-stage amplification. Such a large broadcast system would require 1001 amplifiers and 1000 splitters of $1 \times 100$. Experimental demonstrations of CATV broadcast systems with $10^6$ users are discussed in Chapter 11.

The fiber transmission loss $T$ has been neglected in this power evaluation, as the distance between the most remote station and the central transmitter is small. But in large-scale applications, the maximum distance is likely to be greater than 50 km $(T > 10 \text{ dB}$ with 0.2 dB/km loss coefficient), and the corresponding transmission loss must be compensated. This can be done either by raising the transmitter power level or by introducing additional optical amplifiers along the chain.

While we have considered the EDFA's inherent output power limit, we have not analyzed how this limit is related to the pump power and how, in turn, pump power considerations limit the number of optical channels in the system. An analysis made by I. M. Habbab et al. [91] shows a practical method for the simultaneous optimization of both amplifier design system channel capacity.

Another source of performance limitation is signal distortion. Signal distortion is a consequence of device nonlinearity [14]. If two optical carriers of frequency $\omega_i$ and $\omega_j$ traverse a nonlinear medium, a fraction of their power is converted into tones oscillating at various frequencies. The composite second order (CSO) tones are at frequencies $\omega_i \pm \omega_j$, $2\omega_i$, $2\omega_j$. The tones $2\omega_i$ and $2\omega_j$ represent second harmonic distortion (SHD). The composite triple beat (CTB) tones are at frequencies $2\omega_i \pm \omega_j$, $2\omega_j \pm \omega_i$, $3\omega_i$, $3\omega_j$. The tones $3\omega_i$ and $3\omega_j$ represent third harmonic distortion (THD).

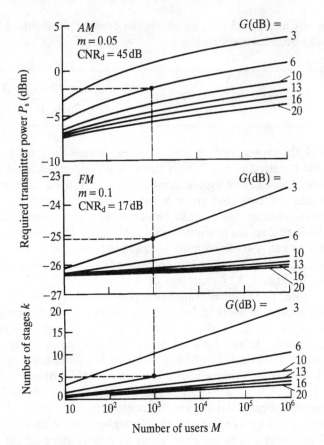

**FIGURE 3.17**   Required transmitter power $P_s$ for analog AM and FM TV broadcast systems with $k$ stage amplification, plotted as a function of maximum number of users $M$ and shown for different amplifier gains $G$. Other parameters are identical to that of Figures 3.14 and 3.15. The bottom graph shows the number of amplification stages $k$ corresponding to number of users $M$ and amplifier gain $G$.

Terms that are not harmonic represent intermodulation distortion. When other channels are turned on, noise may be observed at frequency $\omega_i$ but it neither CSO nor CTB. This noise is called crosstalk or cross-modulation (XM).

By definition, the CSO and CTB at carrier frequency $\omega_m$ are defined as the ratio of the sum of all distortion powers measured at that frequency to the carrier power, i.e.,

$$\mathrm{CSO}(\omega_m) = \frac{\sum\limits_{ij} P(\omega_m = \omega_i \pm \omega_j)}{P(\omega_m)} \tag{3.131}$$

$$\mathrm{CTB}(\omega_m) = \frac{\sum\limits_{ijk} P(\omega_m = \omega_i \pm \omega_j \pm \omega_k)}{P(\omega_m)} \tag{3.132}$$

The quantities CSO, CTB, and XM are expressed in dBc, indicating that the reference is the carrier power. As seen previously with the CNR, the standards of

CSO, CTB, and XM for CATV distribution applications are related to the subjective definition of picture quality. Typical standards at the trunk level are: CSO = CTB − 65 dBc and XM = −52 dBc [92].

In lightwave analog systems, signal distortion can have several causes. A first source of distortion is the nonlinear response of the transmitter (laser diode, external modulator) or of various electronic components in the system, which produce clipping distortion, harmonic distortion, and intermodulation distortion [14]. Fiber dispersion (with standard, or nondispersion-shifted fiber) is also a cause of signal distortion, [90] and [94]. Finally, some distortion also originates from the optical amplifiers. Because EDFAs are characterized by slow gain dynamics (100 $\mu$s to 1 ms), discussed in detail in Chapter 5, nonlinearity induced by gain saturation is a negligible effect at the frequencies of interest. A numerical study by L. K. A. Chen [95] has shown for an EDFA with 10 dB gain and output carrier power of 0 dBm, CSO and CTB are near −60 dBc and −70 dBc at 10 kHz, and drop by −20 dB per tenfold increase in frequency. At video frequencies near 1 MHz, therefore, the theoretical CSO and CTB levels are −100 dBc and −110 dBc, respectively. Thus, gain nonlinearity in the EDFA represents a negligible contribution to distortion in comparison to other sources.

An important cause of CSO distortion is the combined effect of chirping in the laser source and gain nonuniformity, or tilt, in the EDFA, [96]–[98]. The chirp is defined as an instantaneous frequency deviation of the carrier that occurs when semiconductor diodes are directly modulated [14]. For 1.5 $\mu$m laser diode sources with modulation index near 5%, the chirp is typically 2–5 × 10$^{-3}$ nm or 300–600 MHz, [96]–[98]. The EDFA gain tilt is defined as the gain slope $dG(\lambda)/d\lambda$ (expressed in dB/nm) at a given signal wavelength. While the slope is zero at the gain peaks, it can be as large as 0.2 dB/nm in their vicinity [97]. The resulting CSO distortion can be expressed as (C. Y. Kuo et al. [96], K. Kikushima et al. [97]):

$$\text{CSO}(\lambda_\text{s}) = M \frac{m\Delta\lambda_{\max}}{G(\lambda_\text{s})} \left(\frac{dG}{d\lambda}\right)_{\lambda=\lambda_\text{s}} \tag{3.133}$$

where $M$ is the number of carrier components, and $\Delta\lambda_{\max}$ is the maximum wavelength deviation for $m = 100\%$ modulation. The chirp for any modulation index is then equal to $m\Delta\lambda_{\max}$. From this equation, the theoretical CSO level due to the combined effect of chirp and EDFA gain tilt is between −60 dBc and −40 dBc, depending on EDFA gain $G$ and signal wavelength location with respect to the gain peaks, [96] and [97]. A detailed analysis by J. Ohya et al. [98] has shown that chirp/tilt-induced CSO actually varies along the EDFA length and, under certain conditions, the effects of chirp and gain tilt can mutually cancel CSO.

The previous study and basic examples provided have shown the important role that optical amplifiers can play in CATV broadcast systems. The wide bandwidth of EDFAs (25 nm) can accommodate a large number of optical channels (i.e., $n > 100$), each of them able to convey several frequency multiplexed video channels. Applications of EDFAs to CATV broadcast systems, [91] and [93], are described in Chapter 7. A comparison between analog and digital systems for video transmission (W. I. Way [90]) shows the importance of various system design issues to future implementations of optical amplifiers in analog CATV systems, which require careful and progressive network planning.

# CHARACTERISTICS OF ERBIUM-DOPED FIBER AMPLIFIERS

## CHAPTER 4

# CHARACTERISTICS OF ERBIUM-DOPED FIBERS

## INTRODUCTION

The spectroscopy of Er-doped glass fibers plays a fundamental role in the analysis and physical understanding of optical fiber amplifiers. All the important device characteristics of Er-doped fiber amplifiers, i.e., gain spectrum, gain versus pump power and pump wavelength, output saturation power, power conversion efficiency, and noise figure are fundamentally related to spectroscopic properties.

In Chapters 1 and 2, the physics of light amplification and noise in doped fibers are formally described and analyzed without any specification of the physics of the amplifying medium itself. The underlying assumption is that Er:glass can be pumped as a three-level or two-level laser system, characterized by phenomenological, nearly homogeneous absorption and emission line shapes. The results of this analysis, combined with that of Chapter 3 on photodetection of amplified light, are sufficient to a basic understanding of amplifier-based communications systems. A deeper understanding requires insights in physics and spectroscopy specific to Er-doped glass fibers. These explain fundamental limitations, as well as the choice of specific glass materials and pump or signal wavelengths for optimum device performance. Earlier chapters discuss device performance limitations that intrinsically relate to the physics of Er:glass: homogeneous and inhomogeneous gain saturation, minimum noise figure versus pump wavelength, fiber background loss in distributed amplifiers, or gain-tilt-induced second order distortion in analog systems. The slow transient gain dynamics of EDFAs, analyzed in detail in Chapter 5, can also be explained by the relative magnitude of different types of relaxation processes in Er:glass; this is studied by spectroscopy.

The purpose of chapter is to provide a basic understanding and review of the various physical processes that take place in Er:glass. These processes include: radiative and nonradiative atomic decay, laser line broadening, excited state absorption, cooperative energy transfer, and cooperative upconversion. The role played by the glass host composition and Er-doping concentration in the fiber amplifier gain characteristics (cross section spectra, fluorescence lifetime) is also described.

Section 4.1 provides first an introduction of the general properties of laser glass, with a review of the different types of glass hosts for rare earth (RE) ions. The different techniques for the fabrication of RE-doped fibers based on oxide and fluoride glasses, currently the most developed in optical amplifier applications, are then described in Section 4.2.

The energy level structure of Er:glass and its properties are studied in Section 4.3. Various relaxation process (i.e., radiative, nonradiative, and thermal) associated with these energy levels are described in this section, with a review of results from the Judd–Ofelt theory. The effects of homogeneous and inhomogeneous laser line broadening and the Stark level substructure of Er:glass are described in Section 4.4. The effects of glass composition and Er-doping concentration on the laser transition characteristics (emission and absorption line shapes, fluorescence lifetime) are also reviewed in this section.

Theoretical and experimental determinations of emission and absorption cross sections are analyzed in Section 4.5. Section 4.6 review various experimental techniques for the characterization of basic Er-doped fiber parameters: fluorescence lifetime, doping concentration, absorption, and emission cross section spectra.

In Section 4.7, the process of pump excited state absorption is analyzed, both theoretically and experimentally. The ion/ion interaction processes leading to energy transfer and cooperative upconversion are described in Section 4.8. Section 4.9 analyzes the refractive index change occurring near the laser resonance and its effect on fiber dispersion. The effect of pump and signal polarizations and the notion of polarized cross sections are considered in Section 4.10.

## 4.1 CHARACTERISTICS OF LASER GLASS

Following the discovery by E. Snitzer in 1961 of the laser action of neodymium in glass [1], the field of glass lasers has known rapid and important developments. This rapid growth was motivated by the research of laser hosts suitable to high energy and high power applications. The advantages of glass as a laser host, compared to other solid-state materials (i.e., crystalline, polycrystalline, or ceramic), include: optical quality, transparency, low birefringence, high optical damage threshold, thermal shock resistance, weak refractive index nonlinearity, high energy storage and power extraction capacities, variety of possible compositions, size and shape scalability, and low cost of raw materials, [2]–[4].

The first large-scale applications of Nd-doped glass lasers in the 1970s, concerned research in inertial confinement fusion. At Lawrence Livermore Laboratory, Nd:glass disks and rods were used as bulk optics amplifiers to produce giant optical pulses with 10 kJ energy (800 ps) or 20 TW peak power (60 ps) characteristics [3]. Nd-doped glass fibers were also considered as early as 1964 for applications as optical amplifiers (C. J. Koester and E. Snitzer [5]), in anticipation of developments in the newborn field of optical fiber communications [6]. Laser-diode-pumped Nd-doped fiber lasers were also investigated in the mid-1970s as efficient, miniature infrared sources for optical communications (J. Stone and C. A. Burrus, [7] and [8]). Another reason for the investigation of RE-doped glasses was the realization of low loss Faraday rotators of high Verdet constants for application as optical isolators [9].

Most generally, the field of laser glasses have developed, in the same way as other laser materials (e.g. crystals, liquid, and gas). It has several goals: to obtain new laser transition wavelengths for specific applications (e.g. laser surgery, imaging, optical storage, reprography, communications, and atmospheric propagation); to achieve compact or efficient solid-state laser sources with practical pump wavelengths; and finally, to optimize the physical characteristics of laser materials for high power or high energy applications (e.g., thermonuclear fusion, machining, laser surgery, range finding, and defense). The relatively new technology of Re-doped fibers, driven by the need for optical amplifiers in lightwave communications in the mid-1980s, actually represents a renewed interest in the field of laser glasses, a field which has been thoroughly investigated for more than 30 years, and for which there exists abundant literature. We review some basic features and physical properties of RE-doped laser glasses, with particular emphasis on Er-doped glass materials.

Rare earth ions, more specifically bivalent or trivalent lanthanides, have been used as activators in as many as 425 known laser crystals, most of which have an ordered structure [10]. In these crystalline hosts, the most frequent rare earths observed to yield stimulated emission are Nd (neodymium), Ho (Holmium), Er (erbium), and Tm (Thullium) [10]. Trivalent lanthanides are the only ions for which laser oscillation was observed in a glass host, whether in bulk or fiber form, [2] and [11]. The electronic configuration of a trivalent rare earth is $[Xe]4f^{N-1}5s^25p^66s^0$, where [Xe] represents the closed shell electronic configuration of xenon, [12] and [13]. In this ion configuration, one and two electrons are removed from the $4f$ and $6s$ shells, respectively, a consequence of the energetic sequence in which electrons fill up the subshells [12]. On the other hand, the $N-1$ inner electrons of the $4f$ shell are shielded from external fields by the outermost shells $5s$, $5p$. This property causes the $4f \rightarrow 4f$ laser transitions of rare earth solid-state laser materials to exhibit relatively sharp lines, as compared to the case of transition metals, for instance [13]. Another consequence of this shielding effect is that the spectroscopic characteristics of the $4f \rightarrow 4f$ transitions are weakly sensitive to the type of host. In amplifier device applications, the weak perturbations induced by the host actually represent important effects.

The known $4f \rightarrow 4f$ lasing transitions for trivalent rare earths in glass hosts and corresponding energy levels are shown in Figure 4.1, [14]. In the case of praseodymium ($Pr^{3+}$), the recently reported transition near 1.31 $\mu$m is shown in the diagram, [15]–[17], while several other transitions initiating from the $^3P_0$ and $^3P_1$ levels through direct pump excitation [18], or through upconversion processes [19], are also shown in the figure.

A look at this diagram of Figure 4.1 shows that in spite of the relative abundance of energy levels and possible laser transitions in the trivalent rare earth family, only a few elements satisfy three important criteria for optical amplifier applications: (1) the laser transition wavelength should be near 1.31 $\mu$m or 1.55 $\mu$m, corresponding to the transmission windows of optical fiber communications; (2) the pump wavelength, corresponding to a transition starting from the ground level, should be in the near infrared (800 nm to 1500 nm), so as to be compatible with compact semiconductor laser sources; (3) the signal and pump transitions should be free from the completing effect of excited state absorption (ESA), as discussed in Section 4.7. In addition to

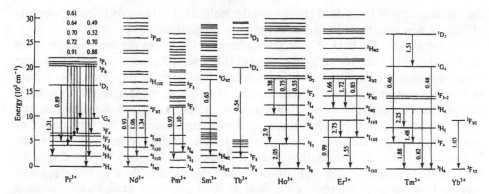

**FIGURE 4.1** Reported $4f \to 4f$ lasing transitions for trivalent rare earths in glass hosts and corresponding energy levels. (Courtesy M. J. Weber, Lawrence Livermore National Laboratory [14]).

criteria (1) to (3), the laser transition should also be characterized by a long fluorescence lifetime, a high branching ratio, and essentially radiative properties.

Rare earths can be associated in pairs, in a donor–acceptor configuration, for instance, ytterbium and erbium [20]. Under certain conditions, energy transfer between the donor and acceptor ions can be achieved; this opens up further possibilities for the pump wavelength (Section 4.8). Figure 4.1 shows that among the trivalent rare earths, only $Pr^{3+}$, $Nd^{3+}$, and $Er^{3+}$ potentially satisfy criteria (1) to (3). The characteristics of fiber amplifiers based on $Pr^{3+}$ and $Nd^{3+}$ doping are briefly discussed in Chapter 5.

We now focus on the physical characteristics of the glass host. By definition, glass is an inorganic product of fusion cooled to a rigid condition without crystallizing, (K. Patek, [13]). The structural organization of glass is well defined at the scale of a few atoms, but is completely random, asymmetric and aperiodic at a larger scale. The glass lattice is built from basic structural units made of *network former* atoms. The most common is the silica tetrahedron $(SiO_4)^{2-}$. Other usual glass formers are $GeO_2$, $P_2O_5$, and $B_2O_3$. In silica glass, for instance, the tetrahedron units are tightly connected by their corners through oxygen atoms (bridging oxygens); these random connections form a three-dimensional, disordered lattice. Other compounds, such as alkali or alkaline earths, can be added to the glass as *network modifiers*. These modifiers can cause former bridging ions to become nonbridging; this breaks the lattice and results in a looser network structure. Network modifiers (such as $Na^+$, $Al^{3+}$) are used to facilitate the incorporation of trivalent rare earth ions in glass, as their size (near 1 angstrom) is substantially greater than that of the basic network formers. The structure of a typical alkali silicate glass is illustrated in Figure 4.2.

The coordination of the rare earth ion by the surrounding network structure can be determined only by theoretical analysis or indirect experimental evidence [13]. Theoretical investigations are based upon crystal field theory, [10] and [13]. However, as in glass hosts, the local crystalline structure of the rare earth site is hypothetical, as opposed to the case of crystalline hosts. Because of the various assumptions and approximations made for the coordination, these investigations generally do not lead to definite conclusions [13]. Indirect experimental confirmation of hypothetical types

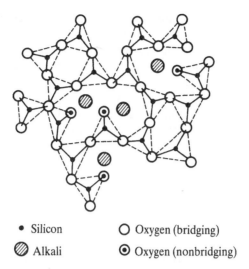

- Silicon       ○ Oxygen (bridging)

▨ Alkali       ◉ Oxygen (nonbridging)

**FIGURE 4.2**  Alkali silicate glass structure based on ($Si^{4+}$, $O^{2-}$) network formers and alkali ion network modifier. The fourth bridging oxygen of the silicon tetrahedron is outside the plane of the figure.

of rare earth coordination is provided by stereochemical reasoning [13], and low temperature spectroscopy [21]. Low temperature spectroscopy determines the Stark level substructure, which can be related to the type of symmetry of the rare earth site (Section 4.4). Changes observed experimentally on cross section line shapes with temperature also enable us to infer the existence of multiple types of rare earth sites [21]. Rare earth coordination is further discussed in Section 4.4.

The variety of glasses that can be made from multiple former and modifier compounds and their relative compositions is not infinite. Indeed, glass forms upon solidification only if certain molar ratios are respected; the range depends upon the number of oxides forming the mixture, [4] and [13]. The different types of laser glasses can be grouped into four main categories: oxide, halide, oxyhalide, and chalcogenides [2].

In the oxide glass category, phosphate glasses activated by $Nd^{3+}$ and $Er^{3+}$ have received much attention because of advantageous properties of thermal stability and chemical strength in high power applications, [11], [22], and [23]. Other oxide glasses such as aluminosilica and germanosilica are generally used as the glass base for the fabrication of Er-doped fibers, due to their compatibility with standard single-mode fibers, as well as other spectroscopic characteristics (see Sections 4.2–4.5).

In the halide glass category, heavy metal fluoride glasses (HMF), based on zirconium fluoride or hafnium fluoride as a network former, and other multicomponent glasses based on various other fluorides, are advantageous for their lower vibrational frequencies [11]. Lower vibrational frequencies reduce intrinsic loss in the near-infrared region [24] and enhance fluorescence properties for laser glass applications [25]. The variety of HMF glasses that can be realized is quite large and is only limited by glass-forming compositional restrictions and considerations of chemical stability. Several other types of HMF glasses exist [11]. Among them, fluoro-zirconate glasses are of interest for $Nd^{3+}$ and $Pr^{3+}$ fiber amplifier applications (see

Section 5.12). These glasses are named by the acronyms ZB, ZBL, ZBA, ZBAN, ZBLA, ZBLAL, and ZBLAN. For instance, the basic composition of ZBLAN is $ZrF_4$–$BaF_2$–$LaF_3$–$AlF_3$–NaF; each fluoride compound has a variable molar percentage. The structure of ZBLAN glasses is not really well known. It is generally assumed that zirconium is a network former surrounded by six to eight fluorine ions, while barium acts as a network modifier [24].

A large variety of glasses are suitable as hosts for rare earths such as $Pr^{3+}$, $Nd^{3+}$, and $Er^{3+}$. However, in device applications where the pump is a low power laser diode, the laser glass must also be put into the form of a fiber waveguide, and therefore must lend itself to several restrictions and constraints related to the fiber manufacturing technology.

## 4.2  FABRICATION OF RE-DOPED FIBERS

The first RE-doped fiber was a rod of $Nd^{3+}$-doped barium crown glass clad with soda lime silicate glass; this device made possible the first observation of laser action in glass (E. Snitzer [1]). Other types of fibers were fabricated by cladding an $Nd^{3+}$-doped silica glass core with a clear undoped glass ($SiO_2$) having lower index (C. J. Koester and E. Snitzer [5], J. Stone and C. A. Burrus [7] and [8]). The refractive index difference between core and cladding was enhanced by incorporation of alumina ($Al_2O_3$) as a codopant in the core, or by surrounding it with a passive glass sleeve of higher index [7]. Other fiber structures used $Nd^{3+}$-doped silica for both core and cladding and germania ($GeO_2$) as an index-raising codopant in the core [26]. More recent types of RE-doped fibers used a core made of $Nd^{3+}$-doped ultraphosphate crystals ($LaNdP_4O_{14}$, $RNdP_5O_{14}$, R = Y, Bi, Gd) and a cladding made of either similar crystals or organosilicon compounds (G. D. Dudko et al. [27]. The process of fiber drawing would convert the resulting crystal rods into a glassy state.

The methods currently used for the fabrication of RE-doped fibers can be divided into three main groups. The first two methods, called vapor phase and liquid phase concern RE-doped fibers based on silica glass. The third method concerns HMF glass fibers, in particular fluorozirconate (ZBLAN) fibers, and consists of a liquid phase, melting/casting process. A comprehensive review of these fabrication methods can be found in [24], [28], and [29]. In this section, we shall briefly describe their principal features.

The first step in each fiber fabrication method is to realize the fiber preform. The preform can be viewed as a macroscopic replica of the fiber to be fabricated. Next, the drawing process melts the rod inside a furnace and pulls from it a glass filament of reduced dimensions. Upon solidifying, this filament becomes the fiber. The process also includes the cladding of the fiber by a polymer jacket, which provides enhanced mechanical strength, [30]–[33]. We shall only discuss here the preform fabrication methods relevant to the case of RE-doped fibers.

The technique of vapor phase deposition is most common in the fabrication of standard single-mode silica fibers. The two principles used are flame hydrolysis and chemical vapor deposition (CVD). Flame hydrolysis achieved the first low loss fibers [34]. In flame hydrolysis halide vapors ($SiCl_4$, $GeCl_4$) are hydrolyzed in an oxygen–hydrogen flame. The resulting products are soots of $SiO_2$ and $GeO_2$, which are deposited onto either a rotating mandrel (OVD, or outside vapor deposition) or a rotating/translating rod (VAD, or vapor axial deposition [35]). In CVD, the soots are deposited inside a quartz tube by reacting the halide vapors with oxygen. With MCVD (modified CVD, [36] and [37]) the reactants are heated by an outside flame. Earlier CVD processes used $SiH_4$ as a reagant, which directly formed silica glass; MCVD was eventually preferred because of faster deposition rates and higher glass purity, [30] and [31]. Another variation of the technique is PCVD (plasma activated CVD [38]), where the reaction is initiated inside the tube by a plasma; this plasma is generated by an external microwave cavity. The different vapor phase techniques and their comparative performances are described in detail in [30]–[33].

Incorporation of rare earth compounds into the silica fiber core required some modifications of the vapor phase techniques. The main reason for the modifications is that the vapor pressure of rare earth halides is generally lower than that of the other reactants used during the core deposition, so the rare earth halides have comparatively lower volatility [28]. Several methods based on MCVD, OVD, and VAD were devised for delivering the rare earth halide vapors to the reaction zone. The main MCVD methods, illustrated in Figure 4.3, include for the source a heated injector [39], a heated chamber [40], a heated sponge [41], or an aerosol [42]. High rare earth concentrations ($>1$ wt%) can be obtained through MCVD by the use of alumina codoping [39] or organic chelates as the rare earth transport gas [43].

VAD methods use a nebulizer for the rare earth chloride, [28] and [29], or other vaporization/impregnation techniques [44]. The OVD method used rare earth organometallic sources of higher vapor pressure, delivered to the reaction zone through heated lines, [28], [29], and [45].

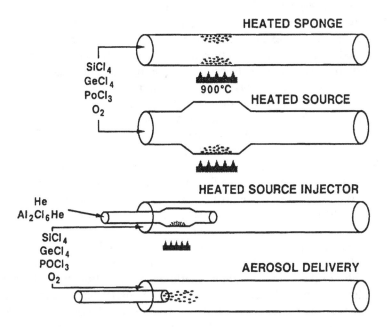

**FIGURE 4.3** Different types of MCVD processes used in the fabrication of RE-doped fibers. (Courtesy J. R. Simpson, AT&T Bell Laboratories).

The rods or boules obtained from the above methods are then sintered into solid glass preforms by traversing a high temperature furnace (OVD, VAD) or by collapse under a high temperature torch, [30]–[33].

The liquid phase fabrication methods for silica fibers use an impregnation of the unsintered rod/boule soots with a solution of rare earth chloride, as first described in [7]. This technique of solution doping, can also be combined with MCVD by iterative steps [46]. The iteration of this process not only makes it possible to yield preforms of high rare earth concentrations, but also permits better control of rare earth doping profile, composition, and concentration, compared to gas phase techniques.

The fabrication method for HMF glass fibers [24], and particularly based on the ZBLAN composition, [25] and [47], is radically different from that applying to silica glass fibers. One of the main characteristics of fluoride glasses is a relatively high viscosity near the melting point, which makes glass formation very slow [24]. The main difficulty is to prevent the formation of crystallites during the process, as the glass transition temperature $T_g$ is very near the crystallisation temperature $T_x$ and the viscosity decreases rapidly above $T_g$. For instance, the temperature difference $T_x - T_g$ is 93°C for ZBLAN, compared with a glass transition temperature of $T_g = 265$°C [24]. Glass formation is generally facilitated when the glass is made from a large number of components. ZBLAN is recognized as being one of the most stable HMF glasses.

The propensity of fluoride glasses to crystallize at temperatures near the melting point makes fiber drawing a very delicate process. Another difficulty is in the realization of a small refractive index difference between core and cladding. The index difference can be made by adding $PbF_2$ to the core (index raising) and $HfF_4$ to the cladding (index depressing) [47]. Any codopant used to raise or lower the refractive index actually alters the glass composition, hence its glass phase characteristics. Furthermore, the thermal expansion coefficients of both core and cladding must be closely matched, and the glass structure uniformized at the core/cladding interface to avoid residual stress, cracking, or crystallization during solidification. Finally, the various compounds of fluoride glasses are highly hygroscopic and reactive to certain crucible materials. To avoid any oxide and hydroxide contamination, the preparation of the preform must be made under a dry and highly controlled atmosphere; the nonreacting crucibles are made of gold, platinum or vitreous carbon [48].

The preform preparation is based on various casting methods. One method casts the glass into a crylindrical crucible. The resulting rod can then be clad with a plastic, such as Teflon FEP, and drawn into a fiber [49]. In the built-in casting method [50], the glass melt is poured into a crucible that is heated at the glass transition temperature. The crucible is then upset, which removes the center portion of the melt. Next, the core glass is poured into the hollow center.

In rotational casting [51], the crucible is partially filled with cladding melt and is spun in the horizontal position to form an inner glass tube. The core melt is poured into this cladding tube. using the above methods, direct fabrication of single-mode fibers is difficult because the core rods cannot be made with a sufficiently small diameter. The jacketing method [52] alleviates this difficulty. It consists of overcladding the preform with a jacketing glass tube. The jacketing tube is also prepared by built-in casting.

The fiber drawing process for ZBLAN fibers is the same as for silica fibers, except for the lower melting temperature of the glass [24]. Comprehensive reviews on techniques and processing parameters for the fabrication of fluoride fibers can be found in [24], [48], [52], and [53].

## 4.3  ENERGY LEVELS OF $Er^{3+}$:glass AND RELAXATION PROCESSES

We consider the energy level structure of $Er^{3+}$:glass, as well as the various relaxation processes originating from these levels. The discrete nature of the energy level structure originates in quantum atomic theory. Some of the results of this theory we briefly outline to provide a basic understanding of the various notations and equations used in $Er^{3+}$:glass spectroscopy.

Section 4.1 showed that the electronic configuration of a trivalent rare earth is $[Xe]4f^{N-1}5s^25p^6$, and that laser transitions take place within the $4f$ shell. In the case of erbium, $N = 12$, so the $4f$ shell of $Er^{3+}$ contains 11 electrons [13]. From quantum atomic theory, each individual electron in a given shell is characterized by four quantum numbers, i.e., $n$, $l$, $m$ and $s$. These numbers define the quantum state of each electron. The principal quantum number ($n = 0, 1, 2 \ldots$) determines the electron's radial probability distribution density. The orbital quantum number ($l$) takes any integer value between 0 and $n - 1$, and determines the electron's total angular momentum, equal to $\sqrt{l(l + 1)}\, \hbar$. By convention, the values $l = 0, 1, 2, 3, 4 \ldots$ are labeled by the letters $s$, $p$, $d$, $f$, $g \ldots$, respectively. Thus, the $4f$ electrons have for two first quantum numbers $n = 4$ and $l = 3$. The quantum number ($m$) determines the orbital momentum orientation, which takes $2l + 1$ possible values from $m = -l$ to $m = +l$. Finally, the spin quantum number ($s$) takes the values $s = 1/2$ or $s = -1/2$; the electron spin is $\pm \hbar/2$. Pauli's exclusion principle requires that in a multielectron atom, two electrons cannot have the same set of quantum numbers.

In a multielectron atom, the total angular momentum $J$ is given by the vector sum of the overall angular momenta $L = \sum l_i$, and overall spin $S = \sum s_i$, where $L$ is an integer and $S$ is an integer or half-integer. As there are several possible ways to obtain a given set of values for $J$, $L$, and $S$, we refer to the collection of quantum states giving $(J, L, S)$ as a term, and the number of these states as the term's multiplicity, equal to $2J + 1$. The number of spin configurations is $2S + 1$, called the spin multiplicity. The above summation rule for angular momenta is referred to as the Russell–Saunders coupling.

By convention, the possible states of a multielectron atom are referred to by the symbol $^{2S+1}L_J$, where $L = 0, 1, 2, 3, 4 \ldots$ corresponds to letters S, P, D, F, G... respectively. Thus, the notation $^4I_{15/2}$ for the ground state of $Er^{3+}$, corresponds to the term $(15/2, 6, 3/2)$, which has a multiplicity $2J + 1 = 8$, and a spin multiplicity $2S + 1 = 4$.

As several quantum states can correspond to the same atomic energy levels (determined by the principal quantum number $n$), the states are said to be degenerate. This degeneracy is actually lifted by three types of weak perturbations related to the electrostatic or Coulomb interaction, the spin–orbit coupling, and the crystal field interaction or Stark effect, [54] and [55]. The first two effects are inherent in the free atom, while the last effect occurs only when the atom is surrounded by an external electric field originating from a crystalline or glass host. In the presence of a magnetic field, additional lifting of degeneracy occurs; this is called the Zeeman effect [13].

The complex study of the energy level structure of rare earth ions taking into account the aforementioned effects was made by B. G. Wybourne [56] and G. H. Dieke [57]. For the $f^N$ electronic configuration, the free atomic states used at the starting point consist of a linear superposition of the russell–Saunders states $|f^N \gamma SLJ\rangle$, where $\gamma$ is a quantum number introduced to distinguish states of identical $S$ and $L$. The effects of Coulomb, spin–orbit and crystal field interactions on these states are then calculated from perturbation and crystal field theories. From the set of perturbed states $|\psi_a\rangle$, we can finally determine the energy eigenvalues (hence the relative energy level positions).

The energy levels corresponding to each possible atomic state $^{2S+1}L_J$ for Er:glass are shown in Figure 4.4 [3]. The figure also shows possible absorption transitions in the visible and near-infrared (pump bands) as well as possible radiative transitions. The approximate wavelengths of radiative transitions that were actually observed in silica and fluorozirconate glass fibers are labeled in the figure in nanometers, [25] and [58–64]. It shows the transition with center wavelength of 1.540 nm, relevant to optical communications, originates from the $^4I_{13/2}$ level and terminates in the $^4I_{15/2}$ ground level. The pump absorption transition near 1.480 nm is justified in the next section, when considering the Stark level substructure of the $^4I_{15/2}$ and $^4I_{13/2}$ levels.

Figure 4.5 shows an experimental absorption spectrum obtained with an alumino-silicate Er-doped fiber [65]. The various absorption bands seen in this spectrum correspond to the absorption transitions shown in Figure 4.4. The broad and intense absorption band near 1.530 nm indicates that Er:glass is a strongly absorbing medium when not activated by any pumping mechanism.

The mere knowledge of the position of energy levels is clearly not sufficient to determine the transition probabilities associated with each pair of levels. Using

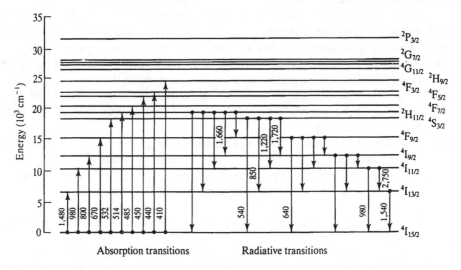

**FIGURE 4.4**   Energy level diagram of Er:glass showing absorption and radiative transitions. The transition wavelengths, given in nanometers, are indicated only for transitions experimentally observed in silicate and fluorozirconate Er-doped fibers.

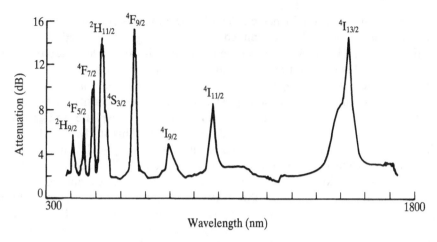

**FIGURE 4.5** Typical absorption spectrum of aluminosilicate Er-doped fiber. The discrete absorption bands correspond to the transitions shown in Figure 4.4. [65].

quantum theory, such transition probabilities can actually be calculated from the quantum states $|\psi_a\rangle$ obtained from the previous analysis. Two types of transitions can be considered, i.e., the electric dipole (ED) transitions or magnetic dipole (MD) transitions. The radiative emission probabilities corresponding to either type of transition can then be calculated through the matrix element of corresponding interaction Hamiltonian operator, taken between the initial and final atomic states.

In the case of electric dipole transitions, such a direct theoretical calculation is subject to numerous and severe difficulties [55]. The theories by B. R. Judd [66] and G. S. Ofelt [67] reported independently in 1962 made the ED problem finally tractable. The fundamental result of the Judd–Ofelt theory is that the oscillator strength $S_{ed}$ of an electric dipole transition taking place between states $|\psi_i\rangle$ and $|\psi_f\rangle$, corresponding to $^{2S+1}L_J$ and $^{2S'+1}L'_{J'}$, respectively, is given by the relation:

$$S_{ed} = \frac{1}{e^2}|\langle\psi_f|\mathcal{H}_{ed}|\psi_i\rangle|^2 = \sum_{k=2,4,6}\Omega_k|\langle f^N\gamma S'L'J'|U^{(k)}|f^N\gamma SLJ\rangle|^2 \qquad (4.1)$$

In Eq. (4.1), $\mathcal{H}_{ed}$ is the ED Hamiltonian, $\Omega_k$ are coefficients reflecting the effect of crystal field, electronic wavefunctions, and energy level separation, and $U^{(k)}$ are reduced tensor operator components reflecting the so-called intermediate coupling approximation, [68] and [69].

The matrix elements of the reduced tensor operators $U^{(k)}$ are constants that are virtually independent of the crystal host, [54] and [68], and their values are tabulated. A list of references where these tables can be found for various rare earth ions is given in [55]. For Er$^{3+}$, tables are given in [52] and [70]–[72].

The three parameters $\Omega_k(k=2, 4, 6)$ are called Judd–Ofelt parameters. The Judd–Ofelt (JO) parameters can be regarded as phenomenological coefficients that reflect the crystalline host influence on the $4f \rightarrow 4f$ radiative transition probabilities. Their values are empirically determined by comparing the theoretical oscillator strength $S_{ed}$ in Eq. (4.1) with the experimental result, through a least-squares fitting algorithm [69]. Instead of the JO parameters, some authors use the intensity parameters $\tau_k$, as related to $\Omega_k$ by $\tau_k = (8\pi^2 m_e c\chi/3hn^2)\Omega_k$, where $m_e$ is the electron mass and $\chi$ a local field correction, [54] and [73]. The intensity parameters $\tau_k$ corresponding to $Er^{3+}$ and other rare earth ions in borate, phosphate, and germanate glass hosts can be found in the experimental study [73]. This study also shows that the intensity parameters can be expressed as the sum $\tau_k = \tau_k' + \tau_k''$, where $\tau_k'$ and $\tau_k''$ are static and vibronic contributions of the host, respectively. The experimental study of how glass composition affects the intensity parameters $\tau_k$ corresponding to a given transition enables us to determine the relative contributions of static and vibronic coupling effects in each type of glass [73]. A comprehensive table of JO parameters for $Er^{3+}$ and other rare earth ions in a variety of crystals, glasses, liquids, and vapors can be found in [74], along with an extensive physical interpretation of the host influence.

The ED transition probability $A_{S'L'J',SLJ}^{ed}$ corresponding to the effect of spontaneous emission from $(S'L'J')$ to $(S, L, J)$ states is given by [54], [68] and [69]:

$$A_{S'L'J',SLJ}^{ed} = \frac{64\pi^4 e^2 n}{3h\langle\lambda\rangle^3} \frac{\chi_{ed}}{2J'+1} S_{ed} \tag{4.2}$$

where $\chi_{ed} = (n^2 + 2)^2/9$ is a local field correction, $n$ is the mean refractive index [69], and $S_{ed}$ is the ED oscillator strength given by Eq. (4.1).

For magnetic dipole transitions the oscillator strength $S_{md}$ is given by [54]:

$$S_{md} = \frac{e\hbar}{2m_e c} |\langle f^N \gamma S'L'J'|(\hat{L} + 2\hat{S})|f^N \gamma SLJ\rangle|^2 \tag{4.3}$$

where $\hat{L} + 2\hat{S}$ is the MD operator. The spontaneous emission probability $A_{S'L'J',SLJ}^{md}$ corresponding to MD transitions is given by [54]:

$$A_{S'L'J',SLJ}^{md} = \frac{64\pi}{3\langle\lambda\rangle^3} \frac{\chi_{md}}{2J'+1} S_{md} \tag{4.4}$$

where $\chi_{md} = n^3$ is the local field correction for MD transitions, and $S_{md}$ is the MD oscillator strength given by Eq. (4.3). In glasses, the contributions of magnetic dipole and electric quadrupole transitions, are weak in comparison to the electric dipole contribution. This fact can be attributed to the low degree of symmetry associated with the rare earth site in the glass medium [54]. For this reason, the total oscillator strength of $RE^{3+}$ : glass is usually approximated by $S \approx S_{ed}$.

The experimental or measured oscillator strength $S_{exp}$ corresponding to ED absorption transitions is given by the relation, [68] and [75]:

$$S_{exp} = \frac{3hcn}{8\pi^3 e^2 \langle\lambda\rangle} \frac{2J+1}{\chi_{ed}} \int_{band} \sigma_a(\lambda)\,d\lambda \tag{4.5}$$

where $\langle\lambda\rangle$ is the mean transition wavelength, $J$ is the orbital momentum of the ground state, $\sigma_a(\lambda) = \alpha(\lambda)/\rho$ is the transition absorption cross section, $\alpha(\lambda)$ the absorption coefficient and $\rho$ the rare earth ion density.

Using the identification $S_{exp} \approx S_{ed}$ in Eqs. (4.2) and (4.5) yields the following relation between the integrated absorption cross section and the spontaneous emission probability $A_{S'L'J',SLJ}^{ed}$:

$$\int_{band} \sigma_a(\lambda)\, d\lambda = \frac{2J' + 1}{2J + 1}\frac{\langle\lambda\rangle^4}{8\pi n^2 c} A_{S'L'J',SLJ}^{ed} \qquad (4.6)$$

If emission is considered instead of absorption, the same equation as (4.5) holds with the substitution $J \rightarrow J'$ and $\sigma_a(\lambda) \rightarrow \sigma_e(\lambda)$, where $\sigma_e(\lambda)$ is the emission cross section. Using the obtained relation with Eq. (4.2) yields:

$$\int_{band} \sigma_e(\lambda)\, d\lambda = \frac{\langle\lambda\rangle^4}{8\pi n^2 c} A_{S'L'J',SLJ}^{ed} \qquad (4.7)$$

Combining Eqs. (4.6) and (4.7) gives the following relation between the absorption and emission cross sections:

$$\sigma_a(\lambda) = \frac{J' + 1/2}{J + 1/2}\sigma_e(\lambda) \qquad (4.8)$$

The quantity $J + 1/2$ represents the degeneracy of the energy level of momentum $J$, as induced by the Stark effect. If $g_1 = J + 1/2$ and $g_2 = J' + 1/2$ are the degeneracies associated with the terminal and upper levels of the transition, respectively, we obtain from Eq. (4.8):

$$\frac{\sigma_e(\lambda)}{\sigma_a(\lambda)} = \frac{g_1}{g_2} \qquad (4.9)$$

The above result, used in many textbooks in the expression of the gain coefficient for multilevel laser systems, [76]–[78], is also discussed in Section 1.3. This result means that the absorption and emission cross section profiles are proportional. But this property is not verified experimentally when considering Er:glass. Section 4.5 contains further discussion of this issue and accurate analysis of cross sections.

We define next the radiative spontaneous emission lifetime $\tau_{S'L'J'}^{rad}$ for the upper state $(S'L'J')$ through [61]:

$$\frac{1}{\tau_{S'L'J'}^{rad}} \equiv A_{tot}^{rad} = \sum_{SLJ} A_{S'L'J',SLJ}^{ed} \qquad (4.10)$$

The above expression means that the radiative or spontaneous emission lifetime of level $(S'L'J')$ is a function of $A_{tot}^{rad}$, which represents the sum of probabilities corresponding to all possible transitions originating from level $(S'L'J')$.

The fluorescence branching ratio $\beta_{S'L'J',SLJ}$, defined by:

$$\beta_{S'L'J',SLJ} = \frac{A^{ed}_{S'L'J',SLJ}}{\displaystyle\sum_{SLJ} A^{ed}_{S'L'J',SLJ}} \tag{4.11}$$

gives the relative probability of spontaneous decay for the particular transition $S'L'J' \rightarrow SLJ$. The fluorescence lifetime $\tau_{S'L'J',SLJ}$, also called radiative lifetime, that is associated with this particular transition, is then defined through:

$$\tau_{S'L'J',SLJ} = \beta_{S'L'J',SLJ} \times \tau^{rad}_{S'L'J'} = \frac{1}{A^{ed}_{S'L'J',SLJ}} \tag{4.12}$$

The effect of spontaneous decay considered is so far attributed to the existence of electric dipole transitions taking place between various energy levels of the rare earth atom. The JO theory predicts the oscillator strength and fluorescence lifetime of these transitions, which are virtually independent of the separation between energy levels. Spectroscopic studies of RE:crystals in the early 1960s, however, showed that fluorescence in various host materials could be observed only for energy separations in excess of $1000 \, \text{cm}^{-1}$ [55]. This revealed the existence of a competing process of nonradiative decay, not explained by the previous theory. Comparing the maximum vibration energy of the crystal lattice with the observed limit of $1000 \, \text{cm}^{-1}$ suggested that the rare earth could nonradiatively decay by excitation of one or several quantized lattice vibrations, or phonons, Figure 4.6.

If the multiphonon decay rate between energy levels $i$ and $j$ is $A^{NR}_{ij}$, and the radiative decay rate of level $i$ is $A^{R}_{ij}$, the lifetime $\tau^{obs}_i$ experimentally observed for this level is given by:

$$\frac{1}{\tau^{obs}_i} = \sum_j (A^{R}_{ij} + A^{NR}_{ij}) \tag{4.13}$$

The radiative quantum efficiency $\eta_{ij}$ of the laser transition $(ij)$ can then be defined as

$$\eta_{ij} = \frac{A^{R}_{ij}}{\displaystyle\sum_j (A^{R}_{ij} + A^{NR}_{ij})} \tag{4.14}$$

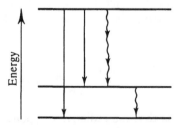

**FIGURE 4.6**  Energy level diagram showing radiative transitions (full lines) and nonradiative transitions corresponding to multiphonon decay (wiggled lines).

If the total nonradiative decay rate is large compared to the total radiative decay date, i.e., $\sum A_{ij}^{NR} \gg \sum A_{ij}^{R}$, fluorescence will not be observed.

The analysis of the process of multiphonon decay in RE:crystals, which could predict the nonradiative rate $A_{ij}^{NR}$, is based on crystal field theory and lattice dynamics. It represents a formidable problem that does not lend itself to a definitive treatment, owing to lack of information on phonon characteristics (frequency, polarization, velocity) and ion/phonons coupling strength [55]. The problem is even less tractable, if not prohibitively difficult, in the case of RE:glass, as the coordination of the RE site is generally unknown and the host structure is amorphous.

But no matter how complex the formal analysis of multiphonon decay, it is possible through intuitive arguments to derive a simple relation that gives the temperature and energy gap dependence of the nonradiative rate, [55], [74], [79], and [80].

The integer number of phons $p$ required to conserve energy in a multiphonon decay between energy levels $i$ and $j$ of energy $E_i$ and $E_j$, respectively, is given by:

$$p = \frac{E_i - E_j}{\hbar \omega} \equiv \frac{\Delta E}{\hbar \omega} \qquad (4.15)$$

where $\Delta E$ is the energy gap between the levels and $\hbar \omega$ is the phonon energy of a given phonon mode. A $p$ order perturbation theory, [55] and [80], shows that the nonradiative rate corresponding to a $p$ phonon decay is dominated by terms involving the factor $(n + 1)^p$, where $n = n(T)$ is the Bose–Einstein occupation number or mean population of the phonon mode, i.e.,

$$n(T) = \frac{1}{\exp\left(\dfrac{\hbar \omega}{k_B T}\right) - 1} \qquad (4.16)$$

The factor $(n + 1)^p$ can be simply explained as resulting from the $p$ time repeated action of the phonon number creation operator $a_\omega^+$ on the phonon eigenstates $|n\rangle$, i.e., the probability of creation of $p$ phonons in a given mode is:

$$|\langle n + p | a_\omega^+ a_\omega^+ \dots a_\omega^+ | n \rangle|^2 = (n + 1)^p \qquad (4.17)$$

The parameter $n$ in Eq. (4.17) can then be substituted by the BE occupation number $n(T)$ defined in Eq. (4.16), in order to account for the temperature dependence. This physical interpretation is only intuitive and does not stem from a rigorous analysis.

We can now express the temperature dependence of the nonradiative rate $A_{ij}^{NR}$ through the relation:

$$A_{ij}^{NR}(T) = A_{ij}^{NR}(0)[n(T) + 1]^p \qquad (4.18)$$

where $A_{ij}^{NR}(0)$ is the decay rate at $T = 0$.

The convergence of the perturbation expansion involved in the multiphonon decay theory requires that the rates satisfy the relation

$$\frac{A_{ij}^{NR}(0, p)}{A_{ij}^{NR}(0, p - 1)} = \varepsilon \ll 1 \qquad (4.19)$$

where $\varepsilon$ is a small phenomenological ion–phonon coupling constant which depends on the host [55], [74], and [80]. The nonradiative rate corresponding to $p$ phonon decay thus decreases as $\varepsilon^p$. Using this result and Eqs. (4.15) and (4.18), we can put the multiphonon rate $A_{ij}^{NR}$ into the form:

$$A_{ij}^{NR}(T, p) = C\varepsilon^p[n(T) + 1]^p = C[n(T) + 1]^p \exp(-\alpha\Delta E) \qquad (4.20)$$

where $\alpha = -\log(\varepsilon)/\hbar\omega$ and $C$ is a host dependent constant.

The result of Eq. (4.20), which predicts both temperature dependence and energy gap dependence of the multiphonon rate $A_{ij}^{NR}(T, p)$ for any rare earth element in a given host, was confirmed experimentally in a variety of crystals and glasses, with an accuracy of two or better, [55] and [80]. The characteristic constants $C$ and $\alpha$ in Eq. (4.20) can be determined by studying the variation of $A_{ij}^{NR}(T, p)$ with energy gap $\Delta E$ for different rare earths, given a glass host at a fixed temperature [80]. The decay rate $A_{ij}^{NR}(T, p)$ of a nonfluorescing level can be determined experimentally by short pulse excitation.

Let energy levels 1 and 2 be characterized by decay rates $W_1$ and $W_2$, respectively, where $W_2$ is uniquely nonradiative, and populations $N_1$ and $N_2$, respectively. We assume now that some of the rare earth atoms are excited to level 2 by a short laser pulse. The transient rate equations for the populations immediately after the pulse excitation take the form:

$$\frac{dN_1}{dt} = W_2 N_2 - W_1 N_1 \qquad (4.21)$$

$$\frac{dN_2}{dt} = -W_2 N_2 \qquad (4.22)$$

Equations (4.21) and (4.22) can be decoupled to give the equation

$$\frac{d^2 N_1}{dt^2} + (W_1 + W_2)\frac{dN_1}{dt} + N_1 W_1 W_2 = 0 \qquad (4.23)$$

of general solution

$$N_1(t) = N_1(0)\{c_1 \exp(-W_1 t) + c_2 \exp(-W_2 t)\} \qquad (4.24)$$

where $c_1, c_2$ are constants. The initial condition $N_1(0) = 0$ gives $c_2 = -c_1$, or

$$N_1(t) = c_1\{\exp(-W_1 t) - \exp(-W_2 t)\} \qquad (4.25)$$

The time dependent fluorescence observed experimentally from level 1 is proportional to $N_1(t)$, and from Eq. (4.25), has a maximum at the time $t_{max}$ given by:

$$t_{max} = \frac{1}{W_2 - W_1}\log\left(\frac{W_2}{W_1}\right) \qquad (4.26)$$

As the rise time constant $t_{max}$ and the fluorescence lifetime $\tau_1 = 1/W_1$ can both be measured experimentally, the nonradiative rate $W_2$ can be deduced from eq. (4.26).

The method of short pulse excitation described above is not the only way to determine nonradiative decay rates. Several alternate methods are outlined in [81]. One of the most practical approaches compares the observed radiative lifetime $\tau_{obs}$ of a given energy level $i$ with the radiative lifetime $\tau_{rad}$ calculated from the JO theory. The nonradiative decay rate can then be deduced through Eq. (4.13), i.e.,

$$A^{NR} = \frac{1}{\tau_{obs}} - \frac{1}{\tau_{rad}} \tag{4.27}$$

with $1/\tau_{rad} = \sum_j A_{ij}^{R}$.

Experimental studies of the multiphonon decay rate dependence with a glass host are reviewed in [74]. The constant $\alpha$ in Eq. (4.20) is observed to vary weakly with most oxide glasses, while the constant $C$ differs by orders of magnitude. In silicate glasses, for instance, $C = 1.4 \times 10^{12}\,\text{s}^{-1}$, $\alpha = 4.7 \times 10^{-3}\,\text{cm}$, and $\hbar\omega = 1100\,\text{cm}^{-1}$ (highest phonon energy). Using these results, we find that *the transition* $^4I_{13/2}-^4I_{15/2}$ *near $\lambda = 1.53\,\mu m$ in Er$^{3+}$:silica is essentially radiative,* corresponding to the case of a negligible multiphonon decay. Indeed, we have for this transition $\Delta E = 1/\lambda(\text{cm}) = 6536\,\text{cm}^{-1}$, $\hbar\omega/k_B T \approx 5.3$ $(T = 300\,\text{K})$, $n(300\,\text{K}) = 0.005$, $p = \Delta E/\hbar\omega \approx 6$, which from Eq. (4.20) gives $A_{ij}^{NR} = 0.067\,\text{s}^{-1}$ (the result weakly depends on $p$, as $n \ll 1$). On the other hand, the radiative decay rate from the $^4I_{13/2}$ level is $A_{21}^{R} \approx 100\,\text{s}^{-1}$, corresponding to a fluorescence lifetime $\tau_{21} = 1/A_{21}^{R} \approx 10\,\text{ms}$ [82]. From Eq. (4.14), these parameters give a radiative quantum efficiency for the laser transition $^4I_{13/2}-^4I_{15/2}$ in Er$^{3+}$:silica of $\eta(^4I_{13/2}) = A_{21}^{R}/(A_{21}^{R} + A_{21}^{NR}) = 99.93\%$, which means that the transition is essentially radiative. This fact was actually confirmed by experimental measurements in oxide glasses, [80] and [83], which gives a conservative estimate $A_{21}^{NR} < 1\,\text{s}^{-1}$, or $\eta(^4I_{13/2}) > 99\%$.

We compare the previous result with the transition $^4I_{11/2}-^4I_{13/2}$ in the same glass host. This transition is characterized by a wavelength near $\lambda = 2.75\,\mu m$, as shown in figure 4.4, which corresponds to the following parameters: $\Delta E = 3636\,\text{cm}^{-1}$ and $p \approx 3$. With the same silica glass host as in the previous example ($C = 1.4 \times 10^{12}\,\text{s}^{-1}$, $\alpha = 4.7 \times 10^{-3}\,\text{cm}$, $\hbar\omega = 1100\,\text{cm}^{-1}$, $T = 300\,\text{K}$, $\hbar\omega/k_B T \approx 5.3$, $n(T) = 0.005$), we obtain from Eq. (4.20) a multiphonon rate $A_{32}^{NR} = 5.53 \times 10^4\,\text{s}^{-1}$. On the other hand, the total radiative decay rate for the $^4I_{11/2}$ level is $A_{31}^{R} + A_{32}^{R} \approx 125\,\text{s}^{-1}$, corresponding to a total radiative lifetime $\tau(^4I_{11/2}) = 1/(A_{31}^{R} + A_{32}^{R}) = 8\,\text{ms}$. The radiative quantum efficiency corresponding to the $^4I_{11/2}$ level in Er$^{3+}$:silica is then $\eta(^4I_{11/2}) = (A_{31}^{R} + A_{32}^{R})/(A_{31}^{R} + A_{32}^{R} + A_{32}^{NR}) = 0.22\%$. In this case, the multiphonon decay is very large in comparison to the radiative decay, and the transition from the $^4I_{11/2}$ level to the $^4I_{13/2}$ level is essentially nonradiative. As the energy levels located above $^4I_{11/2}$ are characterized by smaller energy gaps (see Figure 4.4), the corresponding transitions should be essentially nonradiative too.

Clearly, the effect of multiphonon decay for Er$^{3+}$ in the silica host is very sensitive to energy separation, and it is quite fortuitous that the $^4I_{13/2}-^4I_{15/2}$ transition in Er$^{3+}$:silica turns out to be nearly 100% radiative, while other transitions taking place between other levels (of closer spacing) are essentially nonradiative. This advantage makes Er$^{3+}$:silica glass operate like the ideal three-level laser analyzed in Chapter 1.

For comparison, the parameters corresponding to a ZBLA host are $C = 1.7 \times 10^{10}\,\mathrm{s}^{-1}$, $\alpha = 5.5 \times 10^{-3}\,\mathrm{cm}$, and $\hbar\omega = 500\,\mathrm{cm}^{-1}$ [74]. At $T = 300\,\mathrm{K}$, we find from Eq. (4.20) the multiphonon rates $A_{21}^{NR} = 1.63 \times 10^{-5}\,\mathrm{s}^{-1}$ for the $^{4}I_{13/2}$–$^{4}I_{13/2}$ transition, and $A_{32}^{NR} = 76\,\mathrm{s}^{-1}$ for the $^{4}I_{11/2}$–$^{4}I_{13/2}$ transition. Considering that the radiative rates are $A_{21}^{R} \approx 70\,\mathrm{s}^{-1}$ and $A_{32}^{R} \approx 100\,\mathrm{s}^{-1}$ for the two transitions, respectively [81], the effect of multiphonon decay for the ZLBLA host is much lower than for the silica host. This can be explained by the lower value for the maximum phonon energy $\hbar\omega$, which, for a given energy gap, corresponds to a larger number of required phonons. For the ZBLA host, $C$ is also two orders of magnitude lower, and the dependence with energy gap somewhat higher, in comparison to the silica host. For Er-doped fluorozirconate glasses, the multiphonon decay can be of the same order of magnitude as the radiative decay, so fluorescence is generally observed for a variety of laser lines, [59]–[64], contrary to the case of an Er-doped silica host, where the only fluorescing transition strongly observed is $^{4}I_{13/2}$–$^{4}I_{15/2}$.

The dependence of rare earth multiphonon relaxation rates with the type of glass host (e.g. silicate, phosphate, germanate, borate, tellurite, fluorozirconate) was described in several studies, [74], [80], [81], and [83]. While the nonradiative decay rates of rare earths are found to vary by orders of magnitude with various glass hosts [80], the radiative lifetimes of the laser transitions are found to be nearly independent of the host type. Combining experimental results obtained with $Er^{3+}$:glasses for the radiative lifetimes, [81] and [83], and with rare earths in general for the nonradiative lifetimes [80], it is possible to label the $Er^{3+}$:glass energy levels shown in Figure 4.4 with their respective decay rates, as done in Figure 4.7 [84]. The radiative lifetimes

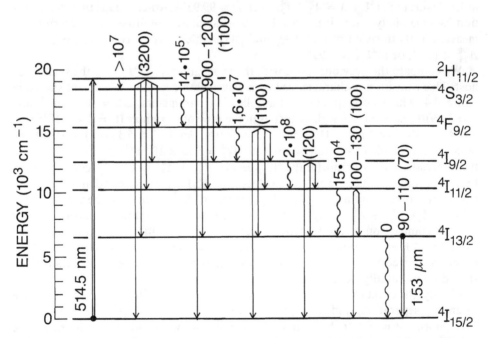

**FIGURE 4.7** Energy level diagram of Er:glass showing ranges of radiative and nonradiative decay rates (in $s^{-1}$) corresponding to various types of glass hosts; the radiative rates shown in parenthesis correspond to ZBLA hosts From [84] © 1989 IEEE.

within parentheses correspond to ZBLA hosts and are shown as indicative values for other hosts where the corresponding information is missing. The data shown in the figure confirm that for glass hosts other than the fluoride type, all transitions except $^4I_{13/2}$–$^4I_{15/2}$ are essentially nonradiative (further discussion in Section 5.12).

## 4.4  LASER LINE BROADENING

The spectral line width $\Delta\omega_{ij}$ of an optical transition $(ij)$ of a free rare earth ion is related to the total decay rates $A_k = A_k^R + A_k^{NR}(k = i, j)$ through [77]:

$$\Delta\omega_{ij} = A_i + A_j \tag{4.28}$$

The line width defined by Eq. (4.28) corresponds to the effect of *lifetime broadening.* Several other effects contribute to increase the transition line width. These can be put into three categories: dephasing broadening, Stark splitting broadening, and inhomogeneous broadening.

In the gas phase, dephasing is caused by collision events occurring between different types of atoms composing the gas mixture (e.g., He–Ne gas), or between the atoms and the tube walls. Dephasing broadening increases linearly with the total gas pressure [77]. In the solid phase, dephasing is caused by high frequency lattice vibrations. The effect on the laser line width is temperature dependent and is called *phonon broadening.* In all cases, the linewidth broadening can be described by the relation $\Delta\omega_{ij} = A_i + A_j + 2/T_{2,ij}$, where $T_{2,ij}$ is an effective dephasing time associated with collison or phonon broadening [77]. Dipolar coupling causes line width broadening at high concentrations. The random effect of dipole–dipole coupling (electric or magnetic) between adjacent atoms in some hosts (e.g. rare earth pentaphosphate) results in a frequency shift of atomic resonances, which can be treated in a way similar to phonon broadening. At room temperature, the effect of dipolar broadening is negligible compared to phonon broadening [77]. The lifetime broadened or phonon broadened laser line resulting from all atomic contributions is said to be homogeneous, when all atoms in the medium host experience the same broadening effects. Density matrix theory in Chapter 1 shows that a homogeneous laser transition is characterized by a Lorentzian spectral line shape, [76]–[78].

Stark splitting of laser transitions is induced by the Stark effect [13]. This is due to the lifting of energy level degeneracy by the host's crystalline field, as shown in Figure 4.8$a$. If the lower and upper energy levels are split into manifolds of $g_1$ and $g_2$ nondegenerate sublevels, respectively, the laser line corresponding to the two main levels, made of the superposition of $g_1 \times g_2$ possible laser transitions, is effectively broadened. As such, Stark splitting broadening is not homogeneous, since each laser transition between the two Stark manifolds has different characteristics. However, the Stark sublevels have small energy gaps and are therefore strongly coupled by the effect of thermalization. Such a strong coupling causes the overall laser line to exhibit homogeneous saturation characteristics, therefore Stark splitting can be viewed indirectly as a cause of homogeneous broadening.

Finally, inhomogeneous broadening stems from the existence of multiple possible atomic sites for a laser activator in a given host. In a crystalline host, differences in sites result from matrix microdefects, which can be isomorphically distributed.

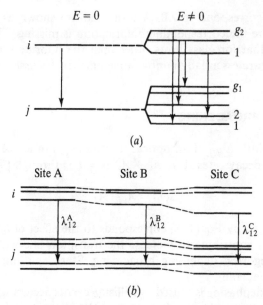

**FIGURE 4.8**   (a) Effect of Stark splitting of energy levels with degeneracies $g_1$ and $g_2$, caused by a crystalline electric field $E$, with possible laser transitions; (b) effect of inhomogeneous broadening, where random field variations from site to site cause changes in the Stark splitting and center wavelengths of laser transitions.

Differences in activator centers can result from symmetry considerations (multicenter system) or charge compensation effects (heterovalent activation) [10]. In a glass host, activator coordination by the crystalline matrix is random, aperiodic, and asymmetric (Section 4.1). In a basic glass matrix, e.g. silica host, the rare earth ion can occupy a single type of site, or activator center, characterized by random coordination. In the case of multicomponent glasses, several types of sites are possible, each corresponding to a different coordination by one or several network modifier neighbors [21]. The term *site* is often used to describe activator centers in glass, regardless of the coordination type. The phonon spectrum corresponding to crystalline vibration modes is random for each site and varies with the type of site, causing random changes in line width phonon broadening. Furthermore, variations of the crystalline electric field from site to site cause a randomization of the Stark splitting effect, as shown in Figure 4.8b. These effects are said to cause inhomogeneous broadening of the laser line. Inhomogeneous broadening can be viewed as the qualitative differences in laser line characteristics (line shape, center frequency) that correspond to any individual atom and to the overall collection of atoms of the medium, respectively. The inhomogeneous line widths generally depend weakly with temperature, and are characterized by Gaussian shapes [10]. For an individual laser transition, the combination of homogeneous and inhomogeneous broadening effects results in a spectral line shape that is between Gaussian and Lorentzian, also called the Voigt profile (Chapter 1).

In this section, we analyze in detail phonon broadening, Stark splitting, and inhomogeneous broadening for the transition $^4I_{13/2}$–$^4I_{15/2}$ in Er:glass.

Investigations of homogeneous and inhomogeneous properties of laser glass started in the early 1970s, [85] and [86]. Several methods have been demonstrated, based for instance on effects of fluorescence line narrowing (FLN), spectral hole burning, and coherent transients.

With the FLN method, fluorescence is generated in the laser glass at cryogenic temperature by a pump source of narrow spectral width. The excitation is site selective, as fluorescence is observed only from atoms that are resonant with the excitation frequency, and the ion concentration is low enough to prevent cross-relaxation effects leading to spectral diffusion. The effect of homogeneous line width narrowing at low temperatures minimizes the overlap between individual Stark transitions, which enables to select single laser transitions in a given family of sites. Due to the high selectivity of the FLN measurement in RE:glass, the observed fluorescence spectrum is narrowed compared to the inhomogeneous line shape, which makes it possible to determine both homogeneous line widths and Stark level energies corresponding to a given subset of rare earth sites (L. A. Riseberg [87]). A nonresonant FLN scheme can also be used, where the excitation and laser transition frequencies are different; the site selective fluorescence is then observed from the Stark levels of lowest energies. A drawback of nonresonant FLN is that resulting spectra exhibit residual inhomogeneous broadening, due to the effect of accidental coincidences, [85] and [88]. But the method does make it possible to determine the Stark level positions. A detailed study of the $^4I_{11/2}-^4I_{15/2}$ and $^4I_{13/2}-^4I_{15/2}$ transitions in a variety of bulk Er:glasses through nonresonant FLN spectroscopy was made by S. Zemon et al. [88].

The method of spectral hole burning (not to be confused with spatial hole burning) is also based on the principle of site selective excitation. Upon intense excitation by a narrow laser source, a spectral hole can be generated in the absorption or fluorescence spectrum due to a variety of effects: saturation, polarization, photochemical, or photophysical, [77] and [86]. In all these cases, the hole is observed only if spectral diffusion, cross-relaxation or atomic rearrangement effects are slow, compared to the stimulated emission rate, or are negligible.

Saturation hole burning (SHB), described below for Er:silica, changes the absorption or stimulated emission rates by selective saturation. In the case of a laser transition with strong inhomogeneous broadening, the hole created in the absorption or fluorescence spectrum has twice the width of the homogeneous transition [77].

Polarization hole burning (POHB) was investigated in Nd:glass by D. Hall et al., [89] and [90]. The effect is based on the fact that each activator site is characterized by a set of three principal axes randomly oriented in space, to each of which can be associated a cross section. These cross sections represents the transition probability corresponding to an excitation field parallel to a given axis. Strong saturation by a polarized source causes anisotropic site selection and results in gain anisotropy or POHB. An experimental measurement of $G_\parallel/G_\perp$, where $G_\parallel$ and $G_\perp$ are the gain in the polarization directions aligned with or orthogonal to the polarization of the saturating signal, makes it possible to determine cross section anisotropy. Given three axis $l$, $m$, $n$ with cross sections $\sigma_p$, $\sigma_q$, $\sigma_r$, respectively, one can assume, for simplicity, $\sigma_q = \sigma_r$ and $\sigma_p > \sigma_q$. The *depolarization ratio* $\mathcal{R}$ is then defined through [90]:

$$\mathcal{R} = \frac{6(\sigma_q/\sigma_p)^2 + 8(\sigma_q/\sigma_p) + 1}{8(\sigma_q/\sigma_p)^2 + 4(\sigma_q/\sigma_p) + 3} \tag{4.29}$$

which represents the ratio of fluorescence intensity polarized along the initial signal to the total fluorescence intensity, as integrated over all possible orientations. When

$\sigma_q = 0$, the laser medium is made of pure dipoles, and $\mathcal{R} = 1/3$. When $\sigma_q = \sigma_p$, the medium is isotropic and the depolarization ratio is $\mathcal{R} = 1$. Measurements through polarized FLN in Nd:glass (ED-2 type) gave a depolarization ratio $\mathcal{R} = 0.89$, corresponding to a cross section anisotropy $\sigma_q/\sigma_p = 0.42$ for this laser material [89]. The observation of polarization hole burning and cross section anisotropy does not mean that the gain in RE:glass depends on the signal polarization. The gain experienced by the saturating signal remains isotropic with respect to any particular orientation. Gain anisotropy induced by polarization hole burning can be only experienced by a weak probe in the presence of this saturating signal. But the anisotropy is subject to a small effect of cross-relaxation at high rare earth concentrations [90], and virtually disappears in the case of a nonpolarization-maintaining medium, such as an optical fiber. Polarization sensitivity is further discussed in Section 4.10 and in Chapter 5.

The effects of photochemical hole burning (PHB) and photophysical hole burning (NPHB, for nonphotochemical hole burning) concern light-induced changes in the activator site due to effects such as: photoionization, photodissociation, anion formation, photoisomerization, and glass network rearrangement, [77] and [86]. These hole burning effects, mostly observed in organic glasses, can last hours or days at cryogenic temperatures (persistent hole burning). NPHB was also measured for rare earths in inorganic glasses, [91]–[95].

Another method for homogeneous line width measurements uses the time domain, through the effects of free induction decay (FID) and photon echoes (PE), [13], [86], [94], and [95]. When a collection of atoms is coherently excited at a laser transition resonance, a macroscopic polarization results, followed by emission of coherent transients (FID for single pulse excitation, PE for multiple pulse excitation). The observation of FID requires that the excitation pulse line width be narrower than the homogeneous line width, which enables site selective measurements, [86], [96], and [97]. In PE experiments, the excitation line width can be broader, which is more practical but loses the advantage of site selectivity. In two-pulse PE experiments, the medium is excited by two short pulses separated by a delay $\tau$. If the pulse delay is less than the atomic dephasing time $T_2$, an echo pulse emitted at time $2\tau$ can be observed. The first pulse creates a macroscopic oscillating dipole, resulting from the superposition of ground and excited state atomic wave functions; the second pulse interchanges the wave function amplitudes, which, after an additional duration $\tau$, restores the initial coherence and results into the echo burst [94]. The excitation can be enhanced over many pulse repetitions (accumulated PE, [95] and [98]). The intensity of the echo pulse is observed to decay in time as $\exp(-2\tau/T)$, which enables the determination of the dephasing time $T_2$, through the relation $1/T_2 = 1/T - 1/2\tau_{21}$, where $\tau_{21}$ is the radiative or fluorescence lifetime of the transition; the homogeneous line width $\Delta v_h$ of the transition is then given by the relation, [86], [95], and [99]:

$$\Delta v_h = \frac{2}{\pi T} = \frac{1}{2\pi\tau_{21}} + \frac{1}{\pi T_2} \tag{4.30}$$

Photon echoes have been proposed for applications in optical memories, optical phase conjugation, and high speed signal processing, e.g., time reversal and convolution (see [100]–[102] and references therein). More generally, PE measurements provide supplementary data for the study of homogeneous line width, which complete SHB and FLN investigations, [94] and [102]. Time domain PE is a more accurate technique for low temperature investigations where SHB and FLN actually reach limits in spectral resolution. A comprehensive review of these experimental techniques and results obtained in various glasses for various trivalent lanthanides is made in [95] and [103]. PE experiments have been carried on since 1983 in single-mode glass fibers with $Nd^{3+}$, [104] and [105], and since 1991, with $Er^{3+}$ [106]. Aside from compatibility with photonic components, the advantages of the fiber geometry for PE experiments are: high intensities can be achieved in the core over long distances; the average separation between ions can be increased by low doping levels, which minimizes parasitic ion–ion interaction effects; PE phase matching is inherent in unidirectionality; heat sinking is efficient at very low temperatures; and the overall cooling apparatus is greatly simplified [105].

The study of the temperature dependence of homogeneous linewidths in RE:glass generally shows two line broadening regimes. At low temperatures (up to 40 K), the homogeneous line width is observed to follow a linear increase law (i.e., $\Delta v_h \approx T^{1+\varepsilon}$); at higher temperatures, the increase is nearly quadratic (i.e., $\Delta v_h \approx T^{2\pm\varepsilon}$), [95] and [103]. The processes that are responsible for the line width broadening with temperature are manifold [107]. The one-phonon direct broadening process is associated with a linear temperature dependence; the two-phonon Raman scattering process is associated with a $T^7$ law at low temperatures (i.e., below the Debye temperature), and with a $T^2$ law at higher temperatures, [107]–[109]. Another Raman type process is the Orbach process, for which a law in $\exp(-a/k_B T)$ is predicted [107]. The mechanism involved at very low temperatures to explain the linear dependence is attributed to glass tunneling dynamics, or coupling with *two-level systems* (TLS) specific to amorphous media, [95] and [110]. The dipole–dipole coupling effect between the ion and a nearby TLS results in the generation of intermediate ground and excited state configurations. Relaxation from these processes produces a shift in the resonance frequency. Such a coupling effect with surrounding TLSs results in spectral diffusion of the laser line, with a broadening $\Delta v \approx A/r_{ij}^3$ ($A$ = coupling constant, $r_{ij}$ = distance between ion and TLS) [110]. A review of various possible theories for TLS interactions in glass is made in [111]. In this section, we shall focus only on the high temperature dependence of the homogeneous line width, since we are interested in the evaluation of this line width for Er:glass at room temperature.

The results from experimental investigation of $Pr^{3+}$ glasses showed that the two-phonon Raman process is dominant over a temperature range of 10–300 K (J. Hegarty and W. M. Yen [112]). Results obtained with $Pr^{3+}$, $Nd^{3+}$, and $Er^{3+}$ show that over this temperature range the line width follows a $T^{1.7}$–$T^{2.2}$ law which depends upon the glass host, the glass composition, and the laser transition considered, [95] and [103]. As in 1990, homogeneous line width data were not available for $Er^{3+}$:glass, a series of SHB experiments were made with different types of Er-doped fibers (E. Desurvire et al. [113]; J. L. Zyskind et al. [114]), whose results are described below.

A basic experimental setup for low temperature SHB measurements of the $^4I_{13/2}$–$^4I_{15/2}$ transition in Er-doped fiber (EDF) is shown in Figure 4.9, [113] and [114]. For these experiments, the pump used to generate fluorescence in the EDF was an argon ion laser operating near $\lambda = 514$ nm or a color center (CC) laser near $\lambda = 1.48\ \mu$m. The saturating signal near $\lambda = 1.53\ \mu$m was a CW polarized, tunable Er-doped fiber laser or CC laser. The pump and signal were combined into the EDF,

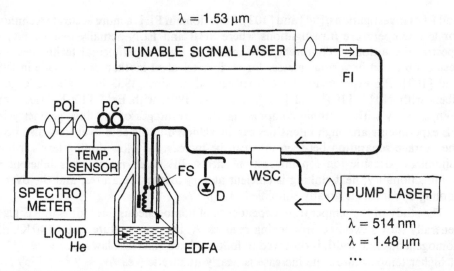

**FIGURE 4.9** Experimental setup for saturated spectral hole burning measurements of $^4I_{13/2}$–$^4I_{15/2}$ transition in Er-doped fibers, [113] and [114]. FI = Faraday optical isolator, WSC = wavelength selective coupler, D = detector, FS = fiber splice, EDFA = Er-doped fiber amplifier, PC = fiber polarization controller, POL = polarizer.

which was placed into the neck of a Dewar tank containing liquid helium ($T = 4.2$ K). In some measurements, the output spectrum was monitored through a polarizer. Two EDF types were investigated, i.e., based on aluminosilicate ($Al_2O_3$–$SiO_2$) and germanosilicate ($GeO_2$–$SiO_2$) glass hosts.

The output spectra are shown in Figure 4.10. Figure 4.10a shows the unpolarized output fluorescence spectrum near $\lambda = 1.535$ $\mu$m of the $GeO_2$ EDF, recorded with our without saturating signal at $T = 75$ K. The saturating signal, shown as a spike in the center, burns a narrow hole in the fluorescence spectrum, indicating that the observed spectral line width is essentially inhomogeneous. The hole characteristics are more clearly observed in the fluorescence spectra of polarization orthogonal with respect to the output signal, Figure 4.10b, c. As the temperature decreases, the hole width, which represents twice the homogeneous transition line width $\Delta \nu_h$ [77], is observed to narrow.

The measured dependence of the hole width $\Delta \lambda_{hole}$ with temperature is plotted in Figure 4.11 for the $Al_2O_3$ EDF and the $GeO_2$ EDF, measured at the peak wavelengths of the fluorescence spectra. As expected, a near-quadratic law is observed for the hole width, i.e., $\Delta \lambda_{hole} \approx T^{1.73}$ for the $Al_2O_3$-EDF and $\Delta \lambda_{hole} \approx T^{1.61}$ for the $GeO_2$ EDF. The break in temperature dependence observed for the $Al_2O_3$ EDF at $T < 20$ K can be attributed to TLS effects and to the limited instrument resolution (0.1 nm). The measured power laws make it possible to extrapolate room temperature values of $\Delta \lambda_{hole} = 23$ nm and $\Delta \lambda_{hole} = 8$ nm for both fiber types, corresponding to homogeneous line widths of $\Delta \lambda_{hom} = 11.5$ nm and $\Delta \lambda_{hom} = 4$ nm, respectively. Near the lowest temperatures ($T < 20$ K) and in the absence of a saturating signal, the observed fluorescence width, which equals the inhomogeneous line width, was measured for the two fibers to be $\Delta \lambda_{inh} = 11.5$ nm and $\Delta \lambda_{inh} = 8$ nm, respectively. Using these homogeneous and inhomogeneous line width values, we can calculate the corre-

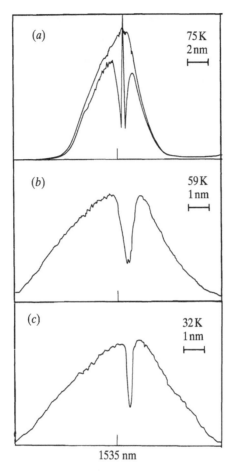

1535 nm

**FIGURE 4.10** (a) Unpolarized output spectra of germanosilicate Er-doped fiber, with saturating signal near $\lambda = 1.535\ \mu$m being on or off, showing hole burning in fluorescence at $T = 75$ K. (b), (c) Polarized spectra with output signal rejection, showing hole width narrowing with temperature [114] © 1990 IEEE.

sponding Voigt profiles from Eq. (1.208), which can be expressed as [113]:

$$V(\lambda) = \int_{-\infty}^{+\infty} \frac{\exp\{-4\log 2 \cdot (\lambda' - \lambda_0)^2 / \Delta\lambda_{\text{inh}}^2\}}{1 + 4(\lambda' - \lambda)^2 / \Delta\lambda_{\text{hom}}^2}\, d\lambda' \qquad (4.31)$$

where $\lambda_0$ is the center wavelength of the transition. In Eq. (4.31), $\Delta\lambda_{\text{inh}}$ represents the inhomogeneous line width, assumed to be temperature independent. Chapter 1 shows the Voigt profile represents the convolution between a homogeneous Lorentzian line shape with a Gaussian distribution representing inhomogeneous broadening. Figure 4.12 plots the emission cross sections determined experimentally at $T = 300$ K, the Voigt profiles calculated from Eq. (4.31), and the experimental values of $\Delta\lambda_{\text{hom}}$ and $\Delta\lambda_{\text{inh}}$ for the two fiber types. The Voigt profiles predicted for $T = 300$ K from the low temperature SHB measurements fit very well the observed cross section spectra. The second peak in the GeO$_2$ EDF near $\lambda = 1.535\ \mu$m was also characterized by

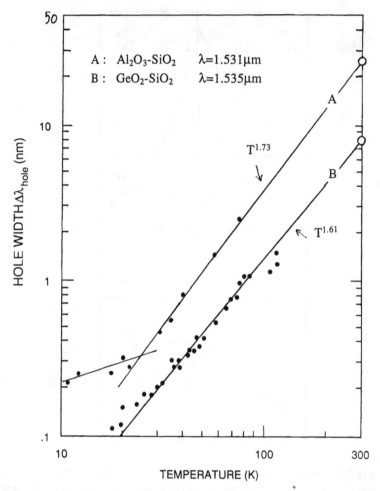

**FIGURE 4.11**   Measured hole width dependence with temperature for aluminosilicate (A) and germanosilicate (B) Er-doped fibers, showing near-quadratic power laws $T^{1.73}$ and $T^{1.61}$, respectively. From [113] and [114] © 1990 IEEE.

SHB. The observed $T^{0.54}$ law [114] reflects the broadening effect of accidental degeneracies near that laser wavelength, rather than a different line broadening mechanism. The good agreement between the predicted Voigt profiles and the observed cross sections for the two fiber types is an indication of the validity of the SHB method.

A comparison of the above results show that for the peak transition of $^{4}I_{13/2}$–$^{4}I_{15/2}$ in Er:glass at $T = 300$ K, homogeneous and inhomogeneous components are in a 1:1 ratio in the $Al_2O_3$ host and in a 1:2 ratio in the $GeO_2$ host. For the $GeO_2$ EDF, very similar values $\Delta\lambda_{hom} = 4$ nm and $\Delta\lambda_{inh} = 7$ nm were also inferred from cross-saturation measurements (R. I. Laming et al. [115]). An independent FLN measurement (S. C. Guy et al. [116] also showed for this glass host the values $\Delta\lambda_{hom} = 3.2$ nm and $\Delta\lambda_{inh} = 6.4$ nm, with 0.5 nm uncertainty, and the same power law, $T^{1.6}$, above $T > 30$ K. The consistency of the results obtained by FLN and SHB confirms the

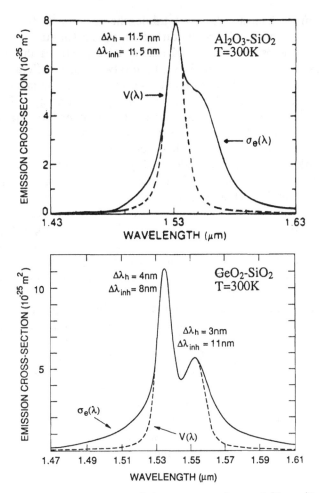

**FIGURE 4.12** Room temperature emission cross sections of $^4I_{13/2}$–$^4I_{15/2}$ transition in aluminosilicate and germanosilicate Er-doped fibers, showing homogeneous and inhomogeneous line widths corresponding to main peaks, and calculated Voigt profiles $V(\lambda)$. From [113] and [114] © 1990 IEEE.

validity of these investigation methods. Further low temperature experiments through SHB, FLN, or PE are necessary to assess possible variations of homogeneous and inhomogeneous line widths across the Er:glass absorption and emission cross section spectra. This data is required for an accurate modeling of inhomogeneous gain saturation in EDFAs (E. Desurvire et al. [117]), which requires modeling inhomogeneous cross sections, as shown in the next section.

The other important cause of line broadening is Stark splitting. The Stark effect, which is induced by the crystalline electric field (also called the ligand field) surrounding the rare earth ion, results in a lifting of degeneracy of the *SLJ* levels, Figure 4.8a. Since in a glass host the ligand field randomly varies from site to site, the Stark splitting between energy levels is also random. This produces site-to-site variations in transition wavelengths and causes inhomogeneous line broadening, Figure 4.8b.

From quantum mechanical rules, the total degeneracy of a $^{2S+1}L_J$ atomic state is $g = 2J + 1$. But for a purely electric perturbation, the maximum number of energy levels, or Stark levels, is $g = J + 1/2$ (Kramers' Rule [15]). In the presence of a magnetic field, each of the $J + 1/2$ levels is split into two sublevels (Kramers doublets) and degeneracy is totally lifted. According to Kramers' Rule, the maximum number of Stark sublevels for $^4I_{15/2}$ and $^4I_{13/2}$ in $Er^{3+}$ are $g_1 = 8$ and $g_2 = 7$, respectively. We refer then to $^4I_{15/2}$ and $^4I_{13/2}$ as ground level and upper level Stark manifolds. The Stark sublevels of manifold $^{2S+1}L_j$ are usually labeled $^{2S+1}\Gamma_n(L)$, where $\Gamma_n$ is the group representation corresponding to the level symmetry.

Because of Stark splitting, there actually exists $g_1 \times g_2$ possible transitions between the Stark manifolds. Experimental observation of 10 individual transitions for $^4F_{3/2}$–$^4I_{9/2}$ in $Nd^{3+}$, indicates that no selection rules prevail between Stark manifolds, and that all laser transitions are possible [14], although not equiprobable since the sublevel populations are weighted by Boltzmann's Law, according to Eq. (1.12). The actual number of Stark sublevels depends upon the symmetry of the ligand field. From group theory, [10] and [14], the number of Stark components for $^4I_{15/2}$ is 5 or 8 for cubic or any lower symmetries, and the corresponding numbers for $^4I_{13/2}$ are 5 or 7. The study of the absorption and emission spectra of RE:glass obtained by FLN reveals the number of Stark components existing in a given manifold $^{2S+1}L_j$, which makes it possible to infer the type of symmetry of the host. In the case of a purely random host structure, corresponding to no symmetry (labeled $C_1$ in symmetry classification), we expect to observe the maximum number of Stark components given by Kramers' Rule. But experimental evidence indicates that there are rare earth sites in glass which are characterized by some degree of symmetry.

Low temperature spectroscopic investigations of Stark levels made with $Yb^{3+}$ and $Er^{3+}$ in various alkali silicate glasses (C. C. Robinson [21]) have revealed the possibility of a site having near-octahedral symmetry $D_3$ and sixfold oxygen coordination, Figure 4.13, [6] and [118]. Charge compensation would require for this

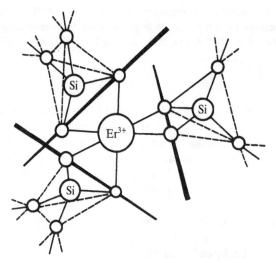

**FIGURE 4.13**   Possible site for $Er^{3+}$ in basic silicate glass. The symmetry is near-octahedral, and distorted into threefold $D_3$ symmetry, as emphasized by the tapered axis [6].

complex three additional monovalent alkali ions or one trivalent ion, e.g., $3Na^+$, $Al^{3+}$, in the rare earth and nonbridging oxygens vicinity.

The distortion of this structure about the three symmetry axes, shown in Figure 4.13 as tapered lines, cause variation of the crystal field and hence inhomogeneous broadening. Using this method, it is possible to calculate the potential energy associated with each angular distortion. The potential energy is then used as a Boltzmann factor to weigh the various contributions to the inhomogeneous line shape, [6] and [118]. The overall structure of the resulting transition line shape (absorption or emission) therefore depends upon both the accidental degeneracies of possible Stark transitions and the statistical weight associated to a given Stark level energy diagram. A calculation based on this model in the case of $^2F_{5/2}-^2F_{7/2}$ transition manifold in $Yb^{3+}$ for both emission and absorption, showed remarkable agreement with the experimental spectra (C. C. Robinson and J. T. Fournier, [118] and [6]).

The existence of a primary site, which is most likely to appear in the random glass structure, does not preclude other types of sites, especially in the case of multicomponent glasses of more complex structure. With $Er^{3+}$ alkali silicate glasses, up to four possible sites were identified (C. C. Robinson [21]). The identification of multiple sites is revealed by observing duplication of features in a given line multiplet, as the concentration of a network modifier is increased. A detailed analysis of the site symmetries in $Er^{3+}$ alkali silicate glasses is given in [21]. The difficulty of extending this type of investigation to any type of Er:glass (e.g., aluminosilicate) comes essentially from the smearing effect of inhomogeneous broadening, from comparatively large homogeneous line widths at low temperatures, from accidental transition degeneracies, and from the possible overlay of site-related spectral features.

Numerical computations through Monte Carlo methods of the transition line shapes make it possible to test various structural models for rare earth sites, (S. A. Brawer and M. J. Weber, [119]–[122]). They have shown that the magnitude of the Stark splitting is not correlated with the degree of rare earth coordination (e.g., sixfold), nor with the proximity of network-modifying ions with the rare earth [123]. Furthermore, the *uniqueness* of the site model for explaining the observed data remains difficult to establish. A main conclusion of these studies is that the actual glass structure cannot be reliably inferred from spectroscopic data, except in simple cases like Yb:silica. Because of the overall complexity of RE:glass, we might wonder whether the exact distributions of sites and laser parameters could ever be known for this system [85].

The phenomenological positions of Stark levels in RE:glass can be accurately determined by low temperature spectroscopy, using resonant or nonresonant FLN techniques, as shown for the case of $Er^{3+}$ in [88]. The line strengths and widths corresponding to each Stark transitions in the FLN absorption of fluorescence spectra can be determined by a multidimensional, nonlinear least-squares fitting method, using Gaussian line shapes, [123] and [124]. This method makes it possible to infer the characteristics of Stark transitions only for a given rare earth site. A reconstruction of the overall inhomogeneous line shape would require a weighting function for all possible sites, which is difficult to obtain from pure spectroscopic investigation. The curve-fitting method is even less conclusive for room temperature line shapes, as the spectral features are smoother, corresponding to more degrees of freedom. While convergence of the nonlinear curve-fitting algorithm can be obtained, there is no uniqueness of solution. The next section uses the curve-fitting method for the room temperature cross section spectra to obtain practical analytical expressions. This curve-fitting, which is not unique, should therefore not be viewed as representing a model for the actual Stark transition line shapes.

The $^4I_{15/2}$ and $^4I_{13/2}$ Stark level positions in aluminosilicate Er-doped fibers can also be approximately determined through an analysis of temperature dependence of absorption and emission spectra (E. Desurvire and J. R. Simpson [125]). In a three-level pumping scheme ($\lambda_p = 514.5$ nm), the spectral narrowing of fluorescence with decreasing temperature, Figure 4.14, is due both to the effect of homogeneous line width narrowing, and the decrease of Boltzmann populations in the Stark manifolds, Eq. (1.12). At the lowest temperature ($T = 4.2$ K), only the lowest energy levels of each manifold are populated, which reduces the number of observed transitions and accidental degeneracies. In these experimental conditions, FLN is not observed, as the homogeneous line width 0.2 nm at $T = 4.2$ K is not resolved in the figure; the 10 nm fluorescence width measured from the figure corresponds to the inhomogeneous width of this Er:glass type. The absence of FLN can be attributed to combined effects of spectral diffusion, spatial diffusion, accidental degeneracies, [87] and [88], and to a relatively large homogeneous line width in this glass host. In spite of the smearing effect of inhomogeneous broadening, the spectral structure changes with temperature that can be observed in Figure 4.14 reveal Stark transition features. The features observed with fluorescence can be correlated with those observed with absorption.

Figure 4.15 shows both absorption and fluorescence spectra recorded for the $Al_2O_3$ EDF at the three temperatures $T = 300$ K, $T = 77$ K (liquid nitrogen) and $T = 4.2$ K (liquid helium) [125]. At $T = 4.2$ K, the absorption spectrum exhibits two peaks (labeled $A_1, A_4$), and three other weakly resolved features ($A_0, A_2, A_3$). The fluorescence spectrum has only two features ($F_1, F_2$). At this temperature, only the levels of lowest energy are occupied in each manifold, so the absorption data show the existence of

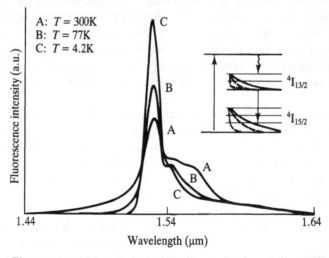

**Figure 4.14**  Fluorescence spectra of aluminosilicate Er-doped fiber recorded for three temperatures (A) $T = 300$, $T = 77$ K, and (C) $T = 4.2$ K. The energy diagram in inset shows the three-level pumping scheme and a sketch of the temperature dependent Boltzmann population distributions within the $^4I_{15/2}$ and $^4I_{13/2}$ manifolds. (a.u. = arbitrary units.)

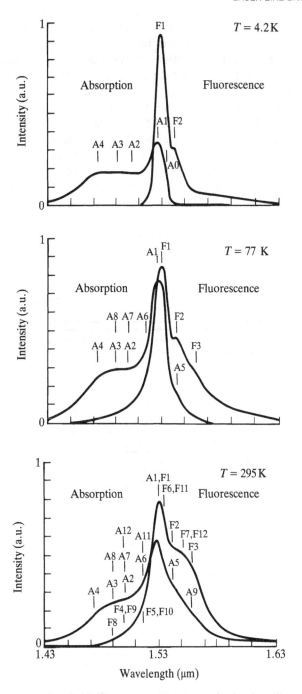

**FIGURE 4.15** Absorption and fluorescence spectra of aluminosilicate Er-doped fiber recorded at temperatures $T = 4.2$ K, 77 K, and 300 K, showing structure changes and positions of Stark transitions. From [125], reprinted with permission from the Optical Society of America, 1990.

four Stark levels in $^4I_{13/2}$ and the fluorescence of two Stark levels in $^4I_{15/2}$; the absorption transitions ($A_1$, $A_2$) are found to correspond to the emission transitions ($F_1$, $F_2$) [125]. The same comparative analysis of spectra obtained at higher temperatures makes it possible to infer the existence of other Stark levels; this yields the energy level diagrams shown in Figure 4.16. This measurement technique enables us to determine energy ranges for five Stark levels in $^4I_{15/2}$ and four Stark levels in $^4I_{13/2}$, between which there exist at least 12 Stark transitions. The lack of resolution does not allow us to draw conclusions about the maximum number of Stark levels. More accurate FLN measurements with $Er^{3+}$ in silicate and fluoride glasses [88] have shown that up to eight Stark levels can be identified in the $^4I_{15/2}$ state, which, for some sites, could be spaced as close together as $20\ cm^{-1}$. This number of Stark levels represents the maximum allowed by Kramers' Rule, which indicates that, as expected, the symmetry of the $Er^{3+}$ site in glass is of a type lower than cubic.

A comparison of Stark energies for different types of glass hosts shows that the total Stark splitting for $^4I_{15/2}$ and $^4I_{13/2}$ varies between $400\ cm^{-1}$ and $500\ cm^{-1}$ (see

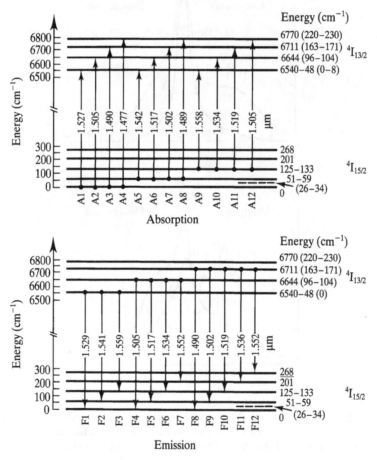

**FIGURE 4.16**    Stark transitions corresponding to absorption and emission aluminosilicate Er-doped inferred from measurements shown in Figure 4.15. From [125], reprinted with permission from the Optical Society of America, 1990.

table in [125]), corresponding to an average Stark level separation of $\Delta\lambda_{\text{Stark}} = 50\text{–}70 \text{ cm}^{-1}$. The range of total Stark splitting $400\text{–}500 \text{ cm}^{-1}$ corresponds approximately to about $2k_{\text{B}}T$, as $k_{\text{b}}T = 208.4 \text{ cm}^{-1}$ at $T = 300 \text{ K}$.

For germanosilicate and aluminosilicate glasses, the homogeneous line widths were found to be $\Delta\lambda_{\text{hom}} = 4 \text{ nm} = 17 \text{ cm}^{-1}$ and $\Delta\lambda_{\text{hom}} = 11.5 \text{ nm} = 49 \text{ cm}^{-1}$, respectively. On the other hand, the inhomogeneous broadenings were found to be $\Delta\lambda_{\text{inh}} = 8 \text{ nm} = 34 \text{ cm}^{-1}$ and $\Delta\lambda_{\text{inh}} = 11.5 \text{ nm} = 49 \text{ cm}^{-1}$, respectively. The additional Zeeman splitting of each Kramers doublet is, in silicate glasses, about $2 \text{ nm}$ or $8 \text{ cm}^{-1}$ [21]. These results indicate that a strong spectral overlap exists from site to site between the many Stark components of the $^4I_{13/2}\text{–}^4I_{15/2}$ transition of $Er^{3+}$ in glass, [88] and [125]. The overlap is the greatest with aluminosilicate hosts, as $\Delta\lambda_{\text{Stark}}/\Delta\lambda_{\text{hom}} \approx \Delta\lambda_{\text{hom}}/\Delta\lambda_{\text{inh}} \approx 1$.

Since the average energy separation between Stark levels corresponds to one or two low energy phonons of the glass matrix, we expect a very fast thermalization effect within each Stark manifold. Thermalization results in a distribution of atomic population densities $N_{nm}$ in each Stark sublevel $(nm)$ given by Boltzmann's Law, Chapter 1, i.e.,

$$N_{nm} = \frac{\exp[-\Delta E_m/k_{\text{B}}T]}{\sum\limits_{m=1}^{g_n} \exp[-\Delta E_m/k_{\text{B}}T]} \bar{N}_n \equiv p_{nm}\bar{N}_n \tag{4.32}$$

where $\Delta E_m = E_m - E_1$ is the energy difference between the Stark sublevel $(nm)$ and the Stark lowest energy $(n, 1)$. In eq. (4.32), $\bar{N}_n$ is the total atomic population of the Stark manifold $n$ at thermal equilibrium. The thermalization rates $A_{\text{NR}}^{\pm}$, corresponding to upward excitation or downward relaxation between Stark sublevels verify the relation, Eqs. (1.11) and (4.32):

$$\frac{A_{\text{NR}}^+}{A_{\text{NR}}^-} = \frac{N_{nm}}{N_{n,m-1}} = \exp\left(-\frac{E_m - E_{m-1}}{k_{\text{B}}T}\right) \tag{4.33}$$

which, by recursion, yields for $N_{nm}$ the Boltzmann distribution (4.32). We can estimate the rates $A_{\text{NR}}^{\pm}$ by assuming an average energy spacing $\Delta E = E_m - E_{m-1} \approx 50\text{–}100 \text{ cm}^{-1}$. For a one-phonon process, the decay rate $A_{\text{NR}}^-$ is given by Eq. (4.20) with $p = 1$. Using the parameters $(C, \alpha)$ corresponding to silica (Section 4.3), we find for $T = 300 \text{ K}$ $A_{\text{NR}}^- \approx (0.2\text{–}0.4 \text{ ps})^{-1}$, and $A_{\text{NR}}^+ \approx (0.2\text{–}0.7 \text{ ps})^{-1}$. Thus, the characteristic thermalization times are theoretically of order $1 \text{ ps}$ or less, which is consistent with experimental data in Nd:glass $(A_{\text{NR}}^{\pm} > (10 \text{ ps})^{-1})$ [90]. In the practical operating regime of the EDFA, the thermalization rates $A_{\text{NR}}^{\pm}$ always overtake the pumping $(\mathscr{R}_{13}, \mathscr{R}_{31})$ and the stimulated emission $(\mathscr{W}_{12}, \mathscr{W}_{21})$ rates, i.e., $A_{\text{NR}}^{\pm} \gg \mathscr{R}_{13}, \mathscr{R}_{31}, \mathscr{W}_{12}, \mathscr{W}_{21} \approx P/\tau P_{\text{sat}} \leq 100/\tau \approx (100 \text{ }\mu\text{s})^{-1}$, where $\tau = 10 \text{ ms}$ is the fluorescence lifetime of the $^4I_{13/2}$ level. In other words, the condition of thermal equilibrium within the Stark manifolds is maintained in any pumping or saturation regimes. Since the relative populations within the Stark manifold are unchanged by pumping or saturation, the spectral characteristics of the $^4I_{13/2}\text{–}^4I_{15/2}$ transition cross sections also remain unchanged. for this reason, we can view the effect of Stark splitting as an additional cause of homogeneous broadening.

Both effects of fast Stark manifold thermalization and strong spectral overlap between Stark components actually explain the *essentially homogeneous saturation behavior of aluminosilicate Er-doped fibers*, discussed in more detail in Chapter 5.

Another important effect of glass host composition is the change in energy separation between the lowest levels of the $^4I_{15/2}$ and $^4I_{13/2}$ manifolds. In glass, this separation varies up to $25\,cm^{-1}$, corresponding to $6\,nm$ [125]. Such a change in energy separation causes a shift in the peak wavelength of the absorption and emission spectra (transitions $A_1$ and $F_1$ in Figure 4.16). The wavelength shift due to codoping was first investigated in Nd:silicate glass (K. Arai et al., [126]). The investigation showed that both alumina ($Al_2O_3$) and phosphorous oxide ($P_2O_5$) codoping result in a shift of fluorescence towards shorter wavelengths; this gives spectra similar to that obtained in multicomponent glasses [126]. This effect results from the modification of the glass network by the codopants, which alters the ligand field in the rare earth sites. Another observed consequence of $Al_2O_3$ and $P_2O_5$ codoping was the suppression of partial microscopic clustering at high rare earth concentrations (Section 4.8). From Raman spectrum analysis, the effect of rare earth ion clustering is alleviated by a shell formation by the codopants around the rare earth. This facilitates rare earth incorporation in the glass matrix, which increases the rare earth solvability [126].

The effect of alumina codoping in $Er^{3+}$-doped fibers on the absorption and emission spectra near $\lambda = 1.5\,\mu m$ [65] is illustrated in Figure 4.17. These curves represent the spectral shapes of the absorption and emission coefficients $\gamma_{ak}$, $\gamma_{ek}$ at wavelength $\lambda_k$, used in Chapter 1 for EDFA theoretical modeling.

In the absence of alumina, the spectra are seen to exhibit two sharp peaks (curves labelled A). The same spectra are observed in germania codoped silica, which indicates

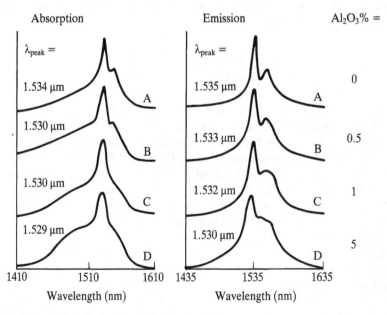

**FIGURE 4.17** Effect of alumina codoping for different molar concentrations on absorption and emission spectra of 1.5 $\mu$m transition in Er:silica glass, showing shift of peak wavelengths $\lambda_{peak}$.

that for this host, $GeO_2$ codoping does not affect the ligand field. As the alumina concentration is increased (curves B, C, D), the spectral features become smoother; in view of previous results, this can be attributed to homogeneous and inhomogeneous line width broadening. The smoothing effect observed between $\lambda = 1.53\,\mu m$ and $\lambda = 1.55\,\mu m$ is advantageous for multichannel amplification, as the gain bandwidth of the EDFA is effectively increased. Another important effect of alumina codoping is the increase of absorption in the short wavelength tail region, relative to the peak absorption. This increase is advantageous for the two-level pumping scheme (Chapter 1), where the pump is near $\lambda_p = 1.48\,\mu m$. This scheme is called two-level since the pump wavelength belongs the signal band. The conditions in which gain can be achieved in such a two-level laser system are analyzed in Chapter 1. The increase of the short wavelength tail is also observed in the emission spectra. The effect of the cross section ratio $\eta_k = \sigma_{ek}/\sigma_{ak} = \gamma_{ek}/\gamma_{ak}$ on the EDFA noise figure is analyzed in Chapter 2. In particular, the maximum inversion cannot be achieved when $\eta_k \neq 0$; this is the case with 1.48 $\mu m$ pumping. As a result, the minimum noise figure is always greater than 3 dB for this pumping scheme, Figure 2.8$b$.

Figure 4.18 shows fluorescence spectra of Er:silica obtained in alumina, germania, and phosphorous codoped glasses, for comparison. Similar comparisons with other types of glasses can be found in [3], [29], and [127]. The effect of line broadening with alumina codoping is clearly observed, while phosphorous codoping represents an intermediate case. Multicomponent glass fibers ($SiO_2$–$Al_2O_3$–$R_2O$–$R'O$, R = Na, Li, R' = Ca, Mg) give different fluorescence features, which consist of two closely spaced peaks of nearly equal magnitude and of overall width narrower than in the $GeO_2$–$SiO_2$ case [128]. Certain compositions of multicomponent glass fibers can exhibit fluorescence features similar to the alumina codoped case, as observed with $SiO_2$–$Al_2O_3$–$Na_2O$–$CaO$ [129]. In HMF hosts, such as ZBLAN, the fluorescence is significantly broader and smoother than observed in $SiO_2$–$Al_2O_3$ fibers, [130] and [131]. The effect of lanthanum codoping yields fluorescence features intermediate between $SiO_2$–$GeO_2$ and $SiO_2$–$Al_2O_3$ fibers (curves B and C in Figure 4.17), with an observed enhancement in gain and saturation power [132].

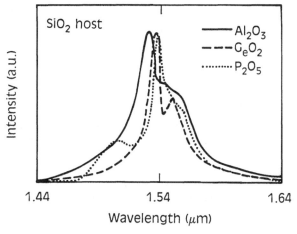

**FIGURE 4.18**   Fluorescence spectra of 1.5 $\mu m$ transition of $Er^{3+}$ for three types of silica glass codopants: alumina ($Al_2O_3$), germania ($GeO_2$), and phosphorous oxide ($P_2O_5$).

There are many reasons to choose a particular glass host composition and stochiometry for $Er^{3+}$ among the many apparent possibilities. First, the glass must lend itself to fiber waveguide fabrication techniques (e.g., MCVD, OVD) and physical constraints (i.e., index difference, deposition rate, expansion coefficients, codopant diffusion, codopant migration, and phase transitions); the fiber background loss must be minimal (low impurity level); the glass must be chemically stable and mechanically resistant. If the above conditions are met, we can choose the type of glass based on the following criteria: relative peak absorption in various pump bands; relative peak emission cross section at the signal wavelength; fast multiphonon decay between pump level and upper laser level; long fluorescence lifetime; smoothness and bandwidth of gain spectrum; gain saturation homogeneity; output saturation power; absence of pump and signal excited state absorption (Section 4.7); solvability at high rare earth concentrations; possibility of sensitization by other rare earth codopants (Section 4.8). Many types of glasses satisfy most of the above conditions. A systematic comparison of the different possibilities is difficult, given the number of degrees of freedom in pump bands, signal wavelengths, and eventually fiber waveguide design (Chapter 5). It is likely that all the possibilities of glass host for EDFAs are not yet exhausted, in spite of the fact that a great variety have already been fully explored. For a state-of-the-art assessment (1993), the reader may refer to comprehensive reviews by W. J. Miniscalco [127] and by B. J. Ainslie [29], which address both Er:glass spectroscopy and fiber fabrication issues.

We now provide some basic relations between various definitions of Er and codopant concentrations generally used in the literature. Let $x(i)$ be the fractional molar concentration (in mol%) of a given oxide $i$ composing the glass, and $Z$ the corresponding molecule mass. We have $Z(Er^{3+}) = 167.3$ g, and for various oxides:

$$Z(SiO_2) = 60.1 \text{ g} \quad Z(Al_2O_3) = 102.0 \text{ g} \quad Z(GeO_2) = 104.6 \text{ g}$$

$$Z(P_2O_5) = 142.0 \text{ g} \quad Z(Er_2O_3) = 382.6 \text{ g}$$

The total weight of one mole of composite glass is then given by:

$$W = x(SiO_2)Z(SiO_2) + x(Al_2O_3)Z(Al_2O_3) + \cdots + x(Er_2O_3)Z(Er_2O_3) \quad (4.34)$$

with

$$\sum_i x(i) = 1 \quad (4.35)$$

Assuming that $D$ is the glass density, and given the Avogadro number $\mathcal{N} = 6.02 \times 10^{23}$, the numbers of $SiO_2$ and $Er_2O_3$ molecules per cubic centimeter (cc) are given by:

$$n(SiO_2) = x(SiO_2)\frac{D\mathcal{N}}{W} \text{ cm}^{-3} \quad (4.36)$$

$$n(Er_2O_3) = x(Er_2O_3)\frac{D\mathcal{N}}{W} \text{ cm}^{-3} \quad (4.37)$$

and the number of $Er^{3+}$ ions per cc or $Er^{3+}$ density is then:

$$\rho(Er^{3+}) = x(Er_2O_3)\frac{2D\mathcal{N}}{W} \text{ ions/cm}^3 \tag{4.38}$$

The ppm mol $Er_2O_3$ concentration (parts per millions in molar concentration) is defined as:

$$\rho(Er_2O_3) = 10^6 \frac{n(Er_2O_3)}{n(SiO_2)} \text{ ppm mol} \tag{4.39}$$

or, using Eqs. (4.36) and (4.37):

$$\rho(Er_2O_3) = \frac{10^6}{x(SiO_2)}\frac{W}{2D\mathcal{N}}\rho(Er^{3+}) \text{ ppm mol} \tag{4.40}$$

Next, the densities of $SiO_2$ and $Er_2O_3$ molecules in gram per cc are given by:

$$\rho(SiO_2) = n(SiO_2)\frac{Z(SiO_2)}{\mathcal{N}} = x(SiO_2)\frac{Z(SiO_2)}{W}D \text{ g/cm}^3 \tag{4.41}$$

$$\rho(Er_2O_3) = n(Er_2O_3)\frac{Z(Er_2O_3)}{\mathcal{N}} = x(Er_2O_3)\frac{Z(Er_2O_3)}{W}D \text{ g/cm}^3 \tag{4.42}$$

The wt% $Er_2O_3$ concentration (percentage of weight in total concentration) is then defined as:

$$\rho(Er_2O_3) = 100\frac{\rho(Er_2O_3)\text{g/cm}^3}{D} \text{ wt\%} \tag{4.43}$$

or, using Eqs. (4.37) and (4.38), and (4.41) and (4.42):

$$\rho(Er_2O_3) = 100\frac{\rho(Er^{3+})\text{cm}^{-3}}{2D\mathcal{N}}Z(Er_2O_3) \text{ wt\%} \tag{4.44}$$

Finally, we can define the ppm wt $Er^{3+}$ concentration (parts per millions of weight in total concentration) through:

$$\rho(Er_2O_3) = 10^6\frac{\rho(Er_2O_3)\text{ g/cm}^3}{D}\frac{2Z(Er^{3+})}{Z(Er_2O_3)} \text{ ppm wt} \tag{4.45}$$

or, using Eqs. (4.38) and (4.42):

$$\rho(Er^{3+}) = 10^6\frac{\rho(Er^{3+})\text{cm}^{-3}}{D}\frac{Z(Er^{3+})}{\mathcal{N}} \text{ ppm wt} \tag{4.46}$$

In most applications, the $Er^{3+}$ and codopant concentrations are small (i.e., $x(SiO_2) \approx 1$, $W \approx Z(SiO_2)$), and the glass density is near $D = 2.86$, which corresponds to pure silica. In this case, assume as a basic example that the $Er^{3+}$ concentration of the fiber core is $\rho(Er^{3+}) = 1.10^{19}$ ions/cm$^3$. The $Er_2O_3$ concentration in ppm is then given by Eq. (4.40), which yields $\rho(Er_2O_3) = 188$ ppm. The $Er_2O_3$ concentration in wt% is given by Eq. (4.44), which yields $\rho(Er_2O_3) = 0.12$ wt%.

Consider next another example, where an $SiO_2$–$Al_2O_3$–$Er_2O_3$ fiber has a core density $D = 2.95$, and molar concentrations $x(SiO_2) \approx 95$ mol%, $x(Al_2O_3) \approx 5$ mol% and $x(Er_2O_3) \approx 10^{-3}$ mol%. What are the erbium ion and erbium oxide densities? Using first Eq. (4.34) we find a molar weight $W = 62.2$ g. Using eqs. (4.38), (4.40) and (4.44), we find then $\rho(Er^{3+}) = 6.0 \times 10^{17}$ ions/cm$^3$, $\rho(Er_2O_3) = 11$ ppm and $\rho(Er_2O_3) = 0.006$ wt%.

## 4.5 DETERMINATION OF TRANSITION CROSS SECTIONS

In this section, we describe the experimental determination and characterization of the absorption and emission cross sections of the $^4I_{13/2}$–$^4I_{15/2}$ laser transition in Er:glass.

Section 4.3 uses the Judd–Ofelt theory to obtain a relation between the transition's oscillator strength and its characteristic radiative lifetime $\tau_{rad} = 1/A^{ed}_{S'L'J',SLJ}$. In most glass hosts, the $^4I_{13/2}$–$^4I_{15/2}$ transition is nearly 100% radiative, i.e., the observed fluorescence lifetime $\tau_{obs}$ is equal to the radiative lifetime $\tau_{rad}$, or $\tau_{obs} = \tau_{rad} \equiv \tau$. Using Eqs. (4.6)–(4.8), we can write:

$$\int_{band} \sigma_e(\lambda) \, d\lambda = \frac{\langle \lambda \rangle^4}{8\pi n^2 c\tau} \tag{4.47}$$

$$\int_{band} \sigma_a(\lambda) \, d\lambda = \frac{g_2}{g_1} \frac{\langle \lambda \rangle^4}{8\pi n^2 c\tau} \tag{4.48}$$

The mean wavelength $\langle \lambda \rangle$ is assumed to be the same for absorption and emission. The integrals in eqs. (4.47) and (4.48) can be put into the form:

$$\sigma_{a,e}^{peak} \int_{band} \frac{\sigma_{a,e}(\lambda)}{\sigma_{a,e}^{peak}} \, d\lambda = \sigma_{a,e}^{peak} \int_{band} \frac{I_{a,e}(\lambda)}{I_{a,e}^{peak}} \, d\lambda \equiv \sigma_{a,e}^{peak} \times \Delta\lambda_{a,e}^{eff} \tag{4.49}$$

where $\sigma_{a,e}^{peak}$ are the peak values of the absorption and emission cross sections, respectively, and $I_{a,e}(\lambda)$ are the experimental absorption and fluorescence spectra with peak values $I_{a,e}^{peak}$. The quantities $\Delta\lambda_{a,e}^{eff}$ in Eq. (4.49) are the effective line widths of the absorption and emission line shapes. Using Eqs. (4.47)–(4.49), we can rewrite:

$$\sigma_e\langle \lambda \rangle = \frac{\langle \lambda \rangle^4}{8\pi n^2 c\tau\Delta\lambda_e^{eff}} \frac{I_{a,e}(\lambda)}{I_e^{peak}} \tag{4.50}$$

$$\sigma_e\langle \lambda \rangle = \frac{g_2}{g_1} \frac{\langle \lambda \rangle^4}{8\pi n^2 c\tau\Delta\lambda_e^{eff}} \frac{I_{a,e}(\lambda)}{I_a^{peak}} \tag{4.51}$$

and, in particular:

$$\sigma_e^{peak} = \frac{\langle\lambda\rangle^4}{8\pi n^2 c\tau\Delta\lambda_e^{eff}} \tag{4.52}$$

$$\sigma_a^{peak} = \frac{g_2}{g_1}\frac{\langle\lambda\rangle^4}{8\pi n^2 c\tau\Delta\lambda_a^{eff}} \tag{4.53}$$

If the line shapes were Gaussian or Lorentzian with full width at half maximum (FWHM) $\Delta\lambda_{FWHM}$, the corresponding effective line widths would be $\Delta\lambda^{eff} = \sqrt{\pi/4\log 2}\,\Delta\lambda_{FWHM} \approx 1.06\Delta\lambda_{FWHM}$ and $\Delta\lambda^{eff} = (\pi/2)\Delta\lambda_{FWHM} \approx 1.57\Delta\lambda_{FWHM}$, respectively, which explains differences in the formulas found in various textbooks [76]–[78] and [133], and throughout the literature. For instance, the peak emission cross section of a Lorentzian line shape is thus, from Eq. (4.52):

$$\sigma_e^{peak} = \frac{\langle\lambda\rangle^4}{4\pi^2 n^2 c\tau\Delta\lambda_{FWHM}} = \frac{\langle\lambda\rangle^2}{4\pi^2 n^2 \tau\Delta\nu_{FWHM}} \tag{4.54}$$

Equations (4.52) and (4.53), which link the peak cross sections $\sigma_{a,e}^{peak}$ to the experimental values of mean wavelength $\langle\lambda\rangle$, refractive index $n$, fluorescence lifetime $\tau$, and effective line widths $\Delta\lambda_{a,e}^{eff}$, is widely used throughout the literature on RE:glass and is sometimes referred to as the *Fuchtbauer–Ladenburg (FL) relation,* [69] and [134]–[137]. The FL relation, which stems from the Judd–Ofelt theory, can also be derived through Einstein's A and B coefficients, [13], [76], [78], and [133]. This derivation is detailed in Appendix L. One of the fundamental results of the Einstein demonstration is the reciprocity between absorption and emission, which can be expressed for multilevel laser systems as:

$$\sigma_a(\lambda) = \frac{g_2}{g_1}\,\sigma_e(\lambda) \tag{4.55}$$

On the other hand, the FL relation predicts a reciprocity which, according to Eqs. (4.52) and (4.53), takes the form:

$$\sigma_a(\lambda) = \frac{g_2}{g_1}\frac{\Delta\lambda_e^{eff}}{\Delta\lambda_a^{eff}}\,\sigma_e(\lambda) \tag{4.56}$$

Equations (4.55) and (4.56) are equivalent when all transitions have equal probabilities, line shapes, and peak wavelengths, which gives $\Delta\lambda_a^{eff} = \Delta\lambda_e^{eff}$.

But the Fuchtbauer–Ladenburg and the Einstein coefficient reciprocity relations can be applied only to the case where the populations of all the Stark sublevels are nearly equal [68]. A second condition, implicit in the derivation of Eqs. (4.6) and (4.7) is that the local field correction factors must be identical for the $J$ and $J'$ manifolds, [135] and [138]. From Eq. (4.33), the first condition is met only when the total Stark splitting energy $\Delta E$ is less than $k_B T$. But experimental evidence with

RE:crystals has shown that within the accuracy of the JO theory, Eqs. (4.6) and (4.7) remain valid when the splitting is somewhat greater than $k_B T$ [68]. The Stark splitting in Er:silica glass is nearly twice $k_B T$; this suggests that the FL relation is inaccurate to some extent. The effect of degeneracy is built into the experimental values of the line widths $\Delta\lambda_a^{eff}$, $\Delta\lambda_e^{eff}$, therefore the ratio $g_1/g_2$ can be overlooked in Eq. (4.53). The inaccuracy was actually confirmed experimentally, [139] and [140], as it was found that the observed cross section ratio $\eta^{peak} = \sigma_e^{peak}/\sigma_a^{peak}$ did not match the value $\eta^{peak} \approx \Delta\lambda_a^{eff}/\Delta\lambda_e^{eff}$ predicted by Eqs. (4.52) and (4.53). Comparative measurements of absorption and gain coefficients in different types of Er-doped fibers have shown that the ratio is weakly dependent of glass composition and is equal to $\eta^{peak} = 0.84-0.94$, while the FL relation predicts a value of $\eta^{peak} = 1.17-1.28$, [142]–[143]. The difference can be essentially attributed to the fact that in presence of substantial Stark splitting (i.e., as large as $2k_B T$), the Stark sublevels are not equally populated, which changes the effective transitions probabilities for each of these levels. Additionally, the effect of inhomogeneous broadening is to weigh each of the transition's probabilities according to the statistics of occurrence of rare earth sites in the glass, which affects the ratio $\Delta\lambda_a^{eff}/\Delta\lambda_e^{eff}$. Another possible cause of the difference is the effect of vibronic interactions, which is observed for instance in Nd:crystals, and results in the enhancement of certain absorption or emission transitions [144]. The effect of vibronics, though considered generally to be negligible in glass hosts, has not been investigated and nor quantified for Er:glass, and could be responsible for additional uncertainty. Finally, the JO theory is known to be somewhat inaccurate (i.e., 10–15%) in the prediction of absolute peak cross sections, which can be attributed to the simplifications used in the model, [11] and [145].

The inadequacy of the FL method for the determination of absolute cross sections of the $^4I_{13/2}-^4I_{15/2}$ transition in Er:glass was clarified in 1989. The accuracy of emission cross section measurements in fibers has improved ever since; this explains the large differences in data found between the earliest and the more recent papers on EDFAs. Systematic experimental and theoretical determinations of absolute cross sections in Er:glasses were reported only in 1991 (W. J. Miniscalco and R. S. Quimby [146]; W. L. Barnes et al. [142]).

A reciprocity relation between absorption and emission cross sections more accurate than Eqs. (4.55) and (4.56) was used in McCumber's theory of phonon-terminated optical masers, [147]–[148]. This reciprocity relation, which we shall refer to as the MC relation, is temperature dependent and is expressed as:

$$\sigma_a(v) = \sigma_e(v)\exp\left\{\frac{h(v-\varepsilon)}{k_B T}\right\}$$

(4.57)

In Eq. (4.57), $h\varepsilon$ represents the net thermodynamical free energy $\Delta F$ required to move one rare earth ion from the ground state to an excited state ($\Delta N_2 = -\Delta N_1 = 1$), while maintaining the lattice temperature $T$ constant, i.e.,

$$h\varepsilon = \Delta F = \left(\frac{\partial F(N_1, N_2, T)}{\partial N_2}\right)_T - \left(\frac{\partial F(N_1, N_2, T)}{\partial N_1}\right)_T$$

(4.58)

The partial derivatives in Eq. (4.58) are the chemical potentials corresponding to the subgroups of ground state ions, population $N_1$, and excited state ions, population $N_2$, [147] and [149]. The function $F(N_1, N_2, T)$ is the total free energy of the system. Assuming that the populations $N_1$, $N_2$ are large, we have:

$$\left(\frac{\partial F(N_1, N_2, T)}{\partial N_1}\right)_T \approx F(N_1 + 1, N_2, T) - F(N_1, N_2, T) \tag{4.59}$$

$$\left(\frac{\partial F(N_1, N_2, T)}{\partial N_2}\right)_T \approx F(N_1, N_2 + 1, T) - F(N_1, N_2, T) \tag{4.60}$$

and

$$h\varepsilon \approx F(N_1, N_2 + 1, T) - F(N_1 + 1, N_2, T) \tag{4.61}$$

which shows that $h\varepsilon$ is the free energy difference between the two ion subgroups. An alternate definition of the free energy $h\varepsilon$ is [147]:

$$h\varepsilon = -k_B T \log\left(\frac{N_2}{N_1}\right) \tag{4.62}$$

where $N_2/N_1$ is the population ratio at thermal equilibrium when the medium is unpumped. If the Stark sublevel populations and absolute energies in each manifold are labeled $N_{1j}$, $E_{1j}$ $(j = 1, \ldots, g_1)$ and $N_{2k}$, $E_{2k}$ $(k = 1, \ldots, g_2)$, Figure 1.2, and have thermal equilibrium values given by Boltzmann's Law, Eq. (1.12), the total populations $N_1$, $N_2$ in each Stark manifold are:

$$N_1 = \sum_{j=1}^{g_1} N_{1j} = N_{11} \sum_{j=1}^{g_1} \exp\left(-\frac{E_{1j} - E_{11}}{k_B T}\right) \tag{4.63}$$

$$N_2 = \sum_{k=1}^{g_2} N_{2k} = N_{21} \sum_{k=1}^{g_2} \exp\left(-\frac{E_{2k} - E_{21}}{k_B T}\right) \tag{4.64}$$

with

$$\frac{N_{21}}{N_{11}} = \exp\left(-\frac{E_{21} - E_{11}}{k_B T}\right) \equiv \exp\left(-\frac{\Delta E_{21}}{k_B T}\right) \tag{4.65}$$

where $\Delta E_{21}$ is the separation between the lowest energy levels of each manifold. Substituting Eqs. (4.63)–(4.65) into Eq. (4.62), we obtain an explicit expression for the free energy [146]:

$$h\varepsilon = -k_B T \log\left\{\exp\left(-\frac{\Delta E_{21}}{k_B T}\right) \frac{1 + \sum_{k=2}^{g_2} \exp\left(-\frac{\delta E_{2k}}{k_B T}\right)}{1 + \sum_{j=2}^{g_1} \exp\left(-\frac{\delta E_{1j}}{k_B T}\right)}\right\} \tag{4.66}$$

where $\delta E_{1j}$, $\delta E_{2k}$ are the energy differences between Stark sublevels with respect to the lowest energy level in the corresponding manifold. The free energy can be calculated from knowledge of the individual Stark sublevel positions. From Eq. (4.57), knowledge of the free energy enables us to determine one cross section spectrum with respect to the other, as well as their ratio at the peak wavelength $\eta^{\text{peak}} = \sigma_e^{\text{peak}}/\sigma_a^{\text{peak}}$.

But if the Stark level energies are not known, it is still possible to assume an average separation $\delta E_1$, $\delta E_2$ for each manifold. We have then $\delta E_{1j} = (j - 1)\delta E_1$ and $\delta E_{2k} = (k - 1)\delta E_2$. This assumption reduces to three (i.e., $\Delta E_{21}$, $\delta E_1$, $\delta E_2$) the number of parameters required for the computation Eq. (4.66), which, using the geometrical sum relation $1 + q + q^2 + \cdots + q^n = (1 - q^{n+1})/(1 - q)$, can be expressed as:

$$
h\varepsilon = -k_{\text{B}} T \log\left\{ \exp\left( -\frac{\Delta E_{21}}{k_{\text{B}} T} \right) \frac{1 - \exp\left( -\dfrac{\delta E_1}{k_{\text{B}} T} \right)}{1 - \exp\left( -\dfrac{\delta E_2}{k_{\text{B}} T} \right)} \frac{1 - \exp\left( -g_2 \dfrac{\delta E_2}{k_{\text{B}} T} \right)}{1 - \exp\left( -g_1 \dfrac{\delta E_1}{k_{\text{B}} T} \right)} \right\} \quad (4.67)
$$

We can now rewrite Eq. (4.57) in the form:

$$
\eta(\lambda) = \frac{\sigma_e(\lambda)}{\sigma_a(\lambda)} = \exp\left[ -\frac{hc}{k_{\text{B}} T} \left( \frac{1}{\lambda} - \frac{1}{\lambda^*} \right) \right] \quad (4.68)
$$

where $\lambda^* = c/\varepsilon$ is the wavelength at which absorption and emission cross sections are equal, according to Eq. (4.57). In particular, the peak cross section ratio is given by:

$$
\boxed{\eta^{\text{peak}} = \frac{\sigma_e^{\text{peak}}}{\sigma_a^{\text{peak}}} = \exp\left[ -\frac{hc}{k_{\text{B}} T} \left( \frac{1}{\lambda_{\text{peak}}} - \frac{1}{\lambda^*} \right) \right]} \quad (4.69)
$$

For the average energy separations $\delta E_1$, $\delta E_2$, values consistent with the experimental absorption and fluorescence spectra can be assumed [146]. The low energy half-width of the emission spectrum is identified as $7\delta E_1$, and similarly, the high energy half-width of the absorption spectrum is identified as $6\delta E_2$. The half-widths are determined by the wavelength separation between the peaks and the point where the spectral intensity is 95% below the peak. These assumptions are based on the fact that the peak wavelengths of both spectra have been identified with the Stark transition bridging the two lowest energy levels of $^4I_{15/2}$ and $^4I_{13/2}$, and are consistent with the energy diagrams shown in Figure 4.16. The energy separation $\Delta E_{21}$ is given by $\Delta E_{21} = hc/\lambda_{\text{peak}}$, and the degeneracy factors are $g_1 = 8$ ($^4I_{15/2}$) and $g_1 = 7$ ($^4I_{13/2}$) (in Er:glass, the peak wavelengths for absorption and emission differ by nearly 1 nm [146], so we must use an average value for $\lambda_{\text{peak}}$). Using these parameters, the value of the free energy $h\varepsilon$ can then be computed from Eq. (4.67).

In a second possible method, we can use the phenomenological values $\lambda_{\text{peak}}$, $\eta^{\text{peak}}$ that are observed experimentally for the peak cross section wavelength and ratios and calculate the free energy $h\varepsilon$, according to Eq. (4.69), i.e.,

$$
h\varepsilon = \frac{hc}{\lambda^*} = \frac{hc}{\lambda_{\text{peak}}} \left\{ 1 + \lambda_{\text{peak}} \frac{k_{\text{B}} T}{hc} \log(\eta^{\text{peak}}) \right\} \quad (4.70)
$$

Equation (4.57) can also be rewritten with Eq. (4.70) in the form:

$$\sigma_{a}(\nu) = \frac{\sigma_{e}(\nu)}{\eta^{\text{peak}}} \exp\left\{ \frac{h(\nu - \nu^{\text{peak}})}{k_{B} T} \right\} \tag{4.71}$$

where $\nu_{\text{peak}} = c/\lambda_{\text{peak}}$ is the peak frequency.

Figure 4.19 shows the experimental absorption and emission cross sections spectra $\sigma_{a}^{\text{exp}}(\lambda)$, $\sigma_{e}^{\text{exp}}(\lambda)$ corresponding to the case of a fluorophosphate glass [146]. It also shows the theoretical emission cross section $\sigma_{e}^{\text{theo}}(\lambda)$ calculated from the MC relation Eq. (4.57) and the experimental cross section $\sigma_{a}^{\text{exp}}(\lambda)$; it uses our earlier method to determine the Stark spacings ($\delta E_1$, $\delta E_2$). Figure 4.19 shows that, for this type of glass, a remarkable agreement exists between the measured and predicted emission cross sections, both in relative peak values and spectral line shapes.

To further illustrate the MC method, we consider three basic types of Er-doped fibers: germanosilicate (Type I), and germano-aluminosilicate (Types II and III); the alumina concentration is greatest in Type III.

The experimental absorption and emission line shapes (normalized to unity) at 1.5 $\mu$m for the three types are shown in Figures 4.20, 4.21, and 4.22, respectively. Each of the six spectra were fitted with Gaussian line shapes $i$ of center wavelength $l_i$, FWHM $d_i$ and peak values $a_i$, respectively, which are also plotted in the figures. The values of parameters ($l_i = \lambda_i$, $d_i = \Delta\lambda_i$, $a_i$) for each of these line shapes are also

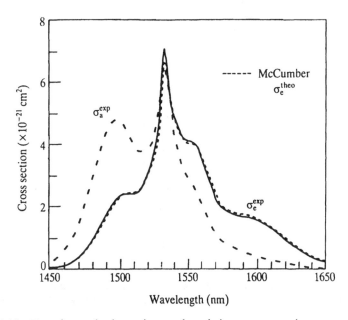

**FIGURE 4.19** Experimental absorption and emission cross section spectra of 1.5 $\mu$m transition of $Er^{3+}$ in a fluorophosphate glass. The emission cross section calculated from the absorption spectrum with McCumber theory is also shown. From [146], reprinted with permission from the Optical Society of America, 1991.

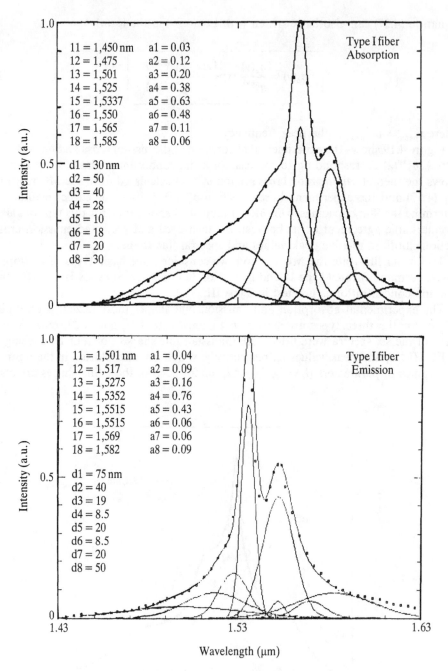

**FIGURE 4.20** Experimental absorption and emission cross section line shapes of 1.5 μm transition of $Er^{3+}$ in germanosilicate glass (Type I). A best fit with eight Gaussian functions (full lines) with peak wavelength ($l_i$), peak intensity ($a_i$), and FWHM ($d_i$) is also shown.

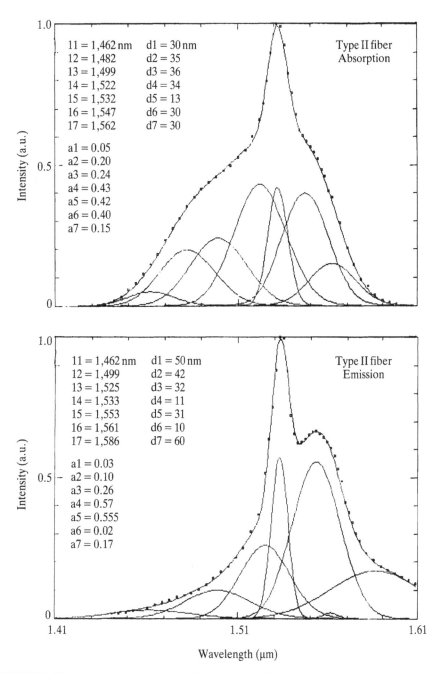

**FIGURE 4.21** Experimental absorption and emission cross section line shapes of 1.5 $\mu$m transition of $Er^{3+}$ in alumino-germanosilicate glass (Type II). A best fit with seven Gaussian functions (full lines) with peak wavelength ($l_i$), peak intensity ($a_i$), and FWHM ($d_i$) is also shown.

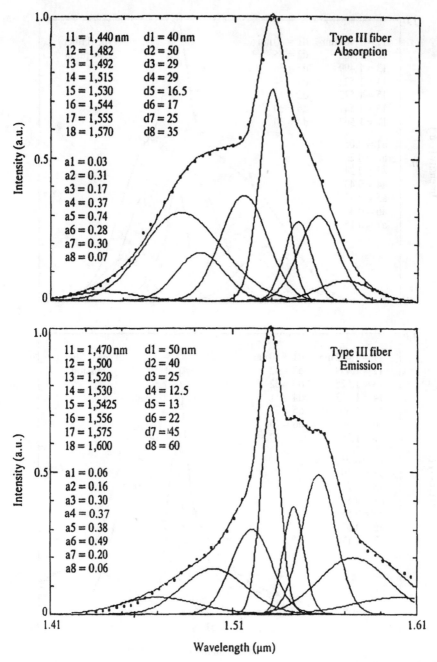

**FIGURE 4.22** Experimental absorption and emission cross section line shapes of 1.5 μm transition of $Er^{3+}$ in alumino-germanosilicate glass (Type III). A best fit with eight Gaussian functions (full lines) with peak wavelength ($l_i$), peak intensity ($a_i$), and FWHM ($d_i$) is also shown.

indicated. The spectra can then be numerically generated for computation purposes through the fitting formula:

$$I(\lambda) = \sum_i a_i \exp\left\{-4\log 2\frac{(\lambda - \lambda_i)^2}{\Delta\lambda_i^2}\right\} \tag{4.72}$$

Such curve-fitting is not unique. More importantly, the Gaussian line shapes of parameters $(l_i, d_i, a_i)$ shown in Figures 4.20, 4.21, and 4.22 represent only fitting functions, not individual Stark transitions. While the overall line shapes are very sensitive to changes in parameters $(l_i, d_i, a_i)$ for some lines $i$, the decomposition is arbitrary. The fitting can always be subjectively improved by small perturbations of these parameters or by adding more lines.

Equation (4.71) can then be used to calculate, for the three fiber types (I, II, III), the emission cross sections from the absorption cross section. The phenomenological or experimental peak ratios are $\eta^{\text{peak}} = 0.84$ for Type I [142], and $\eta^{\text{peak}} = 0.90$ for Types II and III [143], and the peak wavelengths are $\lambda_{\text{peak}} = 1.535\,\mu\text{m}$ (Type I), $\lambda_{\text{peak}} = 1.533\,\mu\text{m}$ (Type II), and $\lambda_{\text{peak}} = 1.531\,\mu\text{m}$ (Type III). The result of this calculation is shown in Figure 4.23, along with the measured cross section spectra, all intensities being normalized in each case to $\sigma_a(\lambda_{\text{peak}})$.

Figure 4.23 shows that a fair agreement exists between the predicted and measured fluorescence line shapes. In particular, a mirror effect between absorption and emission cross section spectra is characteristic of multilevel laser systems. It occurs with Stark split systems, such as RE:glass [6], and phonon terminated systems, such as transition metal fluorides [147]. This mirror image effect is nearly symmetrical with organic dye laser systems, as their vibrational–rotational energy levels are very closely spaced in comparison to RE:glass, [78]–[150]. In RE:glass, the symmetry between absorption and emission line shapes is only relative, due to the comparatively low degeneracy, the difference in Boltzmann distributions within the two level manifolds, the accidental degeneracies of Stark transitions, and the weighting effect of inhomogeneous broadening.

Figure 4.23 shows that the long wavelength sides of the predicted fluorescence line shapes are seen in all three cases to be 13% to 25% higher than the measured line shapes. A possible cause for this error is the uncertainty in the absorption spectrum measurements (e.g., peak wavelength, tails, background loss, coupling to spectrometer), which, with single-mode fibers, are difficult to reproduce accurately. The same type of difference is observed if, on the other hand, absorption spectra are calculated from fluorescence spectra through the same method. In this case, the accuracy of the prediction should be improved, since fluorescence spectra usually contain more reproducible features, but there is no basis for absolute comparison of the resulting absorption spectrum with experiment. These observations suggest that the accuracy of the MC method should be tested more effectively with bulk glass samples, as in [145], where reliable experimental measurements are somewhat easier to achieve.

When the MC method is applied to the three fiber types (I, II, III) for the prediction of the peak cross section ratios $\eta^{\text{peak}}$ using Eq. (4.69) and average Stark level separations $(\delta E_1, \delta E_2)$, the results are also different from experimental values. The predicted ratios are in the range $\eta^{\text{peak}} = 1.17$–$1.45$, somewhat overestimating the experimental values $\eta^{\text{peak}} = 0.84$–$0.9$. But experimental values for aluminosilicate fibers $\eta^{\text{peak}} \approx 1.0$ were

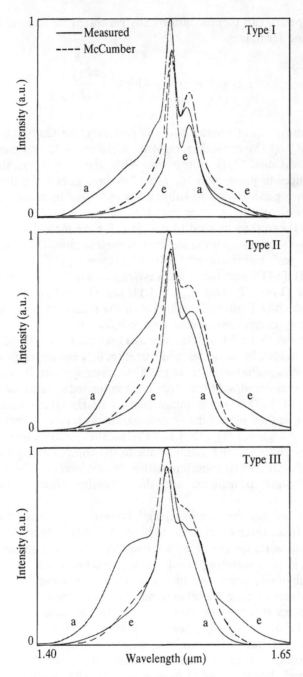

**FIGURE 4.23** Measured absorption and emission cross sections for Types I, II, and III, normalized to the peak absorption values $\sigma_a(\lambda_{peak})$. The dashed curves represent the emission cross sections calculated from the absorption spectra through the McCumber relation, Eq. (4.71).

also reported [151]–[153], which indicates that the difference between experimental and theoretical $\eta^{\text{peak}}$ may not be so important. The major source of uncertainty in the MC method is in the approximation of average Stark spacings ($\delta E_1$, $\delta E_2$) deduced from half-line width measurements. This approximation, apparently quite effective for some glasses [146]), seems too coarse for the three types (I, II, III) considered here and suggests the exact Stark level energies should be used for a more accurate prediction of $\eta^{\text{peak}}$. Some of these energies were determined for Type III fiber (Figure 4.16), but there is no information available regarding the exact number of levels and transitions involved in the absorption or fluorescence line shapes. if the MC method is carried out using the average spacings measured from the Stark levels in Figure 4.16 (i.e., $\delta E_1 \approx 69.2 \, \text{cm}^{-1}$, $\delta E_2 \approx 76.6 \, \text{cm}^{-1}$), instead of the spacings estimated from the half-line widths in Figure 4.22 (i.e., $\delta E_1 \approx 56.8 \, \text{cm}^{-1}$, $\delta E_2 \approx 70.7 \, \text{cm}^{-1}$), the predicted value for the peak ratio becomes $\eta^{\text{peak}} = 1.08$, which is closer to the experimental value ($\eta^{\text{peak}} = 0.9$–$1.0$). These results indicate that the accuracy of the MC method in the prediction of cross sections and relative ratio is sensitive to both experimental errors in spectral line shapes and estimation of the average Stark spacings. But its predictions are qualitatively close to the observed experimental data. There are several methods for the experimental determination of the peak ratio $\eta^{\text{peak}}$ in Er-doped fibers (EDF).

### Cutback method

The cutback method compares the small-signal gain $g_s$ and loss coefficients $\alpha_s$ of the EDF, as measured through a fiber cutback technique (C. R. Giles et al., [154] and [155]). For the gain coefficient measurement, several conditions must be achieved: (1) longitudinally uniform pump distribution, obtained in the high pump and small-signal regimes (i.e., $q \gg 1$, $p \ll 1$); (2) the EDF is pumped as a three-level system, which, along with condition (1), produces full medium inversion; (3) the net signal gain $G$ is relatively small (e.g., $G = 5$–$10 \, \text{dB}$), so that amplified spontaneous emission (ASE) noise and oscillation effects due to fiber-end reflexions in cutback measurements are negligible. From Condition 1, the gain is also unsaturated. A fourth condition is that the glass composition should be uniform across the doped fiber core, independent of the Er-doping distribution.

The absorption measurement can be made using a white light source as the input signal when the pump is turned off. The small-signal gain measurement should use a single-frequency laser, e.g., distributed feedback laser (DFB), with low intensity noise when the pump is turned on. Finally, the pump and the signal should be launched into the same propagation directions. The gain and absorption coefficients $g_s = \log(P_i/P_{i-1})/L_i$, $\alpha_s = \log(P_{i-1}/P_i)/L_i$ can then be determined by cutting small lengths $L_i$ of fiber at the output end and measuring the corresponding ouput signals $P_i$. The procedure must be repeated until the measured values show good reproducibility and consistency. For short fiber lengths, e.g., $L < 2 \, \text{m}$, care must be taken to strip out signal cladding modes, otherwise they can introduce fairly large errors in the output power measurements. Using results derived in Chapter 1, we show in the following how the cross section ratio $\eta^{\text{peak}}$ can be determined from these measurements.

In the unsaturated and ASE noise-free regimes, the small-signal net gain coefficient at fiber coordinate $z$ is $d[\log(p)]/dz$, which can be expressed from Eq. (1.54), with

$p_0 \ll p \ll q$ as:

$$\frac{d\log(p)}{dz} = \rho_0 \sigma_a(\lambda_s) \frac{2}{\omega_s^2} \int_S \frac{\rho(r)}{\rho_0} \psi_s(r) \frac{\eta_s q \psi_p(r) - 1}{q \psi_p(r) + 1} r \, dr \tag{4.73}$$

In this definition, the pump and signal mode envelopes $\psi_{p,s}(r)$ have unity peak. In the high pump regime ($q \gg 1$), $d[\log(p)]/dz$ is equal to the small-signal gain coefficient $g_s$, and Eq. (4.73) takes the form:

$$g_s = \frac{d\log(p)}{dz} = \rho_0 \sigma_e(\lambda_s) \frac{2}{\omega_s^2} \int_S \frac{\rho(r)}{\rho_0} \psi_s(r) r \, dr \tag{4.74}$$

using the relation $\sigma_e(v_s) = \eta_s \sigma_a(v_s)$. On the other hand, in the absence of pumping ($q = 0$), $d[\log(p)]/dz$ is equal to the signal absorption coefficient $-\alpha_s$, and Eq. (4.73) takes the form:

$$\alpha_s = -\frac{d\log(p)}{dz} = \rho_0 \sigma_a(\lambda_s) \frac{2}{\omega_s^2} \int_S \frac{\rho(r)}{\rho_0} \psi_s(r) r \, dr \tag{4.75}$$

Comparing Eqs. (4.74) and (4.75), we obtain at the peak wavelength $\lambda_{peak}$ the ratio:

$$\frac{g_s}{\alpha_s}(\lambda_{peak}) = \frac{\sigma_e(\lambda_{peak})}{\sigma_a(\lambda_{peak})} = \eta^{peak} \tag{4.76}$$

which is independent of the mode and Er-doping distributions. In Eq. (4.73), we implicitly assume that the cross sections $\sigma_{a,e}(\lambda_s)$ are uniform across the fiber core, meaning that the glass composition is uniform. If this is not the case, due to possible diffusion/migration effects of the glass codopants during the fiber fabrication, the ratio of gain to loss coefficients is, instead of Eq. (4.76):

$$\frac{g_s}{\alpha_s}(\lambda_{peak}) = \frac{\int_S \rho(r) \sigma_e(\lambda_{peak}, r) \psi_s(r) r \, dr}{\int_S \rho(r) \sigma_a(\lambda_{peak}, r) \psi_s(r) r \, dr} = \frac{\langle \sigma_e(\lambda_{peak}) \rangle_{r,\psi}}{\langle \sigma_a(\lambda_{peak}) \rangle_{r,\psi}} \tag{4.77}$$

where the brackets indicate an averaging effect due to Er-doping concentration and signal mode distribution.

The difference between Eqs. (4.76) and (4.77), and the possible source of error introduced by glass nonuniformity, makes it important to characterize the concentration profiles of all codopants present in the glass (e.g., $Er^{3+}$, $GeO_2$, $Al_2O_3$). Measurement of glass codopant concentrations can be made in the fiber preform; this is easier than in the fiber itself but the measured values are only indicative. Indeed, large differences between cross section line shapes measured in fiber and preforms are sometimes observed [156]. These differences can be attributed to changes in glass network structure that occur during preform collapse and fiberization, which are produced by various effects of phase transitions, codopant migration or diffusion, and rare earth site rearrangements. Preform/fiber spectroscopic comparisons can be found in the study by K. Dybdal et al. [156]. On the other hand, measurements of codopant profiles show that the core glass composition is generally not uniform (W. L. barnes et al. [142]); the concentration of alumina ($Al_2O_3$) codopant can show a 50% drop from the center to the edge of the core, which, according to Figure 4.17 could be responsible for a radial nonuniformity of cross sections. The results of the cutback method are summarized later on, along with those obtained by other methods. The cutback method is not the only way to determine the absorption and gain coefficients; there are also nondestructive methods such as OTDR (optical time domain reflectometry) and fiber macrobending.

**Net gain–loss method**

Another method for the determination of the cross section ratio $\eta^{peak}$ is based on the measurement of the EDF net gain and loss (W. L. Barnes et al., [142]). From Eq. (4.73), the net fiber gain $G$ at length $L$ is given by:

$$\log G(q, L) = \int_0^L \frac{d \log[p(q, z)]}{dz} dz = \rho_0 \sigma_a(\lambda_s) \int_0^L \hat{\Gamma}(q, z) dz \qquad (4.78)$$

where $\hat{\Gamma}(q, z)$ is the pump power dependent overlap–inversion factor previously defined in Eq. (1.59). For a 3-level system, it takes the form:

$$\hat{\Gamma}(q, z) = 2\pi \int_s \frac{\rho(r)}{\rho_0} \bar{\psi}_s(r) \frac{\eta_s q(z)\psi_p(r) - 1}{q(z)\psi_p(r) + 1} r \, dr \qquad (4.79)$$

In this definition, $\bar{\psi}_s(r)$ represents a normalized signal mode envelope. In the unpumped regime ($q = 0$), $\hat{\Gamma}(q, z)$ reduces to the overlap factor:

$$\hat{\Gamma}(0, z) = -2\pi \int_s \frac{\rho(r)}{\rho_0} \bar{\psi}_s(r) r \, dr \qquad (4.80)$$

while in the high pump regime ($q \gg 1$), $\hat{\Gamma}(q, z)$ reduces to $\hat{\Gamma}(\infty, z) = -\eta_s \hat{\Gamma}(0, z)$ and, from Eq. (4.78) the net gain reaches the maximum value:

$$\log G_{max} \equiv \log G(\infty, L) = \rho_0 \sigma_e(\lambda_s)|\hat{\Gamma}(0, z)|L \qquad (4.81)$$

From Eqs. (4.78)–(4.80), we obtain:

$$\log G(q, L) = \log(G_{max}) \frac{1}{\eta_s L} \int_0^L \frac{\hat{\Gamma}(q, z)}{|\hat{\Gamma}(0, z)|} dz \qquad (4.82)$$

If we consider now short fiber lengths, for which the pump power variation with coordinate $z$ can be neglected, and the case of confined Er-doping, for which the power variation in the doped region can be neglected, Eq. (4.82) with Eqs. (4.79) and

(4.80) reduces to:

$$x(q) = \frac{\log G(q, L)}{\log(G_{\max})} = \frac{q - 1/\eta_s}{q + 1} \tag{4.83}$$

As both maximum gain $G_{\max}$ and unpumped fiber loss $G_{\min} = G(0, L)$ can be measured experimentally, we obtain from Eq. (4.83) the cross section ratio:

$$\eta_s = -\frac{\log(G_{\min})}{\log(G_{\max})} \tag{4.84}$$

which can be measured at the peak wavelength to give $\eta^{peak}$. But the maximum gain $G_{\max}$ can be reached only asymptotically, and some uncertainty remains in its absolute value [142]. To alleviate this difficulty, we measure $x(q)$ as a function of the pump power $\tilde{P}_p = qP_p^{trans}$, where $P_p^{trans}$ is the pump power for which transparency is observed. The unknown values $G_{\max}$ and $\eta_s$ are then determined from a best fit of the curve $x(q)$. Several measurements should be made with different lengths, showing a consistent and reproducible result for $\eta_s$ [142]. The experimental results obtained through this technique are summarized below, along with those obtained from other methods. We consider now various experimental methods that can be used for the accurate determination of *absolute* cross sections in Er:glass fibers.

In the mid-1960s, several methods were proposed for the evaluation of cross sections in bulk Nd:glass, based on measurements of absorption, energy extraction, oscillator gain threshold or the FL relation [134], [136], and [157]–[161]. The absolute accuracy of these methods was estimated to be within 7% at best to within 20% at worst [134]. For methods other than FL, the sources of inaccuracy are, most generally, the uncertainties in rare earth ion density, in measurements of oscillator output energy, in the evaluation of the fraction of lasing ions, and in the detector response. In the case of RE-doped fibers, additional uncertainties are introduced by the determination of mode intensity distribution, RE-doping and glass composition distributions, background loss, power coupling efficiencies, and power and inversion changes along the length; all these parameters are easier to measure with accuracy in bulk materials. The cross sections are preferably determined from the fiber and not from its preform, as discrepancies between the two measurements can be observed [156]. Determination of Er:glass cross sections from fibers is subject to many sources of experimental error; new experimental methods of improved accuracy could be developed in the future. Now we consider various method used for the $^4I_{15/2}$–$^4I_{13/2}$ signal transition and for the various pump transitions in Er-doped fibers.

### Fuchtbauer–Ladenburg method

This method applies the FL formulae in Eqs. (4.50)–(4.53) using the absorption and emission spectral line shapes $I_{a,e}(\lambda)$ determined experimentally (such as shown in Figures 4.20, 4.21, and 4.22) and the observed fluorescence lifetime $\tau$. This method, widely reported in the literature for Nd:glass was also used by several authors in early work on EDFs, [82], [156], and [161]–[163]. Some variations in the early results can be attributed to differences in experimental values for the fluorescence lifetime, to measurement errors in the evanescent tail spectra, and to the inclusion or not of the degeneracy factor $g_2/g_1$ in the FL relation for absorption. The FL method predicts a cross section ratio $\eta^{\text{peak}}$ largely overestimated (i.e., by up to 50%). But the absolute value for the absorption cross section predicted by the FL relation is nearly accurate, therefore it remains a parameter of indicative value.

### Direct absorption cross section measurement method

The direct measurement of the absorption cross section from the fiber (M. P. Singh et al., [164]) requires full knowledge of Er-density $\rho_0$ and Er-doping radial distribution $\rho(r)/\rho_0$ in the core, as well as the intensity distribution of the signal mode $\psi_s(r)$. The fiber absorption coefficient $\alpha_s(\lambda_s)$ can be measured experimentally with good accuracy (e.g., $<1\%$) if the Er concentration or the absorption coefficient is large (e.g., $\alpha_s > 10\,\text{dB/m}$), but the accuracy is reduced with low concentration fibers (e.g., $\alpha_s < 2\,\text{dB/m}$), as power measurements within 0.05 dB error (1%) are more difficult to achieve, and the effect of background loss (not related to $\text{Er}^{3+}$ absorption) could introduce additional error. The fiber absorption coefficient can be measured by fiber cutback. Then the absorption cross section is given from Eq. (4.75) as:

$$\sigma_a(\lambda_s) = \frac{\alpha_s(\lambda_s)}{\rho_0 \dfrac{2}{\omega_s^2} \displaystyle\int_S \frac{\rho(r)}{\rho_0} \psi_s(r) r\, dr} \tag{4.85}$$

This measurement can be made for any transition in the EDF, including one that corresponds to the pump. But for the pump cross section evaluation, we must take care to ensure that all the pump light is propagating in the fundamental mode $\text{LP}_{01}$. For multimode excitation, the overlap integral in Eq. (4.85) is difficult to evaluate, as in experimental conditions the intermodal excitation and its variation with length, as well as the mode radial distributions and relative propagation losses are generally unknown. Multimode excitation makes the relative intensities of the pump absorption bands sensitive to the waveguide parameters ($V$ number, cutoff wavelength) and we need to be careful when interpreting these intensities for pump cross section evaluation.

The emission cross section could also be measured by the same method, using the gain coefficient rather than the absorption coefficient, and Eq. (4.74) instead of Eq. (4.75). But this method amounts to comparing absorption and gain coefficients of the same fiber. Knowledge of the peak absorption cross section $\sigma_a(\lambda_{\text{peak}})$; the coefficient ratio $\eta^{\text{peak}}$, Eq. (4.76); and the fluorescence line shape $I_e(\lambda)$ is sufficient to determine the absolute emission cross section spectrum $\sigma_a(\lambda)$, since these parameters are related through:

$$\sigma_e(\lambda) = \eta^{\text{peak}} \sigma_a(\lambda_{\text{peak}}) \frac{I_e(\lambda)}{I_e^{\text{peak}}} \tag{4.86}$$

For low concentration fibers with long lengths, e.g., $L > 50\,\text{m}$, background loss (or excess loss) which, as in the case of absorption, introduces an additional error in the evaluation of the gain coefficient, can be calibrated by spectral loss measurements away from the 1.5 $\mu$m resonance [165]. With long fibers, OTDR can also be used in both passive and active modes for the simultaneous determination of both absorption and gain coefficients, [166] and [167]. OTDR has two advantages: it is nondestructive and comparatively rapid to implement. Finally, the gain coefficient can also be measured nondestructively by macrobend measurements, where the pump and signal powers can be measured along the fiber by power extraction through macrobending [168]. Further comparative work is needed to assess the relative accuracy of these methods for the determination of the gain coefficient. Experimental results obtained from the direct cross section measurement method are discussed below, along with other data.

### Saturated fluorescence method

Another method for the evaluation of absolute absorption cross-sections uses a measurement of saturated fluorescence (W. L. Barnes et al., [142]). In the absence of an input signal, the output amplified spontaneous emission power, or ASE power, of a length $L$ of EDF is given by integration of Eq. (1.36):

$$
P_{ASE}(L, \lambda_s) = 2P_0 \frac{\displaystyle\int_S \eta_s N_2(r)\bar{\psi}_s(r)r\,dr}{\displaystyle\int_S [\eta_s N_2(r) - N_1(r)]\bar{\psi}_s(r)r\,dr}
$$

$$
\times \left\{ \exp\left\{ 2\pi\sigma_a(\lambda_s)L \int_S [\eta_s N_2(r) - N_1(r)]\bar{\psi}_s(r)r\,dr \right\} - 1 \right\} \quad (4.87)
$$

The fiber length is assumed short, so that longitudinal variations of the populations $N_1$ and $N_2$ are neglected. Assuming that the pump power is high enough so that $N_1 \ll N_2$, Eq. (4.87) can be put into the form:

$$
P_{ASE}(L, \lambda_s) = P_{ASE}^0(\lambda_s) \int_S N_2(r)\bar{\psi}_s(r)r\,dr \quad (4.88)
$$

where $P_{ASE}^0(\lambda_s) = 4\pi L P_0 \sigma_e(\lambda_s)$ is a power spectrum proportional to the emission cross section. Substituting in Eq. (4.88) the expression for the upper-level population $N_2$, Eq. (1.50), while assuming the unsaturated regime ($p \ll q$), we obtain the pump power dependent spectrum:

$$
P_{ASE}(\lambda_s, L, P_p) = P_{ASE}^0(\lambda_s) \int_S \rho(r) \frac{P_p \psi_p(r)}{P_{sat}(\lambda_p) + P_p \psi_p(r)} \bar{\psi}_s(r)r\,dr \quad (4.89)
$$

Equation (4.89) shows that for a very high pump power (i.e., $P_p \gg P_{sat}(\lambda_p)$) the output ASE spectrum, which at low gains is also called the fluorescence spectrum, is proportional to the emission cross section. Thus, the fluorescence spectrum measurement under high pump and with a short fiber length corresponds to a direct measurement of the emission cross section line shape $I_e(\lambda_s)$ (E. Desurvire and J. R. Simpson [82]).

In the saturated fluorescence method [142], the fluorescence spectrum defined in Eq. (4.89) is measured as a function of input pump power $P_p$. For a very high pump power, the output ASE at a given wavelength $\lambda_s$ reaches an asymptotical value $P_{ASE}^{max}$

defined from Eq. (4.89) as:

$$P_{\text{ASE}}^{\max} \equiv P_{\text{ASE}}(\lambda_s, L, \infty) = P_{\text{ASE}}^0(\lambda_s) \int_S \rho(r)\bar{\psi}_s(r)r\, dr \qquad (4.90)$$

If we define then $P_p^*$ as the pump power that generates one half the maximum $P_{\text{ASE}}^{\max}$, we obtain from Eqs. (4.89) and (4.90):

$$\frac{\displaystyle\int_S \frac{\rho(r)}{\rho_0} \frac{P_p^* \psi_p(r)}{P_{\text{sat}}(\lambda_p) + P_p^* \psi_p(r)} \bar{\psi}_s(r)r\, dr}{\displaystyle\int_S \frac{\rho(r)}{\rho_0} \bar{\psi}_s(r)r\, dr} = \frac{1}{2} \qquad (4.91)$$

If the Er-doping distribution is confined near the center of the core so that transverse variations of the pump power can be neglected, we obtain then, from Eqs. (4.91) and (1.45):

$$P_p^* = P_{\text{sat}}(\lambda_p) = \frac{h\nu_p \pi \omega_p^2}{\sigma_a(\lambda_p)\tau} \qquad (4.92)$$

where $\omega_p$ is the experimental pump power mode size, Eq. (1.31), and $\tau$ is the observed fluorescence lifetime. Thus, the pump absorption cross section is experimentally determined from the measurements $(P_p^*, \omega_p, \tau)$ as:

$$\sigma_a(\lambda_p) = \frac{h\nu_p \pi \omega_p^2}{P_p^* \tau} \qquad (4.93)$$

In Eqs. (4.92) and (4.93), it was assumed that the EDF is pumped as a three-level system (for which $\eta_p(\lambda_p) = 0$), but this condition is not truly necessary. The advantage of three-level pumping is that the additional uncertainty in the factor $\eta_p(\lambda_p)$ is removed. In the case where the Er-doping is not confined, and the pump power variation across the core cannot be neglected, Eq. (4.91) can still be solved numerically for $P_{\text{sat}}(\lambda_p)$. This computation requires $P_p^*$, the pump mode envelope $\psi_p(r)$ and the relative Er-doping distribution $\rho(r)/\rho_0$, which can be measured experimentally.

The absolute absorption cross section $\sigma_a(\lambda_s)$ corresponding to the 1.5 $\mu$m transition at $\lambda_s$ can be determined from the experimental values for $\sigma_a(\lambda_p)$ and the ratio $\alpha_s/\alpha_p$ of absorption coefficients at pump and signal wavelengths. Indeed, we obtain from eq. (4.75) expressed for both wavelengths:

$$\sigma_a(\lambda_s) = \sigma_a(\lambda_p) \frac{\alpha_s}{\alpha_p} \frac{\omega_s^2}{\omega_p^2} \frac{\displaystyle\int_S \frac{\rho(r)}{\rho_0} \psi_p(r)r\, dr}{\displaystyle\int_S \frac{\rho(r)}{\rho_0} \psi_s(r)r\, dr} \qquad (4.94)$$

This method requires experimental knowledge of both pump and signal mode envelopes and knowledge of the relative Er-doping distribution. Equations (4.91),

(4.93), and (4.94) do not depend on the absolute or peak erbium concentration $\rho_0$ so only the relative distribution need be known. For the pump absorption coefficient measurement, we need to ensure that all the pump power is propagating in the fundamental mode, as multimode pump excitation could produce large errors in the evaluation of this parameter.

Table 4.1 compares measurements of peak absolute and relative cross sections for the $^4I_{13/2}$–$^4I_{15/2}$ laser transition in Er-doped fibers (I, II, III) as made through the various methods previously described, with theoretical values given by the FL relation. The exact glass compositions are generally not known for results reported with alumino-silicate fibers, so the data were put under Type II or Type III, according to the resemblance of experimental spectra with Figures 4.21 and 4.22.

The table shows that for the three fiber types the measured values for the peak cross section ratios $\eta^{peak}$ agree within 10%. For the absorption cross sections, the experimental values measured through the saturated fluorescence (SF) method are within 10–20% of the theoretical values predicted by the FL relation. The experimental emission cross sections shown in parentheses are calculated from these results and from the measurement of $\eta^{peak}$. The emission cross sections predicted by FL are overestimated by 20–30% in comparison to the experimental values. Finally, the absorption cross section obtained by SF for Type III is 35% higher than the value directly measured. This could be attributed to small differences in the glass composition (as being not revealed by the absorption line shapes) and to the accumulated experimental error associated with the measurements of Er-doping distribution, peak concentration, mode envelopes, and absorption coefficients. Accurate determination of absolute cross sections in Er-doped fibers remains a challenge, and new investigation methods are yet to be developed.

From the experimental data obtained by the SF method and reported in [142] for various fibers, the peak cross sections are observed to decrease in the composition order $SiO_2$–$GeO_2$, $SiO_2$–$Al_2O_3$, $SiO_2$–$GeO_2$–$Al_2O_3$; the maximum decrease represents a factor 35–40% for both absorption and emission. The decrease in peak cross section can be attributed to the broadening effect of homogeneous and inhomogeneous line widths (including Stark splitting), as described in Section 4.4. From the Judd–Ofelt theory (Section 4.3), the oscillator strength for the $^4I_{13/2}$–$^4I_{15/2}$ transition also depends on the glass composition through the JO parameters. Thus, the observed decrease in peak cross section cannot be exclusively attributed to line width broadening. The phenomenological oscillator strength $S_{a,e}$ corresponding to absorption or emission can be expressed as:

$$S_{a,e} = \int \sigma_{a,e}(\lambda)\,d\lambda = \sigma_{a,e}^{peak}\,\Delta\lambda_{a,e}^{eff} \tag{4.95}$$

We can also define the dimensionless transition parameters $T_{a,e}$ through:

$$T_{a,e} \equiv \frac{8\pi n^2 c\tau}{\lambda^4}\,S_{a,e} = \frac{8\pi n^2 c\tau}{\lambda^4}\,\sigma_{a,e}^{peak}\,\Delta\lambda_{a,e}^{eff} \tag{4.96}$$

If the FL relation were to give the exact peak cross sections, the dimensionless transition parameters would be $T_a = g_2/g_1$ and $T_e = 1$). To evaluate how the transition

**TABLE 4.1** Experimental and theoretical cross sections for the $^4I_{13/2}$–$^4I_{15/2}$ transition near $\lambda = 1.5\ \mu$m in Er-doped fibers. (Experimental methods: DI = direct, CB = cut-back, GL = gain-loss, SF = saturated fluorescence; Theoretical method: FL = Fuchtbauer–Ladenburg relation; [*] after Figures 4.20–4.22 with $\tau$ from [142])

| | Experimental cross sections | | | | | Theoretical cross sections (FL) | |
| | Absorption $\sigma_a^{\text{peak}}(10^{-25}\ \text{m}^2)$ | Emission $\sigma_e^{\text{peak}}(10^{-25}\ \text{m}^2)$ | Peak ratio $\eta^{\text{peak}}$ | Method | Ref. | Absorption $\sigma_a^{\text{peak}}(10^{-25}\ \text{m}^2)$ | Emission $\sigma_e^{\text{peak}}(10^{-25}\ \text{m}^2)$ |
| Fiber | | | | | | | |
|---|---|---|---|---|---|---|---|
| Type I | | | | | | | |
| | | | 0.84 | GL | [142] | | |
| | | | 0.90 | CB | [143] | | |
| | | | 0.95 | GL | [152] | | |
| | | | | FL | [142] | 6.2 | 7.9 |
| | | | | FL | [*] | 6.1 | 8.7 |
| | | | | FL | [154] | | 7.5 |
| | | | 0.85 | SF | [142] | | |
| | 7.9 ± 0.2 | (6.7 ± 0.3) | | | | | |
| Type II | | | | | | | |
| | | | 0.90 | CB | [143] | | |
| | | | 0.9–1.0 | GL | [154] | | |
| | | | 0.95 | GL | [152] | | |
| | | | | FL | [154] | | 6.5 |
| | | | | FL | [*] | 5.8 | 6.5 |
| Type III | | | | | | | |
| | | | 0.90 | CB | [143] | | |
| | | | 0.95 | CB | [154] | | |
| | | | 1.0 | GL | [152] | | |
| | | | | FL | [154] | 4.9 | 5.8 |
| | | | | FL | [142] | 5.3 | 5.8 |
| | | | | FL | [*] | | 5.8 |
| | 6.44 | | | DI | [164] | | |
| | 4.7 ± 0.8 | (4.4 ± 1) | 0.93 | SF | [142] | | |

strength changes with glass composition, we can use the experimental data of [142] and calculate the parameters $T_{a,e}$. For Type I, we find $T_a = 1.27 \pm 0.1$ and $T_e = 0.75 \pm 0.05$; for Type III, $T_a = 0.95 \pm 0.21$ and $T_e = 0.76 \pm 0.19$. This indicates that, within the experimental error margins shown, the dimensionless parameters $T_{a,e}$ are not much affected by the glass composition, even if important changes are observed for the peak cross sections, effective line widths, and fluorescence lifetime.

Some confusion and inconsistency exists in the literature regarding cross section data. Indeed, some studies qualify as experimental cross sections values that are actually calculated from the FL relation and calibrated relative to each other by the measured factor $\eta^{peak}$, [152] and [156]. In contrast, truly experimental values are those obtained by the methods previously described, and reported in [142] and [164]. As either absorption or emission cross sections calculated by FL can be used as the absolute reference, and as measured fluorescence lifetimes also vary by 10–20%, there exists a wide range of values for the peak cross sections reported in the literature. Figure 4.24 shows plots of the absorption and emission cross sections of Type I and Type III fibers, whose line shapes correspond to Figures 4.20 and 4.22 and whose peak values correspond to experimental data in [142]. Reference [127] is a comprehensive review of data relative to absolute absorption and emission cross sections of the pump and signal transitions in various Er:glasses.

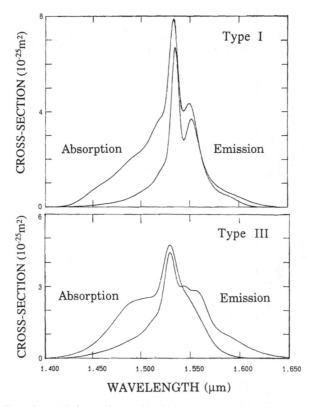

**FIGURE 4.24** Experimental absorption and emission cross-sections for germanosilicate (Type I) and alumino-germanosilicate (Type III) fibers.

Exact knowledge of absolute cross sections and peak Er concentration is not necessary for accurate EDFA modeling. Indeed, in most applications of interest, Er-doping is confined near the center of the fiber core, and transverse intensity variations can be neglected. Chapter 1 shows that in this case, the two required parameters for EDFA modeling are the fiber absorption and gain coefficients, related through $g_s = \eta_s \alpha_s$ and experimentally measurable with good accuracy. Indeed, the normalized rate equations for the pump ($q$) and the signal ($p$) are functions only of these two parameters, Eqs. (1.113)–(1.119). For theory/experiment comparisons, knowledge of the saturation powers at the pump and signal wavelengths, i.e., $P_{sat}(\lambda_p) = P_p/q$ and $P_{sat}(\lambda_s) = P_s/p$, is also necessary. These two parameters, defined by Eq. (1.45), can be determined experimentally through various methods, without knowledge of the absolute cross sections (Section 4.6).

We consider next the issue of *inhomogeneous broadening*. The experimental analysis of Er:glass as a laser system, which was described in the previous sections, has shown that the $^4I_{13/2}$–$^4I_{15/2}$ laser transition is subject to homogeneous and inhomogeneous line broadening. Section 4.4 shows that in aluminosilicate glasses, the two broadening mechanisms are of equal magnitude, while in germanosilicate glasses, inhomogeneity is twice as important than homogeneity.

It is generally accepted that the effect of inhomogeneity appears only in the saturation regime, i.e., when the signal stimulated emission rates $W_{12,21}$ become important in comparison to the pumping rate. While this view is fully accurate in the case of four-level laser systems, it is not applicable to the case of three-level Er:glass. The reason is that the pump and the signal transitions share the $^4I_{15/2}$ ground level. Thus, as $^4I_{15/2}$ is inhomogeneously broadened, a high pumping rate causes inhomogeneous saturation of the pump transition, which must reflect onto the signal absorption cross section. The reason for the cross section change is that saturation by the pump is site selective, which affects the weights of each rare earth site in the statistical averaging of the absorption lines. The effect is more important in the case of two-level pumping, where the pump and the signal transitions share both the $^4I_{15/2}$ ground level and the $^4I_{13/2}$ upper level. The reason why these effects are not generally accounted for in the EDFA modeling is that gain saturation is experimentally observed to be mainly homogeneous. But gain homogeneity in EDFAs is only an approximation; a theoretical description based upon this assumption is not truly comprehensive when the effect of saturation by either pump or signal becomes important.

Modeling of Er:glass as an inhomogeneous gain medium requires knowledge of homogeneous cross sections $\sigma_{a,e}^H(\lambda)$, which at room temperature, cannot be determined experimentally. This is because the spectra of the Stark transitions are smeared by the combined effects of inhomogeneous broadening, of thermal population in the Stark manifolds, of large homogeneous line width, and of accidental transition degeneracies. However, the information obtained at low temperature by SHB or FLN on the relative amounts of inhomogeneity and the homogeneous line width dependence with temperature makes it possible to model the homogeneous cross sections $\sigma_{a,e}^H(\lambda)$ at room temperature (E. Desurvire et al., [117]). This model is based upon an approximation similar to that used in the computation of cross sections in Doppler broadened gases (L. W. Casperson and A. Yariv, [169] and [170]). The assumption is that the distribution of the center frequency of a given laser transition is Gaussian. As a result, the inhomogeneous cross section can be calculated by statistical averaging, using this distribution [133]. In the case of RE:glass, there is no available information about the statistics of rare earth site distribution and transition wavelengths, which is more difficult to model than in a gas. A way to model such statistics is to assign a Boltzmann factor to each rare earth site configuration (degrees of freedom) according to its corresponding potential energy. But this model requires many unverified assumptions about the type of rare earth coordination and the crystalline characteristics of the rare earth site. Certain basic cases (such as $Yb^{3+}$ in silica [46]) are comparatively easier to describe, due to the reduced complexity of the glass network structure and the limited number of Stark transitions involved. In the case of Er:aluminosilicate glass, it is possible that such modeling will always remain speculative, because of the lack of accurate knowledge on the rare earth site coordination and statistics. For this reason, the model described in this section for the computation of homogeneous cross sections should be regarded as tentative, and subject to future possible improvements.

Inhomogeneous gain coefficients from the homogeneous cross sections are derived in Section 1.12. In particular, the homogeneous cross sections $\sigma_{a,e}^{H}(\lambda)$ are determined from the experimental (inhomogeneous) cross sections $\sigma_{a,e}^{I}(\lambda)$ through the inverse Fourier transform relation:

$$\sigma_{a,e}^{H}(\omega) = \mathscr{F}^{-1}\left[\exp\left(\frac{\Delta\omega_{inh}^{2}x^{2}}{16\log(2)}\right) \cdot \mathscr{F}\left[\sigma_{a,e}^{I}(\omega); x\right]; \omega\right] \tag{4.97}$$

where $\Delta\omega_{inh} = 2\pi c\Delta\lambda_{inh}/\lambda^{2}$ is the inhomogeneous bandwidth and $\Delta\lambda_{inh}$ is the inhomogeneous line width. Equation (4.97) assumes that the inhomogeneous distribution is Gaussian, Eq. (123), and that the inhomogeneous cross sections $\sigma_{a,e}^{I}(\lambda)$ are given by convolution integrals, Eqs. (1.205) and (1.206). The deconvolution expressed in Eq. (4.97) has a unique solution, or is meaningful, only when the functions $\sigma_{a,e}^{I}(\lambda)$ and their evanescent tails are well defined analytically. This is not the case with the experimental line shapes $I_{a,e}(\lambda) \approx \sigma_{a,e}^{I}(\lambda)$, but it is possible to fit these line shapes with a superposition of Gaussian functions, as shown in Figures 4.20, 4.21, and 4.22, and in Eq. (4.72). The advantage of using Gaussian functions for this fitting is that the Fourier and inverse Fourier transforms involved in Eq. (4.97) can be calculated analytically. The detailed calculation of Eq. (4.97) is made in Appendix M. In this model, the homogeneous absorption and emission cross sections $\sigma_{a,e}^{H}(\lambda)$ take the form:

$$\sigma_{a,e}^{H}(\lambda) = \sum_{i} a_{i}^{a,e} \frac{\Delta\lambda_{i}}{\sqrt{\Delta\lambda_{i}^{2} - \Delta\lambda_{inh}^{2}}} \exp\left\{-4\log 2 \frac{(\lambda - \lambda_{i})^{2}}{\Delta\lambda_{i}^{2} - \Delta\lambda_{inh}^{2}}\right\} \tag{4.98}$$

where the parameters $(\lambda_{i}, \Delta\lambda_{i}, a_{i})$ correspond to the best fits of experimental line shapes $I_{a,e}(\lambda) \approx \sigma_{a,e}^{I}(\lambda)$ defined in Eq. (4.72), and $\Delta\lambda_{inh}$ is the experimental inhomogeneous line width.

The homogeneous cross sections corresponding to Type III fiber ($\Delta\lambda_{inh} = 11.5$ nm) are shown in Figure 4.25, with the experimental cross sections $\sigma_{a,e}^{I}(\lambda)$ of Figure 4.24, for comparison. The homogeneous cross sections exhibit a detailed structure, which reflects the effect of individual contributions from Stark level transitions. This structure is not an artifact of the Gaussian fitting, but the result of the deconvolution of the experimental line shapes $\sigma_{a,e}^{I}(\lambda)$. Some error is involved in the fitting of smooth details in $\sigma_{a,e}^{I}(\lambda)$, to which the deconvolution operation is very sensitive.

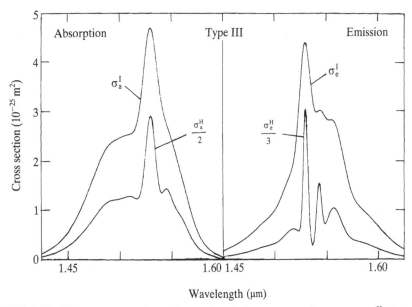

**FIGURE 4.25** Homogeneous absorption and emission cross section spectra $\sigma_{a,e}^{H}(\lambda)$ for Type III fiber, calculated analytically from Eq. (4.98), with corresponding experimental inhomogeneous cross sections $\sigma_{a,e}^{I}(\lambda)$.

Figure 4.25 also shows that, according to the model, the homogeneous line widths are different for each Stark transition, as well as between absorption and emission for a given transition. These homogeneous line width variations could be attributed to the strong spectral overlap existing between adjacent Stark transitions, as well as to the effect of wavelength degeneracies. But these homogeneous cross sections should be interpreted only as representing a simplified description, because of the limitations inherent to the model. The main limitation comes from the assumption that each site is characterized by the same absorption and emission spectra, the center frequencies of which are randomly distributed. As a consequence, it is also assumed that (1) inhomogeneous broadening is uniform with wavelength, i.e., all Stark transitions experience the same inhomogeneous broadening, and (2) the homogeneous line width of a given Stark transition is the same for each site. Experimental results obtained for instance in $Eu^{3+}$, have shown that site dependent variations of single Stark transition homogeneous line width can actually occur; the variation may be up to 50% or more, [85], [171], and [172]. Recent FLN measurements in aluminosilicate Er-doped fibers have also shown large site-to-site variations of homogeneous line width for the Stark transitions (A. M. Brancion et al. [173]). On the other hand, there is no experimental confirmation that the statistics of inhomogeneous line broadening and accidental degeneracies are Gaussian. The randomness of site coordination in the glass network, which generates inhomogeneous broadening, is also a function of potential energy, which has Boltzmann statistics. In the absence of a site coordination model, we cannot assess the validity of any inhomogeneous broadening theory. These observations indicate that while experimental data can provide some basic information (i.e., homogeneous and inhomogeneous line widths, Stark level positions, and site-to-site variations of these parameters), there is not enough information available at present for accurately modeling the inhomogeneous cross sections of the $^4I_{13/2}$–$^4I_{15/2}$ transition in Er:glass. The only possible model is a heuristic description, which could reproduce experimental data of inhomogeneous saturation with better accuracy than the homogeneous model.

A detailed comparison between experiment and homogeneous/inhomogeneous theoretical models is made in Chapter 5, along with further description of spectral gain saturation characteristics in EDFAs.

The temperature dependence of the Er:glass cross sections can be explained by two main effects: the homogeneous line width broadening (proportional to $T^{2-\varepsilon}$ for $T > 20$ K) and the Boltzmann thermal occupation in the Stark manifolds (proportional to $\exp(-\Delta E/k_B T)$). On the other hand, inhomogeneous broadening, which is an effect of the ligand or crystal field, is temperature insensitive [85].

Temperature changes also affect the multiphonon decay between adjacent levels and the intramanifold thermalization rates. In oxide glasses, for instance, the multiphonon rate corresponding to $p = 4$ phonons and an energy gap $\Delta E = 1000$ cm$^{-1}$) is observed to increase by about a 50% from 300 K to 600 K [80]. This result is in good agreement with Eq. (4.20), which predicts that the multiphonon rate follows the temperature dependence:

$$\frac{A_{ij}^{NR}(T, p)}{A_{ij}^{NR}(T', p)} = \left(\frac{n(T) + 1}{n(T') + 1}\right)^p \tag{4.99}$$

For energy levels with strong nonradiative decay, the changes in multiphonon rates affect the level lifetime. As the $^4I_{13/2}$–$^4I_{15/2}$ transition is nearly 100% radiative, the lifetime is not affected by this process. However, the relative change in thermalization rate within the Stark manifolds affect the effective lifetime corresponding to each Stark transition. This change can be evaluated from Eq. (4.99), with the parameter values $p = 1$, $\Delta E = 50$ cm$^{-1}$, to be about 50% for a temperature increase from 250 K to 350 K. This explains the temperature dependence of the $^4I_{13/2}$–$^4I_{15/2}$ transition lifetime observed in Er:glass, which was measured as about 0.01 ms/K between 77 K and 300 K, and 0.003 ms/K between 300 K and 800 K [142]. From these results, we can evaluate that the fluorescence lifetime varies by a factor 2% to 5% within the usual operating range $-40\,°C$ to $+60\,°C$. This temperature change therefore has little impact on the absolute value of the cross sections.

The dependence of cross sections, peak values, and line shapes with temperature can therefore be mainly attributed to changes in homogeneous line width and Boltzmann populations. Given the constant oscillator strength of a Stark transition, line width narrowing corresponds to a proportional increase of peak cross section, and conversely, line width broadening corresponds to decrease in cross section. On the other hand, each Stark transition (i.e., absorption or emission) is weighted by the Boltzmann population of the originating level (i.e., ground or upper level), which causes additional changes in peak cross sections and line shapes. The two effects are illustrated in Figure 4.14, which shows the fluorescence spectrum of Type III fiber recorded at three temperatures. With decreasing temperature, the main Stark transition near $\lambda = 1.53$ μm increases in intensity and narrows, while the contributions of other Stark transitions vanish. The changes in both absorption and emission cross sections corresponding to a $-196\,°C$ to $+85\,°C$ range are shown in Figure 4.26, for

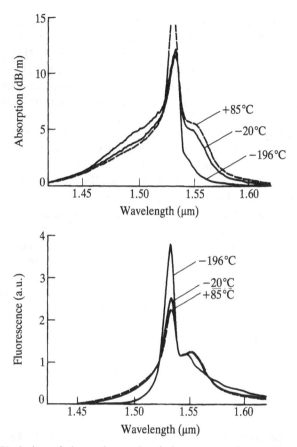

**FIGURE 4.26** Variation of absorption and emission cross section spectra with temperature. From [177], © 1991 IEEE.

a fiber of characteristics similar to Type II (N. Kagi et al. [177]). For the absorption, an increase of temperature results for the cross section in: (1) a decrease at short wavelengths and (2) an increase at long wavelengths. This effect is consistent with an increase of the ground level ($^4I_{15/2}$) Boltzmann population (inset in Figure 4.14). For the emission, the temperature increase causes an increase of the cross section at short wavelengths, contrary to the case of absorption. This fact is also consistent with the increase of Boltzmann population in the upper level ($^4I_{13/2}$). For both absorption and emission, the peak cross sections near $\lambda = 1.53\ \mu m$ decrease with temperature, which reflects the effect of line width broadening. The increase of EDFA gain observed at cryogenic temperatures (E. Desurvire et al. [174], C. A. Millar et al. [175]; and M. Shimizu et al. [176]) is therefore explained by a line width narrowing and a decrease of thermal population in the $^4I_{15/2}$ ground level. At low temperatures (i.e., $T \leqslant 77\,K$), the thermal populations within the ground level and upper level manifolds are small ($\exp(-\Delta E/k_B T) \leqslant 0.4$ for $\Delta E = 50\ cm^{-1}$) and the Stark transitions originate mainly from the two Stark levels of lowest energy. In this configuration, Er:glass operates nearly like a three-level system for short wavelength Stark transitions, while it operates nearly like a four-level system for long wavelength Stark transitions.

Figure 4.26 also shows that in the $-20\,°C$ to $+85\,°C$ temperature range, the cross section changes with respect to the room temperature case ($T = 20\,°C$) are relatively small, i.e., 2–5%. This fact suggest that in this range, the EDFA gain should be weakly sensitive to temperature. Detailed experimental measurements (M. Suyama et al. [178]) have shown that for an EDFA with 20 dB room temperature gain and 1.48 $\mu$m pump, the gain decreases with temperature over a usable range $-40\,°C$ to $+60\,°C$, with a total variation of 40%, or 1.5 dB. This result corresponds to a temperature coefficient $-0.015\,dB/°C$. Other measurements using a 1.48 $\mu$m-pumped EDFA with 35 dB room temperature gain gave a variation of 3.5 dB over a range $-40\,°C$ to $+80\,°C$, corresponding to a temperature coefficient $-0.030\,dB/°C$ (M. Yamada et al. [179]). This study also considered the effect of fiber length. In particular, the investigation shows that for any pump wavelength (e.g., 980 nm or 1.48 $\mu$m), there exists a fiber length for which the gain is nearly temperature insensitive over the usable temperature range. Finally, comparative studies with different pump wavelengths showed that in the case of 980 nm pumping, the effect of fiber length is weaker and the temperature coefficient ($-0.02\,dB/°C$) is always lower than in the case of 1.48 $\mu$m pumping, [177] and [179].

The possibility of achieving temperature insensitivity by an appropriate choice of EDFA length is important for communications systems applications, particularly in terrestrial systems. In such systems, the components usually operate in the $-40\,°C$ to $+60\,°C$ temperature range, with daily and yearly temperature cycles. The possibility of temperature insensitive length also obviates automatic gain control (AGC), such as discussed in Chapter 6, which would otherwise be necessary for EDFA gain regulation.

## 4.6  CHARACTERIZATION OF Er-DOPED FIBER PARAMETERS

We review the different experimental procedures and measurements that make it possible to fully characterize Er-doped fibers (EDF) and determine their basic parameters. The same procedure or types of measurements could also apply to any type of RE-doped fiber, with minor variations. The descriptions of experimental methods are intended for guidance rather than systematic application. Some parameters can be measured in several ways (e.g., mode field diameter, absorption spectrum); we do not describe them all. We concern ourselves only with the experimental determination of the EDF basic parameters, not the EDFA characteristics (e.g., gain versus pump power, output saturation power); these are described in the next chapter.

The experimental characterization of EDFs can be made in two steps. The first concerns measurements made on the fiber preform, the second on the fiber itself. The sequence of measurements described below is illustrated in Figure 4.27 and Figure 4.28, along with the corresponding EDF parameters.

### Preform measurements

***Refractive index profile***  Standard techniques for the measurement of the fiber refractive index profile are outlined in [33] and [180]. This measurement enables to determine an equivalent step index (ESI) fiber having an index difference $\Delta n$ [180]. From this information, one can determine the core diameter $2a$ (hence the outside

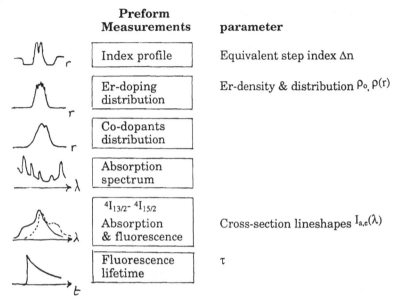

**FIGURE 4.27**  Sequence of measurements in Er-doped fiber preform for EDF basic parameter characterization.

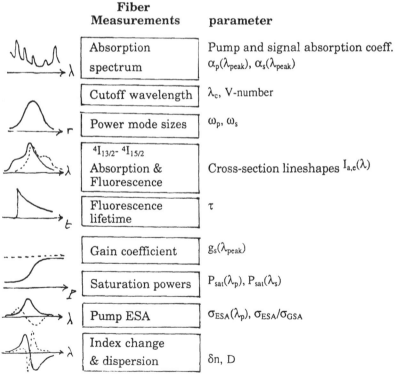

**FIGURE 4.28**  Sequence of measurements in Er-doped fiber for basic parameter characterization.

diameter OD of the fiber to be pulled from the preform) for which a desired cut-off wavelength $\lambda_c$ can be achieved (these parameters are related through $2.405 = (2\pi a/\lambda_c)\sqrt{2n\Delta n}$). The optimal cutoff wavelength $\lambda_c^{opt}$ which maximizes the EDFA gain coefficient can be determined theoretically, as shown in Chapter 5. The cutoff wavelength or core radius that minimizes the mode power size $\omega$ can also be determined from the ESI, using the LP mode or Gaussian formulae, Eqs. (B.5) and (B.7). The actual mode envelopes can also be computed theoretically from the refractive index distribution $n(r)$ by numerical resolution of the scalar wave equation, [33], [180], and [181].

***Erbium-doping distribution***    The normalized distribution $\rho(r)/\rho_0$ of the erbium ions across the fiber core and the absolute Er-concentration $\rho_0$ can be measured through various techniques. For the doping distribution $\rho(r)/\rho_0$, these include: secondary ion mass spectrometry (SIMS), [177] and [182]; X-ray microprobe analysis [142]; electron probe microanalysis (EPMA), [183], and [184]; fluorescence-induced radial scanning, [185] and [186]; Rutherford backscattering spectroscopy (RBS), [156]; and Raman microprobe analysis. On the other hand, the determination of the absolute erbium concentration $\rho_0$ can be made by inductively coupled plasma atomic emission spectroscopy (ICP-ASE), [177] and [187]; by thermal neutron activation (a measure of $^{171}$Er isotope activity) [188]; or by a measurement of the glass absorption coefficient if the absorption cross-section $\sigma_a$ of the Er:glass is known. The absolute Er concentration can also be accurately controlled during fiber fabrication if the solution doping technique is used [189]. The knowledge of the actual Er-doping distribution can be used for accurate EDFA modeling, as described in Chapter 1.

***Codopants distribution***    The knowledge of the radial distributions of the glass codopant (e.g., $GeO_2$, $Al_2O_3$, $P_2O_5$) is important for evaluating the glass homogeneity across the core, which affects the cross section line shapes (see Sections 4.4 and 4.5). The codopant concentrations and distributions can be evaluated by EPMA, [183] and [184], or by Raman microprobe analysis. In the simple case where there is only one index-raising codopant (e.g., $GeO_2$), the concentration and profile of this codopant are given by the refractive index profile $n(r)$, since the two parameters are directly related, [13] and [32].

***Absorption spectrum***    The preform absorption spectrum recorded with a light source from the visible to the near-infrared provides information about the fiber background loss and the strength of various absorption bands of the rare earth (e.g., $Er^{3+}$ and codopants such as $Yb^{3+}$) absorption bands. If the cross sections for the different pump bands are known, the absolute rare earth concentration (as averaged over the preform core) can be evaluated from the absorption spectrum. The rare earth absorption bands are easier to interpret in terms of absolute rare earth concentration when the loss is measured in the preform rather than in the fiber, since the light is unguided.

***$^4I_{13/2}$–$^4I_{15/2}$ absorption and fluorescence spectra***    The absorption spectrum of the $^4I_{13/2}$–$^4I_{15/2}$ transition is made by the same technique as above but with greater accuracy. In the case of low erbium concentration, the background loss spectrum must be subtracted to give the absorption line shape $I_a(\lambda)$. The fluorescence can be generated in a slice of preform using any pump source other than 1.48 $\mu$m (for full medium inversion), and recorded either in the transverse or longitudinal directions. With transverse measurements, the low power level of spontaneous emission requires synchronous detection. With longitudinal measurements, care must be taken that full medium inversion is achieved in the preform core (no ground level reabsorption occurs) and that the medium gain remains small (no amplification of spontaneous emission). These two conditions are met if the fluorescence line shape $I_e(\lambda)$ is observed to be insensitive to the pump power, past a certain level. The drawback of measuring fluorescence and absorption in the preform rather than in the fiber is that it needs a bulk optics spectrometer and analog recording apparatus. On the other hand, using a digital optical spectrum analyzer with optical fiber input makes such measurements directly from the fiber very straightforward. Finally, as differences in absorption and emission line shapes are sometimes observed between preform and fiber, these measurements are indicative only [156].

***Fluorescence lifetime***   The fluorescence lifetime $\tau$ of the $^4I_{13/2}$ level can be measured by pulsed excitation from any pump source. As the lifetime is near 10 ms, the pulse can be generated by chopping a CW pump laser. The fluorescence pulse can be monitored in the longitudinal direction through a high density pump-blocking filter. To ensure that no stimulated emission or spurious lasing effects occur, the $1/e$ characteristic decay time $\tau$ should be independent of pump power, past a certain level. The time analysis of fluorescence decay can also reveal a fast component superimposed on a slowly decaying component. The fast component is attributed to the existence of rare earth clustering, which can occur at high rare earth concentrations (e.g., $> 1$ wt%) as described in Section 4.8. Sizeable differences in fluorescence lifetime can be observed between preform and fibers [156], so the preform measurement is of indicative value. Nonetheless, the preform measurement is important in the detection of unwanted rare earth clustering, or in systematic studies of the effect of glass composition on this clustering [190].

Methods for the determination of the Er:glass cross-sections from knowledge of experimental line shapes, absorption coefficient, absolute Er-concentration, and fluorescence lifetime, are described in Section 4.5.

### Fiber measurements

***Absorption spectrum***   The absorption spectrum can be measured directly from the Er-doped fiber using a white light source. As only very small signal powers ($< -50$ dBm/nm) can be coupled in the single-mode fiber from this source, the measurement requires a very sensitive optical spectrum analyzer (e.g., $-70$ dBm) with signal averaging. The very low signal levels and the few measurable decibels of net fiber absorption can introduce some experimental error in the line shape $I_a(\lambda)$. To determine the absolute absorption coefficient at the pump and signal wavelengths ($\alpha_p(\lambda_p)$, $\alpha_s(\lambda_s)$), monochromatic and intensity-stabilized sources should be used instead of a white light source. repeated cutback measurements of the absorption coefficients in the same fiber ensure good accuracy. Care should be taken to filter the cladding modes, which can introduce large power measurement errors in the case of short fiber lengths (or high Er concentration). See earlier under "Absorption spectrum."

*Cutoff wavelength*   Standard techniques for the determination of the fiber cutoff wavelength $\lambda_c$ are described in [180]. Knowledge of $\lambda_c$ makes it possible to evaluate or numerically compute the pump and signal mode envelopes. This data can be used in turn for accurate EDFA modeling, as described in Chapter 1.

*Power mode sizes*   The determination of mode envelopes and power mode sizes (not to confuse with the mode field diameter, or MFD, used generally in fiber optic technology) can be made through several methods. These methods include: far-field and near-field scanning, transverse offset coupling measurement, spatial filtering, variable aperture measurement, knife edge scan, masking, and harmonic detection [191]. The most straightforward method is the direct monitoring and recording of the fiber mode power envelope with an infrared camera; the size calibration is given by translation of the fiber end.

*${}^4I_{13/2}$–${}^4I_{15/2}$ absorption and fluorescence*   Determination of the absorption and emission line shapes of the transition is straightforward with the fiber. As for the preform, fluorescence can be monitored in either transverse or longitudinal directions. For longitudinal measurements, care must be taken to achieve full inversion in the fiber, with low single-pass gain. The pump source must be at a wavelength different from 1.48 $\mu$m, otherwise complete medium inversion is not possible. The measurement technique for the fluorescence, which leads to the determination of the emission cross section line shape is described in Section 4.5.

*Fluorescence lifetime*   The same comments apply to the fiber as apply to the preform.

*Gain coefficient*   The fiber gain coefficient $g_s$ can be determined by fiber cutback in a regime of full and uniform medium inversion. further description of this measurement technique is given in Section 4.5.

*Saturation powers*   The saturation powers $P_{\text{sat}}(\lambda_{\text{p,s}})$ at the pump and signal wavelengths, defined in Eq. (1.45), can be directly measured in the fiber without knowledge of mode size, fluorescence lifetime, and cross sections. The technique measures the fiber transmission $T = P^{\text{out}}(\lambda)/P^{\text{in}}(\lambda)$ as a function of the input power $P^{\text{in}}(\lambda)$ [192]. An experimental measurement for the wavelength $\lambda = 1.47\,\mu$m is shown in figure 4.29 [193]. The experimental points can be fitted using the analytical solution $T = f[P^{\text{in}}(\lambda), \alpha(\lambda), P_{\text{sat}}(\lambda)]$, [192]–[194] and Section 1.10. The best fit, which uses the experimental value of the absorption coefficient $\alpha(\lambda)$, enables us then to determine the value of the saturation power parameter $P_{\text{sat}}(\lambda)$ at wavelength $\lambda$. This analytical solution is valid only in the case where the Er-doping is confined near the center of the core, i.e., when transverse power variations can be neglected.

In the general case, the fiber transmission $T(\lambda)$ at any wavelength $\lambda$ and fiber coordinate $z$ can be expressed from Eq. (1.56) in the form:

$$\frac{d\log[T(\lambda z)]}{dz} = \frac{d}{dz}\left\{\log\left[\frac{P^{\text{out}}(\lambda z)}{P^{\text{in}}(\lambda)}\right]\right\} = -2\pi\rho_0\sigma_a(\lambda)\int_S \frac{\rho(r)}{\rho_0}\bar{\psi}_s(r)\frac{1}{1+\dfrac{P^{\text{out}}(\lambda)}{P_{\text{sat}}(\lambda)}\psi_s(r)}\,r\,dr$$

$$(4.100)$$

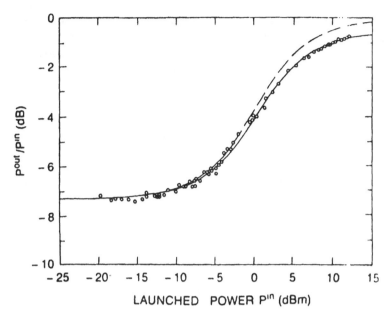

**FIGURE 4.29**  Experimental transmission characteristics of an Er-doped fiber at $\lambda = 1.47\,\mu m$ as a function of input power, used for the determination of the saturation power $P_{sat}(\lambda)$. The dashed curve corresponds to the analytical solution given in Eq. (1.148) with experimental absorption coefficient $\alpha(\lambda) = 1.05\,dB/m$ and best-fit parameter $P_{sat}(\lambda) = 700\,\mu W$. The small difference between the theoretical solution and the experimental points at input powers $P^{in} > 0\,dBm$ is attributed to the effect of $1.47\,\mu m$ power conversion into ASE. From [193] © 1991 IEEE.

or, using Eq. (4.75):

$$\frac{d\log[T(\lambda)]}{dz} = -\alpha(\lambda)\frac{\displaystyle\int_{s}\frac{\rho(r)}{\rho_0}\bar{\psi}_s(r)\frac{1}{1+\dfrac{P^{in}(\lambda)}{P_{sat}(\lambda)T\lambda\psi_s(r)}}r\,dr}{\displaystyle\int_{s}\frac{\rho(r)}{\rho_0}\bar{\psi}_s(r)r\,dr} \tag{4.101}$$

The ASE noise was neglected in Eqs. (4.100)–(4.101). This assumption is valid for short fiber lengths (small gain coefficient) if the wavelength $\lambda$ falls into one of the pump bands (e.g., 980 nm or 1.48 $\mu m$), and for any fiber lengths if $\lambda$ corresponds to a signal in the EDFA gain band. If the mode envelope distribution $\psi_s(r)$ and the relative Er-doping distribution $\rho(r)/\rho_0$ are known, then Eq. (4.101) can be integrated numerically, using the experimental absorption coefficient $\alpha(\lambda)$ and a fitting value for the saturation power $P_{sat}(\lambda)$, given an EDF length $L$ and an input signal power $P^{in}(K)$.

The method described above for the determination of the saturation power $P_{sat}(\lambda)$ is similar to the saturated fluorescence method for the determination of cross sections [142].

It is not necessary to measure the transmission $T(\lambda)$ over a continuous range of wavelengths in order to determine the spectrum $P_{sat}(\lambda)$. If the EDFA is pumped in the $^4I_{15/2}-^4I_{13/2}$ band, and if the emission/absorption cross section ratio $\eta(\lambda)$ is known, only one measurement at $\lambda = \lambda_0$ is theoretically necessary. Indeed, we find from Eq. (1.45) that the saturation power at wavelength $\lambda$ is given by the relation:

$$P_{sat}(\lambda) = P_{sat}(\lambda_0)\frac{\lambda_0}{\lambda}\frac{\sigma_a(\lambda_0)[1 + \eta(\lambda_0)]}{\sigma_a(\lambda)[1 + \eta(\lambda)]}\left(\frac{\omega}{\omega_0}\right)^2 \approx P_{sat}(\lambda_0)\frac{\lambda_0}{\lambda}\frac{I_a(\lambda_0)[1 + \eta(\lambda_0)]}{I_a(\lambda)[1 + \eta(\lambda)]} \qquad (4.102)$$

where $\omega$ and $\omega_0$ are the power mode sizes at $\lambda$ and $\lambda_0$, respectively, which in this case are nearly equal. Equation (4.102) requires only the knowledge of the cross section ratio $\eta(\lambda)$ and the absorption line shape $I_a(\lambda)$. If the EDFA is pumped in bands other than $^4I_{15/2}-^4I_{13/2}$, only one measurement of $P_{sat}(\lambda_0)$ at $\lambda = \lambda_0$ is necessary. If one assumes for instance $\lambda_0 = \lambda_p = 980$ nm, the saturation power at any signal wavelength $\lambda$ is given by the first part of Eq. (4.102) with $\eta(\lambda_p) = 0$. However, the determination of $P_{sat}(\lambda)$ through this equation requires the knowledge of the absolute cross section ratio $\sigma_a(\lambda_0)/\sigma_a(\lambda)$ and mode size ratio $\omega/\omega_0$, in addition to the parameter $\eta(\lambda)$. This knowledge is not straightforward to obtain, so we prefer to systematically determine $P_{sat}(\lambda)$ at both pump and signal wavelengths though separate measurements of $T(\lambda)$.

In the case where the power variations in the fiber transverse plane can be neglected (confined Er-doping), knowledge of the absorption coefficient, cross section ratio and saturation power spectra ($\alpha(\lambda)$, $\eta(\lambda)$, $P_{sat}(\lambda)$) is sufficient for a full theoretical analysis of the EDFA gain and noise spectral characteristics. The corresponding equations (1.113)–(1.121) for the pump and signal powers, or in their normalized form $q = P_p/P_{sat}(\lambda_p)$ and $p = P_s/P_{sat}(\lambda_s)$, are functions only of these three fundamental parameters.

In the intermediate case where the Er-doping is uniform but not confined (step Er-doping), the pump and signal powers are given by Eqs. (1.69)–(1.72). Integration of these equations requires the mode envelopes $\psi_{p,s}(r)$. Integration can be made analytically if a Gaussian envelope approximation is used (Section 1.7). The most accurate approximation uses the experimental power mode sizes $\omega_{p,s}$ for the Gaussian envelopes, instead of the theoretical mode sizes of the ESI fiber obtained from Eq. (B.7). In the most general case, where the Er-doping and mode distributions have arbitrary shapes, the pump and signals are given by the rate equations (1.66)–(1.68), which requires the six parameters: $\alpha(\lambda)$, $\eta(\lambda)$, $\psi_p(r)$, $\psi_s(r)$ and $\rho(r)/\rho_0$. The parameter $\rho_0\sigma_a(\lambda)$ involved in these equations is related to the absorption $\alpha(\lambda)$ through Eq. (4.75).

Finally, for a full characterization of the EDF basic parameters, three other measurements are possible in the doped fiber. The first concerns the effect of pump (or signal) excited state absorption (ESA), the second the effect of gain-induced refractive index changes and dispersion. The techniques for measuring these two effects and determining corresponding parameters are described in Sections 4.7 and 4.9, respectively. The third measurement concerns the Er-doping distribution, which can be evaluated in the fiber by a differential mode-launching technique [195]. The principle of this technique is the following: if the fiber absorption coefficient is experimentally measured at two wavelengths $\lambda_1 < \lambda_c$ and $\lambda_2 > \lambda_c$, such that light at $\lambda_1$ is propagating in the various $LP_{11}$ modes and light at $\lambda_2$ in the $LP_{01}$ mode, the ratio $x$ of the absorption coefficients at these two wavelengths is given by Eq. (4.75):

$$x = \frac{\alpha(\lambda_2)}{\alpha(\lambda_1)} = \frac{\displaystyle\int_S \frac{\rho(r)}{\rho_0} \bar{\psi}_{01}(r, \theta) r \, dr \, d\theta}{\displaystyle\int_S \frac{\rho(r)}{\rho_0} \bar{\psi}_{11}(r, \theta) r \, dr \, d\theta} \tag{4.103}$$

where $\bar{\psi}_{01}$ and $\bar{\psi}_{11}$ are the normalized power mode envelopes corresponding to the $LP_{01}$ and $LP_{11}$ modes, respectively. The degeneracies of the LP modes should be taken into account in Eq. (4.103). The $LP_{11}$ mode distributions depend on the angle $\theta$. The LP mode envelopes can be calculated from the ESI fiber parameters or measured experimentally. If a Gaussian distribution is assumed for the Er-doping profile, i.e., $\rho(r)/\rho_0 = \exp(-r^2/\sigma_{Er}^2)$, where $\sigma_{Er}$ is the $1/e$ half-width of the distribution, it is then possible to compute numerically eq. (4.103) to obtain the ratio $x$ as a function of the $1/e$ half-width $\sigma_{Er}$ [195]. A comparison between the value of $x$ measured experimentally and the theoretical function $x = f(\sigma_{Er})$ makes it possible to determine the parameter $\sigma_{Er}$ [195].

## 4.7  PUMP AND SIGNAL EXCITED STATE ABSORPTION

Excited state absorption (ESA) has been observed to occur in Er-doped fibers in several wavelength bands between 450 nm and 1050 nm (J. R. Armitage et al. [196]; R. I. Laming et al. [197]; S. Zemon et al. [198]).

In the effect of pump ESA, the pump light at frequency $v_p$ is not absorbed from the ground level (1) of the rare earth ion, but from an excited level (2), due to the existence of a third level (3) whose energy gap $\Delta E = E_3 - E_2$ with level 2 happens to closely match the pump photon energy $hv_p$. This process happens only if the ESA cross section overlaps with the ground state absorption (GSA) or pump cross section.

In the effect of signal ESA, the signal light of energy $hv_s$ is absorbed from the metastable level 2 to level 3, due to the same energy gap matching relation, i.e., $\Delta E = E_3 - E_2 \approx hv_s$. Both pump and signal ESA processes therefore result in an excess loss for the pump or the signal, which reduces the fiber gain.

The effect of ESA can occur from any energy level having finite atomic population. In Er:silica glasses, ESA is most likely to initiate from the metastable level $^4I_{13/2}$, as all other levels of higher energy are characterized by fast nonradiative relaxation rates. This is not the case with fluoride glasses, whose energy levels are characterized by relatively long lifetimes.

Figure 4.30 shows an energy level diagram of Er:silica glass with several possible GSA and ESA transitions and associated wavelengths (W. J. Miniscalco [127]). Due to the accidental locations of the energy levels, ESA transition wavelengths are close to each of the GSA transition wavelengths. The cases of signal ESA near 1500 nm (1680 nm ESA transition) and pump ESA initiating from the short-lived $^4I_{11/2}$ are discussed below.

In order for pump or signal ESA to occur at a given wavelength, the corresponding absorption cross sections ($\sigma_{ESA}$) must be important with respect to the GSA cross sections ($\sigma_{GSA}$). We analyze below how the fiber transmission at the pump wavelength $\lambda_p$ is affected by ESA and by the ratio $\delta = \sigma_{ESA}/\sigma_{GSA}$.

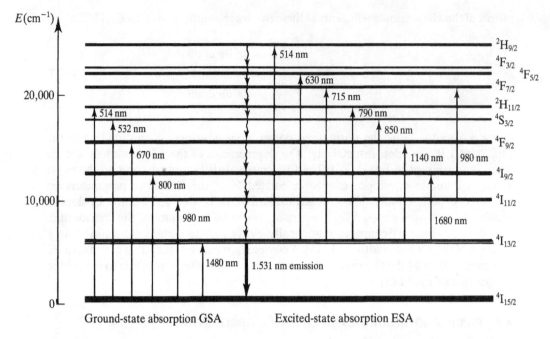

**FIGURE 4.30**   Energy level diagram of Er:glass showing ground state (GSA) and excited state (ESA) transitions, with corresponding peak wavelengths.

Let $N_1, N_2$, and $N_3$ be the atomic populations of the ground (1, or $^4I_{15/2}$), metastable (2, or $^4I_{13/2}$) and higher energy level (3), respectively. From the theory described in Chapter 1 and further developed in Appendix N to include pump ESA, we can write the following rate equation for the normalized pump:

$$\frac{d\log[q(\lambda_p)]}{dz} = 2\pi \int_S \bar{\psi}_p(r)\{\sigma_{GSA}(\lambda_p)[\eta_p N_2(r) - N_1(r)] + \sigma_{ESA}(\lambda_p)[\eta'_p N_3(r) - N_2(r)]\}r\,dr$$

$$(4.104)$$

In the above equation, the parameters $\eta_p$ and $\eta'_p$ correspond to the ratios of emission to absorption cross sections relative to transitions 1–2 and 2–3, respectively. The parameter $d\log[q(\lambda_p)]/dz$ is the fiber transmission coefficient at coordinate $z$. if we assume a fast nonradiative decay from level 3, we have $N_3 \ll N_1, N_2$ and eq. (4.104) takes the form:

$$\frac{d\log[q(\lambda_p)]}{dz} = 2\pi\sigma_{GSA}(\lambda_p) \int_S \bar{\psi}_p(r)[(\eta_p - \delta)N_2(r) - N_1(r)]r\,dr \qquad (4.105)$$

The parameter $\delta = \sigma_{ESA}/\sigma_{GSA}$ introduced in the above equation is meaningful only near the pump bands, i.e., where $\sigma_{GSA}(\lambda_p) \neq 0$. We assume now that the fiber is pumped at a wavelength $\lambda'$ (different from $\lambda_p$), with normalized power $Q = P_p(\lambda')/P_{sat}(\lambda')$, cross section ratio $\eta_p(\lambda') = 0$, and mode envelope $\psi'_p(r)$. Substituting the expressions for the populations $N_1$, $N_2$ in Eq. (4.105) (see Chapter 1), we obtain:

$$\frac{d \log[q(\lambda_p)]}{dz} = -2\pi\rho_0\sigma_{GSA}(\lambda_p) \int_S \frac{\rho(r)}{\rho_0} \bar{\psi}_p(r) \frac{1 + \delta Q \psi'_p(r)}{1 + Q\psi'_p(r)} r\, dr \qquad (4.106)$$

We consider now the two cases where the fiber is unpumped ($Q = 0$) and where the medium inversion is complete ($Q = \infty$). The transmission coefficient at $\lambda_p$ is in these two cases, from Eq. (4.106):

$$\frac{d \log[q(\lambda_p, Q = 0)]}{dz} = -2\pi\rho_0\sigma_{GSA}(\lambda_p) \int_S \frac{\rho(r)}{\rho_0} \bar{\psi}_p(r) r\, dr \equiv -\alpha_{GSA}(\lambda_p) \qquad (4.107)$$

$$\frac{d \log[q(\lambda_p, Q = \infty)]}{dz} = -2\pi\rho_0\sigma_{ESA}(\lambda_p) \int_S \frac{\rho(r)}{\rho_0} \bar{\psi}_p(r) r\, dr \equiv -\alpha_{ESA}(\lambda_p) \qquad (4.108)$$

From Eqs. (4.107) and (4.108), the ESA and GSA absorption cross section ratio $\delta$ is given by the ratio of the corresponding absorption coefficients, i.e.,

$$\delta = \frac{\sigma_{ESA}(\lambda_p)}{\sigma_{GSA}(\lambda_p)} = \frac{\alpha_{ESA}(\lambda_p)}{\alpha_{GSA}(\lambda_p)} \qquad (4.109)$$

The ESA to GSA absorption cross section ratio $\delta$ can therefore be determined experimentally by measuring the absorption coefficient difference:

$$x = \alpha_{ESA}(\lambda_p) - \alpha_{GSA}(\lambda_p) = [\delta(\lambda_p) - 1]\alpha_{GSA}(\lambda_p) \qquad (4.110)$$

The spectrum of the absorption coefficient difference $x$ can be measured with a white light source as a signal probe at $\lambda_p$ and with a pump at $\lambda'$ propagating in the opposite fiber direction [197]. The difference is given by $x = (1/L)\log(T/T')$, where $T$ and $T'$ are the fiber transmissions with the pump on and off, respectively, and $L$ is the fiber length.

Figure 4.31 shows the ESA-induced change in absorption coefficient in the 600 nm to 1050 nm wavelength range, along with the GSA coefficient for reference, as measured in a $GeO_2$–$P_2O_5$ Er-doped silica fiber [127]. From Eq. (4.109), if the value of $x$ is strictly greater than $-\alpha_{GSA}(\lambda_p)$, this means that ESA occurs at this wavelength. From Figure 4.31 ESA occurs near 630 nm, 715 nm, 790 nm, and 850 nm; it is completely absent near 980 nm ($x = -\alpha_{GSA}$). From Figure 4.30, these wavelengths correspond to ESA transitions bridging the levels $^4I_{13/2}$ and $^4F_{3/2-5/2}$, $^4F_{7/2}$, $^2H_{11/2}$ and $^4S_{3/2}$, respectively. The spectra of Figure 4.31 also indicate that there exists a substantial overlap between the GSA and ESA cross section spectra near 650 nm and 800 nm.

Experimental values of the ratio $\delta = \sigma_{ESA}/\sigma_{GSA}$ corresponding to the different pump bands (488 nm, 514.5 nm, 655 nm, and 810 nm) and to various silicate glass compositions, i.e., $GeO_2$, $GeO_2$–$B_2O_3$, $GeO_2$–$P_2O_5$, $Al_2O_3$, can be found in [197]. The measurements show that the ratio $\delta$ decreases in this composition order. At $\lambda_p = 514.5$ nm, for instance, the ratios are $\delta(GeO_2) = 0.95$, $\delta(GeO_2$–$P_2O_5) = 0.55$, and $\delta(Al_2O_3) = 0.50$. The dependence of the parameter $\delta$ on glass composition can be attributed to two factors: (1) the shift in peak wavelengths of the GSA and ESA

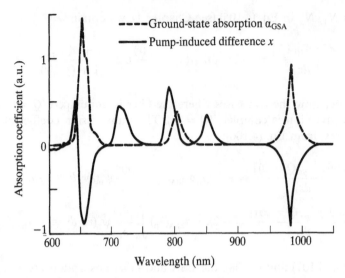

**FIGURE 4.31** Absorption spectrum of Er-doped fiber showing (– – –) ground state absorption coefficient $\alpha_{GSA}$ and (——) pump-induced changes in absorption coefficient due to ESA. From [127] © 1991 IEEE.

transitions relative to each other, which reflects changes in the Stark level positions, and (2) the host dependent changes in peak cross sections for both GSA and ESA absorption. As the position and strength of the ESA bands are observed to vary significantly with the type of host, some optimization can be made if the EDFA must be pumped near these bands. The 810 nm band is of particular interest as GaAs laser diodes (LD) with high output power (>1 W) are available near this wavelength. A systematic study of ESA cross sections corresponding to the two ESA bands near 790 nm and 850 nm (Figure 4.31) was made for various types of hosts, i.e., phosphate, silicate, fluorophosphate (F-P), fluorozirconate, and aluminophosphate (Al-P) silica glasses (S. Zemon et al. [198]). This study showed, in particular, that the 790 nm ESA band is narrower in the F-P host compared to the Al-P host. Because of the small shift (5–10 nm) in center wavelength between ESA and GSA cross sections, the overlap between ESA and GSA bands is reduced in the F-P host, which results in higher EDFA gain. Numerical simulations have shown indeed that for the same pump power near 800 nm (100 mW), the EDFA gain with an F-P host is 10–15 dB higher than with an Al-P host [198].

From Figures 4.4, 4.5, and 4.30, the remaining EDFA pump bands of interest are centered near 514 nm, 532 nm, 670 nm, 980 nm, and 1480 nm. The band near 532 nm is of particular interest since its wavelength matches that of frequency-doubled Nd:YAG lasers, which can be used as compact LD-driven sources for EDFA pumping (M. C. Farries et al. [199]; K. Kannan et al. [200]). Experimental measurements have shown that, while the effect of ESA is significant near 514 nm, it is negligible at 532 nm, [197] and [199]. The band near 670 nm if of relatively limited interest since no commercial high-power LDs are currently available at this wavelength. Further-more, the effect of ESA in this band was measured to represent 15% to 30% of the GSA ($\lambda_p = 655$ nm); the smallest effect ($\delta \approx 0.15$) corresponds to the case of alumino-silicate glass [197].

The experimental measurement shown in Figure 4.31 reveals that pump ESA, initiated from the metastable level $^4I_{13/2}$, is nonexistent near 980 nm. But the energy diagram of Figure 4.30 shows that 980 nm pump ESA can be initiated from the short-lived $^4I_{11/2}$ level; the terminal level is $^4F_{7/2}$. Because the $^4I_{11/2}$ level population is rapidly damped by nonradiative decay, ESA from this level can occur only at high pump power levels.

Experimental evidence of this process is the *upconversion luminescence* ($\lambda = $ 525–550 nm) that can be observed at high 980 nm pump powers (M. G. Seats et al. [201]). This fluorescence corresponds to a radiative transition from the $^4S_{3/2}$ neighboring level of lower energy, Figure 4.4. Detailed measurement of the $^4I_{11/2}$–$^4F_{7/2}$ ESA transition characteristics in Al–P fibers showed that the ESA cross section is very nearly equal to the GSA cross section ($\sigma_{GSA}^{peak} \approx \sigma_{GSA}^{peak} \approx 2, 2.10^{-21}$ cm$^2$), is peaked near 970 nm, and has a broad FWHM of 30 nm (R. S. Quimby et al. [202]). While the ratio of ESA to GSA oscillator strengths is weakly dependent upon glass composition [202], the overlap between ESA and GSA cross section is host dependent. At high pump levels, the effect of pump ESA causes nonlinear pump absorption, which eventually limits the EDFA power conversion efficiency (Chapter 5).

Another possible ESA transition shown in Figure 4.30 has a wavelength near 1140 nm and originates from the metastable level $^4I_{13/2}$ to terminate in the $^4F_{9/2}$ level. Measurements of absorption coefficient changes in germanosilicate fibers (M. C. Farries [203]) have shown that this ESA effect occurs between 1.08 $\mu$m and 1.17 $\mu$m, and is relatively weak (7% change). Since this wavelength range has no overlap with any pump bands and is away from the 1.5 $\mu$m signal band, the effect is of no consequence for the EDFA operation.

The last ESA transition shown in Figure 4.30 is near 1680 nm. Experimental evidence of this effect is the observation of upconversion luminescence ($\lambda = 980$ nm) when the EDFA is pumped near $\lambda = 1.48$ $\mu$m (P. Blixt et al. [204]). This luminescence corresponds to a radiative transition taking place between the levels $^4I_{11/2}$ and $^4I_{15/2}$ and occurring after nonradiative relaxation from the $^4I_{9/2}$ level [204]. A similar process of upconversion luminescence is observed with *cooperative energy transfer* (CET) between two Er$^{3+}$ ions. Cooperative energy transfer is concentration dependent, unlike pump ESA. Using an experimental technique in which CET and ESA can be discriminated, the ESA cross section near $\lambda = 1.48$–1.53 $\mu$m was accurately evaluated to be $\sigma_{ESA} = 5.10^{-26}$ m$^2$ [204]. For a Type III fiber (Figure 4.24), this value correspond to ESA/GSA cross section ratios at the pump and peak signal wavelengths of $\delta(1.48$ $\mu$m$) = 0.25$ and $\delta(1.53$ $\mu$m$) = 0.10$, respectively. This result implies that for an accurate modeling of the EDFA performance, the effects of ESA at 1.48 $\mu$m pump and 1.5 $\mu$m signal wavelengths cannot be neglected. Similar to the case of a 980 nm pump, the pump ESA at 1.48 $\mu$m causes nonlinear absorption in the high power regime, which eventually limits the EDFA power conversion efficiency (see Chapter 5).

The theoretical description of the effect of pump ESA has been described in early work [84], [142], [205], and [206]. For silicate glass hosts, which are characterized by fast nonradiative decay from levels other than $^4I_{13/2}$, it was shown in particular that the only effect of pump ESA is an increase in pump absorption, while the atomic populations $N_1$, $N_2$ remain unchanged [84]. Thus, in a three-level pumping scheme, the signal rate equation derived in Chapter 1 is applicable when pump ESA occurs, while the pump rate equation must be modified. The pump rate equation with ESA is derived in Appendix N. In Appendix N, a general rate equation that includes both effects of pump and signal ESA with bidirectional propagation, and that is applicable to both cases of two- and three-level pumping schemes, is also derived.

Theoretical modeling of the effect of pump ESA in the case of interest where the EDFA is pumped near 800 nm can be found in several studies [198] and [207]–[211]. These studies made it possible to determine optimum EDFA parameters for gain maximization and minimization of noise figure; in particular they show the existence of an optimum pump wavelength detuned from both ESA and GSA absorption peaks. The optimum pump wavelength optimizes the EDFA performance. Chapter 5 discusses numerical evaluations and experimental measurements for 800 nm pumped EDFAs, and compares amplifier performances at various pump bands.

## 4.8 ENERGY TRANSFER AND COOPERATIVE UPCONVERSION

The radiative and nonradiative relaxation processes described so far concern isolated rare earth ions. When the rare earth concentration is sufficiently small, the ions are evenly distributed in the glass matrix and the interionic distance is large, preventing any interaction. But when the concentration is increased above a certain level, the distribution may become uneven and the interionic distance is reduced, which makes possible different effects of ion–ion coupling, and results in cooperative energy transfer (CET). The existence of concentration dependent CET effects that compete with the light amplification process sets an upper limit on rare earth concentration. As a result, the lengths of Er-doped fiber amplifier devices cannot be reduced to arbitrarily small values (e.g., from 1 m to 1 cm) by increasing the Er concentration in inverse proportion (i.e., by 100 times).

The study of rare earth energy transfer in solids began in the 1940s. Three main types of phenomena were investigated, which are named *sensitized luminescence,* *fluorescence quenching,* and *cross-relaxation* [55]. The basic principle of each of these effects is illustrated in Figure 4.32.

In sensitized luminescence (Figure 4.32a), a donor ion of Type 1 (also called sensitizer) transfers its energy to an acceptor ion of Type 2 (also called activator), which is observed to emit fluorescence. Sensitized luminescence with $Er^{3+}$ as the acceptor have been demonstrated for instance with the rare earth pairs Yb–Er, [20] and [212] and Tm–Er [213].

In fluorescence quenching (Figure 4.32b), energy is transferred from the donor to the acceptor, but this latter does not emit fluorescence and relaxes nonradiatively. In this process, the acceptor acts as an energy sink and is sometimes called a deactivator. Since fluorescence from the donor is partially or completely suppressed, the effect is called fluorescence quenching. Three examples of fluorescence quenching of the $^4I_{13/2}$–$^4I_{15/2}$ transition in $Er^{3+}$ are given by the pairs Er–Nd [212], Er–Tm [213], and Er–Ho [214].

In cross-relaxation (Figure 4.32c), energy is transferred from an excited donor to an acceptor ion, thereby promoting the acceptor to a higher energy state. Initially, the acceptor can be either in an excited state or in the ground state, as shown in the figure. The process only requires that both donor and acceptor ions have two energy levels of nearly the same energy gap. The transfer is said to be resonant if the gaps are matched, and phonon-assisted if the mismatch is compensated by the addition of one or two phonons in the process [74]. Cross-relaxation by resonant energy transfer can involve a single type of rare earth ion, which plays both roles of donor and acceptor. For this effect to occur, it is required that the ion has two pairs of equally spaced energy levels. In this case of cross-relaxation involving a single ion type, fluorescence from the ions is partially or completely suppressed, hence the name self-quenching. Possible self-quenching transitions between $Er^{3+}$ ions are shown in

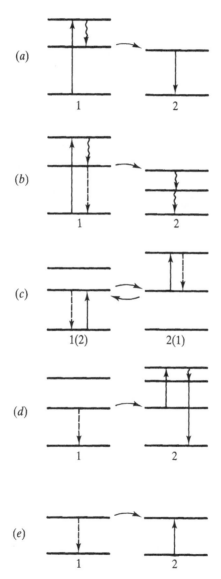

**FIGURE 4.32** Basic processes of cooperative energy transfer (CET) between donor (1) and acceptor (2) RE ions: (*a*) sensitized fluorescence, (*b*) fluorescence quenching, (*c*) cross-relaxation, (*d*) cooperative frequency upconversion, and (*e*) resonant energy migration. The solid arrows indicate radiative absorption or emission transitions, the wiggled arrows represent nonradiative decay and the dashed arrows represent relaxation via CET.

Figure 4.33. Some of these transitions have been identified experimentally in ZBLAN, BYZYT, BATY fluoride Er-doped glasses [214].

Figure 4.32*d* shows cooperative frequency upconversion (CFUC), a special case of cross-relaxation. Initially in an excited state, the acceptor ion is promoted to a higher energy state by CET of energy $hv$. From this state, the acceptor can relax nonradiatively to a lower level then relaxes radiatively to the ground level with the emission of a photon of energy $hv'$, greater than $hv$. As a result, the laser medium is observed to emit fluorescence at an energy greater than that of the incident light, hence the name frequency upconversion.

Frequency upconversion also occurs with pump ESA. But there are two major differences between CFUC luminescence and ESA-induced upconversion luminescence. ESA involves a single ion; CFUC involves a pair of ions. ESA involves the *absorption of photons* from ions in the excited state; CFUC involves a *nonradiative* energy transfer between ions through cross-relaxation.

If the process CET is repeated several times with the same acceptor ion, the frequency of the upconversion fluorescence can take several possible values, which depend upon the number of excitation steps and intermediate nonradiative relaxations in the acceptor. The acceptor acts as a quantum counter, which can be used to convert infrared light signals into the visible domain for applications to low noise photo-detection (N. Bloembergen [215]). For instance, in $BaYF_5:Er^{3+}$ and various other Er-doped fluoride phosphors, [216]–[218], this multistep excitation process involves $Er^{3+}$ in the $^4I_{13/2}$ state as a donor, and $Er^{3+}$ as a single acceptor initially in the $^4I_{13/2}$ state, as illustrated in Figure 4.34. Multistep CET and intermediate nonradiative

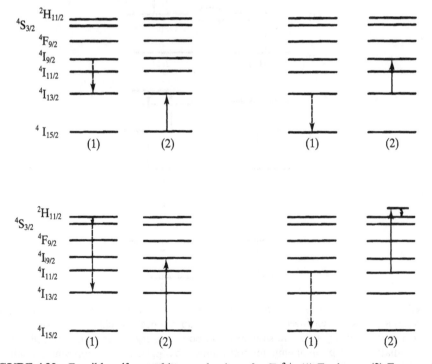

**FIGURE 4.33**   Possible self-quenching mechanisms for $Er^{3+}$. (1) Er-donor, (2) Er-acceptor.

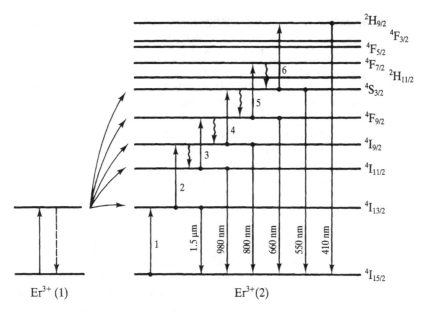

**FIGURE 4.34** Quantum counter process due to multistep Er–Er cooperative energy transfer, resulting in upconversion fluorescence.

relaxations of the $Er^{3+}$ acceptor yield IF–red–blue–green upconversion fluorescence at 1.5 $\mu$m, 980 nm, 800 nm, 660 nm, 550 nm, and 410 nm wavelengths. All these transitions are not simultaneously observed in any host material. In Er-doped phosphors, for instance, only steps 1–2–4–6 in Figure 4.34 are observed [218], while in $BaYF_5:Er^{3+}$ the steps are 1–2–3–5–6 [216]. The probability for each of the steps to occur from a given energy level depends on the corresponding Er–Er CET rate, level and Stark sublevel positions, level lifetime, and level nonradiative decay rate. Additionally, upconversion fluorescence can be strongly reabsorbed by GSA and ESA, as well as by other self-quenching mechanisms such as shown in Figure 4.33. In Er-doped silica fibers, this quantum-counter process is not observed, essentially attributable to the strong nonradiative relaxation effect prevailing in this host at all excited state levels but $^4I_{13/2}$.

Finally, a particular effect of CET is *resonant energy migration*, whereby the relaxation of one donor ion to the ground level promotes a neighboring acceptor ion in the excited state, Figure 4.32e. In turn, the acceptor becomes a donor with respect to another neighboring ions, resulting in the spatial migration and random walk of the excitation energy through the host. This multistep CET effect should not be confused with the radiative energy migration. With radiative energy migration, a photon emitted by one ion upon relaxation is absorbed by a neighboring ion initially in the corresponding lower energy state. as the process repeats itself, energy effectively migrates through the medium from one ion to another in successive emission/absorption steps. This radiative coupling process was implicitly taken into account in the rate equations for the atomic populations ($dN_i/dt$) and the signal light intensity ($dI_s/dz$), Eqs. (1.1), (1.2) and (1.25). Contrary to radiative energy migration, in the four effects previously described (i.e., sensitized luminescence, fluorescence quenching, cross-relaxation, including multistep energy migration, and CFUC) no photon is involved in the energy transfer, and the transfer itself is concentration dependent.

Several other CET effects involving more than two ions are *multi-ion relaxation, cooperative excitation, cooperative luminescence* and *Raman luminescence* [55]. In multi-ion relaxation, the energy $hv$ to be transferred is split between two ions, which are promoted then to excited states of higher energy $hv/2$; this effect was experimentally observed with $LaCl_3 : Ho^{3+}$ [219]. The reverse process is cooperative excitation, where the simultaneous relaxation of two ions, each by an energy amount $hv$, promotes an acceptor ion to an excited state of higher energy $2hv$ [220]. In cooperative luminescence, the simultaneous relaxation of two ions results in the emission of a single photon with energy $2hv$, as observed with $YbPO_4$ [221]. In Raman luminescence, the relaxation of a single ion by an energy amount $hv$, causes the excitation of an acceptor ion to a state of energy $hv'$, with simultaneous emission of a photon at the energy difference $E = hv - hv'$ [222].

Of all the CET effects described above, only three are actually relevant to applications involving Er-doped fiber amplifiers. These effects are sensitized luminescence, cross-relaxation, and CFUC. Sensitized luminescence enables us to pump the EDFA at wavelengths different from that of $Er^{3+}$, or to increase pumping efficiency, which could be advantageous for device considerations. The approach of $Yb^{3+}$-sensitized EDFAs is discussed in Chapter 5. Cross-relaxation and CFUC are detrimental in EDFA applications as they compete with light amplification. The only way to prevent them is to limit the Er-doping concentrations to levels sufficiently low for the interionic average distances to be large.

The physical origins of CET and cross-relaxation between rare earth ions are manifold. They may be multipolar coupling (of electrostatic or magnetic origin) or exchange interaction, also called direct exchange (of S–S coupling origin) [55].

The interaction Hamiltonians of multipolar coupling and direct exchange decrease with interionic distance $R$ according to $1/R^n$ ($n$ is an integer) and $exp(-R/R_B)$ ($R_B$ = Bohr radius), respectively. The electric dipole–dipole (EDD), dipole–quadrupole (EDQ) and quadrupole–quadrupole (EQQ) couplings decrease as $1/R^3$, $1/R^4$ and $1/R^5$, respectively; the magnetic dipole–dipole (MDD) interaction is in $1/R^3$, as in EDD [223]. Since the transition probabilities for each of these effects are given by the square modulus of the corresponding Hamiltonian matrix elements, the probabilities (or CET rates) actually decrease as $1/R^6$ (EDD, MDD), $1/R^8$ (EDQ), $1/R^{10}$ (EQQ), and $exp(-2R/R_B)$ (direct exchange). These abrupt radial dependences suggest EDD and MDD interactions will dominate at large ion separations $R$. For small separations, the effects also differ by their corresponding selection rules, which allow only certain transitions between two *SLJ* states [55]. The selection rules mean that EQQ transitions are more likely than EDQ and EDD at high rare earth concentrations or small separations [224]. Likewise, direct exchange between an ion and its closely spaced neighbor can be important in comparison to EQQ; the effect depends upon the RE/host system under consideration [55].

The first comprehensive theory of sensitized luminescence through multipolar and direct exchange interactions in solids was formulated in the early 1950s by D. L. Dexter [223]. The time dependence of CET by direct exchange was also analyzed in the mid 1960s by M. Inokuti and F. Hirayama [225]. These theories and further developments are reviewed in [55]. The main results are outlined in the following text.

In sensitized luminescence, and CET in general, the characteristic energy transfer quantum yield $\eta_A$ can be defined through:

$$\eta_A = \frac{W_{AB}}{A_A^R + W_{AB}} = \frac{W_{AB}\tau_A}{1 + W_{AB}\tau_A} \qquad (4.111)$$

where $\tau_A = 1/A_A^R$ is the radiative lifetime of donor ion A and $W_{AB}$ the CET rate between donor ion and acceptor ion B. On the other hand, the observed lifetime $\tau_A^{obs}$ of the donor ion is given by:

$$\tau_A^{obs} = \frac{1}{A_A^R + W_{AB}} = \frac{A_A^R \tau_A}{A_A^R + W_{AB}} = (1 - \eta_A)\tau_A \qquad (4.112)$$

Equation (4.112) shows that the observed lifetime $\tau_A^{obs}$ of the donor ion decreases as the CET quantum yield $\eta_A$ increases. The fluorescence of donor ion A is therefore quenched by the effect of CET, and is completely suppressed for $\eta_A = 1$.

The Dexter theory assumes a two-ion pair model, where the donor ion A interacts with the nearest neighboring acceptor ion B. In this model, the quantum yield $\eta_A$ of a given ion pair (A–B) must be averaged over all donor–acceptor separation $R$, using the probability $P(B)$ of finding an acceptor ion B between distance $R$ and $R + dR$, i.e., $P(B)\,dV = \rho_B \exp(-\rho_B V)$ with $V = (4/3)\pi R^3$ and $\rho_B$ = density of ions B. The average quantum yield $\bar{\eta}_A$ is then given by [223]:

$$\bar{\eta}_A = \int_0^\infty \eta_A P(B)\,dV = \int_0^\infty \frac{1}{1 + \dfrac{1}{W_{AB}\tau_A}} \exp(-\rho_B V)\rho_B\,dV = y \int_0^\infty \frac{\exp(-yt)}{1 + t^{n/3}}\,dt \quad (4.113)$$

where $W_{AB}\tau_A = (\gamma_n/V)^{n/3}$, $\gamma_n = 1/\rho_n$, $y = \gamma_n\rho_B = \rho_B/\rho_n$ and $t = V/\gamma_n = \rho_n V$, where $\rho_n$ and $y$ are a critical acceptor density and a reduced acceptor density, respectively. With these definitions, the term $t^{n/3} = (V/\gamma_n)^{n/3} = \text{const.} \times R^n$ in the integrand of Eq. (4.113) accounts for the $1/R^n$ dependence of the multipolar interactions. The exponent $n$ depends upon the type of coupling, i.e., $n = 6$ for EDD, $n = 8$ for EDQ, and $n = 10$ for EQQ, as predicted by the Dexter theory.

The quantum yield $\bar{\eta}_A$ and the corresponding normalized donor luminescence lifetime $\tau_A^{obs}/\tau_A$, defined by Eqs. (4.112) and (4.113), are plotted as functions of the reduced acceptor density $y$ in Figure 4.35 for the three multipolar interactions EDD, EDQ, and EQQ. The quantum yield increases with acceptor density; it reaches 50% near $y = 0.65$, or acceptor concentration $\rho_B = 0.65\rho_n$ [223]. Corresponding to this quantum yield increase is a quenching of the donor luminescence lifetime $\tau_A^{obs}$. The yield weakly depends upon the type of interaction, so the Dexter theory does not make it possible to clearly identify the type of multipolar interaction from experimental data [55].

Another drawback of the two-ion pair model used in the Dexter theory is that the dynamic effect of resonant energy migration is overlooked. As a consequence of the migration, the donors A that are surrounded with a number of acceptors B greater than the average will dominate the A–B energy transfer process, thereby enhancing

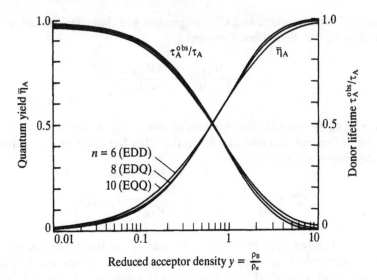

**FIGURE 4.35**   Quantum yield of cooperative energy transfer and normalized donor lumin-escence lifetime predicted by the Dexter theory, as a function of reduced acceptor density, for three types of multipolar interactions: electric dipole–dipole (EDD), electric dipole–quadrupole (EDQ), and electric quadrupole–quadrupole (EQQ).

the short range interactions (EDQ and EQQ) [55]. This enhancement effect is not accounted for in the quantum yield averaging in Eq. (4.113).

In the Inokuti–Hirayama theory, the contribution of the entire donor environment as well as the energy transfer dynamics are taken into account in the determination of the CET yield. According to this model, the time dependent donor population density $\rho_A(t)$ is given by [225]:

$$\rho_A(t) = \rho_A \exp\left(-\frac{t}{\tau_A}\right) \prod_{k=1}^{N} \exp[-tW_{AB}(R_k)] \qquad (4.114)$$

where the CET rate $W_{AB}$ depends upon the distance $R_k$ between the acceptor ion $k$ and the donor ion A, and $N$ represents the total number of acceptor ions B. Let $P(R)$ be the statistical distribution of the A–B interionic distance. The macroscopic average $\bar{\rho}_A(t)$ of the donor density $\rho_A(t)$ can then be expressed through:

$$\bar{\rho}_A(t) = \rho_A \exp\left(-\frac{t}{\tau_A}\right) \times \lim_{N,V \to \infty} \left(\int_V \exp[-tW_{AB}(R)]P(R)\,dV\right)^N \qquad (4.115)$$

The distribution of acceptor ions is assumed random with $P(R)\,dV = 4\pi R^2\,dR/V$ and $V = (4/3)\pi R^3$, we obtain from Eq. (4.115):

$$\bar{\rho}_A(t) = \rho_A \exp\left(-\frac{t}{\tau_A}\right) \times \lim_{N,R \to \infty} \left(\frac{3}{R^3}\int_0^R \exp[-tW_{AB}(R)]R^2\,dR\right)^N \qquad (4.116)$$

To compute the integral in Eq. (4.116), we use the $R$-dependence of multipolar interactions from the Dexter theory, i.e., $W_{AB}\tau_A = (\gamma_n/V)^{n/3} \equiv (R_0/R)^n$, where $R_0$ is the characteristic distance at which the CET rate $W_{AB}$ and the donor radiative rate $\tau_A$ are equal. With the change of variable $z = (t/\tau_A)(R_0/R)^n$, Eq. (4.116) takes the form:

$$\bar{\rho}_A(t) = \rho_A \exp\left(-\frac{t}{\tau_A}\right) \times \lim_{N \to \infty, z \to 0} \left(\frac{3}{n} z^{3/n} \int_z^\infty \frac{\exp(-z)}{z^{1+3/n}} dz\right)^N \tag{4.117}$$

In the limit $z \to 0$, the integral in Eq. (4.117) can be approximated by [225]:

$$\int_z^\infty \frac{e^{-z}}{z^{1+3/n}} dz \approx \frac{n}{3}\left\{z^{-3/n}\left(e^{-z} + \frac{n}{n-3}z\right) - \Gamma\left(1 - \frac{3}{n}\right)\right\} \tag{4.118}$$

and the limit is given by:

$$\lim_{N \to \infty, z \to 0} \left(e^{-z} + \frac{n}{n-3}z - z^{3/n}\Gamma\left(1 - \frac{3}{n}\right)\right)^N \approx \lim_{N \to \infty} \exp\left\{-z^{3/n}\Gamma\left(1 - \frac{3}{n}\right)N\right\}$$

$$\approx \lim_{N \to \infty} \exp\left\{-\Gamma\left(1 - \frac{3}{n}\right)\left(\frac{t}{\tau_A}\right)^{3/n}\frac{1}{\rho_0}\frac{N}{V}\right\} = \exp\left\{-\Gamma\left(1 - \frac{3}{n}\right)\left(\frac{t}{\tau_A}\right)^{3/n}\frac{\rho_B}{\rho_0}\right\} \tag{4.119}$$

In the above equation, the limit of $N/V$ was taken to match the acceptor concentration $\rho_B$ and the critical concentration $\rho_0 = 3/4\pi R_0^3$ was introduced. Substituting Eq. (4.119) into Eq. (4.117), we obtain:

$$\bar{\rho}_A(t, \rho_B) = \rho_A \exp\left\{-\frac{t}{\tau_A} - \Gamma\left(1 - \frac{3}{n}\right)\left(\frac{t}{\tau_A}\right)^{3/n}\frac{\rho_B}{\rho_0}\right\} \tag{4.120}$$

The time dependent donor density obtained in Eq. (4.120) concerns multipolar interactions. The Inokuti–Hirayama theory also makes it possible to derive the density $\bar{\rho}_A(t)$ in the case of direct exchange interaction, using the Dexter CET rate in $\exp(-2R/R_B)$. The donor density for direct exchange is found to be [225]:

$$\bar{\rho}_A(t, \rho_B) = \rho_A \exp\left\{-\frac{t}{\tau_A} - \frac{\rho_B}{\rho_0}\left(\frac{R_B}{2R_0}\right)^3 g\left(\frac{\exp(2R_0 t/R_B)}{\tau_A}\right)\right\} \tag{4.121}$$

with

$$g(z) = 6z \sum_{k=1}^\infty \frac{(-z)^m}{m!(m+1)^4} \tag{4.122}$$

For all types of interactions, the CET quantum yield $\bar{\eta}_A$ is then given by:

$$\bar{\eta}_A = 1 - \frac{\int_0^\infty \bar{P}_A(t, P_B) dt}{\int_0^\infty \bar{P}_A(t, 0) dt} = 1 - \int_0^\infty \bar{P}_A(t, P_B) dt \tag{4.123}$$

and the donor mean luminescence lifetime $\bar{\tau}_A$ by:

$$\bar{\tau}_A = \frac{\displaystyle\int_0^\infty t\bar{\rho}_A(t)\,dt}{\displaystyle\int_0^\infty \bar{\rho}_A(t)\,dt} \tag{4.124}$$

The CET quantum yield $\bar{\eta}_A$ and normalized mean donor luminescence lifetime $\bar{\tau}_A/\tau_A$, defined in Eqs. (4.120), (4.123), and (4.124), are plotted in Figure 4.36 for the three multipolar interactions EDD, EDQ, and EQQ. As the figure shows, the Inokuti–Hirayama theory predicts near the critical donor concentration $\rho_0$ a somewhat larger dependence in the type of multipolar interaction, compared to the predictions of the Dexter theory (Figure 4.35). Near the concentration $\rho_0$, the difference in mean luminescence lifetime is 10–15% between EDD, EDQ, and EQQ interactions.

Experimental investigations of CET in various host materials have shown in several cases remarkable agreement with the Inokuti–Hirayama theory predictions ($\bar{\eta}_A, \bar{\tau}_A$), making it possible to interpret without ambiguity which type of interaction is dominant [55]. For instance, CET in solutions of aromatic molecules (e.g., A = benzophenone, B = naphthalene) is dominated by direct exchange [225]; CET in RE-doped calcium metaphosphate glasses ($Ca(PO_3)_2$ with A–B = $Tb^{3+}$–$Nd^{3+}$ is dominated by EDQ, while direct exchange is inoperative [226]; CET in $YF_3$ with A–B = $Yb^{3+}$–$Ho^{3+}$ is dominated by EDD [227]. In other cases, the analysis is difficult as the theoretical dependence between the various interactions is weak. For instance, in CET measure-

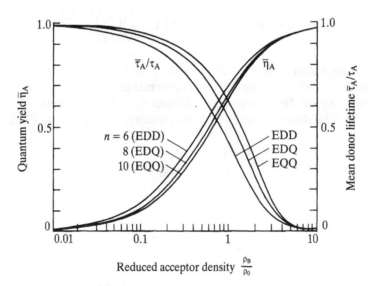

**FIGURE 4.36** Quantum yield of cooperative energy transfer and normalized mean donor luminescence lifetime predicted by the Inokuti–Hirayama theory, as a function of reduced acceptor density, for three types of multipolar interactions: electric dipole–dipole (EDD), electric dipole–quadrupole (EDQ) and electric quadrupole–quadrupole (EQQ).

ments in $Y_2O_3$ with $A$–$B = Yb^{3+}$–$Ho^{3+}$, EDQ provides the best fit, but EDD and EQQ fits are slightly beyond where error bars cannot be unambiguously ruled out [228].

The Inokuti–Hirayama theory has also made it possible to explain the type of CET involved in the effect of self-quenching in Nd:glass. Accurate fits of experimental data through Eq. (4.120) for the time resolved fluorescence in self-quenched Nd:silica glass have demonstrated the predominance of the shorter range interactions EDQ and EQQ [229]. Comparative measurements in Er-doped glasses [230] have shown that in the case of $Er^{3+}$–$Er^{3+}$ self-quenching, the type of dominant CET interaction actually strongly depends upon the type of glass host. Best fits of these experimental data gave $n = 6$ (pure EDD) for a phosphate host, $n = 7$ (EDD and EDQ) for a fluorophosphate host and $n = 8.5$ (EDQ) for a germanate host. The critical $Er^{3+}$ concentrations $\rho_0$ and interaction radii $R_0 = (3/4\pi\rho_0)^{1/3}$ corresponding to these measurements were $\rho_0(10^{21}/cm^3) = 1.16, 1.10, 2.6$ and $R_0(\text{Å}) = 5.9, 6, 4.5$, respectively. These results show that the multipolar interactions occurs at distances near four to six times the $Er^{3+}$ ion diameter [230].

A major consequence of the effect of CET and fluorescence self-quenching of rare earth ions at high concentrations (near or above $\rho_0$) is the loss of their radiative or lasing properties. This detrimental effect is also referred to as *concentration quenching* [3]. Concentration quenching has been thoroughly investigated in the case of Nd:glass, for the purpose of finding an optimum glass host and composition simultaneously to accept large concentrations and exhibit optimal fluorescence characteristics.

The signature of concentration quenching is reduction of fluorescence lifetime with increasing rare earth concentration, with respect to the lifetime observed at low concentrations, according to Eq. (4.120). CET by self-quenching also results in some cases in upconversion fluorescence. An additional effect also observed in concentration-quenched laser materials is *ion clustering*.

The signature of ion clustering is the existence of a fast decay component in the time resolved fluorescence, which immediately follows laser excitation [126]. The fast decay component results from the self-quenching CET of rare earth ion microclusters. These microclusters form randomly in the glass matrix during vitrification when the density of rare earth ions exceeds the number of available sites. The formation of clusters is attributed to an effect of liquid immiscibility in the $RE$–$SiO_2$ system [126]. The rare earth immiscibility results from the limited number of nonbridging oxygens in the host matrix, needed to screen the rare earth ion electric charge. The resulting excess enthalpy is reduced by cluster formation within which nonbridging oxygens are shared by the rare earth ions [231] and [232].

The introduction of network modifiers (such as $Al_2O_3$) and other glass formers (such as $P_2O_5$) to improve rare earth dilution and prevent microclustering was investigated for silica glass hosts with $Nd^{3+}$ (K. Arai et al. [126]) and $Er^{3+}$ (B. J. Ainslie et al. [190]). Previous work on multicomponent fiber fabrication (including RE-doped fibers) by J. R. Simpson and J. B. MacChesney [39] had shown the advantage of $Al_2O_3$ or alumina as a homogeneizing agent for the incorporation of large substituent concentrations (including rare earths). The effect of devitrification, which occurs at high alumina concentrations, can be alleviated by $P_2O_5$ codoping, [39] and [184]. Alumina and phosphorous are also used as index-raising codopants in the fabrication silica fiber waveguides, [30]–[32]. Finally, glass codopants, such as $Al_2O_3$ and $P_2O_5$, can be used to modify the cross section characteristics of the rare earth laser transitions, Section 4.4. For all the above reasons, glass codoping plays a central role in the fabrication and optimization of high-RE concentration fibers.

The study of [126] has shown that microclustering in Nd-doped silica glass disappears when a minimum of alumina codoping with about 10 times the rare earth concentration or phosphorus codoping with about 15 times the rare earth concentration is introduced. The dissolution of microclusters is attributed to the formation of solvation shells surrounding the rare earth ions. In the case of alumina codoping, a possible model for the solvation shell involves two aluminum atoms per rare earth ion [126].

The experimental results reported in [29] and [190] for Er-doped silica fibers have shown that the onset of fast fluorescence decay (or microclustering) is found near Er-doping concentrations of 2.4 wt% in the $GeO_2$ host, while in the $Al_2O_3$ host the onset is near 14 wt%. Therefore, alumina codoping in silica glass makes it possible to incorporate nearly 10 times more rare earth ions than germania codoping. From Eq. (4.44), an Er-doping concentration of 10 wt% corresponds to about $\rho_0 = 1.10^{21}$ ions/cm$^3$. But dissolution of clustering at high rare earth concentrations does not prevent CET between individual (well-coordinated or dissolved) rare earth ions. In the absence of CET effects, a concentration as high as 10 wt% would yield a gain coefficient at full inversion $\Gamma_s \rho_0 \sigma_e = 2.5 \, \text{cm}^{-1}$ ($\Gamma_s = 0.5$, $\sigma_e \approx 5.10^{-21} \, \text{cm}^2$), corresponding to a fiber gain of 11 dB/cm. In this case, high-gain EDGAs ($G = 30$–40 dB) could be realized with fiber lengths of a few centimeters. But the effect of CET prevents this possibility. The results of [190] show that in the alumina host, the $Er^{3+}$ fluorescence lifetime (slow component) decreases at concentrations above 0.15 wt%, corresponding to $\rho_0 = 1.5 \times 10^{19}$ ions/cm$^3$. The gain coefficient corresponding to an optimum EDFA, free from clustering and CET effects, is then $\Gamma_s \rho_0 \sigma_e = 3.5 \, \text{m}^{-1}$, corresponding to a maximum gain of 15 dB/m. A high gain EDFA would therefore require two or three meters of such Er-doped fiber. However, for pumping efficiency considerations, the Er-doping must also be confined near the center of the core, which reduces the overlap factor $\Gamma_s$ (taken here as 0.5) and increases the required length in the same proportion. Chapter 5 reviews EDFA length optimization versus Er-doping concentration, pump wavelength and confinement factor, and experimental results concerning concentration effects in EDFA gain performance.

A detailed experimental study of frequency upconversion by CET in self-quenched Er-doped fibers was made by P. Blixt et al. [204], [233] and [234]. The experiment measures the intensities of 1.5 $\mu$m fluorescence and 980 nm upconversion fluorescence in a 1.48 $\mu$m pumped, high concentration EDFA at various pump powers. As the 1.5 $\mu$m fluorescence is proportional to the upper level population ($N_2$) and the 980 nm upconversion fluorescence to the square of this population ($N_2^2$) [235], a slope of gradient 2 is expected for their intensity ratio. This is consistent with the observed experimental result, a slope of gradient 2.2 [234]. The upconversion fluorescence intensity can be calculated through the rate equation model:

$$\frac{dN_1}{dt} = -\frac{dN_2}{dt} = A_{21}N_2 + W_{21}N_2 - W_{12}N_1 + (1 + 1/m)CN_2^2 = 0 \quad (4.125)$$

In eq. (4.125), $m$ is the branching ratio between the $^4I_{11/2}$–$^4I_{15/2}$ transition ($\lambda = 980$ nm) and the $^4I_{11/2}$–$^4I_{13/2}$ nonradiative transition; $C$ is the two-particle upconversion coefficient; $C$ is concentration independent [235]. A comparison between measured and calculated upconversion fluorescence intensity versus pump power gave best-fit parameters $m = 10^{-4}$ and $C = 10^{-22}$ m$^3$/s, in the case of a Ge/Al/P fiber [234].

Another experimental characterization of Er$^{3+}$ self-quenching in fibers was reported by T. Georges et al. [236]. In this experiment, the EDF transmission function $T = q^{out}/q^{in} = f(q^{in})$ was measured for various Ge/Al/P glass compositions, and the results were fitted with the solution of the equation:

$$\frac{dT}{dz} = \frac{d}{dz} \log[q^{out}(z)/q^{in}] = -\alpha_p \frac{N_1 - N_2 + 2M - M'}{\rho} \qquad (4.126)$$

In this model, CET occurs only between pairs of ions separated by a distance significantly smaller than the average. In Eq. (4.126), $N_1$ and $N_2$ are the ground and upper level populations of ions that are not affected by CET; $M = k\rho$ is the total population of ion pairs ($\rho$ = total Er$^{3+}$ concentration, $k$ = fraction of ion pairs); and $M'$ is the population of pairs in which one of the two ions is in the excited state. At moderately low pump powers, we can assume a negligible number of ion pairs where both ions are in the excited state, thus $2M - M'$ is the number of pairs where both ions are in the ground state. we also have $N_1 + N_2 + 2M = \rho$, and from the steady state rate equations $q = N_2/(N_2 - N_1) = M'/(M - M')$. The number of ions pairs $k$ can then be determined through a best fit of experimental measurements of $T = q^{out}/q^{in}$, as calculated from eq. (4.126) and the above relations. The results show that the fraction of ions pairs increases from $k = 1\%$ to $k = 20\%$ for a tenfold increase of Er$^{3+}$-concentration (i.e., $1.5 \times 10^{18}$ ions/cm$^3$ to $9 \times 10^{18}$ ions/cm$^3$). But the incorporation of alumina and phosphorous oxide codopants for comparable Er$^{3+}$ concentrations was shown to significantly reduce this fraction.

We consider now the important case of ytterbium–erbium sensitized luminescence, an aspect intensively studied and considered for applications in fiber amplifiers.

The first Er:glass laser was a flashlamp-pumped Yb$^{3+}$–Er$^{3+}$ system operating at room temperature (E. Snitzer and R. Woodcock [20]). Laser oscillation of Er$^{3+}$ has been obtained in CaWO$_4$, CaF$_2$, and LaF$_3$ crystals, and Li–Mg–Al–Si glasses, but only at $T = 77$ K, where Er$^{3+}$ operates like a four-level laser system [9]- There are three difficulties with room temperature oscillation of Er:glass: first, as a three-level system with large ground level thermal population, Er$^{3+}$ in bulk glass has a high inversion threshold; second, there is a poor overlap between the flashlamp spectrum and the Er$^{3+}$ absorption bands; third, a large fraction of the flashlamp energy corresponds to pump bands where pump ESA occurs. Room temperature operation of Er:glass lasers could only be achieved by sensitization through Yb$^{3+}$, which has a strong absorption band near 900 nm. The possibility of double sensitization through Nd–Yb–Er or Cr–Yb–Er energy transfers has also been extensively investigated (E. Snitzer, [237] and [9]; J. G. Edwards and J. N. Sandoe [212]; S. G. Lunter et al. [238]).

As recently as 1986, the development of RE-doped single-mode fibers allowed an unsensitized Er:glass laser to be operated at room temperature and in the CW regime (R. J. Mears et al. [58]). There are five reasons for the improvement in pumping efficiency using the fiber laser: the pump intensity is dramatically increased in a single-mode waveguide; the spatial overlap between the pump light and the laser medium is complete; the active interaction length is increased; with a fiber as the laser cavity, the overall cavity loss is reduced; the Er:glass is laser-pumped directly into one of the Er:glass absorption bands. Since the unsensitized Er-fiber lasers can efficiently be pumped, the main reason to consider ytterbium sensitization is the possibility of using alternate pump wavelengths, such as $\lambda = 1.064$ mm or 820–830 nm.

The basic principle of Yb–Er energy transfer through $^2F_{5/2}(Yb) \rightarrow \ ^4I_{11/2}(Er)$ and the principle of double sensitization with Nd–Yb–Er [212] are shown in Figure 4.37. By adding a small amount of neodymium to the Yb–Er glass the laser slope efficiency can be increased up to three times [237]. But at higher Nd concentrations, erbium is quenched by neodymium through the $^4I_{13/2}(Er) \rightarrow \ ^4I_{15/2}(Nd)$ resonant energy transfer. There also exists the possibility of back energy transfer to ytterbium through $^4I_{11/2}(Er) \rightarrow \ ^2F_{5/2}(Yb)$ if the nonradiative relaxation from $^4I_{11/2}$ is not dominant. A detailed experimental and theoretical investigation of the Nd–Yb–Er laser system, leading to the determination of optimum sensitizer concentrations can be found in [212]. With the possibility of laser pumping, as opposed to flashlamp pumping, the doubly sensitized scheme Nd–Yb–Er became rapidly obsolete.

Laser pumping of Yb–Er at $\lambda = 1.064 \ \mu m$ through an Nd:glass laser source was first shown by V. P. Gapontsev et al. [239]–[240]. As recently as 1987–1988, bulk Yb–Er:glass lasers pumped with an Nd:YAG source (D. C. Hanna et al. [241]), and single-mode fiber lasers pumped with Nd:YAG and Nd:YLF sources (M. E. Fermann et al. [242], G. T. Maker and A. I. Ferguson [243]) were demonstrated. The main reason for using an Nd:YAG or an Nd:YLF is that they can be pumped with compact and efficient laser diodes. The drawback is that the corresponding wavelengths ($\lambda = 1.064 \ \mu m$ and $\lambda = 1.053 \ \mu m$, respectively) are located on the steeply falling long wavelength tail of the 920 nm $Yb^{3+}$ absorption band; these two wavelengths also fall into the $^4I_{13/2}$–$^4I_{9/2}$ pump ESA band of $Er^{3+}$ (Figure 4.30). For both reasons, the shorter wavelength of the Nd:YLF is preferable.

The first demonstration of laser-pumped Yb–Er fiber lasers (D. C. Hanna et al. [244]) used pump wavelengths in the 0.8 $\mu m$ region (820–840 nm), closer to the $Yb^{3+}$ absorption peak. The advantage of 0.8 $\mu m$ pumping is that compact GaAlAs laser diodes can be used as the source. But 0.8 $\mu m$ pumping $Yb^{3+}$ suffers from $Er^{3+}$

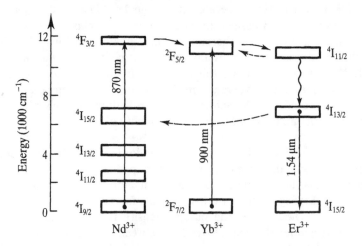

**FIGURE 4.37**   Principle of energy transfer between $Yb^{3+}$ and $Er^{3+}$ (sensitization), and between $Nd^{3+}$, $Yb^{3+}$, and $Er^{3+}$ (double sensitization). Possible back energy transfers from $Er^{3+}$ to $Yb^{3+}$ and to $Nd^{3+}$ are also shown in dashed arrows.

ESA, due to two ESA transitions near 790 nm and 840 nm (Figure 4.31 and Section 4.7). The existence of ESA in $Er^{3+}$ requires the pumping of $Yb^{3+}$ near the optimal wavelength, 830 nm [244]. This is also significantly detuned from the $Yb^{3+}$ absorption peak, 920 nm. A comprehensive theoretical analysis of laser-pumped Yb–Er systems for laser and amplifier applications can be found in the work of A. G. Murzin et al. [245]–[246].

Experimental studies of the Yb–Er energy transfer have shown that the Yb–Er quenching rate $W_{AB}$ is strongly dependent on the type glass host (V. P. Gapontsev et al. [240]). The highest quenching rates ($W_{AB} \approx 10^4 \, s^{-1}$) corresponding to sensitization quantum efficiencies close to unity are obtained in phosphate glasses. the quenching rate dependence on ytterbium concentration is observed to be linear in silicate and fluorophosphate glasses, while it is quadratic in phosphate and borate glasses; the dependence of $W_{AB}$ on erbium concentration is linear for all glasses [240]. Sensitization efficiencies up to 93% have also been obtained in Yb–Er germanate glasses (F. Auzel [230]).

The first application of Yb–Er sensitization to EDFAs used multicomponent phosphate (soft glass) fibers pumped with an Nd:YAG laser (S. G. Grubb et al. [247]). The reason for choosing this host material is its reduced Er–Yb back energy transfer. The reduction of back transfer is due to a stronger effect of multiphonon relaxation from the $^4I_{11/2}$ level [248]. But from fabrication and standard fiber compatibility considerations, silica-based fibers are more advantageous than soft glass fibers. Compositional studies have shown that adding small amounts of phosphorus to silica-based fibers results in a considerable improvement in Yb–Er energy transfer, with efficiencies up to 90%, [249]–[251]. The gain and saturation performance of Nd:YAG- and Nd:YLF-pumped Yb-sensitized EDFAs are reviewed in Chapter 5.

## 4.9 REFRACTIVE INDEX CHANGES AND RESONANT DISPERSION

*Dispersion* is the variation of refractive index with wavelength inside a medium. Refractive index dispersion (also called material dispersion) is caused by the electronic response of the medium to electromagnetic (EM) excitation. Atomic or laser resonances cause additional wavelength dependent and power dependent changes in the refractive index, hence they cause dispersion inside the medium. We consider a simplified model to provide a basic understanding of the origin of refractive index dispersion due to electronic resonances then examine the effect of atomic or laser resonances.

The theory developed by H. A. Lorentz, [30], [78], and [252], describes the response of electrons and nuclei under EM field excitation according to a harmonic oscillator model. In this mode, the Lorentz force (dominated by its Coulomb component) causes the electrons to oscillate about their equilibrium positions; this creates a linear, oscillating polarization. If $x$ represents the electron displacement, then the effective restoring force in the linear regime is given by $-kx$, where $k$ is analogous to a spring constant. The electron's motion is then given by the differential equation:

$$m_e \frac{d^2x}{dt^2} + m_e \Gamma \frac{dx}{dt} + kx = eE(t) \qquad (4.127)$$

where $m_e$ is the electron mass, $E$ is the driving electric field at frequency $\omega$, and $\Gamma$ is a damping coefficient, which accounts for effects of energy dissipation. The solution of eq. (4.127) is:

$$x(t) = \frac{e}{m_e} \frac{1}{\omega_0^2 - \omega^2 - i\Gamma\omega} E(t) \qquad (4.128)$$

where $\omega_0 = \sqrt{k/m_e}$ is a characteristic electronic resonance frequency. The macroscopic polarization density of the medium in the case of one-electron atoms is given by $P(t) = \rho\langle\mu(t)\rangle$ where $\langle\mu(t)\rangle = ex(t)$ is the macroscopic average dipole moment and $\rho$ the atomic density. We can also write $P(t) = \varepsilon_0 n^2 E(t) = \varepsilon_0[1 + \chi(\omega)]E(t)$, where $n$ is the medium refractive index and $\chi(\omega)$ is the medium electric susceptibility [78]. Using these relations, we obtain the medium complex refractive index:

$$n^2(\omega) = 1 + \frac{\rho e^2}{\varepsilon_0 m_e} \frac{1}{\omega_0^2 - \omega^2 - i\Gamma\omega} = 1 + \frac{\rho e^2}{\varepsilon_0 m_e} \frac{\omega_0^2 - \omega^2 + i\Gamma\omega}{(\omega_0^2 - \omega^2)^2 + \Gamma^2\omega^2} \qquad (4.129)$$

Equation (4.129) shows that, away from the electronic resonance, the real part of the refractive index increases monotonically with frequency $\omega$; there is a monotonic decrease with wavelength. In the case of an undoped glass medium, such as fused silica used in standard optical fibers, the electronic resonance is located at a wavelength near 0.1 $\mu$m, [30], [253], and [254]. The real part of the refractive index decays monotonically according to Eq. (4.129) over the wavelength range 0.1–5 $\mu$m. Beyond this range, the existence of lattice absorption resonances cause the refractive index to increase. In bulk and undoped fused silica, the combination of these resonances generates an inflexion point $(d^2n/d\lambda^2 = 0)$ at a wavelength near $\lambda = 1.3 \mu$m, [32], [180], and [255]. In silica fiber waveguides, this inflexion point can be shifted to longer wavelengths (e.g., 1.55 $\mu$m) by controlling the additional effect of waveguide dispersion, not considered here. Textbooks contain a detailed analysis of material and waveguide dispersion in single-mode fibers, [30]–[32] and [180].

On the other hand, Eq. (4.129) that the imaginary part of the refractive index, corresponding to a positive imaginary susceptibility $\chi''$, is maximum at the resonance frequency $\omega_0$. The analysis of plane wave EM propagation outlined in Section 1.11 shows that a positive imaginary susceptibility causes absorption of EM power, i.e., the medium transmission is given by $T = \exp(-\gamma\chi''z)$, where $\gamma$ is a constant. According to Eq. (4.129), the effect of absorption decreases as the EM frequency is detuned from the electronic resonance. The electric susceptibility $\chi = n^2 - 1$ defined by Eq. (4.129) corresponds to the case of a passive medium, such as undoped silica glass.

We consider next the case of a medium doped with activator ions (such as RE-doped silica glass). The electric dipole interaction occurring between the atomic electron clouds and the EM wave can be analyzed through the semiclassical density matrix theory of Chapter 1. The main result of this theory is that the polarization of the activator ions generates a complex atomic susceptibility $\chi = \chi' - i\chi''$, to be distinguished from the host electric susceptibility previously analyzed. Chapter 1 derives the complex atomic susceptibility for a three-level laser system with Stark split energy levels [256] and shows in particular that the real part $\chi'$ of the atomic susceptibility causes a refractive index change $\delta n(\omega)$ relative to the index of the host material. The change in refractive index is given by:

$$\delta n(\omega) = \Gamma_s \frac{1}{2nL} \int_0^L \chi'(\omega, z)\, dz \qquad (4.130)$$

On the other hand, the imaginary part $-\chi''$ causes amplification of EM light with power gain $G(\omega)$:

$$G(\omega) = \exp\left\{ -\Gamma_s \frac{\omega}{nc} \int_0^L \chi''(\omega, z)\, dz \right\} \qquad (4.131)$$

A well-known property of Hilbert transforms [76] relates the real and imaginary parts of the complex atomic susceptibility through the Kramers–Kronig relation (KKR):

$$\chi'(\omega) = \frac{1}{\pi} \text{P.V.} \int_{-\infty}^{+\infty} \frac{\chi''(\omega')}{\omega' - \omega}\, d\omega' \qquad (4.132)$$

Chapter 1 shows that the imaginary part $-\chi''$ is related to the cross-sections $\sigma_{a,e}(\omega)$ and atomic populations densities $N_{1,2}$ through:

$$-\chi''(\omega) = \frac{nc}{\omega} \left\{ \sigma_e(\omega)N_2 - \sigma_a(\omega)N_1 \right\} \qquad (4.133)$$

where $n$ is the refractive index of the host medium.

Using Eqs. (4.132) and (4.133) we can express the real part of the atomic susceptibility in the form:

$$\chi'(\omega) = -\frac{nc}{\omega} \left\{ N_2 \frac{1}{\pi} \text{P.V.} \int_{-\infty}^{+\infty} \frac{\sigma_e(\omega')}{\omega' - \omega}\, d\omega' - N_1 \frac{1}{\pi} \text{P.V.} \int_{-\infty}^{+\infty} \frac{\sigma_a(\omega')}{\omega' - \omega}\, d\omega' \right\} \qquad (4.134)$$

The Hilbert or KKR transform of the cross sections involved in Eq. (4.134) can be performed numerically, using the method described in [257]. Another method [256] fits each experimental cross section through a linear superposition of $N$ Lorentzian line shapes $\mathscr{L}_i(\omega) = a_i[1 + 4(\omega - \omega_i)^2/\Delta\omega_i^2]^{-1}$ $(i = 1, \dots, N)$, for which the KKR transform is given by:

$$-\frac{1}{\pi} \text{P.V.} \int_{-\infty}^{+\infty} \frac{\mathscr{L}_i(\omega')}{\omega' - \omega}\, d\omega' = 2 \frac{\omega - \omega_i}{\Delta\omega_i} \mathscr{L}_i(\omega) \qquad (4.135)$$

Using Eqs. (4.133)–(4.135) and the Lorentzian fitting method, we can write the real and imaginary parts of the atomic susceptibility in the explicit form:

$$\chi'(\omega) = \frac{nc}{\omega} \left\{ N_2 \sum_i 2a_i^e \mathscr{L}_i^e(\omega) \frac{\omega - \omega_i^e}{\Delta\omega_i^e} - N_1 \sum_i 2a_i^a \mathscr{L}_i^a(\omega) \frac{\omega - \omega_i^a}{\Delta\omega_i^a} \right\} \qquad (4.136)$$

$$-\chi''(\omega) = \frac{nc}{\omega} \left\{ N_2 \sum_i a_i^e \mathscr{L}_i^e(\omega) - N_1 \sum_i a_i^a \mathscr{L}_i^a(\omega) \right\} \qquad (4.137)$$

where the superscripts a,e indicate factors relative to absorption and emission cross sections, respectively. In Eqs. (4.136) and (4.137), the factors $a_i^{a,e}$ are in $m^2$, the population densities $N_{1,2}$ are in $m^{-3}$, and the susceptibility $\chi$ is dimensionless.

The real and imaginary parts of the complex atomic susceptibility corresponding to Type III fibers are plotted in Figures 4.38 and 4.39 for different values of the normalized pump power $q = 0$ to 100. In this example, an ideal three-level system with unsaturated gain is assumed, i.e., the population densities are given by

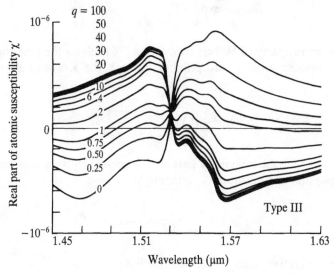

**FIGURE 4.38** Real part $\chi'$ of the complex atomic susceptibility spectrum of aluminosilicate fiber (Type III) obtained from Eq. (4.136) for different values of the normalized pump power $q = 0$ to 100, assuming an $Er^{3+}$ density $\rho_o = 10^{19}$ ions/cm$^3$.

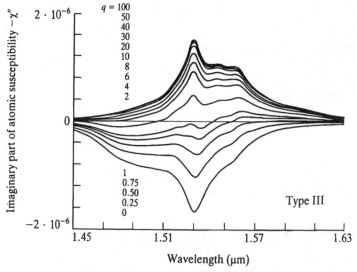

**FIGURE 4.39** Imaginary part $-\chi''$ of the complex atomic susceptibility spectrum of aluminosilicate fiber (Type III) obtained from Eq. (4.137) for different values of the normalized pump power $q = 0$ to 100, assuming an $Er^{3+}$ density $\rho_o = 10^{19}$ ions/cm$^3$.

**TABLE 4.2  Parameters for Lorentzian fitting of absorption and emission cross sections of Type III fiber (unity peak values)**

| | | Absorption | | | Emission | |
|---|---|---|---|---|---|---|
| Number $i$ | Amplitude $a_i^a$ | Center wavelength $\lambda_i^a$(nm) | Line width $\Delta\lambda_i^a$(nm) | Amplitude $a_i^e$ | Center wavelength $\lambda_i^e$(nm) | Line width $\Delta\lambda_i^e$(nm) |
| 1 | 0.22 | 1,479 | 36 | 0.02 | 1,470 | 42 |
| 2 | 0.25 | 1.497 | 36.5 | 0.13 | 1,500 | 39 |
| 3 | 0.22 | 1,515 | 35.5 | 0.23 | 1,520 | 19 |
| 4 | 0.72 | 1,531 | 18 | 0.66 | 1,530 | 12.5 |
| 5 | 0.17 | 1,544 | 16 | 0.38 | 1,545 | 22 |
| 6 | 0.21 | 1,555 | 18.5 | 0.35 | 1,559 | 20 |
| 7 | 0.04 | 1,567 | 18 | 0.11 | 1,575 | 36 |
| 8 | – | – | – | 0.07 | 1,600 | 42 |

$N_1 = \rho_0/(1 + q)$ and $N_2 = \rho_0 q/(1 + q)$, with $\rho_0 = 10^{19}$ Er$^{3+}$/cm$^3$ and the host refractive index $n = 1.46$. In order to compute Eq. (4.138), the absorption and emission cross sections of Type III fibers (Figure 4.22), were fitted with two sets of $N$ Lorentzian functions whose parameters $a_i^{a,e}$ (amplitude), $\lambda_i^{a,e} = 2\pi c/\omega_i^{a,e}$ (center wavelength) and $\Delta\lambda_i^{a,e} = (\lambda_i^{a,e})^2 \Delta\omega_i^{a,e}/c$ (line width), $i = 1,\ldots, N$, are listed in Table 4.2. Cross section fits using Lorentzian functions (involving here seven and eight curves for absorption and emission, respectively) are not unique; they only represent a fair approximation of the experimental line shapes. For simplicity, the sums $\sum \mathscr{L}_i(\omega)$ of the $N$ Lorentzian functions calculated from the parameters in Table 4.2 are chosen to give a peak value of unity for each cross section. The result $\sum \mathscr{L}_i(\omega)$ must therefore be multiplied by the experimental peak cross section values, which are for this fiber $\sigma_a^{peak} = 4.7 \times 10^{-21}$ cm$^2$ and $\sigma_e^{peak} = 4.4 \times 10^{-21}$ cm$^2$ (Table 4.1).

Figures 4.38 and 4.39 show that from the unpumped case ($q = 0$) to the near-complete inversion case ($q = 100$), the sign of both parts of the susceptibility is reversed. For the real part $\chi'$, the spectra are nearly antisymmetrical with respect to a symmetry point at the peak wavelength $\lambda = 1.53$ $\mu$m, while the imaginary part $-\chi''$ changes from negative (unpumped case) to positive (full inversion). For a three-level pumping scheme, the set of curves shown in Figures 4.38 and 4.39 is universal; it corresponds to nearly all possible pump power values; the curves scale linearly with the Er$^{3+}$ concentration $\rho_0$. A set of curves can be obtained for a two-level pumping scheme ($\lambda_p \approx 1.48$ $\mu$m) through the same method. But in this case, the spectrum of $\chi'$ is restricted to signal wavelengths $\lambda_s > \lambda_p$, since the KKR transform is not applicable otherwise (Section 1.11). The Lorentzian-fitting method for the computation of $\chi'$ was also used in [258] in the case of a fully inverted, Er-doped fiber of characteristics similar to Type II.

From Eqs. (4.130) and (4.136), the refractive index change $\delta n(\omega)$ is given by:

$$\delta n(\omega) = \frac{\Gamma_s c}{2\omega}\left\{\sum_i 2a_i^e \mathscr{L}_i^e(\omega) \frac{\omega - \omega_i^e}{\Delta\omega_i^e} \frac{1}{L}\int_0^L N_2(z)\, dz - \sum_i 2a_i^a \mathscr{L}_i^a(\omega)\frac{\omega - \omega_i^a}{\Delta\omega_i^a}\frac{1}{L}\int_0^L N_1(z)\, dz\right\}$$

(4.138)

The above equation shows that the EDFA refractive index change spectrum $\delta n(\omega)$ can be computed from knowledge of the atomic populations $N_{1,2}$, which are functions of the normalized pump power $q(z)$ and the fiber coordinate $z$. Equation (4.138) does not apply to the saturation regime, for in this case the KKR transform is invalid, as discussed in Section 1.11 and [76]. A computation of the refractive index change in the saturation regime would require a full knowledge of the individual Stark transition parameters, as shown in Chapter 1. When the atomic populations $N_{1,2}$ are nearly constant (as in the unpumped of fully inverted regimes), the refractive index change is proportional to the real susceptibility, i.e., $\delta n(\omega) = (\Gamma_s/2n)\chi'(\omega)$. Assuming an overlap factor $\Gamma_s = 0.5$ and $n = 1.46$, we can measure from Figure 4.38 that the range of refractive index change is $-1.5 \times 10^{-7} < \delta n < +1.5 \times 10^{-7}$, corresponding to a maximum absolute difference of $|\delta n| \approx 3 \times 10^{-7}$.

For optical communications applications, the parameter of interest is not the fiber refractive index $n(\lambda)$ but the *material dispersion* $D(\lambda)$, defined through [32], [33], [133], and [180]:

$$D(\lambda) = -\frac{\lambda}{c}\frac{d^2 n(\lambda)}{d\lambda^2} \tag{4.139}$$

In a fiber waveguide, the effect of optical pulse propagation not only depends on material dispersion but also on *waveguide dispersion*, which is not considered here. The total fiber dispersion, also called *chromatic dispersion* or *group velocity dispersion* (GVD), is given by the sum of material and waveguide dispersions. For a detailed analysis of GVD, see for instance [32] and 180]. If the total refractive index is $n(\lambda) = n_H(\lambda) + \delta n(\lambda)$ were $n_H$ is the host refractive index, the total dispersion (overlooking waveguide dispersion) is given by $D(\lambda) = D_H(\lambda) + D_R(\lambda)$, where $D_H$ is the host dispersion in the absence of rare earth dopant. The dispersion $D_R(\lambda)$ induced by the atomic resonance (or RE-doping), also referred to in the literature as *resonant* GVD [258] or *anomalous dispersion* [259], is given from Eqs. (4.130) and (4.139) by:

$$D_R(\lambda) = -\frac{\lambda}{c}\frac{d^2 \delta n(\lambda)}{d\lambda^2} = -\frac{\lambda\Gamma_s}{2n_H c}\frac{1}{L}\int_0^L \frac{d^2\chi'(\lambda, z)}{d\lambda^2}\,dz \tag{4.140}$$

The spectra of the refractive index change $\delta n(\lambda)$ and the resonant dispersion $D_R(\lambda)$ corresponding to Type III fiber, as calculated from Eqs. (4.130) and (4.140) are plotted in Figure 4.40, for the two cases in which the atomic populations are uniform: $q = 0$ (unpumped fiber, or $N_1 = \rho_0$, $N_2 = 0$) and $q = 100$ (near-complete inversion, or $N_1 = 0$, $N_2 = \rho_0$). The other parameters assumed are $\rho_0 = 10^{19}$ $Er^{3+}/cm^3$, $n = 1.46$ and $\Gamma_s = 0.5$. As the figure shows, the index changes are within $\delta n = \pm 1.5 \times 10^{-7}$. This variation actually represents a small perturbation of the host refractive index; indeed, in the case of standard fibers ($a = 4.5$ $\mu$m, NA $= 0.1$) the core cladding index difference is typically $\Delta n = 3.5 \times 10^{-3}$, while for high NA fibers ($a = 1.5$ $\mu$m, NA $= 0.3$) used in EDFA applications, this difference is higher, or $\Delta n = 3.10^{-2}$ [260].

The spectrum of the resonant dispersion $D_R(\lambda)$ shown in Figure 4.40 exhibits several peaks and valleys of comparatively narrow width (5–10 nm), with a main feature near the peak cross section wavelength $\lambda = 1.53\ \mu$m. For the parameters considered in the example of Figure 4.40 ($\rho_0 = 10^{19}\ Er^{3+}/cm^3$, $\Gamma_s = 0.5$) the maximum dispersion variation corresponding to full inversion is $D_R^{max} \approx \pm 30$ fs/m/nm. Figure 4.41 shows experimentally measured resonant dispersion spectra for a multicomponent glass EDFA with an $Er^{3+}$ concentration of $\rho_0 \approx 1, 5.10^{20}$ ions/cm³ (K. Takada et al. [261]). The input pump powers $q = 0.2$, and 36 corresponding to the three curves

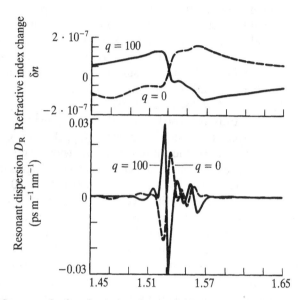

**FIGURE 4.40** Spectra of refractive index change $\delta n(\lambda)$ and resonant dispersion spectra $D_R(\lambda)$ for unpumped fiber ($q = 0$) or fiber with near-complete inversion ($q = 100$), corresponding to Type III fiber. The $Er^{3+}$ density is assumed to be $\rho_o(Er^{3+}) = 10^{19}$ ions/cm³ or $\rho_o(Er_2O_3) = 0.12$ wt%.

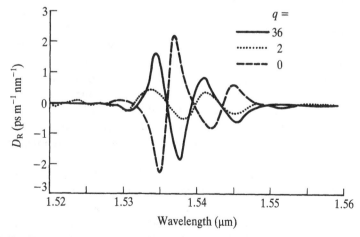

**FIGURE 4.41** Experimental resonant dispersion spectra $D_R(\lambda)$ measured in multicomponent EDFA for three input pump values of $q = 0, 2$ and 36. The $Er^{3+}$ density is $\rho_o(Er^{3+}) \approx 1.5 \times 10^{20}$ ions/cm³ or $\rho_o(Er_2O_3) = 2$ wt%. From [261], reprinted with permission from *Electronic Letters*, 1992.

shown are normalized to a 6 mW threshold. The experimental spectra of $D_R(\lambda)$ are qualitatively very similar to the theoretical spectra shown in Figure 4.40. The difference in number of peaks or valleys is clearly due to changes in the cross section line shapes between the two hosts (i.e., aluminosilicate and multicomponent glasses). The difference in maximum resonant dispersion between the two examples of Figures 4.40 and 4.41 (i.e., $D_R^{max} \approx \pm 0.03$ ps/m/nm and $D_R^{max} \approx \pm 2.5$ ps/m/nm) can be attributed mainly to the difference in $Er^{3+}$ concentration (1:15 ratio) but also to differences in cross section peak values, cross section tail derivatives, and overlap factors.

The maximum resonant dispersion of $D_R^{max} \approx \pm 30$ fs/m/nm found in the theoretical example of Figure 4.40 also corresponds to $D_R^{max} \approx \pm 30$ ps/km/nm, which in absolute value is of the same magnitude as the material dispersion of standard fibers near $\lambda = 1.55$ $\mu$m [31]–[33] and [180]. A similar result ($D_R^{max} \approx \pm 60$ fs/m/nm) was actually measured experimentally in an EDFA with concentration $\rho_0 \approx 1.6 \times 10^{19}$ $Er^{3+}$/cm$^3$ or $\rho_0(Er_2O_3) = 300$ ppm (A. V. Velov et al. [259]).

But EDFAs with concentrations close to $\rho_0 = 10^{19}$ $Er^{3+}$/cm$^3$ cannot be used with kilometer lengths, since given the other parameters $\sigma_e(\lambda_{peak}) = 4.4 \times 10^{-21}$ cm$^2$ and $\Gamma_s = 0.5$, the corresponding gain coefficient is at the peak wavelength $\Gamma_s \rho_0 \sigma_e(\lambda_{peak}) \approx$ 2.2 m$^{-1}$, or 9.5 dB/m. In the case of distributed amplifiers with lengths in the 10–25 km range, the absorption coefficient is $\alpha_s = 2$–10 dB/km $\approx 0.5$–$2.5 \times 10^{-3}$ m$^{-1}$ (Section 2.6), which corresponds to a range in $Er^{3+}$ concentration $\rho_0 \approx 2 \times 10^{15}$–$1 \times 10^{16}$ ions/cm$^3$. For such distributed amplifiers, the resonant dispersion is 1000 to 5000 times smaller than in the previous example, since it scales linearly with $Er^{3+}$ concentration. These considerations show that in EDFA-based communication systems, whether the EDFAs have short lengths and high $Er^{3+}$ concentrations (i.e., $L < 50$ m, $\rho_0 > 10^{19}$ ions/cm$^3$) or long lengths and low $Er^{3+}$ concentrations (i.e., $L > 1$ km, $\rho_0 < 10^{16}$ ions/cm$^3$), the effect of resonant dispersion is negligible compared to that of the material dispersion of the undoped fiber (F. Matera et al. [258], S. C. Fleming and T. J. Whitley [262]). For a detailed analysis of total or chromatic dispersion in EDFAs see for instance [259] and [263].

For short devices such as Er-doped fiber lasers, however, the above examples show that the resonant dispersion can be comparable to the material dispersion of the fiber host, which can produce significant effects, [258], [262], and [263]. The effect of resonant dispersion can also be significant in femtosecond pulse amplification, which results in temporal broadening and distortion (M. Romagnoli et al. [263]). In other types of applications, the enhancement of refractive index dispersion near the EDFA absorption bands can be used for nonlinear optical switching, as demonstrated with two-core fibers (R. A. Betts et al. [264]). Optical switching based on this principle is slow ($<1$ ms) compared to other nonlinearities of electronic origin (e.g., self-phase modulation [265]), owing to the three-level system gain dynamics of Er:glass (Chapter 5).

Because of the potential applications or limitations that resonant dispersion could generate in the field of femtosecond amplification and fiber lasers, the experimental determination of refractive index changes and chromatic dispersion represents an important part of the EDFA characterization. Characterization experiments are described in [261], [262], and [266]. Important effects of glass codopants, such as $GeO_2$, $Al_2O_3$, and $P_2O_5$, on chromatic dispersion in Er-doped fibers are described in [267].

## 4.10  EFFECT OF PUMP AND SIGNAL POLARIZATION

Very important in fiber communication systems is the gain insensitivity of EDFAs to the polarization states (SOPs) of pump and signal. Fundamentally, such SOP insensitivity is an intrinsic property of the Er:glass medium. At the atomic level, the electric dipole transitions of rare earth ions in any type of host have tensor character, i.e., the complex atomic susceptibility $\chi$ is a tensor as opposed to a scalar (Chapter 1). For this reason, the effect of stimulated emission with respect to the signal electric field orientation is anisotropic [77]. In the case of a crystalline host, the rare earth is surrounded with a lattice structure of well-defined symmetry. The static field perturbation induced by the crystal host lifts the degeneracy of the laser transitions (Stark effect), which determines reference axes for each transition's tensor response. This response is either linear along the DC field perturbation axis ($\pi$ transitions), or circularly polarized about it ($\sigma$ transitions). Therefore, and most generally, the response of single, nondegenerate laser transitions is elliptically polarized about the host's reference axis, and therefore it cannot be isotropic [77].

But in the case of a glass host, each rare earth site experiences a crystal field perturbation of random orientation. Because of such randomness, the tensor response of the medium, averaged at the macroscopic level, turns out to be effectively isotropic. Thus, while the response of individual rare earth centers is intrinsically anisotropic, their macroscopic response in a glass is the same for all directions, and the signal gain is therefore isotropic [268].

Despite the property of gain isotropy, there exist three particular effects in RE:glass that are associated with a partially anisotropic response. These effects are *polarized luminescence, polarized laser oscillation*, and *polarization hole burning* (POHB).

In RE:glass, the effect of polarized luminescence corresponds to the selective excitation of a subgroup of rare earth sites in which individual Stark transitions accidentally share the same oscillator characteristics (V. P. Lebedev and A. K. Przhevuskii [268]; P. P. Feofilov [269]). The luminescence (or fluorescence) polarization properties of such a system can be well described theoretically [269]. If $I_\parallel$ and $I_\perp$ are the fluorescence intensities measured in the SOP directions parallel and perpendicular to the direction of linearly polarized pump light, respectively, the fluorescence's degree of polarization $\mathscr{P}$ is defined as

$$\mathscr{P} = \frac{I_\parallel - I_\perp}{I_\parallel + I_\perp} = \frac{1 - \mathscr{R}}{1 + \mathscr{R}} \tag{4.141}$$

where $\mathscr{R} = I_\perp / I_\parallel$ is the depolarization ratio defined in Eq. (4.29). According to such definition, a fully isotropic response (unpolarized fluorescence) would give a degree of polarization $\mathscr{P} = 0$ or $\mathscr{R} = 1$. A maximum value of $\mathscr{P} = 0.5$ or $\mathscr{R} = 1/3$ is obtained when the Stark transitions corresponding to the absorption and the emission are described by the same linear oscillators, [268] and [269], which would be the case if the rare earth ion behaved as a pure dipole. In particular, the condition $\mathscr{P} = 0.5$ occurs in the case of *resonance excitation*, i.e., the emission is observed at the same wavelength as the excitation. For instance, an experimental value $\mathscr{P} = 0.4$ close to the theoretical limit was measured for the single Stark transition $^7F_0 \leftrightarrow {}^5D_0$ at $\lambda = 579$ nm in $Eu^{3+}$:glass [268]. In the case where the LSJ $\leftrightarrow$ L'S'J' transition has several Stark components that strongly overlap, as for $Er^{3+}$ at room temperature (Section 4.4), polarized fluorescence due to resonance excitation is significantly reduced [268]. On the other hand, in the case of nonresonant excitation, a weakly polarized fluorescence (e.g., $|P| \leqslant 0.05$) can generally be observed, which indicates that a partial correlation remains between the excitation and fluorescence SOPs at the macroscopic level. For instance, $\mathscr{P} = 0.04$ was measured in $Er^{3+}$ : tellurite glass for the nonresonant excitation ($^4I_{15/2} \rightarrow {}^2G_{9/2}$, absorption; $^4S_{3/2} \rightarrow {}^4I_{15/2}$, emission) [268].

The effect of polarized laser oscillation has been investigated in Nd- and Er-doped fiber lasers (J. T. Lin et al. [270] and [271]; J. T. Lin and W. A. Gambling [272]). In this effect, the laser output SOP is observed to be related to the input pump SOP. In particular, two independent laser polarization eigenmodes can be selectively excited, each having a different threshold, slope efficiency, and relaxation oscillation frequency [272]. The theoretical model developed by D. W. Hall et al. [273] applied to the case of fiber lasers, [271] and [272], showed that these effects are well explained by the existence of a microscopic stimulated emission cross section anisotropy. Such an isotropy is characterized by a polarized cross section ratio $A = \sigma_q/\sigma_p$.

If, at the atomic level, we assume the emission cross section is different for the three site orientations $\hat{l}$, $\hat{m}$ and $\hat{n}$, the total cross section corresponding to a signal polarized in the direction $\hat{k}$, is given by [273]:

$$\sigma = \sigma_p(\hat{l} \cdot \hat{k})^2 + \sigma_q(\hat{m} \cdot \hat{k})^2 + \sigma_r(\hat{n} \cdot \hat{k})^2 \qquad (4.142)$$

For simplicity, and without loss of generality, we can assume $\sigma_q \approx \sigma_r$ and $\sigma_p > \sigma_q$. As two cross sections are assumed equal, the orientation of a given site is determined by only two spherical angles, corresponding to a differential solid angle $d\Omega$. The effective cross section corresponding to this site in the polarization direction $\hat{k}$ is then given by:

$$\sigma(\Omega, \hat{k}) = \sigma_p\{(\hat{l} \cdot \hat{k})^2 + A(\hat{m} \cdot \hat{k})^2 + A(\hat{n} \cdot \hat{k})^2\} \qquad (4.143)$$

The measurements of slope efficiencies of polarized laser oscillation described in [271] made possible to determine polarized cross section ratios of $A = 0.01$ for Nd-doped fibers, and a higher value of $A = 0.11$ for Er-doped fibers. To date, the dependence of the ratio $A$ with Er:glass composition and the effect of Al/Ge codoping have not been investigated. Chapter 5 shows the degree of polarization of ASE, as measured experimentally under various pump and saturating signal conditions is negligible ($\mathscr{P} \approx 0$). How such an experimental observation can be explained in spite of an apparently high polarized cross section ratio ($A = 0.11$) remains to be fully analyzed.

We consider the effect of polarization hole burning (D. W. Hall et al. [273]; D. W. Hall and M. J. Weber [274]), briefly described in Section 4.4. In the effect of polarization hole burning (POHB, to distinguish from the effect of PHB discussed in Section 4.4), a weak signal probe is experiences gain anisotropy after the RE:glass has been traversed by a short saturating signal pulse. This postpulse gain anisotropy can be explained by selective depopulation of certain classes of ions in the glass. Indeed, the effect of depopulation (i.e., in which excited ions return to the ground level state) is stronger for the class of ions whose principal cross section $\sigma_p$ is oriented in the direction parallel to the saturating signal SOP [273]. Selective depopulation reduces the small-signal gain in the SOP direction parallel to the saturating signal, gain anisotropy is observed, hence the name polarization hole burning. The probability of saturating signal absorption with SOP along direction $\hat{k}$, followed by probe signal stimulated emission of with SOP along direction $\hat{k}'$ orthogonal to $\hat{k}$ is then given by the product $\sigma(\Omega, \hat{k})\sigma(\Omega, \hat{k}')$ integrated over the total solid angle $\Omega$ [273]. The depolarization ratio $\mathscr{R}$ is then given by:

$$\mathscr{R} = \frac{\displaystyle\int_{\Omega} \sigma(\Omega, \hat{k})\sigma(\Omega, \hat{k}') \, d\Omega}{\displaystyle\int_{\Omega} [\sigma(\Omega, \hat{k})]^2 \, d\Omega} \tag{4.144}$$

The calculation of $\mathscr{R}$ in Eq. (4.144) using Eq. (4.143), which is detailed in [273] yields the following result:

$$\mathscr{R} = \frac{6A^2 + 8A + 1}{8A^2 + 4A + 3} \tag{4.145}$$

The above result shows that for a medium made of pure dipoles (i.e., $A = 0$, corresponding to $\sigma_p \neq 0$, $\sigma_q = 0$), the most anisotropic case, the depolarization ratio is $\mathscr{R} = 1/3$, or the degree of polarization defined in Eq. (4.141) is $\mathscr{P} = 0.5$. On the contrary, in the most isotropic case ($A = 1$, corresponding to $\sigma_p = \sigma_q$) the depolarization ratio is $\mathscr{R} = 1$, corresponding to a degree of polarization $\mathscr{P} = 0$. Experimental measurements of the depolarization ratio through resonant FLN gave $\mathscr{R} = 0.89$ or $\mathscr{P} = 0.06$ for Nd:glass (ED-2 type), corresponding to a polarized cross section ratio $A = 0.42$, [273] and [274].

The possibility of POHB in EDFAs does not mean that the gain is polarization sensitive with respect to the saturating signal SOP. The effect of POHB can be produced in any polarization orientation, as there are no priviledged directions in the glass. The gain anisotropy induced by POHB can only be experienced by a probe having nearly the same wavelength as the saturating signal. The effect of POHB, which to date has not been investigated in EDFAs, could be masked by several causes: (1) the multiplicity of Stark transitions contributing to signal gain; (2) the strong spectral overlap between these Stark transitions; (3) the fast thermalization process taking place within each Stark manifold; (4) the cross-relaxation between ions; and (5) the polarization scrambling due to the fiber intrinsic birefringence. Cross-relaxation and scrambling can be suppressed by investigating $Er^{3+}$-doped fibers with low concentrations (i.e., $\rho < 100$ ppm) and short lengths (i.e., $L = 1$ m).

As discussed in Chapter 5, the sensitivity of the EDFA gain with respect to pump and signal polarizations, even in the high saturation regime, is observed to be negligible. This advantageous property of EDFAs was shown both by direct gain measurements and by polarized ASE measurements. However, alternate and highly accurate investigation techniques of polarized fluorescence and POHB could reveal small SOP dependent effects in EDFAs. Recent long-haul transmission experiments have actually shown a related effect of polarization-dependent loss (PDL), as discussed in Section 7.2. Such effects could be analyzed from independent measurements of polarized cross section ratios, such as obtained with Er-doped fiber lasers.

# CHAPTER 5

# GAIN, SATURATION, AND NOISE CHARACTERISTICS OF ERBIUM-DOPED FIBER AMPLIFIERS

## INTRODUCTION

This chapter describes the different characteristics of Er-doped fiber amplifiers (EDFAs). These characteristics are essentially related to the amplifier gain, saturation, and noise properties; all three are generally coupled together. Ideally, the EDFA should yield the highest gain possible (given the pump power available), while having the highest saturation output power and lowest noise possible. The combined EDFA characteristics of gain, saturation power and noise are often referred to in the literature as the EDFA performance.

This chapter shows that the overall EDFA performance is actually limited by several physical effects and device constraints. The most fundamental limitation is due to the energy conservation principle: the maximum signal energy that can be extracted from a fiber amplifier cannot exceed the pump energy that is stored in it. This principle is important in most EDFA applications.

The principle of energy conservation can be expressed in terms of photon flux (number of photons per second). If $\phi_p^{in} = P_p^{in}/h\nu_p$ is the flux of pump photons input to the amplifier, and $\phi_s^{in, out} = P_s^{in, out}/h\nu_s$ the flux of signal photons input and output from the amplifier ($P_p^{in}, P_s^{in}, P_s^{out}$ = input pump, input signal, and output signal powers), we must have:

$$\phi_s^{out} \leqslant \phi_p^{in} + \phi_s^{in} \tag{5.1}$$

The equality in Eq. (5.1) corresponds to a regime where all pump photons are (virtually) converted into signal photons by the amplifier. The inequality in Eq. (5.1) reflects several possible effects: (a) some pump photons can traverse the amplifier without interacting with the activator ions; (b) pump photons can also be lost due to various causes, e.g., background or impurity loss; and (c) a fraction of the pump energy absorbed by the activator ions is lost by spontaneous emission, not converted into signal energy. The inequality is also valid when the three-level laser amplifier is unpumped ($\phi_p^{in} = 0$), in which case the signal flux is attenuated, or $\phi_s^{out} \leqslant \phi_s^{in}$. Equation (5.1) can also be expressed in terms of input/output powers:

$$P_s^{out} \leqslant P_s^{in} + \frac{\lambda_p}{\lambda_s} P_p^{in} \tag{5.2}$$

Equation (5.2) shows that the maximum output signal power extracted from the amplifier depends upon the wavelength ratio $\lambda_p/\lambda_s$. As for any pumping scheme (i.e., two-level or three-level) we have $\lambda_p < \lambda_s$, and as generally we also have $P_s^{in}/P_p^{in} \ll 1$, the *power conversion efficiency* $P_s^{out}/P_p^{in}$ of the amplifier is necessarily less than unity. This result also shows that the conversion efficiency is different for each pump band; the highest value is theoretically obtained with the two-level pumping scheme, where the ratio $\lambda_p/\lambda_s$ is the closest to unity.

Equation (5.2) can also be expressed in terms of amplifier gain $G$. For an amplifier free from spontaneous emission, we have $G = P_s^{out}/P_s^{in}$, and from Eq. (5.2):

$$G \leqslant 1 + \frac{\lambda_p}{\lambda_s} \frac{P_p^{in}}{P_s^{in}} = 1 + \frac{\phi_p^{in}}{\phi_s^{in}} \tag{5.3}$$

Equation (5.3) shows that the upper limit for the gain corresponds approximately to the flux ratio $\phi_p^{in}/\phi_s^{in}$. Thus, the maximum possible amplifier gain corresponds to the case where each pump photon is converted into one signal photon. For a very large input signal power value such that $P_s^{in} \gg (\lambda_p/\lambda_s)P_p^{in}$, the maximum amplifier gain reaches the limit of unity, which corresponds to a condition of medium transparency. This last relation also shows that for achieving a maximum gain $G$, the signal input power or input flux cannot exceed a certain value, i.e., $P_s^{in} \leqslant (\lambda_p/\lambda_s)P_p^{in}/(G-1)$ or $\phi_s^{in} \leqslant \phi_p^{in}/(G-1)$.

The upper limit for the amplifier gain, given by consideration of energy conservation or Eq. (5.3), can be reached only if all pump photons are actually absorbed by the amplifying medium. In actual amplifiers, the absorption of pump photons is limited by the finite number of rare earth ions existing in the medium. For three-level laser systems where the laser transition terminates to the ground level (such as Er:glass), a condition of full medium inversion corresponds to transparency at the pump wavelength or pump absorption bleaching. Under the condition of full inversion, the concentration of rare earth ions in the excited state is $\rho = N_2$ (averaged in the transverse plane to account for mode overlap) and from the theory in Chapter 1 the maximum signal gain corresponding to a three-level laser medium of length $L$ is given by:

$$G = \frac{P_s^{out}}{P_s^{in}} = \exp(\rho \sigma_e L) \tag{5.4}$$

where $\sigma_e$ is the signal emission cross section. The gain expressed in Eq. (5.4) cannot be indefinitely enhanced by increasing the fiber length $L$ or the rare earth concentration $\rho$, due to the principle of energy conservation (5.1), which limits the output signal power $P_s^{out}$, expressed in Eq. (5.2). We conclude that the maximum possible amplifier gain and output signal power are given by the lowest between the two limits given obtained in Eqs. (5.3) and (5.4),

$$
G \leqslant \min\left\{\exp(\rho\sigma_e L), \, 1 + \frac{\lambda_p}{\lambda_s}\frac{P_p^{in}}{P_s^{in}}\right\}
\tag{5.5}
$$

$$
P_s^{out} \leqslant \min\left\{P_s^{in}\exp(\rho\sigma_e L), \, P_s^{in} + \frac{\lambda_p}{\lambda_s}P_p^{in}\right\}
\tag{5.6}
$$

Using neither formalism nor detailed analysis, we have shown that optical amplifier performance in maximum gain and output signal power is limited by a fundamental principle. Therefore, the broad range of EDFA characteristics obtained using different pumping wavelengths, pumping schemes, fiber designs, glass compositions, and various specific parameters, described in this chapter, should be interpreted with the background knowledge of the amplifier's fundamental limits.

In reality, the actual EDFA performance is also limited by secondary physical effects as well as practical technology considerations. Secondary effects include: pump ESA, self-saturation by ASE, concentration quenching, and inhomogeneous broadening. On the other hand, basic technology limitations include: maximum pump power available from laser diodes, control of Er-doping profile, optical isolators extinction ratio, and fiber background loss. These limitations are important in particular applications. Chapter 5 comprehensively reviews the EDFA characteristics of gain, saturation, and noise, and it emphasizes the relation between theoretical predictions from the models described in Chapter 1 and the most important experimental results reported to date.

Section 5.1 outlines the features of current semiconductor laser diodes (LD) and other LD-pumped compact sources, which can be used to pump the EDFA. Other optical components, such as wavelength selective couplers and Faraday isolators, which are generally included in all-fiber EDFA modules, are also briefly described.

In Section 5.2, the gain versus pump power characteristics of EDFAs are reviewed for different possible pump bands. The dependence of EDFA gain with signal power and the EDFA output saturation characteristics are described in Section 5.3. The issue of EDFA power conversion efficiency is addressed in this section, with a comparison between forward, backward, and bidirectional pumping schemes, and results obtained for various pump bands. The effect of inhomogeneous gain saturation is also described theoretically and experimentally.

The ASE noise spectrum and noise figure characteristics of the EDFA are described in Section 5.4. Section 5.5 focusses on the effect of EDFA self-saturation, and outlines a practical method of theoretical analysis not described in Chapter 1. A simplified analysis shows that the discrepancy observed between different pumping schemes can be mainly attributed to the effect of ASE.

Section 5.6 focusses on the optimization of the basic EDFA parameters, i.e., fiber length, fiber cutoff wavelength and core radius, Er-doping concentration, and profile.

Amplifier phase noise is addressed theoretically and experimentally in Section 5.7. Effects of multiple reflections and Rayleigh backscattering in EDFAs are described in Section 5.8.

The transient gain dynamics of the EDFA are described theoretically and experimentally in Section 5.9. The regimes of picosecond and femtosecond pulse amplification, and soliton pulse amplification are described theoretically and experimentally in Sections 5.10 and 5.11, respectively.

Finally, Section 5.12 provides a comparison between EDFAs and other fiber amplifiers, including: Raman, Brillouin, and parametric fiber amplifiers, and other RE-doped fiber amplifiers.

## 5.1 CHARACTERISTICS OF PUMP LASER DIODES AND EDFA-RELATED COMPONENTS

An EDFA module is made of a strand of Er-doped fiber, a pump source, and various optical components. The pump source is the most essential component and has the most elaborate design. The breakthrough in optical amplifier technology by EDFAs and other RE-doped fiber amplifiers (Section 5.12) is not only due to the advantages of the RE:glass fibers as amplifying media, but also to the fact that such media can be pumped with practical, compact, and efficient laser diodes (LD). The improvement in EDFA performance since early investigations (1985–1988) has actually resulted from three main causes: the identification of most suitable pump bands, the optimization of fiber composition and design, and the rapid development of high power, single-transverse mode pump LDs. A measure of this progress can be given by considering the evolution over three years in LD-pumped EDFA performance: in 1988, the first LD-pumped EDFA used for the pump was a 500 mW GaAlAs LD array at 807 nm, of which 15 mW could be coupled into the fiber; a gain of 6 dB was achieved, representing a maximum gain coefficient of 0.4 dB/mW (T. J. Whitley, [1]). In 1990, the use of 980 nm strained quantum well LDs made it possible to achieve gains of 30 dB with only 3 mW pump, corresponding to a maximum gain coefficient of 11 dB/mW (M. Shimizu et al., [2]).

The LD pump sources used in EDFA applications must be single transverse mode (for efficient coupling into single-mode fibers) and are generally multiple longitudinal mode (Fabry–Perot or FP lasers). These sources belong to three main categories, which nearly correspond to the center wavelengths of the three most efficient Er:glass pump bands, [3]–[5]: (1) InGaAsP and multiquantum well (MQW) InGaAs at 1480 nm, (2) strained quantum well (QW) inGaAs at 980 nm, and (3) GaAlAs at 820 nm. Another type of pump source for Yb-sensitized EDFAs (or EYDFAs) is the LD-pumped Nd:YAG (or Nd:YLF) operating at 1.064 $\mu$m or (1.047/1.053 $\mu$m), [6] and [7]. Such sources use high power (e.g., 3 W) GaAlAs laser diode arrays for pumping the Nd crystals. An alternate approach for EYDFA pumping uses double-clad, Nd-doped fiber lasers instead of Nd:crystals [8]. Finally, EYDFAs with a double-clad fiber structure can also be pumped directly with GaAlAs diode arrays [8].

Other possible LD-pumped light sources for the EDFA operate in the visible wavelength range. One is the frequency-doubled LD-pumped Nd:YAG operating at 532 nm, which falls into one EDF absorption band [9]. High power (500 mW) visible GaInP/AlGaInP laser diodes operating at 665 nm, developed for pumping Cr-doped solid-state lasers [10] could also be used for the EDFA. Unlike the 532 nm band, the 665 nm band suffers from the effect of pump ESA and is therefore of limited interest for LD Pumping. The current focus on 820 nm, 980 nm, 1064 nm, and 1480 nm for EDFAs is essentially motivated by considerations of amplifier performance (pumping efficiency), device cost, device reliability (reflecting development history), and the high power (fraction of a watt) currently available from LD sources or LD-pumped sources operating at these wavelengths. In 1992, modules for the three types of high power LDs were available from several manufacturers, with a packaging that includes coupling into standard single-mode fiber.

The first high power LDs for 1480 nm pumping used InGaAsP devices with lengthened cavities (500–800 $\mu$m) for improved heat dissipation; the diode structure was a V-grooved inner stripe on a $p$-substrate (VIPS), characterized by a small leakage current (H. Horikawa et al. [11]; K. Yamada et al. [12]). These pump sources could provide a room temperature CW output of 190 mW [12]. The use of a multiple QW in a double channel, planar buried heterostructure (DCPBH) in InGaAs material made it possible to reduce power absorption by internal loss and bring the LD output power to 170 mW (I. Mito et al. [13]). Current output power performance of MQW–DCPBH 1480 nm lasers with 1800 $\mu$m long cavities is in excess of 250 mW [3]. The schematic structure and $L$–$I$ response of such devices are shown in Figure 5.1. Aging tests conducted for MQW–DCPBH indicated a potential of several 100 thousand hours for 100 mW, CW operation at 20 °C (I. Mito and K. Endo [3]). Another approach for 1480 nm pumped LDs is the strained-layer GaInAsP/GaInAsP MQW separate confinement heterostructure (SCH), for which CW powers near 250 mW have been reported (H. Kamei et al. [14], H. Asano et al. [15]). The performances of the two types of pump LDs and other possible types can eventually be compared not only in terms of maximum output power, but also in terms of epitaxial growth procedures, laser slope efficiency, temperature stability, long-term reliability, and output beam characteristics.

The first pump LDs operating at 980 nm wavelength for EDFA pumping were based on a strained lawer QW InGaAs ridge waveguide structure (S. Uehara et al. [16]). In these devices, a strained single QW was sandwiched between a linearly graded index separate confinement heterostructure (GRIN–SCH), providing output powers near 100 mW, [16] and [17]. Improvements of this approach with single or double QWs with optimized facet coatings have made it possible to achieve CW output powers in the 240–270 mW range (A. Larsson et al. [18]; T. Takeshita et al. [19]). The typical structure and L–I characteristics (with output spectra) of a GRIN–SCH LD pump device are shown in Figure 5.2 [18]; in this example, the cavity length is 600 $\mu$m, and the two laser facets are HR and AR coated, respectively. Aging experiments conducted with these devices have shown stable operation over 5000 hours under severe operating conditions [17]. It can be concluded from this type of testing that strained-layer InGaAs LDs have the degree of long-term reliability required for practical EDFA pumping applications. More recent developments in this field, using strained-layer InGaAs double QW, ridge waveguide lasers with HR/AR coatings, made it possible to achieve 980 nm output powers near 450 mW, with single transverse mode operation up to 320 mW (H. A. Zarem et al. [20]). As in the case of 1480 nm pump LDs, the performances of 980 nm devices must be compared not only in terms of maximum achievable output power, but device fabrication procedure and yield, slope efficiency, temperature stability, device reliability, and output beam characteristics.

In comparison to 980 nm and 1480 nm LD sources, GaAlAs LDs operating near

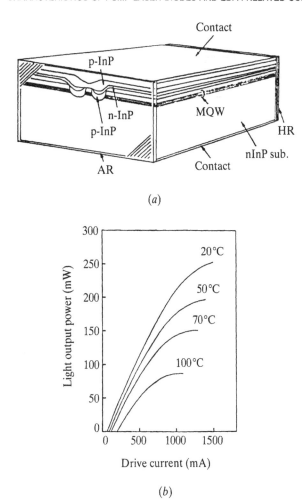

**FIGURE 5.1** (*a*) Basic structure of MQW–DCPBH InGaAs laser diode operating at 1480 nm; (*b*) device power–current response for 1800 $\mu$m cavity length. From [3], reprinted with permission from the Optical Society of America.

820 nm offer the practical advantages: higher output powers (several watts), superior long-term reliability, and relatively lower cost. The main drawback of 810 nm pumping is the effect of pump ESA (Chapter 4), which reduces the power conversion efficiency. In spite of this limitation, the three advantages of available power, realibility, and reduced cost could be conducive to a large-scale commercial development of 820 nm pumped EDFA modules. For instance, EDFAs with 30 or 40 dB gain can be realized using a pump source with one or two commercially available 827 nm GaAsAs LDs that are used for compact disc memory applications (M. Horiguchi et al. [5]). It is possible that future developments in realiability and production yields for 980 nm and 1480 nm LD sources, as driven by the large-scale deployment of long-haul communications systems, could make 810 nm pumping of EDFAs a less competitive approach.

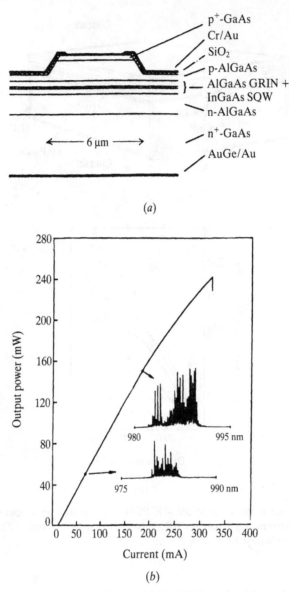

(a)

(b)

**FIGURE 5.2** (a) Basic structure of SQW GRIN–SCH strained-layer InGaAs–GaAs laser diode operating at 980 nm; (b) device power–current response for 600 $\mu$m cavity length. From [18] © 1990 IEEE.

The fourth category of compact pump sources for EYDFAs are diode-pumped lasers (DPL). Such DPLs are made of GaAlAs phased-array LD-pumped Nd:YAG or Nd:YLF, which operate at 1.064 $\mu$m or 1.047/1.053 $\mu$m, [6] and [7], or GaAlAs phased-array LD-pumped Nd-doped fiber lasers operating at 1.053 $\mu$m [8]. The advantage of this approach is that the LD sources used to pump the DPLs are efficient, reliable, and scalable in array size. DPLs also act as brightness converters, as several hundreds of milliwatts of 1.04–1.06 mm laser output can be coupled into a single-mode fiber ($LP_{01}$) from a single, 1 W LD array, [6] and [7]. The first EYDFA based on this pumping scheme used an Nd:YAG DPL (S. G. Grubb et al. [21]). Using a single DPL of this type, gains in excess of 40 dB and output signal powers in excess of 70 mW were demonstrated; with two DPLs, the output signal power could be raised to 130 mW (S. G. Grubb et al. [6]). With a single Nd:YLF, DPL, an EYDFA with 50 dB gain and 290 mW output signal power could also be demonstrated (S. G. Grubb et al. [7]). This last device used a 3 W GaAlAs LD as the DPL pump, which provided 900 mW of 1.06 $\mu$m pump into the doped fiber [7]. The amplified signal power can actually be scaled up by using bidirectional pumping; in this case, it was shown experimentally that 500 mW output signal can be obtained with a 1.5 W pump at 1.06 $\mu$m [7]. These results clearly indicate the great potential of DPL-pumped EYDFA for high power amplifier applications.

Another approach for DPL–EYDFAs uses an LD-array-pumped Nd-fiber lasers for the DPL. In order to achieve a high pumping efficiency from the LD array, the ND.doped fiber has a double-cladding structure; the first cladding surrounding the $Nd^{3+}$-doped core acts as a single-mode waveguide for the 1.06 $\mu$m laser light, while the second or outer cladding enables multimode coupling of the pump array light (E. Snitzer et al. [22]). An improvement of the double-clad fiber design based on a rectangular first cladding made it possible to increase the pump absorption efficiency to 85% (H. Po et al. [23]). The output power characteristics of a 1.053 $\mu$m, 1.2 m long Nd fiber laser pumped with an 808 nm GaAlAs LD-array and based on the improved double-clad fiber design, are shown in Figure 5.3 (J. D. Minelli et al. [8]). The DPL slope efficiency is 45%, which provides 700 mW output at $\lambda = 1.053$ $\mu$m for 1.5 W launched pump power. With this DPL source, an EYDFA gain and output signal power in excess of 100 mW was demonstrated [8].

It was recently demonstrated that EYDFAs with double-clad fiber structures can also be pumped directly with GaAlAs diode arrays (J. D. Minelli et al. [24]). In this demonstration, a 1 W GaAlAs diode array operating at $\lambda = 962$ nm wavelength was used, which made it possible to demonstrate an EYDFA gain and output signal power of 24 dB and 50 mW, respectively. Improvements in fiber geometry and core composition could increase the power conversion efficiency up to 30%, which corresponds to a signal output power in excess of 250 mW [24]. This approach

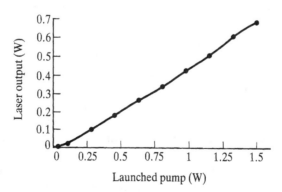

**FIGURE 5.3** Output power at 1.053 $\mu$m versus launched pump power at 808 nm of a double-clad, Nd-doped fiber laser pumped with a GaAlAs phased-array LD. From [8], reprinted with permission from the Optical Society of America, 1992.

obviates an external DPL. However, in order to achieve a fiber-to-fiber amplifier based on this double-clad fiber structure, additional optical components (e.g., optical circulator, dichroic filter) must be included [24].

The output power available from the different LD pump sources described above actually far exceeds (by a factor 10 to 100) the power required to achieve complete Er-doped fiber inversion and high gains ($G = 30$–$50$ dB). But this large power margin is necessary to compensate four types of effects: the pump input coupling loss, the amplifier gain saturation, the pump splitting loss, and the fiber background loss.

The pump input coupling loss is caused the imperfect transmission and limited coupling efficiencies of various optical components located in the pump path. These components include: LD-to-fiber coupling lens/apparatus, pump optical isolator, wavelength selective coupler, coupler taps, and several fiber splices. Recent progress in the design of fiber microlenses has brought the LD-to-fiber coupling efficiency to near 100% (H. M. Presby [25]). On the other hand, a conservative value for the total loss caused by other passive components is 1 dB, corresponding to 80% transmission. These results indicate that, in optimized conditions, a fraction of at least 80% of the LD output power can be used for pumping the EDF.

The second reason for which a large pump power margin is necessary is the effect of EDFA saturation. As the signal power increases, the amplifier gain saturates, which can be compensated by a higher input pump power, according to Eq. (5.3).

In cases where the pump power is shared by several EDFAs (such as in star or broadcast fiber networks, Chapter 7), the pump splitting loss is also compensated for by the available power margin. Previous considerations show that theoretically, 10 to 100 EDFAs could be simultaneously pumped by a single LD (or two LDs, for redundancy and reliability).

In the case of distributed amplifiers and remotely pumped amplifiers (Chapter 6), the effect of fiber background loss can be important; significantly higher pump powers are required. The analysis made in Chapter 1 showed that for a distributed EDFA with 10–25 km length, the pump power ($\lambda = 1480$ nm) required for transparency can vary by a factor two to three, depending on the Er concentration. For 100 km distributed EDFAs, the pump power required for transparency is at least 50 mW, about 10 times higher than the power required for a high gain EDFA.

The above considerations show that a power margin of 10 to 100 for the pump is actually necessary in many practical applications of EDFAs.

We have reviewed the different types of infrared LD sources that can be used to pump EDFAs (or EYDFAs). We consider now the passive optical components that, in addition to the pump source, are necessary to assemble a fiber amplifier module, i.e., a device that can be connected to or inserted into a fiber system. We describe EDFA-related components and EDFA module architectures only for reference, not to suggest an optimum design. Several ways exist to build EDFA modules. Their many applications fall into three categories: power amplifiers, in-line amplifiers, and preamplifiers. Furthermore, the technology of EDFA-related optical components is rapidly evolving, closely following rapid developments in EDFA performance optimization. Choices for specific optical components in EDFA modules are dictated by systems standards and specifications, environmental constraints, and economic considerations. System standards and specifications can largely vary when considering applications such as long-haul, undersea, or terrestrial links; analog or digital video broadcast distribution; and high capacity local area networks.

The three basic EDFA device architectures, corresponding to unidirectional and bidirectional pumping, are illustrated in Figure 5.4. The three layouts include one or two pump LD modules with fiber output (also called fiber pigtail); one or two wavelength selective couplers or combiners (WSC) which, in the literature, are also called wavelength division multiplexing couplers (WDM); and one strand of Er-doped fiber (EDF), or in some cases, of Yb-Codoped EDF. All these optical components have single-mode fiber (SMF) pigtails and are spliced together. Usually, EDFs have smaller cores and higher NAs compared to standard SMFs, and the two fiber splices between dissimilar SMF and EDF must be donw through nonstandard techniques. Finally, these three layouts may be complemented by one or two optical isolators (polarization independent Faraday isolators), required by the possible existence of reflections sources located before or after the EDFA. Without optical isolators, these architectures can also be used for bidirectional amplification (Chapter 6).

A generic architecture for a unidirectionally pumped EDFA module is illustrated in Figure 5.5. In this example, the module includes two optical isolators, and two control loops for pump and signal power stabilization and regulation. Alternate versions of this EDFA module could include optical fiber taps at the input and the output of the EDF for the purpose of monitoring pump and signal powers and spectra. An automatic gain control loop (AGC), can also be implemented by

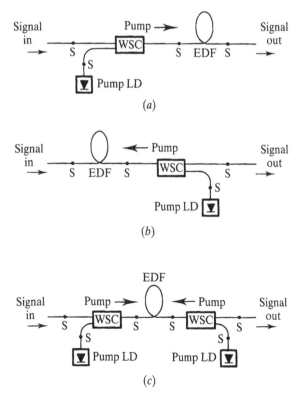

**FIGURE 5.4** Basic EDFA device architectures: (*a*) with unidirectional forward pumping, (*b*) with unidirectional backward pumping, and (*c*) with bidirectional pumping. S: fiber splice, WSC: wavelength selective coupler, EDF: Er-doped fiber.

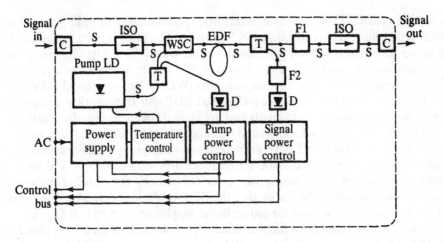

**FIGURE 5.5** Generic example of EDFA module architecture using forward pumping and automatic gain control loop. C: fiber connector, S: fiber splice, ISO: Faraday isolator, WSC: wavelength selective coupler, EDF: Er-doped fiber, T: fiber tap, F1, F2: optical filters, D: photodetector.

monitoring the EDF input and output signal powers, and feeding back the resulting data to the pump module (Chapter 6). Examples of EDFA modules described in the literature and typical performance parameters of commercially available units are described in Section 5.6.

We consider next several issues relevant to passive EDFA-related components: pump-to-fiber coupling efficiency, wavelength selective coupler technologies, SMF to EDF splicing, optical filters, and EDF radiation darkening.

The LD-to-single-mode-fiber coupling efficiency can, with optimized fiber microlens design, approach 100% (H. M. Presby [25]). There exist many possible methods for LD–SMF coupling based on a variety of lenses. These lenses can be spherical, aspherical, or graded index; or assembled as microscope objectives. Microlenses can also be fabricated on SMFs by melting the end of a tapered fiber (H. Kuwahara et al. [26]). Four characteristics are required for ideal LD coupling lenses: a large numerical aperture (NA), a focal length that matches both laser and fiber transverse modes, low spherical aberration, and low surface or Fresnel reflection. A detailed analysis by H. M. Presby et al. shows that the first three conditions cannot be satisfied with spherical lenses; for typical LD and SMF parameters, the intrinsic loss is near $-2.4\,\text{dB}$, corresponding to $\eta = 57\%$ coupling efficiency [25]. This limit is well confirmed by experimental measurements; the best result obtained with spherical microlenses is $\eta = 55\%$ or $-2.5\,\text{dB}$ [27]. Using standard AR-coated microscope objectives with optimized NA, typical coupling efficiencies are usually near $\eta = 50\%$ or $-3\,\text{dB}$. On the other hand, the analysis in [25] shows that the three conditions above are satisfied with aspheric lenses of hyperbolic shape. Fiber microlenses with hyperbolic shape can be realized by a technique of laser micromachining [28]. With additional AR coating, such optimized lenses have a theoretical coupling efficiency near 100%, approached experimentally with the result $\eta = 90\%$ or $-0.45\,\text{dB}$ [25]. A 2–3 dB improvement in LD–SMF coupling loss is important for communications systems, as it increases the power margin, and for LD-pumped EDFA applications.

Another important passive component is the wavelength selective coupler (WSC), also called WDM coupler, used to couple the pump power into the EDF. There are several requirements for an ideal WSC: low excess loss and coupling loss at pump and signal wavelengths, polarization insensitivity, large extinction ratio between the two wavelengths (low crosstalk), low backscatter or reflection, high power damage threshold (for power amplifier applications), large signal bandwidth, temperature stability (terrestrial system applications), compact size, and cost-effectiveness. For forward-pumping, where the WSC is placed on the signal path before the EDF, the WSC insertion loss increases by the same amount as the EDFA noise figure (similarly to a Type C amplifier chain element, Chapter 2). Backscattering and reflections can also generate excess ASE noise in the signal propagation direction and cause BER penalty (Section 5.8).

The four main technologies for the realization of compact, packaged WSCs for EDFA applications are based on: fused–tapered filber couplers (FTFC), evanescent-field polished fiber couplers, integrated optics WDM couplers and bulk optics interference filters (IFs). While many other types of WDM technologies are possible, only these have been considered and developed for the EDFA application market so far.

FTFCs generally use standard SMF; integrated optics WDM couplers and bulk optics IFs require fiber pigtailing, a cause of additional insertion loss. With the IF approach, the effects of imperfect mechanical alignment and beam expansion also cause device insertion loss. A detailed performance comparison between the four aforementioned types of WDC could be rapidly outdated, owing to rapid improvements in related technologies. For instance, an experimental comparison of the FTFC and IF technologies made in 1990 (A. Lord et al. [29]), showed that standard FTFCs exhibit a slight polarization sensitivity (0.5 dB), while there is none in IF couplers. The reason for polarization sensitivity in FTFCs is that many coupling cycles are necessary for achieving a large extinction ratio and high wavelength selectivity [30]. Improved coupler designs, including fiber twist [29], optimized cross sections and cladding refractive index, and optimized fusion parameters [30], have made it possible to reduce FTFC polarization sensitivity in the spectral region of interest ($\lambda_p$, $\lambda_s$) to less than a 0.3 dB level [29].

Of the four WSCs IFs differ in their spectral transmission. The other three are based on the principle of supermode coupling in twin waveguide structures; this generally gives a periodic, raised cosine spectral transmission. In the region of interest, the IF coupler has a step-like spectral transmission [29]. But FTFCs can also be realized with an aperiodic spectral response (K. O. Hill et al. [31]–[33]; F. Gonthier et al. [34]). The advantage of aperiodicity is a flat transmission at the signal wavelength, which can be obtained over a broad spectral range (e.g., 30 nm). A flat transmission with broad range is required in WDM transmission systems, in which several optical channels must be amplified simultaneously.

Finally, the different WSC technologies can be compared on the basis of splice loss and coupling loss, extinction ratio, backscatter, and temperature stability. The comparative performance also depends upon which pump wavelength is considered (e.g., 980 nm or 1480 nm) and the type of standard fiber used. An example of comparative FTFC performance for 980 nm pumping is given in [35].

Power coupling between standard fibers and Er-doped fibers is important. As optimized EDFs usually have smaller mode field diameters (MFD) compared to standard fibers (Section 5.6), SMF–EDF butt coupling or fusion splicing through standard procedures usually yields a significant coupling loss (e.g., > 3 dB). This coupling loss is undesirable for two main reasons: it reduces the pump power launched into the EDF and increases the signal noise figure by the same amount. The reduction or total suppression of coupling loss between fibers of dissimilar MFDs can be achieved by three methods: physical tapering of fiber ends [36] and [37], thermal diffusion tapering, [38]–[43], and multiple splicing of intermediate fibers [44]. Thermal diffusion tapering is based on the principle that fiber core expansion occurs when the fiber is exposed to high temperatures. This expansion is due to the diffusion of index-raising codopants (such as germanium or aluminium) into the cladding, which changes the fiber MFD characteristics. A doubling of MFD can be realized by exposing the high NA fiber end to a heat treatment for a certain duration (e.g., $T = 1300\,^\circ\text{C}$ for 10 hours [40]). But for SMF–EDF coupling, direct splicing by arc fusion is more straightforward. The power coupling efficiency can be monitored at 1530 nm or 1500 nm during splicing, which enables us to optimize the arc splicing parameters (i.e., arc power and fusion time). The splice loss value at $\lambda = 1.53$–$1.54\,\mu\text{m}$ can then be accurately calibrated using a white spectrum source.

A detailed description of arc fusion parameters is given in [42]. Depending upon the type of EDF core glass, the typical SMF–EDF coupling loss obtained through this method is less than 0.5 dB. The loss is usually higher in the EDF–SMF coupling than in the SMF–EDF coupling [42]. By an appropriate control of the arc fusion parameters, the coupling loss can be considerably suppressed. For instance, an SMF–EDF coupling loss of less than 0.1 dB was obtained in both splice directions with an EDF core radius $a = 1.85\,\mu\text{m}$ and numerical aperture NA = 0.42 [42].

Some EDFA applications need to include passive optical filters. In long EDFA chains, they can be used at each amplification stage for reducing ASE noise buildup [44]. In optical preamplifier applications, optical filters are necessary for rejection of ASE noise and minimization of ASE–ASE beat noise [45]. In forward pumped EDFA preamplifiers, they are also necessary to reject (and possibly to recycle) the fiber output pump power, which otherwise could cause excess noise in the detector [46]. Chapter 6 discusses how optical filters can also be used in EDFAs for achieving spectral gain equalization or flattening, [47]–[49]. We consider here only the issue of pump rejection. The pump filtering function alone can be performed with an in-line, narrowband (< 1 nm) IF placed into a beam expander and tuned to the signal wavelength, or by using a standard WSC at the EDFA output. This last solution is required in WDM applications, where the signal spectrum can spread over several nanometers. Other types of filters for pump rejection can be realized based on the principle of doping a fiber section with an absorbing rare earth [50], or by etching a Bragg reflection grating onto the fiber cladding [51]. In particular, the use of chirped Bragg gratings makes it possible to reflect up to 99% of the pump back into the EDFA, while the signal is transmitted over a wide spectral band (> 20 nm) [49]. When considering reflective amplifier configurations, the effect of pump recycling by reflection makes it possible to achieve gain enhancement (Chapter 6).

The effect of Er-doped fiber darkening by ionizing radiation is of importance for system applications in radiation-rich environments (oceanic seabed) or having exposures to brief radiation pulses. Fiber darkening results from color center formation in the glass host and causes both temporary and permanent increases in fiber background loss [52]. In an EDFA, this excess loss affects both pump and signal, and eventually reduces the net amplifier gain. Recent experimental studies of radiation effects on EDFAs have shown that darkening is strongly dependent upon the fiber core composition and is more important in the case of short wavelength pumping (R. B. J. Lewis et al. [53]; G. M. Williams et al. [54]). Both studies have shown that EDFs with Ge–Al codoping have considerable radiation darkening in comparison to standard Ge-doped fibers; the effect is also sensitive to the $Er^{3+}/Al_2O_3$ stochiometry [54]. The study of [53] shows that a radiation exposure of more than 50 Gy causes at least a 1 dB change in EDFA gain, which leads to the conclusion that the cumulative effect of seabed radiation ($< 5$ mGy/year) over a 10 year period is negligible. But for short pulse, intense radiation exposure, EDFAs can be completely disabled (i.e., turned from high gain to lossy fiber devices); the effect is most sensitive when the EDFA is pumped at the shortest wavelengths, 800 nm or 980 nm [54].

## 5.2 GAIN VERSUS PUMP POWER

The gain versus input pump power characteristics are the first and most important measure of the EDFA performance. Throughout this book, we use input pump power to mean the same as launched pump power, which refers to the pump power at the input of the EDF. We must distinguish this notion from absorbed pump power, sometimes used in the literature. Absorbed pump power is the difference $P_p^{abs} = P_p^{in} - P_p^{out}$. For EDFAs, absorbed pump power is generally not relevant, since the amount of unabsorbed pump power due to bleaching can be significant in some cases. Thus, expressing the gain characteristics as a function of absorbed pump power does not reflect how much launched pump power is necessary to achieve the result. Absorbed pump power is relevant to four-level laser systems (e.g., Nd-doped fiber amplifiers), since there is no bleaching effect and in optimized cases, most of the pump is absorbed along the amplifier length, so that $P_p^{abs} \approx P_p^{in}$.

The first gain measurements in Er-doped fibers used as a traveling-wave amplifiers were reported by the University of Southampton in 1987 (R. J. Mears, L. Reekie, I. M. Jauncey, and D. N. Payne, [55] and [56]). The pump was an argon-pumped dye laser operating near $\lambda = 665$ nm, and the maximum gain was $G = 28$ dB for 100 mW absorbed pump power. The same year, AT&T Bell Laboratories reported results obtained with an argon-pumped laser operating at $\lambda = 514.5$ nm (E. Desurvire, J. R. Simpson, and P. C. Becker [57]). In this last experiment, the maximum EDFA gain was $G = 22$ dB for 100 mW launched pump power, which could be increased to $G = 29$ dB by cooling the fiber in liquid nitrogen ($T = 77$ K). Some important aspects of the gain measurements reported in [56] and [57] are shown in Figures 5.6 and 5.7.

Figure 5.6*a* shows the EDFA gain as a function of pump wavelength obtained for different pump powers in the same EDFA length ($L = 3$ m); Figure 5.6*b* shows the EDFA gain as a function of signal wavelength for the maximum pump $P_p^{abs} = 100$ mW [56]. There exists an *optimum pump wavelength* near $\lambda = 665$ nm for which the gain is maximized; this corresponds to the peak absorption wavelength. But at high pumps, the gain is nearly independent of the pump wavelength (over a 30 nm range); this reflects a condition of full population inversion along the fiber. The gain versus pump wavelength characteristics for the 800 nm, 980 nm, and 1480 m, pump bands are discussed later in this section.

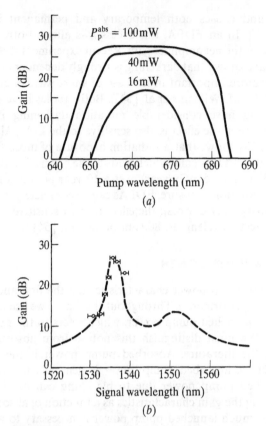

**FIGURE 5.6** Early experimental characterization of EDFA: (*a*) gain versus pump wavelength near $\lambda = 665$ nm, and (*b*) gain spectrum (experimental points) and ASE spectrum (dashed line). From [56], reprinted with permission from *Electronics Letters*, 1987.

Figure 5.6*b* shows by extrapolation from the ASE spectrum that the EDFA gain spectrum is fairly broad (30 nm for $G > 10$ dB), with two peak features near $\lambda = 1535$ nm and $\lambda = 1555$ nm, respectively, specific to Ge-doped EDFs (Type 1). The experimental gain spectrum characteristics of the EDFA are also analyzed later in this section.

The measurements reported in [57] and plotted in Figure 5.7 show the power evolution along the fiber of the pump, signal, and ASE, as well as the EDFA gain versus pump power for different EDFA lengths at the peak signal wavelength $\lambda = 1531$ nm, corresponding to an Al–Ge-doped EDF. In Figure 5.7*a*, the pump and signal power versus length data were fitted with theoretical curves obtained from a basic three-level system model (small adjustments of fiber parameters were necessary for best fits, as the effect of pump ESA was not accounted for in this model). The most important feature of these measurements is the existence of an *optimum length* $L_{opt}$, different for each input pump power, for which the signal gain is maximized. For amplifier lengths $L > L_{opt}$, the signal is reabsorbed along the fiber, as a result of an absence of population inversion in this fiber section, corresponding to a greater population in the $^4I_{15/2}$ ground level. EDFA length optimization is addressed in

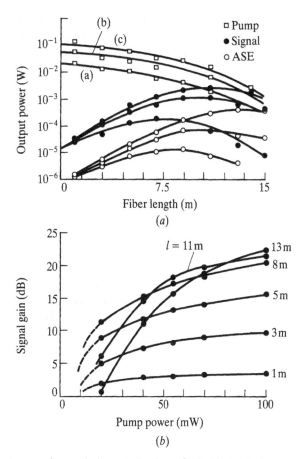

**FIGURE 5.7** Early experimental characterization of EDFA: (a) (□) pump, (●) signal, and (○) ASE powers versus length and (b) signal gain at $\lambda = 1.531\ \mu$m versus input pump power at $\lambda = 514$ nm for different fiber lengths. From [57], reprinted with permission from the Optical Society of America, 1987.

Section 5.6. Figure 5.7b, consistent with the previous observation, shows that the gain versus input pump power characteristics strongly depend upon the EDFA length. As the length is increased, the pump power $P_p^{\text{trans}}$ required for transparency ($G = 0$ dB) also increases. $P_p^{\text{trans}}$ can be interpreted as the input pump power required for bleaching the signal absorption, integrated along the fiber length, up to the transparency level corresponding to a net gain of unity. For a longer fiber, this transparency power must be higher, since the signal absorption is increased. The optimum length and the transparency pump power represents characteristics specific to three-level laser systems such as Er:glass.

We focus now on the physical interpretation of gain versus pump power characteristics, such as shown in Figure 5.7b. A basic analysis of these characteristics does not require the most general theoretical model described in Chapter 1, where the transverse mode and Er-doping distributions, as well as the ASE noise are taken into account. For this analysis, it is sufficient to use the theoretical model of the unsaturated gain regime and confined Er-doping [58]. Section 1.9 yielded a simple analytical expression for the input pump $P_p^{in}$ as a function of EDFA gain $G$, given the set of parameters: fiber length $L$ signal wavelength $\lambda_s$, pump and signal absorption coefficients $\alpha_p$ and $\alpha_s$. In the case of negligible fiber background loss (Section 2.4) this expression takes the form, Eqs. (2.135) and (2.136):

$$P_p^{in} = \alpha_p L \frac{Q_s}{1 - \exp[(Q_s - 1)\alpha_p L]} P_{sat}(\lambda_p) \tag{5.8}$$

with

$$Q_s = \frac{1 + \eta_p}{1 + \eta_s}\left[1 + \frac{\log(G)}{\alpha_s L}\right] \tag{5.9}$$

In Eq. (5.9), the cross section ratios $\eta_{p,s} = \sigma_e(\lambda_{p,s})/\sigma_a(\lambda_{p,s})$ are characteristic parameters of the core Er:glass, as seen in Chapter 4. Equation (5.8) is valid for both forward and backward unidirectional pumping schemes.

We consider now three cases of interest: the unpumped EDFA ($P_p^{in} = 0$), the EDFA at transparency ($P_p^{in} = P_p^{trans}$), and the EDFA with very high input pump power ($P_p^{in} \to \infty$). The first case corresponds to $Q_s = 0$ in Eqs. (5.8) and (5.9), which yields the EDFA gain $G(0) = \exp(-\alpha_s L)$. The net gain $G(0)$ represents the lossy transmission of the unpumped fiber with signal absorption coefficient $\alpha_s$, and corresponds to the case where all $Er^{3+}$ ions in the fiber are in the ground level ($N_1(z) = \rho_0$).

In the case of fiber transparency, we find from Eq. (5.8) after setting $G = 1$ in Eq. (5.9) the transparency pump power $P_p^{trans} = P_p^{in}(G = 1)$ corresponding to the signal wavelength $\lambda_s$:

$$P_p^{trans} = \alpha_p L \frac{\dfrac{1 + \eta_p}{1 + \eta_s}}{1 - \exp\left(-\dfrac{\eta_s - \eta_p}{1 + \eta_s}\alpha_p L\right)} P_{sat}(\lambda_p) \tag{5.10}$$

This last result shows that transparency pump power $P_p^{trans}$ is proportional to the pump saturation power $P_{sat}(\lambda_p)$ and increases with EDFA length. For lengths greater than the characteristic value $L_0 = [(1 + \eta_s)/(\eta_s - \eta_p)]/\alpha_p$, the increase is nearly linear.

Finally, the case of very high pump ($P_p^{in} \to \infty$) corresponds from Eq. (5.8) to $Q_s = 1$, or from Eq. (5.9) to a maximum EDFA gain $G(\infty) = \exp\{[(\eta_s - \eta_p)/(1 + \eta_p)]\alpha_s L\}$. For a three-level pumping scheme, we have $\eta_p = 0$ (Chapter 1), and using Eq. (1.212) for the signal absorption coefficient $\alpha_s$, the maximum gain is then $G(\infty) = \exp(\eta_s \alpha_s L) = \exp(\Gamma_s \rho_0 \sigma_e L)$. This gain value corresponds to the condition of complete and uniform population inversion along the fiber, i.e., $N_2(z) = \rho_0$. For a two-level pumping scheme ($\eta_p \neq 0$), the inversion is necessarily incomplete (Chapter 1), and the maximum gain is somewhat lower than in the three-level pumping case, or

$$G(\infty) = \exp\{[(1 - \eta_p/\eta_s)/(1 + \eta_p)]\Gamma_s \rho_0 \sigma_e L\}.$$

Since this result corresponds to unsaturated conditions, we can conclude that the maximum possible EDFA gain is given by the limit $G(\infty)$, or using the principle of energy conservation:

$$G \leqslant \min\left\{\exp\left(\frac{\eta_s - \eta_p}{1 + \eta_p}\alpha_s L\right), 1 + \frac{\lambda_p}{\lambda_s}\frac{P_p^{in}}{P_s^{in}}\right\} \tag{5.11}$$

The above result is valid for both saturated and unsaturated gain regimes, and represents a generalization of Eq. (5.5), since it includes the case of the two-level pumping scheme.

The above considerations show that in the unsaturated gain regime, the curve $G = f(P_p^{in})$, representing the EDFA gain versus pump power, passes through two points $(0, G(0))$ and $(P_p^{trans}, 1)$, and has an asymptotic value $G(\infty)$ corresponding to $P_p^{in} \rightarrow \infty$. This is also true in the saturated gain regime (including amplifier self-saturation), with the exception of the values of $P_p^{trans}$ and $G(\infty)$, which in the general case must be determined numerically.

Figure 5.8 plots $G = f(P_p^{in})$ curves for different EDFA lengths, obtained from Eq. (5.8) with typical parameters $P_{sat}(\lambda_p) = 1\,\text{mW}$, $\alpha_p = 1\,\text{dB/m}$, $\alpha_s = 2\,\text{dB/m}$, $\eta_p = 0$ and $\eta_s = 0.9$; the two points $(0, G(0))$ and $(P_p^{trans}, G(P_p^{trans}) = 1)$ and the limiting gain value $G(\infty)$, for the case $L = 14\,\text{m}$ are shown. The EDFA gain curve reflects three possible pumping regimes: the underpumped regime $(P_p^{in} < P_p^{trans})$, where the fiber is lossy $(G < 1)$; the incomplete inversion regime $(P_p^{in} \geqslant P_p^{trans})$, where the gain steeply increases with pump power; and the near-complete inversion regime $(P_p^{in} \gg P_p^{trans})$, where the gain approaches the limiting value $G(\infty)$.

The slopes of the tangents to the curves $G = f(P_p^{in})$ passing through the origin of Figure 5.8 are by definition the EDFA *gain coefficients*, expressed in dB/mW units.

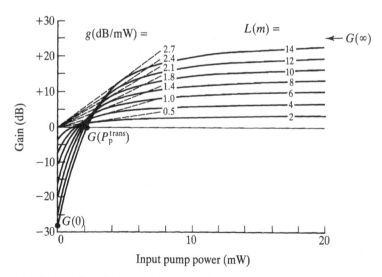

**FIGURE 5.8** Theoretical EDFA gain versus pump power characteristics for different fiber lengths, after Eqs. (5.8)–(5.9). For $L = 14\,\text{m}$, the gains $G(0)$, $G(P_p^{trans})$ and $G(\infty)$ correspond to the unpumped EDFA, to the EDFA pumped at transparency, and to the EDFA with infinite pump, respectively. The slope of the dashed lines shown for each gain curve corresponds to the EDFA gain coefficient. $\alpha_p = 1\,\text{dB/m}$, $\alpha_s = 2\,\text{dB/m}$, $\eta_p = 0$, $\eta_s = 0.9$, and $P_{sat}(\lambda_p) = 1\,\text{mW}$.

The dB/mW gain coefficient was initially defined as the ratio of the maximum gain to the maximum launched pump power [59]. But from Figure 5.8, the highest dB/mW ratio is not obtained at maximum launched pump power, where the gain curve levels off, but at the point defined by the tangent. This is the definition currently adopted in the literature.

The dB/mW gain coefficient is a characteristic device parameter that should not be confused with the usual definition of amplifier gain coefficient ($g_{ek} - g_{ak}$ or $\gamma_{ek} - \gamma_{ak}$) derived in Chapter 1. The latter corresponds to the difference $\sigma_e N_2 - \sigma_a N_1$ integrated over the fiber transverse plane and expressed in $m^{-1}$ units. The dB/mW gain coefficient should not also be confused with the *differential gain coefficient* $dG/dP_p^{in}$, highest for $P_p^{in} = 0$ (see Figure 5.8). In the amplification regime of interest ($G > 0$), the differential gain coefficient is of no indicative value but the slope of the gain curve at the transparency point ($P_p^{trans}$, $G(P_p^{trans}) = 1$) can be used as a reference value, since it increases with the dB/mW gain coefficient. From Eqs. (5.8) and (5.9) with $G = 1$, we find the characteristic value:

$$g^* = \left(\frac{d\log(G)}{dP_p^{in}}\right)_{P_p^{in} = P_p^{trans}} = \frac{\alpha_s}{\alpha_p} \frac{1 + \eta_s}{1 + \eta_p} \frac{1}{P_{sat}(\lambda_p)} \frac{[1 - \exp(-U\alpha_p L)]^2}{1 + \left[\frac{1 + \eta_p}{1 + \eta_s}\alpha_p L - 1\right]\exp(-U\alpha_p L)} \quad (5.12)$$

where $U = (\eta_s - \eta_p)/(1 + \eta_p)$ is a positive parameter. Taking the limit of Eq. (5.12) for $L \to \infty$, we find

$$\lim_{L \to \infty} g^* = \lim_{L \to \infty}\left(\frac{d\log(G)}{dP_p^{in}}\right)_{P_p^{in} = P_p^{trans}} = \frac{\alpha_s}{\alpha_p} \frac{1 + \eta_s}{1 + \eta_p} \frac{1}{P_{sat}(\lambda_p)} \quad (5.13)$$

This last result shows that the differential slope of the gain curve, which in the region $G > 0$ is maximum at fiber transparency, reaches a limit determined by the *intrinsic fiber parameters* $\alpha_p$, $\alpha_s$, $\eta_p$, $\eta_s$ and $P_{sat}(\lambda_p)$. For a given pump wavelength $\lambda_p$, the highest value for $g^*$ is obtained in the case $\eta_p = 0$ (three-level pumping) and at the peak signal wavelength where the product $\alpha_s(1 + \eta_s) \approx \sigma_a(\lambda_s) + \sigma_e(\lambda_s)$ is maximum.

As Figure 5.8 shows, the dB/mW gain coefficient increases with EDFA length. A high gain coefficient corresponds to a steep increase of gain with pump power near the transparency value $P_p^{trans}$. The dB/mW coefficient does not give an indication of the maximum EDFA gain or the pump power required to achieve a given gain level; instead it reflects how rapidly the gain evolves from transparency to its maximum value. The dB/mW coefficient represents a first figure of merit to compare results obtained with different pump bands, Er-concentrations, or fiber designs. It is not advantageous to operate the EDFA in the region of the gain curve where the ratio $G/P_p^{in}$ is optimum or $G/P_p^{in} \approx g(dB/mW)$, as the gain strongly depends on pump power, contrary to the case where $P_p^{in} \gg P_p^{trans}$ (near-complete inversion regime). Equation (5.6) shows another reason for operating the EDFA in the high pump regime or near-complete inversion: a higher saturation power can be achieved.

Actually, the dB/mW gain coefficient does not increase indefinitely with EDFA length, due to the concurrent effect of *amplifier self-saturation*. Indeed, as the maximum gain increases with longer lengths, more amplified spontaneous emission (ASE) power is generated in both fiber directions, which eventually results in gain saturation.

Amplifier self-saturation by ASE is analyzed in detail in Section 5.5. Self-saturation, not accounted for in the model used in the derivation of Eq. (5.8), causes the dB/mW gain coefficient to decrease when the EDFA length is greater than a certain value. This is illustrated in the experimental curves of Figure 5.7b. The optimal EDFA length is $L = 13$ m, but the highest gain coefficient (0.6 dB/mW) is obtained for $L = 8$ m.

The dB/mW coefficient can be used as a figure of merit for comparing the EDFA performance achieved with different pump wavelengths and fiber core compositions (E. A. Snitzer et al. [59], R. I. Laming et al. [60]). During the period 1988–1990, several research groups worked to improve this coefficient for the two pump wavelengths of main potential interest, i.e., 980 nm and 1480 nm, with the goal of achieving efficient LD pumping. For the 980 nm pump band, the gain coefficient was improved in 1989 from 2.2 dB/mW to 3.9 dB/mW (R. I. Laming et al., [60], and in 1990, from 4.9 dB/mW (M. Shimizu et al. [61]) to 10.2–11.0 dB/mW (M. Shimizu et al. [2]; Y. Kimura et al. [62]), which represent the highest values reported to date. For the 1480 nm band, the best coefficient evolved in 1989 from 0.39 dB/mW (M. Nakazawa et al. [63]) to 0.66 dB/mW (P. C. Becker et al. [64]), 2.1 dB/mW (E. Desurvire et al. [65]), and 5.9 dB/mW (J. L. Zyskind et al. [66]), and reached 6.3 dB/mW (T. Kashiwada et al. [67]) in 1991, the highest value reported to date. The results of [2], [67], and [60]–[64], were obtained with practical LD pump sources. The best result to date, 11 dB/mW, obtained with 980 nm LD-pumping represents considerable progress compared to the performance of the first LD-pumped EDFAs, demonstrated in 1988 with an 800 nm GaAlAs phased-array diode (T. J. Whitley [1]), and in 1989 with 980 nm InGaAs and 1480 nm InGaAsP diodes (R. I. Laming et al. [60]; S. Uehara et al. [68]; and M. Nakazawa et al. [63]).

Rapid optimization of the EDFA gain coefficient evolved out of two complementary approaches to Er-doped fiber design: (a) confining the $Er^{3+}$ ions in the center of the fiber core, and (b) reducing the pump mode size (B. J. Ainslie et al. [59], J. R. Armitage [70]). Intuitively, Er-doping confinement concentrates the active ions in the region of the fiber core, where the pump has the highest intensity. A reduction in the pump mode size increases the intensity corresponding to a given pump power level. The reduction in overlap between Er-doping and signal mode due to the doping confinement can then be compensated by the fiber length. Both issues of Er-doping confinement and fiber waveguide optimization are addressed in detail and quantitatively analyzed in Section 5.6.

The highest dB/mW gain coefficients reported up to 1992 for the pump bands at 532 nm, 664 nm, 800 nm, 980 nm, and 1480 nm are shown in relation to the EDFA gain curves plotted in Figure 5.9 (M. M. Choy et al. [72]; M. Horiguchi et al., [5] and [71]; M. Shimizu et al. [2]; and T. Kashiwada et al. [67]). As the figure shows, the optimized EDFA pumped at 980 nm provides a gain of 30 dB with only 2.5 mW pump, and for such a device a near-complete inversion is achieved with about 10 mW pump. With the 1480 nm pump band, an identical performance requires nearly twice as much pump power in each case. With the 800 nm pump band, the pump power required for 30 dB gain is 10 times greater (i.e., $P_p^{in} = 25$ mW), while near-complete inversion is achieved with approximately 50 mW pump [5]. The gain curves corresponding to 820 nm, 980 nm, and 1480 nm pump bands were obtained with a single LD source, while those corresponding to the 532 nm and 665 nm pump bands were obtained with other types of pump sources. The effect of pumping with two

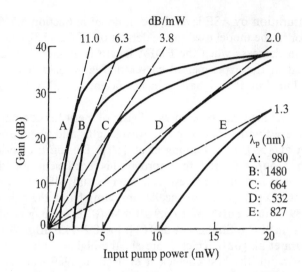

**FIGURE 5.9**    Best experimental EDFA gain versus input pump power characteristics reported to date (1992) using the different pump bands near 532 nm, 820 nm, 665 nm, 980 nm, and 1480 nm; the corresponding dB/mW gain coefficients are also shown. (A) $\lambda_p$ = 980 nm, [2] (B) $\lambda_p$ = 1480 nm, [67] (C) $\lambda_p$ = 664 nm, [71] (D) $\lambda_p$ = 532 nm, [72] (E) $\lambda_p$ = 827 nm, [5].

pump sources (using either bidirectional pumping or two-stage pumping) is discussed below.

The results of Figure 5.9 indicate that high gains can be achieved by pumping the EDFA in any of the erbium pump bands. But the lowest pump power requirements correspond to the cases where the pumping process is free from the concurrent effect of pump ESA (i.e., at pump wavelengths 532 nm, 980 nm, and 1480 nm, Chapter 4), and where the Er-doping can be sufficiently confined with respect to the pump intensity distribution (i.e., at pump wavelengths 980 nm and 1480 nm). All the Er-doped fibers used to obtain these results are of the germanosilicate type, except for the 1480 nm pump for which the fiber is of the alumino-germanosilicate type. Gain measurements were also reported for Er-doped fluorozirconate fibers of ZBLA and ZBLAN composition (T. Sugawa et al. [73]; D. Ronarc'h et al. [74]). The results obtained in [74] show that high gains (31 dB) can be obtained with single-mode ZBLAN EDFAs pumped with 1480 nm LDs. While the pump power required to achieve this result (50 mW) is nearly 10 times higher than for the best silica-based EDFA, a direct performance comparison cannot be made because the fiber designs are different. But, for several reasons, we do not expect the best gain versus pump characteristics and dB/mW coefficients of fluoride glass EDFAs to match those of silica-based EDFAs. First, the peak absorption and emission cross sections are somewhat lower [75]; second, for pump bands other than 1480 nm the nonradiative relaxation rates are also lower (Chapter 4), which increases the pump level lifetime and reduces the pumping efficiency; third, the background loss at pump and signal wavelengths due to glass impurities is difficult to eliminate completely.

The insensitivity of the EDFA gain characteristics to both pump and signal states of polarization (SOP), discussed in Chapter 4, was verified through several experiments. Signal gain insensitivity to SOP was first observed with an EDFA pumped at

$\lambda = 514.5$ nm (C. R. Giles et al. [76]). In these multimode pump conditions, a gain variation of 0.2 dB was measured, corresponding to an output power variation of 4.5% between two orthogonal SOP states. In another experiment, a single-transverse mode pump near 1480 nm and a short fiber length ($L = 1$ m) were used (M. Suyama et al. [77]). A measurement of gain versus polarization angle between pump and signal showed a negligible dependence ($\pm 0.5$ dB about 7 dB gain), which was within experimental error. Another type of experiment measured the ASE output in two orthogonal SOPs under various conditions. In the high gain regime, the ASE spectrum is proportional to the EDFA gain spectrum (Chapter 2). Thus, any effect of gain anisotropy in EDFAs should be observed in the polarized ASE spectra. The first measurement of this type was made for gain hole burning characterization (E. Desurvire et al. [78]), described in detail in Chapter 4. This experiment used a multimode pump at $\lambda = 514.5$ nm and an intense, monochromatic, polarized saturating signal near $\lambda = 1.53$ $\mu$m; no difference between the two polarized ASE spectra could be observed, even at the lowest temperature ($T = 4.2$ K). This property could be advantageously used to measure hole widths in the SOP direction orthogonal to that of the output signal (Chapter 4). The same approach based on the property of ASE isotropy was also used to characterize inhomogeneous broadening in EDFAs (E. Desurvire et al. [79]), and EDFA noise figure in saturated conditions (J. Aspell et al. [80]).

If conditions of near-complete medium inversion could be maintained along the fiber, the total gain experienced by the signal would increase indefinitely with fiber length. But in reality, four factors contribute to reduce the medium inversion in some region of the fiber, thereby preventing such an unlimited signal growth. These factors are: (1) the pump power absorption with length, (2) the gain saturation by the amplified signal, (3) the gain saturation by the ASE, and (4) the gain saturation by laser oscillation.

For a given EDFA length $L$, the effect of pump absorption can always be compensated by increasing the pump power launched into the EDFA, whether in the unidirectional or bidirectional pumping configurations. The result of this is to increase the pumping rate in the region of poor or negative inversion conditions.

The effect of gain saturation by amplified signal occurs in a regime where, in some region of the fiber, the stimulated emission rate induced by the high signal power becomes comparable to the pumping rate, or even takes over (Chapter 1). As before, this effect can be alleviated by increasing the input pump power. EDFA gain saturation can also be suppressed by reducing the input signal power. Upon reducing the signal input below a certain level, the EDFA gain becomes independent of the signal input or output powers, as shown in the next section. This regime is conventionally referred to in the literature as the *unsaturated gain regime* or the *small-signal regime*. As seen previously, the maximum (unsaturated) EDFA gain is given by the expression $G(\infty) = \exp\{[(1 - \eta_p/\eta_s)/(1 + \eta_p)]\rho_0\sigma_e L'\}$, where $L' = \Gamma_s L$ is a reduced effective length accounting for the RE/mode overlap effect. In the ideal, three-level pumping scheme (where complete inversion can be achieved, or $\eta_p = 0$), the highest unsaturated gain is $G(\infty) = \exp(\rho_0\sigma_e L')$. According to this result, since the fiber length $L'$ can in principle be indefinitely increased, there would be no limit to the maximum gain achieveable, if the required pump power could be supplied accordingly. In reality, two limiting effects prevent the EDFA gain from being increased indefinitely: amplifier self–saturation by ASE and laser oscillation.

The physical origin of amplifier self-saturation, analyzed in detail in Section 5.5, can be explained simply through equivalent input noise. Chapter 2 shows that optical amplifiers always generate a certain amount of noise power, called ASE. In the case of fiber amplifiers, the ASE power propagates in both fiber directions (i.e., forward and backward, with respect to the signal propagation direction). The total ASE output obtained at both fiber ends represents the integration over fiber length of the noise power originating initially from spontaneous emission, then amplified along the fiber by stimulated emission. In the unsaturated gain regime, the output ASE power in a given bandwidth $\Delta\nu$ of an amplifier with gain G can be expressed as (Chapter 2):

$$P_{\mathrm{ASE}}^{\pm} = n_{\mathrm{sp}}^{\pm} h\nu_s \Delta\nu(G - 1) = n_{\mathrm{eq}}^{\pm} h\nu_s \Delta\nu G \tag{5.14}$$

where $n_{\mathrm{sp}}^{\pm}$ and $n_{\mathrm{eq}}^{\pm}$ are, by definition, the *spontaneous emission factor* and the *equivalent input noise*, respectively, corresponding to forward and backward propagation directions. Equation (5.14) shows that the generation of ASE noise power is equivalent to the amplification of $n_{\mathrm{sp}}^{\pm}$ or $n_{\mathrm{eq}}^{\pm}$ fictitious input photons. In the high gain regime and complete inversion limit, the equivalent noise input becomes $n_{\mathrm{sp}}^{\pm} \approx n_{\mathrm{eq}}^{\pm} \approx 1$, representing a single photon in bandwidth $\Delta\nu$. In this limit, the ASE power increases linearly with the gain.

As the gain is increased, the stimulated emission by ASE is enhanced to the point where it eventually competes with the pumping rate. This competition occurs in regions located near the fiber ends (where the ASE power is the highest, Section 5.5). In these regions, the medium inversion is reduced and the amplifier gain saturates in the absence of any input signal.

Contrary to the case of saturation by amplified signal, the effect of amplifier self-saturation by ASE cannot be entirely suppressed by increasing the pump input. The reason is that, according to Eq. (5.14), any increase in gain results in an almost proportional increase in ASE. As in the general case, the two effects do not exactly balance out, even when saturated, the amplifier gain can be increased with higher pump power. In this regime, the maximum gain eventually becomes limited by the effect of laser oscillation. Oscillation occurs when the amplifier gain is high enough to compensate for the return loss of any reflecting element in the path of the ASE signal. In this case, the ASE is reflected back and forth into the amplifier, which operates then as a laser. In the regime of laser oscillation, the round-trip gain eventually saturates to a steady state determined by the cavity loss and the input pump conditions [81]. Since the effects of reflection or of backscattering from optical elements located in the signal or ASE paths cannot be totally suppressed, the amplifier gain is ultimately limited by the onset of laser oscillation. The effect of reflections and backscattering, which can be important in high gain EDFAs, is analyzed in Section 5.8.

The power that can be launched into a single-mode fiber end from a pump LD source is limited by device specifications. Bidirectional pumping, Figure 5.4c, enables us to double the amount of pump launched into the EDFA, at the expense of an additional wavelength selective coupler (WSC). Polarization multiplexing two pump LDs at either (or both) EDFA ends, enables us to double (or quadruple) the EDFA input pump. In addition to two WSCs, quadruple pump polarization multiplexing requires two SOP selective couplers or polarization beam splitters (PBS), which can be realized either in bulk optics or in SOP-maintaining fiber optics. Using this approach, a second generation of LD-pumped EDFAs with gains in excess of 45 dB could be demonstrated (Y. Kimura et al. [82]). The notion of *high gain* for optical amplifiers has considerably evolved due to such rapid progress; one way to reflect the state of the art is to use *high gain* for devices with gain $20\,dB < G < 40\,dB$, and *very high gain* for devices with gain of $G > 40\,dB$.

Two other possibilities for achieving very high gain EDFAs, Figure 5.10, are: use of *two-stage amplification* or a *midway isolator*.

Two-stage amplification is equivalent to a two-stage pumping scheme. It cascades two LD-pumped EDFAs, Figure 5.10a. A narrow (e.g., 1 nm) optical bandpass filter placed between the two EDFAs, makes it possible to reduce the effect of amplifier self-saturation, at the expense of amplifier bandwidth. The example of Figure 5.10a corresponds to forward and backward pumping for the first and the second EDF, respectively. Increased gain and output signal power can be obtained by using bidirectional pumping in the second amplification stage. For instance, a gain $G = 51\,dB$ in an LD-pumped EDFA was obtained with the following configuration: forward pumping with two polarization multiplexed LDs in the first stage, bidirectional pumping with four polarization multiplexed LDs in the second stage; the pump power launched into the EDFA from each polarization multiplexed, twin LD pump was 65–80 mW (H. Masuda and A. Takada [83]). More recent developments (1992) in Yb-codoped EDFAs (EYDFA) using LD-array-pumped Nd:YLF sources, have also shown the possibility of achieving gains as high as $G = 51\,dB$ through backward pumping with a single pump device (S. G. Grubb et al. [7]).

Two-stage EDFAs using an optical gate between the two amplifiers (in addition to the optical bandpass filter) can also be implemented for boosting pulsed signals.

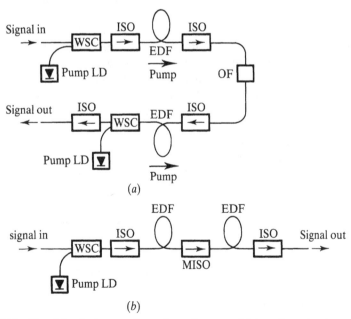

**FIGURE 5.10** Possible device configurations for very high gain EDFAs: (a) two-stage amplification, and (b) use of midway isolator. LD: laser diode, WSC: wavelength selective coupler, ISO: Faraday isolator, EDF: Er-doped fiber, OF: optical filter, MISO: midway Faraday isolator.

The effect of the optical gate is to reduce the self-saturation induced by forward ASE in the second amplification stage. As transient gain saturation is a relatively slow process (Section 5.9), this saturation effect can be effectively suppressed by fast gating (e.g., 1 ns) at a low repetition rate (e.g., 1 MHz). For instance, in a two-stage LD-pumped EDFA having a configuration similar to that of Figure 5.10a, and using for the optical gate a traveling-wave LD amplifier, a gain of 49 dB could be achieved, boosting optical pulses from a DFB laser to a peak power of 105 W (H. Takara et al. [84]).

Self-saturation by backward ASE, which reduces the EDFA gain, can be partially suppressed by using a midway isolator (J. H. Povlsen et al. [85]). Figure 5.10b illustrates this configuration with the forward pumping scheme. The effect of the midway isolator (whose location must be optimized) is to block the backward ASE at some point in the EDF, thus preventing ASE buildup in the input EDF portion and consequent medium saturation. For unidirectional 1480 nm pumped EDFAs, the pump can traverse the isolator with no significant loss (< 1 dB); for unidirectional 980 nm pumped EDFAs, the isolator can be bypassed by the pump through two WSCs placed at both ends. For bidirectional EDFAs, both pump wavelengths can be used, regardless of the isolator transmission characteristics; the backward pump is blocked at the isolator level. The optimal location of the isolator must be determined by numerical analysis (O. Lumholt et al. [86]; S. Yamashita and T. Okoshi [87]). For 1.480 nm bidirectionally pumped EDFAs, it was found both theoretically and experimentally that in the high pump regime, the optimal location for gain enhancement is at 25% of the total EDFA length [86]. An optimal configuration of bidirectional EDFAs with midway isolator that remain to be investigated should use one WSC at each isolator end; in this layout, the counterpropagating pumps could bypass the isolator and be efficiently used throughout the entire EDFA length. The combination of an optical bandpass filter and a midway isolator is the most efficient way to suppress steady state self-saturation in two-stage EDFAs. A further improvement in ASE suppression could be obtained by using a midway polarizer. Such a polarizer would block half of the forward and backward ASE in each EDFA half-section; alternatively, a single-polarization EDFA (SP–EDFA) could be used. SP–EDFAs have been investigated for the amplification of polarized signals (K. Tajima [88]; I. N. Duling III and R. D. Esman [89]), as well as for the realization of narrow line width fiber lasers (K. Iwatsuki et al. [90]). As traveling-wave amplifiers, the drawback of SP–EDFAs is that they are not practical for communications systems based on nonpolarization-maintaining fiber. But in these systems, SP–EDFAs could be implemented in a polarization diversity configuration. In the polarization diversity configuration, the input and output signals are split and recombined through polarization beam splitters; each signal SOP is amplified in a separate SP–EDFA. A major drawback of this approach is SOP dependent gain saturation. As each SP–EDFA has a different and random input signal power, the net signal gain after SOP recombination is also random. The gain, saturation, noise properties, and potential applications of SP–EDFAs still remain to be investigated.

While state-of-the-art technology currently limits the highest EDFA gains to a level near $G = 50$ dB, higher gains (e.g., $G = 60$ dB) could theoretically be achieved. Such an improvement would require multiple stage amplification based on the methods previously described. In addition, higher pump powers (i.e., > 500 mW), optical filters of narrower bandpass (i.e., < 1 nm) and optical isolators of very high return loss (< −60 dB), would be necessary.

We focus now upon the spectral gain characteristics of the EDFA. Next to the gain versus pump power performance, the gain dependence with signal wavelength represents one of the most important EDFA characteristics. The spectral features of the EDFA gain spectrum result from two factors: (a) the particular absorption and emission cross section line shapes of the EDF, and (b) the variation of the gain coefficient along the EDF length, due to nonuniform medium inversion. Thus, in addition to its host dependence, the EDFA gain spectrum depends, in the most general case, upon the fiber length as well as the pump and signal powers. Additionally, it is shown below that the EDFA gain spectrum also depends upon the pumping scheme, i.e., two-level or three-level, since complete medium inversion can be achieved only in the last case.

First, the EDFA gain spectrum is expected to reflect the Er:glass cross section spectral features, as can be shown by simple arguments. Indeed, according to the differential rate equation Eq. (1.24), the transmission spectrum $T(\lambda) = P_s^{out}(\lambda)/P_s^{in}(\lambda)$ of an unsaturated EDFA with length $L$ is given by:

$$T(\lambda) = [\exp\{\rho_0 \sigma_e(\lambda)L\}]^{n_2} \times [\exp\{-\rho_0 \sigma_a(\lambda)L\}]^{n_1} \tag{5.15}$$

In Eq. (5.15), $\rho_0$ is an effective $Er^{3+}$ concentration (averaged over the transverse fiber plane), and $n_1, n_2$ are the fractional ion populations of the ground and upper level, respectively, averaged over the fiber length, according to the definition:

$$n_{1,2}(L) = \frac{1}{\rho_0 L} \int_0^L N_{1,2}(z)\, dz \tag{5.16}$$

According to Eq. (5.15), the EDFA transmission spectrum (or net gain spectrum) can be expressed in decibels as:

$$T(\lambda)_{dB} = 10 \log_{10}[T(\lambda)] = \frac{10\rho_0 L}{\log(10)} \{n_2 \sigma_e(\lambda) - n_1 \sigma_a(\lambda)\} \tag{5.17}$$

The EDFA gain spectrum is proportional to the difference between emission and absorption cross sections $\sigma_a(\lambda)$, $\sigma_e(\lambda)$, weighted by their respective fractional average populations $n_2$, $n_1$. From Eq. (5.16), and in the case of three-level pumping, the fractional populations $n_1, n_2$ have values between zero and unity, with $n_1 + n_2 = 1$. Thus, the dB gain spectrum of the EDFA with three-level pumping has two envelopes represented by the cross section line shapes $\sigma_a(\lambda)$, $\sigma_e(\lambda)$. In the two-level pumping case, complete inversion is not possible, and the maximum value for $n_2$ is $(1 - \eta_p/\eta_s)/(1 + \eta_p)$, as shown in Eq. (1.50). In the two-level pumping case, the dB gain spectrum of the EDFA has two envelopes represented by the cross section line shapes $\sigma_a(\lambda)$ and $\sigma_e(\lambda)[1 - \eta_p/\eta_s(\lambda)]/(1 + \eta_p)$. Thus, the EDFA gain spectrum is also dependent upon whether the pumping scheme is two- or three-level.

In the general case, it is easily verified from the theoretical description of Chapter 1 that with arbitrary Er-doping distributions, and in the regime of saturation by signal or ASE, the result of Eq. (5.17) is also valid with the more general definitions for $n_1, n_2$:

$$n_{1,2}(L) = \frac{2\pi}{\rho_0 L} \int_0^L \int_0^\infty N_{1,2}(r, z)\psi_s(r)r\, dr\, dz \tag{5.18}$$

The theoretical EDFA gain spectra obtained in the unsaturated gain regime can be directly determined from Eqs. (2.125) and (2.169), [58] and [91]. Examples of such spectra obtained for different pump powers in the two-level and three-level pumping schemes are shown in Figures 2.9 and 2.13.

The characteristics of the absorption and emission cross section line shapes $\sigma_a(\lambda)$, $\sigma_e(\lambda)$ of various EDF types are described in Chapter 4. In particular, the cross-sections in germanosilicate EDFs (Type I) exhibit two peaks; cross-sections in aluminosilicate EDFs with high alumina concentrations (Type III) exhibit a single peak with a comparatively broader spectral width (Figure 4.24).

Early spectral gain characterizations of EDFAs have shown the advantage of using aluminosilicate fibers for broadband amplification (C. R. Giles et al. [76]; C. G. Atkins et al. [92]). Figure 5.11 shows experimental EDFA gain spectra (for $G_{dB} \geqslant 0$) obtained in a 10 m long aluminosilicate fiber for different input pump powers (after [92]). The solid line at maximum pump corresponds to a best fit with the parameter values $n_1 = 0.38, n_2 = 0.62$. The EDFA gain bandwidth for $G > 20$ dB is about 40 nm. At maximum pump, the gain spectrum has a peak near $\lambda = 1.53\ \mu$m, a dip near $\lambda = 1.54\ \mu$m, and a nearly flat spectral region of 20 nm width centered near $\lambda = 1.55\ \mu$m.

Figure 5.11 also shows that as the input pump is decreased, the main peak at $\lambda = 1.53\ \mu$m progressively vanishes, while the center of the EDFA gain spectrum shifts towards longer wavelengths near $\lambda = 1.56\ \mu$m. The effect of progressive absorption at $\lambda = 1.53\ \mu$m is clearly due to the fact that the overlap between absorption and emission cross sections is maximum at that wavelength (see Figure 4.24, Type III), which minimizes the difference $n_2\sigma_e(\lambda) - n_1\sigma_a(\lambda)$ in Eq. (5.17). Such pump-induced spectral changes are characteristic of three-level laser systems in which the transition terminates to the ground level. This property was first observed in EDFAs through

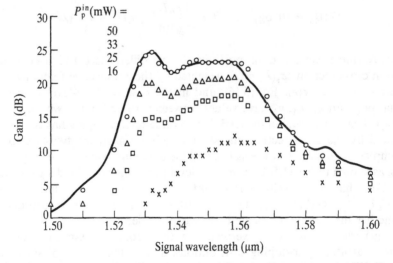

**FIGURE 5.11**   Experimental EDFA gain spectra obtained with a 10 m aluminosilicate fiber for different input pump powers $P_p^{in}$ at $\lambda = 1490$ nm wavelength. The solid line at maximum pump corresponds to a best fit using the average fractional populations $n_1 = 0.38$, $n_2 = 0.62$. From [92], reprinted with permission from *Electronics Letters*, 1989.

the characterization of ASE spectra (E. Desurvire and J. R. Simpson [93]). Another way to interpret these spectral changes is to view the EDFA gain spectrum as being produced by two types of laser systems. The first type, corresponding to the short wavelength region ($\lambda = 1.53\ \mu$m), is a near-three-level system, which has ground level absorption. The second type, corresponding to the long wavelength region ($\lambda = 1.56\ \mu$m), is a near-four-level system, which has a comparatively small ground level absorption. The spectral changes observed in figure 5.11 can then be interpreted in this perspective. Indeed, as the population inversion is decreased, the three-level medium becomes progressively lossy, while the four-level medium becomes progressively transparent, Figure 5.11. The EDFA gain medium can be viewed as a continuous superimposition of laser systems, with pure three-level behavior at short wavelengths and pure four-level behavior at long wavelengths.

The gain spectrum also depends upon the fiber length. This can be shown simply by considering the case of a forward pumped EDFA. Indeed, for a given signal wavelength $\lambda_s$, the local gain coefficient $\gamma(\lambda_s, z) \approx \sigma_e(\lambda_s)N_2(z) - \sigma_a(\lambda_s)N_1(z)$ vanishes at a certain fiber coordinate $z_0$ where the pump power has decayed to the threshold value, Eq. (1.126):

$$P_p(z_0) = \frac{P_{sat}(\lambda_p)}{U(\lambda_p, \lambda_s)} \qquad (5.19)$$

with

$$U(\lambda_p, \lambda_s) = \frac{\eta_s(\lambda_s) - \eta_p(\lambda_p)}{1 + \eta_p(\lambda_p)} \qquad (5.20)$$

The above result corresponds to the case of EDFAs with confined Er-doping and operating in the unsaturated regime. No analytical formula is available to predict the threshold value $P_p(z_0)$ in the general case, but the conclusion remains the same. As the local gain coefficient is negative for $z > z_0$, the EDFA is lossy or absorbing in this region of the fiber. According to Eq. (5.19), the pump power threshold $P_p(z_0)$ depends upon the signal wavelength, so does the coordinate $z_0$, and therefore, the EDFA gain spectrum is fiber length dependent.

The previous considerations also showed that for maximizing the gain at a given signal wavelength, the EDFA length should be $L = z_0$, so that no signal reabsorption occurs. The optimal fiber length depends upon the signal wavelength and is a characteristic feature of EDFAs. EDFA length optimization is addressed in Section 5.6.

So far, we have considered the two pump wavelengths $\lambda = 980$ nm and $\lambda = 1480$ nm, which represent the cases of three-level pumping and two-level pumping, respectively. The conclusions for the three-level pumping case are also relevant to the pump bands of interest at wavelengths near 810 nm and 532 nm. For pumping the EDFA, it is also possible to use two different pump bands simultaneously, e.g., 980 nm and 1480 nm. Er-doped fiber amplifiers pumped in two different bands are referred to in the literature as hybrid (HEDFA). The main interest of 980 nm/1480 nm pumped HEDFAs (J.-M. P. Delavaux et al. [94]) is that they combine both features of (a) the quantum-noise-limited noise figure of the three-level system (980 nm pump) and (b) the quantum-limited power conversion efficiency of the two-level system (1480 nm pump). Another advantage of 980 nm/1480 nm pumped HEDFAs is the relative flatness of their gain spectrum. For instance, a gain bandwidth of 35 nm with 1 dB gain variation was demonstrated [94].

The EDFA gain spectrum can also be extended in the long wavelength region (near $\lambda = 1.6\,\mu m$) by pumping at $\lambda = 1.555\,\mu m$ (J. F. Massicott et al. [95]). Figure 5.12 shows the gain spectrum of an aluminosilicate (Type III) EDFA pumped at $\lambda = 1.555\,\mu m$ [95]. In this experiment, the pump wavelength $\lambda = 1.55\,\mu m$ was found to yield the best compromise between maximum gain and gain bandwidth. A maximum gain of 31 dB at $\lambda = 1.60\,\mu m$ could be achieved with 150 mW input pump at $\lambda = 1.555\,\mu m$, with a 3 dB gain bandwidth of about 35 nm ($\lambda = 1.573 - 1.608\,\mu m$). Hybrid EDFAs can also be implemented using the two pump wavelengths $\lambda = 1.48\,\mu m$ and $\lambda = 1.555\,\mu m$ simultaneously (J. F. Massicott et al. [96]). A first advantage of the hybrid $1.48\,\mu m/1.555\,\mu m$ pump configuration, compared to the single $1.555\,\mu m$ pump configuration, is the flatness of the gain bandwidth; with a proper adjustment of the $1.555\,\mu m$ power, a gain variation of 1 dB about an average 24 dB gain could be achieved over a 35 nm bandwidth [96]. A second advantage of using the HEDFA configuration is the potential to achieve a noise figure close to the limit obtained with the $\lambda = 1.48\,\mu m$ pump. The use of an EDFA pumped at 980 nm or 1480 nm operating in parallel with a $1.48\,\mu m/1.555\,\mu m$ pumped HEDFA would make it possible to extend the usable amplifier bandwidth to about 70 nm ($\lambda = 1.53 - 1.60\,\mu m$).

High gain EDFAs have also been realized with ZBLA or ZBLAN hosts, [73] and [74]. In comparison to the aluminosilicate host, the fluoride materials are characterized by broader and smoother cross section line shapes, which yield somewhat more uniform gain spectra over the 40 nm range considered [73]. The gain spectrum achieved in these fiber materials also exhibits pump- and signal-power dependent changes, as in the case of silicate hosts. Chapter 6 addresses gain flatness in EDFAs, important for broadband or WDM applications.

Section 5.12 compares the EDFA spectral gain characteristics with other types of fiber amplifiers (i.e., Raman, Brillouin, parametric, and other RE-doped fiber amplifiers). We also consider in that section fluorozirconate EDFAs operating at signal wavelengths other than $\lambda = 1.5\,\mu m$ (i.e., $\lambda = 850$ nm and $\lambda = 2.716\,\mu m$).

The experimental gain characteristics previously discussed in this section concerned single-mode EDFAs, i.e., in which the signal always propagates in the fundamental ($LP_{01}$) mode. There are two reasons for using single-mode EDFAs, aside from their

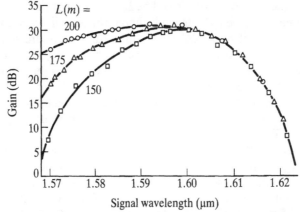

**FIGURE 5.12**   Experimental EDFA gain spectra obtained for different lengths with 140 mW pump power and the long pump wavelength of $\lambda = 1.555\,\mu m$. From [95], reprinted with permission from *Electronics Letters,* 1990.

compatibility with SMF or DSF systems: first, in the single-mode (or near-single-mode) pumping regime, the pump intensity can be maximized, which minimizes the threshold for medium inversion and makes LD pumping possible and practical; second, the ASE noise is also minimized, as spontaneous emission has only two degrees of freedom, which correspond to the two degenerate polarization modes (Chapter 2). The second property occurs because the equivalent input noise $n_{eq}$ of a multimode EDFA having $2N$ nondegenerate modes is at least $2N$ times greater than that of a single-mode EDFA with complete inversion. These two advantages justify the use of single-mode EDFAs but some fiber optic systems (e.g., subscriber loops, fiber sensors) use multimode fibers. The reasons for using multimode fibres are based upon considerations of robustness, simplicity, and cost of the optical components involved (e.g., laser sources, connectors, multiplexers, splitters, and taps). Multimode EDFAs could be implemented in such systems for power loss compensation and performance enhancement. For practical LD pumping, the main drawback of the multimode EDFA is that for a given degree of inversion, a tenfold increase in EDF core size corresponds to a hundredfold increase in required pump power. But as the technology of high power LDs has progressed to the point where output powers in excess of 100 mW are now available from a single device (Section 5.1), LD-pumped, multimode EDFAs are technically possible. For instance, a gain of 18 dB was obtained with about 100 mW pump at $\lambda = 980$ nm in an EDFA supporting theoretically 22 nondegenerate modes (G. Nykolak et al. [97]). In this experiment, the lowest equivalent input noise was measured as $n_{eq} = 21$, compared with $n_{eq} = 1$ for a fully inverted single-mode EDFA.

The EDFA gain versus pump characteristics previously described were obtained with the doped fibers coiled onto reels of relatively large diameters (e.g., 15–30 cm). But in compact packaging applications, the effect of fiber macrobending loss, which could critically affect these characteristics, must be taken into account. Fiber bending not only induces background loss at both pump and signal wavelengths, but also results in the outward displacement of the peak mode intensities from the fiber axis. As a result of an excessive mode displacement effect, the overlap between the pump or signal modes and the Er-doped region is reduced, which degrades the EDFA gain performance. The fiber loss associated with a given bending radius $R$ can be evaluated for both pump and signal operating wavelengths as a function of the waveguide parameters (e.g., $V$ number, cutoff wavelength, core radius, refractive index difference, NA), using the method described by J. Sakai and T. Kimura [98]. Using this method, the bending radius corresponding to a given bending loss can be reduced by increasing the fiber NA and/or increasing the wavelength (i.e., for the pump, choosing $\lambda = 1480$ nm instead of $\lambda = 980$ nm). Considering, for instance, a fiber with NA $= 0.11$, the bending radii causing a 0.100 dB/m loss at $\lambda = 980$ nm and $\lambda = 1480$ nm are found to be about $R = 5$ cm and $R = 2$ cm, respectively; for a fiber with NA $= 0.16$, the corresponding radii are reduced to $R = 2$ cm and $R = 0.5$ cm (A. L. Deus and H. R. D. Sunak [99]). The effect of fiber bending on the EDFA gain performance can be analyzed theoretically by including the corresponding loss coefficient in the EDFA rate equations (M. Ohashi and K. Shiraki [100]). The study of [100] considers for instance an EDFA with 1480 nm pump, NA $= 0.17$, and gain $G = 30$ dB corresponding to cutoff wavelength $\lambda_c = 0.8$ $\mu$m and an absence of bending ($R = \infty$); with a bending radius $R = 1$ cm, the EDFA becomes disabled (lossy); but the theoretical gain drop for the same bending radius is reduced to about 5 dB if the cutoff wavelength is increased to $\lambda_c = 1.0\,\mu$m. This undesirable gain reduction corresponding to small bending radii (e.g., $R = 1$ cm) can be alleviated by using Er-doped fibers with high NAs (e.g., NA $> 0.2$). In this case, the issue of fiber waveguide optimization becomes irrelevant, as the minimum bending radii corresponding to the two relevant operating pump wavelengths of $\lambda = 980$ nm and $\lambda = 1480$ nm are smaller than usual packaging specifications (e.g., $R < 1$ cm).

Now we consider the agreement between theory and experiment in the EDFA gain characteristics. The first EDFA modeling using three-level system rate equations (Chapter 1) was applied to the case of a 514 nm pumped amplifier, for predicting pump and signal versus length measurements, Figure 5.7a. (E. Desurvire et al. [57]). The agreement between theory and experiment was fair, considering that (1) only small adjustments of pump mode area and absorption coefficient were necessary to fit both pump and signal data simultaneously, and (2) the effects of multimode pump propagation, transverse mode distributions, and pump ESA were not taken into account.

The complete theoretical model for EDFAs, taking into account the effect of transverse modes and Er concentration distributions, was first described in three independent papers (J. R. Armitage [101], E. Desurvire and J. R. Simpson [102], A. Bjarklev et al. [103]). Two early papers based on the fundamental analysis of [101]–[103] provided further insights in the modeling of homogeneous gain saturation and pump ESA (E. Desurvire et al. [104]; P. R. Morkel and R. I. Laming [105]). Equations developed in [101] and [103]–[105] include the effect of pump ESA, while those in [102]–[105] include the effect of ASE noise. The studies in [102], [104], and [105] include the first theory/experiment comparisons for the EDFA gain, saturation, and noise characteristics.

The most general equations outlined in [101]–[104] have been widely used in the literature for the theoretical modeling of EDFAs. But there may be uncertainty about which model is the most accurate. Several research groups have since laid claim to more comprehensive theoretical models. But the changes they introduce are not conceptually fundamental and largely consist of improved approximations for numerical solution of the most general rate equations in [101]–[104]. These approximations use LP modes versus Gaussian modes, arbitrary Er-doping profiles versus step-doping profiles, or different definitions for the effective ASE bandwidth. Other improvements could be introduced by the use of accurate spectroscopic data, such as cross section line shapes and peak values for the ground or pump ESA absorption cross sections, which were not available in earlier work. There exists only one basic EDFA model, which lends itself to a wide variety of specific applications and approximations. This EDFA model is applicable to various amplification regimes and fiber designs. Chapter 1 covers the different approximations for which closed-form solutions are obtainable.

Theoretical predictions of EDFA gain characteristics versus pump and signal powers, based on measurable parameters and systematic comparisons with experimental data in the most general cases, have demonstrated the high accuracy of the EDFA model (J. F. Marcerou et al. [106]; B. Pedersen et al. [107]).

Figure 5.13 compares the gain versus pump power characteristics at different EDFA lengths [107]. $G = f(P_p^{in})$ are calculated using the most general rate equations (1.66) and (1.67), which take into account both effects of transverse mode/Er-doping overlap and amplifier self-saturation by ASE. The maximum gain values predicted

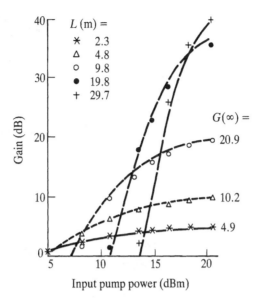

**FIGURE 5.13** Experimental and theoretical gain versus pump power characteristics at $\lambda = 1.53 \,\mu$m obtained with an EDFA pumped near 654 nm for different lengths. The maximum gain values given by $G(\infty) = \exp(\eta_s \alpha_s L)$ with $\eta_s = 1$, $\alpha_s = 2.14$ dB/m are also shown. From [107] © 1990 IEEE.

by $G(\infty) = \exp(\eta_s \alpha_s L)$, which correspond to the horizontal asymptotes of $G = f(P_p^{in})$, are also shown. The formula $G(\infty) = \exp(\eta_s \alpha_s L)$ is valid only for low EDFA gains ($G < 20$ dB) where the ASE does not cause amplifier self-saturation.

In the case of EDFAs with near-confined Er-doping, simplified EDFA models with either numerical or analytical solutions (Chapter 1) have also been used for accurate performance predictions (C. R. Giles et al. [108], A. A. M. Saleh et al. [109]). Figure 5.14 shows for instance the theoretical/experimental characteristics (gain versus pump and dB/mW coefficient) of an EDFA pumped near 980 nm, corresponding to different signal wavelengths and fiber hosts [108]. $G = f(P_p^{in})$ are calculated using the rate equations (1.113)–(1.121), which assumes confined Er-doping and takes into account the effect of self-saturation by ASE. The small discrepancy observed between theory and experiment, which is at maximum 2 dB for gains near 30 dB, can be attributed to the uncertainties in the characterization of the EDF parameters, as well as some error introduced by the confined Er-doping approximation for these fibers. As in Figure 5.13, no adjustable parameters are used for the theoretical calculation of $G = f(P_p^{in})$, which confirms the validity of the EDFA model.

## 5.3 GAIN VERSUS SIGNAL POWER AND AMPLIFIER SATURATION

Section 5.2 concerns EDFAs operated in the small-signal regime, i.e., where the gain is independent of the input signal power. In this section, the EDF gain characteristics versus input signal power are considered. We address power conversion efficiency and homogeneous or inhomogeneous gain saturation.

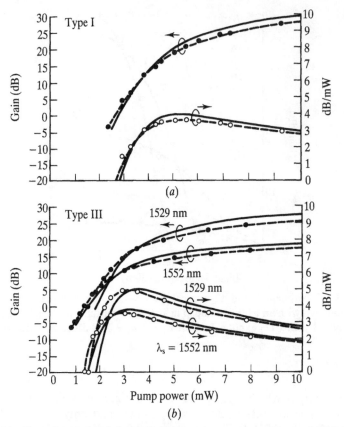

**FIGURE 5.14**  Experimental and theoretical gain versus pump power and dB/mW coefficient characteristics obtained at different signal wavelengths $\lambda_s$ in two EDFA types pumped near 980 nm: (*a*) germanosilicate, Type I, $L = 19.8$ m, $\lambda_p = 982$ nm, $\lambda_s = 1532$ nm (*b*) alumino-germanosilicate, Type III, $L = 24.1$ m, $\lambda_p = 978$ nm. From [108], © 1991 IEEE.

The effect of increasing the input signal power $P_s^{in}$ on the EDFA gain can be characterized by the curves $P_s^{out} = f(P_s^{in})$, $G = f(P_s^{in})$ or $G = f(P_s^{out})$. In the small-signal regime, these curves are linear (i.e., $P_s^{out}(P_s^{in}) = \text{const.} \times P_s^{in}$, $G(P_s^{in}) = G(P_s^{out}) = \text{const.}$). Gain saturation is reached when the EDFA characteristics depart from those linear relations. Figure 5.15 shows experimental measurements of the $G = f(P_s^{in})$ or $G = f(P_s^{out})$ characteristics, obtained at $\lambda = 1.531$ $\mu$m in a 1490 nm pumped EDFA for different input pump powers $P_p^{in}$ and optimum lengths $L_{opt}$ (E. Desurvire et al. [65]). The powers are expressed in decibel-mW or dBm units, according to the definition $P(\text{dBm}) = 10 \log_{10}[P(\text{mW})/P_0]$, with $P_0 = 1$ mW. Figure 5.15*a* shows that the small-signal regime where the gain is constant ($G = G_{max}$) corresponds to input signals with power $P_s^{in} < -30$ dBm or 1 $\mu$W. From Figure 5.15*b*, the corresponding output signal powers, which depend upon the pump power or the maximum gains, range from $P_s^{out} = -5$ dBm to $P_s^{out} = +5$ dBm.

A useful parameter is the EDFA *saturation output power* $P_{sat}^{out}$, defined conventionally as the output power for which the EDFA gain has dropped by $-3$ dB below

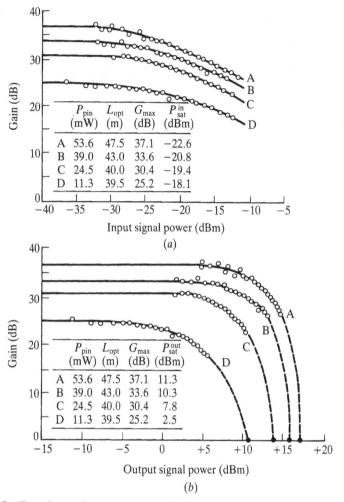

| $P_{pin}$ (mW) | $L_{opt}$ (m) | $G_{max}$ (dB) | $P_{sat}^{in}$ (dBm) |
|---|---|---|---|
| A  53.6 | 47.5 | 37.1 | −22.6 |
| B  39.0 | 43.0 | 33.6 | −20.8 |
| C  24.5 | 40.0 | 30.4 | −19.4 |
| D  11.3 | 39.5 | 25.2 | −18.1 |

Input signal power (dBm)

(a)

| $P_{pin}$ (mW) | $L_{opt}$ (m) | $G_{max}$ (dB) | $P_{sat}^{out}$ (dBm) |
|---|---|---|---|
| A  53.6 | 47.5 | 37.1 | 11.3 |
| B  39.0 | 43.0 | 33.6 | 10.3 |
| C  24.5 | 40.0 | 30.4 | 7.8 |
| D  11.3 | 39.5 | 25.2 | 2.5 |

Output signal power (dBm)

(b)

**FIGURE 5.15**  Experimental measurements of gain versus (a) input or (b) output signal power characteristics, obtained at $\lambda = 1.531\ \mu m$ in a 1490 nm pumped EDFA for different input pump powers $P_p^{in}$ and optimum lengths $L_{opt}$. The points shown in the horizontal axis ($G = 0$ dB) indicate the maximum output signal power that can be extracted from the EDFA, given by the energy conservation limit $P_s^{out}(max) = (\lambda_s/\lambda_p)P_p^{in}$. From [65], reprinted with permission from the Optical Society of America, 1989.

its unsaturated value $G_{max}$. The values of $P_{sat}^{out}$ obtained from Figure 5.15b increase with input pump power and range from $P_{sat}^{out} = +2.5$ dBm to $P_{sat}^{out} = +11.3$ dBm. The power $P_{sat}^{out}$ is also often referred to as saturated output power for 3 dB gain compression. Likewise, we can define an *input saturation power* $P_{sat}^{in}$ for which the gain saturation or compression is $-3$ dB. The relation between input and output saturation power is thus $P_{sat}^{out} = (G_{max}/2)P_{sat}^{in}$ or $P_{sat}^{out}(dBm) = P_{sat}^{in}(dBm) + G_{max}(dB)$ $-3$ dB. Both definitions of input and output saturation powers are relevant and useful EDFA parameters. The two-parameter set $G_{max}$ and $P_{sat}^{in}$ or $P_{sat}^{out}$, which corresponds to a given or available LD pump power, determine the EDFA power dynamic range. Within such a power dynamic range, defined by the relations $P_s^{in} \leqslant P_{sat}^{in}$

or $P_{\mathrm{s}}^{\mathrm{out}} \leqslant P_{\mathrm{sat}}^{\mathrm{out}}$, the amplifier operates near its peak gain performance (conventionally defined by $G_{\max} - 3\,\mathrm{dB} \leqslant G \leqslant G_{\max}$). Many potential applications of EDFAs involve the simultaneous amplification of $N$ optical channels, referred to as wavelength- or frequency-division multiplexing (WDM or FDM), depending upon the channel separation with respect to a 1 nm reference. In such WDM or FDM applications, and assuming equalized gain (Chapter 6), the power dynamic range is then given by $P_{\mathrm{s}}^{\mathrm{in}} \leqslant P_{\mathrm{s}}^{\mathrm{in}}/N$ or $P_{\mathrm{s}}^{\mathrm{out}} \leqslant P_{\mathrm{sat}}^{\mathrm{out}}/N$. The characterization of gain versus input signal power and the determination of $P_{\mathrm{sat}}^{\mathrm{out}}$ are of major importance.

The above definition of $P_{\mathrm{sat}}^{\mathrm{out}}$ should not be confused with the EDFA *saturation power* $P_{\mathrm{sat}}(\lambda)$, defined in Eq. (1.45), which represents an intrinsic EDF parameter independent of experimental conditions such as input/output powers and EDFA length.

Likewise, the definition of $P_{\mathrm{sat}}^{\mathrm{out}}$ should not be confused with the EDFA *saturated output power*, generally referred to as the maximum output signal power than we can achieve under given experimental conditions. In the conditions corresponding to Figure 5.15b, the saturated output powers range from $+5\,\mathrm{dBm}$ to $+15\,\mathrm{dBm}$. The saturated output power increases with the input signal, and is limited by the maximum input pump and signal powers available ($P_{\mathrm{p}}^{\mathrm{in}}(\max) = 53.6\,\mathrm{mW}$ and $P_{\mathrm{s}}^{\mathrm{in}}(\max) = -12\,\mathrm{dBm}$, respectively). For this reason, the saturated output power does not represent a true figure of merit for the amplifier. In the limit of very large input signals ($P_{\mathrm{s}}^{\mathrm{in}} \gg P_{\mathrm{p}}^{\mathrm{in}}$), Eqs. (5.2) and (5.3) show that the maximum saturated output signal power and EDFA gain are $P_{\mathrm{s}}^{\mathrm{out}} \approx P_{\mathrm{s}}^{\mathrm{in}}$ and $G \approx 1$, respectively, which corresponds to a regime of amplifier transparency. According to Eq. (5.15), the fractional populations averaged over the fiber lengths satisfy the relation:

$$\frac{n_1}{n_2} = \frac{\sigma_{\mathrm{e}}(\lambda)}{\sigma_{\mathrm{a}}(\lambda)} = \eta(\lambda) \tag{5.21}$$

Since the fractional population ratio is wavelength independent, Eq. (5.21) implies that EDFA transparency at maximum saturation cannot be spectrally uniform (i.e., $G \neq 1$ for $\lambda' \neq \lambda$).

The maximum signal power extracted from the EDFA, i.e., $P_{\mathrm{s}}^{\mathrm{out}}(\max) = \max(P_{\mathrm{s}}^{\mathrm{out}} - P_{\mathrm{s}}^{\mathrm{in}})$, is determined by the energy conservation limit shown in Eq. (5.2), i.e., $P_{\mathrm{s}}^{\mathrm{out}}(\max) = (\lambda_{\mathrm{s}}/\lambda_{\mathrm{p}})P_{\mathrm{p}}^{\mathrm{in}}$. The values of $P_{\mathrm{s}}^{\mathrm{out}}(\max)$ corresponding to different input pump powers are shown in Figure 5.15 and range from $P_{\mathrm{s}}^{\mathrm{out}}(\max) = 10.4\,\mathrm{dBm}$ to $P_{\mathrm{s}}^{\mathrm{out}}(\max) = 17.2\,\mathrm{dBm}$. The points $(P_{\mathrm{s}}^{\mathrm{out}}(\max), G = 1)$ can be connected to the gain curves $G = f(P_{\mathrm{s}}^{\mathrm{out}})$ if the EDFA gain is defined as $G = (P_{\mathrm{s}}^{\mathrm{out}} - P_{\mathrm{s}}^{\mathrm{in}})/P_{\mathrm{s}}^{\mathrm{in}}$ (dashed lines in Figure 5.15). The two gain definitions $G = P_{\mathrm{s}}^{\mathrm{out}}/P_{\mathrm{s}}^{\mathrm{in}}$ and $G = (P_{\mathrm{s}}^{\mathrm{out}} - P_{\mathrm{s}}^{\mathrm{in}})/P_{\mathrm{s}}^{\mathrm{in}}$ very nearly coincide in the region of interest where the gain compression is less than $-10\,\mathrm{dB}$.

EDFAs that are operated in the saturation regime in order to yield a maximized output signal power are referred to as *power amplifiers*. For power EDFAs, one can define the *power conversion efficiency* (PCE) as the ratio

$$\mathrm{PCE} = \frac{P_{\mathrm{s}}^{\mathrm{out}} - P_{\mathrm{s}}^{\mathrm{in}}}{P_{\mathrm{p}}^{\mathrm{in}}} \tag{5.22}$$

From Eq. (5.2), the maximum value for the PCE is given by:

$$\text{PCE(max)} = \frac{\lambda_p}{\lambda_s} \tag{5.23}$$

As the maximum PCE depends upon both pump and signal wavelengths, it is useful for absolute reference to introduce the *quantum conversion efficiency* (QCE) which is wavelength independent and defined by (B. Pedersen et al. [110]):

$$\text{QCE} = \frac{\phi_s^{out} - \phi_s^{in}}{\phi_p^{in}} = \frac{\lambda_s}{\lambda_p} \text{PCE} \tag{5.24}$$

where $\phi_p^{in}$, $\phi_s^{in}$, and $\phi_s^{out}$ are the input or output pump and signal photon fluxes ($\phi_{p,s}^x = P_{p,s}^x/h\nu_{p,s}$). The maximum possible value for the QCE is unity, which corresponds to the case where all pump photons are effectively converted into signal photons.

Equations (5.23) shows that the highest PCEs are achieved for the longest pump wavelengths, which corresponds to the case where the EDFA is pumped in the $^4I_{15/2}-^4I_{13/2}$ band, i.e., near $\lambda = 1480$ nm, as opposed to other possible bands. A theoretical model providing an explicit analytical form for the PCE as a function of input pump and signal powers is described at the end of this section.

The performance is saturated output signal and PCE of power EDFAs have been experimentally investigated with the two pump wavelengths of 980 nm and 1480 nm (J. F. Massicott et al. [111]; R. I. Laming et al. [112]; and B. Pedersen et al. [113]).

The results obtained in [111] are: PCE = 75.6% (QCE = 79.4%) for 1475 nm pump wavelength, and PCE = 50.0% (QCE = 78.9%) for 980 nm pump wavelength, achieved with a forward pumping configuration. The highest *saturated output powers* achieved in this experiment are: $P_s^{out}(P_p^{in} = 185 \text{ mW}) = 140$ mW or 21.5 dBm for the 1475 nm pump, and $P_s^{out}(P_p^{in} = 1.1 \text{ W}) = 530$ mW or 27.2 dBm for the 980 nm pump. In contrast, the *saturation output power* (corresponding to $-3$ dB gain compression) is $P_{sat}^{out} = 79.5$ mW or $+19$ dBm for the 1475 nm pump [111]. In the experiment described in [113], which used a 980 nm, forward pumped EDFA, a PCE of 56% corresponding to a QCE of 89% was achieved, this result is the highest reported to date (1992).

The effects of fiber length and pumping configuration were investigated for 978 nm pumped EDFAs [112]. Figure 5.16 shows the PCE as a function of EDFA length, for two pump powers and for the two pumping configurations. The most important conclusion of this measurement is that, for any EDFA lengths, *backward pumping yields the highest PCE* [112]. The figure shows that the optimal length at which the PCE is maximized is shorter in the forward pumping case. From the figure, we can derive the following results: at the optimal length $L = 30$ m for backward-pumping, the efficiency is PCE = 55% (QCE = 86.2%); at the optimal length $L = 20$ m for forward pumping, the efficiency is PCE = 45% (QCE = 70.5%). In [112], the PCE is referred to as absolute conversion efficiency, while the main results of the study are interpreted in terms of differential conversion efficiency (DCE). This definition corresponds to DCE = $dP_s^{out}/dP_p^{in}$, which is somewhat higher than the PCE. Unlike the PCE, the DCE does not indicate how much saturated output power can be obtained from the EDFA, since the line $P_s^{out} = f(P_p^{in})$ does not cross the origin. The

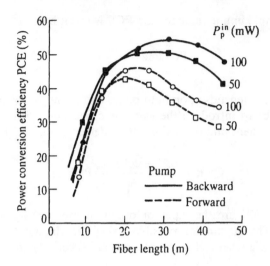

**FIGURE 5.16** Power conversion efficiency PCE versus EDFA length for (– – –) forward and (——) backward pumping, and for two input pump powers at $\lambda = 978$ nm. From [112], reprinted with permission from *Electronics Letters, 1991.*

observed difference between the forward and backward pumping configurations for the PCE can be mainly attributed to the effect of amplifier self-saturation (E. Desurvire [114]). Other secondary causes for the discrepancy, which remain to be fully analyzed, include background loss, signal ESA, and clustering [112]. The effect of self-saturation on the EDFA power conversion efficiency is analyzed theoretically in Section 5.5.

EYDFAs based on Er–Yb-codoped fibers using LD-array-pumped Nd:YAG or Nd:YLF sources near $\lambda = 1.06\ \mu m$ as a pump have also shown a promising potential for power amplifier applications (S. G. Grubb et al., [6] and [7]). For this type of device, the notion of PCE or QCE as referred to the ratio of output signal to LD-array input pump is not relevant for performance comparison, as the power conversion occurs in two stages. Typically, the power conversion efficiency from the LD-array to the $\lambda = 1.06\ \mu m$ output is 45%, corresponding to a quantum efficiency near 60% (Figure 5.3). As the LD-array power can be several watts and is scalable (Section 5.1), the saturated output power of EYDFAs is the only parameter of interest. For instance, with an Nd:YAG-backward-pumped EYDFA, a maximum gain, a saturation output power, and a saturated output power $G_{max} = 42$ dB, $P_{sat}^{out} = 20$ mW ($+13$ dBm), and $P_s^{out} = 71$ mW ($+18.5$ dBm) have been demonstrated; with bidirectional pumping, the output powers were upgraded to $P_{sat}^{out} = 56$ mW ($+17.5$ dBm) and $P_s^{out} = 126$ mW ($+21$ dBm) [6]. These results were improved by using a single Nd:YLF source in the backward pumping configuration, which yielded the maximum gain, saturation, and saturated output powers $G_{max} = 50$ dB, $P_{sat}^{out} = 126$ mW ($+21$ dBm), and $P_s^{out} = 290$ mW ($+24.6$ dBm) [7].

Maximum gain $G_{max}$, saturation output power $P_{sat}^{out}$, and maximum saturated output powers $P_s^{out}$ reported to date (1992) in power EDFAs with different pumping configurations and wavelengths are summarized in Table 5.1. The data ($G_{max}$, $P_{sat}^{out}$) corresponding to table 5.1 are plotted in Figure 5.17. As the figure indicates, the highest gains ($G \geqslant 40$ dB) and saturation output powers ($P_{sat}^{out} \geqslant 10$ dBm) simul-

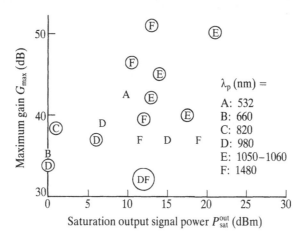

**FIGURE 5.17**   Maximum or unsaturated gains and saturated output signal powers ($-3$ dB gain compression) obtained experimentally for different pump wavelengths, indicated by letters A-F in inset. Circled letters indicate LD pumping. Corresponding values, maximum saturated output powers, pumping configurations, and further information are listed in Table 5.1.

taneously obtained with LD pumping were achieved only with the two pump wavelengths $\lambda_p = 1.05$–$1.06$ $\mu$m and $\lambda_p = 1.48$ $\mu$m, consistent with a PCE proportional to $\lambda_p$, Eq. (5.23). But from Table 5.1, maximum saturated output powers in the range $P_s^{out} = +19$ dBm to $+24.5$ dBm (75 mW to 282 mW) have been obtained for the three wavelengths $\lambda_p = 980$ nm, $\lambda_p = 1.05 - 1.06$ $\mu$m and $\lambda_p = 1.48$ $\mu$m. It is likely that the performance of power EDFAs (as characterized by the three parameters $G_{max}$, $P_{sat}^{out}$, $P_s^{out}$) will still improve in the near future, principally because of the evolution in high power LD technology but also because of fiber design improvements. The issues of reliability and cost of the LD source will also be important factors in the choice of pump wavelength for power EDFA applications.

Now we consider the EDFA gain spectrum changes caused by signal saturation. As discussed in Chapter 4, the Er:glass absorption and emission cross sections do not correspond to a single laser transition, but rather to a superposition of several individual transitions, which initiate from, or terminate in the Stark levels of the ground ($^4I_{15/2}$) and upper ($^4I_{13/2}$) manifolds. Within each Stark manifold, a fast thermalization process results in an equilibrium (or Boltzmann) population distribution. An increase of stimulated emission from a monochromatic signal results in changes in the total manifold populations (i.e., an increase of $N_1$ and a decrease of $N_2$), while the equilibrium distributions remain unchanged (Section 4.4). Since the cross section line shapes are only determined by this equilibrium distribution, and therefore remain also unchanged, the gain coefficient spectrum (i.e., $\sigma_e(\lambda)N_2 - \sigma_a(\lambda)N_1$) has homogeneous saturation characteristics. In the case of a glass host, however, random variations in the Stark effect and the possibility of multiple activator sites cause additional inhomogeneous broadening. The homogeneous and inhomogeneous broadening characteristics of different Er:glass types are discussed qualitatively and quantitatively in Chapter 4. In this section, we consider how these characteristics affects the EDFA gain spectrum during saturation.

TABLE 5.1 Maximum gain $G_{max}$, saturation output power $P_{sat}^{out}$ (3 dB gain compression) and maximum saturated output power $P_s^{out}$ achieved for different pumping wavelengths $\lambda_p$, using forward, backward, and bidirectional pumping configurations

| $\lambda_p$ (nm) | $G_{max}$ (dB) | $P_{sat}^{out}$ (dBm) | $P_s^{out}$ (dBm) | Forward | Backward | Bidirectional | Pump source | Year | Ref. |
|---|---|---|---|---|---|---|---|---|---|
| 532 | 42.5 | 10 | 11 | × | – | – | – | 1990 | [72] |
| 660 | 35 | 0 | 2 | × | – | – | – | 1991 | [71] |
| 827 | 29 | −6 | 2.5 | × | – | × | 1 LD | 1990 | [5] |
| 827 | 38.5 | 1 | 7 | – | – | – | 2 LDs | 1990 | [5] |
| 975 | 36 | 6 | 11 | × | – | – | 1 LD | 1990 | [115] |
| 980 | 34 | 0 | 8 | × | – | – | 2 LDs | 1989 | [68] |
| 980 | 39 | 7 | 13 | × | – | – | – | 1990 | [62] |
| 980 | 36 | 15 | 19 | × | – | – | – | 1991 | [113] |
| 980/1480 | 32 | 12 | 15 | – | × | × | 2 LDs | 1992 | [94] |
| 1050 | 50 | 21 | 24.5 | – | × | – | 1 LD | 1992 | [7] |
| 1064 | 42 | 13 | 18.5 | – | × | – | 1 LD | 1992 | [6] |
| 1064 | 40 | 17.5 | 21 | – | – | × | 2 LDs | 1992 | [6] |
| 1064 | 45 | 14 | 21.5 | – | – | × | 2 LDs | 1991 | [116] |
| 1475 | 37 | 19 | 21.5 | × | – | – | – | 1990 | [111] |
| 1480 | 46.5 | 10.5 | 15 | × | – | – | 4 LDs | 1989 | [82] |
| 1480 | 39.5 | 12 | 15.5 | × | – | – | 1 LD | 1990 | [62] |
| 1480 | 51 | 13 | 15 | × | 2-stage | × | 6 LDs | 1990 | [83] |
| 1490 | 37 | 11.5 | 15 | × | – | – | – | 1989 | [65] |

The homogeneity of the EDFA gain spectrum can be evidenced by the changes of the output ASE (forward direction) in presence of a strong, monochromatic saturating signal. The effect is shown both experimentally and theoretically in Figure 5.18 (E. Desurvire et al. [104]). In this experiment, the wavelength of the saturating signal was $\lambda = 1.5375\,\mu$m, which is close to the center of the EDFA gain spectrum; the small-signal gain was $G(P_s^{in} = 0) = 25$ dB; and the input signal power was varied from 0 to $25\,\mu$W. The succession of spectra obtained for increasing input signal powers reveals two main effects: (a) the gain saturation at $\lambda = 1.5375\,\mu$m, which is evidenced by a sublinear increase of output signal power ($G(P_s^{in} = 25\,\mu$W$) = 21.5$ dB), and (b) the near-uniform decrease of the ASE output power spectrum. The theoretical spectra, which were calculated with the homogeneous gain saturation approximation,

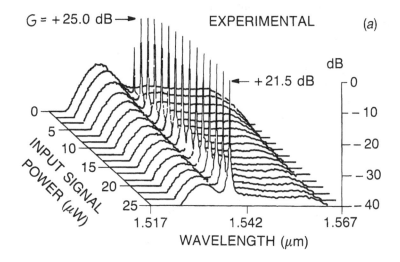

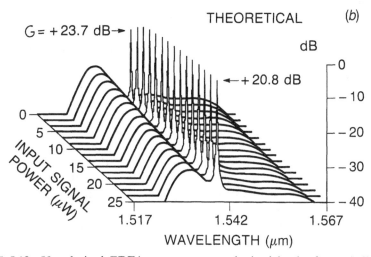

**FIGURE 5.18** Unpolarized EDFA output spectra obtained in the forward direction for different input signal powers at $\lambda = 1.5375\,\mu$m: (*a*) experimental (*b*) theoretical. From [104] © 1989 IEEE.

show fair qualitative and quantitative agreement [104]. Details concerning the method used for this type of calculation and the agreement of theory and experiment are discussed later in this section.

The changes observed in the forward ASE spectrum $P_{ASE}^{forward}(\lambda, L)$ at $z = L$ during saturation approximately follow the changes in the EDFA gain spectrum $G(\lambda, L)$. The changes of $P_{ASE}^{forward}(\lambda, L)$ and $G(\lambda, L)$ are not strictly equal because the ASE power is proportional to $n_{eq}(\lambda, L)G(\lambda, L)$, where $n_{eq}(\lambda, L)$ is the *equivalent input noise factor* (Chapter 2). The equivalent input noise factor is given by the expression

$$n_{eq}(\lambda, L) = \int_0^L \frac{a(\lambda, z)}{G(\lambda, z)} dz \qquad (5.25)$$

where the coefficient $a(\lambda, z)$ is the emission coefficient (i.e., $\sigma_e(\lambda)N_2(z)$) integrated over the fiber transverse plane. Since both numerator and denominator in the integrand in Eq. (5.25) are affected by saturation, so is the equivalent input noise $n_{eq}(\lambda, L)^*$. We analyze the saturation-induced changes in the parameter $n_{eq}(\lambda, L)$ in the next section and show that in the saturation regime of interest (i.e., where the EDFA gain is compressed by not more than 10 dB), the changes in equivalent input noise factor are generally small (e.g., $\Delta n_{eq}(\lambda)/n_{eq}(\lambda) \leqslant 20\%$). In this case, the observed changes in the ASE spectrum approximately correspond to the gain spectrum changes (e.g. within 1 dB accuracy).

Another experimental measurement of ASE power spectrum changes in the saturation regime is shown in Figure 5.19 (E. Desurvire et al. [79]). In this experiment, the output signal is rejected by a polarizer, which makes it possible to accurately measure the ASE in the spectral region near the signal [78]. Unlike in the previous case, the saturating signal is at the peak gain wavelength ($\lambda = 1.531 \ \mu m$) and the input signal power is varied from 1 $\mu$W to 1 mW. The top figure in 5.19 shows the ASE spectra, and the bottom figure shows the ASE power decrease relative to the case where no signal is input in the EDFA ($P_s^{in} = 0$). While EDFA gain homogeneously saturates, the amount of gain compression (reflected by the ASE power change at a specific wavelength) is not spectrally uniform. For the range of input signals considered, a maximum compression of $-20$ dB occurs at the peak gain wavelength, while the effect is nearly 10 times less important ($-10$ dB) at longer wavelengths ($\lambda = 1.55$–$1.56 \ \mu m$). Such uneven gain compression is simply explained by the following: for any change of manifold populations $\Delta N_1 = -\Delta N_2$ induced by saturation, the local gain coefficient decrease $\Delta g(\lambda) \approx \sigma_e(\lambda)\Delta N_2 - \sigma_a(\lambda)\Delta N_1 = [\sigma_e(\lambda) + \sigma_a(\lambda)]\Delta N_2$ is the largest at the wavelength where the sum of the cross sections $\sigma_e(\lambda) + \sigma_a(\lambda)$ is maximum, which occurs at $\lambda = 1.531 \ \mu m$ for aluminosilicate EDFAs.

While the spectral gain saturation effect observed in Figure 5.19 can be essentially attributed to homogeneous cross section characteristics, this measurement does not make it possible to evaluate any effect of inhomogeneous gain saturation due to inhomogeneous broadening. Inhomogeneous gain saturation effects at room temperature in EDFAs have been investigated through several methods. A first method measures the effect of mutual gain saturation in two-signal amplification (K. Inoue

---

* $n_{eq}(\lambda, L)$ is related to the spontaneous emission factor $n_{sp}(\lambda, L)$ through $n_{eq} = n_{sp}(G - 1)/G$, and therefore $n_{eq} \approx n_{sp}$ in the high gain limit. The advantage of using the parameter $n_{eq}$ rather than $n_{sp}$ is that in the limit $G \to 1$, $n_{eq}$ remains finite while $n_{sp}$ tends to infinity (Chapter 2).

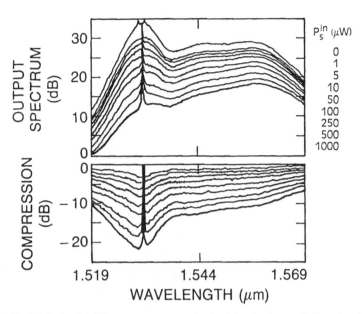

**FIGURE 5.19**   Polarized ASE output spectra obtained in the forward direction for different input signal powers at $\lambda = 1.531 \, \mu m$, showing effects of homogeneous gain saturation (top) with nonuniform compression (bottom). From [79] © 1990 IEEE.

et al. [117]). In this experiment using two input signals of equal powers at wavelengths $\lambda$ and $\lambda'$, the EDFA gain versus total output signal power characteristics (i.e., $G(\lambda) = f[P_s^{out}(\lambda) + P_s^{out}(\lambda')]$) are measured for different wavelength separations $\Delta \lambda = \lambda - \lambda'$. The result is compared to the same characteristics obtained with the second signal at $\lambda'$ turned off (i.e., $G(\lambda) = f[P_s^{out}(\lambda), \, P_s^{out}(\lambda') = 0]$). While the two responses are found to be identical for small wavelength separations (e.g., $\Delta \lambda = 0.5 \, nm$), a discrepancy is observed as the separation is increased. This effect, attributed to inhomogeneity, is not observed in homogeneously broadened laser amplifiers (e.g., semiconductor diode amplifiers), in which mutual gain saturation, for any wavelength pair located within the gain spectrum, is uniquely determined by the total output signal power [117]. To date, the effect of mutual inhomogeneous saturation in EDFAs remains to be analyzed theoretically and fully characterized.

Another method for the investigation of inhomogeneous gain saturation at room temperature uses a monochromatic saturating signal and a weak probe source with broadband spectrum (M. Tachibana et al. [118]). The spectral gain changes $\Delta G(\lambda)$ caused by saturation are recorded for different saturating signal wavelengths. Maximum gain compression is obtained when the saturating signal wavelength $\lambda_s$ matches the peak gain wavelength (e.g., $\lambda = 1.531 \, \mu m$ for aluminosilicate EDFAs); the spectral gain changes obtained with different $\lambda_s$ are similar. Both observations are explained by the homogeneous characteristics of the cross sections, discussed in the example of Figure 5.19. However, an accurate comparison of gain compression curves $\Delta G(\lambda)$ obtained with the same degree of saturation shows small differences (i.e., $< 1 \, dB$), attributed to gain hole burning [118].

The effect of gain hole burning in EDFAs operating at room temperature was also investigated theoretically, using the inhomogeneous rate equation model described in Section 1.12 (E. Desurvire et al. [79]). The main goal of this analysis is to compare the theoretical EDFA characteristics $\Delta G(\lambda) = f[\lambda, P_s^{in}(\lambda_s)]$ predicted by both homogeneous and inhomogeneous models, with respect to experimental data. In the inhomogeneous model, a deconvolution of the Er:glass cross section, based on experimental homogeneous ($\Delta\lambda_{hom}$) and inhomogeneous line widths ($\Delta\lambda_{inh}$ or $\Delta\lambda_i$) can be used to evaluate the homogeneous cross sections $\sigma_{a,e}^H(\lambda)$, as detailed in Section 4.5. The results obtained from both theoretical models for a Type III fiber (aluminosilicate) are shown in Figure 5.20, along with corresponding experimental data previously shown in Figure 5.19 [79]. When no saturation occurs ($P_s^{in} = 0$), the two theoretical models give identical predictions (as expected) and are in good agreement with the experimental data. No fitting parameter was used in either theoretical computation. But in the saturation regime, the shapes of the gain spectra predicted by the two models show small discrepancies, manifest in the gain ripples. At wavelengths far from the saturating signal, the inhomogeneous model predicts a smaller gain compression effect in comparison to the homogeneous model; this is expected. In comparison with experimental data, the homogeneous model is fairly accurate in the regime of moderate saturation ($\Delta G < 10$ dB). But in the regime of deep saturation ($\Delta G > 10$ dB), neither homogeneous nor inhomogeneous models provide accurate predictions. This study does not establish that the current inhomogeneous model is more accurate, but such a model is necessary to predict the mutual gain saturation and hole burning effects previously described. Accurate modeling of inhomogeneous gain saturation is a difficult task, as the homogeneous cross-sections and inhomogeneous probability distribution (Chapter 4) of Er:glass are not known. Consequently, simplifying assumptions must be introduced for these parameters, which may cancel any potential improvement in accuracy. To date, the complex issue of inhomogeneous gain modeling in EDFAs (as well as in other RE-doped fiber amplifiers) remains open to further analysis.

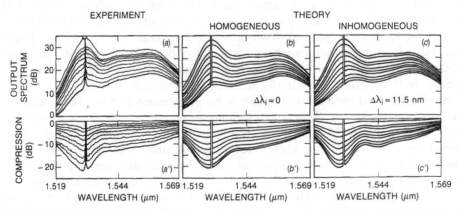

**FIGURE 5.20**    Experimental ($a$) and theoretical ($b, c$) output ASE spectra obtained for different powers of saturating signal at $\lambda = 1.531$ $\mu$m, and corresponding gain compression ($a'$, $b'$, $c'$). The experimental data ($a$) are identical to those of Figure 5.19. The theoretical curves ($b$) and ($c$) are calculated using the homogeneous and inhomogeneous rate equation models, respectively. In the inhomogeneous model, an inhomogeneous line width $\Delta\lambda_i = 11.5$ nm is assumed. From [79] © 1990 IEEE.

The EDFA gain difference between short and long wavelengths is typically 5–10 dB, while the region where the gain is nearly uniform (i.e., 1 dB variation) is about 20 nm wide. An important consequence of nonuniform gain saturation (which is a homogeneous effect) is the flattening of the gain spectrum. This gain flattening effect, which results in a larger usable bandwidth can be obtained by operating the EDFA in the deep saturation regime, to a saturation point where the gain difference between short and long wavelengths is minimized. This regime is illustrated in Figure 5.19 by the gain curve corresponding to $P_s^{in} = 10\,\mu$W. In this example, the maximum variation across the gain spectrum is reduced approximately from 8 dB to 3.5 dB. The EDFA bandwidths are conventionally defined by the region where the gain difference is less than 3 dB or 1 dB, typically, about the average value $\langle G(\lambda) \rangle$. The 1 dB or 3 dB bandwidth can be maximized by using optimized values for the saturating signal power, signal wavelength and fiber length. For instance, 1 dB bandwidths as large as 38 nm ($\langle G(\lambda) \rangle = 13$ dB) were reported for aluminosilicate power EDFAs (R. I. Laming et al. [112]); 3 dB bandwidths of 33 nm ($\langle G(\lambda) \rangle = 21$ dB) were also obtained with germanosilicate power EYDFAs (S. G. Grubb et al. [6]). The effect of gain flattening in power EDFAs or EYDFAs can be advantageously used in WDM applications, which require amplifiers with equalized gains. The main drawback of this approach is the sensitivity of the gain spectrum to random power variations or power drift, which could originate either in the saturating (control) signal or in the WDM channels. Different methods of spectral gain equalization are described in Section 6.4.

In the rest of this section, the theoretical modeling of saturated EDFAs, related computation methods, simplifying assumptions, and agreement with experiment are discussed.

Saturated EDFAs can be modeled through the general rate equations (1.66)–(1.68). The only limiting assumption implied in these equations is gain homogeneity. The assumption of homogeneous gain broadening could be responsible for small differences between theory and experiment when considering the highly saturated regime.

Section 1.6 describes numerical methods for solving Eqs. (1.66)–(1.68). In the general case, solving these $N$ coupled nonlinear differential equations involves three steps: (1) evaluation of boundary conditions, (2) integration of the gain and absorption coefficients across the transverse plane, and (3) integration over the fiber length. Steps 2 and 3 can be performed by usual numerical methods. An example of these methods and related algorithms are given in Appendix C. The evaluation of boundary conditions in Step 1 is by far the most difficult. Consider the case of a power EDFA pumped in the backward direction. The only known boundary conditions are the input signal and pump powers at the fiber ends (i.e., $P_s^{in,+}(0)$, $P_p^{in,-}(L)$ or, in the normalized notation of Chapter 1: $p^+(0)$, $q^-(L)$), while the output powers ($P_s^{out,+}(L)$, $P_p^{out,-}(0)$ or $p^+(L)$, $q^-(0)$) are unknown. If high EDFA gains are considered (e.g., $G > 25$ dB), the effect of self-saturation by ASE cannot be neglected. However, the forward and backward output ASE power spectra ($p_k^{\pm}$, $k = 1 \ldots N$), which are necessary as boundary conditions, are unknown. The most general case involving the maximum of unknown boundary conditions is an EDFA with bidirectional pump, bidirectional signal, and self-saturation. The two main approaches, i.e., the shooting method and the relaxation method that can be used for solving this type of boundary value problem are described in Chapter 1.

The shooting and relaxation methods do not always converge; in some cases they can lead to computational errors and instabilities. For each type of problem, an optimized (and converging) computation algorithm must be determined, which requires trial and error. In the general case, the computation time and memory requirements for solving Eqs. (1.66)–(1.68) are beyond the capacity of standard microcomputers so large mainframe computers are necessary. The required computation time becomes even more prohibitive when input conditions are varied within a certain range (e.g., pump and signal powers or wavelengths) or when the program task is to determine optimal fiber parameters (e.g., length, $V$ number).

Simplifying assumptions and closed-form expressions can be advantageously used for providing a fast and fairly accurate estimate of the EDFA performance. The approximate results can then be used as a guideline in a more accurate computation, in order to narrow the range where optimum parameters are to be found, or as trial boundary values. Examples of approximate computation methods for self-saturated EDFAs are described in Section 5.5. In the rest of this section, we discuss examples of theory/experiment comparisons in the case of saturated EDFAs.

A first example is provided by Figure 5.18, which shows experimental and theoretical EDFA output spectra as a function of input signal power (E. Desurvire et al. [104]). This example consists of an EDFA with forward pump at $\lambda_p = 514.5$ nm, saturation by both by signal and ASE, and pump ESA. The numerical computation used the approximations of a step Er-doping and effective mode sizes for the (multimode) pump and (single-mode) signal, which were evaluated from absorption coefficients. The saturation power $P_{sat}(\lambda_k)$ was also approximated by a single effective saturation power $P_{sat}^*$. The justification for this approximation was an attempt of achieving a best fit to experimental results by using only two adjustable parameters, i.e., $P_{sat}^*$ and $\delta = \sigma_a(\lambda_p, \text{ESA})/\sigma_a(\lambda_p, \text{GSA})$; $\delta$ is the ratio of excited state to ground state absorption cross section at $\lambda_p$. This approximation and the use of adjustable parameters are no longer necessary, as accurate data for the Er:glass cross sections at both signal and pump wavelengths are now available. A resolution of 0.25 nm, corresponding to $N = 200$ rate equations for each ASE spectra was chosen. Equations (1.69) and (1.70) including forward pump, forward ASE + signal and backward ASE were solved iteratively through the relaxation method. The iteration was performed until a convergence near 1% accuracy was reached, this process required about 10 iterations on average, depending on the input signal conditions. The fitting parameters $P_{sat}^* = 280 \, \mu\text{W}$ and $\delta = 1.0$, consistent with current data for Er:glass cross sections, were determined from a best fit to the experimental gain versus input signal power characteristics for two signal wavelengths, Figure 5.21. Good agreement between theory and experiment was then simultaneously achieved for both single-signal and two-signal saturation data. This example corresponds to a first attempt at modeling spectrally resolved, saturated EDFAs, and should not be used as a reference, as spectroscopic data are now available for more accurate modeling. Even in the most accurate treatment, however, the case of multimode pumping, such as at $\lambda_p = 514.5$ nm, requires the use an approximation for the pump intensity distribution.

A comprehensive theory/experiment comparison that would include complete data on EDFA gain and ASE spectra measured as a function of: (1) input pump power, (2) single-signal input power at various wavelengths and (3) two-signal input power with various wavelength spacings, has not yet been reported in the literature. Such a study might represent a final validation of the EDFA model, and in particular of the approximation of homogeneous gain broadening.

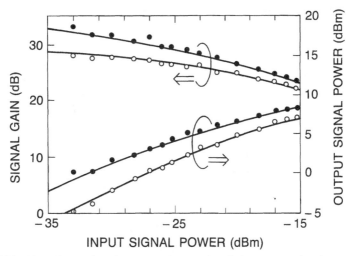

**FIGURE 5.21** Experimental gain versus input signal power at signal wavelengths of $\lambda = 1.531\,\mu$m (closed circles) and $\lambda = 1.5375\,\mu$m (open circles) and corresponding theoretical values (full lines), obtained with an aluminosilicate (Type III) EDFA pumped at $\lambda = 514.5$ nm. From [104] © 1989 IEEE.

The second example of saturated EDFA modeling considered here is based on the analytical model described in Section 1.10. This model enables us to determine numerically the EDFA output power at any wavelength (pump and signal included), with any number of simultaneous optical channels (pump and signal included) with any propagation directions (A. A. Saleh et al. [109]). The restrictive assumptions are: (a) confined Er-doping (no integration in the transverse plane), (b) no pump ESA, and (c) negligible amplifier self-saturation by ASE (no need for spectral resolution and evaluation of boundary conditions). Section 5.5 shows that this model can actually be extended, with further approximations, to include the effect of amplifier self-saturation.

While the model of [109] provides closed-form or analytical solutions, these are found in a transcendental form that must be solved numerically. A computation algorithm for the resolution of such a transcendental equation with arbitrary spectral resolution (or number of optical channels) is described in Section 5.5. Figure 5.22 shows experimental and theoretical results obtained in an EDFA pumped at 1480 nm for two input signal powers at $\lambda = 1.55\,\mu$m. The only fiber parameters used in the theoretical computation are the pump and signal absorption coefficients ($\alpha_{\mathrm{p}}, \alpha_{\mathrm{s}}$) and the saturation powers ($P_{\mathrm{sat}}(\lambda_{\mathrm{p}})$, $P_{\mathrm{sat}}(\lambda_{\mathrm{s}})$), which can be determined experimentally by the transmission measurements $P^{\mathrm{out}}(\lambda_{\mathrm{p,s}}) = f[P^{\mathrm{in}}(\lambda_{\mathrm{p,s}})]$. Note that [109] uses intrinsic saturation powers $\bar{P}_{\mathrm{sat}}(\lambda_{\mathrm{p,s}})$, identical to the definition of photon flux $\bar{P}_{\mathrm{sat}}(\lambda_{\mathrm{p,s}}) = P_{\mathrm{sat}}(\lambda_{\mathrm{p,s}})/h\nu_{\mathrm{p,s}}$. Figure 5.22 shows that the theoretical results based on this low gain, ASE-free model fit the experimental data remarkably well. In the case of an EDFA with unidirectional pump and single-signal saturation, the optimum length for which the gain is maximized can also be determined analytically, as given by Eq. (1.157). This analytical model, which requires comparatively short computation times, permits a rapid evaluation of optimum performance in low gain, power EDFAs.

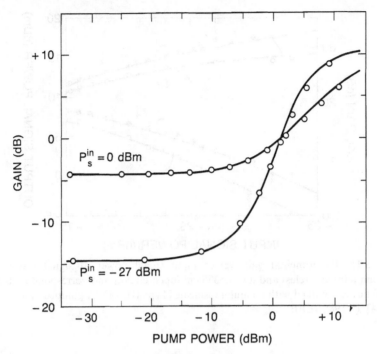

**FIGURE 5.22**   Experimental gain versus input pump power for two input signal powers at $\lambda = 1.55\,\mu$m obtained with a 1480 nm pumped EDFA. The theoretical curves are calculated through an analytical model for low gain, ASE-free EDFAs. From [109] © 1990 IEEE.

Finally, we consider the simplest configuration, an EDFA with unidirectional pump and single-signal saturation. Although this case is straightforward to study through our previous analytical model, another and simpler method is also possible. This method is based upon the graphical resolution of transcendental equation (1.137). The graphical solution method, which enables us to determine the most general characteristics $p_2^{out} = f(p_1^{in}, p_1^{out}, p_2^{in})$, where $p_1$, $p_2$ represent normalized powers at any two wavelengths, is illustrated in Figures 1.5 and 1.6. Implemented with a microcomputer, this graphical method is a succession of one-dimensional array comparisons (i.e., find array indices $I$, $J$ such that $A(I) = B(J)$, then find $J$, $K$ such that $B(J) = C(K)$, etc.). Implementation of such a look-up table search is intrinsically no faster than the usual resolution algorithm for the transcendental equation $f(x) = 0$ (i.e., find index $I$ such that $A(I)$ is nearest to zero, where $A(I)$ is a discrete set of values of $f(x)$). An improvement to the graphical method, which makes it possible to obtain an analytical and explicit expression for the $p_2^{out} = f(p_1^{in}, p_1^{out}, p_2^{in})$ characteristics is described in Appendix O.

Appendix O shows that, in the case of forward pumping, the solution $p_2^{out} = f(p_1^{in}, p_1^{out}, p_2^{in})$ can be put into the form:

$$p_2^{out} = \frac{1}{U_2} 10^{\mathscr{P}(z)} \qquad\qquad (5.26)$$

where $\mathscr{P}(z)$ is an $N$th order polynomial of definition

$$\mathscr{P}(z) = \sum_{i=0}^{N} a_i z^i \tag{5.27}$$

Coefficients $a_i$ are listed in Appendix O; the variable $z$ is defined through:

$$z \equiv z(p_1^{in}, p_1^{out}, p_2^{in}) = \frac{1}{\log(10)} \left\{ U_2 p_2^{in} + \log(U_2 p_2^{in}) + \frac{\alpha_2}{\alpha_1} \left[ U_1(p_1^{in} - p_1^{out}) - \log\left(\frac{p_1^{in}}{p_1^{out}}\right) \right] \right\} \tag{5.28}$$

In eqs. (5.26) and (5.28), the parameters $U_1$ and $U_2$ are defined by $U_1 = (\eta_2 - \eta_1)/(1 + \eta_1)$, $U_2 = (\eta_2 - \eta_1)/(1 + \eta_2)$; $\alpha_1$, $\alpha_2$ are the absorption coefficients at the two wavelengths $\lambda_1$ (pump) and $\lambda_2$ (signal), respectively, and $\eta_i = \sigma_e(\lambda_i)/\sigma_a(\lambda_i)$ are the corresponding cross section ratio. In the case of backward pumping, the same equations (5.26)–(5.28) also apply, with the following substitutions: $p_1^{in} \rightarrow p_1^{out}$ ($=$back-ward pump at $z = 0$), $p_1^{out} \rightarrow p_1^{in}$ ($=$backward pump at $z = L$) and $\alpha_1 \rightarrow -\alpha_1$, which gives an identical definition (5.28) for the variable $z$.

The normalized output signal $p_2^{out}$ defined in eq. (5.26) can be directly plotted as a function of the normalized output pump $p_1^{out}$, given input conditions $p_1^{in}$ and $p_2^{in}$. Figure 5.23 shows a series of plots of $p_2^{out} = f(p_1^{out})$ obtained with forward pumping, input pump $p_1^{in} = 5$ to $p_1^{in} = 50$, input signal $p_2^{in} = 1$, and the following parameters: $\eta_p = 0.26$, $\eta_2 = 0.92$, $\alpha_s/\alpha_p = 3.5$, which are typical for aluminosilicate EDFAs with

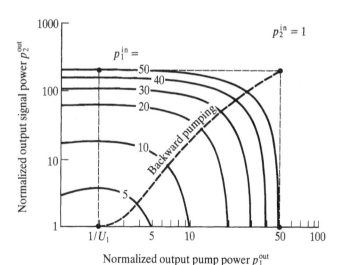

**FIGURE 5.23** Plots of normalized output signal power $p_2^{out}$ versus normalized output pump power $p_1^{out}$ for different input pump powers $p_1^{in}$, obtained in the saturation regime from the explicit analytical solution defined in Eq. (5.26). The example corresponds to the case of forward pumped aluminosilicate EDFAs with pump at $\lambda_1 = 1480$ nm and signal at $\lambda_2 = 1531$ nm ($p_2^{in} = 1$), and with parameters $\eta_p = 0.26$, $\eta_s = 0.92$, $\alpha_s/\alpha_p = 3.5$. The dashed curve corresponds to the same example in the case of backward pumping, with the horizontal axis representing the normalized input pump power at $z = L$.

$\lambda_1 = 1480$ nm and $\lambda_2 = 1531$ nm. As the figure shows, the gain is maximized for the output pump power $p_1^{out} = 1/U_1$, which corresponds to a vanishing gain coefficient at $z = L$. On the other hand, the gain is minimum ($G = 1$) for $p_1^{out} = p_1^{in}$, which corresponds to an EDFA with minimum possible length $L = 0$.

The dashed curve in Figure 5.23 corresponds to plots of $p_2^{out} = f(p_1^{in})$ in the same example, but in the case of backward pumping, with the horizontal axis representing now the normalized input pump power at $z = L$. The forward and backward pumping schemes give identical EDFA gains for identical input pump powers (as expected in the absence of amplifier self-saturation).

The optimal EDFA length $L_{opt}$ for which the saturated gain is maximized (and for which the output pump is $p_1^{out} = 1/U_1$) is given by Eq. (1.157). With the simplified notations used in this section, this equation can be put into the form:

$$L_{opt} = \frac{1}{\alpha_1} \frac{1 + \eta_2}{1 + \eta_1} \left( p_1^{in} - \frac{1 + \eta_1}{\eta_2 - \eta_1} \right) - \frac{1}{\alpha_2} \left\{ \log\left(\frac{p_2^{out}}{p_2^{in}}\right) + p_2^{out} - p_2^{in} \right\} \qquad (5.29)$$

In the previous example, and assuming $\alpha_2 = 1$ dB/m, we find from Eq. (5.29) the optimal lengths $L = 187$ m for $p_1^{in} = 50$ and $L = 138$ m for $p_1^{in} = 20$.

The power conversion efficiency (PCE) is, according to eqs. (5.22) and (5.26):

$$PCE = \frac{P_s^{out} - P_s^{in}}{P_p^{in}} = \frac{p_2^{out} - p_2^{in}}{p_1^{in}} \frac{P_{sat}(\lambda_2)}{P_{sat}(\lambda_1)} = \frac{1}{p_1^{in}} \left\{ \frac{1}{U_2} 10^{\mathscr{P}(z)} - p_2^{in} \right\} \frac{P_{sat}(\lambda_2)}{P_{sat}(\lambda_1)} \qquad (5.30)$$

Equation (5.26) with $p_1^{out} = 1/U_1$ and Eqs. (5.29) and (5.30) are explicit expressions for the maximum saturated output signal power, maximum power conversion efficiency, and corresponding optimal fiber length in unidirectionally pumped EDFAs, given input pump and signal conditions $p_1^{in}$, $p_2^{in}$ and fiber parameters $\eta_1$, $\eta_2$, $\alpha_1$, $\alpha_2$. The result is independent of the pumping scheme (i.e., forward or backward). The effect of EDFA self-saturation actually reduces the PCE from the value predicted by Eq. (5.30). But in the highly saturated regime, self-saturation by ASE is negligible and the highest PCE, achieved with backward pumping, nearly approaches the value predicted by Eq. (5.30). Thus, the explicit expression of Eq. (5.30) can be used for a straightforward prediction of the maximum achievable PCE in EDFAs, given the pump and signal wavelengths ($\lambda_1$, $\lambda_2$), and the six parameters $\eta_1$, $\eta_2$, $\alpha_1$, $\alpha_2$, $P_{sat}(\lambda_1)$, $P_{sat}(\lambda_2)$, which are related to the EDF type and design.

## 5.4 ASE NOISE AND NOISE FIGURE

The amplified spontaneous emission (ASE) noise and the optical noise figure (NF) are the third most important characteristics of EDFAs. The ASE power spectrum closely emulates the gain spectrum, so it provides useful information on the EDFA operating characteristics in various pump and signal power regimes. On the other hand, the NF represents a measure of the SNR degradation from the input to the output of the amplifier. System applications require that a certain level of signal-to-noise ratio (SNR) be achieved at the receiver end; for this reason, the ASE and NF characteristics of EDFAs used in various system locations must be carefully evaluated. Whether EDFAs are used in this system as power amplifiers, in-line repeaters, or preamplifiers, the ASE noise falling into the operating signal bandwidth represents a major parameter according to which the overall system performance (i.e., maximum transmission distance and bit rate) can ultimately be determined.

Chapter 2 gives a fundamental description of the initiation and buildup of ASE in optical amplifiers, and of the signal/ASE photon statistics, from which the NF is defined. This section concerns the experimental characterization of the ASE and NF, as they can be phenomenologically measured in EDFAs. These chararacteristics are then related to the theory developed in Chapter 2. EDFA self-saturation by ASE is analyzed in the next section. The NF definition is based upon a *linear* amplification theory, and therefore, it is not applicable in the saturation (or nonlinear) amplification regime. Chapter 2 addresses nonlinear photon statistics, from which the NF of saturated EDFAs can be evaluated theoretically.

We shall refer to the forward and backward ASE as the noise powers that propagate in the same direction as the amplified signal, or in the opposite direction, respectively. If not specified otherwise, forward and backward ASE refer to the noise powers obtained at the corresponding EDFA ends. A possible experimental setup for the simultaneous measurement of forward and backward ASE as a function of pump power is shown in Figure 5.24 (E. Desurvire et al. [91]). The two ASE spectra are combined via a 3 dB coupler and coupled into the optical spectrum analyzer. Either spectrum can be monitored by blocking one of the two light beams, as shown by the arrows in the figure. Additionally, narrowband (e.g., 1 nm) interference filters can be inserted in the beam paths for accurate measurements of the ASE spectral power density and determination of the equivalent input noise.

Typical experimental ASE spectra obtained from an aluminosilicate fiber (Type III, $L = 1$ m, $\alpha_s = 55$ dB/m, see [104] for other fiber characteristics) for different pump powers at $\lambda = 514.5$ nm ($P_p^{in} = 15$–$205$ mW) are shown in Figure 5.25 in dB scale.

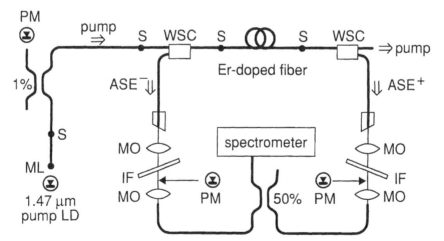

**FIGURE 5.24**  Experimental setup for simultaneous measurements of forward and backward ASE power spectra; ML = fiber microlens, PM = power meter, S = splice, WSC = wavelength selective coupler, MO = microscope objective, IF = interference filter. From [91] © 1991 IEEE.

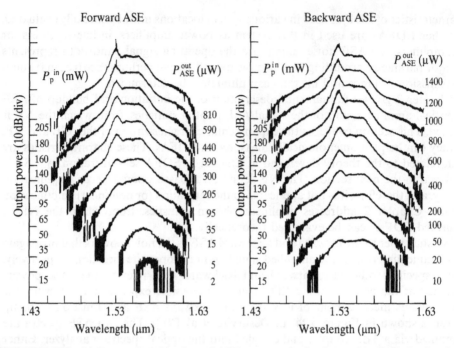

**FIGURE 5.25** (*a*) Forward and backward ASE spectra obtained experimentally in alumino-silicate EDFA for different pump powers $P_p^{in}$. For clarity, the reference level corresponding to each spectra was shifted up in each measurement by 10 dB. The total ASE output powers $P_{ASE}^{out}$ corresponding to each input pump power $P_p^{in}$ are also shown.

For clarity, the reference level of each spectrum is shifted up by 10 dB in each measurement. The total backward ASE power $P_{ASE}^{out}$(backward) is always greater than the forward ASE power, while the ratio $P_{ASE}^{out}$(backward)/$P_{ASE}^{out}$(forward) decreases toward unity as the pump power is increased. Differences between forward and backward ASE spectral shapes are most significant at low pumps. In the regime of low medium inversion, corresponding to low pump powers, ground level absorption dominates at short wavelengths ($\lambda = 1.52$–$1.54$ $\mu$m), resulting in the vanishing of the main peak near $\lambda = 1.53$ $\mu$m. This absorption of the main peak is more important in the forward case than in the backward case. In forward ASE, light is absorbed as it propagates in the direction of the detector; in the backward ASE, light is amplified instead. But in the regime of near-complete inversion corresponding to high pumps, the two ASE spectral shapes become identical, as ground level absorption is bleached.

The changes in ASE spectral shapes observed when the input pump power is varied also depend upon the EDFA length. This feature is illustrated in Figures 5.26 and 5.27, which correspond to EDFA lengths $L = 3.3$ m and $L = 10$ m, respectively (E. Desurvire and J. R. Simpson [93]). In these figures, the forward and backward ASE power spectra are plotted in arbitrary vertical linear scale and are normalized to the same peak value. At low pumps, the effect of ground level absorption and resulting differences between forward and backward ASE spectra are more important in the case of long lengths ($L = 10$ m), as expected. As the pump is increased, forward and backward ASE spectra become nearly identical, for short and long EDFAs; but

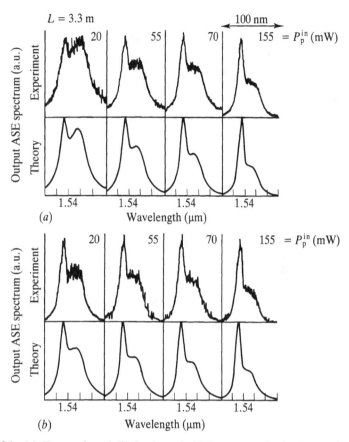

**FIGURE 5.26** (*a*) Forward and (*b*) backward ASE spectra obtained experimentally and theoretically in aluminosilicate EDFA with length $L = 3.3$ m, for different input pump powers $P_p^{in}$ at $\lambda = 514.5$ nm; full horizontal scale per box is 100 nm, vertical scale is arbitrary. From [93] © 1989 IEEE.

at the highest pumps, the power distributions between the main peak near $\lambda = 1.53$ $\mu$m and the second peak near $\lambda = 1.55$ $\mu$m are different for the two lengths.

The theoretical ASE spectra shown in Figures 5.26 and 5.27 are in good qualitative agreement with the experimental spectral. Differences between theoretical and experimental powers found when best matching the ASE spectral shapes, reported in [93], can be attributed to the use of an inaccurate value for the parameter $\eta = \sigma_e/\sigma_a$ (predicted by the Fuchtbauer–Ladenburg relation) and to the fact, that in this early analysis, pump ESA and self-saturation were not accounted for.

A theory/experiment comparison of pump-induced changes in both ASE power spectra, for different fiber lengths, is a stringest test for EDFA modeling. Figure 5.28 shows experimental data of ASE power versus input pump power measured for different EDFA lengths, with theoretical predictions calculated from the general EDFA model, using exact experimental EDF parameters (B. Pedersen et al. [107]). The corresponding gain versus pump characteristics are shown in Figure 5.13. Figures 5.13 and 5.28 show that excellent agreement between theory and experiment for both

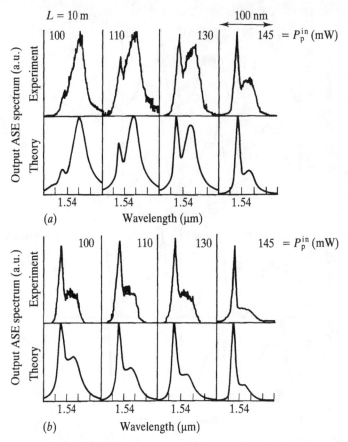

**FIGURE 5.27** (*a*) Forward and (*b*) backward ASE spectra obtained experimentally and theoretically in aluminosilicate EDFA with length $L = 10$ m, for different input pump powers $P_p^{in}$ at $\lambda = 514.5$ nm; full horizontal scale per box is 100 nm, vertical scale is arbitrary. From [93] © 1989 IEEE.

gain and ASE can be obtained when all the EDF parameters are known with good accuracy.

Another EDFA characterization experiment simultaneously measures the forward and backward ASE power generated in a given bandwidth $\Delta\lambda$, centered near two representative wavelengths (e.g., $\lambda = 1.53$ $\mu$m and $\lambda' = 1.55$ $\mu$m), as a function of input pump power. An example of such ASE power measurements obtained in an aluminosilicate EDFA pumped near $\lambda_p = 1480$ nm ($\Delta\lambda = 1$ nm) is shown in Figure 5.29 (E. Desurvire et al. [91]). The solid curves corresponding to each of the four measurements (i.e., forward and backward ASE at $\lambda$ and $\lambda'$) are calculated from the analytical model for unsaturated EDFAs, Eqs. (1.333) and (1.134), using the exact EDF parameters $\alpha_s(\lambda)$, $\alpha_s(\lambda')$, $\eta(\lambda)$, $\eta(\lambda')$, $\eta(\lambda_p)$ and $P_{sat}(\lambda_p)$. Good agreement between theory and experiment is obtained at low pump powers ($P_p^{in} < 5$ mW) for the four measurements. At higher pumps, the observed difference is due to the effect of amplifier self-saturation. The onset of self-saturation observed near $P_p^{in} \approx 5$ mW is consistent

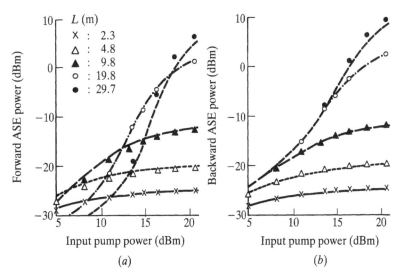

**FIGURE 5.28** Experimental and theoretical ASE powers versus input pump power obtained for different EDFA lengths in (*a*) forward and (*b*) backward directions. From [107] © 1990 IEEE.

with the following measurement: with an input pump of $P_p^{in} = 8\,\text{mW}$, for instance, the total output ASE power measured in each direction is $P_{ASE}^{out} = 70\,\mu\text{W}$, while the saturation power at the peak wavelength is $P_{sat}(\lambda) = 117\,\mu\text{W}$ [91]; since $P_{ASE}^{out}/P_{sat}(\lambda) \approx 0.6$, we expect significant self-saturation at $P_p^{in} = 8\,\text{mW}$, illustrated in Figure 5.29 by the difference between the experiment and the unsaturated gain model.

The simultaneous measurements of EDFA gain $G$ and of the forward ASE power $P_{ASE}^{out}(\text{forward}, \lambda)$ in a narrow bandwidth (e.g., $\Delta\lambda = 1\,\text{nm}$) leads to the experimental determination of the spontaneous emission factor $n_{sp}$, or equivalent input noise factor $n_{eq}$. From the theory developed in Chapter 2, these two parameters are given by:

$$n_{sp}(\lambda) = \frac{1}{G(\lambda) - 1} \frac{P_{ASE}^{out}(\text{forward}, \lambda)}{2h\nu\Delta\nu} = \frac{1}{G(\lambda) - 1} \frac{P_{ASE}^{out}(\text{forward}, \lambda)}{2hc^2\Delta\lambda/\lambda^3} \quad (5.31)$$

$$n_{eq}(\lambda) = \frac{1}{G(\lambda)} \frac{P_{ASE}^{out}(\text{forward}, \lambda)}{2h\nu\Delta\nu} = \frac{1}{G(\lambda)} \frac{P_{ASE}^{out}(\text{forward}, \lambda)}{2hc^2\Delta\lambda/\lambda^3} \quad (5.32)$$

The parameter $n_{eq}$ can be used more generally than the parameter $n_{sp}$, in particular when $G(\lambda) \leqslant 1$, which corresponds to the case of a lossy or transparent EDFA.

Section 2.4, shows that the spontaneous emission factor at signal wavelength $\lambda_k$ decreases with increasing pump power to a minimum limit $n_{sp}^{min}$ given by (E. Desurvire [119]):

$$n_{sp}^{min}(\lambda_p, \lambda_k) = \frac{1}{1 - \dfrac{\eta_p}{\eta_k}} = \frac{1}{1 - \dfrac{\sigma_e(\lambda_p)\sigma_a(\lambda_k)}{\sigma_a(\lambda_p)\sigma_e(\lambda_k)}} \quad (5.33)$$

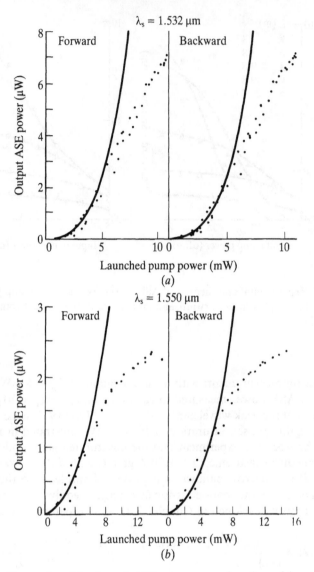

**FIGURE 5.29** Forward and backward ASE power as a function of input pump power, measured experimentally in bandwidth $\Delta\lambda = 1$ nm at (a) $\lambda = 1.532\ \mu$m and (b) $\lambda' = 1.550\ \mu$m wavelengths in a 1480 nm pumped EDFA. Solid lines are calculated from the analytical unsaturated gain model, using exact experimental parameters. From [91] © 1991 IEEE.

In the case of three-level pumping ($\eta_p = 0$ or $\sigma_e(\lambda_p) = 0$), the minimum limit is $n_{sp}^{min}(\lambda_p, \lambda_k) = 1$, while for two-level pumping ($\eta_p \neq 0$), the minimum limit is somewhat greater than unity. Plots of $n_{sp}^{min}(\lambda_p, \lambda_k)$ as a function of pump and signal wavelengths are shown in Figure 2.8a, which corresponds to the case of Type III EDFAs with two-level pumping. The gain and spontaneous emission factor changes with input pump power, experimentally measured in a 1480 nm pumped, Type II EDFA ($L = 15$ m, $\lambda_k = 1.553\ \mu$m, $\alpha_p = 1.0$ dB/m, $\alpha_k = 2.5$ dB/m) are shown in Figure 5.30 (K. Motoshima

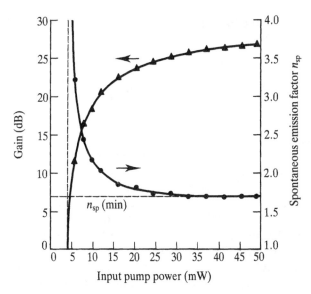

**FIGURE 5.30** Experimental gain and spontaneous emission factor measured as a function of input pump power at $\lambda_p = 1480\,\text{nm}$, showing a minimum spontaneous emission factor limit $n_{sp}^{min} = 1.65$ [120].

[120]). In this example, the minimum spontaneous emission factor is found to be $n_{sp}(\text{min}) = 1.65$, which is 30% higher than the theoretical value $n_{sp}^{min} = 1.25$ predicted by Eq. (5.33). This difference can be attributed to the uncertainties in the experimental parameters $\eta_p$, $\eta_s$, and measurement errors ($<1\,\text{dB}$) in the determination of input/output splice loss and signal gain. This minimum value $n_{sp}(\text{min})$ is reached when the EDFA gain becomes weakly dependent upon pump power and approaches its maximum value ($G = 27\,\text{dB}$), which corresponds to the regime of maximum inversion.

The EDFA optical noise figure spectrum $F_o(\lambda)$ can be determined from measurements of the gain and spontaneous emission factor spectra, i.e., $G(\lambda)$ and $n_{sp}(\lambda)$. Section 2.4 shows the EDFA optical noise figure is given indeed by:

$$F_o(\lambda) = \frac{1 + 2n_{sp}(\lambda)[G(\lambda) - 1]}{G(\lambda)} = \frac{1}{G(\lambda)} + 2n_{eq}(\lambda) \tag{5.34}$$

In Eq. (5.34), the term $1/G$ corresponds to the shot noise and the term proportional to $2n_{sp}$ or $2n_{eq}$ corresponds to the signal–ASE beat noise.

Two assumptions are implicit in eq. (5.34): (a) the input signal has Poisson statistics, and (b) the mean photon number $\langle n_0 \rangle$ of the input signal is large, i.e., $\langle n_0 \rangle \gg n_{sp}$ (Chapter 2). When the input signal statistics differs from Poisson (i.e., $\sigma_0^2 \neq \langle n_0 \rangle$, $\sigma_0^2$ = statistical variance of input signal), and in the general case of arbitrary mean input photon numbers, the optical noise figure is given by (Eq. (2.179) for $k = 1$):

$$F_o = \frac{\langle n_0 \rangle}{\sigma_0^2} \{f_0 - 1 + F_1^{\text{Poisson}}\} \tag{5.35}$$

where $f_0 = \sigma_0^2/\langle n_0 \rangle$ is the Fano factor of the input signal, and $F_1^{\text{Poisson}}$ is defined by (Eq. (2.177) with $i = 1$):

$$F_1^{\text{Poisson}} = \frac{1 + 2n_{\text{eq}}G}{G} + \mathcal{M}\frac{n_{\text{eq}}}{\langle n_0 \rangle}\frac{1 + n_{\text{eq}}G}{G} \tag{5.36}$$

with $\mathcal{M} = 2$ representing the number of polarization modes. In eq. (5.36), the term proportional to $\mathcal{M}n_{\text{eq}}/\langle n_0 \rangle$ corresponds to the ASE–ASE beat noise. Equations (5.34) and (5.35) for the optical noise figure $F_o$ match only if the conditions (a) and (b) are satisfied. If the input signal statistics deviate from Poisson (in which case $f_0 \neq 1$), an excess noise figure $F^{\text{excess}} = (f_0 - 1)/f_0$ is predicted by Eq. (5.35). The total noise figure is given by $F^{\text{excess}} + F_1^{\text{Poisson}}/f_0$, which differs from Eq. (5.34). Equation (5.34) is widely used in the literature for defining the EDFA noise figure but is only an approximation if the input signal statistics deviate from Poisson. Studies of photon statistics in actual InGaAsP laser diodes actually show that these sources do not exhibit purely coherent statistics, as some amount of Gaussian noise is also present (P. L. Liu et al. [121]). The domain of validity for the noise figure approximation (5.34) in system applications using such semiconductor laser sources remains to be fully analyzed.

Equation (5.34) corresponds to the definition of an optical noise figure. Such an optical noise figure could be measured by monitoring the amplified signal with an ideal photodetector having unity quantum efficiency ($\eta = 1$). In the case of actual infrared photodetectors, the quantum efficiency is less than unity and its effect on the amplifier/detector system noise figure must be taken into account. Additionally, it is relevant to define the noise figure with reference to the signal input to the system, as opposed to input to the EDF, which makes it necessary to account for the input coupling loss. In order to obtain a noise figure formula that accounts for both detector quantum efficiency $\eta$ and input coupling loss $\eta_{\text{in}}$, we can use the general result of Eq. (2.179) for a $k$ element amplifier chain. The first element a lossy couply device represented by a noiseless amplifier ($n_{\text{eq1}} = 0$) with input signal $\langle n_0 \rangle$ and gain $G_1 = \eta_{\text{in}} < 1$; the second element is an EDFA with input signal $\langle n_1 \rangle = \eta_{\text{in}}\langle n_0 \rangle$, gain $G_2 = G$, and noise factor $n_{\text{eq2}} = n_{\text{eq}}$; the third element is the photodetector, which is represented by a noise-free amplifier ($n_{\text{eq3}} = 0$) with gain $G_3 = \eta < 1$. The noise figure of a three-element amplifier chain is then given by:

$$F_o = \frac{\langle n_0 \rangle}{\sigma_0^2}\left\{ f_0 - 1 + F_1^{\text{Poisson}} + g_1\frac{F_2^{\text{Poisson}} - 1}{G_1} + g_1 g_2 \frac{F_3^{\text{Poisson}} - 1}{G_1 G_2} \right\} \tag{5.37}$$

with

$$F_i^{\text{Poisson}} = \frac{1 + 2n_{\text{eq}i}G_i}{G_i} + \mathcal{M}\frac{n_{\text{eq}i}}{\langle n_{i-1} \rangle}\frac{1 + n_{\text{eq}i}G_i}{G_i} \tag{5.38}$$

and

$$g_i = \left( 1 + \mathcal{M}\frac{n_{\text{eq}i}}{\langle n_{i-1} \rangle} \right) \tag{5.39}$$

Applying Equations (5.37)–(5.39) to this example, we obtain the following noise figure definition for the EDFA detector system:

$$F_o = \frac{\langle n_0 \rangle}{\sigma_0^2} \left\{ f_0 - 1 + \frac{1}{\eta_{in}} \left[ \frac{1/\eta + 2n_{eq}G}{G} + \mathcal{M} \frac{n_{eq}}{\langle n_0 \rangle} \frac{1/\eta + n_{eq}G}{G} \right] \right\} \qquad (5.40)$$

Equation (5.40) represents the most general definition for noise figure of the EDFA detector system. Note that in this result, the amplified signal $G\langle n_0 \rangle$ is assumed to be large enough so that shot noise and beat noises dominate over the detector's thermal noise. With a Poisson input signal and in the limit of large mean photon number, Eq. (5.40) reduces to the well-known result:

$$F_o \approx \frac{1}{\eta_{in}} \frac{1/\eta + 2n_{eq}G}{G} = \frac{1}{\eta_{in}} \frac{1/\eta + 2n_{sp}(G-1)}{G} \qquad (5.41)$$

which is identical to Eq. (5.34) when $\eta_{in} = \eta = 1$. In the literature, Eq. (5.34) is generally used, instead of Eq. (5.41). The first definition in Eq. (5.34) represents a characteristic EDFA parameter which is of indicative value for device comparisons. On the other hand, the second definition in Eq. (5.41), which applies to the EDFA detector system, can be used for evaluating the receiver sensitivity $\eta_{in}\bar{P}_{in}$ digital direct detection receivers. Section 3.5 shows that the sensitivity is given by, Eq. (3.80):

$$\eta_{in}\bar{P} = 18h\nu_s B_0 \eta_{in}(F_o + \Delta F) \qquad (5.42)$$

where $\Delta F$ is a small correction in the s-ASE beat noise regime. The result of eq. (5.42) is used in Chapter 7 as a theoretical reference for comparing the best sensitivities of direct detection receivers with EDFA preamplifiers.

Equation (5.34) applies only to the linear amplification regime. Indeed, this result stems from a linear analysis of amplification and noise statistics (the STT model of Chapter 2). Nonlinear amplification is analyzed in Section 2.8. The rate equations for the nonlinear amplifier mean and variance can be solved only in two cases: (a) the photon number probability distribution function (PDF) is known a priori, or (b) some decorrelation approximations can be made. In each case, the rate equation system for the mean and variance does not lend itself to analytical solutions and must be solved numerically. As a consequence, no exact formula is available for the prediction of EDFA noise figure in the saturation regime.

We consider now the case of high gain amplifiers (i.e., $G > 20$ dB), which concern most applications of interest (distributed amplifiers, for which $G = 1$ are analyzed in Chapter 6). In the limit $G \gg 1$, the optical noise figure at wavelength $\lambda_s$ defined in Eq. (5.34) reduces to:

$$F_o(\lambda_s) \approx 2n_{sp}(\lambda_s) \approx 2n_{eq}(\lambda_s) \qquad (5.43)$$

From Eqs. (5.33) and (5.43), the minimum noise figure $F_o^{min}$ that can be achieved in the high gain limit is then:

$$F_o^{min}[\lambda_s, G(\lambda_s) \gg 1] = 2n_{sp}^{min} = \frac{2}{1 - \dfrac{\eta_p}{\eta_s}} = \frac{2}{1 - \dfrac{\sigma_e(\lambda_p)\sigma_a(\lambda_s)}{\sigma_a(\lambda_p)\sigma_e(\lambda_s)}} \qquad (5.44)$$

We consider first the case of three-level pumping, for which we have $\sigma_e(\lambda_p) = \eta_p = 0$, which gives an $n_{sp}^{min}$ of unity. The fundamental result of eq. (5.44) shows that in this case, the minimum EDFA noise figure is $F_0^{min} \approx 2$ or 3 dB. Such a limiting value, commonly referred to as the *quantum limit* or signal *spontaneous beat noise limit*, has long been known in the field of semiconductor optical amplifiers (J. C. Simon [122]). In the field of EDFAs, early theoretical analysis has shown that this limit could be very nearly approached (R. Olshansky [123]), which was confirmed experimentally later on. The existence of a 3 dB limit concerns only the high gain regime; in the general case, the noise figure can take any value greater than unity (Chapter 2).

In the case of two-level pumping (pump wavelength near 1480 nm), we expect the minimum achievable noise figure to be somewhat higher than the 3 dB quantum limit, since $\eta_p \neq 0$ in Eq. (5.44). Given the pump and signal wavelengths $\lambda_p$ and $\lambda_s$ in the $^4I_{13/2}$–$^4I_{15/2}$ band, the minimum achievable noise figure is ultimately determined by the cross section ratios $\eta_p$ and $\eta_s$. The McCumber theory of phonon-terminated laser systems shows that the cross section ratio at frequency $v$ is given by the relation (Section 4.5):

$$\eta(v) = \frac{\sigma_e(v)}{\sigma_a(v)} \exp\left\{ -\frac{h(v - \varepsilon)}{k_B T} \right\} \tag{5.45}$$

where $h\varepsilon$ represents a free energy which depends on the host material [124]. Substituting Eq. (5.45) for $\lambda_p$ and $\lambda_s$ into Eq. (5.44) yields (W. J. Miniscalco et al. [125]):

$$F_0^{min}[\lambda_p, \lambda_s, G(\lambda_s) \gg 1] = \frac{2}{1 - \exp\left\{ -\dfrac{h(v_p - v_s)}{k_B T} \right\}} \tag{5.46}$$

The fundamental result of Eq. (5.46) is independent of the host material. Considering the wavelength pairs ($\lambda_p = 1480$ nm, $\lambda_s = 1531$ nm) and ($\lambda_p = 1480$ nm, $\lambda_s = 1550$ nm), we find from Eq. (5.46) the theoretical minimum noise figures at $T = 300$ K of $F_0^{min} = 4.8$ dB and $F_0^{min} = 4.1$ dB, respectively, corresponding to the minimum spontaneous emission factors $n_{sp}^{min} = 1.51$ and $n_{sp}^{min} = 1.30$. These results should not be considered absolute, as the McCumber theory is not 100% accurate for all types of hosts (Chapter 4). Considering, however, that this theory represents a good approximation, we conclude that for EDFAs pumped as two-level systems, the minimum achievable noise figure is 1 dB to 2 dB higher than the 3 dB quantum limit. This important feature of EDFAs is analyzed in more details in Chapter 2. The predicted difference in minimum noise figure between EDFAs pumped as two-level or three-level systems is well confirmed experimentally.

Since the noise figure (NF) is equal to twice the spontaneous emission factor in the high gain regime, Eq. (5.43), we expect a weak spectral dependence over the EDFA gain bandwidth (E. Desurvire [119]). This property is illustrated in Figure 5.31, in the case of an EDFA pumped near 1490 nm (C. R. Giles et al. [126]). The figure shows that over the spectral range where the EDFA gain is greater than 10 dB, the NF is very nearly constant, with a minimum value $F_0 = 3.4$, corresponding to

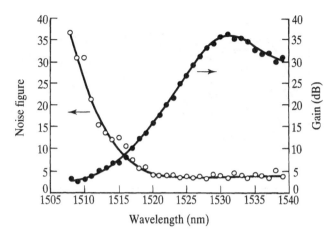

**FIGURE 5.31** Experimental gain and noise figure spectra measured in an EDFA pumped near 1490 nm. From [126] © 1989 IEEE.

$F_o = 5.3$ dB. We also expect the NF to increase as the pump power input to the EDFA is decreased. This property is illustrated in Figure 5.32, showing the gain and NF spectra measured for different input pump powers at $\lambda = 1480$ nm (G. R. Walker [127]). The near-spectral uniformity of NF is achieved as the pump is increased, with a smallest value of 6 dB at signal wavelength $\lambda = 1.57 \, \mu$m. These data include an input coupling loss of $\eta_{in} = 1.5$ dB, so the optical noise figure as referred to the EDF input end is actually $F_o = 4.5$ dB. A description of a computer-controlled measurement technique for such an NF characterization is provided in [127].

An experimental characterization of NF versus input pump power obtained with an EDFA pumped at 980 nm is shown in Figure 5.33 (R. I. Laming and D. N. Payne [128]). Above a certain pump power level for which a high EDFA gain is achieved (here $G \approx 20$ dB for $P_p^{in} = 5$ mW), the NF actually reaches the 3 dB quantum limit predicted by the theory.

The EDFA noise figure can also be determined through a measurement of receiver electrical noise. In this type of measurement, the EDFA output signal is coupled into a photodetector, and the photocurrent noise power spectrum $\sigma^2(f)$ (in $A^2$/Hz) is monitored with an RF spectrum analyzer. The different RF noise components, corresponding to signal and ASE shot noises, signal–ASE and ASE–ASE beat noises, and receiver thermal noise, can be determined individually by measurements where the signal is turned off, the EDFA output blocked, or the EDF bypassed. By subtraction of the obtained noise spectra, it is possible to determine the signal–ASE beat noise component, i.e., $\sigma^2_{s\text{-ASE}}$. The EDFA noise figure is then given by (T. Mukai and Y. Yamamoto [129], C. R. Giles et al. [126]):

$$F_o = \frac{\sigma^2_{s\text{-ASE}}}{2\eta e I_s (G - 1)} \tag{5.47}$$

where $I_s$ is the measured signal photocurrent. Equation (5.47) stems from the analysis of photocurrent noise described in Chapter 3.

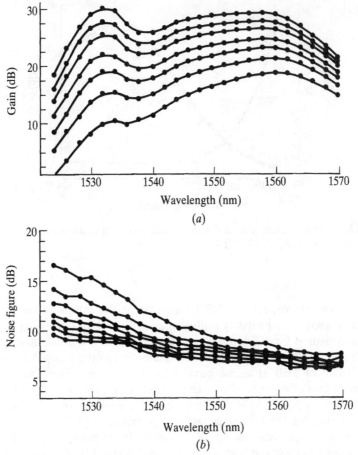

(a)

(b)

**FIGURE 5.32** Experimental gain (a) and noise figure spectra (b) measured in an EDFA pumped near 1480 nm, for different input pump powers $P_p^{in} = 18$ mW to 40 mW. From [127] reprinted with permission from *Electronics Letters*, 1991.

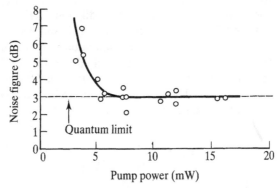

**FIGURE 5.33** Experimental noise figure versus input pump power, as measured in an EDFA pumped at 980 nm, showing a quantum-limited noise regime. From [128] © 1990 IEEE.

The NF can thus be obtained by use of Eq. (5.34), which is based on straightforward optical noise power measurement, or by use of Eq. (5.47), which is based on RF noise power measurement. In the experiment described in [126], the two type of measurements gave $F_o = 5.2$ dB and $F_o = 5.3$ dB, respectively, which shows a good agreement between the two methods.

Through accurate RF noise characterization, the best (or lowest) noise figures, $F_o = 3.2$ dB and $F_o = 4.1$ dB, could be measured in forward pumped EDFAs using 980 nm and 1480 nm pump LDs, respectively (M. Yamada et al. [130]). The minimum spontaneous emission factors corresponding to these NFs are $n_{sp} = 1.05$ for 980 nm pumping and $n_{sp} = 1.29$ for 1480 nm pumping. In this last case, the minimum $n_{sp}$ is apparently very close to the fundamental limit predicted by Eq. (5.33); considering the example of an aluminosilicate EDFA (Type II) with parameters $\eta_p(1480$ nm$) = 0.26$ and $\eta_k(1.545$ nm$) = 1.07$ [131], we find from this equation the minimum value $n_{sp}^{min} = 1.32$, corresponding to an NF of 4.2 dB.

To date, the best NF obtained with EDFAs pumped near 810 nm falls in the range $F_o = 3.9$–$4.2$ dB, which corresponds to a spontaneous emission factor $n_{sp} = 1.26$–$1.58$ dB (Y. Kimura et al. [132]). As the EDFA operates as a three-level system for this pump wavelength, we would expect the NF to reach the 3 dB quantum limit. The main differences between 810 nm pumping and 980 nm pumping are: (1) the effect of pump ESA and (2) a smaller pump absorption coefficient. Both effects contribute to reduce the pumping efficiency, which explains that full medium inversion ($n_{sp} = 1.0$) is hard to achieve with the 810 nm band. The effect of pump ESA is a loss of pump power that does not affect the local population inversion (Section 4.7, Appendix N). We can therefore predict that near-quantum-limited NFs could be achieved at the expense of a large increase of pump power. This can be illustrated by the following consideration: for two EDFAs having the same gain $G = 20$ dB, the best NF, $F_o = 3.2$ dB was achieved with a pump powr of $P_p^{in} = 11$ mW at 980 nm [130], while the best NF, $F_o = 3.9$ dB for 820 nm pumping required a pump power of $P_p^{in} = 100$ mW [132].

A particular pumping configuration uses both 980 nm and 1480 nm pumps, referred to as hybrid pumping (Section 5.2). The advantage of hybrid pumping is to combine both features of quantum-limited noise figure (980 nm pump) and quantum-limited power conversion efficiency (1480 nm pump). A detailed study of the NF and saturation output power characteristics of HEDFAs in various pumping configurations (i.e., backward or bidirectional pumping) has shown that an optimal trade-off is possible (J.-M. Delavaux et al. [94]). The optimum configuration that minimizes the NF ($4.0$ dB $\leqslant F_o \leqslant 5.0$ dB) and maximizes the saturation output power ($P_{sat}^{out} = +12$ dBm) is bidirectional pumping with 980 nm forward pump and 1480 nm backward pump [94]. This result is consistent with the facts that the lowest NF is always achieved with forward pumping, while the highest power conversion efficiency is always achieved with backward pumping (Sections 5.3 and 5.5).

Another pumping configuration of interest for HEDFAs uses both 1480 nm and 1555 nm pumps (J. F. Massicot et al. [96]). The main advantage of such a configuration is to extend the EDFA gain bandwidth in the 1.57–1.61 $\mu$m region (Section 5.2). For signals at 1.575–1.60 $\mu$m wavelengths, the NF corresponding to single forward pumping at 1555 nm is measured to be $F_o = 6.0$–$8.0$ dB, while with hybrid forward pumping, the NF is reduced to $F_o = 3.5$–$4.0$ dB. These results are in close agreement with values predicted by the McCumber theory. Indeed, we find from Eq. (5.46) in this signal wavelength range the values $F_o = 5.4\text{--}7.9\,\text{dB}$ for 1480 nm pumping and $F_o = 3.4\text{--}3.7\,\text{dB}$ for 1555 nm pumping.

In the literature, the NF is usually calculated with reference to the EDF input end, therefore it does not include the effect of input coupling loss due to splices, taps, isolators, and WSC components. In optimized EDFA modules, this loss amounts typically to $\eta_{in} = 0.5\text{--}1.5\,\text{dB}$. Using the NF results reported in [130] and [132], we can then evaluate the best NF of actual EDFA modules to be (approximately) $F_o = 4\text{--}4.5\,\text{dB}$ for 980 nm pumping, $F_o = 4\text{--}5.5\,\text{dB}$ for hybrid pumping and $F_o = 5\text{--}5.5\,\text{dB}$ for 810 nm or 1480 nm pumping.

The lowest EDFA noise figure is always achieved when the pump propagates in the same direction as the signal, which is referred to as forward pumping (R. Olshansky [123]). This important property, previously analyzed in Raman fiber amplifiers or RFAs (E. Desurvire et al., [133] and [134]), is a characteristic feature of all types of end-pumped fiber amplifiers with isotropic gain coefficient. This last property means that the local gain coefficient depends only on the pump intensity and therefore does not depend on pump propagation direction. The pump power launched at either fiber end decreases with propagation distance, due in RFAs to the effect of scattering and in EDFAs to the effect of absorption by $Er^{3+}$ ions. This pump absorption effect means the local gain coefficient is always higher near the fiber end where the pump is launched, which corresponds to a lower spontaneous emission factor. In Section 2.4, a simple demonstration shows that the resulting noise figure is the lowest when both pump and signals propagate in the same direction (referred to as the forward pumping or copropagating scheme). This simple demonstration assumes that the end-pumped fiber amplifier is equivalent to two amplifiers (1 and 2) operating in tandem, which have spontaneous emission factors $n_{sp1}$ and $n_{sp2}$. The pump is launched into amplifier 1 so we have $n_{sp1} < n_{sp2}$. The NF difference between the counter-propagating and copropagating schemes of the tandem amplifier is proportional to $n_{sp2}/n_{sp1} - 1$, Eq. (2.143), which by definition is always positive. As a conclusion, the NF is higher in the counterpropagating case, which is referred to as backward pumping. A real, unidirectionally pumped fiber amplifiers can be viewed as being equivalent to a continuous amplifier chain whose infinitesimal elements have spontaneous emission factors $n_{sp1} < n_{sp2} < \cdots < n_{spk} < \cdots$, which leads to the same conclusion. This demonstration is based on the well-known tandem-amplifier formula (2.141), which for optical amplifiers is accurate only in specific cases (Section 2.5). But in the general case, the conclusion remains identical, as can be shown by numerical integration of the rate equations (R. Olshansky [123]; P. R. Morkel and R. I. Laming [105]; and E. Desurvire, [119] and [135]). The studies of [119] and [135] show that, in the small-signal or linear amplification regime, the NF difference between the two pumping configurations progressively vanishes as the input pump power is increased, which corresponds to a regime of maximum inversion in the fiber [135]. In the case of a large-signal or saturated gain regime, the NF difference between the two pumping configurations is not yet clearly understood, as to date the theory of nonlinear photon statistics (Section 2.8) has not been fully developed. Nonlinear amplification is further discussed in this section.

The phenomenological NF difference between forward and backward pumping is of consequence for the sensitivity of direct detection receivers. In forward pumping, the pump and the signal must be combined together at the amplifier input end through a wavelength selective combiner (WSC), which introduces input coupling loss ($\eta_{in}$).

As a result, while the lowest NF is intrinsically achieved with forward pumping, the input coupling loss due to the WSC increases the effective NF by a penalty of $\delta F = 1/\eta_{in}$. On the other hand, such a penalty is alleviated with backward pumping, but the intrinsic NF of the EDFA is higher. This fact partly explains that record receiver sensitivities are somewhat higher than the quantum limit $\bar{n} = \bar{P}/h\nu_s B_o \approx 40$ photons/bit predicted by Eq. (5.42) for $\eta_{in} = 1$ and $F_o = 3$ dB (Chapters 3 and 7).

Consider the case of saturated amplification. No rigorous definition for the noise figure is available in the case of nonlinear amplification, since the analysis leading to the formula $F_o \approx 2n_{sp}$ stems from a linear theory (Chapter 2). Since the spontaneous emission factor $n_{sp}$ can be measured in either unsaturated or saturated gain regimes, the factor $F_o \approx 2n_{sp}$ still represents a meaningful parameter in both cases. However, the interpretation of this noise figure parameter in the saturated gain regime remains ambiguous. Indeed, the nonlinear photon statistics theory predicts a NF decrease with saturation (Section 2.8), while the parameter $F_o \approx 2n_{sp}$ is experimentally observed to increase. The theoretical NF decrease, which stems from a reduction of photon number fluctuations by the medium nonlinearity, is a well-known feature of saturable absorbers or saturable amplifiers (G. Oliver and C. Bendjaballah, [137]–[139]; N. B. Abraham [140]). This decrease can be intuitively understood in the high signal input limit. Indeed, an optical amplifier with high signal input could be saturated to the point where its net gain is close to unity. As a result of complete gain saturation, the amount of ASE noise generated is negligible, and stimulated emission from the upper level is exactly compensated by absorption from the ground level. It is clear that the signal photon statistics cannot be significantly perturbed by propagation through such a medium. Consequently, the NF of an amplifier saturated to the transparency regime should approach unity, since no SNR degradation is expected to occur. This result can actually be demonstrated rigorously from the nonlinear photon statistics theory, as shown in [140] and discussed in Chapter 2.

In the analysis of noise in saturated EDFAs, a difficulty is introduced by the choice of $n_{sp}$ as a parameter, and not $n_{eq}$. Indeed, $n_{sp}$ tends to infinity as the saturated gain approaches transparency, while $n_{eq}$ remains finite. For this reason, the parameter $F_o \approx 2n_{sp}$ is meaningless in this case. If, on the other hand, the saturated gain remains high (i.e., $G \gg 1$), the two definitions of $n_{sp}$ and $n_{eq}$ are nearly identical ($n_{sp} \approx n_{eq}$), and the parameter $F_o \approx 2n_{sp}$ is meaningful in this case. By consideration of continuity of PDF solutions, we also expect that the linear theory remains valid in the case of moderate gain saturation (e.g., $\Delta G/G \ll 1$). In this case, the parameter $F_o \approx 2n_{sp}$ should represent a good approximation of the true amplifier noise figure. In any case, the functional dependence of the output signal-to-noise ratio (SNR) with the parameter $F_o$ remains to be analyzed for the saturated amplification regime. As this issue is still open to date (1993), we shall use here the tentative denomination of *saturated noise figure* (SNF) for the parameter $F_o \approx 2n_{sp}$, as can be measured in the saturated gain regime, without considering whether this definition necessarily represents the true amplifier NF.

Section 5.3 shows that EDFA saturation can be self-induced by ASE noise, induced by a high input signal, or induced in both ways. We consider first self-saturation and its effect on the SNF.

The near-quantum-limited noise figures close to $F_o = 3.0$ dB were actually achieved with 980 nm pumped EDFAs operating in the small-signal regime with moderate gains (i.e., $G = 20$ dB), for which the effect of saturation is negligible. At higher gains

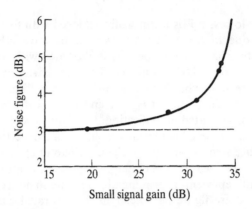

**FIGURE 5.34**   Noise figure $F_o \approx 2n_{sp}$ of forward pumped EDFA measured experimentally for different small-signal gains, corresponding to fiber lengths $L = 7, 12, 17, 25$, and 30 m and input pump power $P_p^{in} = 54$ mW at 980 nm wavelength, showing increase due to self-saturation. From [136] © 1992 IEEE.

(i.e., $G > 25$ dB), a significant increase of the SNF parameter from the 3 dB limit is observed, as shown in Figure 5.34 (R. G. Smart et al. [136]). In this measurement, the pump power is fixed and the EDFA length is increased, as opposed to the case of Figure 5.33. As the data of Figure 5.34 are obtained in the small-signal regime, the SNF increase with gain can be attributed to the effect of EDFA self-saturation.

The effect of self-saturation can be alleviated by using a midway isolator placed at some point in the EDFA (J. H. Povlsen et al. [85]). The function of this isolator is to block the backward ASE, which otherwise saturates the gain at the fiber input end. The effect of midway isolators in very high gain EDFAs on the SNF has been analyzed theoretically in several studies (O. Lumholt et al. [86]; M. N. Zervas et al. [141]; and S. Yamashita and T. Okoshi [87]). These studies show that there exists an optimum isolator location (at 25% to 40% of the total EDFA length) for which the SNF is significantly reduced. This fact was confirmed experimentally. With a 1480 nm pumped EDFA with gain near 40 dB, a 2 dB reduction of SNF could be obtained by this method, to compare with a value of SNF = 7.4 dB achieved without isolator [86]. In the case of 980 nm pumping, an experimental value SNF = 3.1 dB was achieved in an EDFA with 54 dB gain (R. I. Laming et al. [142]). Since the reduced SNF values are close to the minimum noise figures $F_o^{min}$ corresponding to the pump bands, we can conclude that the midway isolator effectively suppresses self-saturation.

We consider next the effect of gain saturation caused by high input signals. Forward ASE noise power (hence, the determination of $n_{sp}$) in the spectral region near the saturating signal wavelength can be measured by two methods. The first method fits a curve to the ASE spectrum, which makes it possible to interpolate the ASE power level near $\lambda_{sat}$ (otherwise masked by the output signal spectrum). A second method, which was initially used for the characterization of low temperature spectral hole burning (E. Desurvire et al. [78]), and extended to room temperature for SNF characterization (J. Aspell et al. [80]), is based on a polarized ASE measurement. In this method, the output signal at $\lambda_{sat}$ is rejected by a polarizing beam splitter, which makes it possible to measure the ASE spectrum in the SOP orthogonal to the signal.

As there is no measurable difference between ASE spectra in the two SOP, we can thus determine the ASE power at $\lambda_{sat}$. The spontaneous emission factor $n_{sp}$ can then be measured as a function of input signal power $P_s^{in}$. A particular feature observed in the $n_{sp} = f(P_s^{in})$ characteristics is a small decrease in a certain signal power range, followed by a rapid increase at higher powers, as shown in Figure 5.35 (J. F. Marcerou et al. [143]). In this example, the pump and signal wavelengths are 1480 nm and 1550 nm, respectively, and the small-signal gain is $G = 39$ dB. The solid lines in the figure are calculated with the general EDFA model. An SNF decrease $\Delta F = 1$ dB with respect to the small-signal value SNF $\approx 6$ dB is observed for input signals near $P_s^{in} = -10$ dBm. For higher input powers, the SNF increases to reach a value near 7 dB at $P_s^{in} = 0$ dBm. The detailed theoretical analysis of [143] shows that the dip in the $n_{sp} = f(P_s^{in})$ characteristics can be attributed to the interplay between self-saturation by backward ASE and signal-induced saturation. The effect of signal-induced saturation is to change the distribution of ASE power along the fiber, which in turn changes the spontaneous emission factor $n_{sp}$. When the saturating signal is sufficiently high to significantly decrease the backward ASE near the fiber input end, a higher medium inversion is achieved in this region. As medium inversion is higher at the EDFA input, a lower SNF is obtained, consistently with the tandem-amplifier rule previously discussed.

The decrease in SNF (or dip effect) induced by signal saturation is observed only at gains that are significantly high for self-saturation to prevail in the small-signal regime (i.e., $G > 30$ dB). The maximum SNF decrease is obtained when the effect of self-saturation is completely suppressed. Therefore, the lowest SNF that can be achieved by this effect must be greater than (or nearly equal to) the minimum value $F_o^{min}$. A detailed experimental measurement of SNF reduction in 980 nm pumped EDFAs confirms that the lowest SNF is always greater than the 3 dB quantum limit (R. G. Smart et al. [136]). The dip effect can therefore be explained by the suppression of self-saturation, followed by purely signal-induced saturation.

The increase of SNF by signal-induced saturation is observed to depend on the initial value of small-signal gain. For EDFAs with small-signal gains near $G = 20$ dB, a gain compression $\Delta G = 5$ dB typically results in a 3 dB increase of SNF, [136] and

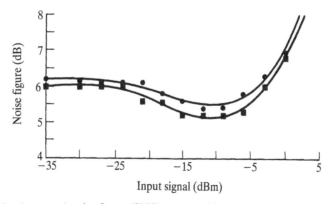

**FIGURE 5.35** Saturated noise figure (SNF) measured in EDFA as a function of signal input power obtained for (●) $P_p^{in} = 21.5$ mW and (■) $P_p^{in} = 35$ mW. From [143], reprinted with permission from the Optical Society of America, 1991.

[144]. In the case where the small-signal gain is near $G = 30\,\text{dB}$ or greater, the SNF remains constant over a broader range of gain compression. In 980 nm pumped EDFAs, an SNF change of $\Delta F \approx \pm 1\,\text{dB}$ was measured for gain compressions up to $\Delta G = 20\,\text{dB}$ [136]; in another experiment, based on RF spectrum measurements, the SNF change was $\Delta F < 1\,\text{dB}$ for gain compressions up to $\Delta G = 23\,\text{dB}$, as shown in figure 5.36 (W. Way et al. [145]).

In [145], a simplified model for the determination of SNF in the signal-induced saturation regime is described. The details of this analysis are outlined in Appendix P, with a generalization to the two-level pumping scheme. The derivation in Appendix P shows that the SNF of saturated EDFAs is given by:

$$\text{SNF} = 2n_{\text{sp}}^{\min}(1 + X) \tag{5.48}$$

where $X$ is an *excess noise factor* due to saturation:

$$X = \frac{G \log G}{G - 1} \frac{1 + \eta_p}{1 + \eta_s} \frac{P_s^{\text{in}}}{P_p^{\text{in}}} \frac{P_{\text{sat}}(\lambda_p)}{P_{\text{sat}}(\lambda_s)} \tag{5.49}$$

The derivation of Eq. (5.49) for the excess noise factor is based upon three important assumptions: (1) the amplifier is free from ASE self-saturation, (2) the pumping regime is such that $P_p^{\text{in}} \gg P_{\text{sat}}(\lambda_p)$, and (3) the pump power distribution along the fiber is nearly constant. Condition (1) is met when the signal-induced saturation takes over self-saturation, and condition (2) is a requirement in power EDFA applications. Condition (3) represents a fair approximation when the fraction of pump power absorbed in the EDFA (i.e., $(P_p^{\text{in}} - P_p^{\text{out}})/P_p^{\text{in}}$) is small; but in most power EDFA applications, this fraction can be large (e.g., 60%), and this approximation could potentially lead to inaccurate predictions.

In the small-signal regime (i.e., $GP_s^{\min}/P_{\text{sat}}(\lambda_s) \ll P_p^{\text{in}}/P_{\text{sat}}(\lambda_p)$), Eq. (5.49) gives $X \ll 1$, and the noise figure in Eq. (5.48) reaches the minimum value $\text{SNF} \approx 2n_{\text{sp}}^{\min}$, which

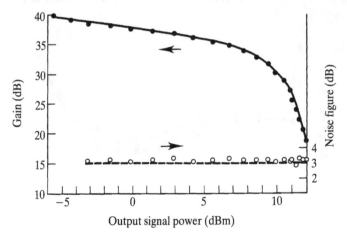

**FIGURE 5.36** Experimental gain and saturated noise figure (SNF) measured as a function of output signal power at $\lambda = 1.531\,\mu\text{m}$ in a 980 nm pumped EDFA. From [145], reprinted with permission from the Optical Society of America, 1990.

corresponds to conditions of maximum inversion. Substituting Eq. (1.45) for the saturation power $P_{sat}(\lambda)$ at the pump and signal wavelengths, we obtain for the excess noise factor:

$$X = \frac{G \log G}{G - 1} \frac{P_s^{in}}{P_p^{in}} \frac{\lambda_s}{\lambda_p} \left(\frac{\omega_p}{\omega_s}\right)^2 \frac{\sigma_a(\lambda_s)}{\sigma_a(\lambda_p)} \qquad (5.50)$$

To check the validity of the excess noise model, we consider the experimental data reported in [145] for a 980 nm pumped EDFA with $\lambda_s = 1.531$ nm. In the highly saturated regime, the following data were measured for pump power, signal power and gain: $P_p^{in} = 60$ mW, $P_p^{out} = 20$ mW, $P_s^{out} = +12$ dBm, and $G = 20$ dB, which correspond to the measurement in Figure 5.36. Assuming typical parameters values $\sigma_a(\lambda_s)/\sigma_a(\lambda_p) \approx 2$ and $\omega_p/\omega_s \approx 0.8$ for the two wavelengths considered, we obtain from Eq. (5.50) the excess noise factor $X \approx 2[\log G/(G - 1)]P_s^{out}/P_p^{in}$ or $X = 0.025$. If in Eq. (5.50) we use $P_p^{out}$ instead of $P_p^{in}$, the excess noise becomes $X = 0.075$. From these two results and from Eq. (5.48) with $n_{sp}^{min} = 1$, we find that the saturated noise figure is given by 3.1 dB $\leqslant$ SNF $\leqslant$ 3.3 dB, corresponding to a maximum SNF change from the quantum limit $\Delta F = 0.3$ dB. This theoretical prediction is found to be in good agreement with the SNF measurement shown in Figure 5.36 [145].

In the case of in the case of EDFAs operating in the regime of quantum-limited power conversion, i.e., $P_s^{out} = (\lambda_p/\lambda_s)P_p^{in}$, SNF predictions from Eqs. (5.49) and (5.50) could be largely underestimated. This is because the approximation made in the model of a constant, high pump power distribution along the fiber, is not valid in this case, as illustrated in Section 5.5). The accuracy of this model could be analyzed by comparing theoretical SNF predictions obtained both from numerical integration and Eqs. (5.50) and (5.51) over a large pump and signal power range. The effect of EDFA saturation on bit-error rate and its observed relation to the increase of SNF [144] are discussed in Section 7.1.

## 5.5  AMPLIFIER SELF-SATURATION

In the previous sections, the importance of the effect of amplifier self-saturation is stressed when considering performance limits in gain, power conversion, and noise figure. As a simple rule, the onset of self-saturation occurs at a gain level for which the total output ASE power $P_{ASE}$ at either fiber end becomes comparable to the saturation power at the peak wavelength $P_{sat}(\lambda_{peak})$. In this regime, the forward and backward ASE power distributions along the fiber are nonlinearly coupled, since both contribute to saturate the gain. The increase of ground level population caused by ASE self-saturation also effects the pump power absorption along the fiber. A self-saturated EDFA with unidirectional pumping can be viewed as a nonlinear system coupling the three waves of pump, forward ASE and backward ASE. Two boundary values constrain the analysis of this system, since the output ASE powers and spectral distributions (i.e., $p_{ASE}^+(z = L, \lambda)$, $p_{ASE}^-(z = 0, \lambda)$) are not known a priori (Section 1.6). In the case of a self-saturated EDFA with bidirectional pumping, the EDFA is a four-wave coupled system with output pump powers and ASE spectra as unknown boundary values (i.e., $q^+(z = L)$, $q^-(z = 0)$, $p_{ASE}^+(z = L, \lambda)$, $p_{ASE}^-(z = 0, \lambda)$). Additionally, the EDFA can be saturated by signals, which increases the complexity of the nonlinear coupling. In this section, we show that the difference in power conversion efficiency observed between forward and backward pumping in saturated EDFAs is actually caused by the concurrent effect of ASE self-saturation (E. Desurvire [114]).

While the case of saturated EDFAs with negligible ASE and multiple pump/signal inputs lends itself to straightforward analytical modeling (A. A. M. Saleh et al. [109]), as described in Section 1.10, the case of self-saturated EDFAs requires a rather complex numerical analysis. We discuss various theoretical models proposed to simplify this analysis and show here that, within some approximations, the model of [109] can be used to include the effect of self-saturation.

We consider first the case of a self-saturated EDFA operating in the small-signal regime. For the purpose of analyzing the basic effect of self-saturation, we need not compute spectrally resolved solutions for the ASE powers. For simplicity and without los of generality, the approximation can be made that the ASE powers are confined within a single spectral channel of width $\Delta \nu_{ASE}$ centered at wavelength $\lambda_{ASE}$, that the Er-doping is confined in the fiber core. We also assume the pump and the signal to be unidirectional. Let $q$, $p$, $p_{ASE}^{\pm}$ be the normalized pump, signal, and ASE powers at $\lambda_p$, $\lambda_s$, $\lambda_A$, respectively, with corresponding absorption coefficients $\alpha_p$, $\alpha_s$, $\alpha_A$. From Eqs. (1.113), (1.114), and (1.117)–(1.119), the rate equations for such a four-wave coupled system take the form [114]:

$$-\frac{dq}{dz} = \pm \alpha_p \frac{U'p + 1 + V'(p_{ASE}^+ + p_{ADE}^-)}{1 + q + p + p_{ASE}^+ + p_{ASE}^-} q \tag{5.51}$$

$$\frac{dp}{dz} = \alpha_s \frac{Uq - 1 + W(p_{ASE}^+ + p_{ASE}^-)}{1 + q + p + p_{ASE}^+ + p_{ASE}^-} p \tag{5.52}$$

$$\frac{dp_{ASE}^{\pm}}{dz} = \pm \alpha_A \frac{(Vq - 1 + W'p)p_{ASE}^{\pm} + \left(\dfrac{q}{1+\eta_p} + \dfrac{p}{1+\eta_s} + \dfrac{p_{ASE}^+ + p_{ASE}^-}{1+\eta_A}\right)\eta_A p_N}{1 + q + p + p_{ASE}^+ + p_{ASE}^-} \tag{5.53}$$

with

$$U = \frac{\eta_s - \eta_p}{1 + \eta_p}, U' = \frac{\eta_s - \eta_p}{1 + \eta_s} \tag{5.54}$$

$$V = \frac{\eta_A - \eta_p}{1 + \eta_p}, V' = \frac{\eta_A - \eta_p}{1 + \eta_A} \tag{5.55}$$

$$W = \frac{\eta_s - \eta_A}{1 + \eta_A}, W' = \frac{\eta_A - \eta_s}{1 + \eta_s} \tag{5.56}$$

and $\eta_x = \sigma_e(\lambda_x)/\sigma_a(\lambda_x)$ $(x = p, s, A)$ and $p_N = 2h\nu_A \Delta \nu_A/P_{sat}(\lambda_A)$. The signal is assumed to propagate in the direction of positive $z$ (referred to as the forward direction). In Eq. (5.51), the cases $(+)$ and $(-)$ correspond to forward and backward pumping, respectively. Equations (5.51)–(5.53) can be solved numerically through the relaxation method (Section 1.6). This method iteratively integrates from $z = 0$ to $z = L$, then from $z = L$ to $z = 0$, and so on, each step using boundary conditions obtained in the previous one. The procedure is stopped when convergence towards stable boundary conditions for the ASE powers (i.e., $P_{ASE}^-(0)$, $p_{ASE}^+(L)$) is finally achieved.

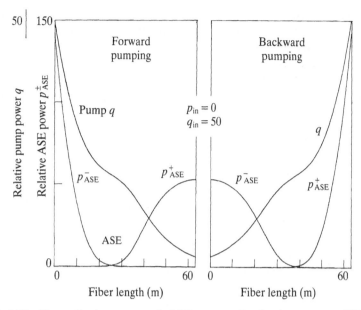

**FIGURE 5.37** Normalized pump and ASE power distributions versus EDFA length, calculated in the forward and backward pumping configurations assuming initial conditions $q_{\text{in}} = 50$, $p_{\text{in}} = 0$. From [114] © 1992 IEEE.

The numerical solutions of coupled Eqs. (5.51) and (5.53) are plotted as functions of the fiber length (coordinate $z$) in Figure 5.37 for forward and backward pumping, assuming no signal input ($p_{\text{in}} = 0$). The parameters chosen in this numerical example are: $\lambda_{\text{p}} = 1480\,\text{nm}$, $\lambda_{\text{s}} = \lambda_{\text{A}} = 1531\,\text{nm}$, $q_{\text{in}} = 50$, $L = 62.5\,\text{m}$, $\alpha_{\text{p}} = 1\,\text{dB/m}$, $\alpha_{\text{s}} = \alpha_{\text{A}} = 3.5\,\text{dB/m}$, $\eta_{\text{p}} = 0.26$, $\eta_{\text{s}} = \eta_{\text{A}} = 0.92$ and $\Delta\nu_{\text{A}} = 1250\,\text{GHz}$ ($\Delta\lambda_{\text{A}} = 10\,\text{nm}$) [114].

The two sets of solutions corresponding to forward or backward pumping are identical by transformation $z \to L - z$, as expected. Obtaining this result represents a valid convergence test for the computing algorithm previously described.

Four observations: (1) the power distribution with length for ASE powers propagating in the same direction as the pump or opposite to the pump are not symmetrical; (2) at the fiber ends, the ASE propagating in the direction opposite to the pump is greater (here by a factor of three) than the copropagating ASE; (3) the forward ($p_{\text{ASE}}^+$) or backward ($p_{\text{ASE}}^-$) ASE powers are negligible over a certain fraction of fiber length $x$, or $L - x$ and consequently, the total ASE power distribution ($p_{\text{ASE}}^+ + p_{\text{ASE}}^-$) reaches a minimum at fiber coordinate $z = x$; (4) near this minimum, the rate of pump power absorption locally decreases, as evidenced by a bump in the pump power distribution.

Concerning the first observation, the qualitative difference in shape or curvatures between the distributions $p_{\text{ASE}}^+(z)$ and $p_{\text{ASE}}^-(z)$ is consistent with that of the gain distributions $G(z)$ in copropagating or counterpropagating pumping [114]. It is seen that in the copropagating case, the ASE distribution reaches a maximum at the fiber output end (i.e., $(dp_{\text{ASE}}^+/dz)_{z=L} = 0$ in forward pumping, for instance), while there is no maximum in the counterpropagating case ($(dp_{\text{ASE}}^-/dz)_{z=L} < 0$ in forward pumping), as also observed with the corresponding gain distributions $G(z)$ [114]. The second observation is a consequence of the property $n_{eq}^- > n_{eq}^+$, i.e., the equivalent input noise is higher in the counterpropagating pump configuration (Section 2.4). This property is also a consequence of the tandem-amplifier rule, which applies to any amplification regime, as seen in the previous section. The third observation shows that the effect of self-saturation is unevenly distributed along the fiber and occurs only in regions near the two fiber ends. The fourth observation shows that self-saturation affects the pump power distribution along the fiber, in particular it results in an increase of the absorbed pump power.

We assume now that the self-saturated EDFA of the previous example is input with a small signal at $z = 0$ ($p_{in} = 0.001$), such that no additional saturation occurs. Figure 5.38 shows the pump, ASE, and signal power distributions for the two pumping configurations. The output signal power $p(L)$ is higher for the backward pumping configuration, corresponding to a 3 dB difference in small-signal gain $G = p(L)/p_{in}$. It can be checked numerically that this difference vanishes when the input pump power is decreased to the point where the pump decay along the fiber is no longer affected by ASE. This operating point can be found by comparing the numerical solution of Eqs. (5.51) calculated while assuming $p_{ASE}^\pm = 0$ (ASE-free amplifier) with that obtained with $p_{ASE}^\pm \neq 0$ (realistic amplifier).

We assume now that the self-saturated EDFA previously considered is input with a higher signal power ($p_{in} = 0.1$), such that signal-induced saturation also occurs. The numerical solutions $q(z)$ and $p(z)$ obtained in the forward and backward pumping configurations are shown in Figure 5.39 (dotted lines). The solid curves correspond to solutions obtained while assuming an ASE-free amplifier, or $p_{ASE}^\pm = 0$. From the pump power distributions, the excess pump absorption caused by ASE self-saturation (power ratio between full and dashed curves) is important in the case of forward

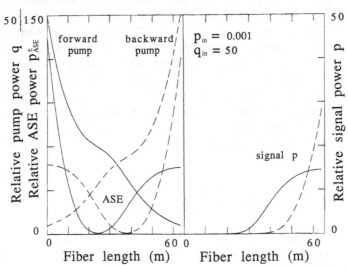

**FIGURE 5.38**  Normalized pump, ASE, and signal power distributions versus EDFA length, calculated in the forward pumping (full lines) and backward pumping (dashed lines) configurations with initial conditions $q_{in} = 50$, $p_{in} = 0.001$; in the small-signal, self-saturated amplification regime, a gain discrepancy of 3 dB is observed between the two pumping configurations. From [114] © 1992 IEEE.

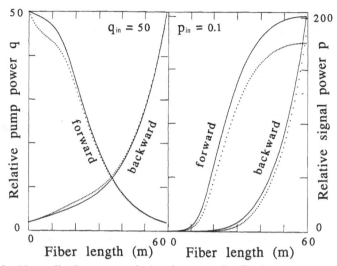

**FIGURE 5.39** Normalized pump and signal power distributions versus EDFA length, calculated in the forward and backward pumping configurations with initial conditions $q_{in} = 50$, $p_{in} = 0.1$, corresponding to a regime where both self-saturation and signal-induced saturation occur (dotted lines); the full lines correspond to numerical solutions obtained through the ASE-free model. From [114] © 1992 IEEE.

pumping, while it is small in the case of backward pumping. As a result, the saturated gain is lower in the case of forward pumping, in this example by about 1 dB. The ASE-free model predicts identical saturated gains for both pumping configurations [114]. Therefore, the gain discrepancy observed between the two configurations cannot be attributed by differences in the way copropagating or counterpropagating pump and signals beams locally saturate the fiber medium, as has been generally thought. The theoretical simulation discussed here shows that the gain difference between forward and backward pumping can be attributed to the effect of amplifier self-saturation, both in the small-signal regime and in the signal-induced saturated regime, as illustrated in Figures 5.38 and 5.39. In the high signal regime where self-saturation is negligible compared to signal-induced saturation, the gain difference between forward and backward pumping vanishes, as predicted by the ASE-free model. It can be concluded from these observations that *the power conversion efficiency (PCE) is saturated EDFAs is the highest with backward pumping, with an upper limit that is given by the ASE-free model description.* This upper limit is therefore given by the explicit solution (5.26), which can be written in the form:

$$ \mathrm{PCE}(x) = \frac{p_{\mathrm{out}} - p_{\mathrm{in}}}{q_{\mathrm{in}}} \frac{P_{\mathrm{sat}}(\lambda_{\mathrm{s}})}{P_{\mathrm{sat}}(\lambda_{\mathrm{p}})} = \frac{P_{\mathrm{sat}}(\lambda_{\mathrm{s}})}{P_{\mathrm{sat}}(\lambda_{\mathrm{p}})} \frac{10^{\mathscr{P}(x)}}{U'} - \frac{P_{\mathrm{s}}^{\mathrm{in}}}{P_{\mathrm{p}}^{\mathrm{in}}} \tag{5.57} $$

with

$$ x \equiv x(q_{\mathrm{in}}, q_{\mathrm{out}}, p_{\mathrm{in}}) = \frac{1}{\log(10)} \left\{ U' p_{\mathrm{in}} + \log(U' p_{\mathrm{in}}) + \frac{\alpha_{\mathrm{s}}}{\alpha_{\mathrm{p}}} \left[ U(q_{\mathrm{in}} - q_{\mathrm{out}}) - \log\left(\frac{q_{\mathrm{in}}}{q_{\mathrm{out}}}\right) \right] \right\} \tag{5.58} $$

where $U'$, $U'$ are defined in Eq. (5.54). the highest PCE is obtained when the saturated gain is maximized, i.e., when the EDFA length $L$ is optimum.*

While the ASE-free model yields the upper limits for the gain and PCE, which are nearly approached in the high-signal saturation regime, it is inadequate to predict these same characteristics in the small-signal self-saturated regime, or any intermediate regimes combining signal-induced and self-induced saturation. In this case, EDFAs can be analyzed through several methods, which are considered now.

The first approach integrates Eqs. (5.51)–(5.53) through the relaxation method, as was done in the previous example. This set of equations corresponds to a simplified description, as ASE is assumed monochromatic with effective spectral width $\Delta\nu_A$. In the general case, the forward and backward ASE spectra must be determined by solving $2k$ differential equations corresponding to ASE powers at wavelength $\lambda_k$ in a spectral bin of width $\delta\nu_A$ (Chapter 1). Such an integration is lengthy, and convergence towards stable solutions can be very slow in some cases, requiring a large number of iterations. More computing time is required if, in addition, the EDFA length must be optimized for maximum gain. Finally, the entire computation must be carried out again each time the initial conditions $q_{in}$, $p_{in}$ are changed.

This difficulty can be alleviated by using the *average power model* (T. G. Hodgkinson, [146] and [147]), detailed in Appendix Q. In this model the amplifier length is divided into a series of elemental sections of length $\Delta L$. In each of these sections, the upper level population $N_2$ is replaced by an effective value that can be evaluated from average signal and ASE powers. Expressions for these average powers, which are similar to that obtained in semiconductor amplifier analysis [148] are defined in Appendix. We can then show that, according to this model, the gain $G_i$ at wavelength $\lambda_i$ is given by:

$$G_i = \exp\left\{\alpha_i\left[(1+\eta_i)\frac{X}{X'}-1\right]\Delta L\right\} \tag{5.59}$$

where $X$ and $X'$ are dummy parameters defined in the Appendix Q. In Eq. (5.59), the gain $G_i$ can be determined by successive iterations, which lead to self-consistent values for $X$ and $X'$. This evaluation procedure must then be carried out sequentially for each section $\Delta L$ over the whole fiber length, both in the forward and the backward directions. The procedure is completed when self-consistent results are obtained between the two directions [147]. The calculation of EDFA gain through this method, using 20 wavelength channels and 20 elemental sections, can be performed with a microcomputer in 20 s, typically; such a computation time is estimated to be 10 to 100 shorter than required by the numerical integration method with a computer

---

* The optimum EDFA length $L = L_{opt}$ is defined in Eq. (5.29), i.e., with the notation used in this section:

$$L_{opt} = \frac{1}{\alpha_p}\frac{1+\eta_s}{1+\eta_p}\left(q_{in}-\frac{1+\eta_p}{\eta_s-\eta_p}\right)-\frac{1}{\alpha_s}\left\{\log\left(\frac{p_{out}}{p_{in}}\right)+p_{out}-p_{in}\right\}$$

In this case, the output pump power is $q_{out}=1/U$, and the maximum PCE is given by Eq. (5.57) with the argument of $\mathscr{P}(x)$ defined by:

$$x(q_{in},p_{in}) = \frac{1}{\log(10)}\left\{U'p_{in}+\log(U'p_{in})+\frac{\alpha_s}{\alpha_p}[Uq_{in}-1-\log(Uq_{in})]\right\}$$

workstation, independently of the number of wavelength channels [147]. As shown in this reference, the method converges rapidly towards stable solutions when the number of wavelength channels and elemental sections are both greater than 20. This model predicts identical signal gains for the forward and backward directions, which is not accurate in the case of realistic self-saturated EDFAs. The reason for such a limitation is that the model is based upon the assumption that in each elemental section, the gains, ASE powers and spontaneous emission factors are the same for both propagation directions. Differences between forward and backward ASE power spectra, shown in [147], are always observed, even when forward and backward signal gains are equal. Such differences in ASE spectra are explained by ground level absorption along the fiber (Section 5.4).

Another approach for a simplified analysis of self-saturated EDFAs is represented by what we shall refer to as the *average inversion model* (T. Georges and E. Delevaque [149]). This model is based upon the property of photon flux conservation:

$$\sum_k \phi_k^{\text{in}} = \sum_k \phi_k^{\text{out}} + \phi_{\text{ASE}}^{\text{tot}} + \phi_{\text{spont}}^{\text{tot}} \tag{5.60}$$

where $\phi_k^{\text{in, out}}$, $\phi_{\text{ASE}}^{\text{tot}}$, and $\phi_{\text{spont}}^{\text{tot}}$ are the input/output signals, the total output ASE (forward + backward) and the total unguided spontaneous emission fluxes, respectively. In this model, the total ASE flux is approximated by:

$$\phi_{\text{ASE}}^{\text{tot}} = 4 \sum_k n_{\text{sp}k}(x)\{G_k(x) - 1\}\delta v \tag{5.61}$$

where $G_k(x)$ is the EDFA gain and $n_{\text{sp}k}(x)$ is an effective spontaneous emission factor; both are functions of the parameter $x$, the fraction of excited ions averaged over the EDFA length [149]. In Eq. (5.61), the factor 4 accounts for the two propagation directions and polarization modes. Using the relations $\phi_k^{\text{out}} = \phi_k^{\text{in}} G_k(x)$ and $\phi_{\text{spont}}^{\text{tot}} = ax$ ($a = $ const.) into Eq. (5.60), a transcendental integral equation is obtained, which can be solved numerically to yield the mean fraction $x$, hence the EDFA gain $G_k(x)$. In [149], it is shown that, without use of any adjustable parameters, the model accurately predicts the EDFA gain spectrum for a broad range of input pump and signal powers. As in the previous model, the gain is independent of the relative propagation directions of the pump and the signal, which is not the case for self-saturated EDFAs.

A third approach that also alleviates the need of numerical integration, is based upon *the equivalent input noise model*. In this model, we assume an a priori value for the equivalent input noise factor $n_{\text{eq}}$, which is spectrally uniform and can be different for the forward and backward directions. Using this assumption, the forward and backward ASE can be treated as equivalent signals in $k$ optical channels, with input powers $n_{\text{eq}}^+ P_{0k}$ and $n_{\text{eq}}^- P_{0k}$ at $z = 0$ and $z = L$, respectively ($P_{0k} = 2hv_k\delta v$). The signal gain $G_k$ can then be calculated using the analytical model for saturated EDFAs with $k$ optical channels, described in Section 1.10 [109].

A basic computer program for the calculation of self-saturated EDFA gain through the equivalent input noise model is given in Appendix R. This program can also be used to solve transcendental equation (1.152) with $n_{\text{eq}}^+ = n_{\text{eq}}^- = 0$. This description corresponds to the model of [109], which applies to the case where ASE self-saturation is negligible (i.e., low gain, saturated EDFAs).

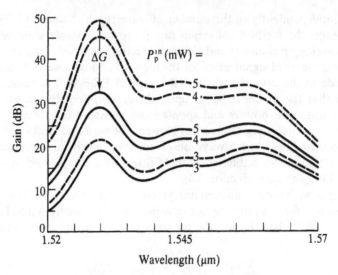

**FIGURE 5.40**   Theoretical gain spectra obtained in the self-saturated regime with the equivalent input noise model, $n_{eq} = 1.02$ (full lines) and with the ASE-free model, $n_{eq} = 0$ (dashed lines), in the case of a 980 nm pumped EDFA with length $L = 50$ m and input pump powers $P_p^{in} = 3, 4$, and 5 mW (see text for parameters). The self-saturated gain regime, obtained by assuming a spontaneous emission factor $n_{eq} = 1.02$, corresponds to a noise figure $F_o = 3.1$ dB.

Figure 5.40 shows small-signal, self-saturated EDFA gain spectra (Type III fiber) calculated for different input pump powers through the equivalent input noise method, assuming $n_{eq} = n_{sp} = 1.02$. For comparison, the gain spectra calculated in the same conditions but with $n_{eq} = n_{sp} = 0$ (ASE noise neglected) are also shown in the figure. In this example, the following parameter values were used: $\lambda_p = 980$ nm, $\lambda_{peak} = 1.531$ $\mu$m, $\eta_p = 0$, $\eta_s(\lambda_{peak}) = 0.9$, $P_{sat}(\lambda_p) = 365$ $\mu$W, $P_{sat}(\lambda_{peak}) = 80$ $\mu$W, and $L = 50$ m. The absorption coefficients $\alpha_{p,s}$ and saturation powers $P_{sat}(\lambda_{p,peak})$ are consistent with the cross section values $\sigma_a(\lambda_p) = 2.1 \times 10^{-25}$ m$^2$, $\sigma_a(\lambda_{peak}) = 5.2 \times 10^{-25}$ m$^2$ [150] and mode sizes of $\omega_p = 1.1$ $\mu$m, $\omega_s = 1.4$ $\mu$m [151], in the approximation of confined Er-doping (see Chapter 1 for definitions). Using a standard microcomputer and the BASIC program in Appendix R, the computation of each gain spectrum with 50 wavelength channels takes less than 1 mn.

The effect of self-saturation is evidenced by a gain reduction $\Delta G$ with respect to the case where ASE is neglected ($n_{sp} = 0$). At the peak wavelength $\lambda_{peak} = 1.531$ $\mu$m for instance, the gain reduction is found to be $\Delta G = 17$ dB for $P_p^{in} = 5$ mW, $\Delta G = 16.5$ dB for $P_p^{in} = 4$ mW, and $\Delta G = 2.7$ dB for $P_p^{in} = 3$ mW; these values correspond to self-saturated gains $G(\lambda_{peak}) = 32.5$ dB, $G(\lambda_{peak}) = 29$ dB, and $G(\lambda_{peak}) = 19$ dB, respectively. Self-saturation therefore becomes important when the peak gain $G(\lambda_{peak})$ is near 20 dB or greater. For EDFA gains greater than 30 dB, ASE-free models are inadequate, as the gains predicted by these models would be far in excess ($\Delta G > 15$ dB) of the actual, self-saturated values. The equivalent input noise model thus makes it possible to predict self-saturated gains in EDFAs using only one parameter, the spontaneous emission factor $n_{sp} \approx n_{eq}$. In the high gain regime ($G > 25$ dB), the spontaneous emission factor is observed to be spectrally uniform and weakly dependent of gain and saturating signal power (Section 5.4), which justifies the assumption $n_{sp}(\lambda_s, P_s^{in}) \approx$ const. made in the model. In order to check whether self-saturation is sensitive to this parameter, the EDFA gain spectra were calculated with the same parameters as in the previous example ($P_p^{in} = 5$ mW) for the different spontaneous emission factors $n_{sp} = 1.0$ to $n_{sp} = 1.6$, corresponding to EDFA noise figures $F_o = 3$ dB to $F_o = 5$ dB. The results, plotted in Figure 5.41, show that a 1 dB noise figure increase ($F_o = 4$ dB) with respect to the quantum limit corresponds to a 1 dB decrease in self-saturated EDFA gain from an initial value of about $G = 33$ dB. Likewise, a 2 dB noise figure increase ($F_o = 5$ dB) with respect to the quantum limit corresponds to a 2.5 dB gain decrease. These results suggest that *when the EDFA noise figure is within 1 dB of the quantum limit, the self-saturated gain is weakly sensitive to the spontaneous emission factor.* Consequently, the equivalent input noise model permits a straightforward and fairly accurate prediction of self-saturated gains in EDFAs, even when the spontaneous emission factor is not accurately known.

As in the two previous models, the equivalent input noise model does not predict gain difference between forward and backward pumping in the self-saturated regime. The reason for this limitation is that in the analytical model of [109], the EDFA gain is determined by a single transcendental equation, i.e., Eq. (1.148):

$$G_k = \exp\left\{-\alpha_k L + \frac{\bar{P}^{in} - \bar{P}^{out}}{P_{sat}(\lambda_k)}\right\}$$

(5.62)

This equation is function of the total input and output photon flux ($\bar{P}^{in}, \bar{P}^{out}$) and not of the relative propagation directions of the optical channels involved. For this reason the equivalent input noise model, based upon Eq. (5.62), does not predict any

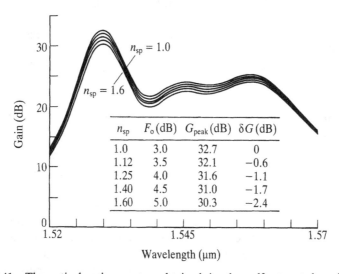

| $n_{sp}$ | $F_o$ (dB) | $G_{peak}$ (dB) | $\delta G$ (dB) |
|------|------|------|------|
| 1.0 | 3.0 | 32.7 | 0 |
| 1.12 | 3.5 | 32.1 | −0.6 |
| 1.25 | 4.0 | 31.6 | −1.1 |
| 1.40 | 4.5 | 31.0 | −1.7 |
| 1.60 | 5.0 | 30.3 | −2.4 |

**FIGURE 5.41** Theoretical gain spectra obtained in the self-saturated regime with the equivalent input noise model, assuming the same parameters as in Figure 5.40 ($\lambda_p = 980$ nm, $L = 50$ m, $P_p^{in} = 5$ mW), and different spontaneous emission factors $n_{sp} = 1.0$–1.6, corresponding to noise figures $F_o = 3.0$–5.0 dB. The peak gains $G_{peak}$ and gain changes $\delta G$ with respect to the case $F_o = 3.0$ dB are also shown.

difference between the forward and backward ASE spectra (given here by $P_{\text{ASE}}(\lambda_k) = 2n_{\text{eq}} h v_k \, \delta v G(\lambda_k)$), unless different equivalent input noise factors ($n_{\text{eq}}^+$, $n_{\text{eq}}^-$) are used for the two directions. The ASE spectrum calculated from the equivalent input noise model thus represents an approximate solution. In the numerical integration method, such a solution can be used for trial boundary conditions for the ASE. For instance, the EDFA rate equations can be integrated from $z = 0$ to $z = L$ using $P_{\text{ASE}}(\lambda_k)$ as boundary condition at $z = 0$ for the backward ASE spectrum. The advantage of this approach is a reduction of the number of iterations required for convergence towards the actual solutions. The equivalent input noise factors corresponding to both propagation directions ($n_{\text{eq}}^+$, $n_{\text{eq}}^-$) can then be corrected at each iteration. This computation algorithm can be viewed as combining both features of the shooting method and the relaxation method, described in Section 1.6. Since the model of [109] can be extended to the case of arbitrary Er-doping profiles (Th. Pfeiffer and H. Bulow [152]), this computation algorithm can be implemented with the most general rate equations of the EDFA model.

In the analysis of self-saturated EDFAs, the main drawbacks of the numerical integration method are the long computing time requirement and, in some cases, the difficulty of reaching algorithm convergence. On the other hand, the main drawback of the three simplified resolution methods for this problem (i.e., based on average power, average inversion, and equivalent input noise models) is the inability to predict gain differences between the forward and backward pumping schemes. A compromise, whose advantages remain to be determined and quantified, combines the two approaches. A current challenge is to develop alternate methods of improved accuracy for solving the EDFAs self-saturation problem.

## 5.6 OPTIMIZATION OF FIBER AMPLIFIER PARAMETERS

In this section, we consider various issues related to the optimization of EDFA parameters. The gain, saturation, and noise characteristics, usually referred to as the EDFA performance, can be optimized by the interplay of several EDFA parameters.

1. The EDFA fiber length $L$
2. The pump absorption band and the pump wavelength $\lambda_p$ within this band
3. The signal wavelength $\lambda_s$
4. The peak erbium concentration $\rho_0$ and concentration profile $\rho(r)$
5. The fiber waveguide characteristics ($\lambda_c$, NA, $a$) and mode envelopes $\psi_{p,s}(r)$.

Characteristics such as cross section peak values and spectra, fluorescence lifetime, fiber background loss, bending loss, and coupling loss also affect the EDFA performance. They are reviewed in earlier sections and are not considered here.

The effect of pump decay along the Er-doped fiber results in nonuniform medium inversion. Since the medium is absorbing when not inverted (a property of three-level systems with laser transition terminating to the ground level), the fiber becomes lossy beyond a certain length. For maximum signal gain, the fiber length must therefore be chosen to some optimized value $L_{\text{opt}}$. This optimum length actually depends on the input pump power, since a longer length of inverted medium can be achieved by a higher pump. This feature is clearly illustrated in Figure 5.7a, which shows both pump and signal power evolution along the fiber, for three different pump inputs. The optimum length also depends on both pump and signal wavelengths, since the pump absorption coefficient and the signal gain coefficient are wavelength dependent. For saturated EDFAs, the optimum length also depends on the signal power, as the effect of saturation is to locally reduce medium inversion. At high gains ($G > 25\,\mathrm{dB}$), self-saturation reduces medium inversion near the fiber ends, an additional factor to account for length optimization. Finally, the optimum length is a function of the peak $Er^{3+}$ concentration ($\rho_0$) as well as of the overlap between the modes and doping profile ($\Gamma_{p,s}$), since both pump absorption and signal gain coefficients are proportional to these two characteristics.

Given an Er-doped fiber with known absorption coefficients $\alpha_{p,s}$ and cross section ratios $\eta_{p,s}$ at given pump and signal wavelengths $\lambda_{p,s}$, how can we determine the optimum length that gives maximum gain?

For a first indicative value of the required fiber length of unsaturated EDFAs, consider forward pumping and assume unsaturated gain with no ESA; the pump decays according to:

$$\frac{dP_p}{dz} = -\alpha_p \frac{P_p}{1 + P_p/P_{sat}(\lambda_p)} \tag{5.63}$$

The solution of Eq. (5.63) gives the pump decay law:

$$P_p^{out} = P_p^{in} - P_{sat}(\lambda_p)\alpha_p L + \log(P_p^{in}) - \log(P_p^{out}) \tag{5.64}$$

Assuming a high input pump, the logarithmic terms in Eq. (5.64) can be neglected, and the pump decay is linear. The fiber length corresponding to input/output pump powers is approximately:

$$L \approx \frac{1}{\alpha_p} \frac{P_p^{in} - P_p^{out}}{P_{sat}(\lambda_p)} \tag{5.65}$$

In Eq. (5.65), the saturation power $P_{sat}(\lambda_p)$ can be determined by measuring the nonlinear transmission $P_p^{out} = f(P_p^{in})$, using any length of EDF having a few dB of pump absorption (Chapter 4). To achieve the conditions for medium inversion over the entire fiber length, the output pump $P_p^{out}$ at $z = L$ must equal the threshold value $P_p(\text{threshold}) = P_{sat}(\lambda_p)/U_{ps}$, with $U_{ps} = (\eta_s - \eta_p)/(1 + \eta_p)$. According to Eq. (5.65), the optimum length is approximately:

$$L_{opt} \approx \frac{1}{\alpha_p}\left\{\frac{P_p^{in}}{P_{sat}(\lambda_p)} - \frac{1 + \eta_p}{\eta_s - \eta_p}\right\} \tag{5.66}$$

To check the validity of Eq. (5.66), consider: $\alpha_p = 0.5\,\mathrm{dB/m}$, $\alpha_s = 1\,\mathrm{dB/m}$, $\eta_p = 0$, $\eta_s = 1$, and $P_{sat}(\lambda_p) = 500\,\mu\mathrm{W}$. For the input pump powers $P_p^{in} = 1\,\mathrm{mW}$, $3.75\,\mathrm{mW}$, $5\,\mathrm{mW}$ corresponding to $P_p^{in}/P_{sat}(\lambda_p) = 5, 7.5, 10$, we obtain from Eq. (5.66) the optimum lengths $L_{opt} = 35\,\mathrm{m}$, $78.5\,\mathrm{m}$, $56.5\,\mathrm{m}$. The actual solutions for pump power and signal gain, calculated from Eqs. (1.113) and (1.114) with ASE neglected, are shown in Figure 5.42. The optimum lengths corresponding to the three input pump powers are

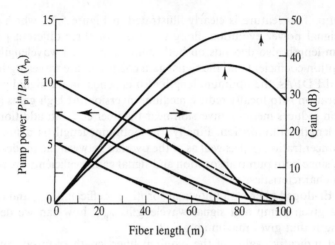

**FIGURE 5.42**  Normalized pump power and small-signal gain distributions along EDFA lengths, calculated for three different input pump powers. Arrows indicate the maximum gains corresponding to optimum EDFA lengths $L_{\text{opt}}$. The dashed lines correspond to the linear pump decay approximation. $\alpha_{\text{p}} = 0.5\,\text{dB/m}$, $\alpha_{\text{s}} = 1\,\text{dB/m}$, $\eta_{\text{p}} = 0$, and $\eta_{\text{s}} = 1$.

$L_{\text{opt}} = 50\,\text{m}$, $73\,\text{m}$, $85\,\text{m}$. These results show that the optimum lengths predicted by Eq. (5.66) are underestimated by less than 25%. The error is due to the difference between the linear law and the actual solutions for the pump decay. Figure 5.42 shows that the optimum lengths are approximately twice the length $L_{1/2}$ at which the pump has decayed to one half its initial value. This observation is consistent with the fact that, according to a quasilinear decay law, the pump approaches zero at length $2L_{1/2}$. Using this rule of thumb and Eq. (5.65), the optimum length is approximately:

$$L_{\text{opt}} \approx 2L_{1/2} \approx \frac{1}{\alpha_{\text{p}}} \frac{P_{\text{p}}^{\text{in}}}{P_{\text{sat}}(\lambda_{\text{p}})} \tag{5.67}$$

The expression for $L_{1/2}$ obtained from Eq. (5.65) is fairly accurate, as the pump decay from $z = 0$ to $z = L_{1/2}$ is well approximated by the linear law, Figure 5.42. Using Eq. (5.67), we find in the previous example $L_{\text{opt}} = 43.5\,\text{m}$, $65\,\text{m}$, $87\,\text{m}$ which, compared to the theoretical values of Figure 5.42, represent a maximum underestimation of 15%. For gains $G > 25\,\text{dB}$, self-saturation somewhat reduces the optimal length. Predictions of $L_{\text{opt}}$ for unsaturated EDFAs obtained by this method are good indicative values. Optimum length can be determined by several experimental methods: cutback, macrobend, and optical time domain reflectometry (OTDR).

In the cutback method, the EDF is input with the highest pump power available and a low signal power; the length is taken 25% longer than the estimated optimum value. The output signal power is measured. The EDF output end is then cut back by small amounts, as long as the measured signal is observed to increase. In each measurement, we take care to avoid end-reflections (e.g., by index matching) and to filter the forward ASE. This is the method that was followed in the measurements of Figure 5.7a. The optimum length thus determined corresponds to the forward pumping configuration. For gains $G < 25$ dB, this optimum length also corresponds to backward pumping. At higher gains, self-saturation reduces the optimum length; as the effect is stronger for forward pumping, the corresponding optimum length is shorter than in the case of backward pumping. The cutback method is less practical for backward pumping, as a low loss splice (with calibrated transmission) for coupling the pump must be made at each measurement, which is quite time-consuming. The main drawback of the cutback technique is the loss of fiber material, which can be important if several optimization tests for different pump and signal wavelengths need to be made.

Other methods for fiber length optimization that are non-destructive measure the signal power distribution along the fiber through macrobending (J. D. Evankow and R. M. Jopson [153]) or through OTDR (D. L. Williams et al. [154]; M. Nakazawa et al. [155]). In the macrobending technique, the entire EDFA length is scanned while passing through a bending point. The low-level signal tapped out from the fiber can then be monitored and discriminated from the pump through a phase sensitive detection apparatus. Polarization control is required as the macrobend coupling coefficient is slightly polarization sensitive [153]. This technique can be used for both forward and backward pumping; in backward pumping, the fiber must be initially cut to a certain length. A direct OTDR measurement of signal transmission through the EDFA is the most straightforward approach. An example of such a measurement obtained for different input pump powers is shown in Figure 5.43, in the case of a 18.2 km long distributed EDFA with forward pumping at $\lambda = 1480$ nm [155]. For input pump powers $P_p^{in} \leqslant 7$ mW the entire fiber is lossy, while transparency and amplification are obtained at higher powers. The optimum length increases with input pump power. The advantage of the macrobend and OTDR methods for the determination of optimum length is to make possible a continuous and nondestructive

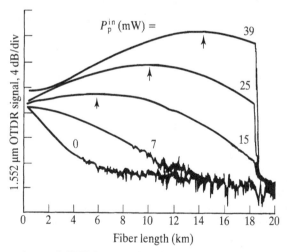

**FIGURE 5.43** Experimental OTDR trace showing 1552 nm signal transmission along the fiber length obtained in an 18.2 km distributed EDFA with 1480 nm forward pumping for different input pump powers; the arrows indicate maximum signal power excursion corresponding to optimum lengths $L_{opt}$. From [155], reprinted with permission from the Optical Society of America, 1990.

acquisition of data with varying pump and signal wavelengths or pump and signal powers.

The optimum EDFA length $L_{opt}$ and corresponding optimum gain $G_{opt}$ can also be determined theoretically, using the various models described in Chapter 1. If the effects of EDFA self-saturation and pump ESA are negligible, several analytic formulae can be used, such as Eq. (1.157) and in several studies (F. F. Rühl, [156]–[158]; M. C. Lin and S. Chi [159]). These formulae differ in their underlying approximations: small-signal, confined Er-doping, or Gaussian mode envelopes. In particular, the study of [159] shows that the analytic formula (1.157) (generalized in this book to include the case of two-level systems) gives accurate predictions of $L_{opt}$ and $G_{opt}$ in EDFAs having Gaussian Er-doping distributions. With a Gaussian doping-width to mode-width ratio $\omega_{Er}/\omega = 0.4$, the predictions for $L_{opt}$ are 23% accurate over a 20 dB range of signal-induced gain compression. These formulae guide experimental measurements and back-up the optimization procedure with realistic predictions.

We consider next the role of the peak $Er^{3+}$ concentration parameter. The maximum gain that can be achieved in an EDF of length $L$ and peak Er concentration $\rho_0$ is given by the value $G(\infty) = \exp(\rho_0 \Gamma_s \sigma_{es} L)$, corresponding to a regime of complete and uniform medium inversion. Using the value of $\sigma_{es} = 4.4 \times 10^{-21}$ cm$^2$ for the peak emission cross section in aluminosilicate fibers (Chapter 4), we find

$$\rho_0(Er^{3+}) \approx 2.3 \times 10^{20} \frac{\log G(\infty)}{\Gamma_s L(cm)} \text{ ions/cm}^3 \qquad (5.68)$$

Using the above result, we find that the Er concentration required to achieve a peak gain of $G = 20$ dB in a 10 m long EDFA with overlap coefficient $\Gamma_s = 0.5$ is $\rho_0 \approx 2.1 \times 10^{18}$ ions/cm$^3$, corresponding to the erbium oxide weight fraction $\rho_0(Er^{3+}) \approx 200$ ppm wt, according to Eq. (4.46) with glass density $D = 2.86$. According to Eq. (5.68), the same gain could theoretically be achieved with shorter lengths by increasing the peak Er concentration in the same proportion, i.e., $\rho_0 \approx 2.1 \times 10^{19}$ ions/cm$^3$ (2000 ppm wt) for $L = 1$ m and $\rho_0 \approx 2.1 \times 10^{20}$ ions/cm$^3$ (20,000 ppm wt) for $L = 1$ cm. But at high ionic concentrations, the effect of cross-relaxation by cooperative energy transfer (CET) reduces the fluorescence lifetime $\tau_{obs}$ (concentration quenching), described in detail in Section 4.8. In the case of silicate glasses, the onset of CET is near concentration $\rho_0 \approx 10^{19}$ ions/cm$^3$ [159]. In the previous example, therefore, EDFAs with lengths $L < 1$ m would be affected by CET. The effect of concentration quenching determines a maximum Er-doping concentration, which corresponds to minimum EDFA length.

The deleterious effect of CET is evidenced by comparing the gain characteristics of EDFAs having different Er concentrations but identical transmission loss $\alpha_s L$ (M. Shimizu et al. [161]; N. Kagi et al. [162]). This effect is illustrated in Figure 5.44. The figure shows the 1552 nm gain versus 1480 nm pump power characteristic obtained in three fibers with signal transmission loss $\alpha_s L = 67.5$ dB and concentrations (by Er weight): $\rho_0 = 77$ ppm, 470 ppm and 970 ppm [161]. In these fibers, the cores were made of Er-doped silica and the cladding of fluorine-doped silica. An Er concentration increase causes a reduction of the dB/mW gain coefficient, which corresponds to a higher pump power for transparency $P_p^{trans}$. In the absence of CET and fiber background loss, the transparency power $P_p^{trans}$ is proportional to the pump transmission $\alpha_p L$, as shown in Eq. (5.10), and therefore is concentration independent

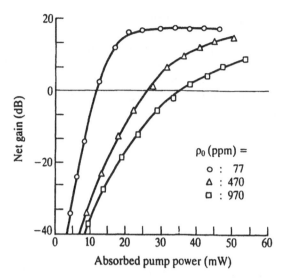

**FIGURE 5.44**  Signal gain at 1552 nm versus 1480 nm pump power for three EDFAs with different Er-doping concentrations of $\rho_0 = 77$ ppm, 470 ppm, 970 ppm and the same unpumped signal transmission $\alpha_s L = 67.5$ dB. From [161] © 1990 IEEE.

for constant transmission. With Eq. (4.46), the concentration $\rho_0 = 470$ ppm wt for which CET is observed corresponds to $\rho_0 \approx 5.10^{18}$ ions/cm$^3$, close to the critical concentration $\rho_0 \approx 10^{19}$ ions/cm$^3$. In comparison, the concentration $\rho_0 = 77$ ppm wt, for which CET is not observed (as evidenced for instance by absence of upconversion fluorescence [162]), corresponds to $\rho_0 \approx 8.10^{17}$ ions/cm$^3$, about one order of magnitude lower than the critical value.

For Er concentrations near $\rho_0 \approx 10^{19}$ ions/cm$^3$, the CET efficiency can also increase due to the additional effect of clustering (Section 4.8). Indeed, when RE clusters are formed in the glass, the mean interionic distance is reduced, which results in CET enhancement. The effect of clustering is host dependent, and can be reduced by alumina codoping. Indeed, nearly an order of magnitude more Er$^{3+}$ ions can be incorporated in the aluminosilicate host, in comparison to the germanosilicate host (B. J. Ainslie et al. [163]). In both types of hosts, a fast component in the fluorescence decay (showing CET) is observed at concentrations near or above 0.15 wt% [163], corresponding to $\rho_0 = 1.5 \times 10^{19}$ ions/cm$^3$. Taking then the value $\rho_0 = 1.10^{18}$ ions/cm$^3$ (or 100 ppm wt) as representing a maximum allowable concentration for CET-free EDFAs, the minimum length required to achieve the gain $G(\infty)$ is, from eq. (5.68):

$$L(m) \approx 2.3 \frac{\log G(\infty)}{\Gamma_s} \qquad (5.69)$$

Using the above formula, we find for $G(\infty) = 30$ dB and $\Gamma_s = 0.5$ a minimum EDFA length $L = 32$ m. This result is independent of the pump wavelength. But in practical applications, it is difficult to achieve complete and uniform inversion along the fiber, due to limits in available pump power. The actual EDFA length required to achieve a gain $G = G(\infty)$ is therefore greater than the value predicted by Eq. (5.69), by an amount which depends upon the pump and signal absorption coefficients, as given by eq. (1.157).

In spite of higher pumping requirements, high concentration EDFAs with lengths $L \leqslant 1$ m offer a significant advantage of compactness. In the case of EDFAs based on fluoride hosts, additional advantages of high concentration devices are the increase in production yield per preform, as well as a reduced effect of background loss. In pulsed fiber laser applications, short and high gain EDF lengths are necessary for achieving high repetition rates (Chapter 6). High gain EDFAs with Er concentration significantly greater than 1000 ppm have been investigated for this purpose. For instance, a 20 dB gain with 60 mW pump at 1480 nm could be achieved in a 50 cm long ZBLAN EDFA with Er concentration of $\rho_0 = 5000$ ppm (Y. Miyajima et al. [164]). Early experiments with 514·6 nm pump used 1 m long EDFAs with Er concentration $\rho_0 = 6500$ ppm; the maximum gain is $G = 35$ dB for 225 mW pump power (E. Desurvire et al. [104]). More recently, a 40 dB gain with 120 mW pump at 980 nm was achieved in a 1 m long aluminosilicate EDFA with 8900 ppm Er concentration, corresponding to $\rho_0 \approx 9.10^{19}$ ions/cm$^3$; the same device with a length reduced to $L = 50$ cm had a 25 dB gain for 80 mW pump (Y. Kimura and M. Nakazawa [165]).

As shown below, the dB/mW gain coefficient can be enhanced by confining the Er$^{3+}$ ions near the center of the fiber core. The effect of this Er-doping confinement is to reduce the overlap factor $\Gamma_s$, and consequently, to increase the required EDFA length. In the Gaussian envelope approximation, the overlap factor corresponding to a step-like doping profile with radius $a_0$ is given by Eq. (1.86), i.e., $\Gamma_s = 1 - \exp(-a_0^2/\omega_s^2)$. In the expression for $\Gamma_s$, the ratio $a_0/\omega_s$ is referred to as the doping confinement factor. The minimum length required for achieving $G(\infty)$ in EDFAs is then given by:

$$L = \frac{\log G(\infty)}{\Gamma_s \rho_0 \sigma_{es}} = \frac{1}{1 - \exp(-a_0^2/\omega_s^2)} \frac{\log G(\infty)}{\rho_0 \sigma_{es}} \tag{5.70}$$

From eq. (2.67), we find that reducing the doping confinement factor from $a_0/\omega_s = 1$ ($\Gamma_s \approx 0.63$) to $a_0/\omega_s = 0.25$ ($\Gamma_s \approx 0.06$) while keeping the same Er concentration, results in a tenfold increase of required length*. When the required EDFA length is increased, due to Er-doping confinement, to the size of one or several kilometers, fiber background loss becomes a source of impairment, particularly at the pump wavelength of 980 nm. Alternatively, the length increase due to doping confinement can be partially compensated by increasing the peak concentration $\rho_0$, so the parameter $\Gamma_s \rho_0$ is maximized. Since the peak concentration cannot be increased beyond the range 100–1000 ppm, it is not possible to simultaneously achieve doping confinement and length minimization. This issue is further addressed when we consider the optimization of Er-doping radius.

Having determined the optimum EDFA lengths assuming fixed pump and signal wavelengths $\lambda_p$, $\lambda_s$, we focus now on pump wavelength optimization, assuming a

---

* This result is not valid in the case where the medium inversion is not uniform along the fiber. In this case, the gain coefficient is given by $\rho_0(\sigma_{es}\Gamma_{es} - \sigma_{as}\Gamma_{as})$, where $\Gamma_{es, as}$ are power dependent overlap integral factors, as shown in Section 1.7. In that section, we show that the overlap integral factors are defined by transcendental functions of the parameter $\Gamma_s = 1 - \exp(-a_0^2/\omega_s^2)$, as shown in Eqs. (1.84)–(1.85).

fixed signal wavelength $\lambda_s$. The three pump bands near 1480 nm, 980 nm, and 810 nm corresponds to: (a) two-level pumping, (b) three-level pumping with negligible pump ESA, (c) three-level pumping with pump ESA.

Consider two-level pumping, which corresponds to the 1480 nm pump band. In this case, the minimum pump power required for local inversion (i.e., $\eta_s N_2 - N_1 > 0$) is given by $P_p = P_{sat}(\lambda_p)/U_{ps} = P_{sat}(\lambda_p)(1 + \eta_p)/(\eta_s - \eta_p)$. From this relation, the required power increases to infinity when the pump wavelength $\lambda_p$ nears the signal wavelength $\lambda_s$, i.e., $\lambda_p \to \lambda_s$. In such a configuration, the EDFA is operated as a true two-level laser system, for which no net gain or inversion can be achieved. Consequently, the pump wavelength must be detuned away from the signal wavelength to maximize $U_{ps}$. But as the pump is detuned towards shorter wavelengths, the pump absorption coefficient decreases. A smaller pump absorption coefficient corresponds to a higher value of the saturation power $P_{sat}(\lambda_p)$, since $P_{sat}(\lambda_p) \approx 1/\sigma_a(\lambda_p) \approx 1/\alpha_p$, thus the required pump power for inversion is higher. The above arguments suggest an optimum wavelength for which the required pump power is minimized. This property is demonstrated by the theoretical plots of Figure 5.45 (E. Desurvire [166]). The plots show the signal gain at $\lambda_s = 1.531$ $\mu$m for a fixed EDFA length ($L = 30$ m) calculated as a function of pump wavelength $\lambda_p$ for different input pump powers $P_p^{in}$. While the signal gain is maximized for $\lambda_p = 1480$ nm, it becomes less sensitive to pump detuning as the pump power is increased. At $P_p^{in} = 10$ mW, the 3 dB gain bandwidth is $\Delta\lambda_p = 29$ nm; at $P_p^{in} = 50$ mW, it increases to $\Delta\lambda_p = 44$ nm. This property can be explained by the fact that at high pumps ($P_p^{in} \gg P_{sat}(\lambda_p)$), the medium inversion at any point in the fiber is weakly sensitive to the pump wavelength. $G(\infty) \to 1$ for $\lambda_p \to \lambda_s = 1.53$ $\mu$m, which corresponds to the true two-level system limit.

The weak gain dependence with pump wavelength predicted at high pumps is confirmed experimentally with both aluminosilicate and germanosilicate EDFAs (J.

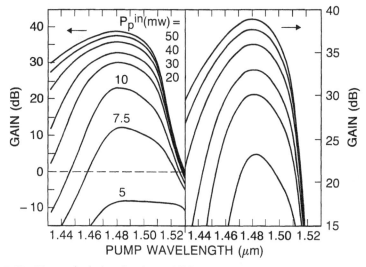

**FIGURE 5.45** Theoretical signal gain at 1531 nm versus pump wavelength near 1480 nm, calculated for a 30 m EDFA with different input pump powers $P_p^{in}$; except for the bottom one, left and right show the same curves but plotted with different vertical scales. From [166] © 1989 IEEE.

L. Zyskind et al. [167]; Y. Kimura et al. [168]). Experimental measurements of $G(\lambda_s) = f(\lambda_p, P_p^{in})$ corresponding to a Type III aluminosilicate EDFA ($L = 55$ m) are shown in Figure 5.46 [167]. In this measurement, the 3 dB gain bandwidth for $\lambda_s^a = 1.531\ \mu$m and $\lambda_s^b = 1.544\ \mu$m is $\Delta\lambda_p = 27$ nm for $P_p^{in} = 10$ mW, and $\Delta\lambda_p = 40$ nm for $P_p^{in} \geqslant 20$ mW, in good qualitative agreement with the results of Figure 5.45. The

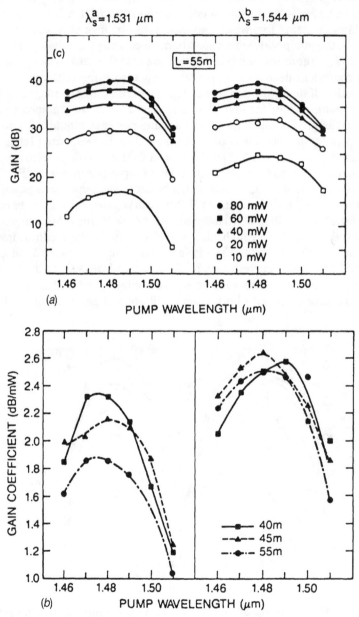

**FIGURE 5.46**    (a) Experimental signal gain at $\lambda_s^a = 1.531\ \mu$m and $\lambda_s^b = 1.544\ \mu$m measured as a function of pump wavelength near 1480 nm, for a 55 m EDFA; (b) corresponding dB/mW gain coefficients for EDFA lengths $L = 40$ m, 45 m, and 55 m. From [167] © 1989 IEEE.

dB/mW gain coefficient, measured for three different lengths as the function of pump wavelength for the two signals at $\lambda_s^a$ and $\lambda_s^b$, is maximum for pump wavelengths falling between $\lambda_p = 1.47\,\mu m$ and $\lambda_p = 1.48\,\mu m$. The highest gain coefficient (2.6 dB/mW in this example) is obtained at the longest signal wavelength ($\lambda_s^b$), for which dependence on EDFA length is also less important. The weak gain dependence withh pump wavelength is important for practical EDFA applications. Indeed, Fabry–Perot LDs with broad multimode spectra (e.g., $\Delta\lambda_p = 20\,nm$) can be used as efficient pump sources. Furthermore, the center pump wavelength specification is fairly broad (i.e., $\lambda_p = 1480\,nm \pm 15\,nm$), which permits high yield production for the pump LDs.

Three-level pumping with negligible pump ESA corresponds in particular to the 980 nm pump band. The first experimental characterization of gain versus pump wavelength near $\lambda_p = 980\,nm$ in germanosilicate and aluminosilicate EDFAs showed that maximum gain is achieved when $\lambda_p$ is at the peak of the pump absorption band (K. Suzuki et al. [169]; P. C. Becker et al. [170]). We expect this, as the saturation power $P_{sat}(\lambda_p) \approx 1/\sigma_a(\lambda_p) \approx 1/\alpha_p$ is minimum at this wavelength, corresponding to a maximum ratio $P_p/P_{sat}(\lambda_p)$. In these initial measurements, the 3 dB gain bandwidths were measured to be $\Delta\lambda_p = 10$–$14\,nm$ and $\Delta\lambda_p = 23\,nm$, depending on the signal wavelength location in the gain band, [169] and [170]. Subsequent experimental characterization has shown that the 3 dB gain bandwidth can actually be extended to $\Delta\lambda_p = 38\,nm$ by using longer and nonoptimized EDFA lengths (R. M. Percival et al. [171]). In this case, a small dip centered at the peak pump absorption wavelength is observed in the curve $G(\lambda_s) = f(\lambda_p, P_p^{in})$, as shown in Figure 5.47 [171]. This dip can be explained by the following: when the fiber length is not optimized, the fiber is lossy over a certain region. The length of this region depends upon the rate at which the pump power is absorbed. As the pump is tuned towards the peak absorption wavelength, the pump absorption rate increases, which increases the length of the lossy region. Consequently, the signal gain drops, which appears as a dip in the tuning curve. From Figure 5.47, the gain dip becomes deeper as the pump power is

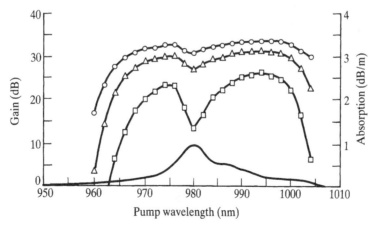

**FIGURE 5.47**  Signal gain at $\lambda = 1536\,nm$ versus pump wavelength near 980 nm for an EDFA with nonoptimized length ($L = 100\,m$), measured with input pump powers $P_p^{in} = 14\,mW$ (bottom) to $P_p^{in} = 23\,mW$ (top); the pump absorption coefficient is also shown. From [171], reprinted with permission from *Electronics Letters, 1991.*

decreased. this confirms that for EDFAs with nonoptimized lengths and in the absence of self-saturation, the dip in the gain tuning curve can be exclusively attributed to the effect of signal reabsorption.

But when self-saturation is present, a similar dip can be observed in the small-signal regime, even with optimized EDFA lengths (B. Pedersen et al. [172]). This can be explained by the effect of self-saturation or more specifically by the effect of backward ASE on pump absorption. Indeed, in the regime of self-saturation backward ASE increases the pump absorption rate near the fiber input end, which results in a higher pump/ASE energy transfer in comparison to the unsaturated case. Pump/ASE energy transfer is maximum when the pump is tuned to the peak absorption wavelength. Theoretical simulations in [172] show that the gain dip becomes deeper with increasing pump, contrary to the previous case (i.e., unsaturated regime, nonoptimized length). This feature, which corresponds to optimized EDFA lengths, clearly indicates that the dip is caused exclusively by self-saturation. We can also view this pump detuning effect as a gain enhancement. Theoretical simulations show that for EDFAs with gains near $G = 35\,\text{dB}$ at the peak pump absorption wavelength ($\lambda_p = 980\,\text{nm}$), the gain enhancement is about 3 dB; the 3 dB gain bandwidth is in this case $\Delta\lambda_p \approx 50\,\text{nm}$ [172]. In the case of power amplifiers, where saturation is primarily induced by the signal, gain enhancement is not observed, and the gain tuning curve $G(\lambda_s) = f(\lambda_p, P_p^{\text{in}})$ exhibits a single peak at the peak pump absorption wavelength, as expected [173].

The saturated noise figure penalties induced by 980 nm pump detuning are different for self-saturation and signal-induced saturation. In the case of self-saturation, the theoretical study of [172] shows that a noise figure penalty $\Delta F = 0.6\,\text{dB}$ incurred per 10 nm of detuning. In the case of signal-induced saturation (power EDFAs), the saturated noise figure penalty is about $\Delta F = 0.3\,\text{dB}$ per 10 nm of detuning [173].

Three-level pumping with pump ESA corresponds, in particular, to the 810 nm pump band (Section 4.7). In the low gain regime, and for nonoptimized EDFA lengths, the same dip effect shown in Figure 5.47 is observed (M. Nakazawa et al. [174]). Additional structure observed in the $G(\lambda_s) = f(\lambda_p, P_p^{\text{in}})$ characteristic is attributed to pump ESA. This structure is an effect of finite overlap between GSA and ESA cross sections and depends on the type of glass host (B. Pedersen et al. [175]). For a $GeO_2$–$P_2O_5$ host, as shown in Figure 4.31, the peak ESA wavelength falls into the short wavelength tail of the GSA. The effect of pump ESA can be reduced by tuning the pump towards the long wavelength tail of the GSA. The structure in the $G(\lambda_s) = f(\lambda_p, P_p^{\text{in}})$ characteristic then depends upon the cross section ratio $\sigma_a^{\text{ESA}}(\lambda_p)/\sigma_a^{\text{GSA}}(\lambda_p)$. Theoretical studies have also shown that the tuning curve $G(\lambda_s) = f(\lambda_p, P_p^{\text{in}})$ is different for unidirectionally pumped and bidirectionally pumped EDFAs (B. Pedersen et al. [176]). The EDFA gain is enhanced with bidirectional pumping (for same total input power), particularly at the short pump wavelengths where ESA is the highest. The tuning curve $G(\lambda_s) = f(\lambda_p, P_p^{\text{in}})$ is also flatter in the case of bidirectional pumping. This can be explained by the following: in bidirectional pumping, the inversion at any point in the fiber is reduced, as the pump power is divided between the two ends, consequently, the inversion dependent ESA is also reduced, which results in the gain enhancement [176]. A comprehensive analysis of EDFAs pumped near 810 nm, taking into account the effects of glass host and pumping configurations, is described in [175]. This analysis shows several important properties: (a) the highest gain is obtained by pumping in the long wavelength tail of the GSA band; (b) fluorophosphate silica hosts can yield the same performance as aluminophosphate silica hosts with a 2–3 dB reduction in required pump power; (c) the effect of ESA on fiber gain can be alleviated by using bidirectional pumping, at the expense of a small noise figure penalty ($\Delta F < 0.5$ dB); (d) independent of pump wavelength, the quantum conversion efficiency (QCE) is nearly doubled when using bidirectional pumping [175].

A comparison of minimum pump power requirements for equal performance in small-signal gain or output signal power, corresponding to optimal pump wavelengths in the three pump bands of 810 nm, 980 nm, and 1480 nm and optimum EDFA lengths, is shown in Figure 5.48 (B. Pedersen et al., [150] and [175]). Where gain or output signal power performance are concerned, the minimum pump power requirement for the 810 nm band is 6–8 dB higher than with the two other bands.

We consider now the effects of Er-doping confinement and fiber waveguide characteristics on EDFA performance. It has long been recognized that confining the $Er^{3+}$ ions near the center of the fiber core results in improved pumping efficiency (B. J. Ainslie et al. [69]; J. R. Armitage [70]). Doping confinement is intuitively justified because pump power rapidly drops at some distance from the center of the fiber core, resulting in reduced medium inversion and possibly signal loss. Thus, the effect of doping the entire core with $Er^{3+}$ ions, or doping both core and cladding, is to yield a poor pumping efficiency. A second intuitive improvement is increased pump intensity near the center of the core. This can be realized by (1) restricting the number of guided modes, i.e., achieving single-mode propagation for the pump, and (2) reducing the spot size of the pump mode. For a given amount of pump power, such an intensity increase causes higher medium inversion in the center of the core. There must be a trade-off between the two approaches, as the effect of confinement is countered by

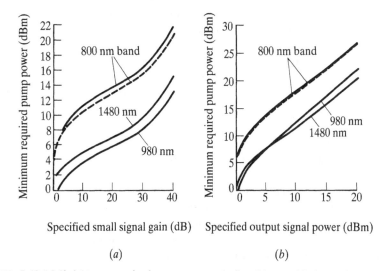

Specified small signal gain (dB)    Specified output signal power (dBm)

(*a*)                                      (*b*)

**FIGURE 5.48** Minimum required pump power for (*a*) specified small-signal gain at $\lambda = 1532$ nm and (*b*) specified output signal power at $\lambda = 1532$ nm for 1 mW signal input, corresponding to aluminophosphate–silica EDFAs with optimum lengths and pump wavelengths in the 810 nm, 980 nm, and 1480 nm bands; solid curves correspond to codirectional pumping and dashed curves to bidirectional pumping. From [150] © 1992 IEEE.

the reduction of the pump mode size. The improvements made by (1) and (2) can be maximized by an appropriate choice of waveguide and doping parameters.

Optimum EDF parameters include the waveguide parameters, i.e., the fiber numerical aperture NA, the fiber core radius $a$ and the cutoff wavelength $\lambda_c = 2\pi a \text{NA}/V_c$, and the Er-doping profile parameters. The effects of Er concentration and fiber length must be considered in relation to these last parameters. We assume that the Er-doping profile is a step-like distribution of characteristic radius $a_0$. EDFAs having other types of Er-doping profiles (e.g. Gaussian, triangular, annular) could also be considered, but the step-distribution is optimum, in view of the previous arguments. The determination of EDFA parameters for which the best performance in gain, saturation, and noise is achieved, is generally referred to in the literature as fiber design optimization. The optimum waveguide parameters found by design optimization exclusively concern the simple case of step-index fibers. The conclusions found in the optimization of step-index fibers are also applicable to graded-index fibbers, although the optimum waveguide parameters for graded-index fibers must be determined by different numerical methods.

Optimization of EDFA became confused by the large number of studies, the variety of models, the profusion of experimental data, and the simplifying approximations. As for EDFA modeling, the basic principles of EDFA design optimization, which have been laid out in the early studies, have become more refined and accurate in subsequent work. It is therefore useful to review the development of important aspects of fiber design optimization from 1988 to 1991.

The first EDFA optimization study, reported in 1988 (J. R. Armitage [70]), considers the case of multimode EDFAs with three-level pumping and includes the effect of pump ESA. The ratio $R = $ (gain per cm)/(absorbed pump power per cm) is used as a figure of merit to conclude that optimum EDFAs should operate with single-mode pump, minimal pump spot size, and confined Er-doping. Subsequent fiber design studies (1990) consider two-level pumping for the determination of optimum $V$ numbers (J. H. Povlsen et al. [177]; N. Kagi et al. [178]), and seek to predict the highest dB/mW performance achieveable with two- and three-level pumping in SMF- and DSF-compatible EDFAs (E. Desurvire et al. [179]). Studies reported in 1991 aim to determine an optimum cutoff wavelength in relation to Er-doping confinement and glass host codopants (M. Ohashi and M. Tsubokawa, [180]–[181]; B. Pedersen et al. [182]–[184]). By determining an optimum cutoff wavelength, the important issue of EDFA input/output coupling loss is overlooked in these last studies. The problem of matching nonstandard EDFs with SMF and DSF fibers was also investigated and solved in the same period.

In the determination of an optimum EDFA design for 1480 nm pumping, the figure of merit used in independent studies [177] and [178] is the pump threshold for transparency corresponding to an infinitesimal EDFA length. Using the actual $LP_{01}$ mode solutions of step-index fibers, the analysis of [177] shows that there exists a range of optimum $V$ number values ($V = 1.4$–$1.5$) for which the 1480 nm pump threshold for transparency at $\lambda = 153$ $\mu$m is minimized. The optimum $V$ number was found to depend on the fiber NA and the doping confinement factor $\varepsilon = a_0/\omega_s$. In the case NA $> 0.1$, the optimum values were found to be $V = 1.4$ for $\varepsilon = a_0/a = 1$ and $V = 1.5$ for $\varepsilon = a_0/\omega_s = 0.2$. Using a Gaussian mode approximation and considering the quasiconfined case (i.e., $\varepsilon = a_0/\omega_s = 1$), the analysis of [178] reaches the same conclusion but with a somewhat higher optimum value $V = 1.6$. The difference between this study and the previous study can be attributed to the inaccuracy of the Gaussian mode approximation.* Reference [178] concludes that the optimum V number value is smaller than needed to minimize the mode spot size at $\lambda = 1480$ nm. This shows that the optimum design corresponds to a trade-off between the effects of Er-doping confinement and pump intensity increase at the center of the fiber core.

The study of [179] considers three basic types of step-index fibers. In the first two types, the signal power mode sizes at $\lambda = 1.5 \mu$m are assumed to be the same as in standard fibers (referred to as SMF) and dispersion-shifted fibers (DSF), while in the third type, a minimum power mode size ($\omega_s = 1.4 \mu$m) obtained with a very high NA (0.3) is assumed. The analysis uses the general EDFA model (Chapter 1), and the exact $LP_{01}$ mode solutions for pumps (980 nm and 1470 nm) and signals (1531 nm and 1544 nm). To account for the effect of self-saturation while keeping the computing time as short as possible, the ASE is assumed to be in a single optical channel with 25 nm bandwidth and $4n_{sp}$ input photons. This last assumption is used in later simplified models for ASE self-saturation, i.e., the average power model, the average inversion model, and the equivalent input noise model (Section 5.5). The issues addressed by this analysis are: (a) considering EDFAs with no coupling loss with SMF and DSF (first two fiber types), what are the highest dB/mW gain coefficient achievable with 980 nm and 1480 nm pumping? How do these results compare with an EDFA of higher NA? (b) How much Er-doping confinement is necessary to approach the best performance? (c) What are the corresponding requirements in Er concentrations and optimum fiber lengths? (d) For 980 nm pumping how is the performance affected by partial excitation of $LP_{11}$ modes?

Numerical simulations in [179] that: (a) for identical Er-doping profiles, the dB/mW coefficients are nearly the same with either pump wavelength i.e. $\approx 1.2$ dB/mW, 3.5 dB/mW, and 8.2 dB/mW for the three fiber types, respectively; (b) for each type, the best dB/mW coefficients are nearly approached with an Er-doping confinement parameter $\varepsilon = a_0/\omega_s \approx 0.25$; (c) the increase in optimal length with Er-doping confinement is nearly quadratic; for a maximum allowable $Er^{3+}$ concentration $\rho_0 = 1 \times 10^{19}$ ions/cm$^3$ (corresponding to the onset of CET), the range of optimum length corresponding to confinement $\varepsilon = a_0/\omega_s \approx 0.25$ is $L_{opt} = 25$–100 m; for lower concentrations, the optimum lengths increase in the same proportion (e.g., $L_{opt} = 250$–1000 m for $\rho_0 = 1 \times 10^{18}$ ions/cm$^3$, etc.), as shown in Figure 5.49; (d) independently of the Er-doping confinement ($\varepsilon = a_0/a \geqslant 0.25$), the EDFA performance is weakly affected by coupling of 980 nm pump into the four $LP_{11}$ modes, provided that at least 40% of the pump power is coupled into the two fundamental $LP_{01}$ modes. In comparison to more recent and accurate experimental data (B. Pedersen et al. [182]), the cross section values used in this study for $Al_2O_3$–$GeO_2$ fibers are 30% underestimated for 980 nm absorption, 99% accurate for 1470 nm or 1500 nm absorptions, and 17% overestimated for 1530 nm emission (the latter error is introduced by the FL relation, Chapter 4). This inaccuracy causes the best dB/mW coefficient to be overestimated for 1480 nm pumping and underestimated for 980 nm pumping. This can explain the fact that the highest gain coefficient of 8.2 dB/mW predicted by this study actually falls between the best results reported to date of

---

* In contrast, the Gaussian *envelope* approximation is fairly accurate, only if we choose the value calculated from Eq. (1.31) for the $1/e$ mode size, using the exact $LP_{01}$ mode solution in the integral computation (Chapter 1, Appendix B).

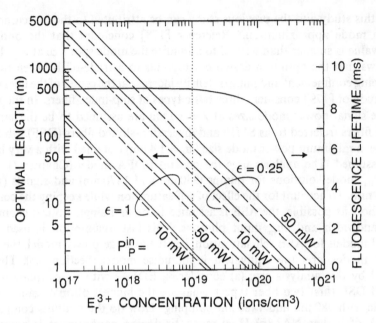

**FIGURE 5.49**   Theoretical optimum length $L_{\text{opt}}$ versus Er-doping concentration $\rho_0$, calculated for SMF-compatible EDFAs pumped near 980 nm (dashed lines) and 1470 nm (full lines), corresponding to different input pump powers $P_p^{\text{in}}$ and Er-doping confinement factors $\varepsilon = a_0/a$; the fluorescence lifetime $\tau$ of typical Er:silica glass is also shown. From [179] © 1990 IEEE.

6.3 dB/mW for 1480 nm pump, and 11.0 dB/mW for 980 nm pump, corresponding to EDFAs with high NA and confined Er-doping (Section 5.2).

The studies of [180] and [181], which are based on a Gaussian mode analysis, concern the determination of an optimum fiber cutoff wavelength for the EDFA, and its relation to Er-doping confinement in the case of 1480 nm pumping. It was found that for quasiconfined doping ($\varepsilon = a_0/a = 1$) this optimum cutoff wavelength is $\lambda_c = 0.8$, independent of the fiber NA; the effect of Er-doping confinement is to shift this optimum wavelength to the value $\lambda_c = 0.9$ ($\varepsilon = a_0/a = 0.2$). Reference [181] contains the same type of investigation for $\alpha$-index profiles (i.e., parabolic for $\alpha = 2$ and triangular for $\alpha = 1$). The optimum cutoff wavelength decreases as the parameter $\alpha$ decreases but the step-index profile ($\alpha = \infty$) with optimum $\lambda_c$ always yields the highest EDFA gain.

Due to the relative inaccuracy of the Gaussian mode approximation for describing the pump and signal power $d$ in the case where $\lambda > \lambda_c$ (Appendix B), the quantitative results of the previous study should be regarded as indicative values only. For an accurate determination of the optimum cutoff wavelength in step-index EDFAs, we must use the exact $LP_{01}$ mode solutions, as in [182]–[184]. The design optimization study of [182], aimed to determine the optimum $\lambda_c$ corresponding to either 980 nm or 1480 nm pumping, using the dB/mW performance as a figure of merit. The effect of glass host codopants (i.e., aluminium and germanium) and the case of $\alpha$-index profiles were also considered. Important conclusions of this comprehensive study are: (a) independent of the fiber NA and host type, the optimum cutoff wavelength in quasiconfined EDFAs ($\varepsilon = a_0/a = 1$) is $\lambda_c = 0.8$ for 980 nm pumping and $\lambda_c = 0.9$ for 1480 nm pumping, Figure 5.50; (b) the step-index profile is the optimum profile with respect to the dB/mW coefficient; (c) germanosilicate EDFAs with 980 nm pumping yield the highest dB/mW coefficients; for fibers with optimum cutoff wavelengths, NA = 0.25, $\varepsilon = a_0/a = 0.2$ and signals at the peak emission wavelength $\lambda_s^{peak}$, the highest theoretical value is 14 dB/mW, compared to 6.7 dB/mW in the aluminosilicate host; (d) in the case of 1480 nm pumping with the same (NA, $\varepsilon$, $\lambda_s^{peak}$) parameters, the best coefficients are 6.0 dB/mW and 4.5 dB/mW for germanosilicate and aluminosilicate hosts, respectively. In the case of germanosilicate EDFAs with NA = 0.4 and $\varepsilon = a_0/a = 1$, this study predicts a coefficient of 23 dB/mW when the pump is at 980 nm, compared to 8 dB/mW when the pump is at 1480 nm [183].

The more detailed study of [183] considers three types of EDFAs with host compositions Ge/silica, Al/Ge/silica and Al/silica (referred to in this book as Types I, II, and III, respectively). The sensitivity of the dB/mW coefficient with signal wavelength was found to increase in the order Type III, II, I. In [184], the overall performance of Type I and Type III EDFAs in dB/mW coefficient, noise figure and

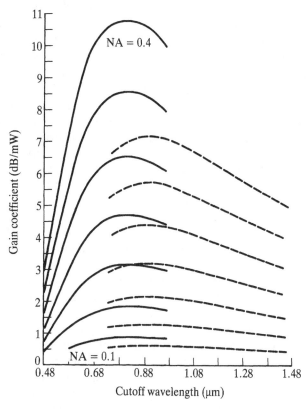

**FIGURE 5.50** Theoretical dB/mW gain coefficient of EDFA versus fiber cutoff wavelength for 980 nm pumping (full lines) and 1480 nm pumping (dashed lines), obtained for different fiber NAs (NA = 0.1–0.4 in steps of 0.05) and assuming uniformly doped core. From [183] © 1991 IEEE.

booster efficiency (or QCE) is compared for both 980 nm and 1480 nm pumping. In addition to conclusions (a)–(d), above, this study shows that: (e) the choice of fiber type (I or III) is significant only for 980 nm pumping when considering optimum dB/mW coefficients, and is less significant when considering optimum QCE; (f) for both fiber types, the noise figure corresponding to optimum dB/mW operating point increases up to 0.5 dB when the numerical aperture increases from NA = 0.1 to NA = 0.4; at this operating point, and independently of NA, the lowest NF is achieved in Type I EDFAs (Ge/silica) with 980 nm pumping, while for 1480 nm pumping Type I yields the highest NF. Observation (f) should not lead us to conclude that Type I EDFAs are preferable in all communications system applications. Indeed, EDFAs are not usually operated at the optimum dB/mW point, as this corresponds to a regime of incomplete inversion (Section 5.2); and Type III EDFAs offer a wider gain bandwidth, important for WDM systems. There is little qualitative difference between EDFAs having gain coefficients of 10 dB/mW and 20 dB/mW, since at some pumping level both reach the same gain limit $G(\infty)$ or a maximum gain determined by self-saturation, whichever is the lowest.

Regarding the optimization of QCE in power EDFAs, the optimum cutoff wavelengths for 980 nm or 1480 nm pumping with quasiconfined doping ($\varepsilon = a_0/a = 1$) are found to be $\lambda_c = 0.8$ and $\lambda_c = 0.9$, respectively [113]. Reference [113] shows that at high pumps (i.e., away from optimum dB/mW point), the effects of NA increase (NA = 0.15 → 0.25) and Er-doping confinement ($\varepsilon = 1 \to 0.15$) is to enhance the QCE by 60% and 20%, respectively, with insignificant effect on the NF.

The best EDFA performance predicted by design optimization is not straightforward to achieve in actual devices. Indeed, a number of factors and limitations related to fiber fabrication contribute to degrade this performance. The diffusion of index-raising codopants during preform collapse reduces the effective NA, and can cause the $Er^{3+}$ ions to form clusters, owing to their low solubility in the glass base matrix. Likewise, the diffusion of $Er^{3+}$ ions into the cladding during collapse reduces the doping uniformity and confinement, which results in increased pump inversion thresholds. Another important source of performance limitation for 980 nm pumping is the effect of fiber background loss (M. N. Zervas et al. [185]). Background loss is generated, in particular, by glass impurities (e.g. OH ions), by high codopant ($GeO_2$, $P_2O_5$) concentrations and by the fiber drawing process. Impurities can be removed by dehydration (other gases, such as $O_2$ and He, are used to prevent GeO formation and to outdiffuse trapped gases), while loss induced by codopants and fiber drawing can be reduced by control of the drawing tension and temperature, [185] and [186]. The experimental study of [185] shows that the dB/mW performance degradation induced by background loss at 980 nm is the most pronounced in low NA fibers. The best dB/mW coefficients that can be achieved experimentally approach 95% of the optimum theoretical values when the pump background loss is less than 1% of the absorption due to the $Er^{3+}$ ions. This illustrates the importance of the fiber fabrication technology, and in particular of the manufacturing process (e.g., VAD, MCVD), for the realization of high performance, optimized EDFAs.

Table 5.2 shows typical performance specifications of various EDFA modules ready for field deployment. This specification list is merely an illustration; it does not suggest any optimum design. The EDFA modules corresponding to these specifications differ by their pump wavelengths, pumping schemes, fiber composition, and most importantly, by the number of components involved (i.e., pump LDs, optical

**TABLE 5.2** **Typical performance specifications of EDFA modules (indicative 1992)** $G_{max}$ = small-signal gain at peak, $\Delta\lambda$ = small-signal gain bandwidth at $-3\,$dB, NF = noise figure, $P_s^{out}$ = saturated output power, SOP = polarization sensitivity, $T$ = operating temperature range

| Pump (nm) | $G_{max}$ (dB) | $\Delta\lambda$ (nm) | $P_s^{out}$ (dBm) | NF (dB) | SOP (dB) | $T$ (°C) | Manufacturer |
|---|---|---|---|---|---|---|---|
| 980 | 38 | 20 | +14 | 4.5 | <0.2 | – | Corning |
| 980 | 35 | 25 | +15 | <4.5 | <0.5 | 0 + 65 | Photonetics |
| 980 | 30 | 30 | +15 | 3.5 | – | −10 + 40 | Pirelli |
| 1064 | 35 | 15 | +10 | 3.5 | – | +10 + 30 | Amoco |
| 1480 | 30 | – | +9 | 7 | <0.8 | −5 + 45 | AT&T |
| 1480 | 35 | – | +14 | 5 | – | +10 + 50 | BT&D |
| 1480 | 30 | 15 | +10 | 7 | <0.5 | 0 + 35 | JDS-Fitel |
| 1480 | 43 | – | +16 | 5.1 | – | −25 + 60 | BT [187] |

isolators). The specifications of these EDFA modules indicate the state of the art in optical components technology (1992) and therefore should not be interpreted as being optimal. In future developments, we may expect improvements in gain bandwidth and flatness, as well as the combination of very high gain ($>45\,$dB), low noise figure ($<5\,$dB), and near-quantum-limited QCE.

## 5.7 AMPLIFIER PHASE NOISE

From a classical description perspective, an optical amplifier can be viewed as a coherent device that enhances the input signal field amplitude $E_s^{in}$ by a factor $\sqrt{G}$. In the case of an ideal amplifier, free from spontaneous emission, the output signal field is then given by $E_s^{out} = \sqrt{G}E_s^{in}\exp(-ink_0L)$, where $L$ is the amplifier length, $n$ the refractive index and $k_0$ the signal wave vector in a vacuum. In this simplified picture, neither amplitude noise nor phase noise are generated by the amplifier. From a quantum description perspective, a noise-free amplifier can be viewed as a photon number multiplier, i.e., the output signal photon number is given by $n^{out} = Gn^{in}$. Chapter 3 shows that even in the absence of spontaneous emission, such a multiplication process (referred to as BD) increases the photon number variance, which is classically equivalent to field amplitude noise. In the case of real amplifiers with spontaneous emission, both semiclassical and quantum descriptions predict the same amplitude noise variance for the output signal. From the two descriptions, the amplitude noise can be viewed either as the result of amplified signal and ASE fields mixing in the detector, or as an intrinsic effect of photon number fluctuations in the amplified light (Chapter 3).

In addition to amplitude noise, *phase noise* can also be introduced in both descriptions. In the quantum description, phase noise be accounted for by use of specific phase operators, as shown in Chapter 3. For the prediction of phase expectation values (i.e., $\langle \widehat{\cos}\,\phi \rangle$, $\langle \widehat{\sin}\,\phi \rangle$) the quantum description requires the use of coherent states, not the photon number states used in Chapter 3, as in this last case the phase angle is completely undetermined [188]. In the semiclassical description, we assume that the amplifier generates an ASE field of expression $E_N(t)\exp\{i\phi(t)\}$, where $E_N(t)$ and $\phi(t)$ are random, time-varying amplitude and phase. Noise results from the coherent superposition of the amplified signal and the ASE field. From the phasor model, Figure 5.51, the in-phase component of the ASE field is responsible for amplitude noise (AM), while the quadrature component is responsible for phase noise (PM). But this physical explanation of phase noise is not complete. Indeed, the amplifying medium is also subject to instantaneous refractive index changes $\delta n(t)$, associated with the complex atomic susceptibility (Section 4.9). The refractive index changes are explained by the statistical fluctuations of the atomic populations ($N_1$ and $N_2$), induced by AM noise in the signal and ASE components. Consequently, an additional phase noise contribution $\delta\phi(t) = k_0\,\delta n(t)\,dz$ is generated in each elemental length $dz$ of the amplifying medium. By analogy, light-induced carrier density fluctuations in semiconductors also cause refractive index changes and phase noise, which results in laser line width enhancement (C. H. Henry [189]).

In this section, we consider the issue of EDFA-generated phase noise, as can be characterized experimentally and analyzed through a basic semiclassical description. This effect should not be confused with the phase noise resulting from Kerr nonlinearity in long-haul transmission systems (Chapter 7). In the case of EDFAs, the possible contributions of statistical refractive index changes to phase noise have not been investigated either theoretically or experimentally. For this reason, we shall consider only the phasor model, in which signal phase noise is assumed to result from the coherent superposition of a deterministic amplified signal field and a random phase ASE field (C. H. Henry [189]; G. J. Cowle [190]; and J. P. Gordon and L. F. Mollenauer [191]).

Let $P_{ASE} = n_{sp}h\nu B_o(G-1)$ be the mean ASE power in optical bandwidth $B_o$, measured at the amplifier output. The mean arrival time of ASE photons at the detector is then given by the reciprocal of the ASE photon flux $\phi_{ASE}$, i.e., $t_{ASE} = 1/\phi_{ASE} = h\nu/P_{ASE}$ or:

$$t_{ASE} = \frac{1}{n_{sp}B_o(G-1)} \tag{5.71}$$

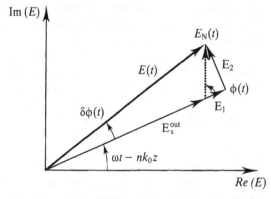

**FIGURE 5.51**  Semiclassical model for ASE-induced phase noise $\delta\phi(t)$. In this phasor diagram, $E_s^{out}$ is the amplitude of the amplified signal; $E_N(t)$ is the ASE field amplitude with in-phase and quadrature components $E_1$, $E_2$; $\phi(t)$ is a random phase angle; and $E(t)$ is the total output field amplitude.

Consider next the signal phase deviation $\delta\phi_1$ generated by a single spontaneous emission event, Figure 5.52. Assuming that a number of signal photons is detected during the time $t_{ASE}$ (i.e., $\langle n_s \rangle \gg 1$), and a random phase angle $\phi(-\pi < \phi < +\pi)$ with respect to the signal phase, the deviation $\delta\phi_1$ for any such event is given by [189]:

$$\tan(\delta\phi_1) \approx \delta\phi_1 = \frac{1}{\sqrt{\langle n_s \rangle}} \sin(\phi) \tag{5.72}$$

Averaging over a large number of events, we obtain the standard phase deviation:

$$\Delta\phi = \sqrt{\langle \delta\phi_1^2 \rangle} = \frac{1}{\sqrt{2\langle n_s \rangle}} \tag{5.73}$$

If $GP_s^{in}$ is the amplified signal power than the number of signal photons detected during $t_{ASE}$ is given by:

$$\langle n_s \rangle = \phi_{sig} t_{ASE} = \frac{GP_s^{in}}{h\nu} \frac{1}{n_{sp} B_o (G-1)} \tag{5.74}$$

From Eqs. (5.73) and (5.74), the standard phase deviation induced by spontaneous emission events during $t_{ASE}$ is then:

$$\boxed{\Delta\phi = \sqrt{\frac{n_{sp} h\nu B_o (G-1)}{2GP_s^{in}}}} \tag{5.75}$$

The above result is the same as obtained in the analysis of optical amplifier chains [191]. In this last analysis, the standard phase deviation is given by $\Delta\phi = 1/\sqrt{2Q}$, where $Q$ is the ratio of path averaged signal power to the ASE noise power accumulated along the chain (Chapter 7). In the case of a single amplifier, this expression reduces to the definition in Eq. (5.75).

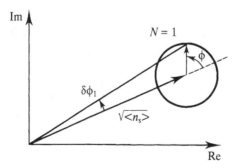

**FIGURE 5.52** Phasor diagram showing phase deviation $\delta\phi_1$ associated with the emission of a single spontaneous photon ($N = 1$) with random phase angle $\phi$, when the mean signal photon number is $\langle n_s \rangle$, corresponding to a mean signal field $\sqrt{\langle n_s \rangle}$.

The phase variance $\Delta\phi^2(T)$ corresponding to time $T$ of a signal source having a Lorentzian spectrum of FWHM $\delta v$ and characterized by Gaussian phase noise is given by (A. Yariv [192]):

$$\Delta\phi^2(T) = 2\pi\delta v T \qquad (5.76)$$

Using Eq. (5.75) and Eq. (5.76) for $T = 1/B_o$, we find the *equivalent line width broadening* (S. Ryu [193]):

$$\boxed{\delta v_{\text{eq}} = \frac{\Delta\phi^2(T)}{2\pi T} = \frac{n_{\text{sp}} h v B_o (G - 1)}{4\pi G P_s^{\text{in}}} B_o} \qquad (5.77)$$

The important conclusion contained in the above result is that the equivalent line width broadening is proportional to the ratio $B_o/\text{SNR}_o$, where $\text{SNR}_o$ is the optical SNR at the amplifier output. This conclusion is intuitively explained by the following: the higher the SNR, the lower the phase angle deviation introduced by ASE, according to Figure 5.51; the broader the bandwidth, the larger the number of randomly phased ASE photons that combine with the signal. Two other conclusions can be derived from Eq. (5.77). First, for gains sufficiently high ($G > 20$ dB), the equivalent line width broadening is independent of gain, as is the case with the SNR. Second, considering the average number of photons per bit $\bar{n} = P_s^{\text{in}}/2hvB_o$ input to the amplifier, the equivalent line width broadening is for high gains:

$$\delta v_{\text{eq}} \approx \frac{1}{8\pi} \frac{n_{\text{sp}}}{\bar{n}} B_o \qquad (5.78)$$

This result shows that in the case of optical preamplifiers ($G \gg 1$, $n_{\text{sp}} \approx 1$) with input signals as low as $\bar{n} = 100$ photons/bit, the line width broadening represents about 0.05% of the bit rate. In the case of an optical amplifier chain with $k$ elements (Type A), the spontaneous emission factor should be replaced by $kn_{\text{sp}}$ (Chapter 3). Thus, for a chain of $k = 100$ elements, the line width increase is about 5% of the bit rate, which represents a small effect. Chapter 7 shows that in long amplifier chains ($L > 100$ km), the dominant source of line width broadening is not the accumulated phase noise generated by the amplifiers, but the Kerr nonlinearity, Section 7.2.

For signals having fairly narrow line widths (e.g., $B_o < 50$ MHz), and in the case $\bar{n} \gg 1$ where Eq. (5.77) is applicable, a negligible broadening effect is predicted. Indeed, assuming $B_o = 30$ MHz, $P_s^{\text{in}} = 1$ $\mu$W, $n_{\text{sp}} = 1$, $\delta v_{\text{eq}} = 25$ Hz and $G = 20$ dB, we obtain from Eq. (5.77) a broadening of $\delta v_{\text{eq}} \approx 10$ Hz. This theoretical prediction contradicts experimental measurements obtained in similar conditions, where the line width broadening is observed to be three orders of magnitude greater, Figure 5.53 (G. J. Cowle et al. [194]). For such a measurement, the experiment used a matched-path Mach–Zehnder interferometer, one arm of which included an EDFA followed by an acousto-optic frequency shifter and a 30 MHz DFB laser source for the signal. The beat spectrum of the signal output was then monitored with an RF spectrum analyzer. This measurement technique effectively deconvolves the input signal spectrum from the broadening component introduced by the amplifier, [190] and [194]. The model developed in [190] assumes that the effect of line width broadening comes from the addition of randomly phased photons only during the time $t_{\text{ASE}}$ (as opposed to the time $1/B_o$). The phase variance of the amplified signal during $t_{\text{ASE}}$ is, according to Eqs. (5.71) and (5.76):

$$\Delta\theta^2(t_{\text{ASE}}) = 2\pi\Delta v_s t_{\text{ASE}} = \frac{2\pi\Delta v_s}{n_{\text{sp}}B_o(G-1)} \tag{5.79}$$

The model shows that the amount of spectral broadening $\delta v$ due to the amplifier relative to the input signal line width $\Delta v_s$ can be approximated by the ratio $\Delta\phi/\Delta\theta$, which from Eqs. (5.75) and (5.79) takes the form:

$$\frac{\delta v}{\Delta v_s} \approx \frac{\Delta\phi}{\Delta\theta} = \sqrt{\frac{m^2 n_{\text{sp}}^2 h v \Delta v_s (G-1)^2}{4\pi G P_s^{\text{in}}}} \tag{5.80}$$

with $m = B_o/\Delta v_s$. According to Eq. (5.80), and compared to Eq. (5.77), the spectral broadening increases like the square root of the gain to optical SNR ratio. Considering signals of relatively narrow line widths, this result is consistent with the experimental measurement shown in Figure 5.53. Indeed, assuming $G = 17$ dB, $n_{\text{sp}} = 1$, $m = 1$, and an input signal at $\lambda = 1.53\,\mu$m with line width $\Delta v_s = 30$ MHz, we obtain from Eq. (5.80) the values $\delta v \approx 35$ kHz for power $P_s^{\text{in}} = 10\,\mu$W and $\delta v \approx 11$ kHz for power $P_s^{\text{in}} = 100\,\mu$W. The amount of broadening and the slow decay in $1/\sqrt{P_s^{\text{in}}}$ predicted by Eq. (5.80) are in fair agreement with the experimental data. At lower signal powers ($P_s^{\text{in}} < 10\,\mu$W) there is a difference between theory and experiment. It is not attributable to the approximation $\langle n_s \rangle \gg 1$ because, for $P_s^{\text{in}} = 1\,\mu$W, the number of signal photons detected during $t_{\text{ASE}}$ is $\langle n_s \rangle \approx 10^5$. This difference is yet to be interpreted.

Another experimental measurement of spectral line width broadening used a delayed self-heterodyne Mach–Zehnder interferometer as the measurement apparatus, and an Er-doped ring fiber laser with $\Delta v_s \leqslant 1.4$ kHz line width as a signal source (H. Okamura and K. Iwatsuki [195]). After passing through a 1480 nm pumped EDFA

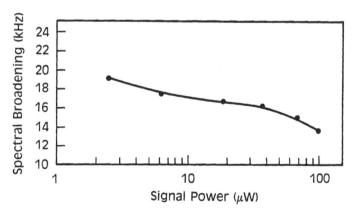

**FIGURE 5.53** Experimental measurement of signal line width broadening $\delta v$ versus input signal power $P_s^{\text{in}}$, obtained with a matched-path Mach–Zehnder interferometer using a DFB laser source of 30 MHz bandwidth and an EDFA with gain $G = 17$ dB. From [194], reprinted with permission from *Electronics Letters*, 1990.

with gains $G = 10–30$ dB ($P_s^{out} = 1$ mW), no signal broadening was observed, consistent with the previous model. Indeed, using Eq. (5.80) and the experimental parameters $\lambda = 1.55$ $\mu$m, $\Delta v_s = 1.4$ kHz, $G = 30$ dB, $n_{sp} \approx 1.6$, $P_s^{in} = 1$ $\mu$W, the maximum line width broadening corresponding to these conditions is $\delta v = 0.3$ Hz, far beyond the resolution of the self-heterodyne measurement apparatus.

The experimental measurements shown in [194]–[195] demonstrate that the effect of phase noise induced by the quadrature component of ASE in EDFAs is not significant, even for very narrow line widths. EDFAs can be used advantageously for coherent detection, either as LO power boosters (ELOS configuration) or as signal preamplifiers (S. Ryu and Y. Horiuchi [196]), which in both cases can enhance receiver sensitivity (Chapter 3, Section 6). Likewise, EDFA-induced phase noise is a negligible source of penalty for coherent systems based on long optical amplifier chains (S. Ryu [193]).

## 5.8 EFFECTS OF REFLECTIONS AND RAYLEIGH BACKSCATTERING

The EDFA performance in gain, saturation, and noise described so far in this chapter assumes total isolation of the amplifier from any external reflection sources. The condition of total isolation from backscattered light is approached experimentally by placing Faraday isolators at both EDFA ends. Without then, a finite ASE (forward and backward) and amplified signal would be fed back into the EDFA, generating an excess noise background and possibly causing amplifier gain saturation and laser oscillation. In actual fiber systems, many discrete reflection sources are present in the light path, such as: connectors, splices, couplers, beam expanders, cleaved fiber ends of taps, optical switches, semiconductor sources, and detector facets. The effect of reflection from discrete components can be suppressed by: index matching, antireflection coating, polarization rotation, mismatched coupling, macrobending loss, optical filters, and isolators. In systems where the EDFAs are separated by long strands of transmission fibers (e.g., $L > 50$ km), the effect of discrete reflections is also alleviated by the fiber return loss (i.e., $< -20$ dB for a 100 km round-trip). But there is a cause of feedback that is distributed over both transmission fiber and EDF which is Rayleigh backscattering (RB).

The effect of internal Rayleigh backscattering in EDFAs can be characterized by backscatter power measurements obtained in the high gain regime (S. L. Hansen et al. [197]). In the experiment described in [197], an EDFA with a small-signal gain $G = 51$ dB ($\lambda = 1.53$ $\mu$m) was used. The measured backscatter power $P_{back}$ was found to be 41 dB higher than the signal input $P_s^{in}$. This backscattering effect is equivalent to having an output mirror of reflectivity $R_{eff}$, defined through the relation $P_{back} = R_{eff} G P_s^{out} = R_{eff} G^2 P_s^{in}$. accounting for component reflections, the effective reflectivity was found in this measurement to be $R_{eff} = -59$ dB. A theoretical simulation of the EDFA with distributed backscattering made it possible to associate the measured reflectivity to a backscattering coefficient value $S = -63$ dB (relative to $1$ m$^{-1}$). This coefficient is nearly $+10$ dB higher than obtained in standard SM fibers, which indicates that Rayleigh scattering is enhanced in EDFAs [197]. A series of measurements obtained with different EDFA types (i.e., aluminosilicate and germanosilicate) followed by a detailed theoretical interpretation of the evaluated scattering coefficients $S$ led to two conclusions: (a) the observed RB enhancement is mainly attributed to the high refractive index difference or NA, which increases the fraction of backscattered power recaptured by the fiber (as compared to standard fibers); and (b) Rayleigh scattering is not influenced by $Er^{3+}$ doping (up to $Er^{3+}$ concentrations 900 ppm), as also observed in $Nd^{3+}$ doped fiber (M. E. Fermann et al. [198]).

The feedback effect induced by internal RB on forward and backward ASE actually limits the maximum gain achievable in EDFAs, determined by conditions of laser oscillation. The maximum gain $G_{max}$ can thus be estimated by the relation $R_{eff} G_{max} = 1$, which corresponds to the oscillation threshold. The Rayleigh reflectivity is approximated by the relation $R_{eff} \approx SL/2 \log(G_{max})$, which L is the EDFA length. The upper gain limit of the different EDFA types investigated in [197] was found to be in the range $G_{max} = 57$–$70$ dB [197]. This range is 5–20 dB above the highest gains reported to date (1992) for amplification regimes limited by ASE self-saturation. The highest EDFA gain achieved to date, using midway isolators at optimum locations to suppress self-saturation is $G = 54$ dB (R. I. Laming et al. [142]; S. L. Hansen et al. [197]), close to the range of maximum gains $G_{max}$. For gains approaching $G = 60$ dB, the effect of discrete reflections originating from fiber splices could be important. This justifies the use of an optical isolator.

A way to increase the RB-limited gain $G_{max}$ is to use EDFAs with lower NAs, at the expense of a reduced dB/mW efficiency [142]. The dB/mW efficiency of EDFAs increases like $(NA)^{\alpha}$ with $\alpha = 1.6$–$1.7$ depending upon the confinement factor [185], whereas the recapture coefficient $S$ increases like $(NA)^2$ (E. Brinkmeyer, [199]–[200]; M. Nakazawa [201]. Thus, decreasing the NA with respect to the optimum value given by design optimization could significantly reduce Rayleigh scattering without significant dB/mW penalty.

We focus next on the effect of reflections in preamplifier/receiver systems, which can be either discrete (optical components) or distributed (RB). In a given fiber path, the interference effect produced by multiple reflections converts signal phase noise into intensity noise. In communications systems, this intensity noise causes BER penalty and in some cases BER floors (J. L. Gimlett et al., [202]–[203]). When optical amplifiers are positioned between the reflection sources, the resulting intensity noise is drastically enhanced, as the reflected signals experience gain over multiple passes. It is therefore important to evaluate the corresponding power penalty for the system.

We consider first the effect of discrete reflections at both amplifier ends (reflectivities $R_1$, $R_2$). Assuming that signal and reflections have matched polarizations and interfere incoherently (as the EDFA length is usually greater than the signal coherence length), the resulting photon statistics variance $\sigma_{ref}^2$ is given by:

$$\sigma_{ref}^2 = \eta^2 G^4 R_{eff}^2 \langle n(0) \rangle^2 \qquad (5.81)$$

where $R_{eff} = R_1 R_2$, $\langle n(0) \rangle = P_s^{in}(1)/h\nu_s B_o$ is the mean EDFA input photon number for 1 symbols, and $\eta$ represents the attenuation between EDFA and receiver [203]. Compare the reflection-induced noise $\sigma_{ref}^2$ with the signal–ASE beat noise: $\sigma_{ref}^2 = 2\eta^2 GN\langle n(0) \rangle \approx \eta^2 G^2 n_{sp}\langle n(0) \rangle$ (Chapter 2). If a Gaussian noise PDF is assumed for 0 and 1 symbols and the decision threshold is placed midway between them, the power penalty at the receiver due to both s–ASE beat noise and reflection-induced noise takes the form (J. L. Gimlett et al. [204]):

$$P(dB) = -5\log\left[1 - 144\left(\frac{n_{sp}}{\bar{n}} + G^2 R_{eff}^2\right)\right] \tag{5.82}$$

where $\bar{n}$ is the average number of photons/bit. For this model, a zero penalty corresponding to BER = $10^{-9}$ is achieved for $\bar{n} = 144n_{sp}$ photons/bit when no reflections occur ($R_{eff} = 0$). Equation (5.82) shows that in the case of large input signals ($\bar{n} \gg n_{sp}$), the penalty is less than 1 dB for $RG < 0.035$. Thus, for an EDFA gain $G = 30$ dB, a penalty $P < 1$ dB is incurred when $R_{eff} < 3 \times 10^{-5}$, corresponding to reflection coefficients at both EDFA ends $R_1 = R_2 < 1.5 \times 10^{-5}$ or $-23$ dB. The study of [204] shows experimental evidence of BER floors occurring in the conditions $GR_{eff} = 0.11$ and $GR_{eff} = 0.08$; we conclude from these measurements and from the above analysis that, for achieving power penalties $P < 1$ dB, EDFA-based systems having discrete reflections should operate in conditions $GR_{eff} < 0.02$.

We consider next the effect of Rayleigh backscattering in EDFA/receiver systems, for direct and coherent digital signal photodetection (J. L. Gimlett et al. [205]; R. K. Staubli et al. [206]). Figure 5.54 schematically shows the path followed by signal reflections, as initiated by RB on both sides of the EDFA at discrete locations ($i = 1, \ldots, N$ and $j = 1, \ldots, M$) in the fiber path. For simplicity, we analyze first the interferometric noise generated by the doubly reflected signal shown in the figure [203]. Let $R_1$ and $R_2$ be the effective Rayleigh mirror reflectivities at locations $i$ and $j$, and $T$ the single-pass transmission from $j$ to $i$. Assuming no modulation, the output signal field at location $i$ is given by:

$$E_0(t) = \sqrt{TP_s^{in}} \exp\{i\omega t + \phi(t)\} \equiv \sqrt{TP_s^{in}}e(t) \tag{5.83}$$

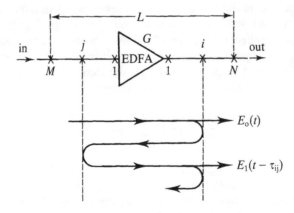

**FIGURE 5.54** Analysis of the effect of Rayleigh backscattering with an EDFA placed between two strands of fiber, showing the paths followed by signal and doubly reflected backscatter and corresponding notations.

where $\phi(t)$ is the signal laser phase noise. The doubly reflected signal field can be expressed as:

$$E_1(t) = \sqrt{R_1 R_2 T^3 P_s^{in}}\, e(t - 2\tau_{12}) \qquad (5.84)$$

where $\tau_{12}$ is the transit time between $i$ and $j$. The contributions of higher order reflections proportional to $(R_1 R_2)$ can be neglected if the reflection coefficients are assumed to be small. In this case, the power detected at the receiver end is given by $P_s(t) = |E_0(t) + E_1(t)|^2$, which with Eqs. (5.83) and (5.84) and neglecting the term in $(R_1 R_2)$, yields:

$$
\begin{aligned}
P_s(t) &= T P_s^{in} + 2T^2 \sqrt{R_1 R_2}\, \mathcal{R}e\{e(t) \cdot e(t - 2\tau_{12})\} \\
&\approx T P_s^{in}\{1 + 2R_{12}\cos[\omega\tau_{12} + \Delta\phi(t, \tau_{12})]\}
\end{aligned}
\qquad (5.85)
$$

with

$$R_{12} = T\sqrt{R_1 R_2}\,\hat{u}(t) \cdot \hat{u}(t - 2\tau_{12}) \qquad (5.86)$$

$$\Delta\phi(t, \tau_{12}) = \phi(t) - \phi(t - 2\tau_{12}) \qquad (5.87)$$

In Eq. (5.86), the unit vectors $\hat{u}(t)$ and $\hat{u}(t - 2\tau_{12})$ are the polarization directions of the transmitted and doubly reflected signals, respectively. They are introduced in order to account for the effect of polarization wandering along the fiber path. Now we model the effect of RB by multiple reflections, as shown in Figure 5.54. We obtain for the output signal power [205]:

$$P_s(t) \approx T P_s^{in}\left\{1 + 2\sum_{i=1}^{N}\sum_{j=1}^{M} R_{ij}\cos[\omega\tau_{ij} + \Delta\phi(t, \tau_{ij})]\right\} \qquad (5.88)$$

In Eq. (5.88), $T = G\exp(-\alpha L)$ is the transmission from point $M$ to point $N$, and the effective reflexion coefficient $R_{ij}$ is given by the expression:

$$R_{ij} = S\alpha_s \Delta L G \exp\{-\alpha(z_i - z_j)\}\hat{u}(t) \cdot \hat{u}(t - 2\tau_{ij}) \qquad (5.89)$$

where $\alpha_s$ is by definition the scattering coefficient, $S$ is the fraction of scattered power recaptured by the fiber, and $\Delta L = L/2N$ is the mean distance over which single scattering events occur. The effect of reflections occurring on one side of the EDFA was neglected, since the corresponding $R_{ij}$ coefficients do not have the factor $G$.

The signal relative intensity noise (RIN) is given by the Fourier transform of the autocovariance function $\langle P_s(t)P_s(t - \tau)\rangle$. Assuming a signal with Lorentzian power spectrum $\mathcal{L}(f)$ and incoherent interference between terms in Eq. (5.88), the RIN is given by [205]:

$$\text{RIN}(f) = \mathcal{L}(f)\sum_{i=1}^{N}\sum_{j=1}^{M}\langle R_{ij}^2\rangle \equiv \mathcal{L}(f)R_{eff}^2 \qquad (5.90)$$

where $R_{eff}$ is an effective mirror reflection coefficient. The double summation in Eq.

(5.90) can be calculated from Eq. (5.89) as follows:

$$R_{\text{eff}}^2 = \sum_{i=1}^{N} \sum_{j=1}^{M} \langle R_{ij}^2 \rangle = \frac{1}{2}(GS\alpha_s)^2 \sum_{i=1}^{N} \sum_{j=1}^{M} \exp\{-2\alpha(z_i - z_j)\}\Delta L^2$$

$$= \frac{1}{2}\left(GS\alpha_s \int_0^{L/2} \exp(-2\alpha z)\, dz\right)^2 = \frac{1}{2}\left(\frac{GS\alpha_s}{2\alpha}[1 - \exp(-\alpha L)]\right)^2 \tag{5.91}$$

where we used $\Delta L \rightarrow dz$ and the property $\langle [\hat{u}(t) \cdot \hat{u}(t - 2\tau_{ij})]^2 \rangle \equiv \langle \cos^2 \theta \rangle = 1/2$, corresponding to random polarization scrambling. From Eqs. (5.90) and (5.91), the RIN can then be put into the form:

$$\boxed{\text{RIN}(f) = \mathscr{L}(f)G^2 R_{\text{eff}}^2 = \mathscr{L}(f)G^2\left(\frac{R_{\text{bs}}}{\sqrt{2}}\right)^2} \tag{5.92}$$

where $R_{\text{bs}}$ is defined as:

$$R_{\text{bs}} = \frac{S\alpha_s}{2\alpha}[1 - \exp(-\alpha L)] \tag{5.93}$$

Comparing the result of Eq. (5.92) with that of [203], we find that the effect of doubly reflected RB is equivalent to a double-pass direct reflection with coefficient $R_{\text{eff}} = R_{\text{bs}}/\sqrt{2}$ [205]. For SM fibers and $\lambda = 1.55\,\mu m$ signals, the ratio $S\alpha_s/2\alpha$, which represents the limit of $R_{\text{bs}}$ at long fiber lengths $L$, is typically $-31\,\text{dB}$ (DSF) to $-34\,\text{dB}$ (SMF).

Assuming Gaussian PDFs for the symbols 0 and 1 with a decision threshold midway between them, the power penalty induced by RB in a direct detection receiver is given by the expression [205]:

$$P(\text{dB}) = -5\log[1 - 144G^2 R_{\text{eff}}^2] = -5\log[1 - 72G^2 R_{\text{bs}}^2] \tag{5.94}$$

In coherent heterodyne detection, the penalty takes the form [206]:

$$P(\text{dB}) = -5\log[1 - 36\xi G^2 R_{\text{eff}}^2] \tag{5.95}$$

with $\xi = 1/3$ for PSK modulation, $\xi = 0.54$ or $\xi = 0.45$ for FSK modulation (index $n = 1$ or $n = 2$), and $\xi = 5/6$ for ASK modulation. For FSK, the modulation index is defined through $n = 2\Delta f/B$, where $2\Delta f$ is the frequency separation between the space and mark symbols and $B$ is the bit rate.

For direct detection systems, Eq. (5.94) shows that for achieving a power penalty $P < 1\,\text{dB}$, we must have $GR_{\text{eff}} < 0.035$. Given the representative value $R_{\text{bs}} = -32\,\text{dB}$, the EDFA gain must be at most $G = 20.6\,\text{dB}$. For coherent detection systems, the same conditions yield $G = 23.2\,\text{dB}$ for PSK, $G = 22.2\,\text{dB}$ or $G = 22.6\,\text{dB}$ for FSK ($n = 1$ or $2$), and $G = 21.2\,\text{dB}$ for ASK. The main conclusion of the two analyses of

[205] and [206] can therefore be stated as follows: *in EDFA chains where no optical isolators are used, the effect of Rayleigh backscattering limits the maximum gain in each amplifier to about 20 dB for direct detection systems and 21–23 dB for coherent systems.*

In AM and FM analog photodetection (Section 3.8) the increase in $G^2$ of the RIN can cause severe degradation of the CNR, resulting in large power penalties. The required signal power corresponding to a desired CNR can be calculated from Eq. (3.127), using Eq. (5.92) to define RIN. In the case of broadcast systems, the distance between EDFAs is limited to relatively short fiber lengths, which reduces somewhat the coefficient $R_{bs}$, as shown in Eq. (5.93). For instance, a fiber length $L = 10$ km with $\alpha = 0.2$ dB/km corresponds to coefficients $R_{bs} = -35$ dB for DSF and $R_{bs} = -38$ dB for SMF. In broadcast systems, EDFAs are used mainly for splitting loss compensation, so $G = 20$ dB at most for each splitting stage. These two considerations show that for analog broadcast systems, the factor $(GR_{eff})^2$ involved in the RIN is at most of the order of $-33$ dB to $-40$ dB. For a complete characterization of CNR degradation in such systems, we should also include the deleterious effects of discrete reflections from splices and optical components (H. Yoshinaga et al. [207]) and of ASE noise accumulation (Chapter 3).

Discrete reflections and distributed Rayleigh backscattering can be modeled with the single parameter $R_{eff}$, representing an effective mirror reflectance. An important consequence of the existence of reflections on both sides of the EDFA, not considered so far, is the increase of the amplifier noise figure. This increase can be explained by the constructive interference of forward and backward ASE, as they are fed back into the amplifier. The case of a fully coherent interference leads to the regime of Fabry–Perot (FP) amplification, as opposed to traveling-wave amplification. In FP amplifiers, gain and ASE spectra exhibit a periodic structure with a characteristic peak spacings $\delta v = c/2nL$, corresponding to the free spectral range of the resonator. In the case of fiber amplifiers, for which lengths are typically greater than 1 m (Section 5.6), this periodicity is less than 100 MHz. Small reflections correspond to a low cavity finesse and consequently to a negligible gain spectrum ripple. But the coherent interference of ASE reflected back into the amplifier results a higher noise background, as shown in the analysis of semiconductor FP amplifiers (E. I. Gordon [208]; Y. Yamamoto, [209]–[211]). This analysis shows that FP effects cause the ASE noise to be enhanced by a certain factor $\chi$, called the *excess noise factor*, of expression:

$$\chi \approx \frac{1 + R_1 G}{1 - R_1} \frac{G - 1}{G} \tag{5.96}$$

where $R_1$ represents the reflectivity of the input side of the amplifier. For high gains and low reflectivities, the excess noise factor is therefore $\chi \approx 1 + R_1 G$. Following the analysis developed in Chapter 2 and taking the factor $\chi$ into account in the signal–spontaneous beat noise component $\sigma_{s-ASE}^2$, the amplifier optical noise figure can be expressed as:

$$F_o = \frac{1 + 2n_{sp}\chi(G - 1)}{G} \tag{5.97}$$

which in the high gain limit $G \gg 1$ takes the form:

$$F_o(G \gg 1) \approx 2n_{sp}\chi = 2n_{sp}(1 + R_1 G)$$    (5.98)

The above result shows that when reflections are present at the input side of the amplifiers, the NF increases by a factor $(1 + R_1 G)$ with respect to the ideal case. Thus, a product $R_1 G = 1$ causes a 3 dB increase in NF. Considering a typical effective Rayleigh backscattering reflectivity $R_1 = R_{bs} = -31$ dB to $-34$ dB for DSF and SMF fibers, respectively, a 3 dB increase in NF is thus predicted for EDFA gains $G = +31$ to $+34$ dB. As the receiver sensitivity $\bar{P}$ for direct detection systems is approximately proportional to the optical noise figure $F_o$, Eq. (3.80). This NF increase also corresponds to a power penalty. It can be confirmed experimentally by measuring the sensitivity degradation obtained when a long strand of fiber (e.g., $L > 10$ km) is placed before the EDFA without optical isolation (N. Henmi et al. [212]). Considering that optical preamplifier/receivers must usually be operated at high gains ($G > 25$ dB) for achieving s–ASE beat-noise-limited detection, the above shows that an optical isolator must be used at the EDFA input to prevent sensitivity degradation by RB.

## 5.9  TRANSIENT GAIN DYNAMICS

This section reviews the EDFA characteristics in the time domain. Such characteristics concern first the transient effects related to the specific gain dynamics of three-level systems. Another issue concerns the amplification of short pulses in the high gain regime, and the related peak power enhancement. Other time domain effects, which are relevant to different regimes of picosecond, femtosecond, and soliton pulse amplification, involve fiber dispersion and nonlinearity. The EDFA characteristics associated with these particular regimes are considered in subsequent sections. In this section, we compare in detail the theoretical analysis of transient gain dynamics in EDFAs with experimental measurements. The most important conclusion of this analysis is that *EDFAs have very slow gain dynamics, i.e., the characteristic saturation and recovery times are, for typical operating conditions, in the range 100 μs to 1 ms. As a result, EDFAs are intrinsically immune to the effect of crosstalk at high data rates,* as confirmed by experimental observation.

The analysis developed in Chapter 1 leads to the general EDFA model. It assumes a steady state amplification regime for which the atomic populations are time invariant. When considering time dependent pump and signals, the model must be developed to predict the associated gain response. We ask four questions. (a) How long does it take to bring an EDFA into steady state conditions from the time the pump is turned on? (b) What is the time dependence of gain saturation in response to signal power changes? (c) How how does it take for the gain to recover to steady state conditions? (d) What are the transient dynamics of two-signal saturation? Automatic gain control (AGC) suppresses such transients and is addressed in Chapter 6.

The theoretical analysis of the transient regime outlined below will be restricted, for simplicity, to the simple case where the transverse power distributions of the pump and the signal(s) can be overlooked (confined or quasiconfined Er-doping). The conclusions of this analysis also apply to the general case, which can be treated by an identical method. Bidirectional pumps and signals, and the regime of self-saturation increase the complexity of two-boundary-value problems and will not be considered here. Likewise, the effects of pump ESA and CET will be overlooked in this analysis, as they are of minor interest in the transient gain regime. As a starting point, we rewrite below the first three equations of this book, which describe the time evolution of the atomic population densities $N_1$, $N_2$, and $N_3$, with the notation to Figure 1.1:

$$\frac{dN_1}{dt} = -R_{13}N_1 + R_{31}N_3 - W_{12}N_1 + W_{21}N_2 + A_{21}N_2 \qquad (5.99)$$

$$\frac{dN_2}{dt} = W_{12}N_1 - W_{21}N_2 - A_{21}N_2 + A_{32}N_3 \qquad (5.100)$$

$$\frac{dN_3}{dt} = R_{13}N_1 - R_{31}N_3 - A_{32}N_3 \qquad (5.101)$$

In the above equations, the time dependence of the populations $N_1$, $N_2$, $N_3$, the pumping rates $R_{13,31}$ and the stimulated emission rates $W_{12,21}$ are not shown for clarity. By conservation of the number of ions, we have $N_3(z, t) = \rho_0 - N_1(z, t) - N_2(z, t)$, and therefore, Eq. (5.101) is redundant. On the other hand, the pump and signal rate equations (forward pumping) take the form:

$$\frac{\partial P_p(z, t)}{\partial z} - \frac{1}{v_g}\frac{\partial P_p(z, t)}{\partial t} = \alpha_p\Gamma_p(N_3 - N_1)P_p(z, t) \qquad (5.102)$$

$$\frac{\partial P_s(z, t)}{\partial z} - \frac{1}{v_g}\frac{\partial P_s(z, t)}{\partial t} = \alpha_s\Gamma_s(\eta_s N_2 - N_1)P_s(z, t) + 2\alpha_s\Gamma_s\eta_s N_2 P_0 \qquad (5.103)$$

where $v_g$ is the group velocity, assumed identical for pump and signal. The effect of two-signal saturation can be analyzed by solving two simultaneous equations of the type (5.103), i.e., one for each signal of power $P_{s1}$ and $P_{s2}$. Likewise, the forward ASE spectrum can be resolved by solving $k$ equations of the type (5.103), i.e., one for each wavelength interval with power $P_{sk}$. If no self-saturation is assumed (as is the case for $G < 25$ dB), the term proportional to $P_0$ in Eq. (5.103) can be neglected.

A first simplification is introduced in Eqs. (5.102) and (5.103) by assuming a priori that the time dependent changes of the atomic populations are slow in comparison to the transit time of the pump and signal through the EDFA. For this reason, the partial derivatives with respect to time can be neglected (we shall not neglect them when we consider time dependent changes associated with dispersion and nonlinearities in the EDFA). The system of partial differential Eqs. (99)–(103) can be integrated numerically through several possible methods [213]. As the system considered has smooth solutions and its boundary values are simple to evaluate, the finite difference method is straightforward to implement. Appendix S details this method for the above equations. The space and time are decomposed into a grid of $M \times N$ discrete bins $\Delta z$ and $\Delta t$, respectively. The space equations are integrated iteratively for each time $t = k\Delta t$ $(k = 1, \ldots, N)$. At each space $(k\Delta z \to (k + 1)\Delta z)$ or time $(k\Delta t \to (k + 1)\Delta t)$

increment, the corresponding variables are evaluated through their first order derivatives. The initial condition of the system for the atomic populations (i.e., $N_1(z, 0)$, $N_2(z, 0)$) is given by the steady state solutions defined in Eqs. (1.8) and (1.9). It is also possible to integrate the system iteratively in time for fixed space coordinates, since the boundary conditions $N_1(0, t)$, $N_2(0, t)$ can also be determined analytically, as shown in Appendix T.

We consider first the case where a saturating pulse having a square shape is input to the EDFA. Figure 5.55 shows experimental measurements and theoretical simulations (calculated as per Appendix S) of the output signal pulse, as obtained in the case of an 514.5 nm pumped EDFA with unsaturated gain near $G = 30$ dB at the peak wavelength $\lambda_s = 1531$ nm (E. Desurvire et al. [214]). In Figure 5.55a, the input signal peak power is kept constant ($P_s^{in}(\lambda_s) = 20 \, \mu W$) while the pump is varied. A power spike, or overshoot, is observed at the leading edge of the output signal pulse, which is followed by steady state gain conditions. The spike is also seen to narrow as the pump power is increased (corresponding to steady state gain increase). In Figure 5.55b, the input signal power is varied while the pump power is kept constant ($P_s^{in} = 210$ mW). As in the previous case, the spike is observed to narrow as the signal power is increased (corresponding to steady state gain decrease). The results of Figures 5.55a,b show that the time constants associated with EDFA transient gain dynamics are functions of both pump and signal powers.

In Figure 5.55c, the output saturating signal pulse is shown along with the output of a weak CW signal at $\lambda_s = 1537$ nm (displayed on a different vertical scale). Following the injection of the saturating input signal into the EDFA, the gain at the probe wavelength is seen to drop (by about 3B), with the same characteristic time as observed in the saturating signal overshoot. As the saturating signal is turned off, the probe gain is seen to recover to its initial value ($G = 30.5$ dB), with a longer characteristic time. The important conclusion of the three types of measurements shown in Figure 5.55 is that in EDFAs, *transient effects of gain saturation and recovery typically occur on a 100 μs–1 ms time scale*. The dynamics of gain saturation and recovery, as predicted by the theory, are seen to be in good qualitative and quantitative agreement with the experiment. For the theoretical computation, an effective input ASE signal with power $P_N^{in} = 5.5 \, \mu W$ to account for self-saturation was used [214]. By this method, and using best-fitting values for the ratios $P_p^{in}/P_{sat}(\lambda_p)$, a good prediction of the time dependent gains could be obtained for all pump and signal power regimes, as shown in Figure 5.55. This agreement is remarkable, considering that the effects of multimode pump propagation and pump ESA were overlooked in the model. The theoretical predictions were obtained with the decay time parameters $\tau_{21} = 10$ ms and $\tau_{32} = 10 \, \mu s$. For the decay time of level 3, the best-fitting value $\tau_{32} = 10 \, \mu s$ is consistent with the smallest nonradiative rate ($A_{32} \approx 15 \times 10^4$ or $\tau_{32} \approx 7 \, \mu s$) observed in Er:silica glass, as shown in the energy level diagram of Figure 4.7.

The same theory/experiment comparison as described above could be done for the transient dynamics of EDFAs pumped near 810 nm and 980 nm. They have not been investigated to date (1992). As EDFA modeling in the case of single-mode pumping is quite accurate, such a comparative study would make it possible to determine with equal accuracy the nonradiative decay time $\tau_{32}$ associated with levels $^4I_{9/2}$ and $^4I_{11/2}$, for different types of glass hosts. This data could then be analyzed and interpreted with the multiphonon decay theory of RE:glass described in Chapter 4. Using the model of [214] with fully resolved spectra, the transient behavior of

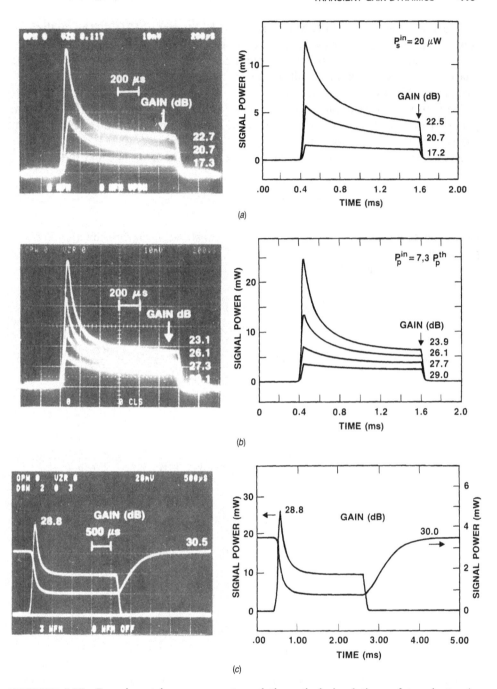

**FIGURE 5.55** Experimental measurements and theoretical simulations of transient gain dynamics in a 514 nm pumped EDFA with a square input signal pulse at wavelength $\lambda_s = 1531$ nm: (a) output signal pulse measured with constant input signal peak power $(P_s^{in}(\lambda_s) = 20\ \mu W)$ while the pump is varied; (b) output signal pulse measured with constant pump power $(P_p^{in} = 210$ mW) while the input signal power is varied; (c) the output saturating signal pulse and output of weak CW signal at $\lambda_s = 1537$ nm (displayed on a different vertical scale), showing transient saturation and recovery. From [214] © 1989 SPIE.

ASE can also be analyzed (Y. Maigron and J. F. Marcerou [215]). The analysis of transient gain requires us to use an empirical or best-fitting value for the parameter $\tau_{32}$.

In the general case, the system of equations (5.99)–(5.103) does not yield simple analytical solutions, making it difficult to physically interpret the observed transient dynamics. This system of equations, however, can be solved exactly at the fiber input end ($z = 0$), in the case of pump and signals with square pulse envelopes (E. Desurvire [215]). This model provides a basic understanding of the transient gain regime in EDFAs. Appendix T shows that the atomic populations $N_i(0, t)$ ($i = 1, 2, 3$) in a three-level medium input with square pulse pump and signals at $t = 0$ take the form:

$$N_i(t) = a_i e^{-t/t_1} + b_i e^{-t/t_2} + c_i \qquad (5.104)$$

where the coefficients $a_i$, $b_i$, $c_i$ and the time constants $t_1$, $t_2$ are functions of the parameters $q = P_p(0, t)/P_{sat}(\lambda_p)$, $p = P_s(0, t)/P_{sat}(\lambda_s)$ and $\tau_{21}$, $\tau_{32}$. If we consider the specific case Er:silica glass, for which $\tau_{32} \ll \tau_{21}$ (Chapter 4), Eq. (5.104) can be simplified:

$$N_1(0, t) = \left( N_1^0 - \frac{1 + W_{21}\tau_{21}}{1 + p + q} \right) e^{-t/t_1} + \frac{1 + W_{21}\tau_{21}}{1 + p + q} \qquad (5.105)$$

$$N_2(0, t) = \left( \frac{1 + W_{21}\tau}{1 + p + q} - N_1^0 \right) e^{-t/t_1} + (N_1^0 + N_2^0 - 1)e^{-t/t_2} + \frac{R\tau + W_{12}\tau}{1 + p + q} \qquad (5.106)$$

$$N_3(0, t) = (1 - N_1^0 - N_2^0)e^{-t/t_2} \qquad (5.107)$$

where the characteristic time constants $t_1$, $t_2$ are now defined by:

$$t_1 = \frac{\tau_{21}}{1 + \dfrac{P_p^{in}(0)}{P_{sat}(\lambda_p)} + \dfrac{P_s^{in}(0)}{P_{sat}(\lambda_s)}} = \frac{\tau_{21}}{1 + R\tau_{21} + (W_{12} + W_{21})\tau_{21}} \equiv \frac{\tau_{21}}{1 + p + q} \qquad (5.108)$$

$$t_2 = \frac{\tau_{32}}{1 + \dfrac{\tau_{32}}{\tau_{21}} \dfrac{P_p^{in}(0)}{P_{sat}(\lambda_p)}} = \frac{\tau_{32}}{1 + \dfrac{\tau_{32}}{\tau_{21}} R\tau_{21}} \equiv \frac{\tau_{32}}{1 + \dfrac{\tau_{32}}{\tau_{21}} q} \qquad (5.109)$$

In Eq. (5.105)–(5.107), the populations are normalized to the peak ion density $\rho_0$, i.e., $N_1 + N_2 + N_3 = 1$, and the populations $N_1^0$, $N_2^0$, $N_3^0$ represent the initial conditions at $t < 0$.

The dynamics of the three-level gain medium can be easily interpreted in several basic cases, using the results of Eqs. (5.105)–(5.109). We consider the following: (a) the pump is turned on with all ions initially in the ground state, (b) the signal is turned on with all ions initially in the ground state, (c) the signal is turned on with ions initially in both ground and excited states. The time dependent populations corresponding to these three cases are then given by:

(a) $\quad N_1(t) = \dfrac{q}{1 + q} e^{-t/t_1} + \dfrac{1}{1 + q}, \; N_2(t) = \dfrac{q}{1 + q} (1 - e^{-t/t_1}), \; t_1 = \tau_{21}/(1 + q)$

$$(5.110)$$

(b)  $N_1(t) = \dfrac{W_{12}\tau_{21}}{1+p} e^{-t/t_1} + \dfrac{1+W_{21}\tau_{21}}{1+p}, N_2(t) = \dfrac{W_{12}\tau_{21}}{1+p}(1 - e^{-t/t_1}), t_1 = \tau_{21}/(1+p)$

$$(5.111)$$

(c)  $N_1(t) = \left(\dfrac{1}{1+q} - \dfrac{1+W_{21}\tau_{21}}{1+p+q}\right) e^{-t/t_1} + \dfrac{1+W_{21}\tau_{21}}{1+p+q}, N_2(t) = 1 - N_1(t),$

$$t_1 = \tau_{21}/(1+p+q) \quad (5.112)$$

In all three cases we have $N_2(t) = 1 - N_1(t)$, so $N_3(t) = 0$. This result stems from the assumption of a rapid decay from level 3, or $\tau_{32} \ll \tau_{21}$. As a consequence, the medium dynamics are function of only one time constant $t_1$, which is given by Eq. (5.108). This conclusion does not apply to all types of Er:glass.* Figure 5.56 shows the time evolution of the ground and upper level populations $N_1(t)$, $N_2(t)$ for the three cases (a), (b), and (c). In (a) and (b), the effect of the excitation pulse (pump or signal) is to decrease $N_1$ and increase $N_2$. However, the steady state populations reached at $T \gg t_1$ are not identical, as indicated by the arrows. For very large pumps ($q \gg 1$), the steady state regime corresponds to a fully inverted medium ($N_1 = 0$, $N_2 = 1$), while in the case of very large signals ($p \gg 1$), this regime corresponds to conditions of equalized populations, according to $N_2 = (W_{12}/W_{21})N_1$. When the excitation pulse is turned off, the populations return to their initial conditions with the characteristic time constant $t_1 = \tau_{21}$. The recovery time ($t_{rec} = \tau_{21}$) is always longer than the saturation time ($t_{sat} = \tau_{21}/(1+q)$ or $t_{sat} = \tau_{21}/(1+p)$).

In (c), the effect of the saturating signal pulse is to increase $N_1$ and decrease $N_2$, which causes gain saturation. The arrows shown in the figure emphasize the competition effect between pump and signal to increase or decrease the medium inversion. The saturation time $t_{sat} = \tau_{21}/(1+p+q)$ corresponding to the on state of the exciting pulse is the shortest of all three cases (a), (b), (c) if identical powers ($p$, $q$) are assumed. The recovery time $t_{rec} = \tau_{21}/(1+q)$ corresponding to the off state decreases with pump power $q$. The ratio of recovery time to saturation time is thus given by:

$$\frac{t_{rec}}{t_{sat}} = \frac{1+p+q}{1+q} \approx 1 + \frac{p}{q} = 1 + \frac{P_{sat}(\lambda_p)}{P_{sat}(\lambda_s)}\frac{P_s}{P_p} > 1 \qquad (5.113)$$

This last result is consistent with the experimental data obtained for the dynamics of saturation and recovery, as shown in Figure 5.55c. From these data, the 10%–90% saturation and recovery times were measured to be $t_{sat} = 110\ \mu s$ and $t_{rec} = 340\ \mu s$, respectively, corresponding to a ratio $t_{rec}/t_{sat} = 3.1$ [214]. The experimental conditions corresponded to the following parameters: $\tau_{21} = 10\ ms$, $G_{max} = 29\ dB$, $q^{in} = 10$, $P_s^{in} = 35\ \mu W$ and $P_{sat}(\lambda_s) \approx 200\ \mu W$. We can then estimate the path averaged powers $\bar{q}$ and $\bar{p}$ through $t_{sat} \approx \tau_{21}\log(9)/(1+\bar{p}+\bar{q})$ and $t_{rec} \approx \tau_{21}\log(9)/(1+\bar{q})$, which gives

---

* Indeed, the case of a fluoride glass host is quite different. Considering the transitions $^4I_{13/2} \to {}^4I_{15/2}$ and $^4I_{11/2} \to {}^4I_{13/2}$ in Er:ZBLA glass, for instance, we have the decay rates $A_{21}^R = 70\ s^{-1}$ and $A_{32}^R = 100\ s^{-1}$, $A_{32}^{NR} = 76\ s^{-1}$ (Section 4.3). The corresponding lifetimes are $\tau_{21} \approx 14\ ms$ and $\tau_{32} \approx 6\ ms$. In this case, the approximation $\tau_{32} \ll \tau_{21}$ does not apply and populations $N_i(0, t)$ are given by the most general result of Eq. (5.104).

$\bar{q} = 64$ and $\bar{p} = 136$. The value of $\bar{q} = 64$ is higher than the value used for best theoretical fit ($q^{in} = 10$), which can be explained by the strong effect of pump ESA, not accounted for in the model. The value $\bar{p} = 136$ gives the effective gain $G = \bar{p}P_{sat}(\lambda_s)/P_s^{in} = 29$ dB, consistent with the experimental value.

The above description shows that the EDFA gain dynamics can be physically explained through a simple analytical model. The solutions obtained from this model

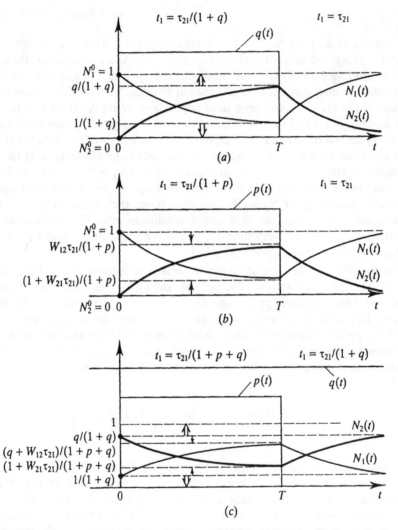

**FIGURE 5.56**    Time evolution of ground ($N_1$) and upper level ($N_2$) populations of a three-level laser system in three basic cases: (*a*) a square pump pulse of duration $T$ and power $q$ is turned on at $t = 0$ with all ions initially in the ground state, (*b*) a square signal pulse of duration $T$ and power $p$ is turned on at $t = 0$ with all ions initially in the ground state, (*c*) a square signal pulse of duration $T$ and power $p$ is turned on at $t = 0$ with ions initially distributed in both ground and excited states, under constant pump $q$. The characteristic time constants ($t_1$) corresponding to the on and off states of the excitation pulses are shown for the three cases (exponential decays not to scale). The clear and dark arrows indicate in which directions the steady state populations move when the pump and the signal powers are increased, respectively.

correspond to the input fiber end ($z = 0$), and therefore do not represent the actual case of EDFAs where pump and signals vary along the fiber. Considering the fact that the EDFA transit time is very short compared to the fluorescence lifetime $\tau_{21}$, (i.e., $t_{trans} = 5$ ns for $L = 1$ m) the EDFA transient dynamics are essentially the same as described in the previous model, with undetermined effective powers and $\bar{q}$ and $\bar{p}$. As in all cases we have $q^{out} < \bar{q} < q^{in}$ and $p^{in} < \bar{p} < p^{out}$ ($p^{out}$ includes ASE), it is possible to give boundary values for the time constant $t_1 = \tau_{21}/(1 + \bar{p} + \bar{q})$ associated with the transient gain. Consider for instance the case of 980 nm pumping in an EDFA having pump mode sizes $\omega_p = 2 \, \mu m$ and $\omega_s = 2.5 \, \mu m$, respectively. With the parameters $\tau_{21} = 10$ ms, $\sigma_a(\lambda_p) = 2.5 \times 10^{-25}$ m$^2$, and $\sigma_a(\lambda_s) + \sigma_e(\lambda_s) = 15 \times 10^{-25}$ m$^2$ [185], we find from Eq. (1.45) the saturation powers $P_{sat}(\lambda_p) \approx 100 \, \mu W$ and $P_{sat}(\lambda_s) \approx 20 \, \mu W$. Assuming an input pump $P_p^{in} = 50$ mW ($q^{in} = 500$) and a maximum signal output $P_s^{out} = 20$ mW ($p^{out} = 1000$), the saturation and recovery times for this EDFA cannot be shorter than the values $t_{sat} = \tau_{21}/(1 + q^{in} + p^{out}) \approx 6 \, \mu s$ and $t_{rec} = \tau_{21}/(1 + q^{in}) \approx 20 \, \mu s$, respectively. This example clearly shows that even in conditions of fairly high pump and signal powers ($q^{in} = 500$, $p^{out} = 1000$) the EDFA gain dynamics remain relatively slow. As the saturation time $t_{sat}$ is the shortest characteristic response, any periodic signal perturbation at frequency $B \gg 1/t_{sat}$ is of no effect on the medium inversion. In the previous example, we have $1/t_{sat} < 0.16$ MHz, which is three to five orders of magnitude smaller than the bit rates used in optical communications ($B = 100$ MHz–10 GHz). For comparison, the characteristic saturation and recovery times in semiconductor amplifiers can be as short as 280 ps (R. M. Jopson et al. [216]), which corresponds to dynamics that are five orders of magnitude faster than in the EDFA. This important result indicates that *in the saturation regime, EDFAs are virtually free from intersymbol interference, patterning, and crosstalk effects.*

The discovery of the slow EDFA gain dynamics and the potential for crosstalk-free amplification has been reported at nearly the same time by four independent groups (E. Desurvire et al. [217]; M. J. Pettitt et al. [218]; R. I. Laming et al. [219]; and Y. Kimura et al. [220]). The study of [217] provided the first measurement of BER in two-signal saturated EDFAs, showing negligible effects of crosstalk at high bit rates (2 Gbit/s). Through AC signal–DC probe measurements, the studies of [218] and [219] showed that the effect of transient gain modulation (referred to as crosstalk) vanishes for modulation frequencies above $f = 1$–10 kHz. The same conclusion applies to pump-induced gain modulation [219]. The study of [220] showed through signal–probe pulse measurements that no transient saturation could be observed in a 50 ps to 5 ns scale. These early reports attributed the absence of crosstalk or gain modulation in EDFAs to the relatively long fluorescence lifetime of Er:glass, i.e., $\tau_{21} = 10$ ms.

A detailed characterization of transient saturation and recovery showed that the characteristic times were in the 100–300 $\mu s$ range, depending on the pump and signal powers (C. R. Giles et al. [221]), as previously shown in Figures 5.55a,b. Thus, the observed dynamics are actually 30 to 100 times faster than the effect of spontaneous decay. This feature could be explained by effecting a numerical integration of the transient equations (5.99)–(5.103) to predict the FM response of the EDFA, which gave remarkable agreement over a 10–100 kHz frequency range [221]. An example of such experimental data with theoretical prediction is shown in Figure 5.57 [214]. This fit requires an empirical value for the lifetime $\tau_{32}$ but is only weakly sensitive to the actual value of $\tau_{32}$, up to a maximum of 100 $\mu s$, consistent with the condition $\tau_{32} \ll \tau_{21}$ prevailing in Er:silica glass (Chapter 4). The drop in interchannel gain

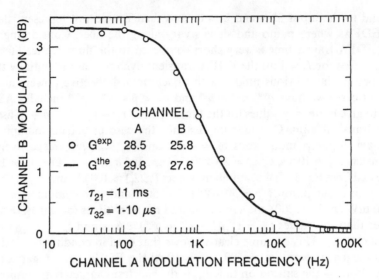

**FIGURE 5.57**  Experimental measurement of interchannel modulation measured in optical channel B as a function of frequency of saturating signal in optical channel A. The theoretical curve was calculated by numerical integration of the transient gain equations (5.99)–(5.103), using the actual experimental parameters and $\tau_{32} = 1$–$10\,\mu s$. From [214] © 1989 SPIE.

modulation near frequencies $f = 1$–$10\,kHz$ can be predicted by the analytical model previously described [215]. Appendix T shows that the relative inversion modulation due to saturation by signals at modulation rate B can be expressed as:

$$\frac{\delta(\Delta N_{12})}{\bar{N}_{12}} = 2\,\frac{(1 - e^{-f/B})(1 - e^{-f'/B})}{e^{-f/B} - e^{-f'/B} + K[1 - e^{-(f+f')/B}]} \qquad (5.114)$$

where the characteristic frequencies $f$, $f'$ are given by

$$f = \frac{1}{2}\frac{1 + p + q}{\tau_{21}},\; f' = \frac{1}{2}\frac{1 + q}{\tau_{21}} \qquad (5.115)$$

and $K$ is a power dependent coefficient defined in the appendix. Plots of the relative inversion modulation $\delta(\Delta N_{12}/\bar{N}_{12})$ as a function of the modulation rate $B$ are shown in Figure 5.58 for different values of the relative powers $p$ and $q$. We assume $W_{21}\tau_{21} = p/(1 + \eta_s) \approx p/2$, corresponding to the peak signal wavelength. In the power regimes $q = 10$, $p \leqslant q$ and $q = 100$, $p \leqslant q$, the relative inversion modulation, less than 0.7 (1.5 dB) at low frequencies, vanishes near 10–100 kHz. In the power regimes $q = 10$, $p = 10q$–$100q$ and $q = 100$, $p = 5q$–$10q$, the relative inversion modulation, which is less than 2.0 (3 dB) at low frequencies, vanishes near 100 kHz–1 MHz. This result shows that at high data rates ($B \gg 1$ Mbit/s), the effect of transient gain dynamics

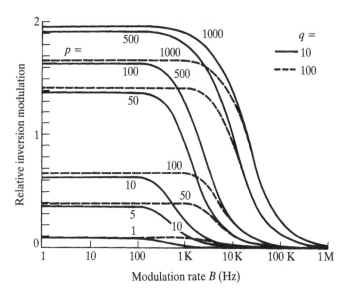

**FIGURE 5.58** Relative inversion modulation $\delta(\Delta N_{12})/\bar{N}_{12}$ as a function of the modulation rate $B$, for relative pump powers (——) $q = 10$ and (– – –) $q = 100$ at different signal powers $p$, plotted from Eq. (5.114).

is negligible, even in the high power regime corresponding to $q = 100$, $p = 1000$, i.e., $P_p^{in} = 100P_{sat}(\lambda_p)$ and $P_s^{out} = 1000P_{sat}(\lambda_s)$.

In the early work of transient gain dynamics characterization, the term *crosstalk* has often been used to designate the time dependent gain modulation $\Delta G(dB) = \delta\Delta N_{12}$. This is not exact. The low frequency gain modulation does not produce intersymbol interference between optical channels. The interference is the actual crosstalk; it would drastically increase the BER, and would cause large power penalties. The first BER measurements in two-signal saturated EDFAs demonstrated the absence of this crosstalk effect (E. Desurvire et al., [104] and [217]). The results of these measurements, obtained at $B = 2$ Gbit/s, are shown in Figure 5.59. A small power penalty ($\Delta P_A^{in} \approx 3$ dB) is incurred by optical channel A when a signal in optical channel B causes a 3 dB gain compression. The observed penalty, not due to crosstalk, can be attributed mainly to: (a) the increase of the EDFA spontaneous emission factor, associated with reduced medium inversion; and (b) the decrease of signal–ASE beat noise, relative to the detector thermal noise. Indeed, these two causes of power penalty have been clearly identified in BER measurements where the EDFA is saturated by CW signals (A. E. Willner et al. [144]), as analyzed in Section 7.1. Since the effect of EDFA gain dynamics vanishes at modulation frequencies $B \gg 1$ MHz, we can conclude that EDFAs are intrinsically free from crosstalk at high signal bit rates.

In spite of the absence of crosstalk, power penalties can yet be generated by low frequency changes in the total power input to the EDFA. In packet-switched communication networks, for instance, such power changes correspond to accidental packet collisions; in WDM systems, these correspond to traffic bursts. In any case, transient gain saturation and recovery cause undesirable fluctuations of received signal power, converted into power penalties. They are countered by automatic gain

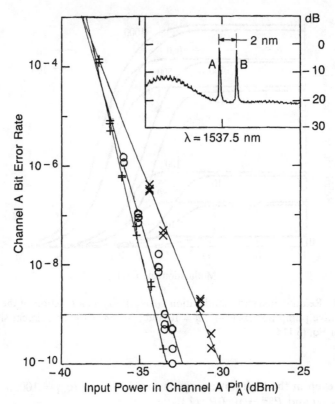

**FIGURE 5.59** BER measurement corresponding to two-signal saturation at 2 Gbit/s in an EDFA pumped at 514.5 nm. The unsaturated gain is $G = 29$ dB in optical channel A (see inset spectrum). The data (+) were measured with channel B turned off. The data (O) and (×) correspond to input powers $P_B^{in} = -32$ dBm and $P_B^{in} = -20$ dBm in channel B, respectively. For $P_B^{in} = -20$ dBm in channel B, a 3 dB gain compression was measured in channel A, causing small (3 dB) power penalty. From [104] © 1989 IEEE.

control (*AGC*). The different techniques for *AGC* in EDFAs are essentially based on the principles of pump modulation, auxiliary signal modulation, or all-optical feedback (Chapter 6).

The major consequence of the slow EDFA gain dynamics is a full suppression of crosstalk in the amplification of high speed signals. But there is another important consequence for short-pulse amplification. Since gain saturation and recovery is slow in EDFAs, the leading and trailing edges of subnanosecond pulses experience identical gains, which results in distortion-free amplification.

## 5.10 PICOSECOND AND FEMTOSECOND PULSE AMPLIFICATION

The possibility of amplifying picosecond or femtosecond pulses through EDFAs is of considerable impact. First, EDFAs can be used to increase the peak power of these pulses by several orders of magnitude, without broadening or distortion. Thus, compact laser sources (such as gain-switched or mode-locked laser diodes) can now be used at high modulation rates with an extended dynamic range. An important application of EDFAs as peak power boosters is the generation of optical solitons in fibers from gain-switched DFB–LDs (K. Iwatsuki et al. [222]). One can measure the progress accomplished considering that soliton sources previous used were Nd:YAG-pumped color center lasers or Raman fiber lasers. The potential of soliton transmission and amplication in fibers has long been demonstrated (L. F. Mollenauer et al., [223] and [224]), but the possibility of generating solitons directly from modulated LDs opened the field of soliton communication systems (K. Iwatsuki et al., [225] and [226]), considered in Chapter 7. Other important applications of EDFAs as peak power boosters are all-optical time domain demultiplexing and switching (Chapter 7). As in the case of soliton generation, these applications are based on nonlinear effects in fibers, previously unobtainable from the limited power of laser diode sources. Thus, EDFAs have opened a new field of practical applications for *nonlinear fiber optics*.

The first experiment of picosecond pulse amplification from LD-generated signals, reported in 1989, used a mode-locked GaInAsP diode as the source and a 532 nm pumped EDFA; with a saturated gain $G = 13$ dB, pulses of width $< 10$ ps and peak power 3.5 W could be generated at a rate of 300 MHz (R. A. Baker et al. [227]). The very first demonstration of picosecond amplification in EDFAs, also reported in 1989, used a CC laser as the signal source (M. Nakazawa et al. [228]). In a subsequent experiment using gain-switched DFB–LD, 9 ps pulses with peak powers of 1 W up to 12 W were obtained at rates of 2 GHz down to 33 MHz (A. Takada et al. [229]). These results were achieved with a two-stage EDFA, which involved one midway bandpass filter and six high power 1480 nm pump LDs. Based on the same scheme, but using only two pump LDs and an optical gate to reduce ASE, this peak power could be increased to 105 W (10 ps, 200 MHz, 1.05 nJ), the highest value reported to date (1992) for picosecond pulses (A. Takara et al. [84]).

Subpicosecond pulse and femtosecond pulse amplification in EDFAs were also demonstrated in 1989–1990 (K. Suzuki et al. [230]; B. J. Ainslie et al. [231]; D. Y. Khrushchev et al. [232]; M. Nakazawa et al. [223]). The characteristics of amplified pulses obtained in these experiments are summarized in Table 5.3. As the table shows, maximum peak powers near 90 kW could be obtained with 80 fs duration [232]. For pulses as short as 80–250 fs, the spectral width is about 30 nm [231] and [232], which is close to the full EDFA gain bandwidth. The amplification regime can be either linear or nonlinear, depending on whether distortion effects are observed in the output pulse's temporal shape or spectrum. At high peak powers, pulse compression from 80–250 fs to 55–60 fs can be observed, which is due to nonlinearity in the EDFA; this nonlinearity is associated with the effects of multiple order soliton excitation [232] or of soliton self-frequency shift (SSFS) [233]. When SSFS is observed, the

**TABLE 5.3  Subpicosecond and femtosecond pulse amplification characteristics**

| Source | Pulse width (fs) | Peak power (W) | Repetition rate (MHz) | Reference |
|---|---|---|---|---|
| CC laser | 700–900 | 140 | – | [230] |
| Dye/YAG/KTP | 250 | 2–4 k | 3.8 | [233] |
| Raman FL | 200 | 50 | 100 | [231] |
| Raman FL | 80 | 91 k | – | [232] |

output spectrum can broaden from 10 nm up to 45 nm, depending on the EDFA gain [233]. For broadband signals, the net power gain is sensitive to the EDFA length, along which different nonlinear regimes can follow one another (D. J. Richardson et al. [234]). A full theoretical analysis including the effects of soliton propagation (with both resonant and third order dispersion), optical amplification and SSFS in EDFAs can be found in [234]. The effect of SSFS during femtosecond pulse amplification is also discussed in the next section, which deals with solitons. Other experiments of picosecond and femtosecond pulse amplification over long EDFA lengths (i.e., 10 km), which rely specifically upon soliton propagation effects, are also described in the next section.

The distortion-free amplification characteristics of saturated EDFAs (obtained in the absence of nonlinear effects) are explained by their intrinsically slow gain dynamics. Thus, for high bit rate, intensity modulated (IM) data sequences, neither intersymbol interference nor patterning effects are expected in the saturation regime, as confirmed experimentally for data rates up to 100 Gbit/s (H. Izadpanah et al. [235]). The study of [235] shows that, in contrast with EDFAs, saturated semiconductor OAs exhibit severe patterning effects at rates of 25–50 Gbit/s, which limits their potential in IM–DD communication systems. Another important application of EDFAs for picosecond signal amplification is the realization of high speed DFB sources with high power and minimal chirp (H. Sundaresan and G. E. Wickens [236]). Indeed, picosecond gain switching in DFBs causes undesirable wavelength chirp, or time dependent wavelength sweep (C. Lin and T. L. Koch [237]). As efficient methods for chirp suppression can result in a significant decrease of LD output power, post-amplification by an EDFA is required. This illustrates the advantage of using EDFAs as external boosters in high speed LD transmitters.

## 5.11 SOLITON PULSE AMPLIFICATION

Before we consider experimental results for the amplication of soliton pulses in EDFAs, we review the properties of soliton propagation in single-mode fibers. A comprehensive introduction to this subject is quite beyond the scope of this book, for solitons actually represent an entire branch of laser optics. For a basic to an in-depth introduction, the reader may refer to various books on photonics (B. E. A. Saleh and M. C. Teich [81]), nonlinear optics (Y. R. Shen [238]), nonlinear fiber optics (G. P. Agrawal [239]) and optical solitons in fibers (A. Hasegawa [240]). Comprehensive descriptions of the complex mathematics of soliton waves and detailed biographies can be found in a 1973 paper by A. C. Scott et al. [241a] (267 references) and in a 1989 paper by Y. S. Kivshar and B. A. Malomed [241b] (564 references). An extensive review of soliton transmission fundamentals and system design considerations can also be found in the 1981 review paper by A. Hasegawa and Y. Kodama [242].

Following a brief description of soliton propagation theory, this section addresses several issues of interest for communication systems, including soliton evolution with amplifier length, characteristics of soliton interactions, effect of ASE noise, and fundamental limits of soliton propagation in very long ($L > 1000$ km) amplifier chains.

What is a soliton? The name of soliton was introduced in 1965 to describe effects in plasma interactions (N. J. Zabusky and M. D. Kruskal [243]) but the concept of solitary waves dates back more than a century and was first related to hydrodynamics

(J. Scott-Russell [244]). In [244], the engineer J. Scott-Russell reports an observation made in the summer of 1834 of a peculiar type of wave generated by the motion of boats in narrow channels. In his account, the wave assumed "the form of a large solitary elevation, a rounded, smooth and well-defined heap of water, which continued its course along the channel apparently without change of form or diminution of speed." The lossless propagation of shallow water waves was first analyzed by Korteveg-de-Vries [241]. In the absence of dissipation, the traveling-wave solution of the KdV equation is an infinite and periodic train of localized humps. The solitary wave, or soliton, corresponds to the case where the separation between the humps is infinite. N. J. Zabusky and M. D. Kruskal observed that the periodic pulse trains obtained from the KdV equation could pass through one another without deformation [243]. The particle-like behavior of these solitary waves solutions therefore justified the name *soliton*.

We consider next basic arguments developed in [241] to explain how solitons are formed. First, the existence of solitary wave solutions in a given physical medium requires either (a) linear propagation with no dispersion, or (b) nonlinear propagation with dispersion. A linear dispersionless medium is described by the well-known propagation equation:

$$\frac{\partial^2 u}{\partial z^2} - \frac{1}{c^2}\frac{\partial^2 u}{\partial t^2} = 0 \qquad (5.116)$$

whose pulse-like solutions $u(z - ct)$ are the envelope of a solitary wave. In a *linear dispersive medium*, the different Fourier components of the wave travel at different speeds, causing the wave energy to spread over time; in this case, no solitary wave solution can exist. The same conclusion applies to a *nonlinear dispersionless medium*, where harmonic generation due to nonlinearity continuously couples the energy of the wave into higher frequency modes, causing continuous changes in the wave envelope. In the case of a *nonlinear dispersive medium*, the Fourier components could be dispersed in such a way to keep the envelope constant or pulsating. In nonlinear media, *solitary waves result from the balance between nonlinearity and dispersion.*

Soliton waves are found in a large variety of physical phenomena. They represent particular solutions of several types of nonlinear differential equations. These include, for instance: the *KdV*, the *sine–Gordon*, the *Hirota* and the *nonlinear Schrödinger equations* [241]. Appendix U summarizes demonstrations made in [240], [242] and [245]. Optical solitons in single-mode fibers are described by the nonlinear Schrödinger equation (NLSE). The canonical form of the NLSE is:

$$\boxed{i\frac{\partial u}{\partial \xi} + \frac{1}{2}\frac{\partial^2 u}{\partial \tau^2} + |u|^2 u - i\Gamma u = 0} \qquad (5.117)$$

In the above, $\xi$, $\tau$ are dimensionless space and time coordinates, which include constants accounting for dispersion and nonlinearity. The dimensionless terms $|u|^2 u$ and $-i\Gamma u$ account for the two effects of *self-phase modulation* (due to the optical Kerr effect) and *amplification or loss* (due to the medium gain and background loss), respectively. In contrast to the KdV soliton, the fiber optical soliton represents the solitary wave envelope of a light wave, as for amplitude modulated signals the pulse represents a wave envelope [240].

The soliton solutions of the NLSE have been known since 1965 in the description of two-dimensional self-focusing and one-dimensional self-phase modulation [241]. In 1973, A. Hasegawa and F. Tappert demonstrated theoretically that picosecond optical solitons could be excited in single-mode fiber waveguides. At anomalous dispersion wavelengths ($k'' < 0$), these solitons would be individual pulses, or bright solitons, and at normal dispersion wavelengths ($k'' > 0$) these would be envelope shocks holes, or dark solitons, [246] and [247]. Considering only the case of bright solitons and a transparent medium ($\Gamma = 0$), two particular solutions of the NLSE, referred to as the $N = 1$ and $N = 2$ solitons, are given by, [248] and [249]:

$$u(\xi, \tau) = \mathrm{sech}(\tau)\exp(i\xi/2) \tag{5.118}$$

$$u(\xi, \tau) = 4\exp(i\xi/2)\,\frac{ch(3\tau) + 3ch(\tau)\exp(4i\xi)}{ch(4\tau) + 4ch(2\tau) + 3\cos(4\xi)} \tag{5.119}$$

where $\mathrm{sech}(\tau) = 1/\cosh(\tau)$, by definition. The two solutions (5.118) and (5.119) correspond to the initial condition $u(0, \tau) = N\,\mathrm{sech}(\tau)$ where $N = 1$ and 2, respectively. The analytical solutions corresponding to higher order $N$-solitons ($N = 3, 4, \ldots$) can be determined through the *inverse scattering method* (J. Satsuma and Y. Yajima [248]; V. E. Zakharov and A. B. Shabat, [250] and [239]), although the computation becomes very fastidious for $N > 2$. The NLSE solutions can also be determined through direct numerical integration of the NLSE. For $N > 4$, the numerical treatment becomes prohibitively lengthy, even for supercomputers such as Cray-1; in this case, numerical implementation of the inverse scattering method remains the most practical approach [251].

As the optical power is proportional to $|u(\xi, \tau)|^2$, Eqs. (5.118) and (5.119) show that the $N = 1$ soliton power envelope is space invariant, while that of the $N = 2$ soliton has a space periodicity $\xi_0 = \pi/2$. The power envelopes $|u(\xi, \tau)|^2$ corresponding to the two solutions are plotted in Figure 5.60. The $N = 2$ soliton undergoes a periodic time compression and side lobe formation which reaches a maximum at each dimensionless space coordinate $\xi = (2k + 1)\pi/4$. A similar pulsation effect with more complex pulse deformation pattern is observed for the $N = 3$ soliton, [248] and [239]. The fundamental value $\xi_0 = \pi/2$ corresponds to a space period (referred to as the *soliton period*) $z_0 = (\pi/2)z_c$, where $z_c$ is a characteristic length (Appendix U). The periodicity in $\xi_0 = \pi/2$ is obeyed by $N$-solitons of any order [250].

The characteristic length $z_c$ is related to the $N = 1$ soliton FWHM ($\Delta T$) through:

$$z_c \approx 0.322\,\frac{2\pi c}{\lambda^2}\frac{\Delta T^2}{D} \tag{5.120}$$

where $D_{s/m^2} = 10^{-6}D_{ps/nm.km}$ is the usual fiber dispersion coefficient. Appendix U also shows that the peak power $P_{peak}$ of the $N = 1$ soliton is given by:

$$P_{peak} = \frac{\lambda A_{eff}}{2\pi n_2 z_c} = 3.09\,\frac{\lambda^3 A_{eff}}{4\pi^2 c n_2}\frac{D}{\Delta T^2} \tag{5.121}$$

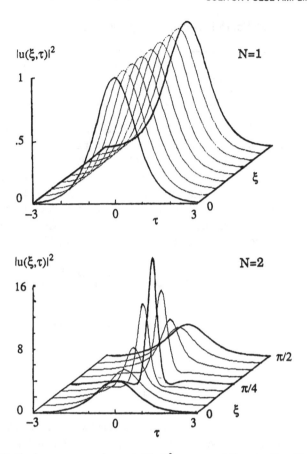

**FIGURE 5.60**  Optical power envelopes $|u(\xi, \tau)|^2$ corresponding to the $N = 1$ and $N = 2$ soliton solutions of the nonlinear Schrödinger equation.

where $n_2$ is the nonlinear refractive index coefficient (defined by the relation $n = n_0 + n_2 I$, $I = P_s/A_{\text{eff}}$), $A_{\text{eff}}$ is the effective mode area. For fused silica fibers, the nonlinear index coefficient is $n_2 = 3.2 \times 10^{-20}$ m$^2$/W [245]. Considering the case of signals at $\lambda = 1.54$ $\mu$m and a fiber mode size comparable to a DSF ($\omega_s$ (field) $\approx 4$ $\mu$m, $A_{\text{eff}} \approx 50$ $\mu$m$^2$), we find from Eq. (5.121) the relation $P_{\text{peak}}(W) \approx 0.153/z_c$(km). On the other hand, the soliton length $z_c$ can be expressed from Eq. (5.120) as $z_c$(km) $\approx 0.256 \times \Delta T_{\text{ps}}^2/D_{\text{ps/nm.km}}$. Thus, for the $N = 1$ soliton in this example, the relation between peak power, FWHM and dispersion is:

$$P_{\text{peak}}(W) \approx 0.597 \frac{D_{\text{ps/nm.km}}}{\Delta T_{\text{ps}}^2} \tag{5.122}$$

Using the above result, we find that near the zero dispersion wavelength with $D \approx 15$ ps/nm.km the $N = 1$ soliton peak power is $P_{\text{peak}} \approx 224$ mW for $\Delta T = 10$ ps and $P_{\text{peak}} \approx 8.9$ mW for $\Delta T = 50$ ps. In the general case, when considering any value for the parameters $(A_{eff}, \lambda)$ Eq. (5.122) can be expressed as:

$$P_{peak}(W) \approx 1.492 \left( \frac{\lambda_{\mu m}}{1.54} \right)^3 \frac{A^{eff}_{\mu m^2}}{50} \frac{D_{ps/nm.km}}{\Delta T^2_{ps}} \tag{5.123}$$

The above analysis has shown that the NLSE has a set of discrete $N$-soliton solutions, which correspond to an initial excitation pulse $u(0, \tau) = N \operatorname{sech}(\tau)$. What are the NLSE solutions for pulses whose initial peak powers and shapes do not match this condition? This important question can be answered by perturbation theory [248]. The theory shows that for an initial excitation $u(0, \tau) = A \operatorname{sech}(\tau)$ with $A = N + \varepsilon$ ($N$ integer, $|\varepsilon| < 1/2$) and width $\Delta T$, the pulse asymptotically evolves along the fiber, with some energy dissipation, into a soliton of order $N$. In particular, the soliton part of the resulting pulse corresponds to an initial excitation, [239], [248], and [249].

$$u(0, \tau) = (N + 2\varepsilon)\operatorname{sech}\left\{ \frac{N + 2\varepsilon}{N} \tau \right\} \tag{5.124}$$

The FWHM, corresponding to the envelope (5.124), obtained at each soliton period $z_0$ is:

$$\Delta T' = \frac{N}{N + 2\varepsilon} \Delta T \tag{5.125}$$

Equation (5.125) shows that for $\varepsilon > 0$, the pulse narrows, while it broadens for $\varepsilon < 0$. The relative change in FWHM, given by $\Delta T'/\Delta T = N/(N + 2\varepsilon) = N/(2A - N)$, is plotted in Figure 5.61. The FWHM change is discontinuous, which provides a useful definition of the order of the soliton (N. J. Doran and K. J. Blow [249]). The change is observed to decrease with order $N$, the broadening effect ($\varepsilon > 0$) being always

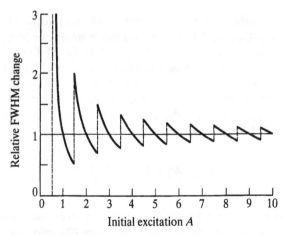

**FIGURE 5.61** Asymptotic change in soliton FWHM relative to input value, corresponding to initial excitation $u(0, \tau) = A \operatorname{sech}(\tau)$, with $A = N + \varepsilon$ ($N$ integer, $|\varepsilon| < 1/2$).

greater than the narrowing effect ($\varepsilon > 0$) for a given order. No soliton forms in the case of an initial excitation $A \leqslant 1/2$, which corresponds to a regime of linear propagation.

The other aspect of the previous question concerns the effect of initial pulse shape. Theoretical simulations show that an input pulse with a Gaussian shape evolves asymptotically towards a soliton (e.g., $u(0, \tau) = \exp(-\tau^2/2)$ evolves into the $N = 1$ soliton), as shown in [239]. The length required to complete the soliton formation can represent several soliton periods ($L = 5z_0$ in the previous example), depending on the initial pulse shape. The same conclusion applies to the case where the initial excitation is sub-Gaussian or super-Gaussian [239]. In fact, any initial pulse shape evolves asymptotically into a set of solitons (V. E. Zakharov and A. B. Shabat, [250]). Taken together, these considerations show that *solitons represent the natural eigenstates of nonlinear pulse propagation in SM fibers*, and they can easily be excited provided the initial peak power is above a certain threshold.

Having analyzed $N$-soliton solution corresponding to a lossless medium, we consider the case where the pulse energy is dissipated along the fiber. The NLSE (5.117) must be solved with the additional term $-i\Gamma u$, with $\Gamma = -z_c \alpha_s/2$, where $\alpha_s$ is the power absorption coefficient. Perturbation theory shows that for small absorption coefficients and short fiber lengths, the $N = 1$ solution (i.e., corresponding to the initial condition $u(0, \tau) = \operatorname{sech}(\tau)$) takes the form (Y. Kodama and A. Hasegawa [252]):

$$u(\xi, \tau) \approx \exp\left\{i\frac{1 - e^{-2\Gamma\xi}}{4\Gamma}\right\}\operatorname{sech}(\tau e^{2\Gamma\xi}) \tag{5.126}$$

The above solution shows that the pulse broadens exponentially, according to $\Delta T' = \Delta T \exp(2\Gamma\xi) = \Delta T \exp(\alpha_s z)$. In order to explain pulse broadening, intuitively consider the case of power dissipation: The balance between nonlinearity and dispersion is progressively broken, thus the transmission becomes linear and dispersive. A comparison between numerical simulations and predictions from Eq. (5.126) for pulse broadening shows that the perturbation solution is accurate up to fiber lengths $\xi = 1/\Gamma$ or $z = 2/\alpha_s$ (K. J. Blow and N. J. Doran, [249] and [253]). For a standard SMF or DSF, the absorption coefficient is $\alpha_s = 0.20$–0.25 dB/km, and this maximum length is 35–43 km, corresponding to a relative power decay and pulse broadening of $-8.7$ dB and $\Delta T'/\Delta T = e^2 \approx 7.5$, respectively. The numerical simulations of [253] show that the asymptotic dispersion of the $N = 1$ soliton is a monotonically increasing function of $\Gamma$, which approaches the regime of linear dispersion as $\Gamma \to \infty$. For any propagation distance, the soliton effect results in a permanent reduction of the dispersion and the reduction decreases with increasing loss.

The first observation of soliton propagation in optical fibers was reported in 1980; with a 700 m long SMF, a 7 ps FWHM pulse at $\lambda = 1.55\,\mu m$ was observed to undergo power dependent narrowing and splitting, in excellent agreement with the $N$-soliton solutions predicted by the NLSE theory (L. F. Mollenauer et al. [254]). One of the first applications of the soliton effect was picosecond pulse compression (L. F. Mollenauer et al. [251]). In a 1983 experiment [251], pulses with 7 ps FWHM could be compressed to 260 fs by using $N = 10$ soliton excitation and fibers shorter than the soliton length $z_c$. One of the most promising applications of solitons envisioned at that time, concerned high-bit rate communication systems. The theoretical maximum bit rate of communication systems based on ASK modulated $N = 1$ solitons, would be 10 to 100 larger than in the best realistic linear system, assuming no signal regeneration (A. Hasegawa and Y. Kodama [242]). But, as in the case of nonregenerative linear systems, the maximum transmission length of soliton systems would be ultimately limited by fiber loss. In 1983–1984, it was suggested that *the performance of soliton systems can be enhanced by using optical amplifiers as a means of in-line regeneration and pulse reshaping* (Y. Kodama and A. Hasegawa [252]).

Considering optical amplification by stimulated Raman scattering (SRS), theoretical studies [252] and [255] showed that solitons can be transmitted without distortion if the fiber is bidirectionally pumped at regular intervals (e.g., every 40 km for 100 mW, 12 ps solitons). The need for a bidirectional and periodic pumping configuration is explained by the effect of pump decay, due to transmission loss depletion by SRS, which results in incomplete soliton loss compensation. The first demonstration of distortion-free soliton amplification by SRS used a 10 km long fiber with backward pumping (L. F. Mollenauer et al. [223]). In the absence of Raman gain, the 10 ps solitons would broaden by a factor 1.5 at the fiber output, while they would recover their initial width with the gain being turned on. The principle of periodic soliton regeneration by SRS amplification over ultralong distances ($L > 1000$ km) was demonstrated in 1988, using a recirculating fiber loop (L. F. Mollenauer and K. Smith [224]).

Because of the effect of pump decay along the fiber length (even in the case of bidirectional pumping), the loss is never exactly canceled over the whole fiber length. The local gain coefficient is either positive of negative, consequently the solitons undergo power excursion. Power excursion is maximum for unidirectional pumping. During periodic amplification, the same argument as in the case of soliton power dissipation applies, i.e. the perturbation is expected to be negligible if the amplification period $L$ is comparable to the soliton period $z_0$. Specific numerical simulations have shown that for bidirectional pumping, the perturbation of the soliton area $\delta S$ is maximum when $L \approx 8z_0$, otherwise it is comparatively small, i.e., when $L \ll z_0$ or $L \gg z_0$ (L. F. Mollenauer et al. [245]). For long amplification periods ($L \gg z_0$) the perturbation is very slow and the dispersion effect can track the power changes. For short amplification periods ($L \ll z_0$) the pulse width remains nearly constant as the dispersion effect is not significant over the distance $L$. The maximum perturbation observed for $L \approx 8z_0$ corresponds to a phase change $2\pi$ in the term $\exp(i\xi/2)$ of the $N = 1$ soliton, and therefore the effect can be interpreted as a resonance between the perturbation period $L$ and the soliton period $z_0$ [245].

Raman fiber amplifiers (RFAs) require relatively high pumping powers ($> 100$ mW) compared to the more recently developed EDFAs (Section 5.12). For this reason, the initial idea of periodic regeneration by RFAs, [223] and [224], was eventually superseded by the EDFA approach. Researchers needed to find out whether regeneration by EDFAs would require distributed amplification (as with RFAs) instead of lumped amplification (Chapter 2). Before we look at EDFA regeneration, we review some soliton amplification experiments using lumped and distributed EDFAs.

Soliton amplification in lumped EDFAs was first reported in 1989 (M. Nakazawa et al. [228]) using initial power boosting in a 3.5 m long EDFA ($G \approx 6$ dB). The $N = 1$ solitons could recover their original FWHM after propagation over a 30 km long DSF; at lower or higher EDFA gains, the solitons would exhibit characteristic

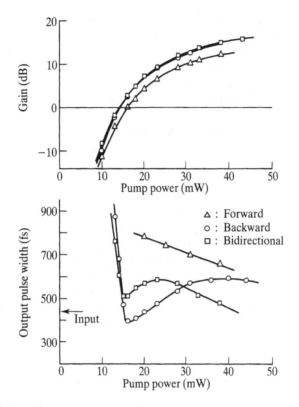

**FIGURE 5.62**   Gain and output FWHM of $N = 1$ solitons at $\lambda = 1.551$ $\mu$m, obtained in an 18 km distributed EDFA pumped in three configurations: forward backward and bidirectional. The input FWHM is $\Delta T = 440$ fs. From [261] © 1992 IEEE.

broadening or pulse splitting, respectively. The same experimental setup demonstrated distortionless transmission over 30 km of a 20 GHz pulse train of 10 ps $N = 1$ solitons (M. Nakazawa et al. [257]). Likewise, subpicosecond ($\Delta T = 700$ fs) soliton amplification followed with 30 km propagation in DSF was also shown (K. Suzuki et al. [230]).

The possibility of distributed amplification over kilometer long, low $Er^{3+}$ concentration DSFs, pumped near 1480 nm, was investigated in 1990 (Chapter 6). The first adiabatic soliton regeneration over DSF–EDFAs ($L = 18$ km) was shown in 1990–1991 for the picosecond regime (M. Nakazawa et al. [258]) and for the femtosecond regime (M. Nakazawa and K. Kurokawa [259]–[261]). In particular, several effects associated with pump power and pumping configuration could be observed with femtosecond solitons. The results obtained for forward, backward, and bidirectional pumping configurations, are shown in Figure 5.62 [261]. At low pump powers, corresponding to lossy transmission, the $N = 1$ solitons broaden (initial $\Delta T = 440$ fs). At higher pumps, corresponding to net pulse amplification, the broadening decreases; and with backward pumping, pulse narrowing (390 fs $< \Delta T < 440$ fs) is even observed. The effects of pulse broadening or narrowing are explained by adiabatic amplification of the $N = 1$ soliton peak power. Indeed, Eq. (5.125), which stems from the perturbation theory of [248], shows that the relative FWHM change associated with $A = 1 + \varepsilon$

solitons is given by $1/(1 + 2\varepsilon)$. For small amplitude gain coefficients $g/2$, the peak amplitude of the soliton envelope becomes $G \approx 1 + g/2$, and the soliton FWHM changes by the factor $1/(1 + g)$ [261]. For lossy transmission $(g < 0)$ the pulse broadens and for net amplification $(g > 0)$ the pulse narrows. But in the case of backward and bidirectional pumping, Figure 5.62 shows regimes of soliton broadening with increasing pump. This broadening is attributed to the effect of soliton self-frequency shift (SSFS).

Discovered in 1986 with RFAs (F. M. Mitschke and L. F. Mollenauer [262]) SSFS is a redshifting of the center frequency of the soliton pulse, induced by self-pumped SRS. The effect occurs for narrow pulses whose spectrum is broad enough to sense SRS. In the SRS process, light at frequency $v_0$ is scattered into a redshifted component at frequency $v'$ (Stokes light) with a certain gain coefficient $g_R(v_0 - v')$. The frequency difference for which the Raman gain coefficient is maximum, referred to as the Stokes or Raman shift, is approximately $\Delta v = 440 \, \text{cm}^{-1}$ in fused silica glass (R. H. Stolen and E. P. Ippen [263]). As the gain coefficient $g_R$ extends down to zero detuning, e.g., $g_R(v_{\text{peak}}/10) \approx g_R(v_{\text{peak}})/10$ (R. H. Stolen et al. [264]), intrapulse Raman scattering can occur with spectral widths $\Delta\lambda \geqslant 40 \, \text{cm}^{-1}$ ($\Delta\lambda \geqslant 10 \, \text{nm}$ at $\lambda = 1.54 \, \mu\text{m}$). As the soliton peak power increases and the pulse narrows, the soliton spectrum broadens and the effect of SSFS becomes significant. According to theory, SSFS varies approximately as the inverse fourth power of the soliton width $\Delta T$ (J. P. Gordon [265]). With a fiber dispersion $D = 15 \, \text{ps/nm.km}$, for instance, the SSFS is given by [265]:

$$\frac{\Delta v_0}{L} \, (\text{THz/km}) \approx 0.0436 \frac{1}{(\Delta T_{\text{ps}})^4} \tag{5.127}$$

For $\Delta T = 1 \, \text{ps}$ pulses, this corresponds to 1 THz for 23 km propagation. In the experiment of [261], whose results are shown in Figure 5.62, the broadening effect observed for pumps corresponds to $P_p^{\text{in}} \geqslant 15 \, \text{mW}$ (backward and bidirectional pumping) corresponds to the onset of SSFS. For $P_p^{\text{in}} = 40 \, \text{mW}$ or $G = 15 \, \text{dB}$, the maximum SSFS was measured to be 5 THz [261]. The broadening effect can be explained by the following: as a shift to longer wavelengths corresponds to a higher dispersion coefficient $D$, a broader pulse width $\Delta T$ is necessary to maintain the $N = 1$ soliton, given the peak power $P_{\text{peak}}$, i.e., $\Delta T \approx \sqrt{D/P_{\text{peak}}}$, according to Eq. (5.121). The difference in pulse broadening observed in Figure 5.62 for the three pumping configurations is explained by the soliton power excursion and related FWHM changes along the fiber. These are not the same for each configuration [261]. SSFS also depends on the soliton input wavelength (K. Kurokawa and M. Nakazawa [267].

Intrapulse Raman scattering and SSFS in femtosecond soliton amplication through EDFAs can be analyzed by incorporating into the NLSE a Raman term proportional to $iu \, \partial|u|^2/\partial t$ (Y. Kodama and A. Hasegawa [266]; I. R. Gabitov et al. [268]; D. J. Richardson et al. [234]; and M. Ding and K. Kikuchi [269]). Effects related to femtosecond soliton amplication in fibers include: soliton collapse and pulse train generation, [268] and [270], and soliton fission, [271] and [272]. In particular, the analysis of [269] shows that by using a distributed EDFA with a linear frequency dependent gain spectrum, compensation of SSFS over a long transmission length (i.e., $L = 30 \, \text{km}$) is theoretically possible.

Nonlinear interaction forces between closely spaced solitons can exist between solitons having either identical or different optical frequencies. Theory shows that adjacent solitons of identical frequencies are subject to attractive or repulsive interaction forces. These depend upon their relative phase, amplitude, and initial separation (J. P. Gordon [273]; K. J. Blow and N. J. Doran [274]; and B. Hermansson and D. Yevick [275]). Interference between the overlapping tails of the two solitons intuitively explain the interaction forces. Depending upon the initial relative phase of the solitons, their tails interfere constructively or destructively. Constructive interference increases the intensity $I(t)$ in the overlapping region, which decreases the chirp (proportional to $dI/dt$) and causes acceleration (or attraction) of the solitons towards each other. The opposite occurs when the two solitons are initially out of phase (F. M. Mitsche and L. F. Mollenauer [276]). For solitons initially in phase ($\Delta\phi = 0$) and separated by $2\delta\tau_0 \gg 1$, the separation at any distance $\xi$ is given by [273]:

$$\exp\{2(\delta\tau - \delta\tau_0)\} = \frac{1 + \cos[4\xi \exp(-\delta\tau_0)]}{2} \tag{5.128}$$

From Eq. (5.128), the soliton separation is found to be periodic with oscillation period:

$$\xi_0 = \frac{\pi}{2}\exp(\delta\tau_0) \tag{5.129}$$

An accurate formula for the oscillation period, valid for any initial separation $2\delta\tau_0$ (C. Desem and P. L. Chu, [277]–[279]) is:

$$\xi_0 = \frac{\pi}{2}\exp(\delta\tau_0)[1 + \exp(-\delta\tau_0)]\frac{\sinh(2\delta\tau_0)}{2\delta\tau_0 + \sinh(2\delta\tau_0)} \tag{5.130}$$

The mutual interaction force between in-phase solitons thus results in periodic attraction, collapse, and repulsion, as shown in the simulations of [278], [279], and [239]. From Eq. (5.129), the collision distance is $L_c = \xi_0 z_c = (2/\pi)\xi_0 z_0 \approx z_0\exp(\delta\tau_0)$. This collision distance and more specifically the ratio $L_c/z_0$, determine the maximum bit rate $B = 0.88/(\delta\tau_0\Delta T)$ allowable in soliton systems [239]. For large initial separations ($\delta\tau_0 \geqslant 10$), the ratio ratio $L_c/z_0$ is very large ($\geqslant 22{,}000$) and the effect of mutual attraction or repulsion can be overlooked. On the other hand, in for short soliton separations (e.g., five pulse widths), soliton interactions can be suppressed through a variety of effects, including: higher order dispersion and nonlinearity, and initial chirp [239]. Soliton collapse can also be avoided by quadrature shifting the phase of consecutive pulses (D. Anderson and M. Lisak [280]) or periodically changing their amplitudes (J. P. Gordon [273]; C. Desem and P. L. Chu [277]; and I. M. Uzumov et al. [281]). Examples of four-soliton interactions with variable consecutive amplitudes and stationary propagation are shown for instance in [281].

The first experimental observtion of single-wavelength soliton interactions in fibers, reported in 1987, confirmed the existence of phase dependent attractive and repulsive forces (F. M. Mitsche and L. F. Mollenauer [276]). Measurements of output versus input pulse separation for $\Delta\phi = 0$ and $\Delta\phi = \pi$ with $\Delta T = 1.1$ ps solitons were in excellent agreement with the theoretical predictions of [276], up to the point where the first pulse collapse was obtained. As the initial separation was reduced, the resulting force was observed to be always repulsive, in agreement with the theory for $\Delta\phi = \pi$ but contrary to the theory for $\Delta\phi = 0$ (where the predicted behavior is oscillatory). The observed discrepancy was attributed to an instability of the attractive phase [273], as triggered by the perturbative effect of SSFS (a self-frequency shift $\delta v$ causes a phase shift $\delta\phi = 2\pi\delta vt$). The estimates in [276] showed that for the attractive phase, the SSFS can be as large as $\pi$, which converts it into a repulsive one.

Experimental measurements in EDFA-based systems have shown that for bit rates up to 20 Gbit/s and propagation distances up to 300 km, single-wavelength soliton interactions are negligible (M. Nakazawa et al. [282]). But at longer distances, random push–pull effects between solitons could be observed, accompanied by pulse broadening and collapse. In particular, it was found that the soliton repulsion force is greater after collapse than the attraction force before collapse, as illustrated by the numerical simulation shown in Figure 5.63 [282], which can be attributed to permanent velocity shifts.

We consider next the case of multiple wavelength soliton interactions, as occurring in WDM soliton systems. In such systems, solitons of different wavelengths travel at different velocities, and therefore periodically pass through each other. In amplifier-based systems, the important issue to be addressed is how multiple collisions between channels could affect the characteristics of the soliton pulses (e.g., relative separation, velocity, width, and phase) and whether undesirable random interferences are generated. The theory of [273] has shown that in the case of a lossless fiber, solitons having different velocities or wavelengths are transparent to each other; the only

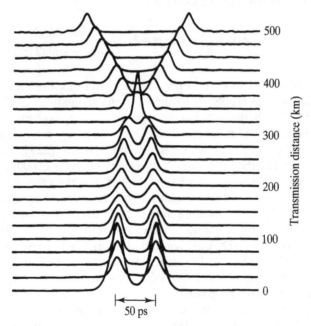

**FIGURE 5.63** Numerical simulation of single-wavelength soliton pair interaction with transmission distance in an EDFA-based system with 25 km amplifier spacing, showing collapse and repulsion near $L = 400$ km. From [282], reprinted with permission from *Electronics Letters,* 1991.

effect of the collision is a slight time advance or retardation for the overtaking or overtaken solitons, respectively. The study of [245] showed that in real systems applications, the net time displacement (and associated standard deviation) caused by hundreds of collisions is at most of the order of a few soliton widths, and therefore is of no consequence.

But in the case of gain-perturbed fibers (where loss is nonuniformly compensated) WDM soliton collisions cause a significant velocity shift, as shown by the numerical simulations in [245]. In one example, these simulations showed that for two solitons with relative velocity $2\Omega$ and colliding at a point in the fiber with gain coefficient $g$, the absolute velocity change $\delta\Omega$ is independent of the initial phase difference and can be empirically described by the relation [245]:

$$\delta\Omega = 1.72g\frac{\sqrt{z_0}}{\Omega^{1.7}} \tag{5.131}$$

where the initial velocity difference $2\Omega$ for frequency or wavelength separation $\delta f$ or $\delta\lambda$ is given by:

$$2\Omega = 2\pi\delta f \times t_c = \frac{3.52}{\pi}\frac{z_0 D}{\Delta T}\delta\lambda \tag{5.132}$$

The definition of $\Omega$ corresponds to a velocity with respect to the center of motion, i.e., the two soliton envelopes are given by $\text{sech}(\tau - \Omega\xi)$ and $\text{sech}(\tau - \Omega\xi)$; therefore, $\Omega$ plays the role of a reciprocal velocity in the $\xi$, $\tau$ space. Additionally, the collision-induced velocity change $\delta\Omega$ in Eq. (5.131) corresponds to a shift in carrier frequency $\Delta f = \delta\Omega/\pi t_c$. The velocity shift is nonpermanent – it occurs only during the collision – but it does set a limit on the maximum number of optical channels.

In the case of a distributed amplifier system, collisions can occur at points of minimum or maximum gain coefficients which, according to Eq. (5.131), correspond to different velocity shifts. In a long amplifier chain, the effects of collision-induced velocity changes are *pulse jitter* or *randomness in pulse arrival time*. The mean change in arrival time $\langle s \rangle$ is zero, as the integral over one period of Eq. (5.131) is zero (transparency condition). On the other hand, assuming a large number of WDM channels and $L = 30–50$ km regeneration spacings in a system with total length $L_{\text{tot}}$, the corresponding deviation is given by [245]:

$$\sigma(s) = \sqrt{\langle s^2 \rangle} \approx 0.306\delta\Omega_c\sqrt{\Omega}\left(\frac{L_{\text{tot}}}{z_0}\right)^{3/2} \tag{5.133}$$

where $\delta\Omega_c$ is the velocity shift associated with collisions occurring at $L/2$ (given by Eq. (5.131) for a gain coefficient $gL/2$). Equation (5.133) makes it possible to calculate the maximum tolerable $\Omega$ corresponding to a given timing jitter $\sigma(s)$, system length $L_{\text{tot}}$, and parameters $z_0$, $g(L/2)$, $\Delta T$ the minimum channel separation $\delta\lambda$ can be determined from $\Omega$ through eq. (5.132), and the maximum number of WDM channels is given by the integer value of the ratio $\Delta\lambda/\delta\lambda$, where $\Delta\lambda$ is the usable amplifier or system bandwidth. Numerical examples of WDM soliton system parameters, based on RFA regeneration, are shown in [245].

The above conclusions applied to the case of WDM systems with periodically distributed amplification. In the case of WDM systems based on periodic lumped amplification, numerical simulations have shown that collisions also result in displacement and deviation in pulse arrival times (L. F. Mollenauer et al. [283]). However, the study in [283] shows that soliton collisions occurring in lumped amplifier chains cause an unbalanced perturbation of soliton velocity, whose result is a net velocity or frequency shift. In this effect, an important parameter is the collision length $L_c$, which is the length over which solitons overlap from and to their half-power points, i.e., $\Delta T/t_c = \Omega\xi_c \equiv \Omega L_c/z_c$, or [283]:

$$L_c = 0.6298\frac{z_0}{\Delta T\delta f} = \frac{2\Delta T}{D\delta\lambda} \tag{5.134}$$

When the collision length matches the amplifier spacing ($L_c/L = 1$), the analysis in [283] shows that the net frequency shift varies sinusoidally with the collision center coordinate $z$, with maxima at $z = (2k + 1)L/2$ and minima at $z = kL$ ($k = 1, \ldots, N$); in this case, maximum frequency shifts are produced by collisions occurring either between two amplifiers or exactly in the amplifiers [283]. The maximum frequency shift (determined by the collision locations giving the maximum effect) goes through a peak value for a ratio $L_c/L$ near 0.5, and becomes rapidly negligible when $L_c/L > 2$. The same conclusion applies when the perturbation is caused by variation in the fiber dispersion parameter $D$ (a particularly important result because $D$ cannot be made constant in real long-haul systems). When the collision length is more than twice the amplifier spacing or any perturbation period, WDM soliton interactions have the same characteristics as those of lossless transmission (i.e., the net result is a time displacement without change of soliton pulse characteristics, as in the unperturbed fiber). The maximum wavelength separation between optical channels $\delta\lambda_{max}$ is given by solving Eq. (5.134) for $L_c \geqslant 2L$, i.e.,

$$\delta\lambda_{max} = \frac{\Delta T}{LD} \tag{5.135}$$

The first experimental measurements of soliton collisions in gain-perturbed fibers were reported in 1990 (P. A. Andrekson et al., [284] and [285]). In these experiments, soliton pulse trains at different wavelengths ($\lambda$, $\lambda'$) were generated by two external cavity, mode-locked LDs. A delay time was used to vary the initial separation between the two pulse trains, thereby controlling the point of collision in the amplifier chain; this chain included a power booster (EDFA1) and an in-line amplifier (EDFA2), according to the sequence EDFA1/35 km-SMF/EDFA2/70 km-SMF. A power envelope topographical map of the soliton pairs at $\lambda$, $\lambda'$ ($\delta\lambda = 0.18$ nm) monitored at the system output ($L = 105$ km) as a function of initial delay is shown in Figure 5.64 [285]. In the absence of interaction, the right-hand soliton should stay at a fixed time location, while the left-hand soliton should shift linearly to the left with increasing delay. The figure shows that when the soliton pulses collide in either amplifier, significant temporal shifts corresponding to attractive interactions occur, as indicated by the arrows. During the collisions, the overtaking pulse (left) is accelerated (moving to the right), while the overtaken pulse (right) is retarded (moving to the left), as per [273]. The comparatively short lengths of the EDFAs mean that collisions actually

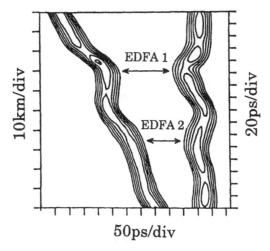

**FIGURE 5.64** Topographical map of power envelopes of two-wavelength soliton pair at the output of a two-element EDFA chain, recorded for different initial time separations; the arrows indicate the delays for which the two solitons pulses collide in the first or the second EDFA. From [285], reprinted with permission from *Electronics Letters, 1990.*

take place before and after crossing the amplifiers; the first and second halves of the collisions are at low and high power levels, respectively [285].

By using a recirculating loop (RL) configuration, two-channel WDM soliton collisions could be investigated in the regime $L_c/L > 2$ over ultralong transmission distances (P. A. Andrekson et al. [286]). The RL included four EDFAs spaced by $L = 30$ km, while the theoretical collision length was $L_c = 450$ km. measurements of soliton output spectra versus time (or transmission length) up to 60 ms (or 12,300 km) showed periodic and nonpermanent spectral shifts of both soliton carrier frequencies, over the interaction length $L_c$ at the three transmission distances $L_{tot} = 1500$ km, 4500 km and 7500 km, where solitons would experience collisions. For wavelength separations of $\delta\lambda = 0.1$–1 nm, the measured frequency shifts were $\Delta f = 0.2$–2 GHz, according to a $1/\delta\lambda$ law. The observation of nonpermanent frequency shifts is consistent with the regime of lossless transmission previously discussed, achieved for $L_c/L > 2$. From the theory of [283], the statistical deviation in arrival time due to WDM soliton collisions over a given system length $L_{tot}$ is, in these conditions [286]:

$$\frac{\sigma(s)}{\Delta T} = 0.1418 \frac{L_{tot} B}{z_0 \delta f} \tag{5.136}$$

where $\delta f$ is the channel frequency separation and $B$ is the bit rate. Consider a numerical example with parameters $D = 2$ ps/nm.km (DSF), $\lambda = 1.54 \mu$m, and $\Delta T = 100$ ps. We find from Eq. (5.120) a soliton period $z_0 = 2000$ km. Now consider a maximum relative jitter $\sigma(s)/\Delta T = 0.25$, a bit rate $B = 1$ Gbit/s, and a total system length $L_{tot} = 10,000$ km. We find from Eq. (5.136) a minimum wavelength separation $\delta\lambda \approx 0.6$ nm, corresponding to 16 optical channels over a 10 nm bandwidth.

During 1992–1993 the complex issues related to soliton interactions in WDM systems and possible countermeasures for transmission capacity enhancement were actively investigated. For instance, it was shown theoretically that the effect of soliton interaction forces can be reduced by bandwidth-limited amplification (Y. Kodama and S. Wabnitz [287]; M. Nakazawa and H. Kubota [288]), although the physical interpretation of this approach is still debated [289]. In the femtosecond amplification regime, recent experiments have shown that the perturbative effect of soliton interactions (in particular with small satellite subpulses) greatly enhances SSFS, explained by its $1/\Delta T^4$ dependence (K. Kurokawa et al., [290] and [291]).

The previous soliton dynamics and mutual interaction effects are produced by the response of the nonlinear medium to the soliton amplitude changes, caused by lumped or distributed amplification. We consider now another fundamental effect which, in very long-haul systems, results from the interaction of solitons with amplified spontaneous emission (ASE).

Following the first experiments of soliton amplification through RFAs in 1985 (L. F. Mollenauer et al. [223]), the question of how solitons and ASE noise interact in the nonlinear fiber medium and possibly limit the transmission distance had to be addressed. In 1986, J. P. Gordon and H. A. Haus showed that the ASE generated along the amplifier chain produces random changes in the soliton's frequency and velocity [292]. Such perturbations causes random walk or timing jitter of the soliton, now referred to as the *Gordon–Haus effect*. The fundamental analysis of [292] gives an expression for the standard deviation $\sigma(s)$ of the soliton arrival time for a lumped amplifier chain of total length $L_{\text{tot}}$, with amplifier spacing $L$ and gain $G = \exp(\alpha_s L)$. In standard (MKSA) units, the deviation can be put into the general form (D. Marcuse, [293] and [294]):

$$\sigma(s) = \left[\frac{1.763 n_2 D h n_{\text{sp}} (G-1) L_{\text{tot}}^3}{9 \Delta T A_{\text{eff}} L Q}\right]^{1/2} \tag{5.137}$$

where the parameter $Q$ is defined by

$$Q = \frac{\alpha_s L}{1 - \exp(-\alpha_s L)} = \frac{G \log G}{G - 1} \tag{5.137}$$

In the case $G - 1 \approx \alpha_s L$ or $Q = 1$ and $n_{\text{sp}} = 1$, eq. (5.137) is identical to Eq. (18) of [292], given in dimensionless units. The parameter $1/Q$, defined from Eq. (5.137), corresponds to the path averaged power of a Type C amplifier chain normalized to the power input, Eqs. (2.184) and (2.186).

What propagation distance is needed to observe the Gordon–Haus effect? Consider a basic numerical example: using the parameters $\alpha_s = 0.25$ dB/km, $D = 3$ ps/nm.km, $A_{\text{eff}} = 35$ $\mu$m$^2$, $n_2 = 3.2 \times 10^{-20}$ m$^2$/W, $n_{\text{sp}} = 1$, and the approximation of short amplifier spacings ($G - 1 \approx \alpha_s L$, $Q = 1$). We find from Eq. (5.137) for $\Delta T = 50$ ps solitons a timing jitter of $\sigma_{\text{ps}} \approx 2.1 \times 10^{-5} L_{\text{tot,km}}^{3/2}$. Thus, for a relative jitter $\sigma = 5$ ps or $\sigma/\Delta T = 10\%$, we find the transmission distance $L_{\text{tot}} \approx 3850$ km.

The first experimental observations of the Gordon–Haus effect, made in 1988, used a recirculating loop (RL) configuration with internal loss compensated by Raman gain (L. F. Mollenauer and K. Smith [224]). In this experiment, the principle of Raman-active RLs, known since 1985 for fiber memory applications (E. Desurvire et al., [295] and [296]), was applied for the first time to study picosecond pulse propagation over thousands of kilometers. Measurements of soliton FWHM and timing jitter as functions of the transmission length up to 10,000 km are shown in Figure 5.65, for input pulses with $\Delta T = 50$ ps (L. F. Mollenauer et al. [297]). The soliton FWHM remains constant for transmission lengths up to $L_{tot} \approx 6000$ km and broadens by 30% near $L_{tot} \approx 10,000$ km. The measured increase in timing jitter $\sigma$, 25% of the initial pulse width $\Delta T$ at the maximum transmission distance, is in very good agreement with the prediction of the Gordon–Haus theory. The experiments described in [224] and [297], reproduced soon after with EDFA-based recirculating loops and more complex error rate measurements, are milestones in the study of soliton propagation in single-mode fibers. We may regard them as providing the first experimental evidence of the potential of soliton regeneration through fiber amplifiers for high speed communications systems.

At the output of the amplifier chain, the solitons are detected in a certain time window $2t_w$. Timing jitter, induced by the Gordon–Haus effect, can prevent the solitons from arriving within this window. If solitons fail to arrive with probability $p$, assuming Gaussian statistics, the condition for $p = 10^{-9}$ is $\sigma(s) \leqslant t_w/6.1$, [292] and [293]. Substituting this condition in Eq. (5.137) gives the maximum allowable system length $L_{tot}^{max}$:

$$L_{tot}^{max} = \left( 0.1372 \frac{t_w^2 \Delta T A_{eff} L Q}{n_2 D h n_{sp}(G-1)} \right)^{1/3} \tag{5.138}$$

Equation (5.138) can be expressed as a function of the soliton data rate $B$ through the general definitions:

$$\kappa = B\Delta T \tag{5.139}$$

$$\kappa_w = Bt_w \tag{5.140}$$

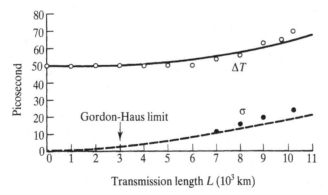

FIGURE 5.65 Evolution of fundamental soliton FWHM ($\Delta T$) and timing jitter ($\sigma$) with transmission length up to 10,000 km, measured in a recirculating fiber loop with internal loss compensated by Raman gain. From [297], reprinted with permission from the Optical Society of America, 1990.

which gives the following condition for the product $BL_{\mathrm{tot}}^{\max}$:

$$BL_{\mathrm{tot}}^{\max} \leqslant \left( 0.1372 \, \frac{\kappa \kappa_{\mathrm{w}}^2 A_{\mathrm{eff}} LQ}{n_2 D h n_{\mathrm{sp}}(G-1)} \right)^{1/3} \tag{5.141}$$

Equation (5.141) expresses the maximum performance of soliton-based communication systems as a bit rate–length product. This result is referred to as the *Gordon–Haus limit*. Consider a numerical example with representative values $\kappa = 0.1$ and $\kappa_{\mathrm{w}} = 1/3$ [292], $\alpha_{\mathrm{s}} = 0.25$ dB/km, $D = 3$ ps/nm.km, $A_{\mathrm{eff}} = 35$ $\mu$m$^2$, $n_2 = 3.2 \times 10^{-20}$ m$^2$/W, $n_{\mathrm{sp}} = 1$. From Eq. (5.141), we find a maximum system capacity $BL_{\mathrm{tot}}^{\max}$(GHz.km) $\approx 938[LQ/(G-1)]^{1/3}$. Assuming first short amplifier spacings or near-distributed amplification ($G - 1 \approx \alpha_{\mathrm{s}}L$, $Q \approx 1$), we have $BL_{\mathrm{tot}}^{\max} \approx 24.3$ THz.km. This system could then transmit soliton data at 10 Gbit/s over 2430 km or 2 Gbit/s over 12,150 km. For comparison, if we assume lumped amplification and amplifier spacings 20–40 km, we find the bit rate–length product $BL_{\mathrm{tot}}^{\max} \approx$ 21–23 THz.km. This corresponds to system lengths 400–1650 km at 2 Gbit/s and 80–330 km at 10 Gbit/s. The reduction in maximum system length associated with amplifier spacing increase is explained by the higher level of path averaged ASE power along the chain (Section 2.6). Numerical simulations of soliton timing jitter for a 9000 km, 5 Gbit/s system are shown in [294]. Experimental data of soliton transmission in very long-haul systems and countermeasures of the Gordon–Haus effect are described in Chapter 7.

The influence of ASE in single- or two-wavelength soliton interaction forces is important for the determination of system performance limits at high bit rates. A full treatment of this problem requires complex numerical simulations. the derivation of practical analytical solutions by perturbation methods (T. Georges and F. Favre, [298]) still represents a challenge.

We consider finally another type of soliton interaction, discovered in amplifier-based systems, which has *long-range* characteristics. It is not related to optical amplification, but it is important in long-haul systems, and deserves to be discussed here. Measurements of single-wavelength soliton pair interactions over $L > 1000$ km paths have revealed the existence of a phase insensitive, periodically attractive or repulsive force, even for initial pulse separations as great as $70 \, \Delta T$ or 4 ns (K. Smith and L. F. Mollenauer [299]). A possible physical explanation is the generation of dispersive wave radiation by fiber birefringence. Colliding with nearby solitons, dispersive waves could cause significant velocity shifts [299]. The effect scales as $D^2$, so its complete suppression could be achieved by using DSFs with low polarization dispersion or a polarization-maintaining fiber. Another mechanism to explain the long-range interaction is the generation of acoustic waves through electrostriction (E. M. Dianov et al., [300]–[302]). In this mechanism, the intense electric field associated with an initial soliton pulse creates a local density perturbation by electrostriction. This perturbation has a compression phase followed by a rarefaction phase. It propagates as an acoustic wave in a direction transverse to the fiber axis, with a 1 ns typical transit time through the core [300]. Reflections of the acoustic wave from the fiber cladding/jacket interface could also produce discrete acoustic mode excitation, whose spectrum depends on the cladding diameter. This transient perturbation causes a periodically damped refractive index modulation of the fiber core ($\delta n(t) \leqslant 1 \times 10^{-14}$ [302]), which produces time dependent velocity shifts in the subsequent soliton pulses. The theory of [301] takes into account the effect of cladding reflection and predicts an oscillatory behavior of the interaction force between soliton pairs with period $\approx 2$ ns, in very good agreement with the experimental data of [299]. Long-range soliton interactions affect pulse timing jitter and BER. The study [302] shows that the jitter caused by electrostrictive interaction increases with the bit rate (due to increased acoustic power in the core), according to:

$$\sigma(s) = 0.0138 D^2 B^{3/2} L_{\text{tot}}^2 \left( 1 - \frac{1.18}{B} \right)^{1/2} \tag{5.142}$$

where $\sigma$ is in ps, $D$ is in ps/nm.km, $L_{\text{tot}}^2$ is in thousands of km, and $B$ in Gbit/s. Since electrostrictive interaction scales as $D^2$, it can be considerably reduced by using low dispersion fibers, similar to reducing dispersive wave radiation. Several issues remain to be investigated: the possible enhancement of electrostriction in WDM systems (consistent with an increase of acoustic power in the core); its dependence on geometrical factors and fluctuations thereof (e.g., fiber OD, core ellipticity); and its possible suppression by appropriate choice of jacket coatings and waveguide geometry [302].

We consider finally the important issue of *soliton pulse generation*. Conventional techniques for short-pulse generation are colliding-pulse mode-locking and fiber pulse compression. Solitons can be directly generated in SMFs from externally pumped or modulated sources by soliton lasers (L. F. Mollenauer and R. H. Stolen [303]), synchronously pumped fiber Raman lasers (E. M. Dianov et al. [304]; J. D. Kafka and T. Baer [305]), and sources based on the effect of modulational instability (A. Hasegawa [306]; K. Tai et al. [307]). Communication systems require practical and compact semiconductor LDs for short-pulse signal sources. To excite $N = 1$ solitons in SMFs, the LD output pulses must be Fourier-transform-limited, free from temporal distortion and spectral chirp, and have enough peak power, as required for solitons of width $\Delta T$. The two basic methods of LD picosecond pulse generation are mode locking (ML) and gain switching (GS). Implementation of ML in semiconductor LDs usually requires external or hybrid elements such as gratings, mirrors, etalons, fiber cavities, gratings, saturable absorbers, or electro-optic switches, whose alignment and adjustment can be environmentally unstable. However, ML can also be implemented in integrated or monolithic devices. Progress in 1990–1991 has shown the possibility of generating transform-limited picosecond and subpicosecond pulses in colliding pulse–mode-locked (CPM) quantum well lasers (M. C. Wu et al., [308] and [309]; Y. K. Chen et al. [310]). The output of CPM–LDs can be boosted by an external EDFA to reach the excitation power required for $N = 1$ soliton excitation [310].

Picosecond pulses can also be generated from LDs by GS, followed by external amplification through a fiber amplifier (K. Iwatsuki et al. [222]). But the generation of transform-limited pulses by GS requires compensation methods to suppress the deleterious effect of frequency chirp. Such methods include: spectral windowing through FP filters (M. Nakazawa et al. [311]) and use of DSF for linear or nonlinear compensation (K. Iwatsuki et al. [312]; H. F. Liu et al. [313]). Soliton LD sources, whether ML or GS, exhibit large FM noise, which in long-haul systems can cause a timing jitter as important as the Gordon–Haus timing jitter (M. Ding and K. Kikuchi [314]).

Another method of soliton generation in SMFs uses the following principle: two CW signals generated by two DFBs at different frequencies and boosted with an EDFA are input to a fiber whose dispersion decreases with length. Nonlinear propagation causes a soliton pulse train to form, its repetition rate determined by the beat frequency (P. V. Mamyshev et al. [315]; E. M. Dianov et al. [316]; and S. V. Chernikov et al. [317]). Finally, a wide range of possibilities for soliton pulse generation is offered by Er-doped fiber lasers (EDFL). LD-pumped EDFLs developed in parallel with EDFAs; they have gained considerable attention over the past few years, justified by their performance, e.g., picosecond and femtosecond pulse generation, single-frequency and continuously tunable sources, their compatibility with EDFAs, and their potential applications for soliton systems. The characteristics of EDFLs, in particular when used as soliton sources, are briefly reviewed in Chapter 6.

## 5.12   OTHER TYPES OF FIBER AMPLIFIERS

In this section, the basic principles and characteristics of various types of fiber amplifiers are reviewed, in comparison with 1.5 $\mu$m EDFAs. We use the name fiber *amplifier* to designate fiber devices for traveling-wave signal amplification. We use the name *fiber laser* to designate signal light sources. The fiber amplifiers investigated to date belong to three main categories:

1. RE-doped fiber amplifiers
2. Scattering fiber amplifiers
3. Parametric or four-wave mixing fiber amplifiers

The first category uses the principle of doping glass fibers with rare earth ions such as: praseodymium ($Pr^{3+}$), neodymium ($Nd^{3+}$), erbium ($Er^{3+}$), thullium ($Tm^{3+}$) and ytterbium ($Yb^{3+}$). Fibers doped with holmium ($Ho^{3+}$) and samarium ($Sm^{3+}$) have also demonstrated as laser sources but not used as amplifiers. The second category of fiber amplifiers is based on the effects of *stimulated Raman* and *Brillouin scattering*. The third category utilizes the effect of *parametric four-wave mixing*, based on third order nonlinearity.

We summarize fundamental similarities and differences between the three types of amplifiers. The physics of scattering and parametric amplifiers is explained by a nonlinear polarization of the silica host (as a dielectric medium). This nonlinear response is third order in electric field strength, as glass can be approximated (to some extent) by a centrosymmetric medium in which second order nonlinearity is absent. In silica glass, the effects of intrinsic nonlinearity are rather weak, and can be observed only by accumulation over relatively long fiber lengths (e.g., 50 m for parametric effect, 1 km for scattering effects) [318]. In RE-doped fibers, the glass is a passive host and the medium polarization responsible for the field interaction is generated by the doping ions; the medium polarization is linear (Chapter 1). As the rare earth concentration can be made large (i.e., up to 10 wt%, limited by unwanted quenching effects, Chapter 4), the fiber lengths can be made relatively short (e.g., 0.5 m). In the three types of amplifiers, the linear or nonlinear polarizations act as an electromagnetic source, which results in the amplification of the input signal field. In the case of RE-doping, signal amplification is due to stimulated emission (accompanied by relaxation of the excited ions to the ground state); in the two other processes, light amplification is due to either stimulated scattering (accompanied by the excitation of molecules into a vibrational state) or to coherent nonlinear wave mixing.

While the rare earth ions in doped amplifiers can be described as two-, three- or four-level laser systems, scattering amplifiers correspond to three-level systems involving glass molecules (e.g. Si–O–Si). In scattering amplifiers, molecules are excited to a virtual level (3) by absorption of a pump photon; they instantly deexcite to level 2, which corresponds to the $n = 1$ vibration state, by emission of a Stokes or signal photon. The molecules then relax to the ground level (1) by annihilation of a vibrational quantum, or phonon. The molecules can also absorb a pump photon from their excited state (2) and relax to the ground level (1) by emission of an anti-Stokes photon. The effect of light amplification by stimulated scattering can be described either quantum mechanically (as a three-level process) or classically (as a wave mixing or parametric process) [319]. The parametric description of scattering corresponds to the nonlinear mixing of the pump with Stokes and anti-Stokes waves (the Stokes wave is amplified, the anti-Stokes absorbed) [319].

In RE-doping amplifiers, the pump and signal band characteristics (i.e., spectrum and center wavelength) are fixed by the rare earth atomic resonances; in scattering amplifiers, the pump can be at any wavelength (i.e., there is no pump absorption band), while the signal gain characteristics are determined by the optical (Raman) or acoustical (Brillouin) phonon spectra. While the pump can be at any wavelength in parametric amplifiers, the Stokes and anti-Stokes gain spectra are determined by optimum *phase matching* conditions. Therefore, RE-doped fiber amplifiers have a gain spectrum with fixed center wavelength and several discrete pump bands, while the other types have gain spectra with variable center wavelengths, as tuned by the pump (scattering and parametric) and phase matching conditions (parametric). In the case of scattering and parametric amplifiers, the difference between pump and peak signal wavelengths is referred to as the Stokes shift. For scattering amplifiers, the Stokes shift is fixed by the medium's phonon spectrum, while for parametric amplifiers it is determined by phase matching. In RE-doped and scattering amplifiers, the gain spectrum characteristics can be modified by the glass codopants (e.g., aluminium, germanium, and phosphorus); for RE-doped fibers, the glass host affects the Stark level positions, the transition homogeneous and inhomogeneous line widths, and the nonradiative decay characteristics; for scattering amplifiers, the host determines the phonon spectrum distribution and the Stokes shift. In RE-doped and Raman fiber amplifiers, pumping can be achieved in either or both propagation directions (i.e., forward or backward with respect to the signal), while pumping must be forward in parametric fiber amplifiers and backward in Brillouin fiber amplifiers, as imposed by phase matching (Brillouin amplifiers are also phase-matched in the forward direction, but with a zero Stokes shift). RE-doped fiber amplifiers are intrinsically polarization insensitive. In contrast, scattering and parametric amplifiers are intrinsically polarization sensitive (i.e., the pump and signal fields must have matched polarizations); but polarization scrambling in long lengths of nonpolarization-maintaining fibers creates an effective polarization insensitivity.

Unlike other amplifiers, ignoring background loss, fibers doped with certain rare earth ions (e.g., $Er^{3+}$, $Tm^{3+}$) are absorbing when unpumped because these ions have a three-level property. In all fiber amplifier types, the gain coefficient is enhanced by reducing the fiber mode size or the effective interaction area, which corresponds to SMFs with high NA and small core. Spontaneous emission, spontaneous scattering, or spontaneous mixing occur in each of the three amplifier types. For RE-doped, Raman and parametric (four-wave mixing) the minimum noise figure achievable in the high gain regime is given by the same quantum limit, 3 dB (Y. Yamamoto and T. Mukai [320]). For Brillouin amplifiers, the minimum spontaneous emission factor is given by $\langle n_v + 1 \rangle$ ($n_v$ = Boltzmann occupation number), the 3 dB quantum limit can be theoretically reached only near $T = 0\,\text{K}$ ($n_v \approx 500$ at $T = 300\,\text{K}$). Now we consider in more detail the gain spectra and pump requirements for interesting examples of all three types.

### RE-doped fiber amplifiers

The energy levels corresponding to several rare earths in glass hosts and approximate transition wavelengths are shown in Figure 4.1.

*Praseodymium-doped fiber amplifiers* (PDFAs), have considerable potential for 1.3 $\mu$m system applications. The first experiments of signal amplification at 1.3 $\mu$m in PDFAs were reported in 1991 by three independent groups (Y. Ohishi et al., [321] and [322]; Y. Durteste et al. [323]; and S. F. Carter et al. [324]). In these experiments, the fiber glass host was of the fluoride type (e.g., ZBLAN or ZBLANP), as opposed to the oxide type (e.g., silica). This fact can be explained by the $Pr^{3+}$ energy level diagram shown in Figure 4.1. The transition of interest originates from level $^1G_4$ and terminates to level $^3H_5$. There are four intermediate levels ($^3F_4$, $^3F_3$, $^3F_2$, $^3H_6$). The highest is $^3F_4$, about 3000 cm$^{-1}$ below $^1G_4$. The relatively small energy gap to the nearest level means that multiphonon relaxation from $^1G_4$ and quenching of the $^1G_4 \rightarrow {}^3H_5$ transition are both highly probable. But in glass hosts with low phonon energy, such as ZBLAN ($\hbar\omega \approx 500$ cm$^{-1}$), the probability of multiphonon decay is reduced (Chapter 4) and fluorescence quenching of the radiative transitions originating from $^1G_4$ can be suppressed. Six radiative transitions originate from level $^1G_4$. Fortuitously, the highest transition probability ($A_R = 197$ s$^{-1}$) corresponds to the 1.3 $\mu$m transition $^1G_4 \rightarrow {}^3H_5$ (for a table of transition probabilities, branching ratios, wavelengths, and cross-sections, see [322]). A comprehensive study of the phonon energies, radiative and nonradiative lifetimes, quantum efficiencies, and fluorescence spectra of $^1G_4$ of $Pr^{3+}$ in a variety of low phonon energy glass hosts can be also found in the work by D. W. Dewak et al. [325]. Two possible pump transitions are $^3H_4 \rightarrow {}^1G_4$ (9833 cm$^{-1}$ or 1.017 $\mu$m) and $^3H_4 \rightarrow {}^1D_2$ (17036 cm$^{-1}$ or 587 nm). $^3H_4 \rightarrow {}^1D_2$ is not advantageous; it is in the visible region (for which high power LDs are not available) and $^1D_2 \rightarrow {}^1G_4$ has a poor branching ratio [326]. The signal transition $^1G_4 \rightarrow {}^3H_5$ is also resonant with the transition $^1G_4 \rightarrow {}^1D_2$, a cause of signal ESA. The early experiments described in [321]–[324] used the $^3H_4 \rightarrow {}^1G_4$ transition with pump wavelength near 1.01 $\mu$m; the first measurements showed that ZBLAN PDFAs have a gain spectrum centered near $\lambda = 1.31$ $\mu$m with 35–40 nm bandwidth; the maximum gains obtained with a few hundreds of mW pump power ($P_p^{\text{in}} \approx 150$–550 mW) from Ti:sapphire laser sources were $G = 7$–15 dB, [321]–[324] and [327].

Improvements in pumping efficiency and reductions in LD pumping wavelength were investigated using $Yb^{3+}$ codoping (Y. Ohishi et al., [328] and [329]). Figure 4.1 shows the excited state level of $Yb^{3+}$ ($^2F_{5/2}$) is isoenergetic with $^1G_4$ of $Pr^{3+}$;

this makes possible $Yb^{3+}-Pr^{3+}$ energy transfer (Chapter 4). The pump transition of $Yb^{3+}$ ($^2F_{7/2} \rightarrow {}^2F_{5/2}$) is resonant with the transition $^1G_4 \rightarrow {}^1D_2$, a cause of pump ESA [326]. Although the absorption band of $Yb^{3+}$ is centered near 1.00 $\mu$m, it peaks near 975 nm and extends to 930 nm [326], amenable to pumping with the high power 980 nm LDs previously developed for EDFAs (M. Miyajima et al. [329]). Before the end of 1991, improvements in fiber design, reduction of coupling, and background loss made it possible to demonstrate efficient pumping and high gain operation ($G = 38$ dB, $P_p^{in} \approx 300$ mW) in uncodoped PDFAs pumped near $\lambda = 1.017\,\mu$m, as shown in Figure 5.66 [329]. LD pumping with $\lambda = 1.017\,\mu$m sources was also demonstrated, [327] and [328]. Experimental characterizations of $Pr^{3+}$ concentration effects, saturation output power, and noise figure can be found in studies by T. Sugawa and Y. Miyajima, [330] and [331], and Y. Ohishi et al., [332] and [333]. The best saturation output power reported in 1991 was $P_{sat}^{out} = +13$ dBm, and the lowest (high gain) noise figure was $F_o = 3.2$ dB [331]. Theoretical models for PDFAs, numerical analysis, and design optimization can be found in the studies by B. Pedersen et al. [334], P. Urquhart [335], and M. Karasek [336]. An example of system characterization (BER, power budget, penalties) using PDFAs as preamplifiers or as power boosters is described in a work by R. Lobbet et al. [337]. A detailed comparison of figures of merit of PDFAs with EDFAs and NDFAs is discussed below (F. F. Rühl [338]).

The first RE-doped fiber amplifiers (demonstrated in 1964–1970) were neodymium-doped fiber amplifiers (NDFAs). They operated near $\lambda = 1.06\,\mu$m signal wavelength (C. J. Koester and E. Snitzer [339]; G. C. Holst and E. Snitzer [340]; and B. Ross and E. Snitzer [341]). Figure 4.1 shows that $Nd^{3+}$ in glass has three transitions originating from the $^4F_{3/2}$ level, at wavelengths near $\lambda = 0.93\,\mu$m, 1.06 $\mu$m, and 1.34 $\mu$m. Researchers hoped the transition $^4F_{3/2} \rightarrow {}^4I_{13/2}$ near $\lambda = 1.34\,\mu$m could be

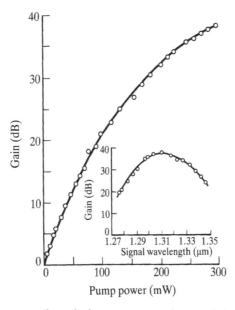

**FIGURE 5.66** Gain versus launched pump power characteristics of $Pr^{3+}$ doped fluoride fiber amplifier pumped at $\lambda = 1.017\,\mu$m; inset is the gain spectrum, centered at $\lambda = 1.310\,\mu$m. From [329], reprinted with permission from *Electronics Letters,* 1991.

used in communications systems but two adverse effects prevent efficient operation of NDFAs in this signal band. Signal ESA occurs as $^4F_{3/2} \rightarrow {}^4I_{13/2}$ is resonant with the group of closely spaced levels $^4G_{7/2}$, $^4G_{9/2}$, and $^2K_{13/2}$. The branching ratio strongly favors the two other transitions ($^4F_{3/2} \rightarrow {}^4I_{11/2}$ and $^4F_{3/2} \rightarrow {}^4I_{9/2}$). Signal ESA is stronger at wavelengths shorter than $\lambda = 1.32\,\mu m$, the region of interest for optical communications. On the other hand, NDFAs do allow pumping near $\lambda = 800\,nm$ ($^4I_{9/2} \rightarrow {}^4F_{5/2}$), a wavelength for which high power GaAs LDs are available. Several studies, driven by the investigation of $Nd^{3+}$ doped fiber lasers, made it possible to determine optimum glass hosts for NDFAs in which the effect of signal ESA is minimized (for comparative reviews of NDFA performance with different glass hosts and compositions, see M. L. Dakss and W. J. Miniscalco [342], S. Zemon et al. [343]). The theoretical analysis of dB/mW pumping efficiency in NDFAs has shown that signal ESA is actually a critical parameter that explains differences in performance for various glass materials (M. L. Dakss and W. J. Miniscalco [344]). In particular, it was observed that in fluoride hosts (ZBLAN, ZBLANP) the ESA cross sections are shifted in wavelength and strength, which makes it possible to operate the NDFA at shorter wavelengths, i.e., near $\lambda = 1.31-1.32\,\mu m$. The first fluoride NDFA was reported in 1988 (M. C. Brierley and C. A. Millar [345]), and short wavelength operation near $\lambda = 1.31-1.32\,\mu m$ was reported in 1990 (Y. Miyajima et al. [346]; M. Brierley et al. [347]). In these early experiments, 150 mW pump power near 795 nm or 496 nm would provide gains at $\lambda = 1.31-1.32\,\mu m$ of $G = 0-3\,dB$ and $G = 0-8\,dB$, respectively, [346] and [347]; for this pump power, 10 dB gains could be obtained only at longer wavelengths, i.e., $\lambda = 1.33-1.34\,\mu m$ [346], T. Sugawa et al. [348]). The poor efficiency of NDFAs operating near $\lambda = 1.3\,\mu m$ is also due to competition with the two other laser transitions at $\lambda = 0.930\,\mu m$ and $\lambda = 1.06\,\mu m$. In fluoride fibers, the strong ASE observed in the high pump regime near $\lambda = 0.880\,\mu m$ and $\lambda = 1.050\,\mu m$ (corresponding to the aforementioned transitions) results in amplifier self-saturation. A way to suppress the unwamted stimulated emission at these two wavelengths (particularly at $1.050\,\mu m$) is to introduce optical blocking filters along the fiber path (M. Øbro et al. [349]). This function can be provided by mode-coupling filters based on mechanical gratings (M. Øbro et al. [350]). Another approach uses the difference in $LP_{01}$ mode sizes between $\lambda = 1.050\,\mu m$ and $\lambda = 1.320\,\mu m$ wavelengths (A. Bjarklev et al. [351]). In this approach, the $Nd^{3+}$ ions are removed from the center of the fiber core (annular doping) to decrease the $\lambda = 1.050\,\mu m$ mode overlap; additionally, the $\lambda = 795\,nm$ pump can be coupled into the $LP_{11}$ mode for improved overlap with the active ions. The theoretical study of [351] shows that with optimum doping profile geometry and waveguide parameters, gains near $G = 15\,dB$ at could be obtained near $\lambda = 1.340\,\mu m$ for 100 mW pump at $\lambda = 795\,nm$. The effect of ND-doping confinement and waveguide optimization for either small-signal or booster amplifier applications of NDFAs has also been investigated theoretically (T. Rasmussen et al. [352]). In the small-signal case, and assuming complete suppression of ASE at $\lambda = 1.050\,\mu m$, the fiber NA should be as high as possible and the optimum cutoff wavelength be chosen close to $\lambda_c = 800\,nm$; in this case, the maximum gain at $\lambda = 1.340\,\mu m$ is near $G = 14\,dB$ with 100 mW pump at $\lambda = 795\,nm$. For booster applications, on the other hand, Nd-doping confinement is shown to improve the QCE up to a value of 80% [352]. The system characterization of fluoride NDFAs used as optical preamplifiers is described by J. E. Pedersen et al [353].

A useful figure of merit for comparing the performance of EDFAs, PDFAs, and NDFAs is the incremental dB gain $g_{inc}$ per additional milliwatt of pump power launched into the fiber (F. F. Rühl [338]). The incremental dB gain $g_{inc}$ corresponding to the three amplifier types is plotted as a function of input pump power in Figure 5.67. The different curves correspond to doped fibers with cutoff wavelength $\lambda_c = 890$ nm, core radius $a = 2\ \mu$m, and confined RE-doping. The pump wavelengths are $\lambda = 800$ nm, $\lambda = 980$ nm, and $\lambda = 1480$ nm for the EDFA (signal wavelength $\lambda = 1536$ nm), $\lambda = 1017$ nm for the PDFA (signal wavelength $\lambda = 1300$ nm) and $\lambda = 795$ nm for the NDFA (signal wavelength $\lambda = 1320$ nm, ASE at $\lambda = 1.050\ \mu$m blocked). As the figure shows, the incremental gain $g_{inc}$ increases for the EDFA (three-level laser system) while it is constant for the PDFA and NDFA (four-level systems). The low pumping efficiencies for PDFAs and NDFAs can be attributed to the short fluorescence lifetimes of their metastable levels, $\tau = 0.1$ ms for $Pr^{3+}(^1G_4)$ and $\tau = 0.5$ ms for $Nd^{3+}(^4F_{3/2})$, compared to the case of EDFAs, $\tau = 10$ ms for $Er^{3+}(^4I_{13/2})$, [338].

Now we consider RE-doped fiber amplifiers that operate at wavelengths different from $\lambda = 1.3$–$1.5\ \mu$m. Investigated to date (1992) are fiber amplifiers based on fluoride hosts (ZBLAN) and fiber amplifiers doped with $Er^{3+}$ and $Tm^{3+}$.

In fluoride EDFAs, the transition $^4I_{11/2} \rightarrow {}^4I_{13/2}$ near $\lambda = 2.75\ \mu$m (Figure 4.1) is of potential interest for ultralow loss communications, provided that the problem of wavelength independent scattering can be overcome (see for instance reviews by D. C. Tran et al. [354] and P. W. France et al. [355]). One of the difficulties of using the $\lambda = 2.75\ \mu$m transition in EDFAs for signal amplification is that the lifetime of the terminal level ($^4I_{13/2}$) is longer than that of the upper level ($^4I_{11/2}$). Methods to reduce the $^4I_{13/2}$ lifetime include quenching with another rare earth codopant (e.g., $Nd^{3+}$), or use of a pump wavelength yielding strong pump ESA. D. Ronarc'h et al., [356] and [357], used a pump wavelength near $\lambda = 642$ nm. The pump ESA corresponding to the $^4I_{13/2} \rightarrow {}^4F_{5/2}$ transition was evidenced by blue upconversion emission (Figure 4.4), and a maximum gain $G = 35$ dB could be obtained with 250 mW

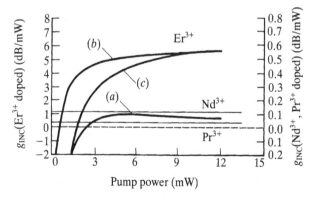

**FIGURE 5.67** Incremental dB gain $g_{inc}$ for 1 mW additional pump power, plotted as a function of input pump power for EDFAs ($\lambda_s = 1536$ nm), PDFAs ($\lambda_s = 1300$ nm) and NDFAs ($\lambda_s = 1320$ nm), assuming identical fiber design (see text). For EDFAs, the pump wavelengths are $\lambda_p = 800$ nm (a), $\lambda_p = 980$ nm (b) and $\lambda_p = 1480$ nm (c). For PDFAs and NDFAs, the pumps are at $\lambda_p = 1017$ nm and $\lambda_p = 795$ nm, respectively. From [338], reprinted with permission from *Electronics Letters*, 1991.

pump power [357]. Another unusual application of fluoride EDFAs is based on *upconversion pumping* near $\lambda = 801$ nm (T. J. Whitley et al. [358]). In this upconversion pumping scheme, the $\lambda = 801$ nm pump first excites the $Er^{3+}$ ions to the $^4I_{9/2}$ level (Figure 4.4); the ions relax (radiatively) to the lower levels $^4I_{11/2}$ and $^4I_{13/2}$; all three levels in fluoride glasses ($^4I_{9/2}$, $^4I_{11/2}$, $^4I_{13/2}$) have long radiative lifetimes (Figure 4.7) so pump ESA can occur at each one; this excites ions to the $^4S_{3/2}$ and higher levels (Figure 4.4). Spontaneous emission from the $^4S_{3/2} \rightarrow {}^4I_{13/2}$ transition, near $\lambda = 850$ nm, can then be observed. As before, signal amplification at this wavelength requires significant pump ESA to deplete the lower level population ($^4I_{13/2}$). In the experiment reported in [358], pump ESA was evidenced by green upconversion fluorescence ($\lambda = 540$ nm, corresponding to $^4S_{3/2} \rightarrow {}^4I_{15/2}$) and signal gains $\lambda = 850$ nm up to $G = 22$ dB could be achieved with 800 mW pump power at $\lambda = 801$ nm. Pumping powers in excess of 1 W are currently available from inexpensive GaAs laser LDs so upconversion pumping EDFAs could be practical devices for amplification of high speed, $\lambda = 850$ nm signals.

Another type of RE-doped fluoride fiber amplifier is based on $Tm^{3+}$ doping (TDFA). For TDFAs, the operating signal wavelengths are $\lambda = 820$ nm and $\lambda = 1480$ nm, corresponding to the $^3H_4 \rightarrow {}^3H_6$ and $^3H_4 \rightarrow {}^3F_4$ transitions, respectively, Figure 4.1 (R. G. Smart et al., [359] and [360]; T. Komukai et al. [361]). Operation at $\lambda = 805$ nm signal wavelength requires pumping in the same band as the signal ($^3H_6 \rightarrow {}^3F_4$) with pump wavelengths near $\lambda = 780$ nm; this corresponds to a two-level pumping scheme similar to 1480 nm pumping in EDFAs (J. N. Carter et al. [362]). A gain $G = 25$ dB at $\lambda = 805$ nm could be demonstrated with only 50 mW pump power at $\lambda = 780$ nm, which reflects the advantages of a high branching ratio in favor of the $^3H_4 \rightarrow {}^3H_6$ transition, as well as the absence of pump or signal ESA [360]. Using an upconversion pumping scheme (transitions $^3H_6 \rightarrow {}^3H_5$ and $^3F_4 \rightarrow {}^3F_2$, $^3F_3$), a gain $G = 25$ dB at $\lambda = 1470$ nm with a 3 dB bandwidth of 40 nm could also be obtained with 450 mW pump at $\lambda = 1064$ nm [361]. This shows *the possibility of extending the fiber operating bandwidth with a transmission window near $\lambda = 1470$ nm.* Such systems could use dispersion-flattened fiber and three types of LD-pumped optical amplifiers: PDFAs ($\lambda_s = 1300$ nm), TDFAs ($\lambda_s = 1470$ nm), and EDFAs ($\lambda_s = 1540$ nm). The $\lambda = 1470$ nm transmission band also could be also used for system monitoring applications [361]. Finally, operation near $\lambda = 1.65$ $\mu$m was also achieved in a 780 nm-pumped *silica-based* TDFA, using the $^3F_4$–$^3H_6$ transition (I. Sankawa et al. [363]).

### Scattering fiber amplifiers

The effects of stimulated Raman and Brillouin scattering in single-mode fibers (SRS and SBS), discovered nearly 20 years ago (E. P. Ippen and R. H. Stolen [364a]; R. H. Stolen and E. P. Ippen [364b]) have received considerable attention. Initially, the interest in SRS and SBS was motivated by the study of new nonlinear interactions (R. H. Stolen [318]); the realization of tunable coherent sources, modulators, and amplifiers in the near infrared (R. H. Stolen [365]; K. O. Hill et al. [366]); and the determination of fundamental limits in the power handling capacity of single-mode fibers (R. G. Smith [367]).

The physical process involved in Raman and Brillouin scattering is the interaction of light with optical phonons (SRS) or with an acoustic wave (SBS), which can be described by both classical and quantum theories (Y. R. Shen and N. Bloembergen [368]; C. L. Tang [369]). Theoretical analysis and modeling of SRS and SBS amplification and noise in single-mode fibers can be found in works by R. G. Smith [367]; J. Auyeung and A. Yariv [370]; R. W. Davis et al. [371]; R. H. Stolen et al., [264] and [372]; K. Mochizuki et al. [373]; D. Cotter [374]; I. Bar-Joseph et al. [375]; and references therein). A review of STS and SBS in single-mode fibers can also be found in G. P. Agrawal [239]. We consider here a simplified description of both effects, which makes possible a straightforward evaluation of power requirements and noise characteristics in signal amplification applications. The rate equations for Raman fiber amplifiers (RFA) and Brillouin fiber amplifiers (BFA) assume the following forms [367]:

$$\pm \frac{dn_s}{dz} = \gamma_e(n_s + 1) - \gamma_a n_s \tag{5.143}$$

In eq. (5.143), forward and backward signal (or Stokes) waves with mean photon number $n_s$ are assumed for SRS and SBS, respectively, and the coefficients $\gamma_e$, $\gamma_a$ correspond to emission and absorption. For SRS, $\gamma_a \equiv \alpha_s$ is the fiber background loss coefficient, and the emission coefficient $\gamma_e$ is defined by:

$$\gamma_e = g_R \frac{P_p}{A_{eff}} \tag{5.144}$$

where $g_R$ is the raman gain coefficient [364b] and $P_p/A_{eff}$ is the pump intensity. $A_{eff}$ is an effective interaction area defined in [318]. For SBS, Eq. (5.143) can be put into the explicit form, N. A. Olsson and J. P. Van der Ziel [376]:

$$-\frac{dn_s}{dz} = K\{(n_s + 1)(n_v + 1)n_p - n_s n_v(n_p + 1)\} - \alpha_s n_s \tag{5.145}$$

In eq. (5.145), the first term in the braces describes the creation of one Stokes photon and one phonon with the annihilation of one pump photon; the second term describes the reverse process. They stem from the action of the operators $a_s^+ a_v^+ a_p$ and $a_s a_v a_p^+$ on the quantum states $|n_s, n_v, n_p\rangle$, where $n_s$, $n_p$ are the signal and pump photon numbers and $n_v$ is the phonon population (A. Yariv [319]). For SBS, the emission and absorption coefficients can be put into the form:

$$\gamma_e = K n_p(n_v + 1) \tag{5.146}$$

$$\gamma_a = K n_v(n_p + 1) + \alpha_s \tag{5.147}$$

In Eqs. (5.146) and (5.147), the constant $K$ involves several characteristics of SBS and is related to the Brillouin gain coefficient $g_B$ through $K n_p \equiv g_B P_p/A_{eff}$. For nonpolarization-maintaining silica fibers, the peak Raman and Brillouin gain coefficients are $g_R \approx (4.7/\lambda_{\mu m}) \times 10^{-14}$ m/W and $g_B \approx 2.2 \times 10^{-11}$ m/W (R. H. Stolen [377]; Y. Aoki [378]; R. W. Tkach et al. [379]). The Raman gain coefficient $g_R$ varies as $1/\lambda$ and increases in $GeO_2$-doped fibers by a factor $(1 + 0.8\Delta)$, where $\Delta$ is the relative index difference expressed in percent [378].

Integration in Eq. (5.143) for a fiber of length $L$, while assuming forward pumping with undepleted pump ($P_p(z) = P_p(0)\exp(-\alpha_p z)$) yields the solution:

$$n_s^{out} = Gn_s^{in} + n_{sp}(G - 1) \tag{5.148}$$

where $n_s^{in}$, $n_s^{out}$ are the input and output signal photon numbers, i.e., $n_s^{in} = n_s(z = 0)$, $n_s^{out} = n_s(z = L)$ for SRS and $n_s^{in} = n_s(z = L)$, $n_s^{out} = n_s(z = 0)$ for SBS. $G$ is the gain defined by:

$$G = \exp\left\{g\frac{P_p(0)}{A_{eff}}\frac{1 - \exp(-\alpha_p L)}{\alpha_p}\right\} \equiv \exp\left\{g\frac{P_p(0)}{A_{eff}}L_{eff}\right\} \tag{5.149}$$

where $L_{eff}$ is an effective interaction length accounting for pump attenuation, and $g = g_R$ or $g = g_B$. In Eq. (5.148), the spontaneous emission factor is defined by $n_{sp} = \{G/(G - 1)\}\int(\gamma_e/G)\,dz$. While an exact analytical expression can be derived for this integral in each fiber propagation direction (E. Desurvire et al. [133]; K. Mochizuki et al. [373]), we shall assume here for simplicity that $n_{sp} \approx \gamma_e/(\gamma_e - \gamma_a)$, a good approximation for low loss fibers ($\alpha_p \approx 0.2\,\text{dB/km}$) with $L < 5$ km. Using this approximation, and Eqs. (5.144), (5.146), and (5.147), we obtain for SRS and SBS:

$$n_{sp}(\text{Raman}) = \frac{\langle n_v + 1\rangle}{1 - \dfrac{\alpha_s A_{eff}}{g_R P_p(0)}} \tag{5.150}$$

$$n_{sp}(\text{Brillouin}) = \frac{\langle n_v + 1\rangle}{1 - \dfrac{\alpha_s A_{eff}}{g_B P_p(0)}} \tag{5.151}$$

In Eq. (5.151), we used approximation $n_p \gg n_v$ and introduced the brackets to remind ourselves that $n_v$ is a mean phonon occupation number).

We consider first the gain defined in Eq. (5.149). Assuming a small core, high NA fiber of effective area $A_{eff} = 20\,\mu\text{m}^2$ ($\omega_{p,s}(\text{field}) \approx 2.5\,\mu\text{m}$), pump and signal wavelengths near $\lambda = 1.5\,\mu\text{m}$ and using the numerical values for the Raman and Brillouin coefficients $g_R$, $g_B$, we obtain the following dB/km gain coefficients:

$$G_{\text{Raman}}(\text{dB/km}) \approx 6.6 \times P_p^{in}(\text{W}) \tag{5.152}$$

$$G_{\text{Brillouin}}(\text{dB/km}) \approx 4.8 \times P_p^{in}(\text{mW}) \tag{5.153}$$

From the above results, to achieve a gain $G = 20$ dB with a 1 km fiber, SRS needs about 3 W pump power while SBS needs about 4 mW. Hence, RFAs require pump powers nearly three orders of magnitude greater than BFAs; this stems from the coefficient ratio $g_R/g_B$. Equation (5.153) also shows that high SBS gains can be achieved with relatively low pump powers (milliwatts), so standard single-mode LDs can be used as pumps.

Equations (5.150) and (5.151) show that in the high pump regime ($P_p(0) \gg A_{eff}/g_{R,B}$), the spontaneous emission factors of RFAs and BFAs reach a minimum value of $n_{sp} \approx \langle n_v + 1\rangle$. The minimum noise figure is $F = 2n_{sp} \approx 2\langle n_v + 1\rangle$. For RFAs, it has been customary in the literature to neglect the phonon contribution $\langle n_v\rangle$ to the noise figure, hence the approximation $F \approx 2 = 3$ dB, which is the quantum limit. For BFAs the minimum noise figure is $F = 2n_{sp} \approx 2\langle n_v + 1\rangle$. The mean photon occupation number $\langle n_v \rangle$ is given by Boltzmann's Law:

$$\langle n_v \rangle = \frac{1}{\exp(h\nu_x/k_B T)} \approx \frac{k_B T}{h\nu_B} \qquad (5.154)$$

where $\nu_x$ ($x = R, B$) is the Raman or Brillouin frequency shift. From Eq. (5.154), we find that for fused silica ($\nu_R \approx 13.2$ THz [379]) ($\nu_B \approx 11$ GHz [379]), the occupation number at $T = 300$ K is $\langle n_v \rangle \approx 0.13$ for SRS and $\langle n_v \rangle \approx 570$ for SBS. Thus, the minimum noise figure of BFAs is nearly 500 times, or 27 dB, higher than the quantum limit obtained in other types of fiber amplifier. More detailed calculations of gain and NF in BFAs, taking into account the effects of pump attenuation and fiber length, can be found in [376] and [379]. These calculations show that, depending on pump power and fiber length, the SBS spontaneous emission factor is typicall $n_{sp} \approx 50$–500.

The spectrum of the Raman gain coefficient $g_R$ corresponding to pure silica fibers is shown in Figure 5.68 [264]. Typical SBS amplified spontaneous emission spectra corresponding to different types of fibers are also shown in Figure 5.69 [380]. SRS and SBS differ considerably by their frequency shifts (i.e., $\nu_R = 440$ cm$^{-1}$ or 13.2 THz for SRS, and $\nu_B \approx 10.7$–11.2 GHz for SBS) and by their gain bandwidths (i.e., $\Delta\nu_R = 240$ cm$^{-1}$ or 7.2 THz for SRS, and $\Delta\nu_B = 15$–50 MHz for SBS). For SBS, the frequency shift is sensitive to the GeO$_2$ concentration in the glass core, as shown in Figure 5.69. A possible way to increase the operating bandwidth of BFAs is pump frequency modulation (N. A. Olsson and J. P. Van der Ziel [376]). The effect of pump modulation is to dither the Brillouin shift (or peak gain frequency) during the signal transit through the fiber, accompanied by a decrease of effective gain. In the experiment described in [376], a peak SBS gain $G = 15$ dB was obtained near $\lambda = 1.5\,\mu$m wavelength in a 30 km standard fiber (core diameter 8 $\mu$m), and the corresponding FWHM was $\Delta\nu_B = 15$ MHz; using pump FM modulation, the SBS gain bandwidth could be increased to $\Delta\nu_B = 150$ MHz, reducing the effective gain to $G = 5$ dB. One

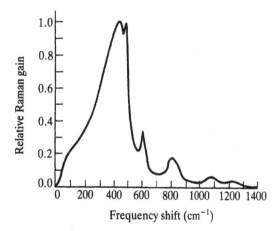

**FIGURE 5.68**    Spectrum of Raman gain coefficient $g_R$ corresponding to pure silica core fibers. From [264], reprinted with permission from the Optical Society of America, 1984.

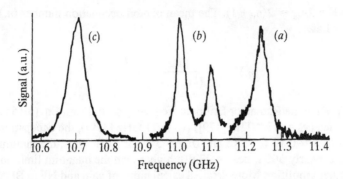

**FIGURE 5.69** Amplified spontaneous emission spectra generated by stimulated Brillouin scattering in different types of fibers: (*a*) silica core fiber, (*b*) depressed cladding fiber, and (*c*) dispersion-shifted fiber. From [380], reprinted with permission from *Electronics Letters*, 1986.

drawback of BFAs is their need for narrow line width pump lasers with precise frequency control. But high SBS gains can be generated in standard fibers using relatively low pump powers (milliwatts). Hence BFAs may have applications in dense FDM systems, where the highly frequency selective SBS process can be used for (tunable) narrowband filtering, selective amplification, demultiplexing, and FSK demodulation (A. R. Chraplyvy and R. W. Tkach [381]; R. W. Tkach et al. [382]). For instance, using pump FM, an SBS bandwidth of 600 MHz with peak gain $G = 30$ dB was achieved with 12 mW LD pump power ($\lambda = 830$ nm) in a 3.5 km fiber; this BFA was used to demodulate and amplify 250 Mb/s FSK signals [382].

The main drawback of Raman fiber amplifiers is their relatively high pump power requirement (1–10 W) for high gain operation. Another drawback is the very fast ($<100$ fs) SRS gain dynamics [372], which in the case of forward pumping, results in pump-to-signal crosstalk. For high gain operation ($G > 20$ dB), the high pump powers required ($>1$ W) must be supplied by a pulsed laser source in the forward pumping configuration (E. Desurvire et al. [383]), which limits the maximum signal bit rate. On the other hand, RFAs offer the advantages of a broad gain bandwidth ($\approx 30$–35 nm) whose peak wavelength is continuously tunable over the 1.3–1.5 $\mu$m region. Furthermore, near-uniform fiber loss compensation and periodic signal regeneration ($L = 30$–50 km) by SRS is also possible with input pump powers less than 100 mW; this is amenable to LD pumping (L. F. Mollenauer [245]; N. Edagawa et al. [384]). For the above reasons and because they have long represented the only approach, RFAs received considerable attention for long-haul communications systems during the years 1985–1988. Reviews of such applications can be found in T. Nakashima et al. [385]; L. F. Mollenauer et al. [245]; G. N. Brown [386]; S. Chi and M. S. Kao [387]; and Y. Aoki, [378] and [388].

Raman fiber amplifiers were overshadowed in the years 1989–1992 by the forceful emergence of RE-doped fiber amplifiers (EDFAs and PDFAs). On the other hand, EDFAs and PDFAs have accelerated the development of high power ($>200$ mW) LDs operating at long wavelengths (e.g., $\lambda = 1.017$ $\mu$m and $\lambda = 1.48$ $\mu$m), which could in turn provide new perspectives for RFA applications. The combination of RE-doped and Raman fiber amplifiers has further potential. For instance, 1.55 $\mu$m pulsed signals boosted by EDFAs can be used to amplify 1.66 $\mu$m signals in RFAs (T. Horiguchi et al. [389]). The 1.6 $\mu$m pulsed signals (whose tunability extends to $\lambda = 1.69\,\mu$m) could then be used in communication systems for live OTDR monitoring applications and for the localization of macro/microbending loss [389].

### Parametric fiber amplifiers

Parametric mixing was investigated in silica optical fibers in 1974–1975 (R. H. Stolen et al. [390]; R. H. Stolen [391]). The main applications of stimulated four-photon mixing (SFPM) in fibers were tunable fiber lasers, optical amplifiers, and frequency upconversion and downconversion. Theoretical analysis and modeling of parametric amplification and frequency conversion by SFPM in single-mode fibers can be found in R. H. Stolen and J. E. Bjorkholm [392]; A. Vatarescu [393]; S. J. Garth [394]; and Y. Chen and A. W. Snyder [395]. A review of SFPM and phase matching in single-mode fibers can also be found in G. P. Agrawal [239].

Traveling-wave amplification by SPFM in fibers was investigated in the years 1980–1985, [396]–[398]. In SMFs, phase matching of the SFPM process required the pump to be near the zero dispersion wavelength ($\lambda = 1.319\,\mu$m). Using a pump at this wavelength with peak power $P_p^{in} = 30$–$70$ W, gains of $G = 30$–$47$ dB at $\lambda = 1.338\,\mu$m signal wavelength could be achieved in a 30 m fiber (K. Washio et al. [396]). In another experiment, phase-matching was achieved for $\lambda = 1.319\,\mu$m pump and $\lambda = 1.57\,\mu$m signals; gains of $G = 20$–$37$ dB could be achieved in a 29 m fiber with pump powers $P_p^{in} = 50$–$120$ W (J. P. Pocholle et al. [398]). Phase matching of SFPM can be achieved in SMFs and DSFs by a variety of techniques, including waveguide design (S. J. Garth [394]; C. Lin et al., [399] and [400]; and J. P. Pocholle et al., [401] and [402]); temperature tuning (J. P. Pocholle et al. [402]); and stress-induced birefringence (K. I. Kitayama and M. Ohashi [403]).

The very high pump powers ($> 10$ W) required for high gain operation of parametric fiber amplifiers are beyond the current capability of near-infrared LD technology. But the use of RE-doped fiber amplifiers (PDFAs and EDFAs) as pump power boosters could generate new applications for LD-pumped SFPM fiber devices, as high gain, wavelength tunable amplifiers. At present, applications of SFPM for traveling-wave amplification are remote. But the possibility of boosting signals generated by LDs through EDFAs has made it possible to use SFPM for practical system applications, where SFPM is used as a means of frequency upconversion and downconversion near the zero dispersion wavelength. For instance, wavelength downconversion by SFPM of FSK signals from $\lambda = 1.555\,\mu$m to $\lambda = 1.547\,\mu$m, using an EDFA as a power booster could be demonstrated (K. Inoue and H. Toba [404]). Another important application of frequency conversion by SFPM is *high speed time division demultiplexing* (P. A. Andrekson et al. [405]). Demultiplexing is based on the generation of a frequency-converted signal by SFPM, occurring when high-rate signal data and low-rate pump pulses coincide in time; the frequency-converted signal represents demultiplexed data at the pump rate. The peak powers required to generate a frequency-converted signal by SFPM through 15 km DSF are between 15 and 35 mW. They can be supplied by prior amplification through an EDFA [405]. Time-domain demultiplexing based on this approach is described further in Section 6.5. Finally, another practical application of SFPM using EDFA boosters concerns *dispersion compensation*, as described in Section 7.2.

The gain versus pump power requirements for Er-doped, Pr-doped, Raman,

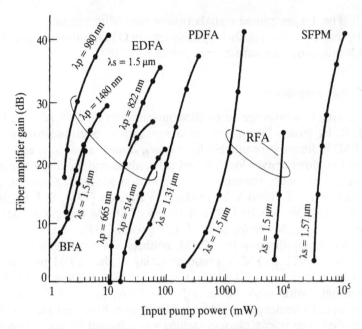

**FIGURE 5.70**   Representative data on gain versus pump power for Er-doped fiber amplifiers (EDFAs), shown for five possible pump bands, Pr-doped fiber amplifiers (PDFA), Raman fiber amplifiers (RFA), Brillouin fiber amplifiers (BFA) and SFPM fiber amplifiers, [2], [56], [57], [329], [376], [383], [402], [406], [407], and [408].

brillouin and SFPM fiber amplifiers (nonpolarization-maintaining fibers) are compared in Figure 5.70. The curves were chosen from representative experimental data reported by: N. A. Olsson and J. P. Van der Ziel [376] for BFAs; R. J. Mears et al. [56], E. Desurvire et al. [57], M. Nakazawa et al. [406], J. L. Zyskind et al. [407], and M. Shimizu et al. [2] for EDFAs; Y. Miyajima et al. [329] for PDFAs; E. Desurvire et al. [383] and K. Nakamura et al. [408] for RFAs; and J. P. Pocholle et al. [402] for SFPM fiber amplifiers. These results chart eight years of progress in the field of fiber amplifiers (1983–1991). This progress led to a reduction of pump power requirement for high gain operation by three to four orders of magnitude. With EDFAs, gains in excess of 30 dB can now be obtained with 3 mW pump power, compared to 100 mW pump power for PDFAs and 1 W pump power for RFAs. As these data were obtained in devices with optimized host materials and fiber design, we can view them as almost ultimate performance limits of fiber amplifiers.

# DEVICE AND SYSTEM APPLICATIONS OF ERBIUM-DOPED FIBER AMPLIFIERS

# CHAPTER 6

# DEVICE APPLICATIONS OF ERBIUM-DOPED FIBER AMPLIFIERS

## INTRODUCTION

This chapter reviews particular EDFA configurations not covered in Chapter 5, in addition to several EDFA-based device applications. We consider two particular EDFA configurations: (1) *distributed* and *remotely pumped* EDFAs; (2) *reflective* and *bidirectional* EDFAs. Distributed and remotely pumped EDFAs have their signal gain distributed over long fiber lengths (i.e., $L > 1$ km) either continuously (distributed EDFAs) or by discrete segments (remotely pumped EDFAs); in each cases, the pump at $\lambda = 1480$ nm, coupled at one or both fiber ends, undergoes propagation loss, which affects the device characteristics (Section 6.1). Reflective and bidirectional fiber amplifiers (Section 6.2) propagate signals in both transmission directions. In reflective EDFAs the signals are amplified in a double-pass scheme by use of a signal end-mirror, while in bidirectional EDFAs the signals are coupled at both fiber ends. Reflective EDFAs include the double-pass pumping configuration (using a pump end-mirror). The characteristics of bidirectionally pumped EDFAs are not considered here, as many examples were discussed in previous chapters. We review duplex transmission links based on various configurations of bidirectional EDFAs. We describe optical time domain reflectometry (OTDR), important for monitoring and fault location in EDFA-based transmission systems.

In Section 6.3 we address *automatic gain control* (AGC) and *automatic power control* (APC) in EDFAs. We review various AGC and APC schemes involving feedforward or feedback loops, as well as their frequency response and dynamic range. Spectral gain equalization in EDFAs, in particular spectral gain flattening, are of central importance for WDM system applications. In Section 6.4 we describe various gain equalization schemes investigated to date.

A second part of this chapter concerns *device applications* of EDFAs. Aside from their usual applications in communication systems as power boosters, in-line amplifiers, and preamplifiers, EDFAs can be implemented in a variety of fiber-based devices for all-optical signal processing applications. *Optically controlled switches and gates* using EDFAs, which have important applications in time domain demultiplexing and time/space/wavelength switching are described in Section 6.5. *Recirculating delay lines*, Section 6.6, are fiber loops that incorporate one or several EDFAs as loss compensators. These active fiber delay lines have been extensively used for the study of ultralong ($L > 10{,}000$ km) signal propagation and nonlinear interactions and may have important applications in fiber networks as optical memories, and in fiber sensors as gyroscopes.

*Fiber lasers* based on EDFAs as intracavity gain elements or Er-doped fiber lasers (EDFLs) have revolutionized the field of active fiber devices. Fiber lasers are very important and the amount of research in this field could fill an entire book. But Section 6.7 confines itself to applications of EDFLs that are relevant to optical communications systems and their main characteristics. These include CW-tunable EDFLs, narrow line width EDFLs, mode-locked EDFLs, and soliton generators.

## 6.1   DISTRIBUTED AND REMOTELY PUMPED FIBER AMPLIFIERS

Much attention has been given to distributed EDFAs for their potential as lossless transmission lines in communication networks and long-haul soliton systems. The concept of distributed amplification was first considered for soliton systems using stimulated Raman amplification in the fiber (A. Hasegawa [1]). Distributed EDFAs, which are based on the same concept, are made of long strands (i.e., $L = 1$–$100$ km) of weakly doped fibers (S. P. Craig-Ryan et al. [2]; J. R. Simpson et al. [3]); these fibers are pumped at a power level which gives net loss compensation or signal transparency ($G = 1$). Because the effect of pump background loss is no longer negligible for long lengths, the pump wavelength is usually chosen near $\lambda = 1480$ nm, the closest possible to minimum absorption. The pump is also absorbed by the $Er^{3+}$ ions along the length, so uniform inversion, to give constant gain coefficient and exact loss compensation at any point in the fiber, is possible only for relatively short fiber lengths ($L < 10$ km) and low doping concentrations. This is true for unidirectional and bidirectional pumping. Thus, in order to cancel the overall fiber loss and achieve transparency over long EDFA lengths, the gain coefficient must vary with fiber coordinate: positive near the highest pump and negative elsewhere. Such variation of gain coefficient causes *signal power excursion*.

The effect of signal power excursion is schematically illustrated in Figure 2.18*b*, for forward, backward, and bidirectional pumping. Figure 6.1 shows experimental OTDR signals obtained in a 55 km bidirectionally pumped DSF–EDFA (D. L. Williams et al. [4]). The two traces in the figure correspond to the small-signal and saturated gain regimes, respectively. There are three regions in this distributed EDFA: from $z = 0$ to $z = 15$ km (approx.), the gain coefficient is positive, causing a maximum small-signal amplification $G = 6$ dB; from $z = 15$ to $z = 40$ km, the gain coefficient becomes negative, causing a net signal loss $\Delta G = 12$ dB; from $z = 40$ to $z = 55$ km, the gain coefficient is again positive, causing the signal power to recover to its initial level. Fiber transparency is therefore relevant only with reference to the fiber output end; before the signal reaches the fiber output end, it undergoes significant power excursion. OTDR signals for forward pumped distributed EDFAs, at different pump powers are shown in Figure 5.43 (M. Nakazawa et al. [5]). At transparency ($P_p^{in} = 25$ mW), the signal power excursion is about $\Delta G = 6$ dB. By using relatively

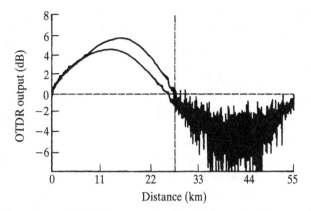

**FIGURE 6.1** Experimental OTDR traces showing 1536 nm signal power distribution along fiber length for a 55 km distributed EDFA with bidirectional pump at $\lambda = 1480$ nm; the top trace corresponds to the small-signal regime, and the bottom trace to the saturated gain regime. From [4], reprinted with permission from *Electronics Letters,* 1991.

low $Er^{3+}$ doping concentrations (e.g., $\rho = 0.05$–$01$ ppm) and relatively short EDFA lengths ($L \leqslant 10$ km), it is possible to significantly reduce the signal power excursion to a regime of near-uniform fiber transparency. For instance a power excursion $\Delta G < 0.2$ dB over $L = 10$ km could be demonstrated in a distributed EDFA with concentration $\rho = 0.06$ ppm (D. L. Williams et al. [6]). In this experiment, the pump was bidirectional with $P_p^{in} \approx 5.5$ mW input at each end. Another possibility for minimizing signal power excursion while using unidirectional pumping requires higher pump power and doping concentrations. In this case, the pump level must be high enough to generate maximum inversion at any point in the fiber. Figure 6.2 shows corresponding OTDR signals obtained for various pump powers in a 10.2 km EDFA

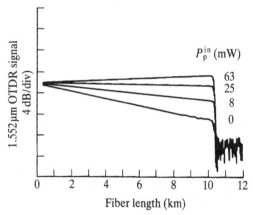

**FIGURE 6.2** Experimental OTDR traces showing 1550 nm signal power distribution along fiber length for a 10.2 km distributed EDFA with forward pump at $\lambda = 1480$ nm, for different input pump powers $P_p^{in}$; there is totally uniform cancellation of fiber loss ($\alpha_s = 0.68$ dB/km) for $P_p^{in} = 25$ mW. From [5], reprinted with permission from the Optical Society of America, 1990.

with $\rho = 0.1$ ppm (M. Nakazawa et al. [5]). Totally uniform loss compensation ($\Delta G \approx 0$) is achieved for $P_p^{in} = 25$ mW. The study of [5] shows that similar results can be obtained for 10 km EDFAs with 980 nm pumping, although the background loss is significantly higher ($\alpha_p' \approx 1.0$ dB/km).

The pump power and Er-doping requirements for achieving transparency in distributed EDFAs with unidirectional or bidirectional pumping can be determined with an exact analytical model (Chapter 2). Examples for 1480 nm pumping are shown in Figures 2.21, 2.24, and 2.25 (E. Desurvire [7]; D. N. Chen and E. Desurvire [8]; and K. Rottwitt et al. [9]). Other theoretical data concerning the effect of fiber background loss on the required pump power and Er-doping concentration on signal power excursion can be found in J. R. Simpson et al. [3].

Fabrication techniques for distributed EDFAs with low doping concentrations and low background loss ($\alpha_s' < 0.5$ dB/km) are based on the MCVD process with volatile halide and solution doping methods (S. P. Craig-Ryan et al. [2]), and seed doping methods (J. R. Simpson et al. [3]). In the seed-doping method, a seed step-index fiber with relatively high Er-doping concentration (e.g., $\rho = 1000$ ppm) is introduced in the deposition tube, prior to collapse. As the fiber is drawn from the collapsed preform, the doping region remains confined near the center of the core, and the average (or effective) concentration ($\rho\Gamma_{p,s}$) can be modified by controlling the fiber diameter. The advantage of the seed method is to separate the core fabrication/deposition process (which can be graded-index for DSF design) from the Er-doping process (which relates to the seed fiber). Aside from the improved flexibility and control of effective Er-doping concentrations, this technique yields a high doping uniformity ($\Delta\rho < 1\%$) in fibres of several kilometers [3]. Glass impurities introduced by the seed fiber or by liquids (in the solution doping technique) can also produce high background loss. Background loss coefficients near the standard value ($\alpha_s' \approx 0.25$ dB/km at $\lambda = 1.55$ $\mu$m) could be achieved with the volatile halide method [2].

The noise figure (NF) characteristics of distributed EDFAs and its dependence on pump power, fiber length, and $Er^{3+}$ concentration are described in detail in Section 2.6. In particular, the theory shows that a maximum noise figure the optimum concentration, which minimizes the pump power required for transparency (whether unidirectional or bidirectional) (D. N. Chen and E. Desurvire [8]). For instance, 1480 nm pumped EDFAs with lengths $L = 10$–25 km (background loss $\alpha_s' \approx 0.5$ dB/km) have maximum noise figures $F_o = 8.5$–9 dB, as illustrated in Figure 2.24. In the ideal case of complete and uniform inversion, obtained with 980 nm pumping, the distributed EDFA noise figure is given by $F_o = 1 + 2\alpha_s'L$, 5.2–8.3 dB in the previous example. The higher NF found for 1480 nm pumping is explained by two causes: (1) the low medium inversion corresponding to the minimum pump power, and (2) the incomplete inversion corresponding to the two-level pumping scheme. The theoretical dependence of NF on signal wavelength is shown in Figure 2.25, which corresponds to a 100 km EDFA bidirectionally pumped at 1480 nm (K. Rottwitt et al. [9]). The NF is concentration dependent; the lowest values correspond to the highest required pump powers and are achieved at long signal wavelengths near $\lambda = 1.55$ $\mu$m. For the range of concentrations investigated in this example, the NF varies between $F_o = 13.5$ dB and $F_o = 18$ dB. Figure 6.3 shows experimental/theoretical gains and noise figures measured as functions of pump power for a 1480 nm bidirectionally pumped EDFA with length $L = 9.3$ km, for signal wavelengths $\lambda = 1.535$ $\mu$m and $\lambda = 1.551$ $\mu$m (K. Rottwitt et al. [10]). The figure shows that near transparency

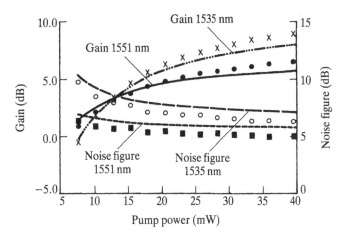

**FIGURE 6.3** Experimental (marks) and theoretical (lines) gain and noise figure versus pump power measured in a 9.3 km distributed EDFA with bidirectional pump at 1480 nm, corresponding to signal wavelengths $\lambda = 1.535\,\mu m$ and $\lambda = 1.551\,\mu m$. From [10], reprinted with permission from the Optical Society of America, 1993.

($G = 0\,dB$), the noise figures are $F_o = 10\,dB$ and $F_o \approx 6\,dB$ for the short and long signal wavelengths, respectively, in agreement with the theoretical predictions. Optimum fiber design considerations for soliton systems based on 1480 nm pumped, distributed EDFAs can be found for instance in the study by K. Rotwitt et al. [11].

For EDFA lengths $L > 10\,km$, the 1480 nm pump can also be amplified by SRS, using an auxiliary pump source (D. M. Patrick et al. [12]). In the experiment described in [12], the auxiliary pump source was an Nd:YAG emitting at $\lambda = 1319\,nm$; its frequency difference with 1480 nm pump falls into the $\Delta\nu = 800\,cm^{-1}$ Raman gain peak (Figure 5.68). The additional effect of SRS results in gain enhancement at the 1.5 $\mu m$ signal wavelength, which extends the range over which fiber transparency can be achieved, or reduces the power requirement at 1480 nm. With an auxiliary pump wavelength $\lambda = 1389\,nm$ corresponding to the peak Raman gain ($\Delta\nu_R = 440\,cm^{-1}$), we can estimate from the data of [12] that the power required for the SRS enhancement is about 50 mW. As the 1389 nm pump power required for SRS enhancement is relatively high in comparison to the 1480 nm pump power required for transparency in the absence of SRS (i.e., $P_p^{in} = 17\,mW$), this approach is not as practical as using a 1480 nm pump source with higher output power.

Recent developments in EDFA-based systems, described in the next chapter, have shown that both linear and soliton data transmission could be achieved over ultralong distances using lumped, closely spaced amplifiers ($L = 5$–$35\,km$). Interest in distributed EDFAs has consequently waned. But it could be revived by developments in fiber networks, where lossless transmission lines in bus or ring topologies (Section 7.5) could be significantly advantageous.

*Remotely pumped* EDFAs are illustrated by Figure 6.4a. A remotely pumped EDFA can be viewed as special case of the distributed EDFA, in which fiber segments are alternately doped and undoped; it can also be viewed as a lumped fiber amplifier chain with its pump coupled at either or both ends of the chain, hence the qualification remotely pumped. This configuration is also called quasidistributed. Its main

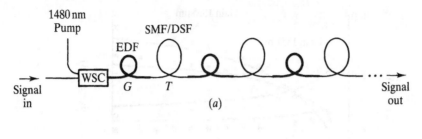

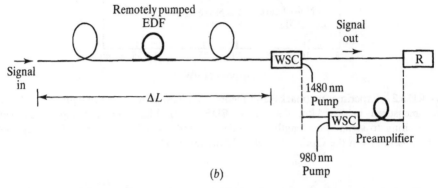

**FIGURE 6.4** Remotely pumped EDFA configurations: (*a*) for lossless transmission lines, (*b*) for transmission length enhancement. Bold lines represent strands of Er-doped fibers (EDF), thin lines represent strands of undoped communication fibers (SMF or DSF); WSC = wavelength selective coupler, R = receiver.

advantage is that several closely spaced EDFAs share the same pump, reducing complexity and cost. Its drawback is that a fraction of the pump power is lost by propagation over the undoped sections (transmission fiber), increasing pump power requirements. As in the case of distributed EDFAs, the goal of remotely pumped EDFAs is to achieve a transparent transmission line with minimal power excursion and large spacing between the pump stations. Remotely pumped EDFAs can also be used to extend the system transmission length prior to photodetection. In either configuration, minimal pump loss through the undoped fiber segments requires 1480 nm pumping.

Remotely pumped EDFAs were first used in recirculating loops for the study of ultralong distance soliton propagation (L. F. Mollenauer et al. [13]). The experiment described in [13] uses a 75 km amplifier chain closed upon itself; the chain is made of the three EDFA/DSF (25 km) elements; transparency is achieved by 1480 nm bidirectional pumping with 50 mW pump power input at each end (i.e., EDFA1 input end and EDFA3 output end). In this configuration, the copropagating pump output from EDFA1 undergoes approximately a 7.5 dB loss through the 25 km DSF strand, before reaching EDFA2, and another 7.5 dB loss before reaching EDFA3, as in the case of the counterpropagating pump. The difference in input pump power between the EDFAs means their lengths must be adjusted so their gains are identical.

Remotely pumped EDFAs can be implemented in lossless transmission lines or as a means to extend the system transmission length, as shown in Figures 6.4*a,b*, respectively. Lossless transmission lines can be used in bus and ring networks, with the advantage that the cost of pump diode components is shared by a large number of end-users. Based on this principle, a transparent fiber bus with an equivalent transmission length of 65 km, a single 1480 nm pump LD with 50 mW launched power and four equally spaced EDFAs could be demonstrated (D. N. Chen et al. [14]). The second type of application concerns transmission distance enhancement (T. Rasmussen et al. [15]). Figure 6.4b show how an in-line EDFA with remote pumping makes it possible to increase the system transmission distance, with the advantage that the pump is located at the receiver end. The study of [15] considers the two configurations shown in Figure 6.4b, including an EDFA preamplifier to extend the system transmission length. For $10^{-9}$ BER, the increases in system transmission distance are $\Delta L = 120$ km and 125 km with and without the preamplifier, respectively; in each case, the power requirement for the remotely pumped EDFA is $P_s^{in} = 50/75$ mW, and its optimum location is 45/50 km away from the receiver [15]. Our two examples illustrate the advantages of remotely pumped EDFAs for terrestrial systems and fiber networks.

## 6.2 REFLECTIVE AND BIDIRECTIONAL FIBER AMPLIFIERS

In *reflective* EDFAs, a device (e.g. mirror, grating) is placed at the EDFA output end to reflect either the pump or the signal; double-passing the pump or signal through the EDFA produces a net gain enhancement. *Bidirectional* EDFAs allow signals to propagate in both fiber directions (bidirectional pumping is analyzed in previous chapters). We describe specific characteristics of gain, saturation, and noise for reflective and bidirectional EDFAs.

Two basic layouts of reflective EDFAs are shown in Figure 6.5b,c,d and compared to the configuration in Figure 6.5a, which corresponds to standard forward pumped EDFAs (S. Nishi et al. [16]; V. Lauridsen et al. [17]). In Figure 6.5b, the pump is reflected back into the EDFA through a dichroic mirror (this can also be achieved through a WSC followed by a broadband mirror); in Figure 6.5c, both pump and signal are reflected through a broadband mirror; Figure 6.5d is similar to Figure 6.5c with additional filtering of ASE. The optical circulators in Figure 6.5c,d configurations circulators could be replaced by 3 dB fiber couplers, which would cause an additional 4 dB round-trip loss and 3 dB noise figure penalty.

The theoretical small-signal gain versus pump power characteristics of Figure 6.5a,b,c configurations and corresponding dB/mW gain coefficients are shown in Figure 6.6, assuming 1480 nm pumping and EDF parameters $L = 14$ m, $\lambda_s = 1.549$ $\mu$m, $\alpha_p = 0.86$ dB/m, $\alpha_s = 1.52$ dB/m (C. R. Giles et al. [18]). In this example, the EDFA gain is relatively small ($G < 25$ dB) so self-saturation by ASE can be neglected. Pump reflection (only) produces an increase of the gain coefficient from 7.5 dB/mW to 12 dB/mW, in comparison to the standard configuration. Signal reflection (only) produces a more dramatic enhancement by double-pass amplification; the gain is nearly twice that of the standard, single-pass configuration. The double-pass gain is slightly lower than twice the signal-pass gain as there is enhanced pump absorption in the double-pass case. Reflection of both pump and signal results in the maximum possible improvement of the gain coefficient, i.e., from 7.5 dB/mW to 22.5 dB/mW. In the above examples, the excess loss introduced by WSC and optical circulators is not accounted for. Calculations that include EDFA self-saturation confirm the even

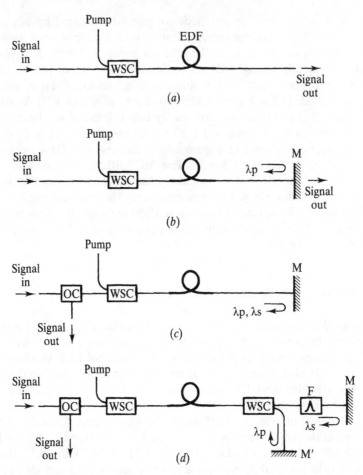

**FIGURE 6.5** Basic EDFA configurations: (a) standard forward pumped EDFA; (b), (c), (d) reflective EDFAs. In (b), a dichroic mirror M is used to reflect the pump and transmit the signal; in (c), both pump and signal are reflected through a broadband mirror M, and the output signal is retrieved via an optical circulator OC; configuration (d) is similar to (c) but uses an additional optical filter F to block the ASE.

greater gain and dB/mW coefficient improvements expected when using the Figure 6.5d configuration with a bandpass filter to prevent ASE feedback (V. Lauridsen et al. [17]).

The study of [17] shows that the noise figure of the other configurations is always higher than the noise figure of the standard configuration. For both 980 nm and 1480 nm pumping, the NF for the Figure 6.5c,d configurations are about 1.5 dB higher than the standard configuration. This accounts for the 1.5 dB excess loss in the optical circulator. Theoretical analysis of EDFAs pumped near 800 nm shows that the Figure 6.5b configuration is more effective than bidirectional pumping (B. Sridar et al. [19]). The study of [19] shows that at relatively low pump powers (i.e., $P_p^{in} = 10$ mW) both gain and NF are improved, while at high pumps (i.e., $P_p^{in} > 100$ mW) only the NF is improved. The improved NF is 1.5 dB lower than the bidirectionally pumped NF [19].

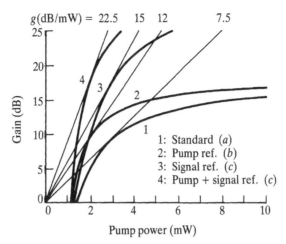

**FIGURE 6.6** Theoretical small-signal gain versus 1480 nm pump power characteristics for EDFA configurations in Figure 6.5, and corresponding dB/mW gain coefficients: (1) standard, forward pumped EDFA, configuration (*a*) (2) EDFA with reflected pump only, configuration (*b*) (3) EDFA with reflected signal only, configuration (*c*) (4) EDFA with reflected pump and signal, configuration (*c*). From [18], reprinted with permission from the Optical Society of America, 1991.

The double-pass pumping configuration in reflective EDFAs can be realized by using different types of pump reflective components at the EDFA output end: (1) a dichroic mirror, and (2) a broadband mirror preceded by a WSC, or a chirped Bragg fiber grating filter (M. C. Farries et al. [20]). Aside from fiber compatibility and low insertion loss, chirped Bragg fiber gratings have high pump reflection coefficients (i.e., 99%) with broad bandwidths amenable to multimode LD pumping [20]. But fiber-etched devices do have the drawback of polarization sensitivity (sometimes compensated by splicing two filters with orthogonal orientations). Finally, EDFA gain enhancement through double-pass pumping can be obtained by frequency-locking the pump LD (H. Masuda et al. [21]). This enhancement, mostly relevant to the case of 980 nm LD pumping, is explained by the EDFA gain dependence in pump wavelength (Figures 5.46 and 5.47). A double-pass pumping configuration with narrowband pump feedback results in pump LD injection locking (the LD has an AR-coated output facet). A coarse tuning of the narrowband filter near the peak pump wavelength results then in EDFA gain maximization, as shown in [21].

*Bidirectional* EDFAs offer the advantages of *transmission bidirectionality, inter-activity, capacity doubling,* and *system monitoring.* By using *fiber pairs,* unidirectional transmission links with optical isolators can be made bidirectional and interactive. A fiber pair doubles the number of components in the system and there is no possibility of system monitoring (i.e., fault location, telemetry) if the two fibers used are independent. Economic and practical considerations may lead to choose single-fiber links with bidirectional propagation, which requires bidirectional EDFAs.

The simplest layout for a bidirectional EDFA is shown in Figure 6.7*a*. It is an EDFA without optical isolators, bidirectionally pumped to achieve identical saturation and noise characteristics in both directions. Signals are input at both ends of the EDFA ends so this configuration is also called *bidirectional signaling.* Section 5.8 shows that for EDFA-based direct detection systems with no optical isolators, the

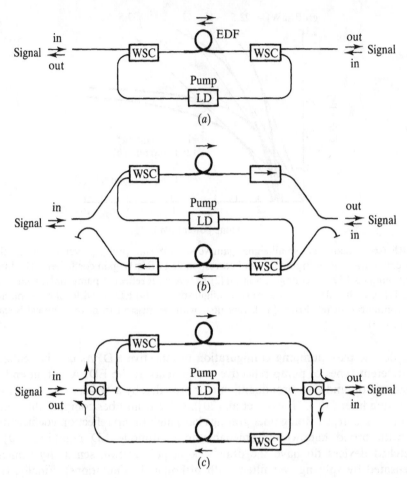

**FIGURE 6.7**   Three basic configurations for bidirectional amplification: (*a*) using a single EDFA, (*b*) using two EDFAs with 3 dB splitters and isolators, and (*c*) using two EDFAs with optical circulators.

effect of reflections (discrete or distributed) is a BER penalty of the form $P(dB) = -5\log[1 - 144G^2R_{eff}^2]$, where $R_{eff}$ is the effective reflection coefficient and $G$ is the EDFA gain. In the case of discrete reflections at each end of the EDFA ($R_1, R_2$), the effective reflection coefficient is $R_{eff} = \sqrt{R_1 R_2}/2$; in the case of Rayleigh backscattering, it is $R_{eff} = R_{bs}/\sqrt{2}$ ($R_{bs} = -31$ dB for DSF and $R_{bs} = -34$ dB for SMF). Thus, assuming that reflections are due exclusively to Rayleigh backscattering, a power penalty $P < 1$ dB, requires that the EDFA gains be limited to less than 18–20 dB. A drawback of using a single EDFA for bidirectional amplification, or bidirectional signaling, is an increased saturation effect in each direction. This enhanced saturation can be compensated at the expense of a higher pumping power. In bidirectional EDFA chains, the ASE is allowed to build up in both propagation directions. This can cause gain saturation in the stages located near the system ends so narrowband optical filtering may be required for long systems. But narrowband filtering restricts the scope for WDM.

At each amplification stage in Figure 6.7b,c, counterpropagating signals are split into separate branches, amplified, then recombined into the trunk (C. W. Barnard et al. [22]). These configurations are based on unidirectional signaling. In Figure 6.7b, splitting/recombining is achieved through 3 dB couplers. Optical isolators are used for unidirectional propagation in the two branches; in this example they include forward pumped EDFAs. A single LD can be used to pump both EDFAs, using a passive 3 dB fiber splitter for the LD output. The 3 dB couplers causes a 4 dB signal loss (which can be compensated by the EDFAs) and a 3 dB noise figure penalty; the required pump power is doubled. In Figure 6.7c, polarization independent optical circulators are used to split and recombine the two counterpropagating signals. Assuming ideal circulators, it has the advantage that no extra loss or noise figure penalties are incurred by the signals and no optical isolators are required. The backward ASE generated by each EDFA is blocked in the fourth port of each circulator. An alternate version of Figure 6.7d is shown in Figure 6.8. In this configuration, similar to that used in an OTDR experiment by Y. H. Cheng et al. [23], the pump is coupled into the upper and lower branches through the two optical circulators. While there is no need for WSC couplers, the pump output from the EDFs must be blocked by two isolators in order to prevent feedback into the LD. An optical circulator costs nearly twice as much as the isolator and coupler it replaces, and circulator insertion loss is currently 1 dB higher than isolator insertion loss [22]. In Figure 6.7b,c and Figure 6.8, the light coming from the upper branch after being backscattered or reflected in the system is then amplified through the lower branch and vice versa. To alleviate corresponding penalties, optical bandpass filters of different center wavelengths can be inserted in each branch. Each filter will pass only one type of signal [22].

Very few experiments of bidirectional (or duplex) transmission through EDFAs have been reported to date (1993). The first experiment used 10 Mbit/subcarrier modulation of one 1480 nm backward pump LD; the EDFA gain is insensitive to pump crosstalk at this modulation rate (Chapter 5), so no penalty was observed for the forward propagating signal (M. Suyama et al. [24]). A long-haul implementation of a duplex system based on this scheme would require amplifiers operating in the 1480 nm signal band. Bidirectional amplification of 1.5 $\mu$m signals based on Figure 6.7a with a single EDFA, was demonstrated later for a 95 km system operating at $B = 622$ Mb/s (J. Haugen et al. [25]). This experiment showed that the largest system

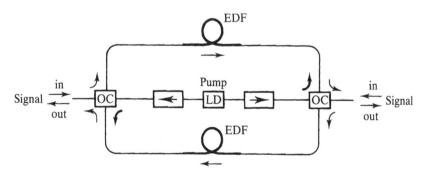

**FIGURE 6.8** Alternate configuration for bidirectional amplification, similar to Figure 6.7c. It uses the optical circulators to combine the pump and the signals into the EDFAs.

gain or power margin improvement (20–25 dB) is achieved when the bidirectional EDFA is placed in the center of the link. The improvement reduces to 7–9 dB when the EDFA is placed near either terminal end. When a bidirectional EDFA is used at either terminal end, it plays the roles of power amplifier and preamplifier. Its gain is most susceptible to saturation and its ASE noise is maximized (see Type A and Type C configurations in Chapter 2). In single-EDFA duplex systems, the optimum EDFA location is at midspan, where ASE noise and saturation effects are reduced. In other experiments, a bidirectional EDFAs (Figure 6.7a) was located at each terminal to obtain power margins 53 dB and 42 dB at bit rates $B = 591$ Mbit/s and $B = 622$ Mbit/s, respectively (K. Kannan and S. Frisken [26]; J. Haugen et al. [27]).

Consider duplex transmission systems based on chains of bidirectional EDFAs. To balance the signal powers input to each amplifier, the EDFAs could be spaced according to Figure 6.9. In this layout, all EDFAs operate as in-line repeaters. The stages nearest to the terminal ends are located at half the amplifier spacing, which ensures equal and minimal input signal powers for each stage. Using Figure 6.7b,c configurations, the input signal powers need not balance at each stage, since copropagating and counterpropagating signals are amplified in separate EDFAs. For Figure 6.7a (bidirectional signaling), the effects of gain and ASE compression must be carefully analyzed at each stage in order to assess the system output SNR (J. Freeman et al. [28]. Further studies are necessary for design optimization and evaluation of bit rate × distance limits of duplex EDF systems based on bidirectional EDFA configurations.

An important application of bidirectional transmission is *system monitoring*. Optical time domain reflectometry (OTDR) is used to supervise EDFA gain characteristics along the link and to locate/prevent possible faults. The dynamic range of OTDR can first be enhanced by using an EDFA power amplifier (L. C. Blank and D. M. Spirit [29]). For unidirectional systems using optical isolators, OTDR

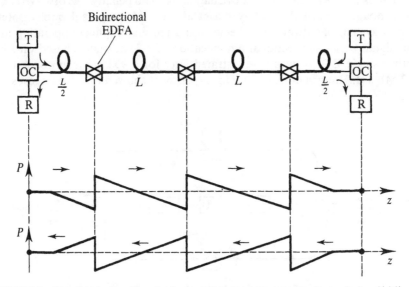

**FIGURE 6.9** Possible layout for a duplex communication system using a chain of bidirectional EDFAs, and corresponding power evolution versus length for the two signals; T = transmitter, R = receiver, OC = optical circulator, L = amplifier spacing.

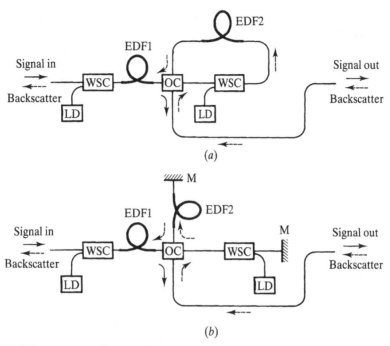

**FIGURE 6.10** Two possible configurations for OTDR implementation in unidirectional EDFA-based transmission systems without optical isolators, using a single optical circulator OC at each stage. EDF1 is used for signal loss compensation. EDF2 is used to compensate additional loss incurred by the backscatter by multiple pass through the OC [31].

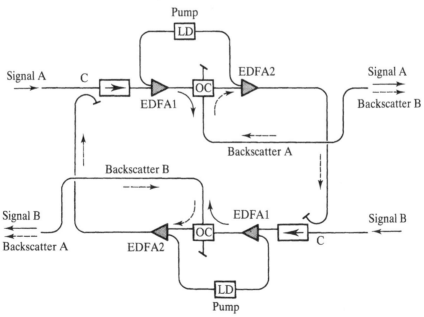

**FIGURE 6.11** Layout for OTDR monitoring in a two-way communication system based on a pair of EDFA chains. In this recursive layout, backscatters generated by signals A and B are carried without attenuation by trunks B and A, respectively. C = coupler, OC = optical circulator [32].

can be implemented at each amplifier stage; this offers the possibility of on-board fault location (J. C. MacKichan et al. [30]). For unidirectional systems without optical isolators, OTDR can be implemented using the in-line EDFA repeater configurations shown in Figure 6.10 (Y. Sato and K. I. Aoyama [31]). A detailed theoretical and experimental study of OTDR based on these configurations is described in [31]. Most generally, OTDR can be implemented in bidirectional EDFA chains based on any of the configurations shown in Figures 6.7 and 6.8, as demonstrated for instance in [23].

Gain saturation and backscattering may fundamentally limit the performance of duplex EDFA-based systems. The only alternative for two-way transmission is the fiber pair configuration, based on unidirectional fiber trunks with optical isolators. How can we ever implement full-span OTDR in a fiber pair configuration? Consider

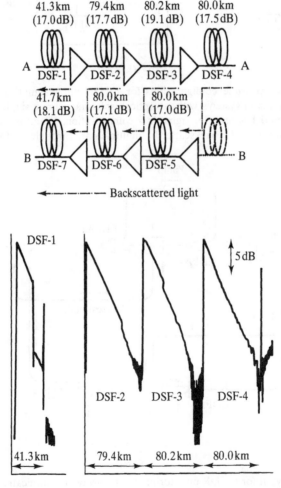

**FIGURE 6.12**   Experimental OTDR trace obtained in a 281 km EDFA chain pair including four DSF sections, based on the recursive layout shown in Figure 6.10. From [32] © 1991 IEEE.

the recursive system layout in Figure 6.11 (Y. Saito and K. I. Aoyama [32]). In this proposed layout, the fiber trunk A (where the modulated signals propagate from left to right) carries the OTDR or backscatter signal corresponding to fiber trunk B (where the modulated signals propagate from right to left), and vice versa. The coupling of backscatter signals from trunks A, B into trunks B, A is performed by two optical circulators. Fiber trunks A, B are truly unidirectional EDFA chains, whether signals propagate from left to right or from right to left. For the modulated signals A and B, the role of EDFA1 is to compensate for propagation loss incurred in the fiber trunk, coupler (c), isolator, and circulator; for the A and B backscatters, the role of EDFA2 is to compensate the loss of the additional circulator crossing. The loss incurred in coupler (c) can be reduced by using WSCs if signals A and B have different wavelengths. An experimental OTDR trace, obtained in a 281 km pair of EDFA chains, including four DSF sections, and based on the previous principle is shown in Figure 6.12 [32].

## 6.3 AUTOMATIC GAIN AND POWER CONTROL

The nonlinear response of EDFAs to large input signals results in undesirable power variations, a cause of BER penalty. In the case of amplifier chains, the effect of gain saturation at any EDFA stage is particularly detrimental, as the chains are designed to operate at exact signal transparency. In WDM systems, the total power resulting from the superposition of several optical channels may randomly vary in time (as in the case of packet-switched networks), which causes unwanted signal power transients and low frequency interchannel modulation (Section 5.9). In EDFA-based systems, the small polarization sensitivity of optical components accumulates along the chain and signal polarization wandering can result in temporary changes in system loss. Furthermore, long-term aging of pump LDs in EDFAs can also result in a gradual increase of system loss. The solution to the above problems is *automatic gain control* (*AGC*). AGC maintains the EDFA gain at a fixed level during transient signal perturbations or changes in system loss. AGC can also perform the important function of *gain linearization.* Some device and system applications (e.g., nonlinear optical switching, fiber networks) require a constant output signal power so output power fluctuations must be suppressed. This is achieved by *automatic power control (APC).* AGC fixes the gain and the output signal power varies; APC fixes the output signal power and the gain varies. We review various schemes for AGC and APC.

Implementation of AGC requires three elementary functions: (1) the detection of signal power variations with respect to some reference level; (2) the generation of an error signal; and (3) the restoration of initial conditions corresponding to zero error. These three functions can be implemented optoelectronically in either feedforward or feedback loops. In either case, the error signal can be used to control the pump power (C. R. Giles et al. [33]), or to control the power of an auxiliary saturating signal (E. Desurvire et al. [34]). We shall refer to the two approaches as *pump-controlled* and *signal-controlled AGCs*, respectively. Pump-controlled AGCs suppress gain saturation by pump increase; signal-controlled AGCs maintain constant the degree of EDFA saturation. Both types produce EDFA gain linearization, i.e., the relation $P_s^{out}/P_s^{in} = G = $ const. holds for any input signal power $P_s^{in}$ falling in a given dynamic range.

Four basic configurations of optoelectronic AGC loops are shown in Figure 6.13. In the feedforward loops of Figure 6.13a,b the error signals are generated by input signal power increase, detected through a fiber tap; the error signals can then be used to increase the pump driving current, Figure 6.13a or to decrease the driving current of the auxiliary signal Figure 6.13b. In the feedback loops of Figure 6.13c,d the error signals are generated by output signal power decrease, detected at the EDFA output through a tap; the error signals can then be used to increase the pump driving current, Figure 6.13c or to decrease the auxiliary signal driving current Figure 6.13d. In the configurations of Figure 6.13b,c the voltage bias $V$ must be adjusted to a level that gives the error signal requested for maximum output power deviation. In the case of feedback loops, the optical bandpass filters F can be centered either at a probe signal wavelength, or at some reference wavelength outside the gain band. At a reference wavelength outside the gain band, the ASE output can be used as a broadband signal probe for the detection of gain fluctuations.

The effect of AGC by pump or signal power control can be simply explained by the analytical model for transient gain dynamics developed in Section 5.9. Although this model gives solutions only for $z = 0$ in the case of square pulse perturbations, these are useful to qualitatively explain the gain dynamics. Consider first the case of

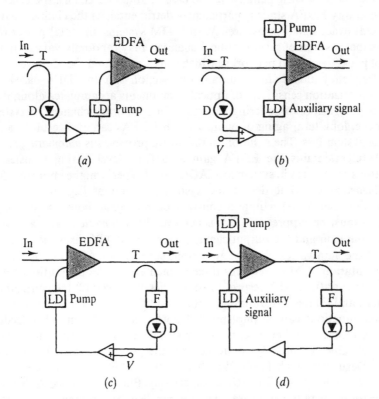

**FIGURE 6.13** Four basic configurations for automatic gain control (AGC) in EDFAs, based on optoelectronic loops: (a) feedforward with pump control, (b) feedforward with auxiliary signal control, (c) feedback with pump control, and (d) feedback with auxiliary signal control. T = tap coupler, D = detector, F = optical bandpass filter.

pump power control. Before the saturating signal pulse excitation, the gain medium is characterized by the steady state ground level population:

$$N_1^0 = \frac{1}{1+q} \tag{6.1}$$

where $q$ is the pump power at $z = 0$ normalized to the saturation power $P_{sat}(\lambda_p)$. After the saturating signal pulse is turned on at $t = 0$, the ground level population at time $t$ becomes, Eq. (5.105):

$$N_1(0, t) = \left( N_1^0 - \frac{1 + W_{21}\tau}{1 + p + q} \right) e^{-t/t_1} + \frac{1 + W_{21}\tau}{1 + p + q} \tag{6.2}$$

where $p$ is the pulse normalized power, $\tau$ is the fluorescence lifetime, $t_1$ is a power dependent time constant defined in Eq. (5.108), and $W_{21}\tau = \eta_s p/(1 + \eta_s)$. The two terms on the right-hand side of Eq. (6.2) are the transient and steady state components of the ground level population $N_1(0, t)$, following a change in input signal power. The transient can be suppressed by instantaneously increasing the pump at $t = 0$ by an amount $\delta q$, such that:

$$N_1^0 = \frac{1 + W_{21}\tau}{1 + p + q + \delta q} \tag{6.3}$$

or, from Eq. (6.1): $\delta q = (1 + q)W_{21}\tau - p = (\eta_s q - 1)p/(1 + \eta_s)$, which from Eq. (6.2) gives the time independent solution $N_1(0, t) = N_1^0$. This shows that it is possible to maintain the ground $N_1$ and upper level $N_2 = \rho - N_1$ populations to their steady state values by appropriate pump power increase. In reality, the populations vary with fiber length, and the exact pump correction needed to compensate for changes in input signal power must be determined by numerical methods (or heuristically in experiments). At present no theory exists to prove whether or not uniform stabilization of atomic populations along the fiber is possible through the pump correction method. The simplified demonstration made above also shows that the pump correction is proportional to the normalized signal power $p = P_s(0)/P_{sat}(\lambda_s)$. Thus, the feedforward AGC configuration, Figure 6.13a, requires knowledge of signal wavelength, hence of $P_{sat}(\lambda_s)$, in order to generate the error signal, based on the detection of input power $P_s(0)$. In the case of WDM systems, the saturating signal wavelength can be random, and a feedforward AGC would require signal spectral analysis, which is not practical. On the other hand, the feedback loop configuration, Figure 6.13c, works independently of the signal wavelength; iterative or trial pump corrections eventually cause convergence toward a null error signal and stable gain.

We consider next the case of signal-controlled AGC. In this configuration, the EDFA is initially saturated by an auxiliary signal input (power $p$), which gives the steady state ground level population:

$$N_1^0 = \frac{1 + W_{21}\tau}{1 + p + q} \tag{6.4}$$

As a saturating pulse of power $p'$ is turned on at $t = 0$, the ground level population becomes at time $t$:

$$N_1(0, t) = \left( N_1^0 - \frac{1 + W_{21}\tau + W'_{21}\tau}{1 + p + p' + q} \right) e^{-t/t_1} + \frac{1 + W_{21}\tau + W'_{21}\tau}{1 + p + p' + q} \qquad (6.5)$$

If the auxiliary signal undergoes a power correction, the transient in Eq. (6.5) can be suppressed and the steady state conditions maintained during the perturbation; the required power correction $-\delta p$ is determined by the condition:

$$N_1^0 = \frac{1 + W_{21}\tau - \delta W_{21}\tau + W'_{21}\tau}{1 + p - \delta p + p' + q} \qquad (6.6)$$

Comparison of Eqs. (6.4) and (6.6) shows that the auxiliary signal power decrease must exactly equal the saturating signal power, i.e., $\delta p = p'$ (and $\delta W_{21} = W'_{21}$). This last relation corresponds to normalized powers, which for actual or measurable powers correspond to $\delta P_s^{comp}/P_{sat}(\lambda_{comp}) = P_s^{sat}/P_{sat}(\lambda_{sat})$. Thus, *cancellation of transient saturation is achieved by keeping constant not the total EDFA input signal power, but the sum of all input powers weighted by their respective saturation powers*, as expected. Our remarks on the pump correction method also apply to the signal correction method: (1) in actual EDFAs, the required signal correction depends upon the gain changes along the fiber and can be determined only numerically or heuristically; (2) there is no proof as to whether transient gain cancellation can be achieved uniformly along the fiber; and (3) the feedforward configuration is inadequate for WDM systems, or randomly varying saturating-signal wavelengths.

Experimental setups for the early demonstration of AGC in EDFAs, according to Figure 6.13a,d are shown in Figures 6.14 and 6.15. The insets shows that AGC

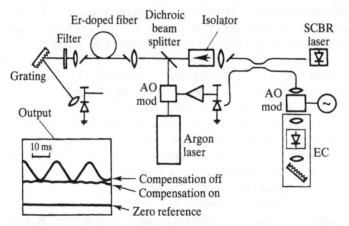

**FIGURE 6.14** Experimental setup for demonstration of AGC through feedforward pump control. In this experiment, the signal probe is provided by an SCBR laser and the saturating signal by an externally modulated external-cavity (EC) laser; an acousto-optic modulator (AO) is used to control the pump power provided by an argon ion laser. The inset shows the output signal probe with the AGC turned on and off. From [33], reprinted with permission from the Optical Society of America, 1989.

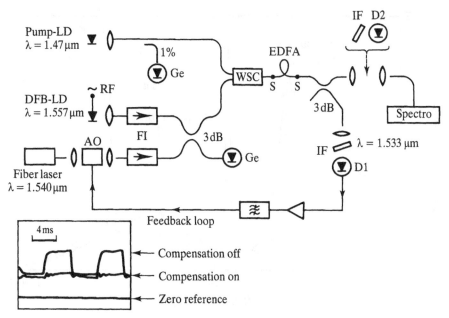

**FIGURE 6.15** Experimental setup for demonstration of AGC through feedback with auxiliary signal power control. In this experiment, the saturating signal is provided by a DFB laser diode, and the auxiliary signal is provided by an Er-doped fiber laser, controlled by an acousto-optic modulator (AO). The inset shows the output ASE, monitored through a bandpass filter, with the AGC turned on and off. From [34] © 1991 IEEE.

effectively suppresses the effect of AC gain modulation otherwise experienced by the signal probe or the ASE, [33] and [34].

Figure 6.16 shows the effect of AGC on the EDFA output spectrum, in the case of a signal-controlled AGC with feedback loop [34]. In this experiment, a saturating signal with $\lambda_{sat} = 1.557\,\mu m$ and modulated at $f = 100\,Hz$, causes the EDFA gain to oscillate. This oscillation causes the spectral gain to vary in time between the two envelopes seen in the figure. The gain oscillations seen in the spectrum are an artifact introduced by the optical spectrum analyzer scanning. Figure 6.16 shows that when the feedback loop is turned on (compensating signal at $\lambda_{comp} = 1.540\,\mu m$), a spectrally uniform stabilization of the EDFA gain results. In this example, the gain is clamped at a level corresponding to the lower spectral gain envelope. The gain is clamped, since any power increase of the output signal probe (here ASE at $\lambda_{ref} = 1.531\,\mu m$) results in a power increase of the compensating signal. Conversely, any detected decrease of the probe results in a decrease of the compensation. This principle of gain clamping can also be implemented in an *all-optical* configuration (M. Zirngibl et al., [35] and [36]).

In high gain EDFAs, signal-controlled AGC requires only a few microwatts of compensating signal power, compared to 10–20 mW compensating pump powers in the case of pump-controlled AGC [34]. But signal-controlled AGC requires an additional (monochromatic) signal source, and the EDFA gain must be reduced to a certain bias level, corresponding to maximum saturation. Pump-controlled AGC requires the EDFA not to operate at maximum pump power, so that extra power

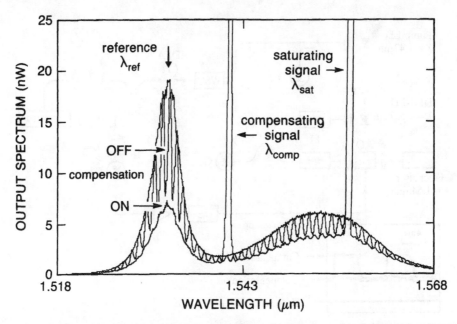

**FIGURE 6.16**  Output spectra showing the effect of AGC through auxiliary signal power control. A 100 HZ saturating signal at $\lambda_{sat} = 1.557\,\mu m$ causes the EDFA gain to oscillate between the two spectral envelopes. As the feedback loop is turned on, the compensating signal at $\lambda_{comp} = 1.540\,\mu m$ uniformly clamps the gain at a bias level which corresponds to the lower envelope. From [34] © 1991 IEEE.

can be available for saturation compensation. The EDFA noise figure is always better in pump-controlled AGC, as it always achieves maximum medium inversion. In the case of relatively small gain saturation, the noise figure increase is negligible (Section 7.1) therefore signal-controlled AGC with low EDFA noise figure can also be realized.

Several experiments have shown that AGC can be effectively implemented using practical components such as laser diodes, fiber couplers, and basic optoelectronic components/circuits. Due to the slow gain dynamics of the EDFA, the AGC loop operates at low frequencies (i.e., DC to 100 kHz), which requires relatively inexpensive components/circuits for implementation. A possible pump-controlled AGC scheme for EDFA-based systems uses *sinusoidal subcarrier modulation* and a minimal number of optical components (A. D. Ellis et al. [37]). In this scheme, a small fraction of the signal power is tapped at the EDFA output; the subcarrier signal, e.g., 10 kHz, is retrieved through a 10 kHz band *AC-coupled* RF receiver. AC-coupling also filters the high frequency components, the ASE and the residual pump output. The error signal, relative to a reference level, is then fed back to the pump for AGC. Another possible scheme for pump feedback AGC uses the total ASE power laterally radiated by the EDFA coil as control signal, which obviates fiber taps (K. Aida and H. Masuda [38]). This method can also be implemented in a two-stage configuration for independent control of gain and noise figure (K. Aida and H. Masuda [39]). In the case where the pump dynamic range is limited, the drawback of pump-controlled AGC is the change in EDFA noise figure with pump power changes: This problem may be alleviated by operating the EDFA at maximum pump and using a variable optical attenuator for controlling the output signal (K. Kinoshita et al. [40]). This method maintains a minimal noise figure.

Figure 6.17 shows the effect of EDFA gain linearization obtained with a basic pump feedback optoelectronic AGC scheme (K. Motoshima et al. [41]). The EDFA gain without AGC is linear for output signal powers up to $P_s^{out} = 0$ dBm, while with AGC the dynamic range for linearity is extended to $P_s^{out} = 11$ dBm, limited by the available LD pump power [41]. As a result of AGC and gain linearization, output square pulses with constant amplitude can be obtained in this power range. Signal-controlled AGC through gain clamping can be entirely optical, as illustrated in Figure 6.18 (M. Zirngibl et al. [35]–[36]). All-optical AGC, Figure 6.18a, feeds back a fraction of output ASE into the EDFA. The EDFA operates as an amplifier for the signal wavelength $\lambda_s$ and as a ring laser oscillating at wavelength $\lambda_F$. Any input signal increase results in a decreased laser feedback, and conversely, any signal decrease results in increased laser feedback. As a result, the single-pass signal gain is clamped by the effect of laser oscillation. The clamping level is controlled by the intracavity attenuator A. Figure 6.18b shows output spectra obtained in the small-signal and large-signal regimes, for feedback on and off; the off spectra show the effect of gain saturation as the signal increases, while the on spectra show that the gain is clamped at a level independent of the input signal power. The advantage of such an AGC scheme is that no electronic components are required, and furthermore, it can be implemented using exclusively all-fiber components.

We consider next the case of *automatic power control* (APC). APC results in signal power equalization and is likely to play an important role in EDFA-based network applications. In these systems, the signal power received at a given node varies as a function of the distance separating the receiving node and the transmitting station. In all-optical packet-switched networks, APC should ideally be performed at every

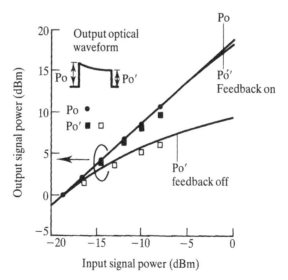

**FIGURE 6.17** Experimental measurement of EDFA gain linearization by pump feedback AGC, in the case of signals with square pulse envelope. From [41], reprinted with permission from the Optical Society of America, 1993.

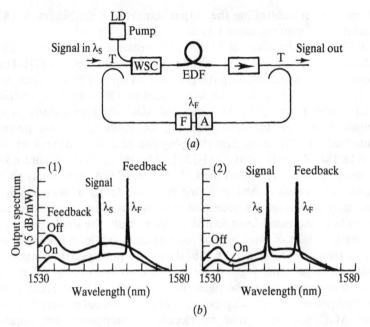

**FIGURE 6.18** (*a*) Basic layout for all-optical implementation of signal feedback AGC in EDFAs (T = fiber tap, A = attenuator, F = optical bandpass filter at $\lambda_F$). (*b*) Output spectra showing the effect of gain clamping with the feedback on: (1) small-signal input, (2) large signal input. From [35], reprinted with permission from the Optical Society of America, 1991.

switching node, in order to equalize signals originating from different stations and maintain a constant power throughput. Likewise, all-optical switches based on *nonlinear loop mirrors* (Section 6.5), require that the signals to be switched have minimal power fluctuations, which can be realized through APC. EDFAs with APC are also referred to as *power-limiting* EDFAs, which is accurate, and sometimes as *gain-limiting* EDFAs, which is not accurate. Indeed, the function of APC is to keep the output signal power constant, which results in a gain decrease with increasing input signal; the maximum gain limit corresponds to the value obtained in the small-signal regime, and is not related to APC. The function of power equalization is achieved when the EDFA gives a constant output signal power for any input signal falling within a given power range. In this respect, a power-limiting EDFA is also a power equalizer.

Several basic methods for implementation of APC in EDFAs are shown in Figure 6.19. The first method (a) uses an optoelectronic pump feedback loop (M. Nishimura et al. [42]). In this method, detected changes in output signal power generate an error signal, which is fed back to the pump (the pump current is decreased if the signal increases, and conversely). This scheme is similar to that of pump feedback AGC, Figure 6.13*c*, except that the error is generated by detection of output signal power fluctuations, as opposed to output probe fluctuations. In order to operate the EDFA at maximum pump (for minimal NF) an optical attenuator A can also be inserted in the loop near the EDFA output. This APC scheme is straightforward to implement with inexpensive optoelectronic components. The optical bandpass filter

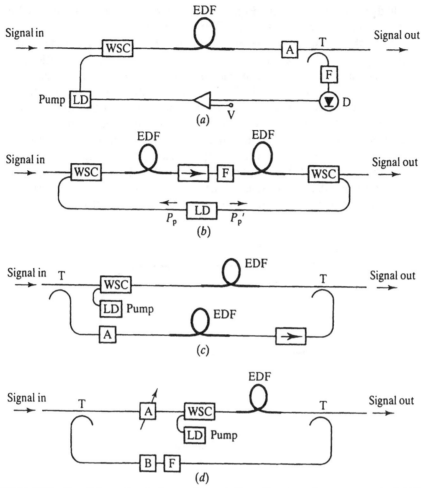

**FIGURE 6.19** Possible methods for implementation of automatic power control (APC) in EDFAs, based on: (*a*) pump feedback loop, (*b*) tandem-EDFAs, (*c*) saturable feedback and (*d*) ring laser. A, B = attenuators, T = fiber tap, D = detector, F = filter, WSC = wavelength selective coupler.

F is not necessary if there is 10 kHz subcarrier modulation with an AC-coupled receiver for the loop detector. In comparison to this approach, all other APC methods described below are passive and entirely optical.

Figure 6.19*b* shows a tandem-EDFA configuration (W. I. Way et al. [43]). This configuration is based on the principle of feedforward saturable gain and can actually include two or three amplification stages. The first stage uses an EDFA with high gain, high output power, and low NF. The last stage uses a low gain, highly saturated EDFA. This operating regime can be achieved by using a single high power pump LD, its output power unequally split between the two EDFAs. Figure 6.20 shows the gain and output power characteristics for the first and the last stage of a three-stage configuration [43]. As the figure shows, the output signal power is limited to the value of $P_s^{out} \approx +13$ dBm, and is constant for input signal powers varying from

$P_s^{in} = -30$ dBm to $P_s^{in} = 0$ dBm. Figure 6.20 shows that such an effect of power equalization is not observed at the first EDFA output.

Figure 6.19c shows a saturable feedback configuration (M. Zirngibl [44]). A fraction of the input signal is fed back into the EDFA, which causes gain saturation, but the feedback signal passes through a saturable absorber, a strand of unpumped Er-doped fiber. An optical isolator prevents the EDFAs output ASE from changing the loss of the unpumped EDF. Nonlinear signal transmission through the unpumped EDF means that the feedback into the EDFA has a negligible effect in the case of weak input signals and a strong effect in the case of large input signals. Thus, output power limiting can be achieved within a certain input signal power range, as shown in the experiment described in [44]. Compared to the tandem-EDFA configuration (feed-forward saturable gain), the advantage of the saturable feedback configuration is that only one EDFA is necessary. The optimal input/output signal characteristics and

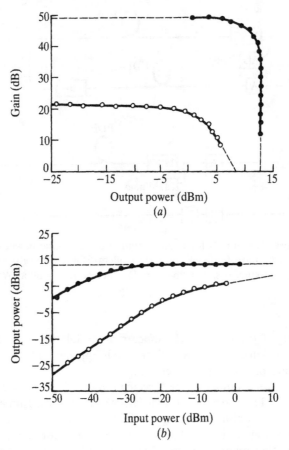

(a)

(b)

**FIGURE 6.20** Experimental gain and output signal power measurements obtained in a three-stage tandem-amplifier APC configuration, showing power-limiting effect: (a) gain versus output signal power, and (b) output signal power versus input signal power, measured at the outputs of the first EDFA (open symbols) and the third EDFA (dark symbols). From [43], reprinted with permission from *Electronics Letters,* 1991.

dynamic ranges of these two APC configurations remain to be fully analyzed and compared.

Figure 6.19d shows another APC scheme based on a ring laser configuration (H. Okamura, [45] and [46]). Instead of compensating EDFA gain changes, this scheme compensates transmission loss changes in a given element in a system. The element with fluctuating loss is represented by the variable attenuator A in Figure 6.19d. To understand loss compensation, first consider the effect of gain saturation induced by laser oscillation in the loop. This clamps the single-pass EDFA gain to a value fixed by the attenuator B. An increase of the loss in B causes an increase of the single-pass EDFA gain. If the attenuation in A is kept constant, this configuration provides gain linearization, i.e., the signal output is proportional to the signal input, as for AGC. On the other hand, if the signal power is obtained. An increase of loss in the loop, induced by the attenuator A, results in a weaker laser feedback and saturation effect, which increases the EDFA gain. The single-pass gain is clamped by the total loop loss from A and B. An increase of loss in attenuator A results in a corresponding increase of the gain, which keeps the output signal power constant. This effect is illustrated by the experimental measurements shown in Figure 6.21 [46]. The transmitted signal is constant for a loss variation in A from 0 dB to 11 dB. Beyond this value, laser oscillation stops and the device response is linear, i.e., the transmission decreases linearly with the attenuation in A. For fixed loop loss and input signal, the output signal power becomes constant past a certain pump threshold value, which corresponds to the onset of laser oscillation [46]. An experimental characterization of the transient dynamics of the ring laser APC configuration is described in [46]. Potential applications of loss-compensating APC are in the field of interferometric fiber sensors, in which unwanted signal loss changes occur at the sensing heads, due to optical phase retardation [45].

The most important applications of APC that have been investigated to date (1993) concern soliton systems (K. Suzuki and M. Nakazawa [47]) and ring networks (W. I. Way et al. [48]), both are based on multiple stage tandem-EDFA configurations. In the case of soliton transmission, experimental measurements have shown that gain

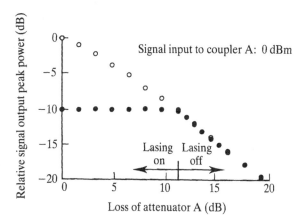

**FIGURE 6.21** Experimental measurement of automatic loss compensation through APC in a ring laser configuration, with feedback on (dark symbols) and off (open circles), showing compensation effect over a 11 dB loss range. From [46] © 1992 IEEE.

saturation in long EDFA chains results in the APC effect, i.e., the soliton output (peak) power is independent of the input (peak) power over a certain dynamic range [47]. On the other hand, the detailed study of [48] shows that when implemented in ring fiber networks, all-optical APC enables power budget increase, component cost decrease, and flexibility improvement.

Due to the slow EDFA gain dynamics, both AGC and APC devices operate at low frequencies (i.e., DC to 100 kHz). In the case of AGC, this fact implies no restrictions, since transient gain perturbations occur only in this frequency range. But in the case of APC, fast signal power changes at frequencies $f > 100$ kGz cannot be controlled. Thus, a high speed TDM sequence (e.g., at $B = 1$ Gbit/s) with peak-to-peak power variations cannot be equalized through EDFA-based APC. On the other hand, EDFA-based APC can process a sequence of optical data packets (e.g., sequence rate $= 10$ kHz, packet length $\Delta T = 100$ $\mu$s) with unequal powers. This is important for fiber networks. For AGC and APC devices based on all-optical loop configurations, a small noise figure increase always occurs when the input signal is large, caused by EDFA saturation. In contrast, AGC and APC devices based on optoelectronic feedback loops with maximum pump, have a minimal NF. This observation is important when considering implementation of AGC and APC in multiple stages, as in the case of EDFA chains or ring networks. Note that in the highly saturated regime, the actual EDFA noise figure is not given by $2n_{sp}$, as this factor becomes infinite near system transparency (Chapter 2). More importantly, the concept of EDFA noise figure has not been completely analyzed and clarified for the regime of EDFA saturation. (Section 2.8).

## 6.4   SPECTRAL GAIN EQUALIZATION AND FLATTENING

Spectral gain equalization is of central importance for WDM system applications of EDFAs. Indeed, as the EDFA gain bandwidth is not spectrally uniform and exhibits some structure or ripple (Chapter 5), gain differences occur between optical channels having large wavelength spacings (e.g., $\Delta\lambda > 1$ nm). In long amplifier chains, even small spectral gain variations (e.g., $\Delta G < 0.5$ dB) can result in large differences in received signal power, causing unacceptably large BER discrepancies between received signals. For some optical channels, complete power extinction can occur at the system output, due to insufficient gain compensation along the amplifier chain. Additionally, the ASE generated in the region of highest gain (i.e., near the peak at $\lambda = 1.53$ $\mu$m) in unequalized EDFAs causes homogeneous gain saturation, which affects WDM channels at longer wavelengths. Finally, gain equalization is desirable in analog systems for suppression of CSO distortion induced by gain tilt (Section 3.8).

Gain equalization means achieving identical gains for a discrete number of optical channels. Gain flattening means achieving a spectrally uniform gain bandwidth. Thus, for two-channel amplification, gain equalization can be achieved without requiring gain flattening. Two-channel gain equalization can be done by simple means: given an EDFA length $L$, for instance, the pump power can be chosen such that the gains in the peak and shoulder regions near $\lambda = 1.53$ $\mu$m and $\lambda = 1.54$ $\mu$m become exactly equal, due to the effect of signal reabsorption at short wavelengths, Figure 5.32a. The same effect can be obtained with a fixed pump power and variable EDFA length. This equalization method has two main drawbacks: (1) the NF is higher at short wavelengths, due to conditions of poor medium inversion (Figure 5.32b); and (2) the effect of gain equalization is removed by gain saturation.

Another gain equalization method places the optical channels at wavelengths giving equal gains under maximum pumping. Figure 6.22 shows typical gain spectra of aluminosilicate EDFAs. The top curve for the small-signal regime shows at least five different wavelengths ($\lambda_1, \ldots, \lambda_5$) for which the EDFA gain is identical. The gain is nearly flat in the spectral region [$\lambda_3, \lambda_4$], which defines a continuum of solutions. But as the gain saturates, simulated in this example by signal power increase at $\lambda_{peak} = 1.531\ \mu m$, the gain spectrum is distorted and the number of equalized channels in the spectral region [$\lambda_1, \lambda_5$] reduces to three. The drawbacks of this approach are the tight wavelength constraint (with reduced tolerance near the peak at $\lambda_{peak} = 1.531\ \mu m$), the inefficient use of the gain bandwidth, and the sensitivity to gain saturation.

The possibility of realizing glass hosts in which the $Er^{3+}$ ions exhibit a uniform gain spectrum seems to be limited. While fluorozirconate glass hosts such as ZBLAN generally provide a gain spectrum more uniform than silicate hosts with high alumina codoping (T. Sugawa et al. [49]; W. J. Miniscalco [50]), their gain characteristics are comparatively poor (Section 5.2). Therefore, optimum gain equalization should be achieved in standard silica hosts through built-in or external compensation means.

The EDFA gain equalization/flattening techniques investigated to date (1993) are based on six different principles: (1) gain clamping with enhanced inhomogeneous saturation, (2) use of passive internal/external filters, (3) use of external active filters, (4) cascading EDFAs with different gain spectra, (5) spatial hole burning in twin-core fiber and (6) adjustment of input signal powers.

The first method uses the *ring laser configuration* shown in Figure 6.18a and described for AGC applications. When the fiber is cooled to a sufficiently low temperature (e.g., $T = 77\ K$), the gain homogeneity decreases (Section 4.4), which results in enhanced inhomogeneous saturation. The enhanced gain inhomogeneity causes the EDFA gain to flatten more uniformly under saturation; a 3 dB gain

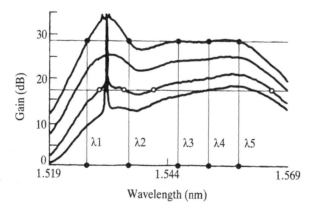

**FIGURE 6.22** Aluminosilicate EDFA gain spectra obtained in the small-signal regime (top curve) and saturated regime for increasing saturating signal at $\lambda = 1.53\ \mu m$ (bottom curves), showing possible choices of WDM channel wavelengths for which gain equalization can be achieved.

bandwidth of 35 nm could be obtained by this method, representing a 10 nm improvement (V. L. da Silva et al. [51]). While gain flattening can be efficiently realized through this approach, its major drawback is the requirement of low temperature cooling, which is not practical for most system applications.

A second method for gain equalization or flattening uses *passive optical filters* (M. Tachibana et al. [52]; M. Wilkinson et al. [53]; R. Kashyap et al. [54]). The principle of the method is to use a band stop or notch filter tuned near the peak gain; the combination of the wavelength dependent filter transmission and EDFA gain results in net spectra gain flattening. For instance, a 3 dB gain bandwidth of 33 nm could be obtained by using a grating filter coupled to fiber cladding modes [52]. Another approach uses a grating filter etched on the surface of a D-fiber section; an example of gain flattening using a D-fiber grating filter at the EDFA output is shown in Figure 6.23 [53]. In this example, the gain variation is less than 0.5 dB over a 30 nm span, and the 3 dB gain bandwidth is as large as 50 nm. The drawbacks of this approach are the D-fiber/SMF coupling loss and the filter polarization sensitivity. A photosensitive blazed grating [54] yields similar results without such drawbacks. Passive filtering of the main gain peak at $\lambda = 1.53$ $\mu$m can also be achieved through a twin-core fiber structure, based on the principle of wavelength selective mode coupling; a 3 dB gain bandwidth of 30 nm could also be demonstrated by this method (G. Grasso et al. [55]). Advantages of passive filters are their simplicity, fiber compatibility, and low excess loss. But passive filters cannot respond to spectral gain changes induced by saturation. Thus, gain equalization through passive optical filtering should be implemented with AGC.

EDFA gain equalization can also be achieved through *active optical filtering*. In integrated acousto-optic tunable filters (AOTF), the transmission at a given wavelength can be controlled by adjusting the center frequency and power of the driving RF signal (D. A. Smith et al. [56]). By driving the AOTF with several RF subcarriers, independent transmission control can be obtained simultaneously for different optical wavelengths (S. F. Su et al. [57]). In this operating mode, the filter transmission spectrum is a comb whose peaks have a relatively narrow width (i.e., 1–2 nm. As a

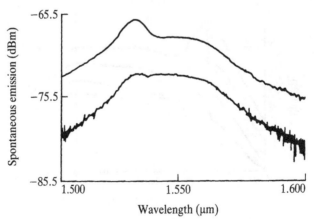

**FIGURE 6.23** Output ASE spectrum obtained after passive filtering through D-fiber grating, showing the effect of EDFA gain flattening (bottom curve). The top curve shows the ASE obtained without filter, for comparison. From [53], reprinted with permission from *Electronics Letters*, 1992.

result, AOTFs also act as efficient blocking filters for ASE [57]. In addition to their polarization insensitivity, AOTFs make gain equalization possible for any set of $N$ discrete wavelengths, as shown in the experiment described in [57]. The individual channels can also be monitored and independently equalized by using subcarrier modulation of the WDM channels and an AC-coupled RF receiver. Each subcarrier associated with a given WDM channel can be used to monitor the corresponding optical power, thus enabling power equalization through RF feedback in the AOTF. Compared to all other methods, the major advantage of AOTF-based gain equalization is the possibility to operate in both small-signal and gain saturation regimes. Furthermore, the method is also applicable to transient gain saturation. Although this has not been demonstrated yet, AOTFs make it thoeretically possible to achieve APC, AGC, and gain equalization simultaneously. The drawbacks of AOTFs are their high insertion loss (<9 dB [58]) and the relatively high RF power required for equalization of a large number of channels (i.e., $P_{RF}^{tot} \approx 1$ W for $N = 10$). Another method for active optical filtering uses integrated-electro-optic Mach-Zehnder interferometers (K. Inoue et al. [58]). The transmission spectrum of such filters is a sinusoid whose period is determined by the path length difference, which can be finely adjusted by control voltages. The transmission period can be made as large as 40 nm, resulting in a smooth loss increase or decrease across the EDFA spectrum [58]. Optimization of control voltages results in gain equalization, as shown in [58]. This equalization method is particularly efficient in the case where the WDM signals are densely spaced (e.g., four channels per nm [58]) and the gain imbalance is a monotonously increasing or decreasing function of wavelength.

Another approach for EDFA gain equalization uses *two-stage amplification with different gain characteristics* (C. R. Giles and D. DiGiovanni [59] and [60]). The principle of this approach can be simply explained by the following. In EDFAs, the signal gain at a given wavelength $\lambda_s$ can be expressed as (Chapter 1):

$$G(\lambda_s) = \exp\{\rho_0 \Gamma_s (\sigma_e^s \hat{N}_2 - \sigma_a^s \hat{N}_1)\} = \exp\{\rho_0 \Gamma_s [(\sigma_e^s + \sigma_a^s)\hat{N}_2 - \sigma_a^s]\} \qquad (6.7)$$

where $\hat{N}_1, \hat{N}_2$ are the atomic populations normalized to the $Er^{3+}$ density and averaged over the fiber length, i.e.,

$$\hat{N}_{1,2} = \frac{1}{\rho_0 L} \int_0^L N_{1,2}(z) \, dz \qquad (6.8)$$

Let $g(\lambda_s) = 10 \log_{10} G(\lambda_s)$ be the dB gain at $\lambda_s$. From Eq. (6.7), it is seen that any change $d\hat{N}_2$ in upper level population (induced by pump power or signal saturation) corresponds to a differential gain change $dg(\lambda_s)/d\hat{N}_2$ given by [60]:

$$\frac{dg(\lambda_s)}{d\hat{N}_2} = 10 \log_{10}(e)\rho_0 \Gamma_s (\sigma_e^s + \sigma_a^s) \qquad (6.9)$$

where $e = \exp(1)$. Considering now two signal wavelengths $\lambda_{s1}$ and $\lambda_{s2}$, we obtain the relation:

$$\frac{dg(\lambda_{s1})}{dg(\lambda_{s2})} = \frac{\sigma_e^{s1} + \sigma_a^{s1}}{\sigma_e^{s2} + \sigma_a^{s2}} \equiv R(\lambda_{s1}, \lambda_{s2}) \qquad (6.10)$$

In the case of a two-stage EDFA made of two sections A and B, we can write for the two signals:

$$dg(\lambda_{s1}) = dg_A(\lambda_{s1}) + dg_B(\lambda_{s1}) \tag{6.11}$$

$$dg(\lambda_{s2}) = R_A(\lambda_{s2}, \lambda_{s1}) \, dg_A(\lambda_{s1}) + R_B(\lambda_{s2}, \lambda_{s1}) \, dg_B(\lambda_{s1}) \tag{6.12}$$

where the subscripts A and B refer to parameters related to the first and the second EDFA sections, respectively. If the two EDFAs are pumped (or saturated) differently, it is always possible to find an operating regime for which $dg_A(\lambda_{s1}) = -dg_B(\lambda_{s1})$, in which case we obtain $dg(\lambda_{s1}) = 0$ for wavelength $\lambda_{s1}$. When this condition is achieved, we obtain $dg(\lambda_{s2}) \neq 0$ for wavelength $\lambda_{s2}$; for two EDFAs of different cross section characteristics we have $R_A(\lambda_{s2}, \lambda_{s1}) \neq R_B(\lambda_{s2}, \lambda_{s1})$ for most wavelengths $\lambda_{s1}, \lambda_{s2}$. Thus, an independent gain adjustment at signal wavelength $\lambda_{s2}$ can be achieved, which provides gain equalization. In the experiment described in [59], the first and second stages included aluminosilicate and germanosilicate EDFAs, respectively, and relative gain compensation of 1 dB between two channels spaced by 2.5 nm could be demonstrated. This equalization method could also be applied to the case where there the number of WDM channels is $N > 2$, through a multiple stage configuration. The possible correction algorithms for a multistage/multichannel implementation of this equalization method could be complex, and the stability of the solution to perturbations (generated by changes in mutual saturation, for instance) remains to be investigated. Finally, optimum equalization algorithms that would be applicable to the general case of an $N$ channel WDM system must be studied.

Another method of EDFA gain equalization is based on the principle of spatial *hole burning* (R. I. Laming et al. [61]). The effect of spatial hole burning, where gain saturation is a function of fiber coordinate, actually occurs in any EDFA operating in the saturation regime. This effect, combined with the wavelength dependence of the gain coefficient, explains the changes of the EDFA gain spectrum with increasing saturation, as shown in Figure 6.22. As saturation is homogeneous, gain compensation can be achieved by controlling the power of one of the input signals. As previously discussed, this compensation method is not advantageous, as only a limited number of WDM channels can be simultaneously equalized. On the other hand, if the effect of spatial hole burning could be decoupled for a given pair of signal wavelengths, then gain saturation would be more inhomogeneous, enabling improved equalization. It is possible to achieve this decoupling using a twin-core EDFA [61]. In a twin-core fiber, the signal intensity is periodically modulated along the fiber length. As the spatial modulation period is approximately proportional to the signal wavelength, two WDM channels become rapidly out of phase, which decouples their corresponding gains and generates effective gain inhomogeneity. The experiment described in [61] showed that for a twin-core EDFA, the maximum equalization rate (expressed as the dB gain change per dB input signal change) was measured to be 0.11 dB/dB, while in conventional EDFAs, this rate is 0.046 dB/dB. This effect of effective saturation inhomogeneity could be used for automatic gain equalization in long EDFA chains [61]. Further characterization and system studies are required to assess the potential of this approach for WDM systems.

We consider finally the equalization method based on *input signal power adjustment* (A. R. Chraplyvy et al. [62]). As opposed to all methods previously described in this section, this approach is based on signal power equalization rather than signal gain equalization. It leaves the EDFA gain characteristics unchanged and scales the input signal power in each channel at $\lambda_{si}$ by a factor inversely proportional to the corresponding gain $G(\lambda_{si})$, i.e.,

$$P_s^{in}(\lambda_{si}) = P_{tot}^{in} \frac{1/G(\lambda_{si})}{\sum_i 1/G(\lambda_{si})} \tag{6.13}$$

In Eq. (6.13), $P_{tot}^{in}$ is the total input power, kept constant through the normalization factor $\sum 1/G(\lambda_{si})$. The power weighting can be done at the system transmitter end through variable optical attenuators, using information on the signal gains $G(\lambda_{si})$ obtained by telemetry from the receiving end [62].

Figure 6.24 shows simulated ASE spectrum and WDM output signal powers computed at the output of a 840 km EDFA chain (70 km amplifier spacing), obtained without equalization for equal input signal powers ($P_s^{in}(\lambda_{si}) \equiv \bar{P}_{in} = -12$ dBm), and with the input signal power adjustment method. The eight WDM wavelengths chosen for this example and corresponding input/output signal powers are listed in the insets. An unequalized system yields a 25 dB output signal power variation across the 14 nm bandwidth used. When the input signal powers are adjusted according to the algorithm of Eq. (6.13), the output power variation is reduced to 4 dB [62]. A second iteration of this algorithm yields the result shown in Figure 6.24b, where the variation is reduced to 0.1 dB (a third iteration yields 0.01 dB variation). The fact that at least two iterations are required for efficient power equalization is explained by the changes in EDFA saturation under different WDM signal power distributions. Suppression of the 25 dB output signal power variation requires a relative input signal power variation of 25 dB. The SNR corresponding to each WDM channel (defined by the ratio of output power to output ASE in 0.2 nm bandwidth) and corresponding theoretical BERs are also shown in the insets. Unbalanced SNRs cause some channels to have unacceptable BERs. To alleviate this, the SNRs can be equalized by an algorithm similar to that of Eq. (6.14), which involves the SNRs corresponding to Figure 6.24a as weighting parameters [62]. In this case, SNRs and BERs are equalized at the expense of output signal power variation. Simultaneous equalization of both output powers and SNRs can be realized by using variable attenuators at the transmitting and receiving ends. SNR is not affected by attenuation at the receiver.

Beside its simplicity and cost-effectiveness, the input signal adjustment method for power equalization in EDFA chains has several advantages: (1) it can be implemented with any type of EDFA (e.g., germanosilicate, aluminosilicate); (2) it can operate under changing saturation conditions; (3) it requires control information only from the receiving terminal; and (4) in theory, the output SNRs are higher than when using gain-equalizing filters (i.e., passive or active), explained by the fact that signal power is rejected by the filtering effect, whereas the total power is maintained in the previous algorithm.

The input signal adjustment method appears to completely solve the problem of power equalization in EDFA-based systems with dispersion-limited lengths or bit rates. This limit is approximately given by the relation $B^2 L_{max}[(\text{Gbit/s})^2 \text{ km}] = 4000(15_{ps/nm\cdot km}/\bar{D}_{ps/nm\cdot km})$, where $\bar{D}$ is the maximum (average) dispersion coefficient

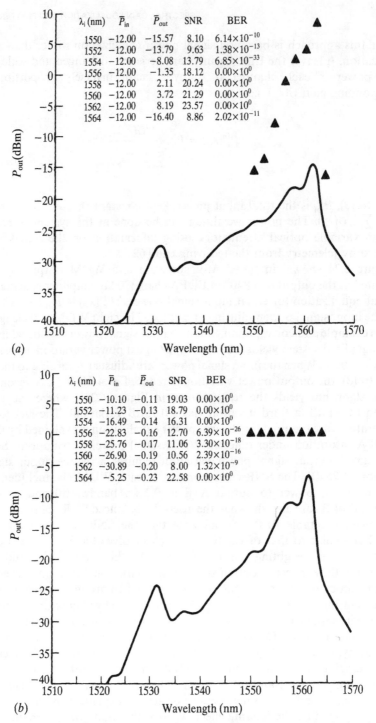

**FIGURE 6.24** Power equalization technique for EDFA-based systems: (*a*) calculated output ASE spectrum (solid line) and WDM output signal powers (▲) obtained without power equalization in an 840 km system (70 km amplifier spacing); (*b*) corresponding output ASE spectrum and WDM output signals obtained with power equalization, according to Eq. (6.13); the insets show the signal wavelengths, input/output signal powers, output SNRs, and theoretical receiver BERs [62] © 1992 IEEE.

for the WDM channels. For instance, the case $\bar{D} = 3_{\text{ps/nm.km}}$ gives $L < 5000\,\text{km}$ for $B = 2\,\text{Gbit/s}$ [63]. Longer transmission lengths and operating bit rates for WDM systems would require compensation techniques for the effects of dispersion and fiber nonlinearities (Chapter 7). Alternatively, nonlinear transmission using optical solitons seems at present to be the most effective solution for very long-haul WDM systems. In this last case, the input signal adjustment method cannot be applied for equalizing the WDM outputs, since the soliton power corresponding to each WDM channel is fixed by fiber dispersion (Section 5.11). For such systems, the elaboration of an effective and practical EDFA gain equalization technique still remains a challenge.

## 6.5  OPTICALLY CONTROLLED SWITCHES AND GATES

The development of LD-pumped EDFAs has opened new perspectives in the field of active fiber devices, in particular for *signal processing* and *switching* applications. The device applications of EDFAs previously discussed in this chapter concerned the signal processing functions of AGC and APC, which due to the slow EDFA gain dynamics, operate in a low-frequency domain (i.e., DC to 100 kHz). EDFAs can also be used as all-optical gates and switches in this operating domain.

High speed signal processing applications require nonlinear device responses in the nanosecond/picosecond time scale. Such a nonlinear response can be generated in standard silica fibers through the *optical Kerr effect*, which causes self- and cross-phase modulation, or SPM and XPM, and *stimulated four-photon mixing* (SFPM). A detailed theoretical and experimental description of SPM, XPM, and SFPM in glass fibers can be found in the book by G. P. Agrawal [64]. As these effects are based on third order nonlinear interactions with very weak susceptibilities (i.e., $\chi^{(3)}_{\text{Kerr}} \approx 6 \times 10^{-15}$ esu, [65] and [66], and $\chi^{(3)}_{\text{SFPM}} \approx 3 \times 10^{-15}$ esu [67], to compare with the case of semiconductor MQWs, where $\chi^{(3)}_{\text{Kerr}} \approx 6 \times 10^{-2}$ esu [68]), the optical powers required to observe them are relatively high for conventional LD sources. For instance, an SPM effect corresponding to a phase shift $\delta\phi = \pi$ would require for an $L = 10\,\text{km}$ DSF fiber a $1.5\,\mu\text{m}$ signal power of $P_s \approx 50\,\text{mW}$ ($\delta\phi = n_2(P_s/A_{\text{eff}})(2\pi/\lambda)L$ with $n_2 = 3.2 \times 10^{-20}\,\text{m}^2/\text{W}$ and $A_{\text{eff}} = 20\,\mu\text{m}^2$). If EDFAs are used as optical power boosters, it is possible to use LD sources to excite these nonlinearities. Thus, a generation of novel active fiber devices combining the advantages of *fast nonlinear response* and *practical LD signal control* has emerged in the wake of EDFA technology. We review the characteristics of the nonlinear optical loop mirror (NOLM) and the SFPM wavelength-converter, active fiber devices based on EDFAs. They can be used for switching and gating (demultiplexing) applications.

We consider first EDFA-based optical gates and switches operating in the *low frequency domain*. Their potential applications are in packet switched networks, where the traffic flow and switching patterns could be remotely controlled by optical signals. Basic examples of $1 \times 2$ and $2 \times 2$ switch configurations are shown in Figure 6.25 (E. Eichen et al. [69]). In both cases, the taps T are $2 \times 2$ passive fiber couplers/combiners. For the switching function, the EDFAs are used as optical gates. When the pump is turned off, the EDFA is absorbing, which effectively blocks the signal (e.g. with $T = -20\,\text{dB}$ transmission loss); on the other hand, turning on the pump generates an EDFA gain that exactly compensates for signal splitting loss (e.g.

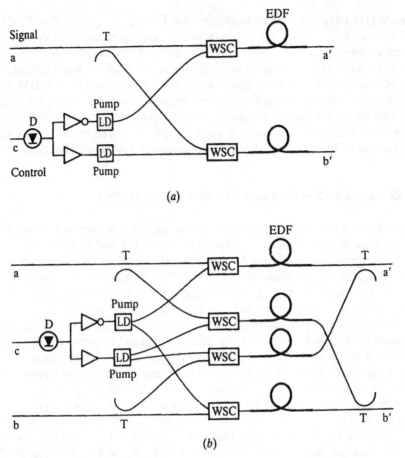

**FIGURE 6.25** Basic configurations of (*a*) 1 × 2 and (*b*) 2 × 2 optical switches, based on the principle of EDFA gating; c = optical control signal, D = detector, T = 3 dB fiber coupler, WDC = wavelength selective coupler.

$G = 3$ dB abd $G = 4$ dB) and other possible excess loss. Figure 6.25*a,b* show the pumps are controlled by a remote optical signal *c*, which can be transmitted through a separate optical fiber. This control signal is converted into an electrical logic signal which either activates or deactivates the pumps. An absence of control signal ($c = 0$) is shown as a bar state (=) in the figure. A nonzero signal is shown as a cross state (×). A self-routing configuration for both switch types could be implemented by conveying the control signal at a different wavelength in the same fiber as the switch input(s) but the long turn-on times of the EDFA (100 $\mu$s–1 ms) would require delaying the input signals by the same amount. This function could be performed by active recirculating fiber delay lines (Section 6.6).

Although the 1 × 2 and 2 × 2 switches described above have intrinsically slow turn-on and turn-off times, they present a certain number of important advantages: (1) they have no insertion loss (fanout = 1), (2) they are polarization insensitive, (3) they generate negligible ASE noise, as the gains are small, (4) they are transparent to bit rate and modulation format, (5) the switching efficiency is near 100% and the crosstalk can be made negligible (e.g., $-30$ to $-40\,\mathrm{dB}$ [69]), (5) with optimized EDFAs, the pump power required for switching is in the $100\,\mu\mathrm{W}$–$1\,\mathrm{mW}$ range, which can be provided by inexpensive GaAs laser diodes at $\lambda = 800\,\mathrm{nm}$.

Another possible implementation of EDFA gates, which has the advantage of being all-optical, is based on a ring laser configuration (M. Zirngibl [70]). Figure 6.26 shows this device uses the same principle as for all-optical signal feedback AGC (Figure 6.18), with the inclusion of an unpumped Er-doped fiber (EDF2) used as an intracavity saturable absorber. Initially, no lasing occurs at $\lambda_F$, due to the high cavity loss; the signal at $\lambda_s$ is transmitted with a net gain determined by EDF1, which corresponds to the on state of the gate. Lasing can then be turned on by injecting a low power control sisgnal at $\lambda_F$ which, after amplification through EDF1, bleaches the loss caused by EDF2. The high saturation effect induced by lasing reduces the gain at $\lambda_s$, which corresponds to the off state of the gate. Once initiated, lasing continues as the control pulse at $\lambda_F$ is turned off. To switch back the gate into the on state, a second control pulse at wavelength $\lambda_R \neq \lambda_F$, $\lambda_s$ is injected in the loop. This second pulse temporarily saturates the gain of EDF1, to a point where lasing is turned off, which causes EDF2 to permanently return to its lossy state. As a result, the gate is reset to the on state. In the experiment described in [70], on/off switching of the gate could be obtained in $10$–$50\,\mathrm{ms}$, using control signal powers of $P(\lambda_F = 1.56\,\mu\mathrm{m}) = 4\,\mu\mathrm{W}$ and $P(\lambda_R = 1.535\,\mu\mathrm{m}) = 140\,\mu\mathrm{W}$, respectively. In spite of a relatively long switching time, this device presents the advantage of a simple, all-optical

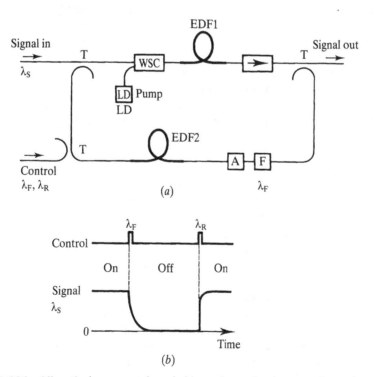

**FIGURE 6.26** All-optical remote gain switching using a ring laser configuration: (*a*) device layout (T = fiber tap, A = attenuator, F = bandpass filter at $\lambda_F$); (*b*) time sequence showing turn-off and turn-on by action of control pulses at $\lambda_F$ and $\lambda_R$, respectively.

and passive configuration. As the required signal power for control is about $100 \, \mu W$, the gate could be remotely switched from long distances; transmission distances of $L > 100 \, km$ for the control signals would require in-line EDFAs.

EDFA gates based on the principle of gain switching can also be realized with faster turn-off responses (i.e., $< 100 \, ns$) by using short saturating signal pulses with high peak power (e.g., $100 \, mW$–$1 \, W$). For instance, 20 dB gain compression or switching could be achieved with a 40 ns response time using a 3 W or 120 nJ control signal pulse (P. Myslinski et al. [71]). For these gates, the gain recovery dynamics are controlled by the pump power (Section 5.9) which, in typical operating conditions, yields recovery times in the millisecond range. The recovery can be made faster by coupling a short pump pulse with high peak power immediately after saturation. Theoretical simulations described in [71] show that for 980 nm pumping, $10 \, \mu s$ recovery is possible with pump pulses having the same energy as the saturating signal pulses; for this pump wavelength, the gain recovery time is limited by the lifetime of the $^4I_{11/2}$ level, i.e., $\tau_{32} \approx 7 \, \mu s$. The gain recovery time could be reduced to a nanosecond scale by using 1480 nm pulses, considering that the associated pump level lifetime is considerably shorter than the lifetime of the $^4I_{11/2}$ level. These results show that nanosecond on/off/on gain switching in EDFAs can be achieved with high energy control pulses. These control pulses could be generated by practical LD-pumped, $Q$-switched or mode-locked fiber lasers (Section 6.7).

We consider next *high speed switching* and *gating* applications of EDFAs that are based on the Kerr nonlinearity. Nonlinear devices based on all-fiber interferometers with Kerr effect have been investigated since the mid-1980s (see [72] and references therein). Among all these devices, the *nonlinear optical loop mirror (NOLM)*, sometimes called the *nonlinear Sagnac interferometer switch (NSIS)*, has received considerable attention owing to its many intrinsic advances and versatility. Before describing how EDFAs can be used in NOLMs, it is useful to briefly outline the main characteristics of the passive version of the device.

A passive NOLM consists of a Sagnac interferometer with imbalanced splitting ratio ($\eta \neq 0.5$), Figure 6.27a (N. J. Doran and D. Wood [73]). In a balanced NOLM ($\eta = 0.5$), the signal is 100% reflected, independent of the input signal power, due to the effect of SPM reciprocity. Reciprocity in nonpolarization-maintaining fibers requires mechanical polarization controllers to align the output polarizations. In the unbalanced NOLM, counterpropagating pulses experience a different SPM effect, so the recombined signal is partially reflected or transmitted by the loop and the response is periodic with the initial power. Pulses having certain peak power values can self-switch through the NOLM. The NOLM gives a similar response in the case of CW input signals, in spite of the additional effect of XPM (K. Otsuka [74]). As most applications of interest concern only short-pulse operation, the effect of XPM between the counterpropagating signals in the NOLM is usually neglected.

For all-optical switching or demultiplexing applications, a single signal pulse must be switched out of a high speed sequence. This function can be performed by coupling the high speed signal at one port of the NOLM, and a single intense signal pulse (pump) at the other port, Figure 6.27b. In this case, switching is due to the effect of XPM between the two types of input pulses. A maximum switching contrast (100%) can be obtained if the NOLM coupler is balanced ($\eta = 0.5$) for the high speed signal and unbalanced ($\eta \neq 0.5$) for the pump; this condition can be achieved if the pump and signal have different wavelengths (M. C. Farries and D. N. Payne [75]).

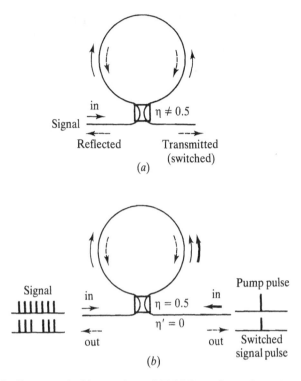

**FIGURE 6.27**  Nonlinear optical loop mirror (NOLM) configurations: (a) unbalanced with coupler coupling ratio $\eta \neq 0.5$; (b) two-wavelength operation for efficient signal pulse switching, using balanced interfometer with signal and pump coupling ratios $\eta = 0.5$ and $\eta' = 0$, respectively.

The NOLM time dependent response to pulsed excitation can be described by coupled nonlinear Schrödinger equations for the pump and the signals. The Kerr nonlinearity has a femtosecond response time (i.e., 2–4 fs [72]), and for all purposes, it can be treated as being instantaneous. A detailed theoretical analysis of NOLMs, which includes the effects of SPM, XPM, and GVD, is the study by M. Jinno [72]. The idea of using optical soliton pulses as natural eigenmodes of nonlinear interferometers or couplers was analyzed and experimentally demonstrated in the late 1980s (N. J. Doran and D. Wood, [73] and [76]; S. Trillo et al. [77]; and K. J. Blow et al. [78]). The NOLM response depends on the instantaneous signal power so we might expect a time dependent excitation produces pulse shaping (i.e., the high power component of the pulse is switched and the low power tails are unswitched). A remarkable feature of soliton pulses however, is that their self-switching characteristics are very similar to the characteristics of CW or square wave excitation [78]. The first soliton switching experiment in NOLM used 412 fs pulses and a 5 km loop; a record-breaking switching energy of 46 pJ was measured, corresponding to approximately 110 W peak power [78]. In contrast, the experiment described in [75], used a pulse pump and a CW signal at different wavelengths, and the pump switching power was 24 W (8.5 nJ). The peak power required for switching in passive NOLMs based on silica fibers generally falls into the range 10–100 W.

The need to decrease the NOLM switching power for practical applications and to obtain 100% contrast (or switching efficiency) has led to the idea of introducing an amplifying element in the loop. This element is placed close to one end of the interferometer and might produce a significant power imbalance between the two counterpropagating signals, hence it reduces the required switching power (M. E. Fermann et al. [79]; D. J. Richardson et al. [80]). The corresponding device is called a NALM, for *nonlinear amplifying loop mirror*. The first experimental implementation of this concept used an $Nd^{3+}$ fiber amplifier with 6 dB gain; a switching power of 900 mW was obtained [79]. By using a 300 m NALM including an EDFA with 46 dB gain, a record-breaking switching power of 200 $\mu$W could be achieved [80]. Figure 6.28 shows the measured input/output signal transmission of the NALM. The maximum signal power of 2 W for 100 $\mu$W input corresponds to a net gain $G = 43$ dB, which accounts for 3 dB loss in the loop. The theoretical transmission $T$ of the NALM, also plotted in the figure, is given by the relation (K. Smith et al. [81]):

$$T = G\{1 - 2\eta(1 - \eta)[1 + \cos(\Delta\phi)]\} \tag{6.14}$$

where $\eta$ is the coupler coupling ratio, $G$ the EDFA gain, and $\Delta\phi$ a power dependent phase given by:

$$\Delta\phi = [(1 - \eta)G - \eta]\frac{2\pi}{\lambda}\delta nL \tag{6.15}$$

where $\delta n = n_2 P_s^{in}/A_{eff}$ is the nonlinear refractive index increase, $P_s^{in}$ the input signal power, and $L$ the loop length. From Eqs. (6.14) and (6.15), the maximum transmission and minimum switching power can be simultaneously achieved as the coupling ratio approaches $\eta = 0.5$. The switching power required for $\Delta\phi = \pi$ in the case $G \gg 1$ and

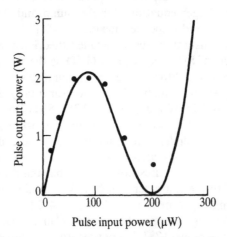

**FIGURE 6.28**  Input/output transmission characteristics of nonlinear amplifying loop mirror (NALM) using EDFA with 43 dB net gain, and theoretical response (full line), showing a switching power of 200 $\mu$W. From [80], reprinted with permission from *Electronics Letters*, 1990.

$\eta = 0.5$, is found from Eq. (6.15) to be $P_s^{in} \approx \lambda A_{eff}/2n_2\,GL$. This result shows that the switching power is reduced by the factor $1/GL$, which explains the performance improvement with respect to passive NOLMs.

In an alternate configuration, the EDFA is placed outside the Sagnac loop, in the path of the pump signal; this active NOLM configuration was used in the first LD-controlled experiments (M. Jinno and T. Matsumoto [82]; B. P. Nelson et al. [83]). The transmission of this device is given by Eqs. (6.14) and (6.15) with $G = 1$; the required switching power is divided by the EDFA gain. In this configuration, the Sagnac interferometer must be unbalanced (i.e., $\eta \neq 0.5$) to exhibit any nonlinear response, otherwise the switching power goes to infinity, according to Eq. (6.15). Thus, the NALM configuration with internal EDFA has superior switching efficiency. But the NOLM configuration with external EDFA has a lower ASE noise output (K. Smith et al. [6.81]). The fact that EDFAs can be used to considerably reduce the switching power makes switching efficiency a less critical issue, and therefore both configurations can be considered as having similar performance.

In addition to being a remarkable nonlinear switching device, the NOLM can also be used for *pulse processing*. Indeed, NOLMs (and their active counterparts, the NALMs) can perform the functions of *pulse shaping*, *cleaning*, *compression*, and *pedestal suppression* (K. Smith et al., [81], [84], and [85]). Pulse processing functions can be used to improve the output characteristics of gain-switched LDs for soliton system applications.

NOLMs can be used for instance to convert arbitrary chirped pulses obtained from a gain-switched DFB laser source at wavelength $\lambda$ into transform limited pulses at another wavelength $\lambda'$ (R. A. Betts et al. [86]). The principle is shown in Figure 6.29. The pump is a gain-switched DFB modulated at bit rate $B$ and emitting at wavelength $\lambda$. It is first passed through an EDFA then launched into the Sagnac loop, near one of the ends, through a fiber coupler of coupling ratio $\eta = 0.7$. For 1 mW (peak) pump power at the input, the (peak) pump power launched into the loop is 500 mW. The signal is a CW DFB at wavelength $\lambda'$. It is launched into the loop through a balanced coupler ($\eta' = 0.5$). The Sagnac loop is made of a 10 km long strand of DSF. When the pump is off, the signal power at $\lambda'$ is completely reflected by the loop, and the NOLM transmission is $T = 0$. When the pump is on, nonreciprocal XPM occurs between pump and signal. The switching effect produces signal pulses at $\lambda'$ at the loop output, while the output pump pulses are blocked by a bandpass filter centered at $\lambda'$. For a certain pump peak power, complete signal switching can be achieved. The signal pulse train obtained at the NOLM output is the faithful counterpart of the input pump pulse train, i.e., it has the same bit rate, bit sequence, and pulse width characteristics. However, each of the signal pulses is transform limited, while the input pump pulses are strongly chirped. This remarkable property was demonstrated in the experiment of [86], where 51 ps pulses could be obtained with a time–bandwidth product $\Delta\nu\Delta t = 0.44$. The presence of the pump coupler ($\eta = 0.7$) into the Sagnac loop effectively imbalances in the NOLM the absence of pump pulses. The NOLM can therefore be operated in the transmitting mode by increasing the signal power. As a result of switching by the pump pulses, dark signal pulses are obtained at the NOLM output (Figure 6.29b). This function corresponds to a local inversion of the input pump pulse train. The transform-limited pulses obtained through this NOLM configuration can be used for the transmission of bright or dark solitons (Section 5.11). Beside the capability of converting chirped data pulses into soliton

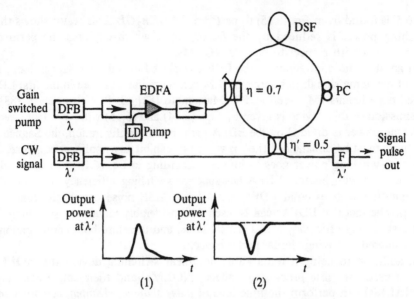

**FIGURE 6.29** Basic layout for conversion of chirped pulses at wavelength $\lambda$ into transform-limited pulses at wavelength $\lambda'$ through NOLM switching (DSF = dispersion-shifted fiber, PC = polarization controller, F = optical bandpass filter). Bright (1) and dark (2) output signal pulses, corresponding to the reflective and transmissive modes of the NOLM, respectively, are also shown.

data pulses, wavelength tunability is also possible, using for instance a two-section DBR laser for the CW signal source. If a second EDFA is used to boost the output signals, complete switching in the NOLM is not required, which makes it possible to reduce the loop length to a few hundred meters. The advantage of incomplete switching is the possibility to exploit additional pulse compression effects [86]. Two other applications could be important in communications systems and fiber networks: (1) wavelength conversion of high speed data, and (2) high speed optical logic for optical data inversion. Another application of this NOLM configuration is the generation of square picosecond pulses with controlable durations (D. U. Noske et al. [87]).

The EDFA-based NOLM configuration where both pump and signal are pulsed can be used to perform three major high speed signal processing functions: (1) *wavelength conversion*, (2) *all-optical signal regeneration*, and (3) *all-optical time-domain demultiplexing*. The function of all-optical regeneration is important in communication systems using short (i.e., 10–100 ps) optical data pulses. In the case of EDFA-based soliton systems, for instance, the accumulated ASE causes timing jitter in the pulse arrival times (Gordon–Haus effect), a source of BER penalty. Likewise, high speed fiber networks based on time division multiple access (TDMA) require low pulse jitter for minimizing pulse separation (guard bands) and optimizing the use of bandwidth. In these systems, amplitude and timing of optical data pulses must be periodically restored using an all-optical regenerator. The NOLM can be used effectively for this purpose, as shown in Figure 6.30 (M. Jinno and M. Abe [88]). The NOLM is input with both signal data (e.g., at $B = 5$ Gbit/s) with wavelength $\lambda$, and clock pulses (e.g., at $f = 5$ GHz) with wavelength $\lambda'$. The clock is generated by

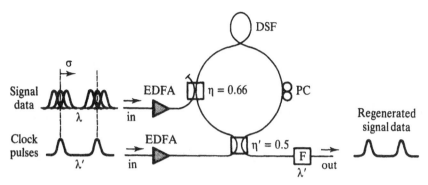

**FIGURE 6.30** Basic layout of an all-optical regenerator based on NALM switching ($\sigma$ = signal timing jitter, DSF = dispersion-shifted fiber, PC = polarization controller, F = optical bandpass filter).

a gain-switched DFB laser. Two EDFAs are used as power boosters for both signal and block. The signal data stream is characterized by timing jitter ($\sigma$) and amplitude fluctuations or AM noise (not shown in the figure). In this NOLM configuration, the signal data act as a pump for the clock, i.e., clock pulses are switched when the input signal is a 1, otherwise the NOLM ouput is zero. The switched clock pulse train exhibits the same modulation characteristics as the input signal pulse train. The switched clock pulses have a constant peak power, in spite of AM noise present in the switching signal pulses. This is due to the NOLM sinusoidal response. Likewise, the switched clock pulses exhibit the same timing characteristics as the input clock pulse-train, in spite of the timing jitter in the switching signal pulses. This property can obtained if the signal pulse width is smaller than the pump/signal walk-off time. This walk-off effect generates a broad switching window for the clock pulses, which explains the absence of timing jitter in the switched output (N. A. Whitaker et al. [89]). Thus, the NOLM can be used as a *LD-controlled all-optical regenerator with both amplitude and timing restoration,* as demonstrated in the 5 Gbit/s experiment of [88].

The important function of *time-domain demultiplexing* can also be performed all-optically with the NOLM (M. Jinno and T. Matsumoto [90], K. J. Blow et al. [91]). In this configuration, the NOLM acts as an optical gate, controlled by clock pulses with low duty cycle and frequency. The clock acts as a pump for the signal input, which is a pulse sequence of high duty cycle. As shown in Figure 6.31, the signal data a bit rate B (wavelength $\lambda$) and the clock pulses at bit rate $B/N$ ($N$ integer) and wavelength $\lambda'$ are input to the NOLM after power amplification through EDFAs. The relative delay between the two pulse trains can be adjusted so that periodic coincidences between pump pulses and signals belonging to a given time slot can be achieved. The relative delay can be controlled by a microwave (or digital) delay line in the synchronization circuit of the pump LD. Any 1 input signal bit is then switched by the clock pulse, while the NOLM output is zero for the 0 bits. In the case of modulated signals, such gating of the high bit rate signal pulse train by the low-frequency clock corresponds to the function of time domain demultiplexing.

Note that the clock and signal wavelengths can be arbitrary, except in high-speed applications where pulse walk-off must be minimized. Two major advantages of such all-optical demultiplexing are (1) its ultrafast response (as limited by walk-off effects),

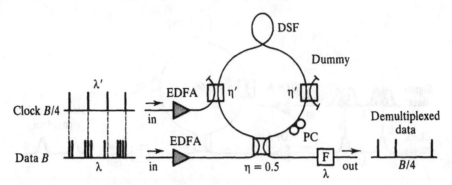

**FIGURE 6.31** Basic layout for all-optical time division demultiplexing, based on NALM switching; in this example, the demultiplexing rate is 1:4 ($B$ = signal bit rate, $B/4$ = clock bit rate, DSF = dispersion-shifted fiber, PC = polarization controller, F = optical bandpass filter).

and (2) both switched and unswitched outputs are retrieved in the optical domain and can therefore be transmitted through fiber systems. Multiple stage demultiplexing is also possible using cascades of NOLMs and EDFAs with unity fanout.

In one early demonstration of all-optical demultiplexing, as described in [91], switching of 492 MHz (unmodulated) signals at $\lambda = 1.53\,\mu m$ could be achieved with a (passive) NOLM, using 100 ps, 10 W clock pulses from an Nd:YAG source at $\lambda = 1.064\,\mu m$. In another experiment, described in [90], switching of 5 Gbit/s signals at $\lambda = 1.54\,\mu m$ could be achieved in a polarization-maintaining NOLM, using 120 ps, 1.8 W clock pulses from an Nd:YAG source at $\lambda = 1.32\,\mu m$. The use of a polarization-maintaining (PM) fiber, as opposed to a standard DSF, is justified by considerations of pump/signal polarization stability with respect to environmental (temperature) changes. The drawback PM–NOLMs is their sensitivity to input signal polarization. All-LD-controlled switching in PM–NALMs could be demonstrated by using EDFAs as clock/signal power boosters (H. Avramopoulos et al. [92]; A. Takada et al. [93]). Polarization insensitive switching could also be achieved in an improved configuration of the PM–NALM including a birefringent fiber polarization compensator (N. A. Whitaker et al. [94]). All-LD-controlled switching in standard DSF NALMs was demonstrated by B. P. Nelson et al. [83]. In their experiment, 20 GHz signals at $\lambda = 1.56\,\mu m$ were switched through 2.5 GHz clock pulses at $\lambda = 1.53\,\mu m$, having 160 mW amplified peak power. The best results to date (1993) for high speed time domain demultiplexing with LD-controlled NALMs are: 40 Gbit/s demultiplexing to 2.5 Gbit/s clock rate, obtained with a PM device (A. Takada et al. [93]), and 64 Gbit/s to 4 Gbit/s clock rate, obtained with a standard DSF device (P. A. Andrekson et al. [95]). The last experiment used the NALM configuration shown in Figure 6.31. A dummy coupler is added to the loop to balance operation in the absence of pump. The results of this experiment are shown in Figure 6.32 for the signal bit rates 16 Gbit/s, 32 Gbit/s, and 64 Gbit/s. The demultiplexed data has a high extinction ratio (14 dB). BER measurements made with the demultiplexed data had low penalties, i.e., 0.6–2.2 dB for 16–64 Gbit/s rates, and showed no error floors [95]. A more recent NALM demultiplexing experiment at 32 Gbit/s included a phase locked loop for clock recovery after 20 km of signal transmission; the measurements showed no power penalty for the 8 Gbit/s demultiplexed signals (S. Kawanishi et al. [96]).

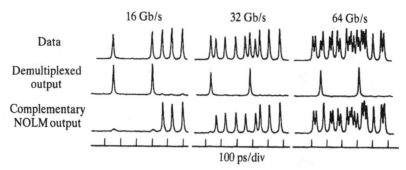

**FIGURE 6.32** Experimental demonstration time domain demultiplexing at very high speed through NOLM, with $f$ = 4 GHz clock rate. Top traces: signal data at $B$ = 4 Gbit/s, 32 Gbit/s, and 64 Gbit/s. Bottom traces: demultiplexed signal outputs at $B$ = 4 Gbit/s and corresponding complementary outputs, representing unswitched or reflected signals. From [95] © 1992 IEEE.

LD-controlled NOLMs have potential applications in high speed communication systems such as TDMA networks. Other potential applications are in high speed optical computing. The NOLM demultiplexer can be used as an AND logic gate. Combined with inverting NOLM gates, the logical $\overline{\text{AND}}$ (or NAND) operation can be be generated. From NAND gates and OR gates, realized through passive power combining, the complete set of logical functions, including the important exclusive OR (or XOR), can be generated. Applications of high speed optical computing are still at a research level and may reveal themselves only during the next decade. But the recent development of practical LD-controlled NOLMs with 100 GHz bandwidth could represent decisive progress in this field.

Before describing other EDFA-based devices for time domain demultiplexing applications, we consider a particular type of PM–NOLM based on the effect of XPM-induced frequency shift using a polarization rotation mirror for the Sagnac loop coupler (T. Morioka et al., [97] and [98]). This application is effectively polarization insensitive. The NOLM siwtches in the *frequency domain* instead of the *time domain*. Instantaneous phase change (or chirp) induced by pump XPM causes the individual signal pulses from a high speed sequence (wavelength $\lambda$) to be frequency shifted according to their relative position within the pump pulse time frame. For a 3 km PM fiber, the maximum chirp is approximately 50 GHz per watt of launched pump power [98]. The output signal pulses (at wavelengths $\lambda + \delta\lambda$, $\lambda + 2\delta\lambda$, $\lambda + 3\delta\lambda$...) can be demultiplexed in the time domain by using a grating at the NOLM output. In the experiment described in [98], polarization independent demultiplexing of a 60 GHz, 4 bit sequence ($\lambda = 1.307\,\mu$m) was demonstrated using 7.5 W pump power from an Nd:YLF laser at $\lambda = 1.313\,\mu$m. The switching power requirement could be potentially reduced to a milliwatt level by using EDFAs or PDFAs as power boosters. This remains to be investigated. Potential applications of this frequency domain switch/demultiplexer NOLM are high speed WDM systems with a wavelength routing capability.

Figure 6.33 shows a high speed time- and frequency-domain switching and gating application of EDFAs based on *stimulated four-photon mixing* (*SFPM*), another fiber nonlinearity. This configuration uses two pulse trains: (1) a signal at frequency $v_s$ with high bit rate $B$ and (2) a pump at frequency $v_p$ with low repetition rate $B/N$

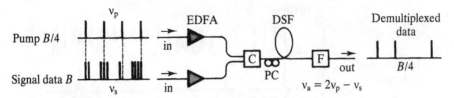

**FIGURE 6.33**  Basic layout for all-optical time division demultiplexing, based on stimulated four-photon mixing (SFPM); in this example, the demultiplexing rate is $1:4$ ($B =$ signal bit rate, $B/4 =$ clock bit rate, DSF $=$ dispersion-shifted fiber, C $=$ fiber coupler, PC $=$ polarization controller, F $=$ optical bandpass filter).

($N$ integer). They are first amplified through EDFAs then coupled into a long ($L > 10$ km) strand of DSF. Due to the fiber third order susceptibility, parametric mixing (of SFPM) occurs along the fiber when pump and 1 signal data pulses are coincident in time. A DSF enables near-phase-matched interaction along the fiber. A mechanical polarization controller is required for optimum polarization alignment and interaction efficiency. The nonlinear mixing between pump and signal along the fiber generates an idler pulse at frequency $v_a = 2v_p - v_s$ (Section 5.12). The idler signal is retrieved at the DSF output end through a narrowband optical filter. Since idler pulses are generated only when the pump coincides with a 1 signal data pulse, this device acts like a logical AND gate. It effects time domain demultiplexing at the pump rate. The two EDFAs boost the pump and signal inputs to a power level where efficient SFPM allows us to use LD sources for the inputs.

As all-LD-controlled implementation of the SFPM demultiplexing gate was demonstrated for 16 Gbit/s and 32 Gbit/s signals with $\lambda = 1.53$ $\mu$m wavelength (P. A. Andrekson et al., [99] and [100]). The pump and signal peak powers required for achieving demultiplexed signals with 8 dB (optical) extinction ratio after 14 km DSF nonlinear transmission was 33 mW and 17 mW, respectively [99]. BER measurements showed a small penalty of 0.9 dB, attributed to crosstalk. This penalty could be reduced by increasing the SFPM conversion efficiency through higher input pump or signal powers. The SFPM conversion efficiency also depends on the wavelength separation between the pump and the fiber zero dispersion point, and exhibits several maxima (K. Inoue and H. Toba [101]). While a conversion efficiency near 100% is theoretically possible, it is difficult to achieve experimentally, due to the variation of the zero dispersion wavelength along the DSF. As defined in units of $P_i^{out}/(P_p^{in})^2 P_s^{in}$, the maximum conversion efficiency that can be obtained with 10 km DSFs, is about $\eta \approx 3 \times 10^{-4}/$mW$^2$; in typical operating conditions, this efficiency corresponds to a signal-to-idler power conversion of $-24$ dB [101].

The principle of demultiplexing through SFPM can also be applied to *picosecond sampling* (P. A. Andrekson [102]). In this application, based on setup of Figure 6.33, the signal frequency is slowly scanned while the pump frequency is fixed. The resulting idler signal exhibits the same modulation as the signal pulse sequence but at a reduced time scale, i.e., a microsecond scale for picosecond pulse sampling. The sampled signal envelope can then be retrieved by detecting the idler through a low pass receiver circuit. By using EDFAs as power boosters, this all-optical sampling principle could be implemented with LDs as practical pump and signal sources; a sampling resolution of 20 ps was achieved with 3 mW pump and signal (average powers) [102].

The variety of device applications reviewed in this section illustrates the great impact EDFAs have had and will likely continue to have in the field of nonlinear fiber optics. The combination of efficient LD-controlled EDFAs with ultrafast fiber nonlinearities has revealed new potential for practical high speed and all-optical signal processing devices.

## 6.6  RECIRCULATING DELAY LINES

The EDFA application of greatest impact in optical fiber communications is undoubtedly the *active recirculating fiber delay line (ARDL)*. ARDLs have been extensively used to evaluate the performance of very long-haul linear transmission systems based on EDFAs and to demonstrate their feasibility. ARDLs have also been used to study the fundamental limits of optical soliton transmission, to discover new interaction effects, to elaborate countermeasures, and to prove the feasibility of nonlinear transmission systems. This process originated from loop experiments using Raman and EDFA-based ARDLs and in the past few years has been followed by more realistic straight-line transmission experiments. Loop and straight-line transmission experiments are described in the next chapter. This section focuses on optical fiber memories, fiber rotation sensors, and signal processing functions relevant to communication and computing systems. We compare earlier ARDL devices, based on Raman fiber amplifiers (RFA) and semiconductor amplifiers (SCA), with EDFA-based devices.

Figure 6.34 shows that a passive recirculating delay line is a closed fiber loop into which an optical signal pulse (width $\Delta T$) is coupled at time $t = 0$. The input signal can be a single optical pulse, a packet of high speed optical data, or a pair of optical soliton pulses. This signal completes a loop circulation in the characteristic transit time $t_{\text{loop}} = nL/c$, where $L$ is the loop length, and $c/n$ the signal group velocity. The loop length $L$ is chosen so that $t_{\text{loop}} > \Delta T$ and the signal pulse does not recombine and interfere with itself. At each recirculation, the signal can be sampled through any tap coupler placed in the loop path, e.g., the input coupler. This results in the periodic output sequence shown in Figure 6.34a. The recirculating delay line thus acts as an optical memory with discrete delays $t_{\text{loop}}, 2t_{\text{loop}}, 3t_{\text{loop}}, \ldots, Nt_{\text{loop}}$. The loop round-trip loss $T$ includes transmission loss, excess loss, and sampling loss. The loop round-trip loss causes the output signal envelope to decay exponentially as $T^N$. This limits the maximum loop delay.

If an optical amplifier (e.g., EDFA, RFA, SCA) is placed within the loop and operated at exact loss compensation ($G = 1/T$), the output signal envelope can be maintained constant, as shown in Figure 6.34b. The loop storage time can be enhanced by a large factor. Since the signal envelope is constant, the maximum storage time is now limited by the effect of recirculating ASE buildup, which causes the signal-to-noise ratio (SNR) to decay, as in the case of optical amplifier chains. For the ring waveguide configuration of Figure 6.34b, the condition $G = 1/T$ represents the threshold of laser oscillation. There are two ways to prevent this oscillation. First, the gain $G$ can be set to a value slightly below the loop loss $T$, i.e., $G = 1/T - \varepsilon$, such that the ring always operates below its oscillation threshold. But this configuration is very sensitive to small fluctuations in round-trip loss, such as induced by polarization wandering, and consequently, is very unstable. The maximum number of recirculations

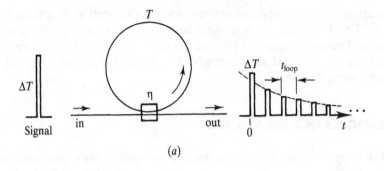

(a)

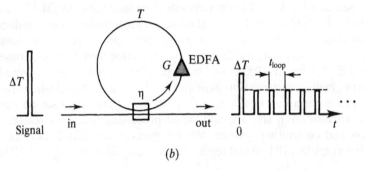

(b)

**FIGURE 6.34** Recirculating fiber delay lines: (a) passive configuration, (b) active configuration, with output signal response for single-pulse excitation (C = fiber coupler with coupling ratio $\eta$, T loop round-trip loss, G = optical amplifier gain, with $G = 1/T$).

$N$ is then limited by the exponential decay of the output signals in $(1 - \varepsilon T)^N$. The second possibility is to turn the gain in the loop on and off for the duration $Nt_{\text{loop}}$, shorter than the time necessary to reach steady state laser oscillation. With EDFA-based ARDLs, this approach requires an additional in-loop switching element (now shown in Figure 6.34b) as the gain cannot be rapidly switched. With RFA-based ARDLs, the gain can be instantaneously switched on and off by using a square pump pulse [103] and [104]. With SCA-based ARDLs, the same function can be performed by square modulating the driving current of the semiconductor device. An additional switching element is not necessary in the case of ARDLs based on RFAs and SCAs.

As a device, the active recirculating delay line can be viewed as being equivalent to a straight-line optical amplifier chain made of a concatenation of elements identical to that of the loop. The basic optical elements in the ARDL periodically encountered by the signal (i.e., coupler, amplifier, transmission fiber) are physically identical. But the basic optical elements in the straight-line amplifier chain are physically different. This is the main difference between the ARDL and the straight-line system. Another difference lies in their transmission characteristics. Straight-line amplifiers can achieve steady state characteristics. ARDL characteristics always transient, because the gain has to be turned on at $t = 0$ and off at $Nt_{\text{loop}}$, due to ASE noise buildup. Furthermore, gain saturation effects are steady state in the straight-line case, whereas they are transient in the ARDL. If we overlook these basic differences and assumes ideal components operating in steady state conditions, we can consider that the active recirculating delay line characteristics are very nearly identical to those of a straight-line amplifier system made of the same periodic elements. Thus, the theory described in Chapter 2 for the study of signal and ASE evolution in optical amplifier chains can be applied to the case of ARDLs. In particular, the three types of amplifier chains A, B and C considered in this analysis correspond to the following cases: in a Type A loop, the EDFA is placed at the loop input end (Figure 6.34); in a Type B look, the EDFA gain is distributed over the whole fiber loop; and in a Type C loop, the EDFA is placed near the loop output end. The additional loss incurred by discrete elements (such as the coupler tap) can be introduced through the transfer matrix formalism developed in Section 2.5.

The first demonstration of an all-optical ARDL, reported in 1985, used stimulated Raman amplification (E. Desurvire et al., [103] and [104]). In the experiment described in [103], loop loss cancellation through SRS was achieved at $\lambda = 1.12\,\mu m$ signal wavelength with 710 mW input pump power at $\lambda = 1.064\,\mu m$, and $N = 130$ signal recirculations could be achieved in a 810 m loop, corresponding to an optical delay of 0.5 ms. The maximum optical delay was limited by a strong amplitude noise generated by incoherent pump recirculations in the loop. Using an improved pump modulation scheme (to prevent pump recirculation) and a dichroic loop coupler (to decrease the pump threshold to a minimum value of 350 mW), the optical delay could be increased to 3 ms, corresponding to $N = 800$ signal recirculations in a 760 m loop [104].

These results demonstrate an actual *optical fiber memory* operation, in which the signal to be stored must be initially coupled into the loop. Other SRS loop devices, in which the initial pulse is generated within the loop, do not correspond to actual optical memories, even if many single-pulse recirculations can be observed. For instance, a synchronously pumped Raman fiber loop could produce $N = 18,250$ signal recirculations after pulse self-initiation, corresponding to a 50 ms delay or a single-pulse propagation over 10,310 km, limited by triggering electronics (E. Desurvire et al. [105]). In this configuration, pulse dispersion was compensated by of all-optical regeneration produced by synchronous pumping. Synchronously pumped SRS loops cannot operate as optical fiber memories for digital signals, as noise pulses would eventually form in the 0 slots. In another SRS loop experiment, pulse self-initiation could be achieved through an internal SRS NOR gate; over $N = 100,000,000$ single-pulse recirculations could be obtained in the 870 m loop, corresponding to more than 10 mn optical delay, or 87 million km propagation (V. I. Belotitskii et al. [106]). In this configuration, pulse dispersion was compensated by the effect of all-optical regeneration produced by the internal NOR gate. This device can operate as a digital optical fiber memory, as the contrast between 1 and 0 pulses is indefinitely preserved [106]. During 1988–1989, SRS-based ARDLs were used to study optical soliton propagation over transmission distances up to 6000 km, corresponding to optical delays of 30 ms (L. F. Mollenauer and K. Smith, [107] and [108]).

The three main drawbacks of SRS-based ARDLs are: (1) their high pump thresholds (300 mW–2.5 W, [104], [106], and [107]), (2) the requirement for a high power source at the Raman-shifted wavelength (i.e., $\lambda_p = 1.45\,\mu m$ for $\lambda_s = 1.55\,\mu m$ signals, Section 5.12), and (3) the high pump-to-signal crosstalk induced by pump intensity noise. Four main advantages of SRS-based ARDLs are: (1) their broad (30 nm) gain bandwidth, with tunable peak wavelength, (2) their low ASE noise (as $n_{sp} = 1$ can be achieved, Section 5.12), (3) the possibility of fast on/off switching, and (4) they can be realized with standard SMF and DSF fibers.

Another type of ARDL, demonstrated in 1987, used a semiconductor optical amplifier for loop loss compensation (N. A. Olsson [109]). Using a 75 km fiber loop and a semiconductor amplifier (SCA) with $G = 37$ dB fiber/fiber gain at $\lambda = 1.5$ $\mu$m, $N = 200$ signal recirculations could be achieved, corresponding to 75 ms delay or 15,000 km propagation distance. Output BERs for 150 Mbit/s signals were also measured, showing less than 0.3 dB penalty for 15,000 km propagation. This loop experiment is the first demonstration of optical data transmission through an ultralong optical amplifier chain. This result could not be achieved in a straight-line system essentially because of the gain polarization sensitivity of SCA devices. This illustrates that loop and straight-line system experiments do not have the same operating constraints, and can in some cases yield quite different BER characteristics.

For SCA-based ARDL applications, the signal polarization can be mechanically controlled. Alternatively, PM fiber loop can be used, then polarization sensitivity is not a device limitation. The main advantage of using SCAs for ARDL applications are their subnanosecond gain dynamics. Their fast response enables on/off memory gating and switching in addition to loss compensation (G. Grosskopf et al. [110]; M. Calzavara et al. [111]). All-optical signal regeneration with signal amplification and retiming is also possible (H. Izadpanah et al. [112]). Alternatively, the functions of loss compensation and gating/switching can be performed in two stages using EDFAs and SCAs, respectively, as shown in [110]. The switching function for on/off operation of the ARDL, can also be performed by integrated acousto-optic or electro-optic devices.

The first demonstration of ARDLs with an RE-doped fiber as a gain element was reported in 1988. It used an LD-pumped $Nd^{3+}$ doped fiber amplifier operating at $\lambda = 1.088$ $\mu$m signal wavelength (P. R. Morkel [113]). In this experiment, the maximum number of signal recirculations in the 34 m long loop was $N = 300$, as limited by the effect of parasitic self-oscillation in the resonator.

The first EDFA-based ARDLs were reported in 1990–1991. They were used to study ultralong distance transmission in linear and nonlinear communication systems (L. F. Mollenauer et al. [114]; D. J. Malyon et al. [115]; and N. S. Bergano et al. [116]). In these experiments, the transmission distances were in 10,000–36,000 km, corresponding to optical delays of 50 ms to 180 ms. For minimum ASE noise accumulation and optimum transmission distance (or storage time), the loop lengths used in these experiments were 35 km, 75 km, and 120 km. The 35 km loop had one EDFA spaced at 35 km. The 75 km loop had three EDFAs spaced at 25 km. And the 120 km loop had four EDFAs spaced at 40 km. In the experiment described in [116], which concerned NRZ signal transmission, $10^{-9}$ BERs were measured for bit rates of 2.4 Gbit/s ($L = 21,000$ km) and 5 Gbit/s ($L = 14,000$ km), respectively. These results essentially impact upon the field of transoceanic communications but they also demonstrate that EDFA-based ARDLs achieve low signal BERs at optical delays of about 100 ms. In another experiment, 2 Gbit/s NRZ signal propagation was studied over distances up to 400,000 km, corresponding to an optical delay of 2 seconds (C. R. Giles et al. [117]). But low BER operation could not be achieved over such distances.

A figure of merit for ARDLs is the product of the maximum optical delay and the signal bit rate, i.e.,

$$M = t_{max}B \tag{6.16}$$

The maximum delay $t_{max}$ can be defined as $t_{max} = N_{max} t_{loop}$, where $N_{max}$ is the number of signal recirculations for which $10^{-9}$ BER is achieved for signals at bit rate $B$ and detected at $t = t_{max}$. For instance, the results previously discussed (i.e., 2.4 Gbit/s, $L = 21,000$ km and 5 Gbit/s, $L = 14,000$ km) delimit a figure of merit range $M = 2.5-3.5 \times 10^8$.

Is there an ultimate limit to the maximum optical delay and figure of merit achievable in ARDLs? Consider the different factors that limit optical signal propagation: SNR degradation due to ASE noise buildup, timing jitter, fiber dispersion, and fiber nonlinearities. The limits due to these effects are different for linear (RZ or NRZ) or nonlinear (soliton) pulse propagation, and become more constraining as the signal bit rate is increased. For a basic ARDL configuration operating in the linear regime and at the exact zero dispersion wavelength of DSFs, the optical delay is first limited by the effect of SNR decay, as in the case of straight-line systems, i.e., $t_{max} = nL_{max}/c$, where $L_{max}$ is the maximum transmission distance giving $10^{-9}$ BER (Chapter 3). At high bit rates (e.g., $B > 1$ Gbit/s), a second limit is dispersion, according to the relation $B^2 L_{max}[(\text{Gbit/s})^2 \text{ km}] = 4000(15_{ps/nm.km}/\bar{D}_{ps/nm.km})$, where $\bar{D}$ is the (average) dispersion coefficient [63]. The dispersion limit corresponds to a maximum optical delay $t_{max} = nL_{max}/c$ or:

$$t_{max}(\text{dispersion}) = 150 \text{ ms} \frac{4 \times 10^6}{B_{GHz}^2} \frac{15_{ps/nm.km}}{\bar{D}_{ps/nm.km}} \approx \frac{20 \text{ ms}}{B_{GHz}^2} \frac{15_{ps/nm.km}}{\bar{D}_{ps/nm.km}} \qquad (6.17)$$

For 10 Gbit/s signals with dispersion $\bar{D} = 0.01$ ps/nm.km, the dispersion-limited optical delay is $t_{max} \approx 300$ ms, corresponding to a figure of merit $M = 3 \times 10^9$. The dispersion limit of ARDLs can effectively be overcome by dispersion compensation techniques.

On the other hand, ARDLs using soliton pulses for recirculating signal data, are intrinsically free from any dispersion limit. In this case, the maximum optical delay is determined by (1) the SNR degradation due to ASE buildup, and (2) the Gordon–Haus effect, which causes timing jitter (Section 5.11). Both effects can be alleviated by two possible methods discussed in Chapter 7: *periodic synchronous modulation* (M. Nakazawa et al., [118]–[120]), or *sliding frequency guiding filters* (L. F. Mollenauer et al. [121]). They correspond to an all-optical regeneration effect, which alleviates SNR decay and timing jitter. Using the periodic synchronous modulation technique, $N = 2,000,000$ recirculations of 10 Gbit/s soliton data in a 510 km EDFA–ARDL could be experimentally demonstrated [118]. In this experiment, the ARDL included 11 EDFAs with 50 km spacing, and one sinusoidally driven amplitude modulator; the total transmission distance was over one million kilometers, corresponding to an optical delay of approximately 5 s. Over this transmission distance, BER measurements showed no power penalty [120]. Transmission distances up to 180 million kilometers were also demonstrated for 10 Gbit/s soliton signals, corresponding to an optical delay of 15 mn [120]. These results clearly show that ARDLs using soliton data and a means of internal all-optical regeneration (e.g., periodic synchronous modulation) can provide virtually unlimited optical delays. A theoretical proof of this unlimited storage time capability is provided in a detailed analysis by H. A. Haus and A. Mecozzi [122]. Consequently, we can conclude that for soliton-based ARDLs with all-optical regeneration, the figure of merit $M$ increases indefinitely in time, i.e., $M = t_{loop} B \times \text{int}(t/t_{loop}) \equiv NBt_{loop}$, where $N$ is the number of signal recirculations obtained at time $t$.

Unlimited storage time capability could also be obtained by using electronic regeneration inside the recirculating loop. Electronic regeneration would detect the signals with zero BER at each loop recirculation, electronically perform the retiming, reshaping, and amplification functions, then convert them back into optical pulses. But this approach presents the same disadvantages as using electronic repeaters for transmission systems: electronic speed limits and fixed bit rate.

All-optical fiber memories may develop considerably in the near future, as a consequence of the current progress in ultralong transmission systems. Having shown that ARDLs can be effectively used as all-optical memories, we discuss potential applications of ARDLs to high speed signal processing in fiber communications and computing networks.

Potential applications for ARDLs concern a large variety of fields:

- Physical measurements and instrumentation
- High speed digital and analog signal processing
- Television frame memories
- Satellite radar altimetry
- Radar and phased-array antennas
- Rotation sensing
- Computer buffer memories
- Photonic packet switching in communication networks

ARDLs can be used as high finesse resonators for optical spectrum analysis; spectral resolutions better than 100 kHz were demonstrated with an EDFA-based device (H. Okamura and K. Iwatsuki [123]). High finesse ring resonators have important applications in rotation sensing (R. E. Meyer et al. [124]; J. T. Kringlebotn [125]). Incorporation of optical amplifiers is known to enhance the finesse of such resonators (J. T. Kringlebotn et al. [126]; H. Okamura and K. Iwatsuki [127]). A finesse as high as $F = 150,000$, corresponding to a round-trip feedback coefficient of 0.99996, could be measured in an EDFA-based ARDL [126].

Digital and analog signal processing functions performed by ARDLs are numerous. They include: bandpass/notch/phase-only filtering of RF signals (B. Moslehi [128]; B. Moslehi and J. W. Goodman [129]); code and word generation, reconfigurable correlation, convolution, bit rate compression, time domain multiplexing, time slot interchanging, frequency translation and matrix–vector multiplication. These are the attributions of fiber optic delay line and lattice signal processing: B. Moslehi et al. [130]; K. P. Jackson et al. [131]; M. Tur et al. [132]; P. R. Prucnal et al. [133] and [134]; and R. I. Macdonald [135]. Figure 6.35 shows the principles of all-optical bit rate compression and frequency translation through ARDLs (S. A. Newton et al. [136]; T. G. Hodgkinson and P. Coppin [137]). Figure 6.35a shows the bit rate $B = 1/T$ of an RZ data sequence is compressed by a factor $T/\tau$ through an ARDL of optical delay $T - \tau$. Considering a bit sequence $(1, 2, 3, \ldots, N)$, the device output consists of packets with the sequence $(1, 12, 123, \ldots, 123 \ldots N)$. The maximum number of bits that can be processed this way to form a compressed sequence is given by $T/\Delta T$, where $\Delta T$ is the RZ pulse width. For this operation, the ARDL active switch must be initially in the cross state $(\eta = 1)$, then turned into an intermediate state (e.g., $\eta = 0.5$) for a maximum duration $T$, and finally turned back into the cross

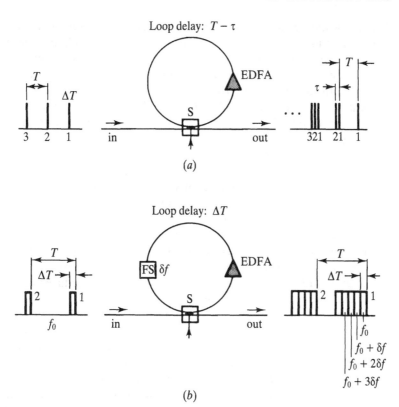

**FIGURE 6.35** Examples of all-optical signal processing through active recirculating delay lines (ARDL): (*a*) bit-rate compression and (*b*) frequency translation (S = active switch, FS = frequency shifter, $B = 1/T$ = bit rate).

state. The compressed sequence $123\ldots N$ must be selected at the output through an optical sampling gate. Figure 6.35*b* shows frequency translation of optical data at carrier frequency $f_0$ achieved through a short ARDL that includes a frequency shifter; the ARDL optical delay $\Delta T$ (nearly) equals the pulse width, which produces the output data packet sequence. The bits within each packet have the same logic value, but are individually frequency-shifted by amounts $\delta f, 2\delta f, \ldots, N\delta f$, respectively. The maximum frequency shift is given by $(T/\Delta T)\delta f$. For this operation, the ARDL active switch operates like the example of Figure 6.35*a*. A high speed optical time domain demultiplexer is necessary to retrieve the output data sequence corresponding to a desired carrier frequency $f_0 + k\delta f$; such a demultiplexing function can be performed through a NALM, as discussed in the previous section. Potential applications of ARDLs as analog devices are broadband radar signal processing (K. P. Jackson et al. [131]) and television frame memories (M. I. Belovolov [138]). Examples of radar signal processing application are satellite radar altimetry, which is used for ocean surface measurements (M. Maignan and J. J. Bernard [139]), and optically steered phased-array antennas (W. W. Ng et al. [140]). Considering TV-frame memories, a complete digital TV picture made of a $1000 \times 1000$ element frame with 8 bit coding could be stored at $B = 10$ Gbit/s in a 160 km ARDL having 0.8 ms delay [138].

Another important application of ARDL concerns *fiber rotation sensors*. Fiber rotation sensors or interferometric fiber optic gyroscopes (IFOG) are based on the principle of detecting rotation-induced phase shifts between two counterpropagating signals in a loop (Sagnac effect). Two active configurations of IFOGs using optical amplifiers are: (1) the amplified fiber ring resonator gyroscope (AFRRG), and (2) the amplified reentrant fiber gyroscope (ARFG). Figure 6.36 shows that the AFRRG configuration uses an ARDL as an active ring resonator; the counterpropagating signals are frequency-shifted and locked to two resonances of the ring; the ring length is modulated by a PZT, which detects rotation-induced frequency splitting between the two signals, proportional to the rotation rate (J. T. Kringlebotn, [6.125] and [6.141]). Theory shows that the AFRRG rotation sensitivity, or minimal error $\delta\Omega$, is determined by the signal–ASE beat noise limit; assuming an ideal device having a ring finesse $F = 3000$ (feedback coefficient 0.999) and a 10 km resonator loop, the calculated sensitivity is $\delta\Omega \approx 10^{-11}$ rad/s, corresponding to $\delta\Omega \approx 2 \times 10^{-6}$ deg/h [125]. This extremely high sensitivity (which corresponds to a full rotation in 21,000 years) is three orders of magnitude better than the best performance of passive IFOG and ring resonator gyros [125]. Finite source coherence length could reduce this sensitivity improvement but was not included in the theory.

The active reentrant fiber gyroscope configuration (ARFG) is shown in Figure 6.37. The principle of this approach is to include an active recirculating relay line (ARDL) within a Sagnac interferometer (G. A. Pavlath and H. J. Shaw [142]). The effect of the ARDL is to effectively increase the Sagnac loop length without transmission loss, which enhances the device sensitivity to rotation. Initially, the active switch is in the cross state ($\eta = 1$); at $t = 0$, the switch is turned for a duration $T$ to an intermediate state with small coupling ratio ($\eta = \varepsilon$). A single signal pulse is also launched into the Sagnac loop at $t = 0$, forming two counterpropagating pulses

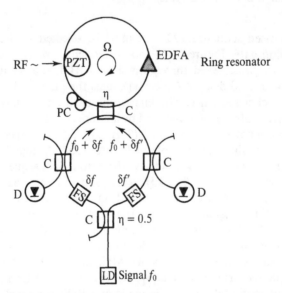

**FIGURE 6.36**    Device configuration of active fiber ring resonator gyro (AFRRG); C = passive couplers with coupling ratio $\eta$, FS = frequency shifter, PZT = piezoelectric transducer, PC = polarization controller, D = detector.

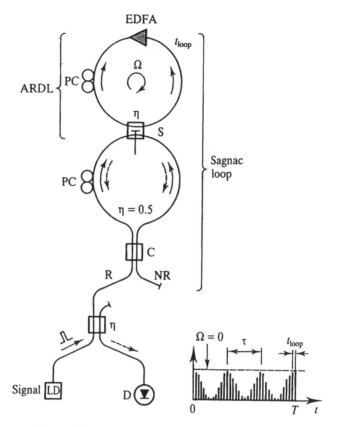

**FIGURE 6.37** Device configuration of active reentrant fiber gyro (ARFG); C = Sagnac loop coupler, R = reciprocal port, NR = nonreciprocal port, S = active switch, PC = polarization controller D = detector.

inside the Sagnac interferometer. A small fraction of these signals is coupled into the ARDL, where they recirculate for the duration $T$. The bidirectional amplification requirement means that no optical isolators are used, which restricts the maximum EDFA gain and the loop loss. At each loop recirculation ($t = Nt_{loop}$), a fraction of the signals exits the ARDL and recombine through the Sagnac loop coupler. The signal detected at the reciprocal port output is a train of pulses with repetition rate $1/t_{loop}$. In the absence of rotation ($\Omega = 0$), the pulse train envelope is constant and has maximum amplitude, consistent with Sagnac interferometer reciprocity. The effect of rotation ($\Omega \neq 0$) is to create a nonreciprocal phase shift $\delta\phi$ between the counterpropagating signals. As the phase shift $\delta\phi$ increases linearly with the loop length, the output waveform is sinusoidally modulated, with a frequency $1/\tau$ proportional to the rotation rate $\Omega$. The rotation rate is therefore determined by a frequency measurement. This measurement can be performed using an RF spectrum analyzer after low pass filtering, or by digitally counting the zero crossings. thus, the ARFG offers the advantage of an intrinsic linear scale factor, with digital frequency readout. Since low rotation rates correspond to long modulation periods, the gyroscope sensitivity is proportional to the maximum optical delay achievable with the ARDL.

The first demonstration of the ARFG, reported in 1988, used a Raman fiber amplifier with optical pump switching (E. Desurvire et al. [143]). A second experiment reported in 1992 was based on an LD-pumped EDFA (D. N. Chen et al. [144]). In both cases, the device rotation sensitivity was limited by short ARDL optical delays (240 $\mu$s and 1.7 ms, respectively). They could not be increased because of pump amplitude noise and excess ASE buildup, [143] and [144]. What is then the maximum rotation sensitivity of ARFGs?

The ARFG sensitivity can be evaluated as follows. After $N$ signal recirculations, the output signal envelope is proportional to $1 + \cos(2\pi f N t_{\text{loop}})$; the modulation frequency $f$ is given by

$$f = \frac{D\Omega}{n\lambda_s} \tag{6.18}$$

where $D$ is the loop diameter, $N$ the fiber refractive index, and $\lambda_s$ the signal wavelength [143]. The minimum time delay $N t_{\text{loop}}$ required for measuring the frequency $f$ is then given by the relation $f N t_{\text{loop}} = 1$. If we match the minimum required time with the maximum optical delay $T_{\text{max}}$ of the ARDL, we obtain from Eq. (6.18) the minimum measurable rotation rate $\Omega_{\text{min}}$:

$$\Omega_{\text{min}} = n \frac{\lambda_s}{D} \frac{1}{N t_{\text{loop}}} \equiv n \frac{\lambda_s}{D} \frac{1}{T_{\text{max}}} \tag{6.19}$$

Considering an ideal device using soliton pulses with in-loop regeneration, the ARDL optical delay $T_{\text{max}}$ is virtually unlimited. Thus, Eq. (6.19) shows that the theoretical ARFG sensitivity can be *indefinitely* increased. But this performance cannot be achieved in a real device. For comparison, we evaluate the theoretical sensitivities of ideal ARFGs with optical delays of 1 ms, 1 s and 1 min. For a device with a loop diameter $D = 10$ cm and operating at $\lambda_s = 1.5$ $\mu$m, the minimum rotation rate is given by $\Omega_{\text{min}}(\text{rad/s}) \approx 0.022/T_{\text{ms}}$, or $\Omega_{\text{min}}(\text{deg/h}) \approx 4.5/T_s$. To achieve a sensitivity comparable to the best IFOG performance ($\Omega_{\text{min}} \approx 10^{-3}$ deg/h), an optical delay of 4500 s, or 1.25 h would be necessary. On the other hand, if the loop diameter $D = 10$ km is used, the required optical delay is reduced to 45 ms. These results show that high sensitivity operation requires a device with relatively large dimensions ($D = 10$ km). This fact suggests that ARFGs could find possible applications in geophysics, astronomy, and cosmology [143]. Nonreciprocal effects, e.g., induced by the earth's magnetic field, could also be sensed through the ARFG.

The ultimate performance of actual (or nonideal) ARFGs remains to be fully investigated, both theoretically and experimentally. A limit on device sensitivity is always imposed by nonreciprocal effects, which cause both random and constant phase errors. For relatively short optical delays (e.g., 100 ms), implementation with soliton signals and all-optical regeneration is not necessary. In this case, the main sources of limitation are the ASE and the fiber Kerr nonlinearity. The main causes of nonreciprocity are SPM and XPM. The nonreciprocity attributed to SPM is caused by (1) unequal power splitting in the Sagnac loop coupler, (2) differences in signal powers before and after amplification, due to nonoptimized location of the amplifier, and (3) small gain differences between the forward and backward propagation directions. For short pulse operation ($\Delta T \ll t_{\text{loop}}$) the nonreciprocity attributed to XPM is caused by the effects of (1) unequal ASE power in both propagation directions, and (2) random ASE power fluctuations, which causes phase noise, as in the Gordon–Haus effect (Section 5.11). Other potential causes of nonreciprocity are (1) the differences in ASE-induced phase noise (Section 5.7), experienced by the signals after amplification, and (2) the effect of amplified incoherent Rayleigh backscattering [145]. These nonreciprocities could in principle be alleviated by using various signal modulation techniques (C. C. Cutler et al. [145]; R. A. Bergh et al. [146]). The above shows that the ARFG is a highly complex system, and significant progress is still to be made before we can observe its best theoretical performance. In the case of nonlinear ARFG implementation (i.e., using soliton signals and all-optical regeneration), the sensitivity may considerably improve, but at the cost of significant device complexity. For this reason alone, the ARFG may never be practical, compared to the AFRRG, for instance, unless it achieves superior performance, according to Eq. (6.19).

At the end, we consider finally a last category of ARDL applications. It concerns high speed communication networks and optical computing. The advantages of ARDLs in this domain have long been known (M. I. Belovolov et al. [147]). Optical fiber memories based on ARDLs have limited physical storage capability (a kilometer of fiber can contain only 5 Kbits at $B = 1$ Gbit/s) and relatively long access time (i.e., 5 $\mu$s per km of loop length). But such devices offer a very high bandwidth (i.e., up to 100 Gbit/s), with the advantage of bit rate and modulation format transparency. Additionally, ARDL memories using bandwidth-equalized amplifiers can support WDM signals. Thus, ARDLs obviate high speed electronic devices for many important functions, including: signal processing (e.g., time division multiplexing, time-slot interchanging, bit rate compression, register shifting) and buffering (e.g., in computer interfaces and photonic switches). Two of the main functions of buffering are to maximize the use of switching capacity and to solve contention between colliding input signals.

For standard RZ or NRZ data, the maximum storage capacity of ARDL memories is limited by fiber dispersion. This capacity is actually independent of the fiber loop length $L$ (C. Q. Maguire and P. R. Prucnal [148]). Consider the maximum dispersion-limited bit rate–length product, given by the relation [63]:

$$B^2 L = \frac{c}{2\bar{D}\lambda_s^2} \tag{6.20}$$

where $\bar{D}$ is the effective (or average) fiber dispersion coefficient. The number of bits $N$ that can be stored in the same length of fiber is given by:

$$N = \frac{n}{c} LB \tag{6.21}$$

Using Eqs. (6.20) and (6.21) we find

$$N = \frac{n}{2\bar{D}\lambda_s^2} \frac{1}{B} \tag{6.22}$$

which is independent of fiber length. For instance, the dispersion-limited capacity of ARDLs at signal wavelength $\lambda_s = 1.5\,\mu$m and bit rate $B = 10$ Gbit/s, with an average dispersion coefficient $\bar{D} \approx 0.005$ ps/nm.km is $N = 6$ Gbits, according to Eq. (6.22). Such a low dispersion coefficient would be very difficult to achieve for long fiber lengths (i.e., $L > 500$ km), due to fluctuations in the zero dispersion wavelength position. With a realistic coefficient of $\bar{D} \approx 1.0$ ps/nm.km, we obtain from Eq. (6.22) a dispersion-limited capacity $N = 30$ Mbits for 10 Gbit/s signals. On the other hand, the loop length must have a limit in size, which restricts the memory capacity according to Eq. (6.21). For instance, a 1000 km loop can store only $N = 50$ Mbits at $B = 10$ Gbit/s, which corresponds to a bit sequence of 5 ms duration. In this example, the maximum storage capacity is slightly in excess of the dispersion limit (i.e., $N = 30$ Mbits), which indicates that maximum capacity could be achieved if the effective dispersion coefficient can be made somewhat lower than the nominal value of $\bar{D} \approx 1.0$ ps/nm.km, or alternatively, if a dispersion compensation technique is used (Chapter 7).

In most applications, the bit sequences or packets to be stored have a relatively small number of bits and are considerably shorter. In ATM networks, standard data packets (or ATM cells) contain 424 bits at 622 Mbit/s, which corresponds, with two 150 ns guard bands, to a $T = 1\,\mu$s duration. For computer interfaces, the cells are even shorter (e.g., 72 bits maximum); with a bit rate $B = 1$ Gbit/s, and accounting for two 15 ns guard bands, the maximum cell length is $T = 100$ ns.

ARDL-based fiber memories for ATM switching have been implemented experimentally with an SCA acting both as gating and amplifying device (M. Calzavara et al. [111]), or with both SCA and EDFA for performing the two functions separately (G. Grosskopf et al. [110]). In either case, the maximum number of packet recirculations was of the order $N = 10$, which corresponds to a practical operating limit.

All-EDFA-based optical fiber shift registers and memories with fast switching capabilities can also be realized in a device configuration that combines a NALM and an ARDL, as shown in Figure 6.38 (N. A. Whitaker et al. [149]; H. Avramopoulos and N. A. Whitaker [150]).

For the shift register application, the probe signal (wavelength $\lambda'$) is turned off. Assume that the device is input with a 1 signal pulse sequence (wavelength $\lambda$). At first, the signal pulses are 100% reflected by the Sagnac loop. A fraction of this signal is tapped by a coupler (C1), amplified through an EDFA and reinjected into the Sagnac loop through a second coupler (C2), which is referred to as the control port. The signal pulses that have reached the control port C2 act then as switching signals for all subsequent pulses. Explicitly, if it takes $N$ bits before the first bit reaches the control port, the loop is 100% reflective for the first $N$ bits, and becomes transmitting (or switches) for the bits following in the sequence. Since the loop is transmitting for the $N$ consecutive bits, no reflected signal is generated during this period. Consequently, the loop switches back to the reflective mode for the next $N$ bits. Thus, the Sagnac loop oscillates with a characteristic period $T$ that corresponds to the transit time in the path C1–C2. In the actual device implementation, the fiber and the couplers are of the PM type. The signal pulses fed back at the control port C2 are coupled in the polarization state orthogonal to that of the signals to be switched. Thus, the control pulses do not appear at the loop output. As shown in [149], this device implements the function of an $N$-bit circulating shift register with inverted feedback and a

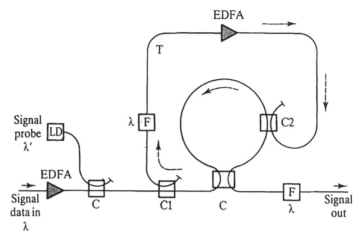

**FIGURE 6.38** Device configuration of an EDFA-based optical shift register and fiber memory using PM-maintaining components (C, C1, C2 = passive PM couplers, F = optical bandpass filter).

recirculating period of 2N bits. Although all signal inputs propagate only twice through the device, the system retains the information of the initial state and therefore acts as a memory [149].

The optical fiber memory application is based on the same configuration as previously described, with the use of a probe signal at wavelength $\lambda'$ and a reduced repetition rate (Figure 6.38). If 1 probe pulses are input to the device at the same rate as the input signal at wavelength $\lambda$, every single bit stored in the loop is read out. On the other hand, bit patterns corresponding to certain sampling rates can be read out from the memory; because of the finite pattern length, the possible sampling rates are given by the divisors of 2N [150].

## 6.7 FIBER LASERS

This section is concerned with applications of EDFAs in *fiber laser* devices. Er-doped fiber lasers (EDFLs) can be viewed as EDFAs operating in the particular regime where coherent oscillation of ASE occurs due to some feedback means. A standard definition could be the following: EDFLs are used as *sources* for coherent light signal generation, while EDFAs are used as *traveling-wave amplifiers* for coherent light signal regeneration. The distinction between EDFAs and EDFLs is not always so clear-cut. For instance, certain AGC applications use a combination of both principles and the devices switch back and forth from EDFL to EDFA regimes (Section 6.3). Likewise, *superfluorescent* fiber laser sources, described below, are based on the principle of double-pass ASE noise amplification. Double-pass or reflective signal amplification is also used in some EDFA applications (Section 6.2). In this case, the ASE output is the output of a superfluorescent EDFL. We have previously described these applications, so in this section we focus exclusively on EDFLs. RE-doped fiber lasers have now become an entire field and EDFLs actually deserve several chapters. We restrict ourselves here to a brief description of the main features of

EDFLs and mention the unique and remarkable properties of some recently investigated device configurations.

Before considering EDFLs, we review some of the basic advantages of RE-doped *fiber lasers* (FLs). All FLs can be pumped with compact efficient and sometimes inexpensive laser diodes. They are compatible with SMFs, DSFs, and fiber optic components used in communications so they have negligible coupling losses. Fiber waveguiding and splicing alleviate any mechanical alignment of parts and provide superior environmental stability. There are many possible laser cavity designs and configurations. Their bulk optics counterparts would sometimes be very cumbersome and unstable. The long fiber lengths (e.g., $L = 1$–50 m) mean the cavity modes are closely spaced (FSR = 2–100 MHz), which makes possible continuous wavelength tuning without mode-hopping.

Because of the high pump intensities available in single-mode fiber waveguides, room temperature laser action can be achieved with three-level laser systems of high inversion thresholds (such as $Er^{3+}$ : glass). The number of possible laser wavelengths and corresponding tunability range are greatly extended by the variety of rare earths that can be used as dopants (singly or in pairs), and the variety of glasses for hosts. FLs based on fluoride glass hosts also offer the possibility of frequency upconversion, making it possible to realize compact visible laser sources with infrared LD pumping. A comprehensive list of RE-doped FL wavelengths reported so far can be found in a review by W. J. Miniscalco [151].

In high power applications, it is possible to use phased-array LDs for the pump, which provide scalable output powers in the 1–10 W range. FLs can therefore be viewed as light converters, from which pump LDs with low coherence characteristics (i.e., longitudinally/transversely multimoded) are used to produce output signals of high spatial and temporal coherence characteristics (i.e., with single longitudinal/spatial mode). In pulsed operation of FLs, peak power outputs up to 10 kW can be achieved through CW LD pumping. Ultrashort pulses ( < 100 fs) can be generated from compact mode-locked FLs. Finally, particular configurations of pulsed infrared FLs can be operated as natural *optical soliton generators*. These practical devices could be deployed in future soliton communication systems (Chapter 7).

The main drawback of RE-doped FLs is their slow gain dynamics ($\mu$s–ms) which removes any possibility of direct internal modulation for high speed data generation. This function can be performed through hybrid technology, using various electro-optic or semiconductor integrated components. Progress in the field of all-fiber switching devices may be driven by interest in RE-doped FLs.

Erbium-doped fiber lasers actually preceded EDFAs. EDFAs have higher pump power requirements. Most experimentalists have also experienced how weak Fresnel reflections at cleaved fiber ends can change an EDFA into an EDFL. The first EDFLs were reported by two independent groups during 1986–1987 (R. J. Mears et al. [152]; A. V. Astakhov et al. [153]). The EDFL described in [153] was made from a 2.5 m multimode EDF coiled around a gas discharge tube. The 90 cm single-mode EDFL described in [152] was pumped with a $\lambda = 514.5$ nm argon ion laser, had a threshold of 30 mW absorbed pump power and a slope efficiency of 0.6%. The EDFL performance has been considerably improved since these experimental results. The study [152] is the first observation of CW lasing of a three-level laser system at room temperature. The first $Q$-switched operation was achieved with an intracavity acousto-optic modulator, and could produce 60 ns pulses with 2 W peak power at

200 Hz repetition rate [152]. Later EDFL improvements (R. J. Mears et al. [154]; P. Myslkinski et al. [155]) could produce $Q$-switched pulse characteristics (30 ns, 120 W, 800 Hz) and (8 ns, 230 W, 1 KHz).

Developments in the field of EDFLs during 1988–1990 have paralleled those of EDFA devices. Among them are the investigation of $Yb^{3+}$ sensitization (D. C. Hanna et al. [156]; W. L. Barnes et al. [157]), and the identification of optimum glass hosts, fiber designs, and pump wavelengths (W. L. Barnes et al. [158]). Slope efficiencies up to 46% could be achieved with 980 nm pumping, compared to 17% with $\lambda = 810$ nm pumped Yb–Er devices [158]. Pumping at $\lambda = 532$ nm with practical frequency-doubled Nd:YAG sources was also investigated (M. C. Farries et al. [159]). Considerable attention was also given to fluorozirconate EDFLs, for which operation at $\lambda = 2.71\ \mu m$, $({}^4I_{11/2}-{}^4I_{13/2}$ transition) is possible (M. C. Brierley and P. W. France [160]). This configuration is unusual, since the ${}^4I_{11/2}-{}^4I_{13/2}$ transition is self-terminated, i.e., the lifetime of the lower level $({}^4I_{13/2})$ is longer than the lifetime of the upper level $({}^4I_{11/2})$; medium inversion can be explained by the effect of pump ESA, which rapidly depopulates the lower level [160]. Fluorozirconate EDFLs operating at $\lambda = 1.6$–$1.7\ \mu m$ $({}^4I_{13/2}-{}^4I_{15/2})$ and $\lambda = 1.0\ \mu m$ $({}^4I_{11/2}-{}^4I_{15/2})$ were also investigated (R. G. Smart et al. [161]; J. Y. Allain et al. [162]). Finally, upconversion pumped fluorozirconate EDFLs with pump at $\lambda = 800$ nm and outputs at $\lambda = 850$ nm or $\lambda = 546$ nm (green) could be demonstrated (C. A. Millar et al. [163]; T. J. Whitley et al. [164]). Recent results (1992) obtained by 970 nm pumping a fluorozirconate EDFL brought the best green output power and slope efficiency to 50 mW and 15%, respectively (J. Y. Allain et al. [165]).

The low pump thresholds of EDFLs made it possible to rapidly demonstrate practical LD pumping for 1.5 $\mu m$ wavelength operation. The first LD-pumped EDFL has a 3 mW oscillation threshold and used a 811 nm AlGaAs pump source (L. Reekie et al. [166]). With an AlGaAs array for pump LD, output powers at $\lambda = 1.5\ \mu m$ could be increased up to 8 mW with 13.5% slope efficiency with respect to launched pump power (R. Wyatt et al. [167]). Similar results were obtained later on with 1480 nm InGaAsP as pump LDs (K. Suzuki et al. [168]). Efficient LD pumping at 810 nm of fluorozirconate EDFLs operating at 2.1 $\mu m$ wavelength could also be demonstrated (R. Allen and L. Esterowitz [169]; H. Yanagita et al. [170]).

A comprehensive review of the basic physics, characteristics, and theoretical modeling of EDFLs are in a recent book edited by M. J. F. Digonnet [171], which contains up to 900 literature references. In particular, [171] provides a description of various fiber device configurations for CW, $Q$-switching, and mode-locking operations. The rest of this section briefly reviews applications of CW tunable EDFLs, single-frequency EDFLs, superfluorescent EDFLs, and mode-locked EDFLs for picosecond/femtosecond pulse and soliton pulse generation.

One of the main advantages of fiber lasers is their very small Fabry–Perot mode spacings, i.e., $\delta v = 2$–$100$ MHz for $L = 1$–$50$ m lengths. This enables continuous laser tuning and in some cases single-longitudinal mode operation. Figure 6.39 shows how *wavelength selection* and *tuning* can be implemented in the EDFL configurations. The simplest tunable EDFL configuration (Figure 6.39a) uses a mechanical grating (R. Wyatt [172]). To alleviate any coupled cavity effects, an antireflection (AR) element must be placed at the fiber end facing the grating. This element can be an index-matching cell, a thin (1–3 mm) glass plate, or a multilayer dielectric AR coating. A mechanical polarization controller is needed because the grating reflectivity (hence

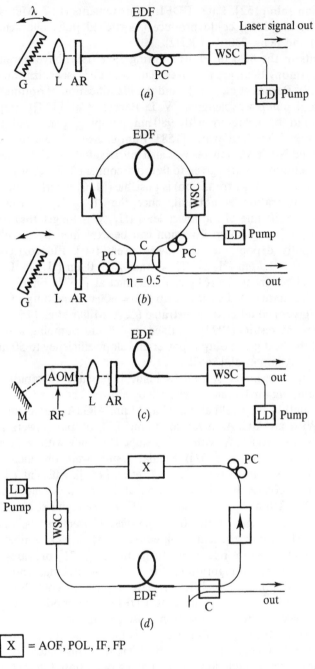

**FIGURE 6.39** Possible configurations of CW-tunable Er-doped fiber lasers: linear cavities with (a) grating, (b) grating and unidirectional loop mirror, (c) acousto-optic modulator; ring cavity (d) with wavelength selective element X, which can be an acousto-optic filter, a polarizer, an interference filter or a Fabry–Perot etalon (G = grating, L = lens, AR = antireflection device, EDF = Er-doped fiber, PC = polarization controller, WSC = wavelength selective coupler, C = fiber coupler, AOM = acousto-optic modulator, AOF = acousto-optic filter, POL = polarizer, IF = interference filter, FP = Fabry–Perot filter).

the cavity loss) is polarization sensitive. With silica-based EDFLs, this configuration permits CW laser operation in the range $\lambda = 1.52–1.58\,\mu m$, representing $60\,nm$ continuous tuning [172]. A possible approach for achieving both wavelength tunability and single-longitudinal mode operation, uses a unidirectional loop mirror (Figure 6.39b) (C. R. O. Cochlain and R. J. Mears [173]). The main drawback of grating-tuned EDFLs is their environmental instability. Instability is caused by random fluctuations in cavity length and signal polarization arising from vibration and thermal effects, which generates intensity and frequency noise.

Figure 6.39c shows an approach using an intracavity acousto-optic modulator (AOM). This permits electronically controlled tuning (P. F. Wysocki et al. [174]). The wavelength selected depends on the deflection angle of the AOM, controlled by driving the modulator at different RF frequencies. In this configuration, the laser output is a single line with 0.3–1.0 nm FWHM [174]. The broad spectrum is explained by the fact that the laser light is frequency shifted (by approximately 150 MHz) at each round-trip through the AOM. The effective cavity loss increases at each round-trip, due to the angular error introduced by the wavelength translation effect. As a result, a limited range of wavelengths and cavity modes can oscillate. If, on the other hand, the AOM is driven by a sinusoidally modulated waveform, the output wavelength sweeps across the tunable range, the maximum excursion depending on the modulation frequency (P. F. Wysocki et al. [175]). The output spectrum of such wavelength-swept EDFL is a single longitudinal mode at any given time [175].

The tunable EDFL configuration shown in Figure 6.39d uses an all-fiber ring laser cavity including exclusively fiber pigtailed optical elements. Wavelength selectivity can be achieved by using for the element X shown in the figure, either an interference filter (K. Iwatsuki et al. [176]) or a polarizing beam splitter (P. D. Humphrey and J. E. Bowers [177]), placed inside a beam expander. In this approach, wavelength tuning is achieved mechanically by tilting the filter or the polarization controller coils, which in this last case is a trial and error procedure. Alternatively, the element X can be an integrated acousto-optic filter based on the principle of TE/TM polarization shifting (D. A. Smith et al. [178]), or a tunable fiber Fabry–Perot etalon (J. L. Zyskind et al. [179]; H. Schmuck et al. [180]). The last two approaches permit electronic tunability and yield single-frequency, narrow line width ($\delta v < 10\,kHz$) outputs. The fiber Fabry–Perot etalon is polarization independent (J. Stone and L. Stulz [181]); the small polarization dependence of the optical isolator used in the ring for traveling-wave operation means a mechanical polarization controller (PC) is required to maximize the EDFL output, [179] and [180]. Another possibility of electronic wavelength tunability, which is totally polarization independent, uses a linear fiber cavity with intracore fiber Bragg reflectors (G. A. Ball and W. W. Morey [182]). In this approach the fiber cavity is made of a short length of EDF (e.g., $L = 10$ cm), which can be controlled by a piezoelectric transducer. The Bragg condition (i.e., $\lambda_B = 2n\Lambda$, $\Lambda$ = grating period) provides sharp wavelength selectivity which, associated with the large longitudinal mode spacing (e.g., FSR = 1 GHz), results in single-mode operation. As the cavity is stretched, continuous tuning is achieved without mode hopping [182]. Other possible configurations of EDFLs can provide multiple wavelength outputs (H. Okamura and K. Iwatsuki [183]; N. Park et al. [184]). This operation can be achieved with a ring laser cavity in which multiple parallel paths are generated through wavelength selective couplers.

We consider next *single-longitudinal mode* and *narrow line width* EDFL configurations operating at fixed signal wavelengths. The first implementation of a narrow line width EDFL used a single evanescent field grating coupler as output mirror, which yielded an FWHM of $\delta\lambda = 0.04$ nm or 5 GHz (I. M. Jauncey et al. [185]). Single-longitudinal mode operation could theoretically be achieved by increasing the cavity's FSR (i.e., shortening the EDF length) to the point where only one cavity mode falls into the grating passband. This aproach was first demonstrated with $Nd^{3+}$ doped FL, where a 5 cm device produced a single-frequency output of 1.3 MHz FWHM (I. M. Jauncey et al. [186]). For EDFLs, the centimeter lengths required for this operation, and the limit in maximum $Er^{3+}$ concentration (Section 5.6) restricts the single-pass gain, which produces very low slope efficiencies. Another narrow-line width implementation of EDFL, in which laser outputs with 620 MHz FWHM could be obtained, used an intracavity etalon air gap (M. S. O'Sullivan et al. [187]).

The first demonstration of single-longitudinal mode operation of EDFLs was based on a Fox–Smith resonator configuration (P. Barnsley et al. [188]). In this configuration, longitudinal mode selection is based on the principle of coupling two laser cavities of different lengths; the lengths are chosen so that the two sets of Fabry–Perot modes coincide only with large frequency periodicity, which results in a broad effective FSR. The main drawback of this approach is the environmental instability of the Fox–Smith resonator, which causes both intensity and frequency noise.

Monolithic and environmentally stable single-mode EDFLs can be realized using short linear fiber cavities in which the mirrors are intracore Bragg reflectors (G. A. Ball et al., [189] and [190]). With a 50 cm device, where gratings were written 10 cm from each end, laser line widths of 47 kHz could be achieved [189]. Although higher performance can be obtained with ring laser configurations, the main advantage of this approach is its (extreme) component simplicity.

Another possible single-mode EDFL configuration uses a unidirectional ring cavity. The cavity is identical to that of Figure 6.39d, without any wavelength selective element X included. Single-mode operation occurs due to an interplay of birefringence and polarization hole burning effects, not fully understood at present (P. R. Morkel et al. [191]). This configuration achieved a line width of 60 kHz [191]. A similar configuration gave a narrower line width of less than 10 kHz. It used a Sagnac-like loop with an external mirror reflector (G. J. Cowle et al. [192]). Finally, a configuration using a unidirectional ring cavity with a polarization-maintaining EDF made it possible to achieve a line width of 1.4 kHz (K. Iwatsuki et al. [176]), the narrowest EDFL line width to date (1993). The performance improvement by use of a PM–EDF is attributed to the suppression of polarization mode competition, a cause of line broadening [176]. A comparable performance (line width less than 5.5 kHz) could be achieved with a standard fiber, ring cavity device that included two cascaded Fabry–Perot etalons (J. L. Zyskind et al. [193]). The advantage of this last approach is the possibility of broadband tunability. With two Fabry–Perot etalons having 800 MHz and 12 GHz FSRs, the device can operate in a single longitudinal mode and be tuned over a comb of discrete wavelengths separated by 12 GHz, or 0.1 nm, from $\lambda = 1.530$ $\mu$m to $\lambda = 1.575$ $\mu$m [193]. Coupled with an external modulator, this device could be used as a tunable transmitter for dense-FDM communication network applications.

The development of *superfluorescent fiber lasers (SLF)* has been motivated by the need for broadband sources for fiber optic gyroscopes (P. F. Wysocki [194]). In IFOG applications, indeed, the use of a broadband signal source alleviates detrimental errors associated with coherent backscattering, Kerr effect and polarization cross-coupling (R. A. Bergh et al. [195]; K. Bohm et al. [196]). The main advantage of SFLs over conventional light-emitting diodes (LED) and superluminescent diodes (SLD) is their higher (fiber coupled) output power and spectral stability with temperature. The highest fiber-coupled powers available from LEDs and SLDs are $10\,\mu\text{W}$ and $1\,\text{mW}$, respectively (N. S. Kwong [197]), to compare with $10\text{--}100\,\text{mW}$ in the case of Er-doped SFLs (P. F. Wysocki [194]; H. Fevrier et al. [198]). The wavelength temperature stability is about $-400\,\text{ppm/}^\circ\text{C}$ for SLDs, compared to $+5\,\text{ppm/}^\circ\text{C}$ for Er-doped SFLs (P. F. Wysocki et al. [199]). For SFLs, the advantage of using Er-doped fibers operating at $\lambda = 1.5\,\mu\text{m}$, rather than Nd-doped fibers operating at $\lambda = 1.06\,\mu\text{m}$, is the minimization of fiber loss in the IFOG. Four possible configurations of Er-doped SFLs are shown in Figure 6.40. The three first configurations consist in single- and double-pass fiber lasers, which use one or two antireflection components at either fiber end. The signal output of the single-pass configuration, Figure 6.40a, corresponds to the ASE noise output of EDFAs. The double-pass configurations, Figure 6.40b,c have a higher output power, due to the doubling of the effective gain length; the output of Figure 6.40b configuration is higher than that of Figure 6.40c for two reasons: the round-trip gain is higher since the WSC is located near the fiber output end, and the backward ASE is always higher than the forward ASE (Section 5.5). Figure 6.40d configuration is based on the principle of post-amplification of a broadband signal though an EDFA; the signal source is either an SLD (N. S. Kwong [197]) or an Er-doped SFL (K. Takada [200]). This device operates in a regime that corresponds both to a superfluorescent fiber laser and a low gain amplifier. The main advantage of this last approach is the possibility of generating an output spectrum that is symmetrical with respect to its center wavelength and has low ripple, contrary to the cases of Figure 6.40a,b,c configurations. When using an SLD signal, the drawback of Figure 6.40d configuration is the temperature dependence of center wavelength. In all cases, the spectral FWHM of the source can be controlled by the launched pump power (or the single-pass EDF gain); the FWHM decreases with increasing pump. For instance, output power characteristics of $5\,\text{mW}$, FWHM $= 10\,\text{nm}$ could be achieved with Figure 6.40a configuration (P. F. Wysocki et al. [201]), or $20\,\text{mW}$, FWHM $= 21\,\text{nm}$ with Figure 6.40d configuration using an SLD signal (N. S. Kwong [197]), and of $150\,\text{mW}$, FWHM $= 1.4\,\text{nm}$ with Figure 6.40d configuration using an Er-doped SLF signal (K. Takada [200]).

The analysis of Er-doped SFLs' output noise characteristics reveals an excess noise component characteristic of thermal light sources (P. R. Morkel et al. [202]). This excess noise corresponds to the ASE–ASE beat noise component of EDFAs (Chapter 3) and is a cause of SNR degradation at the receiver. The threshold at which the excess noise dominates over the shot noise determines the maximum receiver SNR, which depends on the SFL bandwidth. For $1\text{--}50\,\text{nm}$ bandwidths, the maximum SNR falls into a $112\text{--}130\,\text{dB}$ range [202]. For IFOG applications, the minimum rotation rate, determined by the excess noise component, is given by (K. Iwatsuki [203]):

$$\Omega_{\min} = \frac{\lambda^2}{2\pi LD} \sqrt{\frac{B_e c}{\Delta\lambda}} \qquad (6.23)$$

where $L$ and $D$ are the Sagnac loop length and diameter, respectively, $B_e$ the electronic bandwidth, and $\Delta\lambda$ the SFL line width. Thus, the IFOG sensitivity can be increased

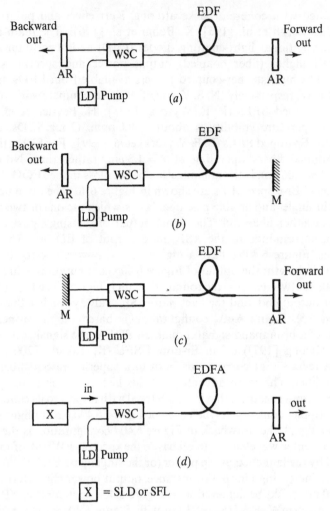

**FIGURE 6.40** Basic device configurations of Er-doped superfluorescent fiber lasers (SFL): (*a*) single-pass, (*b*), (*c*) double-pass, and (*d*) superluminescent LD postamplifier (WSC = wavelength selective coupler, AR = antireflection component, EDF = Er-doped fiber, M = mirror, SLD = superluminescent laser diode).

by using a broadband SLF, as opposed to a conventional LD source. For instance, an IFOG using a 500 m Sagnac loop with 5 cm radius, an Er-doped SFL source with an FWHM of $\Delta\lambda = 2$ nm and an electronic bandwidth $B_e = 1$ Hz has, according to Eq. (6.23) a minimum detectable rotation rate $\Omega_{min} = 5.5 \times 10^{-6}$ rad/s, or 0.57 deg/h, demonstrated experimentally, [203] and [204]. By using an SFL with $\Delta\lambda = 20$ nm FWHM and increasing the loop dimensions to $L = 1$ km and $D = 20$ cm, this sensitivity could be theoretically increased to $\Omega_{min} \approx 2 \times 10^{-2}$ deg/h. In addition to rotation-sensing applications, Er-doped SFLs can also be implemented in high sensitivity, low coherence reflectometry (K. Takada et al. [200]).

We consider next the case of *mode-locked* Er-doped fiber lasers. Mode locking (ML) of EDFLs has been under intense investigation, owing to the possibility of realizing compact practical sources for ultrashort pulse generation, and to the potential of such sources for optical soliton generation. Our description is a brief outline of the main progress made in this field up to 1993. The basic performance of ML–EDFLs can be summarized with the following output characteristics: pulse width, peak power, repetition rate, and time–bandwidth product (TBP). For a soliton pulse with sech$^2$ envelope, the time–bandwidth product is TBP = 0.315.

Since the first demonstration in 1989 of active mode locking of an EDFL, which produced 40 ps pulses (D. C. Hanna et al. [205]), considerable progress was made in cavity design, optimization of intracavity elements, and ML schemes. There are two categories: linear cavity and ring cavity configurations, as shown in Figure 6.41. The ML regime can be achieved actively (by use of AM or FM modulation), passively (by use of saturable absorbers). Figure 6.41$a$ configuration uses an intracavity LiNbO$_3$ phase modulator, which can be made in bulk optics [205] or in integrated optics (K. Smith et al. [206]). A Bragg fiber grating can be incorporated into the cavity for wavelength selection and tuning (R. P. Davey et al. [207]). ML operation requires the phase modulator to be driven at the RF harmonic frequency $f = Nc/2nL$ ($N$ = integer, $L$ = cavity length); as this modulation technique produces two interleaved set of pulses (A. E. Siegman [208]), the most stable operation is actually achieved with modulation $f = (N + 1/2)c/2nL$ [6.207]. Subpicosecond operation (900 fs, 55 W, 480 MHz, TBP = 0.29) could be achieved with this type of device (R. P. Davey et al. [209]). An alternate implementation of Figure 6.41$a$ configuration uses a semiconductor laser diode as a loss modulator (P. G. J. Wigley et al. [210]). A similar approach uses a single 980 nm LD for the simultaneous functions: EDF pumping, ML through loss modulation, and intracavity etalon (E. M. Dianov et al. [211]).

Passive ML devices require a saturable absorber. For linear cavity devices, Figure 6.41$b$, this element can be an MQW semiconductor nonlinear mirror, butt-coupled at the fiber end (W. H. Loh et al. [212]). Passive ML can also be produced through nonlinear polarization evolution in the EDF (M. E. Fermann et al. [213]); in this case, ML of the FL is initiated by a weak intracavity modulation, turned off once the process has started and has been optimized. For linear cavity, passive ML EDFLs, obtained through this approach with an additional prism pair for higher order dispersion compensation, the best performance to date has output characteristics 180 fs, 550 W, 55 MHz, and TBP = 0.31 [213].

The second category of ML–EDFLs is based on active and passive ring cavity configurations, Figure 6.41$c$ and $d$. For active ML devices, the controlling intracavity element can be an integrated LiNbO$_3$ switch or modulator (J. B. Schlager et al. [214]; A. Takada [215]; and K. Smith et al. [206]) or an RF-modulated semiconductor amplifier (D. Burns and W. Sibbett [216]). A tunable filter can also be incorporated into the ring cavity for wavelength selection and tuning (R. P. Davey et al. [217]). Using ring cavities with long fiber lengths (e.g., 5 km), ML pulses with high peak power can be generated at low repetition rates (e.g., 40 kHz); the study [214] describes an experimental ML source with output characteristics 60 ps, 6 W, and 200 kHz. On the other hand, higher repetition rates in the 1–10 GHz range can be obtained at the expense of peak power by harmonic mode locking. For instance, [214] and [217] describe LD-pumped AM ML sources with output characteristics (7.5 ps, 6 mW, 30 GHz, TBP = 0.47) and (21 ps, 25 mW, 14 GHz, TBP = 0.52), respectively.

Dual-wavelength, synchronous ML operation of EDFLs can be achieved by the introduction of a strand of birefringent fiber in the ring cavity (J. B. Schlager et al.

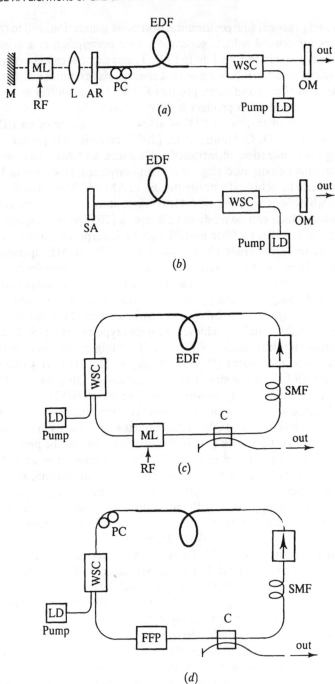

**FIGURE 6.41**   Basic device configurations of mode-locked Er-doped fiber lasers: (*a*) linear cavity with active mode locking (*b*) linear cavity with passive mode locking (*c*) ring cavity with active mode locking, and (*d*) ring cavity with passive mode locking. WSC = wavelength selective coupler, AR = antireflection device, EDF = Er-doped fiber, ML = mode-locking element, PC = polarization controller, M = mirror, OM output mirror, L = lens, SA = saturable absorber, C = fiber coupler, SMF = single-mode fiber, FFP = fiber Fabry–Perot.

[218]). The wavelength separation between the two simultaneous ML outputs is given by $\Delta\lambda = \eta\delta n/\beta$, where $\eta$ is the fraction of the cavity that has birefringence, $\delta n$ is the index difference of the birefringent fiber and $\beta = dn/d\lambda$ is the material phase dispersion [218]. Thus, the wavelength separation can be designed by varying the fraction $\eta$ of the cavity birefringence; with typical parameters, separations of $\Delta\lambda = 1$–2 nm can be achieved with $\eta = 3$–8% [218].

A main drawback of ring ML–EDFLs with long cavities is their output pulse instability. This instability is mainly due to fluctuations of length induced by mechanical vibrations and thermal fluctuations, and secondarily, to polarization wandering. For use in communication systems, ML–EDFLs must produce output pulse trains that are free from amplitude noise and timing jitter, which requires some stabilization means. Stable laser operation is obtained if we can measure very low BERs during external modulation of the output (X. Shan et al. [219]). A possible stabilization technique is to lock the electrical phase of the active ML element to the output pulses; error-free operation of a 100 m ML–EDFL using this technique could be demonstrated at a 2 Gbit/s rate [219]. Another approach uses an all-polarization-maintaining ring cavity with shorter length (H. Takara et al. [220]). For instance, error-free output pulse characteristics 6.9 ps, 90 mW, 20 GHz, TBP = 0.32 at bit rates up to 8 Gbit/s could be demonstrated through this approach [220].

Self-starting, passive ML ring EDFLs can also be realized through a variety of techniques. A device based on a fast intra-cavity saturable absorber, such as a thin MQW semiconductor slab, is described in M. Zirngibl et al., [221]. Its characteristics are 1.2 ps, 10 W, 50 MHz, and TBP = 0.32. A second technique for self-starting passive ML, first demonstrated by L. F. Mollenauer [222], uses the configuration shown in Figure 6.41d. This technique is based on the effect of nonlinear polarization rotation, which plays the role of saturable absorption (C. J. Chen et al. [223]). Pulse stabilization is achieved by the presence of a Fabry–Perot etalon in the cavity, [222] and [223]. A detailed numerical analysis of this device is made in [223]; this analysis shows that the sharpening effects in the time and frequency domains, due to saturable absorption and to the etalon, result in stable optical soliton formation.

In the absence of a Fabry–Perot etalon in the cavity, the ring EDFL shown in Figure 6.41d exhibits several unusual properties (V. J. Matsas et al., [224] and [225]). Two characteristic operating regimes can be observed. At high pump, long square pulses ($> 500$ ps) form at the cavity round-trip frequency. At low pump, the square pulses break to form bunches of soliton pulses with random numbers and spacings, whose patterns repeat at the cavity round-trip frequency. As the pump power is reduced, discrete average power jumps (corresponding to sudden reduction in the number of pulses) can be observed, which follow a power hysteresis, [224] and [225]. In the soliton regime, two different operating modes, which exhibit distinct spectral changes with pump power, can be selected by acting on the polarization state [225]. The self-starting ability and overall stability of the device is also improved by incorporation of a short length of low birefringence fiber (i.e., with beat length $L_{beat} > 2$ m) [225]. This soliton bunching effect was also observed in active ML linear cavity EDFLs (R. P. Davey et al. [226]), and was first discovered in figure-eight lasers (I. N. Duling [227]; D. J. Richardson et al. [228]). The regime where these effects are observed is sometimes referred to as *soliton mode-locking*, to distinguished it from the regime of pure AM or FM mode-locking.

In the soliton mode-locking regime, the laser output power spectrum exhibits narrow sidebands near a broad centerline. Their existence can be attributed to a coupling effect between the amplification period and the soliton period (N. Pandit et al. [229]). Indeed, periodic exponential amplification results in the spatial modulation of fiber nonlinearity; the modulation resonates with the soliton characteristic period and the Fourier decomposition of the perturbation explains the formation of spectral sidebands. Higher order sidebands with larger wavelength separation can also be observed. As shown theoretically and verified experimentally, the separation of $N$th order ($N = 1$) sidebands is proportional to $\sqrt{8Nz_A/z_0 - 1}$, where $z_0$ and $z_A$ are the soliton period and the amplification length, respectively ([229], N. J. Smith et al. [230], D. U. Noske et al. [231]).

ML instability and soliton bunching were first observed using self-starting, passive ML EDFLs based on the figure-eight laser (F8L) configuration. The F8L configuration includes a NALM and an external unidirectional Sagnac loop, connected by their two ports, as shown in Figure 6.42 (D. J. Richardson et al. [228] and [232], I. N. Duling [233]). The output coupler (C') can be placed in either loop. In this device, pulses propagating in the NALM undergo amplification; the high intensity portion of these pulses is self-switched into the external loop (cross coupling ($\times$) in coupler C in Figure 6.42) and reflected back into the NALM. On the other hand, the low intensity portion is transmitted by the NALM (bar coupling ($=$) in coupler C) and blocked by the isolator in the external loop. Such a nonlinear feedback results in the self-starting ML effect, with a repetition rate that can be any harmonic of the loop characteristic frequency $f = c/nL_{loop}$. In the two experiments described in [228] and [233], output soliton pulses as short as 314–320 fs at repetition rates up to 10 GHz could be generated. Subsequent improvements in cavity design made it possible to reduce the output pulse width of the F8L to 290 fs, and finally to 98 fs, the shortest value to date (M. Nakazawa et al., [234] and [235]). The F8L performance reported in [235] was due to several design improvements including: reducing the optical isolator GVD, matching the loop length with twice the soliton period for optimum switching, and including an EDFA in the NALM for adiabatic pulse narrowing (Section 5.11).

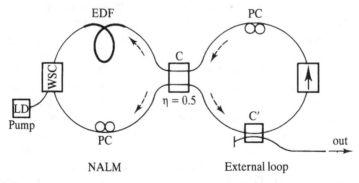

**FIGURE 6.42** Basic device configuration of figure-eight Er-doped fiber laser. NALM = nonlinear active loop mirror, WSC = wavelength selective coupler, EDF = Er-doped fiber, ML = mode-locking element, PC = polarization controller, C = Sagnac loop coupler, C' = fiber output coupler.

In the soliton ML regime of F8Ls, random pulse bunching, energy quantization, and instability of the output pulse trains are observed (D. J. Richardson et al. [236], A. B. Grudinin et al. [237]), as previously discussed in the case of passive ML ring EDFLs. The control of soliton pulse patterns and repetition rate in the F8L remains a challenge [237]. Undoubtedly, the F8L is a very complex nonlinear device for which new developments can be expected.

Other ML–EDFL configurations based on Sagnac loops are also possible. Some use semiconductor amplifiers (SCA) and EDFAs. In a passive ML F8Ls configuration, the SCA is included in the NALM as a saturable absorber (S. J. Frisken et al. [238]); in an active ML, linear cavity configuration, the SCA is included in a NOLM to actively control the mirror reflectivity (C. R. Cochlain et al. [239]). The output characteristics of the device described in [239] are 25 ps, 18 mW, 175 MHz, and TBP = 0.4. With the same active ML linear cavity configuration, the NOLM can also be switched optically by external control pulses, described in Section 6.5 (B. P. Nelson et al. [240]). The output characteristics of the device described in [240] are 8–10 ps, 640 mW, 840 MHz, and TBP = 0.5. Soliton mode locking with repetition rates in the 100 GHz range could also be demonstrated. For instance, FM mode locking in a ring EDFL could produce optical pulses with a substructure of 100 fs pulses at 300 GHz repetition rate (F. Fontana et al. [241]). Soliton pulse trains with tunable repetition rates in the 100 GHz range can also be generated from CW DFB LD sources by nonlinear mixing in a dispersion-tapered fiber (E. M. Dianov et al. [242]; P. V. Mamyshev et al. [243]; and S. V. Chernikov et al. [244]). For this application, EDFAs are used as power boosters to bring the signals into the nonlinear propagation regime.

The many configurations and designs for ML–EDFLs offer great versatility in laser output characteristics. LD-pumped EDFAs have made it possible to realize compact and practical devices. The current challenge is to achieve EDFLs with internal modulation capability, giving error-free output signals. One of the most important application of such sources concerns long-haul soliton communication systems, described in the next chapter.

---

## CHAPTER 7

# SYSTEM APPLICATIONS OF
# ERBIUM-DOPED FIBER AMPLIFIERS

---

## INTRODUCTION

Erbium-doped fiber amplifiers have had a major impact in the field of lightwave communications. This is clearly illustrated by the graph in Figure 7.1, which shows the progress in IM–DD linear system transmission capacity over the past 20 years. The data correspond to the best performance in unrepeatered single-channel transmission, expressed as maximum bit-rate × distance [1]. The growth of system performance, which follows a tenfold capacity increase every four years, has been steadily driven by five major technology advances. The first four are: (1) multimode fiber systems, at of $\lambda = 800$ nm (the first transmission window of early silica fibers), limited by intermodal dispersion and fiber loss; (2) reduced fiber loss and dispersion in single-mode fiber systems at $\lambda = 1.3\ \mu m$; (3) reduced fiber loss and dispersion in single-mode fiber systems at $\lambda = 1.5\ \mu m$; and (4) coherent systems at $\lambda = 1.5\ \mu m$ wavelength, which combine the lowest loss, dispersion and the highest receiver sensitivity achievable. The third and fourth generations of lightwave systems had reached their peak performance by the year 1986. By 1989, a fifth generation was made possible by optical amplifiers, i.e., the EDFAs [2]. The fifth generation system data happen to be consistent with the steady growth rate initiated some 20 years ago.

The 1993 performance of fifth generation lightwave systems is far beyond anything that could be predicted, even dreamt, as little as five years ago. Since the first EDFA-based system experiments reported in 1989, there has been a succession of record-breaking developments. Indeed, the performance data described below speak for themselves.

In 1986, the maximum transmission distance in IM–DD systems was 150 km for 5 Gbit/s signals, and 250 km for 500 Mbit/s signals, corresponding to a maximum bit-rate × distance (system capacity) of 0.75 Tbit.km [1]; in 1987, the best performance in coherent systems was a transmission distance of 202 km at 2 Gbit/s, a system capacity of 0.40 Tbit.km [3].

In early 1993, KDD and AT&T Bell Laboratories could report the successful transmission of 10 Gbit/s signals in a 9000 km system using 274 EDFAs, a performance

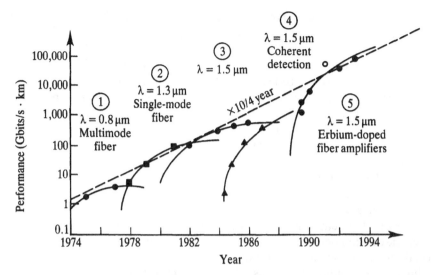

**FIGURE 7.1**   Maximum capacity of IM–DD and coherent lightwave communication systems showing, for a 20 year period, a tenfold increase every four years, through five generations of technology. The data correspond to single-channel, straight-line, unrepeatered transmission. The open circle shown for EDFA-based systems correspond to a loop experiment (2.4 Gbit/s, 21,000 km) distance). The best performance to date (1993) of IM–DD nonsoliton systems is 90 Tbit/s · km, corresponding to the transmission of 10 Gbit/s signals over 9000 km.

of 90 Tbit.km [4]. This is an improvement of 120 times over the previous technology and was obtained only four years after the first experimental EDFA-based systems in 1989 [5] and six years after the first investigation of EDFAs in 1987, [6] and [7]. Such sweeping progress will make it possible to deploy transoceanic links based on EDFA technology across both the Atlantic and the Pacific oceans, as early as 1995–1996 [8].

This considerable progress concerned *linear systems*, which use amplitude modulated signals at the zero dispersion wavelength. During the same period, even more impressive developments occurred in the field of *nonlinear systems* using optical soliton signals. During 1980–1985, many communications experts regarded soliton propagation in fibers as a mere curiosity. Their perspective radically changed with the development of optical amplifiers. Picosecond soliton pulses offered the potential of 50–100 Gbit/s communication. Solitons could propagate in fibers without distortion or fading but not over very long distances. The problem, fiber transmission loss, was elegantly solved by using optical amplifiers based on stimulated Raman scattering and later by using practical EDFAs. Optical amplifiers uniformly or periodically cancelled the fiber loss so the solitons would be able to propagate over longer distances. Unexpectedly the solitons were able to resist the strong perturbations caused by periodic amplification. This was a surprising discovery because soliton propagation requires a delicate balance between fiber nonlinearity and dispersion. It made soliton regeneration by lumped EDFAs all the more practical.

Beginning in 1985, the first studies of soliton propagation over multithousand kilometer distances [9] revealed that soliton transmission is ultimately limited by signal-to-noise ratio degradation and nonlinear interaction with ASE (Gordon–Haus effect). However, new all-optical regeneration techniques to alleviate ASE accumulation in soliton systems were soon discovered. In early 1993, AT&T could then report a demonstration of 10 Gbit/s soliton data transmission over 20,000 km [10]; NTT showed the potential of virtually unlimited transmission distances, with the propagation of 10 Gbit/s data patterns up to 180,000,000 km... Error- and penalty-free propagation was also demonstrated up to 1,000,000 km [11]. Soliton systems, also referred to as nonlinear systems, are now viewed by experts as the natural continuation of the fifth generation of IM–DD communications. Nonlinear systems could to be deployed soon after the year 2000 or even earlier as the sixth generation of lightwave communications.

The considerable progress made over the past five years is not merely the result of discovering the EDFA. Rather, it stems from the convergence of three technologies: RE-doped fiber amplifiers, high power semiconductor lasers, and optical solitons in fibers. The development of high power LDs for practical pump sources immediately followed the development of EDFAs. Developments in amplifier-based soliton systems preceded the development of EDFAs; now EDFAs are accelerating developments in both linear and nonlinear systems. One last factor that explains the successful convergence between the three fields is timeliness.

The invention of a practical optical amplifier (an LD-pumped EDFA) came at the right moment to meet the ever increasing demand for long-haul transoceanic communications, a demand corresponding to a 20–30% capacity increase per year [12]. It could be met in only two ways: by increasing the speed of conventional electronic repeaters (i.e., from 560 Mbit/s to 2.5 Gbit/s) or by developing the optical amplifier technology, led at this time by semiconductor devices. In 1985, neither repeaters nor semiconductor optical amplifiers were ready for implementation in long-haul systems. EDFAs were first investigated in 1986. They initially took the form of cumbersome gas laser or dye laser pumped fiber devices and were considered a curiosity by most system experts. Device researchers were faster to recognize their true potential; they regarded EDFAs as the panacea of future high speed systems. After only a few years of fundamental investigations and design improvements, the LD pumped EDFA turned out to rapidly pass all system qualifications. The fundamental and intrinsic advantages of EDFAs (i.e., low insertion loss, quantum-limited noise, polarization insensitivity, and crosstalk immunity), together with the possibility of practical LD pumping, were responsible for its fast promotion to the level of actual system implementation. Sweeping progress followed.

The impact of EDFAs may have reached its peak in transoceanic systems but developments in other fields are just beginning. Most concern terrestrial systems: all-optical fiber networks, SMF plant upgrade, and analog or digital video signals' distribution. The deployment of the EDFA technology into these fields is far more' complex, owing to the variety of system architectures, protocols, and standards. In the field of local area networks, general consensus has so far proved elusive.

This chapter covers optical preamplification in IM–DD systems in Section 7.1, digital linear systems in Section 7.2, soliton systems in Section 7.3, analog systems in Section 7.4, and local area networks, LANs, in Section 7.5. Fiber nonlinearities and polarization effects in linear systems are described in Section 7.2. Section 7.2 also describes various dispersion compensation techniques, which make it possible to implement EDFAs in the existing system plant based on a standard SMF. EDFA applications in the fields of analog systems and LANs are still at an early research stage. Future developments are expected to occur at a rapid pace. The results described in Sections 7.4 and 7.5 will likely be outdated in the forthcoming years but they can still give an insight into the important role of EDFAs.

## 7.1 EDFA PREAMPLIFIERS

The most fundamental system application of EDFAs concerns *optical preamplification*. Chapter 3 shows that optical preamplification makes it possible to improve the sensitivity of direct detection receivers, when these are limited by thermal noise. With a preamplifier of gain $G$, the received signal power is boosted by a factor $G^2$, while the photodetector noise is increased by signal–ASE beat noise. For signals sufficiently high, this beat noise, also proportional to $G^2$, takes over thermal noise. As a result, the receiver SNR becomes nearly independent of thermal noise. Section 3.3 shows that the highest SNR value that can be achieved in this regime corresponds to the shot noise limit minus 3 dB. Thus, optical preamplification makes it possible to operate direct detection photodetectors at a performance level very close to the ideal shot noise limit. But to achieve this performance requires the optical amplifier to have minimal input coupling loss and a low spontaneous emission factor. The maximum SNR improvement $x$ with respect to the case without optical preamplification is given by:

$$x = \frac{\eta_i}{2\eta n_{eq}} \left( 1 + \frac{2k_B T h v_s}{\eta R e^2} \frac{1}{P_s} \right) \tag{7.1}$$

which is Eq. (3.48) but with the effect of input coupling loss $\eta_i$ taken into account. In this equation, $\eta_i P_s$ is the signal power launched into the optical amplifier and $\eta$ is the detector's quantum efficiency. This result shows that in the absence of thermal noise, or at zero temperature, the factor $x$, corresponding to a detector having unity quantum efficiency, is $x = \eta_i/2n_{eq}$, which in the ideal case ($\eta_i = 1$, $n_{eq} = 1$) represents an SNR degradation of $x = 1/2$ or $-3$ dB. In the realistic case, thermal noise is not negligible, and Eq. (7.1) shows that the factor $x$ can be significantly greater than unity, which means that the SNR is effectively improved by optical preamplification. The improvement increases as $1/P_s$ but it cannot be made arbitrarily large; Chapter 3 shows that the restriction $\langle n_s \rangle = P_s/hv_s B_o \gg n_{eq}$ or $P_s \gg n_{eq} h v_s B_o$ applies to Eq. (7.1). At the low signal powers corresponding to $10^{-9}$ BERs, thermal noise is small compared to signal–ASE beat noise but is not negligible. Therefore, the receiver sensitivity is not determined by the quantum limit, i.e., shot noise minus 3 dB, but by the Johnson noise limit, which is somewhat higher.

In system applications, the receiver sensitivity $\bar{n}$ is, by convention, the average number of photons per bit required for achieving $10^{-9}$ BER, the sensitivity is given by $\bar{n} = \bar{P}_s/hv_s B$, where $B$ is the bit rate. By convention, $\bar{P}_s$ is the signal power actually required at the receiver end; it must include the effect of any input coupling loss. But the receiver sensitivity can also be expressed as $\eta_i \bar{P}_s$ or $\eta_i \bar{n}$, which corresponds to a lower signal level measured at some point within the receiver. For instance, this point can be the EDFA preamplifier input end. In this case, the factor $\eta_i$ corresponds to the splice loss between the transmission fiber output end and the EDFA input end.

In any case, the actual system performance is not given by $\eta_i \bar{P}_s$, but by $\bar{P}_s$. The value of the convention $\eta_i \bar{P}_s$ is to show what the receiver sensitivity would be if the input coupling loss was negligible. Negligible input coupling loss can now be achieved with EDFAs (Section 5.1).

The first experiment in optical preamplification with an RE-doped fiber was made in 1969 (G. C. Holst and E. Snitzer [13]). It used a flashlamp-pumped NDFA operating at $\lambda = 1.06 \,\mu m$ and showed a sensitivity of 4000 photons/pulse for $10^{-3}$ error probability.

An early experimental measurement of RF noise with an EDFA preamplifier made in 1987, showed the potential of achieving a receiver sensitivity of 3470 photons/bit at $B = 140$ Mbit/s (R. J. Mears et al. [6]). For comparison, state-of-the-art Ge-APD receivers had a best sensitivity of $\bar{n} = 642$ photons/bit for $B > 100$ Mbit/s (S. D. Walker and L. C. Blank [14], B. Kasper [15]).

The first actual BER measurements using an EDFA preamplifier receiver were made in 1988–1989 (C. R. Giles et al., [16] and [17]). The experiment described in [16] used an argon-ion-pumped EDFA with 15 dB gain for the amplification of 2 Gbit/s NRZ signals. Figure 7.2 shows the detected signals and corresponding eye diagram. This early measurement is the first proof that EDFAs are able to amplify high speed signals without pulse distortion, intersymbol interference, or patterning effects. It showed that basic error-free detection is possible; moreover, that the EDFA could improve the receiver sensitivity by 6 dB ($\bar{n} = 3850$ photons/bit). The second experiment used a high gain EDFA ($G = 36$ dB) pumped near 1480 nm, and a $p$–$i$–$n$ receiver of higher sensitivity. A BER of $10^{-9}$ at $B = 1.8$ Gbit/s could be achieved with an EDFA input signal of $\eta_i \bar{P}_s = -43$ dBm, corresponding to $\eta_i \bar{n} = 215$ photons/bit. The input splice loss was $\eta_i = 2$ dB, which gave an actual system sensitivity of $\bar{P}_s = -41$ dBm, or $\bar{n} = 341$ photons/bit, a 1 dB improvement with respect to the best sensitivity previously reported for any bit rate with a semiconductor optical preamplifier, i.e., $\bar{n} = 420$ photons/bit (N. A. Olsson and P. Garbinski [18]).

The best receiver sensitivity continued to increase in the following years, and moved closer to the fundamental limit (Chapter 3). The sensitivity improvements could be achieved by reducing the EDFA input coupling loss, by pumping the EDFA at $\lambda = 980$ nm, using two-stage EDFAs, by using optical filters, and by using low noise receivers having optimized electrical bandwidths. The corresponding experimental results, obtained between 1989 and 1993 [19]–[31]), are summarized in Table 7.1. To show the evolution in receiver sensitivity, the relevant photons/bit data and corresponding bit-rates listed in the table are plotted versus time in Figure 7.3. The figure also shows data corresponding to best sensitivities obtained in 1.5 $\mu m$ APD receivers [15]. As the figure indicates, the use of EDFA preamplifiers made it possible to increase the sensitivity of direct detection $p$–$i$–$n$ receivers by a factor of six; the best performance is currently $\bar{n} = 102$ photons/bit for 10 Gbit/s signals (R. I. Laming et al. [30]).

The quantum limit of $\bar{n} = 38$–44 photons/bit shown in Figure 7.3 corresponds to the sensitivity of ideal detectors, calculated from different photon statistics models and listed in Table 3.1. The Johnson noise limit, determined by the minimum detector thermal noise $\sigma_{th}^2 = 4k_B T B_e / R$, is shown for the bit rate range $B = 1$–10 Gbit/s. This limit is calculated from Eq. (3.80) with the parameters: $\lambda = 1.5 \,\mu m$, $B_o = 1$–10 Gbit/s, $B_e = B_o/2$, $G = 30$ dB, $n_{eq} = 1$, $\eta = 0.8$, $\mathcal{M} = 2$, $R = 50 \,\Omega$ and $T = 300$ K. It is found to be of $\bar{n} = 53$ photons/bit for $B = 1$ Gbit/s and 45 photons/bit for $B = 10$ Gbit/s.

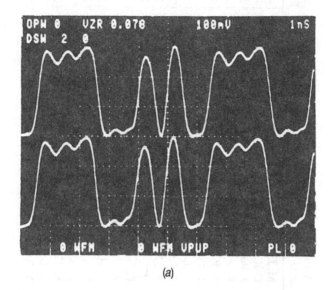

(a)

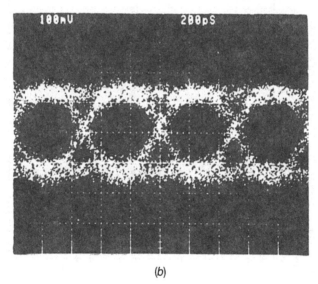

(b)

**FIGURE 7.2**   Early measurement of digital photodetection of 2 Gbit/s signals with a EDFA preamplifier ($G = 15$ dB), showing: (a) received electrical signal with (bottom) and without (top) EDFA, arbitrary vertical scales, and (b) eye diagram corresponding to amplified optical data. From [16] © 1989 IEEE.

Thus, the best sensitivity of $\bar{n} = 102$ photons/bit achieved for 10 Gbit/s signals is only 3.5 dB above the Johnson limit.

We consider next the effect of gain saturation on the sensitivity of EDFA preamplifier receivers. Chapter 2 shows that gain saturation affects the photon statistics of the amplified light and that the linear model is not valid. It is not possible to predict the detector SNR (or the $Q$ factor) without the knowledge of the probability distribution functions (PDFs) corresponding to the nonlinear amplification regime

**TABLE 7.1** EDFA preamplifier direct-detection receiver sensitivities at $10^{-9}$ BER (numbers in parenthesis indicate $\eta_i \bar{n}$ values)

| Year | Photons/bit $\bar{n}$ | Received power $\bar{P}$ (dBm) | Bit rate (Gbit/s) | EDFA pump (nm) | Author et al. | Company | Ref. |
|---|---|---|---|---|---|---|---|
| 1989 | 1382 | −46 | 0.140 | 528 | M. J. Pettitt | STC | [19] |
| 1989 | 341 (215) | −41 | 1.8 | 1490 | C. R. Giles | AT&T | [17] |
| 1990 | 887 | −32.4 | 5 | 1480 | J. Boggis | BTRL | [20] |
| 1990 | 152 | −43 | 0.622 | 980 | P. P. Smith | BTRL | [21] |
| 1990 | 174 | −45.6 | 1.2 | 980 | P. P. Smith | BTRL | [21] |
| 1990 | 152 | −49 | 2.5 | 980 | P. P. Smith | BTRL | [21] |
| 1991 | 606 | −31.1 | 10 | 1480 | L. J. Blair | Hitachi | [22] |
| 1991 | 147 | −37.2 | 10 | 980 | T. Saito | NEC | [23] |
| 1991 | 289 (186) | −40.7 | 2.3 | 1480 | P. M. Gabla | Alcatel-CIT | [24] |
| 1991 | 235 | −33.8 | 10 | 1480 | B. L. Patel | BNR | [25] |
| 1991 | 219 | −41.5 | 2.5 | 980/1480 | C. G. Joergensen | Tech. U. Denmark | [26] |
| 1992 | 193 | −36.0 | 10 | 980 | A. H. Gnauck | AT&T | [27] |
| 1992 | 137 | −43.5 | 2.56 | 980 | A. H. Gnauck | AT&T | [27] |
| 1992 | 124 | −44 | 2.48 | 1480 | Y. K. Park | AT&T | [28] |
| 1992 | 156 (96) | −46.2 | 1.24 | 980 | K. Kannan | OTC | [29] |
| 1992 | 256 (163) | −40.9 | 2.48 | 980 | K. Kannan | OTC | [29] |
| 1992 | **102** | **−38.8** | **10** | 980 | R. I. Laming | U. Southampton | [30] |
| 1993 | 135 | −40.5 | 5 | 980 | Y. K. Park | AT&T | [31] |

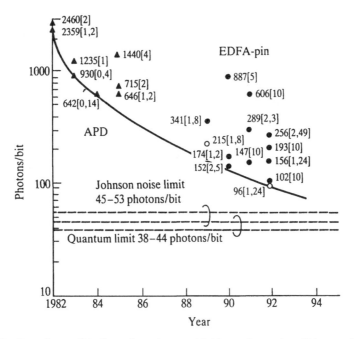

**FIGURE 7.3**    Best photon/bit direct detection sensitivities and associated bit rates (in brackets) reported for 1.5 μm APD receivers and EDFA preamplifier $p–i–n$ receivers (open symbols indicate $\eta_i\bar{n}$ values). The EDFA preamplifier data are listed in Table 7.1.

but it is possible to predict power penalties from the experimental measurement of ASE and RF noise (A. E. Willner and E. Desurvire [32]).

In the saturation regime assume the photocurrent noise variance takes the same canonic form as the linear case, i.e.,

$$\sigma^2 \approx \sigma^2_{\text{s-ASE}} + \sigma^2_{\text{ASE-ASE}} + \sigma^2_{\text{th}} \tag{7.2}$$

where the shot noise contribution is neglected. Assume the signal–ASE beat noise takes the same form as the linear case, i.e., $\sigma^2_{\text{s-ASE}} \approx 2\eta^2 GI_sI_N$, where $I_s$ and $I_N$ are the signal and ASE photocurrents, with $I_s = eP_s/hv_s$ and $I_N = n_{\text{sp}}(G-1)eB_o$, as in Eqs. (3.19) and (3.20). On the other hand, if the light photon statistics are assumed to be described by Gaussian PDFs, the minimal BER is given by the $Q$ factor (Chapter 3):

$$Q = \frac{\sqrt{S(1)} - \sqrt{S(0)}}{\sqrt{\sigma^2(1)} + \sqrt{\sigma^2(0)}} \tag{7.3}$$

where $S = (GI_s)^2$ and $\sigma^2$ are the electrical signal and noise powers associated with either the 0 or the 1 symbols. We assume the condition $Q = 6$; this gives BER $= 10^{-9}$ under unsaturated gain conditions for an amplifier input signal power $P_s$. Under these conditions, the amplifier gain is $G$ and the spontaneous emission factor is $n_{\text{sp}}$. When the amplifier is saturated, the gain decreases to the value $G'$ and the spontaneous emission factor increases to the value $n'_{sp}$, due to the effect of reduced medium inversion. Then the power required for $10^{-9}$ BER is $P'_s$. The BER penalty caused by preamplifier saturation is given by $\mathscr{P} = P'_s/P_s$, the relative amount of signal power increase required for achieving $10^{-9}$ BER in saturation conditions.

Equating the $Q$ factors, obtained from Eq. (7.3) for both saturated and unsaturated conditions, and solving for $\mathscr{P} = P'_s/P_s$ through Eq. (7.2), we obtain the penalty [32]:

$$\mathscr{P}(\text{dB}) = 10 \log_{10}\left\{\frac{G(G'-1)}{G'(G-1)}\right\} + 10 \log_{10}\left\{\frac{n'_{sp}}{n_{sp}}\right\} + 20 \log_{10}\left\{\frac{\sqrt{1+x'}+\sqrt{x'}}{\sqrt{1+x}+\sqrt{x}}\right\}$$

(7.4)

where

$$x = \frac{\sigma^2_{\text{ASE-ASE}} + \sigma^2_{\text{th}}}{\sigma^2_{\text{s-ASE}}}$$

(7.5)

and the primed and unprimed parameters are the saturated and unsaturated conditions, respectively.

The first two terms in Eq. (7.4) are related to an effective decrease in optical SNR for constant input signal, i.e., $\Delta(\text{SNR}) = (G/I_N)_{\text{dB}} - (G'/I'_N)_{\text{dB}}$. The first term in Eq. (7.4) indicates that saturation reduces the gain of both signal and ASE. At high gains, both experience the same power decrease, which does not affect the optical SNR. The first term is negligible. But at high saturation, the spontaneous emission factor increases, so the corresponding SNR change is given by $\Delta(\text{SNR}) = (n'_{sp}/n_{sp})_{\text{dB}}$, which explains the second term in Eq. (7.4). The third term is due to the effect of ASE–ASE and thermal noise background. In the regime where the signal–ASE beat noise dominates, this background noise is negligible and the third term in Eq. (7.4) vanishes, as $x, x' \ll 1$. But at high saturation, the amplified signal power is reduced and the signal–ASE beat noise no longer dominates. The third term thus represents the relative signal power increase necessary to restore the signal–ASE beat-noise-limited regime.

Although the power penalty given by Eq. (7.4) was derived from empirical assumptions, it can be justified *heuristically* by comparison with power penalties, ASE, and RF noises measured experimentally. Figure 7.4 shows experimental BERs measured for incoherent FSK signals at 10 Gbit/s using a saturated, 1480 nm pumped EDFA as a preamplifier [32]. Different saturation conditions with gain compressions $\Delta G = 3$ to 12 dB from the unsaturated value $G = 31$ dB could be obtained by increasing the power of an auxiliary CW saturating signal. It is observed that the power penalty at $10^{-9}$ BER increases with gain saturation. For each regime of saturation, the spontaneous emission factor could be determined from the output ASE. The EDFA gain and the beat noises (s–ASE and ASE–ASE) could also be measured with an RF spectrum analyzer. These data make it possible to predict the power penalty, according to Eq. (7.4). Figure 7.5 shows the power penalty versus gain compression $\Delta G$, determined independently from the BER measurements in Figure 7.4 and the heuristic Eq. (7.4). Figure 7.5 shows that theory and experiment are in good agreement over the 12 dB gain saturation range investigated, and in cases of forward and backward pumping. The power penalty is small, $\mathscr{P} \leqslant 1$ dB, for gain compressions up to $\Delta G = 5$ dB. The maximum penalty $\mathscr{P} \approx 5$ dB, observed for

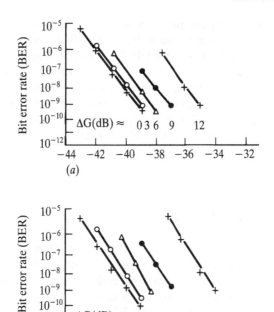

**FIGURE 7.4** Experimental BERs of FSK signals at $B = 1$ Gbit/s, measured for different regimes of EDFA gain saturation (gain compression $\Delta G$), with (a) forward and (b) backward pumping at $\lambda = 1480$ nm; the unsaturated gain is $G = 31$ dB. From [32] © 1991 IEEE.

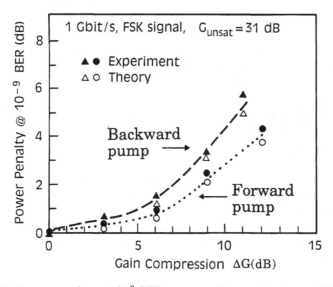

**FIGURE 7.5** Power penalty at $10^{-9}$ BER corresponding to the data of Figure 7.4, with theoretical prediction given heuristically by Eq. (7.4). From [32] © 1991 IEEE.

backward pumping at gain compression $\Delta G = 12\,\text{dB}$, corresponds to a measured spontaneous emission factor increase $n'_{\text{sp}}/n_{\text{sp}} \approx 2.5\,\text{dB}$ and a measured excess noise, given by the third term in Eq. (7.4), of 2.5 dB [32]. For high gain EDFA preamplifiers, the receiver sensitivity is weakly affected by gain saturation up to 5 dB compression. For any degree of saturation, the power penalty can be attributed to two main causes; the increase in spontaneous emission factor, due to reduced medium inversion, and the increase of the thermal and ASE–ASE noises relative to the s–ASE beat noise. A current challenge is to derive a formula similar to Eq. (7.4) from the rigorous nonlinear photon statistics theory of Section 2.8.

## 7.2    DIGITAL LINEAR SYSTEMS

We outline various results obtained in EDFA-based, direct detection or coherent systems but only for long-haul transmission, where no electronic repeaters are used between the transmitting and the receiving ends. In the literature, the expressions *unrepeatered transmission, repeaterless transmission*, and *nonregenerative transmission* are generally used to designate such systems. Lightwave systems based on architectures other than trunk (e.g., broadcast, ring) are considered in Sections 7.4 and 7.5.

Linear system refers to RZ or NRZ digital data transmission, in which dispersion is not compensated for by nonlinearity. Nonlinear systems use the principle of optical soliton transmission (Section 7.3). Fiber nonlinearities affect such linear systems; SPM, through the optical Kerr effect, is a cause of significant signal chirp. In WDM transmission systems, which operate near the zero dispersion wavelength, quasiphase-matched SFPM also occurs, resulting in undesirable crosstalk. We consider the detrimental effects of various fiber nonlinearities (i.e., SPM, XPM, SFPM, SRS, and SBS). And we review techniques of dispersion compensation, which make it possible to use standard SMF systems with signals at 1.5 $\mu$m. We give separate descriptions of system experiments related to either single-channel or multiple channel (WDM) transmission. We cover direct detection systems before coherent detection systems. But first of all we consider the possible architectures of unrepeatered trunk systems, Figure 7.6, and provide numerical examples of their associated transmission capacity.

Figure 7.6*a* shows a standard passive transmission line; the capacity of this system is either loss-limited or dispersion-limited. In the case of direct detection systems considered here, the maximum transmission distance at $B = 10$ Gbit/s is about 40 km for SMF and 135 km for DSF, which represent the dispersion and loss limits, respectively; these values are obtained by assuming a maximum transmitter power $P = 1$ mW at $\lambda = 1.5\,\mu$m, a receiver sensitivity $\bar{n} = 300$ photons/bit and a fiber loss $\alpha_{\text{s}} = 0.25$ dB/km [1]. The corresponding system capacities are then 400 Gbit/s.km and 1.35 Tbit/s.km, respectively. Figure 7.6*b* shows a system with an EDFA preamplifier/receiver. As before, the maximum transmission capacity is determined by the receiver sensitivity $\bar{n}$ (photons/bit) for the loss limit, and by the fiber dispersion coefficient $D$ for the dispersion limit. If dispersion is negligible, the maximum transmission distance is given by:

$$L = \frac{1}{\alpha_{\text{s}}}(P_0 - P_{\text{rec}} - M) \tag{7.6}$$

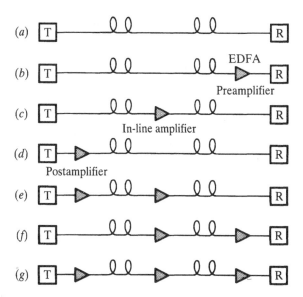

**FIGURE 7.6** Various architectures for nonregenerative trunk systems: (*a*) passive, (*b*) with EDFA preamplifier, (*c*) with in-line EDFAs, (*d*) with EDFA postamplifier, (*e*) Type A chain, (*f*) Type C chain, and (*g*) with preamplifier, in-line and postamplifier EDFAs.

where $L$ is in km, $\alpha_s$ is the fiber loss coefficient in dB/km ($\alpha_s \approx 0.25$ dB/km at $\lambda = 1.55\,\mu$m), $P_0$ is the signal power launched into the system or the transmitter output power in dBm, $P_{\text{rec}} = \{\bar{n}h\nu_s B\}_{\text{dBm}}$ is the EDFA preamplifier/receiver sensitivity, and $M$ is a system margin (e.g., $M = 3$ dB). Thus, for signals at $B = 10$ Gbit/s with $\bar{n} = 100$ photons/bit, $P_0 = 0$ dBm and no margin, the maximum transmission distance is 155 km, corresponding to a capacity of 1.55 Tbit/s.km.

Figure 7.6*c* shows a single (or multiple) in-line amplifier. Generally, the gain $G$ of each EDFA is set to a level that exactly compensates the signal loss incurred in the preceding fiber section. The calculation of maximum capacity in this system, determined by the receiver SNR (or $Q$ factor), can be done using the theory of Chapter 3.

Figure 7.6*d* includes at the transmitter a single EDFA power booster, also called a postamplifier. As before, the maximum transmission capacity can be evaluated from results in Chapter 3. Fairly high signal powers (e.g., $+20$ dBm) can be generated at the EDFA postamplifier output. But the amount of signal power that can be launched into the system is limited by fiber nonlinearities, hence the transmission distance is also limited.

Figure 7.6*e,f* shows architectures that correspond to Type A and Type C amplifier chains. Their characteristics are analyzed in Chapters 2 and 3. In Type A systems, the average signal power is higher than in Type C systems. This makes Type A systems more sensitive to fiber nonlinearity limitations but gives them a greater transmission distance than Type C systems, as their SNR degradation along the chain is slower. In the limit of small amplifier spacings (i.e., $L_{\text{amp}} < 1/\alpha_s$), Figure 7.6*e,f* architectures perform as a distributed amplifier chain and are nearly identical. Figure 7.6*g* uses a combination of power amplifier, preamplifier, and in-line amplifiers. This system architecture can be analyzed using the same model as Type C and considering the postamplifier as part of the transmitter. This approximation neglects the ASE noise contribution of the postamplifier but is accurate if the corresponding gain is highly saturated. For a fixed input signal power, figure 7.6g gives the longest possible transmission distance.

Let us give some numerical examples of transmission distance and capacity based on the Figure 7.6 architectures. Chapter 3 shows that the receiver sensitivities ($10^{-9}$ BER) in a $k$ element amplifier chain of Type A or Type C are given by:

$$\bar{P}(k, \text{Type A}) = 18 h v_s B \left\{ \frac{1}{\eta} + 2 k n_{eq} + \frac{1}{6} \sqrt{ 2 k^2 n_{eq}^2 \left( \frac{4 B_o}{B} - 1 \right) + \frac{8 k n_{eq} B_o}{\eta B} + \frac{4 \sigma_{th}^2}{e^2 \eta^2 B^2} } \right\} \quad (7.7)$$

$$\bar{P}(k, \text{Type C}) = 18 h v_s B \left\{ \frac{1}{\eta} + 2 k G n_{eq} + \frac{1}{6} \sqrt{ 2 k^2 G^2 n_{eq}^2 \left( \frac{4 B_o}{B} - 1 \right) + \frac{8 k G n_{eq} B_o}{\eta B} + \frac{4 \sigma_{th}^2}{e^2 \eta^2 B^2} } \right\} \quad (7.8)$$

where $B$ is the bit rate, $B_o$ the optical filter passband, $\eta$ the detector quantum efficiency, $n_{eq}$ the equivalent input noise factor, $G$ the EDFA gain and $\sigma_{th}^2$ the receiver thermal noise. The above results assume that all EDFAs in the chain operate in the linear gain regime and have identical parameters. Likewise, the effect of fiber nonlinearities and other causes of power penalties are overlooked. The first two terms in braces in Eqs. (7.7) and (7.8) are due to signal shot noise and s–ASE beat noise, respectively. The three terms under the square root are due to ASE–ASE beat noise, ASE shot noise, and thermal noise, respectively. For simplicity, we assume that the ASE–ASE beat noise dominates over the other contributions under the square root, which is the case when the number of EDFAs is large ($k \gg 1$). As a numerical example, we assume that the optical filter passband is $B_o = 125$ GHz (corresponding to $\Delta\lambda = 1$ nm at $\lambda = 1.5\,\mu$m), the equivalent input noise $n_{eq} = 1$, and a large number of EDFAs such that $k \gg 1/2\eta$. Under these assumptions, we obtain from Eqs. (7.7) and (7.8):

$$\bar{P}(k, \text{Type A}) \approx k \times 36 h v_s B \left( 1 + \sqrt{ \frac{6.9}{B_{GHz}} } \right) \quad (7.9)$$

$$\bar{P}(k, \text{Type C}) \approx G \times \bar{P}(k, \text{Type A}) \quad (7.10)$$

The above result shows that the required power for a Type C chain is $G$ times the required power for a Type A chain. A Type C chain has the same input power requirement if a postamplifier of gain $G$ is used at the transmitter. Equations (7.9) and (7.10) can be used in different ways.

Assume first a fixed transmission distance $L_{tot} = 9000$ km (transpacific link). With a fiber loss coefficient $\alpha_s = 0.25$ dB/km, such distance corresponds to a total transmission loss of $T = 2250$ dB. The number $k$ of EDFAs required in the chain is then the integer value of the ratio $x = T/G$. For a 40 km amplifier spacing, we have $G = 10$ dB and $k = 225$. For signals at bit rate $B = 10$ Gbit/s, Eq. (7.9) gives $\bar{P} = -17$ dBm for Type A chains and $\bar{P} = -7$ dBm for Type C chains.

Another way to use Eqs. (7.9) and (7.10) is to assume that the transmitter output power (with no postamplifier) is fixed at a maximum $\bar{P}(\text{max})$. Consider the amplifier spacing $L = G/\alpha_s$ ($G$ in dB and $\alpha_s$ in dB/km). The maximum transmission distance

$L_{tot} = kL$ for a Type A chain is then given by:

$$L_{tot}(\text{Type A}) = \frac{G(\text{dB})}{\alpha_s(\text{dB/km})} \frac{\bar{P}(\text{max})}{36hv_sB\left(1 + \sqrt{\dfrac{6.9}{B_{GHz}}}\right)} \qquad (7.11)$$

Assuming that for linear operation of the EDFAs the maximum input power is $\bar{P}(\text{max}) = -20\,\text{dBm}$ or $10\,\mu\text{W}$, we find for 10 Gbit/s signals the maximum distance $L_{tot}(\text{Type A}) = G(\text{dB}) \times 470\,\text{km}$, equal to 4700 km for $G = 10\,\text{dB}$ and 9400 km for $G = 20\,\text{dB}$.

These examples use SNR degradation to determine the maximum capacity of EDFA-based trunk systems. Their simple formulas account for various system parameters, e.g., EDFA gain and spacing, number of EDFAs, fiber loss, EDFA equivalent input noise, input signal power, bit rate, optical filter passband, receiver thermal noise, and quantum efficiency. But the maximum system capacity is not only determined by the minimum received signal power $\bar{P}(k)$ (according to the above theory), but also by various power penalties. These penalties are caused by several impairments, among them: amplifier saturation, chromatic and polarization dispersion, chirp, phase noise, backscattering, and fiber nonlinearities. A detailed performance evaluation of EDFA-based trunk systems and discussion of these effects is beyond the scope of this book but can be found in: K. Hagimoto et al. [33]; C. R. Giles and E. Desurvire [34]; K. Nakagawa et al. [35]; and K. Inoue et al. [36].

The maximum transmission distance and bit rate performance of single-channel direct detection systems in early 1993 is plotted in Figure 7.7. The three different

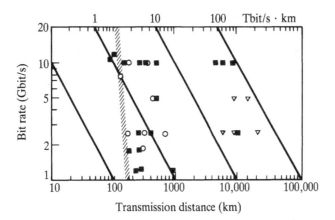

**FIGURE 7.7** Maximum transmission distance and bit rates of direct detection, nonregenerative EDFA-based linear systems reported to date (early 1993), corresponding to data shown in Table 7.2. Square and triangle symbols correspond to straight-line and loop transmission experiments, respectively, using DSF fiber (zero dispersion at $\lambda = 1.5\,\mu\text{m}$); circle symbols correspond to straight-line transmission in SMF fiber (zero dispersion at $\lambda = 1.3\,\mu\text{m}$). The lines indicate constant system capacities of 0.1, 1, 10, and 100 Tbit/s.km. The shaded region corresponds to the maximum capacity range of nonregenerative DSF systems without optical amplifiers, assuming a transmitter power of 0 dBm, a fiber loss coefficient of $\alpha_s = 0.25\,\text{dB/km}$, and receiver sensitivities of $\bar{n} = 300\text{--}1000$ photons/bit.

**TABLE 7.2  Direct-detection, single-channel linear transmission system experiments**

| Year | # | Bit rate (Gbit/s) | Length (km) | Capacity (Tbit.km/s) | Signal λ (μm) | Modulation [a] | Fiber type [b] | Number of EDFAs [c] | Launched signal power (dBm) | Received signal power (dBm) [d] | Author et al. | Company | Ref. |
|---|---|---|---|---|---|---|---|---|---|---|---|---|---|
| 1989 | 1 | 1.8 | 212 | 0.38 | 1.552 | ASK | DSF | 1 | +11.1 | −38.1 | K. Hagimoto | NTT | [37] |
| | 2 | 1.1 | 94 | 1.03 | 1.540 | ASK | DSF | 1 | 0 | −18.5 | J. L. Gimlett | Bellcore | [38] |
| | 3 | 1.2 | 218 | 0.26 | 1.536 | ASK | DSF | 1 | −1.5 | −36.5 | N. Edagawa | KDD | [39] |
| | 4 | 1.8 | 250 | 0.45 | 1.553 | ASK | DSF | 1 | +12.2 | −37.8 | K. Hagimoto | NTT | [40] |
| | 5 | 1.2 | 267 | 0.32 | 1.536 | ASK-ex | SMF | 2 | – | −35.5 | N. Edagawa | KDD | [41] |
| | 6 | 1.8 | 210 | 0.38 | 1.554 | ASK | DSF | 1 | −0.2 | −31 | A. Takada | NTT | [42] |
| | 7 | 10 | 161 | 1.61 | 1.535 | ASK-ex | DSF | 1 | +8.6 | −20 | K. Hagimoto | NTT | [43] |
| | 8 | 12 | 100 | 1.2 | 1.536 | ASK-ex | DSF | 1 | +1.7 | −20 | H. Nishimoto | Fujitsu | [44] |
| | 9 | 11 | 151 | 1.66 | 1.536 | ASK | DSF | 1 | 0 | −17.6 | M. Z. Iqbal | Bellcore | [45] |
| 1990 | 10 | 1.2 | 904 | 1.08 | 1.536 | ASK | DSF | 11 | +8.7 | −32.5 | N. Edagawa | KDD | [46] |
| | 11 | 10 | 375 | 3.75 | 1.536 | ASK-ex | DSF | 5 | +5 | −22.7 | K. Nakagawa | NTT | [73] |
| | 12 | 5 | 523 | 2.61 | 1.536 | ASK-ex | DSF | 5 | +5 | −32.3 | K. Nakagawa | NTT | [73] |
| | 13 | 17 | 150 | 2.55 | 1.536 | ASK-ex | SMF | 1 | +9.8 | −24.8 | K. Hagimoto | NTT | [76] |
| | 14 | 5 | 200 | 1.0 | 1.536 | ASK-ex | DSF | 3 | | −30 | N. Henmi | NEC | [77] |
| | 15 | 11 | 260 | 2.86 | 1.536 | ASK | DSF | 2 | | −18 | A. Righetti | Pirelli | [47] |
| | 16 | 2.4 | 250 | 0.60 | 1.537 | ASK [e] | SMF | 0 | +20.4 | −32.5 | E. G. Bryant | BTRL | [48] |
| | 17 | 2.4 | 710 | 1.70 | 1.536 | ASK-ex | SMF | 9 | +7.7 | −31.5 | N. Edagawa | KDD | [49] |
| | 18 | 2.488 | 132 | 0.33 | 1.527 | ASK [e] | SMF | 0 | +8.3 | −33.5 | E. G. Bryant | BTRL | [71] |
| | 19 | 1.8 | 308 | 0.55 | 1.553 | ASK-ex | SMF | 1 | +16 | −40 | K. Aida | NTT | [72] |
| | 20 | 1.7 | 177 | 0.30 | 1.5 | FSK | DSF | 2 | | −34 | D. A. Fishman | AT&T | [50] |
| | 21 | 10 | 505 | 5.05 | 1.552 | ASK-ex | DSF | 5 | +11 | −18 | K. Hagimoto | NTT | [33] |
| | 22 | 5 | 505 | 2.52 | 1.536 | ASK-ex | DSF | 5 | +5 | −26 | K. Hagimoto | NTT | [33] |
| | 23 | 2.5 | 160 | 0.40 | 1.541 | ASK | SMF | 0 | +8.5 | −30.3 | C. Y. Kuo | AT&T | [51] |

| | | | | | | | | | | | | |
|---|---|---|---|---|---|---|---|---|---|---|---|---|
| 22 | 1991 | 10 | 100 | 1.0 | 1.536 | ASK-ex | **SMF** | 1 | — | -22 | N. Henmi | NEC | [74] |
| 23 | | 5 | **9000** | 45 | 1.5 | ASK-ex | DSF-LOOP | — | — | -24 | N. S. Bergano | AT&T | [52-53] |
| | | 5 | **14,300** | 71.5 | 1.5 | ASK-ex | DSF-LOOP | — | — | — | N. S. Bergano | AT&T | [52-53] |
| | | 2.4 | **21,000** | 50.4 | 1.5 | ASK-ex | DSF-LOOP | — | — | — | N. S. Bergano | AT&T | [52-53] |
| 24 | | 10 | 309 | 3.09 | 1.552 | ASK-ex | DSF | 4 | +10 | -29.3 | K. Nakagawa | NTT | [67] |
| 25 | | 2.5 | **9000** | 22.5 | 1.550 | ASK-ex | DSF-LOOP | — | — | — | S. Yamamoto | KDD | [70] |
| 26 | | 2.5 | **6000** | 15 | 1.545 | ASK-ex | DSF-LOOP | — | — | -24.5 | T. Widdowson | BT | [54] |
| 27 | | 8 | 130 | 1.04 | 1.53 | ASK-ex | **SMF** | 1 | 0 | -31 | A. H. Gnauck | AT&T | [55]a |
| 28 | | 2.5 | **1316** | 3.29 | 1.553 | ASK-ex | DSF | 26 | +12 | -34 | P. M. Gabla | Alcatel-CIT | [55]b |
| 29 | 1992 | 10 | 250 | 2.50 | 1.53 | ASK-ex | DSF | 1 | +12.2 | -36.0 | H. Gnauck | AT&T | [27] |
| | | 2.5 | 250 | 0.62 | 1.53 | ASK-ex | DSF | 1 | +12.2 | -43.5 | H. Gnauck | AT&T | [27] |
| 30 | | 2.48 | 318 | 0.79 | 1.558 | ASK-ex | **SMF** | 1 | +16 | -41 | Y. Park | AT&T | [28] |
| 31 | | 10 | **4500** | 45.5 | 1.558 | ASK-ex | DSF | 138 | — | -25 | H. Taga | KDD | [56] |
| 32 | | 10 | 150 | 1.5 | 1.545 | ASK | SMF + DC | 3 | — | -31 | J. M. Dugan | Corning | [75] |
| 33 | | 5 | 450 | 2.25 | 1.5 | ASK-ex | **SMF** | 9 | +17.9 | -29.5 | A. D. Ellis | BT | [57] |
| 34 | | 2.5 | 301 | 0.75 | 1.558 | ASK [e] | SMF | 1 | +6 | -41 | C. G. Joergensen | Tech. U. of Denmark | [78] |
| 35 | | 2.48 | 10,073 | 25 | 1.552 | ASK-ex | DSF | 199 | +5 | -28 | T. Imai | NTT | [58] |
| 36 | | 2.5 | 4520 | 11.3 | 1.552 | ASK-ex | DSF | 48 | +3 | -27.5 | S. Saito | NTT | [79] |
| 37 | | 5 | **9000** | 45 | 1.5 | ASK-ex | DSF | 274 | +15.5 | -32.5 | N. S. Bergano | AT&T + KDD | [68] |
| 38 | | 5 | 226 | 1.13 | 1.557 | ASK-ex | **SMF** | 1 | — | -38 | Y. K. Park | AT&T | [69] |
| 39 | | 2.5 | 110 | 0.27 | 1.550 | ASK | SMF + DC | 1 | — | -29 | H. Izadpanah | Bellcore | [59] |
| 40 | | 2.48 | 357(341) | 0.88 | 1.554 | ASK-ex | (SMF) | 2 | +15.4 | -47.2 | P. M. Gabla | Alcatel-CIT | [60] |
| | | 0.622 | 401(365) | 0.25 | 1.554 | ASK-ex | (SMF) | 2 | +16.2 | -50.3 | P. M. Gabla | Alcatel-CIT | [60] |
| 41 | | 10 | **6000** | 60 | 1.552 | ASK-ex | DSF | 119 | +2 | -27 | M. Murakami | NTT | [61] |
| 42 | 1993 | 5 | **4500** | 22.5 | 1.5 | ASK-ex | DSF | 138 | — | -30 | S. Yamamoto | KDD | [62] |
| 43 | | 10 | **9000** | 90 | 1.559 | ASK-ex | DSF | 274 | +2 | -21 | H. Taga | KDD + AT&T | [63] |
| 44 | | 100 | 50 | 5 | 1.556 | ASK-ex | DSF | 0 | — | -23 | S. Kawanishi | NTT | [64] |
| 45 | | 10 | 360 | 3.6 | 1.543 | ASK-ex | **SMF** | 2 | — | -27 | R. Jopson | AT&T | [65] |
| | | 2.5 | 400 | 0.8 | 1.543 | ASK-ex | SMF | 2 | — | -31.4 | R. Jopson | AT&T | [65] |
| 46 | | 10 | 215 | 2.15 | 1.548 | ASK-ex | DSF | 1 | +15 | -32.5 | B. L. Patel | BNR | [66] |
| | | 10 | 180 | 1.8 | 1.548 | ASK-ex | **SMF + DC** | 1 | +13.5 | -28.2 | B. L. Patel | BNR | [66] |

([a]: ASK = amplitude-shift-keying, FSK = frequency-shift-keying, ex = external modulation, [b] **SMF** = single-mode fiber with $\lambda_o = 1.3$ μm, DSF = single-mode fiber with $\lambda_o = 1.55$ μm, DC = dispersion compensation fiber, [c]: not including post-amplifiers, [d]: at $10^{-9}$ BER, [e]: self-homodyne FSK transmitter).

symbols correspond to *straight-line* and *loop* transmission with DSF, and *straight-line* transmission with SMF. The shaded region corresponds to the maximum capacity range of nonregenerative DSF systems without optical amplifiers, (given by the relation $L(\text{km}) = (\bar{n}hv_s B)_{\text{dBm}}/\alpha_s(\text{dB/km})$, assuming a maximum transmitter power of 0 dBm, a fiber loss coefficient $\alpha_s = 0.25$ dB/km, and a direct detection receiver sensitivity range $\bar{n} = 300$–1000 photons/bit. The current capacity of EDFA-based systems is two to three orders of magnitude better than conventional loss-limited systems, whose capacity falls within the shaded region.

The spread of experimental points in Figure 7.7 reflects the considerable progress made since the first long-haul, high bit rate system experiments reported in 1989. This progress is also shown in Table 7.2, the year-to-year results of 46 system experiments where $10^{-9}$ BER was measured after some transmission length. In addition to bit rate, length, and system capacity, the table indicates some experimental parameters and references for each experiment, [37]–[79].

All experiments in Table 7.2 with transmission length greater than $L_{\text{tot}} = 1000$ km used transmitters with *external modulation*. Direct modulation produces considerable chirp or line width broadening ($\alpha$-parameter = 4–6), while the effect is negligible with external modulation ($\alpha < 0.1$) [33]. After transmission through a dispersive fiber, the signal phase noise is converted into intensity noise, a cause of power penalty (N. Edagawa [80]). In very long-haul systems ($L_{\text{tot}} = 10,000$ km), the total chromatic dispersion is near 10,000 ps/nm (10 ns/nm) or higher, which shows that transmitters with fairly narrow line widths are required (N. Edagawa et al. [49]). Another advantage of external modulation is the broad bandwidth available ($B > 20$ GHz for Ti : LiNbO$_3$ Mach–Zenhder intensity modulators), and bit rate flexibility with constant bias conditions (K. Hagimoto et al. [33]). Large variations in launched and received powers can be attributed to differences in system architecture, Figure 7.6, in receiver sensitivities, in amplifier spacings, in EDFA spontaneous emission factors, as well as various causes of power penalties. We shall first fully discuss the case of DSF-based systems, and consider only later on the case of SMF-based systems, which use specific dispersion compensation techniques.

As shown in Table 7.2, a first generation of EDFA-based linear systems having transmission capacities greater than 1 Tbit/s.km (i.e., $B \approx 10$ Gbit/s and $L \geqslant 100$ km, or $B \geqslant 1$ Gbit/s and $L \approx 1000$ km), was first demonstrated in the period 1989–1990 (J. L. Gimlett et al. [38]; K. Hagimoto et al., [33] and [43]; H. Nishimoto et al. [44]; M. Z. Iqbal et al. [45]; N. Edagawa et al., [46] and [49]; and K. Nakagawa et al. [73]). The experimental layout of the first system with length near 1000 km (i.e., $L_{\text{tot}} = 904$ km), which used 11 in-line LD-pumped EDFAs at $\lambda = 1480$ nm and 1.2 Gbit/s signals, is shown in Figure 7.8 [46]. In this experiment, the penalty with respect to the back-to-back sensitivity was only 0.6 dB.

A second generation of EDF-based linear systems having transmission capacities in the range of 10–70 Tbit/s.km could be demonstrated in 1991 through *recirculating loop experiments*. The first of these experiments used a 120 km loop including four EDFAs spaced at 40 km, Figure 7.9 (N. S. Bergano et al., [52] and [53]). Successful transmission of 2.4 Gbit/s and 5 Gbit/s NRZ signals could be obtained over 21,000 km and 14,300 km, respectively. As no BER floor was observed, we can extrapolate these results for transatlantic distances ($L_{\text{tot}} = 6400$ km), which gives a BER lower than $10^{-18}$ at both bit rates [53]. Electrical signals corresponding to pseudorandom NRZ data after 9000 km loop transmission and measured at $10^{-9}$ BER are shown in Figure

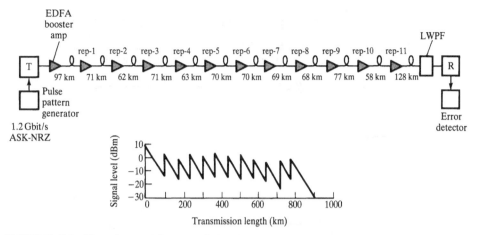

**FIGURE 7.8** Experimental layout of the first EDFA-based system demonstration with transmission length near 1000 km, and corresponding signal level diagram. $10^{-9}$ BER, $P_{rec} = -33$ dBm, penalty at 904 km $\sim 0.6$ dB. From [46], reprinted with permission from *Electronics Letters*, 1990.

7.10, along with the corresponding input signals. There is no effect of pulse broadening or distortion, while intensity noise and jitter effects are relatively small (note the higher intensity noise in the 1 symbols, which reflects the additional effect of s–ASE beat noise).

A third generation of EDF-based linear systems having transmission capacities in the range 20–90 Tbit/s.km and transmission lengths in the range 4500–10,000 km was demonstrated in 1992–1993 through actual *straight-line systems*. Compared to loop experiments, these system demonstrations use a considerable amount of equipment, considering both the fiber length involved and the number of LD-pumped EDFA modules (140 to 275). By early 1993, the maximum system length achieved was $L_{tot} = 10,073$, corresponding to a bit rate of 2.5 Gbit/s (T. Imai et al. [58]); the maximum system capacity achieved was 90 Tbit/s.km, corresponding to a 9000 km system operating at $B = 10$ Gbit/s (H. Taga et al. [63]). These record performances have been surpassed only by soliton systems.

Could the capacity of single-channel, linear direct detection systems still increase in the future? Most certainly yes, considering the possibilities for system and device improvements: reduction of overall chromatic dispersion, use of shorter amplifier spacings, reduction of system excess loss, use of optimum modulation format, and optimization of transmitter and receiver characteristics. It is also possible to increase the bit rate (or the transmission distance) by implementing *error-correcting codes* (ECCs). The two main error-control strategies are called automatic repeat request (ARQ) and forward error correction (FEC) (A. B. Carlson [81]). With FEC, all errors can be corrected (BER $\ll 10^{-15}$) if the uncorrected line BER is less than $10^{-4}$ but the encoding operation requires 14% redundancy in the transmission rate. This means that 2.5 Gbit/s coded signals must be transmitted at a rate $B = 2.85$ Gbit/s (P. M. Gabla et al. [60]). Thus, transoceanic communications based using ECCs could, in practice, operate at BERs higher than $10^{-9}$, while virtually no errors would be incurred at the receiving end. The maximum achievable bit rate capacity offered by EDFA-based linear systems with ECCs remains to be evaluated.

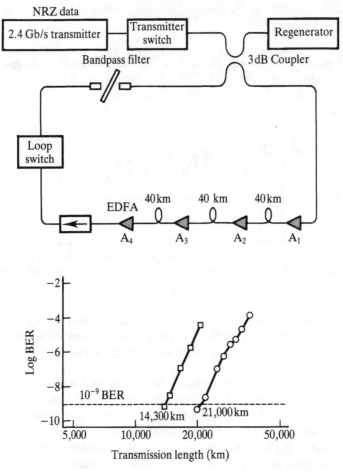

**FIGURE 7.9** Experimental layout of the first EDFA-based recirculating loop system demonstrating the possibility of transmission lengths of 14,300 km and 21,000 km at $B = 5$ Gbit/s and 2.4 Gbit/s, respectively; the measured BER versus transmission length up to 36,300 km is also shown. From [52], reprinted with permission from the Optical Society of America, 1991.

System capacity is increased by multiplexing optical data in the time, wavelength, or frequency domains. In time division multiplexing (TDM), transmitters with picosecond outputs are required and the modulation format is generally ASK; the RZ format is for tolerance to timing errors and jitter. Demultiplexing in TDM systems can be done by all-optical fiber devices (Section 6.5). In wavelength division multiplexing (WDM), several optical channels with different wavelengths, modulated in ASK or FSK, are combined in the same fiber; the wavelength spacing between channels is of the order of 1 nm. WDM systems require transmitters and receivers with wavelength agility (i.e., rapidly and digitally wavelength tunable or addressable); they also require optical amplifiers with flat gain passband. In the case of frequency division multiplexing (FDM), the optical channel separation is of the order of 1–10 GHz; FDM systems, which use FSK modulation, require transmitters with narrow line widths, low chirp and absolute frequency control; demultiplexing requires

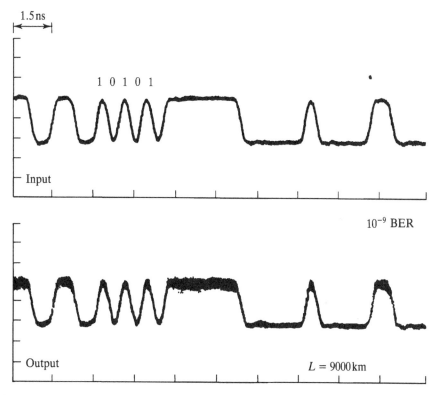

**FIGURE 7.10** Input and output pseudorandom NRZ signals at 2.4 Gbit/s, corresponding to a 9000 km loop transmission experiment, measured at $10^{-9}$ BER. (Courtesy N. S. Bergano, AT&T Bell Laboratories).

high resolution, tunable optical filters. In long-haul linear systems, the use of dense-TDM is limited by chromatic dispersion and fiber nonlinearities (SBS, SRS). The limitations by nonlinearities is explained by the fact that in the case of short pulses, a given number of photons/bit requires higher peak power. The use of TDM and FDM in long-haul linear systems is severely limited by three factors: gain ripple and finite bandwidth of EDFAs, fiber chromatic dispersion, and fiber nonlinearity (SFPM). Near the zero dispersion wavelength, fiber nonlinearity produces near-phase-matched wave mixing between different channels, which for very long interaction lengths can result in significant crosstalk (A. R. Chraplyvy [82]).

Table 7-3 lists for EDFA-based, TDM, WDM, and FDM linear systems with direct detection, reported during 1989 to early 1993, [83]–[90]. For each experiment, the table indicates the number of channels, the total (or aggregate) bit rate, the corresponding capacity, the channel separation and the modulation format. The maximum WDM system capacity reported is 45 Tbit/s.km, achieved in a two-channel, 4550 km link (H. Taga et al. [90]). This result is one half the best performance achieved in single-channel systems (Table 7.2). This experiment showed that fluctuations in the zero dispersion wavelength and EDFA self-filtering could reduce the buildup of SFPM side modes. For both channels, the power penalty measured at $10^{-9}$ BER was about 1 dB [90]. Another two-channel WDM experiment showed that for 1111 km

**Table 7.3  Direct-detection, multiple–channel linear transmission system experiments**

| | Year | Number of channels | Total bit rate (Gbit/s) | System length (km) | Total capacity (Tbit.km/s) | Channel separation Δt, Δλ or Δf | Modulation and system type [a] | Fiber type [b] | Number of EDFAs [c] | Launched signal power (dBm) [d] | Received signal power (dBm) [e] | Author et al. | Company | Ref. |
|---|---|---|---|---|---|---|---|---|---|---|---|---|---|---|
| 1 | 1989 | 16 | 9.95 | 0 | – | 5 GHz | ASK-FDM | – | 0 | – | –38 | H. Toba | NTT | [83][36] |
| 2 | 1990 | 4 | 9.6 | 459 | 4.40 | 2 nm | ASK-WDM | SMF | 6 | +12 | –30, –33 | H. Taga | KDD | [84] |
| 3 | | 4 | 6.8 | 70 | 0.47 | 4.3 nm | FSK-WDM | SMF | 1 | +3.5 | –33.5 | D. A. Fishman | AT&T | [50] |
| 4 | 1991 | 4 | 20 | 205 | 4.1 | 50 ps | ASK-TDM | DSF | 4 | +7 | –19.7 | G. E. Wickens | BTRL | [85] |
| 5 | 1992 | 4 | 20 | 137 | 2.74 | 50 ps | ASK-TDM | DSF | 1 | +7.5 | –15.8 | D. M. Spirit | BT | [86] |
| 6 | | 2 | 5 | 1,111 | 5.55 | 3.1 nm | ASK-WDM | DSF | 21 | +8 | –34 | P. M. Gabla | Alcatel-CIT | [86] |
| 7 | 1993 | 4 | 6.8 | 100 | 0.68 | 30 GHz | FSK-FDM | DSF | 1 | –0.9 | –35 | W. Y. Guo | Telecom. Lab. Taiwan | [88] |
| 8 | | 100 | 62.2 | (500)[f] | (31.1)[f] | 10 GHz | FSK-FDM | – | 6 | –10 | –37 | H. Toba | NTT | [89] |
| 9 | | 2 | 10 | 4,550 | 45 | 2 nm | ASK-WDM | DSF | 138 | – | –14, –15 | H. Taga | KDD | [90] |

[a]: TDM = time-domain multiplexing, WDM = wavelength-domain multiplexing, FDM = frequency-domain multiplexing, ASK = amplitude-shift-keying, FSK = frequency-shift-keying, [b] SMF = single mode fiber with $\lambda_o$ = 1.3 $\mu$m, DSF = single-mode fiber with $\lambda_o$ = 1.55 $\mu$m, [c]: not including post-amplifiers, [d]: including all channels [e]: per channel at $10^{-9}$ BER, [f]: potential performance.

EDF-based systems, the wavelength range over which signals can be tuned with less than 1 dB power penalty is 15 nm (P. M. Gabla et al. [87]). The 100 FDM channels experiment shown in Table 7.3 demonstrated gain equalization in a six-stage EDFA chain over an 8 nm bandwidth, using two in-line Mach–Zehnder filters (H. Toba et al. [89]). Transmission loss between the EDFAs was simulated by variable attenuators; output power equalization with 2.3 dB maximum deviation could be achieved with a 98 dB gain-compensated loss. This demonstrates the possibility of transmission over 500 km distance [89]. With an aggregate bit rate of 62 Gbit/s, the potential capacity of this system is thus 31 Gbit/s.km. But this result overlooks the potential effect of SFPM-induced crosstalk in an actual SMF- or DSF-based system. It remains to be investigated. The maximum channel capacity of such dense-FDM systems is also limited by the finite bandwidth of gain-equalized EDFAs, as well as crosstalk effects in filtering/demultiplexing (K. Inoue et al. [36], A. E. Willner [91]). Concerning dense-TDM systems, limitations imposed by chromatic dispersion and fiber nonlinearities suggest the use of soliton signals.

Considerable interest focuses on SMF-based systems because they form most of the fiber plant in the United States and many other countries. In the United States, the SMF plant represents more than 5 million miles. Operation at 1.55 $\mu$m signal wavelength and the deployment of EDFs would make it possible to upgrade the bit rate × distance capacity of this fiber network. But signal transmission at 1.55 $\mu$m in SMFs requires some means of *dispersion compensation (DC)*, as the zero dispersion wavelength in these fibers is at 1.3 $\mu$m. Fiber chromatic dispersion can be optically equalized by passive components such as reflective Fabry–Perot filters (A. H. Gnauck et al. [92]), two-mode fibers (C. D. Poole et al. [93]), or actively with an external modulator by signal predistortion or prechirping (N. Henmi et al. [94]). Another approach for DC uses at the receiver end a passive strand of fiber which has a very high negative dispersion coefficient (J. M. Dugan et al. [75]; H. Izadpanah et al. [59]). Experiments [59] and [75] achieved equalization with *dispersion-compensating fibers (DCF)* having coefficients $D = -45$ ps/nm.km and $D = -65$ ps/nm.km, respectively, at $\lambda = 1.55$ $\mu$m. The DC fiber length required depends on the SMF transmission distance. A 150 km SMF corresponds to nearly 2250 ps/nm total chromatic dispersion; using a DC fiber with coefficient $D = -45$ ps/nm.km, the required DCF length is therefore $L = 2250/45 = 50$ km. Because the absorption coefficient of DCFs can be as high as $\alpha_s = 0.48$ dB/km at $\lambda = 1.55$ $\mu$m [75], EDFAs are also required for loss compensation. In the previous example, the loss incurred in the DCF would be $-24$ dB. The experiment described in [59] used a single-stage DC followed by two EDFAs and 100 km of SMF. The experiment described in [75], used a three-stage DC with three EDFAs, in the sequence DC/EDFA/SMF/EDFA/DC/EDFA/DC, achieving compensation of 150 km SMF dispersion. Various parameters corresponding to SMF–DC– EDFA system experiments are summarized in Table 7.2. By early 1993, the highest capacity demonstrated through DC was 1.8 Tbit/s/km, corresponding to a 180 km SMF transmission distance at $B = 10$ Gbit/s (B. L. Patel et al. [66]).

Another possible approach for passive chromatic dispersion equalization uses *midsystem spectral inversion by SFPM* (R. M. Jopson et al. [65]). This approach, proposed nearly 15 years ago as a means of compensating optical signal dispersion, is based on the effect of nonlinear phase conjugation (A. Yariv et al. [95]). In this effect, a chirped signal at center frequency $\omega_s$ is mixed with a parametric pump at

$\omega'_s$, which produces a phase conjugate (idler) signal at frequency $\omega''_s = 2\omega'_s - \omega_s$. The idler is also chirped, but with a time reversed spectrum. This property can be explained as follows. Let $\omega_{sb} = \omega_s + \delta\omega$ the high frequency (blue-shifted) component of the incident signal pulse; parametric mixing instantaneously produces an idler component at frequency $\omega''_{s1} = 2\omega'_s - \omega_{sb}$. Consider next the low frequency (red-shifted) component $\omega_{sr} = \omega_s - \delta\omega$ of the signal pulse; the corresponding idler component has a frequency $\omega''_{s2} = 2\omega'_s - \omega_{sr}$. From these relations, we obtain $\omega''_{s1} = \omega''_s - \delta\omega$ and $\omega''_{s2} = \omega''_s + \delta\omega$, which shows that the chirp in the idler pulse is reversed with respect to that of the incident signal pulse. Figure 7.11 shows the experimental setup used to demonstrate this principle in the case of SMF systems [65]. The parametric mixing apparatus A is placed midway in the system. After propagation through distance $L/2$ in the SMF, the signal pulses at $\lambda = 1542.7$ nm incident in A have been chirped by chromatic dispersion; the redshifted component is in the trailing edge. The signal is then combined in A with a pump at $\lambda' = 1546.7$ nm, generated by an LD followed by an EDFA booster. Near-phase-matched parametric mixing occurs next in a 20 km DSF. The output idler wavelength is given by the relation $1/\lambda'' = 2/\lambda' - 1/\lambda$, or $\lambda'' = 1550.5$ nm. An interference filter tuned at $\lambda''$ and placed at the output of A rejects the unwanted components. An EDFA is used to boost the idler before propagation over the second SMF half of the system. As the idler pulse chirp is the reverse of that of the signal input to A, the idler pulse narrows back to the initial pulse width. The compensation is independent of the system length, as long as the phase conjugation apparatus A is placed midway in the system. This technique made it possible to demonstrate total SMF transmission distances up to 360 km and 400 km at $B = 10$ Gbit/s and $B = 2.5$ Gbit/s, respectively, with negligible power penalty [65]. The first result corresponds to a record capacity for 1.5 $\mu$m SMF systems of 3.6 Tbit/s·km.

We consider next the case of EDFA-based systems based on *coherent detection*. Section 3.6 analyzes the sensitivity of coherent receivers with optical preamplifiers for various modulation formats and heterodyning schemes. Section 3.7 analyzes the

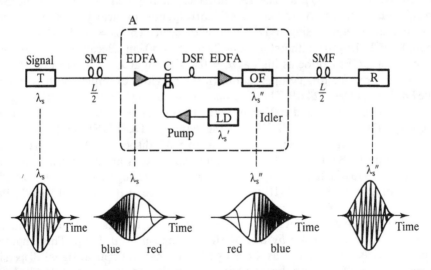

**FIGURE 7.11**   Dispersion compensation in SMF systems through mid-distance nonlinear optical phase conjugation; electric field amplitudes of the signal and idler are also shown [65].

case of coherent detection with optical amplifier chains. It shows that in coherent systems, the primary uses of EDFAs are as power boosters and in-line amplifiers. A numerical comparison of EDFA-based systems with either direct detection or coherent heterodyne detection shows that, for transmission lengths up to 10,000 km, the power penalties are similar, Figure 3.11 and 3.12. But for coherent systems, the back-to-back receiver sensitivities are higher, i.e., 20–25 dB. Thus, coherent systems can operate at much lower average signals powers, which reduces the deleterious effect of fiber nonlinearities. On the other hand, coherent systems are far more sensitive to such effects. In coherent PSK and FSK systems, a major source of impairment is SPM; in WDM applications, XPM and SFPM could also limit system performance. Aside from potential limitations due to fiber nonlinearities, a main advantage of coherent detection for EDFA-based systems is that narrowband ASE noise filtering is intrinsically achieved by LO mixing (T. J. Whitley et al. [96]). Interest in the field of EDFA-based coherent systems is motivated by the possibility of upgrading the capacity (or increasing the transmission distance) of currently existing unrepeatered links that use coherent receivers. A description of coherent lightwave communication systems and their performance prior to 1989, can be found in a review by R. A. Linke and A. H. Gnauck [97].

The distance and bit rate performance of EDFA-based heterodyne coherent systems, reported through the years 1989 to early 1993, is shown in Figure 7.12. The best performance reported to date for homodyne and heterodyne coherent systems without EDFAs is also shown for comparison (R. A. Linke and A. H. Gnauck [97]; J. M. Kahn et al. [98]; and A. H. Gnauck et al. [99]). EDFAs made it possible to increase the capacity of coherent systems by more than one order of magnitude. Relevant parameters and additional data corresponding to these system experiments are listed in Table 7.4, [100]–[113]. For EDFA-based coherent systems, the longest

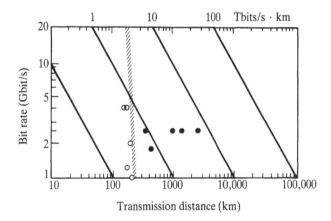

**FIGURE 7.12**  Maximum transmission distance and bit rates of coherent detection, non-regenerative EDFA-based linear systems reported to date (early 1993), corresponding to data shown in Table 7.4. The lines indicate constant system capacities of 0.1, 1, 10, and 100 Tbit/s.km. The shaded region corresponds to the maximum capacity range of nonregenerative coherent systems without optical amplifiers, assuming a transmitter power of 0 dBm, a fiber loss coefficient of $\alpha_s = 0.25$ dB/km, and receiver sensitivities of $\bar{n} = 9$–36 photons/bit. The open symbols correspond to best performance obtained in coherent systems without amplifiers.

**Table 7.4  Coherent transmission system experiments**

| | Year | Number of channels | Total bit rate (Gbit/s) | System length (km) | Total capacity (Tbit.km/s) | Signal λ (nm) | Modulation and system type [a] | Fiber type [b] | Number of EDFAs [c] | Launched signal power (dBm) [d] | Received signal power (dBm) [e] | Penalty (dB) [f] | Author et al. | Company | Ref. |
|---|---|---|---|---|---|---|---|---|---|---|---|---|---|---|---|
| 1 | 1989 | 1 | 0.565 | 0 | – | 1.536 | DPSK | – | 1 | – | –38 | –0.4 | T. J. Whitley | BTRL | [95] |
| 2 | 1990 | 1 | 2.5 | 400 | 1.0 | 1.537 | FSK | SMF | 2 | +5.1 | –26.3 | – | S. Saito | NTT | [100] |
| 3 | | 1 | 2.5 | 351 | 0.88 | 1.554 | CPFSK | SMF | 0 | +18.8 | –45 | –0.3 | T. Sugie | NTT | [101] |
| 4 | | 1 | 0.622 | 200 | 0.12 | 1.543 | DPSK | SMF | 1 | –5.3 | –48.1 | – | M.J.Creaner | BTRL | [102] |
| | | 1 | 0.622 | 200 | 0.12 | 1.543 | FSK | SMF | 1 | –1.3 | –41.1 | – | M.J.Creaner | BTRL | [102] |
| 5 | | 1 | 2.5 | **2223** | 5.56 | 1.554 | FSK | DSF | 25 | +8.8 | –42.0 | –4.2 | S. Saito | NTT | [103–104] |
| 6 | | 1(2) | 0.622 | 264 | 0.16(0.32) | 1.545 | DPSK | SMF | 2 | –8.3 | –48.1 | – | M.J.Creaner | BTRL | [105] |
| | | 1(2) | 0.622 | 264 | 0.16(0.32) | 1.543 | FSK | SMF | 2 | –4.3 | –44.1 | – | M.J.Creaner | BTRL | [105] |
| 7 | | 1 | 0.560 | **1028** | 0.57 | 1.536 | FSK | SMF | 10 | –1.5 | –42.4 | 0 | S. Ryu | KDD | [106] |
| 8 | | 1 | 0.565 | 218 | 0.12 | 1.532 | DPSK | SMF | 0 | +12.2 | –50.7 | 0 | B. Clesca | Alcatel-CIT | [107] |
| 9 | | 1 | 2.5 | **1360** | 3.4 | 1.554 | CPFSK | DSF | 16 | –12.3 | –35 | 10 | S. Saito | NTT | [112] |
| 10 | | 1 | 1.7 | 419 | 0.71 | 1.536 | FSK | SMF | 2 | +8.2 | –39 | –1 | Y. K. Park | AT&T | [108] |
| 11 | 1991 | 1(100) [g] | 0.622 | 300 | 0.18(18.6) [g] | 1.556 | DPSK | SMF | 4 | –15 | –47 | – | G. R. Walker | BTRL | [109] |
| 12 | | 1 | 2.5 | 364 | 0.91 | 1.554 | CPFSK | SMF | 0 | +19.1 | –45.8 | –1 | T. Sugie | NTT | [110] |
| 13 | 1992 | 1 | 2.5 | 310 | 0.77 | 1.552 | CPFSK | SMF | 0 | +18 | –42 | –1.9 | N. Ohkawa | NTT | [113] |
| 14 | | 1 | 1.19 | 195 | 0.23 | 1.548 | CPFSK | SMF | 0 | +13 | –41.8 | – | S. Ryu | KDD | [111] |
| 15 | | 1 | 2.5 | **2500** | 6.25 | 1.552 | CPFSK | DSF | 24 | +6 | –39.8 | –6.7 | S. Saito | NTT | [79] |

[a]: ASK, FSK, PSK = amplitude-, frequency-, phase-shift-keying, WDM = wavelength-domain multiplexing, FDM = frequency-domain multiplexing, [b] SMF = single-mode fiber with $\lambda_o$ = 1.3 $\mu$m, DSF = single-mode fiber with $\lambda_o$ = 1.55 $\mu$m, [c]: not including post-amplifiers, [d]: including all channels [e]: per channel at $10^{-9}$ BER, [f]: at $10^{-9}$ BER, [g]: potential.

transmission distance achieved to date is $L_{tot} = 2500$ km and the highest bit rate for any distance is $B = 2.5$ Gbit/s, a performance lower than for direct detection systems. It reflects that long-haul coherent systems suffer from greater impairments, as shown experimentally (S. Saito et al. [79]). In coherent systems, the main source of impairment is SPM nonlinearity, to which direct detection systems are basically immune [79]. Two other sources of phase noise that affect the performance of coherent systems are the ASE noise quadrature component (Section 5.7) and ASE-noise-induced XPM nonlinearity.

Concerning SMF-based systems with transmission distances in the 300–400 km range, a comparison between Tables 7.3 and 7.4 shows that, as expected, coherent detection provides greater receiver sensitivities and power margins. The longest transmission distances achieved to date in coherent SMF-based systems are 419 km ($B = 1.7$ Gbit/s) and 1028 km ($B = 0.56$ Gbit/s, [106] and [108]. In direct detection SMF-based systems, the maximum distances are 450 km ($B = 5$ Gbit/s) and 710 km ($B = 2.4$ Gbit/s), [49] and [57], which show a capacity advantage (2.25 Tbit/s.km).

We discuss now the effect of *fiber nonlinearities*. In silica fibers, they belong to four basic types: stimulated Raman ((SRS) and Brillouin (SBS) scattering, the optical Kerr effect, and stimulated four-photon mixing (SFPM). The effects of SRS and SBS cause frequency conversion (hence signal excess attenuation), the Kerr effect causes SPM and XPM (hence signal phase noise), and SFPM causes frequency mixing (hence WDM crosstalk). A description of these effects and related system limitations can be found in the review by A. R. Chraplyvy [82].

Consider first SRS. For an $N$-channel WDM system at $1.55\,\mu$m wavelength, a penalty of less than 1 dB is incurred if the maximum input power $P_s$ per channel verifies (A. Chraplyvy, [82] and [116]):

$$NP_s \times (N - 1)\Delta\lambda < 4.10^3 \left(\frac{\omega_{\mu m}}{4}\right)^2 \text{mW.nm} \tag{7.12}$$

where $\omega$ is the signal field mode size and $\Delta\lambda$ the channel separation. For instance, assuming an equalized EDFA gain bandwidth of 10 nm, $N = 10$ channels, the maximum power per channel in DSF ($\omega$(field) 3.2 $\mu$m) is found to be $P_s = 2.7$ mW.

Consider next SBS. The backscattered Stokes power depends on the signal modulation format and the bit rate [82]. Indeed, the SBS gain coefficient is given by:

$$g = g_B \left\{ \frac{1}{2} - \frac{B}{4\Delta\nu_B} \left[ 1 - \exp\left(-\frac{\Delta\nu_B}{B}\right) \right] \right\} \tag{7.13}$$

for ASK (NRZ) and FSK, while for PSK (suppressed carrier modulation) it is given by:

$$g = g_B \left\{ 1 - \frac{B}{\Delta\nu_B} \left[ 1 - \exp\left(-\frac{\Delta\nu_B}{B}\right) \right] \right\} \tag{7.14}$$

In Eqs. (7.13) and (7.14), $g_B$ and $\Delta\nu_B$ are the Brillouin gain coefficient and bandwidth, respectively and (Section 5.12). As for silica fibers the Brillouin gain bandwidth is $\Delta\nu_B \leqslant 50$ MHz, which gives $B/\Delta\nu_B \geqslant 20$, the effective SBS coefficient $g$ at bit rate $B = 1$ Gbit/s is reduced at least by a factor of 4 (ASK, FSK) to a factor of 40 (PSK), in comparison to the CW, unmodulated case ($g = g_B$). Chapter 5 shows that in the case of DSF, the unmodulated SBS gain is about 5 dB per km and mW of signal power, Eq. (5.153). Thus, for ASK signals at $B = 1$ Gbit/s, the gain is reduced to 1.25 dB/km/mW, which over long transmission distances, corresponds to significant SBS. But complete SBS suppression can be achieved by concatenating fiber segments having different Brillouin shifts (T. Sugie et al. [110]; T. Sugie [114]). The length of each segment is chosen such that SBS power conversion for each stage remains negligible.

Consider next the optical Kerr effect. In the case of SPM and XPM, the phase noise $\delta\phi$ induced by random signal power fluctuations $\delta P$ is given by:

$$\delta\phi = \frac{5}{6}\frac{2\pi n_2}{\lambda}\frac{L_{int}}{A_{eff}}\delta P \tag{7.15}$$

where $n_2 \approx 3.10^{-20}$ m²/W for SPM and $n_2 \approx 6.10^{-20}$ m²/W for XPM [82], $L_{int}$ is the effective interaction length and $A_{eff}$ is the effective area ($A_{eff} = \pi\omega^2$(field)). The factor 5/6 reflects the effect of polarization scrambling. Indeed, the nonlinear refractive index change for circular polarization is two thirds the change for linear polarization; the average coefficient is therefore $(5/6)n_2$ (R. H. Stolen and C. Lin [115]). The effect of XPM is twice that of SPM, i.e., $\delta\phi_{XPM} = 2\delta\phi_{SPM}$. In lossy or passive fibers, the effective interaction length is given by $L_{int} = [1 - \exp(-\alpha_s L)]/\alpha_s$, where $L$ is the propagation distance (for short pulses at different wavelengths, the effect of walk-off must also be taken into account in the determination of $L_{int}$). For long transmission lengths (i.e., $L \gg 1/\alpha_s$), we have $L_{int} \approx 1/\alpha_s$. For signals at $\lambda = 1.55$ μm with absorption coefficient $\alpha_s = 0.25$ dB/km, the SPM-induced phase noise reduces from Eq. (7.15) to:

$$\delta\phi_{SPM} \approx 0.035\left(\frac{4}{\omega_{\mu m}}\right)^2 \delta P_{mW} \tag{7.16}$$

which is the result shown in [82]. In EDFA-based systems with $k$ amplification stages, the accumulated phase noise due to SPM is $k\delta\phi_{SPM}$. For instance, a DSF-based system ($\omega \approx 3.3$ μm) with $k = 100$ stages would produce an accumulated SPM phase noise $k\delta\phi_{SPM} = \pi$ for a signal power fluctuation $\delta P = 0.5$ mW. In WDM systems, the phase noise $\delta\phi_{tot}$ produced by both SPM and XPM in a given channel is defined by the quadrature sum (A. R. Chraplyvy et al. [116]):

$$\delta\phi_{tot} = \sqrt{\delta\phi_{SPM}^2 + \sum_{i=2}^{N}\delta\phi_{XPM}^2(i)} \tag{7.17}$$

where $\delta\phi_{XPM}(i)$ is the XPM phase noise induced from channel $i$. Equation (7.17) neglects the Raman contribution to the nonlinear index [116] and any pulse walk-off effects. Assuming that all channels have the same intensity noise $\delta P(i)$, and using the property $\delta\phi_{XPM} = 2\delta\phi_{SPM}$, we obtain from Eq. (7.17):

$$\delta\phi_{tot} \approx 2\sqrt{N-1}\,\delta\phi_{SPM} = 0.07\left(\frac{4}{\omega_{\mu m}}\right)^2\sqrt{N-1}\,\delta P_{mW} \tag{7.18}$$

This result shows that the phase noise induced by XPM increases only like the square root of the number of channels [82]. In the previous example ($k = 100$, DSF system) and assuming $N = 10$ channels, a total phase noise $k\delta\phi = \pi$ is produced by power fluctuations $\delta P = 0.05\,\text{mW}$ in each channel. For EDFA-based systems, ASE noise is present in each channel and is the main source of power fluctuation and XPM. We need to carefully evaluate the effect of ASE-induced XPM.

Considering the case of a single-channel system, the ASE-induced XPM phase noise, caused only by the in-phase component of ASE, is given by the expression (J. P. Gordon and L. F. Mollenauer [117]):

$$\langle \delta\phi^2_{\text{ASE-XPM}} \rangle = \frac{2}{3} n_{\text{sp}} h\nu B F_{\text{path}} \alpha_s L_{\text{tot}} \left( \frac{2\pi n_2 L_{\text{tot}}}{\lambda A_{\text{eff}}} \right)^2 P_{\text{ave}} \tag{7.19}$$

where $F_{\text{path}}$ is the penalty factor function, Eq. (2.204), defined by:

$$F_{\text{path}}(G) = \frac{1}{G} \left( \frac{G-1}{\log G} \right)^2 \tag{7.20}$$

and $P_{\text{ave}}$ is the path averaged signal power along the chain, given by:

$$P_{\text{ave}} = P_s^{\text{out}} \frac{G-1}{G \log G} \tag{7.21}$$

In Eqs. (7.19)–(7.21), $B$ is the bit rate, $L_{\text{tot}}$ is the total system length, $G$ is the amplifier gain in each segment of the chain, with $\alpha_s L = \log G$ ($L$ = amplifier spacing), and $P_s^{\text{in}}$ is the signal power input to the system ($P_s^{\text{out}} = GP_s^{\text{in}}$). The phase noise defined in Eq. (7.19) corresponds to an equivalent line width broadening of (A. Yariv [118]):

$$\delta\nu_{\text{eq}}(\text{ASE–XPM}) = \langle \delta\phi^2_{\text{ASE-XPM}} \rangle \frac{B}{2\pi} \tag{7.22}$$

Section 5.7 derives the signal phase noise $\langle \delta\phi^2_{\text{EDFA}} \rangle$ induced by the quadrature component of ASE in each EDFA. The corresponding equivalent line width broadening is given by Eq. (5.77) multiplied by the number of EDFAs ($k = L_{\text{tot}}/L$). With the definitions of Eqs. (7.20) and (7.21), this line width broadening can be put into the form:

$$\delta\nu_{\text{eq}}(\text{EDFA}) = k\langle \delta\phi^2_{\text{EDFA}} \rangle \frac{B}{2\pi} = \frac{n_{\text{sp}} h\nu B^2}{4\pi P_{\text{ave}}} F_{\text{path}} \alpha_s L_{\text{tot}} \tag{7.23}$$

Considering that the in-phase and quadrature components of ASE are statistically independent, the total line width broadening is then given by the sum $\delta\nu_{\text{eq}}(\text{ASE–XPM}) + \delta\nu_{\text{eq}}(\text{EDFA})$. From Eqs. (7.19), (7.22), and (7.23), this takes the final form (S. Ryu [119]):

$$\delta\nu_{\text{eq}}(\text{total}) = \frac{n_{\text{sp}} h\nu B^2}{4\pi} F_{\text{path}} \alpha_s L_{\text{tot}} \left\{ \frac{1}{P_{\text{ave}}} + \frac{4}{3} \left( \frac{2\pi n_2 L_{\text{tot}}}{\lambda A_{\text{eff}}} \right)^2 P_{\text{ave}} \right\} \tag{7.24}$$

For sufficiently high average powers $P_{\text{ave}}$, ASE-induced XPM is the dominant cause of line width broadening. The effect increases as the third power of the total system length $L_{\text{tot}}$, or the number of amplifiers $k = L_{\text{tot}}/L$. A numerical application of Eq. (7.24) shows that for a system with length $L_{\text{tot}} = 10,000$ km, amplifier spacing $L = 100$ km, 1480 nm pumped EDFAs ($n_{\text{sp}} = 1.6$), bit rate $B = 2.5$ Gbit/s and EDFA output power $P_s^{\text{out}} = +10$ dBm, the line width broadening is about 1 GHz [119]. In CPFSK coherent systems, this effect is responsible for large power penalties, as shown in the detailed study of [119]. Theoretical and experimental studies have also shown that the Kerr nonlinearity produces a line shape deformation of the signal spectrum (M. Murakami and S. Saito, [120] and [121]; S. Ryu [122]). The resulting spectrum is the sum of two Lorentzians; one which corresponds to the original signal, the other corresponds to the broadband phase noise induced by Kerr nonlinearity. This deformation is a cause of power penalty for coherent systems and also for direct detection systems that use narrowband optical filtering [121]. Deleterious spectral changes are also produced by the interplay between Kerr nonlinearity and chromatic dispersion [122].

Consider finally parametric wave mixing or stimulated four-photon mixing (SFPM), which occurs in WDM systems. Wave mixing is explained by the nonlinear response of an optical medium under intense field excitation $E$, which takes the form of the third order nonlinear polarization:

$$P_{\text{NL}} = \chi^{(3)} EEE \qquad (7.25)$$

where $\chi^{(3)}$ is the third-order nonlinear susceptibility. If only one signal field excitation is involved, i.e., $E(\omega, t) = E_0 \exp(i\omega t) + E_0^* \exp(-i\omega t)$, the tensor product in Eq. (7.25) produces oscillating terms at frequencies $\omega$ and $3\omega$ (third harmonic). When two signal fields at frequencies $\omega_1$ and $\omega_2$ are present, i.e., $E(\omega, t) = E_1(\omega_1, r) + E_2(\omega_2, t)$, the third order product generates new frequency components or tones, which take the form $2\omega_1 - \omega_2$ and $2\omega_2 - \omega_1$, as shown in Figure 7.13a. These two tones, which result from the nonlinear wave mixing of the two incident signals, form sidebands located at a distance corresponding to the signal frequency separation $\Delta\omega$. If three fields are mixed, nine new components are formed, which have frequencies $\omega_i + \omega_j - \omega_k$ ($i, j, k = 1, 2, 3$), as shown in Figure 7.13b. If the three signals are not equally spaced, i.e., $\omega_2 - \omega_1 \neq \omega_3 - \omega_2$, there are nine different tones, at frequencies distinct from the signal frequencies. Figure 7.13c shows the case where the three signals are equally spaced, i.e., $\omega_2 - \omega_1 = \omega_3 - \omega_2 = \Delta\omega$. Two pairs of tones have degenerate frequencies, while three tones have the frequencies of the input signals. Thus, nonlinear parametric mixing causes not only *power dissipation* (or excess attenuation), due to the generation of sidebands, but also results in channel interference or *crosstalk*. This crosstalk can also occur in the two-channel case. Considering Fig. 7.13a, the four waves at $2\omega_1 - \omega_2$, $\omega_1$, $\omega_2$ and $2\omega_2 - \omega_1$ can mix again to form new tones at $2\omega_1 - (2\omega_1 - \omega_2) = \omega_2$ and $2\omega_2 - (2\omega_2 - \omega_1) = \omega_1$, which correspond to the two initial signals. The effect of crosstalk is explained by the fact that such tones appear only when the two signals are simultaneously turned on (1 symbols in ASK), which results in the duplication of one channel information into the other, as a noise background. Figure 7.13c shows for instance that crosstalk occurs in the channel at $\omega_1$ only when there is a 1 symbol in both channels at $\omega_2$ and $\omega_3$, and so on.

The effect of crosstalk induced by SFPM is illustrated in Figure 7.14 (K. Inoue

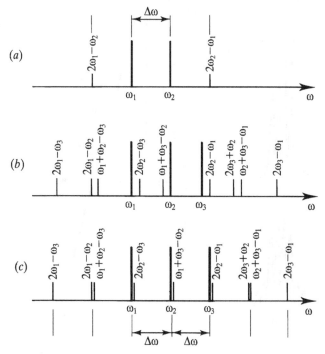

**FIGURE 7.13** Effect of parametric wave mixing showing the generation of multiple sideband tones in the nonlinear interaction of: (a) two signals with frequency separation $\Delta\omega$, (b) three signals with unequal frequency separation, and (c) three signals with equal frequency separation $\Delta\omega$, where some tones have degeneracies.

and H. Toba [123]). Figure 7.14a shows the input signal spectrum with four FSK modulated channels near $\lambda = 1.55 \, \mu m$ with 10 GHz separation. The side modes seen on the left are an artifact of spectrum analyzer misalignment and do not actually exist. After EDFA postamplification, the input power per channel was between 5 mW and 8 mW. The output spectrum obtained after 26 km DSF transmission, shown in Figure 7.14b, exhibits the characteristic sidebands produced by SFPM. Figure 7.14c shows the BER degradation or crosstalk, as measured when all four channels are modulated. A small power penalty is also observed when the three other channels are not modulated. This feature can be explained by the excess shot noise produced by the SFPM tones falling into the measured channel [123].

SFPM is a phase dependent interaction. Indeed, for efficient power conversion into the third order tones, the nonlinear polarization shown in Eq. (7.25) must be in phase with the corresponding fields. For the tone at frequency $2\omega_2 - \omega_1$, the phase matching condition is:

$$\beta(2\omega_2 - \omega_1) \approx 2\beta(\omega_2) - \beta(\omega_1) \Leftrightarrow \Delta\beta(\omega_2, \omega_2, \omega_1) = \beta(2\omega_2 - \omega_1) - 2\beta(\omega_2) - \beta(\omega_1) \approx 0$$

$$(7.26)$$

where $\beta(\omega_i)$ is the propagation constant at frequency $\omega_i$. Equation (7.26) means that the tone field at frequency $\Omega = 2\omega_2 - \omega_1$ (with phasor $\exp\{i[\Omega t - \beta(\Omega)z]\}$) is at any

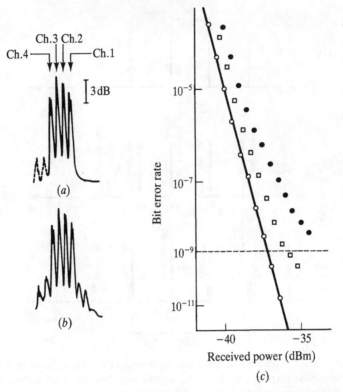

**FIGURE 7.14** Effect of parametric wave mixing between four FSK channels: (*a*) input spectrum, showing the four modulated channels, (*b*) output spectrum after 26 km transmission through DSF, showing sidebands, and (*c*) BER measurements before transmission (open circles), after transmission with all other channels modulated (closed circles), and after transmission with all other channels unmodulated (squares). From [123] © 1991 IEEE.

fiber coordinate $z$ in phase with the nonlinear polarization at the same frequency (with phasor $\exp\{i\Omega t - i[2\beta(\omega_2) - \beta(\omega_1)]z\}$. For a fiber length $L$ with loss coefficient $\alpha_s$, the power conversion efficiency $\eta_{\text{SFPM}}$ of the SFPM process takes the form (K. Inoue [124])

$$\eta_{\text{SFPM}} = \frac{\alpha_s^2}{\alpha_s^2 + \Delta\beta^2}\left\{1 + 4\frac{\exp(-\alpha_s L)\sin^2(\Delta\beta L/2)}{[1 - \exp(-\alpha_s L)]^2}\right\} \tag{7.27}$$

where $\Delta\beta$ is the phase mismatch between the interacting waves. The conversion efficiency is $\eta = 100\%$ for $\Delta\beta = 0$. The power generated in a third order tone at frequency $\omega_i + \omega_j - \omega_k$ with degeneracy $d$ ($d = 3$ for $i = j$, $d = 6$ for $i \neq j$) is then given by the expression [124]:

$$P_s^{\text{out}}(\omega_i + \omega_j - \omega_k) = \eta_{\text{SFPM}}^{ijk}\frac{1024\pi^6}{n^4\lambda^2 c^2}\left(\frac{L_{\text{eff}}}{A_{\text{eff}}}\right)^2 (d\chi^{(3)})^2\exp(-\alpha_s L)P_s^{\text{in}}(\omega_i)P_s^{\text{in}}(\omega_j)P_s^{\text{in}}(\omega_k)$$

$$\tag{7.28}$$

where $\eta_{\text{SFPM}}^{ijk}$ is the conversion efficiency corresponding to the phase mismatch $\Delta\beta(\omega_i, \omega_j, \omega_k)$, $P_s^{\text{in}}(\omega_m)$ is the signal power at frequency $\omega_m$ ($m = i, j, k$), and $L_{\text{eff}} = [1 - \exp(-\alpha_s L)]/\alpha_s$ is the effective interaction length (for the definition and derivation of the effective interaction area $A_{\text{eff}}$ of SFPM in single-mode fibers, see R. H. Stolen and J. E. Bjorkholm [125]). In silica fibers, the SFPM susceptibility is $\chi^{(3)} \approx 5.10^{-14}$ esu (K. O. Hill et al. [126]). The theory leading to these results does not take into account the effects of power depletion in the source channels (parametric approximation) and of fiber polarization scrambling.

SFPM nonlinearity is nearly phase matched near the zero dispersion wavelength. Thus, for closely spaced WDM systems with DSF, high SFPM conversion efficiencies can be obtained over a relatively long coherence length $2\pi/\Delta\beta$ (which represents a spatial oscillation period $\eta_{\text{SFPM}}^{ijk}$). Detailed studies of phase matching in DSF are: S. J. Garth [127]; R. A. Sammut and S. J. Garth [128]; N. Shibata et al., [129] and [130]; M. W. Maeda et al. [131]; and references therein. The phase mismatch $\Delta\beta(\omega_i, \omega_j, \omega_k)$ can be expressed as a function of the fiber chromatic dispersion coefficient $D(\lambda)$ through [129]:

$$\Delta\beta(\omega_i, \omega_j, \omega_k) = \frac{2\pi\lambda^2}{c} \Delta f_{ik}\Delta f_{jk}\left\{D(\lambda) + (\Delta f_{ik} + \Delta f_{jk})\frac{\lambda^2}{2c}\frac{\partial D(\lambda)}{\partial\lambda}\right\} \qquad (7.29)$$

where $\Delta f_{lm} = |f_l - f_m|$. Detailed theoretical and experimental studies of coherent FDM system power and capacity limitations, caused by near-phase-matched SFPM, can be found in [130] and [131].

In the case of optical amplifier chains based on DSF transmission, the interaction length of near-phase-matched SFPM can be extended to several thousand kilometers, which greatly enhances its otherwise weak effect. The SFPM conversion efficiency in optical amplifier chains (Type C) takes the form (K. Inoue [124]):

$$\eta_{\text{SFPM}} = \frac{1}{k^2}\frac{\alpha_s^2}{\alpha_s^2 + \Delta\beta^2}\frac{\sin^2(k\Delta\beta L/2)}{\sin^2(\Delta\beta L/2)}\left\{1 + 4\frac{\exp(-\alpha_s L)\sin^2(\Delta\beta L/2)}{[1 - \exp(-\alpha_s L)]^2}\right\} \qquad (7.30)$$

where $k$ is the number of optical amplifiers in the chain ($L$ = amplifier spacing). The power generated in the third order tone is given by Eq. (7.28) multiplied by $k^2$. When the interaction is phase-matched ($\Delta\beta = 0$ and $\eta_{\text{SFPM}} = 1$), the absolute SFPM power conversion is $k^2$ times greater than in a system of length $L$ without amplifiers. But as the channel separation is increased, the condition $\Delta\beta = 0$ is not realized and the conversion efficiency rapidly drops with weak oscillations; with the parameters $D = 2$ ps/nm.km, $\lambda = 1.55$ $\mu$m, $\alpha_s = 0.2$ dB/km, $k = 10$, and $L = 50$ km, the conversion efficiency is about 50% for a channel spacing of 10 GHz, 10% for a channel spacing of 50 GHz, and less than 1% for a channel spacing of 100 GHz [124].

The detailed theoretical investigation of non-soliton pulse propagation in long-haul WDM systems leads to three major conclusions: (1) in a system made of DSF with uniform dispersion characteristics, catastrophic buildup of SFPM occurs when one WDM channel is exactly at the zero dispersion wavelength $\lambda_0$; (2) transient collisions of pulses at different wavelengths away from $\lambda_0$ generate spurious SFPM tones, but these tones decay as the pulses separate; (3) a two-channel, 2.5 Gbit/s WDM system 7500 km long is feasible, provided the signals are not closer than 0.5 nm to $\lambda_0$ and are spaced by 2–3 nm (D. Marcuse et al. [132]). The effect of channel spacing on SFPM efficiency was investigated in several WDM system experiments (D. A. Cleland et al., [133] and [134]. Several studies also show that SFPM can be reduced through an adequate choice of the amplifier spacing (D. G. Schadt [135]; C. Bungarzeanu [136]) and of the power assigned to each channel, which depends on the amplifier spacing and the wavelength spacing (C. Kurtzke and K. Petermann [137]).

A proposed method for efficient SFPM suppression is to use WDM signals having unequal frequency separations (F. Forghieri et al. [138]). This approach is illustrated in Figure 7.15, which shows theoretical simulations for a 10-channel WDM system operating at $B = 10$ Gbit/s ($L = 50$ km, $L_{tot} = 500$ km) [138]. In Figure 7.15a, the channels are equally spaced by $\Delta f = 125$ GHz or $\Delta \lambda = 1$ nm; the corresponding output spectrum at $L_{tot} = 500$ km exhibits a strong SFPM conversion, with the generation of multiple third order products or tones. The occurrence of these tones in each channel causes a high level of crosstalk, as shown by the closure of the simulated eye diagram. If, on the other hand, the channel frequencies are distributed with unequal and optimized spacings, Figure 7.15b, the output spectrum clears from SFPM tones and the eye diagram is almost completely opened [138]. This SFPM

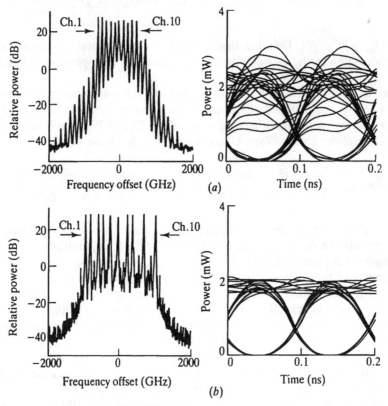

**FIGURE 7.15**  Suppression of SFPM in EDFA-based WDM systems: (a) theoretical output spectrum in a 500 km system with 10 equally spaced channels (125 GHz), and corresponding eye diagram; (b) theoretical output spectrum in the same system with 10 unequally spaced channels, and corresponding eye diagram. From [138], reprinted with permission from the Optical Society of America, 1993.

suppression techniques requires a larger bandwidth (16 nm in this example), within the EDFA capacity. A second requirement concerns the stability of the channel absolute frequencies, which must be controlled within 1 GHz; this control can be achieved by locking the channels to the modes of a suitable Fabry–Perot filter [138].

The previous overview of system limitations due to fiber nonlinearities illustrates the actual complexity of EDFA-based transmission systems. This brief description, however, would not be complete without mentioning potential limitations related to *signal polarization.*

A first limitation is caused by the effect of *polarization mode dispersion (PMD).* In single-mode fibers, the effect of PMD is caused by unwanted birefringence, due to internal /external stress and core ellipticity. In the case of DSF, the PMD coefficient is typically 0.1 ps$\sqrt{\text{km}}$ (Y. Namihira et al. [139]). Additional PMD is caused by various optical components in the system, such as optical isolators, wavelength selective couplers, filters, and EDFs. An experimental measurement made in a 4564 km 138-EDFA system has shown that the total PMD coefficient of the system varies with length between 0.2 ps$\sqrt{\text{km}}$ and 0.3 ps/$\sqrt{\text{km}}$ (Y. Namihira et al. [140]). This result makes it possible to estimate the total dispersion (or differential group delay DGD) of a 10,000 km system to be 20–30 ps. This dispersion effect, is of negligible consequence for bit rates near 1–5 Gbit/s but could cause power penalties at bit rates near 10 Gbit/s or higher. In the system described in [139], each of the 138 EDFAs had an optical isolator and a wavelength selective coupler with low DGD characteristics of 0.79 ps and 0.002 ps, respectively. Other measurements have shown, however, that PMD can largely vary between commercially available components. In EDFs, the PMD coefficient can vary between 0.75 ps/$\sqrt{\text{km}}$ and 7.5 ps/$\sqrt{\text{km}}$, one to two orders of magnitude greater than that of DSF; this large PMD is attributed to an increased birefringence effect in the drawing process of small-core fibers (A. Galtarossa and M. Schiano [141]). In optical isolators, the measured DGD varies from less than 0.1 ps to more than 9 ps [141]. This shows that a single isolator can have a DGD equivalent to that of 8000 km DSF. Significant system limitations could result when using optical components and EDFA modules that are not PMD optimized.

Another detrimental effect in EDFA-based systems is *polarization dependent loss (PDL).* Experimental measurements have shown that a PDL as low as 0.125 dB per amplifier span results in BER floors and severe reduction in transmission distances, with failure at $L_{tot} = 6000$ km (D. J. Malyon et al. [142]). The time-evolution of the overall system PDL (fading) causes BER fluctuations, observed in a 4500 km experimental system (S. Yamamoto et al. [62]). This fading effect is not attributed to fluctuations in received signal power, but in output SNR [62]. Another cause of fading is *polarization hole burning* Section 4.10. Recent measurements obtained in long-haul EDFA-based systems have shown that a polarized signal probe experiences higher gain in the polarization direction where the signal power is the lowest, regardless of its orientation (M. G. Taylor [143]). In an actual 3100 km system, the effect was measured to correspond to an active PDL of 0.07 dB per amplifying section [143]. The consequence of active PDL is SNR fading. SNR fading can be explained by the fact that in each EDFA, the gain is adjusted for exact signal loss compensation therefore the ASE in the polarization transverse to the signal experiences higher throughput gain. The time dependent changes in the signal polarization cause SNR fluctuations. A solution to this problem is signal polarization modulation using acousto-optic modulators [143].

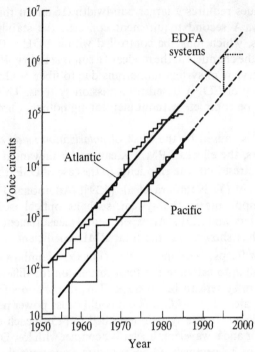

**FIGURE 7.16** Growth in voice circuit capacity of transatlantic and transpacific systems. The projected capacity of EDFA-based systems, to be deployed in 1995–1996, is also shown. From [144] ©1990 IEEE.

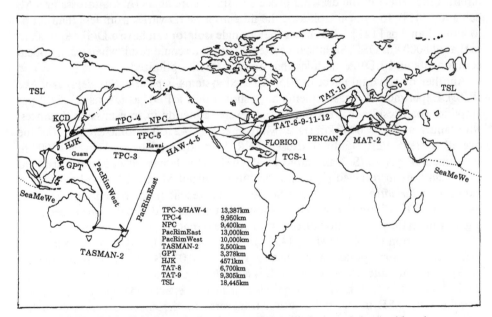

**FIGURE 7.17** Main lightwave systems currently installed around the world or in progress. The transatlantic and transpacific links TAT-12 and TPC-5, to be deployed in 1995–1996, will be nonregenerative EDFA-based systems.

The transmission capacity of long-haul, nonregenerative EDFA-based linear transmission systems is inherently limited by fundamental physical principles, e.g. SNR degradation, fiber nonlinearities. But their performance largely exceeds the performance of current electronic repeater technology. As the traffic demand in both transatlantic and transpacific communications steadily increases at a yearly rate of 25% (Figure 7.16 [144]), the technology of nonregenerative linear systems has appeared in time to meet it.

Figure 7.17 is a map of the main lightwave links currently installed around the world or in progress, [144] and [145]. As the inset in the figure shows, the transmission distances of the longest links fall in the 6700–18,500 km range. This corresponds to the capability of EDFA-based systems, with bit rates between 2.5 Gbit/s and 10 Gbit/s. The deployment of the first transoceanic EDFA-based systems in the Atlantic and the Pacific oceans (TAT-12 and TPC-5) is planned for 1995 and 1996. At bit rates $B = 2.0$ Gbit/s, such systems will have a minimum capacity of at least 120,000 voice circuits per fiber pair [146]. Increasing the line bit rates to $B = 10$ Gbit/s, the system capacity could then be upgraded to more than 600,000 voice circuits.

## 7.3  SOLITON SYSTEMS

We review the performance characteristics of *soliton systems*, also referred to as *nonlinear systems*. They are based on direct detection with optical soliton pulses as RZ data. Compared to linear systems, soliton systems have a greater potential capacity, both in bit rate and transmission distance. Actually, certain all-optical regeneration techniques that can be implemented only with solitons make it possible to limit the effect of SNR degradation. In soliton systems, therefore, the transmission distances can be increased indefinitely. We refer to soliton systems that use these techniques as *active*. In contrast, *passive* soliton systems use the same components as linear direct detection systems, except for the transmitter.

The maximum transmission distances and bit rates achieved by early 1993 in all types of soliton systems, including active or passive, DSF or SMF, single-channel or WDM, and for which BER $\leqslant 10^{-9}$ are plotted in Figure 7.18. Experimental parameters for passive soliton experiments are summarized in Table 7.5, [147]–[177].

The results form two groups: straight-line systems with bit-rates ranging from 4 Gbit/s to 80 Gbit/s and transmission distances up to 1200 km; and loop experiments with bit rates ranging from 2 Gbit/s to 20 Gbit/s and transmission distances from 4200 km up to $10^6$ km. In all cases, bit rates greater than 10 Gbit/s correspond to WDM, TDM, or WDM–PDM experiments. This does not imply any potential limitations for high speed, single channel soliton systems; it can be explained by the lack of BER test sets operating at such rates. Likewise, the shorter transmission distances reported for straight-line systems can also be explained by limitations in available fiber lengths, rather than in potential system performance. The shaded area is the Gordon–Haus limit, assuming DSF with $L = 40$ km amplifier spacing and a dispersion coefficient range $\bar{D} = 0.5$–5 ps/nm.km. This limit is calculated from Eq. (5.141) wih the following parameters: $\kappa = B\Delta T = 0.1$, $\kappa_w = Bt_w = 1/3$, $A_{eff} = 35$ $\mu$m², $n_2 = 3.2 \times 10^{-20}$ m²/W, and $n_{sp} = 1.7$. The single-channel data fall below or near this limit, except for the point corresponding to a 5 Gbit/s, 15,000 km loop experiment

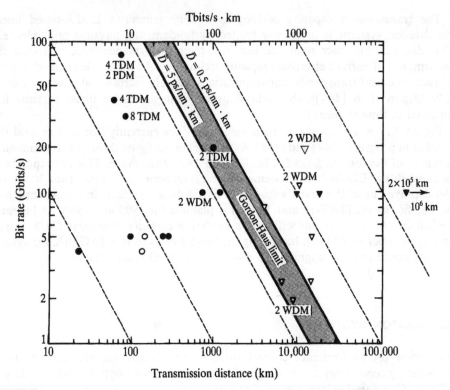

**FIGURE 7.18**   Maximum transmission distances and total bit rates of soliton systems reported with BER $\leqslant 10^{-9}$; circles correspond to straight-line transmission (open symbol = SMF, closed symbol = DSF), while triangles correspond to loop experiments (open symbol = passive system, closed symbol = active system); in the case of WDM, TDM and PDM experiments, the number of channels is also indicated, [147]–[177]. The dashed lines indicate constant system capacity. The shaded area corresponds to the Gordon–Haus limit, assuming DSF with 40 km amplifier spacing and a range of dispersion coefficient of $\bar{D} = 0.5$–5 ps/nm.km.

(L. F. Mollenauer et al. [173]); this difference corresponds to a factor of two in the allowed bit rate and can be explained by the use of different parameter values for $\kappa$ and $\kappa_w$, as well as a shorter amplifier spacing ($L = 27$ km).

So far, the highest capacity achieved in straight-line soliton systems is 37 Tbit/s.km, which corresponds to a transmission distance of 1850 km at a total WDM bit rate $B = 20$ Gbit/s (M. Nakazawa et al. [160]). In loop experiments and in the case of passive soliton systems, on the other hand, the highest capacity achieved to date is 110 Tbit/s.km, which corresponds to a transmission distance of 11,000 km at a total WDM bit rate $B = 10$ Gbit/s (L. F. Mollenauer et al. [173]). Table 7.5 shows that since the first BER measurements in soliton systems in 1990 (K. Iwatsuki et al., [148] and [151]), the transmission capacity has increased by nearly two thousand times. In *active* soliton systems, the notion of maximum system capacity becomes meaningless, as the transmission capacity has increased by nearly two thousand times. In active soliton systems, the notion of maximum system capacity becomes meaningless, as the transmission distance is virtually unlimited. Indeed, the experimental measurement of $10^{-9}$ BER at $B = 10$ Gbit/s for $L_{tot} = 10^6$ km (M. Nakazawa et al. [176]),

**TABLE 7.5 Passive soliton system experiments**

| | Year | Number of channels [a] | Total bit rate (Gbit/s) [b] | System length (km) [c] | Total capacity (Tbit.km/s) | Pulse width (ps) [d] | Dispersion coeff. $\bar{D}$ (ps/nm.km) | Signal wavelength (nm) [e] | Channel separation (nm) | Number of EDFAs [f] | Launched signal power (mW) [g] | Received signal power (dBm) [h] | Author et al. | Company | Ref. |
|---|---|---|---|---|---|---|---|---|---|---|---|---|---|---|---|
| 1 | 1988 | 1 | 33 GHz | 4,000(SMF)-LOOP | – | 55 | 17 | 1.600 | – | 0(Raman) | – | – | L. F. Mollenauer | AT&T | [165] |
| 2 | 1989 | 1 | 0.100 GHz | 32 | – | 9 | 2.7 | 1.535 | – | 1 | 40–190 | – | M. Nakazawa | NTT | [147] |
| | | 2(WDM) | 0.100 GHz | 32 | – | 9(13) | 2.7(3.8) | 1.535 | 17 | 1 | 19(61) | – | M. Nakazawa | NTT | [147] |
| 3 | | 1 | 0.100 GHz | 6,000(SMF)-LOOP | – | 50 | 17 | 1.600 | – | 0(Raman) | – | – | L. F. Mollenauer | AT&T | [166] |
| 4 | | 1 | 2.8 | 23 | 0.064 | 15.6 | 4 | 1.55 | – | 0(Raman) | 18.2 | –13 | K. Iwatsuki | NTT | [148] |
| 5 | 1990 | 1 | 5 | 100 | 0.50 | 20 | 2.35 | 1.545 | – | 3 | 6–10 | BER < $10^{-9}$ | M. Nakazawa | NTT | [149] |
| 6 | | 1 | 5 | 250 | 1.25 | 20 | 2.2 | 1.545 | – | 9 | 3.5–5.5 | BER < $10^{-9}$ | K. Suzuki | NTT | [150] |
| 7 | | 1 | 3.6 | 23 | 0.083 | 15.6 | 4 | 1.55 | – | 0(Raman) | 43.6 | –23 | K. Iwatsuki | NTT | [151] |
| 8 | | 1 | 4 | 136(SMF) | 0.54 | 80 | 16.9 | 1.559 | – | 3 | 44 | BER = $10^{-11}$ | N. A. Olsson | AT&T | [152] |
| 9 | | 1 | 20 GHz | 70 | – | 5.7 | 3.6 | 1.55 | – | 0(Raman) | 353 | – | M. Nakazawa | NTT | [170] |
| 10 | | 1 | 20 GHz | 200 | – | 20 | 2.4 | 1.552 | – | 7 | 4–6 | – | M. Nakazawa | NTT | [153] |
| 11 | | 2(WDM) | 1 GHz | 106(SMF) | – | 70 | 16 | 1.559 | 0.18 | 2 | 55 | – | P. A. Andrekson | AT&T | [168] |
| 12 | | 1 | 0.200 GHz | 10,000-LOOP | – | 50 | 1.38 | 1.532 | – | 3 | 0.6 | – | L. F. Mollenauer | AT&T | [169] |
| 13 | 1991 | 1 | 1.2 GHz | 12,000-LOOP | – | 55 | 1.38 | 1.532 | – | 3 | 1.5 | – | L. F. Mollenauer | AT&T | [154] |
| 14 | | 2(WDM) | 2 GHz | 9,000-LOOP | – | – | 2.4 | 1.56 | 0.52 | 4 | – | BER = $10^{-9}$ | N. A. Olsson | AT&T | [155] |
| 15 | | 1 | 10 | 1,000 | 10 | 45 | 2.1 | 1.552 | – | 20 | 0.3–1.3 | – | E. Yamada | NTT | [156] |
| 16 | | 8(TDM) | 32 | 90 | 2.9 | 15 | 1 | 1.53 | – | 4 | 34 | –24.5 | P. A. Andrekson | AT&T | [158] |
| 17 | | 1 | 5 | 150(SMF) | 0.75 | 40 | 18 | 1.554 | – | 1 | – | –29.0 | Y. Sunohara | NEC | [171] |
| 18 | | 1 | 20 | 500 | 10 | 18 | 1.8 | 1.552 | – | 20 | 3.0–6.1 | – | M. Nakazawa | NTT | [167] |
| 19 | | 1 | 2.5 | 14,000-LOOP | 35 | 50 | 1.36 | 1.555 | – | 3 | – | BER = $10^{-10}$ | L. F. Mollenauer | AT&T | [157] |
| 20 | 1992 | 2(WDM) | 10 | 678 | 6.8 | 26 | 0.7 | 1.552 | 0.4 | 8 | 7.0 | –6.5 | A. D. Ellis | BT | [172] |
| 21 | | 1 | 5 | 15,000-LOOP | 75 | 40 | 0.7 | 1.555 | – | 3 | – | BER = $10^{-10}$ | L. F. Mollenauer | AT&T | [173] |
| | | 2(WDM) | 10 | 11,000-LOOP | 110 | 40 | 0.7 | 1.555 | 0.36 | 3 | – | BER = $10^{-10}$ | L. F. Mollenauer | AT&T | [173] |
| 22 | | 1 | 10 | 1,200 | 12 | 30 | 0.5 | 1.552 | – | 23 | 0.85 | –29.4 | M. Nakazawa | NTT | [159] |
| 23 | | 2(TDM) | 20 | 1,020 | 20.4 | 12 | 0.4 | .1552 | – | 20 | 4 | –30.0 | M. Nakazawa | NTT | [160] |
| 24 | | 1 | 2.5 | 6,400-LOOP | 16 | 60 | 1.1 | 1.551 | – | 4 | 4.2 | BER = $10^{-10}$ | H. Taga | KDD | [161] |
| 25 | | 1 | 8.2 GHz | 15,000-LOOP | – | 20 | 0.7 | 1.548 | – | 3 | – | – | P. B. Hansen | AT&T | [174] |
| 26 | | 4(TDM) | 40 | 65 | 2.6 | 6.8 | 0.11–0.51 | 1.5 | – | 5 | – | –27.2 | K. Iwatsuki | NTT | [162] |
| 27 | | 1 | 5 | 3,000 | 15 | 35 | 0.4 | 1.558 | – | 92 | 4.5 | –15.2 | H. Taga | KDD | [163] |
| 28 | 1993 | 8(TDM + PDM) | 80 | 80 | 6.4 | 6.8 | 0.11–0.17 | 1.55 | – | 4 | – | –23.5 | K. Iwatsuki | NTT | [164] |
| 29 | | 1 | 8.2 | 4,200-LOOP | 34.4 | 20 | 0.7 | 1.548 | – | 4 | – | BER = $10^{-9}$ | C. R. Giles | AT&T | [175a] |
| 30 | | 2(TDM) | 20 | 1850 | 37 | 12 | 0.4 | 1.552 | – | 37 | 4.0 | –27 | M. Nakazawa | NTT | [175b] |
| | | 4(TDM + PDM) | 40 | 750 | 30 | 12 | 0.4 | 1.552 | – | 15 | 4.0 | –28 | M. Nakazawa | NTT | [175b] |

[a]: WDM, TDM, PDM = wavelength-, time-, polarization-division multiplexing; [b]: unmodulated signals if indicated in GHz; [c]: DSF except when SMF indicated and straight-line except when LOOP indicated; [d]: at system input; [e]: shortest wavelength of all channels; [f]: in-line or in-loop, not including post-amplifiers; [g]: peak power per channel; [h]: mean power per channel at $10^{-9}$ BER.

corresponds to a bit rate–length product of 10,000 Tbit/s.km. The experiment described in [176] shows there is actually no limit in transmission distance, so the only relevant parameter to consider is the bit rate. For this reason, soliton system data are not shown in Figure 7.1, which considers only results achieved in *linear* systems. Soliton systems can thus be viewed as part of a sixth generation of technology, which will likely appear in the wake of nonsoliton systems.

In view of the progress in linear systems, we can expect a steady increase in the performance of straight-line soliton systems during forthcoming years. The loop experiment data (summarized in Figure 7.18 and Table 7.5), show that the capacity potential of straight-line soliton systems is at least of 200 Tbit/s.km, which corresponds to a two-channel WDM at a total rate $B = 20$ Gbit/s with transmission length $L_{tot} = 10,000$ km. This performance can potentially be increased using a four-channel WDM and EDFA gain equalization. Alternatively, *polarization division multiplexing (PDM)* can be used (S. G. Evangelides et al. [178]). The PDM technique time multiplexes two soliton channels having orthogonal polarization states. Theory and experiment confirm that soliton pairs conserve their polarization orthogonality over transoceanic distances, in spite of the perturbations caused by fiber birefringence and ASE [178]. At the receiving end, the soliton channels can be demultiplexed through a polarization diversity apparatus, which includes automatic polarization control and tracking ([178]; F. Heismann and M. S. Whalen [179]). Another important property of soliton PDM is the reduction of soliton–soliton interactions (Section 5.11). For single-channel soliton systems with transoceanic distances, PDM makes it possible to reduce soliton pulse spacings and therefore to enhance the channel bit rate [178].

A discussion of the maximum theoretical performance of passive soliton systems is beyond the scope of this book. Section 5.11 comments several important references and further insights may be gained from: H. Kubota and M. Nakazawa, [180] and [181]; J. P. Gordon and L. F. Mollenauer [182]; L. F. Mollenauer et al. [183]; and J. D. Moores [184]. But the field of long-haul soliton systems is evolving at a rapid pace. New effects and techniques are discovered or demonstrated every year. The aforementioned studies are best used as a guide to fundamental considerations as their conclusions already may have been reinterpreted. This last observation is justified by the latest developments in *active soliton systems*, based on the principle of nonlinear all-optical regeneration described below. Before considering nonlinear all-optical regeneration in soliton systems, we discuss first to what extent signal regeneration can improve the performance of soliton systems.

Chapter 5 shows that the maximum transmission distance in passive soliton systems is primarily limited by two effects: SNR degradation and timing jitter. SNR degradation is due to ASE accumulation along the system; timing jitter (or the Gordon–Haus effect) is due to random frequency changes induced by ASE phase noise on soliton pulses. In passive amplifier chains (Chapter 2), ASE accumulation is unavoidable. This makes infinite transmission distance for concatenated amplifier chains an impossible notion (Chapter 3). Not only is it true for passive systems, it also applies when active elements perform periodic signal processing. These elements can be electronic or all-optical repeaters, whose global function is known as 3R, for pulse retiming, reshaping, and regeneration. EDFAs represent all-optical 1R repeaters, for which the only function is power regeneration. The other functions, including retiming, can also be performed all-optically through a variety of semiconductor and fiber devices (C. R. Giles et al.

[185]; M. Jinno and T. Matsumoto [186]; and M. Jinno and M. Abe [187]). Can the system transmission distance be indefinitely increased by using 3R repeaters? Communication theory, [81] and [188], shows that the bit-error rate of a $k$ element system using 3R repeaters is given by BER = $k\{p(1|0) + p(0|1)\}/2$, where $p(x|y)$ is the conditional error probability of an individual segment. The BER always increases in 3R systems, although the rate of increase can be effectively reduced by small signal power corrections, to which the probabilities $p(x|y)$ are sensitive [81]. In any case, the theoretical transmission distance of an even ideal 3R system cannot be infinite.

In-line retiming in soliton systems could alleviate limitations in transmission distance due to the Gordon–Haus effect. A remarkable property of soliton systems is that retiming alone can also suppress SNR degradation, contrary to the case of linear systems. The principle of this effect, which is known as the *guiding center soliton,* is to periodically and synchronously modulate the soliton pulse train with a sinusoid (M. Nakazawa et al., [189], [190] and [191]). This modulation removes one half the ASE noise present between soliton pulses and not synchronous with them. The overall suppression of ASE noise can then be explained by the following: the signal emerging from the modulator is a superposition of soliton pulses and ASE noise pulses. The ASE noise pulses are dispersive so they vanish with propagation distance. For the soliton pulses, the opposite applies. If synchronous modulation and optical filtering is repeated at intervals such that the ASE buildup in the selected spectral window is effectively countered by this dispersive effect, a constant SNR can be achieved for the soliton pulses, while negligible ASE accumulation occurs in the empty bit periods. This property can be demonstrated formally through the following [190]. Let $T_m$ be the modulator transfer function for the ASE and $G_{ex}$ the excess gain required to maintain the control. The total noise power after an infinite number of stages can be expressed as:

$$P_N^{tot} = \{\ldots(((P_N T_m G_{ex} + P_N)T_m G_{ex} + P_N)T_m G_{ex} + P_N) + \ldots\} = P_N \sum_{k=0}^{\infty} (T_m G_{ex})^k \quad (7.31)$$

In Eq. (7.31), $P_N$ represents the ASE noise power, as generated over the EDFA chain between each modulator. The modulator transfer function is given by:

$$T_m = \frac{P_N^{out}}{P_N^{in}} = \frac{1}{T}\int_{-T/2}^{T/2}\left\{\alpha + (1-\alpha)\cos^2\left(\pi\frac{t}{T}\right)\right\}dt = \frac{1+\alpha}{2} \quad (7.32)$$

where $1/\alpha$ is the modulator extinction ratio and $T = 1/B$ the bit rate. An extinction of $-20\,dB$ corresponds to $\alpha = 0.01$, and the transmission is $T_m = 0.5$, meaning that one half the ASE is blocked by the modulator. The other half of the ASE is dispersed in the following fiber section, to form a CW background when reaching the next modulator. This effect justifies Eq. (7.31). Since the excess gain is $G_{ex} = 1 + \delta g$, where $\delta g$ is small compared to unity, the factor $T_m G_{ex}$ is smaller than unity, consequently the total ASE noise given by the series in Eq. (7.31) converges to:

$$P_N^{tot} = P_N \frac{1}{1 - T_m G_{ex}} \approx P_N \frac{1}{1 - (1 + \delta G)/2} \approx P_N \frac{3 + \delta G}{2} \quad (7.33)$$

independent of the transmission length. $G_{ex}$ must correspond to the loss $T_s$ experienced by the soliton through the modulation and filtering. A Gaussian envelope approximation shows that for the soliton pulse this loss takes the form [191]:

$$T_s = \alpha + \frac{1-\alpha}{2}\left\{1 + \exp\left(-\frac{\pi^2}{4\log 2}\frac{\Delta T^2}{T^2}\right)\right\} = \frac{1}{G_{ex}} \quad (7.34)$$

The above calculations shows that ASE can be effectively suppressed by synchronous modulation, provided that the soliton timing does not fluctuate between each modulation stage. Timing stabilization is actually achieved at each stage by the optical filter. The filter cuts the wings of the soliton spectrum. This causes pulse broadening. Broadening and synchronous modulation effectively recenter the pulse in the timing window ([190]; K. J. Blow et al. [192]).

The principle of synchronous modulation for ASE and timing jitter suppression was confirmed theoretically, through the nonlinear Schrödinger equation (NLSE) formalism, Section 5.11, and experimentally through recirculating loop experiments (M. Nakazawa et al., [189] and [190]). Using a 509 km loop that included a single intensity modulator and an optical filter, propagation of 10 Gbit/s soliton signals over *one million kilometers* was first demonstrated [189]. Figure 7.19 shows more recent results, in which the propagation distance was increased to 180 million kilometers (M. Nakazawa et al. [176]). This distance corresponds to approximately 15 mn optical delay in the loop. The soliton data patterns emerge from the loop

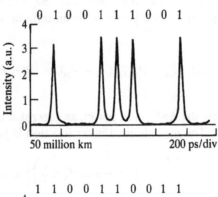

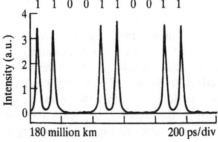

**FIGURE 7.19** Soliton data patterns at $B = 10$ Gbit/s measured after 50 and 180 million kilometer transmission in a 509 km recirculating loop, which included a synchronous modulator and optical filtering From [176], reprinted with permission from the Optical Society of America, 1993.

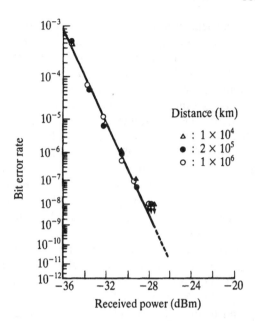

**FIGURE 7.20** BER measurements at $B = 10$ Gbit/s ($2^9 - 1$ PRBs) obtained in loop experiment with synchronous modulation and filtering for distances up to $10^6$ km, demonstrating penalty-free soliton transmission. From [176], reprinted with permission from the Optical Society of America, 1993.

virtually undistorted and free from ASE noise. BER measurements using pseudorandom bit patterns, Figure 7.20, confirms that for distances up to $10^6$ km, the soliton transmission is free from any penalty. Computer simulations have revealed that synchronous modulation suppresses soliton interaction forces [190].

*The transmission distance in active soliton systems with synchronous modulation is virtually unlimited.* But for implementation in actual straight-line systems, the main issue is the effect of PDL in the modulators, more difficult to control than in recirculating loops. Finally, the application of synchronous modulation to multiplexed data must be investigated. PDM requires polarization independent modulators operating at twice the channel bit rate. WDM channels have to be demultiplexed before modulation then recombined after modulation. This is because the arrival times depend on wavelength. Each of the regenerator modules must have build-in clock recovery. A 9000 km transoceanic system requires approximately 18 such regenerators, each including as many modulators as the number of WDM channels. Practical implementation of such a system must take into account important factors such as reliability, component complexity, and cost-effectiveness.

A second type of active soliton regeneration is based on the principle of *sliding frequency guiding filters.* Two recent theoretical studies have shown that the Gordon–Haus effect could be suppressed by using narrowband optical filters periodically distributed along the link (A. Mecozzi and J. D. Moores [193]; Y. Kodama and A. Hasegawa [194]). The principle of this technique is to guide the soliton center frequency, which effectively reduces pulse random walk. But because the optical filters cut the wings of the soliton spectrum, extra gain is required at each stage to compensate for the corresponding loss. While the frequency guiding filter technique proves to be efficient in jitter suppression, the extra gain required causes the ASE noise to grow exponentially. The exponential growth of the ASE rapidly degrades the SNR and eventually limits the transmission distance. A solution to this problem, proposed by L. F. Mollenauer et al., [195] and [196], is to gradually translate the center frequency of each optical filter with increasing distance, hence the name sliding frequency guiding filters. Filtering reduces jitter in the same way as before but the gradual frequency displacement in each filter impedes the growth of ASE. Theory shows that the soliton pulses emerge at each filtering stage with a mean frequency lagging behind the filter peak [195]. The important difference in the sliding frequency guiding filter approach is that *the transmission system is opaque to ASE noise while it is transparent to solitons.* This remarkable property could not be achieved with nonsoliton pulses, since the system would be opaque to them.

The required frequency sliding rate for effective Gordon–Haus jitter suppression is relatively small. Indeed, [195], shows that for solitons with width $\Delta T = 20$ ps, a tenfold (twentyfold) reduction of the soliton jitter can be achieved at a transmission distance of 10,000 km (20,000 km) when the average sliding rate is $-5.6$ MHz/km or $-5.6$ GHz/Mm. For a 10,000 km system, the soliton's center frequency is therefore downshifted by 56 GHz, only 3.5 times the soliton FWHM (given by $\Delta \nu \Delta T = 0.315$). Downsliding (as opposed to upsliding) keeps the frequency lag to a minimum, as shown by numerical simulations [195].

The sliding frequency guiding filter technique was implemented experimentally in a recirculating loop containing three EDFAs and three ramp-driven tunable filters with a sliding rate of 6 GHz/Mm (L. F. Mollenauer et al. [177]). The eye diagram corresponding to 10,000 km transmission at 10 Gbit/s ($\Delta T = 18$ ps) is shown in Figure 7.21. The output soliton pulses (corresponding to 1 symbols in the diagram) are remarkably free from amplitude noise and jitter. The observed jitter, consistent with the theoretical prediction $\sigma = 2$ ps in these conditions [195], indicates that the bit rate could probably be increased to 20 Gbit/s without significant eye closure.

Figure 7.22 shows the BERs measured for single-channel and two-channel WDM experiments, using pseudorandom bit patterns [177]. Error-free transmission is achieved at 21,000 km and 13,000 km, respectively, which correspond to system capacities of 210 Tbit/s.km and 130 Tbit/s.km. These results are matched only by the synchronous modulation results in Figure 7.18.

To call the sliding frequency filtering technique a form of *active* soliton control is

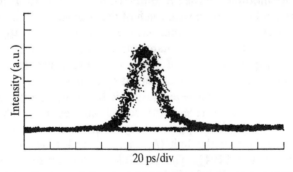

20 ps/div

**FIGURE 7.21**  Eye diagram of $B = 10$ Gbit/s soliton data, obtained after 10,000 km transmission through a recirculating loop including sliding frequency guiding filters. From [177], reprinted with permission from the Optical Society of America, 1993.

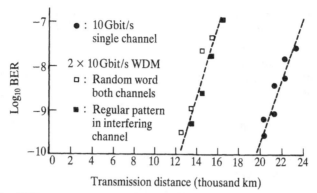

**FIGURE 7.22**    BER measurements at $B = 10$ Gbit/s obtained in loop experiment with sliding frequency guiding filters, showing error-free soliton transmission over 21,000 km for single channel and over 13,000 km for two-channel WDM; $2^{14}$ bit word always pseudorandom in measured channel. From [177], reprinted with permission from the Optical Society of America, 1993.

justified because the signal center frequency is altered in the process. In passive soliton systems the center frequency is not altered. The loop experiment required active filters to effect frequency sliding but it can be implemented using *passive* optical components, unlike the synchronous modulation technique. An additional advantage of passive filtering is polarization insensitivity, realized in Fabry–Perot etalons (J. Stone and L. W. Stultz [198]). Finally, the sliding frequency guiding filter technique is fully compatible with WDM soliton systems [196]. In this case, the guiding filters are Fabry–Perot etalons whose FSR matches the channel separation. An additional advantage of frequency periodic filtering is the observed reduction of WDM soliton interactions forces [196]. These forces cause random time and frequency shifts between channels (Section 5.11). Due to the filter polarization insensitivity, the technique is also applicable to PDM and PDM–WDM soliton systems. But for practical, straight-line system implementation, automatic control of both the absolute peak frequencies and FSR of the guiding filters must be achieved. Ideally, the control should be internal to each amplifier stage, independent of the rest of the system. Whether such a function could be realized with strictly passive components, without any form of external control or feedback, remains to be investigated.

Sliding frequency filtering makes it possible to reduce the soliton timing jitter and to maintain the ASE to a fixed background level at the expense of a small carrier frequency shift. If the sliding rate is kept constant (e.g., $-5$ GHz/Mm), the soliton frequency falls outside the equalized EDFA gain bandwidth after some transmission distance; a way to prevent this effect is to reverse the frequency sliding after some transmission distance (e.g., at $+5$ GHz/Mm), and so on. But this system is not free from timing jitter, unlike synchronous modulation.

To summarize, the two state-of-the-art strategies for enhancing the performance of soliton systems are based on time domain modulation and filtering (with fixed frequency) and frequency domain modulation (with shifting frequency). The first approach is free from timing jitter so it offers the possibility of virtually unlimited transmission distances at bit rates potentially up to 20 Gbit/s. But WDM may be difficult or possibly impractical to implement. Frequency domain modulation offers the potential of 20 Gbit/s transmission over one half the earth circumference

(20,000 km) and is compatible with WDM and PDM. Assuming negligible soliton interactions, a ten-channel WDM–PDM implementation of a system based on frequency domain modulation would provide a total bit rate capacity $B = 200$ Gbit/s for intercontinental distances. Assuming that present transoceanic communications needs are currently met using systems operating at $B = 5$ Gbit/s, the 25% annual growth in traffic demand (Figure 7.16) suggests that $B = 20, 50, 100,$ and $200$ Gbit/s will be required in nearly 6, 10, 13, and 17 years. Such a projection assumes a constant demand over the period considered. It is possible that this demand will actually increase, as driven by the deployment of earth-size communication networks, Figure 7.17. Soliton systems might in this case be implemented within a shorter year span.

Aside from long-haul communications applications, active soliton transmission could be used to investigate new types of nonlinearities over ultralong fiber distances, e.g., related to $\chi^{(5)}$ or higher order interactions. Practical soliton sources can now be realized with monolithic LDs or LD-pumped EDFLs and recirculating fiber loops can be implemented with standard research equipment. So there may be considerable progress in this field.

## 7.4  ANALOG SYSTEMS

The impact of EDFAs has not been limited to the digital communication systems. Analog lightwave transmission through EDFA-based fiber systems is likely to play an increasing role in the future, considering important applications in video signal distribution (e.g., CATV and HDTV), subcarrier multiplexed lightwave networks and analog data transport in electromagnetically perturbed environments. The impact of EDFAs on analog systems is explained by the same reasons given for their impact on digital systems: (1) as postamplifiers they increase the power budget of analog transmission links, (2) as preamplifiers they increase the sensitivity of analog receivers, and (3) as in-line amplifiers they increase the system transmission distance. Other potential advantages of EDFAs as passive repeaters lie in their low noise figure and low intrinsic nonlinearity (CSO, CTB, and XM) (P. M. Gabla et al. [199]).

Chapter 3 analyzes the signal power requirements of AM and FM analog modulation schemes. It consider numerical examples with parameters and standards used in video distribution, and refers to theoretical studies of system optimization and limitations due to various sources of distortion. This section reviews experimental results obtained in video broadcast systems based on lightwave networks.

The generic layout of video broadcast systems, generally referred to as common antenna television (CATV), is shown in Figure 7.23. Video signals can actually originate from different sources, including satellite, terrestrial microwave, broadcasts, and local distribution. North American Standard (NTSC) video signals use either AM–VSB (vestigial sideband) modulation or FM modulation. FM signals can also carry high definition TV channels (HDTV). Digital video signals or data can also be fed into the system. In the central or receiving station, the analog video signals corresponding to individual TV channels are subcarrier multiplexed (SCM) and fed to single-frequency LDs operating at carrier wavelengths $\lambda_1 \ldots \lambda_n$. Digital signals are also converted into a TDM lightwave carrier at wavelength $\lambda_p$. All optical signals are then multiplexed through an $m \times m$ star coupler (Section 7.5) and amplified

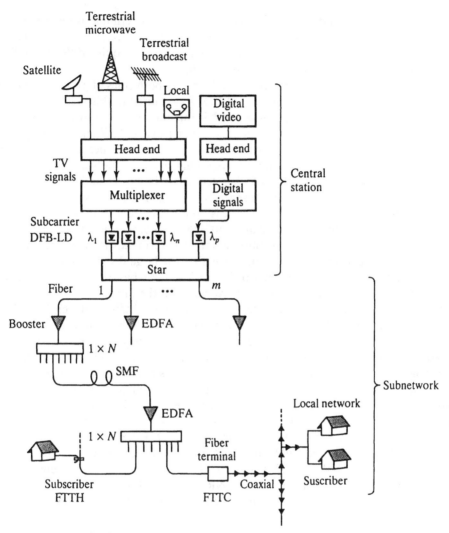

**FIGURE 7.23**  Basic layout of CATV broadcast system using EDFA-based fiber network (see text for description).

through an EDFA power booster. The central station has $m$ fiber outputs, which feed as many independent subnetworks.

The optical multiplex fed into one subnetwork is first split through a $1 \times N$ tree coupler, then transmitted along a strand of SMF (5–25 km long). Power boosting and subsequent splitting can be repeated a certain number of times. *Wavelength routing* can be used to split the carriers into different branches of the subnetworks. At the receiving end, the fiber can reach individual homes (FTTH); in this case, the home receiver must perform both optical and RF demultiplexing. Alternatively, the fiber can be sent to a receiving terminal located near the curb of a residential area (FTTC); this terminal performs optical demultiplexing and directs the multiplexed video channels to individual homes through coaxial lines with RF repeaters. The number of end-users that can be serviced from each fiber terminal is determined by the power budget of the coaxial network and by cost-effectiveness. For FTTH and FTTC, the component complexity and system costs are quite different. FTTH inherently provides the greatest bandwidth to the end-user.

In the CATV system described above, EDFAs play both roles of postamplifier and splitting loss compensator. The number of EDFAs required in this system is $mN^{k-1}$, where $m$ is the number of subnetworks and $k$ is the number of $1 \times N$ splitting stages with loss compensation. The number of fiber terminals is $mN^k$. When $1 \times 16$ splitters are used, a three-stage splitting provides 4096 fiber terminals per subnetwork and each subnetwork requires 256 EDFAs. Assuming a CATV network with $m = 12$ optical carriers, the total number of fiber terminals is 49,152 and the number of EDFAs is 3072. In comparison to transoceanic nonregenerative links, the number of EDFAs required in such a CATV network is about one order of magnitude greater. While a transoceanic link can service up to 500,000 users simultaneously (see previous section), the CATV system has 10 times fewer fiber terminals (50,000) for 10 times more EDFAs (3000). This observation shows that the large distribution potential of EDFA-based CATV systems may not be fully utilized due basic economic considerations, unless EDFAs become relatively inexpensive in the future.

Various experimental results and relevant parameters obtained during 1989–1992 in analog AM and FM video transmission systems are summarized in Table 7.6, [199]–[218]. Results concerning digital video transmission, based on the same broadcast principle, are also shown. Most experiments used only one optical carrier. Experiment #6 (H. E. Tohme et al. [204]) used a single DFB laser operating at $\lambda = 1.536\,\mu m$ to generate an optical multiplex containing 10 FM–NTSC video channels, three FM–HDTV video channels and one digital signal at $B = 622$ Mbit/s. The rate $B = 622$ Mbit/s corresponds to a transmission capacity of 24 compressed HDTV signals at $B = 155$ Mbit/s (H. Giaggoni and D. J. LeGall [219]). By using a single EDFA booster in the middle of a splitting tree network similar to that shown in Figure 7.23, a CNR of 12.7–17 dB could be achieved at the receiving end, corresponding to a standard-compatible SNR in excess of 51 dB. A fiber network of $1 \times 16 \times 8 \times 16 \times 4$ splittings has the potential to service as many as 8192 fiber terminals [204].

Other analog video transmission experiments used WDM. Experiment #8 used $m = 16$ optical channels spaced by 2 nm intervals, Figure 7.24 (W. I. Way et al. [205]). The spectrum recorded at the EDFA booster output shows the 16 channels, spread over a 34 nm bandwidth, with characteristic ASE background. The maximum gain difference between channels is 7 dB. Six DFB lasers corresponding to channels 1, 7, 9, 11, 14 and 15 were modulated each at $B = 622$ Mbit/s, which corresponds to an equivalent capacity of 144 HDTV signals. The remaining 10 lasers were each modulated with 10 SCM–FM video signals, corresponding to a total capacity of 100 FM video channels. The tree network used in this experiment was based on a $1 \times 8 \times 8 \times 4$ splitting; a single EDFA was used after the first $1 \times 8$ splitter. The number of fiber terminals per subnetwork is thus 256, and the total number of terminals is $m \times 256$ or 4096. The total number of EDFAs required to complete this CATV network is $m \times 8 = 128$. Because of the gain differences between channels, the receiver sensitivities for CNR = 16.5 dB in each FM channel varied by about 2 dB. This experiment shows that for some system applications, it is possible to fully use the EDFA bandwidth without gain equalization.

**Table 7.6  Analog and digital video transmission system experiments**

| No. | Year | Number of channels [a] Electrical | Number of channels [a] Optical | Modulation type [b] | Modulation index (%) | Number of EDFAs [c] | Power budget (dB) | Receiver CNR (dB) [d] | Receiver weighted SNR (dB) | Number of terminals [e] | Author et al. | Company | Ref. |
|---|---|---|---|---|---|---|---|---|---|---|---|---|---|
| 1 | 1989 | 19 | 1 | AM-VSB | 5 | 1 | 15 | 40–46 | – | – | W. I. Way | Bellcore | [200] |
| 2 | 1990 | 160 | 16 | FM | – | 1 (16) | 9.2 | 16.5 | – | 16 | W. I. Way | Bellcore | [201] |
| 3 | | 11 | 1 | FM | – | 6 | 41 | 30 | 60 | 12,500 | K. Kikushima | NTT | [210] |
| 4 | | 4 | 1 | FM | 4 | 1 | – | – | – | – | E. Eichen | GTE | [202] |
| 5 | | 32 | 9 | ASK(2.2 Gb/s, PAL) | | 1 (1) | – | $10^{-9}$ BER | – | 7,023 | A. M. Hill | BTRL | [203] |
| | | 8 | | ASK(155 b/s, PAL) | | | | $10^{-9}$ BER | | | | | |
| 6 | | 1 | 1 | ASK(622 Mb/s) | 30.5 | 1 (128) | 49.5 | $10^{-9}$ BER | – | 8,192 | H. E. Tohme | Bellcore | [204] |
| | | 10 | | FM-NTSC | 9 | | | 12.7 | | | | | |
| | | 3 | | FM-HDTV | 21 | | | 17 | | | | | |
| 7 | | 40 | 1 | AM-FDM | 3 | 4 (4,097) | – | | 45 | 65,536 | K. Kikushima | NTT | [211] |
| | | 7 | | FM-NTSC | 0.3 | | | | 55 | | | | |
| | | | | FM-HDTV | 0.3 | | | | 33 | | | | |
| 8 | | 6 | 16 | ASK(622 Mb/s) | 15–25 | 1 (128) | 36 | $10^{-9}$ BER | – | 4,096 | W. I. Way | Bellcore | [205] |
| | | 100 | | FM | | | 36 | 16.5 | | | | | |
| 9 | | 384 | 12 | ASK(2.2 Gb/s, PAL) | | 2 (1,029) | – | $10^{-9}$ BER | – | 39.5 M | A. M. Hill | BTRL | [206] |
| | | | | ASK(155 b/s, PAL) | | | | $10^{-9}$ BER | | | | | |
| 10 | 1991 | 35 | 1 | AM-VSB | 7.2 | 1 | 16 | 48 | – | – | P. M. Gabla | Alcatel-CIT | [199] |
| | | 12 | | FM-st | | | | | | | | | |
| 11 | | 40 | | AM-VSB | 4 | 1 | | 48–52 | – | – | D. R. Huber | General Inst. | [212] |
| 12 | | 20 | | AM-VSB | | 1 | | 51 | – | – | D. R. Huber | General Inst. | [213] |
| 13 | | 35 | 2 | AM-VSB | 5.2 | 1 | | 47.6 | – | – | R. Heidelman | Alcatel-SEL | [214] |
| | | 30 | | FM-st | | | | | | | | | |
| | | 1 | | ASK(2.5 Gb/s) | | | | $10^{-9}$ BER | | | | | |
| 14 | | 40 | 1 | AM-VSB | 5 | 1 | 14 | 51 | – | – | M. Shigematsu | Sumitomo | [215] |
| 15 | | 40 | 1 | AM-VSB | 7.75 | 1 | 21 | 48 | – | – | G. R. Joyce | GTE | [216] |
| 16 | 1992 | 60 | 1 (1.5 µm) | AM-VSB | 3 | 1 | – | 48.7 | – | – | M. R. Phillips | AT&T | [217] |
| | | 52 | 1 (1.3 µm) | AM-VSB | 3 | 0 | – | 51.4 | – | – | | | |
| 17 | | 30 | 1 | AM-VSB | 6 | 3 | 45 | 48 | – | – | P. M. Gabla | Alcatel-CIT | [207] |
| 18 | | 32 | 1 | AM-VSB | 6 | 3 (64) | 46 | 48 | – | 1,024 | B. Clesca | Alcatel-CIT | [218] |
| 19 | | 35 | 1 | AM-VSB | 6.7 | 4 (131,072) | 89 | – | 45 | 4.4 M | H. Bülow | Alcatel-SEL | [208] |
| | | 12 | | FM-st | | | | | | | | | |
| 20 | | 35 | 1 | AM-VSB | 6.7 | 3 (8,192) | 73 | – | 45 | 0.524 M | H. Bülow | Alcatel-SEL | [209] |
| | | 12 | | FM-st | | | | | | | | | |

[a]: electrical (RF) and optical (λ = 1.5 µm) channels simultaneously transmitted in fiber; [b]: AM-VSB = amplitude-modulation vestigial-sideband video, FM = frequency-modulation video, FM-st = FM-stereo frequency-modulation [c]: used in the experiment as post and in-line amplifiers (number required in full broadcast system is shown in parenthesis); [d]: or BER for ASK, when applicable; [e]: in broadcast system, M = million.

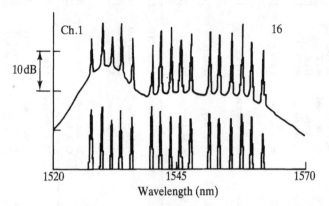

**FIGURE 7.24**   Input (bottom) and output (top) spectrum in a 16-channel WDM CATV broadcast network experiment. Optical channels #1, 7, 9, 11, 14, and 15 are modulated with 622 Mbit/s signals, corresponding to 24 HDTV video channels; the 10 other optical channels are modulated with 10 subcarrier FMTV signals, corresponding to 100 FM video channels. From [205] © 1990 IEEE.

As a matter of fact, the potential capacity of WDM CATV broadcast systems based on EDFAs, expressed as the number of video channels and end-users, is actually *enormous*. The high number of video channels is due primarily to the wide bandwidth of the EDFAs and secondarily to the even larger bandwidth of optical fibers. The high number of potential end-users is a consequence of geometric progression. Indeed, in a transparent $1 \times N \times \cdots \times N$ tree network, where the splitting loss is compensated at each stage $k$ by EDFAs, the number of terminals increases as $N^k$. The maximum number of splittings is determined by the minimum CNR required at the receiving end (Section 3.8). Because the EDFAs have low noise figures and insertion losses, the number of splittings can be quite large. This property was demonstrated in Experiment #9. (A. M. Hill et al. [206]). The experiment used $m = 12$ optical carriers containing each 32 digital video channels at $B = 70$ Mbit/s (standard PAL), corresponding to a total rate $B = 2.2$ Gbit/s per carrier, and a total of 384 video channels. The experimental layout is shown in Figure 7.25. The tree network comprises $4 \times 3 \times 7^6 \times 4 \times 7$ splittings, which corresponds to 39,530,064 potential end-users. The number of EDFAs required to complete the network is $4 \times 3 \times 7^3 = 4116$, one EDFA per 9604 users. In such a large implementation, the EDFA component cost/user becomes negligible. The network includes 30 km of SMF, which corresponds to an area 2800 km$^2$. Larger areas could be covered by the CATV network with longer SMF distances, additional in-line EDFAs and improved receiver sensitivities.

In the above experiment, the BER measurements showed an error floor at BER $= 10^{-7}$ but the PAL channels showed no visible sign of degradation. This feature illustrates that digital video systems have a higher tolerance to BER degradation, and can therefore operate at error rates somewhat higher than in standard transmission systems. The notion of required CNR is subjective, as it corresponds to the human appreciation of image quality. Chapter 3 shows that the CATV standards are CNR $= +50$ to $+55$ dB for AM TV; pictures with CNR $= +25$ dB are clearly objectionable, but pictures with CNR $= +45$ dB are still judged excellent (K. Simon [220]).

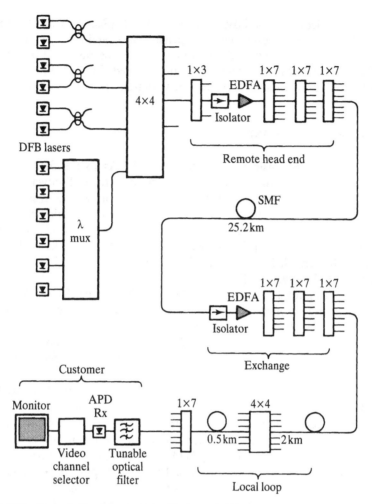

**FIGURE 7.25** Experimental layout of a 12-channel WDM CATV network, with a potential of broadcasting 384 video channels to 39,530,064 end-users. In the complete network, the total number of EDFAs is 4116, representing one EDFA per 9604 users. From [206] © 1990 IEEE.

Could optical solitons be used in future digital CATV systems? This question becomes relevant when considering potential developments in FTTH and integrated digital networks (see the review by P. E. Green [222]). First, soliton signals offer the highest bandwidth achievable in lightwave systems. Second, for transmission distances limited to 100–500 km, soliton interactions and jitter are negligible (e.g., $1/39,530,064 = -76$ dB loss, which corresponds to 304 km fiber transmission). Very high bit rates could be achieved while using a single optical carrier (e.g., $B = 100$ Gbit/s), obtained through passive TDM. In broadcast networks, $N = 1$ solitons would not be perturbed by transmission loss, splitting loss, and power boosting, as the peak power would be nearly constant at each point in the network. The techniques of synchronous modulation or frequency guiding filters would make possible to achieve virtually noiseless transmission. High speed demultiplexing could be performed by all-fiber,

LD-controlled devices such as those based on NALMs (Section 6.5). Due to the low noise figure of the transmission system, inexpensive baseband receivers with relatively low sensitivities could be used at the terminals. While the potential performance of such a soliton system clearly surpasses any current needs for CATV, this perspective may rapidly change with the emergence of new digital HDTV services.

## 7.5  LOCAL AREA NETWORKS

In this last section, we consider applications of EDFAs to the field of local area networks (LANs). As a detailed review of the current status of lightwave (or optical, or photonic) networks is far beyond the scope of this book, we will restrict ourselves to basic considerations. We will not outline possible trends in this field, since both communication concepts and hardware technologies in LANs are evolving at a rapid pace. The multiple issues of optical networks are covered in books by P. E. Green [222]; S. E. Miller and I. P. Kaminow [223]; and J. Walrand [224]. Some recent reviews and tutorial papers are [225]–[231].

Before considering applications of EDFAs to LANs, we briefly recall a few basic concepts and introduce some of the acronyms most frequently used in the literature. First, a LAN can be defined most generally as: *an information transport system which provides connectivity among distributed entities within a structure or a geographic area* (S. D. Personick [232]). A more detailed definition distinguishes the *local area network* (LAN), which extends over a few kilometers, from the *metropolitain area* and *wide area networks (MAN, WAN),* which extend to 100 km and 1000 km ranges, respectively. A description of existing types of LANs, MANs, WANs, as well as corresponding architectures, protocols, communication layers, and standards, can be found in [222]. In the case of lightwave networks, another important distinction exists between *passive optical networks (PON)* and *all-optical networks (AON).* In PONs, the optical fiber plays the role of a broadband, passive interconnect; the functions of switching and control are played by electronics, whether at terminals or at various network nodes. A multiple access, passive star network is a representative PON [223]. Another example is given by the CATV broadcast system described in the previous section. Thus, PONs can contain EDFAs as passive in-line boosters or repeaters. The PON connectivity between terminals is not necessarily fixed by the network physical path topology. For instance, if the PON contains WDM couplers, the connectivity can be controlled from the terminals, through electronic wavelength switching [227]. Transceiver tunability and passive WDM couplers are not even required with multihop protocols (A. S. Acampora [233]). In all-optical networks (AONs), the role of electronics is reduced to basic control functions at the terminals, while that of optics is emphasized for a greater use of the fiber bandwidth. A possible definition for AONs is that all-optical devices (called here AODs) are used exclusively between any two nodes and terminals. For instance, an $N \times N$ all-optical switch with input/output ends connected to $N$ terminals is a basic example of AON. Thus, an AON is ideally an uninterrupted mesh of reconfigurable optical paths, which includes AODs.

AOD remains an elusive notion. It escapes any simple definitions such as "three- or four-port photonic coupler," but the device concept can be conveyed by several examples. Examples of fiber-based AODs are: $2 \times 2$ couplers, Fabry–Perot etalons, fiber Bragg gratings, fiber mach–Zehnder interferometers, recirculating delay lines, NALMs, and EDFAs. EDFAs can be considered as performing an AOD function, even if they are based on hybrid optoelectronic technology. EDFAs can also be implemented to perform *active* and *optically controlled functions*, described in detail in Chapter 6. While there is much progress in the field of active AODs (as based on fiber or semiconductor devices), the advent of AONs is yet far out of sight. This fact can be explained by the very high degree of complexity involved in the issues of LAN management and control. Compared with their electronic or hybrid counterparts, how competitive, are all-optical LANs in simplicity, capacity, efficiency, and cost? An answer to this question is quite beyond the scope of this book. We can only infer here that, in all likelyhood, the fields of active AODs and AONs will eventually meet in the future. Other reasonable inferences are that *optical amplifiers* (i.e., EDFAs, PDFAs, and SCAs) will be key elements in future AON technology and likely constituents of all-optical signal processors.

Considering that the field of EDFAs is only a few years old, it is not surprising that there are very few studies dedicated to their implementation in LANs, aside from CATV applications discussed in the previous section. These studies expectedly concern mostly device and system applications in PONs. We briefly review some of the main results and conclusions of these studies, grouped according to LAN topology: broadcast, bus, ring, and star.

*Broadcast topology* experiments have demonstrated the potential of EDFAs as in-line repeaters for system capacity enhancement. For instance, a 16-channel WDM broadcast LAN with a 39.8 Gbit/s aggregate bit rate, a 527 km range, and 43,787,856 potential end-users was demonstrated with 10 EDFAs (D. S. Forrester et al. [234]). Conceptually, this network represents a long-haul extension of the CATV system shown in Figure 7.25. Another type of EDFA-based broadcast LAN used 128 channels (10 GHz spacings) with polarization diversity heterodyne receivers (H. Toba et al. [235]). The novelty is that the trunk lines are bidirectional so some of the optical carriers are used for interactive channel selection. A detailed theoretical analysis of FDM broadcast LAN performance with direct and coherent detection terminals can be found in a study by H. Toba et al. [236].

Distributed or lumped EDFAs fit naturally in many types of LANs having *bus topologies*. The capacity of such LANs (bit rate and number of stations) is strongly limited by excess loss, tapping loss, and transmission loss. EDFAs can be used for in-line loss compensation along the bus. They can even be used as the transmission medium in distributed amplification (Section 6.1). Another advantage of implementing optical amplifiers is their potential to increase the number of stations and the bus length, otherwise limited by the power budget. A distributed Er-doped fiber bus network, 850 m long with five stations (50% taps) was investigated by T. J. Whitley et al. [237]. The experiment showed that bus transparency, in which all tap outputs have nearly identical powers, could be achieved with 25 mW pumping at 980 nm from the bus fiber end. Theoretical optimization studies showed that EDFAs make it possible to increase the number of bus stations by several hundred times, taking into account limiting effects of gain saturation and ASE noise accumulation (K. Liu and R. Ramaswami, [238] and [239]). The study of [239] shows that for a fixed tapping loss $\alpha$, the number of stations is maximized when there is only one station between each EDFA repeater. On the other hand, for a fixed EDFA gain, the maximum number of station increases with the tapping loss $\alpha$; this can be explained by the fact that a greater tapping loss corresponds to a higher signal input into the bus, and a slower ASE growth [239]. Another approach for EDFA-based LANs uses remote pumping (D. N. Chen et al. [240]). In the remote pumping configuration (Section 6.1), the pump power at $\lambda = 1480$ nm is coupled into the bus at a some location. Signal loss compensation is achieved by splicing lumped Er-doped fiber sections between each station. The EDF lengths are calculated to provide equal gains; the length differences compensate the effect of pump decay along the bus. The advantage of this approach is the reduction in component cost, dominated by the EDFA pump diodes for the passive bus. A single pump can be used to achieve bus transparency without changing the number and the location of the EDFAs along the bus. The experiment described in [240] showed that a four-station transparent bus ($\alpha = 50\%$), 6.5 km long, can be implemented with three EDF sections and a single 50 mW pump LD. A theoretical analysis, which takes into account gain saturation and ASE accumulation effects, shows that with a 250 mW pump, the bus would support up to 40 stations with BER $\leqslant 10^{-9}$ at each receiver. The maximum transmission capacity of distributed, lumped, or remotely pumped bus WDM LANs remains to be comprehensively studied to take into account limitations by EDFA gain ripple.

LANs with *ring topologies* using lumped or distributed amplification have also been investigated theoretically (E. Goldstein et al., [241] and [242]; A. F. Elrefaie and S. Zaidi [243]) and experimentally (W. I. Way et al. [244]). The main difference between the bus and the ring LANs is that ring LANs allow the ASE to accumulate and recirculate indefinitely. One way to avoid laser oscillation, is to achieve conditions where the gain integrated along the loop circumference $g$ is kept slightly below the loop loss $\alpha$. Simulations in [241] show that even in the case where the loop gain is marginally smaller than the loss (i.e., $g = 0.99\alpha$), the total accumulated ASE power converges toward a very small value, a few microwatts for a ring with a 10 km circumference. An SNR analysis shows, for instance, that a 350 km ring with $g = 0.90\alpha$ could support up to 450 stations at a bit rate $B = 2.5$ Gbit/s [241]. Another theoretical study, which concerned unidirectional WDM rings, showed that a 15-channel system with a central hub station could support up to 11 nodes at a bit rate $B = 10$ Gbit/s [242]. Because of the uneven EDFA gain spectrum, such a performance can be achieved by proper ordering of the station wavelengths, assigning the wavelengths of highest gains to the nodes that are the most remote from the central hub [243]. Loop instability with transient gain saturation is an important source of impairment not studied so far. Indeed, weak gain transients could be produced by random changes in the power distribution between optical channels, which would make the loop depart from the steady state condition $g = 0.90\alpha$, and possibly cause spurious oscillations. To alleviate this effect, the EDFAs can be pumped in a near 100% inversion regime, in which case no saturation occurs, or AGC can be implemented at each node. The slow response time of AGC (0.1–1 ms) limits the loop circumference to 20–200 km. EDFAs with automatic power control (APC), can also be implemented in SONET self-healing ring LANs [244]. The principle of self-healing can be implemented in a bidirectional, four-ring system made of two working pairs ($w_1$, $w_2$) and two backup pairs ($b_1$, $b_2$). In the event of a fiber cut occurring between any stations, the network is automatically switched at these two stations to form two new ring paths ($w_1$, $b_1$) and ($w_2$, $b_2$). The drawback of the approach is the costly duplication of add–drop multiplexer (ADM) components in the backup system. In [244], it is shown that EDFAs with APC can alleviate the need for this duplication;

in this case, the function of APC is to keep the signal received at each node at a constant level whenever the ring is switched to the healing mode.

Finally, the performance of LANs with *star topologies* can be enhanced with EDFAs. A first application is cancellation of loss in passive star couplers (A. E. Willer et al. [245]; H. M. Presby and C. R. Giles [246]). The principle of this application is illustrated in Figure 7.26. A passive $N \times N$ star coupler distributes any signal input (power $P_0$) into $N$ equal outputs (power $P_0/N$), as shown in Figure 7.26a. In addition to the $1/N$ transmission loss, the signal undergoes excess loss due the components forming the star coupler. The star coupler is based on a perfect shuffle interconnection algorithm, which utilizes basic $2 \times 2$ couplers. In this algorithm, the number of $2 \times 2$ couplers to be traversed by any signal is given by $k = \log_2 N$ [223]. Thus, if $\gamma$ is the $2 \times 2$ coupler excess loss including splice loss), expressed in dB, the total loss of the star is given by $10 \log_{10} N + \gamma \log_2 N$. This loss can be compensated at the input or the output of the star through $N$ EDFAs. One advantage of the star is that it can also be used to distribute the pump power to each amplifier (at the expense of one input port). In this case, loss cancellation can be achieved by splicing $N$ strands of EDF at each of the star inputs or outputs, as shown in Figure 7.26b. Backward pumping is similar to a Type A chain element (Chapter 2), where amplification

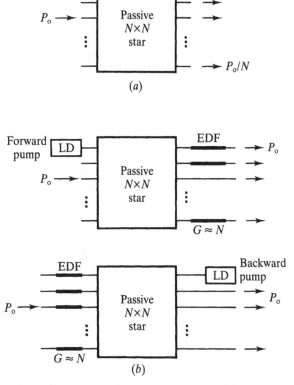

(a)

(b)

**FIGURE 7.26** Principle of (a) passive $N \times N$ star coupler with $1/N$ transmission loss, (b) lossless or transparent $N \times N$ star coupler with forward or backward EDF pumping.

precedes loss. It therefore has the lowest noise figure (assuming identical spontaneous emission factors for both pumping schemes). The practical advantage of this approach is that a transparent star coupler requires only one pump LD and small lengths of highly doped EDFs. With $N \times N$ couplers built from specially designed, wavelength insensitive $2 \times 2$ couplers, the pump could be supplied by an inexpensive $\lambda = 800$ nm GaAlAs diode. As large-size $N \times N$ star couplers can be built from basic $m \times m$ and $n \times n$ star subunits $(mn = N)$, EDFs can also be spliced between each of these subunits. Increased pump power requirements can be met by coupling additional pump LDs at the $N \times N$ coupler inputs. It is also possible to reduce the number of required EDFs through a more complex connection diagram, involving various $m \times m$ subunits (M. I. Irshid and M. Kavehrad [247]). Regardless of the locations of EDFs, the pump power can always be increased above the transparency level in order to provide net signal gain. In this case, the device operates both as a star coupler and as a postamplifier, which can be advantageously used for transmission loss compensation. Star couplers with net gain can also be implemented in certain LAN configurations.

This principle of star coupler transparency through EDFAs was demonstrated experimentally with a $64 \times 64$ fiber based star coupler, using a single pump LD at 1480 nm [245]. Another demonstration was made with an integrated $19 \times 19$ silica waveguide coupler with EDFs spliced at the output ports [246]. As this type of coupler is nearly wavelength independent, efficient 980 nm pumping with optimized EDFs can be used. In the experiment of [246], the 17 dB transmission loss of the coupler could be compensated by 5.5 mW pump at 980 nm. The coupler excess loss (4 dB) means that the pump power actually required at the input is not $19 \times 5.5 \approx$ 105 mW, but 200 mW. One advantage of the integrated version is the flexibility in number of pump ports. A drawback of integrated stars is their limited connectivity when considering large-scale implementation in complex mesh networks. Indeed, each integrated star must be fiber pigtailed to be connected to another integrated star, while connectivity is straightforward in fiber-based stars.

The advantageous connectivity of fiber-based star couplers is illustrated by the reflective star LAN layout shown in Figure 7.27 (Y. K. Chen [248]). This LAN bidirectionally connects 64 terminals through four nodes or concentrators. Each of the nodes is built from two $8 \times 8$ lossless stars. Each node has one pump LD, the output of which is fed to each of the two $8 \times 8$ star pump inputs. Figure 7.28 shows how the $8 \times 8$ lossless stars can be build from basic $2 \times 2$ couplers. Figure 7.28$a$ shows that the $8 \times 8$ star layout includes two $4 \times 4$ stars with pump inputs and four $2 \times 2$ couplers. For $8 \times 8$ star transparency, the $4 \times 4$ stars must have a net gain of 3 dB. Figure 7.28$b$ shows how the $4 \times 4$ stars can be implemented. In this configuration, the $4 \times 4$ star includes six $2 \times 2$ couplers, one WSC and a strand of EDF. Assuming negligible excess loss, the total gain required is $10 \log_{10}(2^4) + 3$ dB, or 15 dB. This gain is sufficiently low to alleviate the neeed for optical isolators, which is important, since all stars are bidirectional. Another possible configuration uses a standard $4 \times 4$ passive star and requires a lower gain. This configuration uses four EDFs spliced at the $4 \times 4$ star output ports; pumping the EDFs requires four WSCs and one $1 \times 4$ pump splitter. Because of the reduced number of signal splittings, the required gain is reduced to $10 \log_{10}(2^2) + 3$ dB, or 7 dB. But this configuration requires more fiber components than the configuration in Figure 7.28$b$. On the other hand, the lower gain yields a lower ASE noise and noise figure, which obviates bandpass filters. Finally, this 64-station LAN requires only four pump diodes, which is cost-effective.

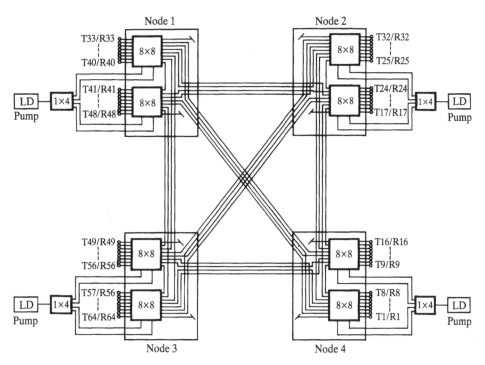

**FIGURE 7.27** Reflective star LAN with 64 terminals, using four lossless nodes; the nodes include two 8 × 8 transparent star couplers and one pump LD. From [248] © 1992 IEEE.

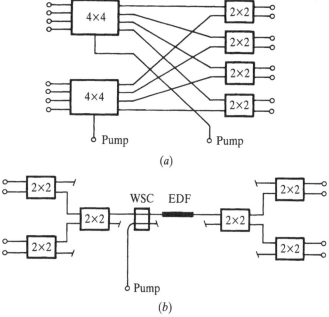

**FIGURE 7.28** Detailed layout of the 8 × 8 star couplers used in the reflective LAN of Figure 7.27: (a) internal layout showing two 4 × 4 stars with 3 dB net gain and four 2 × 2 couplers; (b) detail of the 4 × 4 star, showing six 2 × 2 couplers and one EDF with pump input port.

Only a few experiments involving EDFA-based star LANs have been demonstrated so far. System capacity enhancement was shown by using EDFAs as preamplifiers in an FSK–FDMA star LAN (A. E. Willner et al. [249]), or as a central power boosters in an ASK–TDMA star LAN (S. Culverhouse et al. [250]).

The reflective star LAN in Figure 7.27 is a good example of interactive AON. The possibility of achieving lossless $N \times N$ star couplers could have a considerable impact on the design of future AONs. Complex architectures that were previously considered impractical, because of high intrinsic loss, could now be implemented on large scales, with minimal optical component cost. The accumulated ASE noise is relatively low in such systems, due to the low excess gain required in each star subcouplers and the relatively small number of amplification stages required. The small number of stages is a direct consequence of using perfect shuffle as a connection algorithm. Lossless star LANs could therefore have the potential of connecting as many as 1000 to 10,000 users. Communication protocols of TDM and FDM networks with passive star topology allow us to use only a fraction of the available capacity because of unresolved contention between data packets (I. Habbab et al. [251]). This could be alleviated by increasing the end-user bit rate and using novel collision-correcting algorithms. A large increase in the user's bit rate is made possible by several new factors: the availability of optical amplifiers with wide bandwidth, the availability of practical picosecond GS–LD sources with high output powers, and the possibilities offered by passive TDM and all-optical time domain demultiplexing (Chapter 6).

While the impact of EDFAs in the field of long-haul communication systems is confirmed to a point where it could almost be regarded as past history, the potential of EDFAs in LANs has been only partially explored. It is likely that the impact will be equally significant, though spread over a longer period of time, as the demand for gigabit networks still belongs to the future.

Currently, most of the LAN research splits into two categories: control, management, and protocols (the communication level); and photonic components, system layout, and power budget (the physical level). The content of technical papers shows there is little overlap between the two. Maybe this partly explains why progress in all-optical LANs is apparently slow. Another reason may be that these two newly emerged domains have not had enough time to converge. Both sides continue to progress on their own but the development of AONs strongly depends on convergence.

The same way new protocols must be tested at the communication level, novel EDFA-based architectures must be investigated at the physical level. The high cost of photonic components make large-scale implementation difficult and testbed experiments prohibitive. But in both cases, testing can be achieved through intensive computer simulations. EDFA-based LAN simulations can be extremely time-consuming, due to the large number of differential equations involved, as well as to the multiplicity of physical parameters. As a matter of fact, the determination of optimum parameters could be more complex than in EDFA design, due to ever changing system configurations and operating conditions. In the first phase, the simulation of complex LAN topologies could be greatly facilitated by using various EDFA analytical models (Chapter 1). These models make it possible to rapidly predict the EDFA operating conditions (i.e., gain, saturation and noise), with fair accuracy. Using a mainframe computer, or possibly a workstation, this approach could permit real-time LAN design and simulation. Optimum locations of EDFAs and related parameters could be rapidly determined, using interactive software similar to software for IC design. Once the feasibility of a proposed system is established at the physical level (though BER, SNR, or CNR estimates), the simulation can be continued at the communications level; data from the physical level can also be refined through comprehensive modeling. Alternatively, novel LAN architectures that have proven effective at the communication level could be tried at the physical level, with the possibility of interactive optimization. This approach is likely to facilitate a rapid convergence between communication network and physical network realities. And in order to meet future demand, rapid convergence is a current challenge.

# APPENDICES

# APPENDIX A

# RATE EQUATIONS FOR STARK SPLIT THREE-LEVEL LASER SYSTEMS (SEE SECTION 1.2)

From Figure 2.1, corresponding to a Stark split, three-level laser system, the following rate equations can be derived. First, for the ground level manifold 1:

$$\frac{dN_{11}}{dt} = -A_{NR}^{+}N_{11} + A_{NR}^{-}N_{12} + \sum_{k} A_{k1}N_{2k} - \sum_{l} R_{l1}(N_{11} - N_{3l}) + \sum_{k} W_{k1}(N_{2k} - N_{11})$$

(A.1)

$$\frac{dN_{1j}}{dt} = -(A_{NR}^{+} + A_{NR}^{-})N_{1j} + A_{NR}^{-}N_{1,j+1} + A_{NR}^{+}N_{1,j-1}$$

$$+ \sum_{k} A_{kj}N_{2k} - \sum_{l} R_{lj}(N_{1j} - N_{3l}) + \sum_{k} W_{kj}(N_{2k} - N_{1j}); \; 1 < j < g_1 \quad \text{(A.2)}$$

$$\frac{dN_{1,g1}}{dt} = -A_{NR}^{-}N_{1,g1} + A_{NR}^{+}N_{1,g1-1} + \sum_{k} A_{k,g1}N_{2k} - \sum_{l} R_{l,g1}(N_{1,g1} - N_{3l})$$

$$+ \sum_{k} W_{k,g1}(N_{2k} - N_{1,g1}) \quad \text{(A.3)}$$

For the metastable-level manifold 2:

$$\frac{dN_{21}}{dt} = -A_{NR}^{+}N_{21} + A_{NR}^{-}N_{22} - \sum_{j} A_{j1} \; N_{21} - \sum_{j} W_{1j}(N_{21} - N_{1j}) \quad \text{(A.4)}$$

$$\frac{dN_{2k}}{dt} = -(A_{NR}^{+} + A_{NR}^{-})N_{2k} + A_{NR}^{-}N_{2,k+1} + A_{NR}^{+}N_{2,k-1}$$

$$- \sum_{j} A_{kj}N_{2k} - \sum_{j} W_{kj}(N_{2k} - N_{1j}); \; 1 < k < g_2 \quad \text{(A.5)}$$

$$\frac{dN_{2,g2}}{dt} = A_{32}^{NR}N_{31} - A_{NR}^{-}N_{2,g2} + A_{NR}^{+}N_{2,g2-1} - \sum_{j} A_{g2,j}N_{2,g2} - \sum_{k}(W_{g2} - N_{1j}) \quad \text{(A.6)}$$

For the pump level manifold 3:

$$\frac{dN_{31}}{dt} = -A_{32}^{NR}N_{31} - A_{NR}^{+}N_{31} + A_{NR}^{-}N_{32} + \sum_k A_{k1}N_{2k} + \sum_j R_{1j}(N_{1j} - N_{31}) \quad (A.7)$$

$$\frac{dN_{3l}}{dt} = -(A_{NR}^{+} + A_{NR}^{-})N_{3l} + A_{NR}^{-}N_{3,l+1} + A_{NR}^{+}N_{3,l-1} + \sum_l R_{lj}(N_{1j} - N_{3l}); \quad 1 < l < g_3$$

$$(A.8)$$

$$\frac{dN_{3,g3}}{dt} = -A_{NR}^{-}N_{3,g3} + A_{NR}^{+}N_{3,g3-1} + \sum_j R_{g3,j}(N_{1j} - N_{3,g3}) \quad (A.9)$$

# APPENDIX B

# COMPARISON OF LP$_{01}$ BESSEL SOLUTION AND GAUSSIAN APPROXIMATION FOR THE FUNDAMENTAL FIBER MODE ENVELOPE (SEE SECTIONS 1.3 AND 1.7)

For graded-index fibers with core radius $a$ and numerical aperture NA, the normalized frequency, or $V$ number is defined at wavelength $\lambda$ by $V = 2\pi a\text{NA}/\lambda$ [1]. The LP$_{01}$ solution for the fundamental mode envelope is given by:

$$\psi_s(r < a) = J_0^2\left( U \frac{r}{a} \right) \tag{B.1}$$

$$\psi_s(r > a) = K_0^2\left( W \frac{r}{a} \right) \frac{J_0^2(U)}{K_0^2(W)} \tag{B.2}$$

which correspond to the core $(r < a)$ and the cladding $(r > a)$ regions, respectively. The parameters $U$ and $W$ are related to $V$ through $W^2 = V^2 - U^2$, where $U$ can be approximated by [2]:

$$U = \frac{(1 + \sqrt{2})V}{1 + (4 + V^4)^{0.25}} \tag{B.3}$$

The mode envelope $\psi_s(r)$ has a unity peak value, i.e., $\psi_s(0) = 1$. Defining the mode radius $\omega_s$ corresponding to the Bessel solution through, Eq. (1.131):

$$\omega_s = \left\{ 2 \int_S^r \psi_s(r) r \, dr \right\}^{1/2} \tag{B.4}$$

we obtain from Eqs. (B.1) and (B.2)

$$\omega_s(\text{Bessel}) = a \frac{V K_1(W)}{U K_0(W)} J_0(U) \tag{B.5}$$

In the Gaussian approximation, the mode envelope is defined instead through:

$$\psi_s(r) = \exp\left(-\frac{r^2}{\omega_s^2}\right) \tag{B.6}$$

where $\omega_s$ is the $1/e$ mode radius. The $1/e$ mode radius for which the power coupling efficiency into the $LP_{01}$ mode solution is maximized is given by [2]:

$$\omega_s(\text{Gauss}) = \frac{a}{\sqrt{2}}\left(0.65 + \frac{1.619}{V^{1.5}} + \frac{2.879}{V^6}\right) \tag{B.7}$$

In the following, we compare the intensity distributions corresponding to the exact $LP_{01}$ mode or exact Bessel solution, (B.1) and (B.2), and to the Gaussian mode approximation, (B.6) and (B.7). As an example, we choose two representative wavelengths, $\lambda = 980$ nm and $\lambda = 1530$ nm, corresponding to one typical choice of pump and signal wavelengths in an Er-doped fiber amplifiers. The fiber is chosen to have a cutoff wavelength exactly at $\lambda_c = 980$ nm, which can be realized for instance with a core radius $a = 1.875\ \mu m$ and a numerical aperture of $NA = 0.2$.

Figure B.1 shows the calculated mode envelopes $\bar{\psi}_s(r) = \psi_s(r)/\pi\omega_s^2$ at the two wavelengths, for both Bessel and Gaussian modes, corresponding to this fiber, as a function of distance $r$ to the fiber axis. The same total mode power is assumed, and both envelopes are normalized so that $\int \bar{\psi}_s(r)r\,dr\,d\theta = 1$.

Figure B.1 shows that at the cutoff wavelength ($\lambda = \lambda_c = 980$ nm), the exact and Gaussian intensity distributions are very similar and the corresponding error in actual intensity across the core remains relatively small. The Gaussian approximation predicts a peak intensity at $r = 0$ in excess of 3% of the exact value.

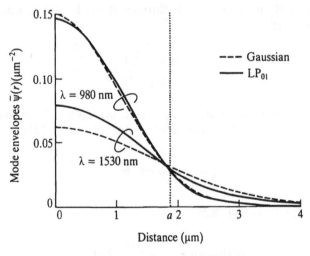

**FIGURE B.1** Comparison between (——) exact Bessel mode intensity distribution ($LP_{01}$) and (– – –) Gaussian mode approximation versus core distance $r$, for the two wavelengths $\lambda = 980$ nm and $\lambda = 1530$ nm. The fiber cutoff wavelength is $\lambda_c = 980$ nm.

On the other hand, at wavelengths well above cutoff, e.g., $\lambda = 1530$ nm, the peak intensity predicted by the Gaussian approximation is much below the actual value, representing an underestimation of 22%. The difference is due to the faster decay of the Bessel solution in the cladding, compared to the Gaussian mode.

Thus, the difference in intensity distributions between the exact Bessel mode solution and the approximated Gaussian mode solution is increasing with the ratio $\lambda/\lambda_c$, as discussed in [1]. This difference is not a real problem for passive fiber waveguides, since the important parameter generally considered is the total power launched into the mode. But it may introduce significant errors in the modeling of doped waveguides, as the mode/dopant interaction is an effect of local intensity rather than of power launched into the mode.

A second type of approximation uses a Gaussian envelope with $1/e$ mode size identical to the Bessel mode size $\omega_s$ defined in Eqs. (B.4) and (B.5). We refer to this approximation as the Gaussian *envelope* approximation, in contrast to the Gaussian *mode* approximation. Figure B.1 shows that the difference in envelope (and corresponding intensities) between Gaussian and Bessel modes can be large at wavelengths above cutoff. In the Gaussian envelope approximation, the Bessel envelope is matched with a Gaussian envelope with $1/e$ size given by $\omega_s$(Bessel). Figure B.2 shows plots of both mode envelopes for the two wavelengths, $\lambda = 980$ nm and $\lambda = 1530$ nm, corresponding to the same fiber waveguide chosen for the previous example. Matching between the two types of mode envelopes is improved, especially at long wavelengths. The absolute difference between the envelopes, represents a maximum absolute error of 5% for any radius $r$, and vanishes near the peak ($r=0$) at any wavelength. As a result, the Gaussian envelope approximation enables a better match of the exact

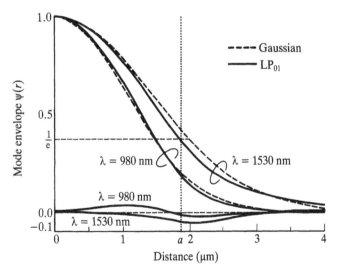

**FIGURE B.2**   Comparison between (———) exact Bessel mode intensity distribution ($LP_{01}$) and (– – –) Gaussian envelope approximation versus core distance $r$, for the two wavelengths $\lambda = 980$ nm and $\lambda = 1530$ nm. The Gaussian envelope has a $1/e$ radius equal to the mode size $\omega_s$ of the exact Bessel envelope, Eqs. (B.4) and (B.5). The fiber cutoff wavelength is $\lambda_c = 980$ nm. The figure also shows plots of the absolute difference between Bessel and Gaussian envelopes versus distance $r$ for the two wavelengths.

intensity distribution, compared to the Gaussian mode approximation. While the Gaussian mode approximation provides the best match with total power launched into the mode, the Gaussian envelope approximation provides the best match with the intensity distribution of the mode at identical powers.

## REFERENCES

[1]   L. Jeunhomme, *Single-mode fiber optics, principles and applications*, second edition, Marcel Dekker, New York, 1990

[2]   D. Gloge, "Weakly guiding fibers," *Applied Optics*, vol 10, no 10, 2252 (1971)

# EXAMPLE OF PROGRAM ORGANIZATION AND SUBROUTINES FOR NUMERICAL INTEGRATION OF GENERAL RATE EQUATION (1.68) OF SECTION 1.6

We give a basic program organization and examples of FORTRAN program subroutines for numerical integration of Eq. (1.68). These can be used as the basis of a complete program for the numerical simulation of Er-doped fiber amplifiers. We use the Runge–Kutta algorithm to integrate Eq. (1.68), but alternate methods may prove more effective in certain applications. The algorithms and basic organization of the subroutines are merely examples. They can serve as guidelines for an initial start in the conception and realization of a more comprehensive program. Many modifications or improvements in their organization can be made after several trial runs.

The example chosen here corresponds to the case of an Er fiber amplifier pumped near $\lambda_p = 1480\,\text{nm}$ in the forward direction, with an input signal power spectrum made of different channels located at arbitrary wavelengths $\lambda_k$. The program computes the output power spectra in the forward and backward directions, corresponding to the solution of Eq. (1.68), for any fiber waveguide and Er-doping profile. For simplicity, the pump and signal mode envelopes are assumed equal.

The source program should include the definition of the EDF parameters (waveguide parameters, cross-section spectra, Er-doping profile), the initial input power conditions (spectra of pump and signals), the declaration of different integration parameters (number of resolved points in fiber length and power spectrum). It should also include a certain number of output statements for monitoring the development of the computation during the run (number of points left to compute, convergence of iteration procedure). We shall not detail here such a complete program organization, which is straightforward for any programmer to do. However, we outline here some important steps and portions of the source program, along with the definitions of the parameters used in the subroutines detailed below.

First, the declaration of main parameter in the source program can follow the sequence:

### Integer and real parameters

| | |
|---|---|
| NPOINT | : number of longitudinal points |
| NRAD | : number of radial points |
| NSPEC | : number of spectral points |
| XLENG | : maximum fiber length $L$ (in m) |
| XRAD | : maximum fiber radius $r$ (in $\mu$m) |
| A | : mode area $\pi\omega^2$ (in $\mu$m$^2$) |
| DZ = XLENG/NPOINT | : increment $dz$ |
| DR = XRAD/NRAD | : increment $dr$ |
| PI | : number $\pi$ |

### Real arrays

| | |
|---|---|
| PSI | : mode envelopes with unity peak $\psi(r)$ |
| RO | : Er-doping density profile (in m$^{-3}$ units) |
| SIGA | : absorption cross section spectrum (in m$^2$) |
| SIGE | : emission cross section spectrum (in m$^2$) |
| ETA | : emission to absorption cross section ratio $\eta$ |
| PSAT | : saturation power spectrum $P_{sat}(\lambda)$ (in mW) |
| GE | : emission coefficient (in m$^{-1}$ units) |
| GA | : absorption coefficient (in m$^{-1}$ units) |
| Z | : fiber coordinate $z$ (in m) |
| R | : radial coordinate $r$ (in $\mu$m) |
| PFOR, FF | : normalized forward power spectrum $p^+$ |
| PBAK, FB | : normalized backward power spectrum $p^-$ |
| PO | : normalized equivalent input noise power spectrum |

The source program must include the following steps:

- Input Er-doped fiber waveguide parameters
- Input pump and signal powers (in mW)
- Calculation of mode envelope solution PSI (averaged for all wavelengths)
- Determination of mode area $A$ by surface integration of mode envelope PSI
- Acquisition of experimental cross section datafiles SIGA, SIGE from disk
- Generation of spectrum ETA
- Generation of spectrum PSAT
- Generation of normalized spectrum PO
- Generation of normalized input spectra PFOR, PBAK
- Declaration of common variables:

```
COMMON/a/NSPEC,DZ,GE,GA,PO
```

- Generation of fiber longitudinal and radial coordinate arrays:

```
      DO 50 L=1,NPOINT
      Z(L)=L*DZ
 50   CONTINUE
      DO 51 K=1,NRAD
R(K)=K*DR
 51   CONTINUE
```

- Setting input conditions:

```
      DO 60 I=1,NSPEC
      FF(I)=PFOR(I)
      FB(I)=PBAK(I)
      QF(I)=0
      QB(I)=0
 60   CONTINUE
```

- Fiber integration loop (for integration in backward direction change DZ into −DZ):

```
   DO 1000 L=1,NPOINT
   C    Calculation of absorption and emission coefficients
        DO 500 I=1,NSPEC
   C    Calculation of normalized power sums in Eq. (1.68)
        S1=0
        S2=0
        S3=0
        DO 150 M=1,NSPEC
        S1=S1+FF(M)+FB(M)
        S2=S2+(FF(M)+FB(M))/(1+ETA(M))
        S3=S3+(FF(M)+FM(M))*ETA(M)/(1+ETA(M))
   150  CONTINUE
        SE=0
        SA=0
   C    Integrate coefficients over the fiber transverse plane
        DO 200 K=1,NRAD-1
        E1=RO(K)*(PSI(K)/A)*(ETA(I)*S2*PSI(K))/
                                        (1+S1*PSI(K))
        E2=RO(K+1)*(PSI(K+1)/A)*(ETA(I)*S2*PSI(K+1))/
                                        (1+S1*PSI(K+1))
        A1=RO(K)*(PSI(K)/A)*(1+S3*PSI(K))/(1+S1*PSI(K))
        A2=RO(K+1)*(PSI(K+1)/A)*(1+S3*PSI(K+1))/
                                        (1+S1*PSI(K+1))
        SE=SE+DR*(E1*R(K)*+E2*R(K+1))/2
        SA=SA+DR*(A1*R(K)+A2*R(K+1))/2
```

```
 200  CONTINUE
      GE(I)=SIGE(I)*SE*2*PI
      GA(I)=SIGA(I)*SA*2*PI
 500  CONTINUE
C     Integration over fiber length from z to z+dz
      CALL EQDIF(FF,FB,QF,QB)
1000  CONTINUE
C     Save output spectra
      DO 600 I=1,NSPEC
      PFOR(I)=FF(I)
      PBAK(I)=FB(I)
 600  CONTINUE
```

- Compare output solutions at $z = L$ to solutions from previous trials
- Iterate fiber integration loop until convergence is reached
- Determine signal gain, total ASE power, pump output power, etc.
- Store data files Z, PFOR, PBAK, and other results on disk
- Display results
- Plot various data on screen
- Iterate the above program with modified values for parameters such as pump power, signal powers, signal wavelengths, fiber length, waveguide parameters, Er concentration, etc.

```
1100  STOP
      END
      SUBROUTINE EQDIF(FF,FB,QF,QB)
C  1  Corresponds to differential equation system of Eq. (1.68)
      COMMON/a/NSPEC,DZ,GE,GA,PO
      INTEGER NSPEC
      REAL DZ,SF,SB,RF,RB
      REAL FF(200),FBB(200),QF(200),QB(200)
      REAL PO(200),GE(200),GA(200)
      REAL GF(200),GB(200)
      DO 100 I=1,NSPEC
C  1  ------------------------------------------------
      SF=DZ*(GE(I)*(FF(I)+PO(I))-GA(I)*FF(I))
      SB=-DZ*(GE(I)*(FB(I)+PO(I))-GA(I)*FB(I))
      RF=(SF-2*QF(I))/2
      RB=(SB-2*QB(I))/2
      GF(I)=FF(I)+RF
      GB(I)=FB(I)+RB
      QF(I)=QF(I)+3*RF-SF/2
      QB(I)=QB(I)+3*RB-SB/2
C  2  ------------------------------------------------
      SF=DZ*(GE(I)*(GF(I)+PO(I))-GA(I)*GF(I))
      SB=-DZ*(GE(I)*(GB(I)+PO(I))-GA(I)*GB(I))
      RF=(SF-QF(I))/2
      RB=(SB-QB(I))/2
```

```
      FF(I)=GF(I)+RF
      FB(I)=GB(I)+RB
      QF(I)=QF(I)+3*RF-0.2928*SF
      QB(I)=QB(I)+3*RB-0.2928*SB
C  3  ------------------------------------------------
      SF=DZ*(GE(I)*(FF(I)+PO(I))-GA(I)*FF(I))
      SB=-DZ*(GE(I)*(FB(I)+PO(I))-GA(I)*FB(I))
      RF=1.5*(SF-QF(I))
      RB=1.5*(SB-QB(I))
      GF(I)=FF(I)+RF
      GB(I)=FB(I)+RB
      QF(I)=QF(I)+3*RF-1.707*SF
      QB(I)=QB(I)+3*RB-1.707*SB
C  4  ------------------------------------------------
      SF=DZ*(GE(I)*(GF(I)+PO(I))-GA(I)*GF(I))
      SB=-DZ*(GE(I)*(GB(I)+PO(I))-GA(I)*GB(I))
      RF=0.166*(SF-2*QF(I))
      RB=0.166*(SB-2*QB(I))
      FF(I)=GF(I)+RF
      FB(I)=GB(I)+RB
      QF(I)=QF(I)+3*RF-SF/2
      QB(I)=QB(I)+3*RB-SB/2
100   CONTINUE
      RETURN
      END
C     ------------------------------------------------
```

# EMISSION AND ABSORPTION COEFFICIENTS FOR THREE-LEVEL SYSTEMS WITH GAUSSIAN ENVELOPE APPROXIMATION (SEE SECTION 1.7)

We derive approximate closed-form expressions for the absorption and emission coefficients $g_{ap}$, $g_{ek}$, $g_{ak}$ defined in Eqs. (1.74)–(1.76) corresponding to a three-level system, with Gaussian envelope approximation and step-like dopant distribution with radius $a_0$. The pump and signal mode radii are $\omega_p$ and $\omega_s$, respectively. We introduce first the following definitions:

$$\alpha = \left(\frac{\omega_p}{\omega_s}\right)^2 \tag{D.1}$$

$$\Gamma_{p,s} = 1 - \exp\left(-\frac{a_0^2}{\omega_{p,s}^2}\right) \tag{D.2}$$

$$\kappa = \sum_j (p_j^+ + p_j^-) \tag{D.3}$$

$$\mathscr{D}_e = \frac{1}{\kappa}\sum_j \frac{p_j^+ + p_j^-}{1 + \eta_j} \tag{D.4}$$

$$\mathscr{D}_a = \frac{1}{\kappa}\sum_j \frac{\eta_j}{1 + \eta_j}(p_j^+ + p_j^-) \tag{D.5}$$

Using the above equations and Eqs. (1.80) and (1.81) for the mode envelopes and the dopant density distribution we obtain, after a change of variables, from eqs. (1.74)–(1.76):

$$g_{ap}(z) = \rho_0 \sigma_{ap} \int_{1-\Gamma_p}^{1} \frac{1 + \kappa \mathscr{D}_a x^\alpha}{1 + (q^+ + q^-)x + \kappa x^\alpha}\, dx \equiv \rho_0 \sigma_{ap} \Gamma_{ap}(z, q^{\pm}, p_j^{\pm}, \alpha) \tag{D.6}$$

$$g_{ek}(z) = \eta_k \rho_0 \sigma_{ak} \int_{1-\Gamma_s}^{1} \frac{\kappa \mathscr{D}_e x + (q^+ + q^-)x^{1/\alpha}}{1 + \kappa x + (q^+ + q^-)x^{1/\alpha}} \, dx \equiv \eta_k \rho_0 \sigma_{ak} \Gamma_{ek}(z, q^\pm, p_j^\pm, \alpha) \quad (D.7)$$

$$g_{ak}(z) = \rho_0 \sigma_{ak} \int_{1-\Gamma_s}^{1} \frac{1 + \kappa \mathscr{D}_a x}{1 + \kappa x + (q^+ + q^-)x^{1/\alpha}} \, dx \equiv \rho_0 \sigma_{ak} \Gamma_{ak}(z, q^\pm, p_j^\pm, \alpha) \quad (D.8)$$

As $\alpha$ is not generally an integer, it is not possible to reduce Eqs. (D.6)–(D.8) to integrated, closed-form expressions for the power dependent overlap integrals $\Gamma_{ap}$ and $\Gamma_{ek,ak}$. The overlap integrals $\Gamma_{ep}(z, 0, p_j^\pm, 1)$ and $\Gamma_{ek,ak}(z, 0, p_j^\pm, 1)$, corresponding to the two-level system case (the pump is included in the $p_j^\pm$ spectrum), can be expressed as integrated, closed-form expressions, as shown in Section 1.7. But for three-level systems, we can make the approximation $\alpha \approx 1$ in the integrands in order to yield elementary integrals; this approximation is justified a posteriori. We obtain then from Eqs. (D.6)–(D.8):

$$\Gamma_{ap}(z, q^\pm, p_j^\pm, 1) \approx \Gamma_p \left\{ \mathscr{D}_a' + \frac{\mathscr{D}_a' - 1}{\kappa' \Gamma_p} \log\left(1 - \frac{\kappa'}{1 + \kappa'} \Gamma_p\right) \right\} \quad (D.9)$$

$$\Gamma_{ek}(z, q^\pm, p_j^\pm, 1) \approx \Gamma_s \left\{ \mathscr{D}_e' + \frac{\mathscr{D}_e'}{\kappa' \Gamma_s} \log\left(1 - \frac{\kappa'}{1 + \kappa'} \Gamma_s\right) \right\} \quad (D.10)$$

$$\Gamma_{ak}(z, q^\pm, p_j^\pm, 1) \approx \Gamma_s \left\{ \mathscr{D}_a' + \frac{\mathscr{D}_a' - 1}{\kappa' \Gamma_s} \log\left(1 - \frac{\kappa'}{1 + \kappa'} \Gamma_s\right) \right\} \quad (D.11)$$

with

$$\kappa' = q^+ + q^- + \sum_j (p_j^+ + p_j^-) \quad (D.12)$$

$$\mathscr{D}_e' = \frac{1}{\kappa'} \left\{ q^+ + q^- + \sum_j \frac{p_j^+ + p_j^-}{1 + \eta_j} \right\} \quad (D.13)$$

$$\mathscr{D}_a' = \frac{1}{\kappa'} \sum_j \frac{\eta_j(p_j^+ + p_j^-)}{1 + \eta_j} \quad (D.14)$$

How accurate are Eqs. (D.9)–(D.11) for the power dependent overlap integrals with respect to their exact expressions (D.6)–(D.8) with $\alpha \neq 1$?

To answer this question, we can study the coefficients changes with pump and signal power as a function of the ratio $a_0/\omega_s$ (doping confinement factor). We choose the same fiber waveguide as in Appendix B, i.e., with cutoff wavelength $\lambda_c = 980$ nm, core radius $a = 1.875\ \mu$m, pump at $\lambda_p = 980$ nm and signal at $\lambda_s = 1530$ nm. For this waveguide, the exact mode radii for the pump and signal are $\omega_p = 1.475\ \mu$m and $\omega_p = 2.00\ \mu$m, respectively, corresponding to $\alpha = 0.544$. For simplicity, and without loss of generality, we can assume $\eta_j = 1$, which gives $\mathscr{D}_e = \mathscr{D}_a = 1/2$. Let $q = q^+ + q^-$ and $p = \Sigma(p_j^+ + p_j^-)$, which gives $\kappa' = p + q$, $\mathscr{D}_e' = (q + p/2)\ \kappa'$ and $\mathscr{D}_a' = p/(2\kappa')$ in Eqs. (D.12)–(D.14).

The power dependent overlap integrals obtained from Eqs. (D.6)–(D.8), exact forms, and Eqs. (D.9)–(D.11), approximate forms, for the pump absorption and the signal

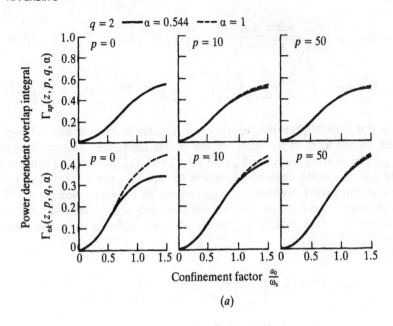

(a)

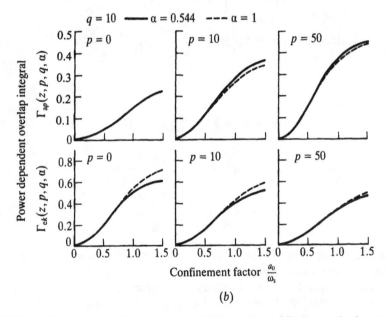

(b)

**FIGURE D.1** Power dependent overlap integral factors $\Gamma_{ap}$ and $\Gamma_{ek}$ versus doping confinement factor $a_0/\omega_s$, for pump at $\lambda_p = 980$ nm and signal at $\lambda_s = 1530$ nm, respectively, showing a comparison between exact formula, $\alpha = 0.544$ (full line) and approximate formula, $\alpha = 1$ (dashed line). The factors are plotted for different pump power $q$ and signal power $p$: in Figure D.1a, $q = 2$, $p = 0, 10, 50$ and in Figure D.1b $q = 10$, $p = 0, 10, 50$. The fiber waveguide parameters are identical to those of the example in Appendix B. The parameter $\alpha$ is the square of the mode size ratio $\omega_p/\omega_s$.

emission coefficients are plotted in Figures D.1$a$,$b$ as functions of the confinement factor $a_0/\omega_s$, for different values of $q$ and $p$. The figures show that for $p = 0$ or $p \ll q$ (small-signal or unsaturated gain regime), both exact and approximate forms coincide for $\Gamma_{ap}$, as the dependence in $\alpha$ in Eq. (D.6) disappears. Likewise, for $p = 50$ or $p > q$ (high signal or highly saturated gain regime), the approximate formula for $\Gamma_{ek}$ is very nearly accurate, due to the same reason in Eq. (D.7). A maximum difference of 15% (for $a_0/\omega_s < 1$) is observed at low pump ($q = 2$, Figure D.1$a$). This case is not of interest as it corresponds to a locally underpumped fiber with no signal. The difference reduces to a maximum of 6% (for $a_0/\omega_s < 1$) for higher pump ($q = 10$) and signal powers $p \leqslant 50$. The same observations apply for $\Gamma_{ek}$.

These plots show that for confinement factors of unity or less ($a_0/\omega_s \leqslant 1$), the approximate expressions for the power dependent overlap integrals are accurate within 6%, depending on the pump-to-signal power ratio $p/q$. The formula are 100% accurate at high pump, i.e., $p \gg q$, corresponding to the unsaturated gain regime, and 96% accurate at high signals, i.e., $q \gg p$, corresponding to the highly saturated gain regime.

---

# ANALYTICAL SOLUTIONS FOR PUMP AND SIGNAL + ASE IN THE UNSATURATED GAIN REGIME, FOR UNIDIRECTIONAL AND BIDIRECTIONAL PUMPING (SEE SECTION 1.9)

---

We derive approximate closed-form expressions corresponding to analytical solutions of Eqs. (1.125) and (1.126). The pump and signal + ASE Eqs. (1.125) and (1.126) form the coupled nonlinear system:

$$\frac{dq^+}{dz} = -\alpha_p \left( \frac{1}{1 + q^+ + q^-} + \varepsilon_p \right) q^+ \tag{E.1}$$

$$\frac{dq^-}{dz} = \alpha_p \left( \frac{1}{1 + q^+ + q^-} + \varepsilon_p \right) q^- \tag{E.2}$$

$$\frac{dp_k^+}{dz} = \alpha_k \frac{1}{1 + q^+ + q^-} \left[ \left\{ \frac{\eta_k - \eta_p}{1 + \eta_p}(q^+ + q^-) - 1 - \varepsilon_k(1 + q^+ + q^-) \right\} p_k^+ \right.$$
$$\left. + \frac{\eta_k}{1 + \eta_p}(q^+ + q^-)2p_{0k} \right] \tag{E.3}$$

$$\frac{dp_k^-}{dz} = -\alpha_k \frac{1}{1 + q^+ + q^-} \left[ \left\{ \frac{\eta_k - \eta_p}{1 + \eta_p}(q^+ + q^-) - 1 - \varepsilon_k(1 + q^+ + q^-) \right\} p_k^- \right.$$
$$\left. + \frac{\eta_k}{1 + \eta_p}(q^+ + q^-)2p_{0k} \right] \tag{E.4}$$

In the above equations, we have introduced the parameters $\varepsilon_{p,k} = \alpha'_{p,k}/\alpha_{p,k}$ corresponding to the ratio of fiber background loss to ionic absorption for the pump and signal wavelengths at $\lambda_p$ and $\lambda_k$, respectively.

We solve Eqs. (E.1)–(E.4) by first decoupling the two pump Eqs. (E.1) and (E.2). We multiply them by $q^-$ and $q^+$, respectively, then sum the result. We obtain $d(q^+ q^-)/dz = 0$, or:

$$q^+(z)q^-(z) = \text{const.} = q^+(0)q^-(0) = q^+(L)q^-(L) \equiv a \tag{E.5}$$

where $L$ is the total fiber length. This result enables us to decouple the forward and backward pump equations, (E.1) and (E.2), by substituting $q^-(z) = q^+(L)q^-(L)/q^+(z)$ in the $dq^+/dz$ equation (E.1). For simplicity, we use the notation $q^\pm(0) = q_0^\pm$ and $q^\pm(L) = q_L^\pm$. After integration in Eq. (E.1), we find a transcendental relation for the forward pump solution:

$$\log\left(\frac{q_0^+}{q_L^+}\right) + \frac{1}{\varepsilon_p}\int_{q_0^+}^{q_L^+}\frac{dx}{a + bx + x^2} = \alpha'_p L \qquad (E.6)$$

with $a = q_L^+ q_L^-$ and $b = (1 + \varepsilon_p)/\varepsilon_p$.

The analytical expression of the integral in Eq. (E.6) depends upon the sign of the parameter $\Delta = 4a - b^2$ [1], a function of both forward output and backward input pumps at $z = L$, since $a = q_L^+ q_L^- = q_0^+ q_0^-$. In the case of unidirectional pumping, we have $q_L^- = 0$, or $a = 0$, and the parameter $\Delta$ is always negative. But in the general case of bidirectional pumping, as the forward output pump $q_L^+$ is not known a priori for given forward and backward input pumps and fiber length, the sign of $\Delta$ can only be known a posteriori. In the following, we first consider the most general case of bidirectional pumping followed by the simpler case of unidirectional (forward) pumping.

Depending of the sign of $\Delta$, the integral in Eq. (E.6) takes the following forms [1]:

$$I(y) = \int^y\frac{dx}{a + bx + x^2} = \begin{cases} \dfrac{-2}{\sqrt{-\Delta}}\operatorname{artanh}\left(\dfrac{b + 2y}{\sqrt{-\Delta}}\right); & \Delta < 0 \\[3mm] \dfrac{-2}{b + 2y} & ; \Delta = 0 \\[3mm] \dfrac{-2}{\sqrt{\Delta}}\operatorname{arctg}\left(\dfrac{b + 2y}{\sqrt{\Delta}}\right) & ; \Delta > 0 \end{cases} \qquad (E.7)$$

The case $\Delta = 0$ is of limited interest, as it is too restrictive. In this case, the forward output pump power $q_L^+$ at $z = L$ is exactly $q_L^+ = b^2/4q_L^- = (1 + \alpha_p/\alpha'_p)^2/4q_L^-$. This condition is very unlikely to occur, as the input pump power conditions $q_0^+, q_L^-$ are then bound to take specific values, determined by the solution of the following transcendental equation obtained from Eqs. (E.6) and (E.7):

$$\log\left(\frac{4}{b^2}q_0^+ q_L^-\right) + \frac{2}{b\varepsilon_p}\frac{1 - \dfrac{4}{b^2}q_0^+ q_L^-}{1 + \dfrac{4}{b^2}q_0^+ q_L^- + \dfrac{2}{b}(q_0^+ + q_L^-)} = \alpha'_p L \qquad (E.8)$$

The condition $\Delta = 0$ is thus overly restrictive and unlikely to occur in most cases of interest, as confirmed later in this analysis.

Considering now the case $\Delta \neq 0$, we obtain from Eqs. (E.6) and (E.7) the two possible equations for the forward output pump $q_L^+$:

$$\log\left(\frac{q_L^+}{q_0^+}\right) - \frac{1}{\varepsilon_p\sqrt{-\Delta}}\log\left\{\frac{b(q_0^+ + q_L^+) + 2q_L^+(q_0^+ + q_L^-) - (q_0^+ - q_L^+)\sqrt{-\Delta}}{b(q_0^+ + q_L^+) + 2q_L^+(q_0^+ + q_L^-) + (q_0^+ - q_L^+)\sqrt{-\Delta}}\right\}$$

$$= -\alpha'_p L; \Delta < 0 \quad (E.9)$$

$$\log\left(\frac{q_L^+}{q_0^+}\right) + \frac{2}{\varepsilon_p\sqrt{\Delta}}\, \text{arctg}\left\{\frac{(q_0^+ - q_L^+)\sqrt{\Delta}}{b(q_0^+ + q_L^+) + 2q_L^+(q_0^+ + q_L^-)}\right\} = -\alpha_p' L; \ \Delta > 0 \quad \text{(E.10)}$$

with $\Delta = 4q_L^+ q_L^- - b^2$. Both Eqs. (E.9) and (E.10) are transcendental and solving them explicitly for $q_L^+$ requires a numerical method. But we need not actually solve them to obtain an explicit expression of the fiber gain as a function of the input power conditions, [2] and [3], as shown below.

We focus now on the signal + ASE equations (E.3) and (E.4). For simplicity of notation, let $Q = q^+ + q^- \equiv q^+ + a/q^+$ be the total pump power at coordinate $z$. We take the ratio of eqs. (E.3) and (E.4) to eq. (E.1), which yields a $z$-independent equation:

$$\frac{dp_k^\pm}{dq^+} = \pm\frac{\alpha_k}{\alpha_p}\frac{1}{q^+(1 + \varepsilon_p + \varepsilon_p Q)}\left[\left\{\frac{\eta_p - \eta_k}{1 + \eta_p}Q + 1 + (1 + Q)\varepsilon_k\right\}p_k^\pm - \frac{\eta_k}{1 + \eta_p}2p_{0k}Q\right]$$

$$\text{(E.11)}$$

Equation (E.11) is linear in $p_k^\pm$ and can be straightforwardly integrated to give for instance:

$$p_k^+(q_L^+) = G_k^+(q_L^+)\left\{p_k^+(q_0^+) + \frac{\alpha_k}{\varepsilon_p\alpha_p}\frac{\eta_k}{1 + \eta_p}2p_{0k}\int_{q_L^+}^{q_0^+}\frac{a + x^2}{x(a + bx + x^2)G_k^+(x)}\, dx\right\}\text{(E.12)}$$

with the forward gain $G_k^+(x)$ at $\lambda_k$ being defined by:

$$G_k^+(x) = \exp\left\{\frac{\alpha_k}{\alpha_p}\frac{1 + \varepsilon_k}{\varepsilon_p}\int_{q_0^+}^{x}\frac{C(a + x^2) + x}{x(a + bx + x^2)}\, dx\right\} \quad \text{(E.13)}$$

with

$$C = \frac{\dfrac{\eta_p - \eta_k}{1 + \eta_p} + \varepsilon_k}{1 + \varepsilon_k} \quad \text{(E.14)}$$

Let $G_k$ be the EDFA gain at $\lambda_k$ and at $z = L$, i.e., $G_k \equiv G_k^+(q_L^+)$. An elementary calculation in Eq. (E.13) yields:

$$\log G_k = \frac{\alpha_k}{\alpha_p}\frac{1 + \varepsilon_k}{\varepsilon_p}\left\{C\log\left(\frac{q_L^+}{q_0^+}\right) + (1 - bC)\int_{q_0^+}^{q_L^+}\frac{dx}{a + bx + x^2}\right\}$$

$$= \frac{\alpha_k}{\alpha_p}\frac{1 + \varepsilon_k}{\varepsilon_p}\left\{C\log\left(\frac{q_L^+}{q_0^+}\right) + (1 - bC)[I(q_L^+) - I(q_0^+)]\right\} \quad \text{(E.15)}$$

where $I(y)$ is defined by the integral in Eq. (E.7).

Combining Eqs. (E.6) and (E.15), we can eliminate the unknown integral term $I(y)$ to obtain, after elementary algebra, a simple relation linking the forward output pump $q_L^+$ to the forward input pump $q_0^+$ and the gain $G_k$ [3]:

$$\boxed{q_L^+ = q_0^+ \exp(-\mathscr{A}_{pk}L)} \tag{E.16}$$

with

$$\mathscr{A}_{pk} = \alpha_p \frac{1 + \eta_p}{1 + \eta_k} \left\{ (1 + \varepsilon_k)(1 - bC)\varepsilon_p - \frac{\log G_k}{\alpha_k L} \right\} \tag{E.17}$$

By definition, we call the parameter $\mathscr{A}_{pk}$ the *gain dependent pump absorption coefficient*. The backward output pump $q_0^-$ at $z = 0$ is given by the relation $q_0^- = a/q_0^+ = q_L^+ q_L^-/q_0^+$, so we obtain for $q_0^-$ an expression similar to Eq. (1.16), i.e., $q_0^- = q_L^- \exp(-\mathscr{A}_{pk}L)$.

Since we now have an explicit expression for the forward output pump power $q_L^+$, we can express the parameter $\Delta$ as:

$$\boxed{\Delta = 4q_0^+ q_L^- \exp(-\mathscr{A}_{pk}L) - b^2} \tag{E.18}$$

In the general case of bidirectional pumping, the sign of $\Delta$ can then be determined from the value of the gain $G_k$, as we obtain from Eqs. (E.17) and (E.18)

$$\begin{aligned} \Delta > 0 &\Leftrightarrow G_k > \exp(U_k) \\ \Delta < 0 &\Leftrightarrow G_k < \exp(U_k) \end{aligned} \tag{E.19}$$

with

$$U_k = \alpha_k(1 + \varepsilon_k)(1 - bC)\varepsilon_p L + \frac{\alpha_k}{\alpha_p} \frac{1 + \eta_k}{1 + \eta_p} \log\left(\frac{4q_0^+ q_L^-}{b^2}\right) \tag{E.20}$$

It can be verified that in practical cases, the value of the exponent $U_k$ is high ($U_k > 10$), while the unsaturated gain value is comparatively low ($G_k < 100$), so the condition $G_k > \exp(U_k)$, corresponding to the case $\Delta > 0$, is generally not applicable. This can be checked by taking typical values as an example: $\eta_k = 1$, $\eta_p = 0$, $\alpha_k/\alpha_p = 2$, $\varepsilon_k = \varepsilon_p = 0.5$ to $0.0001$ and $q_0^+ = q_L^- = 5$ to $50$, which gives $U_k > 10$, or $\Delta > 0 \Leftrightarrow G_k > 43\,\mathrm{dB}$). For bidirectional pumping, we shall therefore consider that, in all practical cases of interest, assuming an unsaturated gain regime, the condition $\Delta < 0$ is verified. For unidirectional pumping, $\Delta = -b^2$ and the condition $\Delta < 0$ holds.

Consider the transcendental pump equation (E.9) for $\Delta < 0$, and using relation (E.16), we can obtain, after elementary transformation, an implicit equation linking the amplifier gain $G_k$ (contained in the definition of the parameter $\mathscr{A}_{pk}$) to the input pump powers $q_0^+$, $q_L^-$:

$$q_0^+ + q_L^- = \frac{b}{2} \frac{e^{\mathscr{A}_{pk}L}}{e^{-wL} - 1} \left\{ (1 + e^{-\mathscr{A}_{pk}L})(1 - e^{-wL}) - (1 - e^{-\mathscr{A}_{pk}L})(1 + e^{-wL}) \sqrt{1 - 4\frac{q_0^+ q_L^-}{b^2}} \right\} \tag{E.21}$$

with

$$w = b\varepsilon_p \sqrt{1 - 4\frac{q_0^+ q_L^-}{b^2}} \, (\mathscr{A}_{pk} - \alpha_p'). \tag{E.22}$$

We have thus obtained in Eq. (E.21) a relation between the input pump power conditions $q_0^+, q_L^-$ for a given fiber gain $G_k$, length $L$, and all the basic fiber parameters contained in the definition of $\mathscr{A}_{pk}$ and $b$. When the forward $q_0^+$ and backward $q_L^-$ input pump powers are nonzero, the factor $\exp(-wL)$ makes Eq. (E.21) transcendental. It must therefore be solved numerically. When one of the input pump powers is zero (unidirectional pumping), Eq. (E.21) can be solved exactly.

Consider forward pumping ($q_0^+ \neq 0$, $q_L^- = 0$). From Eqs (E.21) and (E.22), we find the input pump power $q_0^+$ [2]:

$$\boxed{q_0^+ = be^{BL}\frac{e^{AL} - 1}{e^{BL} - 1}} \tag{E.23}$$

with

$$A = \mathscr{A}_{pk} - B = \varepsilon_p \alpha_p \frac{1 + \eta_p}{1 + \eta_k}\left\{1 + \varepsilon_k + \frac{\log G_k}{\alpha_k L}\right\} \tag{E.24}$$

$$B = w(q_L^- = 0) = b\varepsilon_p(\mathscr{A}_{pk} - \alpha_p') = -\alpha_p(1 + \varepsilon_p)\frac{1 + \eta_p}{1 + \eta_k}\left\{(1 + \varepsilon_k)C + \frac{\log G_k}{\alpha_k L}\right\} \tag{E.25}$$

where $C$ is defined in Eq. (E.14). Note that the expression for $B$ in [2] should read as Eq. (E.25).

We have thus obtained with Eq. (E.23) an explicit relation between input pump power and gain at any signal wavelength $\lambda_k$, in the simplest case of unsaturated gain and forward pumping, i.e., $q_0^+ = f(G_k)$. The symmetry of eqs. (E.21) and (E.22) in their dependence on $q_0^+$ and $q_L^-$ means that eq. (E.23) also describes the backward unidirectional pump power $q_L^-$. From Eq. (E.16) and (E.24), we find that the output pump can be also expressed as $q_L^+ = q_0^+ \exp\{-(A + B)L\}$.

In the simplest case of unidirectional pumping, the amplified spontaneous emission (ASE) power spectra can also be derived under the form of explicit analytical solutions, which are functions only of the input and output pump powers [4].

In order to show this, we rewrite Eq. (E.12) with $a = 0$, assuming no input signal ($p_k^+(q_0^+) = 0$), which gives the forward ASE power at $z = L$, i.e., $p_{k,\,ASE}^+(q_L^+)$, and we derive a similar equation for the backward ASE power at $z = 0$, i.e., $p_{k,\,ASE}^-(q_0^+)$:

$$p_k^+(q_L^+) = 2p_{0k}\frac{\alpha_k}{\varepsilon_p \alpha_p}\frac{\eta_k}{1 + \eta_p}G_k^+(q_L^+)\int_{q_L^+}^{q_0^+}\frac{dx}{(b + x)G_k^+(x)} \tag{E.26}$$

and

$$p_k^-(q_0^+) = 2p_{0k}\frac{\alpha_k}{\varepsilon_p \alpha_p}\frac{\eta_k}{1 + \eta_p}G_k^-(q_0^+)\int_{q_L^+}^{q_0^+}\frac{dx}{(b + x)G_k^-(x)} \tag{E.27}$$

with

$$G_k^+(x) = \exp\left\{\frac{\alpha_k}{\alpha_p}\frac{1+\varepsilon_k}{\varepsilon_p}\int_{q_0^+}^x \frac{Cx+1}{x(b+x)}\,dx\right\} \tag{E.28}$$

$$G_k^-(x) = \exp\left\{-\frac{\alpha_k}{\alpha_p}\frac{1+\varepsilon_k}{\varepsilon_p}\int_x^{q_L^+} \frac{Cx+1}{x(b+x)}\,dx\right\} \tag{E.29}$$

Integration in equation (E.28) and (E.29) yields:

$$G_k^+(x) = \exp\left\{\frac{\alpha_k}{\alpha_p}\frac{1+\varepsilon_k}{1+\varepsilon_p}\left[(bC-1)\log\left(\frac{b+x}{b+q_0^+}\right)-\log\left(\frac{q_0^+}{x}\right)\right]\right\} \tag{E.30}$$

$$G_k^-(x) = \exp\left\{-\frac{\alpha_k}{\alpha_p}\frac{1+\varepsilon_k}{1+\varepsilon_p}\left[(bC-1)\log\left(\frac{b+x}{b+q_L^+}\right)-\log\left(\frac{q_L^+}{x}\right)\right]\right\} \tag{E.31}$$

Finally, substitution of Eqs. (E.30) and (E.31) into Eqs. (E.26) and (E.27) gives:

$$p_{k,\text{ASE}}^+(q_L^+) = 2p_{0k}\frac{\alpha_k}{\alpha_p}\frac{\eta_k}{1+\eta_p}I^+(q_0^+,q_L^+)\cdot\exp\left\{\frac{\alpha_k}{\alpha_p}\frac{1+\varepsilon_k}{1+\varepsilon_p}\left[(bC-1)\log(b+q_L^+)+\log(q_L^+)\right]\right\} \tag{E.32}$$

$$p_{k,\text{ASE}}^-(q_0^+) = 2p_{0k}\frac{\alpha_k}{\alpha_p}\frac{\eta_k}{1+\eta_p}I^-(q_0^+,q_L^+)\cdot\exp\left\{-\frac{\alpha_k}{\alpha_p}\frac{1+\varepsilon_k}{1+\varepsilon_p}\left[(bC-1)\log(b+q_0^+)+\log(q_0^+)\right]\right\} \tag{E.33}$$

with

$$I^+(q_0^+,q_L^+) = \frac{1}{\varepsilon_p}\int_{q_L^+}^{q_0^+} dx\,\frac{X^{-u_{pk}}}{(b+X)^{1+u_{pk}(bC-1)}} \tag{E.34}$$

$$I^-(q_0^+,q_L^+) = \frac{1}{\varepsilon_p}\int_{q_L^+}^{q_0^+} dx\,\frac{X^{u_{pk}}}{(b+X)^{1-u_{pk}(bC-1)}} \tag{E.35}$$

and

$$u_{pk} = \frac{\alpha_k}{\alpha_p}\frac{1+\varepsilon_k}{1+\varepsilon_p} \tag{E.36}$$

Equations (E.32) and (E.33) explicitly define the forward and backward ASE output power spectra (i.e., the power distribution versus wavelength $\lambda_k$) as a function of the input and output (forward) pump powers. Both pump powers are related to the fiber gain $G_k$ through Eqs. (E.16) and (E.23). therefore, the ASE power spectra is explicitly determined as function of the fiber gain $G_k$ only. It can also be explicitly determined as a function of the input pump power $q_0^+$ only, by taking the numerical inverse of Eq. (E.23) for $q_0^+ = f(G_k)$. The integrals involved in Eqs. (E.25) and (E.36) must be computed numerically, as in the general case, they have no analytical form.

In the case of negligible fiber background loss, for which $\varepsilon_{p,k} \to 0$ or $b \to \infty$, as considered in the study [4], we obtain from the above results, Eqs. (E.32)–(E.35)

$$p_{k,\text{ASE}}^+(q_L^+) = 2p_{0k}\frac{\alpha_k}{\alpha_p}\frac{\eta_k}{1+\eta_p}I^+(q_0^+, q_L^+)\exp\left\{\frac{\alpha_k}{\alpha_p}\left[\log(q_L^+) - \frac{\eta_k - \eta_p}{1+\eta_p}q_L^+\right]\right\} \quad \text{(E.37)}$$

$$p_{k,\text{ASE}}^-(q_0^+) = 2p_{0k}\frac{\alpha_k}{\alpha_p}\frac{\eta_k}{1+\eta_p}I^-(q_0^+, q_L^+)\exp\left\{-\frac{\alpha_k}{\alpha_p}\left[\log(q_0^+) - \frac{\eta_k - \eta_p}{1+\eta_p}q_0^+\right]\right\} \quad \text{(E.38)}$$

with

$$I^\pm(q_0^+, q_L^+) = \int_{q_L^+}^{q_0^+} dx\, \frac{\exp\left(\pm\dfrac{\alpha_k}{\alpha_p}\dfrac{\eta_k - \eta_p}{1+\eta_p}x\right)}{X^{\pm\alpha_k/\alpha_p}} \quad \text{(E.39)}$$

The integrals involved in Eq. (E.39) can be approximated by a rapidly converging series, obtained through a Taylor expansion of the exponential term in the integrands.

## REFERENCES

[1]  I. S. Gradshteyn and I. M. Ryzhik, *Table of Integrals, Series, and Products*, corrected and enlarged edition, Academic Press, New York, 1980

[2]  E. Desurvire, "Analysis of distributed erbium-doped fiber amplifiers with fiber background loss," *IEEE Photonics Technol. Lett.*, vol 3, no 2, 127 (1991)

[3]  D. Chen and E. Desurvire, "Noise performance evaluation of distributed erbium-doped fiber amplifiers with bidirectional pumping at 1.48 $\mu$m," *IEEE Photonics Technol. Lett.*, vol 3, no 7, 625 (1991)

[4]  E. Desurvire, M. Zirngibl, H. M. Presby and D. DiGiovanni, "Characterization and modeling of amplified spontaneous emission in unsaturated erbium-doped fiber amplifiers," *IEEE Photonics Technol. Lett.*, vol 3, no 2, 127 (1991)

## APPENDIX F

# DENSITY MATRIX DESCRIPTION OF STARK SPLIT THREE-LEVEL LASER SYSTEMS (SEE SECTION 1.11)

We consider the Stark split, three-level laser system shown in Figure 1.2. By convention, the subscripts $j$, $k$, and $l$ in the density matrix elements will refer to the manifolds 1 (ground manifold), 2 (upper manifold) and 3 (pump manifold). The Heisenberg equation of motion for the density matrix elements $\rho_{mn}$ is, from Eq. (1.159):

$$\frac{d\rho_{mn}}{dt} = -i\omega_{mn}\rho_{mn} + \frac{i}{\hbar} E(z,t) \sum_k \{\mu_{mk}\rho_{kn} - \rho_{mk}\mu_{kn}\} \tag{F.1}$$

where the indices $m$ and $n$ can take the values $(j1 \ldots jN, N = g_1)$, $(k1 \ldots kN, N = g_2)$, $(l1 \ldots lN, N = g_3)$. Following the same approach as in the derivation of equations (1.160) and (1.161) and similarly to the derivation of the general rate equations made in Appendix A, we obtain for the diagonal elements $\rho_{mm}$ corresponding to the ground manifold 1:

$$\frac{d\rho_{j1,j1}}{dt} = \frac{i}{\hbar} E(z,t) \sum_p \{\mu_{j1,p}\rho_{p,j1} - \rho_{j1,p}\mu_{p,j1}\} - \sum_l R_{l,j1}\{\rho_{j1,j1} - \rho_{ll}\} + \sum_k A_{k,j1}\rho_{kk}$$
$$- A_{NR}^+\rho_{j1,j1} + A_{NR}^-\rho_{j2,j2} \tag{F.2}$$

$$\frac{d\rho_{jj}}{dt} = \frac{i}{\hbar} E(z,t) \sum_p \{\mu_{jp}\rho_{pj} - \rho_{jp}\mu_{pj}\} - \sum_l R_{lj}\{\rho_{jj} - \rho_{ll}\} + \sum_k A_{kj}\rho_{kk}$$
$$- (A_{NR}^+ + A_{NR}^-)\rho_{jj} + A_{NR}^-\rho_{j+1,j+1} + A_{NR}^+\rho_{j-1,j-1} \tag{F.3}$$

$$\frac{d\rho_{jN,jN}}{dt} = \frac{i}{\hbar} E(z,t) \sum_p \{\mu_{jN,p}\rho_{p,jN} - \rho_{jN,p}\mu_{p,jN}\} - \sum_l R_{l,jN}\{\rho_{jN,jN} - \rho_{ll}\} + \sum_k A_{k,jN}\rho_{kk}$$
$$- A_{NR}^-\rho_{jN,jN} + A_{NR}^+\rho_{jN-1,jN-1} \tag{F.4}$$

In order to simplify the above equations, several properties can be used: (1) all diagonal elements of the dipole operator $\mu_{mn}$ must vanish, i.e., $\mu_{mm} = 0$; (2) we assume

**607**

that no radiative transitions take place within each manifold, i.e., $\mu_{jp,jq} = 0$, $\mu_{kp,kq} = 0$ and $\mu_{lp,lq} = 0$; (3) the off-diagonal dipole operator elements corresponding to radiative transitions between manifolds 1 and 2 are symmetric, i.e., $\mu_{kj} = \mu_{jk}$; (4) we assume that thermal equilibrium is maintained within each manifold.

If we define the quantities $\bar{\rho}_{11}$, $\bar{\rho}_{22}$, $\bar{\rho}_{33}$ through:

$$\bar{\rho}_{11} = \sum_j \rho_{jj} \; ; \; \bar{\rho}_{22} = \sum_k \rho_{kk} \; ; \; \bar{\rho}_{33} = \sum_l \rho_{ll} \tag{F.5}$$

then the thermal equilibrium condition gives:

$$\rho_{jj} = p_{1j}\bar{\rho}_{11} \; ; \; \rho_{kk} = p_{2k}\bar{\rho}_{22} \; ; \; \rho_{ll} = p_{3l}\bar{\rho}_{33} \tag{F.6}$$

The quantities $\bar{\rho}_{11}$, $\bar{\rho}_{22}$, $\bar{\rho}_{33}$ correspond to the total probabilities of finding an atom in an energy state corresponding to the manifolds 1, 2, or 3. The terms $p_{1j}$, $p_{2k}$, and $p_{3l}$ are the Boltzmann distributions of manifolds 1, 2, and 3 with sublevels $j$, $k$, and $l$, defined in Eq. (1.12). Using the above property, and summing Eqs. (F.2)–(F.4) we find:

$$\frac{d\bar{\rho}_{11}}{dt} = \sum_j \frac{d\rho_{jj}}{dt} = \frac{i}{\hbar} E(z,t) \sum_j \sum_k (\mu_{jk}\rho_{kj} - \rho_{jk}\mu_{kj}) - \sum_j \sum_l R_{lj}(p_{1j}\bar{\rho}_{11} - p_{3l}\bar{\rho}_{33})$$

$$+ \sum_j \sum_k A_{kj} p_{2k}\bar{\rho}_{22}$$

$$= \frac{i}{\hbar} E(z,t) \sum_j \sum_k \mu_{kj}(\rho_{kj} - \rho_{jk}) - \mathscr{R}_{13}\bar{\rho}_{11} - \mathscr{R}_{31}\bar{\rho}_{33} + \mathscr{A}_{21}\bar{\rho}_{22} \tag{F.7}$$

In Eq. (F.7) we have introduced the same definitions of the total pumping rates and spontaneous emission rate as used in Section 1.2:

$$\mathscr{R}_{13} = \sum_j \sum_l R_{lj} p_{1j} \tag{F.8}$$

$$\mathscr{R}_{31} = \sum_j \sum_l R_{lj} p_{3l} \tag{F.9}$$

$$\mathscr{A}_{21} = \sum_j \sum_k A_{kj} p_{2k} \tag{F.10}$$

Following the same analysis, the evolution equations for the quantities $\bar{\rho}_{22}$, $\bar{\rho}_{33}$, are easily established:

$$\frac{d\bar{\rho}_{22}}{dt} = -\frac{i}{\hbar} E(z,t) \sum_j \sum_k \mu_{kj}(\rho_{kj} - \rho_{jk}) + \mathscr{A}_{21}\bar{\rho}_{22} + \mathscr{A}_{32} \tag{F.11}$$

$$\frac{d\bar{\rho}_{33}}{dt} = -\mathscr{A}_{32} + \mathscr{R}_{13}\bar{\rho}_{11} - \mathscr{R}_{31}\bar{\rho}_{33} \tag{F.12}$$

with by definition $\mathscr{A}_{32} = A_{32}^{NR} p_{31} \bar{\rho}_{33}$. As the sum of all probabilities $\rho_{mm}$ is unity, or $\bar{\rho}_{11} + \bar{\rho}_{22} + \bar{\rho}_{33} = 1$, the equation for $d\bar{\rho}_{33}/dt$ can be derived also from Eqs. (F.7) and (F.11), and $\bar{\rho}_{33}$ can be eliminated from these two equations.

From Eq. (F.1), the off-diagonal elements $\rho_{jk}$, are given by:

$$
\begin{aligned}
\frac{d\rho_{jk}}{dt} &= -i\omega_{jk}\rho_{jk} + \frac{i}{\hbar} E(z,t) \sum_p (\mu_{jp}\rho_{pk} - \rho_{jp}\mu_{pk}) - A'_{jk}\rho_{jk} \\
&= -i\omega_{jk}\rho_{jk} + \frac{i}{\hbar} E(z,t) \sum_{k'} (\mu_{jk'}\rho_{k'k} - \rho_{jk'}\mu_{k'k}) \\
&\quad + \frac{i}{\hbar} E(z,t) \sum_{j'} (\mu_{jj'}\rho_{j'k} - \rho_{jj'}\mu_{j'k}) - A'_{jk}\rho_{jk} \\
&= i\omega_{kj}\rho_{jk} + \frac{i}{\hbar} E(z,t) \sum_{k'} \mu_{k'j}\rho_{k'k} - \frac{i}{\hbar} E(z,t) \sum_{j'} \mu_{kj'}\rho_{jj'} - A'_{jk}\rho_{jk}
\end{aligned} \tag{F.13}
$$

with $\omega_{kj} = (E_k - E_j)/\hbar$. In developing Eq. (F.13), we have used the property $\mu_{jj'} = \mu_{kk'} = 0$, meaning that no radiative transitions take place within each manifold. We have also introduced the damping term $A'_{jk}$, which ensures that, in absence of external field, the off-diagonal matrix elements vanish, as imposed by equilibrium conditions. We use then the other property according to which $\rho_{jj'} = \rho_{kk'} = 0$ if $j \neq j'$, $k \neq k'$, which means that (thermal) equilibrium is maintained between the Stark sublevels, causing the corresponding off-diagonal density matrix elements to vanish. The nonvanishing elements $\rho_{jj}$ and $\rho_{kk}$ are functions of $\bar{\rho}_{11}$, $\bar{\rho}_{22}$, according to Eq. (F.6), and we finally get:

$$
\frac{d\rho_{jk}}{dt} = i\omega_{kj}\rho_{jk} + \frac{i}{\hbar} E(z,t)\mu_{kj}\{p_{2k}\bar{\rho}_{22} - p_{1j}\bar{\rho}_{11}\} - A'_{jk}\rho_{jk} \tag{F.14}
$$

The next step consists in eliminating the fast time dependence of $\rho_{jk}$ through the substitution $\rho_{jk} = \tilde{\sigma}_{jk} \exp(i\omega t)$ and $E(z,t) = E_0[\exp(i\omega t) + cc.]$, which gives, keeping only the synchronous terms in Eq. (F.14):

$$
\frac{d\tilde{\sigma}_{jk}}{dt} = i(\omega_{kj} - \omega)\tilde{\sigma}_{jk} + i\Omega_{kj}p_{2k}\bar{\rho}_{22} - i\Omega_{kj}p_{1j}\bar{\rho}_{11} - A'_{jk}\rho_{jk} \tag{F.15}
$$

with the following definition for the precession frequencies:

$$
\Omega_{kj} = \frac{E_0}{2\hbar} \mu_{kj} \tag{F.16}
$$

Taking then the difference between Eqs. (F.11) and (F.7), we obtain and equation for the population inversion probability $\bar{\rho}_{22} - \bar{\rho}_{11}$:

$$
\begin{aligned}
\frac{d(\bar{\rho}_{22} - \bar{\rho}_{11})}{dt} &= -2i \sum_{jk} \Omega_{kj}(\tilde{\sigma}_{kj} - \tilde{\sigma}_{jk}) + \bar{\rho}_{11}(\mathscr{R}_{13} + \mathscr{R}_{31} + \mathscr{A}_{21}) \\
&\quad + \bar{\rho}_{22}(\mathscr{R}_{13} - 2\mathscr{A}_{21} - \mathscr{A}_{32}) + \mathscr{A}_{32} - \mathscr{R}_{31}
\end{aligned} \tag{F.17}
$$

We assume first the steady state regime for $\rho_{33}$, which gives $d(\rho_{11} + \rho_{22})/dt = 0$, or from Eqs. (F.7) and (F.11):

$$\bar{\rho}_{22} = 1 - \bar{\rho}_{11} \frac{\mathscr{R}_{13} + \mathscr{R}_{31} + \mathscr{A}_{32}}{\mathscr{R}_{31} + \mathscr{A}_{32}} \tag{F.18}$$

Using this last result, Eq. (F.17) can be put under the form:

$$\frac{d(\bar{\rho}_{22} - \bar{\rho}_{11})}{dt} = -2i \sum_{jk} \Omega_{kj}(\tilde{\sigma}_{kj} - \tilde{\sigma}_{jk}) + \frac{1}{T}\left\{(\bar{\rho}_{22} - \bar{\rho}_{11}) - \frac{\mathscr{R}_{13}\tau - 1}{\mathscr{R}_{13}\tau + 1}\right\} \tag{F.19}$$

where $\tau = 1/\mathscr{A}_{21}$ is the total or effective fluorescence lifetime of the upper level manifold 2, and $T$ is a time constant defined through $T = \tau_{21}/(\mathscr{R}_{13}\tau_{21} + 1)$. This time constant represents the characteristic time for the difference $\bar{\rho}_{22} - \bar{\rho}_{11}$, representing the population inversion probability to reach its equilibrium value $\Delta\rho_{eq} = (\mathscr{R}_{13}\tau - 1)/(\mathscr{R}_{13}\tau + 1)$ as the external electric field $E$ (contained in $\Omega_{jk}$) is turned off. This regime can be called thermal equilibrium, as the Stark sublevel populations follow Boltzmann's Law, according to Eqs. (F.5) and (F.6). This regime of thermal equilibrium corresponds to the case of an atomic system driven by an external pump source (contained in $\mathscr{R}_{13}$). Consistently, if all external sources are turned off ($\mathscr{R}_{13} = \Omega_{jk} = 0$), Eq. (F.19) shows that the equilibrium value of $\bar{\rho}_{22} - \bar{\rho}_{11}$ is $\Delta\rho_{eq} = -1$, which means, using relation (F.18), i.e., $\bar{\rho}_{11} + \bar{\rho}_{22} = 1$, that $\bar{\rho}_{11} = 1$, $\bar{\rho}_{22} = 0$. Thus, as external sources are turned off, the new regime of thermal equilibrium predicted by Eq. (F.19) corresponds to a situation where all the atoms return to their ground state; the Stark sublevel occupation fits the Boltzmann distribution. Consistently, this equilibrium is reached in the characteristic time constant $T = \tau$, which is the fluorescence lifetime of the upper level manifold. Equation (F.19) thus represents the generalization to the case of a Stark-split three-level system of the result obtained in Eq. (1.165) for the basic three-level system.

We consider next the steady state regime for which $d\tilde{\sigma}_{jk}/dt = d(\bar{\rho}_{22} - \bar{\rho}_{11})/dt = 0$ in Eqs. (F.15) and (F.19). With the first condition, we obtain:

$$\tilde{\sigma}_{jk}(\omega) = 2i \frac{\Omega_{kj}}{\Delta\omega_{kj}}(p_{2k}\bar{\rho}_{22} - p_{1j}\bar{\rho}_{11})\left(1 + 2i \frac{\omega_{kj} - \omega}{\Delta\omega_{kj}}\right)\mathscr{L}_{kj} \tag{F.20}$$

where the Lorentzian function is:

$$\mathscr{L}_{kj} = \frac{1}{1 + 4\left(\dfrac{\omega_{kj} - \omega}{\Delta\omega_{kj}}\right)^2} \tag{F.21}$$

and $\Delta\omega_{kj} = 2A_{jk}^r$ is the line width (FWHM) of the transition with center frequency $\omega_{kj}$. Using the property $\tilde{\sigma}_{jk} = \tilde{\sigma}_{kj}^*$ (the density matrix is a Hermitian operator), we obtain from eq. (F.20):

$$\tilde{\sigma}_{kj} - \tilde{\sigma}_{jk} = -4i \frac{\Omega_{kj}}{\Delta\omega_{kj}}(p_{2k}\bar{\rho}_{22} - \bar{\rho}_{1j}\bar{\rho}_{11})\mathscr{L}_{kj} \tag{F.22}$$

We substitute Eq. (F.22) into Eq. (F.17) with $d(\bar{\rho}_{22} - \bar{\rho}_{11})/dt = 0$ to obtain:

$$\bar{\rho}_{22}\left\{1 + 8T\sum_{jk} p_{2k}\frac{\Omega_{kj}^2}{\Delta\omega_{kj}}\mathcal{L}_{kj}\right\} - \bar{\rho}_{11}\left\{1 + 8T\sum_{jk} p_{1j}\frac{\Omega_{kj}^2}{\Delta\omega_{kj}}\mathcal{L}_{kj}\right\} = \Delta\rho_{eq} \quad \text{(F.23)}$$

We use then a property of Er:glass according to which the nonradiative decay from manifold 3 is generally much greater than the pumping rates, i.e., $\mathcal{A}_{32} \gg \mathcal{R}_{13}, \mathcal{R}_{31}$; in which case we obtain from Eq. (F.18) the simple relation $\bar{\rho}_{11} + \bar{\rho}_{22} = 1$. This relation expresses that the probability of atomic occupation of manifold 3 is negligible, due to the damping effect of the nonradiative decay of rate $\mathcal{A}_{32}$. Using this relation and eq. (F.23), we can obtain the steady state expressions of $\bar{\rho}_{11}, \bar{\rho}_{22}$ (total probabilities of occupation of manifolds 1 and 2):

$$\bar{\rho}_{11} = \frac{1 - \Delta\rho_{eq} + 8T\sum_{jk} p_{2k}\dfrac{\Omega_{kj}^2}{\Delta\omega_{kj}}\mathcal{L}_{kj}}{2 + 8T\sum_{jk}(p_{2k} + p_{1j})\dfrac{\Omega_{kj}^2}{\Delta\omega_{kj}}\mathcal{L}_{kj}} \quad \text{(F.24)}$$

$$\bar{\rho}_{22} = \frac{1 + \Delta\rho_{eq} + 8T\sum_{jk} p_{1j}\dfrac{\Omega_{kj}^2}{\Delta\omega_{kj}}\mathcal{L}_{kj}}{2 + 8T\sum_{jk}(p_{2k} + p_{1j})\dfrac{\Omega_{kj}^2}{\Delta\omega_{kj}}\mathcal{L}_{kj}} \quad \text{(F.25)}$$

With Eqs. (F.24) and (F.25), we can now express the off-diagonal element $\tilde{\sigma}_{jk}$ through Eq. (F.20):

$$\tilde{\sigma}_{jk} = 2\frac{\Omega_{kj}}{\Delta\omega_{kj}}\left(i - 2\frac{\omega_{kj} - \omega}{\Delta\omega_{kj}}\right)\mathcal{L}_{kj}\{p_{2k}\bar{\rho}_{22} - p_{1j}\bar{\rho}_{11}\} \quad \text{(F.26)}$$

We can introduce at this point a certain number of definitions. First, the cross section of the laser transition centered at frequency $\omega_{kj}$ is given by, similarly to Eq. (1.169):

$$\sigma_{kj}(\omega) = \mathcal{L}_{kj}\sigma_{kj}^{\text{peak}} = \mathcal{L}_{kj}\frac{\lambda_{kj}^2}{2\pi n^2 \tau_{kj}\Delta\omega_{kj}} \quad \text{(F.27)}$$

where $\lambda_{kj} = 2\pi c/\omega_{kj}$ is the $(kj)$ transition wavelength and $\tau_{kj} = 1/A_{kj}$ is the associated fluorescence lifetime. using then the definition of the dipole moment operator element $\mu_{kj}$:

$$\mu_{kj}^2 = \frac{\pi\varepsilon_0\hbar c^3}{n\tau_{kj}\omega_{kj}^2} \quad \text{(F.28)}$$

and converting $E_0^2$ into an intensity though $cn\varepsilon_0 E_0^2/2 = P_s/A$ ($A$ = effective area). we obtain from Eqs. (F.16), (F.27), and (F.28)

$$4\frac{\Omega_{kj}^2\mathcal{L}_{kj}}{\Delta\omega_{kj}} = \frac{P_s}{A}\frac{\sigma_{kj}(\omega)}{h\nu_{kj}} \quad \text{(F.29)}$$

with $v_{kj} = 2\pi/\lambda_{kj}$. We define next the total absorption and emission cross sections $\sigma_{a,e}(\omega)$ through:

$$\frac{\sigma_a(\omega)}{h\bar{v}} = \sum_{jk} \frac{\sigma_{kj}(\omega)}{hv_{kj}} p_{1j} \tag{F.30}$$

$$\frac{\sigma_e(\omega)}{h\bar{v}} = \sum_{jk} \frac{\sigma_{kj}(\omega)}{hv_{kj}} p_{2j} \tag{F.31}$$

where $\bar{v}$ is an average transition frequency. With these different results and definitions, we can rewrite $\bar{\rho}_{11}$, $\bar{\rho}_{22}$, $\tilde{\sigma}_{jk}$ in Eqs. (F.24)–(F.31) through:

$$\bar{\rho}_{11} = \frac{1 + \dfrac{P_s}{A} \sigma_e(\omega_s)\tau}{1 + \mathcal{R}_{13}\tau + \dfrac{P_s}{h\bar{v}A}[\sigma_e(\omega_s) + \sigma_a(\omega_s)]\tau} \tag{F.32}$$

$$\bar{\rho}_{22} = \frac{\mathcal{R}_{13}\tau + \dfrac{P_s}{A} \sigma_a(\omega_s)\tau}{1 + \mathcal{R}_{13}\tau + \dfrac{P_s}{h\bar{v}A}[\sigma_e(\omega_s) + \sigma_a(\omega_s)]\tau} \tag{F.33}$$

$$\tilde{\sigma}_{jk}(\omega) = nc\frac{\varepsilon_0 E_0}{2}\frac{\sigma_{kj}(\omega)}{\mu_{kj}\omega_{kj}}\left(i - 2\frac{\omega_{kj} - \omega}{\Delta\omega_{kj}}\right)$$

$$\times \frac{\mathcal{R}_{13}\tau p_{2k} - p_{1j} + \dfrac{P_s}{h\bar{v}A}[p_{2k}\sigma_a(\omega_s) - p_{1j}\sigma_e(\omega_s)]\tau}{1 + \mathcal{R}_{13}\tau + \dfrac{P_s}{h\bar{v}A}[\sigma_e(\omega_s) + \sigma_a(\omega_s)]\tau} \tag{F.34}$$

In Eqs. (F.32) and (F.33) we have used the symbol $\omega_s$ for the fixed frequency of the saturating signal, not $\omega$, which is a variable. This means that the quantities $\bar{\rho}_{11}$, $\bar{\rho}_{22}$ are functions only of the signal frequency $\omega_s$ and are therefore fixed parameters in Eq. (F.26) defining $\tilde{\sigma}_{jk}(\omega)$.

The final step is to calculate the macroscopic polarization $P$, which is related to the complex atomic susceptibility $\chi$ through $P = \rho\langle\mu\rangle = \mathcal{R}e[\chi\varepsilon_0 E_0 \exp(i\omega t)]$ ($\rho$ here is the ionic dopant density). Using a fundamental property of the density matrix $\hat{\rho}$, we calculate the macroscopic (average) dipole moment $\langle\mu\rangle$ through:

$$\langle\mu\rangle = \text{tr}(\hat{\rho}\mu) = \sum_{jk} \rho_{jk}\mu_{kj} = \sum_{jk} \mu_{kj}\{\tilde{\sigma}_{jk}\exp(i\omega t) + \tilde{\sigma}_{kj}\exp(-i\omega t)\} \tag{F.35}$$

Defining the complex atomic susceptibility as a sum of real and imaginary parts, i.e., $\chi = \chi' - i\chi''$, we find from Eq. (F.35):

$$\chi' = \frac{2}{\varepsilon_0 E_0}\rho \sum_{jk} \mathcal{R}e(\mu_{kj}\tilde{\sigma}_{jk}) \tag{F.36}$$

$$\chi'' = -\frac{2}{\varepsilon_0 E_0} \rho \sum_{jk} \mathscr{Im}(\mu_{kj} \tilde{\sigma}_{jk}) \tag{F.37}$$

and finally from Eqs. (F.30), (F.31), and (F.34):

$$\chi' = 2nc\rho \sum_{jk} \frac{\sigma_{kj}(\omega)}{\omega_{kj}} \frac{\omega - \omega_{kj}}{\Delta\omega_{kj}} \frac{\mathscr{R}_{13}\tau p_{2k} - p_{1j} + \dfrac{P_\mathrm{s}}{h\bar{v}A}[p_{2k}\sigma_\mathrm{a}(\omega_\mathrm{s}) - p_{1j}\sigma_\mathrm{e}(\omega_\mathrm{s})]\tau}{1 + \mathscr{R}_{13}\tau + \dfrac{P_\mathrm{s}}{h\bar{v}A}[\sigma_\mathrm{e}(\omega_\mathrm{s}) + \sigma_\mathrm{a}(\omega_\mathrm{s})]\tau} \tag{F.38}$$

$$\chi'' = -nc\rho \sum_{jk} \frac{\sigma_{kj}(\omega)}{\omega_{kj}} \frac{\mathscr{R}_{13}\tau p_{2k} - p_{1j} + \dfrac{P_\mathrm{s}}{h\bar{v}A}[p_{2k}\sigma_\mathrm{a}(\omega_\mathrm{s}) - p_{1j}\sigma_\mathrm{e}(\omega_\mathrm{s})]\tau}{1 + \mathscr{R}_{13}\tau + \dfrac{P_\mathrm{s}}{h\bar{v}A}[\sigma_\mathrm{e}(\omega_\mathrm{s}) + \sigma_\mathrm{a}(\omega_\mathrm{s})]\tau}$$

$$= -\frac{nc}{2\pi\bar{v}}\rho \frac{\mathscr{R}_{13}\tau\sigma_\mathrm{e}(\omega) - p_{1j}\sigma_\mathrm{a}(\omega) + \dfrac{P_\mathrm{s}}{h\bar{v}A}\displaystyle\sum_{jk}\frac{\sigma_{kj}(\omega)}{\omega_{kj}}[p_{2k}\sigma_\mathrm{a}(\omega_\mathrm{s}) - p_{1j}\sigma_\mathrm{e}(\omega_\mathrm{s})]\tau}{1 + \mathscr{R}_{13}\tau + \dfrac{P_\mathrm{s}}{h\bar{v}A}[\sigma_\mathrm{e}(\omega_\mathrm{s}) + \sigma_\mathrm{a}(\omega_\mathrm{s})]\tau} \tag{F.39}$$

or

$$\chi'' = -\frac{nc}{2\pi\bar{v}}\rho \frac{\mathscr{R}_{13}\tau\sigma_\mathrm{e}(\omega) - \sigma_\mathrm{a}(\omega) + \dfrac{P_\mathrm{s}}{h\bar{v}A}(\sigma_\mathrm{e}(\omega)\sigma_\mathrm{a}(\omega_\mathrm{s}) - \sigma_\mathrm{a}(\omega)\sigma_\mathrm{e}(\omega_\mathrm{s}))\tau}{1 + \mathscr{R}_{13}\tau + \dfrac{P_\mathrm{s}}{h\bar{v}A}[\sigma_\mathrm{e}(\omega) + \sigma_\mathrm{a}(\omega)]\tau} \tag{F.40}$$

If now we define a complex atomic susceptibility $\chi_{jk}$ corresponding to the individual transition $(jk)$ as:

$$\chi_{jk} = -nc\rho \frac{\sigma_{kj}(\omega)}{\omega_{kj}} \left(2\frac{\omega_{kj} - \omega}{\Delta\omega_{kj}} - i\right) \frac{\mathscr{R}_{13}\tau p_{2k} - p_{1j} + \dfrac{P_\mathrm{s}}{h\bar{v}A}[p_{2k}\sigma_\mathrm{a}(\omega_\mathrm{s}) - p_{1j}\sigma_\mathrm{e}(\omega_\mathrm{s})]\tau}{1 + \mathscr{R}_{13}\tau + \dfrac{P_\mathrm{s}}{h\bar{v}A}[\sigma_\mathrm{e}(\omega_\mathrm{s}) + \sigma_\mathrm{a}(\omega_\mathrm{s})]\tau}$$

$$\tag{F.41}$$

we find that from Eqs. (F.36), (F.37), (F.38), and (F.40) that the total atomic susceptibility $\chi$ is given by:

$$\chi = \sum_{jk} \chi_{jk} \tag{F.42}$$

---

# RESOLUTION OF THE AMPLIFIER PGF DIFFERENTIAL EQUATION IN THE LINEAR GAIN REGIME (SEE SECTION 2.3)

---

From Chapter 2, the probability generating function $F$ corresponding to the amplifier linear gain regime satisfies the following partial differential equation, Eq. (2.39):

$$\frac{\partial F}{\partial z} = (x - 1)\left\{(ax - b)\frac{\partial F}{\partial x} + aF\right\} \tag{G.1}$$

The above has the form of the Lagrange linear equation [1]:

$$\xi\frac{\partial F}{\partial x} + \eta\frac{\partial F}{\partial z} = \zeta \tag{G.2}$$

with $\zeta = (x - 1)aF$. The characteristic equation of Eq. (G.2) is $dx/\xi = dz/\eta = dF/\zeta$, or from Eq. (G.1), with the substitution $y = 1/(1 - x)$:

$$dz = \frac{dy}{(a - b)y - a} = -y\frac{dF}{aF} \tag{G.3}$$

We solve first the left part of Eq. (G.3) to obtain:

$$y(z) = G(z)y(0) - N(z) \tag{G.4}$$

with

$$G(z) = \exp\left\{\int_0^z [a(z') - b(z')]\,dz'\right\} \tag{G.5}$$

and

$$N(z) = G(z)\int_0^z \frac{a(z')}{G(z')}\,dz' \tag{G.6}$$

$G(z)$ and $N(z)$ have the same definitions as in Eqs. (2.30) and (2.31). They are the amplifier gain and output ASE noise photon number, respectively.

The right-hand part of Eq. (G.3) can be expressed through Eq. (G.4) in the form:

$$\frac{1}{F}\frac{dF}{dz} = -\frac{a}{G}\frac{1}{y(0) - N/G} = \frac{d}{dz}\log\{y(0) - N/G\} \tag{G.7}$$

using the property $a/G = d(N/G)/dz$, which stems from Eq. (G.6). The solution of Eq. (G.7) is

$$F(z) = \text{const.}\{y(0) - N/G\} \tag{G.8}$$

The constant in Eq. (G.8) is given by the initial conditions at $z = 0$. As $N(0) = 0$ and $G(0) = 1$, from Eqs. (G.5) and (G.6), we find const. $= F(x_0, 0)/y(0)$, with $x_0 = 1 - 1/y(0)$. using the PGF definition (2.34) for the initial conditions, i.e., $F(x_0, 0) = \sum_m x_0^m P_m(0)$, where $P_m(0)$ is the probability of having $m$ input photons at $z = 0$, the solution (G.8) using eq. (G.4) takes the form:

$$F(x, z) = \frac{F(x_0, 0)}{y(0)}\{y(0) - N/G\} = \sum_m x_0^m P_m(0)\frac{G}{y + N}\left(\frac{y + N}{G} - N/G\right)$$

$$= \frac{y}{y + N}\sum_m \left(1 - \frac{G}{y + N}\right)^m P_m(0) = \frac{1}{1 + N(1 - x)}\sum_m \left[\frac{1 + (N - G)(1 - x)}{1 + N(1 - x)}\right]^m P_m(0) \tag{G.9}$$

We must now express the right-hand side of Eq. (G.9) as a series in $x^m$. We first rewrite the terms involving $x$ in Eq. (G.9) as:

$$\frac{1}{1 + N(1 - x)}\left[\frac{1 + (N - G)(1 - x)}{1 + N(1 - x)}\right]^m = \frac{(1 + N - G)^m}{(1 + N)^{m+1}}\frac{1}{1 - Bx}\left(1 + \frac{1 + Ax}{1 - Bx}\right)^m \tag{G.10}$$

where $A = G/(1 + N)(1 + N - G)$ and $B = N/(1 + N)$. We use the binomial formula:

$$(1 + C)^m = \sum_{j=0}^{m}\frac{m!}{j!(m - j)!}C^j \tag{G.11}$$

with $C = (1 + Ax)/(1 - Bx)$ and the Taylor series for $[1/(1 - Bx)]^j$

$$\left(\frac{1}{1 - Bx}\right)^j = \sum_{p=0}^{\infty}\frac{B^p x^p}{p!}\frac{(j + p - 1)!}{(j - 1)!} \tag{G.12}$$

to express Eq. (G.9) as

$$F(x, z) = \sum_{m=0}^{\infty} P_m(0)\frac{(1 + N - G)^m}{(1 + N)^{m+1}}\sum_{j=0}^{m}\frac{m!}{j!(m - j)!}(Ax)^j \sum_{p=0}^{\infty}\frac{(j + p)!}{p!j!}(Bx)^p \tag{G.13}$$

Let $n = j + p$ in Eq. (G.13). We obtain:

$$F(x, z) = \sum_{m=0}^{\infty} P_m(0) \sum_{n=0}^{\infty} \frac{(1 + N - G)^m N^n}{(1 + N)^{m+n+1}} \sum_{j=0}^{m \text{ or } n} \frac{m! n!}{j! j! (m-j)! (n-j)!} \left[ \frac{G}{N(1+N-G)} \right]^j x^n$$

(G.14)

The summation over index $j$ is carried on up to $j = m$ for $m \leqslant n$ and up to $j = n$ for $n \leqslant m$; this stems from the condition $p \geqslant 0$. We rewrite Eq. (G.14) in the form:

$$F(x, z) = \sum_m x^m P_m(z)$$

(G.15)

where $P_m(z)$ is the photon probability distribution at coordinate $z$ and is defined by:

$$P_m(z) = \sum_{n=0}^{\infty} P_n(0) \frac{(1 + N - G)^n N^m}{(1 + N)^{m+n+1}} \sum_{j=0}^{m \text{ or } n} \frac{m! n!}{j! j! (m-j)! (n-j)!} \left[ \frac{G}{N(1+N-G)} \right]^j$$

(G.16)

The above equation is one of the fundamental results of the STT analysis [2]. The term multiplying $P_n(0)$ in the sum in Eq. (G.16) is called $P_{n,m}$ in this analysis. The function $P_{n,m}(z)$ corresponds to the *probability of finding m output photons assuming that n photons are input to the amplifier*. The probability of having $n$ input photons is $P_n(0)$, so the output probability distribution $P_m(z)$ is:

$$P_m(z) = \sum_n P_{m,n}(z) P_n(0)$$

(G.17)

with

$$P_{n,m}(z) = \frac{(1 + N - G)^n N^m}{(1 + N)^{m+n+1}} \sum_{j=0}^{\min(m, n)} \frac{m! n!}{j! j! (m-j)! (n-j)!} \left[ \frac{G}{N(1+N-G)} \right]^j$$

(G.18)

Considering now the case where the coefficients $a$ and $b$ are constant with coordinate $z$, we have from Eqs. (G.5) and (G.6), $G(z) = \exp[(a-b)z]$ and $N(z) = a[G(z)-1]/(a-b)$. Substituting them in Eq. (G.18) yields the corresponding probability $P_{n,m}(z)$:

$$P_{n,m}(z) = \frac{(a-b)a^m b^n (G-1)^{n+m}}{(aG-b)^{m+n+1}} \sum_{j=0}^{m \text{ or } n} \frac{m! n!}{j! j! (m-j)! (n-j)!} \left[ \frac{(a-b)^2 G}{ab(G-1)^2} \right]^j$$

(G.19)

which is exactly Eq. (30) of [2].

The functions $P_{n,m}(z)$ defined in Eq. (G.18) for the general case, correspond to Jacobi polynomials. The Jacobi polynomials are defined by [3]:

$$P_N^{(\alpha, \beta)}(X) = \frac{(X-1)^N}{2^N} \sum_{j=0}^{N} \binom{N+\alpha}{j} \binom{N+\beta}{N-j} \left( \frac{X+1}{X-1} \right)^j$$

(G.20)

Identification of Eq. (G.18) with Eq. (G.20) for the two cases $m < n$ and $n < m$ yields:

$$P_{n,m}(z) = \frac{(N + 1 - G)^{n-m}}{(N + 1)^{n+1}} (G - N)^m P_m^{(n-m, 0)}(X) \; ; \; m < n \qquad (G.21)$$

$$P_{n,m}(z) = \frac{N^{m-n}}{(N + 1)^{m+1}} (G - N)^n P_n^{(m-n, 0)}(X) \; ; \; n < m \qquad (G.22)$$

with

$$X = \frac{1}{N - G}\left(G\frac{N - 1}{N + 1} - N\right) \qquad (G.23)$$

A case of interest considered in [2] is that of large gain $G$ and output photon number $m$, i.e., $G \gg 1$ and $m \gg n$, assuming $G/m$ remains finite. In this case, the following approximations can be made:

$$(aG - b)^{m+1} = (aG)^{m+1}\left(1 - \frac{b}{aG}\right)^{m+1} \approx (aG)^{m+1}\left(1 - m\frac{b}{aG}\right) \approx (aG)^{m+1}\exp\left(-m\frac{b}{aG}\right)$$

$$(G.24)$$

$$(G - 1)^{m+n} = (G - 1)^n G^m\left(1 - \frac{1}{G}\right)^m \approx (G - 1)^n G^m\left(1 - m\frac{1}{G}\right) \approx (G - 1)^n G^m \exp\left(-\frac{m}{G}\right)$$

$$(G.25)$$

Substituted in Eq. (G.19), corresponding to the case of constant coefficients $a$ and $b$, for $m > n$, these approximations yield:

$$P_{n,m}(z) = \frac{a - b}{aG}\left(\frac{b}{a}\right)^n \exp\left(-m\frac{a - b}{aG}\right) \sum_{j=0}^{n} \frac{m!\,n!}{j!\,j!\,(m - j)!\,(n - j)!}\left[\frac{(a - b)^2}{abG}\right]^j \quad (G.26)$$

As we have $m \gg n > j$, we can make the additional approximation $m!/(m - j)! \approx m^j$. Using the definition of the Laguerre polynomials $L_n(X)$ [3]:

$$L_n(X) = \sum_{j=0}^{n} \frac{n!}{j!\,j!\,(n - j)!}(-X)^j \qquad (G.27)$$

we obtain finally, from Eq. (G.26) the result in [2]:

$$P_{n,m}(z) = \frac{a - b}{aG}\left(\frac{b}{a}\right)^n \exp\left(-m\frac{a - b}{aG}\right) L_n\left[-m\frac{(a - b)^2}{abG}\right] \qquad (G.28)$$

In the general case where the coefficients $a$ and $b$ depend on coordinate $z$, we find from Eqs. (G.18) and (G.27) that $P_{n,m}$ takes the form:

$$P_{n,m}(z) = \frac{(1 + N - G)^n N^m}{(1 + N)^{m+n+1}} L_n\left[-m\frac{G}{N(1 + N - G)}\right] \qquad (G.29)$$

In the case of full medium inversion for all coordinates $z$, we have $b \to 0$ and $N \approx (G - 1)$, which results in a singularity in the arguments of the Laguerre polynomials in Eqs. (G.28) and (G.29). But, using Eq. (G.27) for the Laguerre polynomials, we observe that:

$$A^n L_n\left(-\frac{X}{A}\right) = \sum_{j=0}^{n} \frac{n!}{j!\,j!\,(n-j)!} X^j A^{n-j} = \frac{X^n}{n!} \qquad (G.30)$$

when $A \to 0$, therefore the functions $P_{n,m}$ in Eqs. (G.28) and (G.29) remain finite in both cases where $A = b = 0$ or $A = 1 + N - G = 0$.

A different treatment for the resolution of the PGF (G.1) is described in [4]. In the following, we outline the demonstration leading to the PGF solution using notation consistent with the previous analysis, and apply it to the general case.

Following the approach in [4], we integrate Eq. (G.1) along a certain path $x(z)$ from an initial point $(x_0, z = 0)$ to an arbitrary point $(x(z), z)$. The function $x(z)$ is defined through

$$\frac{dx}{dz} = -(x - 1)(ax - b) \qquad (G.31)$$

We obtain then

$$\frac{dF(x, z)}{dz} = \frac{\partial F}{\partial z} + \frac{\partial F}{\partial x}\frac{dx}{dz} = -(x - 1)(ax - b)\frac{\partial F}{\partial x} + \frac{\partial F}{\partial z} \qquad (G.32)$$

Substituting Eq. (G.32) into Eq. (G.1) yields the basic first order linear equation:

$$\frac{dF}{dz} = a(x - 1)F \qquad (G.33)$$

with solution:

$$F(x, z) = F(x(0), 0)\exp\left\{\int_0^z [x(z') - 1]a(z')\,dz'\right\} \qquad (G.34)$$

To obtain next an expression for the path $x(z)$, we rewrite Eq. (G.31) in the form:

$$\frac{d(x - 1)}{dz} = -a(x - 1)^2 + (b - a)(x - 1) \qquad (G.35)$$

Introducing the function

$$h(z) = \frac{1}{G(z)} = \exp\left\{\int_0^z [b(z') - a(z')]\,dz'\right\} \qquad (G.36)$$

and noting that $b - a = (dh/dz)/h$, we can put Eq. (G.35) into the form:

$$\frac{d}{dz}\left(\frac{h}{x - 1}\right) = ah \qquad (G.37)$$

whose solution is

$$\frac{h(z')}{x(z') - 1} - \frac{h(z)}{x(z) - 1} = k(z') - k(z) \tag{G.38}$$

with

$$k(z) = \frac{N(z)}{G(z)} = \int_0^z h(z')a(z')\,dz' \tag{G.39}$$

From Eq. (G.38), the path $x(z)$ is then defined by:

$$x(z') - 1 = \frac{h(z')[x(z) - 1]}{h(z) + [k(z') - k(z)][x(z) - 1]} \tag{G.40}$$

and in particular, using the properties $h(0) = 1$, $k(0) = 0$:

$$x(0) = 1 + \frac{x(z) - 1}{h(z) - k(z)[x(z) - 1]} \tag{G.41}$$

The PGF solution in Eq. (G.34) is then, from Eq. (G.40):

$$F(x, z) = F[X(x, z; 0), 0]\exp\left\{\int_0^z [X(x, z; z') - 1]a(z')\,dz'\right\} \tag{G.42}$$

with

$$X(x, z; z') = 1 + \frac{h(z')(x - 1)}{h(z) + [k(z') - k(z)](x - 1)} \tag{G.43}$$

and

$$X(x, z; 0) = 1 + \frac{x - 1}{h(z) - k(z)(x - 1)} \tag{G.44}$$

which is the result obtained in [4]. It is possible, however, to perform the integration in Eq. (G.42) in the most general case, where the coefficients $a$ and $b$ depend on fiber coordinate $z$, as shown in the following. Using Eqs. (G.36) and (G.39) for the functions $h(z)$ and $k(z)$, we first rewrite Eqs. (G.43) and (G.44) in the form:

$$X(x, z; z') = 1 + \frac{(x - 1)G(z)}{G(z') - (x - 1)[N(z)G(z') - N(z')G(z)]} \tag{G.45}$$

and

$$X(x, z; 0) = 1 + \frac{(x - 1)G(z)}{1 - (x - 1)N(z)} \tag{G.46}$$

Using Eq. (G.45), the exponential term in Eq. (G.42) becomes:

$$F_1(x, z) = \exp\left\{ \int_0^z [X(x, z; z') - 1]a(z')\, dz' \right\}$$

$$= \exp\left\{ \frac{(x-1)G(z)}{1-(x-1)N(z)} \int_0^z \frac{a(z')}{G(z')} \frac{1}{1 + \dfrac{(x-1)G(z)}{1-(x-1)N(z)} \dfrac{N(z')}{G(z')}}\, dz' \right\} \qquad (G.47)$$

Using the property

$$\frac{d}{dz}\left\{ \frac{N(z)}{G(z)} \right\} = \frac{a(z)}{G(z)} \qquad (G.48)$$

which stems from Eq. (G.6), and $G(0) = 1$, $N(0) = 0$, we obtain from Eq. (G.47):

$$F_1(x, z) = \exp\left\{ \frac{(x-1)G(z)}{1-(x-1)N(z)} \int_0^{N(z)/G(z)} dU \frac{1}{1 + \dfrac{(x-1)G(z)}{1-(x-1)N(z)} U} \right\} = \frac{1}{1-(x-1)N(z)}$$

$$(G.49)$$

Eq. (G.42) can thus be expressed explicitly as function of $x$, $z$, $G(z)$ and $N(z)$, see Eqs. (2.49) to (2.51).

The photon statistics first and second moments $\langle n(z) \rangle$ and $\langle n^2(z) \rangle$ corresponding to the output PGF are given by:

$$\langle n(z) \rangle = \left( \frac{\partial F(x, z)}{\partial x} \right)_{x=1} \qquad (G.50)$$

$$\langle n^2(z) \rangle = \left( \frac{\partial^2 F(x, z)}{\partial x^2} \right)_{x=1} + \langle n(z) \rangle \qquad (G.51)$$

We have then from Eqs. (G.42), (G.50), and (G.51):

$$\langle n(z) \rangle = \frac{\partial F(X, 0)}{\partial X}\left( \frac{\partial X(x, z; 0)}{\partial x} \right)_{x=1} + \left( \frac{\partial F_1(x, z)}{\partial x} \right)_{x=1} \qquad (G.52)$$

$$\langle n^2(z) \rangle = \left( \frac{\partial^2 F(X, 0)}{\partial x\, \partial X} \right)_{x=1}\left( \frac{\partial X(x, z; 0)}{\partial x} \right)_{x=1} + \frac{\partial F(X, 0)}{\partial X}\left( \frac{\partial^2 X(x, z; 0)}{\partial x^2} \right)_{x=1}$$

$$+ \frac{\partial F(X, 0)}{\partial X}\left( \frac{\partial X(x, z; 0)}{\partial x} \frac{\partial F_1(x, z)}{\partial x} \right)_{x=1} + \left( \frac{\partial^2 F_1(x, z)}{\partial x^2} \right)_{x=1}$$

$$+ \left( \frac{\partial F_1(x, z)}{\partial x} \right)^2_{x=1} + \langle n(z) \rangle \qquad (G.53)$$

where we have used the properties $F[X(1, z; 0), 0] = 1$ and $F_1(1, z) = 1$. We have by definition:

$$\frac{\partial F(X, 0)}{\partial X} = \langle n(0) \rangle \tag{G.54}$$

$$\frac{\partial^2 F(X, 0)}{\partial X^2} = \langle n^2(0) \rangle - \langle n(0) \rangle \tag{G.55}$$

where $\langle n(0) \rangle$ and $\langle n^2(0) \rangle$ are the first and second moments at the input of the amplifier, respectively. We find from Eqs. (G.49) and (G.52)–(G.55):

$$\langle n(z) \rangle = G(z)\langle n(0) \rangle + N(z) \tag{G.56}$$

$$\langle n^2(z) \rangle = G^2(z)[\langle n^2(0) \rangle - \langle n(0) \rangle] + 4G(z)N(z)\langle n(0) \rangle + G(z)\langle n(0) \rangle + 2N^2(z) + N(z) \tag{G.57}$$

## REFERENCES

[1]   E. L. Ince, *Ordinary Differential Equations*, Dover Publications, New York, 1944

[2]   K. Shimoda, H. Takahasi, and C. H. Townes, "Fluctuations in amplification of quanta with application to maser amplifiers," *J. Phys. Soc. Japan*, vol 12, no 6, 686 (1957)

[3]   M. Abramovitz and I. A. Stegun, *Handbook of Mathematical Functions*, Dover Publications, New York, 1972, p775

[4]   P. Diament and M. C. Teich, "Evolution of the statistical properties of photons passed through a traveling-wave laser amplifier," *IEEE J. Quantum Electron.*, vol 28, no 5, 132 (1992)

# APPENDIX H

## CALCULATION OF THE OUTPUT NOISE AND VARIANCE OF LUMPED AMPLIFIER CHAINS (SEE SECTION 2.5)

For amplifier chains of Types A and C, the vector equations (2.165) and (2.166) are:

$$\begin{pmatrix} \langle n_k \rangle \\ \sigma_k^2 \end{pmatrix}_{\text{Type A}} = [\hat{T}\hat{M}]^k \begin{pmatrix} \langle n_0 \rangle \\ \sigma_0^2 \end{pmatrix} + \sum_{j=1}^{k} [\hat{T}\hat{M}]^{j-1}\hat{T}\mathbf{N} \tag{H.1}$$

$$\begin{pmatrix} \langle n_k \rangle \\ \sigma_k^2 \end{pmatrix}_{\text{Type C}} = [\hat{M}\hat{T}]^k \begin{pmatrix} \langle n_0 \rangle \\ \sigma_0^2 \end{pmatrix} + \sum_{j=1}^{k} [\hat{M}\hat{T}]^{j-1}\mathbf{N} \tag{H.2}$$

Using Eqs. (2.160) and (2.161) to define the matrices $\hat{T}$ and $\hat{M}$, and the property of geometrical series:

$$\sum_{j=0}^{k-1} Q^j = \frac{1 - Q^k}{1 - Q} \tag{H.3}$$

we find:

$$[\hat{T}\hat{M}]^k = (GT)^k \begin{pmatrix} 1 & 0 \\ 1 + 2NTx_k - (GT)^k & (GT)^k \end{pmatrix} \tag{H.4}$$

$$[\hat{M}\hat{T}]^k = (GT)^k \begin{pmatrix} 1 & 0 \\ 1 + 2Nx_k - (GT)^k & (GT)^k \end{pmatrix} \tag{H.5}$$

$$\sum_{j=1}^{k} [\hat{T}\hat{M}]^{j-1} = x_k \begin{pmatrix} 1 & 0 \\ \dfrac{1 - GT + 2TN}{1 - (GT)^2} GT[1 - (GT)^{k-1}] & \dfrac{1 + (GT)^k}{1 + GT} \end{pmatrix} \tag{H.6}$$

$$\sum_{j=1}^{k} [\hat{M}\hat{T}]^{j-1} = x_k \begin{pmatrix} 1 & 0 \\ \dfrac{1 - GT + 2N}{1 - (GT)^2} GT[1 - (GT)^{k-1}] & \dfrac{1 + (GT)^k}{1 + GT} \end{pmatrix} \tag{H.7}$$

with

$$x_k = \frac{1 - (GT)^k}{1 - GT} \tag{H.8}$$

Substituting the results of Eqs. (H.4)–(H.7) into Eqs. (H.1) and (H.2) with definition (2.162) for the noise vector **N** yields:

$$\left( \begin{array}{l} \langle n_k \rangle = (GT)^k \langle n_0 \rangle + \mathcal{M} TNx_k \\[4pt] \quad \sigma_k^2 = (GT)^k [\sigma_0^2 - \langle n_0 \rangle] + (GT)^k [1 + 2TNx_k]\langle n_0 \rangle \\[4pt] \quad + \dfrac{\mathcal{M} TNx_k}{1 + GT} \left\{ \dfrac{1 - GT + 2TN}{1 - GT} GT[1 - (GT)^{k-1}] + (1 + TN)[1 + (GT)^k] \right\} \end{array} \right)_{\text{Type A}}$$

$$\tag{H.9}$$

and

$$\left( \begin{array}{l} \langle n_k \rangle = (GT)^k \langle n_0 \rangle + \mathcal{M} Nx_k \\[4pt] \quad \sigma_k^2 = (GT)^k [\sigma_0^2 - \langle n_0 \rangle] + (GT)^k [1 + 2Nx_k]\langle n_0 \rangle \\[4pt] \quad + \dfrac{\mathcal{M} Nx_k}{1 + GT} \left\{ \dfrac{1 - GT + 2N}{1 - GT} GT[1 - (GT)^{k-1}] + (1 + N)[1 + (GT)^k] \right\} \end{array} \right)_{\text{Type C}}$$

$$\tag{H.9}$$

# APPENDIX I

# DERIVATION OF A GENERAL FORMULA FOR THE OPTICAL NOISE FIGURE OF AMPLIFIER CHAINS (SEE SECTION 2.5)

We consider a chain of optical amplifiers, each characterized by gains and equivalent input noise factors $G_i$ and $n_{eqi}$, respectively. Any amplifier can be an attenuating element with gain $G_i = T_i < 1$ and noise factor $n_{eqi} = 0$, which represents a strand of lossy fiber; this does not affect the generality of the result.

The optical noise figure for each amplifier $i$ is defined by Eq. (2.115) as $F_i = \text{SNR}_{in}(i)/\text{SNR}_{out}(i)$, where $\text{SNR}_{in}(i) = \langle n_{i-1}(\text{signal})\rangle^2/\sigma_{i-1}^2$ and $\text{SNR}_{out}(i) = \langle n_i(\text{signal})\rangle^2/\sigma_i^2$ are the input and output signal-to-noise ratios; $\langle n_i(\text{signal})\rangle = G_i\langle n_{i-1}(\text{signal})\rangle = G_i G_{i-1}\ldots G_1\langle n_0(\text{signal})\rangle$ is the amplified signal. The optical noise figure of amplifier $i$ is thus:

$$F_i = \frac{\sigma_i^2}{G_i^2\sigma_{i-1}^2} \tag{I.1}$$

According to the above definitions, the optical noise figure $F'$ of the amplifier chain with $k$ elements is then given by:

$$F' = \frac{\langle n_0(\text{signal})\rangle^2}{\sigma_0^2}\frac{\sigma_k^2}{\langle n_k(\text{signal})\rangle^2} = \frac{\sigma_1^2}{G_1^2\sigma_0^2}\cdots\frac{\sigma_{k-1}^2}{G_{k-1}^2\sigma_{k-2}^2}\frac{\sigma_k^2}{G_k^2\sigma_{k-1}^2} = F_1\ldots F_{k-1}F_k \tag{I.2}$$

Using Eqs. (2.75) and (2.123), the variance of amplifier $i$ is given by

$$\sigma_i^2 = G_i^2(\sigma_{i-1}^2 - \langle n_{i-1}\rangle) + G_i\langle n_{i-1}\rangle(1 + 2n_{eqi}G_i) + \mathcal{M}n_{eqi}G_i(1 + n_{eqi}G_i) \tag{I.3}$$

and the optical noise figure in Eq. (I.1) takes the form:

$$F_i = \frac{\langle n_{i-1}\rangle}{\sigma_{i-1}^2}\left\{f_{i-1} - 1 + \frac{1 + 2n_{eqi}G_i}{G_i} + \mathcal{M}\frac{n_{eqi}}{\langle n_{i-1}\rangle}\frac{1 + n_{eqi}G_i}{G_i}\right\} \tag{I.4}$$

where $f_i$ is the Fano factor:

$$f_i = \frac{\sigma_i^2}{\langle n_i \rangle} \tag{I.5}$$

In the case where the signal input to the amplifier $i$ has Poisson statistics, we have $\sigma_i^2 = \langle n_i \rangle$ and $f_i = 1$. We define an optical noise figure $F_i^{\text{Poisson}}$, where the input signal has Poisson statistics, through:

$$F_i^{\text{Poisson}} = \frac{1 + 2n_{\text{eq}i}G_i}{G_i} + \mathcal{M} \frac{n_{\text{eq}i}}{\langle n_{i-1} \rangle} \frac{1 + n_{\text{eq}i}G_i}{G_i} \tag{I.6}$$

so that the optical noise figure in the general case of Eq. (I.4) takes the form:

$$F_i = \frac{\langle n_{i-1} \rangle}{\sigma_{i-1}^2} \{f_{i-1} - 1 + F_i^{\text{Poisson}}\} \tag{I.7}$$

In order to derive an explicit formula for the optical noise figure $F'$ of the amplifier chain, defined in Eq. (I.2), we must consider first the simple case of a chain of two amplifiers in tandem. Using Eqs. (2.73), (2.75), and (2.123), we find for the output mean, output variance, and noise figure of each amplifier:

$$\langle n_1 \rangle = G_1 \langle n_0 \rangle + \mathcal{M} n_{\text{eq}1} G_1 \tag{I.8}$$

$$\langle n_2 \rangle = G_2 \langle n_1 \rangle + \mathcal{M} n_{\text{eq}2} G_2 \tag{I.9}$$

$$\sigma_1^2 = G_1^2(\sigma_0^2 - \langle n_0 \rangle) + G_1 \langle n_0 \rangle (1 + 2n_{\text{eq}1} G_1) + \mathcal{M} n_{\text{eq}1} G_1 (1 + n_{\text{eq}1} G_1) \tag{I.10}$$

$$\sigma_2^2 = G_2^2(\sigma_1^2 - \langle n_1 \rangle) + G_2 \langle n_1 \rangle (1 + 2n_{\text{eq}2} G_2) + \mathcal{M} n_{\text{eq}2} G_2 (1 + n_{\text{eq}2} G_2) \tag{I.11}$$

$$F_1 = \frac{\langle n_0 \rangle}{\sigma_0^2} \{f_0 - 1 + F_1^{\text{Poisson}}\} \tag{I.12}$$

$$F_2 = \frac{\langle n_1 \rangle}{\sigma_1^2} \{f_1 - 1 + F_2^{\text{Poisson}}\} \tag{I.13}$$

From Eqs. (I.2), (I.12), and (I.13), the optical noise figure $F'$ of the chain is then

$$F' = F_1 F_2 = \frac{\langle n_0 \rangle}{\sigma_0^2} \frac{\langle n_1 \rangle}{\sigma_1^2} (f_0 - 1 + F_1^{\text{Poisson}})(f_1 - 1 + F_2^{\text{Poisson}}) \tag{I.14}$$

From Eqs. (I.8) and (I.10), after straightforward algebra, we find:

$$\frac{1}{f_1} = \frac{\langle n_1 \rangle}{\sigma_1^2} = \frac{1 + \mathcal{M} \dfrac{n_{\text{eq}1}}{\langle n_0 \rangle}}{G_1(f_0 - 1 + F_1^{\text{Poisson}})} \tag{I.15}$$

and we obtain from (I.14) and (I.15) the explicit formula:

$$F' = \frac{\langle n_0 \rangle}{\sigma_0^2} \left\{ f_0 - 1 + F_1^{\text{Poisson}} + \left( 1 + \mathcal{M} \frac{n_{\text{eq}1}}{\langle n_0 \rangle} \right) \frac{F_2^{\text{Poisson}} - 1}{G_1} \right\} \qquad (I.16)$$

For an input signal with Poisson statistics, the noise figure of the tandem amplifier chain reduces to:

$$F' = F_1^{\text{Poisson}} + \frac{F_2^{\text{Poisson}} - 1}{G_1} + \mathcal{M} \frac{n_{\text{eq}1}}{\langle n_0 \rangle} \frac{F_2^{\text{Poisson}} - 1}{G_1} \qquad (I.17)$$

Finally, assuming an input signal large enough so that $\langle n_0 \rangle G_1 \gg n_{\text{eq}1}$, we find

$$F' \approx F_1^{\text{Poisson}} + \frac{F_2^{\text{Poisson}} - 1}{G_1} \qquad (I.18)$$

which is the usual result found in the literature and is used in Section 2.5. Equation (I.16) is different from the result in [1], which neglects the excess noise terms proportional to $\sigma_{i-1}^2 - \langle n_{i-1} \rangle$ in the expression of the amplifier variance $\sigma_i^2$. This is equivalent to assuming that the signal input to each amplifier has Poisson statistics, in which case each amplifier would be characterized by the noise figure $F_i^{\text{Poisson}}$(called $F$ in [1]); this approximation is not accurate, although the high-signal limit is identical to Eq. (I.18).

The procedure described above makes it possible to derive a general formula for the optical noise figure of a chain of $k$ amplifiers. We find:

$$F' = \frac{\langle n_0 \rangle}{\sigma_0^2} \left\{ f_0 - 1 + F_1^{\text{Poisson}} + g_1 \frac{F_2^{\text{Poisson}} - 1}{G_1} + g_1 g_2 \frac{F_3^{\text{Poisson}} - 1}{G_1 G_2} \right.$$

$$\left. + g_1 g_2 \dots g_{k-1} \frac{F_k^{\text{Poisson}} - 1}{G_1 G_2 \dots G_{k-1}} \right\} \qquad (I.19)$$

with

$$g_i = \left( 1 + \mathcal{M} \frac{n_{\text{eq}i}}{\langle n_{i-1} \rangle} \right) \qquad (I.20)$$

and

$$\langle n_i \rangle = G_1 G_2 \dots G_i \langle n_0 \rangle + \mathcal{M} \{ G_1 G_2 \dots G_i n_{\text{eq}1} + G_2 \dots G_i n_{\text{eq}2} + \dots + G_i n_{\text{eq}i} \} \qquad (I.21)$$

## REFERENCES

[1]  T. Okoshi, "Exact noise figure formulas for optical amplifiers and amplifier-fiber cascaded chains," in *Proc. Topical Meeting on Optical Amplifiers and Applications, 1990*, postdeadline paper Pdp11, Technical Digest Series vol 13, Optical Society of America, Washington DC.

# DERIVATION OF THE NONLINEAR PHOTON STATISTICS MASTER EQUATION AND MOMENT EQUATIONS FOR TWO- OR THREE-LEVEL LASER SYSTEMS (SEE SECTION 2.8)

We consider the steady state populations $N_1$, $N_2$ generalized to the cases of two- or three-level laser systems, defined in Eqs. (1.49) and (1.50), while assuming that the doping concentration is confined in the fiber core. These populations take the form:

$$N_1 = \rho \frac{1 + \dfrac{\eta_p}{1 + \eta_p} \dfrac{P_p}{P_{sat}(\nu_p)} + \dfrac{\eta_s}{1 + \eta_s} \dfrac{P_s}{P_{sat}(\nu_s)}}{1 + \dfrac{P_p}{P_{sat}(\nu_p)} + \dfrac{P_s}{P_{sat}(\nu_s)}} \tag{J.1}$$

$$N_2 = \rho \frac{\dfrac{1}{1 + \eta_p} \dfrac{P_p}{P_{sat}(\nu_p)} + \dfrac{1}{1 + \eta_s} \dfrac{P_s}{P_{sat}(\nu_s)}}{1 + \dfrac{P_p}{P_{sat}(\nu_p)} + \dfrac{P_s}{P_{sat}(\nu_s)}} \tag{J.2}$$

where the saturation power $P_{sat}(\nu_{p,s})$ for the pump and the signal are defined in Eq. (1.45). The three-level system expressions correspond to the case $\eta_p = 0$. Introducing the definition $q = P_p/P_{sat}(\nu_p)$ in Eqs. (J.1) and (J.2) and factorizing $(1 + q)$ yields:

$$N_1 = \rho \frac{1}{1 + q} \frac{1 + \dfrac{\eta_p}{1 + \eta_p} q + \dfrac{(1 + q)\eta_s}{1 + \eta_s} \dfrac{P_s}{(1 + q)P_{sat}(\nu_s)}}{1 + \dfrac{P_s}{(1 + q)P_{sat}(\nu_s)}} \tag{J.3}$$

$$N_2 = \rho \frac{1}{1 + q} \frac{\dfrac{1}{1 + \eta_p} q + \dfrac{1 + q}{1 + \eta_s} \dfrac{P_s}{(1 + q)P_{sat}(\nu_s)}}{1 + \dfrac{P_s}{(1 + q)P_{sat}(\nu_s)}} \tag{J.4}$$

**627**

We define now a saturation photon number $n_{sat}$ and the saturation coefficient $s = 1/n_{sat}$ through:

$$n_{sat} = \frac{1}{s} = (1 + q) \frac{P_{sat}(v_s)}{hv_s B_0} = (1 + q) \frac{\pi \omega_s^2}{\sigma_a(v_s)(1 + \eta_s) B_0 \tau} \qquad (J.5)$$

where $B_0$ is the signal optical bandwidth. The signal photon number is then $n = P_s/hv_s B_0$, and we have $P_s/P_{sat}(v_s) = sn$. The emission and absorption coefficients are given by $a = \Gamma_s \sigma_e N_2 = \eta_s \alpha_s N_2/\rho$ and $b = \Gamma_s \sigma_a N_1 = \eta_s \alpha_s N_1/\rho$, where $\Gamma_s$ is the overlap factor, Eq. (1.86) and $\alpha_s$ is the signal absorption coefficient. Using Eqs. (J.3) and (J.4), and after some algebra, the emission and absorption coefficients can be put into the form:

$$a_n = \alpha_s \frac{\eta_s}{1 + \eta_s} + \alpha_s \frac{\eta_s}{1 + \eta_s} \frac{Uq - 1}{1 + q} f_n \equiv A_0 + A_1 f_n \qquad (J.6)$$

$$b_n = \alpha_s \frac{\eta_s}{1 + \eta_s} - \alpha_s \frac{1}{1 + \eta_s} \frac{Uq - 1}{1 + q} f_n \equiv B_0 + B_1 f_n \qquad (J.7)$$

where a subscript $n$ is introduced to indicate the dependence of coefficients on signal photon number $n$, $U = (\eta_s - \eta_p)/(1 + \eta_p)$ and $f_n = 1/(1 + sn)$ is the saturation function.

We consider next the photon statistics master equation (2.72), in which we substitute now the $n$-dependent emission and absorption coefficients previously defined:

$$\frac{dP_n}{dz} = a_{n-1+\mathcal{M}}(n - 1 + \mathcal{M})P_{n-1} - a_{n+\mathcal{M}}(n + \mathcal{M})P_n + b_{n+1}(n + 1)P_{n+1} - b_n n P_n$$

$$= A_0\{(n - 1 + \mathcal{M})P_{n-1} - (n + \mathcal{M})P_n\} + B_0\{(n + 1)P_{n+1} - nP_n\}$$

$$+ A_1\{f_{n-1+\mathcal{M}}(n - 1 + \mathcal{M})P_{n-1} - f_{n+\mathcal{M}}(n + \mathcal{M})P_n\}$$

$$+ B_1\{f_{n+1}(n + 1)P_{n+1} - f_n n P_n\} \qquad (J.8)$$

The substitution of the nonlinear coefficients $a_n$ and $b_n$ into the master equation is heuristic, not based upon a rigorous demonstration. It is justified only by analogy with the case of four-level systems, for which a similar transformation of the master equation into its nonlinear form is based upon a rigorous theory (Section 2.8). For simplicity, we shall consider only the single-mode case ($\mathcal{M} = 1$) which, as far as the functions $f_{n-1+\mathcal{M}}$ and $f_{n+\mathcal{M}}$ in Eq. (J.8) are concerned, is equivalent to assuming that the effect of saturation by all the ASE modes is negligible in comparison to the saturation caused by the amplified signal, i.e., $\mathcal{M} \ll n$. Under the single-mode assumption, the nonlinear photon statistics master equation takes the form:

$$\frac{dP_n}{dz} = -\{A_0(n + 1) + B_0 n\}P_n + A_0 n P_{n-1} + B_0(n + 1)P_{n+1}$$

$$- \{A_1 f_{n+1}(n + 1) + B_1 f_n n\}P_n + A_1 f_n n P_{n-1} + B_1 f_{n+1}(n + 1)P_{n+1} \qquad (J.9)$$

The first and second order moments $\langle n \rangle$, $\langle n^2 \rangle$ are given from Eq. (J.9) by $\langle n \rangle = \sum n P_n$ and $\langle n^2 \rangle = \sum n^2 P_n$, which, with the usual translation property $\sum g(n) P_{n \pm 1} = \sum g[n - (\pm 1)] P_n$ yields, after summation:

$$\frac{d\langle n \rangle}{dz} = A_0 + A_1 \left\langle \frac{n+1}{1 + s(n+1)} \right\rangle - B_1 \left\langle \frac{n}{1 + sn} \right\rangle \tag{J.10}$$

$$\frac{d\langle n^2 \rangle}{dz} = A_0 + (3A_0 + B_0)\langle n \rangle + A_1 \left\langle \frac{2n^2 + 3n + 1}{1 + s(n+1)} \right\rangle - B_1 \left\langle \frac{2n^2 - n}{1 + sn} \right\rangle \tag{J.11}$$

The same procedure as above, but using Eq. (J.8) for the multimode case $\mathcal{M} > 1$ yields:

$$\frac{d\langle n \rangle}{dz} = \mathcal{M} A_0 + A_1 \left\langle \frac{n + \mathcal{M}}{1 + s(n + \mathcal{M})} \right\rangle - B_1 \left\langle \frac{n}{1 + s(n - 1 + \mathcal{M})} \right\rangle \tag{J.12}$$

$$\frac{d\langle n^2 \rangle}{dz} = \mathcal{M} A_0 + [(2\mathcal{M} + 1)A_0 + B_0]\langle n \rangle + A_1 \left\langle \frac{(n + \mathcal{M})(2n + 1)}{1 + s(n + \mathcal{M})} \right\rangle$$
$$- B_1 \left\langle \frac{2n^2 - n}{1 + s(n - 1 + \mathcal{M})} \right\rangle \tag{J.13}$$

We focus now onto the single-mode case $\mathcal{M} = 1$. The right-hand side brackets in Eqs. (J.10) and (J.11) can be evaluated through the decorrelation Taylor expansion in Eq. (2.39), i.e.,

$$\left\langle \frac{n^j}{1 + sn} \right\rangle = \frac{\langle n \rangle^j}{1 + s\langle n \rangle} + \sum_k \frac{\langle (n - \langle n \rangle)^k \rangle}{k!} \left[ \frac{d^k}{dn^k} \left( \frac{n^j}{1 + sn} \right) \right]_{\langle n \rangle}$$

$$= \frac{\langle n \rangle^j}{1 + s\langle n \rangle} + s^2 \frac{\sigma^2}{2!} \left[ \frac{1}{s^2} \frac{d^2}{dn^2} \left( \frac{n^j}{1 + sn} \right) \right]_{\langle n \rangle}$$

$$+ s^3 \frac{\langle (n - \langle n \rangle)^3 \rangle}{3!} \left[ \frac{1}{s^3} \frac{d^3}{dn^3} \left( \frac{n^j}{1 + sn} \right) \right]_{\langle n \rangle} + \cdots \tag{J.14}$$

and

$$\left\langle \frac{n^j}{1 + s(n+1)} \right\rangle = \frac{\langle n \rangle^j}{1 + s\langle n \rangle} + \sum_k \frac{\langle (n - \langle n \rangle)^k \rangle}{k!} \left[ \frac{d^k}{dn^k} \left( \frac{n^j}{1 + s(n+1)} \right) \right]_{\langle n \rangle}$$

$$= \frac{\langle n \rangle^j}{1 + s\langle n \rangle} + s^2 \frac{\sigma^2}{2!} \left[ \frac{1}{s^2} \frac{d^2}{dn^2} \left( \frac{n^j}{1 + s(n+1)} \right) \right]_{\langle n \rangle}$$

$$+ s^3 \frac{\langle (n - \langle n \rangle)^3 \rangle}{3!} \left[ \frac{1}{s^3} \frac{d^3}{dn^3} \left( \frac{n^j}{1 + s(n+1)} \right) \right]_{\langle n \rangle} + \cdots \tag{J.15}$$

where we use the approximation

$$\frac{1}{1 + s(\langle n \rangle + 1)} \approx \frac{1}{1 + s\langle n \rangle} \tag{J.16}$$

Equation (J.16) assumes large photon numbers $n \gg 1$, which is the case during saturation as $n$ is of the order of $1/s = n_{sat} \gg 1$. In the linear regime, this approximation does not affect the result because $s(n+1) \ll 1$ and $sn \ll 1$. Using Eqs. (J.14) and (J.15) in Eqs. (J.10) and (J.11) yields:

$$\frac{d\langle n \rangle}{dz} = A_0 + \frac{(A_1 - B_1)\langle n \rangle + A_1}{1 + s\langle n \rangle} - sf^3_{\langle n \rangle}(A_1 - B_1)\sigma^2$$

$$+ s^2 f^4_{\langle n \rangle}(A_1 - B_1)\langle (n - \langle n \rangle)^3 \rangle + \cdots \tag{J.17}$$

$$\frac{d\langle n^2 \rangle}{dz} = A_0 + 4A_0\langle n \rangle + \frac{2(A_1 - B_1)\langle n \rangle^2 + (3A_1 + B_1)\langle n \rangle + A_1}{1 + s\langle n \rangle}$$

$$+ (2 + s)f^3_{\langle n \rangle}(A_1 - B_1)\sigma^2 - s(2 + s)f^4_{\langle n \rangle}(A_1 - B_1)\langle (n - \langle n \rangle)^3 \rangle + \cdots \tag{J.18}$$

Numerical solution of Eqs. (J.17) and (J.18) can be made up to the second order $k = 2$, assuming for instance an input signal with Poisson statistics (mean $\langle n_0 \rangle$ and $\langle n_0^2 \rangle = \langle n_0 \rangle^2 + \langle n_0 \rangle$). For higher orders $k$ of the Taylor series, the knowledge of moments $\langle n^k \rangle$ is required, and additional equations $d\langle n^k \rangle/dz$ must be solved simultaneously. If the photon number $\langle n \rangle$ is large (e.g., $\langle n \rangle > 10^5$), and the saturation function $f_{\langle n \rangle}$ is near unity ($f_{\langle n \rangle} \lesssim 1$), the convergence of the series (J.14) and (J.15) is not rapid and the computation of a large number of moments $\langle n^k \rangle$ may be required.

# APPENDIX K

## SEMICLASSICAL DETERMINATION OF NOISE POWER SPECTRAL DENSITY IN AMPLIFIED LIGHT PHOTODETECTION (SEE SECTION 3.2)

The instantaneous detector photocurrent is given by:

$$i(t) = \frac{2e}{hv_s} GP_s^{in} \cos^2(2\pi v_s t) + \frac{4e}{hv_s} \sqrt{GP_s^{in} Nhv_s \delta v} \sum_{k=-M}^{M} \cos\{2\pi(v_s + k\delta v)t + \varphi_k\}\cos(2\pi v_s t)$$

$$+ 2eN\delta v \left[ \sum_{k=-M}^{M} \cos\{2\pi(v_s + k\delta v)t + \varphi_k\} \right]^2 \tag{K.1}$$

The time average of the first term in Eq. (K.1), i.e., $I_s = eGP_s^{in}/hv_s$, corresponds to the mean signal photocurrent with DC frequency. The second and third terms in Eq. (K.1) correspond to signal–ASE beat noise and ASE–ASE beat noise, respectively.

We look first at the signal–ASE beat noise component (s–ASE), which can be rewritten as:

$$i_{s\text{-}ASE}(t) = \frac{4e}{hv_s} \sqrt{GP_s^{in} Nhv_s \, \delta v} \sum_{k=-M}^{M} \cos\{2\pi(v_s + k\delta v)t + \varphi_k\}\cos(2\pi v_s t)$$

$$= \frac{2e}{hv_s} \sqrt{GP_s^{in} Nhv_s \, \delta v} \sum_{k=-M}^{M} \left[\cos\{2\pi(2v_s + k\delta v)t + \varphi_k\} + \cos(2\pi k\delta v t + \varphi_k)\right]$$

$$\tag{K.2}$$

While the time average of the s–ASE current $\langle i_{\text{s-ASE}}(t) \rangle$ is zero, the corresponding electrical power is given by:

$$\langle (i_{\text{s-ASE}})^2 \rangle = \left( \frac{2e}{h\nu_{\text{s}}} \right)^2 GP_{\text{s}}^{\text{in}} N h\nu_{\text{s}} \delta\nu$$

$$\times \sum_{k=-M}^{M} \sum_{l=-M}^{M} \left[ \langle \cos\{2\pi(2\nu_{\text{s}} + k\delta\nu)t + \varphi_k\} \cos(2\pi l\delta\nu t + \varphi_l) \rangle \right.$$

$$+ \langle \cos\{2\pi(2\nu_{\text{s}} + l\delta\nu)t + \varphi_l\} \cos(2\pi k\delta\nu t + \varphi_k) \rangle$$

$$+ \langle \cos\{2\pi(2\nu_{\text{s}} + k\delta\nu)t + \varphi_k\} \cos\{2\pi(2\nu_{\text{s}} + l\delta\nu)t + \varphi_l\} \rangle$$

$$\left. + \langle \cos(2\pi k\delta\nu t + \varphi_k) \cos(2\pi l\delta\nu t + \varphi_l) \rangle \right]$$

$$= \sum_{k=-M}^{M} \left[ 2\langle \cos\{2\pi(2\nu_{\text{s}} + k\delta\nu)t + \varphi_k\} \cos(2\pi k\delta\nu t + \varphi_k) \rangle \right.$$

$$\left. + \langle \cos^2\{2\pi(2\nu_{\text{s}} + k\delta\nu)t + \varphi_k\} \rangle + \langle \cos^2(2\pi k\delta\nu t + \varphi_k) \rangle \right] \qquad \text{(K.3)}$$

In Eq. (K.3), the time average of cross-terms containing products of cosines with random phases $\varphi_k \neq \varphi_l$ vanish. The three remaining terms correspond to components near twice the optical frequency $(2\nu_{\text{s}} \pm k\delta\nu)$, whose power spectra fall outside the electronic bandwidth, and components $\langle \cos^2(2\pi k\delta\nu t + \varphi_k) \rangle$ whose uniform power spectrum falls into the baseband frequency interval $[0, B_{\text{o}}/2]$. This contribution yields:

$$\langle (i_{\text{s-ASE}})^2 \rangle = \frac{4e^2}{h\nu_{\text{s}}} GP_{\text{s}}^{\text{in}} N \sum_{k=-M}^{M} \langle \cos^2(2\pi k\delta\nu t + \varphi_k) \rangle$$

$$= \frac{4e^2}{h\nu_{\text{s}}} GP_{\text{s}}^{\text{in}} N \delta\nu 2M \frac{1}{2} = \frac{4e^2}{h\nu_{\text{s}}} GP_{\text{s}}^{\text{in}} N M \delta\nu \qquad \text{(K.4)}$$

As the mean s–ASE photocurrent is zero, the power defined in Eq. (K.4) also corresponds to a noise variance. The s–ASE power spectrum is uniform over $[0, B_{\text{o}}/2]$, and has a spectral density:

$$\hat{\sigma}_{\text{s-ASE}}^2(f) = \frac{4e^2}{h\nu_{\text{s}}} GP_{\text{s}}^{\text{in}} N = 4GI_{\text{s}}I_{\text{N}} \frac{1}{B_{\text{o}}} \qquad \text{(K.5)}$$

The total s–ASE noise power in the electronic bandwidth $B_{\text{e}}$ is then:

$$\sigma_{\text{s-ASE}}^2 = 4GI_{\text{s}}I_{\text{N}} \frac{B_{\text{e}}}{B_{\text{o}}} \qquad \text{(K.6)}$$

We focus next on the ASE–ASE beat noise term in Eq. (K.1), which can be rewritten as:

$$i_{\text{ASE-ASE}}(t) = 2eN\delta\nu \sum_{k=-M}^{M} \sum_{l=-M}^{M} \cos\{2\pi(\nu_{\text{s}} + k\delta\nu)t + \varphi_k\} \cos\{2\pi(\nu_{\text{s}} + l\delta\nu)t + \varphi_l\}$$

$$= eN\delta\nu \sum_{k=-M}^{M} \sum_{l=-M}^{M} \left[ \cos\{2\pi[2\nu_{\text{s}} + (k+1)\delta\nu]t + \varphi_k + \varphi_l\} \right.$$

$$\left. + \cos\{2\pi(k-1)\delta\nu t + \varphi_k - \varphi_l\} \right] \qquad \text{(K.7)}$$

The terms oscillating at twice the optical frequency in Eq. (K.7) yield a zero time average photocurrent, while the terms with frequencies $(k - 1)\delta v$ yield nonzero contributions for $k = 1$, corresponding to the mean DC photocurrent:

$$I_N = eN\delta v \sum_{k=1}^{2M} \sum_{l=1}^{2M} \langle \cos\{2\pi(k - 1)\delta vt + \varphi_k - \varphi_l\}\rangle = eN\delta v.2M = NeB_o \quad \text{(K.8)}$$

defined in Eq. (3.20).

Considering next only the relevant terms in Eq. (K.7) which oscillate at frequencies $(k - 1)\delta v$, the corresponding noise power is given by:

$$\langle (i_{\text{ASE–ASE}})^2 \rangle = (eN\delta v)^2 \left\langle \left[ \sum_{k=1}^{2M} \sum_{l \neq k}^{2M} \cos\{2\pi(k - 1)\delta vt + \varphi_k - \varphi_l\}\rangle \right]^2 \right\rangle$$

$$= 2(eN\delta v)^2 \sum_{k=1}^{2M} \sum_{l \neq k}^{2M} \langle \cos^2\{2\pi(k - 1)\delta vt + \varphi_k - \varphi_l\}\rangle \quad \text{(K.9)}$$

In Eq. (K.9), we used the property that all cross-terms having unrelated random phases have zero time average. Recursion verifies that the double sum in Eq. (K.8) contains $2M - 1$ terms at $\delta v$, $2M - 2$ terms at $2\delta v, \ldots$, 1 term at $(2M - 1)\delta v$, and no term at $2M\delta v$, and the same number of corresponding terms with negative frequencies and opposite phases. The spectral density takes the value $2.2M(eN\delta v)^2/\delta v = 2e^2 N^2 B_o$ near DC, and decreases linearly with frequency to vanish at $v = B_o$. The ASE–ASE spectral density is thus:

$$\hat{\sigma}^2_{\text{ASE–ASE}}(f) = 2e^2 N^2 B_o \left( 1 - \frac{f}{B_o} \right) = \frac{2I_N^2}{B_o} \left( 1 - \frac{f}{B_o} \right) \quad \text{(K.10)}$$

The total ASE–ASE noise power falling into the electronic bandwidth $B_e$ is then given by:

$$\sigma^2_{\text{ASE–ASE}} = \frac{2I_N^2}{B_o} \int_0^{B_e} \left( 1 - \frac{f}{B_o} \right) df = I_N^2 \frac{B_e}{B_0^2} (2B_o - B_e) \quad \text{(K.11)}$$

# DERIVATION OF THE ABSORPTION AND EMISSION CROSS SECTIONS THROUGH EINSTEIN'S *A* AND *B* COEFFICIENTS (SECTION 4.5)

Consider a two-level system with ground and upper level atomic populations $N_1$ and $N_2$, respectively. In thermal equilibrium, these populations are related through the Boltzmann factor:

$$\frac{N_2}{N_1} = \frac{g_2}{g_1} \exp\left(-\frac{h\nu}{k_B T}\right) \tag{L.1}$$

where $h\nu = \Delta E$ is the energy difference between the two atomic levels. The populations also follow the rate equations:

$$\frac{dN_1}{dt} = AN_2 - W_{12}N_1 + W_{21}N_2 \tag{L.2}$$

$$\frac{dN_2}{dt} = -AN_2 + W_{12}N_1 - W_{21}N_2 \tag{L.3}$$

where $A$, $W_{12}$, $W_{21}$ are the spontaneous emission rate, the absorption rate, and the stimulated emission rate, respectively. At thermal equilibrium, absorption and emission exactly balance, so $dN_1/dt = dN_2/dt = 0$. From Eqs. (L.1) and (L.2) we obtain then

$$\frac{N_2}{N_1} = \frac{W_{12}}{A + W_{21}} = \frac{g_2}{g_1} \exp\left(-\frac{h\nu}{k_B T}\right) \tag{L.4}$$

Einstein assumed that the induced transition rates $W_{12}$, $W_{21}$ are related to the thermal radiation energy density $\rho(\nu, T)$ through:

$$W_{12} = B_{12}\rho(\nu, T) \tag{L.5}$$

$$W_{21} = B_{21}\rho(\nu, T) \tag{L.6}$$

Combining Eqs. (L.4), (L.5), and (L.6) we obtain:

$$\rho(v, T) = \frac{g_2}{g_1} \frac{A}{B_{12} \exp\left(-\dfrac{hv}{k_B T}\right) - \dfrac{g_2}{g_1} B_{21}} \tag{L.7}$$

On the other hand, the thermal radiation energy density $\rho(v, T)$ can be put into the form:

$$\rho(v, T) = M(v)hvn(T) \tag{L.8}$$

where $M(v)$ is the density of radiation modes and $n(T)$ the Bose–Einstein occupation number:

$$n(T) = \frac{1}{\exp\left(\dfrac{hv}{k_B T}\right) - 1} \tag{L.9}$$

The density of radiation modes $M(v)$ can be determined by the following. Consider a cubic cavity of volume $V = d^3$ and refractive index $n$. The radiation mode wavelength must verify the boundary condition $q\lambda/2n = d$, where $q$ is an integer. The modes wave vectors are then given by $k = 2\pi n/\lambda = q\pi/d$. The wave vector difference between two adjacent modes is then $\pi/d$, corresponding to an elementary volume in $k$-space $(\pi/d)^3$. The density of such elementary volumes in a sphere of radius $k$ in $k$-space is:

$$\frac{N}{V} = \frac{1}{V} \frac{\frac{4}{3}\pi k^3}{(\pi/d)^3} = \frac{4}{3} \frac{8\pi n^3 v^3}{c^3} \tag{L.10}$$

The spectral density of radiation modes with wave vector $k$ is then given by

$$M(v) = 2 \cdot \frac{1}{8} \frac{\partial}{\partial v}\left(\frac{N}{V}\right) = \frac{8\pi n^3 v^2}{c^3} \tag{L.11}$$

where the factor $1/8$ takes into account the fact that the wave vectors $(k)$ and $(-k)$ correspond to the same radiation mode, so only the positive octant of the $k$-sphere must be considered, and the factor 2 accounts for the two possible polarizations. From Eqs. (L.8), (L.9), and (L.11), we obtain:

$$\rho(v, T) = M(v)hvn(T) = \frac{8\pi h n^3 v^3}{c^3} \frac{1}{\exp\left(\dfrac{hv}{k_B T}\right) - 1} \tag{L.12}$$

and finally, from Eq. (L.7):

$$\frac{8\pi h n^3 v^3}{c^3} \frac{1}{\exp\left(\dfrac{hv}{k_B T}\right) - 1} = \frac{g_2}{g_1} \frac{A}{B_{12} \exp\left(-\dfrac{hv}{k_B T}\right) - \dfrac{g_2}{g_1} B_{21}} \tag{L.13}$$

Equation (L.13) is satisfied for any temperature only if

$$B_{12} = \frac{g_2}{g_1} B_{21} \tag{L.14}$$

and

$$B_{21} = \frac{c^3}{8\pi h n^3 v^3} A \tag{L.15}$$

The coefficients $A$ and $B_{21}$ are called Einstein's A and B coefficients.

Using this result, the stimulated emission rates $W_{12}$, $W_{21}$ can then be written as:

$$W_{12} = \frac{g_2}{g_1} W_{21} \tag{L.16}$$

$$W_{21} = \frac{c^3}{8\pi c h n^3 v^3} \rho(v, T) \tag{L.17}$$

where $\tau = 1/A$ is the radiative lifetime. Equation (L.17) gives the stimulated emission rate induced by thermal light. In the case of a monochromatic excitation of intensity $I_s(v)$, the corresponding radiation density $\rho(v, T)$ is given the relation $I_s(v) = (c/n)\rho(v)$, and the stimulated emission rate in Eq. (L.17) takes the form:

$$W_{21} = \frac{\lambda^2}{8\pi n^2 \tau} \frac{I_s(v)}{hv} g(v) \tag{L.18}$$

where $\lambda = c/n$ is the wavelength in a vacuum and $g(v)$ is a normalized line shape which represents the probability of emission near the resonance of frequency $v$. We define next the transition emission cross section through:

$$\sigma_e(v) = \frac{\lambda^2}{8\pi n^2 \tau} g(v) \tag{L.19}$$

and the stimulated emission coefficient is now defined as:

$$W_{21} = \frac{I_s(v)}{hv} \sigma_e(v) \tag{L.20}$$

The peak emission cross section $\sigma_e^{peak}$ can then be obtained by integration of (L.19):

$$\int_0^{+\infty} \sigma_e(v)\, dv = \frac{c}{\lambda^2} \int_0^{+\infty} \sigma_e(\lambda)\, d\lambda = \sigma_e^{peak} \frac{c\Delta\lambda_e}{\lambda^2} = \frac{\lambda^2}{8\pi n^2 \tau} \int_0^{+\infty} g(v)\, dv = \frac{\lambda^2}{8\pi n^2 \tau} \tag{L.21}$$

or

$$\boxed{\sigma_e^{peak} = \frac{\lambda^4}{8\pi c n^2 \tau \Delta\lambda_e}} \tag{L.22}$$

From Eqs. (L.16) and (L.20), the absorption rate is then defined as:

$$W_{12} = \frac{g_2}{g_1} \frac{I_s(v)}{hv} \sigma_e(v) = \frac{I_s(v)}{hv} \sigma_a(v) \tag{L.23}$$

where $\sigma_a(v) = g_2/g_1 \sigma_e(v)$ is the absorption cross section.

# CALCULATION OF HOMOGENEOUS ABSORPTION AND EMISSION CROSS SECTIONS BY DECONVOLUTION OF EXPERIMENTAL CROSS SECTIONS (SEE SECTION 4.5)

The homogeneous absorption and emission cross sections $\sigma_{e,a}^{H}(\omega)$ are defined by the inverse Fourier transform relation:

$$\sigma_{e,a}^{H}(\omega) = \mathscr{F}^{-1}\left[ \exp\left\{\frac{\Delta\omega_{inh}^2 x^2}{16\log 2}\right\} \cdot \mathscr{F}\left[\sigma_{e,a}^{I}(\omega); x\right]; \omega \right] \tag{M.1}$$

The inhomogeneous or experimental cross-sections $\sigma_{a,e}^{I}(\lambda)$ are given by a best fit with Gaussian functions:

$$\sigma_{a,e}^{I}(\omega) = \sum_k a_k^{a,e} \exp\left\{ -4\log 2 \frac{(\omega - \omega_k)^2}{\Delta\omega_k^2} \right\} \tag{M.2}$$

For the computation of Eq. (M.1), we use first two properties of the Fourier transforms:

$$\mathscr{F}[\exp\{-a^2(\omega - \omega_k)^2; x] = \exp\{i\omega_k x\}\mathscr{F}[\exp\{-a^2\omega^2\}; x] \tag{M.3}$$

$$\mathscr{F}[\exp\{-a^2\omega^2\}; x] = \frac{1}{a\sqrt{2}} \exp\left\{ -\frac{x^2}{4a^2} \right\} \tag{M.4}$$

and the property of linearity for this transformation, which gives:

$$\mathscr{F}[\sigma_{a,e}^{I}(\omega); x] = \sum_k a_k^{a,e} \frac{\Delta\omega_k}{2\sqrt{2\log 2}} \exp\{-i\omega_k x\}\exp\left\{ -\frac{\Delta\omega_k^2 x^2}{16\log 2} \right\} \tag{M.5}$$

From Eqs. (M.1) and (M.5), the homogeneous cross sections are then given by the expression:

$$\sigma_{a,e}^{H}(\omega) = \sum_k a_k^{a,e} \frac{\Delta\omega_k}{2\sqrt{2\log 2}} \mathscr{F}^{-1}\left[ \exp\left\{ -\frac{(\Delta\omega_k^2 - \Delta\omega_{inh}^2)x^2}{16\log 2} - i\omega_k x \right\}; \omega \right] \tag{M.6}$$

The argument in the exponential function in Eq. (M.6) can be expressed in the form $b^2 - (ax - b)^2$, where $a$ and $b$ are constants defined through:

$$a = \frac{\Delta\omega_{inh}\sqrt{\beta_k^2 - 1}}{4\sqrt{\log 2}} \tag{M.7}$$

$$b = i\,\frac{2\omega_k\sqrt{\log 2}}{\Delta\omega_{inh}\sqrt{\beta_k^2 - 1}} \tag{M.8}$$

and

$$\beta_k = \frac{\Delta\omega_k}{\Delta\omega_{inh}} \tag{M.9}$$

In Eqs. (M.7) and (M.8) it was assumed that $\beta_k > 1$, or $\Delta\omega_k > \Delta\omega_{inh}$. If any of the Gaussian fitting functions has a width smaller or equal to the inhomogeneous broadening, the convolution equations (1.205) and (1.206) do not apply. Now we use the translation property of Fourier transforms:

$$\mathscr{F}^{-1}[\exp\{-(ax - b)^2\}; \omega] = \exp\left\{-i\frac{b}{a}\omega\right\}\mathscr{F}^{-1}[\exp\{-a^2x^2\}; \omega] \tag{M.10}$$

and the inverse relation:

$$\mathscr{F}^{-1}[\exp\{-a^2x^2\}; \omega] = \frac{1}{a\sqrt{2}}\exp\left\{-\frac{\omega^2}{4a^2}\right\} \tag{M.11}$$

Substituting the result of Eqs. (M.10) and (M.11) into Eq. (M.6) yields:

$$\sigma_{a,e}^{H}(\omega) = \sum_k a_k^{a,e}\,\frac{\Delta\omega_k}{\sqrt{\Delta\omega_k^2 - \Delta\omega_{inh}^2}}\exp\left\{-4\log 2\,\frac{(\omega - \omega_k)^2}{\Delta\omega_k^2 - \Delta\omega_{inh}^2}\right\} \tag{M.12}$$

Equation (M.12) coincides with Eq. (M.2) in the case where there is no inhomogeneous broadening, $\Delta\omega_{inh} = 0$, and $\sigma_{a,e}^{H}(\omega) = \sigma_{a,e}^{I}(\omega)$. Equation (M.12) can also be rewritten as a function of wavelength in the form:

$$\sigma_{a,e}^{H}(\lambda) = \sum_i a_i^{a,e}\,\frac{\Delta\lambda_i}{\sqrt{\Delta\lambda_i^2 - \Delta\lambda_{inh}^2}}\exp\left\{-4\log 2\,\frac{(\lambda - \lambda_i)^2}{\Delta\lambda_i^2 - \Delta\lambda_{inh}^2}\right\} \tag{M.13}$$

# APPENDIX N

# RATE EQUATIONS FOR THREE-LEVEL SYSTEMS WITH PUMP EXCITED STATE ABSORPTION (SEE SECTION 4.7)

Consider the energy level diagram of Figure N.1. Let $R$ and $R'$ be the pumping rates corresponding to ground state absorption (GSA) and excited state absorption (ESA) which take place between levels 1 and 3, and 2 and 4, respectively. For simplicity, we assume $R_{13} = R_{31} = R$ and $R_{24} = R_{42} = R'$. The nonradiative decay rates from levels 3 and 4 are $A_{32}$ and $A_{43}$, respectively, while the radiative decay rate from the metastable level, level 2, is $A_{21}$; the stimulated emission rates are $W_{12}$ and $W_{21}$. The rate equations for the populations $N_1$, $N_2$, $N_3$, and $N_4$ of levels 1, 2, 3, and 4 are:

$$\frac{dN_1}{dt} = A_{21}N_2 + W_{21}N_2 - W_{12}N_1 - R(N_1 - N_3) \tag{N.1}$$

$$\frac{dN_2}{dt} = A_{32}N_3 - A_{21}N_2 - W_{21}N_2 + W_{12}N_1 - R'(N_2 - N_4) \tag{N.2}$$

$$\frac{dN_3}{dt} = -A_{32}N_3 + A_{43}N_4 + R(N_1 - N_3) \tag{N.3}$$

$$\frac{dN_4}{dt} = -A_{43}N_4 + R'(N_2 - N_4) \tag{N.4}$$

with $N_1 + N_2 + N_3 + N_4 = \rho(r)$. We introduce the following definitions:

$$\alpha = R + A_{32} \tag{N.5}$$

$$\beta = W_{21} + A_{21} + R' \tag{N.6}$$

$$\kappa = \frac{1}{1 + A_{43}/R'} \tag{N.7}$$

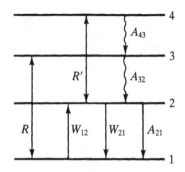

**FIGURE N.1** Energy level diagram corresponding to the analysis of pump excited state absorption.

The steady state solutions $(dN_i/dt = 0, i = 1, \ldots, 4)$ of the rate equations (N.1)–(N.4) become:

$$N_1 = \rho \frac{\alpha\beta - \kappa(\alpha R' + A_{32}A_{43})}{\beta(R + \alpha) + (\alpha W_{12} + RA_{32})(1 + \kappa) + \kappa[A_{43}(W_{12} - A_{32}) - R'(R + \alpha)]} \quad \text{(N.8)}$$

$$N_2 = \rho \frac{\alpha W_{12} + RA_{32}}{\beta(R + \alpha) + (\alpha W_{12} + RA_{32})(1 + \kappa) + \kappa[A_{43}(W_{12} - A_{32}) - R'(R + \alpha)]} \quad \text{(N.9)}$$

$$N_3 = \rho \frac{R\beta + \kappa(A_{43}W_{12} - RR')}{\beta(R + \alpha) + (\alpha W_{12} + RA_{32})(1 + \kappa) + \kappa[A_{43}(W_{12} - A_{32}) - R'(R + \alpha)]} \quad \text{(N.10)}$$

and $N_4 = \kappa N_2$. We introduce next:

$$\tau = 1/A_{21} \quad \text{(N.11)}$$

$$\varepsilon = A_{21}/A_{32} \quad \text{(N.12)}$$

$$\varepsilon' = A_{21}/A_{43} \quad \text{(N.13)}$$

$$R'' = \frac{R'}{1 + \varepsilon'R'\tau} \quad \text{(N.14)}$$

and Eqs. (N.8)–(N.10) can be written as:

$$N_1 = \rho \frac{(1 + \varepsilon R\tau)[1 + W_{21}\tau + R'\tau(1 - \kappa)] - R''\tau}{(1 + 2\varepsilon R\tau)[1 + W_{21}\tau + R'\tau(1 - \kappa)] + (1 + \kappa)[(1 + \varepsilon R\tau)W_{12}\tau + R\tau] + R''\tau(\varepsilon W_{12}\tau - 1)} \quad \text{(N.15)}$$

$$N_2 = \rho \frac{(1 + \varepsilon R\tau)W_{12}\tau + R\tau}{(1 + 2\varepsilon R\tau)[1 + W_{21}\tau + R'\tau(1 - \kappa)] + (1 + \kappa)[(1 + \varepsilon R\tau)W_{12}\tau + R\tau] + R''\tau(\varepsilon W_{12}\tau - 1)} \quad \text{(N.16)}$$

$$N_3 = \rho \frac{\varepsilon R\tau[1 + W_{21}\tau + R'\tau(1 - \kappa)] + \varepsilon R'\tau}{(1 + 2\varepsilon R\tau)[1 + W_{21}\tau + R'\tau(1 - \kappa)] + (1 + \kappa)[(1 + \varepsilon R\tau)W_{12}\tau + R\tau] + R''\tau(\varepsilon W_{12}\tau - 1)} \quad \text{(N.17)}$$

We consider now the case of silicate glass where the spontaneous decay rate $A_{21}$ is negligible compared to the nonradiative decay rates, i.e., $A_{21}/A_{32} \ll 1$ and $A_{21}/A_{43} \ll 1$. In this case, we have $\varepsilon \approx \varepsilon' \approx \kappa \approx 0$ and, consequently, $N_3 \approx N_4 \approx 0$. This result shows that because of fast nonradiative decay, levels 3 and 4 are very weakly populated. With these approximations, Eqs. (N.15) and (N.16) become:

$$N_1 = \rho \frac{1 + W_{21}\tau}{1 + W_{12}\tau + W_{21}\tau + R\tau} \tag{N.18}$$

$$N_2 = \rho \frac{R\tau + W_{12}\tau}{1 + W_{12}\tau + W_{21}\tau + R\tau} \tag{N.19}$$

The above result shows that in the presence of pump ESA, the populations of levels 1 and 2 are the same as in the absence of ESA. That no change of population occurs, in spite of pump absorption from level 2, is due to fast nonradiative decay from levels 4 and 3 to level 2.

We introduce into Eqs. (N.18) and (N.19) the explicit definitions (1.64) and (1.65) of the stimulated emission rates in the form:

$$W_{12}\tau(r) = \sum_k \frac{1}{1 + \eta_k} \frac{P_s(\lambda_k)}{P_{sat}(\lambda_k)} \psi_{sk}(r) \tag{N.20}$$

$$W_{21}\tau(r) = \sum_k \frac{\eta_k}{1 + \eta_k} \frac{P_s(\lambda_k)}{P_{sat}(\lambda_k)} \psi_{sk}(r) \tag{N.21}$$

where $P_s(\lambda_k)$, $P_{sat}(\lambda_k)$, $\eta_k$, $\psi_{sk}(r)$ have the usual definitions given in Chapter 1. In order to obtain formulae that are valid for both two-level and three-level pumping schemes, we can set $R = 0$ in Eqs. (N.18) and (N.19) and include the pumping rate in the sums in Eqs. (N.20) and (N.21). The case where $\eta_p = 0$ for $\lambda_k = \lambda_p$ in the sums corresponds to the three-level pumping scheme, while the case $\eta_p \neq 0$ corresponds to the two-level pumping scheme.

We assume now that there is only one pump at $\lambda_k = \lambda_p$ and one signal at $\lambda_k = \lambda_s$, with corresponding mode envelopes $\psi_p(r)$ and $\psi_s(r)$, respectively. Using Eqs. (N.20) and (N.21) in Eqs. (N.18) and (N.19), and $P_s(\lambda_p)/P_{sat}(\lambda_p) = q$, $P_s(\lambda_s)/P_{sat}(\lambda_s) = p$, we obtain:

$$N_1(r) = \rho(r) \frac{1 + \dfrac{\eta_p}{1 + \eta_p} q\psi_p(r) + \dfrac{\eta_s}{1 + \eta_s} p\psi_s(r)}{1 + q\psi_p(r) + p\psi_s(r)} \tag{N.22}$$

$$N_2(r) = \rho(r) \frac{\dfrac{1}{1 + \eta_p} q\psi_p(r) + \dfrac{1}{1 + \eta_s} p\psi_s(r)}{1 + q\psi_p(r) + p\psi_s(r)} \tag{N.23}$$

For a three-level pumping scheme (unidirectional pump) with pump ESA occurring between levels 2 and 4, the pump rate equation given by:

$$\frac{1}{q} \frac{dq}{dz} = 2\pi \int_S \bar{\psi}_p(r)\{\sigma_{GSA}(\lambda_p)[\eta_p N_3(r) - N_1(r)] + \sigma_{ESA}(\lambda_p)[\eta_p' N_4(r) - N_2(r)]\}r\,dr \tag{N.24}$$

while for a two-level pumping scheme with pump ESA occurring between levels 2 and 3, it is given by:

$$\frac{1}{q}\frac{dq}{dz} = 2\pi \int_S \bar{\psi}_p(r)\{\sigma_{GSA}(\lambda_p)[\eta_p N_2(r) - N_1(r)] + \sigma_{ESA}(\lambda_p)[\eta'_p N_3(r) - N_2(r)]\} r\, dr \quad (N.25)$$

where $\sigma_{GSA}(\lambda_p) = \sigma_{13}(\lambda_p)$, $\sigma_{ESA}(\lambda_p) = \sigma_{24}(\lambda_p)$, $\eta_p = \sigma_{31}(\lambda_p)/\sigma_{13}(\lambda_p)$ and $\eta'_p = \sigma_{42}(\lambda_p)/\sigma_{24}(\lambda_p)$ for the three-level pumping scheme, and $\sigma_{GSA}(\lambda_p) = \sigma_{12}(\lambda_p)$, $\sigma_{ESA}(\lambda_p) = \sigma_{23}(\lambda_p)$, $\eta_p = \sigma_{21}(\lambda_p)/\sigma_{12}(\lambda_p)$ and $\eta'_p = \sigma_{32}(\lambda_p)/\sigma_{23}(\lambda_p)$ for the two-level pumping scheme.

We use next the result $N_3 \approx N_4 \approx 0$ and substitute Eqs. (N.22) and (N.23) into Eq. (N.24) to obtain the pump rate equation for the three-level pumping scheme:

$$\frac{1}{q}\frac{dq}{dz} = -2\pi \int_S \bar{\psi}_p(r)\{\sigma_{GSA}(\lambda_p)N_1(r) + \sigma_{ESA}(\lambda_p)N_2(r)\} r\, dr$$

$$= -2\pi\rho_0\sigma_{GSA}(\lambda_p)\int_S \frac{\rho(r)}{\rho_0}\bar{\psi}_p(r)\frac{1 + \delta q\psi_p(r) + \dfrac{\eta_s + \delta}{1 + \eta_s}p\psi_s(r)}{1 + q\psi_p(r) + p\psi_s(r)} r\, dr \quad (N.26)$$

with $\delta = \sigma_{ESA}(\lambda_p)/\sigma_{GSA}(\lambda_p)$. On the other hand, the pump rate equation for the three-level pumping scheme is, from Eq. (N.25):

$$\frac{1}{q}\frac{dq}{dz} = 2\pi \int_S \bar{\psi}_p(r)\{\sigma_{GSA}(\lambda_p)[\eta_p N_2(r) - N_1(r)] - \sigma_{ESA}(\lambda_p)N_2(r)\} r\, dr$$

$$= -2\pi\rho_0\sigma_{GSA}(\lambda_p)\int_S \frac{\rho(r)}{\rho_0}\bar{\psi}_p(r)\frac{1 + \dfrac{\delta}{1 + \eta_p} q\psi_p(r) + \dfrac{\eta_s - \eta_p + \delta}{1 + \eta_s}p\psi_s(r)}{1 + q\psi_p(r) + p\psi_s(r)} r\, dr \quad (N.27)$$

Note that Eq. (N.27) reduces to Eq. (N.26) in the case $\eta_p = 0$ (three-level pumping scheme).

Equation (N.27) can be generalized for any wavelength $\lambda_k$ to include effects of pump and signal ESA:

$$\frac{1}{p_k}\frac{dp_k}{dz} = 2\pi\rho_0\sigma_{GSA}(\lambda_k)\int_S \frac{\rho(r)}{\rho_0}\bar{\psi}_{sk}(r)\frac{\displaystyle\sum_j \frac{\eta_k - \eta_j - \delta_k}{1 + \eta_j}p_j\psi_{sj}(r) - 1}{1 + \displaystyle\sum_j p_j\psi_{sj}(r)} r\, dr \quad (N.28)$$

with $\delta_k = \sigma_{ESA}(\lambda_k)/\sigma_{GSA}(\lambda_k)$. The ASE noise, which for simplicity is overlooked in Eq. (N.28), can be introduced through the additional term $2p_{0k}\sigma_{GSA}(\lambda_k)\eta_k N_2(r)/p_k$ in the integrand of this equation. The most general equation, which includes bidirectional pump and signals with ASE noise takes the form:

$$\frac{dp_k^{\pm}}{dz} = \pm 2\pi\rho_0\sigma_{GSA}(\lambda_k) \times \int_S \frac{\rho(r)}{\rho_0}\bar{\psi}_{sk}(r)$$

$$\times \frac{(\eta_k - \delta_k)\left[\displaystyle\sum_j \frac{(p_j^+ + p_j^-)}{1 + \eta_j}\psi_{sj}(r)\right](p_k^{\pm} + 2p_{0k}) - \left[1 + \displaystyle\sum_j \eta_j\frac{(p_j^+ + p_j^-)}{1 + \eta_j}\psi_{sj}(r)\right]p_k^{\pm}}{1 + \displaystyle\sum_j (p_j^+ + p_j^-)\psi_{sj}(r)} r\, dr \quad (N.29)$$

---

# DETERMINATION OF EXPLICIT ANALYTICAL SOLUTION FOR A LOW GAIN, UNIDIRECTIONALLY PUMPED EDFA WITH SINGLE-SIGNAL SATURATION (SEE SECTION 5.3)

---

The transcendental equation (1.140) giving $p_2^{\text{out}} = f(p_1^{\text{in}}, p_2^{\text{in}})$ in the case of forward pumping can be written in the form (Section 1.10):

$$y = h(U_2 p_2^{\text{out}}) = \mathscr{C}' g(U_1 p_1^{\text{out}}) \tag{O.1}$$

where

$$h(x) = x \exp(x) \tag{O.2}$$

$$g(x) = \{x \exp(-x)\}^{\alpha_2/\alpha_1} \tag{O.3}$$

$$\mathscr{C}' = U_2 p_2^{\text{in}}(U_1 p_1^{\text{in}})^{-\alpha_2/\alpha_1} \cdot \exp\left(U_2 p_2^{\text{in}} + \frac{\alpha_2}{\alpha_1} U_1 p_1^{\text{in}}\right) \tag{O.4}$$

and $U_1 = (\eta_2 - \eta_1)/(1 + \eta_1)$, $U_2 = (\eta_2 - \eta_1)/(1 + \eta_2)$ with $\alpha_1$, $\alpha_2$ being the absorption coefficients at the two wavelengths and $\eta_i = \sigma_e(\lambda_i)/\sigma_a(\lambda_i)$ the corresponding cross section ratios. The explicit solution of Eq. (O.1) is given by

$$p_2^{\text{out}} = \frac{1}{U_2} h^{-1}(y) = \frac{1}{U_2} h^{-1}[\mathscr{C}' \cdot g(U_1 p_1^{\text{out}})] \tag{O.5}$$

where $h^{-1}(y)$ is the inverse function of $h(x)$. Equation (O.5) can be explicitly expressed through an accurate polynomial fit of the inverse function $h^{-1}(y)$. In order to obtain a polynomial fit valid over a large range of values for the variable $y = h(x)$, it is convenient to take first the decimal logarithm of Eq. (O.5), and convert it into the form:

$$p_2^{\text{out}} = \frac{1}{U_2} 10^{\mathscr{P}(z)} \tag{O.6}$$

with

$$z = \log_{10}(y) = \frac{1}{\log(10)} \left\{ U_2 p_2^{in} + \log(U_2 p_2^{in}) + \frac{\alpha_2}{\alpha_1} \left[ U_1(p_1^{in} - p_1^{out}) - \log\left(\frac{p_1^{in}}{p_1^{out}}\right) \right] \right\} \quad (O.7)$$

In Eq. (O.6) the function $\mathscr{P}(z) = \log_{10}(x)$ is the inverse function of $z = \log_{10}(y) = \log_{10}(x) \cdot \exp[\log_{10}(x)]$. This inverse function can be expressed through an $N$th order polynomial fit of expression:

$$\mathscr{P}(z) = \sum_{i=0}^{N} a_i z^i \quad (O.8)$$

The coefficients $a_i(i = 0, \ldots, N)$ in Eq. (O.7) can be determined from an arbitrary set of values of $\{\log_{10}(x_i), z_i\}$ $(i = 0, \ldots, N)$ by solving through standard numerical methods the matrix equation (shown here for $N = 2$):

$$\begin{pmatrix} 1 & z_0 & z_0^2 \\ 1 & z_1 & z_1^2 \\ 1 & z_2 & z_2^2 \end{pmatrix} \begin{pmatrix} a_0 \\ a_1 \\ a_2 \end{pmatrix} = \begin{pmatrix} \log_{10}(x_0) \\ \log_{10}(x_1) \\ \log_{10}(x_2) \end{pmatrix} \quad (O.9)$$

Because of the rapid growth of the function $z = \log_{10}(y)$, the polynomial fit $\mathscr{P}(z)$ must be calculated for two different intervals for the variable $z$. For instance, the fit can be made over the two intervals ($z = -3$ to $z = 4$) and ($z = 4$ to $z = 40$), which are found to cover the range of interest for the variable $y$. The corresponding coefficients are listed in Table O.1.

**TABLE O.1**    Definition of coefficients for the polynomial $\mathscr{P}(z)$

| $-3 \leqslant z \leqslant 4$ | $4 \leqslant z \leqslant 40$ |
|---|---|
| $a_0 = -0.2496093$ | $a_0 = 0.1818821$ |
| $a_1 = 0.6433086$ | $a_1 = 0.2487224$ |
| $a_2 = -0.1564108$ | $a_2 = -2.571578 \times 10^{-2}$ |
| $a_3 = 1.273030 \times 10^{-3}$ | $a_3 = 1.871872 \times 10^{-3}$ |
| $a_4 = 1.002236 \times 10^{-2}$ | $a_4 = -8.959120 \times 10^{-5}$ |
| $a_5 = -1.354737 \times 10^{-3}$ | $a_5 = 2.746533 \times 10^{-6}$ |
| $a_6 = -3.2217997 \times 10^{-4}$ | $a_6 = -5.153686 \times 10^{-8}$ |
| $a_7 = 9.992659 \times 10^{-5}$ | $a_7 = 5.355500 \times 10^{-10}$ |
| $a_8 = -9.265364 \times 10^{-6}$ | $a_8 = -2.347970 \times 10^{-12}$ |
| $a_9 = 2.937424 \times 10^{-7}$ | $a_9 = 0$ |

# APPENDIX P

# DETERMINATION OF EDFA EXCESS NOISE FACTOR IN THE SIGNAL-INDUCED SATURATION REGIME (SEE SECTION 5.4)

We consider the rate equation (1.114) for the normalized signal at wavelength $\lambda_k$ propagating in the forward direction, which applies to the case of confined Er-doping:

$$\frac{dp_k^+}{dz} = (\gamma_{ek} - \gamma_{ak})p_k^+ + 2\gamma_{ek}p_{0k} \qquad (P.1)$$

In Eq. (P.1), the emission and absorption coefficients are defined by Eqs. (1.115) and (1.116), i.e.,

$$\gamma_{ek} = \alpha_k \frac{\sum_j \frac{\eta_k}{1 + \eta_j}(p_j^+ + p_j^-)}{1 + \sum_j (p_j^+ + p_j^-)} \qquad (P.2)$$

$$\gamma_{ak} = \alpha_k \frac{1 + \sum_j \frac{\eta_j}{1 + \eta_j}(p_j^+ + p_j^-)}{1 + \sum_j (p_j^+ + p_j^-)} \qquad (P.3)$$

We assume now that each signal and pump are at one single wavelength. For simplicity, let $p_k^+ \equiv p$ and $p_{j \neq k}^+ + p_{j \neq k}^- \equiv q$; the latter represents the normalized pump power, whether unidirectional or bidirectional. The notation can be clarified by using $\gamma_{ek} \equiv \gamma_{es}$, $\gamma_{ak} \equiv \gamma_{as}$, $\alpha_k \equiv \alpha_s$, $\eta_{j \neq k} \equiv \eta_p$, $\eta_k \equiv \eta_s$ and $p_{0k} \equiv p_0$. Then the coefficients in Eqs. (P.2) and (P.3) the form:

$$\gamma_{es} = \alpha_s \frac{\frac{\eta_s}{1 + \eta_p}q + \frac{\eta_s}{1 + \eta_s}p}{1 + p + q} \qquad (P.4)$$

$$\gamma_{as} = \alpha_s \frac{1 + \frac{\eta_p}{1 + \eta_p}q + \frac{\eta_s}{1 + \eta_s}p}{1 + p + q} \qquad (P.5)$$

The general solution of Eq. (P.1) can be written as:

$$p(L) = G(L)p(0) + 2p_0 G(L) \int_0^L \frac{\gamma_{es}}{G(z)} \, dz \qquad (P.6)$$

with the gain $G(z)$ defined by:

$$G(z) = \exp\left\{ \int_0^z (\gamma_{es} - \gamma_{as}) \, dz \right\} \qquad (P.7)$$

The second term in Eq. (P.6) is the forward ASE. This term can be written in the form $2p_0 n_{sp}[G(L) - 1]$, where $n_{sp}$ is the spontaneous emission factor, i.e.,

$$n_{sp} = \frac{G(L)}{G(L) - 1} \int_0^L \frac{\gamma_{es}}{G(z)} \, dz \qquad (P.8)$$

From Eq. (P.7), we obtain $dG = (\gamma_{es} - \gamma_{as})G \, dz$, which can be substituted into Eq. (P.8) along with definitions in Eqs. (P.4) and (P.5) to yield:

$$n_{sp} = \frac{G(L)}{G(L) - 1} \int_1^{G(L)} \frac{\dfrac{\eta_s}{1 + \eta_p} q + \dfrac{\eta_s}{1 + \eta_s} p}{\dfrac{\eta_s - \eta_p}{1 + \eta_p} q - 1} \frac{dG}{G^2} \qquad (P.9)$$

In order to compute the integral in Eq. (P.9), we approximate the pump power in the fiber as almost constant with gain $G$ (i.e., $q(G) = \text{const.} = q_{in}$). Additionally, we assume a high pump regime, i.e., $q \gg 1$. With these two assumptions, and using the relation $p(G) = Gp(1) = Gp_{in}$, we obtain from Eq. (P.9):

$$
\begin{aligned}
n_{sp} &= \frac{G(L)}{G(L) - 1} \frac{\eta_s}{\eta_s - \eta_p} \int_1^{G(L)} \left( 1 + G \frac{1 + \eta_p}{1 + \eta_s} \frac{p_{in}}{q_{in}} \right) \frac{dG}{G^2} \\
&= \frac{1}{1 - \dfrac{\eta_p}{\eta_s}} \left( 1 + \frac{1 + \eta_p}{1 + \eta_s} \frac{p_{in}}{q_{in}} \frac{G \log G}{G - 1} \right) \equiv n_{sp}^{min}(1 + X)
\end{aligned} \qquad (P.10)
$$

where $G = G(L)$ is the saturated gain, $n_{sp}^{min} = 1/(1 - \eta_p/\eta_s)$ is the minimum spontaneous emission factor, and $X$ is the excess noise factor, defined as:

$$X = \frac{G \log G}{G - 1} \frac{1 + \eta_p}{1 + \eta_s} \frac{p_{in}}{q_{in}} = \frac{G \log G}{G - 1} \frac{1 + \eta_p}{1 + \eta_s} \frac{P_s^{in}}{P_p^{in}} \frac{P_{sat}(\lambda_p)}{P_{sat}(\lambda_s)} \qquad (P.11)$$

The saturated noise figure is then given by:

$$SNF = 2n_{sp} = 2n_{sp}^{min}(1 + X) \qquad (P.12)$$

---

# AVERAGE POWER ANALYSIS FOR SELF-SATURATED EDFAs (SEE SECTION 5.5)

---

We consider the rate equation (1.36) for the power $P_k$ at wavelength $\lambda_k$:

$$\frac{dP_k(\lambda_k)}{dz} = \sigma_a(\lambda_k)2\pi \int_S \{\eta(\lambda_k)N_2(r)[P_k(\lambda_k) + 2P_{0k}] - N_1(r)P_k(\lambda_k)\}\bar{\psi}_k(r)r\,dr \quad (Q.1)$$

where $P_{0k} = h\nu_k\delta\nu_k$ is the power of one photon in frequency bin $\delta\nu_k$. For simplicity, the effect of fiber background loss is neglected in Eq. (Q.1); this effect can be accounted for by introducing a factor $-\alpha'_k P_k$ ($\alpha'_k =$ background loss coefficient) in the right-hand side of Eq. (Q.1) and the following equations. Assuming confined Er-doping with concentration $\rho_0$ (Section 1.8) and normalizing $P_k$ to the saturation power $P_{\text{sat}}(\lambda_k)$, i.e., $p_k = P_k(\lambda_k)/P_{\text{sat}}(\lambda_k)$, we obtain from Eq. (Q.1)

$$\frac{dp_k}{dz} = \frac{\alpha_k}{\rho_0} \{\eta_k(N_2 - N_1)p_k + 2p_{0k}N_2\} = \alpha_k\left\{(1 + \eta_k)\frac{N_2}{\rho_0} - 1\right\}p_k + 2\alpha_k p_{0k}N_2 \quad (Q.2)$$

The relation $N_1 + N_2 = \rho_0$ is used in Eq. (Q.2). The upper level population $N_2$ is given by Eq. (1.50). This expression can be generalized to $k$-wavelength channels (including pump) and applied to the case of confined Er-doping to take the form:

$$N_2 = \rho_0 \frac{\sum_k \frac{1}{1 + \eta_k}p_k}{1 + \sum_k p_k} \quad (Q.3)$$

The principle of the average power method is to replace the terms $p_k$ in Eq. (Q.3) by the average quantities:

$$p_k \to \langle p_k \rangle = p_k^{\pm}(\text{in})\frac{G_k - 1}{\log G_k} \quad (Q.4)$$

for terms corresponding to signals, and by the average quantities

$$p_k \rightarrow \langle p'_k \rangle = 4n_{spk}P_{0k}\left(\frac{G_k - 1}{\log G_k} - 1\right) \tag{Q.5}$$

for terms corresponding to ASE. The factor 4 accounts for the two ASE powers in the two polarization modes. In Eqs. (Q.4) and (Q.5), $G_k$ is the gain at wavelength $\lambda_k$, $P_k^\pm(\text{in})$ is all powers at this wavelength input in each direction, and $n_{spk}$ is the corresponding spontaneous emission factor, defined by:

$$n_{spk} = \frac{\eta_k N_2/\rho_0}{(1 + \eta_k)N_2/\rho_0 - 1} \tag{Q.6}$$

With approximations (Q.4) and (Q.5), the upper level population in Eq. (Q.3) becomes the average quantity:

$$N_2 \approx \rho_0 \frac{\sum_k \frac{1}{1 + \eta_k}(\langle p_k \rangle + \langle p'_k \rangle)}{1 + \sum_k (\langle p_k \rangle + \langle p'_k \rangle)} \equiv \rho_0 \frac{X}{X'} \tag{Q.7}$$

with

$$X = \sum_k \frac{1}{1 + \eta_k}(\langle p_k \rangle + \langle p'_k \rangle) \tag{Q.8}$$

and

$$X' = 1 + \sum_k (\langle p_k \rangle + \langle p'_k \rangle) \tag{Q.9}$$

Integration of Eq. (Q.2) from $z = 0$ to $z = \Delta L$ with Eq. (Q.7) yields the gain at wavelength $\lambda_i$:

$$G_i = \exp\left\{\alpha_i\left[(1 + \eta_i)\frac{X}{X'} - 1\right]\Delta L\right\} \tag{Q.10}$$

The gain $G_i$ can be evaluated by first estimating the dummy parameters $X$ and $X'$, and adjusting $X'$ until Eqs. (Q.9) and (Q.10) give self-consistent results. The parameter $X$ can then be adjusted iteratively until Eqs. (Q.8) and (Q.10) are consistent. The output signal and ASE powers for this section are given by:

$$p_i^\pm(\text{out}) = p_i^\pm(\text{in})G_i \tag{Q.11}$$

$$p_i^\pm(\text{out, ASE}) = 2n_{spi}p_{0i}(G_{i-1} - 1) \tag{Q.12}$$

The above procedure must be repeated sequentially for each amplifier section over the entire amplifier length $L$, in both forward and backward directions, until both give self-consistent results [1]. While the appendix and the study [1] differ in some of their definitions and notation the calculation procedure is the same. Equation (Q.11) predicts identical gains in the forward and backward directions, which is only an approximation in the case of actual self-saturated EDFAs.

## REFERENCE

[1]   T. G. Hodgkinson, "Improved average power analysis technique for erbium-doped fiber amplifiers," *IEEE Photonics Lett.*, vol 4, no 11, 1273 (1992)

# APPENDIX R

# A COMPUTER PROGRAM FOR THE DESCRIPTION OF AMPLIFIER SELF-SATURATION THROUGH THE EQUIVALENT INPUT NOISE MODEL (SEE SECTION 5.5)

This appendix outlines a program example for the calculation of gain and ASE spectra in the amplifier self-saturation regime, using the equivalent input noise model (Section 5.5). This program, written here in BASIC (owing to its simplicity), solves trascendental equation (1.152) and generates output gain and ASE spectra. The algorithms and organization of the subprograms outlined here are merely examples and are not necessarily optimal. However, we consider here quite a simple type of problem and there is actually little need for such optimization. This program is quite straightforward and rapid to run with a small computer.

The transcendental equation (1.152) takes the form $F(U) = 0$, i.e.,

$$F(U) = U - \sum_j a_j \exp(-b_j U) = 0 \tag{R.1}$$

where $U = \bar{P}^{out}$ is the total output flux and

$$a_j = \bar{P}_j^{in} \exp(-\alpha_j L + b_j \bar{P}^{in}) \tag{R.2}$$

$$b_j = \frac{1}{\bar{P}_{sat}(\lambda_j)} \tag{R.3}$$

$$\bar{P}^{in,out} = \sum_k \bar{P}_k^{in,out} \tag{R.4}$$

where $\bar{P}_k = P_k/h\nu_k$ and $\bar{P}_{sat}(\lambda_k) = P_{sat}(\lambda_k)/h\nu_k$ are the photon flux associated with the powers $P_k$ and $P_{sat}(\lambda_k)$, $\alpha_j$ are the fiber absorption coefficient at $\lambda_j$, and $L$ is the total fiber length. The quantity $\bar{P}^{in}$ is the total flux input to the amplifier. The ASE in both forward and backward directions is modeled by a broadband, equivalent input noise of total flux:

$$\bar{P}^{in}_{ASE, k} = 4n_{eq}\delta\nu \tag{R.5}$$

**651**

where $n_{eq}$ is the equivalent input noise factor ($n_{eq} = n_{sp}$ for $G \gg 1$), and $\delta v$ is a frequency bin width. The factor 4 accounts for two polarization modes and two propagation directions. In this example $n_{eq}$ is chosen to be the same for both propagation directions. The gain $G_k$ can then be calculated for all wavelengths $\lambda_k$ through Eq. (1.148) and the solution $U$ of Eq. (R.1):

$$G_k = \frac{\bar{P}_k^{\text{out}}}{\bar{P}_k^{\text{in}}} = \exp\left\{-\alpha_k L + \frac{\bar{P}^{\text{in}} - U}{\bar{P}_{\text{sat}}(\lambda_k)}\right\} \tag{R.6}$$

The source program should include the definition of the absorption coefficient and saturation power spectra $\alpha_k(\lambda_k)$ and $P_{\text{sat}}(\lambda_k)$ corresponding to the Er-doped fiber, the initial input power conditions (spectra of pump and signals), the declaration of different parameters (number of resolved points in the search of solution $U$ and for the power spectra). It should also include a certain number of output statements for monitoring the development of the computation during the run (number of points left to compute, optional convergence tests of iteration procedure) and for storing the data on disk. We shall not detail here such a complete program organization, which is quite straightforward. We outline here only the important steps and relevant parts of the program, along with the definitions of the parameters used.

The source program starts with a declaration of the main parameters:

### Integer and real parameters

NUPOINT : number of points for search of solution $U$

NSPEC     : number of spectral points

L            : fiber length $L$ (in m)

APL       : pump absorption coefficient times $L$

PPIN      : input pump power flux $\bar{P}_p^{\text{in}}$

DNU      : frequency bin width $\delta v$

NSP      : spontaneous emission factor $n_{sp}$

PSATP    : pump saturation flux $P_{\text{sat}}(\lambda_p)$

HNUS     : signal photon energy $hv_s$

### Real arrays

U          : parameter $U$

F          : function $F$ in Eq. (R.1)

LAM      : wavelength $\lambda_k$

ALPHAL   : absorption coefficient spectrum times $L$

PKIN      : input flux spectrum $\bar{P}_k^{\text{in}}(\lambda_k)$

PSAT      : saturation flux spectrum $\bar{P}_{\text{sat}}(\lambda_k)$

G          : gain spectrum $G_k$

ASE       : output ASE power spectrum in dBm

The source program must also include the following steps:

- Input pump and signal powers (in mW) from program user
- Acquisition of cross section datafiles SIGA, SIGE (fiber absorption and emission cross sections) from disk
- Generation of spectrum PSAT
- Conversion of all powers into photon flux (normalizing all fluxes by $10^{16}$ prevents possible overflow)
- Setting input conditions. For instance, if no input signals other than ASE equivalent input noises are assumed:

```
        HNUS = (6.62E-34)*(3E+8)/(1.55E-6)
        PO = 4*NSP*DNU
REM   --- (ASE input noise only) ---
        FOR K = 1 TO NSPEC
        PKIN(K) = PO
        NEXT I
REM   --- Total input flux ---
        PIN = PPIN
        FOR K = 1 TO NSPEC
        PIN = PIN + PKIN(K)
        NEXT K
```

- Solving transcendental equation (R.1):

    This can be done in two loops: first, a coarse search of solution with UMIN = 0, UMAX = PIN, representing the search interval, and NUPOINT = 50; as the function F(I) increases monotonously from I = 1 to I = NUPOINT, the root is localized at I = IROOT when F(IROOT-1)*F(IROOT) < 0, meaning that F has crossed the U axis; the actual solution is near U(IROOT); redefine then the search interval, with UMIN = U(IROOT-1), UMAX = U(IROOT+1), with a finer resolution NUPOINT = 70, and restart the loop; as F can now be approximated by line, i.e., F(U) = AU+B, the solution of F(U) = 0 is USOL = -B/A.

```
100   PRINT"First loop"
        COUNTER = 0
        FMIN = 0    :    FMAX = 0
        UMIN = 0    :    UMAX = 0
100   DU = (UMAX-UMIN)/NUPOINT
        U(1) = UMIN + DU
        SUM = U(1) - (PIN*EXP(-APL+(PIN-U(1))/PSATP)
        FOR K = 1 TO NSPEC
        SUM = SUM - PKIN(K)*EXP(-ALPHAL(K)+(PIN-U(K))/PSAT(K))
        NEXT K
        F(1) = SUM
        FOR I = 2 TO NUPOINT
        U(I) = UMIN + I*DU
        SUM = U(I) - PPIN*EXP(-APL+(PIN-U(I))/PSATP)
```

```
        FOR K=1 TO NSPEC
        SUM=SUM-PKIN(K)*EXP(-ALPHAL(K)+(PIN-U(I))/PSAT(K))
        NEXT K
        F(I)=SUM
        IF F(I-1)*F(I)<0 THEN IROOT=I
        NEXT I
REM  --- Narrow search interval ---
        UMIN=U(IROOT-1)
        UMAX=U(IROOT+1)
        COUNTER=COUNTER+1
        IF COUNTER<2 THEN GOTO 100
        A=(F(IROOT)-F(IROOT-1))/(U(IROOT)-U(IROOT-1))
        B=F(IROOT)-A*U(IROOT)
        USOL=-B/A
        POUT=USOL
```

The choice of NUPOINT, UMIN, and UMAX can be optimized as a function of the input conditions. At low input pump powers, for instance, the output flux POUT is small compared to PIN, and UMIN can be chosen to be equal to PIN/10 instead of PIN. This choice enables a rapid determination of IROOT with the fewest number of points NUPOINT.

- Determine signal gain, total ASE power, pump output power, etc.

```
        FOR K=1 TO NSPEC
REM  ---Determine gain spectrum in dB ---
        GAIN=EXP(-ALPHAL(K)+(PIN-POUT)/PSAT(K))
        G(K)=4.34*LOG(GAIN)
REM  ---Determine ASE spectrum in dBm ---
        ASE(K)=4.34*LOG(GAIN*P0*HNUS*1.5+19)
        NEXT K
```

In the above definition of the dBm ASE spectra, the factor $10^{19} = 10^{16}/10^{-3}$ comes from the fact that all fluxes were normalized to $10^{16}$ photons, and that the power reference is 1 mW.

- Store datafiles for $G$, ASE, and other parameters on external disk
- Display results
- Plot gain and ASE spectra on screen
- Iterate the above program with modified values for initial parameters, e.g., pump power, signal powers, signal wavelengths, and fiber length
- Close program

```
STOP
END
```

# APPENDIX S

## FINITE DIFFERENCE RESOLUTION METHOD FOR TRANSIENT GAIN DYNAMICS IN EDFAs (SEE SECTION 5.9)

A straightforward numerical integration method for the partial differential equation system (5.99)–(5.103) with $\partial P_\mathrm{p}/\partial t = \partial P_\mathrm{s}/\partial t = 0$, uses the so-called forward Euler algorithm. Let the fiber coordinate and the time be defined by $z = m\Delta z$ and $t = n\Delta t$, respectively. The atomic populations and powers are defined through $N_i(z, t) = N_i^{m,n}$ ($i = 1, 2, 3$) and $P_x(z, t) = P_x^{m,n}$ ($x = \mathrm{p, s}$). In the forward difference (or forward Euler) algorithm, the atomic populations at time $(n + 1)\Delta t$ and the powers at coordinate $(m + 1)\Delta z$ are given by the following relations:

$$N_i^{m,n+1} = N_i^{m,n} + \Delta t \frac{\partial N_i^{m,n}}{\partial t} \tag{S.1}$$

$$P_x^{m+1,n} = P_x^{m,n} + \Delta z \frac{\partial P_x^{m,n}}{\partial z} \tag{S.2}$$

Using Eqs. (1.1) and (1.2), and the definitions:

$$R_{13} = R_{31} = \frac{1}{\tau_{21}} \frac{P_\mathrm{p}(z, t)}{P_\mathrm{sat}(\lambda_\mathrm{p})} \equiv \frac{1}{\tau_{21}} \frac{P_\mathrm{p}^{m,n}}{P_\mathrm{sat}^\mathrm{p}} \tag{S.3}$$

$$W_{12} = \sum_s \frac{1}{1 + \eta_s} \frac{1}{\tau_{21}} \frac{P_\mathrm{s}(z, t)}{P_\mathrm{sat}(\lambda_\mathrm{s})} \equiv \sum_s \frac{1}{1 + \eta_s} \frac{1}{\tau_{21}} \frac{P_\mathrm{s}^{m,n}}{P_\mathrm{sat}^\mathrm{s}} \tag{S.4}$$

$$W_{21} = \sum_s \frac{\eta_s}{1 + \eta_s} \frac{1}{\tau_{21}} \frac{P_\mathrm{s}(z, t)}{P_\mathrm{sat}(\lambda_\mathrm{s})} \equiv \sum_s \frac{\eta_s}{1 + \eta_s} \frac{1}{\tau_{21}} \frac{P_\mathrm{s}^{m,n}}{P_\mathrm{sat}^\mathrm{s}} \tag{S.5}$$

and Eq. (S.1) for the populations of levels 1 and 2 takes the form:

$$N_1^{m,n+1} = N_1^{m,n} + \frac{\Delta t}{\tau_{21}} \left\{ \frac{P_\mathrm{p}^{m,n}}{P_\mathrm{sat}^\mathrm{p}} (N_3^{m,n} - N_1^{m,n}) + \sum_s \frac{\eta_s}{1 + \eta_s} \frac{P_\mathrm{s}^{m,n}}{P_\mathrm{sat}^\mathrm{s}} N_2^{m,n} - \sum_s \frac{1}{1 + \eta_s} \frac{P_\mathrm{s}^{m,n}}{P_\mathrm{sat}^\mathrm{s}} N_1^{m,n} \right\} \tag{S.6}$$

$$N_2^{m,n+1} = N_2^{m,n} + \frac{\Delta t}{\tau_{21}} \left\{ \sum_s \frac{1}{1+\eta_s} \frac{P_s^{m,n}}{P_{\text{sat}}^s} N_1^{m,n} - \sum_s \frac{\eta_s}{1+\eta_s} \frac{P_s^{m,n}}{P_{\text{sat}}^s} N_2^{m,n} + \frac{\tau_{21}}{\tau_{32}} N_3^{m,n} \right\} \quad \text{(S.7)}$$

with $\tau_{32} = 1/A_{32}$. Using Eqs. (5.102), (5.103), and (S.2), the pump and signal rate equations can be expressed as follows:

$$P_p^{m+1,n} = P_p^{m,n} + \Delta z \alpha_p \Gamma_p (N_3^{m,n} - N_1^{m,n}) P_p^{m,n} \quad \text{(S.8)}$$

$$P_s^{m+1,n} = P_s^{m,n} + \Delta z \alpha_s \Gamma_s (\eta_s N_2^{m,n} - N_1^{m,n}) P_s^{m,n} \quad \text{(S.9)}$$

Multiple signals (s) are assumed in Eqs. (S.4) and (S.5), and therefore there are many equations of the type (S.9) to solve. In eqs. (S.6), (S.7), and (S.8), the population $N_3^{m,n}$ is given by the conservation relation $N_3^{m,n} = \rho_0 - N_1^{m,n} - N_2^{m,n}$. Consider the case of a single saturating signal. The summation symbols in Eqs. (S.4)–(S.7) must be overlooked. If, on the other hand, a two-level pumping scheme is assumed, the terms containing $P_p^{m,n}$ and $N_3^{m,n}$ in Eqs. (S.6) (S.7) must be set to zero. In this two-level pumping case, the pump and signal equations are the same as in Eqs. (S.8) and (S.9), with the substitution ($N_3^{m,n} \rightarrow N_2^{m,n}$ in Eq. (S.8).

The initial conditions necessary to solve Eqs. (S.6)–(S.9) are given by the input powers at $z = m = 0$ (i.e., $P_p^{0,n}$, $P_s^{0,n}$), and the steady state populations in Eqs. (1.8) and (1.9):

$$N_1^{m,0} = \rho_0 \frac{(1 + W_{21}^0 \tau_{21})(1 + R_{13}^0 \tau_{32})}{(1 + W_{21}^0 \tau_{21})(1 + 2R_{13}^0 \tau_{32}) + W_{12}^0 \tau_{21}(1 + R_{13}^0 \tau_{32}) + R_{13}^0 \tau_{21}} \quad \text{(S.10)}$$

$$N_2^{m,0} = \rho_0 \frac{R_{13}^0 \tau_{21} + W_{12}^0 \tau_{21}(1 + R_{13}^0 \tau_{32})}{(1 + W_{21}^0 \tau_{21})(1 + 2R_{13}^0 \tau_{32}) + W_{12}^0 \tau_{21}(1 + R_{13}^0 \tau_{32}) + R_{13}^0 \tau_{21}} \quad \text{(S.11)}$$

with by definition:

$$R_{13}^0 = R_{31}^0 = \frac{1}{\tau_{21}} \frac{P_p^{m,0}}{P_{\text{sat}}^p} \quad \text{(S.12)}$$

$$W_{12}^0 = \sum_s \frac{1}{1+\eta_s} \frac{1}{\tau_{21}} \frac{P_s^{m,0}}{P_{\text{sat}}^s} \quad \text{(S.13)}$$

$$W_{21}^0 = \sum_s \frac{\eta_s}{1+\eta_s} \frac{1}{\tau_{21}} \frac{P_s^{m,0}}{P_{\text{sat}}^s} \quad \text{(S.14)}$$

The power distributions at $t = n = 0$ (i.e., $P_p^{m,0}$, $P_s^{m,0}$) and coordinate $z = m\Delta z$, which are required to evaluate the steady state parameters defined in Eqs. (S.10)–(S.14), must be first calculated by solving iteratively Eqs. (S.8) and (S.9) for $n = 0$, using Eqs. (S.10) and (S.11) and starting with the boundary values $P_p^{0,0}$, $P_s^{0,0}$, i.e.,

$$P_p^{m+1,0} = P_p^{m,0} + \Delta z \alpha_p \Gamma_p (N_3^{m,0} - N_1^{m,0}) P_p^{m,0} \quad \text{(S.15)}$$

$$P_s^{m+1,0} = P_s^{m,0} + \Delta z \alpha_s \Gamma_s (\eta_s N_2^{m,0} - N_1^{m,0}) P_s^{m,0} \quad \text{(S.16)}$$

However complex this procedure may appear, it is relatively straightforward to implement. In the case of single- or two-signal saturation, numerical integration can be performed with a microcomputer. If a larger number of equations are considered in order to resolve ASE spectra, the increase in computation time requires the use of a large mainframe computer. In the regime of self-saturation where each of these equations must be integrated iteratively at each time step $t = n\Delta t$, the computation time is increased by at least a factor of 10. Convergence is not always achieved in the self-saturation problem, as discussed in Chapter 1.

In order to avoid computational instabilities and divergence in the implementation of the forward difference algorithm, care must be taken that the boundary conditions $P_p^{0,n}$, $P_s^{0,n}$ smoothly and continuously change over time. the choice of proper integration step in the time domain ($\Delta t$) is also very important. As a trial value, $\Delta t$ can be initially chosen as one tenth of the oscillation period (or characteristic time evolution) of the pump or the signal. The integration step can be divided by two, and the new solutions compared to the previous ones. We may consider that good convergence is achieved if, for instance, the relative difference between the solutions obtained with $\Delta t$ and $\Delta t/2$ steps is less than 1%. In the case where the oscillation frequency of the input signal is varied (as in a frequency sweep), such a convergence test must be done repeatedly. Other finite difference algorithms are possible; some have greater convergence properties. For instance, the central difference algorithm uses for a function $F(z, t)$ the following definitions:

$$F^{m+1,n} = F^{m-1,n} + 2\Delta z \frac{\partial F^{m,n}}{\partial z} \tag{S.17}$$

$$F^{m,n+1} = F^{m,n-1} + 2\Delta t \frac{\partial F^{m,n}}{\partial t} \tag{S.18}$$

which converge in $(\Delta z)^2$ and $(\Delta t)^2$, as opposed to $\Delta z$ and $\Delta t$ for the forward difference algorithm. However, the central difference algorithm produces computational instabilities when solving equations of the type (S.6)–(S.9) with actual EDFA parameters. On the other hand, the forward difference algorithm is observed to be free from such a problem, when properly implemented.

## APPENDIX T

---

# ANALYTICAL SOLUTIONS FOR TRANSIENT GAIN DYNAMICS IN EDFAs (SEE SECTION 5.9)

---

The time dependent atomic populations of energy levels 1 and 2 are given by Eqs. (5.99) and (5.100). Assuming $R_{13} = R_{31} \equiv R$, they can be put under the form:

$$\frac{dN_1}{dt} = -(R + W_{12})N_1 + (W_{21} + A_{21})N_2 + RN_3 \tag{T.1}$$

$$\frac{dN_2}{dt} = W_{12}N_1 - (W_{21} + A_{21})N_2 + A_{32}N_3 \tag{T.2}$$

Considering normalized populations $N_i/\rho_0$ for simplicity, and using then the conservation relation $N_3 = 1 - N_1 - N_2$, Eqs. (T.1) and (T.2) become:

$$\frac{dN_1}{dt} = -(2R + W_{12})N_1 + (W_{21} - R + A_{21})N_2 + R \tag{T.3}$$

$$\frac{dN_2}{dt} = (W_{12} - A_{32})N_1 - (W_{21} + A_{21} + A_{32})N_2 + A_{32} \tag{T.4}$$

Taking the derivative of eq. (T.4) while assuming $W_{12}$, $W_{21}$ are constant in time, and eliminating terms in $N_1$ by substitution of $dN_1/dt$ and $N_1$ from Eqs. (T.3) and (T.4) into the result, we obtain the second order equation:

$$\frac{d^2N_2}{dt^2} + B\frac{dN_2}{dt} + CN_2 = D \tag{T.5}$$

with

$$B = 2R + W_{12} + W_{21} + A_{21} + A_{32} \tag{T.6}$$

$$C = R(2W_{21} + W_{12} + 2A_{21} + A_{32}) + A_{32}(W_{12} + W_{21} + A_{21}) \tag{T.7}$$

$$D = R(W_{12} + A_{32}) + W_{12}A_{32} \tag{T.8}$$

The general solution of Eq. (T.5) takes the form:

$$N_2(t) = \mathscr{A}\exp(\omega_1 t) + \mathscr{B}\exp(\omega_2 t) + \frac{\exp(\omega_1 t)}{\omega_1 - \omega_2}\int_0^t \exp(-\omega_1 t')D\,dt'$$

$$- \frac{\exp(\omega_2 t)}{\omega_1 - \omega_2}\int_0^t \exp(-\omega_2 t')D\,dt' \tag{T.9}$$

with

$$\omega_{1,2} = \frac{-B \pm \sqrt{\Delta}}{2},\ \Delta = B^2 - 4C \tag{T.10}$$

Assuming $D$ to be constant, Eq. (T.9) becomes:

$$N_2(t) = \mathscr{A}\exp(\omega_1 t) + \mathscr{B}\exp(\omega_2 t) + \frac{D}{\sqrt{\Delta}}\left\{\frac{\exp(\omega_1 t) - 1}{\omega_1} - \frac{\exp(\omega_2 t) - 1}{\omega_2}\right\} \tag{T.11}$$

The constants $\mathscr{A}$ and $\mathscr{B}$ in Eq. (T.11) are determined from the initial conditions:

$$N_2(0) \equiv N_2^0 = \mathscr{A} + \mathscr{B} \tag{T.12}$$

$$\left(\frac{dN_2}{dt}\right)_{t=0} \equiv \dot{N}_2^0 = (W_{12} - A_{32})N_1^0 - (W_{21} + A_{21} + A_{32})N_2^0 + A_{32} = \mathscr{A}\omega_1 + \mathscr{B}\omega_2 \tag{T.13}$$

which gives:

$$\mathscr{A} = \frac{\dot{N}_2^0 - \omega_2 N_2^0}{\sqrt{\Delta}},\ \mathscr{B} = \frac{\omega_1 N_2^0 - \dot{N}_2^0}{\sqrt{\Delta}} \tag{T.14}$$

On the other hand, the ground level population $N_1(t)$ is given by Eq. (T.4)

$$N_1(t) = \frac{1}{W_{12} - A_{32}}\left\{\frac{dN_2}{dt} + (W_{21} + A_{21} + A_{32})N_2 - A_{32}\right\} \tag{T.15}$$

which, after using eq. (T.11) for $N_2(t)$ yields:

$$N_1(t) = \frac{1}{W_{12} - A_{32}}\left\{\mathscr{A}\omega_1\exp(\omega_1 t) + \mathscr{B}\omega_2\exp(\omega_2 t) + \frac{D}{\sqrt{\Delta}}[\exp(\omega_1 t) - \exp(\omega_2 t)\right.$$

$$+ (W_{21} + A_{21} + A_{32})\left\{\mathscr{A}\exp(\omega_1 t) + \mathscr{B}\exp(\omega_2 t)\right.$$

$$\left.+ \frac{D}{\sqrt{\Delta}}\left[\frac{\exp(\omega_1 t) - 1}{\omega_1} - \frac{\exp(\omega_2 t) - 1}{\omega_2}\right]\right\} \tag{T.16}$$

The atomic populations $N_i(t)(i = 1, \ldots, 3)$ can be put into the general form:

$$\boxed{N_i(t) = a_i e^{\omega_1 t} + b_i e^{\omega_2 t} + c_i} \tag{T.17}$$

with

$$a_1 = \frac{1}{W_{12} - A_{32}}\left\{\mathscr{A}\omega_1 + \frac{D}{\sqrt{\Delta}} + (W_{21} + A_{21} + A_{32})\left(\mathscr{A} + \frac{D}{\omega_1\sqrt{\Delta}}\right)\right\} \tag{T.18}$$

$$b_1 = \frac{1}{W_{12} - A_{32}}\left\{\mathscr{B}\omega_2 - \frac{D}{\sqrt{\Delta}} + (W_{21} + A_{21} + A_{32})\left(\mathscr{B} - \frac{D}{\omega_2\sqrt{\Delta}}\right)\right\} \tag{T.19}$$

$$c_1 = \frac{1}{W_{12} - A_{32}}\left\{\frac{D}{\omega_1\omega_2}(W_{21} + A_{21} + A_{32}) - A_{32}\right\} \tag{T.20}$$

$$a_2 = \mathscr{A} + \frac{D}{\omega_1\sqrt{\Delta}}, b_2 = \mathscr{B} - \frac{D}{\omega_2\sqrt{\Delta}}, c_2 = \frac{D}{\omega_1\omega_2} \tag{T.21}$$

$$a_3 = -(a_1 + a_2), b_3 = -(b_1 + b_2), c_3 = 1 - c_1 - c_2 \tag{T.22}$$

We now determine explicit expressions for the characteristic frequencies $\omega_1$, $\omega_2$ defined in Eq. (T.10). We first introduce $\tau_{21} \equiv \tau = 1/A_{21}$ and the parameter:

$$\varepsilon = \frac{A_{21}}{A_{32}} \tag{T.23}$$

Using the definitions in Eqs. (T.6) and (T.7) we find from Eq. (T.10):

$$\Delta = B^2 - 4C = A_{32}^2[1 + \varepsilon(1 + 2R\tau + W_{12}\tau + W_{21}\tau)]^2$$
$$- 4A_{32}^2\{\varepsilon(R\tau + W_{12}\tau + W_{21}\tau) + \varepsilon^2 R\tau(2 + W_{12}\tau + 2W_{21}\tau)\} \tag{T.24}$$

Neglecting the terms in $\varepsilon^2$ in Eq. (T.24), as $\varepsilon = A_{21}/A_{32} \ll 1$ for Er:silica glass, we obtain:

$$\Delta \approx A_{32}^2[1 - 2\varepsilon(1 + W_{12}\tau + W_{21}\tau)] \tag{T.25}$$

Considering again that $\varepsilon \ll 1$, the square root of $\Delta$ in Eq. (T.25) is approximately:

$$\sqrt{\Delta} \approx A_{32}[1 - \varepsilon(1 + W_{12}\tau + W_{21}\tau)] \tag{T.26}$$

Substituting Eq. (T.26) into Eq. (T.10) we obtain after straightforward algebra:

$$\omega_1 \approx -A_{21}(1 + W_{12}\tau + W_{21}\tau) \tag{T.27}$$

$$\omega_2 \approx -A_{32}(1 + \varepsilon R\tau) \tag{T.28}$$

Using now the relation $\tau_{32} = 1/A_{32}$ and the definitions

$$W_{12}\tau + W_{21}\tau = \frac{P_s^{in}(0)}{P_{sat}(\lambda_s)} \equiv p \tag{T.29}$$

$$R\tau = \frac{P_p^{in}(0)}{P_{sat}(\lambda_p)} \equiv q \tag{T.30}$$

the characteristic time constants $t_1 = -1/\omega_1$, $t_2 = -1/\omega_2$ with eqs. (T.27)–(T.30) take the form:

$$t_1 = \frac{\tau_{21}}{1 + \dfrac{P_p^{in}(0)}{P_{sat}(\lambda_p)} + \dfrac{P_s^{in}(0)}{P_{sat}(\lambda_s)}} \equiv \frac{\tau_{21}}{1 + p + q} \tag{T.31}$$

$$t_2 = \frac{\tau_{32}}{1 + \dfrac{\tau_{32}}{\tau_{21}}\dfrac{P_p^{in}(0)}{P_{sat}(\lambda_p)}} \equiv \frac{\tau_{32}}{1 + \dfrac{\tau_{32}}{\tau_{21}} q} \tag{T.32}$$

and the general solution (T.17) for the atomic populations can be expressed finally as:

$$\boxed{N_i(t) = a_i e^{-t/t_1} + b_i e^{-t/t_2} + c_i} \tag{T.33}$$

Now we express the coefficients $a_i$, $b_i$, $c_i$ in explicit form, making use of the approximation $\varepsilon \ll 1$. With this approximation, we obtain:

$$\mathcal{A} \approx 1 - N_1^0, \quad \mathcal{B} \approx N_1^0 + N_2^0 - 1, \quad \frac{D}{\sqrt{\Delta}} \approx A_{21}(R\tau + W_{12}\tau) \tag{T.34}$$

and for the coefficients $a_i$, $b_i$, $c_i$:

$$a_1 \approx N_1^0 - \frac{1 + W_{21}\tau}{1 + p + q}, \quad b_1 \approx 0, \quad c_1 \approx \frac{1 + W_{21}\tau}{1 + p + q} \tag{T.35}$$

$$a_2 \approx \frac{1 + W_{21}\tau}{1 + p + q} - N_1^0, \quad b_2 \approx N_1^0 + N_2^0 - 1 = -N_3^0, \quad c_2 \approx \frac{R\tau + W_{12}\tau}{1 + p + q} \tag{T.36}$$

$$a_3 \approx 0, \quad b_3 \approx 1 - N_1^0 - N_2^0 = N_3^0, \quad c_3 \approx 0 \tag{T.37}$$

Using finally Eqs. (T.33)–(T.37), the populations $N_i(0, t)$ can be expressed explicitly as:

$$N_1(0, t) = \left(N_1^0 - \frac{1 + W_{21}\tau}{1 + p + q}\right)e^{-t/t_1} + \frac{1 + W_{21}\tau}{1 + p + q} \tag{T.38}$$

$$N_2(0, t) = \left(\frac{1 + W_{21}\tau}{1 + p + q} - N_1^0\right)e^{-t/t_1} - N_3^0 e^{-t/t_2} + \frac{R\tau + W_{12}\tau}{1 + p + q} \tag{T.39}$$

$$N_3(0, t) = (1 - N_1^0 - N_2^0)e^{-t/t_2} = N_3^0 e^{-t/t_2} \tag{T.40}$$

We consider next the case of an EDFA input with a periodic signal sequence made of square pulses with 50% duty cycle. The ground level population $N_1(t)$ can be calculated iteratively from Eq. (T.38), with the input signal alternatively turned on and off for durations $\Delta T$. From the first on/off sequence, we obtain:

$$N_1(\Delta T) = (N_1^0 - U)P + U \tag{T.41}$$

$$N_1(2\Delta T) = [N_1(\Delta T) - U']P' + U' = PP'N_1^0 + P'U(1 - P) + U'(1 - P') \tag{T.42}$$

with

$$P = \exp(-\Delta T/t_1), \; P' = \exp(-\Delta T/t_1') \tag{T.43}$$

$$t_1 = \frac{\tau_{21}}{1 + p + q}, \; t_1' = \frac{\tau_{21}}{1 + q} \tag{T.44}$$

$$U = \frac{1 + W_{21}\tau}{1 + p + q}, \; U' = \frac{1}{1 + p + q} \tag{T.45}$$

Iteration shows that after $2n$ and $2n + 1$ such on/off periods, the population $N_1(t)$ is given by:

$$N_1(2n\Delta T) = (PP')^n N_1^0 + [P'U(1 - P) + U'(1 - P')]S_{n-1} \tag{T.46}$$

$$N_1[(2n + 1)\Delta T] = P(PP')^n N_1^0 + U(1 - P)S_n + PU'(1 - P')S_{n-1} \tag{T.47}$$

with

$$S_n = 1 + PP' + \cdots + (PP')^n \tag{T.48}$$

For large $n$, the series in Eq. (T.48) can be approximated by $S_n = 1/(1 - PP')$. The population inversion corresponding to each period is given by $\Delta N_{12} = N_2 - N_1 = 1 - 2N_1$. The difference in inversion $\delta\Delta N_{12}$ between the on and off periods of the saturating pulse is given by:

$$\delta\Delta N_{12} = 2\{N_1[2n\Delta T] - N_1[(2n + 1)\Delta T]\} \tag{T.49}$$

Using Eqs. (T.46) and (T.47) while taking the limit $n \to \infty$, we obtain from Eq. (T.49):

$$\delta\Delta N_{12} = 2\frac{(1 - P)(1 - P')}{1 - PP'}(U - U') \tag{T.50}$$

On the other hand, the average inversion $\Delta\bar{N}_{12}$ corresponding to a full on/off period is given by:

$$\Delta\bar{N}_{12} = \frac{\Delta N_{12}[2n\Delta T] - \Delta N_{12}[(2n + 1)\Delta T]}{2}$$

$$= 1 - N_1[2n\Delta T] - N_1[(2n + 1)\Delta T] \tag{T.51}$$

Using Eqs. (T.46) and (T.47), while taking the limit $n \rightarrow \infty$, we obtain from Eq. (T.51):

$$\Delta \bar{N}_{12} = \frac{(1 - PP')(1 - U - U') + (P - P')(U - U')}{1 - PP'} \tag{T.52}$$

The relative change in inversion is therefore, from Eqs. (T.50) and (T.52):

$$\frac{\delta(\Delta N_{12})}{\bar{N}_{12}} = 2 \frac{(1 - P)(1 - P')}{P - P' + K(1 - PP')} \tag{T.53}$$

with $K = (1 - U - U')/(U - U')$. Using the characteristic frequencies $f = 1/2t_1$, $f' = 1/2t_1'$ and the on/off modulation rate $B = 1/2\Delta T$, Eq. (T.53) can be put into the form:

$$\boxed{\frac{\delta(\Delta N_{12})}{\bar{N}_{12}} = 2 \frac{(1 - e^{-f/B})(1 - e^{-f'/B})}{e^{-f/B} - e^{-f'/B} + K[1 - e^{-(f+f')/B}]}} \tag{T.54}$$

# DERIVATION OF THE NONLINEAR SCHRÖDINGER EQUATION (SEE SECTION 5.11)

Consider an electromagnetic wave represented by the function:

$$U(z, t) = u(z, t)\exp\{i(\omega_s t - k_s z)\} \tag{U.1}$$

where $u(z, t)$ is a dimensionless complex amplitude representing the slowly varying envelope of the wave, and $\omega_s$ is the wave central frequency ($k_s = n\omega_s/c$). The wave intensity is given by $I = I_c|u(z, t)|^2$, where $I_c$ is a constant. The optical Kerr effect increases the refractive index by the quantity $\delta n = n_2 I$, where $n_2$ is the nonlinear refractive index coefficient. We assume then that near the central frequency of the wave, the following dispersion relation holds:

$$k(\omega) = k_s + (\omega - \omega_s)\left(\frac{\partial k}{\partial \omega}\right)_{\omega_s} + \frac{(\omega - \omega_s)^2}{2}\left(\frac{\partial^2 k}{\partial \omega^2}\right)_{\omega_s} + \frac{(\omega - \omega_s)^3}{3!}\left(\frac{\partial^3 k}{\partial \omega^3}\right)_{\omega_s} + \cdots$$

$$+ k_2 I + I(\omega - \omega_s)\left(\frac{\partial k_2}{\partial \omega}\right)_{\omega_s} + \cdots \tag{U.2}$$

Equation (U.2) is a Taylor development of the wave vector near $\omega_s$, with the addition of the effect of nonlinearity $\delta k = k_2 I$ with $k_2 = n_2\omega_s/c$. In this equation, the derivatives $k'$ and $k''$ are related to the group velocity $v_g$ through $v_g = \partial \omega/\partial k$, or using eq. (U.2):

$$\frac{1}{v_g} = \frac{\partial k}{\partial \omega} = \left(\frac{\partial k}{\partial \omega}\right)_{\omega_s} + (\omega - \omega_s)\left(\frac{\partial^2 k}{\partial \omega^2}\right)_{\omega_s} + \frac{(\omega - \omega_s)^2}{2}\left(\frac{\partial^3 k}{\partial \omega^3}\right)_{\omega_s} + \cdots + I\left(\frac{\partial k_2}{\partial \omega}\right)_{\omega_s} + \cdots$$

$$\equiv k' + (\omega - \omega_s)k'' + \frac{(\omega - \omega_s)^2}{2}k''' + \cdots + Ik_2' + \cdots \tag{U.3}$$

The fiber dispersion coefficient $D$ is related to $k''$ by the usual definition:

$$D = -\frac{2\pi c}{\lambda^2} k'' \tag{U.4}$$

At the frequencies of interest for solitons in single-mode fibers, the terms in $k''$ and $k_2 I$ in Eq. (U.2) have comparable magnitudes, while the higher order terms in $k'''$ and $Ik'_2$ are small perturbations. Thus, eq. (U.2) can be approximated by:

$$\Delta k \approx \Delta\omega k' + \frac{\Delta\omega^2}{2} k'' + k_2 I \tag{U.5}$$

with $\Delta k = k - k_s$ and $\Delta\omega = \omega - \omega_s$. From eq. (U.3), we find:

$$k'' \approx \frac{\partial}{\partial\omega}\left(\frac{1}{v_g}\right) = -\frac{1}{v_g^2}\frac{\partial v_g}{\partial\omega} \tag{U.6}$$

which shows that $k''$ is the dispersion of the wave's group velocity.

We consider now the Fourier transform of the envelope function:

$$\tilde{u}(\Delta k, \Delta\omega) = \int u(z, t)\exp\{i(\Delta\omega t - \Delta kz)\}\, d(\Delta\omega)\, d(\Delta k) \tag{U.7}$$

and the inverse transform:

$$u(z, t) = \frac{1}{(2\pi)^2} \int \tilde{u}(\Delta k. \Delta\omega)\exp\{-i(\Delta\omega t - \Delta kz)\}\, dt\, dz \tag{U.8}$$

Equations (U.7) and (U.8) show that the quantities $\partial u/\partial z = i\Delta k u$ and $\partial u/\partial t = -i\Delta\omega u$ are the Fourier transforms of $i\Delta k\tilde{u}$ and $-i\Delta\omega\tilde{u}$, respectively. Thus $\Delta k$, $\Delta\omega$ can be put into the form of operators $-i\partial/\partial z$, $i\partial/\partial t$, and Eq. (U.5) can be expressed in operator form:

$$-i\frac{\partial}{\partial z} \approx ik'\frac{\partial}{\partial t} - \frac{i}{2}k''\frac{\partial^2}{\partial z^2} + k_2 I \tag{U.9}$$

Applying q. (U.9) to the wave envelope $u(z, t)$ and using $I = I_c|u(z, t)|^2$, we obtain:

$$i\frac{\partial u}{\partial z} + ik'\frac{\partial u}{\partial t} - \frac{k''}{2}\frac{\partial^2 u}{\partial t^2} + k_2 I_c|u|^2 u \approx 0 \tag{U.10}$$

Equation (U.10) can be transformed to correspond to a retarded time frame and be made dimensionless through the following substitutions:

$$\tau = \frac{1}{t_c}(t - k'z) \tag{U.11}$$

$$\xi = \frac{z}{z_c} \tag{U.12}$$

where $t_c$, $z_c$ are constants with dimensions of time and space, respectively. We have thus:

$$\frac{\partial u}{\partial z} = -\frac{k'}{t_c}\frac{\partial u}{\partial \tau} + \frac{1}{z_c}\frac{\partial u}{\partial \xi} \tag{U.13}$$

$$\frac{\partial u}{\partial t} = \frac{1}{t_c}\frac{\partial u}{\partial \tau}, \frac{\partial^2 u}{\partial t^2} = \frac{1}{t_c^2}\frac{\partial^2 u}{\partial \tau^2} \tag{U.14}$$

and from Eq. (U.10):

$$-i\frac{t_c^2}{k''z_c}\frac{\partial u}{\partial \xi} + \frac{1}{2}\frac{\partial^2 u}{\partial \tau^2} - \frac{t_c^2}{k''}k_2 I_c|u|^2 u = 0 \tag{U.15}$$

The constants $t_c$, $z_c$ can then be arbitrarily defined by:

$$\frac{t_c^2}{z_c} = -k'' \tag{U.16}$$

and

$$z_c = \frac{1}{k_2 I_c} \tag{U.17}$$

With the above definitions, Eq. (U.15) takes the canonical form:

$$i\frac{\partial u}{\partial \xi} + \frac{1}{2}\frac{\partial^2 u}{\partial \tau^2} + |u|^2 u = 0 \tag{U.18}$$

The above equation is similar to the Schrödinger equation:

$$i\frac{\partial \psi}{\partial t} + \frac{\partial^2 \psi}{\partial x^2} + V\psi = 0 \tag{U.19}$$

where $(\xi, \tau)$ plays the role of $(t, x)$ and the potential $V$ is represented by the wave intensity $|\psi|^2$, which contributes to a nonlinear term. For this reason, Eq. (U.18) is called the nonlinear Schrödinger equation. In the case where the medium has gain or loss, a term $-i\Gamma u$ with $\Gamma = z_c g/2$ must be added to the left-hand side of Eq. (U.18), where $g$ is the net power gain coefficient:

$$\boxed{i\frac{\partial u}{\partial \xi} + \frac{1}{2}\frac{\partial^2 u}{\partial \tau^2} + |u|^2 u - i\Gamma u = 0} \tag{U.20}$$

It is easily checked that with such a definition for $\Gamma$, Eq. (U.20) transforms back into:

$$\frac{\partial u}{\partial z} + k'\frac{\partial u}{\partial t} + i\frac{k''}{2}\frac{\partial^2 u}{\partial t^2} - ik_2 I_c|u|^2 u = \frac{g}{2}u \tag{U.21}$$

whose term in $g/2$ indeed reflects the effect of amplitude gain or loss.

In the case of a transparent medium ($\Gamma = 0$) and assuming the initial condition $u(0, \tau) = N \operatorname{sech}(\tau)$ (with $\operatorname{sech}(\tau) = 1/\cosh(\tau)$), the soliton solutions of Eq. (U.18) or Eq. (U.20) for $N = 1$ and $N = 2$ can be put into the form [U.1] and [U.2]:

$$u(\xi, \tau) = \operatorname{sech}(\tau)\exp(i\xi/2) \tag{U.22}$$

$$u(\xi, \tau) = 4 \exp(-i\xi/2) \frac{ch(3\tau) + 3ch(\tau)\exp(-4i\xi)}{ch(4\tau) + 4ch(2\tau) + 3 \cos(4\xi)} \tag{U.23}$$

The soliton intensity is given by $(\xi, \tau) = I_c|u(\xi, \tau)|^2$, so the $N = 1$ or fundamental solution is space invariant $(I(\xi, \tau) = I_c \operatorname{sech}^2(\tau))$, while the $N = 2$ solution has space periodicity $\xi_0 = \pi/2$. We can introduce the soliton period through:

$$z_0 = \frac{\pi}{2} z_c = \frac{\pi}{2k_2 I_c} \tag{U.24}$$

where $z_c$ is by definition the characteristic length. The peak power $P_{peak} = A_{eff} I_{peak}$ of the fundamental soliton is reached at $\tau = 0$ and is given by:

$$P_{peak} = A_{eff} I_c |u(\xi, 0)|^2 = \frac{A_{eff}}{k_2 z_c} = \frac{\lambda A_{eff}}{4n_2 z_0} \tag{U.25}$$

The fundamental soliton FWHM $(\Delta T)$ is found from the relation $\operatorname{sech}^2(\tau) = 1/2$ with $\tau = \Delta T/t_c$, which yields:

$$\frac{\Delta T}{t_c} = 2 \cosh^{-1}(\sqrt{2}) = 1.762747 \tag{U.27}$$

Using Eq. (U.27) and Eqs. (U.4), (U.16), and (U.25), we find the characteristic length and the peak power:

$$\boxed{z_c \approx 0.322 \frac{2\pi c}{\lambda^2} \frac{\Delta T^2}{D}} \tag{U.28}$$

$$\boxed{P_{peak} = \frac{\lambda A_{eff}}{2\pi n_2 z_c} = 3.09 \frac{\lambda^3 A_{eff}}{4\pi^2 c n_2} \frac{D}{\Delta T^2}} \tag{U.29}$$

In Eq. (U.29), the effective mode interaction area $A_{eff}$ is given by [5.245]:

$$A_{eff} = \frac{[\int\int I_s(r, \theta)\, r dr d\theta]^2}{\int\int I_s^2(r, \theta)\, r dr d\theta} \tag{U.30}$$

where $I_s(r, \theta)$ is the signal intensity distribution, Eq. (1.28). With Gaussian mode approximation $(I_s(r) = I_{max} \exp[-r^2/\omega_s^2])$, this area reduces to $A_{eff} = \pi\omega_s^2$ (field), where $\omega_s$ (field) $= \sqrt{2}\omega_s$ (power).

## REFERENCES

[1]  J. Satsuma and Y. Yajima, "Initial value problems of one-dimensional self-modulation of nonlinear waves in dispersive media," in supplement to *Progress in Theoretical Physics*, no 55, 284 (1974)

[2]  N. J. Doran and K. J. Blow, "Solitons in optical communications," *IEEE J. Quantum Electron.*, vol 19, no 12, 1883 (1983)

# REFERENCES

## CHAPTER 1

[1.1] A. L. Schawlow and C. H. Townes, "Infrared and optical masers," *Phys. Rev.*, vol 112, no 6, 1940 (1958)

[1.2] K. Shimoda, H. Takahasi, and C. H. Townes, "Fluctuations in amplification of quanta with application to maser amplifiers," *J. Phys. Soc Japan*, vol 12, no 6, 686 (1957)

[1.3] A. Yariv and J. P. Gordon, "The laser," in *Proc. IRE*, January 1963, p4

[1.4] H. A. Haus and J. A. Mullen, "Quantum noise in linear amplifiers," *Phys. Rev.*, vol 128, no 5, 2407 (1962)

[1.5] C. J. Koester and E. A. Snitzer, "Amplification in a fiber laser," *Applied Optics*, vol 3, no 10, 1182 (1964)

[1.6] K. Kubodera and K. Otsuka, "Single-transverse mode LiNd $P_4O_{12}$ slab waveguide laser," *J. Applied Phys.*, vol 50, 653 (1979)

[1.7] K. Kubodera and K. Otsuka, "Laser performance of a glass-clad $LiNdP_4O_{12}$ rectangular waveguide," *J. Applied Phys.*, vol 50, 653 (1979)

[1.8] M. J. F. Digonnet and C. J. Gaeta, "Theoretical analysis of optical fiber laser amplifiers and oscillators," *Applied Optics*, vol 24, no 3, 333 (1985)

[1.9] M. J. F. Digonnet, "Theory of superfluoresecent fiber lasers," *IEEE J. Lightwave Technol.*, vol 4, no 11, 1631 (1986)

[1.10] R. J. Mears, L. Reekie, S. B. Poole, and D. N. Payne, "Low-threshold, tunable cw and Q-switched fibre laser operating at 1.55 $\mu$m," *Electron. Lett.*, vol 22, no 3, 159 (1986)

[1.11] R. J. Mears, L. Reekie, I. M. Jauncey, and D. N. Payne, "Low-noise erbium-doped fibre amplifier operating at 1.54 $\mu$m," *Electron. Lett.*, vol 23, no 19, 1026 (1987)

[1.12] E. Desurvire, J. R. Simpson, and P. C. Becker, "High-gain erbium-doped traveling-wave fiber amplifier," *Optics Lett.*, vol 12, no 11, 888 (1987)

[1.13] R. Olshansky, "Noise figure for erbium-doped optical fibre amplifiers," *Electron. Lett.*, vol 24, no 22, 1363 (1988)

[1.14]  J. R. Armitage, "Three-level fiber laser amplifier: a theoretical model," *Applied Optics*, vol 27, no 23, 4831 (1988)

[1.15]  E. Desurvire and J. R. Simpson, "Amplification of spontaneous emission in erbium-doped single-mode fibers," *IEEE J. Lightwave Technology*, vol 7, no 5, 835 (1989)

[1.16]  A. Bjarklev, S. L. Hanse, and J. H. Povlsen, "Large-signal modeling of an erbium-doped fiber amplifier," *Proc. SPIE Conference on Fiber Laser Sources and Amplifiers*, vol 1171, 118 (1989)

[1.17]  E. Desurvire, C. R. Giles and and J. R. Simpson, "Gain saturation effects in high-speed, multichannel erbium-doped fiber amplifiers at $\lambda = 1.53$ $\mu$m," *IEEE J. Lightwave Technology*, vol 7, no 12, 2095 (1989)

[1.18]  P. R. Morkel and R. I. Laming, "Theoretical modeling of erbium-doped fiber amplifiers with excited-state absorption," *Optics Lett.*, vol 14, no 19, 1062 (1989)

[1.19]  E. Desurvire, "Analysis of transient gain saturation and recovery in erbium-doped fiber amplifiers," *IEEE Photonics Technol. Lett.*, vol 1, no 8, 196 (1989)

[1.20]  E. Desurvire, "Analysis of erbium-doped fiber amplifiers pumped in the $^{4}I_{13/2}$-$^{4}I_{15/2}$ band," *IEEE Photonics Technol. Lett.*, vol 1, no 10, 293 (1989)

[1.21]  E. Desurvire, "Analysis of noise figure spectral distribution in erbium-doped fiber amplifiers pumped near 980 and 1480 nm," *Applied Optics*, vol 29, no 21, 3118 (1990)

[1.22]  J. F. Marcerou, H. A. Fevrier, J. Ramos, J. C. Auge, and P. Bousselet, "General theoretical approach describing the complete behavior of the erbium-doped fiber amplifier," in *Proc. SPIE Conference on Fiber Laser Sources and Amplifiers*, vol 1373, 168 (1990)

[1.23]  M. Peroni and M. Tamburrini, "Gain in erbium-doped fiber amplifiers: a simple analytical solution for the rate equations," *Optics Lett.*, vol 15, no 15, 842 (1990)

[1.24]  A. A. M. Saleh, R. M. Jopson, J. D. Evankow and J. Aspell, "Modeling of gain in erbium-doped fiber amplifiers," *IEEE Photonics Technol. Lett.*, vol 2, no 10, 714 (1990)

[1.25]  E. Desurvire, J. W. Sulhoff, J. L. Zyskind, and J. R. Simpson, "Study of spectral dependence of gain saturation and effect of inhomogeneous broadening in erbium-doped aluminosilicate fibre amplifiers," *IEEE Photonics Technol. Lett.*, vol 2, no 9, 653 (1990)

[1.26]  M. J. F. Digonnet, "Closed-form expressions for the gain in three- and four-level laser fibres," *IEEE J. Quantum Electron.* vol 26, no 10, 1788 (1990)

[1.27]  E. Desurvire, "Study of the complex atomic susceptibility of erbium-doped fiber amplifiers," *IEEE J. Lightwight Technol.*, vol 8, no 10, 1517 (1990)

[1.28]  E. Desurvire, J. L. Zyskind, and C. R. Giles, "Design optimization for efficient erbium-doped fiber amplifiers," *IEEE J. Lightwave Technol.*, vol 8, no 11, 1730 (1990)

[1.29]  B. Pedersen, K. Dybdal, C. D. Hansen, A. Bjarklev, J. H. Povlsen, H. Vendeltorp-Pommer, and C. C. Larsen, "Detailed theoretical and experimental investigation of high-gain erbium-doped fiber amplifier," *IEEE Photonics Technol. Lett.*, vol 2, no 12, 863 (1990)

[1.30]  A. Yariv, *Quantum Electronics,* second edition, John Wiley, New York, 1975

[1.31]  B. E. A. Saleh and M. C. Teich, *Fundamentals of Photonics,* John Wiley, New York, 1991.

[1.32]  A. E. Siegman, *Lasers,* University Science Books, Mill Valley, CA, 1986

[1.33]  L. Jeunhomme, *Single-mode Fiber Optics, Principles and Applications*, second edition, Marcel Dekker, New York, 1990

[1.34]  D. Gloge, "Weakly guiding fibers," *Applied Optics*, vol 10, no 10, 2252 (1971)

[1.35]  D. Marcuse, "Gaussian approximation of the fundamental modes of graded-index fibers," *J. Opt. Soc. Am.* vol 68, no 1, 103 (1978)

[1.36] F. F. Rühl, "Accurate analytical formula for gain-optimized EDFAs," *Electron. Lett.*, vol 28, no 3, 312 (1992)

[1.37] H. Kogelnick and A. Yariv, "Considerations of noise and schemes for its reduction in laser amplifiers," *Proc. IEEE*, February 1964, p165

[1.38] W. H. Press, B. P. Flannery, S. A. Teukolsky, and W. T. Vetterling, *Numerical Recipes, The Art of Scientific Computing,* Cambridge University Press, New York, 1986

[1.39] C. R. Giles and E. Desurvire, "Modeling erbium-doped fiber amplifiers," *IEEE J. Lightwave Technol.*, vol 9, no 2, 271 (1991)

[1.40] E. Desurvire, "Analysis of distributed erbium-doped fiber amplifiers with fiber background loss," *IEEE Photonics Technol. Lett.*, vol 3, no 7, 625 (1991)

[1.41] D. Chen and E. Desurvire, "Noise performance evaluation of distributed erbium-doped fiber amplifiers with bidirectional pumping at 1.48 $\mu m$," *IEEE Photonics Technol. Lett.*, vol 4, no 1, 52 (1992)

[1.42] E. Desurvire, M. Zirngibl, H. M. Presby, and D. DiGiovanni, "Characterization and modeling of amplified spontaneous emission in unsaturated erbium-doped fiber amplifiers," *IEEE Photonics Technol. Lett.*, vol 3, no 2, 127 (1991)

[1.43] H. T. Davis, *Introduction to Nonlinear Differential and Integral Equations,* Dover Publications, New York, 1962

[1.44] Th.Pfeiffer and H. Bülow, "Analytical gain equation for erbium-doped fiber amplifiers including mode field profiles and dopant distributions," *IEEE Photonics Technol. Lett.*, vol 4, no 5, 449 (1992)

[1.45] E. Desurvire, "Modeling self-saturation in erbium-doped fiber amplifiers: a rapid evaluation method," Center for Telecommunications Research, Columbia University, internal report 210692, June 1992

[1.46] P. Meystre and M. Sargent III, *Elements of Quantum Optics,* second edition. Springer-Verlag, New York, 1991

[1.47] D. Marcuse, *Principles of Quantum Electronics,* Academic Press, New York, 1980

[1.48] A. A. Kaminskii, *Laser Crystals,* second edition, Springer series in optical sciences, Springer-Verlag, New York, 1990

[1.49] P. W. Milonni and J. H. Eberly, *Lasers,* John Wiley, New York, 1988

[1.50] M. Abramowitz and I. A. Stegun, *Handbook of Mathematical Functions,* Dover Publications, New York, 1972

[1.51] I. S. Gradshteyn and I. M. Ryzhik, *Table of Integrals, Series and Products,* Academic Press, New York, 1980

## CHAPTER 2

[2.1] D. Marcuse, *Principles of Quantum Electronics,* Academic Press, New York, 1980

[2.2] A. Yariv, *Quantum Electronics,* second edition, John Wiley, New York, 1975

[2.3] R. Loudon, *The Quantum Theory of Light,* second edition, Oxford University Press, New York, 1983

[2.4] B. E. A. Saleh and M. C. Teich, *Fundamentals of Photonics,* John Wiley, New York, 1991.

[2.5] M. C. Teich and B. E. A. Saleh, "Squeezed states of light," *Quantum Optics*, vol 1, 153 (1989)

[2.6] B. Yurke, "The apropriateness of squeezed light for long-distance communications," in *Proc. Topical Meeting on Optical Amplifiers and their Applications*, 1991, paper FA1,

Technical digest series vol 13, 180, Optical Society of America, Washington DC, 1991

[2.7] Y. Yamamoto and T. Mukai, "Fundamentals of optical amplifiers," *Optical and Quantum Electron.*, vol 21, 1 (1989)

[2.8] H. Heffner, "The fundamental noise limit of linear amplifiers," *Proc. IRE*, July 1962, p1604

[2.9] S. O. Rice, "Statistical properties of a sine-wave plus random noise," *Bell Syst. Tech. J.*, vol 27, 109 (1948)

[2.10] H. A. Haus and J. A. Mullen, "Quantum noise in linear amplifiers," *Phys. Rev.*, vol 128, no 5, 2407 (1962)

[2.11] M. W. P. Strandberg, "Inherent noise of quantum-mechanical amplifiers," *Phys. Rev.*, vol 106, no 4, 617 (1957)

[2.12] H. A. Haus and J. A. Mullen, "Noise in optical maser amplifiers," in *Proc. Symposium On Optical Masers*, Polytechnic Institute of Brooklyn, April 16-19, 1963

[2.13] R. J. Glauber, "Coherent and incoherent states of the radiation field," *Phys. Rev.*, vol 131, no 6, 2766 (1963)

[2.14] C. Cohen-Tannoudji, J. Dupont-Roc, and G. Grynberg, *Photons and Atoms,* John Wiley, New York, 1989

[2.15] B. W. Shore, *The Theory of Coherent Atomic Excitation,* vols 1 and 2, John Wiley, New York, 1990

[2.16] W. H. Louisell, A. Yariv, and A. E. Siegman, "Quantum fluctuations and noise in parametric processes," *Phys. Rev.*, vol 124, 1646 (1961)

[2.17] M. Lax, "Quantum noise, IV. Quantum theory of noise sources," *Phys. Rev.*, vol 145, no 1, 110 (1966)

[2.18] M. Lax, "Quantum noise. VII. The rate equations and amplitude noise in lasers," *IEEE J. Quantum Electron.*, vol 3, no 2, 37 (1967)

[2.19] M. Lax and W. H. Louisell, "Quantum noise. IX. Quantum Fokker–Planck solution for laser noise," *IEEE J. Quantum Electron.*, vol 3, no 2, 47 (1967)

[2.20] M. Lax and W. H. Louisell, "Quantum noise. XII. Density-operator treatment of field and population fluctuations," *Phys. Rev.*, vol 185, no 2, 568 (1969)

[2.21] M. O. Scully and W. E. Lamb, "Quantum theory of an optical maser. I. General theory," *Phys. Rev.*, vol 159, no 2, 208 (1967)

[2.22] Y. K. Wang and W. E. Lamb, "Quantum theory of an optical maser. VI. transient behavior," *Phys. Rev. A*, vol 8, no 2, 866 (1973)

[2.23] M. Sargent, M. O. Scully, and W. E. Lamb, Laser physics, Addison-Wesley, London, 1974

[2.24] H. Haken, *Laser Theory,* Springer-Verlag, New York, 1984

[2.25] P. Meystre and M. Sargent III, *Elements of Quantum Optics,* second edition, Springer-Verlag, New York, 1991

[2.26] J. Perina, *Quantum Statistics of Linear and Nonlinear Phenomena*, D. Reidel Publishing Co., Kluwer Academic Publishers, Boston, 1984

[2.27] K. Shimoda, H. Takahasi, and C. H. Townes, "Fluctuations in amplification of quanta with application to maser amplifiers," *J. Phys. Soc Japan*, vol 12, no 6, 686 (1957)

[2.28] E. B. Rockower, N. B. Abraham, and S. R. Smith, "Evolution of the quantum statistics of light," *Phys. Rev. A*, vol 17, no 3, 1100 (1978)

[2.29] P. Diament and M. C. Teich, "Evolution of the statistical properties of photons passed through a traveling-wave laser amplifier," *IEEE J. Quantum Electron.*, vol 28, no 5, 132 (1992), and references therein

[2.30]  S. Karlin and H. M. Taylor, *A first Course in Stochastic Processes*, second edition, Academic Press, New York, 1975

[2.31]  M. Abramovitz and I. A. Stegun, *Handbook of Mathematical Functions*, Dover Publications, New York, 1972, chapters 15 and 22

[2.32]  Y. Y. Yamamoto, "Noise and error rate performance of semiconductor laser amplifiers in PCM-IM optical transmission systems," *IEEE J. Quantum Electron.*, vol 16, no 10, 1073 (1980)

[2.33]  J. A. Arnaud, "Enhancement of optical receiver sensitivities by amplification of the carrier," *IEEE J. Quantum Electron.*, vol 4, no 11, 893 (1968)

[2.34]  S. Ruiz-Moreno, G. Junyent, J. R. Usandigaza, and A. Caldaza, "Resolution of moment equations in a nonlinear optical amplifier," *Electron. Lett.*, vol 23, no 1, 15 (1987)

[2.35]  M. Sargent III, M. O. Scully, and W. E. Lamb, "Buildup of laser oscillations from quantum noise," *Applied Optics*, vol 9, no 11, 2423 (1970)

[2.36]  J. Perina, "Superposition of coherent and incoherent fields," *Phys. Lett.*, vol 24A, 333 (1967)

[2.37]  M. C. Teich and W. J. McGill, "Neural counting and photon counting in the presence of dead time," *Phys. Rev. Lett.*, vol 36, 754 (1976)

[2.38]  R. Loudon and T. J. Shepherd, "Properties of the optical quantum amplifier," *Optica Acta*, vol 31, 1243 (1984)

[2.39]  J. P. Gordon, L. R. Walker, and W. H. Louisell, "Quantum statistics of masers and attenuators," *Phys. Rev.*, vol 130, no 2, 806 (1963)

[2.40]  E. Desurvire, "Spectral noise figure of $Er^{3+}$-doped fiber amplifiers," *IEEE Photonics Technol. Lett.*, vol 2, no 3, 208 (1990)

[2.41]  C. R. Giles and E. Desurvire, "Modeling erbium-doped fiber amplifiers," *IEEE J. Lightwave Technol.*, vol 9, no 2, 271 (1991)

[2.42]  E. Desurvire, M. J. F. Digonnet, and H. J. Shaw, "Theory and implementation of a Raman active fiber delay line," *IEEE J. Lightwave Technol.* vol 4, no 4, 426 (1986)

[2.43]  E. Desurvire, M. Zimgibl, H. M. Presby, and D. DiGiovanni, "Characterization and modeling of amplified spontaneous emission in unsaturated erbium-doped fiber amplifiers," *IEEE Photonics Technol. Lett.*, vol 3, no 2, 127 (1991)

[2.44]  E. Desurvire, "Analysis of noise figure spectral distribution in erbium-doped fiber amplifiers pumped near 980 and 1480 nm," *Applied Optics*, vol 29, no 21, 3118 (1990)

[2.45]  R. Olshansky, "Noise figure for erbium-doped optical fibre amplifiers," *Electron. Lett.*, vol 24, no 22, 1363 (1988)

[2.46]  E. Desurvire, M. Tur, and H. J. Shaw, "Signal-to-noise ratio in Raman active fiber systems: application to recirculating delay lines," *IEEE J. Lightwave Technol.*, vol 4, no 5, 560 (1986)

[2.47]  P. Urquhart and T. J. Whitley, "Long span fiber amplifiers," *Applied Optics*, vol 29, no 24, 3503 (1990)

[2.48]  G. R. Walker, D. M. Spirit, D. L. Williams, and S. T. Davey, "Noise performance of distributed fiber amplifiers," *Electron. Lett.*, vol 27, no 15, 1390 (1991)

[2.49]  D. Chen and E. Desurvire, "Noise performance evaluation of distributed erbium-doped fiber amplifiers with bidirectional pumping at 1.48 μm," *IEEE Photonics Technol. Lett.*, vol 4, no 1, 52 (1992)

[2.50]  E. Desurvire, "Analysis of distributed erbium-doped fiber amplifiers with fiber background loss," *IEEE Photonics Technol. Lett.*, vol 3, no 7, 625 (1991)

[2.51]  K. Kikuchi, "Generalized formula for optical amplifier noise and its application to erbium-doped fibre amplifiers," *Electron. Lett.*, vol 26, no 22, 1851 (1990)

[2.52] A. Yariv, "Signal-to-noise ratio considerations in fiber links with periodic or distributed amplification," *Optics Lett.*, vol 15, no 19, 1064 (1990)

[2.53] R. H. Stolen, "Nonlinear properties of optical fibers", in *Optical Fiber Telecommunications,* edited by S. E. Miller and A. G. Chynoweth, Academic Press, New York, 1979

[2.54] Y. Aoki, "Fibre Raman amplifier properties for applications to long-distance optical communications," *Optical and Quantum Electron.*, vol 21, 89 (1989)

[2.55] K. Rottwitt, J. H. Povlsen, A. Bjarklev, O. Lumholt, B. Pedersen, and T. Rasmussen, "Optimum signal wavelength for a distributed erbium-doped fiber amplifier," *IEEE Photonics Technol. Lett.*, vol 4, no 7, 714 (1992)

[2.56] S. Wen and S. Chi, "Distributed erbium-doped fiber amplifiers with stimulated Raman scattering," *IEEE Photonics Technol. Lett.*, vol 4, no 2, 189 (1992)

[2.57] J. P. Gordon and L. F. Mollenauer, "Effects of fiber nonlinearities and amplifier spacing on ultra-long distance transmission," *IEEE J. Lightwave Technol.*, vol 9, no 2, 170 (1991)

[2.58] J. P. Gordon and H. A. Haus, "Random walk of coherently amplified solitons in optical fiber transmission," *Optics Lett.*, vol 11, no 10, 665 (1986)

[2.59] K. Rottwitt, A. Bjarklev, O. Lumholt, J. H. Povlsen, and T. P. Rasmussen, "Design of long distance distributed erbium-doped fiber amplifier," *Electron. Lett.*, vol 28, no 3, 287 (1992)

[2.60] T. Li and M. C. Teich, "Performance of a lightwave system incorporating a cascade of erbium-doped fiber amplifiers," *Opt. Comm.*, vol 91, nos 1–2, 41 (1992)

[2.61] M. C. Teich and B. E. A. Saleh, "Effect of random deletion and additive noise on bunched and antibunched photon-counting statistics," *Optics Lett.*, vol 7, no 8, 365 (1982)

[2.62] E. Desurvire, "Erbium-doped fiber amplifiers basic physics and characteristics," in *Rare-earth-doped Fibers and Devices,* edited by M. J. Digonnet, Marcel Dekker, New York, 1992

[2.63] G. Oliver and C. Bendjaballah, "Statistical properties of coherent radiation in a nonlinear optical amplifier," *Phys. Rev. A*, vol 22, no 2, 630 (1980)

[2.64] C. Bendjaballah and G. Oliver, "Comparison of statistical properties of two models for a saturated laser-light amplifier," *Phys. Rev. A*, vol 22, no 6, 2726 (1980)

[2.65] N. B. Abraham, "Quantum theory of a saturable optical amplifier," *Phys. Rev. A*, vol 21, no 5, 1595 (1980)

[2.66] U. Mohr, "Photon statistics in saturated $m$-photon amplification and attenuation of coherent light," *Opt. Comm.*, vol 41, no 1, 21 (1982)

[2.67] C. Bendjaballah and G. Oliver, "Detection of coherent light after nonlinear amplification," *IEEE Trans. Aerospace and Electron. Syst.*, vol 17, no 5, 620 (1981)

[2.68] A. E. Willner and E. Desurvire, "Effect of gain saturation on receiver sensitivity in 1Gbit/s multichannel FSK direct-detection systems using erbium-doped fiber pre-amplifiers," *IEEE Photonics Technol. Lett.*, vol 3, no 3, 259 (1991)

[2.69] M. Sargent III, M. O. Scully, and W. E. Lamb, "Buildup of laser oscillations from quantum noise," *Applied Optics*, vol 9, no 11, 2423 (1970). A movie showing the transient buildup up to the saturated gain regime of the laser photon statistics can be seen by flipping through the pages of this journal issue.

[2.70] F. A. Vorobev and R. I. Sokolovskii, *Opt. Spectrosc.* (USSR), vol 36, 625 (1974)

[2.71] W. H. Press, B. P. Flannery, S. A. Teukolsky, and W. T. Vetterling, *Numerical Recipes, The Art of Scientific Computing,* Cambridge University Press, New York, 1986

## CHAPTER 3

[3.1]   D. Marcuse, *Principles of Quantum Electronics*, Academic Press, New York 1980

[3.2]   R. Loudon, *The Quantum Theory of Light*, second edition, Oxford University Press, New York, 1983

[3.3]   B. E. A. Saleh and M. C. Teich, *Fundamentals of Photonics*, John Wiley, New York, 1991

[3.4]   J. Perina, *Quantum Statistics of Linear and Nonlinear Phenomena*, D. Reidel Publishing Co., Kluwer Academics Publishers, Boston, 1984

[3.5]   M. C. Teich and T. Li, "The retinal rod as a chemical photomultiplier," *J. Visual comm. and image representation*, vol 1, no 1, 104 (1990)

[3.6]   C. E. Shannon, "A mathematical theory of communication," *Bell Syst. Tech. J.*, vol 27, 379–423/623–656 (1948)

[3.7]   D. Gabor, "Communication theory and physics," *Phil. Mag.*, vol 41, 1161 (1950)

[3.8]   J. P. Gordon, "Quantum effect in communication systems," *Proc. IRE*, vol. 50, 1898 (1962)

[3.9]   H. Takahasi, "Information theory and quantum mechanical channels," in *Advances in Communications Systems*, edited by Balakrishnan, vol 1, 117, Academic Press, 1965.

[3.10]  O. Hirota and H. Tsushima, "Quantum communication theory and its applications," *Trans. IEICE*, vol 72, no 5, 460 (1989)

[3.11]  I. Garett, "Towards the fundamental limits of optical fiber communications," *J. Lightwave technol.*, vol 1, no 1, 131 (1983)

[3.12]  J. Gowar, *Optical Communications Systems*, Prentice-Hall, London, 1984

[3.13]  J. M. Senior, *Optical Fiber Communications, Principles and Practice*, Prentice-Hall, London, 1985

[3.14]  G. Keiser, *Optical Fiber Communications*, second edition, McGraw-Hill, New York, 1991

[3.15]  P. E. Green, *Fiber-Optic Networks*, Prentice-Hall, London 1993

[3.16]  S. E. Miller and I. P. Kaminow, *Optical fiber telecommunications II*, Academic Press, New York, 1988

[3.17]  S. M. Sze, *Semiconductor Devices, Physics and Technology*, John Wiley, New York, 1985

[3.18]  M. O. Scully and W. E. Lamb, "Quantum theory of an optical maser III. Theory of photoelectron counting statistics," *Phys. Rev.*, vol 179, no 2, 368 (1969)

[3.19]  K. Shimoda, H. Takahasi, and C. H. Townes, "Fluctuations in amplification of quanta with application to maser amplifiers," *J. Phys. Soc Japan*, vol 12, no 6, 686 (1957)

[3.20]  R. Loudon and T. J. Shepherd, "Properties of the optical quantum amplifier," *Optica Acta*, vol 31, no 11, 1243 (1984)

[3.21]  M. C. Teich and B. E. A. Saleh, "Effect of random deletion and additive noise on bunched and antibunched photon-counting statistics," *Optics Lett.*, vol 7, no 8, 365 (1982)

[3.22]  N. A. Olsson, "Lightwave systems with optical amplifiers," *IEEE J. Lightwave technol.* vol 7, no 7, 1071 (1989)

[3.23]  P. L. Kelley and W. H. Kleiner, *Phys. Rev. A*, vol 136, 316 (1964)

[3.24]  L. Mandel, *Proc. Phys. Soc*, vol 72, 1037 (1958); *Proc. Phys. Soc.*, vol 74, 233 (1959)

[3.25]  T. J. Shepherd, A model for photodetection of single-mode cavity radiation," *Optica Acta*, vol 28, no 4, 567 (1981)

[3.26]  W. B. Davenport and W. L. Root, *An Introduction to the Theory of Random Signals and Noise,* Chapter 12, McGraw-Hall, New York, 1985

[3.27]  R. C. Steele and G. R. Walker, "High-sensitivity FSK signal detection with an erbium-doped fiber preamplifier and Fabry–Perot etalon demodulation," *IEEE Photonics Technol. Lett.,* vol 2, no 10, 753 (1990)

[3.28]  R. C. Steele, G. R. Walker, and N. G. Walker, "Sensitivity of optically preamplified receivers with optical filtering," *IEEE Photonics technol. Lett.,* vol 3, no 6, 545 (1991)

[3.29]  O. K. Tonguz and L. G. Kazovsky, "Theory of direct-detection lightwave receivers using optical amplifiers," *IEEE J. Lightwave Technol.,* vol 9, no 2, 174 (1991)

[3.30]  L. G. Kazovsky and O. K. Tonguz, "Sensitivity of direction-detection receivers using optical preamplifiers," *IEEE Photonics Technol. Lett.,* vol 3, no 1, 53 (1991)

[3.31]  P. M. Lane, C. R. Medeiros, and J. J. O'Reilly, "Moment based optimization of optically preamplified systems in presence of jitter," in *Proc. Topical Meeting on Optical Amplifiers and Applications,* paper ThC3, 119 (1992), Optical Society of America, Washington DC, 1992

[3.32]  Y. E. Dallal and S. Shamai, "Analytical techniques assessing the performance of optical amplifiers," in *Proc. Topical Meeting on Optical Amplifiers and Applications,* paper ThC5, 127 (1992), Optical Society of America, Washington DC, 1992

[3.33]  S. D. Personick, "Receiver design for digital fiber-optic communications systems, I," *Bell Syst. Tech. J,* vol 52, no 6, 843 (1973); S. D. Personick, "Receiver design for digital fiber-optic communications systems, II," *Bell Syst. Tech. J,* vol 52, no 6, 875 (1973)

[3.34]  Y. Yamamoto, "Noise and error rate performance of semiconductor laser amplifiers in PCM-IM optical transmission systems," *IEEE J. Quantum Electron.,* vol 16, no 10, 1073 (1980)

[3.35]  T. Mukai, Y. Yamamoto, and T. Kimura, "S/N and error rate performance in AlGaAs semiconductor laser preamplifiers and linear repeater systems," *IEEE J. Quantum Electron.* vol 18, no 10, 1560 (1982)

[3.36]  H. Steinberg, "The use of a laser amplifier in a laser communication system," in *Proc. IEEEE (correspondence),* 943 (1963)

[3.37]  H. Steinberg, "Signal detection with a laser amplifier," in *Proc. IEEE (correspondence),* 28 (1964)

[3.38]  H. Kogelnick and A. Yariv, "Considerations of noise and schemes for its reduction in laser amplifiers," *Proc. IEEE,* 165 (1964)

[3.39]  F. Arams and M. Wang, "Infrared laser preamplifier system," in *Proc. IEEE (correspondence),* vol 53, 329 (1965)

[3.40]  W. Bridges and G. Picus, "Gas laser preamplifier performance," *Applied Optics,* vol 3, 1189 (1964)

[3.41]  J. A. Arnaud, "Enhancement of optical receiver sensitivies by amplification of the carrier," *J. Quantum Electron.,* vol 4, no 11, 893 (1968)

[3.42]  C. J. Koester and E. A. Snitzer, "Amplification in a fiber laser," *Applied Optics,* vol 3, no 10, 1182 (1964)

[3.43]  H. A. Haus and J. A. Mullen, "Quantum noise in linear amplifiers," *Phys. Rev.,* vol 128, no 5, 2407 (1962)

[3.44]  S. D. Personick, "Applications for quantum amplifiers in simple digital optical communication systems," *Bell Syst. Tech. J.,* vol 52, no 1, 117 (1973)

[3.45]  J. C. Simon, Semiconductor laser amplifiers for single-mode optical fiber communications," *J. Opt. Comm.,* vol 4, no 2, 51 (1982)

[3.46]  Y. Yamamoto, "Receiver performance evaluation of various digital optical modulatioñdemodulation systems in the 0.5–10 mm wavelength region," *IEEE J. Quantum Electron.*, vol 16, 1251 (1980)

[3.47]   S. Ryu, S. Yamamoto, H. Taga, N. Edagawa, Y. Yoshida, and H. Wakabayashi, "Long-haul coherent optical fiber communication systems using optical amplifiers," *IEEE J. Lightwave Technol.*, vol 9, no 2, 251 (1991)

[3.48]   K. Inoue, H. Toba, and K. Nosu, "Multichannel amplification utilizing an $Er^{3+}$ doped fiber amplifier," *IEEE J. Lightwave Technol.*, vol 9, no 3, 368 (1991)

[3.49]   D. Marcuse, "Derivation of analytical expressions for the bit-error probability in lightwave systems with optical amplifiers," *IEEE J. Lightwave Technol.*, vol 8, 1816 (1991)

[3.50]   D. Marcuse, "Calculation of bit-error probability for a lightwave system with optical amplifiers and post-detection Gaussian noise," *IEEE J. Lightwave Technol.*, vol 9, no 4, 505 (1991)

[3.51]   D. Marcuse, "Bit-error rate of lightwave systems at the zero-dispersion wavelength," *IEEE J. Lightwave Technol.*, vol 9, no 10, 1330 (1991)

[3.52]   P. A. Humblet and M. Azizoglu, "On the bit error rate of lightwave systems with optical amplifiers," *IEEE J. Lightwave Technol.*, vol 9, no 11, 1576 (1991)

[3.53]   G. Jacobsen, "Multichannel system design using optical preamplifiers and accounting for the effects of phase noise, amplifier noise, and receiver noise," *IEEE J. Lightwave Technol.*, vol 10, no 3, 367 (1991)

[3.54]   R. S. Fyath and J. J. O'Reilly, "Comprehensive moment-generating function characrization of optically preamplified receivers," *Proc. IEE J*, vol 137, no 6, 391 (1990)

[3.55]   R. S. Fyath and J. J. O'Reilly, "Accurate assessment of sensitivity and optimum threshold for semiconductor laser preamplified optical receivers," *Proc. IEE J*, vol 138, no 3, 221 (1991)

[3.56]   J. Zhou and S. D. Walker, "Application of the Edgeworth series expansion method to optically preamplified receiver bit-error-ratio calculation," in *Proc. Topical Meeting on Optical Amplifiers and Application*, paper ThC4, 123 (1992), Optical Society of America, Washington DC, 1992

[3.57]   C. Bendjaballah and M. Charbit, "Detection of optical signals in generalized Gaussian noise," *IEEE Trans. Comm.*, vol 30, no 2, 367 (1982)

[3.58]   C. Bendjaballah and K. Hassan, "Probability of detecting a coherent optical signal in thermal noise," *J. Opt. Soc. Am.*, vol 73, no 12, 1840 (1983)

[3.59]   C. Bendjaballah and G. Oliver, "Detection of coherent light after nonlinear amplification," *IEEE Trans. Aerospace and Electron. Syst.*, vol 17, no 5, 620 (1981)

[3.60]   Y. Yamamoto, "AM and FM quantum noise in semiconductor lasers – part I: theoretical analysis," *IEEE J. Quantum Electron.*, vol 19, no 1, 34 (1983) Y. Yamamoto, "AM and FM quantum noise in semiconductor lasers – part II: comparison of theoretical and experimental results for AlGaAs lasers," *IEEE J. Quantum Electron.*, vol 19, no 1, 47 (1983)

[3.61]   T. Li and M. C. Teich, "Bit-error rate for a lightwave communication system incorporating and erbium-doped fibre amplifier," *Electron. Lett.*, vol 27, no 7, 598 (1991)

[3.62]   T. Li and M. C. Teich, "Performance of a lightwave system incorporating a cascade of erbium-doped fiber amplifiers," *Opt. Comm.*, vol 91, no 1–2, 41 (1992)

[3.63]   M. Abramowitz and I. A. Stegun, *Handbook of Mathematical Functions,* Dover Publications, New York, 1972

[3.64]   M. Schwartz, *Information, Transmission, Modulation, and Noise,* Chapter 5, McGraw-Hill, New York, 1980

[3.65]  S. O. Rice, "Mathematical analysis of random noise," *Bell Syst. Tech. J.*, vol 23, 282 (1944)

[3.66]  P. S. Henry, "Error-rate performance of optical amplifiers," *Proc. Conference on Optical Fiber Communications, OFC'89*, paper THK3, p170, Optical Society of America, Washington DC, 1989

[3.67]  P. S. Henry, "Lightwave primer," *IEEE J. Quantum Electron.*, vol 21, no 12, 1862 (1985)

[3.68]  S. K. Morshnev and A. V. Fantsesson, "Coherent fiber-optic communications," *Sov. J. Quantum Electron.*, vol 15, no 9, 1183 (1985)

[3.69]  R. A. Linke and A. H. Gnauck, "High-capacity coherent lightwave systems," *IEEE J. Lightwave Technol.*, vol 6, no 11, 1750 (1988)

[3.70]  S. Ryu and Y. Horiuchi, "Use of an optical amplifier in a coherent receiver," *IEEE Photonics Technol. Lett.*, vol 3, no 7, 663 (1991)

[3.71]  B. Glance, G. Eisenstein, P. J. Fitzgerald, K. J. Pollock, and G. Raybon, "Sensitivity of an optical heterodyne receiver in presence of an optical preamplifier," *Electron. Lett.*, vol 24, no 19, 1229 (1988)

[3.72]  S. Saito, T. Imai, and T. Ito, "An over 2200-km coherent transmission experiment at 2.5 Gb/s using erbium-doped-fiber in-line amplifiers," *IEEE J. Lightwave Technology*, vol 9, no 2, 161 (1991)

[3.73]  S. Ryu, S. Yamamoto, H. Taga, N. Edagawa, Y. Yoshida, and H. Wakabayashi, "Long-haul coherent optical fiber communication systems using optical amplifiers," *IEEE J. Lightwave Technol.*, vol 9, no 2, 251 (1991)

[3.74]  K. Inoue, H. Toba, and K. Nosu, "Multichannel amplification utilizing an $Er^{3+}$-doped fiber amplifier," *IEEE Photonics Technol. Lett.*, vol 9, no 3, 368 (1991)

[3.75]  H. P. Yuen and V. W. S. Chan, "Noise in homodyne and heterodyne detection," *Optics Lett.*, vol 8, no 3, 177 (1983)

[3.76]  J. H. Shapiro and S. S. Wagner, "Phase and amplitude uncertainties in heterodyne detection," *IEEE J. Quantum Electronics*, vol 20, no 7, 803 (1984)

[3.77]  K. S. Shanmugan, *Digital and Analog Communication Systems*, John Wiley, New York, 1979

[3.78]  H. Taub and D. L. Schilling, *Principles of Communication Systems*, second edition, McGraw-Hill, New York, 1986

[3.79]  T. Okoshi and K. Kikuchi, *Coherent Optical Fiber Communications*, Kluwer Academic, Boston, 1986

[3.80]  A. B. Carlson, *Communication Systems*, third edition, McGraw-Hill, New York, 1990

[3.81]  J. Salz, "Coherent lightwave communications," *AT&T Tech. J.*, vol 64, 2153 (1985)

[3.82]  L. G. Kazovsky, "Optical heterodyning vs. optical homodyning: a comparison," *J. Opt. Comm.*, vol 6, 18 (1985)

[3.83]  J. Salz, "Modulation and detection for coherent lightwave communications," *IEEE Comm. Mag.*, vol 24, 38 (1986)

[3.84]  T. G. Hodgkison, "Receiver analysis for synchronous coherent optical fiber transmission systems," *IEEE J. Lightwave Technol.*, vol 5, 573 (1987)

[3.85]  L. G. Kazovsky and O. K. Tonguz, "ASK and FSK coherent lightwave systems: a simplified approximate analysis," *IEEE J. Lightwave Technol.*, vol 8, 338 (1990)

[3.86]  K. Hagimoto, S. Nishi, and K. Nakagawa, "An optical bit-rate flexible transmission system with 5 Tbit/s.km capacity employing multiple in-line erbium-doped fiber amplifiers," *IEEE J. Lightwave Technol.*, vol 8, no 9, 1387 (1990)

[3.87] C. R. Giles and E. Desurvire, "Propagation of signal and noise in concatenated erbium-doped fiber optical amplifiers," *IEEE J. Lightwave Technol.*, vol 9, no 2, 147 (1991)

[3.88] G. R. Walker, N. G. Walker, R. C. Steele, M. J. Creaner, and M. C. Brain, "Erbium-doped fiber amplifier cascade for multichannel coherent optical transmission," *IEEE J. Lightwave Technol.*, vol 9, no 2, 182 (1991)

[3.89] S. Saito, M. Murakami, A. Naka, Y. Fukada, T. Imai, M. Aiki, and T. Ito, "In-line amplifier transmission experiments over 4500 km at 2.5 Gbit/s," *IEEE J. Lightwave Technol.*, vol 10, no 8, 1117 (1992)

[3.90] W. I. Way, "Optical fiber amplifiers for multichannel video transmission and distribution," in *Proc. Topical Meeting on Optical Amplifiers and Applications, 1991*, paper FB-1, p206, Optical Society of America, Washington DC, 1991

[3.91] I. M. Habbab and L. J. Cimini, "Optimized performance of erbium-doped fiber amplifiers in subcarrier multiplexed lightwave AM-VSB CATV systems," *IEEE J. Lightwave Technol.*, vol 9, no 10, 1321 (1991)

[3.92] K. Simon, *Technical Handbook for CATV Systems,* third edition, General Instruments, 1986

[3.93] A. A. M. Saleh, "Fundamental limit on the number of channels in subcarrier-multiplexing lightwave CATV systems," *Electron. Lett.*, vol 25, no 12, 776 (1989)

[3.94] E. E. Bergmann, C. Y. Kuo, and S. Y. Huang, "Dispersion induced composite second order distortion at 1.5 $\mu m$", *IEEE Photonics Technol. Lett.*, vol 3, no 1, 59 (1991)

[3.95] Lian-Kuan Allen Chen, *Highly Linear Lightwave Video Distribution System,* PhD thesis, Columbia University, New York, 1992, chapter 4

[3.96] C. Y. Kuo and E. E. Bergmann, "Erbium-doped fiber amplifier second-order distortion in analog links and electronic compensation," *IEEE Photonics Technol. Lett.*, vol 3, no 9, 829 (1991)

[3.97] K. Kikushima and H. Yoshinaga, "Distortion due to gain tilt of erbium-doped fiber amplifiers," *IEEE Photonics Technol. Lett.*, vol 3, no 10, 945 (1991)

[3.98] J. Ohya, H. Sato, and T. Fujita, "Second-order distortion generated by amplification of intensity-modulated signals with chirping in erbium-doped fibers," *IEEE Photonics Technol. Lett.*, vol 4, no 9, 1000 (1991)

## CHAPTER 4

[4.1] E. Snitzer, "Optical maser action of $Nd^{3+}$ in a barium crown glass," *Phys. Rev. Lett.*, vol 7, 444 (1961)

[4.2] M. J. Weber, "Science and technology of laser glass", *J. Non-Cryst. Solids*, 1 (1990)

[4.3] S. E. Stokowski, "Glass lasers," in *Handbook of Laser Science and Technology*, vol 1, edited by M. J. Weber, CRC Press

[4.4] C. F. Rapp, "Laser glasses," in *Handbook of Laser Science and Technology*, vol 5, edited by M. J. Weber, CRC Press

[4.5] C. J. Koester and E. Snitzer, "Amplification in a fiber laser," *Applied Optics*, vol 3, no 10, 1182 (1964)

[4.6] E. Snitzer, "Lasers and glass technology," *Ceramic Bulletin*, vol 52, no 6, 516 (1973)

[4.7] J. Stone and C. A. Burrus, "Neodymium-doped silica lasers in end-pumped fiber geometry," *Applied Phys. Lett.*, vol 23, no 7, 388 (1973)

[4.8]  J. Stone and C. A. Burrus, "Neodymium-doped fiber lasers: room-temperature cw operation with an injection laser pump," *Applied Optics*, vol 13, no 6, 1256 (1974)

[4.9]  E. Snitzer, "Glass lasers," *Applied Optics*, vol 5, no 10, 1487 (1966)

[4.10]  A. A. Kaminskii, *Laser Crystals,* Springer series in optical sciences, vol 14, Springer-Verlag, New York, 1990

[4.11]  D. W. Hall and M. J. Weber, "Glass lasers," in *Handbook of Laser Science and Technology*, supplement 1, CRC Press, Boca Raton, FL

[4.12]  H. Haken and H. C. Wolf, *Atomic and Quantum Physics,* second edition, Springer-Verlag, New York 1987, chapters 5, 15, and 19

[4.13]  K. Patek, *Glass Lasers*, CRC Press, Butterworth, London, 1970

[4.14]  M. J. Weber, private communication

[4.15]  Y. Okoshi, T. Kanamori, T. Kitagawa, T. Takahashi, E. Snitzer, and G. H. Siegel, "$Pr^{3+}$-doped fluoride fiber amplifier operating at 1.31 $\mu$m," *Optics Lett.*, vol 16, no 22, 1747 (1991)

[4.16]  R. M. Percival, M. W. Phillips, D. C. Hanna, and A. C. Tropper, "Characterization of spontaneous and stimulated emission from praseodymium ($Pr^{3+}$) ions doped into a silica-based monomode optical fiber," *IEEE J. Quantum Electron.*, vol 25, no 10, 2119 (1989)

[4.17]  B. Pedersen, W. J. Miniscalco, and R. S. Quimby, "Optimization of $Pr^{3+}$:ZBLAN fiber amplifiers," *IEEE Photonics Technol. Lett.*, vol 4, no 5, 446 (1992)

[4.18]  J. Y. Allain, M. Monerie, and H. Poignant, "Tunable cw lasing around 610, 635, 695, 715, 885 and 910 nm in praseodymium-doped fluorozirconate fibre," *Electron. Lett.*, vol 27, no 2, 189 (1991)

[4.19]  Y. Okoshi, T. Kanamori, and S. Takahashi, "$Pr^{3+}$-doped fluoride single-mode fiber laser," *IEEE Photonics Technol. Lett.*, vol 3, no 8, 688 (1991)

[4.20]  E. Snitzer and R. Woodcock, "$Yb^{3+}$–$Er^{3+}$ glass laser," *Applied Phys. Lett.*, vol 6, no 3, 45 (1965)

[4.21]  C. C. Robinson, "Multiple sites for $Er^{3+}$ in alkali silicate glasses (I). Evidence of four sites for $Er^{3+}$," *J. Non-Cryst. Solids*, vol 15, 1 (1974); "Multiple sites for $Er^{3+}$ in alkali silicate glasses (II). Evidence of four sites for $Er^{3+}$," *J. Non-Cryst. Solids,* vol 15, 11 (1974)

[4.22]  N. E. Alekseiev, V. P. Gapontsev, M. E. Zhabotinskii, V. B. Kravchenko, and Y. P. Rudnitskii, in *Laser Phosphate Glasses,* edited by M. E. Zhabotinskii, Nauka, Moscow, 1980. Translation by Berkeley Scientific, UCRL TRANS-11817 for Lawrence Livermore Laboratory, 1983

[4.23]  S. E. Stokowski, R. A. Saroyan, and M. J. Weber, in *Nd-doped Laser Glass Spectroscopic and Physical Properties,* Lawrence Livermore National laboratory, Livermore, CA, 1981

[4.24]  P. W. France, S. F. Carter, M. W. Moore, and C. R. Day, "Progress in fluoride fibres for optical communications," *Br. Telecom Technol. J.*, vol 5, no 2, 28 (1987)

[4.25]  S. T. Davey and P. W. France, "Rare-earth-doped fluorozirconate glasses for fibre devices," *Br. Telecom Technol. J.*, vol 7, no 1, 58 (1989)

[4.26]  A. V. Astakhov, M. M. Butusov, and S. L. Galkin, "Characteristics of laser effects in active fiber waveguides," *Opt. Spectrosc.* (USSR), vol 59, no 4, 551 (1985)

[4.27]  G. D. Dudko, R. S. Shelevich, A. N. Izotov, V. N. Tabrin, B. N. Litvin, and L. G. Bebikh, "Generation of light in optical fibers made of glasses formed from rare-earth ultraphosphate crystals," *Sov. J. Quantum Electron.*, vol 12, no 9, 1210 (1982)

[4.28]  J. R. Simpson, "Fabrication of rare-earth doped glass fibres," in *Proc. SPIE*

*Conference on Fiber Laser Sources and Amplifiers*, vol 1171, p2 (1989); J. R. Simpson, "Rare earth doped fiber fabrication: techniques and physical properties", in *Rare Earth Doped Fiber Lasers and Amplifiers*, edited by M. J. F. Digonnet, Marcel Dekker, New York, 1993

[4.29] B. J. Ainslie, "A review of the fabrication and properties of erbium-doped fibers for optical amplifiers," *IEEE J. Lightwave Technol.*, vol 9, no 2, 220 (1991)

[4.30] J. Gowar, *Optical Communications Systems*, Prentice-Hall, London, 1984

[4.31] J. M. Senior, *Optical Fiber Communications, Principles and Practice*, Prentice-Hall, London, 1985

[4.32] G. Keiser, *Optical Fiber Communications*, second edition, McGraw-Hill, New York 1991

[4.33] S. E. Miller and I. P. Kaminow, *Optical Fiber Telecommunications II*, Academic Press, New York, 1988

[4.34] F. P. Kapron, D. B. Keck, and R. D. Maurer, "Radiation losses in optical waveguides," *Applied Phys. Lett.*, vol 10, 423 (1970)

[4.35] S. Sudo, M. Kawachi, M. Edahiro, T. Izawa, T. Shoida, and H. Gotoh, "Low-OH-content optical fiber fabricated by vapor-phase axial deposition method," *Electron. Lett.*, vol 14, no 17, 534 (1978)

[4.36] W. G. French, J. B. MacChesney, P. B. O'Conner and G. W. Tasker, "Optical waveguides with very low losses," *Bell Syst. Tech. J.*, vol 53, 951 (1974)

[4.37] D. N. Payne and W. A. Gambling, "New silica-based low-loss optical fibres," *Electron. Lett.*, vol 10, no 15, 289 (1974)

[4.38] D. Kuppers and J. Koenigs, "Preform fabrication by deposition of thousands of layers with the aid of plasma-activated CVD," in *Proc. Second Conference on Optical Fiber Communications, OFC'76*, p 49, Optical Society of America, Washington DC, 1976

[4.39] J. R. Simpson and J. B. MacChesney, "Alternate dopants for silicate waveguides," in *Proc. Optical Fiber Communications Conference*, OFC'82, paper TuCC5, Optical Society of America, Washington DC, 1982; J. R. Simpson and J. B. MacChesney, *Electron. Lett.*, vol 19, 261 (1983)

[4.40] S. B. Poole, D. N. Payne, M. E. Fermann, "Fabrication of low-loss optical fibres containing rare-earth ions," *Electron. Lett.*, vol 21, no 17, 737 (1985); S. B. Poole, D. N. Payne, R. J. Mears, M. E. Fermann, and R. I. Laming, "Fabrication and characterization of low-loss optical fibers containing rare-earth ions," *IEEE J. Lightwave Technol.*, vol 4, no 7, 870 (1986)

[4.41] B. J. Ainslie, S. P. Craig, and S. T. Davey, "The fabrication and optical properties of $Nd^{3+}$ in silica-based fibres," *Material Lett.*, vol 5, no 4, 143 (1987)

[4.42] T. F. Morse, L. Reinhart, A. Kilian, W. Risen, and J. W. Cipolla, "Aerosol-doping technique for MCVD and OVD," in *Proc. SPIE Conference on Fiber Laser Sources and Amplifiers*, vol 1171, 72 (1989)

[4.43] R. P. Tumminelli, B. C. MacCollum, and E. Snitzer, "Fabrication of high-concentration rare-earth doped optical fibers using chelates," *IEEE J. Lightwave Technol.*, vol 8, no 11, 1680 (1990)

[4.44] M. Shimizu, F. Hanawa, H. Suda, and M. Horiguchi, "Transmission loss characteristics of Nd-doped silica single-mode fibers fabricated by the VAD method," *Japanese J. Applied Phys.*, vol 28, no 3, 476 (1989)

[4.45] P. L. Bocko, "Rare-earth doped optical fibres by the outside vapour deposition process," in *Proc. Optical Fiber Communications Conference, OFC'89*, paper TuG2, Optical Society of America, Washington DC, 1989

[4.46]   J. E. Townsend, S. B. Poole, and D. N. Payne, "Solution-doping technique for fabrication of rare-earth doped optical fibers," *Electron. Lett.*, vol 23, no 7, 329 (1987)

[4.47]   P. W. France and M. C. Brierley, Fluoride glass fiber lasers and amplifiers," in *Proc. SPIE Conference on Fiber Laser sources and Amplifiers*, vol 1171, 65 (1989)

[4.48]   D. C. Tran, G. H. Sigel, and B. Bendow, "Heavy metal fluoride glasses and fibers: a review," *IEEE J. Lightwave Technol.*, vol 2, no 5, 566 (1984)

[4.49]   S. Mitachi, S. Shibata, and T. Manabe, "Teflon FEP-clad fluoride glass fibres," *Electron. Lett.*, vol 17, 128 (1981)

[4.50]   S. Mitachi, T. Miyashita, and T. Kanamori, "Fluoride-glass-cladded optical fibres for mid-infrared ray transmission," *Electron. Lett.*, vol 17, 591 (1981)

[4.51]   D. C. Tran, C. F. Fischer, and G. H. Siegel, "Fluoride glass preforms prepared by a rotational casting process," *Electron. Lett.*, vol 18, 659 (1982)

[4.52]   Y. Ohishi, S. Mitachi, and S. Takahashi, "Fabrication of fluoride glass single-mode fibers," *IEEE J. Lightwave Technol.*, vol 3, 593 (1984)

[4.53]   P. W. France, M. G. Drexhage, J. M. Parker, M. W. Moore, S. F. Carter, and J. V. Wright, *Fluoride Glass Optical Fibers*, Blackie, London, 1990

[4.54]   R. Reisfeld, "Radiative and non-radiative transitions of rare-earth ions in glass," in *Structure and Bonding*, no 22, Springer-Verlag, New York

[4.55]   L. A. Riseberg and M. J. Weber, "Relaxation phenomena in rare-earth luminescence," in *Progress in Optics*, vol XIV, edited by E. Wolf, North-Holland, 1976

[4.56]   B. G. Wybourne, *Spectroscopic Properties of Rare-earths*, John Wiley, New York, 1965

[4.57]   G. H. Dieke, *Spectra and Energy Levels of Rare-earth Ions in Crystals*, John Wiley, New York, 1968

[4.58]   R. J. Mears, L. Reekie, S. B. Poole, and D. N. Payne, "Low-threshold tunable cw and Q-switched fibre laser operating at 1.55 $\mu$m," *Electron. Lett.*, vol 22, no 3, 159 (1986)

[4.59]   M. C. Brierley and P. W. France, "Continuous-wave lasing at 2.7 $\mu$m in an erbium-doped fluorozirconate fibre," *Electron. Lett.*, vol 24, no 15, 935 (1988)

[4.60]   C. A. Millar, M. C. Brierley, M. H. Hunt, and S. F. Carter, "Efficient up-conversion pumping at 800 nm of an erbium-doped fluoride fibre laser operating at 850 nm," *Electron. Lett.*, vol 26, no 22, 1871 (1990)

[4.61]   R. G. Smart, J. N. Carter, D. C. Hanna, and A. C. Tropper, "Erbium doped fluorozirconate fibre laser operating at 1.66 and 1.72 $\mu$m," *Electron. Lett.*, vol 26, no 10, 649 (1990)

[4.62]   R. Allen, L. Esterowitz, and R. J. Ginther, "Diode-pumped single-mode fluorozirconate fiber laser from the $^4I_{11/2} \rightarrow {}^4I_{13/2}$ transition in erbium," *Applied Phys. Lett.*, vol 56, no 17, 1635 (1990)

[4.63]   T. J. Whitley, C. A. Millar, R. Wyatt, M. C. Brierley, and D. Szebesta, "Upconversion pumped green lasing in erbium-doped fluorozirconate fibre," *Electron. Lett.*, vol 27, no 20, 1785 (1991)

[4.64]   J. Y. Allain, M. Monerie, and H. Poignant, "Tunable green upconversion erbium fibre laser," *Electron. Lett.*, vol 28, no 2, 111 (1992)

[4.65]   E. Desurvire, "Erbium-doped fiber amplifiers: basic physics and theoretical modeling," *Int. J. High-Speed electron.*, vol 2, no 1–2, 89 (1991)

[4.66]   B. R. Judd, "Optical absorption intensities of rare-earth ions," *Phys, Rev.*, vol 127, no 3, 750 (1962)

[4.67]   G. S. Ofelt, "Intensities of crystal spectra of rare-earth ions," *J. Chem. Physics.*, vol 37, no 3, 511 (1962)

[4.68]   W. F. Krupke, "Radiative transition probabilities within the $4f^3$ ground configuration of Nd:YAG," *IEEE J. Quantum Electron.*, vol 7, no 4, 153 (1971)

[4.69]   R. R. Jacobs and M. J. Weber, "Dependence of the $^4F_{3/2} \rightarrow {}^4I_{11/2}$ induced-emission cross-section for $Nd^{3+}$ on glass composition," *IEEE J. Quantum Electron.*, vol 12, no 2, 102 (1976)

[4.70]   M. J. Weber, *Phys. Rev*, vol 157, 262 (1967)

[4.71]   J. R. Chamberlain, A. C. Everitt, and J. W. Orton, *Phys. Rev.*, *J. Phys. C1*, 157 (1968)

[4.72]   W. T. Carnall, P. R. Fields, and K. Rajnak, "Electronic energy levels in the trivalent lanthanide aquo ions I. $Pr^{3+}$, $Nd^{3+}$, $Pm^{3+}$, $Sm^{3+}$, $Dy^{3+}$, $Ho^{3+}$, $Er^{3+}$ and $Tm^{3+}$," *J. Chem. Phys.*, vol 49, 4424 (1968)

[4.73]   R. Reisfeld and Y. Eckstein, "Intensity parameters of $Tm^{3+}$ and $Er^{3+}$ in borate, phosphate and germanate glasses," *Solid-State Commun.*, vol 13, 265 (1973)

[4.74]   R. Reisfeld and C. K. Jorgensen, "Excited-state phenomena in vitreous materials," in *Handbook on the Physics and Chemistry of Rare-earths*, edited by K. A. Gschneider and L. Eyring editors, Elsevier Science Publishers, 1987

[4.75]   W. F. Krupke, "Optical absorption and fluorescence intensities in several rare-earth-doped $Y_2O_3$ and $LaF_3$ single crystals," *Phys. Rev.*, vol 145, 325 (1966)

[4.76]   A. Yariv, *Quantum Electronics*, second edition, John Wiley, New York, 1975, Chapter 8

[4.77]   A. E. Siegman, *Lasers*, University Science Books, Mill Valley, CA, 1986, Chapters 3, 7, and 30

[4.78]   P. W. Milonni and J. H. Eberly, *Lasers*, John Wiley, New York, 1988, Chapters 2, 10, and 13

[4.79]   L. A. Riseberg and H. W. Moos, *Phys. Rev*, vol 174, 429 (1968)

[4.80]   C. B. Layne, W. H. Lowdermilk, and M. J. Weber, "Multiphonon relaxation of rare-earth ions in oxide glasses," *Phys. Rev. B*, vol 16, no 1, 10 (1977); C. B. Layne, W. H. Lowdermilk, and M. J. Weber, "Nonradiative relaxation of rare-earth ions in silicate laser glass," *IEEE J. Quantum Electron.*, vol 11, 798 (1975)

[4.81]   M. D. Shinn, W. A. Sibley, M. G. Drexhage, and R. N. Brown, "Optical transitions of $Er^{3+}$ ions in fluorozirconate glasses," *Phys. Rev. B*, vol 27, no 11, 6635 (1983)

[4.82]   E. Desurvire and J. R. Simpson, "Amplification of spontaneous emission in erbium-doped-single-mode fibers," *IEEE J. Lightwave Technol.*, vol 5, no 5, 835 (1989)

[4.83]   Zheng Haixing and Gan Fuxi, "Investigation of glasses doped with $Er^{3+}$ ions used as laser materials," *Chinese Physics*, vol 6, no 4, 978 (1986)

[4.84]   E. Desurvire, C. R. Giles and J. R. Simpson, "Gain saturation effects in high-speed, multichannel erbium-doped fiber amplifiers at $\lambda = 1.53$ $\mu m$," *IEEE J. Lightwave Technol.*, vol 7, no 12, 2095 (1989)

[4.85]   M. J. Weber, "Fluorescence and glass lasers," *J. Non-Cryst. Solids*, vol 47, no 1, 117 (1982)

[4.86]   M. J. Weber, "Laser spectroscopy of glasses," *Ceramic Bulletin*, vol 64, no 11, 1439 (1985)

[4.87]   L. A. Riseberg, "Laser-induced fluorescence line-narrowing spectroscopy of glass: Nd," *Phys. Rev. A*, vol. 7, no 2, 671 (1973)

[4.88]   S. Zemon, G. Lambert, W. J. Miniscalco, L. J. Andrews, and B. T. Hall, "Characterization of $Er^{3+}$-doped glasses by fluorescence line narrowing," in *Proc. SPIE Conference on Fiber Laser Sources and Amplifiers*, vol 1171, 219 (1989): S. Zemon, G. Lambert, L. J. Andrews, W. J. Miniscalco, B. T. Hall, T. Wei, and R. C. Folweiler "Characterization of $Er^{3+}$-doped glasses by fluorescence line narrowing," *J. Applied Phys.*, May 1991

[4.89] D. W. Hall and M. J. Weber, "Polarized fluorescence line-narrowing measurements of laser glasses: evidence of stimulated emission cross-section anisotropy," *Appl. Phys. Lett.*, vol 42, 157 (1983)

[4.90] D. W. Hall, R. A. Haas, W. F. Krupke, and M. J. Weber, "Spectral and polarization hole burning in neodymium glass lasers," *IEEE J. Quantum Electron.* vol 19, no 11, 1704 (1983)

[4.91] R. M. MacFarlane and R. M. Shelby, "Measurement of optical dephasing in $Eu^{3+}$ and $Pr^{3+}$-doped silicate glasses by spectral hole-burning," *Opt. Comm.*, vol 45, no 1, 46 (1983)

[4.92] R. M. MacFarlane and R. M. Shelby, "Measurement of optical dephasing by spectral hole-burning in rare-earth-doped inorganic glasses," in *Coherence and Energy Transfer in Glasses*, edited by P. A. Fleury and B. Golding, Plenum, New York, 1984

[4.93] W. S. Brocklesby, B. Golding and J. R. Simpson, "Absorption fluctuations and persistent spectral hole-burning in a $Nd^{3+}$-doped glass waveguide," *Phys. Rev. Lett.*, vol 63, no 17, 1833 (1989)

[4.94] M. Sargent. M. O. Scully, and W. E. Lamb, *Laser Physics,* Addison-Wesley, London, 1974, chapter 13

[4.95] R. M. MacFarlane and R. M. Shelby, "Homogeneous line broadening of optical transitions of ions and molecules in glasses," *J. Luminescence*, vol 36, 179 (1987)

[4.96] A. Z. Genak, R. M. MacFarlane, and R. G. Brewer, *Phys. Rev. Lett.*, vol 37, 1078 (1976)

[4.97] R. M. Shelby and R. M. MacFarlane, *Opt. Comm.*, vol 27, 399 (1978)

[4.98] R. M. Shelby, "Measurement of optical homogeneous linewidths in a glass with picosecond accumulated photon echoes," *Optics Lett.*, vol 8, no 2, 88 (1983)

[4.99] M. M. Broer, B. Golding, W. H. Haemmerle, and J. R. Simpson, "Low-temperature dephasing of rare-earth ions in inorganic glasses," *Phys. Rev. B*, vol 33, no 6, 4160 (1986)

[4.100] N. W. Carlson, L. J. Rothberg, A. G. Yodh, W. R. Babbitt, and T. W. Mossberg, "Storage and time reversal of light pulses using photon echoes," *Optics Lett.*, vol 8, no 9, 483 (1983)

[4.101] M. Mitsunaga, K. I. Kubodera, and H. Kanbe, "Effects of hyperfine structures on an optical stimulated echo memory device," *Optics Lett.*, vol 11, no 5, 339 (1986)

[4.102] M. K. Kim and R. Kachru, "Storage and phase conjugation of multiple images using backward-stimulated echoes in $Pr^{3+}:LaF_3$," *Optics Lett.*, vol 12, no 8, 593 (1987)

[4.103] W. M. Yen and R. T. Brundage, "Fluorescence line narrowing in inorganic glasses: linewidth measurements," *J. Luminescence*, vol 36, 209 (1987)

[4.104] J. Hegarty, M. M. Broer, B. Golding, J. R. Simpson, and J. B. MacChesney, "Photon echoes below 1 K in a $Nd^{3+}$-doped glass fiber," *Phys. Rev. Lett.*, vol 51, no 22, 2033 (1983)

[4.105] M. M. Broer and B. Golding, "Coherent transients in active optical fibers," *J. Opt. Soc. Am. B*, vol 3, no 4, 523 (1986)

[4.106] V. L. DaSilva, Y. Silberberg, J. P. Heritage, E. W. Chase, M. A. Saifi, and M. J. Andrejco, "Femtosecond accumulated photon echo in Er-doped fibers," *Optics Lett.*, vol 16, no 17, 1340 (1991)

[4.107] J. M. Pellegrino, W. M. Yen, and M. J. Weber, "Composition dependence of $Nd^{3+}$ homogeneous linewidths in glasses," *J. Appl. Phys.*, vol 51, no 12, 6332 (1980)

[4.108] W. M. Yen, W. C. Scott, and A. L. Schalow, *Phys. Rev.* no 136, A271 (1964)

[4.109] R. Flach, D. S. Hamilton, P. M. Selzer, and W. M. Yen, *Phys. Rev.* B15, 1248 (1977)

[4.110] W. S. Brocklesby, B. Golding, and J. R. Simpson, "Absorption fluctuations and persistent spectral hole-burning in a $Nd^{3+}$-doped glass waveguide," *Phys. Rev. Lett.*, vol 63, no 17, 1833 (1989)

[4.111]   R. Silbey and K. Kassner, "Theoretical studies of homogenneous linewidths of optical transitions in glasses," *J. Luminescence*, vol 36, 283 (1987)

[4.112]   J. Hegarty and W. M. Yen, "Optical homogeneous linewidths of $Pr^{3+}$ in $BeF_2$ and $GeO_2$ glasses," *Phys. Rev. Lett.*, vol 43, no 15, 1126 (1979)

[4.113]   E. Desurvive, J. L. Zyskind, and J. R. Simpson, "Spectral gain hole-burning at 1.53 $\mu$m in erbium-doped fiber amplifiers," *IEEE Photonics Technol. Lett.*, vol 2, no 4, 246 (1990)

[4.114]   J. L. Zyskind, E. Desurvire, J. W. Sulhoff, and D. DiGiovanni, "Determination of homogeneous linewidth by spectral gain hole-burning in an erbium-doped fiber amplifier with $GeO_2$–$SiO_2$ core," *IEEE Photonics Technol. Lett.*, vol 2, no 12, 869 (1990)

[4.115]   R. I. Laming, L. Reekie, P. R. Morkel, and D. N. Payne, "Multichannel crosstalk and pump noise characterization of Er-doped fibre amplifier pumped at 980 nm," *Electron. Lett.*, vol 25, no 7, 455 (1989)

[4.116]   S. C. Guy, R. A. Minasian, S. B. Poole, and M. G. Seats, "Fluorescence line narrowing in erbium-doped fiber amplifiers," in *Proc. Optical Fiber Communications Conference, OFC'90*, paper FA2, p194, Optical Society of America, Washington DC, 1990; M. G. Seats, S. C. Guy, R. A. Minasian, and S. B. Poole, "Stress fluctuations and optical coherence of erbium in doped fibers," in *Proc. Optical Fiber Communica-ons Conference, OFC'92*, paper TuL1, p64, Optical Society of America, Washington DC, 1992

[4.117]   E. Desurvire, J. W. Sulhoff, J. L. Zyskind, and J. R. Simpson, "Study of spectral dependence of gain situation and effect of inhomogeneous broadening in erbium-doped aluminosilicate fiber amplifiers," *IEEE Photonics Technol. Lett.*, vol 2, no 9, 653 (1990)

[4.118]   C. C. Robinson and J. T. Fournier, *J. Phys. Chem. Solids*, vol 31, 895 (1970); J. T. Fournier and R. H. Bartram, "Inhomogeneous broadening of the optical spectra of $Yb^{3+}$ in phosphate glass," *J. Phys. Chem. Solids*, vol 31, 2615 (1970)

[4.119]   S. A. Brawer and M. J. Weber, *Phys. Rev. Lett*, vol 45, 460 (1980)

[4.120]   S. A. Brawer and M. J. Weber, "Theoretical study of the structure and optical properties of rare-earth doped $BeF_2$ glass," *J. Non-Cryst. Solids*, vol 38–39, 9 (1980)

[4.121]   S. A. Brawer and M. J. Weber, "Neodymium fluorescence in glass: comparison with computer simulation of glass structure," *J. Luminescence*, vol 24–25, 115 (1981)

[4.122]   M. J. Weber and S. A. Brawer, "Investigations of glass structure using fluorescence line narrowing and molecular dynamics simulations," *J. Phys. Coll. C9*, vol 43, 291 (1982)

[4.123]   T. F. Belliveau and D. J. Simkin, "On the coordination environment of rare earth ions in oxide glasses," *J. Non-Cryst. Solids*, vol 110, 127 (1989)

[4.124]   T. F. Belliveau, PhD thesis, McGill University, Montreal, Canada (1988)

[4.125]   E. Desurvire and J. R. Simpson, "Evaluation of $^4I_{15/2}$ and $^4I_{13/2}$ Stark-level energies in erbium-doped aluminosilicate glass fibers," *Optics Lett.*, vol 15, no 10, 547 (1990)

[4.126]   K. Arai, H. Namikawa, K. Kumata and T. Honda, "Aluminum or phosphorus codoping effects on the fluorescence and structural properties of neodymium-doped silica glass," *J. Applied Phys.*, vol 59, no 10, 3430 (1986)

[4.127]   W. J. Miniscalco, "Erbium-doped glasses for fiber amplifiers at 1500 nm," *IEEE J. Lightwave Technol.*, vol 9, no 2, 234 (1991); W. J. Miniscalco, "Optical and electronic properties of rare earth ions in glasses," in *Rare earth doped fiber lasers and amplifiers*, edited by M. J. F. Digonnet, Marcel Dekker, New York, 1993

[4.128]   M. Yamada, M. Shimizu, M. Horiguchi, M. Okayasu, and E. Sugita, "Gain characteristics of an $Er^{3+}$-doped multicomponent glass single-mode optical fiber," *IEEE Photonics Technol. Lett.*, vol 2, no 9, 656 (1990)

[4.129]   M. Ohashi and K. Shiraki, "$Er^{3+}$-doped multicomponent glass core fibre amplifier pumped at 1.48 $\mu$m," *Electron. Lett.*, vol 27, no 23, 2143 (1991)

[4.130]   Y. Miyajima, T. Sugawa, and K. Komukai, "20 dB gain at 155 $\mu$m wavelength in 50 cm long $Er^{3+}$-doped fluoride fibre amplifier," *Electron. Lett.*, vol 26, no 18, 1527 (1990)

[4.131]   D. Ronarc'h, M. Guibert, H. Ibrahim, M. Monerie, H. Poignant, and A. Tromeur, "30 dB optical net gain at 1.543 $\mu$m in $Er^{3+}$ doped fluoride fibre pumped around 1.48 $\mu$m," *Electron Lett.*, vol 27, no 11, 908 (1991)

[4.132]   M. Nakazawa and Y. Kimura, "Lanthanum co-doped erbium fibre amplifier," *Electron. Lett.*, vol 27, no 12, 1065 (1991)

[4.133]   B. E. A. Saleh and M. C. Teich, *Fundamentals of photonics*, John Wiley, New York, 1991

[4.134]   J. G. Edwards, "Measurement of the cross-section for stimulated emission in neodymium-doped glass from the output of a free-running oscillator," *Br. J. Phys. Phys D*, ser 2, vol 1, 449 (1968)

[4.135]   W. F. Krupke, "Induced-emission cross-sections in neodymium laser glasses," *IEEE J. Quantum Electron.* vol 10, no 4, 450 (1974)

[4.136]   J. N. Sandoe, P. H. Sarkies, and S. Parke, "Variation of $Er^{3+}$ cross section for stimulated emission with glass composition," *J. Phys. D Applied Phys.*, vol 5, 1788 (1972)

[4.137]   M. D. Shinn, W. F. Krupke, R. W. Solarz, and T. A. Kirchoff, "Spectroscopic and laser properties of $Pm^{3+}$," *IEEE J. Quantum Electron.* vol 24, no 6, 1100 (1988)

[4.138]   W. B. Fowler and D. L. Dexter, "Relation between absorption and emission probabilities in luminescent centers in ionic solids," *Phys. Rev.*, vol 128, 2154 (1962)

[4.139]   W. L. Barnes, R. I. Laming, P. R. Morkel, and E. J. Tarbox, "Absorption-emission cross-section ratio for $Er^{3+}$-doped fibers at 1.5 $\mu$m," in *Proc. Conference on Lasers and Electro-Optics, CLEO '90*, paper JTUA3, p50, Optical Society of America, Washington DC. 1990

[4.140]   P. R. Morkel and R. I. Laming, "Theoretical modeling of erbium-doped fiber with excited-state absorption," *Optics Lett.*, vol 14, no 19, 1062 (1989)

[4.141]   K. Dybdal, N. Bjerre, J. E. Pedersen, and C. C. Larsen, *Proc. Soc. Photo-Opt. Instrum. Eng.* no 1171, 209 (1989)

[4.142]   W. L. Barnes, R. I. Laming, E. J. Tarbox, and P. R. Morkel, "Absorption and emission cross-sections of $Er^{3+}$ doped silica fibers," *IEEE J. Quantum Electron.* vol 27, no 4, 1004 (1991)

[4.143]   C. R. Giles and E. Desurvire, "Modeling erbium-doped fiber amplifiers," *IEEE J. Lightwave Technol.*, vol 9, no 2, 271 (1991)

[4.144]   B. F. Aull and H. P. Jenssen, "Vibronic interactions in Nd:YAG resulting in nonreciprocity of absorption and stimulated emission cross-sections," *IEEE J. Quantum Electron.*, vol 18, no 5, 925 (1982)

[4.145]   J. A. Caird, A. J. Ramponi, and P. R. Staver, "Quantum efficiency and excited-state relaxation dynamics in neodymium-doped phosphate laser glasses," *Phys. Rev. B*, 1989

[4.146]   W. J. Miniscalco and R. S. Quimby, "General procedure for the analysis of cross-sections," *Optics Lett.*, vol 16, no 4, 258 (1991)

[4.147]  D. E. McCumber, "Theory of phonon-terminated optical masers," *Phys. Rev.*, vol 134, no 2A, 299 (1964)

[4.148]  A. Einstein, *Z. Physik*, vol 18, 121 (1917)

[4.149]  L. Landau and E. Lifchitz, *Statistical Physics,* second edition, Addison-Wesley, Reading MA, 1967, chapter 9

[4.150]  B. B. Snavely, "Flashlamp-excited organic dye lasers," *Proc. IEEE*, vol 57, no 8, 1374 (1969)

[4.151]  B. Pedersen, K. Dybdal, C. D. Hansen, A. Bjarklev, J. H. Povlsen, H. Vendeltorp-Pommer, and C. C. Larsen, "Detailed theoretical and experimental investigation of a high-gain erbium-doped fiber amplifier," *IEEE Photonics Technol. Lett.*, vol 2, no 12, 863 (1990)

[4.152]  B. Pedersen, A. Bjarklev, O. Lumholt, and J. H. Povlsen, "Detailed design analysis of erbium-doped fiber amplifiers," *IEEE Photonics Technol. Lett.*, vol 3, no 6, 548 (1991)

[4.153]  B. Pedersen, A. Bjarklev, J. H. Povlsen, K. Dybdal, and C. C. Larsen, "The design of erbium-doped fiber amplfiers," *IEEE J. Lightwave Technol.*, vol 9, no 9, 1105 (1991)

[4.154]  C. R. Giles, C. A. Burrus, D. DiGiovanni, N. K. Dutta, and G. Raybon. "Characterization of erbium-doped fibers and application to modeling 980-nm and 1480-nm pumped amplifiers," *IEEE Photonics Technol. Lett.*, vol 3, no 4, 363 (1991)

[4.155]  C. R. Giles and D. DiGiovanni, "Spectral dependence of gain and noise in erbium-doped fiber amplifiers," *IEEE Photonics Technol. Lett.*, vol 2, no 11, 797 (1990)

[4.156]  K. Dybdal, N. Bjerre, J. E. Pedersen, and C. C. Larsen, "Spectroscopic properties of Er-doped silica fibers and performs," in *Proc. SPIE Conference on Fiber Laser Sources and Amplifiers*, vol 1171, 209 (1989)

[4.157]  P. Mauer, "Amplification coefficient of neodymium-doped glass at 1.06 micron," *Applied Optics*, vol 3, 433 (1964)

[4.158]  D. W. Harper, "Assessment of neodymium optical maser glass," *Phys. Chem. Glasses*, vol 5, 11 (1964)

[4.159]  C. G. Young and J. W. Kantorski, "Saturation operation and gain coefficient of a neodymium-glass amplifier," *Applied Optics*, vol 4, 1675 (1965)

[4.160]  J. G. Edwards, "An accurate carbon core calorimeter for pulsed lasers," *J. Sci. Instrum.*, vol 44, 835 (1967)

[4.161]  P. H. Sarkies, J. N. Sandoe, and S. Parke, "Variation of $Nd^{3+}$ cross section for stimulated emission with glass composition," *J. Phys. D Appl. Phys.*, vol 4, 1642 (1971)

[4.162]  C. G. Atkins, J. F. Massicott, J. R. Armitage, R. Wyatt, B. J. Ainslie, and S. P. Craig-Ryan, "High-gain, broad spectral bandwidth erbium-doped fibre amplifier pumped near 1.5 $\mu$m," *Electron. Lett.*, vol 25, no 14, 910 (1989)

[4.163]  E. Desurvire, "Analysis of erbium-doped fiber amplifiers pumped in the $^4I_{15/2}$–$^4I_{13/2}$ band," *IEEE Photonics Technol. Lett.*, vol 1, no 10, 293 (1989)

[4.164]  M. P. Singh, D. W. Oblas, J. O. Reese, W. J. Miniscalco, and T. Wei, "Measurement of spectral dependence of absorption cross-section for erbium-doped single-mode optical fiber," in *Technical Digest of Symposium on Optical Fiber Measurements, 1990*, p93, NIST special publication no 792, 1990

[4.165]  S. P. Craig-Ryan, B. J. Ainslie, and C. A. Millar, "Fabrication of long lengths of low excess loss erbium-doped optical fibre," *Electron. Lett.*, vol 26, no 3, 185 (1990)

[4.166]  D. L. Williams, S. T. Davey, D. M. Spirit, and B. J. Ainslie, "Transmission over 10km or erbium doped fibre with ultralow signal power excursion," *Electron. Lett.*, vol 26, no 18, 1517 (1990)

[4.167]  M. Nakazawa, Y. Kimura, and K. Suzuki, "Gain-distribution measurements along an ultralong erbium-doped fiber amplifier using optical-time-domain reflectometry," *Optics Lett.*, vol 15, no 21, 1200 (1990)

[4.168] J. D. Evankow and R. M. Jopson, "Nondestructive measurement of length dependence of gain characteristics in fiber amplifiers," *IEEE Photonics Technol. Lett.*, vol 3, no 11, 993 (1991)

[4.169] L. W. Casperson and A. Yariv, "Spectral narrowing in high-gain lasers," *IEEE J. Quantum Electron.*, vol 8, no 2, 80 (1972)

[4.170] L. W. Casperson, "Threshold characteristics of mirrorless lasers," *J. Applied Phys.*, vol 48, no 1, 256 (1977)

[4.171] P. Avouris, A. Campion, and M. A. El-Syed, "Variations in homogeneous fluorescence linewidth and electron-phonon coupling within an inhomogeneous spectral profile," *J. Chem. Phys.*, vol 67, no 7, 3397 (1977)

[4.172] J. R. Morgan, E. P. Chock, W. D. Hopewell, M. A. El-Sayed, and R. Orbach, "Origin of homogeneous and inhomogeneous line widths of the $^5D_0$-$^7F_0$ transition of $Eu^{3+}$ in amorphous solids," *J. Chem. Phys.*, vol 85, 747 (1981)

[4.173] A. M. Brancion, B. Jacquier, J. C. Gacon, C. LeSergent and J. F. Marcerou, "Inhomogeneous line broadening of optical transitions in $Er^{3+}$ and $Nd^{3+}$ doped preforms and fibers," in *Proc. SPIE Conference on Fiber Laser Sources and Amplifiers II*, vol 1373, 9 (1990)

[4.174] E. Desurvire, J. R. Simpson, and P. C. Becker, "High-gain erbium-doped traveling-wave fiber amplifier," *Optics Lett.*, vol 12, no 11, 888 (1987)

[4.175] C. A. Millar, T. J. Whitley, and S. C. Fleming, "Thermal properties of an erbium-doped fibre amplifier," *IEEE Proc. J.*, vol 137, no 3, 155 (1990)

[4.176] M. Shimizu, M. Yamada, M. Horiguchi, and E. Sugita, "Gain characteristics of erbium-doped single-mode fiber amplifiers operated at liquid-nigrogen temperature," *Applied Phys. Lett.*, vol 55, no 23, 2273 (1990)

[4.177] N. Kagi, A. Oyobe, and K. Nakamura, "Temperature dependence of the gain in erbium-doped fibers," *IEEE J. Lightwave Technol.*, vol 9, no 2, 261 (1991)

[4.178] M. Suyama, R. I. Laming, and D. N. Payne, "Temperature dependent gain and noise characteristics of a 1480 nm-pumped erbium-doped fibre amplifier," *Electron. Lett.*, vol 26, no 21, 1756 (1990)

[4.179] M. Yamada, M. Shimizu, M. Okayasu, and M. Horiguchi, "Temperature insensitive $Er^{3+}$ doped optical fibre amplifiers," *Electron. Lett.*, vol 26, no 20, 1649 (1990)

[4.180] L. Jeunhomme, *Single-mode Fiber Optics, Principles and Applications*, second edition, Marcel Dekker, New York, 1990

[4.181] A. W. Snyder and J. D. Love, *Optical Waveguide Theory*, Chapman and Hall, New York, 1983

[4.182] N. Kagi, A. Oyobe, and K. Nakamura, "Gain characteristics of $Er^{3+}$-doped fiber with a quasi-confined structure," *IEEE J. Lightwave Technol.*, vol 8, no 9, 1319 (1990)

[4.183] T. Kashiwada, M. Shigematsu, T. Kougo, H. Kanamori, and M. Nishimura, "Erbium-doped fiber amplifier pumped at 1.48 $\mu$m with extremely high efficiency," *IEEE Photonics Technol. Lett.*, vol 3, no 8, 721 (1991)

[4.184] B. J. Ainslie, S. P. Craig, S. T. Davey, and B. Wakefield, "The fabrication, assessment and optical properties of high-concentration $Nd^{3+}$ and $Er^{3+}$-doped silica fibres," *Material Lett.*, vol 6, no 5–6, 139 (1988)

[4.185] J. F. Marcerou, B. Jacquier, A. M. Brancion, J. C. Gacon, H. Fevrier, and J. Auge, "Rare earth concentration and localization in $Nd^{3+}$- and $Er^{3+}$-doped fibre amplifiers," *J. Luminescence*, vol 45, 108 (1990)

[4.186] J. F. Marcerou, H. Fevrier, J. Auge, J. Ramos, and A. Dursin, "Feasibility demonstration of low pump power operation for 1.48 $\mu$m diode-pumped erbium-doped fibre amplifier module," *Electron. Lett.*, vol 26, no 15, 1102 (1990)

[4.187] M. Shimizu, M. Yamada, M. Horiguchi, and E. Sugita, "Concentration effects on optical amplification characteristics of erbium-doped silica single-mode fibers," *IEEE Photonics Technol. Lett.*, vol 2, no 1, 43 (1990)

[4.188] D. W. Oblas, F. Pink, M. P. Singh, J. Connolly, D. Dugger, and T. Wei, in *Digest of Materials Research Society Annual Meeting*, paper J6.3, Materials Research Society, Pittsburgh, PA, 1992

[4.189] J. E. Townsend, S. B. Poole, and D. N. Payne, "Solution-doping technique for fabrication of rare-earth-doped optical fibres," *Electron. Lett.*, vol 23, no 7, 329 (1987)

[4.190] B. J. Ainslie, S. P. Craig-Ryan, S. T. Davey, J. R. Armitage, C. G. Atkins, and R. Wyatt, "Optical analysis of erbium-doped fibres for efficient lasers and amplifiers," in *Proc. Seventh International Conference on Integrated Optics and Optical Fiber Communications, IOOC '89*, paper 20A3-2, p22, Kobe, Japan, 1989

[4.191] M. Artiglia, G. Coppa, P. DiVita, M. Potenza, and A. Sharma, "Mode field diameter measurements in single-mode optical fibres," *IEEE J. Lightwave Technol.*, vol 7, no 8, 1139 (1989)

[4.192] A. A. M. Saleh, R. M. Jopson, J. D. Evankow, and J. Aspell, "Modeling of gain in erbium-doped fiber amplifiers," *IEEE Photonics Technol. Lett.*, vol 2, no 10, 714 (1990)

[4.193] E. Desurvire, M. Zirngibl, H. M. Presby, and D. DiGiovanni, "Characterization and modeling of amplified spontaneous emission in erbium-doped fiber amplifiers," *IEEE Photonics Technol. Lett.*, vol 3, no 2, 127 (1991)

[4.194] M. Peroni and M. Tamburrini, "Gain in erbium-doped fiber amplifiers: a simple analytical solution for the rate equations," *Optics Lett.*, vol 15, no 15, 842 (1990)

[4.195] A. M. Vengsargar, D. DiGiovanni, W. A. Reed, K. W. Quoi, and K. L. Walker, "Measurement of erbium confinement in optical fibers: a differential mode-launching technique," *Optics Lett.*, vol 17, no 18, 1277 (1992)

[4.196] J. R. Armitage, C. G. Atkins, R. Wyatt, B. J. Ainslie, and S. P. Craig, "Spectroscopic studies of $Er^{3+}$-doped single-mode silica fiber," in *Proc. Conference on Lasers and Electro-Optics*, CLEO'88, paper TUM27, Optical Society of America, Washington DC, 1988; C. G. Atkins, J. R. Armitage, R. Wyatt, B. J. Ainslie, and S. P. Craig-Ryan, "Pump excited-state absorption in $Er^{3+}$-doped optical fibers," *Opt. Comm.*, vol vol 73, no 3, 217 (1989)

[4.197] R. I. Laming, S. B. Poole, and E. J. Tarbox, "Pump excited-state absorption in erbium-doped fibers," *Optics Lett.*, vol 13, no 12, 1084 (1988)

[4.198] S. Zemon, G. Lambert, W. J. Miniscalco, R. W. Davies, B. T. Hall, R. C. Folweiler, T. Wei, L. J. Andrews, and M. P. Singh, "Excited state cross sections for Er-doped glasses," in *Proc. SPIE Conference on Fiber Laser Sources and Amplifiers*, vol 1373, 21 (1990)

[4.199] M. C. Farries, P. R. Morkel, R. I. Laming, T. A. Birks, D. N. Payne, and E. J. Tarbox, "Operation of erbium-doped fiber amplifiers and lasers with frequency-doubled Nd:YAG lasers," *IEEE J. Lightwave Technol.*, vol 7, no 10, 1473 (1989)

[4.200] K. Kannan, S. Friesken, and P. S. Atherton. "Characteristics of an erbium-doped fiber amplifier pumped by a frequency doubled diode-pumped Nd:YAG laser," *IEEE Photonics Technol. Lett.*, vol 3, no 2, 124 (1991)

[4.201] M. G. Seats, P. A. Krug, G. R. Atkins, S. C. Guy, and S. B. Poole, "Non-linear excited state absorption in $Er^{3+}$-doped fibre with high power 980 nm pumping," in *Proc. Topical Meeting on Optical Amplifiers and applications, 1991*, paper WD2, p48, Optical Society of America, Washington DC, 1991

[4.202] R. S. Quimby, W. J. Miniscalco and B. T. Thomson, "Excited state absorption at 980nm in erbium-doped silica glass," in *Proc. Topical Meeting on Optical Amplifiers and Applications, 1992*, paper WE3, p67, Optical Society of America, Washington DC, 1992

[4.203] M. C. Farries, "Excited-state absorption and gain in erbium-fiber amplifiers between 1.05 and 1.35 $\mu$m," *IEEE Photonics Technol. Lett.*, vol 3, no 7, 619 (1991)

[4.204] P. Blixt, J. Nilsson, J. Babonas, and B. Jaskorzynska, "Excited-state absorption at 1.5 $\mu$m in $Er^{3+}$-doped fiber amplifiers," in *Proc. Topical Meeting on Optical Amplifiers and Applications, 1992*, paper WE2, p63, Optical Society of America, Washington DC, 1992

[4.205] A. Bjarklev, S. L. Hansen, and J. H. Povlsen, "Large-signal modeling of an erbium-doped fiber amplifier," in *Proc. SPIE Conference on Fiber Laser Sources and Amplifiers*, vol 1171, 118 (1989)

[4.206] M. Monerie, T. Georges, P. L. Francois, J. Y. Allain, and D. Neveux, "Ground-state and excited-state absorption in rare-earth doped optical fibres," *Electron. Lett.*, vol 26, no 5, 320 (1990)

[4.207] S. Zemon, B. Pedersen, G. Lambert, W. J. Miniscalco, L. J. Andrews, R. W. Davies, and T. Wei, "Excited-state absorption cross-sections in the 800 nm band for Er-doped Al/P silica fibers: measurements and amplifier modeling," *IEEE Photonics Technol. Lett.*, vol 3, no 7, 621 (1991)

[4.208] B. Pedersen, S. Zemon, and W. J. Miniscalco, "Erbium-doped fibres pumped in 800nm band," *Electron. Lett.*, vol 27, no 14, 1295 (1991)

[4.209] S. P. Bastien and H. R. D. Sunak, "Comparison of forward and bidirectional pumping at 805 and 819 nm in erbium-doped silica fiber amplifiers," *IEEE Photonics Technol. Lett.*, vol 3, no 5, 456 (1991)

[4.210] S. P. Bastien and H. R. D. Sunak, "Pump wavelength independent gain with bidirectionally pumped aluminosilicate erbium-doped fiber amplifiers in the 800 nm pump band," *IEEE Photonics Technol. Lett.*, vol 3, no 9, 825 (1991)

[4.211] S. P. Bastien and H. R. D. Sunak, "Analysis of the performance expected in fluorophosphate erbium-doped fiber amplifiers with the 800 nm pump band," *IEEE Photonics Technol. Lett.*, vol 3, no 12, 1088 (1991)

[4.212] J. G. Edwards and J. N. Sandoe, "A theoretical study of the Nd:Yb:Er glass laser," *J. Phys. D, Appl. Phys.*, vol 7, 1078 (1974)

[4.213] R. Reisfeld and Y. Eckstein, "Energy transfer between $Tm^{3+}$ and $Er^{3+}$ in borate and phosphate glasses," *J. Non-Cryst. Solids*, vol 11, 261 (1973)

[4.214] B. Moine, A. Brenier, and C. Pedrini, "Fluorescence dynamics of $Er^{3+}$ and ions and energy transfer in some fluoride glasses," *IEEE J. Quantum Electron.*, vol 25, no 1, 88 (1989)

[4.215] N. Bloembergen, "Solid-state infrared quantum counters," *Phys. Rev. Lett.*, vol 2, 84 (1959)

[4.216] L. F. Johnson, H. J. Guggenheim, T. C. Rich, and F. W. Ostermayer, "Infrared-to-visible conversion by rare-earth ions in crystals," *J. Applied Phys.*, vol 43, 1125 (1972)

[4.217] F. E. Auzel, "Materials and devices using double-pumped phosphors with energy transfer," *Proc. IEEE*, vol 61, no 6, 758 (1973)

[4.218] J. P. Van der Ziel, L. G. Van Uitert, W. H. Grodkiewicz, and R. M. Mikulyak, "1.5 $\mu$m infrared excitation of visible luminescence in $Y_{1-x}Er_xF_3$ and $Y_{1-x-y}Er_x Tm_yF_3$ via resonant energy transfer," *J. Applied Phys.*, vol 60, no 12, 4262 (1986)

[4.219] J. F. Porter, "Energy transfer processes in $LaCl_3$: $Ho^{3+}$: three ion process," *Bull. Am. Phys. Soc.*, vol 13, 102 (1968)

[4.220]  P. P. Feofilov and V. V. Ovsyankin, "Cooperative luminescence od solids," *Applied Optics*, vol 6, 1828 (1967)

[4.221]  E. Nakazawa and S. Shionoya, "Cooperative luminescence in YbPO$_4$," *Phys. Rev. Lett.*, vol 25, 1710 (1970)

[4.222]  P. P. Feofilov and A. K. Trotimov, "Raman luminescence of Yb$_2$O$_3$-Gd," *Opt. Spectrosc.*, no 27, 291 (1967)

[4.223]  D. L. Dexter, "A theory of sensitized luminescence in solids," *J. Chem. Phys.*, vol 21, no 5, 836 (1953)

[4.224]  J. D. Axe and P. F. Weller, "Fluorescence and energy transfer in Y$_2$O$_3$:Eu," *J. Chem. Phys.*, vol 40, 3066 (1964)

[4.225]  M. Inokuti and F. Hirayama, "Influence of energy transfer by the exchange mechanism on donor luminescence," *J. Chem. Phys.*, vol 43, no 6, 1978 (1965)

[4.226]  E. Nakazawa and S. Shionoya, "Energy transfer between trivalent rare-earth ions in inorganic solids," *J. Chem. Phys.*, vol 47, no 9, 3211 (1967)

[4.227]  R. K. Watts and H. J. Richter, "Diffusion and transfer of optical excitation in YF$_3$:Yb, Ho," *Phys. Rev. B6*, 1584 (1972)

[4.228]  N. S. Yamada, S. Shionoya, and T. Kushida, *J. Phys. Soc. Japan*, vol 32, 1577 (1972)

[4.229]  J. Chrysochoos, "Nature of the interaction forces associated with the concentration fluorescence quenching of Nd in silicate glasses," *J. Chem. Phys.*, vol 61, no 11, 4596 (1974)

[4.230]  F. Auzel, "Application des transfers d'énergie résonnants à l'effet laser de verre dopé avec Er$^{3+}$, *Ann. Telecomm.* vol 24, no 9–10 (1969)

[4.231]  B. E. Warren and A. G. Pincus, "Atomic consideration of immiscibility in glass systems," *J. Am. Ceram. Soc.*, vol 23, 301 (1940)

[4.232]  F. P. Glasser, I. Warshaw, and R. Roy., "Liquid immiscibility in silicate systems," *Phys. Chem. Glass*, vol 1, 39 (1960)

[4.233]  P. Blixt, J. Nilsson, T. Carlnas, and B. Jaskorzynska, "Amplification reduction in fiber amplifiers due to up-conversion at Er$^{3+}$-concentrations below 1000 ppm," in *Proc. Topical Meeting on Optical Amplifiers and Applications,* paper WD3, p 52, Optical Society of America, Washington DC, 1992

[4.234]  P. Blixt, J. Nilsson, T. Carlnas, and B. Jaskorzynska, "Concentration-dependent upconversion in Er$^{3+}$-doped fibers: experiments and modeling," *IEEE Photonics Technol. Lett.*, vol 3, no 11, 996 (1991)

[4.235]  W. Q. Shi, M. Bass, and M. Birnbaum, "Effects of energy transfer among ions on the fluorescence decay and lasing properties of heavily doped Er:Y$_3$Al$_5$O$_{12}$," *J. Opt. Soc. Am. B*, vol 7, 1456 (1990)

[4.236]  T. Georges, E. Delevaque, M. Monerie, P. Lamouler, and J. F. Bayon, "Pair-induced quenching in erbium-doped silicate fibers," in *Proc. Topical Meeting on Optical Amplifiers and Applications 1992*, paper WE4, p71, Optical Society of America, Washington DC, 1992; E. Delevaque, T. Georges, M. Monerie, P. Lamouler, and J. F. Bayon, "Modeling of pair-induced quenching in erbium-doped silicate fibers," *IEEE Photonics Technol. Lett.*, vol 5, no 1, 73 (1993)

[4.237]  E. Snitzer, "Phosphate glass Er$^{3+}$ laser," *IEEE J. Quantum Electron.*, May 1968, p360

[4.238]  S. G. Lunter, A. G. Murzin, M. N. Tolstoi, Y. K. Fedorov, and V. A. Fromzel, "Energy parameters utilizing erbium glasses sensitized with ytterbium and chromm," *Sov. J. Quantum Electron.*, vol 145, no 1, 66 (1984)

[4.239]  V. P. Gapontsev, M. E. Zhabotinskii, A. A. Izyneev, V. B. Kravchenko, and Y. P. Rudnitskii, "Effective 1.059→1.54 $\mu$m stimulated emission conversion," *JETP Lett.*, vol 18, 251 (1973)

[4.240]  V. P. Gapontsev, S. M. Matitsin, A. A. Isyneev, and V. B. Kravchenko, "Erbium glass lasers and their applications," *Optics and Laser Technol.*, August 1982, p189

[4.241]  D. C. Hanna, A. Kazer, and D. P. Shepherd, "A 1.54 $\mu$m Er glass laser pumped by a 1.064 mm Nd:YAG laser," *Opt. Comm.*, vol 63, no 6, 417 (1987)

[4.242]  M. E. Fermann, D. C. Hanna, D. P. Shepherd, P. J. Suni, and J. E. Townsend, "Efficient operation of an Yb-sensitized Er fibre laser at 1.56 $\mu$m," *Electron. Lett.*, vol 24, no 18, 1135 (1988)

[4.243]  G. T. Maker and A. I. Ferguson, "1.56 $\mu$m Yb-sensitized Er fiber laser pumped by diode-pumped Nd:YAG and ND:YLF lasers," *Electron. Lett.*, vol 24, no 18, 1160 (1988)

[4.244]  D. C. Hanna, R. M. Percival, I. R. Perry, R. G. Smart, and A. C. Tropper, "Efficient operation of an Yb-sensitized Er fibre laser pumped in 0.8 $\mu$m region," *Electron. Lett.*, vol 24, no 17, 1068 (1988)

[4.245]  A. G. Murzin, D. S. Prilezhaev, and V. A. Fromzel, "Some features of laser excitation of Ytterbium-erbium glasses," *Sov. J. Quantum Electron.*, vol 15, no 3, 349 (1985)

[4.246]  A. G. Murzin and V. A. Fromzel, "Maximum gains of laser-pumped glasses activated with $Yb^{3+}$ and $Er^{3+}$ ions," *Sov. J. Quantum Electron.*, vol 11, no 3, 304 (1981)

[4.247]  S. G. Grubb, R. S. Cannon, T. W. Windhorn, S. W. Vendetta, P. A. Leilababy, D. W. Anton, K. L. Sweeney, W. L. Barnes, E. R. Taylor, and J. E. Townsend, "High-power sensitized erbium optical fibre amplifiers," in *Proc. Conference on Optical Fiber Communications, OFC'91*, paper PD7, Optical Society of America, Washington DC, 1991

[4.248]  V. P. Gapontsev, S. M. Matitsin and A. A. Isyneev, "Channels of energy losses in erbium laser glasses in the stimulated emission process," *Opt. Comm.*, vol 46, 226 (1983)

[4.249]  J. E. Townsend, W. L. Barnes, and K. P. Jedrezejewski, "$Yb^{3+}$ sensitized $Er^{3+}$-doped silica optical fibre with ultrahigh transfer efficiency and gain," *Electron. Lett.* vol 27, no 21, 1958 (1991)

[4.250]  S. G. Grubb, W. H. Humer, R. S. Cannon, S. W. Vendetta, K. L. Sweeney, P. A. Leilabady, M. R. Keur, J. G. Kwasegroch, T. C. Munks, and D. W. Anthon, "+24.6 dBm output power Er/Yb codoped optical amplifier pumped by diode-pumped Nd:YLF laser," *Electron Lett.* vol 28, no 13, 1275 (1992)

[4.251]  S. G. Grubb, W. F. Humer, R. S. Cannon, T. H. Windhorn, S. W. Vendetta, K. L. Sweeney, P. A. Leilabady, W. L. Barnes, K. P. Jedrzejewski, and J. E. Townsend, "+21 dBm erbium power amplifier pumped by a diode-pumped Nd:YAG laser," *IEEE Photonics technol. Lett.*, vol 4, no 6, 553 (1992)

[4.252]  M. Born and E. Wolf, *Principles of optics*, Pergamon Press, New York, 1980 chapter 2

[4.253]  H. Rawson, *Properties and Applications of Glass,* Elsevier Scientific Publishing, New York, 1980

[4.254]  I. Fanderlik, *Optical Properties of Glass,* Elsevier Scientific Publishing, New York, 1983

[4.255]  J. W. Fleming, "Material dispersion in lightguide glasses," *Electron. Lett.*, vol 14, 326 (1978)

[4.256]  E. Desurvire, "Study of the complex atomic susceptibility of erbium-doped fiber amplifiers," *IEEE J. Lightwave Technol.*, vol 8, no 10, 1517 (1990)

[4.257] A. R. Chraplyvy, D. Marcuse, and P. S. Henry, "Carrier-induced phase noise in angle-modulated optical fiber systems," *IEEE J. Lightwave Technol.*, vol 2, no 1, 6 (1984)

[4.258] F. Matera, M. Romagnoli, M. Settembre, and M. Tamburrini, "Evaluation of chromatic dispersion in erbium-doped fibre amplifiers," *Electron. Lett.*, vol 27, no 20, 1867 (1991)

[4.259] A. V. Belov, A. A. Voloshinskaya, V. A. Semenov, and A. V. Chicolini, "Anomalous dispersion at 1.53 $\mu$m in erbium-doped single-mode fibres," *Sov. Lightwave Comm.*, vol 1, 255 (1991)

[4.260] E. Desurvire, J. L. Zyskind, and C. R. Giles, "Design optimization for efficient erbium-doped fiber amplifiers," *IEEE J. Lightwave Technol.*, vol 8, no 11, 1730 (1990)

[4.261] K. Takada, T. Kitagawa, K. Hattori, M. Yamada, M. Horiguchi, and R. K. Hickernell, "Direct dispersion measurement of highly doped optical amplifiers using a low coherence reflectometer coupled with dispersive Fourier spectroscopy," *Electron. Lett.*, vol 28, no 20, 1889 (1992)

[4.262] S. C. Fleming and T. J. Whitley, "Measurement of pump-induced refractive index change in erbium doped fibre amplifier," *Electron. Lett.*, vol 27, no 21, 1959 (1991)

[4.263] M. Romagnoli, F. S. Locati, F. Matera, M. Settembre, M. Tamburrini, and S. Wabnitz, "Role of pump-induced dispersion on femtosecond soliton amplification in erbium-doped fibers," *Optics Lett.*, vol 17, no 13, 923 (1992); erratum, *Optics Lett.*, vol 17, no 23, 1721 (1992)

[4.264] R. A. Betts, T. Tjugiarto, Y. L. Xue, and P. L. Chu, "Nonlinear refractive index in erbium-doped optical fiber: theory and experiment," *IEEE J. Quantum Electron.*, vol 27, no 4, 908 (1991)

[4.265] G. P. Agrawal, *Nonlinear Fiber Optics*, Academic Press, New York, 1989

[4.266] I. N. Duling III, "Dispersion in rare-earth-doped fibers," *Optics Lett.*, vol 16, no 24, 1947 (1991)

[4.267] B. Deutsch and T. Pfeiffer, "Chromatic dispersion of erbium-doped silica fibres," *Electron. Lett.*, vol 28, no 3, 303 (1992)

[4.268] V. P. Lebedev and A. K. Przhevuskii, "Polarized luminescence of rare-earth activated glasses," *Sov. Phys. Solid State*, vol 19, no 8, 1389 (1977)

[4.269] P. P. Feofilov, *The Physical Basis of Polarized Emission*, Consultants Bureau, New York, 1961

[4.270] J. T. Lin, P. R. Morkel, and D. N. Payne, "Polarization effects in fibre lasers," in *Proc. European Conference on Optical Communications, ECOC '87*, vol I, 109 (1987)

[4.271] J. T. Lin, W. A. Gampling, and D. N. Payne, "Modeling of polarization effects in fiber lasers," in *Proc. Conference on Lasers and Electro-Optics, CLEO '89*, paper TUJ25, p90, Optical Society of America, Washington DC, 1989

[4.272] J. T Lin and W. A. Gambling, "Polarization effects in fibre lasers: phenomena, theory and applications," in *Proc. SPIE Conference on Fiber Laser Sources And Amplifiers II*, vol 1373, 42 (1990)

[4.273] D. W. Hall, R. A. Hass, W. F. Krupke, and M. J. Weber, "Spectral and Polarization hole burning in neodymium glass lasers," *IEEE J. Quantum Electronics*, vol 19, no 11, 1704 (1983)

[4.274] D. W. Hall and M. J. Weber, "Polarized fluorescence line-narrowing measurements of Nd laser glasses: evidence of stimulated emission cross-section anisotropy," *Applied Phys. Lett.*, vol 42, 157 (1983)

## CHAPTER 5

[5.1] T. J. Whitley, "Laser diode pumped operation of $Er^{3+}$-doped fibre amplifier," *Electron. Lett.*, vol 24, no 25, 1537 (1988)

[5.2] M. Shimizu, M. Yamada, M. Horiguchi, T. Takeshita, and M. Oyasu, "Erbium-doped fibre amplifiers with an extremely high gain coefficient of 11.0 dB/mW," *Electron. Lett.*, vol 26, no 20, 1641 (1990)

[5.3] I. Mito and K. Endo, "1.48 $\mu$m and 0.98 $\mu$m high-power laser diodes for erbium-doped fiber amplifiers," in *Proc. Topical Meeting of Optical Amplifiers and Applications, 1991*, paper WC1, p22, Optical Society of America, Washington DC, 1991

[5.4] K. Suzuki, Y. Kimura and M. Nakazawa, "High gain $Er^{3+}$-doped fibre amplifier pumped by 820nm GaAlAs laser diodes," *Electron. Lett.*, vol 26, no 13, 948 (1990)

[5.5] M. Horiguchi, M. Shimizu, M. Yamada, K. Yoshino, and H. Hanafusa, "Highly efficient optical fibre amplifier pumped by a 0.8 $\mu$m band laser diode," *Electron. Lett.*, vol 26, no 21, 1758 (1990)

[5.6] S. G. Grubb, W. F. Humer, R. S. Cannon, T. H. Windhorn, S. W. Vendetta, K. L. Sweeney, P. A. Leilabady, W. L. Barnes, K. P. Jedrezejewski, and J. E. Townsend, " +21 dBm erbium power amplifier pumped by a diode-pumped Nd:YAG laser," *IEEE Photonics Technol. Lett.*, vol 4, no 6, 553 (1992)

[5.7] S. G. Grubb, W. H. Humer, R. S. Cannon, S. W. Vendetta, K. L. Sweeney, P. A. Leilabady, M. R. Keur, J. G. Kwasegroch, T. C. Munks, and D. W. Anthon, " +24.6 dBm output power Er/Yb codoped optical amplifier pumped by diode-pumped Nd:YLF laser," *Electron. Lett.*, vol 28, no 13, 1275 (1992)

[5.8] J. D. Minelli, R. I. Laming, J. E. Townsend, W. L. Barnes, E. R. Taylor, K. P. Jedrezejewski, and D. N. Payne, "High gain power amplifier tandem-pumped by a 3 W multistripe diode," in *Proc. Conference on Optical Fiber Communications, OFC'92*, paper TuG2, p32, Optical Society of America, Washington DC, 1992

[5.9] K. Kannan, S. Friesken, and P. S. Atherton, "Characteristics of an erbium-doped fiber amplifier pumped by a frequency doubled diode-pumped Nd:YAG laser," *IEEE Photonics Techno. Lett.*, vol 3, no 2, 124 (1991)

[5.10] H. B. Serreze, Y. C. Chen, R. G. Waters, and C. M. Harding, "Very-low threshold, high-power GaInP/AlGaInP visible laser diodes," in *Proc. Conference on Lasers and Electro-Optics, CLEO'91*, paper CPDP1, p571, Optical Society of America, Washington DC, 1991

[5.11] H. Horikawa, S. Oshiba, A. Matoba, and Y. Kawai, "V-groove inner-stripe laser diodes on a p-type substrate operating over 100 mW at 1.5 $\mu$m wavelength," *Applied Phys. Lett.*, vol 50, no 7, 374 (1987)

[5.12] K. Yamada, S. Oshiba, T. Kunji, Y. Ogawa, and T. Kamijoh, "More than 3000 hours stable cw operation of 1.48 $\mu$m LD for EDFA pumping source," in *Proc. Topical Meeting of Optical Amplifiers and Applications, 1990*, paper WA3, p214, Optical Society of America, Washington DC, 1990

[5.13] I. Mito, H. Yamazaki, H. Yamada, T. Sazaki, S. Takano, Y. Aoki and M. Kitamura, "170 mW output power cw operation in 1.48–1.51 $\mu$m InGaAs MQW DCPBH-LD," in *Proc. Conference on Integrated Optics and Optical Fiber Communications, IOOC'89*, paper 20PDB-12, Kobe, Japan, 1989

[5.14] H. Kamei, M. Yoshimura, H. Kobayashi, N. Tatoh, and H. Hayashi, "High-power operation of 1.48 $\mu$m GaInAsP/GaInAsP strained-layer multiple quantum well lasers," in *Proc. Topical Meeting of Optical Amplifiers and Applications, 1991*, paper WC3, p30, Optical Society of America, Washington DC, 1991

[5.15] H. Asano, S. Takano, M. Kawaradani, M. Kitamura, and I. Mito, "1.48 $\mu$m high-power InGaAs/InGaAsP MQW LD's for Er-doped fiber amplifiers," *IEEE Photonics Technol. Lett.*, vol 3, no 5, 415 (1991)

[5.16] S. Uehara, M. Horiguchi, T. Takeshita, M. Okayasu, M. Yamada, M. Shimizu, U. Kogure, and K. Oe, "0.98 $\mu$m strained-quantum-well lasers for erbium-doped fiber optical amplifiers," in *Proc. Conference on Integrated Optics and Optical Fiber Communications, IOOC'89*, paper 20PDB-11, Kobe, Japan, 1989

[5.17] M. Okayasu, M. Fukuda, T. Takeshita, and S. Uehara, "Stable operation (over 5000h) of high-power 0.98um InGaAs–GaAs strained quantum well ridge waveguide lasers for pumping $Er^{3+}$-doped fiber amplifiers," *IEEE Photonics Technol. Lett.*, vol 2, no 10, 689 (1990)

[5.18] A. Larsson, S. Forouhar, J. Cody, R. J. Lang, and P. A. Andrekson, "A 980 nm pseudomorphic single quantum well laser for pumping erbium-doped optical fiber amplifiers," *IEEE Photonics Technol. Lett.*, vol 2, no 8, 540 (1990)

[5.19] T. Takeshita, M. Okayasu and S. Uehara, "High-power operation in 0.98 $\mu$m strained-layer InGaAs-GaAs single-quantum well ridge waveguide lasers," *IEEE Photonics Technol. Lett.*, vol 2, no 12, 849 (1990)

[5.20] H. A. Zarem, J. Paslaski, M. Mittlestein, J. Ungar, and I. Ury, "High power fiber coupled strained layer InGaAs lasers emitting at 980 nm," in *Proc. Topical Meeting of Optical Amplifiers and Applications, 1991*, paper PdP4, Optical Society of America, Washington DC, 1991

[5.21] S. G. Grubb, R. S. Cannon, T. W. Windhorn, S. W. Vendetta, P. A. Leilabady, D. W. Anthon, K. L. Sweeney, W. L. Barnes, E. R. Taylor, and J. E. Townsend, "High-power sensitized erbium optical fiber amplifier," in *Proc. Conference on Optical Fiber Communications, OFC'91*, paper PD7, Optical Society of America, Washington DC, 1991

[5.22] E. Snitzer, H. Po, F. Hakimi, R. Tumminelli, and B. C. McCollum, "Double-clad, offset core Nd fiber laser," in *Proc. Conference on Optical Fiber Sensors, OFS'88*, paper PD5, Optical Society of America, Washington DC, 1988

[5.23] H. Po, E. Snitzer, R. Tumminelli, L. Zenteno, F. Hakimi, N. M. Cho, and T. Haw, "Double clad high brightness Nd fiber laser pumped by GaAlAs phased array," in *Proc. Conference on Optical Fiber Communications, OFC'89*, paper PD7, Optical Society of America, Washington DC, 1989

[5.24] J. D. Minelli, W. L. Barnes, R. I. Laming, P. R. Morkel, J. E. Townsend, S. G. Grubb, and D. N. Payne, "$Er^{3+}/Yb^{3+}$ co-doped power amplifier pumped by a 1 W diode array," in *Proc. Topical Meeting of Optical Amplifiers and Applications, 1992*, paper Pd2, Optical Society of America, Washington DC, 1992

[5.25] H. M. Presby, "Near 100% efficient fiber microlenses," in *Proc. Conference on Optical Fiber Communications, OFC'92*, paper PD24, Optical Society of America, Washington DC, 1992; H. M. Presby and C. A. Edwards, "Near 100% efficient fibre microlenses," *Electron. Lett.*, vol 28, no 6, 582 (1992)

[5.26] H. Kuwahara, M. Sazaki, and N. Tokoyo, "Efficient coupling from semiconductor lasers into single-mode fibers with tapered hemispherical ends," *Applied Optics*, vol 19, no 15, 2578 (1980)

[5.27] H. Ghafoori-Shiraz and T. Asano, "Microlens for coupling a semiconductor laser to a single-mode fiber," *Optics Lett.*, vol 11, no 8, 537 (1986)

[5.28] H. M. Presby, A. F. Benner, and C. A. Edwards, "Laser micromachining of efficient fiber microlenses," *Applied Optics.*, vol 29, no 18, 2692 (1990)

[5.29] A. Lord, I. J. Wilkinson, A. Ellis, D. Cleland, R. A. Garnham, and W. A. Stallard,

"Comparison of WDM coupler technologies for use in erbium doped fibre amplifier systems," *Electron. Lett.*, vol 26, no 13, 900 (1990)

[5.30] J. D. Minelli and M. Suyama, "Wavelength combining fused-taper couplers with low sensitivity to polarization for use with 1480 nm-pumped erbium-doped fibre amplifiers," *Electron. Lett.*, vol 26, no 8, 523 (1990)

[5.31] K. O. Hill, F. Bilodeau, B. Malo, and D. C. Johson, "WDM all-fiber compound devices; bimodal fiber narrowband tap and equal-arm dissimilar fiber unbalanced Mach–Zehnder interferometer," in *Proc. Conference on Optical Fiber Communications OFC'90*, paper WM2, p99, Optical Society of America, Washington DC, 1990

[5.32] K. O. Hill, B. Malo, D. C. Johson, and F. Bilodeau, "A novel low-loss in-line bi-modal fiber tap: wavelength-selective properties," *IEEE Photonics Technol. Lett.*, vol 2, no 7, 484 (1990)

[5.33] K. O. Hill, F. Bilodeau, B. Malo, and D. C. Johnson, "Comment on $2 \times 2$ multiplexing couplers for all-fibre 1.55 $\mu$m amplifiers and lasers," *Electron. Lett.*, vol 27, no 9, 786 (1991)

[5.34] F. Gonthier, D. Ricard, S. Lacroix, and J. Bures, "$2 \times 2$" multiplexing couplers for all-fibre 1.55 $\mu$m amplifiers and lasers," *Electron. Lett.*, vol 27, no 1, 42 (1991)

[5.35] D. W. Hall, C. M. Truesdale, D. L. Weidman, M. E. Vance, and L. J. Button. "Wavelength division multiplexers for 980 nm pumping of erbium-doped fiber optical amplifiers," *Proc. Topical Meeting of Optical Amplifiers and Applications, 1990*, paper WA5, Optical Society of America, Washington DC, 1990

[5.36] K. P. Jedrzejewski, F. Martinez, J. D. Minelli, C. D. Hussey, and F. P. Payne, "Tapered-beam expander for single-mode opticalfiber gap devices," *Electron. Lett.*, vol 22, 105 (1986)

[5.37] N. Amitay, H. M. Presby, F. V. Dimarcello, and K. T. Nelson. "Single-mode optical fiber tapers for self-align beam expansion," *Electron. Lett.*, vol 22, 702 (1986)

[5.38] D. B. Mortimore and J. V. Wright, "Low-loss joints between dissimilar fibres by tapering diffusion splices," *Electron. Lett.*, vol 22, 318 (1986)

[5.39] J. S. Harper, C. P. Botham, and S. Hornung, "Tapers in single-mode fiber by controlled core diffusion," *Electron. Lett.*, vol 24, 245 (1988)

[5.40] K. Shiraishi, Y. Azawa, and S. Kawakami, "Beam expanding fiber using thermal diffusion of the dopant," *IEEE J. Lightwave Technol.*, vol 8, no 8, 1151 (1990)

[5.41] H. Hanafusa, M. Horiguchi, and J. Noda, "Thermally diffused expanded core fibers for low-loss and inexpensive photonic components," *Electron. Lett.*, vol 27, 1968 (1991)

[5.42] H. Y. Tam, "Simple fusion splicing technique for reducing splicing loss between standard single-mode fibres and erbium-doped fibre," *Electron. Lett.*, vol 27, no 17, 1597 (1991)

[5.43] Y. Ando and H. Hanafusa, "Low-loss optical connector between dissimilar single-mode fibers using local core expansion technique by thermal diffusion," *IEEE Photonics Technol. Lett.*, vol 4, no 9, 1028 (1992)

[5.44] M. J. Holmes, F. P. Payne, and D. M. Spirit, "Matching fibres for low-loss coupling into fibre amplifiers," *Electron. Lett.*, vol 26, no 25, 2102 (1990)

[5.44] C. R. Giles and E. Desurvire, "Propagation of signal and noise in concatenated erbium-coped fiber optical amplifiers," *IEEE J. Lightwave Technol.*, vol 9, no 2, 147 (1991)

[5.45] A. E. Willner, E. Desurvire, H. M. Presby, C. A. Edwards, and J. R. Simpson, "Use of LD-pumped erbium-doped fiber preamplifiers with optimal noise filtering in a FDMA-FSK 1 Gb/s star network," *IEEE Photonics Technol. Lett.*, vol 2, no 9, 669 (1990)

[5.46]   M. C. Farries, C. M. Ragdale, and D. C. Reid, "Broadband chirped fibre Bragg filters for pump rejection and recycling in erbium-doped fibre amplifiers," *Electron. Lett.*, vol 28, no 5, 497 (1992)

[5.47]   M. Tachibana, R. I. Laming, P. R. Morkel, and D. N. Payne, "Erbium-doped fiber amplifier with flattened gain spectrum," *IEEE Photonics Technol. Lett.*, vol 3, no 2, 118 (1991)

[5.48]   K. Inoue, T. Kominato, and H. Toba, "Tunable gain equalization using a Mach–Zenhder optical filter in multistage fiber amplifiers," *IEEE Photonics Technol. Lett.*, vol 3, no 8, 718 (1991)

[5.49]   M. Wilkinson, A. Bebbington, S. A. Cassidy, and P. McKee, "D-fibre filter for erbium gain spectrum flattening," *Electron Lett.*, vol 28, no 2, 131 (1992)

[5.50]   M. C. Farries, J. E. Townsend, and S. B. Poole, "Very high rejection optical fibre fiber filters," *Electron Lett.*, vol 22, 1126 (1986)

[5.51]   I. Bennion, D. C. J. Reid, C. J. Rowe, and W. J. Stewart, "High-reflectivity monomode-fibre grating filters," *Electron. Lett.*, vol 22, 341 (1986)

[5.52]   R. H. West, "A local view of radiation effects in optical fibres," *IEEE J. Lightwave Technol.*, vol 6, no 2, 155 (1988)

[5.53]   R. B. J. Lewis, E. S. R. Sikora, J. V. Wright, R. H. West, and S. Dowling, "Investigation of effects of gamma radiation on erbium doped fibre amplifiers," *Electron. Lett.*, vol 28, no 17, 1589 (1992)

[5.54]   G. M. Williams, M. A. Putnam, C. G. Askins, M. E. Gingerich and E. J. Friebele, "Radiation effects in erbium-doped optical fibres," *Electron. Lett.*, vol 28,, no 19, 1816 (1992)

[5.55]   R. J. Mears, L. Reekie, I. M. Jauncey and D. N. Payne, "High-gain rare-earth-doped fiber amplifier at 1.54 μm," in *Proc. Conference on Optical Fiber Communications and Conference on Integrated Optics and Optical Communications, OFC/IOOC '87*, paper WI2, p167, Optical Society of America, Washington DC, 1987

[5.56]   R. J. Mears, L. Reekie, I. M. Jauncey, and D. N. Payne, "Low-noise erbium-doped fibre amplifier operating at 1.54 μm," *Electron. Lett.*, vol 23, no 19, 1026 (1987)

[5.57]   E. Desurvire, J. R. Simpson, and P. C. Becker, "High-gain erbium-doped traveling-wave fiber amplier," *Optics Lett.*, vol 12, no 11, 888 (1987)

[5.58]   E. Desurvire, "Analysis of distributed erbium-doped fiber amplifiers with fiber background loss," *IEEE Photonics technol. Lett.*, vol 3, no 7, 625 (1991)

[5.59]   E. A. Snitzer, H. Po., F. Hakimi, R. Tuminelli, and B. C. MacCollum, "Erbium fiber laser amplifiers at 1.55 μm with pump at 1.49 μm and Yb-Sensitized Er oscillators," in *Proc. Conference on Optical Fiber Communications, OFC '88,* paper PD2, Optical Society of America, Washington DC, 1988

[5.60]   R. I. Laming, M. C. Farries, P. R. Morkel, L. Reekie, D. N. Payne, P. L. Scrivener, F. Fontana, and A. Righetti, "Efficient pump wavelengths of erbium-doped fibre optical amplifier," *Electron. Lett.*, vol 25, no 1, 12 (1989); R. I. Laming, V. Shah, L. Curtis, R. S. Vodhanel, F. J. Favire, W. L. Barnes, J. D. Minelly, D. P. Bour, and E. J. Tarbox, "Highly efficient 978 nm diode pumped erbium-doped fibre amplifier with 24 dB gain," in *Proc. Conference on Integrated Optics and Optical Communications, IOOC '89*, paper 20PDA-4, Kobe, Japan, 1989; R. S. Vodhanel, R. I. Laming, V. Shah, L. Curtis, D. P. Bour, W. L. Barnes, J. D. Minelly, E. J. Tarbox, and F. J. Favire, "Highly efficient 978 nm diode pumped erbium-doped fibre amplifier with 24 dB gain," *Electron. Lett.*, vol 25, no 20, 1386 (1989)

[5.61]   M. Shimizu, M. Horiguchi, M. Yamada, M. Okayasu, T. Takeshita, I. Nishi, S. Uehara, J. Noda, and E. Sugita, "Highly efficient integrated optical fiber amplifier module pumped by a 0.98 μm laser diode," *Electron. Lett.*, vol 26, no 8, 498 (1990)

[5.62]  Y. Kimura, M. Nakazawa, and K. Suzuki, "Ultra-efficient erbium-doped fiber amplifier," *Applied Phys. Lett.*, vol 57, no 25, 2635 (1990)

[5.63]  M. Nakazawa, Y. Kimura, and K. Suzuki, "Efficient $Er^{3+}$-doped optical fiber amplifier pumped by a 1.48 $\mu$m InGaAsP laser diode," *Applied Phys. Lett.*, vol 54, no 4, 295 (1989)

[5.64]  P. C. Becker, J. R. Simpson, N. A. Olsson, and N. K. Dutta, "High-gain and high-efficiency diode laser pumped fiber amplifier at 1.56 $\mu$m," *IEEE Photonics Technol. Lett.*, vol 1, no 9, 267 (1989)

[5.65]  E. Desurvire, C. R. Giles, J. R. Simpson, and J. L. Zyskind, "Efficient erbium-doped fiber amplifier at a 1.53 $\mu$m wavelength with a high output saturation power," *Optics Lett.*, vol 14, no 22, 1266 (1989)

[5.66]  J. L. Zyskind, D. J. DiGiovanni, J. W. Sulhoff, P. C. Becker, and C. H. Brito Cruz, "High performance erbium-doped fiber amplifier pumped at 1.48 $\mu$m and 0.97 $\mu$m," in *Proc. Topical Meeting of Optical Amplifiers and Applications, 1990*, paper PdP6, Optical Society of America, Washington DC, 1990

[5.67]  T. Kashiwada, M. Shigematsu, T. Kougo, H. Kanamori and M. Nishimura, "Erbium-doped fiber amplifier pumped at 1.48um with extremely high efficiency," *IEEE Photonics Technol. Lett.*, vol 3, no 8, 721 (1991)

[5.68]  S. Uehara, M. Horiguchi, T. Takeshita, M. Okayasu, M. Yamada, M. Shimizu, O, Kogure, and K. Oe, "0.98-$\mu$m InGaAs strained quantum well lasers for erbium-doped fiber optical amplifiers," in *Proc. Conference of Integrated Optics and Optical Communications, IOOC'89*, paper 20PB-11, Kobe, Japan, 1989; M. Yamada, M. Shimizu, T. Takeshita, M. Okayasu, M. Horiguchi, S. Uehara, and E. Sugita, "$Er^{3+}$-doped fiber amplifier pumped by 0.98 $\mu$m laser diodes," *IEEE Photonics Technol. Lett.*, vol 1, no 12, 422 (1989)

[5.69]  B. J. Ainslie, J. R. Armitage, S. P. Carig, and B. Wakefield, "Fabrication and optimization of the erbium distribution in silica based doped fibres," in *Proc. European Conference on Optical Communications, ECOC'88*, Brighton, 1988; IEE Conference Publication 292, part 1, 1988, p62.

[5.70]  J. R. Armitage, "Three-level fiber laser amplifier: a theoretical model," *Applied Optics*, vol 27, no 23, 4831 (1988)

[5.71]  M. Horiguchi, M. Shimizu, M. Yamada, K. Yoshino, and H. Hanafusa, "Highly efficient Er-doped fibre amplifiers pumped in 660 nm band," *Electron. Lett.*, vol 27, no 25, 2319 (1991)

[5.72]  M. M. Choy, C. Y. Chen, M. Andrejco, M. Saifi, and C. Lin, "A high-gain, high-output saturation power erbium-doped fiber amplifier pumped at 532 nm," *IEEE Photonics Technol. Lett.*, vol 2, no 1, 38 (1990)

[5.73]  T. Sugawa, T. Kokumai, and Y. Miyajima, "Optical amplification in $Er^{3+}$-doped single-mode fluoride fibers," *IEEE Photonics Technol. Lett.*, vol 2, no 7, 475 (1990)

[5.74]  D. Ronarc'h M. Guibert, H. Ibrahim, M. Monerie, H. Poignant, and A. Tromeur, "30 dB optical net gain at 1.543 $\mu$m in $Er^{3+}$ doped fluoride fibre pumped around 1.48 $\mu$m," *Electron. Lett.*, vol 27, no 11, 908 (1991)

[5.75]  W. J. Miniscalco, "Erbium-doped glasses for fiber amplifiers at 1500 nm," *IEEE J. Lightwave Technol.*, vol 9, no 2, 234 (1991)

[5.76]  C. R. Giles, E. Desurvire, J. R. Talman, J. R. Simpson, and P. C. Becker, "2-Gbit/s signal amplification at $\lambda = 1.53$ $\mu$m in an erbium-doped single-mode fiber amplifier," *IEEE J. Lightwave Technol.*, vol 7, no 4, 651 (1989)

[5.77]  M. Suyama, K. Nakamura, S. Kashiwa, and H. Kuwahara, "Polarization-independent gain in $Er^{3+}$-doped fiber amplifier under single-mode pumping," in *Proc. Conference on Integrated Optics and Optical Communications, IOOC'89,* paper 20A4-3, Kobe, Japan, 1989

[5.78]   E. Desurvire, J. L. Zyskind, and J. R. Simpson, "Spectral gain hole-burning at 1.53 um in erbium-doped fiber amplifiers," *IEEE Photonics Technol. Lett.,* vol 2, no 4, 246 (1990)

[5.79]   E. Desurvire, J. W. Sulhoff, J. L. Zyskind, and J. R. Simpson, "Study of spectral dependence of gain saturation and effect of inhomogeneous broadening in erbium-doped aluminosilicate fiber amplifiers," *IEEE Photonics Technol. Lett.,* vol 2, no 9, 653 (1990)

[5.80]   J. Aspell, J. F. Federici, B. M. Nyman, D. L. Wilson, and D. S. Shenk, "Accurate noise figure measurements of erbium-doped fiber amplifiers in saturation condition," in *Proc. Conference on Optical Fiber Communications, OFC'92,* paper ThA4, p189, Optical Society of America, Washington DC, 1992

[5.81]   B. E. A. Saleh and M. C. Teich, *Fundamentals of Photonics,* John Wiley, New York, 1991

[5.82]   Y. Kimura, K. Susuki, and N. Nakazawa, "46.5 dB gain in $Er^{3+}$-doped fibre amplifier pumped by 1.48 $\mu$m GaInAsP laser diodes," *Electron. Lett.,* vol 25, no 24, 1656 (1989)

[5.83]   H. Masuda and A. Takada, "High gain two-stage amplification with erbium-doped fibre amplifier," *Electron. Lett.,* vol 26, no 10, 661 (1990)

[5.84]   H. Takara, A. Takada, and M. Saruwatari, "A highly efficient two stage $Er^{3+}$-doped optical fiber amplifier employing an optical gate to effectively reduce ASE," *IEEE Photonics Technol. Lett.,* vol 4, no 3, 241 (1992)

[5.85]   J. H. Povlsen, A. Bjarklev, O. Lumholt, H. Vendeltorp-Pommer, and K. Rottwitt, "Optimizing gain and noise performance of EDFA's with insertion of a filter or an isolator," in *Proc. Conference SPIE OE/FIBERS'91,* no 1581-11, 1991

[5.86]   O. Lumholt, K. Schlusler, A. Bjarklev, S. Dahls-Pedersen, J. H. Povlsen, T. Rasmussen, and K. Rottwitt, "Optimum position of isolators withing erbium-doped fibers," *IEEE Photonics Technol. Lett.,* vol 4, no 6, 568 (1992)

[5.87]   S. Yamashita and T. Okoshi, "Performance improvement and optimization of fiber amplifier with a midway isolator," *IEEE Photonics Technol. Lett.,* vol 4, no 11, 1276 (1992)

[5.88]   K. Tajima, "$Er^{3+}$-doped single-polarization optical fibres," *Electron. Lett.,* vol 26, no 18, 1498 (1990)

[5.89]   I. N. Duling III and R. D. Esman, "Single-polarization fibre amplifier," *Electron. Lett.,* vol 28, no 12, 1126 (1992)

[5.90]   K. Iwatsuki, H. Okamura, and M. Saruwatari, "Wavelength tunable single frequency and single polarization Er-doped fiber ring laser with 1.4 kHz linewidth," *Electron. Lett.,* vol 26, no 24, 2033 (1990)

[5.91]   E. Desurvire, M. Zirngibl, H. M. Presby, and D. DiGiovanni, "Characterization and modeling of amplified spontaneous emission in unsaturated erbium-doped fiber amplifiers," *IEEE Photonics Technol. Lett.,* vol 3, no 2, 127 (1991)

[5.92]   C. G. Atkins, J. F. Massicott, J. R. Armitage, R. Wyatt, B. J. Ainslie, and S. P. Craig-Ryan, "High-gain, broad spectral bandwidth erbium-doped fibre amplifier pumped near 1.5 $\mu$m," *Electron. Lett.,* vol 25, no 14, 910 (1989)

[5.93]   E. Desurvire and J. R. Simpson, "Amplification of spontaneous emission in erbium-doped single-mode fibers," *IEEE J. Lightwave Technol.,* vol 7, no 5, 835 (1989)

[5.94]   J. M. P. Delavaux, C. F. Flores, R. E. Tench, T. C. Pleiss, T. W. Cline, D. J. DiGiovanni, J. Federici., C. R. Giles, H. Presby, J. S. Major, and W. J. Gignac, "Hybrid Er-doped fibre amplifiers at 980–1480 nm for long distance optical communication," *Electron. Lett.,* vol 28, no 17, 1642 (1992)

[5.95]   J. F. Massicott, J. R. Armitage, R. Wyatt, B. J. Ainslie, and S. P. Craig-Ryan, "High gain, broadband, 1.6 $\mu$m $Er^{3+}$ doped silica fibre amplifier," *Electron. Lett.*, vol 26, no 20, 1645 (1990)

[5.96]   J. F. Massicott, R. Wyatt, and B. J. Ainslie, "Low noise operation of $Er^{3+}$ doped silica fibre amplifier around 1.6 $\mu$m," *Electron. Lett.*, vol 28, no 20, 1924 (1992)

[5.97]   G. Nykolak, S. A. Kramer, J. R. Simpson, D. DiGiovanni, C. R. Giles, and H. M. Presby, "An erbium-doped multimode optical fiber amplifier," *IEEE Photonics Technol. Lett.*, vol 3, no 12, 1079 (1991)

[5.98]   J. Sakai and T. Kimura, "Splicing and bending loss of single-mode optical fibres," *Applied Optics*, vol 17, 3653 (1978)

[5.99]   A. L. Deus and H. R. D. Sunak, "Design considerations for minimizing macrobending loss in erbium-doped fiber amplifiers," *IEEE Photonics Technol. Lett.*, vol 3, no 1, 50 (1991)

[5.100]  M. Ohashi and K. Shiraki, "Bending loss effect on signal gain in an $Er^{3+}$-doped fiber amplifier," *IEEE Photonics Technol. Lett.*, vol 4, no 2, 192 (1992)

[5.101]  J. R. Armitage, "Three-level fiber laser amplifier: a theoretical model," *Applied Optics*, vol 27, no 23, 4831 (1988)

[5.102]  E. Desurvire and J. R. Simpson, "Amplification of spontaneous emission in erbium-doped single-mode fibers," *IEEE J. Lightwave Technol.*, vol 7, no 5, 835 (1989)

[5.103]  A. Bjarklev, S. L. Hansen, and J. H. Povlsen, "Large-signal modeling of an erbium-doped fiber amplifier," in *Proc. SPIE Conference on Fiber Laser Sources and Amplifiers*, vol 1171, 118 (1989)

[5.104]  E. Desurvire, C. R. Giles and J. R. Simpson, "Gain saturation effects in high-speed, multichannel erbium-doped fiber amplifiers at $\lambda = 1.53 \mu$m," *IEEE J. Lightwave Technol.*, vol 7, no 12, 2095 (1989)

[5.105]  P. R. Morkel and R. I. Laming, "Theoretical modeling of erbium-doped fiber with excited-state absorption," *Optics Lett.*, vol 14, no 19, 1062 (1989)

[5.106]  J. F. Marcerou, H. A. Fevrier, J. Ramos, J. C. Auge, and P. Bousselet, "General theoretical approach describing describing the complete behavior of the erbium-doped fiber amplifier," in *Proc. SPIE Conference on Fiber Laser Sources and Amplifiers*, vol 1373, 168 (1990)

[5.107]  B. Pedersen, K. Dybdal, C. D. Hansen, A. Bjarklev, J. H. Povlsen, H. Vendeltorp-Pommer, and C. C. Larsen, "Detailed theoretical and experimental investigation of a high-gain erbium-doped fiber amplifier," *IEEE Photonics Technol. Lett.*, vol 2, no 12, 863 (1990)

[5.108]  C. R. Giles, C. A. Burrus, C. A. Burrus, D. DiGiovanni, N. K. Dutta, and G. Raybon. "Characterization of erbium-doped fibers and application to modeling 980 nm and 1480 nm pumped amplifiers," *IEEE Photonics Technol. Lett.*, vol 3, no 4, 363 (1991)

[5.109]  A. A. M. Saleh, R. M. Jopson, J. D. Evankow, and J. Aspell, "Modeling of gain in erbium-doped fiber amplifiers," *IEEE Photonics Technol. Lett.*, vol 2, no 10, 714 (1990)

[5.110]  B. Pedersen, M. L. Dakss, and W. J. Miniscalco, "Conversion efficiency and noise in erbium-doped fiber power amplifiers," in *Proc. Topical Meeting of Optical Amplifiers and Applications, 1991*, paper ThE3, p170, Optical Society of America, Washington DC, 1991

[5.111]  J. F. Massicott, R. Wyatt, B. J. Ainslie, and S. P. Craig-Ryan, "Efficient, high power, high gain, $Er^{3+}$ doped silica fibre amplifiers," *Electron. Lett.*, vol 26, no 14, 1038 (1990)

[5.112]  R. I. Laming, J. E. Townsend, D. N. Payne, F. Meli, G. Grasso, and E. J. Tarbox, "High-power erbium-doped-fiber amplifiers operating in the saturated regime," *IEEE Photonics Technol. Lett.*, vol 3, no 3, 253 (1991)

[5.113]  B. Pedersen, M. L. Dakss, B. A. Thompson, W. J. Miniscalco, T. Wei, and L. J.

Andrews, "Experimental and theoretical analysis of efficient erbium-doped fiber power amplifiers," *IEEE Photonics Technol.*, vol 3, no 12, 1085 (1991)

[5.114] E. Desurvire, "Analysis of gain difference between forward- and backward-pumped erbium-doped fiber amplifiers in the saturation regime," *IEEE Photonics Technol. Lett.*, vol 4, no 7, 711 (1992)

[5.115] A. Lidgard, J. R. Simpson, and P. C. Becker, "Output saturation characteristics of erbium-doped fiber amplifiers pumped at 975 nm," *Applied Phys. Lett.*, vol 56, no 26, 2607 (1990)

[5.116] J. E. Townsend, W. L. Barnes, and K. P. Jedrzejewski, "$Yb^{3+}$ sensitized $Er^{3+}$-doped silica optical fibre with ultrahigh transfer efficiency and gain," *Electron. Lett.*, vol 27, no 21, 1958 (1991)

[5.117] K. Inoue, H. Toba, N. Shibata, K. Iwatsuki, A. Takada, and M. Shimizu, "Mutual signal gain saturation in $Er^{3+}$-doped fibre amplifier around 1.54 $\mu$m wavelength," *Electron. Lett.*, vol 25, no 9, 594 (1989)

[5.118] M. Tachibana, R. I. Laming, P. R. Morkel, and D. N. Payne, "Spectral gain cross saturation and hole-burning in wideband erbium-doped fibre amplifiers," in *Proc. Topical Meeting of Optical Amplifiers and Applications, 1991*, paper ThB1, p104, Optical Society of America, Washington DC, 1991

[5.119] E. Desurvire, "Spectral noise figure of $Er^{3+}$-doped fiber amplifiers," *IEEE Photonics Technol. Lett.*, vol 2, no, 208 (1990)

[5.120] K. Motoshima, L. M. Leba, D. N. Chen, M. M. Downs, T. Li and E. Desurvire, "Dynamic compensation of transient gain saturation in erbium-doped fiber amplifiers by pump feedback control," *IEEE Photonics Technology Lett.* vol 5, no 12, 1423 (1993)

[5.121] P. L. Liu, L. E. Fencil, J. S. Ko, I. P. Kaminow, T. P. Lee, and C. A. Burrus, "Amplitude fluctuations and photon statistics of InGaAsP injection lasers," *IEEE J. Quantum Electron.*, vol 19, no 9, 1348 (1983)

[5.122] J. C. Simon, "Semiconductor laser amplifiers for single-mode optical fiber communications," *J. Opt. Comm.*, vol 4, no 2, 51 (1982)

[5.123] R. Olshansky, "Noise figure for erbium-doped optical fibre amplifiers," *Electron. Lett.*, vol 24, no 22, 1363 (1988)

[5.124] D. E. McCumber, "Theory of phonon-terminated optical masers," *Phys. Rev.*, vol 134, no 2A, 299 (1964)

[5.125] W. J. Miniscalco, B. A. Thompson, M. L. Dakss, S. A. Zemon, and L. J. Andrews, "The measurement and analysis of cross-sections for rare-earth-doped glasses," in *Proc. SPIE Conference on Fibre Laser Sources and Amplifiers III*, vol 1581, 80 (1991)

[5.126] C. R. Giles, E. Desurvire, J. L. Zyskind, and J. R. Simpson, "Noise performance of erbium-doped fiber amplifier pumped at 1.49 $\mu$m and application to signal pre-amplification at 1.8 Gbit/s," *IEEE Photonics Technol. Lett.*, vol 1, no 11, 367 (1989)

[5.127] G. R. Walker, "Gain and noise characterization of erbium-doped fiber amplifiers," *Electron. Lett.*, vol 27, no 9, 744 (1991)

[5.128] R. I. Laming and D. N. Payne, "Noise characteristics of erbium-doped fiber amplifier pumped at 980 nm," *IEEE Photonics Technol. Lett.*, vol 2, no 6, 418 (1990)

[5.129] T. Mukai and Y. Yamamoto, "Noise in AlGaAs semiconductor laster amplifier," *IEEE J. Quantum Electron.*, vol 18, 564 (1982)

[5.130] M. Yamada, M. Shimizu, M. Okayasu, T. Takeshita, M. Horiguchi, Y. Tachikawa, and E. Sugita, "Noise characteristics of Er-doped fiber amplifiers pumped by 0.98 and 1.48 $\mu$m laser diodes," *IEEE Photonics Technol. Lett.*, vol 2, no 3, 205 (1990)

[5.131] C. R. Giles and E. Desurvire, "Modeling erbium-doped fiber amplifiers," *IEEE J. Lightwave Technol.*, vol 9, no 2, 271 (1991)

[5.132] Y. Kimura, K. Suzuki and M. Nakazawa, "Noise figure characteristics of $Er^{3+}$-doped fibre amplifier pumped in 0.8 $\mu m$ band," *Electron. Lett.*, vol 27, no 2, 146 (1991)

[5.133] E. Desurvire, M. J. F. Digonnet, and H. J. Shaw, "Theory and implementation of a Raman active fiber delay line," *IEEE J. Lightwave Technol.*, vol 4, no 4, 426 (1986)

[5.134] E. Desurvire, M. Tur, and H. J. Shaw, "Signal-to-noise in Raman active fiber systems: application to recirculating delay lines," *IEEE J. Lightwave Technol.*, vol 4, no 5, 560 (1986)

[5.135] E. Desurvire, "Analysis of noise figure spectral distribution in erbium-doped fiber amplifiers pumped near 980 and 1480 nm," *Applied Optics.*, vol 29, no 21, 3118 (1990)

[5.136] R. G. Smart, J. L. Zyskind, W. J. Sulhoff, and D. J. DiGiovanni, "An investigation of the noise figure and conversion efficiency of 0.98 $\mu m$ pumped erbium-doped fiber amplifiers under saturated conditions," *IEEE Photonics Technol. Lett.*, vol 4, no 11, 1261 (1992)

[5.137] G. Oliver and C. Bendjaballah, "Statistical properties of coherent radiation in a nonlinear optical amplifier," *Phys. Rev. A*, vol 22, no 2, 630 (1980)

[5.138] C. Bendjaballah and G. Oliver, "Comparison of statistical properties of two models for saturated laser-light amplifier," *Phys. Rev. A*, vol 22, no 6, 2726 (1980)

[5.139] C. Bendjaballah and G. Oliver, "Detection of coherent light after nonlinear amplification," *IEEE Trans. Aerospace and Electron. Syst.*, vol 17, no 5, 620 (1981)

[5.140] N. B. Abraham, "Quantum theory of a saturable optical amplifier," *Phys. Rev. A*, vol 21, no 5, 1595 (1980)

[5.141] M. N. Zervas, R. I. Laming, and D. N. Payne, "Efficient erbium-doped fibre amplifier with integral isolator," in *Proc. Topical Meeting of Optical Amplifiers and Applications, 1992*, paper FB2, p162, Optical Society of America, Washington DC, 1992

[5.142] R. I. Laming, A. H. Gnauck, C. R. Giles, M. N. Zervas, and D. N. Payne, "High sensivity optical pre-amplifier at 10 Gbit/s employing a low noise composite EDFA with 46 dB gain," in *Proc. Topical Meeting of Optical Amplifiers and Applications, 1992*, postdeadline paper PD13, Optical Society of America, Washington DC, 1992; R. I. Laming, M. N. Zervas and D. N. Payne, "Erbium-doped fiber amplifier with 54 dB gain and 3.1 dB noise figure," *IEEE Photonics Technol. Lett.*, vol 4, no 12, 1345 (1992)

[5.143] J. F. Marcerou, H. Fevrier, J. Herve, and J. Auge, "Noise characteristics of the EDFA in gain saturation regions," in *Proc. Topical Meeting of Optical Amplifiers and Applications, 1991*, paper ThE1, p162, Optical Society of America, Washington DC, 1991

[5.144] A. E. Willner and E. Desurvire, "Effect of gain saturation on receiver sensitivity in 1 Gbit/s multichannel FSK direct-detection systems using erbium-doped fiber preplifiers," *IEEE Photonics Technol. Lett.*, vol 3, no 3, 259 (1991)

[5.145] W. Way, A. C. VonLehman, M. J. Andrejco, M. A. Saifi, and C. Lin, "Noise figure of gain-saturated erbium-doped fiber amplifier pumped at 980 nm," in *Proc. Topical Meeting of Optical Amplifiers and Applications, 1990*, paper TuB3, p134, Optical Society of America, Washington DC, 1990

[5.146] T. G. Hodgkinson, "Average power analysis technique for erbium-doped fiber amplifiers," *IEEE Photonics Lett.*, vol 3, no 12, 1082 (1991)

[5.147] T. G. Hodgkinson, "Improved average power analysis technique for erbium-doped fiber amplifiers," *IEEE Photonics Lett.*, vol 4, no 11, 1273 (1992)

[5.148] M. J. Adams, J. V. Collins, and I. D. Henning, "Analysis of semiconductor laser amplifiers," *IEE Proc. J*, vol 132, 58 (1985)

[5.149] T. Georges and E. Delevaque, "Analytic modeling of high-gain erbium-doped fiber amplifiers," *Optics Lett.*, vol 17, no 16, 1113 (1992)

[5.150]  B. Pedersen, B. A. Thomson, S. Zemon, W. J. Miniscalco, and T. Wei, "Power requirements for erbium-doped fiber amplifiers pumped in the 800, 980 and 1480 nm bands," *IEEE Photonics Technol. Lett.*, vol 4, no 1, 46 (1992)

[5.151]  E. Desurvire, J. L. Zyskind and C. R. Giles, "Design optimization for efficient erbium-doped fiber amplifiers," *IEEE J. Lightwave Technol.*, vol 8, no 11, 1730 (1990); erratum: *IEEE J. Lightwave Technol.*, vol 9, no 6, 809 (1991)

[5.152]  Th.Pfeiffer and H. Bülow, "Analytical gain equation for erbium-doped fiber amplifiers including mode field profiles and dopant distributions," *IEEE Photonics Technol. Lett.*, vol 4, no 5, 449 (1992)

[5.153]  J. D. Evankow and R. M. Jopson, "Nondestructive measurement of length dependence of gain characteristics in fiber amplifiers," *IEEE Photonics Technol. Lett.*, vol 3, no 11, 993 (1991)

[5.154]  D. L. Williams, S. T. Davey, D. M. Spirit, and B. J. Ainslie, "Transmission over 10 km or erbium doped fibre with ultralow signal power excursion," *Electron. Lett.*, vol 26, no 18, 1517 (1990)

[5.155]  M. Nakazawa, Y. Kimura, and K. Suzuki, "Gain-distribution measurements along an ultralong erbium-doped fiber amplifier using optical-time-domain reflectometry," *Optics Lett.*, vol 15, no 21, 1200 (1990)

[5.156]  F. F. Rühl, "Prediction of optimum fibre lengths for erbium doped fibre amplifiers," *Electron. Lett.*, vol 27, no 9, 769 (1991)

[5.157]  F. F. Rühl, "Calculation of optimum fibre lengths for EDFAs at arbitrary pump wavelengths," *Electron. Lett.*, vol 27, no 16, 1443 (1991)

[5.158]  F. F. Rühl, "Accurate analytical formulas for gain-optimized EDFAs," *Electron. Lett.*, vol 28, no 3, 312 (1992)

[5.159]  Min-Chuan Lin and Sien Chi, "The gain and optimal length in the erbium-doped fiber amplifiers with 1480 nm pumping," *IEEE Photonics Technol. Lett.*, vol 4, no 4, 354 (1992)

[5.160]  J. G. Edwards and J. N. Sandoe, "A theoretical study of the Nd:Yb:Er glass laser," *J. Phys. D, Appl. Phys.*, vol 7, 1078 (1974)

[5.161]  M. Shimizu, M. Yamada, M. Horiguchi, and E. Sugita, "Concentration effects on optical amplification characteristics of erbium-doped silica single-mode fibers," *IEEE Photonics Technol. Lett.*, vol 2, no 1, 43 (1990)

[5.162]  N. Kagi, A. Oyobe, and K. Nakamura, "Efficient optical amplifier using a low-concentration erbium-doped fiber," *IEEE Photonics Technol. Lett.*, vol 2, no 8, 559 (1990)

[5.163]  B. J. Ainslie, S. P. Craig-Ryan, S. T. Davey, J. R. Armitage, C. G. Atkins, and R. Wyatt, "Optical analysis of erbium-doped fibres for efficient lasers and amplifiers," in *Proc. Seventh International Conference on Integrated Optics and Optical Fiber Communications, IOOC'89*, paper 20A3-2, p22, Kobe, Japan, 1989

[5.164]  Y. Miyajima, T. Sugawa, and K. Komukai, "20 dB gain at 1.55 $\mu$m wavelength in 50 cm long $Er^{3+}$-doped fluoride fibre amplifier," *Electron. Lett.*, vol 26, no 18, 1527 (1990)

[5.165]  Y. Kimura and M. Nakazawa, "Gain characteristics of erbium-doped fibre amplifiers with high erbium concentration," *Electron. Lett.*, vol 28, no 15, 1420 (1992)

[5.166]  E. Desurvire, "Analysis of erbium-doped fiber amplifiers pumped in the $^4I_{13/2}$–$^4I_{15/2}$ band," *IEEE Photonics Technol. Lett.*, vol 1, no 10, 293 (1989)

[5.167]  J. L. Zyskind, C. R. Giles, E. Desurvire, and J. R. Simpson, "Optimal pump wavelength in the $^4I_{15/2}$–$^4I_{13/2}$ absorption band for efficient $Er^{3+}$-doped fiber amplifiers," *IEEE*

*Photonics technol. Lett.*, vol 1, no 12, 428 (1989)

[5.168]  Y. Kimura, K. Suzuki, and M. Nakazawa, "Pump wavelength dependence of the gain factor in 1.48 $\mu$m pumped $Er^{3+}$-doped fiber amplifiers," *Applied Phys. Lett.*, vol 56, no 17, 1611 (1990)

[5.169]  K. Suzuki, Y. Kimura, and M. Nakazawa, "Pumping wavelength dependence on gain factor of 0.98 $\mu$m pumped $Er^{3+}$ fiber amplifier," *Applied Phys. Lett.*, vol 55, no 25, 2573 (1989)

[5.170]  P. C. Becker, A. Lidgard, J. R. Simpson, and N. A. Olsson, "Erbium-doped fiber amplifier pumped in the 950–1000 nm region," *IEEE Photonics Technol. Lett.*, vol 2, no 1, 35 (1990)

[5.171]  R. M. Percival, S. Cole, D. M. Cooper, S. P. Craig-Ryan, A. D. Ellis, C. J. Rowe, and W. A. Stallard. "Erbium-doped fibre amplifier with constant gain for pump between 966 and 1004 nm," *Electron. Lett.*, vol 27, no 14, 1266 (1991)

[5.172]  B. Pedersen, J. Chirravuri, and W. J. Miniscalco, "Gain and noise properties of small-signal erbium-doped fiber amplifiers pumped in the 980-nm band," *IEEE Photonics Technol. Lett.*, vol 4, no 6, 556 (1992)

[5.173]  B. Pedersen, J. Chirravuri, and W. J. Miniscalco, "Gain and noise penalty for detuned 980 nm pumping of erbium-doped fiber power amplifiers," *IEEE Photonics Technol. Lett.*, vol 4, no 4, 351 (1992)

[5.174]  M. Nakazawa, Y. Kimura, and K. Suzuki, "High gain erbium fibre amplifier pumped by 800 nm band," *Electron. Lett.*, vol 26, no 8, 548 (1990)

[5.175]  B. Pedersen, W. J. Miniscalco, and A. Zemon, "Evaluation of the 800 nm pump band for erbium-doped fiber amplifiers," *IEEE J. Lightwave Technol.*, vol 10, no 8, 1041 (1992)

[5.176]  B. Pedersen, S. Zemon, and W. J. Miniscalco, "Erbium-doped fibres pumped in 800 nm band," *Electron. Lett.*, vol 27, no 14, 1295 (1991)

[5.177]  J. H. Povlsen, A. Bjarklev, B. Pedersen, H. Vendeltorp-Pommer, and K. Rottwitt, "Optimum design of erbium fibre amplifiers pumped with sources emitting at 1480 nm," *Electron. Lett.*, vol 26, no 17, 1429 (1990)

[5.178]  N. Kagi, A. Oyobe, and K. Nakamura, "Gain characteristics of $Er^{3+}$-doped fiber with a quasi-confined structure," *IEEE J. Lightwave Technol.*, vol 8, no 9, 1319 (1990)

[5.179]  E. Desurvire, J. L. Zyskind, and C. R. Giles, "Design optimization for efficient erbium-doped fiber amplifiers," *IEEE J. Lightwave Technol.*, vol 8, no 11, 1730 (1990)

[5.180]  M. Ohashi and M. Tsubokawa, "Optimumm parameter design of $Er^{3+}$-doped fibers for optical amplifiers," *IEEE Photonics Technol. Lett.*, vol 3, no 2, 121 (1991)

[5.181]  M. Ohashi, "Design considerations for an $Er^{3+}$-doped fiber amplifier," *IEEE J. Lightwave Technol.*, vol 9, no 9, 1099 (1991)

[5.182]  B. Pedersen, A. Bjarklev, and J. H. Povlsen, "Design of erbium doped fibre amplifiers for 980 nm or 1480 nm pumping," *Electron. Lett.*, vol 27, no 3, 255 (1991)

[5.183]  B. Pedersen, A. Bjarklev, O. Lumholt, and J. H. Povlsen, "Detailed design analysis of erbium-doped fiber amplifiers," *IEEE Photonics Technol. Lett.*, vol 3, no 6, 548 (1991)

[5.184]  B. Pedersen, A. Bjarklev, J. H. Povlsen, K. Dybdal, and C. C. Larsen, "The design of erbium-doped fiber amplifiers," *IEEE J. Lightwave Technol.*, vol 9, no 9, 1105 (1991)

[5.185]  M. N. Zervas, R. I. Laming, J. E. Townsend, and D. N. Payne, "Design and fabrication of high gain-efficiency erbium-doped fiber amplifiers," *IEEE Photonics Technol. Lett.*, vol 4, no 12, 1342 (1992)

[5.186]  B. J. Ainslie, K. J. Beales, C. R. Day, and J. D. Rush, "Interplay of design parameters and fabrication conditions on the performance of monomode fibers made by MCVD,"

*IEEE J. Quantum Electron.*, vol 17, no 6, 854 (1981)

[5.187] P. Garner, P. P. Smyth, P. Eardley, K. H. Cameron, S. M. Webster, M. A. Collins, N. D. Harvey, D. G. Peker, and M. Fake, "High performance optical fiber amplifier designed for field deployment," in *Proc. Conference on Optical Fiber Communications, OFC'92*, paper ThA6, p192, Optical Society of America, Washington DC, 1992

[5.188] R. Loudon, *The Quantum Theory of Light*, second edition, Oxford University Press, New York, 1983

[5.189] C. H. Henry, "Theory of the linewidth of semiconductor lasers," *IEEE J. Quantum Electron.*, vol 18, no 2, 259 (1982)

[5.190] G. J. Cowle, PhD thesis, University of Southampton, 1990, chapter 6

[5.191] J. P. Gordon and L. F. Mollenauer, "Phase noise in photonic communication systems using linear amplifiers," *Optics Lett.*, vol 15, no 23, 1351 (1990)

[5.192] A. Yariv, *Optical Electronics*, third edition, Holt, Reinehart and Winston, New York, 1985

[5.193] S. Ryu, "Signal linewidth broadening due to nonlinear Kerr effect in longhaul coherent systems using cascaded optical amplifiers," *IEEE J. Lightwave Technol.*, vol 10, no 10, 1450 (1992)

[5.194] G. J. Cowle, P. R. Morkel, R. I. Laming, and D. N. Payne, "Spectral broadening due to fibre amplifier phase noise," *Electron. Lett.*, vol 26, no 7, 424 (1990)

[5.195] H. Okamura and K. Iwatsuki, "Spectral linewidth broadening in Er-doped fibre amplifiers with 1.4 KHz linewidth light source," *Electron. Lett.*, vol 26, no 23, 1965 (1990)

[5.196] S. Ryu and Y. Horiuchi, "Use of an optical amplifier in a coherent receiver," *IEEE Photonics Technol. Lett.*, vol 3, no 7, 663 (1991)

[5.197] S. L. Hansen, K. Dybdal, and C. C. Larsen, "Gain limit in erbium-doped fiber amplifiers due to internal Rayleigh backscattering," *IEEE Photonics Technol. Lett.*, vol 4, no 6, 559 (1992)

[5.198] M. E. Fermann, S. B. Poole, D. N. Payne, and G. Martinez, "Comparative measurement of Rayleigh scattering in single-mode optical fibers based on an OTDR technique," *IEEE J. Lightwave Technol.*, vol 6, 545 (1988)

[5.199] E. Brinkmeyer, "Backscattering in single-mode fiber," *Electron. Lett.*, vol 16, 329 (1980)

[5.200] E. Brinkmeyer, "Analysis of the backscattering method for single-mode fibers," *J. Opt. Soc. Am.*, vol 70, 1010 (1980)

[5.201] M. Nakazawa, "Rayleigh backscattering theory for single-mode optical fibers," *J. Opt. Soc. Am.*, vol 73, no 9, 1175 (1983)

[5.202] J. L. Gimlett, J. Young, R. E. Spicer, and N. K. Cheung, "Degradation in Gbit/s DFB laser transmission systems due to phase-to-intensity noise conversion by multiple reflection point," *Electron. Lett.*, vol 24, 406 (1988)

[5.203] J. L. Gimlett and N. K. Cheung, "Effects of phase-to-intensity noise conversion by multiple reflections on gigabit-per-second DFB laser transmission systems," *IEEE J. Lightwave Technol.*, vol 7, no 1, 888 (1989)

[5.204] J. L. Gimlett, M. Z. Iqbal, L. Curtis, N. K. Cheung, A. Righetti, F. Fontana, and G. Grasso, "Impact of multiple reflection noise in Gbit/s lightwave systems with optical fibre amplifiers," *Electron. Lett.*, vol 25, no 20, 1393 (1989)

[5.205] J. L. Gimlett, M. Z. Iqbal, N. K. Cheung, A. Righetti, F. Fontana, and G. Grasso, "Observation of Rayleigh scattering mirrors in lightwave systems with optical amplifiers," *IEEE Photonics TEchnol. Lett.*, vol 2, no 3, 211 (1990)

[5.206] R. K. Staubli, P. Gysel, and R. U. Hofstetter, "Power penalties due to multiple

Rayleigh backscattering in coherent transmission systems using in-line optical amplifiers," *IEEE Photonics Technol. Lett.*, vol 2, no 12, 872 (1990)

[5.207]   H. Yoshinaga, K. Kikushima, and E. Yoneda, "Influence of reflected light on erbium-doped fiber amplifiers for optical AM video signal transmission systems," *IEEE Photonics Technol. Lett.*, vol 10, no 8, no 8, 1132 (1992)

[5.208]   E. I. Gordon, "Optical maser oscillators and noise," *Bell Syst. Tech. J.*, vol 43, 507 (1964)

[5.209]   Y. Yamamoto, "Characteristics of AlGaAs Fabry-Perot cavity type laser amplifiers," *IEEE J. Quantum ELectron.*, vol 16, no 10, 1047 (1980)

[5.210]   Y. Yamamoto, "Noise and error rate performance of semiconductor laser amplifiers in PCM-IM optical transmission systems,," *IEEE J. Quantum Electron.*, vol 16, no 10, 1073 (1980)

[5.211]   Y. Yamamoto, "S/N and error rate performance in AlGaAs semiconductor laser preamplifiers and linear repeater systems," *IEEE J. Quantum Electron.*, vol 18, no 10, 1560 (1982)

[5.212]   N. Henmi, S. Fujita, Y. Sunohara, and M. Shikada, "Rayleigh scattering influence on performance of 10 Gbit/s optical receiver with Er-doped optical fiber preamplifier," *IEEE Photonics Technol. Lett.*, vol 2, no 4, 277 (1990)

[5.213]   W. H. Press, B. P. Flannery, S. A. Teukolsky, and W. T. Vetterling, *Numerical Recipes, The Art of Scientific Computing,* Cambridge University Press, New York, 1986

[5.214]   E. Desurvire, C. R. Giles, and J. R. Simpson, "Gain dynamics of erbium-doped fiber amplifiers," in *Proc. SPIE Conference on Fiber Laser Sources and Amplifiers,* vol 1171, 103 (1989)

[5.215]   E. Desurvire, "Analysis of transient gain saturation and recovery in erbium-doped fibre amplifiers," *IEEE Photonics Technol. Lett.*, vol 1, no 8, 196 (1989)

[5.216]   R. M. Jopson, T. E. Darcie, K. T. Gayliard, R. T. Ku, R. E. Tench, T. C. Rice, and N. A. Olsson, "Measurement of carrier-density mediated intermodulation distortion in an optical amplifier," *Electron. Lett.*, vol 23, no 25, 1394 (1987)

[5.217]   E. Desurvire, C. R. Giles, and J. R. Simpson, "Saturation-induced crosstalk in high-speed erbium-doped fiber amplifiers at $\lambda = 1.53\ \mu m$," in *Proc. Conference on Optical Fiber Communications, OFC'89,* paper TuG7, p25, Optical Society of America, Washington DC. 1989

[5.218]   M. J. Pettitt, A. Hadjifotiou and R. A. Baker, "Crosstalk in erbium-doped fibre amplifier," *Electron. Lett.*, vol 25, no 6, 416 (1989)

[5.219]   R. I. Laming, L. Reekie, P. R. Morkel and D. N. Payne, "Multichannel crosstalk and pump noise in an $Er^{3+}$-doped fibre amplifier pumped at 980 nm," *Electron. Lett.*, vol 25, no 7, 455 (1989); R. I. Laming, L. Reekie, P. R. Morkel and D. N. Payne, "Multichannel crosstalk and pump noise in an $Er^{3+}$-doped fibre amplifier pumped at 980 nm," in *Proc. Conference on Integrated Optics and Optical Communications, IOOC'89,* Kobe, Japan, 1989, paper 20A4-2.

[5.220]   Y. Kimura, K. Suzuki, and M. Nakazawa, "Gain dynamics of an $Er^{3+}$-doped fiber amplifier," in *Proc. Conference on Integrated Optics and Optical Communications, IOOC'89,* Kobe, Japan, 1989, paper 20A4-6

[5.221]   C. R. Giles, E. Desurvire, and J. R. Simpson, "Transient gain and crosstalk in erbium-doped fiber amplifiers," *Optics Lett.*, vol 14, no 16, 880 (1989); C. R. Giles, E. Desurvire and J. R. Simpson, "Transient gain and crosstalk effects in erbium-doped fiber amplifiers," in *Proc. Conference on Integrated Optics and Optical Communications, IOOC'89,* Koke, Japan, 1989, paper 20A4-4

[5.222]   K. Iwatsuki, A. Takada, and M. Saruwatari, "Optical soliton propagation using

3 GHz gain-switched 1.3 $\mu$m laser diodes," *Electron. Lett.*, vol 24, no 25, 1572 (1988)

[5.223]   L. F. Mollenauer, R. H. Stolen, and M. N. Islam, "Experimental demonstration of soliton propagation in long fibers: loss compensated by Raman gain," *Optics Lett.*, vol 10, no 5, 229 (1985)

[5.224]   L. F. Mollenauer and K. Smith, "Demonstration of soliton transmission over more than 4000 km in fiber with loss periodically compensated by Raman gain," *Optics Lett.*, vol 13, no 8, 675 (1988)

[5.225]   K. Iwatsuki, S. Nishi, M. Saruwatari, and M. Shimizu, "2.8 Gbit/s optical soliton transmission employing all-optical laser diodes," *Electron. Lett.*, vol 26, no 1, 1 (1990)

[5.226]   K. Iwatsuki, S. Nishi, and K. Nakagawa, "3.6 Gb/s all-laser diode optical soliton transmission," *IEEE Photonics Technol. Lett.*, vol 2, no 5, 255 (1990)

[5.227]   R. A. Baker, K. C. Byron, D. Burns, and W. Sibett, "Amplification of mode-locked semiconductor diode laser pulses in erbium-doped fibre amplifier," *Electron. Lett.,"* *Electron. Lett.*, vol 25, no 17, 1131 (1989)

[5.228]   M. Nakazawa, Y. Kimura, and K. Suzuki, "Soliton amplification and transmission with $Er^{3+}$-doped fibre repeater by InGaAsP laser diode," *Electron. Lett.*, vol 25, no 3, 199 (1989)

[5.229]   A. Takada, K. Iwatsuki, and M. Saruwatari, "Picosecond laser diode pulse amplification up to 12 W by laser diode pumped erbium-doped fiber," *IEEE Photonics Technol. Lett.*, vol 2, no 2, 122 (1990)

[5.230]   K. Suzuki, Y, Kimura, and M. Nakazawa, "Subpicosecond soliton amplification and transmission using $Er^{3+}$-doped fibers pumped by InGaAsP laser diodes," *Optics Lett.*, vol 14, no 16, 865 (1989)

[5.231]   B. J. Ainslie, K. J. Blow, A. S. Gouveia-Neto, P. G. J. Wigley, A. S. B. Sombra, and J. R. Taylor, "Femtosecond soliton amplification in erbium-doped silica fibre," *Electron. Lett.*, vol 26, no 3, 186 (1990)

[5.232]   D. Y. Khrushchev, A. B. Grudinin, E. M. Dianov, D. V. Korobkin, V. A. Semenov, and A. M. Prokhorov, "Amplification of femtosecond pulses in $Er^{3+}$-doped single-mode optical fibres," *Electron. Lett.*, vol 26, no 7, 456 (1990)

[5.233]   M. Nakazawa, K. Kurokawa, H. Kubota, K. Suzuki, and Y. Kimura, "Femtosecond erbium-doped optical fiber amplifier," *Applied Phys. Lett.*, vol 57, no 7, 653 (1990)

[5.234]   D. J. Richardson, V. V. Afanasjev, A. B. Grudinin, and D. N. Payne, "Amplification of femtosecond pulses in a passive all-fiber soliton source," *Optics Lett.*, vol 17, no 22, 1597 (1992)

[5.235]   H. Izadpanah, D. Chen, C. Lin, M. A. Saifi, W. I. Way, A. Yi-Yan, and J. L. Gimlett, "Distortion-free amplification of high-speed test patterns up to 100 Gbit/s with erbium-doped fibre amplifiers," *Electron. Lett.*, vol 27, no 3, 196 (1991)

[5.236]   H. Sundaresan and G. E. Wickens, "Very high amplitude, minimal chirp optical pulse generation at 1.55 $\mu$m using multicontact DFB's and an erbium-doped fibre amplifier," *Electron. Lett.*, vol 26, no 11, 725 (1990)

[5.237]   C. Lin and T. L. Koch, "Chirping in 1.55 $\mu$m vapour-phase-transport distributed feedback (VPTDFB) semiconductor lasers under picosecond gain switching and 4 GHz modulation," *Electron. Lett.*, vol 21, 958 (1985)

[5.238]   Y. R. Shen, *The Principles of Nonlinear Optics,* Wiley-Interscience, New York, 1984

[5.239]   G. P. Agrawal, *Nonlinear Fiber Optics,* Academic Press, New York, 1989, Chapter 5

[5.240]   A. Hasegawa, *Optical Solitons in Fibers,* second edition, Springer-Verlag, New York, 1990

[5.241a]   A. C. Scott, F. Y. F. Chu, and D. W. McLaughlin, "The soliton: a new concept in applied science," *Proc. IEEE*, vol 61, no 10, 1443 (1973)

[5.241b] Y. S. Kivshar and B. A. Malomed, "Dynamics of solitons in nearly integrable systems," *Reviews of Mod. Phys.*, vol 61, no 4, 763–915 (1989)

[5.242] A. Hasegawa and Y. Kodama, "Signal transmission by optical solitons in monomode fibers," *Proc. IEEE*, vol 69, no 9, 1145 (1981)

[5.243] N. J. Zabusky and M. D. Kruskal, "Interaction of solitons in a collisionless plasma and the recurrence of initial states," *Phys. Rev. Lett.*, vol 15, 240 (1965)

[5.244] J. Scott-Russell, "Report on waves," *Proc. Roy. Soc. Edinburgh*, 319 (1844)

[5.245] L. F. Mollenauer, J. P. Gordon, and M. N. Islam, "Soliton propagation in long fibers with periodically compensated loss," *IEEE J. Quantum Electron.*, vol 22, no 1, 157 (1986)

[5.246] A. Hasegawa and F. Tappert, "Transmission of stationary optical pulses in dispersive dielectric fibers I. Anomalous dispersion," *Applied Phys. Lett.*, vol 23, no 3, 142 (1973)

[5.247] A. Hasegawa and F. Tappert, "Transmission of stationary optical pulses in dispersive dielectric fibers II. Normal dispersion," *Applied Phys. Lett.*, vol 23, no 4, 171 (1973)

[5.248] J. Satsuma and Y. Yajima, "Initial value problems of one-dimensional self-modulation of nonlinear waves in dispersive media," in supplement of *Progress in Theoretical Physics*, no 55, 284 (1974)

[5.249] N. J. Doran and K. J. Blow, "Solitons in optical communications," *IEEE J. Quantum Electron.*, vol 19, no 12, 1883 (1983)

[5.250] V. E. Zakharov and A. B. Shabat, "Exact theory of two-dimensional self-focusing and one-dimensional self-modulation of waves in nonlinear media," *Sov. Phys. JETP*, vol 34, no 1, 62 (1972)

[5.251] L. F. Mollenauer, R. H. Stolen, J. P. Gordon, and W. J. Tomlinson, "Extreme picosecond pulse narrowing by means of soliton effect in single-mode optical fibers," *Optics Lett.*, vol 8, no 5, 289 (1983)

[5.252] Y. Kodama and A. Hasegawa, "Amplification and reshaping of optical solitons in glass fiber II," *Optics Lett.*, vol 7, 339 (1982)

[5.253] K. J. Blow and N. J. Doran, "The asymptotic dispersion of solitons pulses in lossy fibres," *Opt. Comm.*, vol 52, no 5, 367 (1985)

[5.254] L. F. Mollenauer, R. H. Stolen, and J. P. Gordon, "Experimental observation of picosecond pulse narrowing and solitons in optical fibers," *Phys. Rev. Lett.*, vol 45, no 13, 1095 (1980)

[5.255] A. Hasegawa, "Numerical study of optical soliton transmission amplified periodically by the stimulated Raman process," *Applied Optics*, vol 23, 3302 (1984)

[5.256] L. F. Mollenauer, R. H. Stolen, and M. N. Islam, "Experimental demonstration of soliton propagation in long fibers: loss compensated by Raman gain," *Optics Lett.*, vol 10, no 5, 229 (1985)

[5.257] M. Nakazawa, K. Suzuki, and Y. Kimura, "20 GHz soliton amplification and transmission with an $Er^{3+}$-doped fiber," *Optics Lett.*, vol 14, no 19, 1065 (1989)

[5.258] M. Nakazawa, Y. Kimura, and K. Suzuki, "Ultralong dispersion-shifted erbium-doped fiber amplifier and its application to soliton transmission," *IEEE J. Quantum Electron.*, vol 26, 2103 (1990)

[5.259] M. Nakazawa and K. Kurokawa, "Femtosecond soliton transmission in 18 km-long dispersion-shifted, distributed erbium-doped fibre amplifier," *Electron. Lett.*, vol 27, no 15, 1369 (1991)

[5.260] K. Kurokawa and M. Nakazawa, "Femtosecond soliton transmission in 18km erbium-doped fibre amplifier with different pumping configurations," *Electron. Lett.*, vol 27, no 19, 1765 (1991)

[5.261] K. Kurokawa and M. Nakazawa, "Femtosecond soliton transmission characteristics in an ultralong erbium-doped fiber amplifier with different pumping configurations," *IEEE J. Quantum Electron.*, vol 28, no 9, 1922 (1992)

[5.262] F. M. Mitschke and L. F. Mollenauer, "Discovery of the soliton self-frequency shift," *Optics Lett.*, vol 11, no 10, 659 (1986)

[5.263] R. H. Stolen and E. P. Ippen, "Raman gain in glass optical waveguides," *Applied Phys. Lett.*, vol 22, 276 (1973)

[5.264] R. H. Stolen, C. Lee and R. K. Jain, "Development of the stimulated Raman spectrum in single-mode silica fibers," *J. Opt. Soc. Am. B*, vol 1, no 4, 652 (1984)

[5.265] J. P. Gordon, "Theory of the soliton self-frequency shift", *Optics Lett.*, vol 11, no 10, 662 (1986)

[5.266] Y. Kodama and A. Hasegawa, "Nonlinear pulse propagation in a monomode dielectric guide," *IEEE J. Quantum Electron.*, vol 23, 510 (1987)

[5.267] K. Kurokawa and M. Nakazawa, "Wavelength-dependent amplification characteristics of femtosecond erbium-doped optical fiber amplifiers," *Applied Phys. Lett.*, vol 58, no 25, 2871 (1991)

[5.268] I. R. Gabitov, M. Romagnoli, and S. Wabnitz, "Femtosecond soliton collapse and coherent pulse train generation in erbium-doped fiber amplifiers," *Applied Phys. Lett.*, vol 59, no 15, 1811 (1991)

[5.269] M. Ding and K. Kikuchi, "Analysis of soliton transmission in optical fibers with the soliton self-frequency shift being compensated by distributed frequency dependent gain," *IEEE Photonics Technol. Lett.*, vol 4, no 5, 497 (1992)

[5.270] G. P. Agrawal, "Amplification of ultrashort solitons in erbium-doped fiber amplifiers," *IEEE Photonics Technol. Lett.*, vol 2, no 12, 875 (1990)

[5.271] A. Hasegawa, K. Tai, and N. Bekki, "Fission of optical solitons induced by stimulated Raman effect," in *Proc. International Quantum Electronics Conference, IQEC '88*, paper ThE3, Optical Society of America, Washington DC, 1988

[5.272] Y. Kodama and K. Nozaki, "Soliton interactions in optical fibers," *Optics Lett.*, vol 12, no 12, 1038 (1987)

[5.273] J. P. Gordon, "Interaction forces among solitons in optical fibers," *Optics Lett.*, vol 8, no 11, 596 (1983)

[5.274] K. J. Blow and N. J. Doran, "Bandwidth limits of nonlinear (soliton) optical communication systems," *Electron. Lett.*, vol 19, 429 (1983)

[5.275] B. Hermansson and D. Yevick, "Numerical investigation of soliton interaction," *Electron. Lett.*, vol 19, 570 (1983)

[5.276] F. M. Mitsche and L. F. Mollenauer, "Experimental observation of interaction forces between solitons in optical fibers," *Optics Lett.*, vol 12, no 5, 355 (1987)

[5.277] C. Desem and P. L. Chu, "Reducing soliton interaction in single-mode optical fibers," *IEE Proc. J.*, vol 134, 145 (1987)

[5.278] C. Desem and P. L. Chu, "Soliton interaction in the presence of loss and periodic amplification in optical fibers", *Optics Lett.*, vol 12, 349 (1987)

[5.279] C. Desem and P. L. Chu, "Soliton propagation in the presence of source chirping and mutual interaction in single-mode optical fibres," *Electron. Lett.*, vol 23, 260 (1987)

[5.280] D. Anderson and M. Lisak, "Bandwidth limits due to mutual pulse interaction in optical soliton communication systems," *Optics Lett.*, vol 11, no 3, 174 (1986)

[5.281] I. M. Uzumov, V. D. Stoev, and T. I. Tzoleva, "*N*-soliton interaction in trains of unequal soliton pulses in optical fibers," *Optics Lett.*, vol 17, no 20, 1417 (1992)

[5.282] M. Nakazawa, K. Suzuki, E. Yamada, and H. Kubota, "Observation of nonlinear interactions in 20 Gbit/s soliton transmission over 500 km using erbium-doped fibre amplifiers," *Electron. Lett.*, vol 27, no 18, 1662 (1991)

[5.283] L. F. Mollenauer, S. G. Evangelides, and J. P. Gordon, "Wavelength division multiplexing with solitons in ultra-long distance transmission using lumped amplifiers," *IEEE J. Lightwave Technol.*, vol 9, no 3, 362 (1991)

[5.284] P. A. Andrekson, N. A. Olsson, P. C. Becker, J. R. Simpson, T. Tanbun-Ek, R. A. Logan, and K. W. Wecht, "Observation of multiple wavelength soliton collisions in optical systems with fiber amplifiers," *Applied Phys. Lett.*, vol 57, no 17, 1715 (1990)

[5.285] P. A. Andrekson, N. A. Olsson, J. R. Simpson, T. Tanbun-Ek, R. A. Logan, P. C. Becker, and K. W. Wecht, "Soliton collision interaction force dependence on wavelength separation in fibre amplifier based systems," *Electron. Lett.*, vol 26, no 18, 1499 (1990)

[5.286] P. A. Andrekson, N. A. Olsson, J. R. Simpson, T. Tanbun-Ek, R. A. Logan, and K. W. Wecht, "Observation of collision induced temporary soliton carrier frequency shifts in ultra-long fiber transmission systems," *IEEE J. Lightwave Technol.*, vol 9, no 9, 1132 (1991)

[5.287] Y. Kodama and S. Wabnitz, "Reduction of soliton interaction forces by bandwidth limited amplification," *Electron. Lett.*, vol 27, 1931 (1991)

[5.288] M. Nakazawa and H. Kubota, "Physical interpretation of reduction of soliton interaction forces by bandwidth limited amplification," *Electron. Lett.*, vol 28, no 10, 958 (1992)

[5.289] Y. Kodama and S. Wabnitz, Comment on "Physical interpretation of reduction of soliton interaction forces by bandwidth limited amplification," and reply by M. Nakazawa and H. Kubota, *Electron, Lett.*, vol 29, no 2, 226 (1993)

[5.290] K. Kurokawa, H. Kubota, and M. Nakazawa, "Soliton self-frequency shift accelerated by femtosecond soliton interaction," *Electron. Lett.*, vol 28, no 22, 2052 (1992)

[5.291] K. Kurokawa, H. Kubota, and M. Nakazawa, "Significant modification of femtosecond soliton interaction in gain medium by small subpulses," *Electron. Lett.*, vol 28, no 25, 2334 (1992)

[5.292] J. P. Gordon and H. A. Haus, "Random walk of coherently amplified solitons in optical fiber transmission," *Optics Lett.*, vol 11, no 10, 665 (1986)

[5.293] D. Marcuse, "An alternative derivation of the Gordon–Haus effect," *IEEE J. Lightwave Technol.*, vol 10, no 2, 273 (1992)

[5.294] D. Marcuse, "Simulations to demonstrate the Gordon–Haus effect," *Optics Lett.*, vol 17, no 1, 35 (1992)

[5.295] E. Desurvire, "Raman amplification of recirculating pulses in a reentrant fiber loop," *Optics Lett.*, vol 10, no 2, 83 (1985)

[5.296] E. Desurvire, "Theory and implementation of a Raman active fiber delay line," *IEEE J. Lightwave Technol.*, vol 4, no 4, 426 (1986)

[5.297] L. F. Mollenauer, M. J. Neubelt, S. G. Evangelides, J. P. Gordon, J. R. Simpson and L. G. Cohen, "Experimental study of soliton transmission over more than 10,000 km in dispersion-shifted fiber," *Optics Lett.*, vol 15, no 21, 1203 (1990)

[5.298] T. Georges and F. Favre, "Influence of soliton interaction on amplifier-noise induced jitter: a first-order analytical solution," *Optics Lett.*, vol 16, no 21, 1656 (1991)

[5.299] K. Smith and L. F. Mollenauer, "Experimental observation of soliton interaction over long fiber paths: discovery of a long-range interaction," *Optics Lett.*, vol 14, no 22, 1284 (1989)

[5.300] E. M. Dianov, A. V. Luchnikov, A. N. Pilipetskii, and A. N. Starodumov, "Electro-riction mechanism of soliton interaction in optical fibers," *Optics Lett.*, vol 15, no 6, 314 (1990)

[5.301] E. M. Dianov, A. V. Luchnikov, A. N. Pilipetskii, and A. N. Starodumov, "Long range interaction of soliton pulse trains in single-mode fibre," *Sov. Lightwave Comm.,* vol I, 37 (1991)

[5.302] E. M. Dianov, A. V. Luchnikov, A. N. Pilipetskii, and A. M. Prokhorov, "Long-range interaction of solitons in ultra-long communication systems," *Sov. Lightwave Comm.,* vol I, 235 (1991)

[5.303] L. F. Mollenauer and R. H. Stolen, "The soliton laser," *Optics Lett.,* vol 9, no 1, 13 (1984)

[5.304] E. M. Dianov, A. M. Prokhorov, and V. N. Serkin, "Dynamics of ultrashort-pulse generation by Raman fiber lasers: cascade self-mode-locking, optical pulsons and solitons," *Optics Lett.,* vol 11, no 3, 168 (1986)

[5.305] J. D. Kafka and T. Baer, "Fiber Raman soliton laser pumped by a Nd:YAG laser," *Optics Lett.,* vol 12, no 3, 181 (1987)

[5.306] A. Hasegawa, "Generation of a train of soliton pulses by induced modulational instability in optical fibers," *Optics Lett.,* vol 9, no 7, 288 (1984)

[5.307] K. Tai, A. Tomita, J. L. Jewell, and A. Hasegawa, "Generation of subpicosecond solitonlike optical pulses at 0.3 THz repetion rate by induced modulational instability," *Applied Phys. Lett.,* vol 49, no 5, 236 (1986)

[5.308] M. C. Wu, Y. K. Chen, T. Tanbun-Ek, R. A. Logan, M. A. Chin, and G. Raybon, "Transform-limited 1.4 ps optical pulses from a monolithic colliding-pulse mode-locked quantum well laser," *Applied Phys. Lett.,* vol 57, 759 (1990)

[5.309] M. C. Wu, Y. Chen, T. Tanbun-Ek, R. A. Logan, and M. A. Chin, "Tunable monolithic colliding-pulse mode-locked quantum well lasers," *IEEE Photonics Technol. Lett.,* vol 3, no 10, 874 (1991)

[5.310] Y. K. Chen, M. C. Wu, T. Tanbun-Ek, R. A. Logan, and M. A. Chin, "Subpicosecond monolithic colliding-pulse mode-locked multiple quantum well laser," *Applied Phys. Lett.,* vol 58, 1253 (1991)

[5.311] M. Nakazawa, K. Suzuki, and Y. Kimura, "Transform-limited pulse generation in the gigahertz region from a gain-switched distributed feedback laser diode using spectral windowing," *Optics Lett.,* vol 15, 715 (1990)

[5.312] K. Iwatsuki, S. Nishi, M. Saruwatari, and K. Nakagawa, *Optics Lett.,* vol 2, 507 (1990)

[5.313] H. F. Liu, S. Oshiba, Y. Ogawa, and Y. Kawai, "Method of generating nearly transform-limited pulses from gain-switched distributed-feedback laser diodes and its application to soliton transmission," *Optics Lett.,* vol 17, no 1, 64 (1992)

[5.314] M. Ding and K. Kikuchi, "Limits of long-distance soliton transmission in optical fibers with laser diodes as pulse sources," *IEEE Photonics Technol. Lett.,* vol 4, no 6, 667 (1992)

[5.315] P. V. Mamyshev, S. V. Chernikov, and E. M. Dianov, "Generation of fundamental soliton trains for high bit rate optical fiber communications," *IEEE J. Quantum Electron.,* vol 27, 2347 (1991)

[5.316] E. M. Dianov, P. V. Mamyshev, A. M. Prokhorov, and S. V. Chernikov, "Generation of a train of fundamental solitons at high repetition rate in optical fiber," *Optics Lett.,* vol 14, 1008 (1989)

[5.317] S. V. Chernikov, J. R. Taylor, P. V. Mamyshev, and E. M. Dianov, "Generation of soliton pulse train in optical fibre using two cw single-mode diode lasers," *Electron. Lett.,* vol 28, no 10, 931 (1992)

[5.318] R. H. Stolen, "Nonlinear properties of optical fibers," in *Optical Fiber Telecommunications,* edited by S. E. Miller and A. G. Chynoweth, Academic Press, New York, 1979

[5.319] A. Yariv, *Quantum Electronics,* second edition, John Wiley, New York, 1975

[5.320] Y. Yamamoto and T. Mukai, "Fundamentals of optical amplifiers," *Optical and*

*Quantum Electron.*, vol 21, 1 (1989)

[5.321]  Y. Ohishi, T. Kanamori, T. Kitagawa, and S. Takahashi, "$Pr^{3+}$-doped fiber amplifier operating at 1.3 $\mu$m," in *Proc. Conference on Optical Fiber Communications, OFC'91*, paper PD2, p10, Optical Society of America, Washington DC, 1991

[5.322]  Y. Ohishi, T. Kanamori, T. Kitagawa, S. Takahashi, E. Snitzer, and G. H. Sigel, "$Pr^{3+}$-doped fluoride fiber amplifier operating at 1.3 $\mu$m," *Optics Lett.*, vol 16, no 22, 1747 (1991)

[5.323]  Y. Durteste, M. Monerie, J. Y. Allain, and H. Poignant, "Amplification and lasing at 1.3 $\mu$m in praseodymium-doped fluorozirconate fibres," *Electron. Lett.*, vol 27, no 8, 626 (1991)

[5.324]  S. F. Carter, D. Szebesta, S. T. Davey, R. Wyatt, M. C. Brierley, and P. W. France, "Amplification and lasing at 1.3um in a $Pr^{3+}$-doped single-mode fluorozirconate fibre," *Electron. Lett.*, vol 27, no 8, 628 (1991)

[5.325]  D. W. Dewak, R. S. Deol, J. Wang, G. Wylangowski, J. A. Mederios Neto, B. N. Samson, R. I. Laming, W. S. Brocklesby, D. N. Payne, A. Jha, M. Poulain, S. Otero, S. Surinach, and M. D. Baro, "Low phonon energy glasses for efficient 1.3 $\mu$m optical fibre amplifiers," *Electron. Lett.*, vol 29, no 2, 237 (1993)

[5.326]  Y. Ohishi, T. Kanamori, T. Nishi, S. Takahashi, and E. Snitzer, "Gain characteristics of $Pr^{3+}$–$Yb^{3+}$ codoped fluoride fiber for 1.3 $\mu$m amplification," *IEEE Photonics Technol. Lett.*, vol 3, no 11, 990 (1991)

[5.327]  M. Yamada, M. Shimizu, Y. Ohishi, J. Temmyo, M. Wada, T. Kanamori, M. Horiguchi, and S. Takahashi, "15.1 dB gain $Pr^{3+}$-doped fluoride fiber amplifier pumped by high-power laser diode modules," *IEEE Photonics Technol. Lett.*, vol 4, no 9, 994 (1992)

[5.328]  Y. Ohishi, T. Kanamori, J. Temmyo, M. Wada, M. Yamada, M. Shimizu, K. Yoshino, H. Hanafusa, M. Horiguchi, and S. Takahashi, "Laser diode pumped $Pr^{3+}$-doped and $Pr^{3+}$–$Yb^{3+}$-codoped fluoride fibre amplifiers operating at 1.3 $\mu$m," *Electron. Lett.*, vol 27, no 22, 1995 (1991)

[5.329]  Y. Miyajima, T. Sugawa, and Y. Fukasaku, "38.2 dB amplification at 1.31 $\mu$m and possibility of 0.98 $\mu$m pumping in $Pr^{3+}$-doped fluoride fibre," *Electron. Lett.*, vol 27, no 19, 1706 (1991)

[5.330]  T. Sugawa and Y. Miyajima, "Gain and output-saturation limit evaluation in $Pr^{3+}$-doped fluoride fiber amplifier operating in the 1.3 $\mu$m band," *IEEE Photonics Technol. Lett.*, vol 3, no 7, 616 (1991)

[5.331]  T. Sugawa and Y. Miyajima, "Noise characteristics of $Pr^{3+}$-doped fluoride fiber amplifiers," *Electron. Lett.*, vol 28, no 3, 246 (1992)

[5.322]  Y. Ohishi, T. Kanamori, T. Nishi, and S. Takahashi, "A high gain, high output saturation power $Pr^{3+}$-doped fluoride fiber amplifier operating at 1.3 $\mu$m," *IEEE Photonics Technol. Lett.*, vol 3, no 8, 715 (1991)

[5.333]  Y. Ohishi, T. Kanamori, T. Nishi, S. Takahashi, and E. Snitzer, "Concentration effect of gain of $Pr^{3+}$-doped fluoride fiber for 1.3 $\mu$m amplification," *IEEE Photonics Technol. Lett.*, vol 4, no 12, 1338 (1992)

[5.334]  B. Pedersen, W. J. Miniscalco, and R. S. Quimby, "Optimization of $Pr^{3+}$:ZBLAN fiber amplifiers," *IEEE Photonics Technol. Lett.*, vol 4, no 5, 446 (1992)

[5.335]  P. Urquhart, "Praseodymium-doped fiber amplifiers: theory of 1.3 $\mu$m operation," *IEEE J. Quantum Electron.*, vol 28, no 10, 1962 (1992)

[5.336]  M. Karasek, "Numerical analysis of $Pr^{3+}$-doped fluoride fiber amplifier," *IEEE Photonics Technol. Lett.*, vol 4, no 11, 1266 (1992)

[5.337]  R. Lobbett, R. Wyatt, P. Eardley, T. J. Whitley, P. Smyth, D. Szebesta, S. F. Carter,

S. T. Davey, C. A. Millar, and M. C. Brierley, "System characterization of high gain and high saturated output power, $Pr^{3+}$-doped fluorozirconate fibre amplifier at 1.3 $\mu$m," *Electron. Lett.*, vol 27, no 16, 1472 (1991)

[5.338] F. F. Rühl, "Figures of merit for doped fibre amplifiers in 1300 nm and 1550 nm windows," *Electron. Lett.*, vol 27, no 18, 1605 (1991)

[5.339] C. J. Koester and E. Snitzer, "Amplification in a fiber laser," *Optics Lett*, vol 3, no 10, 1182 (1964)

[5.340] G. C. Holst and E. Snitzer, "Detection with a fiber laser preamplifier at 1.06 $\mu$m," *IEEE J. Quantum Electron.*, vol 5, 319 (1969)

[5.341] B. Ross and E. Snitzer, "Optical amplification of 1.06$\mu$ $InAs_{1-x}P_x$ injection-laser emission," *IEEE J. Quantum Electron.*, vol 6, no 6, 361 (1970)

[5.342] M. L. Dakss and W. J. Miniscalco, "A large-signal model and signal/noise ratio analysis for $Nd^{3+}$-doped fiber amplifiers at 1.3 $\mu$m," in Fiber laser sources and amplifiers II, *SPIE Proc.*, vol 1373, 1990

[5.343] S. Zemon, B. Pedersen, G. Lambert, W. J. Miniscalco, B. T. Hall, R. C. Folweiler, B. A. Thompson, and L. J. Andrews, "Excited-state-absorption cross-sections and amplifier modeling in the 1300-nm region for Nd-doped glasses," *IEEE Photonics Technol. Lett.*, vol 4, no 3, 244 (1992)

[5.344] M. L. Dakss and W. J. Miniscalco, "Fundamental limits on $Nd^{3+}$-doped fiber amplifier performance at 1.3 $\mu$m," *IEEE Photonics Technol. Lett.*, vol 2, no 9, 650 (1990)

[5.345] M. C. Brierley and C. A. Millar, "Amplification and lasing at 1350 nm in a neodymium doped fluorozirconate fibre," *Electron. Lett.*, vol 24, no 7, 438 (1988)

[5.346] Y. Miyajima, T. Komukai, and T. Sugawa, "1.31–1.36 $\mu$m optical amplification in $Nd^{3+}$-doped fluorozirconate fibre," *Electron. Lett.*, vol 26, no 3, 194 (1990)

[5.347] M. Brierley, S. Carter, P. France, and J. E. Pedersen, "Amplification in the 1300 nm telecommunications window in a Nd-doped fluoride fibre," *Electron. Lett.*, vol 26, no 5, 329 (1990)

[5.348] T. Sugawa, Y. Miyajima, and T. Komukai, "10 dB gain and high saturation power in a $Nd^{3+}$-doped fluorozirconate fibre amplifier," *Electron. Lett.*, vol 26, no 24, 2042 (1990)

[5.349] M. Øbro, B. Pedersen, A. Bjarklev, J. H. Povlsen and J. E. Pedersen, "Highly improved fibre amplifier for operation around 1300 nm," *Electron. Lett.*, vol 27, no 5, 470 (1991)

[5.350] M. Øbro, J. E. Pedersen, and M. C. Brierley, "Gain enhancement in $Nd^{3+}$-doped ZBLAN fibre amplifier using mode coupling filter," *Electron. Lett.*, vol 28, no 1, 99 (1992)

[5.351] A. Bjarklev, T. Rasmussen, J. H. Povlsen, O. Lumholt, K. Rottwitt, S. Dahl-Petersen, and C. C. Larsen, "9 dB improvement of 1300 nm optical amplifier by amplified spontaneous emission suppressing fibre design," *Electron. Lett.*, vol 27, no 19, 1701 (1991)

[5.352] T. Rasmussen, A. Bjarklev, O. Lumholt, M. Øbro, B. Pedersen, J. H. Povlsen, and K. Rottwitt, "Optimum design of Nd-doped fiber optical amplifiers," *IEEE Photonics Technol. Lett.*, vol 4, no 1, 49 (1992)

[5.353] J. E. Pedersen, M. C. Brierley, and R. A. Lobbett, "Noise characterization of a neodymium-doped fluoride fiber amplifier and its performance in a 2.4 Gb/s system," *IEEE Photonics Technol. Lett.*, vol 2, no 10, 750 (1990)

[5.354] D. C. Tran, G. H. Sigel, and B. Bendow, "Heavy metal fluoride glasses and fibers: a review," *IEEE J. Lightwave Technol.*, vol 2, no 5, 566 (1984)

[5.355] P. W. France, S. F. Carter, M. W. Moore, and C. R. Day, "Progress in fluoride abres for optical communications," *Br. Telecom Technol J.*, vol 5, no 2, 28 (1987)

[5.356]  D. Ronarc'h, J. Y. Allain, M. Guibert, M. Monerie, and H. Poignant, "Erbium-doped fluoride fibre optical amplifier operating around 2.75 $\mu$m," *Electron. Lett.*, vol 26, no 13, 903 (1990)

[5.357]  D. Ronarc'h, M. Guibert, F. Auzel, D. Mechenin, J. Y. Allain, and H. Poignant, "35 dB optical gain at 2.716 $\mu$m in erbium doped ZBLAN fibre pumped at 0.642 $\mu$m," *Electron. Lett.*, vol 27, no 6, 511 (1991)

[5.358]  T. J. Whitley, C. A. Millar, M. C. Brierley, and S. F. Carter, "23 dB gain upconversion pumped erbium doped fibre amplifier operating at 850 nm," *Electron. Lett.*, vol 27, no 2, 184 (1991)

[5.359]  R. G. Smart, J. N. Carter, A. C. Tropper, D. C. Hanna, S. F. Carter, and D. Szebesta, "20 dB gain thulium-doped fluorozirconate fibre amplifier operating at around 0.8 $\mu$m," *Electron. Lett.*, vol 27, no 13, 1123 (1991)

[5.360]  R. G. Smart, A. C. Tropper, D. C. Hanna, J. N. Carter, S. T. Davey, S. F. Carter, and D. Szebesta, "High efficiency, low threshold amplification and lasing at 0.8 $\mu$m in monomode $Tm^{3+}$-doped fluorozirconate fibre," *Electron. Lett.*, vol 28, no 1, 58 (1992)

[5.361]  T. Komukai, T. Yamamoto, T. Sugawa, and Y. Miyajima, "1.47 $\mu$m band $Tm^{3+}$ doped fluoride fibre amplifier using a 1.064 $\mu$m upconversion pumping scheme," *Electron. Lett.*, vol 29, no 1, 110 (1993)

[5.362]  J. N. Carter, R. G. Smart, D. C. Hanna, and A. C. Tropper, "Lasing and amplification in the 0.8 $\mu$m region in thulium doped fluorozirconate fibres," *Electron. Lett.*, vol 26, no 21, 1759 (1990)

[5.363]  I. Sankawa, H. Izumita, S-I. Furukawa and K. Ishihara, "An optical fiber amplifier for wide-band wavelength range around 1.65 $\mu$m," *IEEE Photonics Technology Lett.*, vol 2, no 6, 422 (1990)

[5.364a]  E. P. Ippen and R. H. Stolen, "Stimulated Brillouin scattering in optical fibers," *Applied Phys. Lett.*, vol 21, no 11, 539 (1972)

[5.364b]  R. H. Stolen and E. P. Ippen, "Raman gain in glass optical waveguides," *Applied Phys. Lett.*, vol 22, no 6, 276 (1973)

[5.365]  R. H. Stolen, "Fiber Raman lasers," *Fiber and Integrated Optics*, vol 3, no 1, 21 (1980)

[5.366]  K. O. Hill, B. S. Kawazaki, and D. C. Johnson, "CW Brillouin fiber laser," *Applied Phys. Lett.*, vol 28, no 10, 608 (1976)

[5.367]  R. G. Smith, "Optical power handling capacity of low loss optical fibers as determined by stimulated Raman and Brillouin scattering," *Applied Optics.*, vol 11, no 11, 2489 (1972)

[5.368]  Y. R. Shen and N. Bloembergen, "Theory of stimulated Brillouin and Raman scattering," *Phys. Rev.*, vol 137, no 6A, 1787 (1965)

[5.369]  C. L. Tang, "Saturation and spectral characteristics of the Stokes emission in the stimulated Brillouin process," *J. Applied Phys.*, vol 37, no 8, 2945 (1966)

[5.370]  J. Auyeung and A. Yariv, "Spontaneous and stimulated Raman scattering in low loss fibers," *IEEE J. Quantum Electron.*, vol 14, no 5, 347 (1978)

[5.371]  R. W. Davies, P. Melman, W. H. Nelson, M. L. Dakss, and B. M. Foley, "Output moments and photon statistics in fiber Raman amplification," *IEEE J. Lightwave Technol.*, vol 5, no 8, 1068 (1987)

[5.372]  R. H. Stolen, C. Lee, and R.K, "Raman response function of silica-core fibers," *J. Opt. Soc. Am. B.*, vol 6, no 6, 1159 (1989)

[5.373]  K. Mochizuki, N. Edagawa, and Y. Iwamoto, "Amplified spontaneous Raman scattering in fiber Raman amplifiers," *IEEE J. Lightwave Technol.*, vol 4, no 9, 1328 (1986)

[5.374] D. Cotter, "Stimulated Brillouin scattering in monomode optical fiber," *J. Opt. Comm.*, vol 4, no 1, 10 (1983)

[5.375] I. Bar-Joseph, A. A. Friesem, E. Lichtman, and R. G. Waarts, "Steady and relaxation oscillations of stimulated Brillouin scattering in single-mode optical fibers," *J. Opt. Soc. Am.*, vol 2, no 10, 1606 (1985)

[5.376] N. A. Olsson and J. P. Van der Ziel, "Characteristics of a semiconductor laser pumped Brillouin amplifier with electronically controlled bandwidth," *IEEE J. Lightwave Technol.*, vol 5, no 1, 147 (1987)

[5.377] R. H. Stolen, "Polarization effects in fiber Raman and Brillouin lasers," *IEEE J. Quantum Electron.*, vol 15, no 10, 1157 (1979)

[5.378] Y. Aoki, "Fibre Raman fiber amplifier properties for applications to long-distance optical communications," *Optical and Quantum Electron.*, vol 21, 89 (1989)

[5.379] R. W. Tkach and A. R. Chraplyvy, "Fibre Brillouin amplifiers," *IEEE J. Quantum Electron.*, vol 21, 105 (1989)

[5.380] R. W. Tkach, A. R. Chraplyvy, and R. M. Derosier, "Spontaneous Brillouin scattering for single-mode optical fibre characterization," *Electron. Lett.*, vol 22, no 19, 1011 (1986)

[5.381] A. R. Chraplyvy and R. W. Tkach, "Narrowband tunable optical filter for channel selection in densely packed WDM systems," *Electron. Lett.*, vol 22, no 20, 1084 (1986)

[5.382] R. W. Tkach, A. R. Chraplyvy, R. M. Derosier, and H. T. Shang, "Optical demodulation and amplification of FSK signals of FSK signals using AlGaAs lasers," *Electron. Lett.*, vol 24, no 5, 260 (1988)

[5.383] E. Desurvire, M. Papuchon, J. P. Pocholle, J. Raffy, and D. Ostrowsky, "High-gain optical amplification of laser diode signal by Raman scattering in single-mode fibres," *Electron. Lett.*, vol 19, no 19, 751 (1983)

[5.384] N. Edagawa, K. Mochizuki, and Y. Iwamoto, "Simultaneous amplification of wavelength-division-multiplexed signals by a highly efficient fibre Raman amplifier pumped by high-power semiconductor lasers," *Electron. Lett.*, vol 23, no 5, 196 (1987)

[5.385] T. Nakashima, S. Seikai, M. Nakazawa, and Y. Negishi, "Theoretical limit of repeater spacing in an optical transmission line utilizing Raman amplification," *IEEE J. Lightwave Technol.*, vol 4, no 8, 1267 (1986)

[5.386] G. N. Brown, "Raman fibre amplifier characteristics for use in optical transmission systems," *Br. Telecom Technol. J.*, vol 5, no 2, 45 (1987)

[5.387] S. Chi and M. S. Kao, "Bidirectional optical fiber transmission systems using Raman amplification," *IEEE J. Lightwave Technol.*, vol 6, no 2, 312 (1988)

[5.388] Y. Aoki, "Properties of fiber Raman amplifiers and their applicability to digital optical communications systems," *IEEE J. Lightwave Technol.*, vol 6, no 7, 1225 (1988)

[5.389] T. Horiguchi, T. Sato, and Y. Koyamada, "Stimulated Raman amplification of 1.6 $\mu m$ band pulsed light in optical fibers," *IEEE Photonics Technol. Lett.*, vol 4, no 1, 64 (1992)

[5.390] R. H. Stolen, J. E. Bjorkholm, and A. Ashkin, "Phase-matched three-wave mixing in silica fiber optical waveguides," *Applied Phys. Lett.*, vol 24, 308 (1974)

[5.391] R. H. Stolen, "Phase-matched-stimulated four-photon mixing in silica-fiber wave-guides," *IEEE J. Quantum Electron.*, vol 11, no 3, 100 (1975)

[5.392] R. H. Stolen and J. E. Bjorkholm, "Parametric amplification and frequency conversion in optical fibers," *IEEE J. Quantum Electron.*, vol 18, no 7, 1062 (1982)

[5.393] A. Vatarescu, "Light conversion in nonlinear monomode optical fibers," *IEEE J. Lightwave Technol.*, vol 5, no 12, 1652 (1987)

[5.394] S. J. Garth, "Phase matching the stimulated four-photon mixing process on single-mode fibers operating in the 1.55 $\mu m$ region," *Optics Lett.*, vol 13, no 12, 1117 (1988)

[5.395] Y. Chen and A. W. Snyder, "Four-photon parametric mixing in optical fibers: effect of pump depletion," *Optics Lett.*, vol 14, no 1, 87 (1989)

[5.396] K. Washio, K. Inoue, and S. Kishida, "Efficient large-frequency-shifted three wave mixing in low dispersion wavelength region in a single-mode optical fiber ," *Electron. Lett.*, vol 16, 658 (1980)

[5.397] M. Ohashi, K. Kitayama, Y. Ishida, and N. Uchida, "Phase-matched light amplification by three-wave mixing process in a birefringent fibre due to externally applied stress," *Applied Phys. Lett.*, vol 41, 1111 (1982)

[5.398] J. P. Pocholle, J. Raffy, M. Papuchon, and E. Desurvire, "Raman and four photon mixing amplification in single mode fibers," *Optical Engineering*, vol 24, no 4, 600 (1985)

[5.399] C. Lin, W. A. Reed, A. D. Pearson, and H. T. Shang, "Phase matching in the minimum chromatic dispersion region of single-mode fibers for stimulated four-photon mixing," *Optics Lett.*, vol 10, 493 (1981)

[5.400] C. Lin, W. A. Reed, A. D. Pearson, H. T. Shang, and P. F. Glodis, "Designing single mode fibres for near-IR (1.1–1.7 $\mu$m) frequency generation by phase-matched four-photon mixing in the minimum chromatic dispersion region," *Electron. Lett.*, vol 18, no 2, 87 (1982)

[5.401] J. P. Pocholle, M. Papuchon, J. Raffy, and E. Desurvire, "Nonlinear optical amplification in single-mode fibers: potential applications to optical communication systems," in *Proc. Conference on Lasers and Electro-Optics, CLEO'84*, paper FR1, p266, Optical Society of America, Washington DC, 1984

[5.402] J. P. Pocholle, M. Papuchon, J. Raffy, and E. Desurvire, "Nonlinearities and optical amplification in single-mode fibers," in *Revue Technique of Thomson-CSF*, vol 22, no 2, 187 (1990)

[5.403] K. I. Kitayama and M. Ohashi, "Frequency tuning for stimulated four-photon mixing by bending-induced birefringence in a single-mode fiber," *Applied Phys. Lett.*, vol 41, no 7, 619 (1982)

[5.404] K. Inoue and H. Toba, "Wavelength conversion experiment using fiber four-wave mixing," *IEEE Photonics Technol. Lett.*, vol 4, no 1, 69 (1992)

[5.405] P. A. Andrekson, N. A. Olsson, J. R. Simpson, T. Tanbun-Ek, R. A. Logan, and M. Haner, "16 Gbit/s all-optical demultiplexing using four-wave mixing," *Electron. Lett.*, vol 27, no 11, 922 (1991)

[5.406] M. Nakazawa, Y. Kimura, E. Yoshida, and K. Suzuki, "Efficient erbium-doped fibre amplifier pumped at 820 nm," *Electron. Lett.*, vol 26, no 23, 1936 (1990)

[5.407] J. L. Zyskind, D. J. DiGiovanni, J. W. Sulhoff, P. C. Becker, and C. H. Brito-Cruz, "High performance erbium-doped fiber amplifier pumped at 1.48 $\mu$m and 0.97 $\mu$m," in *Proc. Topical Meeting on Optical Amplifiers and Applications*, paper PDP6, 1990, Optical Society of America, Washington DC, 1990

[5.408] K. Nakamura, M. Kimura, S. Yoshida, T. Hidaka, and M. Mitsuhashi, "Raman amplification of 1.50 $\mu$m laser diode light in a low fiber loss region ," *IEEE J. Lightwave Technol.*, vol 2, no 4, 379 (1984)

## CHAPTER 6

[6.1] A. Hasegawa, "Amplification and reshaping of optical solitons in a glass fiber IV: use of the stimulated Raman process," *Optics Lett.*, vol 8, no 12, 650 (1983)

[6.2] S. P. Craig-Ryan, B. J. Ainslie and C. A. Millar, "Fabrication of long lengths of low excess loss erbium-doped optical fibre," *Electron. Lett.*, vol 26, no 3, 185 (1990)

[6.3]   J. R. Simpson, H. T. Shang, L. F. Mollenauer, N. A. Olsson, P. C. Becker, K. S. Kranz, P. J. Lemaire and M. J. Neubelt, "Performance of a distributed erbium-doped dispersion-shifted fiber amplifier," *IEEE J. Lightwave Technol.*, vol 9, no 2, 228 (1991); J. R. Simpson, L. F. Mollenauer, K. S. Kranz, P. J. Lemaire, N. A. Olsson, H. T. Shang, and P. C. Becker, "A distributed erbium-doped fiber amplifier," in *Proc. Conference on Optical Fiber Communications, OFC '90*, paper PD19, Optical Society of America, Washington DC, 1990

[6.4]   D. L. Williams, S. T. Davey, D. M. Spirit, G. R. Walker, and B. J. Ainslie, "Transmission and saturation performance of dispersion-shifted distributed erbium fibre," *Electron. Lett.*, vol 27, no 10, 812 (1991)

[6.5]   M. Nakazawa, Y. Kimura, and K. Suzuki, "Gain-distribution measurements along an ultralong erbium-doped fiber amplifier using optical-time-domain reflectometry," *Optics Lett.*, vol 15, no 21, 1200 (1990)

[6.6]   D. L. Williams, S. T. Davey, D. M. Spirit, and B. J. Ainslie, "Transmission over 10 km or erbium doped fibre with ultralow signal power excursion," *Electron. Lett.*, vol 26, no 18, 1517 (1990)

[6.7]   E. Desurvire, "Analysis of distributed erbium-doped fiber amplifiers with fiber background loss," *IEEE Photonics Technol. Lett.*, vol 3, no 7, 625 (1991)

[6.8]   D. N. Chen and E. Desurvire, "Noise performance evaluation of distributed erbium-doped fiber amplifiers with bidirectional pumping at 1.48 µm," *IEEE Photonics Technol. Lett.*, vol 4, no 1, 52 (1992)

[6.9]   K. Rottwitt, J. H. Povlsen, A. Bjarklev, O. Lumholt, B. Pedersen, and T. Rasmussen, "Optimum signal wavelength for a distributed erbium-doped fiber amplifier," *IEEE Photonics Technol. Lett.*, vol 4, no 7, 714 (1992)

[6.10]  K. Rottwitt, S. L. Hansen, J. H. Povlsen, P. Thorsen, K. Dybdal, A. Bjarklev, and C. C. Larsen, "Experimental analysis of signal wavelength for a distributed erbium-doped fiber," in *Proc. Conference on Optical Fiber Communications, OFC '93*, paper TuL4, p55, Optical Society of America, Washington DC, 1993

[6.11]  K. Rottwitt, A. Bjarklev, O. Lumholt, J. H. Povlsen, and T. R. Rasmussen, "Design of long distance distributed erbium-doped fibre amplifier," *Electron. Lett.*, vol 28, no 3, 287 (1992)

[6.12]  D. M. Patrick, D. M. Spirit, and D. L. Williams, "Gain enhancement in distributed erbium-doped fibre amplifier by Raman amplification of 1480 nm pump radiation," *Electron. Lett.*, vol 28, no 13, 1260 (1992)

[6.13]  L. F. Mollenauer, M. J. Neubelt, S. G. Evangelides, J. P. Gordon, J. R. Simpson, and L. G. Cohen, "Experimental study of soliton transmission over more than 10,000 km in dispersion-shifted fiber," *Optics Lett.*, vol 15, no 21, 1203 (1990)

[6.14]  D. N. Chen, K. Motoshima and E. Desurvire, "Transparent optical bus using a chain of remotely pumped erbium-doped fiber amplifiers," in *Proc. Conference on Optical Fiber Communications, OFC '93*, paper WI10, p144, Optical Society of America, Washington DC, 1993; D. N. Chen, K. Motoshima and E. Desurvire, "A transparent optical bus using a chain of remotely pumped erbium-doped fiber amplifiers," *IEEE Photonics Technol. Lett.*, vol 5, no 3, 351 (1993)

[6.15]  T. Rasmussen, A. Bjarklev, J. H. Povlsen, O. Lumholt, and K. Rottwitt, "Transmission length improvement in 2.5 Gbit/s direct detection system with Er-doped fibre amplifiers by efficient remote pumping," *Electron. Lett.*, vol 27, no 17, 1537 (1991)

[6.16]  S. Nishi, K. Aida, and K. Nakagawa, "Highly efficient configuration of erbium-doped fibre amplifiers," in *Proc. European Conference on Optical Communications, ECOC '90*, vol 1, 99 (1990)

[6.17]  V. Lauridsen, R. Tadayoni, A. Bjarklev, J. H. Povlsen and B. Pedersen, "Gain and noise performance of fibre amplifiers operating in new pump configurations," *Electron. Lett.*, vol 27, no 4, 327 (1991)

[6.18]   C. R. Giles, J. Stone, L. W. Stulz, K. Walker, and C. A. Burrus, "Gain enhancement in reflected-pump erbium-doped fiber amplifiers," in *Proc. Topical Meeting on Optical Amplifiers and Applications, 1991*, paper ThD2, Technical Digest Series, vol 13, 148, Optical Society of America, Washington DC, 1991

[6.19]   B. Sridar, S. P. Bastien, and H. D. Sunak, "Erbium-doped fiber power amplifiers with pump reflecting mirrors in the 800 nm band," *IEEE Photonics Technol. Lett.*, vol 4, no 8, 917 (1992)

[6.20]   M. C. Farries, C. M. Ragdale, and D. C. J. Reid, "Broadband chirped Bragg filter for pump rejection and recycling in erbium-doped fibre amplifiers," *Electron. Lett.*, vol 28, no 5, 487 (1992)

[6.21]   H. Masuda, K. Aida, K. Nakagawa, K. Yoshino, M. Wada, and J. Temmyo, "Pump-wavelength-locked erbium-doped fibre amplifier employing novel external cavity for 0.98 μm laser diode," *Electron. Lett.*, vol 28, no 20, 1855 (1992)

[6.22]   C. W. Barnard, J. Chrotowski, and M. Kaverad, "Bidirectional fiber amplifiers," *IEEE Photonics Technol. Lett.*, vol 4, no 8, 911 (1992)

[6.23]   Y. Y. Cheng, N. Kagi, A. Oyobe, and K. Nakamura, "Novel fibre amplifier configuration suitable for bidirectional system," *Electron. Lett.*, vol 28, no 6, 559 (1992)

[6.24]   M. Suyama, S. Watanabe, I. Yokota, and H. Kuwahara, "Bidirectional transmission scheme using intensity modulation of 1.48 mm pump laser diode for erbium-doped fibre amplifiers," *Electron. Lett.*, vol 27, no 1, 89 (1991)

[6.25]   J. Haugen, J. Freeman, and J. Conradi, "Bidirectional transmission at OC-12 with erbium-doped fiber amplifiers," in *Proc. Conference on Optical Fiber Communications, OFC'92*, paper ThK5, p247, Optical Society of America, Washington DC, 1992; J. Haugen, J. Freeman, and J. Conradi, "Bidirectional transmission at 622 Mb/s utilizing erbium-doped fiber amplifiers," *IEEE Photonics Technol. Lett.*, vol 4, no 8, 913 (1992)

[6.26]   K. Kannan and S. Frisken, "Unrepeatered bidirectional transmission system over a single fiber using optical fiber amplifiers," *IEEE Photonics Technol. Lett.*, vol 5, no 1, 76 (1993)

[6.27]   J. Haugen, J. Freeman and J. Conradi, "Full-duplex bidirectional transmission at 622 Mb/s with two erbium-doped fiber amplifiers," in *Proc. Conference on Optical Fiber Communications, OFC'93*, paper TuI6, p42, Optical Society of America, Washington DC, 1993

[6.28]   J. Freeman, C. J. Chung, and J. Conradi, "Gain compression due to bidirectional signaling in erbium-doped fiber amplifiers," in *Proc. Conference on Optical Fiber Communications, OFC'93*, paper ThF3, p181, Optical Society of America, Washington DC, 1993

[6.29]   L. C. Blank and D. M. Spirit, "OTDR performance enhancement through erbium fibre amplification," *Electron. Lett.*, vol 25, no 25, 1693 (1989)

[6.30]   J. C. MacKichan, J. A. Kitchen, and C. W. Pitt, "Innovative approach to interspan fibre break location in fibre amplifier repeatered communication systems," *Electron. Lett.*, vol 28, no 7, 626 (1992)

[6.31]   Y. Sato and K. I. Aoyama, "Optical time domain reflectometry in optical transmission lines containing in-line Er-doped fiber amplifiers," *IEEE Photonics Technol. Lett.*, vol 10, no 1, 78 (1992)

[6.32]   Y. Saito and K. I. Aoyama, "OTDR in optical transmission systems using Er-doped fiber amplifiers containing optical circulators," *IEEE Photonics Technol. Lett.*, vol 3, no 11, 1001 (1991)

[6.33]   C. R. Giles, E. Desurvire, and J. R. Simpson, "Transient gain and crosstalk in erbium-doped fiber amplifiers," *Optics Lett.*, vol 14, no 16, 880 (1989)

[6.34]   E. Desurvire, M. Zirngibl, H. M. Presby, and D. DiGiovanni, "Dynamic gain compensation in saturated erbium-doped fiber amplifiers," *IEEE Photonics Technol. Lett.*, vol 3, no 5, 453 (1991)

[6.35]   M. Zirngibl, P. B. Hansen, G. Raybon, B. Glance, and E. Desurvire, "Passive gain control by all-optical feedback loop in erbium-doped fiber amplifiers," in *Proc. Conference on Optical Fiber Communications*, OFC'91, postdeadine paper PD21, p100, Optical Society of America, Washington DC, 1991

[6.36]   M. Zirngibl, "Gain control in erbium-doped fibre amplifiers by an all-optical feedback loop," *Electron. Lett.*, vol 27, no 7, 560 (1991)

[6.37]   A. D. Ellis, R. M. Percival, A. Lord, and W. A. Stallard, "Automatic gain control in cascaded erbium-doped fibre amplifier systems," *Electron. Lett.*, vol 27, no 3, 193 (1991)

[6.38]   K. Aida and H. Masuda, "Automatic gain control of erbium-doped fiber amplifier by detecting spontaneous emission power along the fiber," in *Proc. Topical Meeting on Optical Amplifiers and Applications*, 1991, paper FE3, Technical Digest Series, vol 13, 276, Optical Society of America, Washington DC, 1991

[6.39]   K. Aida and H. Masuda, "Independent control of noise figure and gain of a two-stage erbium-doped fiber amplifier by detecting local spontaeous emission," in *Proc. Topical Meeting on Optical Amplifiers and Applications, 1992,,* paper FB1, Technical Digest Series, vol 17, 158, Optical Society of America, Washington DC, 1992

[6.40]   K. Kinoshita, M. Suyama, and H. Kuwahara, "Novel configuration for low noise Er-doped fiber optical amplifiers during automatic-level-controlled operation," in *Proc. Topical Meeting Meeting on Optical Amplifiers and Applications, 1992*, paper FB3, Technical Digest Series, vol 17, 166, Optical Society of America, Washington DC, 1992

[6.41]   K. Motoshima, D. N. Chen, L. M. Leba, M. M. Downs, T. Li, and E. Desurvire, "Dynamic compensation of transient gain saturation in erbium-doped fiber amplifiers by pump feedback control," in *Proc. Conference on Optical Fiber Communications, OFC'93*, paper TuI5, p40, Optical Society of America, Washington DC, 1993

[6.42]   M. Nishimura, T. Kashiwada, and M. Shigematsu, "Low noise figure erbium-doped fiber amplifiers with gain adjustment by pump power control," in *Proc. Topical Meeting on Optical Amplifiers and Applications, 1992*, paper FB4, Technical Digest Series, vol 17, 170, Optical Society of America, Washington DC, 1992

[6.43]   W. I. Way, D. Chen, M. A. Siafi, M. J. Andrejco, A. Yi-Yan, A. VonLehman, and C. Lin, "High gain limiting erbium-doped fibre amplifier with over 30 dB dynamic range," *Electron. Lett.*, vol 27, no 3, 211 (1991)

[6.44]   M. Zirngibl, "An optical power equalizer based on one Er-doped fiber amplifier," *IEEE Photonics Technol. Lett.*, vol 4, no 4, 357 (1992)

[6.45]   H. Okamura, "Automatic optical-loss compensation with Er-doped fibre amplifier," *Electron. Lett.*, vol 27, no 23, 2155 (1991)

[6.46]   H. Okamura, "Automatic optical loss compensation with Erbium-doped fiber amplifier," *IEEE J. Lightwave Technol.*, vol 10, no 8, 1110 (1992)

[6.47]   K. Suzuki and M. Nakazawa, "Automatic optical soliton control using cascaded $Er^{3+}$-doped fibre amplifiers," *Electron. Lett.*, vol 26, no 14, 1032 (1990)

[6.48]   W. I. Way, T. H. Wu, A. Yi-Yan, M. Andrejco, and C. Lin, "Optical power limiting amplifier and its applications in a SONET self-healing ring network," *IEEE J. Lightwave Technol.*, vol 10, no 2, 206 (1992)

[6.49]   T. Sugawa, T. Kokumai, and Y. Miyajima, "Optical amplification in $Er^{3+}$-doped single-mode fluoride fibers," *IEEE Photonics Technol. Lett.*, vol 2, no 7, 475 (1990)

[6.50]   W. J. Miniscalco, "Erbium-doped glasses for fiber amplifiers at 1500 nm," *IEEE J. Lightwave Technol.*, vol 9, no 2, 234 (1991)

[6.51]   V. L. da Silva, "Automatic gain flattening in Er-doped fiber-amplifiers," in *Proc. Conference on Optical Fiber Communications, OFC'93*, paper ThD2, p174, Optical Society of America, Washington DC, 1993

[6.52]   M. Tachibana, R. I. Laming, P. R. Morkel, and D. N. Payne, "Erbium-doped fiber amplifier with flattened gain spectrum," *IEEE Photonics Technol. Lett.*, vol 3, no 2, 118 (1991)

[6.53]   M. Wilkinson, A. Bebbington, S. A. Cassidy, and P. McKee, "D-fibre filter for erbium gain spectrum flattening," *Electron. Lett.*, vol 28, no 2, 131 (1992)

[6.54]   R. Kashyap, R. Wyatt, and R. J. Campbell, "Wideband gain flattened erbium fibre amplifier using a photosensitive fibre blazed grating," *Electron. Lett.*, vol 29, no 2, 154 (1993)

[6.55]   G. Grasso, F. Fontana, A. Righetti, P. Scrivener, P. Turner, and P. Maton, "980 nm diode-pumped Er-doped fiber optical amplifiers with high gain-bandwidth product," in *Proc. Conference on Optical Fiber Communications, OFC'91*, paper FA3, p195, Optical Society of America, Washington DC, 1991

[6.56]   D. A. Smith, J. E. Baran, K. W. Cheung, and J. J. Johnson, "Polarization-independent acoustically-tunable optical filters," *Applied Phys. Lett.*, vol 56, 209 (1990)

[6.57]   S. F. Su, R. Olshansky, G. Joyce, D. A. Smith, and J. E. Baran, "Gain equalization in multiwavelength lightwave systems using acousto-optic tunable filters," *IEEE Photonics Technol. Lett.*, vol 4, no 3, 269 (1992)

[6.58]   K. Inoue, T. Kominato, and H. Toba, "Tunable gain equalization using a Mach–Zenhder optical filter in multistage fiber amplifiers," *IEEE Photonics Technol. Lett.*, vol 3, no 8, 718 (1991)

[6.59]   C. R. Giles and D. DiGiovanni, "Dynamic gain equalization in two-stage fiber amplifiers," *IEEE Photonics Technol. Lett.*, vol 2, no 12, 866 (1990)

[6.60]   C. R. Giles and D. DiGiovanni, "Spectral dependence of gain and noise in erbium-doped fiber amplifiers," *IEEE Photonics Technol. Lett.*, vol 2, no 11, 797 (1990)

[6.61]   R. I. Laming, J. D. Minelli, L. Long, and M. N. Zervas, "Erbium-doped-fiber amplifier with passive spectral-gain equalization," in *Proc. Conference on Optical Fiber Comnications, OFC'93*, paper ThD3, p175, Optical Society of America, Washington DC, 1993

[6.62]   A. R. Chraplyvy, J. A. Nagel, and R. W. Tkach, "Equalization in amplified WDM lightwave transmission systems," *IEEE Photonics Technol. Lett.*, vol 4, no 8, 920 (1992)

[6.63]   S. E. Miller and I. P. Kaminow, *Optical fiber telecommunications II*, Academic Press, New York, 1988, chapter 21

[6.64]   G. P. Agrawal, *Nonlinear Fiber Optics,* Academic Press, New York, 1989, Chapters 4, 7, and 10

[6.65]   J. M. Dziedzic, R. H. Stolen, and A. Ashkin, "Optical Kerr effect in long fibers," *Applied Optics*, vol 20, no 8, 1403 (1981)

[6.66]   R. H. Stolen, "Nonlinear properties of optical fibers, in *Optical fiber Telecommunications*, edited by S. E. Miller and A. G. Chynoweth, Academic Press, New York, 1979

[6.67]   R. H. Stolen and J. E. Bjorkholm, "Parametric amplification and frequency conversion in optical fibers," *IEEE J. Quantum Electron.*, vol 18, no 7, 1062 (1982)

[6.68]   D. S. Chemla, D. A. B. Miller, and P. W. Smith, "Nonlinear optical properties of GaAs/GaAlAs multiple quantum well material: phenomena and applications," *Optical Engineering*, vol 24, no 4, 556 (1985); D. A. B. Miller, D. S. Chemla, P. W. Smith, A. C. Gossard, and W. Wiegmann, "Nonlinear optics with a diode-laser light source,"

*Optics Lett.*, vol 8, no 9, 477 (1983)

[6.69] E. Eichen, W. J. Miniscalco, J. McCabe, and T. Wei, "Lossless, 2 × 2, all-fiber optical routing switch," in *Proc. Conference on Optical Fiber Communications, OFC'90*, paper PD-20, Optical Society of America, Washington DC, 1990

[6.70] M. Zirngibl, "All-optical remote gain switching in Er-doped fibre amplifiers," *Electron. Lett.*, vol 27, no 13, 1164 (1991)

[6.71] P. Myslinski, G. Cheney, J. Chrostowski, B. Sytett, and J. Glinski, "Nanosecond all-optical gain switching of an erbium-doped fibre amplifiers," in *Proc. Second IEEE International Workshop on Photonic Networks, Components, and Applications (Photonics'92)*, paper 1.10; P. Myslinski, C. Barnard, G. Cheney, J. Chrostowski, B. Sytett, and J. Glinski, "Nanosecond all-optical gain switching of an erbium-doped fibre amplifier," *Opt. Comm.*, vol 97, 340, 1993

[6.72] M. Jinno, "Effects of group velocity dispersion on self/cross phase modulation in a nonlinear Sagnac interferometer switch," *IEEE J. Lightwave Technol.*, vol 10, no 8, 1167 (1992)

[6.73] N. J. Doran and D. Wood, "Nonlinear-optical loop mirror," *Optics Lett.*, vol 13, no 1, 56 (1988)

[6.74] K. Otsuka, "Nonlinear antiresonant ring interferometer," *Optics Lett.*, vol 8, 471 (1983)

[6.75] M. C. Farries and D. N. Payne, "Optical fiber switch employing a Sagnac interferometer," *Applied Phys. Lett.*, vol 55, no 1, 25 (1989)

[6.76] N. J. Doran and D. Wood, "A soliton processing element for all-optical switching and logic," *J. Opt. Soc. Am. B*, vol 4, 1843 (1987)

[6.77] S. Trillo, S. Wabnitz, E. M. Wright, and G. I. Stegemann, "Soliton switching in fibre nonlinear direction couplers," *Optics Lett.*, vol 13, 672 (1988)

[6.78] K. J. Blow, N. J. Doran, and B. K. Nayar, "Experimental demonstration of optical soliton switching in an all-fibre nonlinear Sagnac interferometer," *Optics Lett.*, vol 14, no 14, 754 (1989)

[6.79] M. E. Fermann, F. Haberl, M. Hofer, and H. Hochreiter, "Nonlinear amplifying loop mirror," *Optics Lett.*, vol 15, no 13, 752 (1990)

[6.80] D. J. Richardson, R. I. Laming, and D. N. Payne, "Very low threshold Sagnac switch incorporating an erbium doped fibre amplifier," *Electron. Lett.*, vol 26, no 21, 1779 (1990)

[6.81] K. Smith, E. J. Greer, N. J. Doran, D. M. Bird, and K. H. Cameron, "Pulse amplification and shaping using a nonlinear loop mirror that incorporates a saturable gain," *Optics Lett.*, vol 17, no 6, 409 (1992)

[6.82] M. Jinno and T. Matsumoto, "Demonstration of laser-diode-pumped ultrafast all-optical switching in a nonlinear Sagnac interferometer," *Electron. Lett.*, vol 27, no 1, 75 (1991)

[6.83] B. P. Nelson, K. J. Blow, P. D. Constantine, N. J. Doran, J. K. Lucek, I. W. Marshall, and K. Smith, "All-optical Gbit/s switching using nonlinear optical loop mirror," *Electron. Lett.*, vol 27, no 9, 704 (1991)

[6.84] K. Smith, N. J. Doran, and P. G. J. Wigley, "Pulse shaping, compression and pedestal suppression employing a nonlinear-optical loop mirror," *Optics Lett.*, vol 15, 1294 (1990)

[6.85] K. Smith, E. J. Greer and N. J. Doran, "Square pulse amplification using nonlinear loop mirror incorporating saturable gain," *Electron. Lett.*, vol 27, no 22, 2046 (1991)

[6.86] R. A. Betts, J. W. Lears, S. J. Frisken, and P. S. Atherton, "Generation of transform limited optical pulses using all-optical gate," *Electron. Lett.*, vol 28, no 11, 1035 (1992)

[6.87] D. U. Noske, N. Pandit, and J. R. Taylor, "Picosecond square pulse generation using nonlinear fibre loop mirror," *Electron. Lett.*, vol 28, no 10, 908 (1992)

[6.88] M. Jinno and M. Abe, "All-optical regenerator based on nonlinear fibre Sagnac interferometer," *Electron. Lett.*, vol 28, no 14, 1350 (1992)

[6.89] N. A. Whitaker, H. Avramopoulos, P. M. W. French, M. C. Gabriel, R. E. LaMarche, D. J. DiGiovanni, and H. M. Presby, "All-optical arbitrary demultiplexing at 2.5 Gbit/s with tolerance to timing jitter," *Optics Lett.*, vol 16, no 23, 1838 (1991)

[6.90] M. Jinno and T. Matsumoto, "Ultrafast, low power, and highly stable all-optical switching in an all polarization maintaining fiber Sagnac interferometer," *IEEE Photonics Technol. Lett.*, vol 2, no 5, 349 (1990)

[6.91] K. J. Blow, N. J. Doran, and B. P. Nelson, "Demonstration of the nonlinear fibre loop mirror as an ultrafast all-optical demultiplexer," *Electron. Lett.*, vol 26, 962 (1990)

[6.92] H. Avramopoulos, P. M. W. French, M. C. Gabriel, H. H. Houh, N. A. Whitaker, and T. Morse, "Complete switching in a three-terminal Sagnac switch," *IEEE Photonics Technol. Lett.*, vol 3, no 3, 235 (1991)

[6.93] A. Takada, K. Aida, and M. Jinno, "Demultiplexing of a 40 Gbit/s optical signal to 2.5 Gbit/s using a nonlinear fiber loop mirror driven by amplified, gain-switched laser diode pulses," in *Proc. Conference on Optical Fiber Communications, OFC '91*, paper TuN3, p50, Optical Society of America, Washington DC, 1991

[6.94a] N. A. Whitaker, P. M. W. French, M. C. Gabriel, and H. Avramopoulos, "Polarization-independent all-optical switching," *IEEE Photonics Technol. Lett.*, vol 4, no 3, 260 (1992)

[6.94b] K. Uchiyama, H. Takara, S. Kawanishi, T. Morioka and M. Saruwatari, "Ultrafast polarization-independent all-optical switching using a polarization diversity scheme in the nonlinear optical loop mirror," *Electron. Lett.*, vol 28, no 20, 1864 (1992)

[6.95] P. A. Andrekson, N. A. Olsson, J. R. Simpson, D. J. DiGiovanni, P. A. Morton, T. Tanbun-Ek, R. A. Logan, and K. W. Wecht, "64 Gbit/s All-optical demultiplexing with the nonlinear optical-loop mirror," *IEEE Photonics Technol. Lett.*, vol 4, no 6, 644 (1992)

[6.96] S. Kawanishi, H. Takara, K. Uchiyama, and M. Saruwatari, "4 × 8 Gbit/s optical TDM transmission and all-optical demultiplexing experiment with clock recovery by new phase-lock-loop technique," in *Proc. Conference on Optical Fiber Communications, OFC '93*, paper WC4, p91, Optical Society of America, Washington DC, 1993

[6.97] T. Morioka, H. Takada, H. Saruwatari, and M. Saruwatari, "Ultrafast reflective optical; Kerr demultiplexer using a polarization rotation mirror," *Electron. Lett.*, vol 28, 521 (1992)

[6.98] T. Morioka, K. Mori, and M. Saruwatari, "Ultrafast polarization-independent optical demultiplexer using optical carrier frequency shift through crossphase modulation," *Electron. Lett.*, vol 28, no 11, 1070 (1992)

[6.99] P. A. Andrekson, N. A. Olsson, J. R. Simpson, T. Tanbun-Ek, R. A. Logan, and M. Haner, "16 Gbit/s all-optical demultiplexing using four-wave mixing," *Electron. Lett.*, vol 27, no 11, 922 (1991)

[6.100] P. A. Andrekson, N. A. Olsson, M. Haner, J. R. Simpson, T. Tanbun-Ek, R. A. Logan, D. Coblentz, H. M. Presby, and K. W. Wecht, "32 Gbit/s optical soliton transmission over 90 km," *IEEE Photonics Technol. Lett.*, vol 4, no 1, 76 (1992)

[6.101] K. Inoue and H. Toba, "Wavelength conversion experiment using fiber four-wave mixing," *IEEE Photonics Technol. Lett.*, vol 4, no 1, 69 (1992)

[6.102] P. A. Andrekson, "Picosecond optical sampling using four-wave mixing," *Electron. Lett.*, vol 27, no 16, 1440 (1991)

[6.103] E. Desurvire, M. J. F. Digonnet and H. J. Shaw, "Raman amplification of recirculating pulses in a reentrant fiber loop ," *Optics Lett.*, vol 10, no 2, 83 (1985)

[6.104]  E. Desurvire, M. J. F. Digonnet and H. J. Shaw, "Theory and implementation of a Raman active fiber delay line," *IEEE J. Lightwave Technol.*, vol 4, no 4, 426 (1986)

[6.105]  E. Desurvire, A. Imamoglu and H. J. Shaw, "Low-threshold synchronously pumped all-fiber ring Raman laser," *IEEE J. Lightwave Technol.*, vol 5, no 1, 89 (1987)

[6.106]  V. I. Belotitskii, E. A. Kuzin, M. P. Petrov and V. V. Spirin, "Demonstration of over 100 million round trips in recirculating fibre loop with all-optical regeneration," *Electron. Lett.*, vol 29, no 1, 49 (1993)

[6.107]  L. F. Mollenauer and K. Smith, "Demonstration of soliton transmission over more than 4000 km in fiber with loss periodically compensated by Raman gain," *Optics Lett.*, vol 13, no 8, 675 (1988)

[6.108]  L. F. Mollenauer and K. Smith, "Soliton transmission over more than 6000 km in fiber with loss periodically compensated by Raman gain," in *Proc. European Conference on Optical Communications, ECOC'89*, paper TuA5, p71

[6.109]  N. A. Olsson, "15,000 km fibre-optic transmission using a linear repeater," *Electron. Lett.*, vol 23, no 12, 659 (1987)

[6.110]  G. Grosskopf, L. Kuller, R. Ludwig, W. Pieper, R. Schnabel, and H. G. Weber, "Optical amplifiers operating in a random access fiber loop memory," in *Proc. Topical Meeting on Optical Amplifiers and Applications, 1991*, paper FE6, Technical Digest Series, vol 13, 288, Optical Society of America, Washington DC, 1991

[6.111]  M. Calzavara, P. Gambini, M. Puelo, B. Bostica, P. Cinato, and E. Vezzoni, "Optical-fiber-loop memroy for multiwavelength packet buffering in ATM switching applica-tions," in *Proc. Conference on Optical Fiber Communications, OFC'93*, paper TuE3, p19, Optical Society of America, Washington DC, 1993

[6.112]  H. Izadpanah, A. Elrefaie, W. Sessa, C. Lin, S. Tsuji, and H. Inoue, "A multi-Gb/s self-synchronized optical regenerator using a 1.55 μm traveling-wave semiconductor optical amplifier," in *Proc. Topical Meeting on Optical Amplifiers and Applications, 1990*, paper WC5, Technical Digest Series, vol 13, 276, Optical Society of America, Washington DC, 1990

[6.113]  P. R. Morkel, "All-fibre, diode-pumped recirculating-ring delay line," *Electron. Lett.*, vol 24, no 10, 608 (1988)

[6.114]  L. F. Mollenauer, M. J. Neubelt, S. G. Evangelides, J. P. Gordon, J. R. Simpson, and L. G. Cohen, "Experimental study of soliton transmission over more than 10,000 km in dispersion-shifted fiber," *Optics Lett.*, vol 15, no 21, 1203 (1990)

[6.115]  D. J. Malyon, T. Widdowson, E. G. Bryant, S. F. Carter, J. V. Wright, and W. A. Stallard, "Demonstration of optical pulse propagation over 10,000 km of fibre using recriculating loop," *Electron. Lett.*, vol 27, no 2, 120 (1991)

[6.116]  N. S. Bergano, J. Aspell, C. R. Davidson, P. R. Trischitta, B. M. Nyman, and F. W. Kerfoot, "A 9,000 km 5 Gb/s and 21,000 km 2.4G bit/s feasibility demonstration of transoceanic EDFA systems using a circulating loop," in *Proc. Conference on Optical Fiber Communications, OFC'91*, postdeadline paper PD13, Optical Society of America, Washington DC, 1991; N. S. Bergano, J. Aspell, C. R. Davidson, P. R. Trischitta, B. M. Nyman, and F. W. Kerfoot, "Bit error rate measurements of 14,000 km 5 Gbit/s fibre amplifier transmission system using circulating loop," *Electron. Lett.*, vol 27, no 21, 1889 (1991)

[6.117]  C. R. Giles, J. M. Kahn, S. K. Korotky, J. J. Veselka, C. A. Burrus, J. Perino, and H. M. Presby, "Polarization effects on ultralong distance signal transmission in amplified optical-fiber loops," *IEEE Photonics Technol. Lett.*, vol 3, no 10, 948 (1991)

[6.118]  M. Nakazawa, E. Yamada, H. Kubota, and K. Suzuki, "10 Gbit/s soliton data transmission over one million kilometres," *Electron. Lett.*, vol 27, no 14, 1270 (1991)

[6.119]  M. Nakazawa, H. Kubota, E. Yamada, and K. Suzuki, "Infinite-distance soliton transmission with soliton controls in time and frequency domains," *Electron. Lett.*, vol 28, no 12, 1099 (1992)

[6.120]  M. Nakazawa, K. Suzuki, E. Yamada, H. Kubota, Y. Kimura, and M. Takaya, "Experimental demonstration of soliton data transmission over unlimited distances with soliton control in time and frequency domains," in *Proc. Conference on Optical Fiber Communications, OFC '93*, postdeadline paper PD7, Optical Society of America, Washington DC, 1993

[6.121]  L. F. Mollenauer, J. P. Gordon, and S. G. Evangelides, "The sliding-frequency guiding filter: and improved form of soliton jitter control," *Optics Lett.*, vol 17, no 22, 1575 (1992)

[6.122]  H. A. Haus and A. Mecozzi, "Long-term storage of a bit stream of solitons," *Optics Lett.*, vol 17, no 21, 1500 (1992)

[6.123]  H. Okamura and K. Iwatsuki, "Er-doped fibre ring resonator applied to optical spectrum analyser with less than 100 kHz resolution," *Electron. Lett.*, vol 27, no 12, 1047 (1991)

[6.124]  R. E. Meyer, S. Ezekiel, D. W. Stowe, and V. J. Tekippe, "Passive fiber-optic ring resonator for rotation sensing," *Optics Lett.*, vol 8, 644 (1983)

[6.125]  J. T. Kringlebotn, "Amplified fiber ring resonator gyro," *IEEE Photonics Technol. Lett.*, vol 4, no 10, 1180 (1992)

[6.126]  J. T. Kringlebotn, P. R. Morkel, C. N. Pannell, D. N. Payne, and R. I. Laming, "Amplified fibre delay line with 27,000 recirculations," *Electron. Lett.*, vol 28, no 2, 201 (1992)

[6.127]  H. Okamura and K. Iwatsuki, "A finesse-enhanced Er-doped fiber ring resonator," *IEEE J. Lightwave Technol.*, vol 9, 1554 (1991)

[6.128]  B. Moslehi, "Fibre-optic filters employing optical amplifiers to provide design flexibility," *Electron. Lett.*, vol 28, no 3, 226 (1992)

[6.129]  B. Moslehi and J. W. Goodman, "Novel amplified fiber-optic recirculating delay line processor," *IEEE J. Lightwave Technol.*, vol 10, no 8, 1142 (1992)

[6.130]  B. Moslehi, J. W. Goodman, M. Tur, and H. J. Shaw, "Fiber-optic signal processing," *Proc. IEEE*, vol 72, no 7, 909 (1984)

[6.131]  K. P. Jackson, S. A. Newton, B. Moslehi, M. Tur, C. C. Cutler, J. W. Goodman, and H. J. Shaw, "Optical fiber delay-line signal processing," *IEEE Trans. Microwave Theory and Techniques*, vol MTT-33, no 3, 193 (1985)

[6.132]  M. Tur, J. W. Goodman, B. Moslehi, J. E. Bowers, and H. J. Shaw, "Fiber-optic signal processor with applications to matrix-vector multiplication and lattice filtering," *Optics Lett.*, vol 7, no 9, 463 (1982)

[6.133]  P. R. Prucnal, M. A. Santoro, and S. K. Sehgal, "Ultrafast all-optical synchronous multiple-access fiber networks," *IEEE J. Selected Areas In Communications*, vol SAC-4, no 9, 1484 (1986)

[6.134]  P. R. Prucnal, M. A. Santoro, and T. R. Fan, "Spread spectrum fiber-optic local area network using optical processing," *IEEE J. Lightwave Technol.*, vol 4, no 5, 547 (1986)

[6.135]  R. I. Macdonald, "Switched optical delay line signal processors," *IEEE J. Lightwave Technol.*, vol 5, no 6, 856 (1987)

[6.136]  S. A. Newton, R. S. Howland, K. P. Jackson, and H. J. Shaw, "High-speed pulse-train generation using single-mode-fibre recirculating delay lines," *Electron. Lett.*, vol 19, no 19, 756 (1983)

[6.137]  T. G. Hodgkinson and P. Coppin, "Optical domain frequency translation of a modulated optical carrier," *Electron. Lett.*, vol 26, no 16, 1262 (1990)

[6.138] M. I. Belovolov, Y. M. Dianov, and V. I. Karpov, "Fiber-optic memories," in *Proc. of the Institute of General Physics, Russian Academy of Sciences*, vol. 5, edited by A. M. Prokhorov, p159, 1985

[6.139] M. Maignan and J. J. Bernard, "Wideband 150 μs optical delay line for satellite altimetric radar checking," *Electron. Lett.*, vol 24, no 14, 902 (1988)

[6.140] W. W. Ng, A. A. Watson, G. L. Tangonan, J. J. Lee, I. L. Newberg, and N. Bernstein, "The first demonstration of an optically steered microwave phased-array antenna using true-time-delay," *IEEE J. Lightwave Technol.*, vol 9, no 9, 1124 (1991)

[6.141] J. T. Kringlebotn, *Fiber-optic Recirculating-ring Delay Lines and Ring Resonators with Doped Fiber Amplifiers*, Dr. In. Thesis (NTH 1992:53), University of Trondheim, Norwegian Institute of Technology, 1992

[6.142] G. A. Pavlath and H. J. Shaw, "Reentrant fiber-optic rotation sensors," in *Fiber Optic Rotation Sensors and Related Technologies*, edited by S. Ezekiel and H. J. Arditty, Springer-Verlag, 1982

[6.143] E. Desurvire, B. Y. Kim, K. Fesler, and H. J. Shaw, "Reentrant fiber Raman gyroscope," *IEEE J. Lightwave Technol.*, vol 6, no 4, 481 (1988)

[6.144] D. N. Chen, K. Motoshima, M. M. Downs, and E. Desurvire, "Reentrant Sagnac fiber gyroscope with a recirculating delay line using an erbium-doped fiber amplifier," *IEEE Photonics Technol. Lett.*, vol 4, no 7, 813 (1992)

[6.145] C. C. Cutler, S. A. Newton, and H. J. Shaw, "Limitation of rotation sensing by scattering," *Optics Lett.*, vol 5, no 11, 488 (1980)

[6.146] R. A. Bergh, H. C. Lefevre, and H. J. Shaw, "Compensation of the optical Kerr effect in fiber-optic gyroscopes," *Optics Lett.*, vol 7, no 6, 282 (1982)

[6.147] M. I. Belovolov, E. M. Dianov, V. I. Karpov, V. N. Protopopov, and V. N. Serkin, "Fiber optic dynamic memory for fast signal-processing and optical computing," in *Proc. SPIE Conference on Optical Computing*, vol 963, 90 (1988)

[6.148] C. Q. Maguire and P. R. Prucnal, "High-density optical storage using optical delay lines," *SPIE Conference on Medical Imaging III*, vol 1093, 571 (1989)

[6.149] N. A. Whitaker, M. C. Gabriel, H. Avramopoulos, and A. Huang, "All-optical, all-fiber circulating shift register with an inverter," *Optics Lett.*, vol 16, no 24, 1999 (1991)

[6.150] H. Avramopoulos and N. A. Whitaker, "Addressable fiber-loop memory," *Optics Lett.*, vol 18, no 1, 22 (1993)

[6.151] W. J. Miniscalco, "Optical and electronic properties of rare earth ions in glasses," in *Rare Earth Doped Fiber Lasers and Amplifiers*, edited by M. J. F. Digonnet, Marcel Dekker, New York, 1993

[6.152] R. J. Mears, L. Reekie, S. B. Poole, and D. N. Payne, "Low-threshold tunable cw and Q-switched fibre laser operating at 1.55 μm," *Electron. Lett.*, vol 22, no 3, 159 (1986)

[6.153] A. V. Astakhov, M. M. Butusov, S. L. Galkin, N. V. Ermakova, and Y. K. Fedorov, "Fiber laser with 1.54 μm radiation wavelength," *Opt. Spectrosc* (USSR), vol 62, no 1, 140 (1987)

[6.154] R. J. Mears, L. Reekie, and I. M. Jauncey, "High power tunable erbium-doped fiber laser operating at 1.55 μm," in *Proc. Conference on Lasers and Electr-Optics, CLEO '87*, paper WD3, Optical Society of America, Washington DC, 1987

[6.155] P. Myslkinski, J. Chrostowski, J. A. Koningstein, and J. R. Simpson, "High power Q-switched erbium-doped fiber laser," *IEEE J. Quantum Electron.*, vol 28, no 1, 371 (1992)

[6.156] D. C. Hanna, R. M. percival, I. R. Perry, R. G. Smart, and A. C. Tropper, "Efficient operation of an Yb-sensitized Er fibre laser pumped in 0.8 μm region," *Electron. Lett.,* vol 24, no 17, 1068 (1988)

[6.157] W. L. Barnes, S. B. Poole, J. E. Townsend, L. Reekie, D. J. Taylor, and D. N. Payne, "$Er^{3+}$–$Yb^{3+}$ and $Er^{3+}$ doped fiber lasers," *IEEE J. Lightwave Technol.,* vol 7, no 10, 1461 (1989)

[6.158] W. L. Barnes, P. R. Morkel, L. Reekie, and D. N. Payne, "High-quantum efficiency $Er^{3+}$ fiber lasers pumped at 980 nm," *Optics Lett.,* vol 14, no 18, 1002 (1989)

[6.159] M. C. Farries, P. R. Morkel, R. I. Laming, T. A. Birks, D. N. Payne, and E. J. Tarbox, "Operation of erbium-doped fiber amplifiers and lasers with frequency-doubled Nd:YAG lasers," *IEEE J. Lightwave Technol.,* vol 7, no 10, 1473 (1989)

[6.160] M. C. Brierley and P. W. France, "Continuous wave lasing at 2.7 μm in an erbium-doped fluorozirconate fibre," *Electron. Lett.,* vol 24, no 15, 935 (1988)

[6.161] R. G. Smart, J. N. Carter, D. C. Hanna, and A. C. Tropper, "Erbium doped fluorozirconate fibre laser operating at 1.66 and 1.72 μm," *Electron. Lett.,* vol 26, no 10, 649 (1990)

[6.162] J. Y. Allain, M. Monerie, and H. Poignant, "Lasing at 1.00 μm in erbium-doped fluorozirconate fibers," *Electron. Lett.,* vol 25, no 5, 318 (1989)

[6.163] C. A. Millar, M. C. Brierley, M. H. Hunt, and S. F. Carter, "Efficient up-conversion pumping at 800 nm of an erbium-doped fluoride fibre laser operating at 850 nm," *Electron. Lett.,* vol 26, no 22, 1871 (1990)

[6.164] T. J. Whitley, C. A. Millar, R. Wyatt, M. C. Brierley, and D. Szebesta, "Upconversion pumped green lasing in erbium-doped fluorozirconate fibre," *Electron. Lett.,* vol 27, no 20, 1785 (1991)

[6.165] J. Y. Allain, M. Monerie, and H. Poignant, "Tunable green upconversion erbium fibre laser," *Electron. Lett.,* vol 28, no 2, 111 (1992)

[6.166] L. Reekie, I. M. Jauncey, S. B. Poole, and D. N. Payne, "Diode-laser-pumped operation of an $Er^{3+}$-doped single-mode fibre laser," *Electron. Lett.,* vol 23, no 20, 1076 (1987)

[6.167] R. Wyatt, B. J. Ainslie, and S. P. Craig, "Efficient operation of array-pumped $Er^{3+}$ doped silica fibre laser at 1.5 μm," *Electron. Lett.,* vol 24, no 22, 1362 (1988)

[6.168] K. Suzuki, Y. Kimura, and M. Nakazawa, "An 8mW cw $Er^{3+}$-doped fiber laser pumped by 1.46 μm InGaAsP laser diodes," *Japanese J. of Applied Phys.,* vol 28, no 6, 1000 (1989)

[6.169] R. Allen and L. Esterowitz, "Diode-pumped single-mode fluorozirconate fiber laser from the $^4I_{11/2} \rightarrow {}^4I_{13/2}$ transition in erbium," *Applied Phys. Lett.,* vol 56, no 17, 1635 (1990)

[6.170] H. Yanagita, I. Masuda, T. Yamashita, and H. Toratani, "Diode laser pumped $Er^{3+}$ fibre laser operation between 2.7–2.8 μm," *Electron. Lett.,* vol 26, no 22, 1836 (1990)

[6.171] M. J. F. Digonnet, editor, *Rare Earth Doped Fiber Lasers and Amplifiers,* Marcel Dekker, New York, 1993

[6.172] R. Wyatt, "High-power broadly tunable erbium-doped silica fibre laser," *Electron. Lett.,* vol 25, no 22, 1498 (1989)

[6.173] C. R. O. Cochlain and R. J. Mears, "Broadband tunable single-frequency diode-pumped erbium-doped fibre laser," *Electron. Lett.,* vol 28, no 2, 124 (1992)

[6.174] P. F. Wysocki, M. J. F. Digonnet, and B. Y. Kim, "Electronically tunable, 1.55 μm erbium-doped fiber laser," *Optics Lett.,* vol 15, no 5, 273 (1990)

[6.175] P. F. Wysocki, M. J. F. Digonnet, and B. Y. Kim, "Broad-spectrum, wavelength-swept, erbium-doped fiber laser at 1.55 μm," *Optics Lett.,* vol 15, no 16, 879 (1990)

[6.176] K. Iwatsuki, H. Okamura, and M. Saruwwatari, "Wavelength-tunable single-frequency and single-polarization Er-doped fibre ring-laser with 1.4 kHz linewidth," *Electron. Lett.*, vol 26, no 24, 2033 (1990)

[6.177]  P. D. Humphrey and J. E. Bowers, "Fiber birefringence tuning technique for an erbium-doped fiber ring lasers," *IEEE Photonics Technol. Lett.*, vol 5, no 1, 32 (1993)

[6.178]  D. A. Smith, M. W. Maeda, J. J. Johnson, J. S. Patel, M. A. Saifi, and A. VonLehman, "Acoustically tuned erbium-doped fiber ring laser," *Optics Lett.*, vol 16, no 6, 367 (1991)

[6.179]  J. L. Zyskind, J. W. Sulhoff, J. Stone, D. DiGiovanni, L. W. Stultz, H. M. Presby, A. Piccirili, and P. E. Pramayon, "Electrically tunable, diode-pumped erbium-doped fibre ring laser with fibre Fabry–Perot etalon," *Electron. Lett.*, vol 27, no 21, 1950 (1991)

[6.180]  H. Schmuck, T. Pfeiffer, and G. Veith, "Widely tunable narrow linewidth erbium-doped fibre ring laser," *Electron. Lett.*, vol 27, no 23, 2117 (1991)

[6.181]  J. Stone and L. Stulz, "Pigtailed high-finesse tunable fibre Fabry–Perot interferometers with large, medium and small free spectral range," *Electron. Lett.*, vol 23, 781 (1987)

[6.182]  G. A. Ball and W. W. Morey, "Continuously tunable single-mode erbium fiber laser," *Optics Lett.*, vol 17, no 6, 421 (1992)

[6.183]  H. Okamura and K. Iwatsuki, "Simultaneous oscillation of wavelength-tunable, singlemode lasers using an Er-doped fibre amplifier," *Electron. Lett.*, vol 28, no 5, 461 (1992)

[6.184]  N. Park, J. W. Dawson, and K. J. Vahala, "Multiple wavelength operation of an erbium-doped fiber laser," *IEEE Photonics Technol. Lett.*, vol 4, no 6, 540 (1992)

[6.185]  I. M. Jauncey, L. Reekie, R. J. Mears, and C. J. Rowe, "Narrow-linewidth fiber laser operating at 1.55 $\mu$m," *Optics Lett.*, vol 12, no 3, 164 (1987)

[6.186]  I. M. Jauncey, L. Reekie, J. E. Townsend, and D. N. Payne, "Single-longitudinal-mode operation of an $Nd^{3+}$-doped fibre laser," *Electron. Lett.*, vol 24, no 1, 24 (1988)

[6.187]  M. S. O'Sullivan, J. Chrostowski, E. Desurvire, and J. R. Simpson, "High-power narrow-linewidth $Er^{3+}$-doped fiber laser," *Optics Lett.*, vol 14, no 9, 438 (1989)

[6.188]  P. Barnsley, P. Urquhart, C. Millar, and M. Brierley, "Fiber Fox–Smith resonators: application to single-longitudinal-mode operation of fiber lasers," *J. Opt. Soc. Am. A*, vol 5, no 8, 1339 (1988)

[6.189]  G. A. Ball, W. W. Morey, and W. H. Glenn, "Standing-wave monomode erbium fiber laser," *IEEE Photonics Technol. Lett.*, vol 3, no 7, 613 (1991)

[6.190]  G. A. Ball and W. H. Glenn, "Design of a single-mode linear-cavity erbium fiber laser utilizing Bragg reflectors," *IEEE J. Lightwave Technol.*, vol 10, no 10, 1338 (1992)

[6.191]  P. R. Morkel, G. J. Cowle, and R. I. Laming, "Travelling-wave erbium fibre ring laser with 60 kHz linewidth," *Electron. Lett.*, vol 26, no 10, 632 (1990)

[6.192]  G. J. Cowle, D. N. Payne, and D. Reid, "Single-frequency travelling-wave erbium-doped fibre loop laser," *Electron. Lett.*, vol 27, no 3, 229 (1991)

[6.193]  J. L. Zyskind, J. W. Sulhoff, Y. Sun, J. Stone, L. W. Stulz, G. T. Harvey, D. DiGiovanni, H. M. Presby, A. Piccirilli, U. Koren, and R. M. Jopson, "Singlemode diode-pumped tunable erbium-doped fibre laser with linewidth less than 5.5 kHz," *Electron. Lett.*, vol 27, no 23, 2148 (1991)

[6.194]  P. F. Wysocki, "Broadband operation of erbium- and neodymium-doped fiber sources," in *Rare Earth Doped Fiber Lasers and Amplifiers*, edited by M. J. F. Digonnet, Marcel Dekker, New York, 1993

[6.195]  R. A. Bergh, B. Culshaw, C. C. Cutler, H. C. Lefevre, and H. J. Shaw, "Source statistics and the Kerr effect in fibre-optic gyroscopes," *Optics Lett.*, vol 7, 563 (1982)

[6.196]  K. Bohm, P. Marten, K. Petermann, E. Weidel, and R. Ulrich, "Low-drift fibre gyro using a superluminescent diode," *Electron. Lett.*, vol 17, 352 (1981)

[6.197]  N. S. Kwong, "High-power, broadband 1550 nm light source by tandem combination of a superluminescent diode and an Er-doped fiber amplifier," *IEEE Photonics Technol. Lett.*, vol 4, no 9, 996 (1992)

[6.198]  H. Fevrier, J. F. Marcerou, P. Bousselet, J. Auge, and M. Jurczyszyn, "High power, compact 1.48 $\mu$m diode-pumped broadband superfluorescent fibre source at 1.55 $\mu$m," *Electron. Lett.*, vol 27, no 3, 261 (1991)

[6.199]  P. F. Wysocki, M. J. F. Digonnet, and B. Y. Kim, "Wavelength stability of a high-output, broadband, Er-doped superfluorescent fiber source pumped near 980 nm," *Optics Lett.*, vol 16, no 12, 961 (1991)

[6.200]  K. Takada, T. Kitagawa, M. Shimizu, and H. Horiguchi, "High-sensitivity low coherence reflectometer using erbium-doped superfluorescent fibre source and erbium-doped power amplifier," *Electron. Lett.*, vol 29, no 4, 365 (1993)

[6.201]  P. F. Wysocki, M. J. F. Digonnet, and B. Y. Kim, "Spectral characteristics of high-power 1.5 $\mu$m broad-band superluminescent fiber sources," *IEEE Photonics Technol. Lett.*, vol 2, no 3, 178 (1990)

[6.202]  P. R. Morkel, R. I. Laming, and D. N. Payne, "Noise characteristics of high-power doped fibre superluminescent sources," *Electron. Lett.*, vol 26, no 2, 96 (1990)

[6.203]  K. Iwatsuki, "Excess noise reduction in fiber gyroscope using broad spectrum linewidth Er-doped superfluorescent fiber laser," *IEEE Photonics Technol. Lett.*, vol 3, no 3, 281 (1991)

[6.204]  K. Iwatsuki, "Long-term bias stability of all-panda fibre gyroscope with Er-doped superfluorescent fibre laser," *Electron. Lett.*, vol 27, no 12, 1092 (1991)

[6.205]  D. C. Hanna, A. Kazer, M. W. Phillips, D. P. Shepherd, and P. J. Suni, "Active mode-locking of an Yb:Er fibre laser," *Electron. Lett.*, vol 25, no 2, 95 (1989)

[6.206]  K. Smith, E. J. Greer, R. Wyatt, P. Wheatley, N. J. Doran, and M. Lawrence, "Totally integrated erbium fibre soliton laser pumped by laser diode," *Electron. Lett.*, vol 27, no 3, 244 (1991)

[6.207]  R. P. Davey, R. P. E. Fleming, K. Smith, R. Kashyap, and J. R. Armitage, "Mode-locked erbium fibre laser with wavelength selection by means of a fibre Bragg grating reflector," *Electron. Lett.*, vol 27, no 22, 2087 (1991)

[6.208]  A. E. Siegman, *Lasers,* University Science Books, Mill Valley, CA, 1986, chapter 7

[6.209]  R. P. Davey, N. Langford, and A. I. Fergusin, "Subpicosecond pulse generation from erbium-doped fibre laser," *Electron. Lett.*, vol 27, no 9, 726 (1991)

[6.210]  P. G. J. Wigley, A. V. Babushkin, J. I. Vukusik, and J. R. Taylor, "Active mode-locking of an erbium-doped fiber laser using an intracavity laser diode device," *IEEE Photonics Technol. Lett.*, vol 2, no 8, 543 (1990)

[6.211]  E. M. Dianov, T. R. Martirosian, O. G. Okhotnikov, V. M. Paranov, and A. M. Prokhorov, "Diode-pumped drive-current-controlled mode-locked fiber laser," in *Proc. Conference on Optical Fiber Communications, OFC '93,* paper FA1, p236, Optical Society of America, Washington DC, 1993

[6.212]  W. H. Loh, D. Atkinson, P. R. Morkel, M. Hopkinson, A. RIvers, A. J. Seeds, and D. N. Payne, "Passively mode-locked $Er^{3+}$ fiber laser using a semiconductor nonlinear mirror," *IEEE Photonics Technol. Lett.*, vol 5, no 1, 35 (1993)

[6.213]  M. E. Fermann, M. J. Andrejco, Y. Silberberg, and A. M. Weiner, "Generation of pulses shorter than 200 fs from a passively mode-locked Er-fiber laser," *Optics Lett.*, vol 18, no 1, 48 (1993)

[6.214]  J. B. Schlager, Y. Yamabayashi, D. L. Franzen, and R. I. Juneau, ," *IEEE Photonics*

*Technol. Lett.*, vol 1, no 9, 264 (1989)

[6.215] A. Takada, "30 GHz picosecond pulse generation from actively mode-locked erbium-doped fibre laser," *Electron. Lett.*, vol 26, no 3, 216 (1990)

[6.216] D. Burns and W. Sibbett, "Controlled amplifier modelocked $Er^{3+}$ fibre ring laser," *Electron. Lett.*, vol 26, no 8, 505 (1990)

[6.217] R. P. Davey, K. Smith, and A. McGuire, "High-speed, mode-locked, tunable integrated erbium fibre laser," *Electron. Lett.*, vol 28, no 5, 482 (1992)

[6.218] J. B. Schlager, S. Kawanishi, and M. Saruwatari, "Dual wavelength pulse generation using mode-locked erbium-doped fibre ring laser," *Electron. Lett.*, vol 27, no 22, 2072 (1991)

[6.219] X. Shan, D. Cleland, and A. Ellis, "Stabilizing Er fibre soliton laser with pulse phase locking," *Electron. Lett.*, vol 28, no 2, 182 (1992)

[6.220] H. Takara, S. Kawanishi, M. Saruwatari, and K. Noguchi, "Generation of highly stable 20 GHz transform-limited optical pulses from actively mode-locked $Er^{3+}$-doped fibre lasers with an all-polarization-maintaining ring cavity," *Electron. Lett.*, vol 28, no 22, 2095 (1992)

[6.221] M. Zirngibl, L. W. Stulz, J. Stone, J. Hugi, D. DiGiovanni, and P. B. Hansen, "1.2 ps pulses from passively mode-locked laser diode pumped Er-doped fibre ring laser," *Electron. Lett.*, vol 27, no 19, 1734 (1991)

[6.222] L. F. Mollenauer, "Solitons in ultra long distance transmission," in *Proc. Conference on Optical Fiber Communications, OFC '91*, tutorial paper WE1, p200, Optical Society of America, Washington DC, 1991; G. T. Harvey and L. F. Mollenauer, "Harmonically mode-locked fiber ring laser with an internal Fabry-Perot stabilizer for soliton transmission," *Optics Lett.*, vol 18, no 2, 107 (1993)

[6.223] C. J. Chen, P. K. A. Wai, and C. R. Menyuk, "Soliton fiber ring laser," *Optics Lett.*, vol 17, no 6, 417 (1992)

[6.224] V. J. Matsas, T. P. Newson, D. J. Richardson, and D. N. Payne, "Self-starting passively mode-locked fibre ring soliton laser exploiting nonlinear polarization rotation," *Electron. Lett.*, vol 28, no 15, 1391 (1992)

[6.225] V. J. Matsas, D. J. Richardson, T. P. Newson, and D. N. Payne, "Characterization of a self-starting, passively mode-locked fiber ring laser that exploits nonlinear polarization evolution," *Optics Lett.*, vol 18, no 5, 358 (1993)

[6.226] R. P. Davey, N. Langford, and A. I. Ferguson, "Interacting solitons in erbium doped fibre laser," *Electron. Lett.*, vol 27, no 14, 1257 (1991)

[6.227] I. N. Duling, "All-fiber ring soliton laser mode-locked with a nonlinear loop mirror," *Optics Lett.*, vol 16, no 8, 539 (1991)

[6.228] D. J. Richardson, R. I. Laming, D. N. Payne, M. W. Phillips, and V. J. Matsas, "320 fs soliton generation with passively mode-locked erbium fibre laser," *Electron. Lett.*, vol 27, no 9, 730 (1991)

[6.229] N. Pandit, D. U. Noske, S. M. J. Kelly, and J. R. Taylor, "Characteristic instability of fibre loop soliton lasers," *Electron. Lett.*, vol 28, no 5, 455 (1992)

[6.230] N. J. Smith, K. J. Blow, and I. Andonovic, "Side-band generation through perturbations to the average soliton mode," *IEEE J. Lightwave* technology (to be published, 1993)

[6.231] D. U. Noske, N. Pandit, and J. R. Taylor, "Source of spectral and temporal instability in soliton fiber lasers," *Optics Lett.*, vol 17, no 21, 1515 (1992)

[6.232] D. J. Richardson, R. I. Laming, D. N. Payne, V. J. Matsas, and M. W. Phillips, "Self-starting, passively mode-locked erbium fibre ring laser based on the amplifying Sagnac switch," *Electron. Lett.*, vol 27, no 6, 542 (1991)

[6.233]  I. N. Duling, "Subpicosecond all-fibre erbium laser," *Electron. Lett.*, vol 27, no 6, 544 (1991)

[6.234]  M. Nakazawa, E. Yoshida, and Y. Kimura, "low threshold, 290 fs erbium-doped fibre laser with a nonlinear amplifying loop mirror pumped by InGaAsP laser diodes," *Applied Phys. Lett.*, vol 59, 2073 (1991)

[6.235]  M. Nakazawa, E. Yoshida, and Y. Kimura, "Generation of 98 fs optical pulses directly from an erbium-doped fibre ring laser at 1.57 $\mu$m," *Electron. Lett.*, vol 29, no 1, 63 (1993)

[6.236]  D. J. Richardson, R. I. Laming, D. N. Payne, V. J. Matsas, and M. W. Phillips, "Pulse repetition rates in passive, self-starting, femtosecond soliton fibre laser," *Electron. Lett.*, vol 27, no 16, 1451 (1991)

[6.237]  A. B. Grudinin, D. J. Richardson, and D. N. Payne, "Energy quantization in figure eight fibre laser," *Electron. Lett.*, vol 28, no 1, 67 (1992)

[6.238]  S. J. Frisken, C. A. Telford, R. A. Betts, and P. S. Atherton, "Passively mode-locked erbium-doped fibre laser with nonlinear fibre mirror," *Electron. Lett.*, vol 27, no 10, 887 (1991)

[6.239]  C. R. Cochlain, R. J. Mears and G. Sherlock, "Low threshold tunable soliton source," *IEEE Photonics Technol. Lett.*, vol 5, no 1, 25 (1993)

[6.240]  B. P. Nelson, K. Smith, and K. J. Blow, "Mode-locked erbium fibre laser using all-optical nonlinear loop mirror," *Electron. Lett.*, vol 28, no 7, 656 (1992)

[6.241]  F. Fontana, G. Grasso, N. Manfredini, M. Romagnoli, and B. Daino, "Generation of sequences of ultrashort pulses in erbium doped fibre single ring lasers," *Electron. Lett.*, vol 28, no 14, 1291 (1992)

[6.242]  E. M. Dianov, P. V. Mamyshev, A. M. Prokhorov, and S. V. Chernikov, "Generation of a train of fundamental solitons at high repetition rate in optical fiber," *Optics Lett.*, vol 14, 1008 (1989)

[6.243]  P. V. Mamyshev, S. V. Chernikov, and E. M. Dianov, "Generation of fundamental soliton trains for high bit rate optical fiber communications," *IEEE J. Quantum Electron.*, vol 27, 2347 (1991)

[6.244]  S. V. Chernikov, J. R. Taylor, P. V. Mamyshev, and E. M. Dianov, "Generation of soliton pulse train in optical fibre using two cw single-mode diode lasers," *Electron. Lett.*, vol 28, no 10, 931 (1992)

[6.245]  S. V. Chernikov, D. J. Richardson, R. I. Laming, E. M. Dianov, and D. N. Payne, "70 Gbit/s fibre based source of fundamental solitons at 1550 nm," *Electron. Lett.*, vol 28, no 13, 1210 (1992)

## CHAPTER 7

[7.1]  P. S. Henry, R. A. Linke, and A. H. Gnauck, "Introduction to lightwave systems," in *Optical Fiber Telecommunications II*, edited by S. E. Miller and I. P. Kaminow, Academic Press, New York, 1988

[7.2]  E. Desurvire, "Lightwave communications: the fifth generation," *Scientific American*, vol 266, no 1, 114 (1992)

[7.3]  R. A. Linke, "Coherent lightwave communications," in *Proc. Conference on Optical Fiber Communications, OFC '88*, tutorial paper TuE1, p1, Optical Society of America, Washington DC, 1988

[7.4]  H. Taga, N. Edagawa, H. Tanaka, M. Suzuki, S. Yamamoto, H. Wakabayashi, N. S. Bergano, C. R. Davidson, G. M. Homsey, D. J. Kalmus, P. R. Trischitta, D. A. Gray, and R. L. Maybach, "10 Gbit/s, 9,000 km IM-DD transmission experiments using 27'

Er-doped fiber amplifiers," in *Proc. Conference on Optical Fiber Communications, OFC'93*, paper PD1, Optical Society of America, Washington DC, 1993

[7.5]   *Proc. Conference on Optical Fiber Communications, OFC'89*, Optical Society of America, Washington DC, 1989; *Proc. Conference on Integrated Optics and Optical Fiber Communication, IOOC'89*, Kobe, Japan, 1989

[7.6]   R. J. Mears, L. Reekie, I. M. Jauncey, and D. N. Payne, "High-gain rare-earth-doped fiber amplifier at 1.54 μm," in *Proc. Conference on Optical Fiber Communications and Conference on Integrated Optics and Optical Communications, OFC/IOOC'87*, paper WI2, p167, Optical Society of America, Washington DC, 1987; R. J. Mears, L. Reekie, I. M. Jauncey, and D. N. Payne, "Low-noise erbium-doped fibre amplifier operating at 1.54 μm," *Electron. Lett.*, vol 23, no 19, 1026 (1987)

[7.7]   E. Desurvire, J. R. Simpson, and P. C. Becker, "High-gain erbium-doped traveling-wave fiber amplifier," *Optics Lett.*, vol 12, no 11, 888 (1987)

[7.8]   N. S. Bergano, "Undersea lightwave systems using erbium-doped fiber amplifiers," *Optics and Photonics News*, vol 4, no 1, 8 (1993)

[7.9]   L. F. Mollenauer and K. Smith, "Demonstration of soliton transmission over more than 4000 km in fiber with loss periodically compensated by Raman gain," *Optics Lett.*, vol 13, no 8, 675 (1988)

[7.10]  L. F. Mollenauer, E. Lichtman, M. J. Neubelt, and G. T. Harvey, "Demonstration, using sliding-frequency guiding filters, of error-free soliton transmission over more than 20,000 km at 10 Gbit/s, single channel, and over more than 13,000 km at 20 Gbit/s in a two-channel WDM," in *Proc. Conference on Optical Fiber Communications and Conference on Integrated Optics and Optical Communications, OFC/IOOC'93*, paper PD8, Optical Society of America, Washington DC, 1993

[7.11]  M. Nakazawa, K. Suzuki, E. Yamada, H. Kubota, Y. Kimura, and M. Takaya, "Experimental demonstration of soliton data transmission over unlimited distances with soliton control in time and frequency domains," in *Proc. Conference on Integrated Optics and Optical Communications, OFC/IOOC'93*, paper PD7, Optical Society of America, Washington DC, 1993

[7.12]  P. K. Runge, "Undersea lightwave systems," *Optics and Photonics News*, vol 1, no 11, 12 (1990)

[7.13]  G. C. Holst and E. Snitzer, "Detection with a fiber laser preamplifier at 1.06 μm," *IEEE J. Quantum Electron.*, vol 5, 319 (1969); G. C. Holst and E. Snitzer, "Detection with a fiber laser preamplifier at 1.06μm," in *Proc. Optical Society of America*, spring meeting, paper ThF19, p506, 1969

[7.14]  S. D. Walker and L. C. Blank "Ge APD/GaAs FET/op-amp transimpedance optical receiver design having minimum noise and intersymbol interference characteristics," *Electron. Lett.*, vol 20, 808 (1984)

[7.15]  B. Kasper, "Receiver design," in *Optical Fiber Telecommunications II*, edited by S. E. Miller and I. P. Kaminow, Academic Press, New York, 1988

[7.16]  C. R. Giles, E. Desurvire, J. R. Talman, J. R. Simpson, and P. C. Becker, "2-Gbit/s signal amplification at λ = 1.53 μm in an erbium-doped single-mode fiber amplifier," *IEEE J. Lightwave Technol.*, vol 7, no 4, 651 (1989)

[7.17]  C. R. Giles, E. Desurvire, J. L. Zyskind, and J. R. Simpson, "Noise performance of erbium-doped fiber amplifier pumped at 1.49 μm, and application to signal preamplication at 1.8 Gbit/s," *IEEE Photonics Technol. Lett.*, vol 1, no 11, 367 (1989)

[7.18]  N. A. Olsson and P. Garbinski, "High sensitivity direct-detection receiver with 1.5 μm optical preamplifier," *Electron. Lett.*, vol 22, 114 (1986)

[7.19]  M. J. Pettitt, and A. Hadjifotiou, "System performance of optical fibre preamplifier," *Electron. Lett.*, vol 25, no 4, 273 (1989)

[7.20] J. Boggis, A. Lord, M. Violas, D. Heatley, and W. A. Stallard, "Broadband, high sensivity erbium amplifier receiver for operation over wide wavelength range," *Electron. Lett.*, vol 26, no 8, 532 (1990)

[7.21] P. P. Smyth, R. Wyatt, A. Fidler, P. Eardley, A. Sayles, and S. Craig-Ryan, "152 photons per bit detection at 622 Mbit/s to 2.5 Gbit/s using an erbium fibre preamplifier," *Electron. Lett.*, vol 26, no 19, 1604 (1990)

[7.22] L. T. Blair and H. Nakano, "High sensitivity 10 Gbit/s optical receiver using two cascaded EDFA preamplifiers," *Electron. Lett.*, vol 27, no 10, 835 (1991)

[7.23] T. Saito, Y. Sunohara, K. Fukagai, S. Ishikawa, N. Henmi, S. Fujita, and Y. Aoki, "High receiver sensitivity at 10 Gbit/s using an Er-doped fiber preamplifier pumped with a 0.98 $\mu$m laser diode," *IEEE Photonics Technol. Lett.*, vol 3, no 6, 551 (1991)

[7.24] P. M. Gabla, E. Leclerc, and C. Coeurjoli, "Practical implementation of a highly sensitive receiver using an erbium-doped fiber preamplifier," *IEEE Photonics Technol. Lett.*, vol 3, no 8, 727 (1991)

[7.25] B. L. Patel, E. M. Kimber, M. G. Taylor, A. N. Robinson, I. Hardcastle, A. Hadjifotiou, S. J. Wilson, R. Keys, and J. E. Righton, "High performance 10 Gbit/s optical transmission system using erbium-doped fibre preamplifier," *Electron. Lett.*, vol 27, no 23, 2179 (1991)

[7.26] C. G. Joergensen, B. Mikkelsen, T. Durhuus, K. E. Stubkjaer, J. A. Van den Berk, C. F. Pedersen, and C. C. Larsen, "High sensitivity fibre preamplifier receiver for multichannel applications," *Electron. Lett.*, vol 27, no 23, 2153 (1991)

[7.27] A. H. Gnauck and C. R. Giles, "2.5 and 10 Gb/s transmission experiments using a 137 photons/bit erbium-fiber preamplifier receiver," *IEEE Photonics Technol. Lett.*, vol 4, no 1, 80 (1992)

[7.28] Y. K. Park, S. W. Granlund, T. W. Cline, L. D. Tzeng, J. S. French, J-M. P. Delavaux, R. E. Tench, S. K. Korotky, J. J. Veselka, and D. J. DiGiovanni, "2.488 Gb/s–318 km repeaterless transmission using erbium-doped fiber amplifiers in a direct-detection system," *IEEE Photonics Technol. Lett.*, vol 4, no 2, 179 (1992)

[7.29] K. Kannan, A. Bartos, and P. S. Atherton, "High-sensitivity receiver optical pre-amplifiers," *IEEE Photonics Technol. Lett.*, vol 4, no 3, 272 (1992)

[7.30] R. I. Laming, A. H. Gnauck, C. R. Giles, M. N. Zervas, and D. N. Payne, "High-sensitivity two-stage erbium-doped fiber preamplifier at 10 Gb/s," *IEEE Photonics Technol. Lett.*, vol 4, no 12, 1348 (1992)

[7.31] Y. K. Park, J-M. P. Delavaux, O. Mizuhara, L. D. Tzeng, T. V. Nguyen, M. L. Kao, P. D. Yeates, S. W. Granlund, and J. Stone, "5 Gbit/s optical preamplifier receiver with 135 photons/bit usable receiver sensitivity," in *Proc. Conference on Optical Fiber Communications and Conference on Integrated Optics and Optical Communications, OFC/IOOC'93*, paper TuD4, p16, Optical Society of America, Washington DC, 1993

[7.32] A. E. Willner and E. Desurvire, "Effect of gain saturation on receiver sensitivity in 1 Gb/s multichannel FSK direct-detection systems using erbium-doped fiber pre-amplifiers," *IEEE Photonics Technol. Lett.*, vol 3, no 3, 259 (1991)

[7.33] K. Hagimoto, S. Nishi, and K. Nakagawa, "An optical bit-rate flexible transmission system with 5 Tb/s.km capacity employing multiple in-line erbium-doped fiber amplifiers," *IEEE J. Lightwave Technol.*, vol 8, no 9, 1387 (1990)

[7.34] C. R. Giles and E. Desurvire, "Propagation of signal and noise in concatenated erbium-doped fiber optical amplifiers," *IEEE J. Lightwave Technol.*, vol 9, no 2, 147 (1991)

[7.35] K. Nakagawa, S. Nishi, K. Aida, and E. Yoneda, "Trunk and distribution network application of erbium-doped fiber amplifier," *IEEE J. Lightwave Technol.*, vol 9, no 2, 198 (1991)

[7.36] K. Inoue, H. Toba, and K. Nosu, "Multichannel amplification utilizing an Er-doped fiber amplifier," *IEEE J. Lightwave Technol.*, vol 9, no 3, 368 (1991)

[7.37]  K. Hagimoto, K. Iwatsuki, A. Takada, M. Nakazawa, M. Saruwatari, K. Aida, K. Nakagawa, and M. Horiguchi, "A 212 km non-repeated transmission experiment at 1.8 Gb/s using LD pumped Er-doped fiber amplifiers in an IM/direct-detection repeater system," in *Proc. Conference on Optical Fiber Communications, OFC'89*, paper PD15, Optical Society of America, Washington DC, 1989

[7.38]  J. L. Gimlett, M. Z. Iqbal, C. E. Zah, J. Young, L. Curtis, R. Spicer, C. Caneau, F. Favire, S. G. Menocal, N. Andreakis, T. P. Lee, N. K. Cheung, and S. Tsujii, "A 94 km, 11 Gb/s NRZ transmission experiment using a 1540 nm DFB laser with an optical amplifier and a OIN/HEMT receiver," in *Proc. Conference on Optical Fiber Communications, OFC'89*, paper PD16, Optical Society of America, Washington DC, 1989

[7.39]  N. Edagawa, K. Mochizuki, and H. Wakabayahsi, "1.2 Gbit/s, 218 km transmission experiment using inline Er-doped optical fibre amplifier," *Electron. Lett.*, vol 25, no 5, 363 (1989)

[7.40]  K. Hagimoto, K. Iwatsuki, A. Takada, M. Nakazawa, M. Saruwatari, K. Aida, and K. Nakagawa, "250 km nonrepeated transmission experiment at 1.8 Gb/s using LD pumped Er-doped fibre amplifiers in IM/direct detection system," *Electron. Lett.*, vol 25, no 10, 662 (1989)

[7.41]  N. Edagawa, M. Suzuki, K. Mochizuki, and H. Wakabayashi, "276 km, 1.2 Gbit/s optical transmission experiment using two in-line LD-pumped Er-doped optical fibre amplifiers and an electroabsorption modulator," in *Proc. Conference on Integrated Optics and Optical Communications, IOOC'89*, paper 21B4-1, Kobe, Japan, 1989

[7.42]  A. Takada, K. Hagimoto, K. Iwatsuki, K. Aida, K. Nakagawa, and M. Shimizu, "1.8 Gbit/s transmission over 210 km using an Er-doped fiber laser amplifier with 20 dB repeater gain in a direct detection system," in *Proc. Conference on Integrated Optics and Optical Communications, IOOC'89*, paper 21B3-3, Kobe, Japan, 1989

[7.43]  K. Hagimoto, Y. Miyagawa, Y. Miyamoto, M. Ohhashi, M. Ohhata, K. Aida, and K. Nakagawa, "A 10 Gbit/s long-span fiber transmission experiment employing optical amplification technique and monolithic IC technology," in *Proc. Conference on Integrated Optics and Optical Communications, IOOC'89*, paper 20PDA-6, Kobe, Japan, 1989

[7.44]  H. Nishimoto, I. Yokota, M. Suyama, T. Okiyama, M. Sino, T. Horimatsu, H. Kuwahara, and T. Touge, "Transmission of 12 Gb/s over 100 km using an LD-pumped erbium-doped fiber amplifier and a Ti:LiNbO$_3$ Mach–Zehnder modulator," in *Proc. Conference on Integrated Optics and Optical Communications, IOOC'89*, paper 20PDA-8, Kobe, Japan, 1989

[7.45]  M. Z. Iqbal, J. L. Gimlett, M. M. Choy, A. Yi-Yan, M. J. Andrejco, L. Curtis, M. A. Saifi, C. Lin, and N. K. Cheung, "An 11 Gbit/s, 151 km transmission experiment employing a 1480 nm pumped erbium-doped in-line fiber amplifier," *IEEE Photonics Technol. Lett.*, vol 1, no 10, 334 (1989)

[7.46]  N. Edagawa, Y. Yoshida, H. Taga, S. Yamamoto, K. Mochizuki, and H. Wakabayashi, "904 km, 1.2 Gbit/s non-regenerative optical fibre transmission experiment using 12 Er-doped fibre amplifiers," *Electron. Lett.*, vol 26, no 1, 66 (1990)

[7.47]  A. Righetti, F. Fontana, G. Delrosso, G. Grasso, M. Z. Iqbal, J. L. Gimlett, R. D. Standley, J. Young, N. K. Cheung, and E. J. Tarbox, "11 Gbit/s, 260 km transmission experiment using a directly-modulated 1536 nm DFB laser with two Er-doped fibre amplifiers and clock recovery," *Electron. Lett.*, vol 26, no 5, 330 (1990)

[7.48]  E. G. Bryant, S. F. Carter, A. D. Ellis, W. A. Stallard, J. V. Wright, and R. Wyatt, "Unrepeatered 2.4 Gbit/s transmission experiment over 250 km of step-index fibre using erbium power amplifier," *Electron. Lett.*, vol 26, no 8, 528 (1990)

[7.49]  N. Edagawa, Y. Yoshida, H. Taga, S. Yamamoto, and H. Wakabayashi, "12,300 ps/nm, 2.4 b/s nonregenerative optical fiber transmission experiment and effect of transmitter phase noise," *IEEE Photonics Technol. Lett.*, vol 2, no 4, 274 (1990)

[7.50]  D. A. Fishman, J. A. Nagel, T. W. Cline, R. E. Tench, T. C. Pleiss, T. Miller, D. G. Coult, M. A. Milbrodt, P. D. Yeates, A. Chraplyvy, R. Tkach, A. B. Piccirilli, J. R. Simpson, and C. M. Miller, "A high capacity noncoherent FSK lightwave field experiment using Er-doped fiber optical amplifiers," *IEEE Photonics Technol. Lett.*, vol 2, no 9, 662 (1990)

[7.51]  C. Y. Kuo, M. L. Kao, J. S. French, R. E. Tench, and T. W. Cline, "1.55 μm, 2.5 Gb/s direct detection repeaterless transmission of 160 km nondispersion shifted fiber," *IEEE Photonics Technol. Lett.*, vol 2, no 12, 911 (1990)

[7.52]  N. S. Bergano, J. Aspell, C. R. Davidson, P. R. Trischitta, B. M. Nyman, and F. W. Kerfoot, "A 9,000 km 5 Gb/s and 21,000 km 2.4 Gb/s feasibility demonstration of transoceanic EDFA systems using circulating loop," *Proc. Conference on Optical Fiber Communications, OFC'91*, paper PD13, Optical Society of America, Washington DC, 1991

[7.53]  N. S. Bergano, J. Aspell, C. R. Davidson, P. R. Trischitta, B. M. Nyman, and F. W. Kerfoot, "Bit error rate measurements of 14,000 km 5 Gbit/s fibre-amplifier transmission system using circulating loop," *Electron. Lett.*, vol 27, no 21, 1889 (1991)

[7.54]  T. Widdowson and D. J. Malyon, "Error ratio measurements over transoceanic distances using recirculating loop," *Electron. Lett.*, vol 27, no 24, 2201 (1991)

[7.55]a  A. H. Gnauck, C. R. Giles, L. J. Cimini, J. Stone, L. W. Stult, S. K. Korotky, and J. J. Veselka, "8 Gb/s–130 km transmission experiment using Er-doped fiber preamplifier and optical dispersion equalization," *IEEE Photonics Technol. Lett.*, vol 3, no 12, 1147 (1991)

[7.55]b  P. M. Gabla, O. Scaramucci, J. O. Frorud, G. Bassier, J. B. Leroy, V. Havard, and E. leclerc, "1316 km, 2.5 Gbits/s IM-DD penalty free transmission through 26 in-line erbium-doped fiber amplifiers," in *Proc. European Conference on Optical Communications, ECOC '91, and conference on Integrated Optics and Optical Communications, IOOC '91*, paper PDP-4, 1991

[7.56]  H. Taga, N. Edagawa, Y. Yoshida, S. Yamamoto, M. Suzuki, and H. Wakabayashi, "10 Gbit/s, 4500 km transmission experiment using 138 cascaded Er-doped fiber amplifiers," in *Proc. Conference on Optical Fiber Communications, OFC'92*, paper PD12, Optical Society of America, Washington DC, 1992

[7.57]  A. D. Ellis, S. J. Pycock, D. A. Cleland, and C. H. F. Sturrock, "Dispersion compensation in 450 km transmission system employing standard fibre," *Electron. Lett.*, vol 28, no 10, 954 (1992)

[7.58]  T. Imai, M. Murakami, Y. Fukada, M. Aiki, and T. Ito, "2.5 Gbit/s, 10,073 km straight line transmission system experiment using 199 Er-doped fibre amplifiers," *Electron. Lett.*, vol 28, no 16, 1484 (1992)

[7.59]  H. Izadpanah, C. Lin, J. L. Gimlett, A. J. Antos, D. W. Hall, and D. K. Smith, "Dispersion compensation in 1310 nm-optimized SMFs using optical equalizer fibre, EDFAs and 1310/1550 nm WDM," *Electron. Lett.*, vol 28, no 15, 1469 (1991); H. Izadpanah, C. Lin, J. L. Gimlett, H. Johnson, W. Way, and P. Kaiser, "Dispersion compensation for upgrading inter-office networks built with 1310 nm-optimized SMFs using an equalizer fiber, EDFAs and 1310/1550 nm WDM," in *Proc. Conference on Optical Fiber Communications OFC '92*, paper PD15, Optical Society of America, Washington DC, 1992

[7.60]  P. M. Gabla, J. L. Pamart, R. Uhel, E. Leclerc, J. O. Frorud, F. X. Ollivier, and S.

Borderieux, "401 km, 622 Mb/s and 347 km, 2.488 Gb/s IM/DD repeaterless transmission experiments using erbium-doped fiber amplifiers and error correcting code," *IEEE Photonics Technol. Lett.*, vol 4, no 10, 1148 (1992)

[7.61]　M. Murakami, T. Kataoka, T. Imai, K. Hagimoto, and M. Aiki, "10 Gbit/s, 6000 km transmission experiment using erbium-doped fibre in-line amplifiers," *Electron. Lett.*, vol 28, no 24, 2254 (1992)

[7.62]　S. Yamamoto, N. Edagawa, H. Taga, Y. Yoshida, and H. Wakabayashi, "Observation of BER degradation due to fading in long-distance optical amplifier system," *Electron. Lett.*, vol 29, no 2, 209 (1993)

[7.63]　H. Taga, N. Edagawa, H. Tanaka, M. Suzuki, S. Yamamoto, and H. Wakabayashi, "10 Gbit/s, 9,000 km IM-DD transmission experiments using 274 Er-doped fiber amplifier repeaters," in *Proc. Conference on Optical Fiber Communications and Conference on Integrated Optics and Optical Communications OFC/IOOC'93*, paper PD1, Optical Society of America, Washington DC, 1993

[7.64]　S. Kawanishi, H. Takara, K. Uchiyama, T. Kitoh, and M. Saruwatari, "100 Gbit/s, 50 km optical transmission employing all-optical multi/demultiplexing and PLL timing extraction," in *Proc. Conference on Optical Fiber Communications* and *Conference on Integrated Optics and Optical Communications OFC/IOOC'93*, paper PD2, Optical Society of America, Washington DC, 1993

[7.65]　R. M. Jopson, A. H. Gnauck, and R. M. Derosier, "10 Gb/s 360 km transmission over normal-dispersion fiber using mid-system spectral inversion," in *Proc. Conference on Optical Fiber Communications and Conference on Integrated Optics and Optical Communications OFC/IOOC'93*, paper PD3, Optical Society of America, Washington DC, 1993

[7.66]　B. L. Patel, E. M. Kimber, N. E. Jolley, and A. Hadjifotiou, "Repeaterless transmission at 10 Gb/s over 215 km of dispersion shifted fibre, and 180 km of standard fibre," in *Proc. Conference on Optical Fiber Communications and Conference on Integrated Optics and Optical Communications OFC/IOOC'93*, paper PD6, Optical Society of America, Washington DC, 1993

[7.67]　K. Nakagawa, K. Hagimoto, S. Nishi, and K. I. Aoyama, "A bit-rate flexible transmission field trial over 300 km installed cables employing optical fiber amplifiers," in *Proc. Topical Meeting on Optical Amplifiers and Applications*, 1991, paper PdP11, Optical Society of America, Washington DC, 1991

[7.68]　N. S. Bergano, C. R. Davidson, G. M. Homsey, D. J. Kalmus, P. R. Trischitta, J. Aspell, D. A. Gray, R. L. Maybach, S. Yamamoto, H. Taga, N. Edagawa, Y. Yoshida, Y. Horiuchi, T. Kawazawa, Y. Namihira, and S. Akiba, "9000 km, 5 Gbit/s NRZ transmission experiment using 274 erbium-doped fiber amplifiers," in *Proc. Topical Meeting on Optical Amplifiers and Applications, 1992*, paper PD11, Optical Society of America, Washington DC, 1992

[7.69]　Y. K. Park, O. Mizuhara, L. D. Tzeng, J-M. P. Delavaux, T. V. Nguyen, M-L. Kao, and P. D. Yeates, "A 5 Gbit/s repeaterless transmission system using optical amplifiers," in *Proc. Topical Meeting on Optical Amplifiers and Applications, 1992*, paper PD14,, Optical Society of America, Washington DC, 1992

[7.70]　S. Yamamoto, H. Taga, Y. Yoshida, and H. Wakabayashi, "Characteristics of single-carrier fiber-optics transmission systems using optical amplifiers," in *Proc. Topical Meeting on Optical Amplifiers and Applications, 1992*, paper ThA4, p90, Optical Society of America, Washington DC, 1991

[7.71]　E. G. Bryant, S. F. Carter, R. B. J. Lewis, D. M. Spirit, T. Widdowson, and J. V. Wright, "Field demonstration of FSK transmission at 1.488 Gbit/s over a 132 km submarine cable using an erbium power amplifier," in *Proc. Topical Meeting on Optical Amplifiers and Applications, 1990*, paper TuC2, p152, Optical Society of America,

Washington DC, 1990

[7.72] K. Aida, H. Masuda, and A. Takada, "Long-span IM/DD transmission system experiment using high-efficiency EDF amplifiers," in *Proc. Topical Meeting on Optical Amplifiers and Applications, 1990*, paper TuC5, p164, Optical Society of America, Washington DC, 1990

[7.73] K. Nakagawa, K. Hagimoto, and S. Nishi, "Optical bit rate flexible transmission experiment over 10 Gbit/s, 375 km employing in-line fiber amplifiers," in *Proc. Conference on Optical Fiber Communications OFC'90*, paper WC2, p56, Optical Society of America, Washington DC, 1990

[7.74] N. Henmi, T. Saito,, M. Yamaguchi, and S. Fujita, "10 Gbit/s, 100 km normal fiber transmission experiment employing a modified prechirp technique," in *Proc. Conference on Optical Fiber Communications OFC'91*, paper TuO2, p54, Optical Society of America, Washington DC, 1991

[7.75] J. M. Dugan, A. J. Price, M. Ramadan, D. L. Wolf, E. F. Murphy, A. J. Antos, D. K. Smith, and D. W. Hall, "All-optical, fiber-based 1550 nm dispersion compensation in a 10 Gbit/s, 150 km transmission experiment over 1310 nm optimized fiber," in *Proc. Conference on Optical Fiber Communications OFC'92*, paper PD14, Optical Society of America, Washington DC, 1992

[7.76] K. Hagimoto, Y. Miyamoto, T. Kataoka, K. Kawano, and M. Ohhata, "A 17 Gb/s long-span fiber transmission experiment using a low-moise broadband receiver with optical amplification and equalization," in *Proc. Topical Meeting on Optical Amplifiers and Applications, 1990*, paper TuA2, p100, Optical Society of America, Washington DC, 1990

[7.77] N. Henmi, T. Saitoh, S. Fujita, M. Yamaguchi, H. Asano, I. Mito, and M. Shikada, "Dispersion compensation by prechirp technique in multigigabit optical amplifier repeater systems," in *Proc. Topical Meeting on Optical Amplifiers and Applications, 1990*, paper TuA4, p108, Optical Society of America, Washington DC, 1990

[7.78] C. G. Joergensen, B. Mikkelsen, C. F. Pedersen, K. Schusler, O. Frolish, C. C. Larsen, J. A. Berk, T. Durhuus, and K. E. Stubkjaer, "Refined amplifier design for unrepeated 2.5 Gbit/s transmission over 59 dB of fibre loss," in *Proc. Topical Meeting on Optical Amplifiers and Applications, 1992*, paper FC3, p192, Optical Society of America, Washington DC, 1992

[7.79] S. Saito, M. Murakami, A. Naka, Y. Kukada, T. Imai, M. Aiki, and T. Ito, "In-line amplifier transmission experiments over 4500 km at 2.5 Gbit/s," *IEEE J. Lightwave Technol.*, vol 10, no 8, 1117 (1992)

[7.80] N. Edagawa, "Applications of fiber amplifiers to telecommunications systems," in *Rare earth Doped Fiber Lasers and Amplifiers*, edited by M. J. F. Digonnet, Marcel Dekker, New York, 1993

[7.81] A. B. Carlson, *Communications Systems*, third edition, McGraw-Hill, New York, 1968, chapters 11 and 13

[7.82] A. R. Chraplyvy, "Limitations on lightwave communications imposed by optical-fiber nonlinearities," *IEEE J. Lightwave Technol.*, vol 8, no 10, 1548 (1990)

[7.83] H. Toba, K. Inoue, N. Shibata, K. Nosu, K. Iwatsuki, N. Takato, and M. Shimizu, "16-channel optical FDM distribution/transmission experiment utilizing Er-doped fibre amplifier," *Electron. Lett.*, vol 25, no 14, 885 (1989)

[7.84] H. Taga, Y. Yoshida, N. Edagawa, S. Yamamoto, and H. Wakabayashi, "459 km, 2.4 Gbit/s 4 wavelength multiplexing optical fiber transmission experiment using 6 Er-doped fiber amplifiers," in *Proc. Conference on Optical Fiber Communications OFC'90*, paper PD9, Optical Society of America, Washington DC, 1990; H. Taga, Y. Yoshida, N. Edagawa, S. Yamamoto, and H. Wakabayashi, "459 km, 2.4 Gbit/s four wavelength multiplexing optical fibre transmission experiment using six Er-doped fiber amplifiers," *Electron. Lett.*, vol 26, no 8, 500 (1990)

[7.85]   G. E. Wickens, D. M. Spirit, and L. C. Blank, "20 Gbit/s, 205 km optical time division multiplexed transmission system," *Electron. Lett.*, vol 27, no 11, 973 (1991)

[7.86]   D. M. Spirit, G. E. Wickens, T. Widdowson, G. R. Walker, D. L. Williams, and L. C. Blank, "137 km, 4 × 5 Gbit/s optical time division multiplexed unrepeatered system with distributed erbium fibre preamplifier," *Electron. Lett.*, vol 28, no 13, 1218 (1992)

[7.87]   P. M. Gabla, J. O. Frorud, E. Leclerc, S. Gauchard, and V. Havad, "1111 km, two-channel IM-DD transmission experiment at 2.5 Gb/s through 21 in-line erbium-doped fiber amplifiers," *IEEE Photonics Technol. Lett.*, vol 4, no 7, 717 (1992)

[7.88]   W. Y. Guo and Y. K. Chen, "High-speed bidirectional four-channel optical FDM-NCFSK transmission using an Er-doped fiber amplified," *IEEE Photonics Technol. Lett.*, vol 5, no 2, 232 (1993)

[7.89]   H. Toba, K. Nakanishi, K. Oda, K. Inoue, and T. Kominato, "A 100-channel optical FDM six-stage in-line amplifier system employing tunable gain equalizers," *IEEE Photonics Technol. Lett.*, vol 5, no 2, 248 (1993)

[7.90]   H. Taga, N. Edagawa, S. Yamamoto, Y. Yoshida, Y. Horiuchi, T. Kawazawa, and H. Wakabayashi, "Over 4,500 km IM-DD 2-channel WDM transmission experiments at 5 Gbit/s using 138 in-line Er-doped fiber amplifiers," in *Proc. Conference on Optical Fiber Communications OFC'93*, paper PD4, Optical Society of America, Washington DC, 1993

[7.91]   A. E. Willner, "SNR analysis of crosstalk and filtering effects in an amplified multichannel direct-detection dense-WDM system," *IEEE Photonics Technol. Lett.*, vol 4, no 2, 186 (1992)

[7.92]   A. H. Gnauck, L. J. Cimini, J. Stone, and L. W. Stultz, "Optical equalization of fiber chromatic dispersion in a 5 Gbit/s transmission system," in *Proc. Conference on Optical Fiber Communications OFC'90*, paper PD7, Optical Society of America, Washington DC, 1990

[7.93]   C. D. Poole, K. T. Nelson, J. M. Wiesenfeld, and A. McCormick, "Broad-band dispersion compensation using the higher-order spatial mode in a two-mode fiber," in *Proc. Conference on Optical Fiber Communications OFC'92*, paper PD13, Optical Society of America, Washington DC, 1992

[7.94]   N. Henmi, S. Fujita, T. Saito, M. Yamaguchi, M. Shikada, and J. Namiki, "A novel dispersion compensation technique for multigigabit transmission with normal optical fiber at 1.5 $\mu$m wavelength," in *Proc. Conference on Optical Fiber Communications OFC'90*, paper PD8, Optical Society of America, Washington DC, 1990

[7.95]   A. Yariv, D. Fekete, and D. M. Pepper, "Compensation for channel dispersion by nonlinear optical phase conjugation," *Optics Lett.*, vol 4, no 2, 52 (1979)

[7.96]   T. J. Whitley, M. J. Creaner, R. C. Steele, M. C. Brain, and C. A. Millar, "Laser diode pumped Er-doped fiber amplifier in a 565 Mbit/s DPSK coherent transmission experiment," *IEEE Photonics Technol. Lett.*, vol 1, no 12, 425 (1989)

[7.97]   R. A. Linke and A. H. Gnauck, "High-capacity coherent lightwave systems," *IEEE J. Lightwave Technol.*, vol 6, no 11, 1750 (1988)

[7.98]   J. M. Kahn, A. H. Gnauck, J. J. Veselka, S. K. Korotky, and B. L. Kasper, "4 Gb/s PSK homodyne transmission system using phase-locked semiconductor lasers," *IEEE Photonics Technol. Lett.*, vol 2, no 4, 285 (1990)

[7.99]   A. H. Gnauck, K. C. Reichmann, J. M. Kahn, S. K. Korotky, J. J. Veselka, and T. L. Koch, "4 Gb/s heterodyne transmission experiments using ASK, FSK and DPSK modulation," *IEEE Photonics Technol. Lett.*, vol 2, no 12, 908 (1990)

[7.100]  S. Saito, T. Imai, T. Sugie, N. Ohkawa, Y. Ichihashi, and T. Ito, "2.5 Gbit/s 400 km coherent transmission experiment using two in-line erbium-doped fiber amplifiers," in *Proc. Conference on Optical Fiber Communications OFC'90*, paper WC3, p57, Optical Society of America, Washington DC, 1990

[7.101]  T. Sugie, T. Imai, and T. Ito, "Over 350 km CPFSK repeaterless transmission at 2.5 Gbit/s employing high-output power erbium-doped fibre amplifiers," *Electron. Lett.*, vol 26, no 19, 1577 (1990)

[7.102]  M. J. Creaner, D. Spirit, G. R. Walker, N. G. Walker, R. C. Steele, J. Mellis, S. Al Chalabi, W. Hale, I. Sturgess, M. Rutherford, D. Trivett, and M. Brain, "Field demonstration of coherent transmission system with diode-pumped erbium-doped fibre amplifiers," *Electron. Lett.*, vol 26, no 7, 442 (1990)

[7.103]  S. Saito, T. Imai, T. Sugie, N. Ohkawa, Y. Ichihashi, and T. Ito, "Coherent transmission experiment over 2,223 km at 2.5 Gbit/s using erbium-doped fibre amplifiers," *Electron. Lett.*, vol 26, no 10, 669 (1990); S. Saito, T. Imai, T. Sugie, N. Ohkawa, Y. Ichihashi and T. Ito, "An over 2,200 coherent transmission experiment at 2.5 Gbit/s using erbium-doped fiber amplifiers," in *Proc. Conference on Optical Fiber Communications OFC'90*, paper PD3, Optical Society of America, Washington DC, 1990

[7.104]  S. Saito, T. Imai, and T. Ito, "An over 2200 km coherent transmission experiment at 2.5 Gb/s using erbium-doped-fiber in-line amplifiers," *IEEE J. Lightwave Technol.*, vol 9, no 2, 161 (1991)

[7.105]  M. J. Creaner, R. C. Steele, D. Spirit, G. R. Walker, N. G. Walker, J. Mellis, S. Al Chalabi, W. Hale, I. Sturgess, M. Rutherford, D. Trivett, and M. Brain "Field demonstration of two channel coherent transmission with a diode-pumped fibre amplifier repeater," *Electron. Lett.*, vol 26, no 19, 1621 (1990)

[7.106]  S. Ryu, N. Edagawa, Y. Yoshida, and H. Wakabayashi, "Over 1000 km FSK heterodyne transmission system experiment using erbium-doped optical fiber amplifiers and conventional single-mode fibers," *IEEE Photonics Technol. Lett.*, vol 2, no 6, 428 (1990)

[7.107]  B. Clesca, J. Auge, B. Biotteau, P. Bousselet, A. Dursin, C. Clergeaud, P. Kretzmeyer, V. Lemaire, O. Gautheron, G. Grandpierre, E. Leclerc, and P. Gabla, "Repeaterless transmission with 62.9 dB power budget using a highly efficient erbium-doped fibre amplifier module," *Electron. Lett.*, vol 26, no 18, 1426 (1990); J. Auge, B. Clesca, B. Biotteau, P. Bousselet, A. Dursin, C. Clergeaud, P. Kretzmeyer, V. Lemaire, O. Gautheron, G. Grandpierre, E. Leclerc, and P. Gabla, "Repeaterless transmission with 62.9 dB power budget using a highly efficient erbium-doped fiber amplifier module," in *Proc. Topical Meeting on Optical Amplifiers and Applications, 1990*, paper TuC3, p156, Optical Society of America, Washington DC, 1990

[7.108]  Y. K. Park, J-M. P. Delavaux, R. E. Tench, and T. W. Cline, "1.7 Gb/s–419 km transmission experiment using a shelf-mounted FSK coherent system and packaged fiber amplifier modules," *IEEE Photonics Technol. Lett.*, vol 2, no 12, 917 (1990); Y. K. Park, J-M. P. Delavaux, R. E. Tench, and T. W. Cline, "1.7 Gb/s–419 km transmission experiment using a shelf-mounted FSK coherent system and packaged fiber amplifier modules," in *Proc. Topical Meeting on Optical Amplifiers and Applications*, 1990, paper TuC4, p160, Optical Society of America, Washington DC, 1990; Y. K. Park, "1.7 Gb/s coherent optical transmission field trial," in *Proc. Conference on Optical Fiber Communications OFC'91*, paper TuH3, p30, Optical Society of America, Washington DC, 1991

[7.109]  G. R. Walker, N. G. Walker, R. C. Steele, M. J. Creaner, and M. C. Brain, "Erbium-doped fiber amplifier cascade for multichannel coherent optical transmission," *IEEE J. Lightwave Technol.*, vol 9, no 2, 182 (1991)

[7.110]  T. Sugie, N. Ohkawa, T. Imai, and T. Ito, "A novel repeaterless CPFSK coherent lightwave system employing an optical booster amplifier," *IEEE J. Lightwave*

*Technol.*, vol 9, no 9, 1178 (1991)

[7.111] S. Ryu, T. Miyazaki, T. Kawazawa, Y. Namihira, and H. Wakabayashi, "Field demonstration of 195 km-long coherent unrepeatered submarine cable system using optical booster amplifier," *Electron. Lett.*, vol 28, no 21, 1965 (1992)

[7.112] S. Saito, T. Imai, A. Naka, and T. Ito, "Transmission performance of coherent in-line-amplifier systems," in *Proc. Topical Meeting on Optical Amplifiers and Applications, 1990*, paper TuA5, p112, Optical Society of America, Washington DC, 1990

[7.113] N. Ohkawa, Y. Hayashi, H. Fushimi, and D. Yanai, "2.4 Gbit/s CPFSK booster amplifier/polarization diversity receiver for a repeaterless optical transmission system," in *Proc. Conference on Optical Fiber Communications OFC'92*, paper FC4, p287, Optical Society of America, Washington DC, 1992

[7.114] T. Sugie, "Suppression of SBS by discontinuous Brillouin frequency shifted fibre in CPFSK coherent lightwave system with booster amplifier," *Electron. Lett.*, vol 27, no 14, 1231 (1991)

[7.115] R. H. Stolen and C. Lin, "Self-phase modulation in silica optical fibers," *Phys. Rev.*, vol 17, no 4, 1448 (1978)

[7.116] A. R. Chraplyvy, D. Marcuse and P. S. Henry, "Carrier-induced phase noise in angle-modulated optical fiber systems," *IEEE J. Lightwave Technol.*, vol 2, no 1, 6 (1984)

[7.117] J. P. Gordon and L. F. Mollenauer, "Phase noise in photonic communication systems using linear amplifiers," *Optics Lett.*, vol 15, no 23, 1351 (1990)

[7.118] A. Yariv, *Optical Electronics*, third edition, Holt, Rinehart and Winston, New York, 1985

[7.119] S. Ryu, "Signal linewidth broadening due to nonlinear Kerr effect in long-haul coherent systems using cascaded optical amplifiers," *IEEE J. Lightwave Technol.*, vol 10, no 10, 1450 (1992)

[7.120] M. Murakami and S. Saito, "Deformation of carrier field spectrum due to fiber-nonlinearity-induced phase noise in lightwave systems with optical amplifiers," *Electron. Lett.*, vol 28, no 21, 1987 (1992)

[7.121] M. Murakami and S. Saito, "Evolution of field spectrum due to fibre-nonlinearity-induced phase noise in in-line optical amplifier system," *IEEE Photonics Technol. Lett.*, vol 4, no 11, 1269 (1992)

[7.122] S. Ryu, "Change of field spectrum of signal light due to fibre nonlinearities and chromatic dispersion in long-haul coherent systems using in-line optical amplifiers," *Electron. Lett.*, vol 28, no 24, 2212 (1992)

[7.123] K. Inoue and H. Toba, "Error-rate degradation due to fiber four-wave mixing in four-channel FSK direct-detection transmission," *IEEE Photonics Technol. Lett.*, vol 3, no 1, 77 (1991)

[7.124] K. Inoue, "Phase mismatching characteristic of four-wave mixing in fiber lines with multistage amplifiers," *Optics Lett.*, vol 17, no 11, 810 (1992)

[7.125] R. H. Stolen and J. E. Bjorkholm, "Parametric amplification and frequency conversion in optical fiber," *IEEE J. Quantum Electron.*, vol 18, no 7, 1062 (1982)

[7.126] K. O. Hill, D. C. Johnson, B. S. Kawazaki, and I. R. MacDonald, "CW three- wave mixing in single-mode optical fibers," *J. Applied Phys.*, vol 49, 5098 (1978)

[7.127] S. J. Garth, "Phase matching the stimulated four-photon mixing process on single-mode fibers operating in the 1.55 $\mu$m region," *Optics Lett.*, vol 13, no 12, 1117 (1988)

[7.128] R. A. Sammut and S. J. Garth, "Multiple-frequency generation on single-mode optical fibers," *J. Opt. Soc. Am. B*, vol 6, no 9, 1732 (1989)

[7.129] N. Shibata, R-P. Braun, and R. G. Waarts, "Phase-mismatch dependence of efficiency of wave generation through four-wave mixing in a single-mode optical fiber," *IEEE J. Quantum Electron.*, vol 23, 1205 (1987)

[7.130] N. Shibata, K. Nosu, K. Iwashita, and Y. Azuma, "Transmission limitations due to fiber nonlinearities in optical FDM systems" *IEEE J. Selected Areas in Communications*, vol 8, no 6, 1068 (1990)

[7.131] M. W. Maeda, W. B. Sessa, W. I. Way, A. Yi-Yan, L. Curtis, R. Spicer, and R. I. Laming, "The effect of four-wave mixing in fibers on optical frequency-division multiplexed signals," *IEEE J. Lightwave Technol.*, vol 8, no 9, 1402 (1990)

[7.132] D. Marcuse, A. R. Chraplyvy, and R. W. Tkach, "Effect of fiber nonlinearity on long-distance transmission," *IEEE J. Lightwave Technol.*, vol 9, no 1, 121 (1991)

[7.133] D. A. Cleland, X. Y. Gu, J. D. Cox, and A. D. Ellis, "Limitations of WDM transmission over 560 km due to degenerate four wave mixing," *Electron. Lett.*, vol 28, no 3, 307 (1992)

[7.134] D. A. Cleland, A. D. Ellis, and C. H. F. Sturrock, "Precise modelling of four wave mixing products over 400 km of step-index fibre," *Electron. Lett.*, vol 28, no 12, 1171 (1992)

[7.135] D. G. Schadt, "Effect of amplifier spacing on four-wave mixing in multichannel coherent communications," *Electron. Lett.*, vol 27, no 20, 1805 (1991)

[7.136] C. Bungarzeanu, "Influence of optical amplifier location on four-wave mixing induced crosstalk," *Electron. Lett.*, vol 29, no 1, 43 (1993)

[7.137] C. Kurtzke and K. Petermann, "Impact of fiber four-photon mixing on the design of *n*-channel megameter optical communication systems," in *Proc. Conference on Optical Fiber Communications OFC'93*, paper FC3, p251, Optical Society of America, Washington DC, 1993

[7.138] F. Forghieri, R. W. Tkach, A. R. Chraplyvy, and D. Marcuse, "Reduction of four-wave mixing crosstalk in WDM systems using unequally spaced channels," in *Proc. Conference on Optical Fiber Communications OFC'93*, paper FC4, p252, Optical Society of America, Washington DC, 1993

[7.139] Y. Nahimira, T. Kawazawa, and H. Wakabayashi, "Polarization mode dispersion measurements in 1520 km EDFA system," *Electron. Lett.*, vol 28, no 9, 881 (1992)

[7.140] Y. Namihira, T. Kawazawa, and H. Wakabayashi, "Polarization mode dispersion measurements in 4564 km EDFA system," *Electron. Lett.*, vol 29, no 1, 32 (1993)

[7.141] A. Galtarossa and M. Schiano, "Complete characterization of polarization mode dispersion in erbium-doped optical amplifiers," *Electron. Lett.*, vol 28, no 23, 2143 (1992)

[7.142] D. J. Malyon, T. Widdowson, and A. Lord, "Assessment of the polarization loss dependence of transoceanic systems using a recirculating loop," *Electron. Lett.*, vol 29, no 2, 207 (1993)

[7.143] M. G. Taylor, "Observation of new polarization dependence effect in long-haul optically amplified systems," in *Proc. Conference on Optical Fiber Communications OFC'93*, paper PD5, Optical Society of America, Washington DC, 1993

[7.144] K. Amano and Y. Iwamoto, "Optical fiber submarine cable systems," *IEEE J. Lightwave Technol.*, vol 8, no 4, 595 (1990)

[7.145] G. G. Koudriavtsev, L. E. Varakine, and P. S. Kourakov, "La ligne transovietique (TSL) de communications a fibres optiques et l'anneau planetaire de telecommunications numeriques," *J. des Telecommunications*, vol 57, no 10, 678 (1990)

[7.146] N. S. Bergano, "Optical amplifier undersea lightwave transmission systems," in *Proc.*

*Conference on Optical Fiber Communications OFC'92*, paper WF4, p117, Optical Society of America, Washington DC, 1992

[7.147]   M. Nakazawa, Y. Kimura, K. Suzuki, and H. Kubota, "Wavelength multiple soliton amplification and transmission with an Er-doped optical fiber," *J. Applied Phys.*, vol 66, no 7, 2803 (1989); M. Nakazawa, Y. Kimura, and K. Suzuki, "Soliton amplification and transmission with an Er-doped fiber repeater pumped by InGaAsP laser diodes," in *Proc. Conference on Optical Fiber Communications OFC'89*, paper PD2, Optical Society of America, Washington DC, 1989

[7.148]   K. Iwatsuki, S. Nishi, M. Saruwatari, and M. Shimizu, "2.8 Gbit/s optical soliton transmission employing all laser diodes," *Electron. Lett.*, vol 26, no 1, 1 (1990); K. Iwatsuki, S. Nishi, M. Saruwatari, and M. Shimizu, "2.8 Gbit/s error-free optical soliton transmission employing all laser diodes," in *Proc. Conference on Integrated Optics and Optical Communications IOOC'89*, paper 20PDA-1, Kobe, Japan, 1989

[7.149]   M. Nakazawa, K. Suzuki, and Y. Kimura, "3.2-5 Gb/s, 100 km error-free soliton transmissions with erbium amplifiers and repeaters," *IEEE Photonics Technol. Lett.*, vol 2, no 3, 216 (1990)

[7.150]   K. Suzuki, M. Nakazawa, E. Yamada, and Y. Kimura, "5 Gbit/s, 250 km error-free soliton transmission with Er-doped fibre amplifiers and repeaters," *Electron. Lett.*, vol 26, no 8, 551 (1990); M. Nakazawa, K. Suzuki, E. Yamada, and Y. Kimura, "Distortion-free single-pass soliton communication over 250 km using multiple Er-doped optical repeaters," in *Proc. Conference on Optical Fiber Communications OFC'90*, paper PD5, Optical Society of America, Washington DC, 1990

[7.151]   K. Iwatsuki, S. Nishi, and K. Nakagawa, "3.6 Gb/s all laser diode optical soliton transmission," *IEEE Photonics Technol. Lett.*, vol 2, no 5, 355 (1990)

[7.152]   N. A. Olsson, P. A. Andrekson, P. C. Becker, J. R. Simpson, T. Tanbun-Ek, R. A. Logan, H. Presby, and K. Wecht, "4 Gb/s soliton data transmission over 136km using erbium-doped fiber amplifiers," *IEEE Photonics Technol. Lett.*, vol 2, no 5, 358 (1990); N. A. Olsson, P. A. Andrekson, P. C. Becker, J. R. Simpson, T. Tanbun-Ek, R. A. Logan, H. Presby, and K. Wecht, "4 Gb/s soliton data transmission experiment," in *Proc. Conference on Optical Fiber Communications OFC'90*, paper PD4, Optical Society of America, Washington DC, 1990

[7.153]   M. Nakazawa, K. Suzuki, E. Yamada and Y. Kimura, "20 Gbit/s soliton transmission over 200 km using erbium-doped fibre repeaters," *Electron. Lett.*, vol 26, no 19, 1592 (1990)

[7.154]   L. F. Mollenauer, B. M. Nyman, M. J. Neubelt, G. Raybon, and S. G. Evangelides, "Demonstration of soliton transmission at 2.4 Gbit/s over 12,000 km," *Electron. Lett.*, vol 27, no 2, 178 (1991)

[7.155]   N. A. Olsson, P. A. Andrekson, J. R. Simpson, T. Tanbun-Ek, R. A. Logan, and K. Wecht, "Bit-error-rate investigation of two-channel soliton propagation over more than 10,000 km," *Electron. Lett.*, vol 27, no 9, 695 (1991); N. A. Olsson, P. A. Andrekson, J. R. Simpson, T. Tanbun-Ek, R. A. Logan, and K. Wecht, "Two-channel soliton pulse propagation over more than 9,000 km with bit-error-rate," in *Proc. Conference on Optical Fiber Communica-tions OFC'91*, paper PD1, Optical Society of America, Washington DC, 1991

[7.156]   E. Yamada, K. Suzuki, and M. Nakazawa, "10 Gbit/s single-pass transmission over 1000 km," *Electron. Lett.*, vol 27, no 14, 1289 (1991)

[7.157]   L. F. Mollenauer, M. J. Neubelt, M. Haner, E. Lichtman, S. G. Evangelides and B. M. Nyman, "Demonstration of error-free soliton transmission at 2.5 Gbit/s over more than 14,000 km," *Electron. Lett.*, vol 27, no 22, 2055 (1991)

[7.158]   P. A. Andrekson, N. A. Olsson, M. Haner, J. R. Simpson, T. Tanbun-Ek, R. A.

Logan, D. Coblentz, H. M. Presby and K. W. Wecht, "32 Gb/s optical soliton data transmission over 90 km," *IEEE Photonics Technol. Lett.*, vol 4, no 1, 76 (1992); P. A. Andrekson, N. A. Olsson, M. Haner, J. R. Simpson, T. Tanbun-Ek, R. A. Logan, D. Coblentz, H. M. Presby, and K. W. Wecht, "Soliton data transmission at 32Gb/s over 90km," *Proc. Topical Meeting on Optical Amplifiers and Applications, 1991*, paper PdP2, vol 13, Optical Society of America, 1991

[7.159] M. Nakazawa, K. Suzuki, E. Yamada, H. Kubota and Y. Kimura, "10 Gbit/s, 1200 km error-free soliton data transmission using erbium-doped fibre amplifiers," *Electron. Lett.*, vol 28, no 9, 817 (1992); M. Nakazawa, K. Suzuki, E. Yamada, H. Kubota, and Y. Kimura, "10 Gbit/s–1200 km single-pass soliton data transmission using erbium-doped fiber amplifiers," in *Proc. Conference on Optical Fiber Communications OFC'92*, paper PD11, p355, Optical Society of America, Washington DC, 1992

[7.160] M. Nakazawa, K. Suzuki, and E. Yamada, "20 Gbit/s, 1020 km penalty-free soliton data transmission using erbium-doped fibre amplifiers," *Electron. Lett.*, vol 28, 11, 1046 (1992)

[7.161] H. Taga, M. Suzuki, H. Tanaka, Y. Yoshida, N. Edagawa, S. Yamamoto, and H. Wakabayashi, "Bit error rate measurement of 2.5 Gbit/s data modulated solitons generated by InGaAsP EA modulator using a circulating loop," *Electron. Lett.*, vol 28, no 13, 1280 (1992)

[7.162] K. Iwatsuki, K. Suzuki, S. Nishi and M. Saruwatari, "40 Gbit/s optical soliton transmission over 65 km," *Electron. Lett.*, vol 28, no 19, 1821 (1992)

[7.163] H. Taga, M. Suzuki, N. Edagawa, Y. Yoshida, S. Yamamoto, S. Akiba and H. Wakabayashi, "5 Gbit/s optical soliton transmission experiment over 3000 km employing 91 cascaded Er-doped fibre amplifier repeaters," *Electron. Lett.*, vol 28, no 24, 2247 (1992)

[7.164] K. Iwatsuki, K. Suzuki, S. Nishi, and M. Saruwatari, "80 Gb/s optical soliton transmission over 80 km with time/polarization division multiplexing," *IEEE Photonics Technol. Lett.*, vol 5, no 2, 245 (1993)

[7.165] L. F. Mollenauer and K. Smith, "Demonstration of soliton transmission over more than 4000 km in fiber with loss periodically compensated by Raman gain," *Optics Lett.*, vol 13, no 8, 675 (1988)

[7.166] L. F. Mollenauer and K. Smith, "Soliton transmission over more than 6000km in fiber with loss periodically compensated by Raman gain," in *Proc. European Conference on Optical Communications ECOC'89*, paper TuA5, p71, 1989

[7.167] M. Nakazawa, K. Suzuki, E. Yamada, and H. Kubota, "Observation of nonlinear interactions in 20 Gbit/s soliton transmission over 500 km using erbium-doped fibre amplifiers," *Electron. Lett.*, vol 27, no 18, 1662 (1991)

[7.168] P. A. Andrekson, N. A. Olsson, P. C. Becker, J. R. Simpson, T. Tanbun-Ek, R. A. Logan, and K. W. Wecht, "Observation of multiple wavelength soliton collisions in optical systems with fiber amplifiers," *Applied Phys. Lett.*, vol 57, no 17, 1715 (1990)

[7.169] L. F. Mollenauer, M. J. Neubelt, S. G. Evangelides, J. P. Gordon, J. R. Simpson, and L. G. Cohen, "Experimental study of soliton transmission over more than 10,000 km in dispersion-shifted fiber," *Optics Lett.*, vol 15, no 21, 1203 (1990)

[7.170] M. Nakazawa, K. Suzuki, S. Nishi, and M. Saruwatari, "20 Gb/s optical soliton data transmission over 70 km using distributed fiber Raman amplifiers," in *Proc. Topical Meeting on Optical Amplifiers and Applications, 1990*, paper PdP4, vol 13, Optical Society of America, 1990

[7.171] Y. Sunohara, Y. Aoki, S. Nakaya, T. Saito, T. Koyama, and S. Fujita, "Quasi-soliton demonstration for over-dispersion-limited distance transmission at 5 Gb/s with Er-doped optical fiber amplifiers," in *Proc. Topical Meeting on Optical Amplifiers and*

*Applications, 1991,* paper PdP3, vol 13, Optical Society of America, 1991

[7.172]   A. D. Ellis, D. A. Cleland, J. D. Cox, and W. A. Stallard. "Two-channel soliton transmission over 678 km," in *Proc. Conference on Optical Fiber Communications OFC'92,* paper WC3, p108, Optical Society of America, Washington DC, 1992

[7.173]   L. F. Mollenauer, E. Lichtman, G. T. Harvey, M. J. Neubelt, and B. M. Nyman, "Demonstration of error-free soliton transmission over more than 15,000 km at 5 Gbit/s, single-channel and over 11,000 km at 10 Gbit/s in a two-channel WDM," in *Proc. Conference on Optical Fiber Communications OFC'92,* paper PD10, p351, Optical Society of America, Washington DC, 1992

[7.174]   P. B. Hansen, C. R. Giles, G. Raybon, A. H. Gnack, U. Koren. B. I. Miller, M. G. Young, M-D. Chein, R. C. Alferness, and C. A. Burrus, "15,000 km, 8.2 GHz soliton propagation in a fiber loop using 980 nm dioed pumped amplifiers and a mode-locked semiconductor laser source," in *Proc. Topical Meeting on Optical Amplifiers and Applications, 1992,* paper FC2, vol 17, 188, Optical Society of America, 1992

[7.175a]   C. R. Giles, P. B. Hansen, S. G. Evangelides, G. Raybon, U. Koren, B. I. Miller, M. G. Young, M. A. Newkirk, J-M. P. Delavaux, S. K. Korotky, J. J. Veselka, and C. A. Burrus, "Soliton transmission over 4200 km by using a mode-locked monolithic external-cavity laser as a soliton source," in *Proc. Conference on Optical Fiber Communications OFC'93,* paper WC2, p88, Optical Society of America, Washington DC, 1993

[7.175b]   M. Nakazawa, K. Suzuki, E. Yamada, H. Kubota, M. Takaya and Y. Kimura, "20 Gbit/s–1850 km and 40 Gbit/s–750 km soliton data transmissions using erbium-doped fiber amplifiers," in *Proc. Conference on Optical Fiber Communications OFC'93,* paper PD9, Optical Society of America, Washington DC, 1993

[7.176]   M. Nakazawa, K. Suzuki, E. Yamada, H. Kubota, Y. Kimura, and M. Takaya, "Experimental demonstration of soliton data transmission over unlimited distances with soliton control in time and frequency domains," in *Proc. Conference on Optical Fiber Communications OFC'93,* paper PD7, Optical Society of America, Washington DC, 1993

[7.177]   L. F. Mollenauer, E. Lichtman, M. J. Neubelt, and G. T. Harvey, "Demonstration, using sliding-frequency-guiding filters, of error-free soliton transmission over more than 20,000 km at 10 Gbit/s, single-channel, and over more than 13,000 km at 20Gbit/s in a two-channel WDM," in *Proc. Conference on Optical Fiber Communications, OFC'93,* paper PD8, Optical Society of America, Washington DC, 1993

[7.178]   S. G. Evangelides, L. F. Mollenauer, J. P. Gordon, and N. S. Bergano, "Polarization multiplexing with solitons," *IEEE J. Lightwave Technol.,* vol 10, no 1, 28 (1992)

[7.179]   F. Heismann and M. S. Whalen, "Broadband, reset-free, automatic polarization controller," *Electron. Lett.,* vol 27, 377 (1991)

[7.180]   H. Kubota and M. Nakazawa, "Long-distance optical soliton transmission with lumped amplifiers," *IEEE J. Quantum Electron.,* vol 26, no 4, 692 (1990)

[7.181]   H. Kubota and M. Nakazawa, "Maximum transmission capacity of a soliton communication system with lumped amplifiers," *Electron. Lett.,* vol 26, 18, 1454 (1990)

[7.182]   J. P. Gordon and L. F. Mollenauer, "Effects of fiber nonlinearities and amplifier spacing on ultra-long distance transmission," *IEEE J. Lightwave Technol.,* vol 9, no. 2, 170 (1991)

[7.183]   L. F. Mollenauer, S. G. Evangelides, and H. A. Haus, "Long-distance soliton propagation using lumped amplifiers and dispersion-shifted fiber," *IEEE J. Lightwave Technol.,* vol 9, no 2, 194 (1991)

[7.184]   J. D. Moores, "Ultra-long distance wavelength-division-multiplexed soliton transmission using inhomogeneously broadened fiber amplifiers," *IEEE J. Lightwave*

[7.184]  J. D. Moores, "Ultra-long distance wavelength-division-multiplexed soliton trans-mission using inhomogeneously broadened fiber amplifiers," *IEEE J. Lightwave Technol.*, vol 10, no 4, 482 (1992)

[7.185]  C. R. Giles, T. Li, T. H. Wood, C. A. Burrus, and D. A. B. Miller, "All-optical regenerator," *Electron. Lett.*, vol 24, no 14, 848 (1988)

[7.186]  M. Jinno and T. Matsumoto, "All-optical timing extraction using a 1.5 $\mu$m self-pulsating multielectrode DFB-LD," *Electron. Lett.*, vol 24, no 23, 1426 (1988)

[7.187]  M. Jinno and M. Abe, "All-optical regenerator based on nonlinear fibre Sagnac interferometer," *Electron. Lett.*, vol 28, no 14, 1350 (1992)

[7.188]  M. Schwartz, *Information, Transmission, Modulation and Noise*, McGraw-Hill, New York, Chapter 5, 1980

[7.189]  M. Nakazawa, E. Yamada, H. Kubota, and K. Suzuki, "10 Gbit/s soliton data transmis-sion over one million kilometres," *Electron. Lett.*, vol 27, no 14, 1270 (1991)

[7.190]  M. Nakazawa, E. Yamada, H. Kubota, and K. Suzuki, "Infinite-distance soliton trans-mission with soliton controls in time and frequency domains," *Electron. Lett.*, vol 28, no 12, 1099 (1992)

[7.191]  H. Kubota and M. Nakazawa, "Soliton transmission control in time and frequency do-mains," *IEEE J. Quant. Electron.*, vol 29, no 7, 2189 (1993)

[7.192]  K. J. Blow, N. J. Doran, and D. Wood, "Generation and stabilization of short soliton pulses in the amplified Schrödinger equation," *J. Opt. Soc. Am, B.* vol 5, 381 (1988)

[7.193]  A. Mecozzi, J. D. Moores, H. A. Haus, and Y. Lai, "Soliton transmission control," *Optics Lett.*, vol 16, no 23, 1841 (1991)

[7.194]  Y. Kodama and A. Hasegawa, "Generation of asymptotically stable optical solitons and suppression of the Gordon–Haus effect," *Optics Lett.*, vol 17, no 1, 31 (1992)

[7.195]  L. F. Mollenauer, J. P. Gordon, and S. G. Evangelides, "The sliding-frequency guiding filter: an improved form of soliton jitter control," *Optics Lett.*, vol 17, no 22, 1575 (1992)

[7.196]  L. F. Mollenauer, "Introduction to solitons," in *Proc. Conference on Optical Fiber Communications, OFC'93*, tutorial paper TuP, Optical Society of America, Washing-ton DC, 1993

[7.198]  J. Stone and L. W. Stultz, "Pigtailed, high-finesse tunable fiber Fabry–Perot inter-ferometers with large, medium and small free spectral range," *Electron. Lett.*, vol 23, 781 (1987)

[7.199]  P. M. Gabla, V. Lemaire, H. Krimmel, J. Otterback, J. Auge, and A. Dursin, "35 AM-VSB TV channels distribution with high signal quality using 1480 nm diode-pumped erbium-doped fiber postamplifiers," *IEEE Photonics Technol. Lett.*, vol 3, no 1, 56 (1991)

[7.200]  W. I. Way, M. M. Choy, A. Yi-Yan, M. J. Andrejco, M. Saifi, and C. Lin, "Multichannel AM-VSB television signal transmission using an erbium-doped optical fiber power amplifier," *IEEE Photonics Technol. Lett.*, vol 1, no 10, 343 (1989); W. I. Way, M. M. Choy, A. Yi-Yan, M. J. Andrejco, M. Saifi, and C. Lin, "Multi-channel AM-VSB television signal transmission using an erbium-doped optical fiber power amplifier, in *Proc. Conference on Integrated Optics and Optical Communications IOOC '89*, paper 20PDA-10, Kobe, Japan, 1989

[7.201]  W. I. Way, M. W. Maeda, A. Yi-Yan, M. J. ANdrejco, M. M. Choy, M. Saifi, and C. Lin, "160-channel FM-video transmission using optical FM/FDM and subcarrier multiplexing and an erbium-doped optical fibre amplifier," *Electron. Lett.*, vol 26, no 2, 139 (1990)

[7.202]  E. Eichen, J. McCabe, W. J. Miniscalco, R. Olshansky, and T. Wei, "FM microwave

F. Massicott, R. A. Lobbett, P. Smith, and T. G. Hodgkinson, "7203-user WDM broadcast network employing one erbium-doped fibre power amplifier," *Electron. Lett.*, vol 26, no 9, 605 (1990)

[7.204] H. E. Tohme, C. N. Lo, and M. A. Saifi, "Simultaneous transmission and amplification of 622 Mbit/s, 10FM NTSC video channels and 3FM HDTV components using a high-gain erbium doped fibre amplifiers," *Electron. Lett.*, vol 26, no 16, 1280 (1990)

[7.205] W. I. Way, S. S. Wagner, M. M. Choy, C. Lin, R. C. Menendez, H. Tohme, A. Yi-Yan, A. C. Von Lehman, R. E. Spicer, M. Andrejco, M. A. Saifi, and H. L. Lemberg, "Simultaneous distribution of multichannel analog and digital video channels to multiple terminals using high-density WDM and a broad-band in-line erbium-doped fiber amplifier," *IEEE Photonics Technol. Lett.*, vol 2, no 9, 665 (1990); W. I. Way, S. S. Wagner, M. M. Choy, C. Lin, R. C. Menendez, H. Tohme, A. Yi-Yan, A. C. Von Lehman, R. E. Spicer, M. Andrejco, M. A. Saifi, and H. L. Lemberg, "Distribution of 100 FM-TV channels and six 622 Mbit/s channels to 4096 terminals using high-density WDM and a broad-band in-line erbium-doped fiber amplifier," in *Proc. Conference on Optical Fiber Communications, OFC'90*, paper PD21, Optical Society of America, Washington DC, 1990

[7.206] A. M. Hill, R. Wyatt, J. F. Massicott, K. . Blyth, D. S. Forrester, R. A. Lobbett, P. J. Smith, and D. B. Payne, "39.5 million-way WDM broadcast network employing two-stage of erbium-doped fibre amplifiers," *Electron. Lett.*, vol 26, no 22, 1882 (1990); A. M. Hill, D. B. Payne, R. Wyatt, J. F. Massicott, K. J. Blyth, D. S. Forrester, R. A. Lobbett, and P. J. Smith, "High-capacity broadcasting over an optical network," in *Proc. Conference on Optical Fiber Communications, OFC'91*, paper ThD3, p 153 Optical Society of America, Washington DC, 1991

[7.207] P. M. Gabla, C. Bastide, Y. Cretin, P. Bousselet, A. Pitel, and J. P. Blondel, "45 dB power budget in a 30-channel AM-VSB distribution system with three cascaded erbium-doped fiber amplifiers," *IEEE Photonics Technol. Lett.*, vol 4, no 5, 510 (1990)

[7.208] H. Bülow, R. Fritschi, R. Heidemann, B. Junginger, H. G. Krimmel and J. Otterback, "4.2 million subscribers with cascaded EDFA-distribution system for 35 TV AM-VSB channels," *Electron. Lett.*, vol 28, no 19, 1836 (1992)

[7.209] H. Bülow, R. Fritschi, R. Heidemann, B. Junginger, H. G. Krimmel and J. Otterbach, "Analog video distribution system with three cascaded 980 nm single-pumped EDFA's and 73 dB power budget," *IEEE Photonics Technol. Lett.*, vol 4, no 11, 1287 (1992)

[7.210] K. Kikushima, E. Yoneda, and K. Aoyama, "6-stage erbium fiber amplifiers for analog AM and FM-FDM video distribution systems," in *Proc. Conference on Optical Fiber Communications OFC'90*, paper PD22, Optical Society of America, Washington DC, 1990

[7.211] K. Kikushima, E. Yoneda, K. Suto and H. Yoshinaga, "Simultaneous distribution of AM/FM FDM TV signals to 65,536 subscribers using a 4 stage cascade EDFA," in *Proc. Topical Meeting on Optical Amplifiers and Applications, 1990*, paper WB1, vol 13, 232, Optical Society of America, 1990

[7.212] D. R. Huber, "40-channel VSB-AM CATV link utilizing a high-power erbium amplifier," in *Proc. Conference on Optical Fiber Communications OFC'91*, paper ThD3, p153, Optical Society of America, Washington DC, 1991

[7.213] D. R. Huber and Y. S. Trisno, "20 channel VSB-AM CATV link utilizing an external modulator, erbium laser and a high power erbium amplifier," in *Proc. Conference on Optical Fiber Communications OFC'91*, paper PD16, Optical Society of America, Washington DC, 1991

[7.214] R. Heidelman, B. Junginger, H. Krimmel, J. Otterbach, D. Schlump, and B. Wedding, "Simultaneous distribution of analogue AM-TV and multigigabit HDTV with optical *on Optical Fiber Communications OFC'91*, paper PD16, Optical Society of America, Washington DC, 1991

[7.214]  R. Heidelman, B. Junginger, H. Krimmel, J. Otterbach, D. Schlump, and B. Wedding, "Simultaneous distribution of analogue AM-TV and multigigabit HDTV with optical amplifier," in *Proc. Topical Meeting on Optical Amplifiers and Applications, 1991*, paper FB2, vol 13, 210, Optical Society of America, 1991

[7.215]  M. Shigematsu, M. Nishimura, T. Okita, and K. Izuka, "A high power and low noise transmitter for AM-VSB transmission using erbium-doped fiber post amplifier," in *Proc. Topical Meeting on Optical Amplifiers and Applications, 1991*, paper FB3, vol 13, 214, Optical Society of America, 1991

[7.216]  G. R. Joyce, R. Olshansky, R. Childs, and T. Wei, "A 40 channel AM-VSB distribution system with a 21 dB link budget," in *Proc. Topical Meeting on Optical Amplifiers and Applications, 1991*, paper FB4, vol 13, 218, Optical Society of America, 1991

[7.217]  M. R. Phillips, A. H. Gnauck, T. E. Darcie, N. J. Frigo, G. E. Bodeep, and E. A. Pitman, "112 channel WDM split-band CATV system," in *Proc. Conference on Optical Fiber Communications, OFC'92*, paper PD6, Optical Society of America, Washington DC, 1992

[7.218]  B. Clesca, J-P. Blondel, P. Bousselet, J-P. Hebert, J-L. Beylat, F. Brillouet, L. Sniadower, and Y. Cretin, "32 channel, 48 dB CNR and 46 dB budget AM-VSB transmission experiment with field-ready post-amplifiers," in *Proc. Topical Meeting on Optical Amplifiers and Applications, 1992*, paper ThA4, vol 17, 91, Optical Society of America, 1992

[7.219]  H. Giaggoni and D. J. LeGall, "Digital video transmission and coding for the broadband ISDN," *IEEE Trans. Consumer Electron.*, vol 34, 16 (1988)

[7.220]  K. Simon, *Technical Handbook for CATV Systems,* third edition, General Instruments, 1986

[7.222]  P. E. Green, *Fiber Optic Networks*, Prentice-Hall, New York, 1993, Chapter 5

[7.223]  P. E. White and L. S. Smoot, "Optical fibers in loop distribution systems," in *Optical Fiber Telecommunications II*, edited by S. E. Miller and I. P. Kaminow, Academic Press, New York, 1988

[7.224]  J. Walrand, *Communication networks: a first course*, Irwin, Boston, 1991

[7.225]  I. P. Kaminow, "Non-coherent photonic frequency-multiplexed access networks," *IEEE Network*, March 1989, p4

[7.226]  K. W. Lu, M. I. Eiger, and H. L. Lemberg, "System and cost analyses of broad-band fiber loop architectures," *IEEE J. Selected Areas in Communications*, vol 8, no 6, 1058 (1990)

[7.227]  C. A. Brackett, "Dense wavelength division multiplexing networks: principles and applications," *IEEE J. Selected Areas in Communications*, vol 8, no 6, 948 (1990)

[7.228]  R. E. Wagner, "Multichannel lightwave networks," in *Proc. Conference on Optical Fiber Communications OFC'91*, tutorial paper TuD1, Optical Society of America, Washington DC, 1991

[7.229]  J. O. Limb, "Multiple access high-speed network," in *Proc. Conference on Optical Fiber Communications OFC'92*, tutorial paper WJ, Optical Society of America, Washington DC, 1992

[7.230]  R. Ramaswami, "Multiwavelength lightwave networks for computer communication," *IEEE Communications Magazine*, February 1993, p18

[7.231]  B. Alexander et al., "A precompetitive consortium on wide-band all-optical networks," *IEEE J. Lightwave Technology*, vol 11, no 5–6, 714 (1993)

43.8 million-way WDM broadcast network with 527 km range," *Electron. Lett.*, vol 27, no 22, 2051 (1991)

[7.235] H. Toba, K. Oda, K. Nakanishi, K. Nosu, K. Kato, and Y. Hibino, "128-channel OFDM information-distribution system with interactive video channels based on OA-OFDMA," in *Proc. Conference on Optical Fiber Communications, OFC'93*, paper TuN2, p66, Optical Society of America, Washington DC, 1993

[7.236] H. Toba, K. Oda, K. Nosu, and N. Takato, "Factors affecting the design of optical FDM information distribution systems," *IEEE J. Selected Areas in Communication*, vol 8, no 6, 965 (1990)

[7.237] T. J. Whitley, C. A. Millart, S. P. Craig-Ryan, and P. Urquhart, "Demonstration of a distributed optical fibre amplifier BUS network," in *Proc. Topical Meeting on Optical Amplifiers and Applications*, 1990, paper WB2, vol 13, 236, Optical Society of America, 1990

[7.238] K. Liu and R. Ramaswami, "Analysis of optical bus networks using doped-fiber amplifiers," in *Proc. IEEE/LEOS topical meeting on optical multiaccess networks*, 1989

[7.239] R. Ramaswami and K. Liu, "Analysis of multiple-channel optical bus networks using doped fiber amplifiers," in *Proc. Conference on Optical Fiber Communications OFC'91*. paper FE5, p219, Optical Society of America, Washington DC, 1991

[7.240] D. N. Chen, K. Motoshima, and E. Desurvire, "A transparent optical bus using a chain of remotely pumped erbium-doped fiber amplifiers," *IEEE Photonics Technol. Lett.*, vol 5, no 3, 351 (1993)

[7.241] E. Goldstein, "Optical ring networks with distributed amplification," *IEEE Photonics Technol. Lett.*, vol 3, no 4, 390 (1991)

[7.242] E. L. Goldstein, A. F. Elrefaie, N. Jackman, and S. Zaidi, "Multiwavelength cascades in unidirectional interoffice ring networks," in *Proc. Conference on Optical Fiber Communications OFC'93*, paper TuJ3, p44, Optical Society of America, Washington DC, 1993

[7.243] A. F. Elrefaie and S. Zaidi, "Fibre amplifier in closed-ring WDM networks," *Electron. Lett.*, vol 28, no 25, 2340 (1992)

[7.244] W. I. Way, T. H. Wu, A. Yi-Yan, M. Andrejco, and C. Lin, "Optical power limiting amplifier and its applications in a SONET self-healing ring network," *IEEE J. Lightwave Technol.*, vol 10, no 2, 206 (1992)

[7.245] A. E. Willner, A. A. M. Saleh, H. M. Presby, D. DiGiovanni, and C. A. Edwards, "Star couplers with gain using multiple erbium-doped fibers pumped with a single laser," *IEEE Photonics Technol. Lett.*, vol 3, no 3, 250 (1991)

[7.246] H. M. Presby and C. R. Giles, "Amplified integrated star couplers with zero loss," *IEEE Photonics Technol. Lett.*, vol 3, no 8, 724 (1991)

[7.247] M. I. Irshid and M. Kavehrad, "Star couplers with gain using fiber amplifiers," *IEEE Photonics Technol. Lett.*, vol 4, no 1, 58 (1992)

[7.248] Y. K. Chen, "Amplified distributed reflective optical star couplers," *IEEE Photonics Technol. Lett.*, vol 4, no 6, 570 (1992)

[7.249] A. E. Willner, E. Desurvire, H. M. Presby, C. A. Edwards, and J. R. Simpson, "Use of LD-pumped erbium-doped fiber preamplifiers with optimal noise filtering in a FDMA-FSK 1G bit/s star networks," *IEEE Photonics Technol. Lett.*, vol 2, no 9, 669 (1990)

[7.250] S. Culverhouse, R. A. Lobbett, and P. J. Smith, "Optically amplified TDMA distributive switch network with 2.488 Gbit/s capacity offering interconnection to over 1000 customers at 2 Mbit/s," *Electron. Lett.*, vol 28, no 17, 1672 (1992)

[7.251] I. Habbab, M. Kavehrad, and C. E. W. Sundberg, "Protocols for very high-speed optical fiber local networks using a passive star topology," *IEEE J. Lightwave Technol.*, vol 5, no 12, 1782 (1987)

FDMA-FSK 1G bit/s star networks," *IEEE Photonics Technol. Lett.*, vol 2, no 9, 669 (1990)

[7.250]  S. Culverhouse, R. A. Lobbett, and P. J. Smith, "Optically amplified TDMA distributive switch network with 2.488 Gbit/s capacity offering interconnection to over 1000 customers at 2 Mbit/s," *Electron. Lett.*, vol 28, no 17, 1672 (1992)

[7.251]  I. Habbab, M. Kavehrad, and C. E. W. Sundberg, "Protocols for very high-speed optical fiber local networks using a passive star topology," *IEEE J. Lightwave Technol.*, vol 5, no 12, 1782 (1987)

# INDEX

Boldface page numbers indicate an emphasis of the subject.

A and B coefficients, *see* Einstein's A and B
    coefficients
Absorption:
    bleaching, 21, 307, 319–320, 356, 489
    excited-state, *see* Excited-state absorption (ESA)
    ground-level, *see* Ground-level absorption (GSA)
    saturable, 148, 152, 369, 439, 478, 519, 521,
      523
Absorption coefficient:
    asymptotical value, 127
    distribution, 114
    gain-dependent pump, 38
    inhomogeneous, 64
    ionic, or related to $Er^{3+}$-doping, 21–22, 28–29,
      35–36, 37, 128
    ionic/background ratio, 127
    measurement, 259
    minimum, 128
    related to background loss, 26
    spectrum, 216–217, 272–273
Absorption length, 125, 128
Acceptor, 282, 284–285
Accumulated photon echoes, 228
Acousto-optic tunable filter (AOFT), *see* Filter
Activator ions, 59, 211, 226, 282, 296
Activator site, *see* Site
Adiabatic decoupling, 141
All-optical signal regeneration, 494–495, 502–503,
    508, 526, 559, 562–568
Alumina, *see* Glass
Aluminosilicate glass, *see* Er:glass
Alumino-germano silicate glass, *see* Er:glass

Amplified spontaneous emission (ASE):
    accumulation or buildup, 192–194, 409,
      499–500, 502–503, 508, 526, 562–564, 566,
      575–576, 580
    affecting photon statistics, 93
    analytical solution, **39, 104**, 129, 600–606
    determined from photon statistics, 76–77
    difference between forward and backward, 104,
      109, 355–360
    effective bandwidth, 336, 374, 378, 395
    equivalent-noise power, 169, 190, 412
    experimental characterization, 54–360
    filtering, *see* Optical filtering
    from blackbody modes, 16
    in distributed EDFA's, 129
    in Raman fiber amplifiers, 126
    in semiclassical photodetection, 164
    in signal rate equation, 16
    noise background, 76, 98
    path-averaged, 134–135
    polarization filtering, *see* Polarizer
    polarized, 327, 370
    power, 260, 324, 328
    related to phase noise, 399–404
    timing jitter induced by, *see* Timing
    spectrum or noise spectrum, 308, 333, 346–348,
      354–360
Amplifier:
    bandwidth, 70
    bulk-optics, 208
    cascades or chains, *see* Chains
    design, *see* Fiber